MODERN CHEMISTRY

HOLT, RINEHART AND WINSTON

Harcourt Brace & Company

Austin • New York • Atlanta • San Francisco • Boston • Dallas • Toronto • London

Authors

RAYMOND E. DAVIS, PH.D.
Distinguished Teaching Professor
Department of Chemistry and Biochemistry
The University of Texas at Austin
Austin, TX

H. CLARK METCALFE
Former Chemistry Teacher and Science Department Chair
Wilkinsburg, PA

JOHN E. WILLIAMS
Former Chemistry Teacher and Science Department Chair
Indianapolis, IN

JOSEPH F. CASTKA
Former Adjunct Associate Professor
C. S. Post College
Long Island, NY

Contributing Writers

Seth Madej
Associate Producer
Pulse of the Planet radio series
Jim Metzner Productions, Inc.
Yorktown Heights, NY

Jim Metzner
Executive Producer
Pulse of the Planet radio series
Yorktown Heights, NY

Jay A. Young, Ph.D.
Chemical Safety Consultant
Silver Spring, MD

Cover: Portion of the periodic table superimposed on a scanning tunneling electron micrograph of a protein, photo by Digital Instruments/Fran Heyl Associates

Portions of this work were published in previous editions.

Acknowledgments appear on page v, which is an extension of the copyright page.

Printed in the United States of America

ISBN 0-03-051122-4 234567 41 02 01 00 99 98

Reviewers

George F. Atkinson, Ph.D.
Professor of Chemistry
Department of Chemistry
University of Waterloo
Waterloo, Ontario, Canada

G. Lynn Carlson, Ph.D.
Department of Chemistry
University of Wisconsin–Parkside
Kenosha, WI

Doris I. Lewis, Ph.D.
Professor of Chemistry
Suffolk University
Boston, MA

Daniel B. Murphy, Ph.D.
Professor Emeritus of Chemistry
Department of Chemistry
Herbert H. Lehman College
The City University of New York
Bronx, NY

R. Thomas Myers, Ph.D.
Professor Emeritus of Chemistry
Kent State University
Kent, OH

Keith B. Oldham, Ph.D.
Professor of Chemistry
Trent University
Peterborough, Ontario, Canada

Charles Scaife, Ph.D.
Professor of Chemistry
Union College
Schenectady, NY

David C. Taylor, Ph.D.
Professor of Chemistry
Department of Chemistry
Slippery Rock University
Slippery Rock, PA

Richard S. Treptow, Ph.D.
Professor of Chemistry
Department of Chemistry and Physics
Chicago State University
Chicago, IL

Martin VanDyke, Ph.D.
Consultant
Chem-Safe Environmental Services
Denver, CO

Laverne Weidler, Ph.D.
Professor of Chemistry
Science and Engineering
Black Hawk College
Kewanee, IL

Charles M. Wynn, Sr., Ph.D.
Professor of Chemistry
Department of Physical Sciences
Eastern Connecticut State University
Willimantic, CT

aff Credits

Executive Editor
Ellen Standafer

Managing Editor
William Wahlgren

Project Editors
Patrick Earvolino
Erin Edwards
Derika Hatcher
Cecelia Schneider
Valerie Uzcategui

Copyeditors
Steve Oelenberger, *Copyediting Supervisor*
Amy Daniewicz
Denise Haney

Book Design
Christine Schueler, *Design Manager*
Jennifer Dix
Cathy Jenevein

Image Services
Greg Geisler, *Art Director*
Elaine Tate
Sherry France
Ian Christopher
Linda Wilburn

Photo Research
Peggy Cooper, *Photo Research Manager*
Jeannie Taylor
Caroline Robbins

Photo Studio
Sam Dudgeon, *Senior Staff Photographer*
Victoria Smith

Media Design
Joe Melomo, *Design Manager*

Cover, PE and ATE Design and Production
The Quarasan Group

Production
Mimi Stockdell, *Senior Production Manager*
Jessica Wyatt
Sara Carroll-Downs

Manufacturing
Michael Roche

Media
Nancy Hargis
Susan Mussey

New Media
Armin Gutzmer, *Manager, Training and Technical Support*
Cathy Kuhles

Editorial Permissions
Amy Minor

Acknowledgments

For permission to reprint copyrighted material, grateful acknowledgment is made to the following sources:

Doubleday, a division of Bantam Doubleday Dell Publishing Group, Inc.: From *Science Matters: Achieving Scientific Literacy* by Robert M. Hazen and James Trefil. Copyright © 1991 by Robert M. Hazen and James Trefil.

HarperCollins Publishers, Inc.: Two figures and excerpt from *The Periodic Kingdom* (Science Masters Series) by Peter Atkins. Copyright © 1995 by P. W. Atkins.

Henry Holt and Company, Inc.: From "Bound for Glory" by Roald Hoffman from *Scientific American: Triumph of Discovery*. Copyright © 1995 by Henry Holt & Co., Inc. and Scientific American, Inc.

The McGraw-Hill Companies: From "Philosophy" from *A Concise Introduction to Philosophy* by William H. Halverson. Copyright © 1967, 1972, 1976, 1981 by Random House, Inc.

John W. Moore: From "Faraday's Contribution to Electrolytic Solution Theory" by Ollin J. Drennan from *Journal of Chemical Education,* vol. 42, no. 12, December, 1965, pp. 679–681. Copyright © 1965 by Division of Chemical Education, American Chemical Society. From "Acid and Water: A Socratic Dialogue" by David Todd from *Journal of Chemical Education,* vol. 70, no. 12, December, 1993, page 1022. Copyright © 1993 by Division of Chemical Education, American Chemical Society. From "The Chemical Adventures of Sherlock Holmes: The Hound of Henry Armitage" by Thomas G. Waddel and Thomas R. Rybolt from *Journal of Chemical Education,* vol. 71, no. 12, December, 1994, pp. 1049–1051. Copyright © 1994 by Division of Chemical Education, American Chemical Society.

The Philisophical Library: Quote by Albert Einstein as quoted in a personal memoir of William Miller, an editor, in *Life* magazine, May 2, 1955. Copyright © 1955 by The Philisophical Library, New York.

Schocken Books, distributed by Pantheon Books, a division of Random House, Inc.: From "Travels with C" from *The Periodic Table* by Primo Levi, translated by Raymond Rosenthal. Translation copyright © 1984 by Schocken Books, Inc.

Smithsonian Institution Press: From *Chemistry Imagined: Reflections on Science* by Roald Hoffman and Vivian Torrence. Copyright © 1993 by Roald Hoffman and Vivian Torrence.

Texas Christian University Press: From "Chemistry and the Plastic and Graphic Arts: Creating and Caring for Works of Art" by Jonathon E. Ericson from *The Central Science: Essays on the Uses of Chemistry,* edited by George B. Kauffman and H. Harry Szmant. Copyright © 1984 by George B. Kauffman and H. Harry Szmant.

University of California Press: From "Exile in Stockholm" from *Lise Meitner: A Life in Physics* by Ruth Lewin Sime. Copyright © 1996 by The Regents of the University of California.

Other Sources Cited:

Quote by Louis Pasteur from *Pasteur and Modern Science* by René J. Dubos. Science Tech Publishers, Madison, WI, 1988.

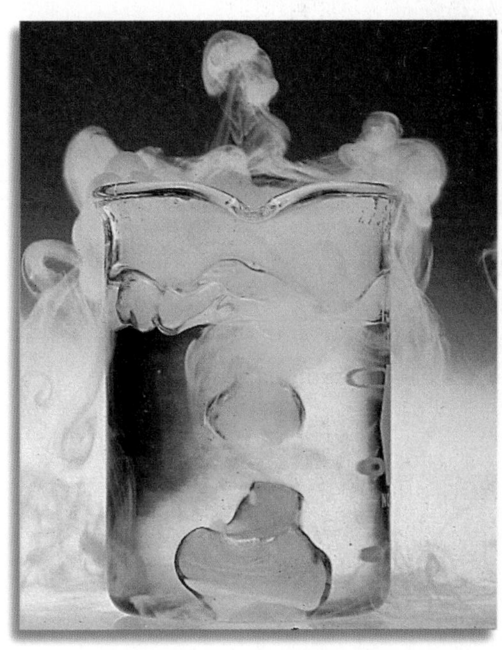

Contents

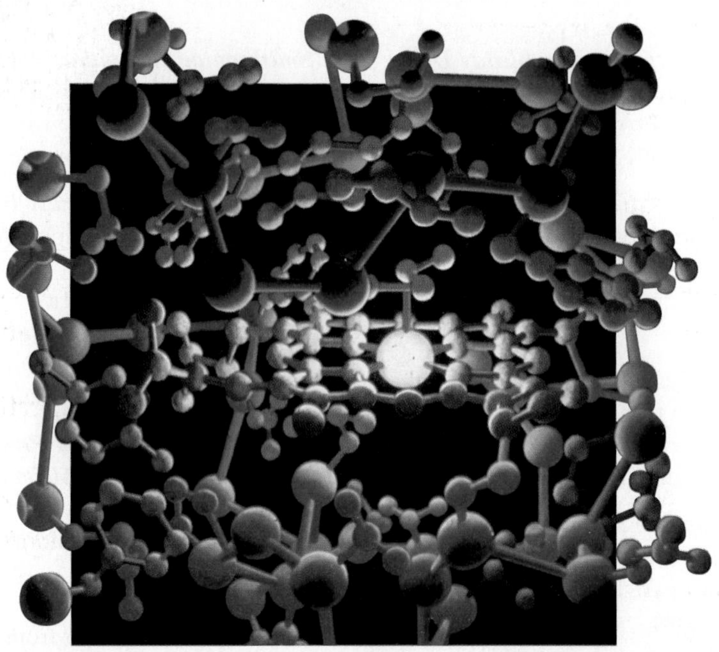

UNIT 1 *Introduction to Chemistry and Matter* 2

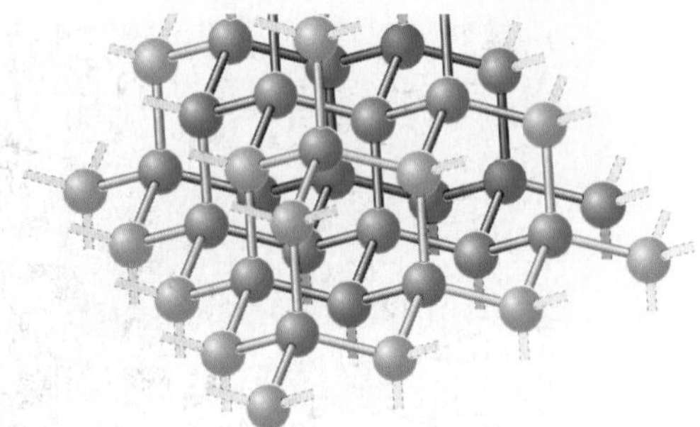

UNIT 2 *Organization of Matter* 62

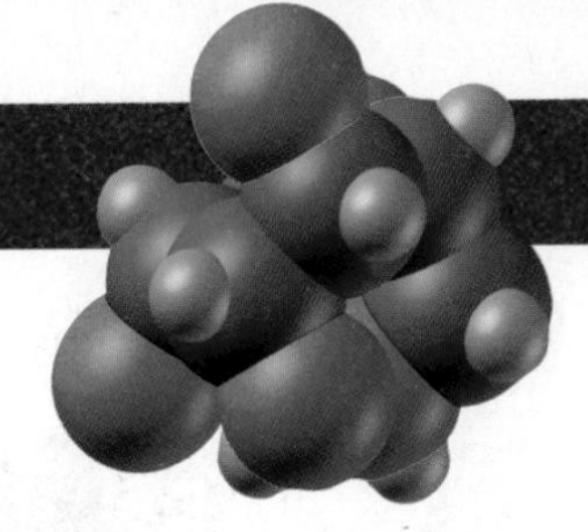

CHAPTER 3

Atoms: The Building Blocks of Matter

CHAPTER 4

Arrangement of Electrons in Atoms

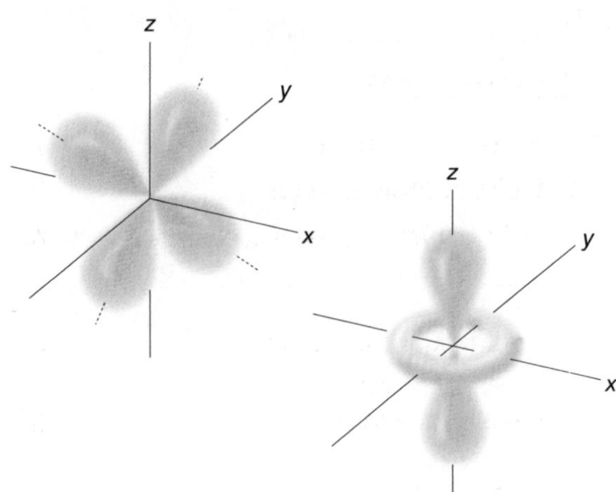

CHAPTER 5

The Periodic Law

CHAPTER 6

Chemical Bonding

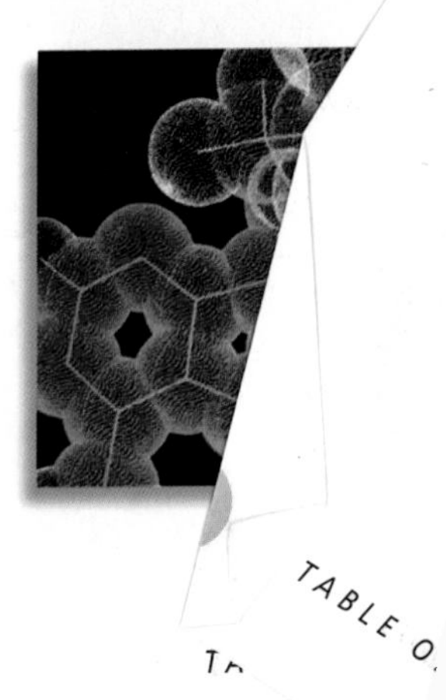

TABLE O

Tr

CHAPTER 7

Chemical Formulas and Chemical Compounds

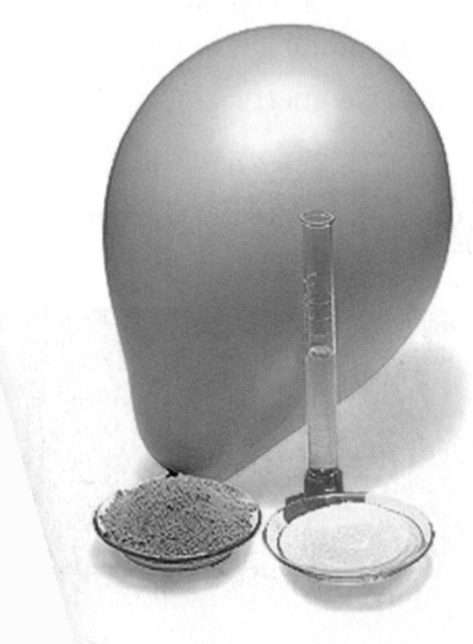

CHAPTER 8

Chemical Equations and Reactions

CHAPTER 9

Stoichiometry

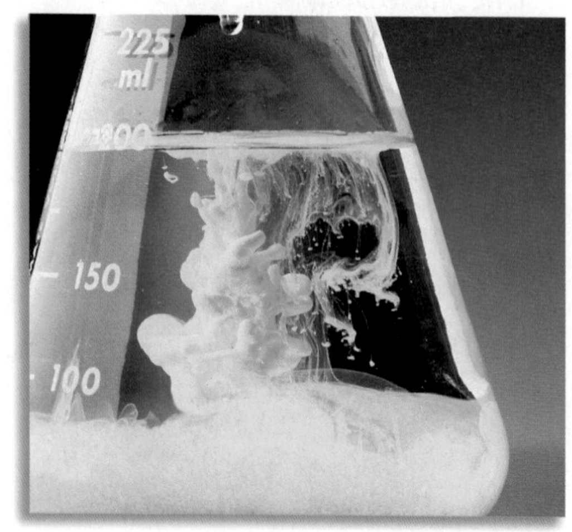

UNIT 4 *Phases of Matter* 300

CHAPTER 10

Physical Characteristics of Gases

CHAPTER 11

Molecular Composition of Gases

CHAPTER 12

Liquids and Solids

CHAPTER 15

Acids and Bases

CHAPTER 13

Solutions

CHAPTER 16

Acid-Base Titration and pH

CHAPTER 14

Ions in Aqueous Solutions and Colligative Properties

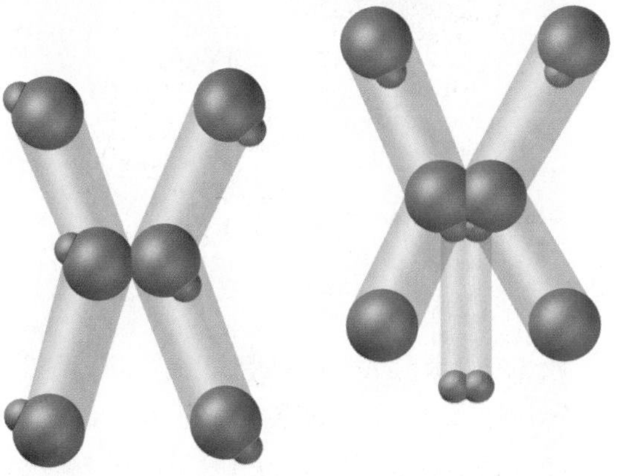

CHAPTER 17

Reaction Energy and Reaction Kinetics

CHAPTER 18

Chemical Equilibrium

CHAPTER 19

Oxidation-Reduction Reactions

UNIT 7 *Organic and Nuclear Chemistry* 622

CHAPTER 20

Carbon and Hydrocarbons

CHAPTER 21

Other Organic Compounds

CHAPTER 22

Nuclear Chemistry

Laboratory Experiments 784

Reference Section 894

Features

RESEARCH NOTES

CHEMICAL COMMENTARY

GREAT DISCOVERIES

DESKTOP INVESTIGATIONS

UNIT 1

Introduction to Chemistry and Matter

CHAPTERS

WILLIAM H. HALVERSON, ON THE DIFFERENCE BETWEEN SCIENCE AND THE GADGETS OF SCIENCE

It is a grave mistake to think of science as having to do primarily with test tubes, microscopes, cyclotrons, and the like. One should not confuse the gadgets of science with science itself. "Science" is simply the collective name for the totality of human efforts to achieve a systematic understanding of the physical universe through disciplined inquiry. The purpose of the gadgets is merely to assist in these efforts.

(From *A Concise Introduction to Philosophy*)

CHAPTER 1

Matter and Change

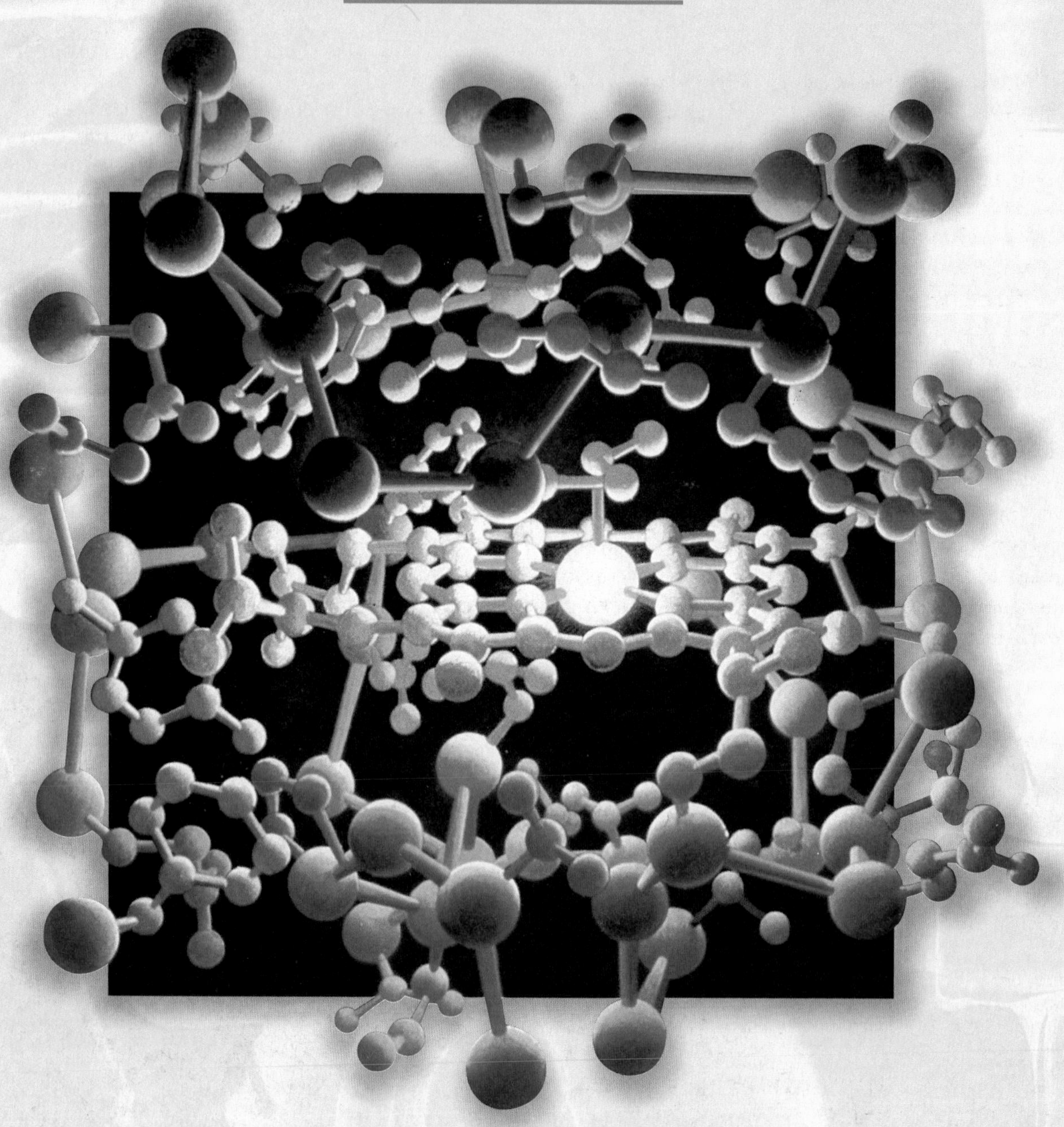

Chemistry is central to all of the sciences.

Chemistry Is a Physical Science

Objectives

- Define *chemistry.*
- List examples of the branches of chemistry.
- Compare and contrast basic research, applied research, and technological development.

The natural sciences were once divided into two broad categories: the biological sciences and the physical sciences. Living things are the main focus of the biological sciences. The physical sciences focus mainly on nonliving things. However, because we now know that both living and nonliving matter have a chemical structure, chemistry is central to all the sciences, and there are no longer distinct divisions between the biological and physical sciences.

Chemistry *is the study of the composition, structure, and properties of matter and the changes it undergoes.* Chemistry deals with questions such as, What is that material made of? What is its makeup and internal arrangement? How does it behave and change when heated, cooled, or mixed with other materials and why does this behavior occur? Chemists answer these kinds of questions in their daily work.

Instruments are routinely used in chemistry to extend our ability to observe and make measurements. Instruments make it possible, for example, to look at microstructures—things too tiny to be seen with the unaided eye. The scanning electron microscope reveals tiny structures by beaming particles called electrons at materials. When the electrons hit a material, they scatter and produce a pattern that shows the material's microstructure. Invisible rays called X rays can also be used to

FIGURE 1-1 A balance (a) is an instrument used to measure the mass of materials. A sample of DNA placed in a scanning electron microscope produces an image (b) showing the contours of the DNA's surface.

(a)

(b)

"look at" microstructures. The patterns that appear, called X-ray diffraction patterns, can be analyzed to reveal the arrangement of atoms, molecules, or other particles that make up the material. By learning about microstructures, chemists can explain the behavior of macrostructures—the visible things all around you.

Branches of Chemistry

Chemistry includes many different branches of study and research. The following are six main areas, or branches, of study. But like the biological and physical sciences, these branches often overlap.

1. *Organic chemistry*—the study of most carbon-containing compounds
2. *Inorganic chemistry*—the study of all substances not classified as organic, mainly those compounds that do not contain carbon
3. *Physical chemistry*—the study of the properties, changes, and relationships between energy and matter
4. *Analytical chemistry*—the identification of the components and composition of materials
5. *Biochemistry*—the study of substances and processes occurring in living things
6. *Theoretical chemistry*—the use of mathematics and computers to design and predict the properties of new compounds

In all areas of chemistry, scientists work with chemicals. *A* **chemical** *is any substance that has a definite composition.* For example, consider the material called sucrose, or cane sugar. It has a definite composition in terms of the atoms that compose it. It is produced by certain plants, which use carbon dioxide and water in the chemical process of photosynthesis. Sucrose is a chemical. Carbon dioxide, water, and countless other substances are chemicals as well.

Knowing the properties of chemicals allows chemists to find suitable uses for them. For example, researchers have synthesized new substances, such as artificial sweeteners and synthetic fibers. The reactions used to make these chemicals are carried out on a large scale to make new products such as sweeteners and fabrics available for consumers.

Basic Research

Basic research is carried out for the sake of increasing knowledge, such as how and why a specific reaction occurs and what the properties of a substance are. Chance discoveries can be the result of basic research. The properties of Teflon, for example, were first discovered by accident. A researcher named Roy Plunkett was puzzled by the fact that a gas cylinder used for an experiment appeared to be empty even though the measured mass of the cylinder clearly indicated there was something inside. Plunkett cut the cylinder open and found a solid white material. Through basic research, Plunkett's research team determined the nonstick properties, chemical structure, and chemical composition of the new material.

Applied Research

Applied research is generally carried out to solve a problem. For example, when refrigerants escape into the upper atmosphere, they damage the ozone layer, which helps block harmful ultraviolet rays from reaching the surface of Earth. In response to concerns that this could pose potential health problems, chemists have developed new compounds to replace refrigerants. In applied research, the researchers are driven not by simple curiosity or a desire to know but by a desire to solve a specific problem.

FIGURE 1-2 The chemical structure of the material in an optical fiber gives it the property of total internal reflection. This property, which allows these fibers to carry light, was discovered through basic and applied research. The use of this property to build telecommunications networks by sending data on light pulses is the technological development of fiber optics.

Technological Development

Technological development typically involves the production and use of products that improve our quality of life. Examples include computers, catalytic converters for cars, and biodegradable materials.

Technological applications often lag far behind the basic discoveries that are eventually used in the technologies. For example, nonstick cookware, a technological application, was developed well after the accidental discovery of Teflon. When it was later discovered that the Teflon coating on cookware often peeled off, a new problem had to be solved. Using applied research, scientists were then able to improve the bond between the Teflon and the metal surface of the cookware so that it did not peel.

Basic research, applied research, and technological development often overlap. Discoveries made in basic research may trigger ideas for applications that can result in new technologies. For example, knowledge of crystals and the behavior of light was gained from basic research, and this knowledge was used to develop lasers. It was then discovered that pulses of light from lasers can be sent through optical fibers. Today, information, such as telephone messages and cable television signals, can now be carried quickly over long distances using fiber optics.

SECTION REVIEW

1. Define *chemistry*.
2. Name the six branches of study in chemistry.
3. Compare and contrast basic research, applied research, and technological development.

Modern Alchemy

HISTORICAL PERSPECTIVE

Merely a hundred years ago, chemists were still debating the validity of John Dalton's atomic theory. Few, however, challenged the notion that the elements were unchangeable. Near the beginning of the twentieth century, the discovery of some new elements and the strange radiation they emitted established the connection between atoms and the elements while resurrecting an ancient notion long discarded by science.

Before Chemistry

Until the chemical revolution of the seventeenth and eighteenth centuries, most theories about matter were based on the ideas of the ancient Greek philosopher Aristotle. He postulated that all matter consisted of four elements: earth, water, air, and fire. In turn, each of these elements exhibited two of four fundamental properties: moistness, dryness, coldness, and hotness. By altering these basic properties, Aristotle claimed, the elements could be transformed, or transmuted, into one another.

The interior of an alchemist's laboratory is depicted by artist Eugene Isabey.

The practical pursuit of transmutation became known as alchemy, and for more than 1,500 years investigators searched in vain for alchemical methods that would transform common metals such as mercury and lead into precious gold. Then, in the seventeenth century, chemists began to question Aristotle's assumptions. They defined an element as a material that can't be broken down into simpler substances, and with no evidence to support the possibility of transmutation of the modern elements, alchemy fell into ill repute. By the nineteenth century, the subject was generally disregarded by scientists.

Strange Rays

In 1896, French scientist Henri Becquerel discovered that the element uranium gave off a strange, invisible radiation. The report of these "uranic rays" caught the attention of a young chemist by the name of Marie Curie, who decided to dedicate her studies to the mysterious radiation, which she later named radioactivity.

Working with her husband, Pierre, Marie began to test various substances for radioactivity. Analyzing a mineral composite called pitchblende, a known source of uranium, she was startled to find that the composite's level of radioactivity was greater than that of a similar amount of pure uranium. This meant that another radioactive material besides uranium was present in the pitchblende.

After months of tedious work, Marie had isolated two new radioactive elements, which she named polonium and radium, from pitchblende. Curie later won the Nobel Prize for her discovery, but at the time, she was troubled by the seemingly constant energy source of the radioactive process that had led her to the new elements. As she wrote in 1900:

> *The emission of the uranic rays is very constant . . . The uranium shows no appreciable change of state, no visible chemical transformation, it remains, in appearance at least, the same as ever, the source of the energy it discharges remains undetectable. . . .*

The Revival of Transmutation

While Marie Curie was making her momentous discoveries, other scientists were establishing that the chemical elements were actually different types of atoms. The connection between this emerging theory of matter and radioactivity was made by the famous scientist Ernest Rutherford.

In 1902, Rutherford and his assistant, Frederick Soddy, reported that the radioactivity of a sealed sample of thorium (a known element determined to be radioactive by the Curies) had actually *increased* over time. This increase was accompanied by the simultaneous evolution of a radioactive gas. The two investigators began to question whether radioactive elements were as stable as elements were supposed to be. After further studies, Rutherford and Soddy presented their shocking explanation:

> *The cause and nature of radioactivity is at once an atomic phenomenon and the accompaniment of a chemical change in which new kinds of matter are produced. The two considerations force us to the conclusion that radioactivity is a manifestation of subatomic chemical change.*

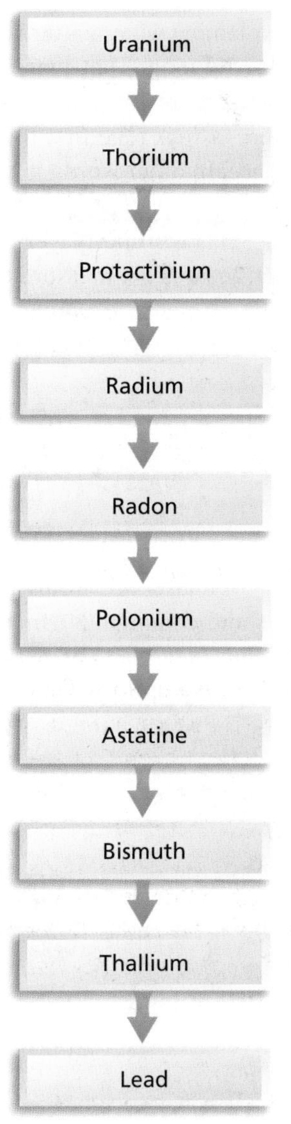

A simplified decay diagram for uranium

Rutherford and Soddy's "subatomic chemical change" was nothing less than transmutation. Nature had turned out to be an alchemist!

The New Alchemy Explained

Rutherford later showed that the spontaneous transmutation, or decay, of radioactive atoms involves the emission of nuclear particles or high-energy waves known as gamma rays, or both. The atom's nucleus is consequently reconfigured, changing the atom into that of another element.

Thus, the source of energy in Marie Curie's uranium sample came from within the radioactive atoms. And although it seemed as if the energy output was constant, the sources of the radiation were continually changing as one radioactive element decayed into another, which in turn decayed into another, and so on.

Years later, it would be shown that the transmutation of certain elements could be deliberately initiated by bombarding nuclei with accelerated subatomic particles. But the goal of modern alchemy is no longer gold, as Soddy remarked in 1917:

> *If man ever achieves this control over Nature, it is quite certain that the last thing he would want to do would be to turn lead or mercury into gold*—for the sake of gold. *The energy that would be liberated, if the control of these sub-atomic processes were possible . . . would far exceed in importance and value the gold.*

SECTION 1-2

Matter and Its Properties

Objectives

- Distinguish between the physical properties and chemical properties of matter.
- Classify changes of matter as physical or chemical.
- Explain the gas, liquid, and solid states in terms of particles.
- Distinguish between a mixture and a pure substance.

Look around you. You can see a variety of objects—books, desks, chairs, and perhaps trees or buildings outside. All those things are made up of matter, but exactly what is matter? What characteristics, or properties, make matter what it is? In this section, you will learn the answers to these questions.

Explaining what matter is involves finding properties that all matter has in common. That may seem difficult, given that matter takes so many different forms. For the moment, just consider one example of matter—a rock. The first thing you might notice is that the rock takes up space. In other words, it has *volume*. Volume is the amount of three-dimensional space an object occupies. All matter has volume. All matter also has a property called mass. **Mass** *is a measure of the amount of matter*. Mass is the measurement you make using a balance. **Matter** can thus be defined as *anything that has mass and takes up space*. These two properties are the general properties of all matter.

Basic Building Blocks of Matter

You know that matter comes in many forms. The fundamental building blocks of matter are atoms and molecules. These particles make up elements and compounds. *An* **atom** *is the smallest unit of an element that maintains the properties of that element. An* **element** *is a pure substance made of only one kind of atom.* Carbon, hydrogen, and oxygen are elements. They each contain only one kind of atom.

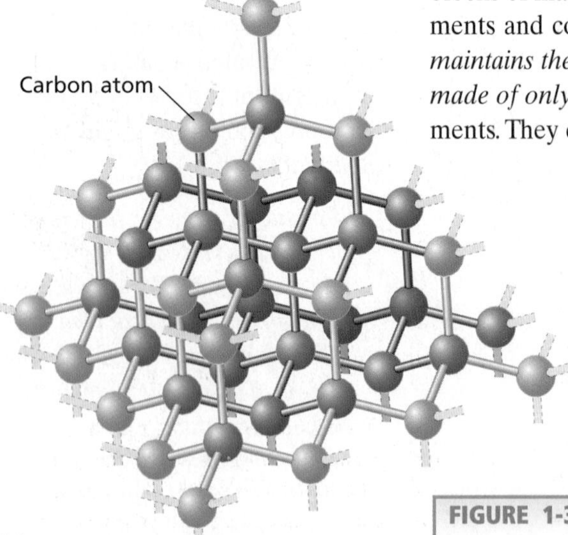

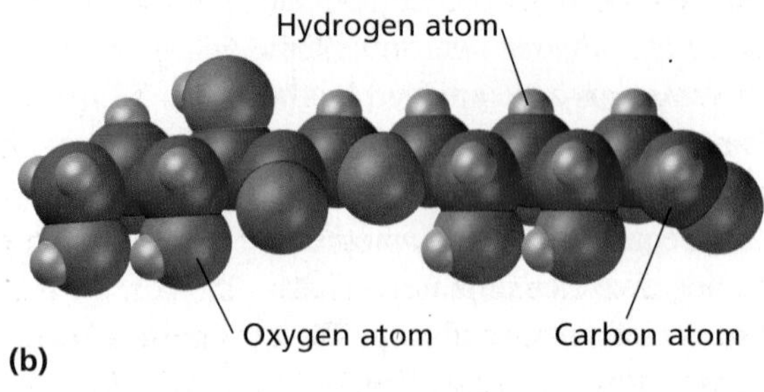

FIGURE 1-3 Both elements and compounds are made of atoms, as shown in these models of (a) diamond and (b) sucrose (table sugar).

A **compound** *is a substance that is made from the atoms of two or more elements that are chemically bonded.* Many compounds consist of molecules. Water is an example of a compound. It is made of two elements, hydrogen and oxygen. The atoms of hydrogen and oxygen are chemically bonded to form a water molecule. You will learn more about the particles that make up compounds when you study chemical bonding in Chapter 6. For now, you can think of a *molecule* as the smallest unit of an element or compound that retains all of the properties of that element or compound.

Properties and Changes in Matter

Every substance, whether it is an element or a compound, has characteristic properties. Chemists use properties to distinguish between substances and to separate them. Most chemical investigations are related to or depend on the properties of substances.

A property may be a characteristic that defines an entire group of substances. That property can be used to classify an unknown substance as a member of that group. For example, one large group of elements is the metals. The distinguishing property of metals is that they conduct electricity well. Therefore, if an unknown element is tested and found to conduct electricity well, it is a metal.

Properties also define subgroups of substances. These properties can then be used to classify substances as belonging to one subgroup or another. For example, members of the large group of organic compounds called sugars can be classified into one of two subgroups: reducing sugars or nonreducing sugars. A reagent called Fehling's solution can be added to an unknown sugar to determine the subgroup to which the sugar belongs. *Reagent* is another name for any chemical used in a chemical reaction. If the Fehling's solution reacts and forms a red solid, the sugar is a reducing sugar. If it does not, the sugar is a nonreducing sugar.

Properties can help reveal the identity of an unknown substance. However, conclusive identification usually cannot be made based on only one property. Comparisons of several properties can be used together to establish the identity of an unknown. Properties are either intensive or extensive. **Extensive properties** *depend on the amount of matter that is present.* Such properties include volume, mass, and the amount of energy in a substance. In contrast, **intensive properties** *do not depend on the amount of matter present.* Such properties include the melting point, boiling point, density, and ability to conduct electricity and heat. These properties are the same for a given substance regardless of how much of the substance is present. Properties can also be grouped into two general types: physical properties and chemical properties.

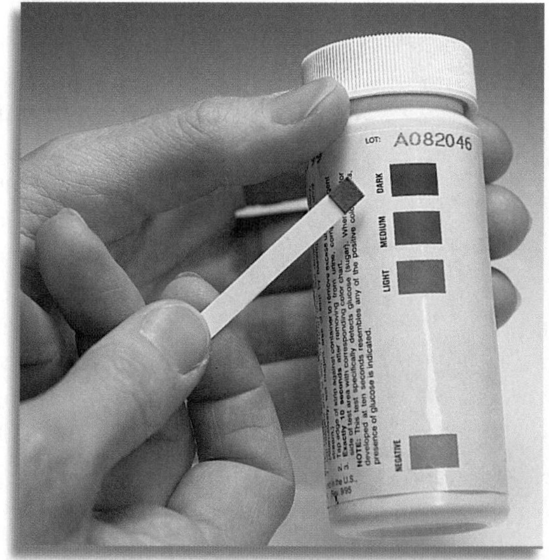

FIGURE 1-4 Benedict's solution is used to test for the presence of sugar in urine. The test strip is dipped into the sample. The test strip is matched to a color scale to determine the sugar level in the urine.

Physical Properties and Physical Changes

A **physical property** *is a characteristic that can be observed or measured without changing the identity of the substance.* Physical properties describe the substance itself, rather than describing how it can change into other substances. Examples of physical properties are melting point and boiling point. Those points are, respectively, the temperature at which a substance melts from solid to liquid and the temperature at which it boils from liquid to gas. For example, water melts from ice to liquid at 0°C (273 K or 32°F). Liquid water boils to vapor at 100°C (373 K or 212°F).

A change in a substance that does not involve a change in the identity of the substance is called a **physical change.** Examples of physical changes include grinding, cutting, melting, and boiling a material. These types of changes do not change the identity of the substance present.

Melting and boiling are part of an important class of physical changes called changes of state. As the name suggests, *a* **change of state** *is a physical change of a substance from one state to another.* The three common states of matter are solid, liquid, and gas. *Matter in the* **solid** *state has definite volume and definite shape.* For example, a piece of quartz or coal keeps its size and its shape, regardless of the container it is in. Solids have this characteristic because the particles in them are packed together in relatively fixed positions. The particles vibrate about a fixed point. *Matter in the* **liquid** *state has a definite volume but an indefinite shape;* a liquid assumes the shape of its container. For example, a given quantity of liquid water takes up a definite amount of space, but the water takes the shape of its container. Liquids have this characteristic because the particles in them are close together but can flow around one another. *Matter in the* **gas** *state has neither definite volume nor definite shape.* For example, a given quantity of helium expands to fill any size container and takes the shape of the container. All gases have this characteristic because they are composed of particles that are at great distances from one another compared with the particles of liquids and solids.

FIGURE 1-5 Water boils at 100°C no matter how much water is in the container. Boiling point is an intensive property.

Melting, the change from solid to liquid, is an example of a change of state. Boiling is a change of state from liquid to gas. Freezing, the opposite of melting, is the change from a liquid to a solid. A change of state does not affect the identity of the substance. For example, when ice melts to liquid water or when liquid water boils to form water vapor, the same substance, water, is still present, as shown in Figure 1-6. The water has simply changed state, but it has not turned into a different compound. Only the relative distances and interactions between the particles that make up water have changed.

Chemical Properties and Chemical Changes

Physical properties can be observed without changing the identity of the substance, but properties of the second type—chemical properties—cannot. *A* **chemical property** *relates to a substance's ability to undergo changes that transform it into different substances.* Chemical properties are easiest to see when substances react to form new substances.

FIGURE 1-6 Models for water in three states. The molecules are close together in the solid and liquid states but far apart in the gas state. The molecules in the solid state are relatively fixed in position, but those in the liquid and gas states can flow around each other.

For example, the ability of charcoal (carbon) to burn in air is a chemical property. When charcoal burns, it combines with oxygen in air to become a new substance, carbon dioxide gas. After the chemical change, the original substances, carbon and oxygen, are no longer present. A different substance with different properties has been formed. Other examples of chemical properties include the ability of iron to rust by combining with oxygen in air and the ability of silver to tarnish by combining with sulfur.

A change in which one or more substances are converted into different substances is called a **chemical change** *or* **chemical reaction.** *The substances that react in a chemical change are called the* **reactants.** *The substances that are formed by the chemical change are called the* **products.** In the case of burning charcoal, carbon and oxygen are the reactants in a combustion, or burning, reaction. Carbon dioxide is the product. The chemical change can be described as follows:

Carbon plus oxygen yields (or forms) carbon dioxide.

Arrows and plus signs can be substituted for the words *yields* and *plus,* respectively:

$$\text{carbon} + \text{oxygen} \longrightarrow \text{carbon dioxide}$$

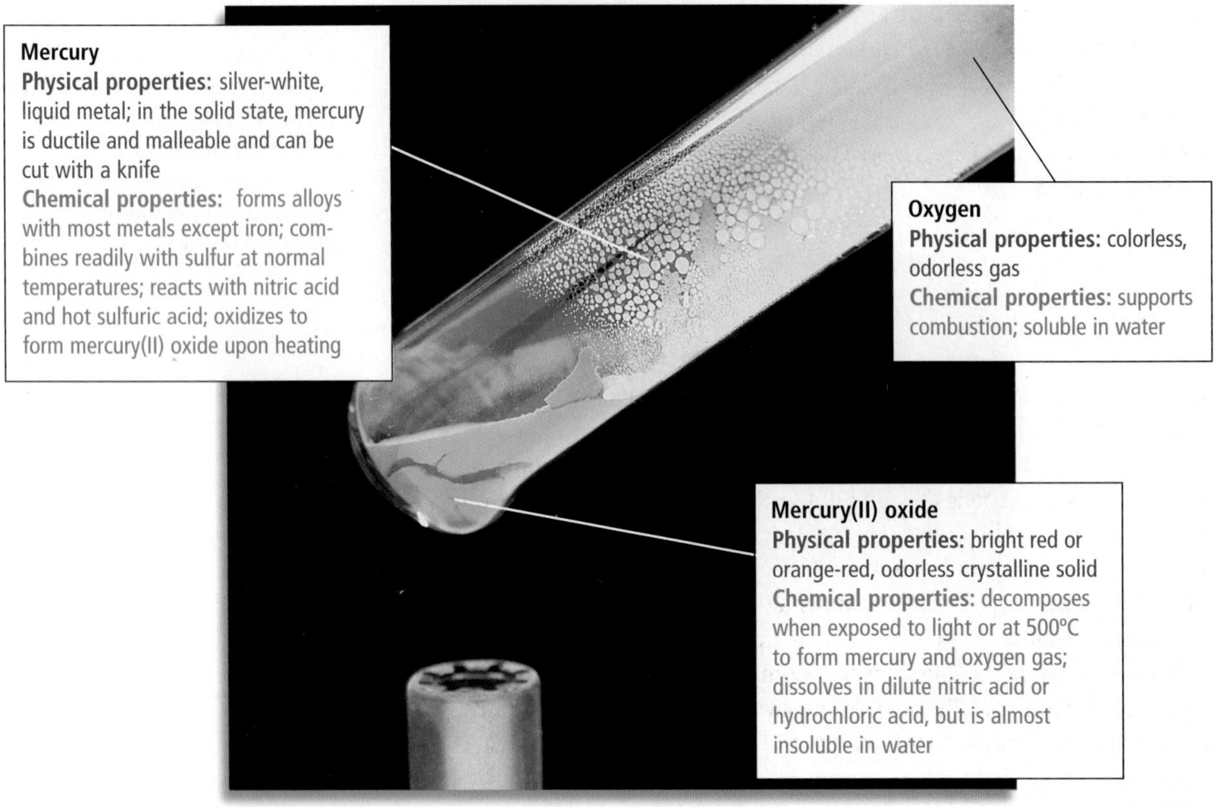

Mercury
Physical properties: silver-white, liquid metal; in the solid state, mercury is ductile and malleable and can be cut with a knife
Chemical properties: forms alloys with most metals except iron; combines readily with sulfur at normal temperatures; reacts with nitric acid and hot sulfuric acid; oxidizes to form mercury(II) oxide upon heating

Oxygen
Physical properties: colorless, odorless gas
Chemical properties: supports combustion; soluble in water

Mercury(II) oxide
Physical properties: bright red or orange-red, odorless crystalline solid
Chemical properties: decomposes when exposed to light or at 500°C to form mercury and oxygen gas; dissolves in dilute nitric acid or hydrochloric acid, but is almost insoluble in water

FIGURE 1-7 When mercury(II) oxide is heated, it decomposes to form oxygen gas and mercury (which can be seen on the side of the test tube). Decomposition is a chemical change that can be observed by comparing the properties of mercury(II) oxide, mercury, and oxygen.

The decomposition of the mercury compound shown in Figure 1-7 can be expressed as follows:

$$\text{mercury(II) oxide} \longrightarrow \text{mercury} + \text{oxygen}$$

Chemical changes and reactions, such as combustion and decomposition, form products whose properties differ greatly from those of the reactants. However, chemical changes do not affect the total amount of matter present before and after a reaction. The amount of matter, and therefore the total mass, remains the same.

Energy and Changes in Matter

When physical or chemical changes occur, energy is almost always involved. The energy can take several different forms, such as heat or light. Sometimes heat provides enough energy to cause a physical change, as in the melting of ice, and sometimes heat provides enough energy to cause a chemical change, as in the decomposition of water vapor to form oxygen gas and hydrogen gas. But the boundary between physical and chemical changes isn't always so clear. For example, while most chemists would consider the dissolving of sucrose in water to be a physical change, many chemists would consider the dissolving of table salt in water to be a chemical change. As you learn more about the structure of matter, you will better understand why the boundaries between chemical and physical changes can be confusing.

Although energy can be absorbed or released in a change, it is not destroyed or created. It simply assumes a different form. This is the law of conservation of energy. Accounting for all the energy present before and after a change is not a simple process. But scientists who have done such experimentation are confident that the total amount of energy remains the same.

Classification of Matter

The variety of forms in which matter exists is enormous. However, all matter can be classified into one of two groups: pure substances or mixtures. A pure substance can be an element or compound. The composition of a pure substance is the same throughout and does not vary from sample to sample. Mixtures, in contrast, contain more than one substance. They can vary in composition and properties from sample to sample and sometimes from one part of a sample to another part of the same sample. All matter, whether it is a pure substance or a mixture, can be classified in terms of uniformity of composition and properties of a given sample. Figure 1-8 illustrates the overall classification of matter into elements, compounds, and mixtures.

Mixtures

You deal with mixtures every day. Nearly every object around you, including most things you eat and drink and even the air you breathe, is a mixture. *A* **mixture** *is a blend of two or more kinds of matter, each*

FIGURE 1-8 This classification scheme for matter shows the relationships among mixtures, compounds, and elements.

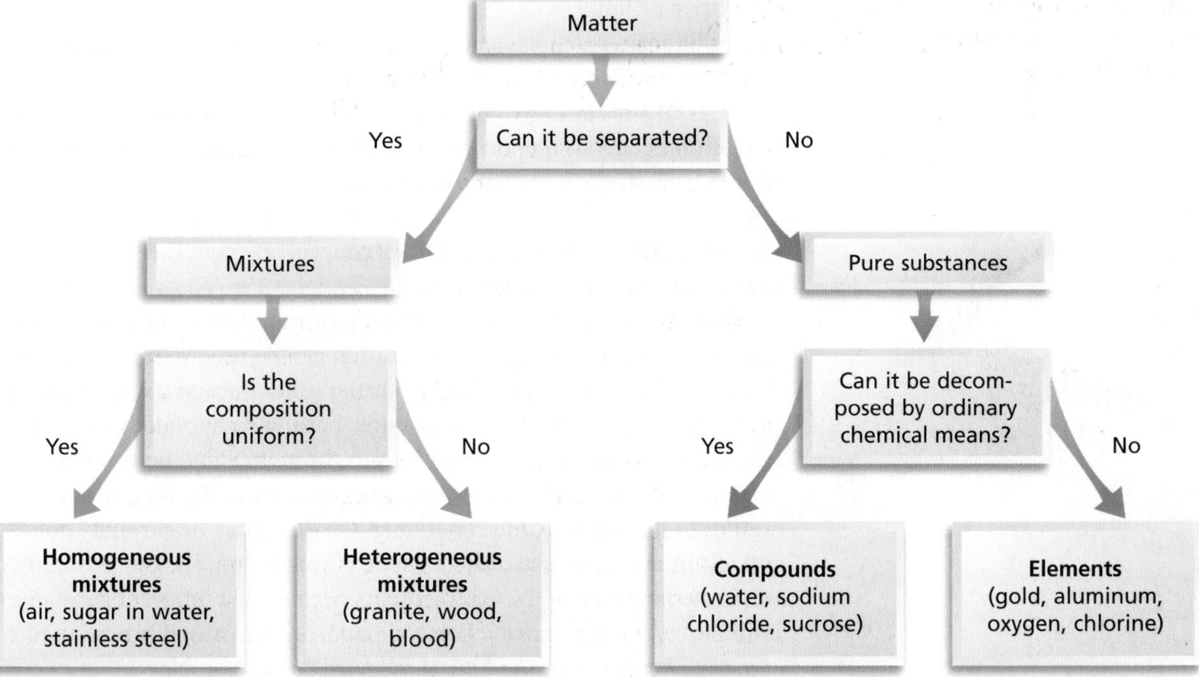

(a)

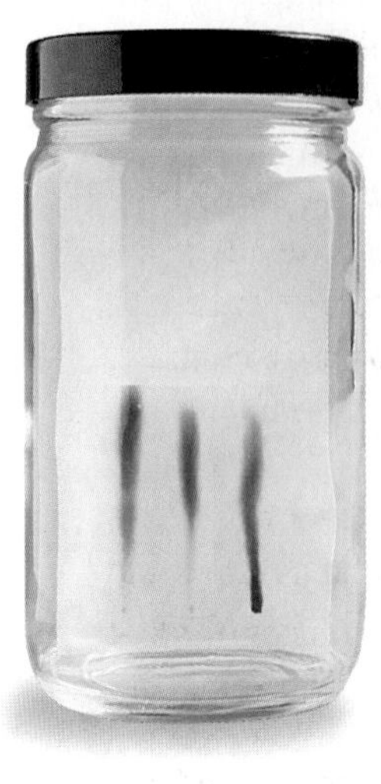
(b)

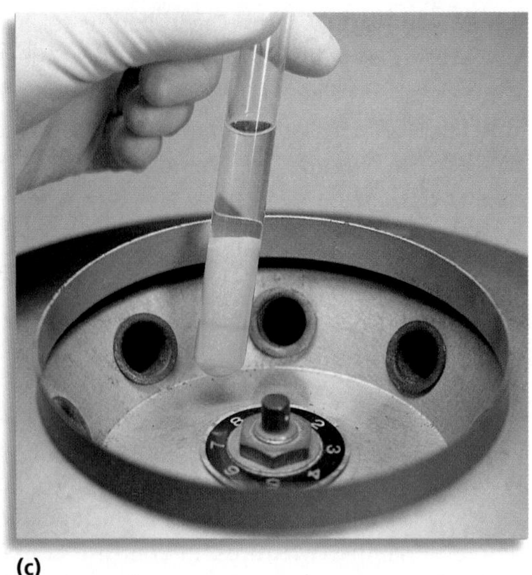
(c)

FIGURE 1-9 (a) Barium chromate can be separated from the solution in the beaker using filtration. (b) The components of an ink can be separated using paper chromatography. (c) A centrifuge can be used to separate certain solid components. The centrifuge spins rapidly, which pushes the solids to the bottom of the test tube.

of which retains its own identity and properties. The parts, or components, of a mixture are simply mixed together physically and can usually be separated. As a result, the properties of a mixture are a combination of the properties of its components. Because mixtures can contain various amounts of different substances, a mixture's composition must be specified. This is often done in terms of percentage by mass or by volume. For example, a mixture might be 5% sodium chloride and 95% water by mass.

Some mixtures are uniform in composition; that is, they are said to be *homogeneous*. They have the same proportion of components throughout. Homogeneous mixtures are also called solutions. A salt-water solution is an example of such a mixture. Other mixtures are not uniform throughout; that is, they are *heterogeneous*. For example, in a mixture of clay and water, heavier clay particles concentrate near the bottom of the container.

Some mixtures can be separated by filtration or vaporized to separate the different components. Filtration can be used to separate a mixture of solid barium chromate from the other substances, as shown in the beaker in Figure 1-9(a). The yellow barium compound is trapped by the filter paper, but the solution passes through. Some liquid-solid mixtures can be decanted to separate the liquid from the solid. If the solid settles to the bottom of the container, the liquid can be carefully poured off. Another technique, called paper chromatography, can be used to separate mixtures of dyes or pigments because the different substances will move at different rates on the paper. A centrifuge can be used to separate some solid-liquid mixtures, such as those in blood.

Pure Substances

In contrast to a mixture, a pure substance is homogenous as a single entity. *A* **pure substance** *has a fixed composition and differs from a mixture in the following ways:*

1. *Every sample of a given pure substance has exactly the same characteristic properties.* All samples of a pure substance have the same characteristic physical and chemical properties. These properties are so specific that they can be used to identify the substance. In contrast, the properties of a mixture depend on the relative amounts of the mixture's components.
2. *Every sample of a given pure substance has exactly the same composition.* Unlike mixtures, all samples of a pure substance have the same makeup. For example, pure water is always 11.2% hydrogen and 88.8% oxygen by mass.

Pure substances are either compounds or elements. A compound can be decomposed, or broken down, into two or more simpler compounds or elements by a chemical change. Water is a compound made of hydrogen and oxygen chemically bonded to form a single substance. Water can be broken down into hydrogen and oxygen through a chemical reaction called electrolysis, as shown in Figure 1-10(a).

Sucrose is made of carbon, hydrogen, and oxygen. Sucrose breaks down to form the other substances shown in Figure 1-10(b). Under intense heating, sucrose breaks down to produce carbon and water.

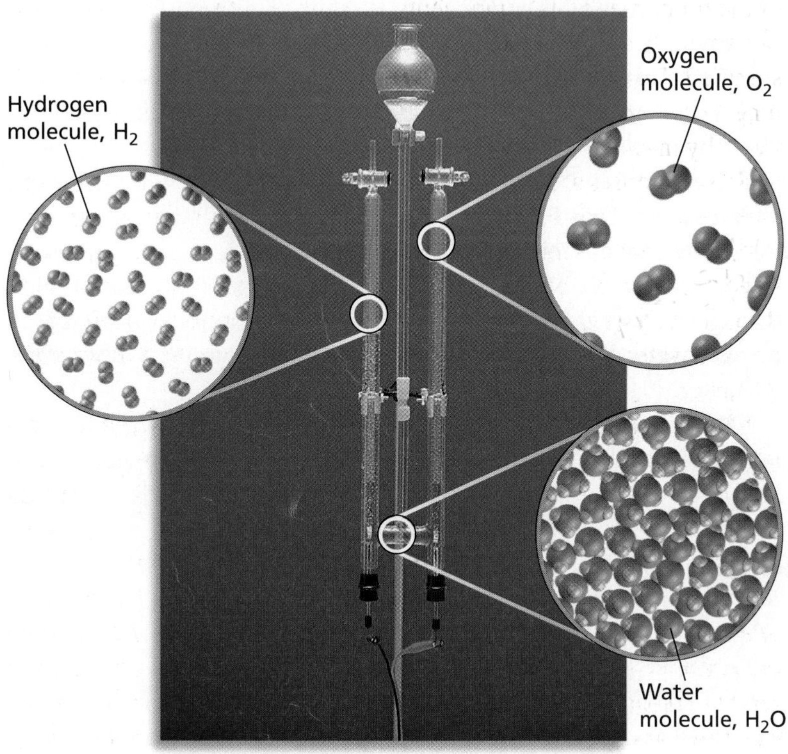

(a)

(b)

FIGURE 1-10 (a) Passing an electric current through water causes the compound to break down into the elements hydrogen and oxygen, which differ in composition from water. (b) When sucrose is heated, it caramelizes. When it is heated to a high temperature, it breaks down completely into carbon and water.

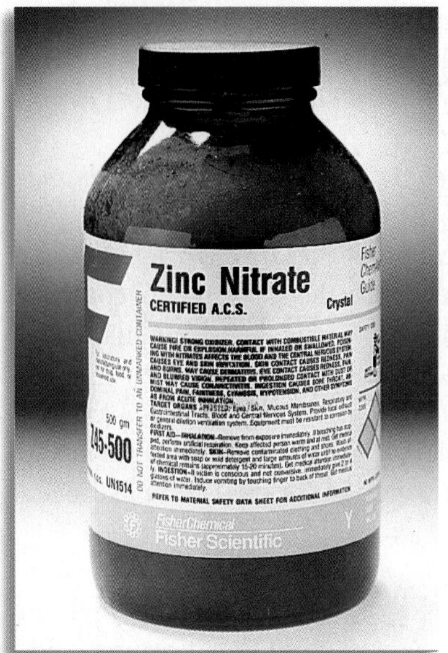

$Zn(NO_3)_2 \cdot 6H_2O$ F.W. 297.47

Certificate of Actual Lot Analysis

Acidity (as HNO_3)	0.008%
Alkalies and Earths	0.02%
Chloride (Cl)	0.005%
Insoluble Matter	0.001%
Iron (Fe)	0.0002%
Lead (Pb)	0.001%
Phosphate (PO_4)	0.0002%
Sulfate (SO_4)	0.002%

Store separately from and avoid contact with combustible materials. Keep container closed and in a cool, dry place. Avoid contact with skin, eyes and clothing.

LOT NO. 917356

FL-02-0588 CAS 10196-18-6

FIGURE 1-11 The labeling on a reagent bottle lists the grade of the reagent and the percentages of impurities for that grade. What grade is this chemical?

TABLE 1-1 *Some Grades of Chemical Purity*

Increasing purity (↑)
Primary standard reagents
ACS (American Chemical Society–specified reagents)
USP (United States Pharmacopoeia standards)
CP (chemically pure; purer than technical grade)
NF (National Formulary specifications)
FCC (Food Chemical Code specifications)
Technical (industrial chemicals)

Laboratory Chemicals and Purity

The chemicals in laboratories are generally treated as if they are pure. However, all chemicals have some impurities. Chemical grades of purity are listed in Table 1-1. The purity ranking of the grades can vary when agencies differ in their standards. For some chemicals, the USP grade may specify higher purity than the CP grade. For other chemicals, the opposite may be true. However, the primary standard reagent grade is always purer than the technical grade for the same chemical. Chemists need to be aware of the kinds of impurities in a reagent because these impurities could affect the results of a reaction. For example, the chemical label shown in Figure 1-11 shows the impurities for that grade. The chemical manufacturer must ensure that the standards set for that reagent by the American Chemical Society are met.

SECTION REVIEW

1. a. How do physical properties and chemical properties differ?
   b. Give an example of each.
2. Classify each of the following as either a physical change or a chemical change.
   a. tearing a sheet of paper
   b. melting a piece of wax
   c. burning a log
3. You are given a sample of matter to examine. How do you decide whether the sample is a solid, liquid, or gas?
4. Contrast mixtures with pure substances.

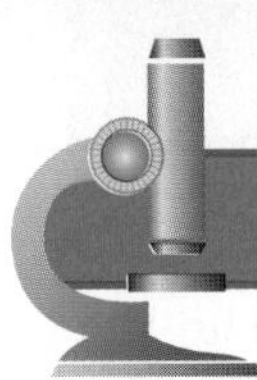

Secrets of the Cremona Violins

What are the most beautiful sounding of all violins? Most professionals will pick the instruments created in Cremona, Italy, following the Renaissance. At that time, Antonio Stradivari, the Guarneri family, and other designers created instruments of extraordinary sound that have yet to be matched. The craftsmen were notoriously secretive about their techniques, but, based on 20 years of research, Dr. Joseph Nagyvary, a professor of biochemistry at Texas A&M University, thinks he has discovered the key to the violins' sound hidden in the chemistry of their materials.

According to Dr. Nagyvary, Stradivarius instruments are nearly free of the shrill, high-pitched noises produced by modern violins. Generally, violin makers attribute this to the design of the instrument, but Dr. Nagyvary traces it to a different source. In Stradivari's day, wood for the violins was transported by floating it down a river from the mountains to Venice, where it was stored in sea water. Dr. Nagyvary first theorized that the soaking process could have removed ingredients from the wood that made it inherently noisy. His experiments revealed that microbes and minerals also permeated the wood, making their own contribution to the mellow musical sound.

Dr. Nagyvary and his violin

Dr. Nagyvary found other clues to the sound of the violins in the work of Renaissance alchemists. Aside from trying to turn lead into gold, alchemists made many useful experiments, including investigating different chemical means of preserving wood in musical instruments. Attempting to duplicate their techniques and to reproduce the effects of sea water, Dr. Nagyvary soaks all his wood in a "secret" solution. One of his favorite ingredients is a cherry-and-plum puree, which contains an enzyme called pectinase. The pectinase softens the wood, making it resonate more freely.

"The other key factor in a violin's sound," says Dr. Nagyvary, "is the finish, which is the filler and the varnish covering the instrument. Most modern finishes are made from rubbery materials, which limit the vibrations of the wood." Modern analysis has revealed that the Cremona finish was different: it was a brittle mineral microcomposite of a very sophisticated nature. According to historical accounts, all violin makers, including Stradivari, procured their varnishes from the local drugstore chemist, and they didn't even know what they were using! Dr. Nagyvary thinks this unknown and unsung drugstore chemist could have been the major factor behind the masterpieces for which violin makers like Stradivari received exclusive credit. By now, Dr. Nagyvary and his co-workers have identified most of the key ingredients of the Cremona finish.

Many new violins made from the treated wood and replicated finish have been made, and their sound has been analyzed by modern signal analyzers. These violins have been favorably compared with authentic Stradivari violins.

A number of expert violinists have praised the sound of Dr. Nagyvary's instruments, but some violin makers remain skeptical of the chemist's claims. They insist that it takes many years to reveal just how good a violin is. In the meantime, most everyone agrees that the art and science of violin making are still epitomized by the instruments of Cremona.

SECTION 1-3

Elements

Objectives

- Use a periodic table to name elements, given their symbols.
- Use a periodic table to write the symbols of elements, given their names.
- Describe the arrangement of the periodic table.
- List the characteristics that distinguish metals, nonmetals, and metalloids.

As you have read, elements are pure substances that cannot be decomposed by chemical changes. The elements serve as the building blocks of matter. Each element has characteristic properties. The elements are organized into groups based on similar chemical properties. This organization of elements is the *periodic table,* which is shown in Figure 1-12 on the next page.

Introduction to the Periodic Table

Each small square of the periodic table shows the name of one element and the letter symbol for the element. For example, the first square, at the upper left, represents element 1, hydrogen, which has the atomic symbol H.

As you look through the table, you will see many familiar elements, including iron, sodium, neon, silver, copper, aluminum, sulfur, and lead. You can often relate the symbols to the English names of the elements. Though some symbols are derived from the element's older name, which was often in Latin, wolfram comes from the German name for tungsten. Table 1-2 lists some of those elements.

TABLE 1-2 *Elements with Symbols Based on Older Names*

Modern name	Symbol	Older name
Antimony	Sb	stibium
Copper	Cu	cuprum
Gold	Au	aurum
Iron	Fe	ferrum
Lead	Pb	plumbum
Mercury	Hg	hydrargyrum
Potassium	K	kalium
Silver	Ag	argentum
Sodium	Na	natrium
Tin	Sn	stannum
Tungsten	W	wolfram

Periodic Table

Metals | Nonmetals | Metalloids | Noble Gases

Period	Group 1	Group 2	Group 3	Group 4	Group 5	Group 6	Group 7	Group 8	Group 9	Group 10	Group 11	Group 12	Group 13	Group 14	Group 15	Group 16	Group 17	Group 18
1	1 H																	2 He
2	3 Li	4 Be											5 B	6 C	7 N	8 O	9 F	10 Ne
3	11 Na	12 Mg											13 Al	14 Si	15 P	16 S	17 Cl	18 Ar
4	19 K	20 Ca	21 Sc	22 Ti	23 V	24 Cr	25 Mn	26 Fe	27 Co	28 Ni	29 Cu	30 Zn	31 Ga	32 Ge	33 As	34 Se	35 Br	36 Kr
5	37 Rb	38 Sr	39 Y	40 Zr	41 Nb	42 Mo	43 Tc	44 Ru	45 Rh	46 Pd	47 Ag	48 Cd	49 In	50 Sn	51 Sb	52 Te	53 I	54 Xe
6	55 Cs	56 Ba	57 La	72 Hf	73 Ta	74 W	75 Re	76 Os	77 Ir	78 Pt	79 Au	80 Hg	81 Tl	82 Pb	83 Bi	84 Po	85 At	86 Rn
7	87 Fr	88 Ra	89 Ac	104 Rf	105 Db	106 Sg	107 Bh	108 Hs	109 Mt									

58 Ce	59 Pr	60 Nd	61 Pm	62 Sm	63 Eu	64 Gd	65 Tb	66 Dy	67 Ho	68 Er	69 Tm	70 Yb	71 Lu
90 Th	91 Pa	92 U	93 Np	94 Pu	95 Am	96 Cm	97 Bk	98 Cf	99 Es	100 Fm	101 Md	102 No	103 Lr

FIGURE 1-12 The periodic table of elements. The names of the elements can be found on Table A-6 in the appendix.

The vertical columns of the periodic table are called **groups,** or **families.** Notice that they are numbered from 1 to 18 from left to right. Each group contains elements with similar chemical properties. For example, the elements in Group 2 are beryllium, magnesium, calcium, strontium, barium, and radium. All of these elements are reactive metals with similar abilities to bond to other kinds of atoms. The two major categories of elements are metals and nonmetals. Metalloids and noble gases are really nonmetals.

The horizontal rows of elements in the periodic table are called **periods.** Physical and chemical properties change somewhat regularly across a period. Elements that are close to each other in the same period tend to be more similar than elements that are farther apart. For example, in Period 2, the elements lithium and beryllium, in Groups 1 and 2, respectively, are somewhat similar in properties. However, they are very different from fluorine, the Period-2 element in Group 17.

The two sets of elements placed below the periodic table make up what are called the lanthanide series and the actinide series. These metallic elements fit into the table just after elements 57 and 89. They are placed below the table to keep the table from being too wide.

There is a section in the back of this book called the *Elements Handbook,* which covers some representative elements in greater detail. You will use information from the handbook to complete the questions in the Handbook Search sections in the chapter reviews.

Types of Elements

The periodic table is broadly divided into two main sections: metals and nonmetals. As you can see in Figure 1-12, the metals are at the left and in the center of the table. The nonmetals are toward the right. The elements along the dividing line show characteristics of both metals and nonmetals.

Metals

Some of the properties of metals may be familiar to you. For example, you can recognize metals by their shininess, or metallic luster. Perhaps the most important characteristic property of metals is the ease with which they conduct heat and electricity. Thus, *a* **metal** *is an element that is a good conductor of heat and electricity*.

At room temperature, most metals are solids. Most metals also have the property of *malleability,* that is, they can be hammered or rolled into thin sheets. Metals also tend to be *ductile,* which means that they can be drawn into a fine wire. Metals behave this way because they have high *tensile strength,* the ability to resist breaking when pulled.

Although all metals conduct electricity well, metals also have very diverse properties. Mercury is a liquid at room temperature, whereas tungsten has the highest melting point of any element. The metals in Group 1 are so soft that they can be cut with a knife, yet others, like chromium, are very hard. Some metals, such as manganese and bismuth, are very brittle, yet others, such as iron and copper, are very malleable and ductile. Most metals have a silvery or grayish white *luster*. Two exceptions are gold and copper, which are yellow and reddish brown, respectively. Figure 1-13 shows examples of metals.

FIGURE 1-13 (a) Gold has a low reactivity, which is why it may be found in nature in relatively pure form. (b) Copper is used in wiring because it is ductile and is an excellent conductor of electricity. (c) Aluminum is malleable. It can be rolled into foil that is used for wrapping food.

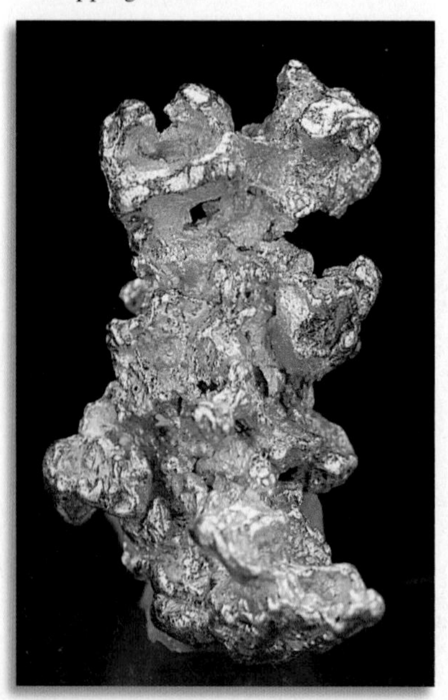

(a)

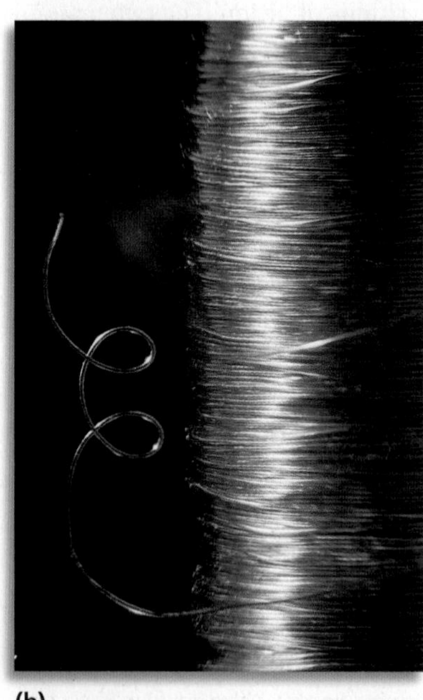

(b)

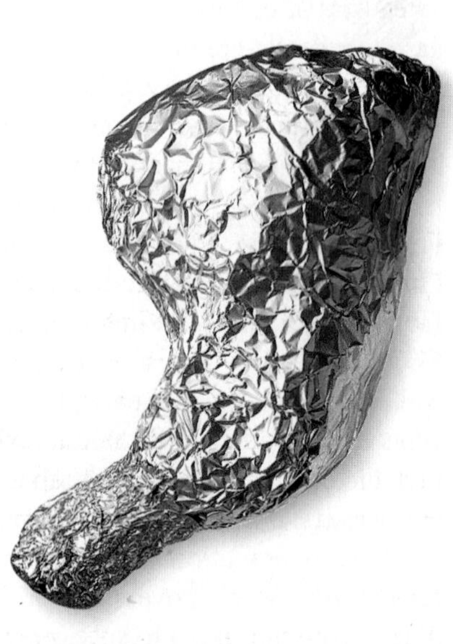

(c)

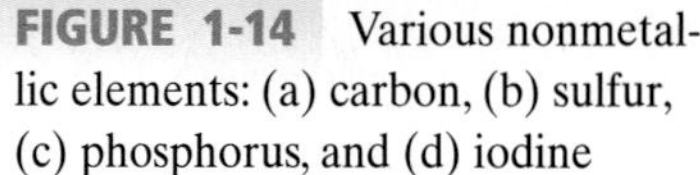

FIGURE 1-14 Various nonmetallic elements: (a) carbon, (b) sulfur, (c) phosphorus, and (d) iodine

Copper: A Representative Metal

Copper has a characteristic reddish color and a metallic luster. It is found naturally in minerals such as chalcopyrite and malachite. Pure copper melts at 1083°C and boils at 2567°C. It can be readily drawn into fine wire, pressed into thin sheets, and formed into tubing. Copper conducts electricity with little loss of energy.

Copper remains unchanged in pure, dry air at room temperature. When heated, it reacts with oxygen in air. It also reacts with sulfur and the elements in Group 17 of the periodic table. The green coating on a piece of weathered copper comes from the reaction of copper with oxygen, carbon dioxide, and sulfur compounds. Copper is an essential mineral in the human diet.

Nonmetals

Many nonmetals are gases at room temperature. These include nitrogen, oxygen, fluorine, and chlorine. One nonmetal, bromine, is a liquid. The solid nonmetals include carbon, phosphorus, selenium, sulfur, and iodine. These solids tend to be brittle rather than malleable and ductile. Some nonmetals are illustrated in Figure 1-14.

Low conductivity can be used to define nonmetals. *A* **nonmetal** *is an element that is a poor conductor of heat and electricity.* If you look at Figure 1-12, you will see that there are fewer nonmetals than metals.

Phosphorus: A Representative Nonmetal

Phosphorus is one of five solid nonmetals. Pure phosphorus is known in two common forms. Red phosphorus is a dark red powder that melts at 597°C. White phosphorus is a waxy solid that melts at 44°C. Because it ignites in air, white phosphorus is stored underwater.

Phosphorus is too reactive to exist in pure form in nature. It is present in huge quantities in phosphate rock, where it is combined with oxygen and calcium. All living things contain phosphorus.

FIGURE 1-15 Selenium is a nonmetal, though it looks metallic.

Metalloids

A stair-step line separates the metals from the nonmetals on the periodic table. The elements in the area of this line are often referred to as

FIGURE 1-16 The noble gases helium, neon, argon, krypton, and xenon are all used to make lighted signs of various colors.

metalloids. *A* **metalloid** *is an element that has some characteristics of metals and some characteristics of nonmetals.* All metalloids are solids at room temperature. They tend to be less malleable than metals but not as brittle as nonmetals. Some metalloids, such as antimony, have a somewhat metallic luster.

Metalloids tend to be semiconductors of electricity. That is, their ability to conduct electricity is intermediate between that of metals and that of nonmetals. Metalloids are used in the semiconducting materials found in desktop computers, hand-held calculators, digital watches, televisions, and radios.

Noble Gases

The elements in Group 18 of the periodic table are the noble gases. These elements are generally unreactive. In fact, prior to 1962 no noble gas compounds had been identified. That year, the first noble gas compound, xenon tetrafluoride, was prepared. Their low reactivity sets noble gases apart from the other families of elements. Group-18 elements are gases at room temperature. Neon, argon, krypton, and xenon are all used in lighting. Helium is used in party balloons and weather balloons because it is less dense than air.

SECTION REVIEW

1. Use the periodic table on the inside back cover to write the names for the elements that have the following symbols: O, S, Cu, Ag.
2. Use the periodic table to write the symbols for the following elements: iron, nitrogen, calcium, mercury.
3. Which elements are most likely to undergo the same kinds of reactions, those in a group or those in a period?
4. Describe the main differences between metals, nonmetals, and metalloids.

CHAPTER 1 REVIEW

CHAPTER SUMMARY

1-1

- Chemistry is the study of the composition, structure, and properties of matter and its changes.
- Chemistry is classified as physical science. Six areas of study in chemistry are organic chemistry, inorganic chemistry, physical chemistry, analytical chemistry, biochemistry, and theoretical chemistry.
- A chemical is any substance that has a definite composition or is used or produced in a chemical process.
- Basic research is carried out for the sake of increasing knowledge. Applied research is carried out to solve practical problems. Technological development involves the use of existing knowledge to make life easier or more convenient.

Vocabulary

chemical (6)
chemistry (5)

1-2

- All matter has mass and takes up space. Mass is one measure of the amount of matter.
- An element is composed of one kind of atom. Compounds are made from two or more elements. Any pure compound has fixed proportions of elements.
- All substances have characteristic properties that enable chemists to tell the substances apart and to separate them.
- The physical properties of a substance can be observed or measured without changing the identity of the substance. Physical changes do not involve changes in identity.
- The three major states of matter are solid, liquid, and gas. The particles in these states differ in proximity to one another and ease of flow. Changes of state, such as melting and boiling, are physical changes.
- Chemical properties refer to a substance's ability to undergo changes that alter its composition and identity. Chemical changes, or chemical reactions, involve changes in identity.
- Energy changes accompany physical and chemical changes. Energy may be released or absorbed, but it is neither created nor destroyed.
- Matter can be classified into mixtures and pure substances. Pure substances differ from mixtures in that they have a definite composition that does not vary. Solutions are homogeneous mixtures.

Vocabulary

atom (10)
change of state (12)
chemical change (13)
chemical property (12)
chemical reaction (13)
compound (11)
element (10)
extensive property (11)
gas (12)
intensive property (11)
liquid (12)
mass (10)
matter (10)
mixture (15)
physical change (12)
physical property (12)
product (13)
pure substance (17)
reactant (13)
solid (12)

1-3

- Each element has a unique symbol. The periodic table shows the elements organized by their chemical properties. Columns on the table represent groups or families of elements with similar chemical properties. Properties vary across the rows, or periods.
- The elements can be classified as metals, nonmetals, metalloids, and noble gases. These classes occupy different areas of the periodic table. Metals tend to be shiny, malleable, ductile, and good conductors. Nonmetals tend to be brittle and poor conductors. Metalloids are intermediate in properties between metals and nonmetals, and they tend to be semiconductors of electricity. The noble gases are generally unreactive elements.

Vocabulary

family (21)
group (21)
metal (22)
metalloid (24)
nonmetal (23)
period (21)

REVIEWING CONCEPTS

1. What is chemistry? (1-1)
2. What branch of chemistry is most concerned with the study of carbon compounds. (1-1)
3. What is meant by the word *chemical*, as used by scientists? (1-1)
4. Briefly describe the differences between basic research, applied research, and technological development. Provide an example of each. (1-1)
5. a. What is mass?
   b. What is volume? (1-2)
6. How does the composition of a pure compound differ from that of a mixture? (1-2)
7. a. Define *property*.
   b. How are properties useful in classifying materials? (1-2)
8. What is the difference between extensive properties and intensive properties? (1-2)
9. a. Define *physical property*.
   b. List two examples of physical properties. (1-2)
10. a. Define *chemical property*.
    b. List two examples of chemical properties. (1-2)
11. Distinguish between a *physical change* and a *chemical change*. (1-2)
12. a. Name the three common states of matter.
    b. How does a solid differ from a liquid?
    c. How does a liquid differ from a gas?
    d. How is a liquid similar to a gas? (1-2)
13. What is meant by a change in state? (1-2)
14. How are elements organized in the periodic table? (1-3)
15. Compare the properties of metals, nonmetals, metalloids, and noble gases. (1-3)
16. In which of the six branches of chemistry would a scientist be working if he or she were doing the following: (1-1)
    a. designing new compounds using computer modeling
    b. investigating energy relationships for various reactions
    c. comparing properties of alcohols with those of sugars
    d. studying reactions that occur during the digestion of food
    e. carrying out tests to identify unknown substances
17. Identify the reactants and products in the following reaction: (1-2)
    potassium + water $\longrightarrow$
    potassium hydroxide + hydrogen
18. Suppose element *X* is a poor conductor of electricity and breaks when hit with a hammer. Element *Z* is a good conductor of electricity and heat. In what area of the periodic table does each element most likely belong? (1-3)
19. Identify each of the following as either a physical change or a chemical change. Explain your answers. (1-2)
    a. A piece of wood is sawed in half.
    b. Milk turns sour.
    c. Melted butter solidifies in the refrigerator.
20. Use the periodic table to write the names of the elements that have the following symbols, and identify each as a metal, nonmetal, metalloid, or noble gas. (1-3)
    a. K c. Si e. Hg
    b. Ag d. Na f. He
21. An unknown element is shiny and is found to be a good conductor of electricity. What other properties would you predict the element to have? (1-3)
22. Identify each of the following as an example of either basic research, applied research, or technological development: (1-1)
    a. A new type of refrigerant is developed that is less damaging to the environment.
    b. A new element is synthesized in a particle accelerator.
    c. A computer chip is redesigned to increase the speed of the computer.
23. Use the periodic table to identify the group numbers and period numbers of the following elements: (1-3)
    a. carbon, C d. barium, Ba
    b. argon, Ar e. iodine, I
    c. chromium, Cr f. gold, Au

24. a. Suppose different parts of a sample material have different compositions. What can you conclude about the material? (1-2)

b. Suppose different parts of a sample have the same composition. What can you conclude about the material? Explain your answer. (1-2)

TECHNOLOGY & LEARNING

25. Graphing Calculator Graphing Tabular Data
The graphing calculator can be programmed to graph ordered pairs of data, such as temperature versus time. In this problem you will learn how to create a table of data. Then you will learn how to program the calculator to plot the data.

A. Create lists L_1 and L_2.
Keystrokes for creating lists: [STAT] [4] [2nd] [L1] [,] [2nd] [L2] [ENTER] [STAT] [1]. Enter the time data in L_1, and enter the temperature data in L_2.

Time (min)	Temperature (°C)
1	28.3
3	32.2
5	36.5
8	40.1
10	42.7
13	45.8

B. Program the calculator to graph.
Keystrokes for naming the program: [PRGM] [▶] [▶] [ENTER]
Name the program: [G] [R] [A] [P] [H] [ENTER]
Program keystrokes: [2nd] [STAT PLOT] [1] [2nd] [STAT PLOT] [▶] [1] [,] [2nd] [L1] [,] [2nd] [L2] [,] [2nd] [STAT PLOT] [▶] [▶] [1] [)] [2nd] [:] [ZOOM] [9] [.]

C. Check the display.
Your screen display should look like the following. If it does not, start again.
Plot1(Scatter, L1,L2,): ZoomStat
Press [2nd] [QUIT] to exit the program editor.

D. Run the program.
a. Press [PRGM]. Select GRAPH and press [ENTER] [ENTER]. The calculator will provide a graph of the data.
b. Press [TRACE] and [▶] to move from one data point to the next. The coordinates of the data point will be shown at the bottom of the screen.

c. Use the GRAPH program with the data set shown below. Refer back to creating L_1 and L_2. Use L_1 for time data and L_2 for temperature data. Now use the program to graph these data.

Time (min)	Temperature (°C)
1	23.3
3	21.2
5	19.5
8	18.1
10	15.7
13	14.8

HANDBOOK SEARCH

26. Review the information on trace elements in the *Elements Handbook* in the back of this text.
a. What are the functions of trace elements in the body?
b. What transition metal plays an important role in oxygen transport throughout the body?
c. What two Group 1 elements are part of the electrolyte balance in the body?

RESEARCH & WRITING

27. Research any current technological product of your choosing. Find out about its manufacture and uses. Also find out about the basic research and applied research that made its development possible.

ALTERNATIVE ASSESSMENT

28. Make a list of all the changes that you see around you involving matter during a one-hour period. Note whether each change seems to be a physical change or a chemical change. Give reasons for your answers.

29. Make a concept map using at least 15 terms from the vocabulary lists. An introduction to concept mapping is found in Appendix B of this book.

CHAPTER 2

Measurements and Calculations

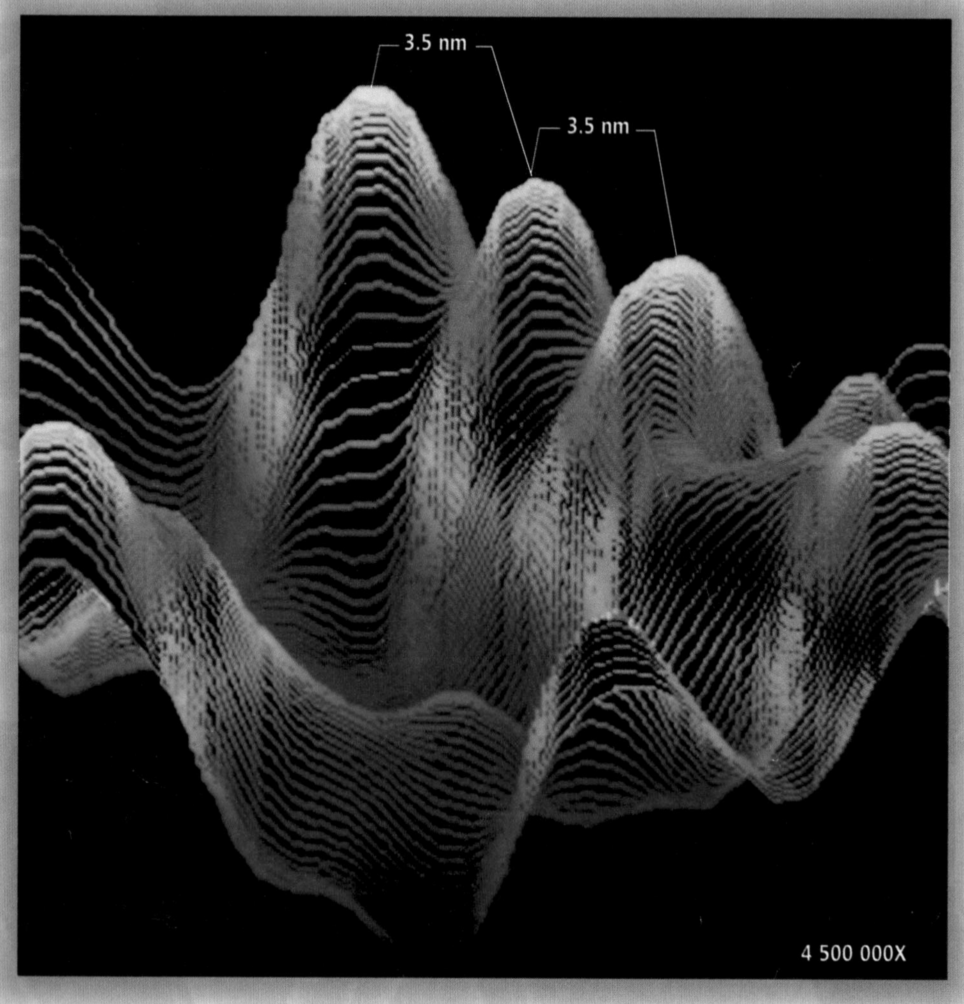

Measurements provide quantitative information.

Scientific Method

SECTION 2-1

OBJECTIVES

- Describe the purpose of the scientific method.
- Distinguish between qualitative and quantitative observations.
- Describe the differences between hypotheses, theories, and models.

Sometimes progress in science comes about through accidental discoveries. However, most scientific advances result from carefully planned investigations. The process researchers use to carry out their investigations is often called the scientific method. *The* **scientific method** *is a logical approach to solving problems by observing and collecting data, formulating hypotheses, testing hypotheses, and formulating theories that are supported by data.*

Observing and Collecting Data

Observing is the use of the senses to obtain information. Observation often involves making measurements and collecting data. The data may be descriptive (qualitative) or numerical (quantitative) in nature. Numerical information, such as the fact that a sample of copper ore has a mass of 25.7 grams, is *quantitative*. Non-numerical information, such as the fact that the sky is blue, is *qualitative*.

Experimenting involves carrying out a procedure under controlled conditions to make observations and collect data. To learn more about matter, chemists study systems. *A* **system** *is a specific portion of matter in a given region of space that has been selected for study during an experiment or observation.* When you observe a reaction in a test tube, the test tube and its contents form a system.

FIGURE 2-1 These students are designing an experiment to determine how to get the largest volume of popped corn from a fixed number of kernels. They think that the volume is likely to increase as the moisture in the kernels increases. Their experiment will involve soaking some kernels in water and observing whether the volume of the popped corn is greater than that of corn popped from kernels that have not been soaked.

FIGURE 2-2 A graph of data can show relationships between two variables. In this case the graph shows data collected during an experiment to determine the effect of phosphorus fertilizer compounds on plant growth. The following is one possible hypothesis: *If* phosphorus stimulates corn-plant growth, *then* corn plants treated with a soluble phosphorus compound should grow faster, under the same conditions, than corn plants that are not treated.

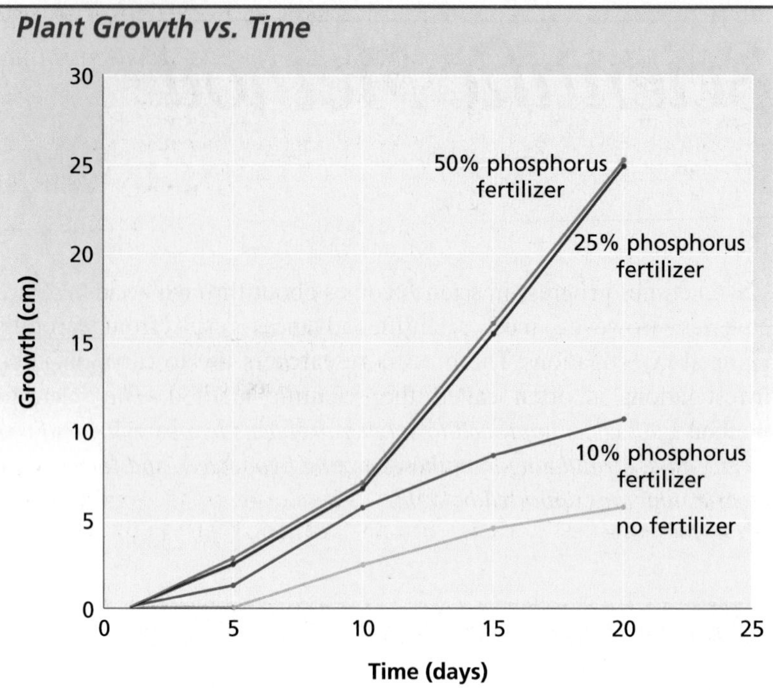

Formulating Hypotheses

As scientists examine and compare the data from their own experiments, they attempt to find relationships and patterns—in other words, they make generalizations based on the data. Generalizations are statements that apply to a range of information. To make generalizations, data are sometimes organized in tables and analyzed using statistics or other mathematical techniques, often with the aid of graphs and a computer.

Scientists use generalizations about the data to formulate a **hypothesis,** *or testable statement.* The hypothesis serves as a basis for making predictions and for carrying out further experiments. Hypotheses are often drafted as "if-then" statements. The "then" part of the hypothesis is a prediction that is the basis for testing by experiment. Figure 2-2 shows data collected to test a hypothesis.

Testing Hypotheses

Testing a hypothesis requires experimentation that provides data to support or refute a hypothesis or theory. Do the data in Figure 2-2 support the hypothesis? If testing reveals that the predictions were not correct, the generalizations on which the predictions were based must be discarded or modified. One of the most difficult yet most important aspects of science is rejecting a hypothesis that is not supported by data.

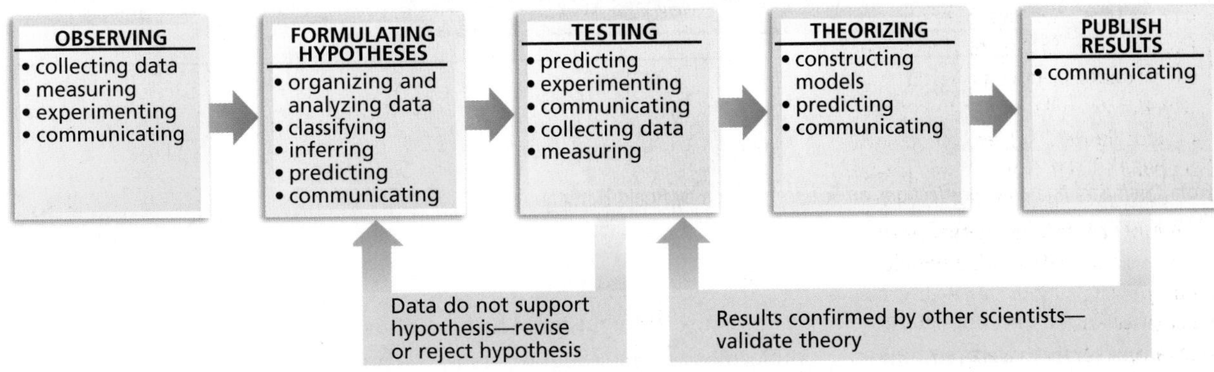

FIGURE 2-3 The scientific method is not a stepwise process. Scientists may repeat steps many times before there is sufficient evidence to formulate a theory. You can see that each stage represents a number of different activities.

Theorizing

When the data from experiments show that the predictions of the hypothesis are successful, scientists typically try to explain the phenomena they are studying by constructing a model. *A* **model** *in science is more than a physical object; it is often an explanation of how phenomena occur and how data or events are related.* Models may be visual, verbal, or mathematical. One of the most important models in chemistry is the atomic model of matter, which explains that matter is composed of tiny particles called atoms.

If a model successfully explains many phenomena, it may become part of a theory. The atomic model is a part of the atomic theory, which you will study in Chapter 3. *A* **theory** *is a broad generalization that explains a body of facts or phenomena.* Theories are considered successful if they can predict the results of many new experiments. Examples of the important theories you will study in chemistry are kinetic-molecular theory and collision theory. Figure 2-3 shows where theory fits in the scheme of the scientific method.

SECTION REVIEW

1. What is the scientific method?
2. Which of the following are quantitative?
   a. the liquid floats on water
   b. the metal is malleable
   c. the liquid has a temperature of 55.6°C
3. How do hypotheses and theories differ?
4. How are models related to theories and hypotheses?
5. What are the components of the system in the graduated cylinder shown on page 38?

CHEMICAL COMMENTARY

Chemistry's Holy Grail

From *Chemistry Imagined, Reflections on Science,* written by Roald Hoffmann and illustrated by Vivian Torrence

In other fields the important questions seem to be breathtakingly simple: How does the brain work? Let's land a manned space craft on Mars . . . What is the "cure" for cancer? The outsider romanticizes, to be sure. But what about chemistry, where is the Holy Grail of the molecular science? . . .

To search for the Holy Grail, 150 of King Arthur's knights committed their hearts and resources. Translated to modern times, a like group effort demands some gigantic, or at least intricately expensive, machinery. Big science, in other words: a supercollider to search the innards, a space telescope to probe the outer fringes, a genome project to map human heredity. None of these is typical of chemistry, where from the beginning small groupings of people, working with relatively cheap cookware, have transformed a wondrous variety of matter.

Chemistry is an intermediate science. Its universe is defined not by reduction to a few elementary particles, or even to the hundred or so elements, but by a reaching out to the infinities of molecules that can be synthesized. A registry of new molecules contains over ten million man-made entries. A small fraction of these is of natural origin, though millions are waiting to be analyzed. And millions more are lost in species that our ecological pressure extinguishes. Most molecules are man- and woman-made. The beauty I would claim for chemistry is that of richness and complexity, the realm of the possible. There is no end to the range of structure and function that molecules exhibit. . . .

Roald Hoffmann

There is no Holy Grail in chemistry. Yes, we would like to have a magic machine that separates the most awful mixture, purifies every component to 99.4% purity (or better if we pay more) and determines the precise arrangement of atoms in space in each molecule. Yes, we'd like to know in complete detail the resistance of a molecule to every twist, bend, stretch, rock, and roll. And, yes, we certainly must espy the secret, rapid motions molecules undergo in their most intimate transformations. And, most of all, most fundamental to the science of transformations, we desire control—ways to synthesize to order, in a short time, using cheap materials, in one pot, any molecule in the world.

The secret of the Holy Grail is that it is to be found not in the consummation but in the search. Imagine every woman rejuvenated, every man saved, all ills, physical and mental, cured, all humanity perfect, and, of course, at peace. What a dull world! . . .

If the grand desires of chemistry were achieved—to know what one has, how things happen on the molecular scale, how to create molecules with absolute control—chemistry would simply vanish. To come to terms with complexity and the never-ending search, to find joy and beauty in the plain thing, the small step—that is the grail.

Reading for Meaning

What does Hoffmann mean when he says that chemistry is an intermediate science?

Read Further

Hoffmann states that, most of all, chemists desire ways to control chemical reactions. Research three ways that chemists control reactions, and write a paragraph describing each.

Units of Measurement

SECTION 2-2

Objectives

- Distinguish between a quantity, a unit, and a measurement standard.
- Name SI units for length, mass, time, volume, and density.
- Distinguish between mass and weight.
- Perform density calculations.
- Transform a statement of equality to a conversion factor.

Measurements are quantitative information. A measurement is more than just a number, even in everyday life. Suppose a chef were to write a recipe listing quantities such as 1 salt, 3 sugar, and 2 flour. The cooks could not use the recipe without more information. They would need to know whether the number 3 represented teaspoons, tablespoons, cups, ounces, grams, or some other unit for sugar.

Measurements *represent* quantities. *A* **quantity** *is something that has magnitude, size, or amount.* A quantity is not the same as a measurement. For example, the quantity represented by a teaspoon is volume. The teaspoon is a unit of measurement, while volume is a quantity. A teaspoon is a measurement standard in this country. Units of measurement compare what is to be measured with a previously defined size. Nearly every measurement is a number plus a unit. The choice of unit depends on the quantity being measured.

Many centuries ago, people sometimes marked off distances in the number of foot lengths it took to cover the distance. But this system was unsatisfactory because the number of foot lengths used to express a distance varied with the size of the measurer's foot. Once there was agreement on a standard for foot length, confusion as to the real length was eliminated. It no longer mattered who made the measurement, as long as the standard measuring unit was correctly applied.

SI Measurement

Scientists all over the world have agreed on a single measurement system called *Le Système International d'Unités*, abbreviated **SI.** This system was adopted in 1960 by the General Conference on Weights and Measures. SI has seven base units, and most other units are derived from these seven. Some non-SI units are still commonly used by chemists and are also used in this book.

SI units are defined in terms of standards of measurement. The standards are objects or natural phenomena that are of constant value, easy to preserve and reproduce, and practical in size. International organizations monitor the defining process. In the United States, the National Institute of Standards and Technology plays the main role in maintaining standards and setting style conventions. For example, numbers are written in a form that is agreed upon internationally. The number seventy-five thousand is written 75 000, not 75,000, because the comma is used in other countries to represent a decimal point.

TABLE 2-1 SI Base Units

Quantity	Quantity symbol	Unit name	Unit abbreviation	Defined standard
Length	l	meter	m	the length of the path traveled by light in a vacuum during a time interval of 1/299 792 458 of a second
Mass	m	kilogram	kg	the unit of mass equal to the mass of the international prototype of the kilogram
Time	t	second	s	the duration of 9 192 631 770 periods of the radiation corresponding to the transition between the two hyperfine levels of the ground state of the cesium-133 atom
Temperature	T	kelvin	K	the fraction 1/273.16 of the thermodynamic temperature of the triple point of water
Amount of substance	n	mole	mol	the amount of substance of a system which contains as many elementary entities as there are atoms in 0.012 kilogram of carbon-12
Electric current	I	ampere	A	the constant current which, if maintained in two straight parallel conductors of infinite length, of negligible circular cross section, and placed 1 meter apart in vacuum, would produce between these conductors a force equal to 2×10^{-7} newton per meter of length
Luminous intensity	I_v	candela	cd	the luminous intensity, in a given direction, of a source that emits monochromatic radiation of frequency 540×10^{12} hertz and that has a radiant intensity in that direction of 1/683 watt per steradian

SI Base Units

The seven SI base units and their standard abbreviated symbols are listed in Table 2-1. All the other SI units can be derived from the fundamental units.

Prefixes added to the names of SI base units are used to represent quantities that are larger or smaller than the base units. Table 2-2 lists SI prefixes using units of length as examples. For example, the prefix *centi-*, abbreviated c, represents an exponential factor of 10^{-2}, which equals 1/100. Thus, 1 centimeter, 1 cm, equals 0.01 m, or 1/100 of a meter.

Mass

As you learned in Chapter 1, mass is a measure of the quantity of matter. The SI standard unit for mass is the kilogram. The standard for mass defined in Table 2-1 is used to calibrate balances all over the world.

TABLE 2-2 *SI Prefixes*

Prefix	Unit abbreviation	Exponential factor	Meaning	Example
Tera	T	10^{12}	1 000 000 000 000	1 terameter (Tm) = 1×10^{12} m
Giga	G	10^{9}	1 000 000 000	1 gigometer (Gm) = 1×10^{9} m
Mega	M	10^{6}	1 000 000	1 megameter (Mm) = 1×10^{6} m
Kilo	k	10^{3}	1000	1 kilometer (km) = 1000 m
Hecto	h	10^{2}	100	1 hectometer (hm) = 100 m
Deka	da	10^{1}	10	1 decameter (dam) = 10 m
		$\mathbf{10^{0}}$	**1**	**1 meter (m)**
Deci	d	10^{-1}	1/10	1 decimeter (dm) = 0.1 m
Centi	c	10^{-2}	1/100	1 centimeter (cm) = 0.01 m
Milli	m	10^{-3}	1/1000	1 millimeter (mm) = 0.001 m
Micro	μ	10^{-6}	1/1 000 000	1 micrometer (μm) = 1×10^{-6} m
Nano	n	10^{-9}	1/1 000 000 000	1 nanometer (nm) = 1×10^{-9} m
Pico	p	10^{-12}	1/1 000 000 000 000	1 picometer (pm) = 1×10^{-12} m
Femto	f	10^{-15}	1/1 000 000 000 000 000	1 femtometer (fm) = 1×10^{-15} m
Atto	a	10^{-18}	1/1 000 000 000 000 000 000	1 attometer (am) = 1×10^{-18} m

The mass of a typical textbook is about 1 kg. The gram, g, which is 1/1000 of a kilogram, is more useful for measuring masses of small objects, such as flasks and beakers. For even smaller objects, such as tiny quantities of chemicals, the milligram, mg, is often used. One milligram is 1/1000 of a gram, or 1/1 000 000 of a kilogram.

Mass is often confused with weight because people often express the weight of an object in grams. Mass is determined by comparing the mass of an object with a set of standard masses that are part of the balance. **Weight** *is a measure of the gravitational pull on matter.* Unlike weight, mass does not depend on such an attraction. Mass is measured on instruments such as a balance, and weight is typically measured on a spring scale. Taking weight measurements involves reading the amount that an object pulls down on a spring. As the force of Earth's gravity on an object increases, the object's weight increases. The weight of an object on the moon is about one-sixth of its weight on Earth.

Length

The SI standard unit for length is the meter. A distance of 1 m is about the width of an average doorway. To express longer distances, the kilometer, km, is used. One kilometer equals 1000 m. Road signs in the United States sometimes show distances in kilometers as well as miles. The kilometer is the unit used to express highway distances in most other countries of the world. To express shorter distances, the centimeter

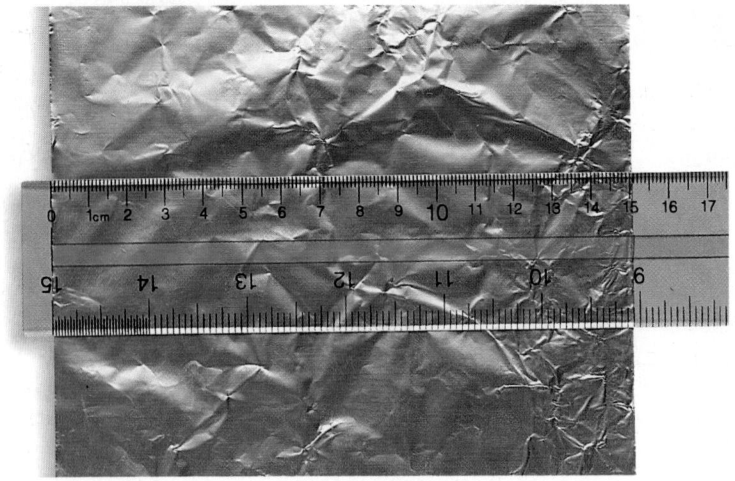

FIGURE 2-4 The meter is the SI unit of length, but the centimeter is often used to measure smaller distances. What is the width in cm of the rectangular piece of aluminum foil shown?

is often used. From Table 2-2, you can see that one centimeter equals 1/100 of a meter. The width of this book is just over 20 cm.

Derived SI Units

Many SI units are combinations of the quantities shown in Table 2-1. *Combinations of SI base units form* **derived units.** Some derived units are shown in Table 2-3.

Derived units are produced by multiplying or dividing standard units. For example, area, a derived unit, is length times width. If both length and width are expressed in meters, the area unit equals meters times meters, or square meters, abbreviated m^2. The last column of

TABLE 2-3 *Derived SI Units*

Quantity	Quantity symbol	Unit	Unit abbreviation	Derivation
Area	A	square meter	m^2	length × width
Volume	V	cubic meter	m^3	length × width × height
Density	D	kilogram per cubic meter	$\frac{kg}{m^3}$	$\frac{\text{mass}}{\text{volume}}$
Molar mass	M	kilograms per mole	$\frac{kg}{mol}$	$\frac{\text{mass}}{\text{amount of substance}}$
Concentration	c	moles per liter	M	$\frac{\text{amount of substance}}{\text{volume}}$
Molar volume	V_m	cubic meters per mole	$\frac{m^3}{mol}$	$\frac{\text{volume}}{\text{amount of substance}}$
Energy	E	joule	J	force × length

Table 2-3 shows the combination of fundamental units used to obtain derived units.

Some combination units are given their own names. For example, pressure expressed in base units is the following.

$$kg/m \cdot s^2$$

The name *pascal,* Pa, is given to this combination. You will learn more about pressure in Chapter 10. Prefixes can also be added to express derived units. Area can be expressed as cm^2, square centimeters, or mm^2, square millimeters.

FIGURE 2-5 The speed that registers on a speedometer represents distance traveled per hour and is expressed in the derived units kilometers per hour or miles per hour.

Volume

Volume *is the amount of space occupied by an object.* The derived SI unit of volume is cubic meters, m^3. One cubic meter is equal to the volume of a cube whose edges are 1 m long. Such a large unit is inconvenient for expressing the volume of materials in a chemistry laboratory. Instead, a smaller unit, the cubic centimeter, cm^3, is often used. There are 100 centimeters in a meter, so a cubic meter contains 1 000 000 cm^3.

$$1.000\ m^3 = (100\ cm)(100\ cm)(100\ cm) = 1\ 000\ 000\ cm^3$$

When chemists measure the volumes of liquids and gases, they often use a non-SI unit called the liter. The liter is equivalent to one cubic decimeter. Thus, a liter, L, is also equivalent to 1000 cm^3. Another non-SI unit, the milliliter, mL, is used for smaller volumes. There are 1000 mL in 1 L. Because there are also 1000 cm^3 in a liter, the two units—milliliter and cubic centimeter—are interchangeable.

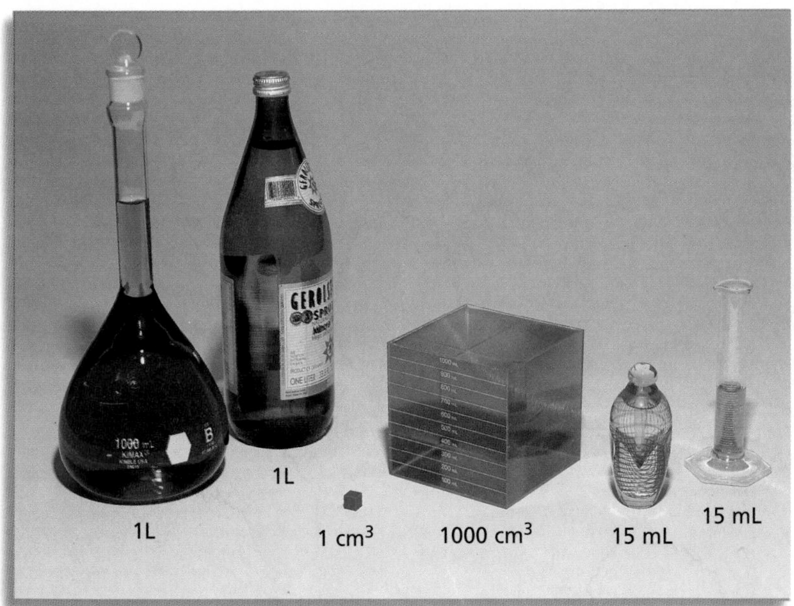

FIGURE 2-6 The relationships between various volumes are shown here. One liter contains 1000 mL of liquid, and 1 mL is equivalent to 1 cm^3. A small perfume bottle contains about 15 mL of liquid. The volumetric flask and graduated cylinder are used for measuring liquid volumes in the lab.

FIGURE 2-7 Density is the ratio of mass to volume. Both water and copper shot float on mercury because mercury is more dense.

Density

An object made of cork feels lighter than a lead object of the same size. What you are actually comparing in such cases is how massive objects are compared with their size. This property is called density. **Density** *is the ratio of mass to volume, or mass divided by volume*. Mathematically, the relationship for density can be written in the following way.

$$density = \frac{mass}{volume} \text{ or } D = \frac{m}{V}$$

The quantity m is mass, V is volume, and D is density.

The SI unit for density is derived from the base units for mass and volume—the kilogram and the cubic meter, respectively—and can be expressed as kilograms per cubic meter, kg/m^3. This unit is inconveniently large for the density measurements you will make in the laboratory. You will often see density expressed in grams per cubic centimeter, g/cm^3, or grams per milliliter, g/mL. The densities of gases are generally reported either in kilograms per cubic meter, kg/m^3, or in grams per liter, g/L.

Density is a characteristic physical property of a substance. It does not depend on the size of the sample because as the sample's mass increases, its volume increases proportionately, and the ratio of mass to volume is constant. Therefore, density can be used as one property to help identify a substance. Table 2-4 shows the densities of some common materials. As you can see, cork has a density of only 0.24 g/cm^3, which is less than the density of liquid water. Because cork is less dense than water, it floats on water. Lead, on the other hand, has a density of 11.35 g/cm^3. The density of lead is greater than that of water, so lead sinks in water.

Note that Table 2-4 specifies the temperatures at which the densities were measured. That is because density varies with temperature. Most objects expand as temperature increases, thereby increasing in volume. Because density is mass divided by volume, density usually decreases with increasing temperature.

TABLE 2-4 *Densities of Some Familiar Materials*

Solids	Density at 20°C (g/cm^3)	Liquids	Density at 20°C (g/mL)
Cork	0.24*	gasoline	0.67*
Butter	0.86	alcohol, ethyl	0.791
Ice	0.92†	kerosene	0.82
Sucrose	1.59	turpentine	0.87
Bone	1.85*	water	0.998
Diamond	3.26*	sea water	1.025**
Copper	8.92	milk	1.031*
Lead	11.35	mercury	13.6

† measured at 0°C
* typical density
** measured at 15°C

DESKTOP INVESTIGATION

Density of Pennies

MATERIALS

- **balance**
- **100 mL graduated cylinder**
- **40 pennies dated before 1982**
- **40 pennies dated after 1982**
- **water**

PROCEDURE

1. Using the balance, determine the mass of the 40 pennies minted prior to 1982. Repeat this measurement two more times. Average the results of the three trials to determine the average mass of the pennies.
2. Repeat step 1 with the 40 pennies minted after 1982.
3. Pour about 50 mL of water into the 100 mL graduated cylinder. Record the exact volume of the water. Add the 40 pennies minted before 1982. Record the volume of the water and pennies. Repeat this process two more times. Determine the volume of the pennies for each trial. Average the results of those trials to determine the average volume of the pennies.
4. Repeat step 3 with the 40 pennies minted after 1982.
5. Review your data for any large differences between trials that could increase the error of your results. Repeat those measurements.
6. Use the average volume and average mass to calculate the average density for each group of pennies.
7. Compare the average calculated densities with the density of the copper listed in Table 2-4.

DISCUSSION

1. Why is it best to use the results of three trials rather than a single trial for determining the density?
2. How did the densities of the two groups of pennies compare? How do you account for any difference?
3. Use the results of this investigation to formulate a hypothesis about the composition of the two groups of pennies. How could you test your hypothesis?

SAMPLE PROBLEM 2-1

A sample of aluminum metal has a mass of 8.4 g. The volume of the sample is 3.1 cm³. Calculate the density of aluminum.

SOLUTION **Given:** mass (m) = 8.4 g
volume (V) = 3.1 cm^3
Unknown: density (D)

$$density = \frac{mass}{volume} = \frac{8.4\ g}{3.1\ cm^3} = 2.7\ g/cm^3$$

PRACTICE

1. What is the density of a block of marble that occupies 310 cm^3 and has a mass of 853 g? *Answer* 2.75 g/cm^3
2. Diamond has a density of 3.26 g/cm^3. What is the mass of a diamond that has a volume of 0.35 cm^3? *Answer* 1.14 g
3. What is the volume of a sample of liquid mercury that has a mass of 76.2 g, given that the density of mercury is 13.6 g/mL? *Answer* 5.60 mL

Conversion Factors

A **conversion factor** *is a ratio derived from the equality between two different units that can be used to convert from one unit to the other.* For example, suppose you want to know how many quarters there are in a certain number of dollars. To figure out the answer, you need to know how quarters and dollars are related. There are four quarters per dollar and one dollar for every four quarters. Those facts can be expressed as ratios in three conversion factors.

$$\frac{4 \text{ quarters}}{1 \text{ dollar}} = 1 \qquad \frac{1 \text{ dollar}}{4 \text{ quarters}} = 1 \qquad \frac{0.25 \text{ dollar}}{1 \text{ quarter}} = 1$$

Notice that each conversion factor equals 1. That is because the two quantities divided in any conversion factor are equivalent to each other—as in this case, where 4 quarters equal 1 dollar. Because conversion factors are equal to 1, they can be multiplied by other factors in equations without changing the validity of the equations. When you want to use a conversion factor to change a unit in a problem, you can set up the problem in the following way.

$$\text{quantity sought} = \text{quantity given} \times \text{conversion factor}$$

For example, to determine the number of quarters in 12 dollars, you would carry out the unit conversion that allows you to change from dollars to quarters.

$$\text{number of quarters} = 12 \text{ dollars} \times \text{conversion factor}$$

Next you would have to decide which conversion factors give you an answer in the desired unit. In this case, you have dollars and you want quarters. To obtain quarters, you must divide the quantity by dollars. Therefore, the conversion factor in this case must have dollars in the denominator. That factor is 4 quarters/1 dollar. Thus, you would set up the calculation as follows.

$$? \text{ quarters} = 12 \text{ dollars} \times \text{conversion factor}$$

$$= 12 \cancel{\text{dollars}} \times \frac{4 \text{ quarters}}{1 \cancel{\text{dollar}}} = 48 \text{ quarters}$$

Notice that the dollars have divided out, leaving an answer in the desired unit—quarters.

Suppose you had guessed wrong and used 1 dollar/4 quarters when choosing which of the two conversion factors to use. You would have an answer with entirely inappropriate units.

$$? \text{ quarters} = 12 \text{ dollars} \times \frac{1 \text{ dollar}}{4 \text{ quarters}} = \frac{3 \text{ dollars}^2}{\text{quarter}}$$

You will work many problems in this book. It is always best to begin with an idea of the units you will need in your final answer. When working through the Sample Problems, keep track of the units needed for the unknown quantity. Check your final answer against what you've written as the unknown quantity.

Deriving Conversion Factors

You can derive conversion factors if you know the relationship between the unit you have and the unit you want. For example, from the fact that *deci-* means "1/10," you know that there is 1/10 of a meter per decimeter and that each meter must have 10 decimeters. Thus, from the equality

$$1 \text{ m} = 10 \text{ dm}$$

you can write the following conversion factors relating meters and decimeters.

$$\frac{1 \text{ m}}{10 \text{ dm}} \text{ and } \frac{0.1 \text{ m}}{\text{dm*}} \text{ and } \frac{10 \text{ dm}}{\text{m}}$$

The following sample problem illustrates an example of deriving conversion factors to make a unit conversion.

SAMPLE PROBLEM 2-2

Express a mass of 5.712 grams in milligrams and in kilograms.

SOLUTION

Given: 5.712 g

Unknown: mass in mg and kg

The expression that relates grams to milligrams is

$$1 \text{ g} = 1000 \text{ mg}$$

The possible conversion factors that can be written from this expression are

$$\frac{1000 \text{ mg}}{\text{g}} \text{ and } \frac{1 \text{ g}}{1000 \text{ mg}}$$

*In this book, when there is no digit shown in the denominator, you can assume the value is 1.

To derive an answer in mg, you'll need to multiply 5.712 g by 1000 mg/g.

$$5.712\ \cancel{g} \times \frac{1000\ mg}{\cancel{g}} = 5712\ mg$$

This answer makes sense because milligrams is a smaller unit than grams and, therefore, there should be more of them.

The kilogram problem is solved similarly.

$$1\ kg = 1000\ g$$

Conversion factors representing this expression are

$$\frac{1\ kg}{1000\ g} \text{ and } \frac{1000\ g}{kg}$$

To derive an answer in kg, you'll need to multiply 5.712 g by 1 kg/1000 g.

$$5.712\ \cancel{g} \times \frac{1\ kg}{1000\ \cancel{g}} = 0.005712\ kg$$

The answer makes sense because kilograms is a larger unit than grams and, therefore, there should be fewer of them.

PRACTICE

1. Express a length of 16.45 m in centimeters and in kilometers.

Answer
1645 cm, 0.01645 km

2. Express a mass of 0.014 mg in grams.

Answer
0.000 014 g

SECTION REVIEW

1. Why are standards needed for measured quantities?

2. Label each of the following measurements by the quantity each represents.

a. 5.0 g/mL
b. 37 s
c. 47 J
d. 39.56 g
e. 25.3 cm^3
f. 325 ms
g. 500 m^2
h. 30.23 mL
i. 2.7 mg
j. 0.005 L
k. 2000.5 kg
l. 63.5 km/h

3. Complete the following conversions.

a. 10.5 g = ____ kg
b. 1.57 km = ____ m
c. 3.54 μg = ____ g
d. 3.5 mol = ____ μmol
e. 1.2 L = ____ mL
f. 358 cm^3 = ____ m^3
g. 548.6 mL = ____ cm^3

4. Write conversion factors to represent the following equalities.

a. 1 m^3 = 1 000 000 cm^3
b. 1 in. = 2.54 cm
c. 1 μg = 0.000 001 g
d. 1 Mm = 1 000 000 m

5. a. What is the density of an 84.7 g sample of an unknown substance if the sample occupies 49.6 cm^3?

b. What volume would be occupied by 7.75 g of this same substance?

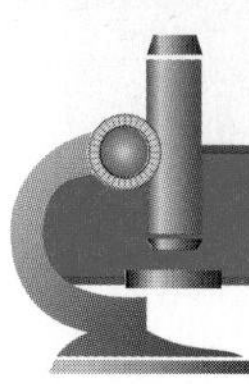

Roadside Pollution Detector

Dr. Donald Stedman, a chemist at the University of Denver, has developed a device that monitors exhaust emissions on highways. It pinpoints those cars that are dirtying the atmosphere.

The pollution detector sits on the side of a highway and shines a beam of infrared light across the road. After the beam passes through a car's exhaust fumes, it strikes a rotating mirror on the other side of the highway, which reflects the light onto four different sensors. These sensors detect changes in the infrared beam, and then each sensor uses that information to make different measurements. One detector gauges the amount of carbon dioxide in the exhaust. The second calculates the amount of carbon monoxide. A third sensor measures the amount of hydrocarbons, which contribute to the production of smog. To ensure accurate measurements, the fourth sensor measures a reference beam.

A carbon monoxide rating of higher than 4.5% gets a poor rating on the display.

A car driving down the highway will break the infrared beam, which signals the detector to store the measurement of the air directly in front of the vehicle. Then an exhaust reading is taken after the car passes for half a second to ensure that the beam measures data from the middle of the exhaust fumes. At the same instant, a video camera captures an image of the automobile. In this way, the detector can match emissions measurements with specific cars.

This technology is an efficient way of measuring the air pollution caused by automobiles because it can monitor a thousand cars in an hour.

Stedman put the detector into action on a highway exit ramp in Denver. The device gives every car that drives by an emissions rating and automatically displays the rating on a nearby billboard. If less than 1.3% of the car's exhaust is carbon monoxide, it earns a "good" rating. A rating of less than 4.5% carbon monoxide receives a "fair" rating. A rating higher than 4.5% is a "poor" rating. Stedman has found that the billboard not only informs people that their cars are polluters but also motivates the drivers to get their cars fixed.

Stedman has determined that only a small percentage of cars are responsible for automobile pollution. In fact, half of all the pollution from automobiles is created by about 10% of the cars on the road.

Stedman adds that the drivers will share the economic benefits of cleaning up their act. "If you have a gross-polluting car," he says, "you will save the amount of money that the repair might cost you in your fuel economy in a couple of years because you get a tremendous 10 to 15% fuel-economy improvement by fixing a gross-polluting car."

SECTION 2-3

OBJECTIVES

- Distinguish between accuracy and precision.
- Determine the number of significant figures in measurements.
- Perform mathematical operations involving significant figures.
- Convert measurements into scientific notation.
- Distinguish between inversely and directly proportional relationships.

Using Scientific Measurements

If you have ever measured something several times, you know that the results can vary. In science, for a reported measurement to be useful, there must be some indication of its reliability or uncertainty.

Accuracy and Precision

The terms *accuracy* and *precision* mean the same thing to most people. However, in science their meanings are quite distinct. **Accuracy** *refers to the closeness of measurements to the correct or accepted value of the quantity measured.* **Precision** *refers to the closeness of a set of measurements of the same quantity made in the same way.* Thus, measured values that are accurate are close to the accepted value. Measured values that are precise are close to one another but not necessarily close to the accepted value.

Figure 2-8 should help you visualize the difference between precision and accuracy. A set of darts thrown separately at a dartboard may land in various positions, relative to the bull's-eye and to one another. The

FIGURE 2-8 The sizes and locations of the areas covered by thrown darts illustrate the difference between precision and accuracy.

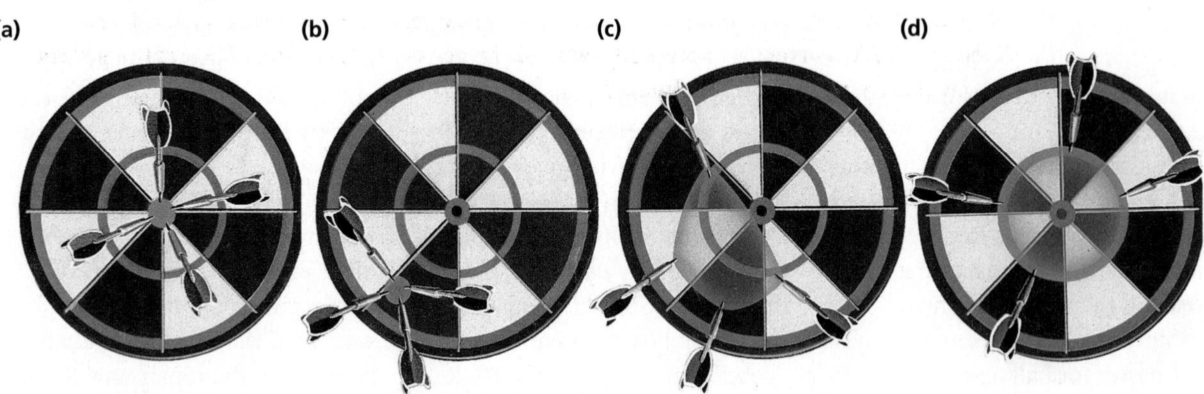

(a) Darts within small area = High precision. Area covered on bull's-eye = High accuracy

(b) Darts within small area = High precision. Area far from bull's-eye = Low accuracy

(c) Darts within large area = Low precision. Area far from bull's-eye = Low accuracy

(d) Darts within large area = Low precision. Area centered around bull's-eye = High accuracy (on average)

closer the darts land to the bull's-eye, the more accurately they were thrown. The closer they land to one another, the more precisely they were thrown. Thus, the set of results shown in Figure 2-8(a) is both accurate and precise because the darts are close to the bull's-eye and close to each other. In Figure 2-8(b), the set of results is inaccurate but precise because the darts are far from the bull's-eye but close to each other. In Figure 2-8(c), the set of results is both inaccurate and imprecise because the darts are far from the bull's-eye and far from each other. Notice also that the darts are not evenly distributed around the bull's-eye, so the set, even considered on average, is inaccurate. In Figure 2-8(d), the set on average is accurate compared with the third case, but it is imprecise. That is because the darts are distributed evenly around the bull's-eye but are far from each other.

Percent Error

The accuracy of an individual value or of an average experimental value can be compared quantitatively with the correct or accepted value by calculating the percent error. **Percent error** *is calculated by subtracting the experimental value from the accepted value, dividing the difference by the accepted value, and then multiplying by 100.*

$$\textit{Percent error} = \frac{\textit{Value}_{accepted} - \textit{Value}_{experimental}}{\textit{Value}_{accepted}} \times 100$$

Percent error has a positive value if the accepted value is greater than the experimental value. It has a negative value if the accepted value is less than the experimental value. The following sample problem illustrates the concept of percent error.

SAMPLE PROBLEM 2-3

A student measures the mass and volume of a substance and calculates its density as 1.40 g/mL. The correct, or accepted, value of the density is 1.36 g/mL. What is the percent error of the student's measurement?

SOLUTION

$$\textit{Percent error} = \frac{\textit{Value}_{accepted} - \textit{Value}_{experimental}}{\textit{Value}_{accepted}} \times 100$$

$$= \frac{1.36\ \cancel{g/mL} - 1.40\ \cancel{g/mL}}{1.36\ \cancel{g/mL}} \times 100 = -2.9\%$$

PRACTICE

1. What is the percent error for a mass measurement of 17.7 g, given that the correct value is 21.2 g? *Answer* 17%

2. A volume is measured experimentally as 4.26 mL. What is the percent error, given that the correct value is 4.15 mL? *Answer* –2.7%

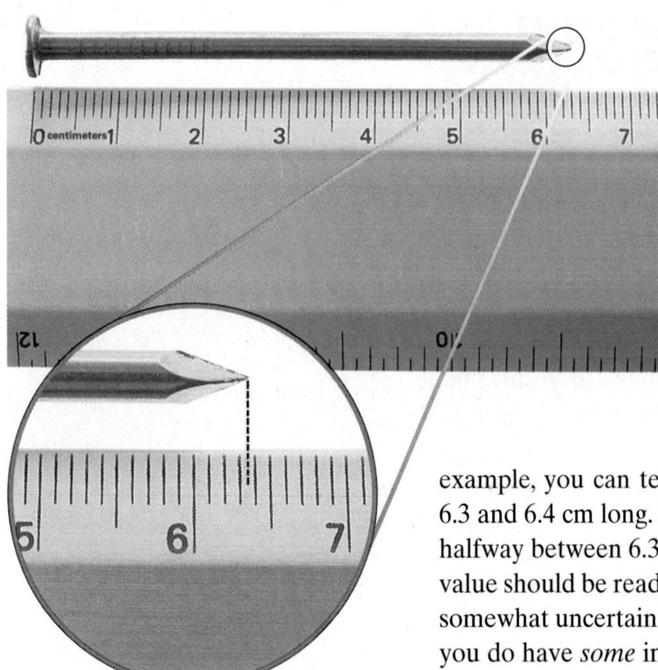

FIGURE 2-9 What value should be recorded for the length of this nail?

Error in Measurement

Some error or uncertainty always exists in any measurement. The skill of the measurer places limits on the reliability of results. The conditions of measurement also affect the outcome. The measuring instruments themselves place limitations on precision. Some balances can be read more precisely than others. The same is true of rulers, graduated cylinders, and other measuring devices.

When you use a properly calibrated measuring device, you can be almost certain of a particular number of digits in a reading. For example, you can tell that the nail in Figure 2-9 is definitely between 6.3 and 6.4 cm long. Looking more closely, you can see that the value is halfway between 6.3 and 6.4 cm. However, it is hard to tell whether the value should be read as 6.35 cm or 6.36 cm. The hundredths place is thus somewhat uncertain. Simply leaving it out would be misleading because you do have *some* indication of the value's likely range. Therefore, you would estimate the value to the final questionable digit, perhaps reporting the length of the nail as 6.36 cm. You might include a plus-or-minus value to express the range, for example, 6.36 cm ± 0.01 cm.

Significant Figures

In science, measured values are reported in terms of significant figures. **Significant figures** *in a measurement consist of all the digits known with certainty plus one final digit, which is somewhat uncertain or is estimated.* For example, in the reported nail length of 6.36 cm discussed above, the last digit, 6, is uncertain. All the digits, including the uncertain one, are significant, however. All contain information and are included in the reported value. Thus, the term *significant* does not mean *certain.* In any correctly reported measured value, the final digit is significant but not certain. Insignificant digits are never reported. As a chemistry student, you will need to use and recognize significant figures when you work with measured quantities and report your results, and when you evaluate measurements reported by others.

Determining the Number of Significant Figures

When you look at a measured quantity, you need to determine which digits are significant. That process is very easy if the number has no zeros because all the digits shown are significant. For example, if you see a number reported as 3.95, all three digits are significant. The significance of zeros in a number depends on their location, however. You need to learn and follow several rules involving zeros. After you have studied the rules in Table 2-5, use them to express the answers in the sample problem that follows.

TABLE 2-5 *Rules for Determining Significant Zeros*

Rule	Examples
1. Zeros appearing between nonzero digits are significant.	a. 40.7 L has three significant figures. b. 87 009 km has five significant figures.
2. Zeros appearing in front of nonzero digits are not significant.	a. 0.095 897 m has five significant figures. b. 0.0009 kg has one significant figure.
3. Zeros at the end of a number and to the right of a decimal are significant.	a. 85.00 g has four significant figures. b. 9.000 000 000 mm has 10 significant figures.
4. Zeros at the end of a number but to the left of a decimal may or may not be significant. If such a zero has been measured or is the first estimated digit, it is significant. On the other hand, if the zero has not been measured or estimated but is just a placeholder, it is not significant. A decimal placed after the zeros indicates that they are significant.	a. 2000 m may contain from one to four significant figures, depending on how many zeros are placeholders. **For measurements given in this text, assume that 2000 m has one significant figure.** b. 2000. m contains four significant figures, indicated by the presence of the decimal point.

SAMPLE PROBLEM 2-4

How many significant figures are in each of the following measurements?

a. 28.6 g
b. 3440. cm
c. 910 m
d. 0.046 04 L
e. 0.006 700 0 kg

SOLUTION Determine the number of significant figures in each measurement using the rules listed in Table 2-5.

a. 28.6 g
There are no zeros, so all three digits are significant.

b. 3440. cm
By rule 4, the zero is significant because it is immediately followed by a decimal point; there are 4 significant figures.

c. 910 m
By rule 4, the zero is not significant; there are 2 significant figures.

d. 0.046 04 L
By rule 2, the first two zeros are not significant; by rule 1, the third zero is significant; there are 4 significant figures.

e. 0.006 700 0 kg
By rule 2, the first three zeros are not significant; by rule 3, the last three zeros are significant; there are 5 significant figures.

PRACTICE

1. Determine the number of significant figures in each of the following.
   a. 804.05 g
   b. 0.014 403 0 km
   c. 1002 m
   d. 400 mL
   e. 30 000. cm
   f. 0.000 625 000 kg

2. Suppose the value "seven thousand centimeters" is reported to you. How should the number be expressed if it is intended to contain the following:
   a. 1 significant figure
   b. 4 significant figures
   c. 6 significant figures

Answer

1. a. 5
   b. 6
   c. 4
   d. 1
   e. 5
   f. 6

2. a. 7000 cm
   b. 7000. cm
   c. 7000.00 cm

Rounding

When you perform calculations involving measurements, you need to know how to handle significant figures. This is especially true when you are using a calculator to carry out mathematical operations. The answers given on a calculator can be derived results with more digits than are justified by the measurements.

Suppose you used a calculator to divide a measured value of 154 g by a measured value of 327 mL. Each of these values has three significant figures. The calculator would show a numerical answer of 0.470948012. The answer contains digits not justified by the measurements used to calculate it. Such an answer has to be rounded off to make its degree of certainty match that in the original measurements. The answer should be 0.471 g/mL.

The rules for rounding are shown in Table 2-6. The extent of rounding required in a given case depends on whether the numbers are being added, subtracted, multiplied, or divided.

TABLE 2-6 *Rules for Rounding Numbers*

If the digit following the last digit to be retained is:	then the last digit should:	Example (rounded to three significant figures)
greater than 5	be increased by 1	42.68 g $\longrightarrow$ 42.7 g
less than 5	stay the same	17.32 m $\longrightarrow$ 17.3 m
5, followed by nonzero digit(s)	be increased by 1	2.7851 cm $\longrightarrow$ 2.79 cm
5, not followed by nonzero digit(s), and preceded by an odd digit	be increased by 1	4.635 kg $\longrightarrow$ 4.64 kg (because 3 is odd)
5, not followed by nonzero digit(s), and the preceding significant digit is even	stay the same	78.65 mL $\longrightarrow$ 78.6 mL (because 6 is even)

Addition or Subtraction with Significant Figures

Consider two mass measurements, 25.1 g and 2.03 g. The first measurement, 25.1 g, has one digit to the right of the decimal point, in the tenths place. There is no information on possible values for the hundredths place. That place is simply blank and cannot be assumed to be zero. The other measurement, 2.03 g, has two digits to the right of the decimal point. It provides information up to and including the hundredths place.

Suppose you were asked to add the two measurements. Simply carrying out the addition would result in an answer of 25.1 g + 2.03 g = 27.13 g. That answer suggests there is certainty all the way to the hundredths place. However, that result is not justified because the hundredths place in 25.1 g is completely unknown. The answer must be adjusted to reflect the uncertainty in the numbers added.

When adding or subtracting decimals, the answer must have the same number of digits to the right of the decimal point as there are in the measurement having the fewest digits to the right of the decimal point. When working with whole numbers, the answer should be rounded so that the final digit is in the same place as the leftmost uncertain digit. Comparing the two values 25.1 g and 2.03 g, the measurement with the fewest digits to the right of the decimal point is 25.1 g. It has only one such digit. Following the rule, the answer must be rounded so that it has no more than one digit to the right of the decimal point. It should therefore be rounded to 27.1 g.

Multiplication and Division with Significant Figures

Suppose you calculated the density of an object that has a mass of 3.05 g and a volume of 8.47 mL. The following division on a calculator will give a value of 0.360094451.

$$density = \frac{mass}{volume} = \frac{3.05\ \text{g}}{8.47\ \text{mL}} = 0.360094451\ \text{g/mL}$$

The answer must be rounded to the correct number of significant figures. The values of mass and volume used to obtain the answer have only three significant figures each. The degree of certainty in the calculated result is not justified. *For multiplication or division, the answer can have no more significant figures than are in the measurement with the fewest number of significant figures.* In the calculation just described, the answer, 0.360094451 g/mL, would be rounded to three significant figures to match the significant figures in 8.47 mL and 3.05 g. The answer would thus be 0.360 g/mL.

SAMPLE PROBLEM 2-5

Carry out the following calculations. Express each answer to the correct number of significant figures.

a. 5.44 m – 2.6103 m

b. 2.4 g/mL × 15.82 mL

SOLUTION Carry out each mathematical operation. Follow the rules in Table 2-5 and Table 2-6 for determining significant figures and for rounding.

a. The answer is rounded to 2.83 m (for subtraction there should be two digits to the right of the decimal point, to match 5.44 m).

b. The answer is rounded to 38 g (for multiplication there should be two significant figures in the answer, to match 2.4 g/mL).

PRACTICE

1. What is the sum of 2.099 g and 0.05681 g? *Answer* 2.156 g

2. Calculate the quantity 87.3 cm – 1.655 cm. *Answer* 85.6 cm

3. Calculate the area of a crystal surface that measures 1.34 µm by 0.7488 µm. (Hint: Recall that *area* = *length* × *width* and is measured in square units.) *Answer* 1.00 μm^2

4. Polycarbonate plastic has a density of 1.2 g/cm^3. A photo frame is constructed from two 3.0 mm sheets of polycarbonate. Each sheet measures 28 cm by 22 cm. What is the mass of the photo frame? *Answer* 440 g

Conversion Factors and Significant Figures

Earlier in this chapter, you learned how conversion factors are used to change one unit to another. Such conversion factors are typically exact. That is, there is no uncertainty in them. For example, there are exactly 100 cm in a meter. If you were to use the conversion factor 100 cm/m to change meters to centimeters, the 100 would not limit the degree of certainty in the answer. Thus, 4.608 m could be converted to centimeters as follows.

$$4.608 \text{ m} \times \frac{100 \text{ cm}}{\text{m}} = 460.8 \text{ cm}$$

The answer still has four significant figures. Because the conversion factor is considered exact, the answer would not be rounded. Most exact conversion factors are defined, rather than measured, quantities. Counted numbers also produce unlimited conversion factors. For example, if you counted that there are 10 test tubes for every student, that would produce an exact conversion factor of 10 test tubes/student. There is no uncertainty in that factor.

Scientific Notation

In **scientific notation,** *numbers are written in the form* M × 10^n, *where the factor* M *is a number greater than or equal to 1 but less than 10 and* n *is a whole number.* For example, to write the quantity 65 000 km in scientific notation and show the first two digits as significant, you would write the following.

$$6.5 \times 10^4 \text{ km}$$

Writing the M factor as 6.5 shows that there are exactly two significant figures. If, instead, you intended the first three digits in 65 000 to be significant, you would write 6.50×10^4 km. When numbers are written in scientific notation, only the significant figures are shown.

Suppose you are expressing a very small quantity, such as the length of a flu virus. In ordinary notation this length could be 0.000 12 mm. That length can be expressed in scientific notation as follows.

$$0.000\ 12 \text{ mm} = 1.2 \times 10^{-4} \text{ mm}$$

move four places to the right and
multiply the number by 10^{-4}

1. Determine M by moving the decimal point in the original number to the left or the right so that only one nonzero digit remains to the left of the decimal point.
2. Determine n by counting the number of places that you moved the decimal point. If you moved it to the left, n is positive. If you moved it to the right, n is negative.

Mathematical Operations Using Scientific Notation

1. *Addition and subtraction* These operations can be performed only if the values have the same exponent. If they do not, adjustments must be made to the values so that the exponents are equal. Once the exponents are equal, the M factors can be added or subtracted. The exponent of the answer can remain the same, or it may then require adjustment if the M factor of the answer has more than one digit to the left of the decimal point. Consider the example of the addition of 4.2×10^4 kg and 7.9×10^3 kg.

   We can make both exponents either 3 or 4. The following solutions are possible.

$$\begin{array}{r l} 4.2\ \ \times 10^4 \text{ kg} & \\ +0.79 \times 10^4 \text{ kg} & \\ \hline 4.99 \times 10^4 \text{ kg} & \text{rounded to } 5.0 \times 10^4 \text{ kg} \end{array}$$

or

$$\begin{array}{r l} 7.9\ \ \times 10^3 \text{ kg} & \\ +42\ \ \ \ \times 10^3 \text{ kg} & \\ \hline 49.9\ \ \times 10^3 \text{ kg} & = 4.99 \times 10^4 \text{ kg rounded to } 5.0 \times 10^4 \text{ kg} \end{array}$$

   Note that the units remain kg throughout.

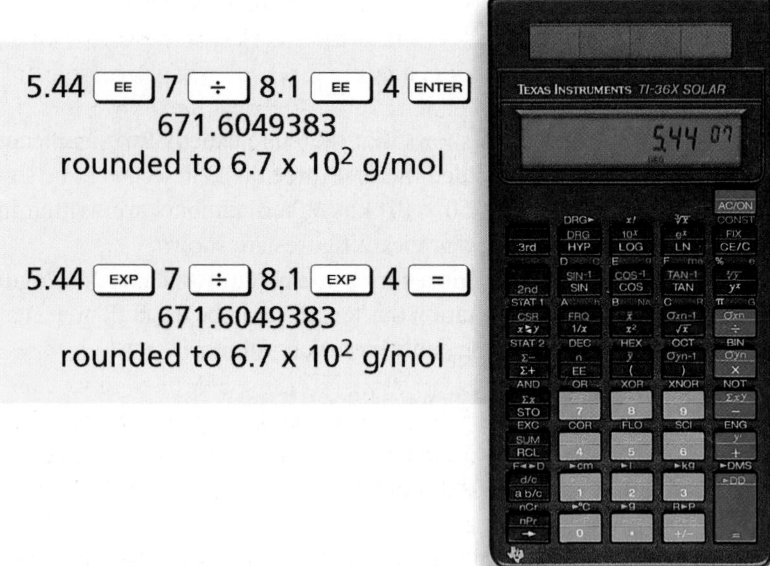

FIGURE 2-10 When you use a scientific calculator to work problems in scientific notation, don't forget to express the value on the display to the correct number of significant figures and show the units when you write the final answer.

2. *Multiplication* The *M* factors are multiplied, and the exponents are added algebraically.

   Consider the multiplication of 5.23×10^6 μm by 7.1×10^{-2} μm.

$$
\begin{aligned}
(5.23 \times 10^6\ \mu\text{m})(7.1 \times 10^{-2}\ \mu\text{m}) &= (5.23 \times 7.1)(10^6 \times 10^{-2}) \\
&= 37.133 \times 10^4\ \mu\text{m}^2 \text{ (adjust to two significant digits)} \\
&= 3.7 \times 10^5\ \mu\text{m}^2
\end{aligned}
$$

   Note that when length measurements are multiplied, the result is area. The unit is now μm^2.

3. *Division* The *M* factors are divided, and the exponent of the denominator is subtracted from that of the numerator. The calculator keystrokes for this problem are shown in Figure 2-10.

$$
\begin{aligned}
\frac{5.44 \times 10^7\ \text{g}}{8.1 \times 10^4\ \text{mol}} &= \frac{5.44}{8.1} \times 10^{7-4}\ \text{g/mol} \\
&= 0.6716049383 \times 10^3 \text{ (adjust to two significant figures)} \\
&= 6.7 \times 10^2\ \text{g/mol}
\end{aligned}
$$

   Note that the unit for the answer is the ratio of grams to moles.

Using Sample Problems

Learning to analyze and solve such problems requires practice and a logical approach. In this section, you will review a process that can help you analyze problems effectively. Most Sample Problems in this book are organized by four basic steps to guide your thinking in how to work out the solution to a problem.

Analyze

The first step in solving a quantitative word problem is to read the problem carefully at least twice and to analyze the information in it. Note any important descriptive terms that clarify or add meaning to the problem. Identify and list the data given in the problem. Also identify the unknown—the quantity you are asked to find.

Plan

The second step is to develop a plan for solving the problem. The plan should show how the information given is to be used to find the unknown. In the process, reread the problem to make sure you have gathered all the necessary information. It is often helpful to draw a picture that represents the problem. For example, if you were asked to determine the volume of a crystal given its dimensions, you could draw a representation of the crystal and label the dimensions. This drawing would help you visualize the problem.

Decide which conversion factors, mathematical formulas, or chemical principles you will need to solve the problem. Your plan might suggest a single calculation or a series of them involving different conversion factors. Once you understand how you need to proceed, you may wish to sketch out the route you will take, using arrows to point the way from one stage of the solution to the next. Sometimes you will need data that are not actually part of the problem statement. For instance, you'll often use data from the periodic table.

Compute

The third step involves substituting the data and necessary conversion factors into the plan you have developed. At this stage you calculate the answer, cancel units, and round the result to the correct number of significant figures. It is very important to have a plan worked out in step 2 before you start using the calculator. All too often, students start multiplying or dividing values given in the problem before they really understand what they need to do to get an answer.

Evaluate

Examine your answer to determine whether it is reasonable. Use the following methods, when appropriate, to carry out the evaluation.

1. Check to see that the units are correct. If they are not, look over the setup. Are the conversion factors correct?
2. Make an estimate of the expected answer. Use simpler, rounded numbers to do so. Compare the estimate with your actual result. The two should be similar.
3. Check the order of magnitude in your answer. Does it seem reasonable compared with the values given in the problem? If you calculated the density of vegetable oil and got a value of 54.9 g/mL, you would know that something is wrong. Oil floats on water; therefore, its density is less than water, so the value obtained should be less than 1.0 g/mL.
4. Be sure that the answer given for any problem is expressed using the correct number of significant figures.

Look over the following quantitative Sample Problems. Notice how the four-step approach is used in each, and then apply the approach yourself in solving the practice problems that follow.

SAMPLE PROBLEM 2-6

Calculate the volume of a sample of aluminum that has a mass of 3.057 kg. The density of aluminum is 2.70 g/cm³.

SOLUTION

1 ANALYZE **Given:** mass = 3.057 g, density = 2.70 g/cm³
Unknown: volume of aluminum

2 PLAN The density unit in the problem is g/cm³, and the mass given in the problem is expressed in kg. Therefore, in addition to using the density equation, you will need a conversion factor representing the relationship between grams and kilograms.

$$1000 \text{ g} = 1 \text{ kg}$$

Also, rearrange the density equation to solve for volume.

$$density = \frac{mass}{volume} \quad \text{or} \quad D = \frac{m}{V}$$

$$V = \frac{m}{D}$$

3 COMPUTE

$$V = \frac{3.057 \text{ kg}}{2.70 \text{ g/cm}^3} \times \frac{1000 \text{ g}}{\text{kg}} = 1132.222\ldots \text{cm}^3 \text{ (calculator answer)}$$

The answer should be rounded to three significant figures.

$$V = 1.13 \times 10^3 \text{ cm}^3$$

4 EVALUATE The unit of volume, cm³, is correct. An order-of-magnitude estimate would put the answer at over 1000 cm³.

$$\frac{3}{2} \times 1000$$

The correct number of significant digits is three, to match the number of significant figures in 2.70 g/cm³.

PRACTICE

1. What is the volume of a sample of helium that has a mass of 1.73×10^{-3} g, given that the density is 0.178 47 g/L? *Answer* 9.69 mL
2. What is the density of a piece of metal that has a mass of 6.25×10^5 g and is 92.5 cm × 47.3 cm × 85.4 cm? *Answer* 1.67 g/cm³
3. How many millimeters are there in 5.12×10^5 kilometers? *Answer* 5.12×10^{11} mm
4. A clock gains 0.020 second per minute. How many seconds will the clock gain in exactly six months, assuming exactly 30 days per month? *Answer* 5.2×10^3 s

Direct Proportions

Two quantities are **directly proportional** *to each other if dividing one by the other gives a constant value.* For example, if the masses and volumes of different samples of aluminum are measured, the masses and volumes will be directly proportional to each other. As the masses of the samples increase, their volumes increase at the same rate, as you can see from the data in Table 2-7. Doubling the mass doubles the volume. Halving the mass halves the volume.

When two variables, x and y, are directly proportional to each other, the relationship can be expressed as $y \propto x$, which is read as "y is *proportional* to x." The general equation for a directly proportional relationship between the two variables can be written as follows.

$$\frac{y}{x} = k$$

The value of k is a constant called the proportionality constant. Written in this form, the equation expresses an important fact about direct proportion: the ratio between the variables remains constant. Note that using the mass and volume values in Table 2-7 gives a mass-volume ratio that is relatively constant (neglecting measurement error). The equation can be rearranged into the following form.

$$y = kx$$

The equation $y = kx$ may look familiar to you. It is the equation for a straight line. If two variables related in this way are graphed versus one another, a straight line, or linear plot that passes through the origin (0,0), results. The data for aluminum from Table 2-7 are graphed in Figure 2-11. The mass and volume of a pure substance are directly proportional to each other. Consider mass to be y and volume to be x. The constant ratio, k, for the two variables is density. The slope of the line reflects the constant density, or mass-volume ratio, of aluminum, which

TABLE 2-7 *Mass-Volume Data for Aluminum at 20°C*

Mass (g)	Volume (cm^3)	$\frac{m}{V}$ (g/cm^3)
54.4	20.1	2.70
65.7	24.15	2.72
83.5	30.9	2.70
97.2	35.8	2.71
105.7	39.1	2.70

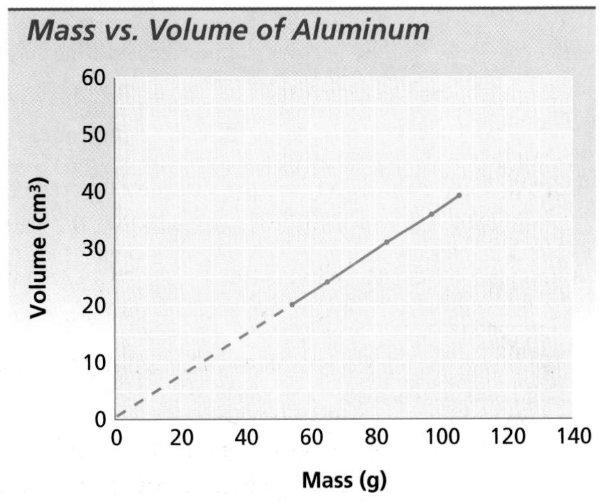

FIGURE 2-11 The graph of mass versus volume shows a relationship of direct proportion. Notice that the line is extrapolated to pass through the origin.

is 2.70 g/cm^3 at 20°C. Notice also that the plotted line passes through the origin. All directly proportional relationships produce linear graphs that pass through the origin.

Inverse Proportions

Two quantities are **inversely proportional** *to each other if their product is constant.* An example of an inversely proportional relationship is that between speed of travel and the time required to cover a fixed distance. The greater the speed, the less time that is needed to go a certain fixed distance. Doubling the speed cuts the required time in half. Halving the speed doubles the required time.

When two variables, x and y, are inversely proportional to each other, the relationship can be expressed as follows.

$$y \propto \frac{1}{x}$$

This is read "y is *proportional* to 1 divided by x." The general equation for an inversely proportional relationship between the two variables can be written in the following form.

$$xy = k$$

In the equation, k is the proportionality constant. If x increases, y must decrease to keep the product constant.

A graph of variables that are inversely proportional produces a curve called a hyperbola. Such a graph is illustrated in Figure 2-12. When the temperature of the gas is kept constant, the volume (V) of the gas sample decreases as the pressure (P) increases. Look at the data in Table 2-8. Note that $P \times V$ gives a reasonably constant value. The graph of this data is shown in Figure 2-12.

TABLE 2-8 *Pressure-Volume Data for Nitrogen at Constant Temperature*

Pressure (kPa)	Volume (cm^3)	$P \times V$
100	500	50 000
150	333	49 500
200	250	50 000
250	200	50 000
300	166	49 800
350	143	50 500
400	125	50 000
450	110	49 500

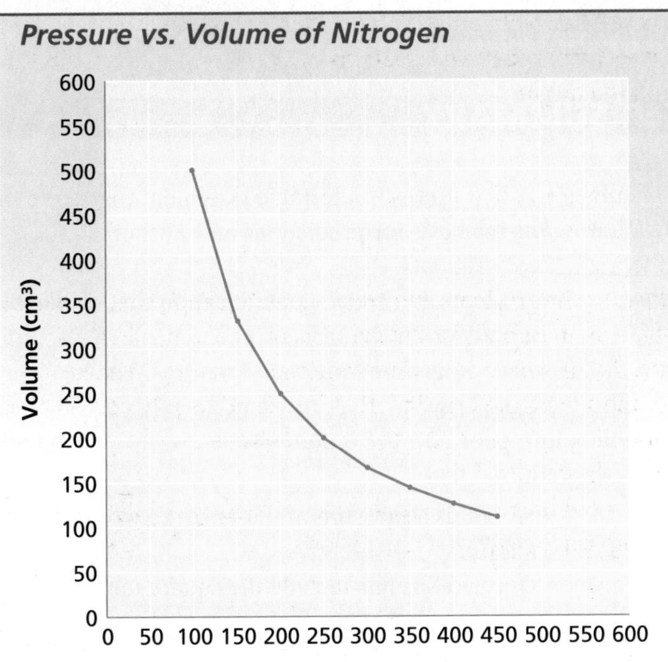

FIGURE 2-12 The graph of pressure versus volume shows an inversely proportional relationship. Note the difference between the shape of this graph and that of the graph in Figure 2-11.

SECTION REVIEW

1. The density of copper is listed as 8.94 g/cm^3. Two students each make three density determinations of samples of the substance. Student A's results are 7.3 g/mL, 9.4 g/mL, and 8.3 g/cm^3. Student B's results are 8.4 g/cm^3, 8.8 g/cm^3, and 8.0 g/cm^3. Compare the two sets of results in terms of precision and accuracy.
2. How many significant figures are there in each of the following measured values?
   a. 6.002 cm
   b. 0.0020 m
   c. 10.0500 g
   d. 7000 kg
   e. 7000. kg
3. Round 2.6765 to two significant figures.
4. Carry out the following calculations.
   a. 52.13 g + 1.7502 g
   b. 12 m × 6.41 m
   c. $\frac{16.25 \text{ g}}{5.1442 \text{ mL}}$
5. Perform the following operations. Express each answer in scientific notation.
   a. $(1.54 \times 10^{-2} \text{ g}) + (2.86 \times 10^{-1} \text{ g})$
   b. $(7.023 \times 10^{9} \text{ g}) - (6.62 \times 10^{7} \text{ g})$
   c. $(8.99 \times 10^{-4} \text{ m}) \times (3.57 \times 10^{4} \text{ m})$
   d. $\frac{2.17 \times 10^{-3} \text{ g}}{5.022 \times 10^{4} \text{ mL}}$
6. Write the following measurements in scientific notation.
   a. 560 000
   b. 33 400
   c. 0.000 4120
7. A student measures the mass of a beaker filled with corn oil. The mass reading averages 215.6 g. The mass of the beaker is 110.4 g.
   a. What is the mass of the corn oil?
   b. What is the density of the corn oil if its volume is 114 cm^3?
8. Calculate the mass of a sample of gold that occupies 5.0×10^{-3} cm^3. The density of gold is 19.3 g/cm^3.
9. What is the difference between a graph representing data that are directly proportional and a graph of data that are inversely proportional?

CHAPTER 2 REVIEW

CHAPTER SUMMARY

2-1

- The scientific method is a logical approach to solving problems that lend themselves to investigation.
- The processes of observing, generalizing, theorizing, and testing are aspects of the scientific method.
- A hypothesis is a testable statement that serves as the basis for predictions and further experiments.
- A theory is a broad generalization that explains a body of known facts or phenomena.

Vocabulary

hypothesis (30) · scientific method (29) · system (29) · theory (31) · model (31)

2-2

- The result of nearly every measurement is a number and a unit.
- The SI system of measurement is used in science. It has seven base units: the meter (length), kilogram (mass), second (time), kelvin (temperature), mole (quantity of substance), ampere (electric current), and candela (luminous intensity).
- Weight is a measure of the gravitational pull on matter.
- Derived SI units include the square meter (area) and the cubic meter (volume).
- Density is the ratio of mass to volume.
- Conversion factors are used to convert from one unit to another.

Vocabulary

conversion factor (40) · derived unit (36) · SI (33) · weight (35) · density (38) · quantity (33) · volume (37)

2-3

- Accuracy refers to the closeness of a measurement to the correct or accepted value. Precision refers to the closeness of values for a set of measurements.
- The measurement average is the sum of a group of measurements divided by the total number of measurements.
- Percent error is the difference between the accepted and the experimental value, divided by the accepted value, then multiplied by 100.
- The significant figures in a number consist of all digits known with certainty plus one final digit, which is uncertain or estimated. A set of logical rules must be followed to determine the number of significant figures in numbers containing zeros.
- After addition or subtraction, the answer should be rounded so that it has no more digits to the right of the decimal point than there are in the measurement with the smallest number of digits to the right of the decimal point. After multiplication or division, the answer should be rounded so that it has no more significant figures than there are in the measurement with the fewest number of significant figures.
- Exact conversion factors are completely certain and do not limit the number of digits in a calculation.
- A number written in scientific notation is of the form $M \times 10^n$, where M is greater than or equal to 1 but less than 10 and n is an integer.
- Two quantities are directly proportional to each other if dividing one by the other gives a constant value. The graphs of variables related in this way are straight lines that pass through the origin.
- Two quantities are inversely proportional to each other if their product has a constant value. The graphs of variables related in this way are hyperbolas.

Vocabulary

accuracy (44) · inverse proportion (56) · precision (44) · significant figure (46) · direct proportion (55) · percent error (45) · scientific notation (50)

REVIEWING CONCEPTS

1. How does quantitative information differ from qualitative information? (2-1)
2. What is a hypothesis? (2-1)
3. a. What is a model in the scientific sense?
   b. How does a model differ from a theory? (2-1)
4. Why is it important for a measurement system to have an international standard? (2-2)
5. How does a quantity differ from a unit? Use two examples to explain the difference. (2-2)
6. List the seven SI base units and the quantities they represent. (2-2)
7. What is the numerical equivalent of each of the following SI prefixes?
   a. kilo-
   b. centi-
   c. mega-
   d. micro-
   e. milli- (2-2)
8. Identify the SI unit that would be most appropriate for expressing the length of the following.
   a. width of a gymnasium
   b. length of a finger
   c. distance between your town and the closest border of the next state
   d. length of a bacterial cell (2-2)
9. Identify the SI unit that would be most appropriate for measuring the mass of each of the following objects.
   a. table
   b. coin
   c. a 250 mL beaker (2-2)
10. Explain why the second is not defined by the length of the day. (2-2)
11. a. What is a derived unit?
    b. What is the SI derived unit for area? (2-2)
12. a. List two SI derived units for volume.
    b. List two non-SI units for volume, and explain how they relate to the cubic centimeter. (2-2)
13. a. Why are the units used to express the densities of gases different from those used to express the densities of solids or liquids?
    b. Name two units for density.
    c. Why is the temperature at which a density is measured usually specified? (2-2)
14. a. Which of the solids listed in Table 2-4 will float on water?
    b. Which of the liquids will sink in milk?
15. a. Define *conversion factor*.
    b. Explain how conversion factors are used. (2-2)
16. Contrast accuracy and precision. (2-3)
17. a. Write the equation that is used to calculate percent error.
    b. Under what condition will percent error be negative? (2-3)
18. How is the average for a set of values calculated?
19. What is meant by a mass measurement expressed in this form: 4.6 g ± 0.2 g?
20. Suppose a graduated cylinder were not correctly calibrated. How would this affect the results of a measurement? How would it affect the results of a calculation using this measurement?
21. Round each of the following measurements to the number of significant figures indicated.
    a. 67.029 g to three significant figures
    b. 0.15 L to one significant figure
    c. 52.8005 mg to five significant figures
    d. 3.174 97 mol to three significant figures (2-3)
22. State the rules governing the number of significant figures that result from each of the following operations.
    a. addition and subtraction
    b. multiplication and division (2-3)
23. What is the general form for writing numbers in scientific notation? (2-3)
24. a. State the general equation for quantities that are directly proportional.
    b. For two directly proportional quantities, what happens to one variable when the other increases? (2-3)
25. a. State the general equation for quantities that are inversely proportional.
    b. For two inversely proportional quantities, what happens to one variable when the other increases? (2-3)

PROBLEMS

Volume and Density

26. What is the volume, in cubic meters, of an object that is 0.25 m long, 6.1 m wide, and 4.9 m high?

27. Find the density of a material, given that a 5.03 g sample occupies 3.24 mL. (Hint: See Sample Problem 2-1.)

28. What is the mass of a sample of material that has a volume of 55.1 cm^3 and a density of 6.72 g/cm^3?

29. A sample of a substance that has a density of 0.824 g/mL has a mass of 0.451 g. Calculate the volume of the sample.

Conversion Factors

30. How many grams are there in 882 μg? (Hint: See Sample Problem 2-2.)

31. Calculate the number of mL in 0.603 L.

32. a. Find the number of km in 92.25 m.
b. Convert the answer in km to cm.

Percent Error

33. A student measures the mass of a sample as 9.67 g. Calculate the percent error, given that the correct mass is 9.82 g. (Hint: See Sample Problem 2-3.)

34. A handbook gives the density of calcium as 1.54 g/cm^3. What is the percent error of a density calculation of 1.25 g/cm^3 based on lab measurements?

35. What is the percent error of a length measurement of 0.229 cm if the correct value is 0.225 cm?

Significant Figures

36. How many significant figures are there in each of the following measurements? (Hint: See Sample Problem 2-4.)
a. 0.4004 mL
b. 6000 g
c. 1.000 30 km
d. 400. mm

37. Calculate the sum of 6.078 g and 0.3329 g.

38. Subtract 7.11 cm from 8.2 cm. (Hint: See Sample Problem 2-5.)

39. What is the product of 0.8102 m and 3.44 m?

40. Divide 94.20 g by 3.167 22 mL.

Scientific Notation

41. Write the following numbers in scientific notation.
a. 0.000 673 0
b. 50 000.0
c. 0.000 003 010

42. The following numbers are in scientific notation. Write them in ordinary notation.
a. 7.050×10^{-3} g
b. $4.000\ 05 \times 10^{7}$ mg
c. $2.350\ 0 \times 10^{4}$ mL

43. Perform the following operations. Express each answer in scientific notation and with the correct number of significant figures.
a. $(8.205 \times 10^{4}\ \text{mg}) + (7.266 \times 10^{2}\ \text{mg})$
b. $(3.5011 \times 10^{-2}\ \text{kg}) - (9.2448 \times 10^{-3}\ \text{kg})$
c. $(8.61 \times 10^{5}\ \text{m})(7.9304 \times 10^{-8}\ \text{m}^2)$
d. $\dfrac{6.124\ 33 \times 10^{6}\ \text{m}^3}{7.15 \times 10^{-3}\ \text{m}}$

44. What quantities are represented by each of the answers in item 43?

45. A sample of a certain material has a mass of 2.03×10^{-3} g. Calculate the volume of the sample, given that the density is 9.133×10^{-1} g/cm^3. Use the four-step method in solving the problem. (Hint: See Sample Problem 2-6.)

MIXED REVIEW

46. A man finds that he has a mass of 100.6 kg. He goes on a diet, and several months later he finds that he has a mass of 96.4 kg. Express each number in scientific notation, and calculate the number of kilograms the man has lost by dieting.

47. A large office building is 1.07×10^{2} m long, 31 m wide, and 4.25×10^{2} m high. What is its volume?

48. An object is found to have a mass of 57.6 g. Find the object's density, given that its volume is 40.25 cm^3.

49. A student measures the mass of some sucrose as 0.947 mg. Convert that quantity to grams and to kilograms.

50. A student calculates the density of iron as 6.80 g/cm^3 using lab data for mass and volume. A handbook reveals that the correct value is 7.86 g/cm^3. What is the percent error?

HANDBOOK SEARCH

51. Find the table of properties for Group 1 elements in the *Elements Handbook*. Calculate the volume of a single atom of each element listed in the table using the equation for the volume of a sphere.

$$\frac{4}{3}\pi r^3$$

52. Using the volume of a sodium atom that you calculated in item 51, calculate the number of sodium atoms in a block of sodium with the following measurements.

3.0 cm × 5.0 cm × 5.0 cm

53. a. If the block described in item 52 has a measured mass of 75.5 g, calculate its density.

b. Compare your calculated density with the value in the properties table for Group 1 elements. Calculate the percent error for your density determination.

RESEARCH & WRITING

54. Find out how the metric system, which was once a standard for measurement, differs from SI. Why was it necessary to change to SI?

55. Find out what ISO 9000 standards are. How do they affect industry on an international level?

ALTERNATIVE ASSESSMENT

56. Performance Obtain three metal samples from your teacher. Determine the mass and volume of each sample. Calculate the density of each metal from your measurement data. (Hint: Consider using the water displacement technique to measure the volume of your samples.)

57. Using the data from the Nutrition Facts label below, answer the following.

a. Use the data given on the label for grams of fat and Calories from fat to construct a conversion factor with the units Calories per gram.

b. Calculate the mass in kilograms of 20 servings of the food.

c. Calculate the mass of protein in micrograms for one serving of the food.

d. What is the correct number of significant figures for the answer in item a? Why?

Nutrition Facts

Serving Size ¾ cup (30g)
Servings Per Container About 14

Amount Per Serving	Corn Crunch	with ½ cup skim milk
Calories	120	160
Calories from Fat	15	20
	% **Daily Value****	
Total Fat 2g*	**3%**	**3%**
Saturated Fat 0g	**0%**	**0%**
Cholesterol 0mg	**0%**	**1%**
Sodium 160mg	**7%**	**9%**
Potassium 65mg	**2%**	**8%**
Total Carbohydrate 25g	**8%**	**10%**
Dietary Fiber 3g		
Sugars 3g		
Other Carbohydrate 11g		
Protein 2g		

*Amount in Cereal. A serving of cereal plus skim milk provides 2g fat, less 5mg cholesterol, 220mg sodium, 270mg potassium, 31g carbohydrate (19g sugars) and 6g protein.

**Percent Daily Values are based on a 2,000 calorie diet. Your daily values may be higher or lower depending on your calorie needs:

	Calories	2,000	2,500
Total Fat	Less than	65g	80g
Sat Fat	Less than	20g	25g
Cholesterol	Less than	300mg	300mg
Sodium	Less than	2,400mg	2,400mg
Potassium		3,500mg	3,500mg
Total Carbohydrate		300g	375g
Dietary Fiber		25g	30g

UNIT 2

Organization of Matter

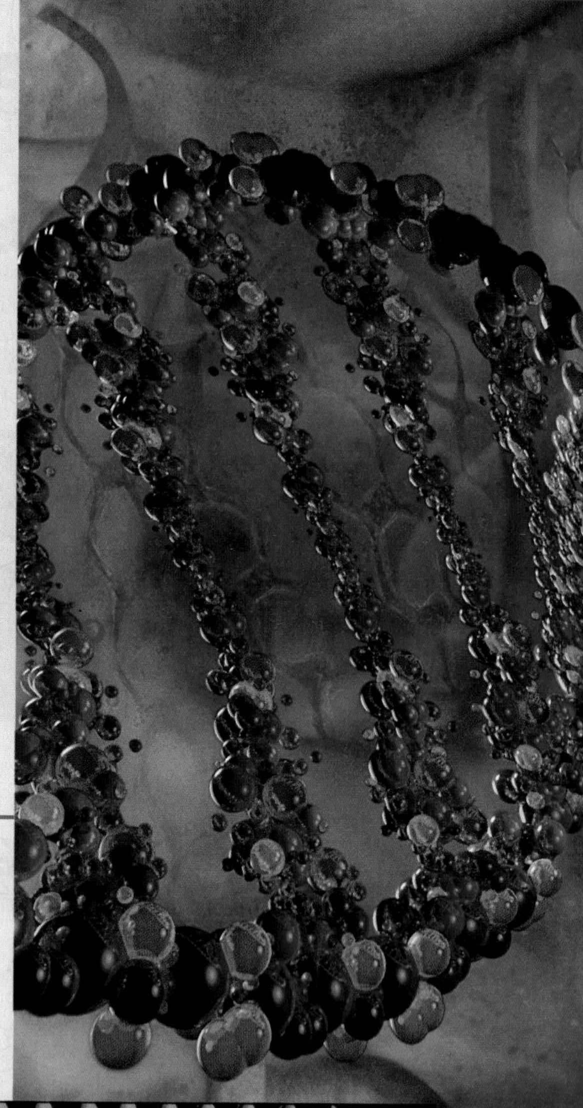

CHAPTERS

ROALD HOFFMANN, ON THE INDIRECT NATURE OF ATOMIC STUDY

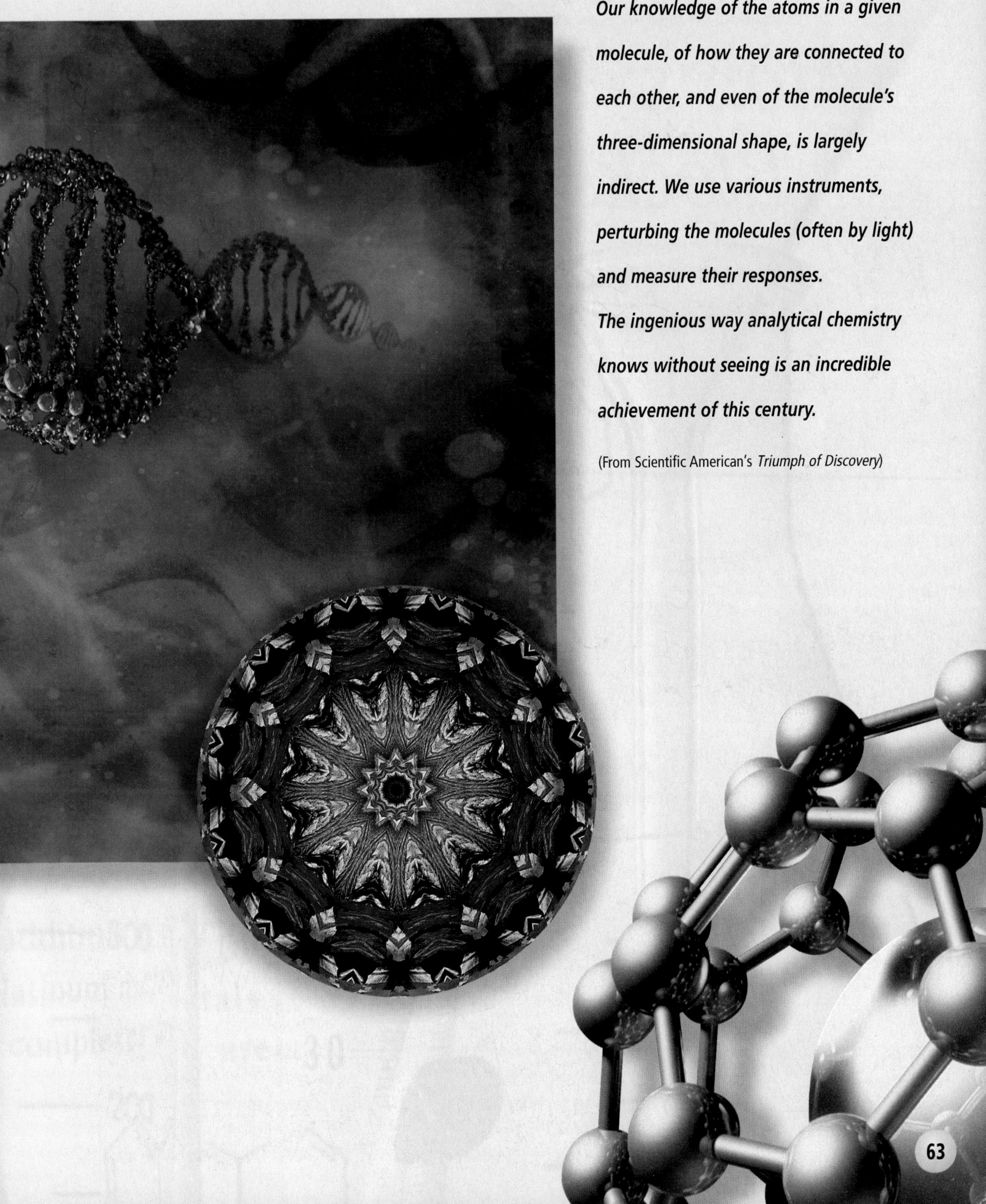

Our knowledge of the atoms in a given molecule, of how they are connected to each other, and even of the molecule's three-dimensional shape, is largely indirect. We use various instruments, perturbing the molecules (often by light) and measure their responses.

The ingenious way analytical chemistry knows without seeing is an incredible achievement of this century.

(From Scientific American's *Triumph of Discovery*)

CHAPTER 3

Atoms: The Building Blocks of Matter

An atom is the smallest particle of an element that retains the chemical properties of that element.

SECTION 3-1

The Atom: From Philosophical Idea to Scientific Theory

Objectives

- Explain the law of conservation of mass, the law of definite proportions, and the law of multiple proportions.
- Summarize the five essential points of Dalton's atomic theory.
- Explain the relationship between Dalton's atomic theory and the law of conservation of mass, the law of definite proportions, and the law of multiple proportions.

When you crush a lump of sugar, you can see that it is made up of many smaller particles of sugar. You may grind these particles into a very fine powder, but each tiny piece is still sugar. Now suppose you dissolve the sugar in water. The tiny particles seem to disappear completely. Even if you look at the sugar-water solution through a powerful microscope, you cannot see any sugar particles. Yet if you were to taste the solution, you'd know that the sugar is still there. Observations like these led early philosophers to ponder the fundamental nature of matter. Is it continuous and infinitely divisible, or is it divisible only until a basic, invisible particle that cannot be divided further is reached?

The particle theory of matter was supported as early as 400 B.C. by certain Greek thinkers, such as Democritus. He called nature's basic particle an *atom,* based on the Greek word meaning "indivisible." Aristotle was part of the generation that succeeded Democritus. His ideas had a lasting impact on Western civilization, and he did not believe in atoms. He thought that all matter was continuous, and his opinion was accepted for nearly 2000 years. Neither the view of Aristotle nor that of Democritus was supported by experimental evidence, so each remained speculation until the eighteenth century. Then scientists began to gather evidence favoring the atomic theory of matter.

Foundations of Atomic Theory

Virtually all chemists in the late 1700s accepted the modern definition of an element as a substance that cannot be further broken down by ordinary chemical means. It was also clear that elements combine to form compounds that have different physical and chemical properties than those of the elements that form them. There was great controversy, however, as to whether elements always combine in the same ratio when forming a particular compound.

The transformation of a substance or substances into one or more new substances is known as a *chemical reaction*. In the 1790s, the study of matter was revolutionized by a new emphasis on the quantitative

FIGURE 3-1 Each of the salt crystals shown here contains exactly 39.34% sodium and 60.66% chlorine by mass.

analysis of chemical reactions. Aided by improved balances, investigators began to accurately measure the masses of the elements and compounds they were studying. This lead to the discovery of several basic laws. One of these laws was the **law of conservation of mass,** *which states that mass is neither destroyed nor created during ordinary chemical or physical reactions.* This discovery was soon followed by the assertion that, regardless of where or how a pure chemical compound is prepared, it is composed of a fixed proportion of elements. For example, sodium chloride, also known as ordinary table salt, *always* consists of 39.34% by mass of the element sodium, Na, and 60.66% by mass of the element chlorine, Cl. *The fact that a chemical compound contains the same elements in exactly the same proportions by mass regardless of the size of the sample or source of the compound is known as the* **law of definite proportions.**

It was also known that two elements sometimes combine to form more than one compound. For example, the elements carbon and oxygen form two compounds, carbon dioxide and carbon monoxide. Consider examples of each of these compounds, each containing 1.0 g of carbon. In carbon dioxide, 2.66 g of oxygen combine with 1.0 g of carbon. In carbon monoxide, 1.33 g of oxygen combine with 1.0 g of carbon. The ratio of the masses of oxygen in these two compounds is exactly 2.66 to 1.33, or 2 to 1. This illustrates the **law of multiple proportions:** *If two or more different compounds are composed of the same two elements, then the ratio of the masses of the second element combined with a certain mass of the first element is always a ratio of small whole numbers.*

Dalton's Atomic Theory

In 1808, an English schoolteacher named John Dalton proposed an explanation for the law of conservation of mass, the law of definite proportions, and the law of multiple proportions. He reasoned that elements were composed of atoms and that only whole numbers of atoms can combine to form compounds. His theory can be summed up by the following statements.

1. All matter is composed of extremely small particles called atoms.
2. Atoms of a given element are identical in size, mass, and other properties; atoms of different elements differ in size, mass, and other properties.
3. Atoms cannot be subdivided, created, or destroyed.
4. Atoms of different elements combine in simple whole-number ratios to form chemical compounds.
5. In chemical reactions, atoms are combined, separated, or rearranged.

According to Dalton's atomic theory, the law of conservation of mass is explained by the fact that chemical reactions involve merely the combination, separation, or rearrangement of atoms and that during these processes atoms are not subdivided, created, or destroyed. This

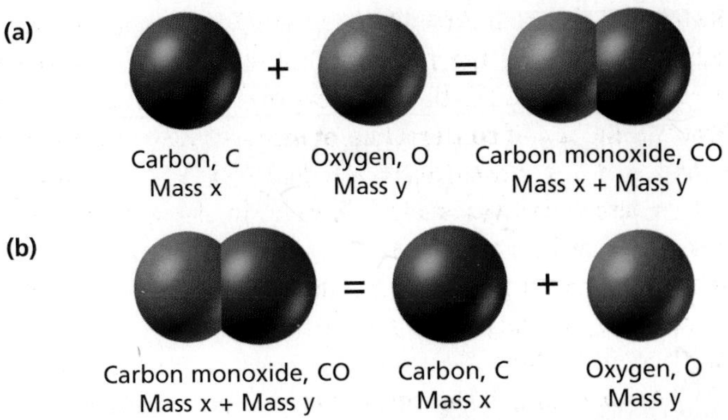

(a)

Carbon, C
Mass x

Oxygen, O
Mass y

Carbon monoxide, CO
Mass x + Mass y

(b)

Carbon monoxide, CO
Mass x + Mass y

Carbon, C
Mass x

Oxygen, O
Mass y

FIGURE 3-2 (a) An atom of carbon, C, and an atom of oxygen, O, can combine chemically to form a molecule of carbon monoxide, CO. The mass of the CO molecule is equal to the mass of the C atom plus the mass of the O atom. (b) The reverse holds true in a reaction in which a CO molecule is broken down into its elements.

idea is illustrated in Figure 3-2 for the formation of carbon monoxide from carbon and oxygen.

The law of definite proportions, on the other hand, results from the fact that a given chemical compound is always composed of the same combination of atoms (see Figure 3-3). As for the law of multiple proportions, in the case of the carbon oxides, the 2-to-1 ratio of oxygen masses results because carbon dioxide always contains twice as many atoms of oxygen (per atom of carbon) as does carbon monoxide. This can also be seen in Figure 3-3.

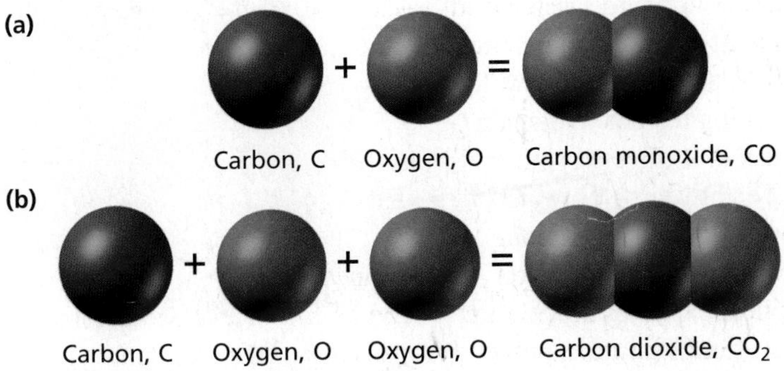

(a)

Carbon, C

Oxygen, O

Carbon monoxide, CO

(b)

Carbon, C

Oxygen, O

Oxygen, O

Carbon dioxide, CO_2

FIGURE 3-3 (a) CO molecules are always composed of one C atom and one O atom. (b) CO_2 molecules are always composed of one C atom and two O atoms. Note that a molecule of carbon dioxide contains twice as many oxygen atoms as does a molecule of carbon monoxide.

Modern Atomic Theory

By relating atoms to the measurable property of mass, Dalton turned Democritus's *idea* into a *scientific theory* that could be tested by experiment. But not all aspects of Dalton's atomic theory have proven to be correct. For example, today we know that atoms are divisible into even smaller particles (although the law of conservation of mass still holds true for chemical reactions). And, as you will see in Section 3-3, we know that a given element can have atoms with different masses. Atomic theory has not been discarded, however. Instead, it has been modified to explain the new observations. The important concepts that (1) all matter is composed of atoms and that (2) atoms of any one element differ in properties from atoms of another element remain unchanged.

Travels with C

From "Travels with C" by Primo Levi in *Creation to Chaos*

It was to carbon, the element of life, that my first literary dream was turned—and now I want to tell the story of a single atom of carbon.

My fictional character lies, for hundreds of millions of years, bound to three atoms of oxygen and one of calcium, in the form of limestone. (It already has behind it a very long cosmic history, but that we shall ignore.) Time does not exist for it, or exists only in the form of sluggish daily or seasonal variations in temperature. Its existence, whose monotony cannot be conceived of without horror, is an alternation of hots and colds.

The limestone ledge of which the atom forms a part lies within reach of man and his pickax. At any moment—which I, as narrator, decide out of pure caprice to be the year of 1840—a blow of the pickax detached the limestone and sent it on its way to the lime furnace, where it was plunged into the world of things that change. The atom of carbon was roasted until it separated from the limestone's calcium, which remained, so to speak, with its feet on the ground and went on to meet a less brilliant destiny. Still clinging firmly to two of its three companions, our fictional character issued from the chimney and rode the path of the air. Its story, which once was immobile, now took wing.

Accompanied by two oxygen atoms (red), the carbon atom (green) took to the air.

The atom was caught by the wind, flung down onto the earth, lifted ten kilometers high. It was breathed in by a falcon, but did not penetrate the bird's rich blood and was exhaled. It dissolved three times in the sea, once in the water of a cascading torrent, and again was expelled. It traveled with the wind for eight years—now high, now low, on the sea and among the clouds, over forests, deserts and limitless expanses of ice. Finally, it stumbled into capture and the organic adventure.

The year was 1848. The atom of carbon, accompanied by its two satellites of oxygen, which maintained it in a gaseous state, was borne by the wind along a row of vines. It had the good fortune to brush against a leaf, penetrate it, and be nailed there by a ray of the sun. On entering the leaf, it collided with other innumerable molecules of nitrogen and oxygen. It adhered to a large and complicated molecule that activated it, and simultaneously it received the decisive message from the sky, in the flashing form of a packet of solar light: in an instant, like an insect caught by a spider, the carbon atom was separated from its oxygen, combined with hydrogen, and finally inserted in a chain of life. All this happened swiftly, in silence, at the temperature and pressure of the atmosphere.

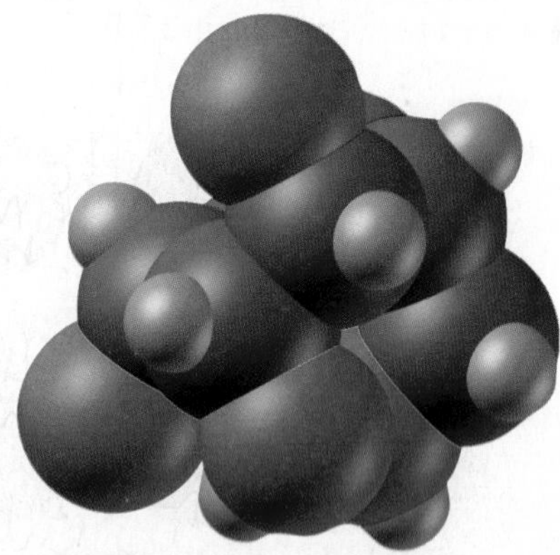

Once inside the leaf, the carbon atom joined other carbon atoms (green), as well as hydrogen (blue) and oxygen (red) atoms, to form this molecule essential to life.

Reading for Meaning

Name the various compounds that the carbon atom was a component of during the course of Levi's story.

Read Further

As a component of one particular compound, Levi's carbon atom is breathed in and exhaled by a falcon. Why was it unlikely for the carbon atom to have been taken into the bird's bloodstream?

DESKTOP INVESTIGATION

Constructing a Model

MATERIALS

- **can covered by a sock sealed with tape**
- **one or more objects that fit in the container**
- **metric ruler**
- **balance**

QUESTION

How can you construct a model of an unknown object by (1) making inferences about an object that is in a closed container and (2) touching the object without seeing it?

PROCEDURE

1. Your teacher will provide you with a can that is covered by a sock sealed with tape. Without unsealing the container, try to determine the number of objects inside the can as well as the mass, shape, size, composition, and texture of each. To do this, you may carefully tilt or shake the can. Record your observations in a data table.
2. Remove the tape from the top of the sock. Do *not* look inside the can. Put one hand through the opening, and make the same observations as in step 1 by handling the objects. To make more-accurate estimations, practice estimating the sizes and masses of some known objects outside the can. Then compare your estimates of these objects with actual measurements using a metric ruler and a balance.

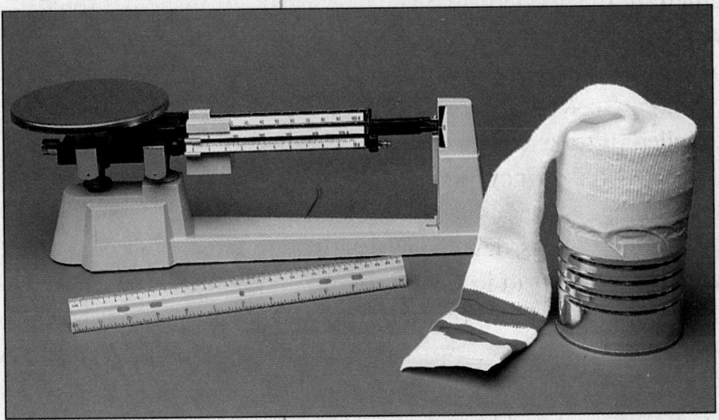

DISCUSSION

1. Scientists often use more than one method to gather data. How was this illustrated in the investigation?
2. Of the observations you made, which were qualitative and which were quantitative?
3. Using the data you gathered, draw a model of the unknown object(s) and write a brief summary of your conclusions.

SECTION REVIEW

1. Describe the major contributions of each of the following to the modern theory of the atom:
   a. Democritus
   b. John Dalton
2. List the five essential points of Dalton's atomic theory.
3. Using Dalton's theory, explain the law of conservation of mass, the law of definite proportions, and the law of multiple proportions.

SECTION 3-2

The Structure of the Atom

Objectives

- Summarize the observed properties of cathode rays that led to the discovery of the electron.
- Summarize the experiment carried out by Rutherford and his co-workers that led to the discovery of the nucleus.
- List the properties of protons, neutrons, and electrons.
- Define *atom.*

Although John Dalton thought atoms were indivisible, investigators in the late 1800s proved otherwise. As scientific advances allowed a deeper exploration of matter, it became clear that atoms are actually composed of several basic types of smaller particles and that the number and arrangement of these particles within an atom determine that atom's chemical properties. Today we define an **atom** *as the smallest particle of an element that retains the chemical properties of that element.*

All atoms consist of two regions. The *nucleus* is a very small region located near the center of an atom. In every atom the nucleus contains at least one positively charged particle called a *proton* and usually one or more neutral particles called *neutrons.* Surrounding the nucleus is a region occupied by negatively charged particles called *electrons.* This region is very large compared with the size of the nucleus. Protons, neutrons, and electrons are often referred to as *subatomic particles.*

Discovery of the Electron

The first discovery of a subatomic particle resulted from investigations into the relationship between electricity and matter. In the late 1800s, many experiments were performed in which electric current was passed through various gases at low pressures. (Gases at atmospheric pressure don't conduct electricity well.) These experiments were carried out in glass tubes like the one shown in Figure 3-4. Such tubes are known as *cathode-ray tubes.*

FIGURE 3-4 A simple cathode-ray tube. Particles pass through the tube from the *cathode,* the metal disk connected to the negative terminal of the voltage source, to the *anode,* the metal disk connected to the positive terminal.

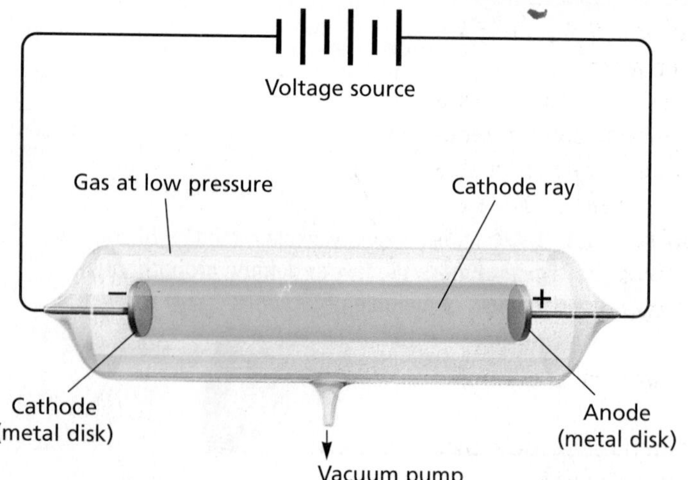

Cathode Rays and Electrons

Investigators noticed that when current was passed through a cathode-ray tube, the surface of the tube directly opposite the cathode glowed. They hypothesized that the glow was caused by a stream of particles, which they called a cathode ray. The ray traveled from the cathode to the anode when current was passed through the tube. Experiments devised to test this hypothesis revealed the following observations.

1. An object placed between the cathode and the opposite end of the tube cast a shadow on the glass.
2. A paddle wheel placed on rails between the electrodes rolled along the rails from the cathode toward the anode (see Figure 3-5).

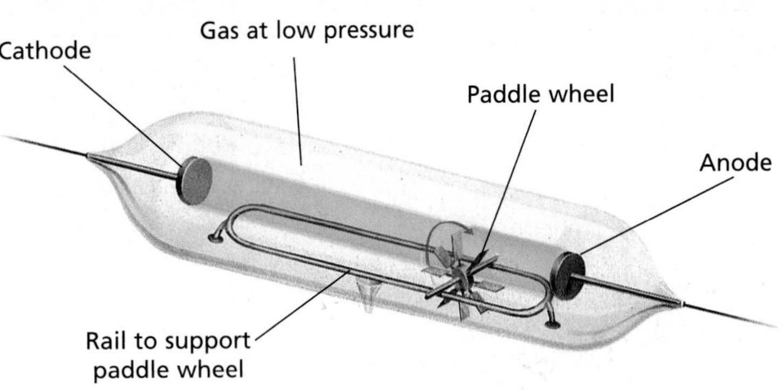

FIGURE 3-5 A paddle wheel placed in the path of the cathode ray moves away from the cathode and toward the anode. The movement of the wheel led scientists to conclude that cathode rays have mass.

These facts supported the existence of a cathode ray. Furthermore, the paddle-wheel experiment showed that a cathode ray had sufficient mass to set the wheel in motion.

Additional experiments provided more information.

3. Cathode rays were deflected by a magnetic field in the same manner as a wire carrying electric current, which was known to have a negative charge.
4. The rays were deflected away from a negatively charged object.

These observations led to the hypothesis that the particles that compose cathode rays are negatively charged. This hypothesis was strongly supported by a series of experiments carried out in 1897 by the English physicist Joseph John Thomson. In one investigation, he was able to measure the ratio of the charge of cathode-ray particles to their mass. He found that this ratio was always the same, regardless of the metal used to make the cathode or the nature of the gas inside the cathode-ray tube. Thomson concluded that all cathode rays are composed of identical negatively charged particles, which were later named electrons.

Charge and Mass of the Electron

Thomson's experiment revealed that the electron has a very large charge for its tiny mass. In 1909, experiments conducted by the American physicist Robert A. Millikan showed that the mass of the electron is in fact about one two-thousandth the mass of the simplest type of hydrogen atom, which is the smallest atom known. More-accurate experiments conducted since then indicate that the electron has a mass of 9.109×10^{-31}kg, or 1/1837 the mass of the simplest type of hydrogen atom.

Millikan's experiments also confirmed that the electron carries a negative electric charge. And because cathode rays have identical properties regardless of the element used to produce them, it was concluded that electrons are present in atoms of all elements. Thus, cathode-ray experiments provided evidence that atoms are divisible and that one of the atom's basic constituents is the negatively charged electron.

Based on what was learned about electrons, two other inferences were made about atomic structure.

1. Because atoms are electrically neutral, they must contain a positive charge to balance the negative electrons.
2. Because electrons have so much less mass than atoms, atoms must contain other particles that account for most of their mass.

Discovery of the Atomic Nucleus

More detail of the atom's structure was provided in 1911 by New Zealander Ernest Rutherford and his associates Hans Geiger and Ernest Marsden. The scientists bombarded a thin, gold foil with fast-moving *alpha particles,* which are positively charged particles with about four times the mass of a hydrogen atom. Geiger and Marsden assumed that mass and charge were uniformly distributed throughout the atoms of the gold foil. So they expected the alpha particles to pass through with only a slight deflection. And for the vast majority of the particles, this was the case. However, when the scientists checked for the possibility of wide-angle deflections, they were shocked to find that roughly 1 in 8000 of the alpha particles had actually been redirected back toward the source (see Figure 3-6). As Rutherford later exclaimed, it was "as if you had fired a 15-inch [artillery] shell at a piece of tissue paper and it came back and hit you."

After thinking about the startling result for two years, Rutherford finally came up with an explanation. He reasoned that the rebounded alpha particles must have experienced some powerful force within the atom. And he figured that the source of this force must occupy a very small amount of space because so few of the total number of alpha particles had been affected by it. He concluded that the force must be caused by a very densely packed bundle of matter with a positive electric charge. Rutherford called this positive bundle of matter the nucleus (see Figure 3-7).

Rutherford had discovered that the volume of a nucleus was very small compared with the total volume of an atom. In fact, if the nucleus were the size of a marble, then the size of the atom would be about the size of a football field. But where were the electrons? Although he had no supporting evidence, Rutherford suggested that the electrons surrounded the positively charged nucleus like planets around the sun. He could not explain, however, what kept the electrons in motion around the nucleus.

FIGURE 3-6 (a) Geiger and Marsden bombarded a thin piece of gold foil with a narrow beam of alpha particles. (b) Some of the particles were redirected by the gold foil back toward their source.

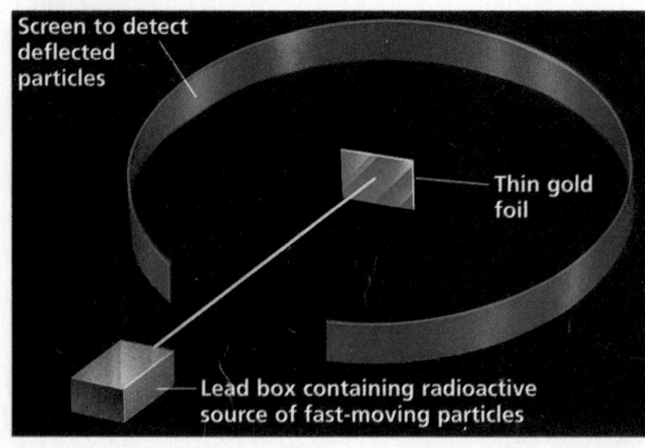

(a)

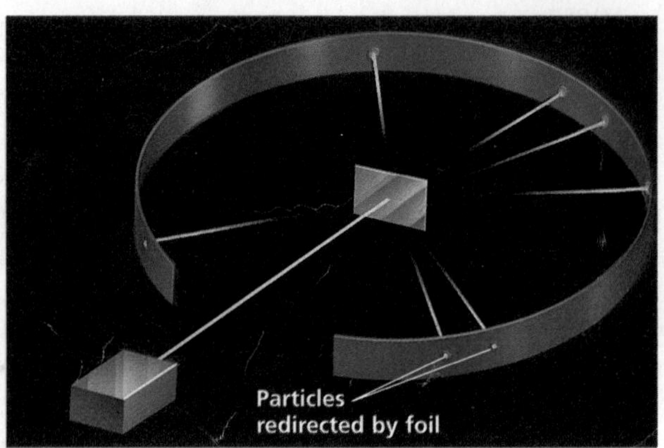

(b)

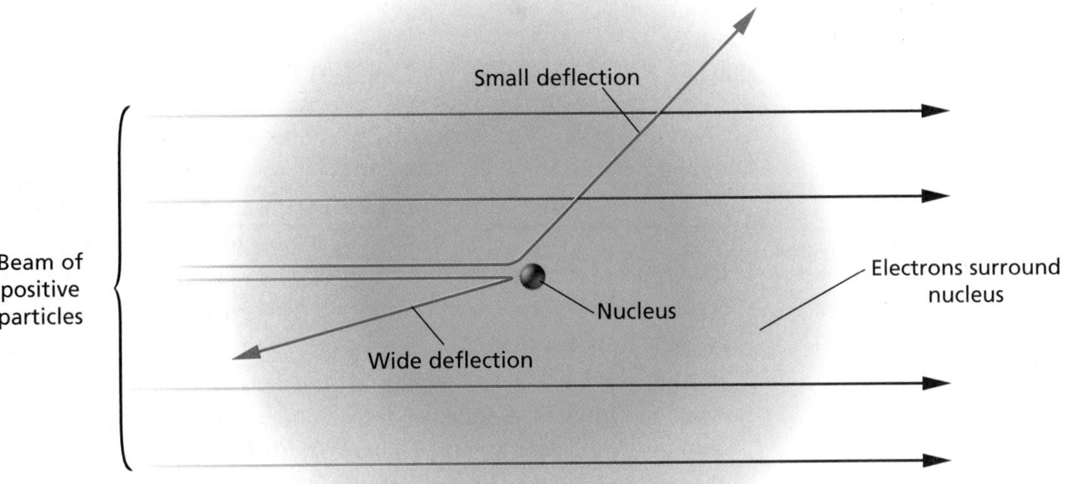

FIGURE 3-7 Rutherford reasoned that each atom in the gold foil contained a small, dense, positively charged nucleus surrounded by electrons. A small number of the alpha particles directed toward the foil were deflected by the tiny nucleus (red arrows). Most of the particles passed through undisturbed (black arrows).

Composition of the Atomic Nucleus

Except for the nucleus of the simplest type of hydrogen atom (discussed in the next section), all atomic nuclei are made of two kinds of particles, protons and neutrons. A proton has a positive charge equal in magnitude to the negative charge of an electron. Atoms are electrically neutral because they contain equal numbers of protons and electrons. A neutron is electrically neutral.

The simplest hydrogen atom consists of a single-proton nucleus with a single electron moving about it. A proton has a mass of 1.673×10^{-27} kg, which is 1836 times greater than the mass of an electron and 1836/1837, or virtually all, of the mass of the simplest hydrogen atom. All atoms besides the simplest hydrogen atom also contain neutrons. The mass of a neutron is 1.675×10^{-27} kg—slightly larger than that of a proton.

The nuclei of atoms of different elements differ in the number of protons they contain and therefore in the amount of positive charge they possess. Thus, the number of protons in an atom's nucleus determines that atom's identity. Physicists have identified other subatomic particles, but particles other than electrons, protons, and neutrons have little effect on the chemical properties of matter. Table 3-1 on page 74 summarizes the properties of electrons, protons, and neutrons.

Forces in the Nucleus

Generally, particles that have the same electric charge repel one another. Therefore, we would expect a nucleus with more than one proton to be unstable. However, when two protons are extremely close to each other, there is a strong attraction between them. In fact, more than 100

TABLE 3-1 ***Properties of Subatomic Particles***

Particle	Symbols	Relative electric charge	Mass number	Relative mass (amu*)	Actual mass (kg)
Electron	$e^-, {}_{-1}^{0}e$	–1	0	0.000 5486	9.109×10^{-31}
Proton	$p^+, {}_{1}^{1}\mathrm{H}$	+1	1	1.007 276	1.673×10^{-27}
Neutron	$n^\circ, {}_{0}^{1}n$	0	1	1.008 665	1.675×10^{-27}

*1 amu (atomic mass unit) = $1.660\,540 \times 10^{-27}$ kg (see page 78)

protons can exist close together in a nucleus. A similar attraction exists when neutrons are very close to each other, or when protons and neutrons are very close together. *These short-range proton-neutron, proton-proton, and neutron-neutron forces hold the nuclear particles together and are referred to as* **nuclear forces.**

The Sizes of Atoms

It is convenient to think of the region occupied by the electrons as an electron cloud—a cloud of negative charge. The radius of an atom is the distance from the center of the nucleus to the outer portion of this electron cloud. Because atomic radii are so small, they are expressed using a unit that is more convenient for the sizes of atoms. This unit is the picometer. The abbreviation for the picometer is pm (1 pm = 10^{-12} m = 10^{-10} cm). To get an idea of how small a picometer is, consider that 1 cm is the same fractional part of 10^3 km (about 600 mi) as 100 pm is of 1 cm. Atomic radii range from about 40 to 270 pm. By contrast, the nuclei of atoms have much smaller radii, about 0.001 pm. Nuclei also have incredibly high densities, about 2×10^8 metric tons/cm^3.

SECTION REVIEW

1. Define each of the following:
   a. atom
   b. electron
   c. nucleus
   d. proton
   e. neutron
2. Describe one conclusion made by each of the following scientists that led to the development of the current atomic theory:
   a. Thomson
   b. Millikan
   c. Rutherford
3. Compare and contrast the three types of subatomic particles in terms of location in the atom, mass, and relative charge.
4. Why is the cathode-ray tube in Figure 3-4 connected to a vacuum pump?
5. Label the charge on the following in a cathode-ray tube. State the reasons for your answers.
   a. anode
   b. cathode

Counting Atoms

SECTION 3-3

OBJECTIVES

- Explain what isotopes are.
- Define *atomic number* and *mass number*, and describe how they apply to isotopes.
- Given the identity of a nuclide, determine its number of protons, neutrons, and electrons.
- Define *mole* in terms of Avogadro's number, and define *molar mass*.
- Solve problems involving mass in grams, amount in moles, and number of atoms of an element.

Consider neon, Ne, the gas used in many illuminated signs. Neon is a minor component of the atmosphere. In fact, dry air contains only about 0.002% neon. And yet there are about 5×10^{17} atoms of neon present in each breath you inhale. In most experiments, atoms are much too small to be measured individually. Chemists can analyze atoms quantitatively, however, by knowing fundamental properties of the atoms of each element. In this section you will be introduced to some of the basic properties of atoms. You will then discover how to use this information to count the number of atoms of an element in a sample with a known mass. You will also become familiar with the *mole,* a special unit used by chemists to express amounts of particles, such as atoms and molecules.

Atomic Number

All atoms are composed of the same basic particles. Yet all atoms are not the same. Atoms of different elements have different numbers of protons. Atoms of the same element all have the same number of protons. *The* **atomic number** *(Z) of an element is the number of protons in the nucleus of each atom of that element.*

Turn to the inside back cover of this textbook. In the periodic table shown, an element's atomic number is indicated above its symbol. Notice that the elements are placed in order of increasing atomic number. At the top left of the table is hydrogen, H, which has atomic number 1. Atoms of the element hydrogen have one proton in the nucleus. Next in order is helium, He, which has two protons in each nucleus. Lithium, Li, has three protons; beryllium, Be, has four protons; and so on.

The atomic number identifies an element. If you want to know which element has atomic number 47, for example, look at the periodic table. You can see that it is silver, Ag. All silver atoms contain 47 protons in their nuclei. Because atoms are neutral, we know from the atomic number that all silver atoms must also contain 47 electrons.

3
Li
Lithium
6.941
[He]$2s^1$

FIGURE 3-8 The atomic number in this periodic-table entry reveals that an atom of lithium has three protons in its nucleus.

Isotopes

The simplest atoms are those of hydrogen. All hydrogen atoms contain only one proton. However, like many naturally occurring elements, hydrogen atoms can contain different numbers of neutrons.

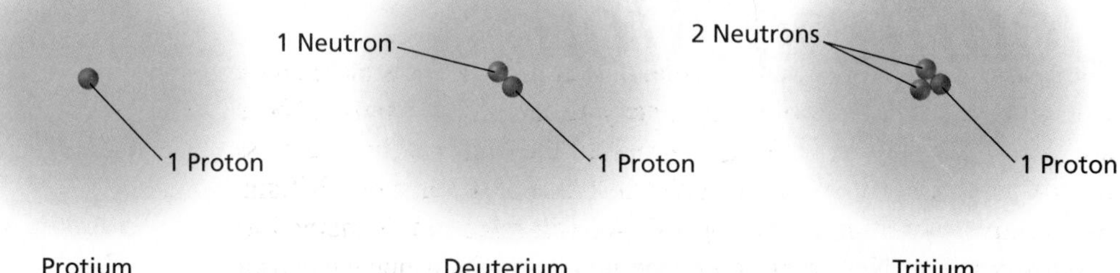

FIGURE 3-9 The nuclei of different isotopes of the same element have the same number of protons but different numbers of neutrons. This is illustrated above by the three isotopes of hydrogen.

Three types of hydrogen atoms are known. The most common type of hydrogen is sometimes called *protium.* It accounts for 99.985% of the hydrogen atoms found on Earth. The nucleus of a protium atom consists of one proton only, and it has one electron moving about it. There are two other known forms of hydrogen. One is called *deuterium,* which accounts for 0.015% of Earth's hydrogen atoms. Each deuterium atom has a nucleus containing one proton and one neutron. The third form of hydrogen is known as *tritium,* which is radioactive. It exists in very small amounts in nature, but it can be prepared artificially. Each tritium atom contains one proton, two neutrons, and one electron.

Protium, deuterium, and tritium are isotopes of hydrogen. **Isotopes** *are atoms of the same element that have different masses.* The isotopes of a particular element all have the same number of protons and electrons but different numbers of neutrons. In all three isotopes of hydrogen, the positive charge of the single proton is balanced by the negative charge of the electron. Most of the elements consist of mixtures of isotopes. Tin has 10 stable isotopes, for example, the most of any element.

Mass Number

Identifying an isotope requires knowing both the name or atomic number of the element and the mass of the isotope. *The* **mass number** *is the total number of protons and neutrons in the nucleus of an isotope.* The three isotopes of hydrogen described earlier have mass numbers 1, 2, and 3, as shown in Table 3-2.

TABLE 3-2 *Mass Numbers of Hydrogen Isotopes*

	Atomic number (number of protons)	Number of neutrons	Mass number
Protium	1	0	$1 + 0 = 1$
Deuterium	1	1	$1 + 1 = 2$
Tritium	1	2	$1 + 2 = 3$

Designating Isotopes

The isotopes of hydrogen are unusual in that they have distinct names. Isotopes are usually identified by specifying their mass number. There are two methods for specifying isotopes. In the first method, the mass number is written with a hyphen after the name of the element. Tritium, for example, is written as hydrogen-3. We will refer to this method as *hyphen notation.* The uranium isotope used as fuel for nuclear power plants has a mass number of 235 and is therefore known as uranium-235. The second method shows the composition of a nucleus as the isotope's *nuclear symbol.* For example, uranium-235 is written as $^{235}_{92}U$. The superscript indicates the mass number and the subscript indicates the atomic number. The number of neutrons is found by subtracting the atomic number from the mass number.

mass number − atomic number = number of neutrons
235 (protons + neutrons) − 92 protons = 143 neutrons

Thus, a uranium-235 nucleus contains 92 protons and 143 neutrons.

Table 3-3 gives the names, symbols, and compositions of the isotopes of hydrogen and helium. **Nuclide** *is a general term for any isotope of any element.* We could say that Table 3-3 lists the compositions of five different nuclides.

TABLE 3-3 *Isotopes of Hydrogen and Helium*

Isotope	Nuclear symbol	Number of protons	Number of electrons	Number of neutrons
Hydrogen-1 (protium)	$^{1}_{1}H$	1	1	0
Hydrogen-2 (deuterium)	$^{2}_{1}H$	1	1	1
Hydrogen-3 (tritium)	$^{3}_{1}H$	1	1	2
Helium-3	$^{3}_{2}He$	2	2	1
Helium-4	$^{4}_{2}He$	2	2	2

SAMPLE PROBLEM 3-1

How many protons, electrons, and neutrons are there in an atom of chlorine-37?

SOLUTION

1 *ANALYZE*

Given: name and mass number of chlorine-37
Unknown: numbers of protons, electrons, and neutrons

2 *PLAN*

atomic number = number of protons = number of electrons
mass number = number of neutrons + number of protons

3 *COMPUTE* The mass number of chlorine-37 is 37. Consulting the periodic table reveals that chlorine's atomic number is 17. The number of neutrons can be found by subtracting the atomic number from the mass number.

mass number of chlorine-37 – atomic number of chlorine =
number of neutrons in chlorine-37

mass number – atomic number = 37 (protons plus neutrons) – 17 protons
= 20 neutrons

An atom of chlorine-37 contains 17 electrons, 17 protons, and 20 neutrons.

4 *EVALUATE* The number of protons in a neutral atom equals the number of electrons. And the sum of the protons and neutrons equals the given mass number.

PRACTICE

1. How many protons, electrons, and neutrons are in an atom of bromine-80?
   Answer
   35 protons, 35 electrons, 45 neutrons

2. Write the nuclear symbol for carbon-13.
   Answer
   $^{13}_{6}C$

3. Write the hyphen notation for the element that contains 15 electrons and 15 neutrons.
   Answer
   phosphorus-30

Relative Atomic Masses

Masses of atoms measured in grams are very small. An atom of oxygen-16, for example, has a mass of 2.657×10^{-23} g. For most chemical calculations it is more convenient to use *relative* atomic masses. As you read in Chapter 2, scientists use standards of measurement that are constant and are the same everywhere. In order to set up a relative scale of atomic mass, one atom has been arbitrarily chosen as the standard and assigned a relative mass value. The masses of all other atoms are expressed in relation to this defined standard.

The standard used by scientists to govern units of atomic mass is the carbon-12 nuclide. It has been arbitrarily assigned a mass of exactly 12 atomic mass units, or 12 amu. *One* **atomic mass unit,** *or 1 amu, is exactly 1/12 the mass of a carbon-12 atom, or 1.660 540 × 10^{-27} kg.* The atomic mass of any nuclide is determined by comparing it with the mass of the carbon-12 atom. The hydrogen-1 atom has an atomic mass of *about* 1/12 that of the carbon-12 atom, or about 1 amu. The precise value of the atomic mass of a hydrogen-1 atom is 1.007 825 amu. An oxygen-16 atom has about 16/12 (or 4/3) the mass of a carbon-12 atom. Careful measurements show the atomic mass of oxygen-16 to be 15.994 915 amu. The mass of a magnesium-24 atom is found to be slightly less than twice that of a carbon-12 atom. Its atomic mass is 23.985 042 amu.

Some additional examples of the atomic masses of the naturally occurring isotopes of several elements are given in Table 3-4 on page 80. Isotopes of an element may occur naturally, or they may be made in the laboratory *(artificial isotopes)*. *Although isotopes have different masses, they do not differ significantly in their chemical behavior.*

The masses of subatomic particles can also be expressed on the atomic mass scale (see Table 3-1). The mass of the electron is 0.000 5486 amu, that of the proton is 1.007 276 amu, and that of the neutron is 1.008 665 amu. Note that the proton and neutron masses are close to but not equal to 1 amu. You have learned that the mass number is the total number of protons and neutrons in the nucleus of an atom. You can now see that the mass number and relative atomic mass of a given nuclide are quite close to each other. They are not identical because the proton and neutron masses deviate slightly from 1 amu and the atomic masses include electrons. Also, as you will read in Chapter 22, a small amount of mass is changed to energy in the creation of a nucleus from protons and neutrons.

Average Atomic Masses of Elements

Most elements occur naturally as mixtures of isotopes, as indicated in Table 3-4. The percentage of each isotope in the naturally occurring element on Earth is nearly always the same, no matter where the element is found. The percentage at which each of an element's isotopes occurs in nature is taken into account when calculating the element's average atomic mass. **Average atomic mass** *is the weighted average of the atomic masses of the naturally occurring isotopes of an element.*

The following is a simple example of how to calculate a *weighted average.* Suppose you have a box containing two sizes of marbles. If 25% of the marbles have masses of 2.00 g each and 75% have masses of 3.00 g each, how is the weighted average calculated? You could count the marbles, calculate the total mass of the mixture, and divide by the total number of marbles. If you had 100 marbles, the calculations would be as follows.

$$25 \text{ marbles} \times 2.00 \text{ g} = 50 \text{ g}$$
$$75 \text{ marbles} \times 3.00 \text{ g} = 225 \text{ g}$$

Adding these masses gives the total mass of the marbles.

$$50 \text{ g} + 225 \text{ g} = 275 \text{ g}$$

Dividing the total mass by 100 gives an average marble mass of 2.75 g.

A simpler method is to multiply the mass of each marble by the decimal fraction representing its percentage in the mixture. Then add the products.

$$25\% = 0.25 \qquad 75\% = 0.75$$
$$(2.00 \text{ g} \times 0.25) + (3.00 \text{ g} \times 0.75) = 2.75 \text{ g}$$

TABLE 3-4 Atomic Masses and Abundances of Several Naturally Occurring Isotopes

Isotope	Mass number	Percentage natural abundance	Atomic mass (amu)	Average atomic mass of element (amu)
Hydrogen-1	1	99.985	1.007 825	1.007 94
Hydrogen-2	2	0.015	2.014 102	
Carbon-12	12	98.90	12 (by definition)	12.0111
Carbon-13	13	1.10	13.003 355	
Carbon-14	14	trace	14.003 242	
Oxygen-16	16	99.762	15.994 915	15.9994
Oxygen-17	17	0.038	16.999 131	
Oxygen-18	18	0.200	17.999 160	
Copper-63	63	69.17	62.929 599	63.546
Copper-65	65	30.83	64.927 793	
Cesium-133	133	100	132.905 429	132.905
Uranium-234	234	0.005	234.040 947	238.029
Uranium-235	235	0.720	235.043 924	
Uranium-238	238	99.275	238.050 784	

Calculating Average Atomic Mass

The average atomic mass of an element depends on both the mass and the relative abundance of each of the element's isotopes. For example, naturally occurring copper consists of 69.17% copper-63, which has an atomic mass of 62.929 598 amu, and 30.83% copper-65, which has an atomic mass of 64.927 793 amu. The average atomic mass of copper can be calculated by multiplying the atomic mass of each isotope by its relative abundance (expressed in decimal form) and adding the results.

$$0.6917 \times 62.929\ 599 \text{ amu} + 0.3083 \times 64.927\ 793 \text{ amu} = 63.55 \text{ amu}$$

The calculated average atomic mass of naturally occurring copper is 63.55 amu.

The average atomic mass is included for the elements listed in Table 3-4. As illustrated in the table, most atomic masses are known to four or more significant figures. *In this book, an element's atomic mass is usually rounded to two decimal places before it is used in a calculation.*

Relating Mass to Numbers of Atoms

The relative atomic mass scale makes it possible to know how many atoms of an element are present in a sample of the element with a measurable mass. Three very important concepts—the mole, Avogadro's number, and molar mass—provide the basis for relating masses in grams to numbers of atoms.

The Mole

The mole is the SI unit for amount of substance. *A* **mole** *(abbreviated* mol*) is the amount of a substance that contains as many particles as there are atoms in exactly 12 g of carbon-12.* The mole is a counting unit, just like a dozen is. We don't usually order 12 or 24 ears of corn; we order one dozen or two dozen. Similarly, a chemist may want 1 mol of carbon, or 2 mol of iron, or 2.567 mol of calcium. In the sections that follow, you will see how the mole relates to masses of atoms and compounds.

Avogadro's Number

The number of particles in a mole has been experimentally determined in a number of ways. The best modern value is $6.022\ 1367 \times 10^{23}$. This means that exactly 12 g of carbon-12 contains $6.022\ 1367 \times 10^{23}$ carbon-12 atoms. The number of particles in a mole is known as Avogadro's number, named for the nineteenth-century Italian scientist Amedeo Avogadro, whose ideas were crucial in explaining the relationship between mass and numbers of atoms. **Avogadro's number**—*$6.022\ 1367 \times 10^{23}$—is the number of particles in exactly one mole of a pure substance.* For most purposes, Avogadro's number is rounded to 6.022×10^{23}.

To get a sense of how large Avogadro's number is, consider the following: If every person living on Earth (5 billion people) worked to count the atoms in one mole of an element, and if each person counted continuously at a rate of one atom per second, it would take about 4 million years for all the atoms to be counted.

Molar Mass

An alternative definition of *mole* is the amount of a substance that contains Avogadro's number of particles. Can you figure out the approximate mass of one mole of helium atoms? You know that a mole of carbon-12 atoms has a mass of exactly 12 g and that a carbon-12 atom has an atomic mass of 12 amu. The atomic mass of a helium atom is 4.00 amu, which is about one-third the mass of a carbon-12 atom. It follows that a mole of helium atoms will have about one-third the mass of a mole of carbon-12 atoms. Thus, one mole of helium has a mass of about 4.00 g.

The mass of one mole of a pure substance is called the **molar mass** *of that substance.* Molar mass is usually written in units of g/mol. The molar mass of an element is numerically equal to the atomic mass of the element in atomic mass units (which can be found in the periodic table). For example, the molar mass of lithium, Li, is 6.94 g/mol, while the molar mass of mercury, Hg, is 200.59 g/mol (rounding each value to two decimal places).

A molar mass of an element contains one mole of atoms. For example, 4.00 g of helium, 6.94 g of lithium, and 200.59 g of mercury all contain a mole of atoms. Figure 3-10 shows molar masses of three common elements.

Gram/Mole Conversions

Chemists use molar mass as a conversion factor in chemical calculations. For example, the molar mass of helium is 4.00 g He/mol He. To

(a)

(b)

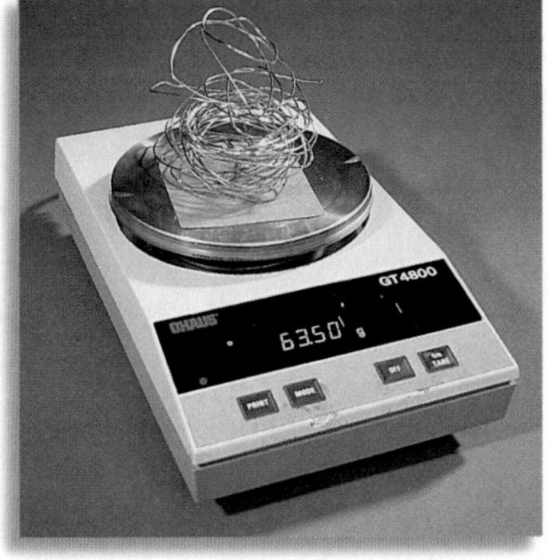

(c)

FIGURE 3-10 Shown is approximately one molar mass of each of three elements: (a) carbon (as graphite), (b) iron (nails), and (c) copper (wire).

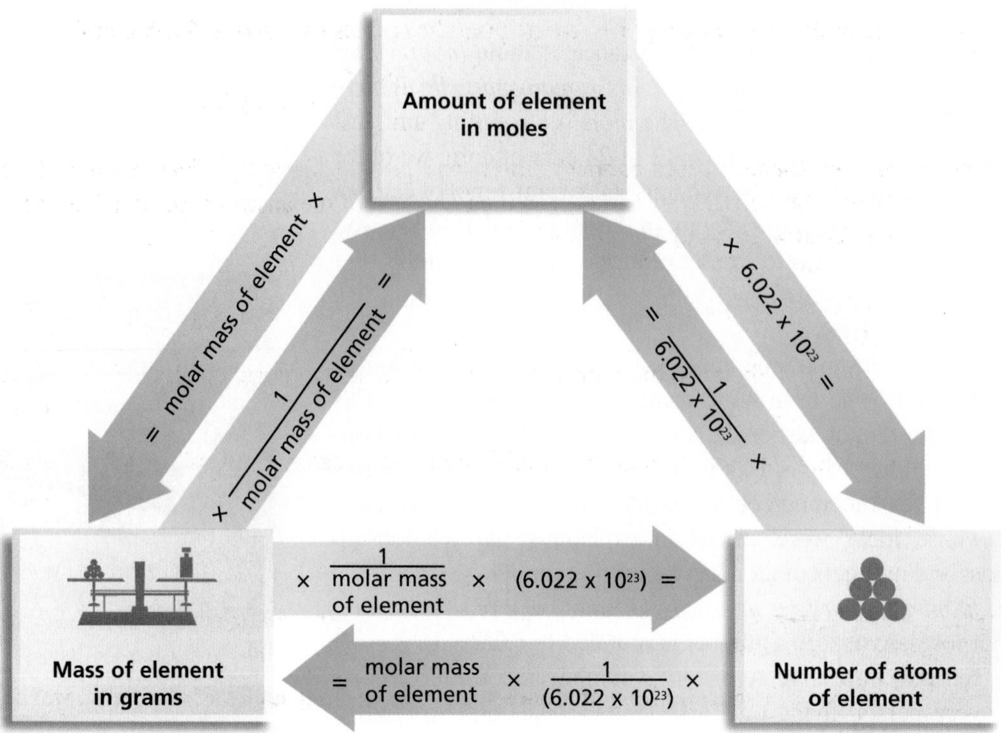

FIGURE 3-11 The diagram shows the relationship between mass in grams, amount in moles, and number of atoms of an element in a sample.

find how many grams of helium there are in two moles of helium, multiply by the molar mass.

$$2.00 \cancel{\text{mol He}} \times \frac{4.00 \text{ g He}}{\cancel{\text{mol He}}} = 8.00 \text{ g He}$$

Figure 3-11 shows how to use molar mass, moles, and Avogadro's number to relate mass in grams, amount in moles, and number of atoms of an element.

SAMPLE PROBLEM 3-2

What is the mass in grams of 3.50 mol of the element copper, Cu?

SOLUTION

1 *ANALYZE* **Given:** 3.50 mol Cu
Unknown: mass of Cu in grams

2 *PLAN*

$$\text{amount of Cu in moles} \longrightarrow \text{mass of Cu in grams}$$

According to Figure 3-11, the mass of an element in grams can be calculated by multiplying the amount of the element in moles by the element's molar mass.

$$\text{moles Cu} \times \frac{\text{grams Cu}}{\text{moles Cu}} = \text{grams Cu}$$

3 COMPUTE The molar mass of copper from the periodic table is rounded to 63.55 g/mol.

$$3.50 \cancel{\text{mol Cu}} \times \frac{63.55 \text{ g Cu}}{\cancel{\text{mol Cu}}} = 222 \text{ g Cu}$$

4 EVALUATE Because the amount of copper in moles was given to three significant figures, the answer was rounded to three significant figures. The size of the answer is reasonable because it is somewhat more than 3.5 times 60.

PRACTICE

1. What is the mass in grams of 2.25 mol of the element iron, Fe? *Answer* 126 g Fe
2. What is the mass in grams of 0.375 mol of the element potassium, K? *Answer* 14.7 g K

SAMPLE PROBLEM 3-3

A chemist produced 11.9 g of aluminum, Al. How many moles of aluminum were produced?

SOLUTION

1 ANALYZE **Given:** 11.9 g Al
Unknown: amount of Al in moles

2 PLAN

$$\text{mass of Al in grams} \longrightarrow \text{amount of Al in moles}$$

As shown in Figure 3-11, amount in moles can be obtained by *dividing* mass in grams by molar mass, which is mathematically the same as *multiplying* mass in grams by the *reciprocal* of molar mass.

$$\text{grams Al} \times \frac{\text{moles Al}}{\text{grams Al}} = \text{moles Al}$$

3 COMPUTE The molar mass of aluminum from the periodic table is rounded to 26.98 g/mol.

$$11.9 \cancel{\text{g Al}} \times \frac{\text{mol Al}}{26.98 \cancel{\text{g Al}}} = 0.441 \text{ mol Al}$$

4 EVALUATE The answer is correctly given to three significant figures. The answer is reasonable because 11.9 g is somewhat less than half of 26.98 g.

PRACTICE

1. How many moles of calcium, Ca, are in 5.00 g of calcium? *Answer* 0.125 mol Ca
2. How many moles of gold, Au, are in 3.60×10^{-10} g of gold? *Answer* 1.83×10^{-12} mol Au

Conversions with Avogadro's Number

Figure 3-11 shows that Avogadro's number can be used to find the number of atoms of an element from the amount in moles or to find the amount of an element in moles from the number of atoms. While these types of problems are less common in chemistry than converting between amount in moles and mass in grams, they are useful in demonstrating the meaning of Avogadro's number. Note that in these calculations, Avogadro's number is expressed in units of atoms per mole.

SAMPLE PROBLEM 3-4

How many moles of silver, Ag, are in 3.01×10^{23} atoms of silver?

SOLUTION

1 *ANALYZE*

Given: 3.01×10^{23} atoms of Ag
Unknown: amount of Ag in moles

2 *PLAN*

$$\text{number of atoms of Ag} \longrightarrow \text{amount of Ag in moles}$$

From Figure 3-11, we know that number of atoms is converted to amount in moles by dividing by Avogadro's number. This is equivalent to multiplying numbers of atoms by the reciprocal of Avogrado's number.

$$\text{Ag atoms} \times \frac{\text{moles Ag}}{\text{Avogadro's number of Ag atoms}} = \text{moles Ag}$$

3 *COMPUTE*

$$3.01 \times 10^{23}\ \cancel{\text{Ag atoms}} \times \frac{\text{mol Ag}}{6.022 \times 10^{23}\ \cancel{\text{Ag atoms}}} = 0.500\ \text{mol Ag}$$

4 *EVALUATE*

The answer is correct—units cancel correctly and the number of atoms is exactly one-half of Avogadro's number.

PRACTICE

1. How many moles of lead, Pb, are in 1.50×10^{12} atoms of lead?
   Answer 2.49×10^{-12} mol Pb
2. How many moles of tin, Sn, are in 2500 atoms of tin?
   Answer 4.2×10^{-21} mol Sn
3. How many atoms of aluminum, Al, are in 2.75 mol of aluminum?
   Answer 1.66×10^{24} atoms Al

SAMPLE PROBLEM 3-5

What is the mass in grams of 1.20×10^{8} atoms of copper, Cu?

SOLUTION

1 *ANALYZE*

Given: 1.20×10^{8} atoms of Cu
Unknown: mass of Cu in grams

2 PLAN

number of atoms of Cu ⟶ amount of Cu in moles ⟶ mass of Cu in grams

As indicated in Figure 3-11, the given number of atoms must first be converted to amount in moles by dividing by Avogadro's number. Amount in moles is then multiplied by molar mass to yield mass in grams.

$$\text{Cu atoms} \times \frac{\text{moles Cu}}{\text{Avogadro's number of Cu atoms}} \times \frac{\text{grams Cu}}{\text{moles Cu}} = \text{grams Cu}$$

3 COMPUTE

The molar mass of copper from the periodic table is rounded to 63.55 g/mol.

$$1.20 \times 10^8 \cancel{\text{Cu atoms}} \times \frac{1 \cancel{\text{mol Cu}}}{6.022 \times 10^{23} \cancel{\text{Cu atoms}}} \times \frac{63.55 \text{ g Cu}}{\cancel{\text{mol Cu}}} = 1.27 \times 10^{-14} \text{ g Cu}$$

4 EVALUATE

Units cancel correctly to give the answer in grams. The size of the answer is reasonable—10^8 has been divided by about 10^{24}.

PRACTICE

1. What is the mass in grams of 7.5×10^{15} atoms of nickel, Ni? — *Answer* 7.3×10^{-7} g Ni

2. How many atoms of sulfur, S, are in 4.00 g of sulfur? — *Answer* 7.51×10^{-22} atoms S

3. What mass of gold, Au, contains the same number of atoms as 9.0 g of aluminum, Al? — *Answer* 66 g Au

SECTION REVIEW

1. Define each of the following:
   a. atomic number
   b. mass number
   c. relative atomic mass
   d. average atomic mass
   e. mole
   f. Avogadro's number
   g. molar mass
   h. isotope

2. Determine the number of protons, electrons, and neutrons in each of the following isotopes:
   a. sodium-23
   b. calcium-40
   c. $^{64}_{29}\text{Cu}$
   d. $^{108}_{47}\text{Ag}$

3. Write the nuclear symbol and hyphen notation for each of the following isotopes:
   a. mass number of 28 and atomic number of 14
   b. 26 protons and 30 neutrons
   c. 56 electrons and 82 neutrons

4. To two decimal places, what is the relative atomic mass and the molar mass of the element potassium, K?

5. Determine the mass in grams of the following:
   a. 2.00 mol N
   b. 3.01×10^{23} atoms Cl

6. Determine the amount in moles of the following:
   a. 12.15 g Mg
   b. 1.50×10^{23} atoms F

7. Determine the number of atoms in the following:
   a. 2.50 mol Zn
   b. 1.50 g C

CHAPTER 3 REVIEW

CHAPTER SUMMARY

3-1

- The *idea* of atoms has been around since the time of the ancient Greeks. In the nineteenth century, John Dalton proposed a *scientific theory* of atoms that can still be used to explain properties of many chemicals today.
- When elements react to form compounds, they combine in fixed proportions by mass.
- Matter and its mass cannot be created or destroyed in chemical reactions.
- The mass ratio of the atoms of the elements that make up a given compound is always the same, regardless of how much of the compound there is or how it was formed.
- If two or more different compounds are composed of the same two elements, then the ratio of the masses of the second element combined with a certain mass of the first element can be expressed as a ratio of small whole numbers.

Vocabulary

law of conservation of mass (66) law of definite proportions (66) law of multiple proportions (66)

3-2

- Cathode-ray tubes supplied evidence of the existence of electrons, which are negatively charged subatomic particles that have relatively little mass.
- Rutherford found evidence for the existence of the atomic nucleus—a positively charged, very dense core within the atom—by bombarding metal foil with a beam of positively charged particles.
- Atomic nuclei are composed of protons, which have an electric charge of +1, and (in all but one case) neutrons, which have no electric charge.
- Isotopes of an element differ by the number of neutrons in their nuclei.
- Atomic nuclei have radii of about 0.001 pm (pm = picometers; 1 pm = 10^{-12} m), while atoms have radii of about 40–270 pm.

Vocabulary

atom (70) nuclear forces (74)

3-3

- The atomic number of an element is equal to the number of protons in the nucleus of an atom of that element.
- The mass number is equal to the total number of protons and neutrons in the nucleus of an atom of that element.
- The relative atomic mass unit (amu) is based on the carbon-12 atom and is a convenient unit for measuring the mass of atoms. It equals $1.660\,540 \times 10^{-24}$ kg.
- The average atomic mass of an element is found by calculating the weighted average of the atomic masses of the naturally occurring isotopes of the element.
- Avogadro's number is equal to approximately $6.022\,137 \times 10^{23}$. It is equal to the number of atoms in exactly 12 g of carbon-12. A sample that contains a number of particles equal to Avogadro's number contains a mole of those particles.
- The molar mass of an element is the mass of one mole of atoms of that element.

Vocabulary

atomic mass unit (78) Avogadro's number (81) mass number (76) mole (81)
atomic number (75) isotope (76) molar mass (81) nuclide (77)
average atomic mass (79)

REVIEWING CONCEPTS

1. Explain each of the following in terms of Dalton's atomic theory:
   a. the law of conservation of mass
   b. the law of definite proportions
   c. the law of multiple proportions (3-1)
2. According to the law of conservation of mass, if element A has an atomic mass of 2 mass units and element B has an atomic mass of 3 mass units, what mass would be expected for compound AB? for compound A_2B_3? (3-1)
3. a. What is an atom?
   b. What two regions make up all atoms? (3-2)
4. Describe at least four properties of electrons that were determined based on the experiments of Thomson and Millikan. (3-2)
5. Summarize Rutherford's model of an atom, and explain how he developed this model based on the results of his famous gold-foil experiment. (3-2)
6. What one number identifies an element? (3-2)
7. a. What are isotopes?
   b. How are the isotopes of a particular element alike?
   c. How are they different? (3-3)
8. Copy and complete the following table concerning the three isotopes of silicon, Si. (Hint: See Sample Problem 3-1.) (3-3)

Isotope	Number of protons	Number of electrons	Number of neutrons
Si-28			
Si-29			
Si-30			

9. a. What is the atomic number of an element?
   b. What is the mass number of an isotope?
   c. In the nuclear symbol for deuterium, ${}^{2}_{1}H$, identify the atomic number and the mass number. (3-3)
10. What is a nuclide? (3-3)
11. Use the periodic table and the information that follows to write the hyphen notation for each isotope described.
    a. atomic number = 2, mass number = 4
    b. atomic number = 8, mass number = 16
    c. atomic number = 19, mass number = 39 (3-3)
12. a. What nuclide is used as the standard in the relative scale for atomic masses?
    b. What is its assigned atomic mass? (3-3)
13. What is the atomic mass of an atom if its mass is approximately equal to the following?
    a. $\frac{1}{3}$ that of carbon-12
    b. 4.5 times as much as carbon-12 (3-3)
14. a. What is the definition of a mole?
    b. What is the abbreviation for mole?
    c. How many particles are in one mole?
    d. What name is given to the number of particles in a mole? (3-3)
15. a. What is the molar mass of an element?
    b. To two decimal places, write the molar masses of carbon, neon, iron, and uranium. (3-3)
16. Suppose you have a sample of an element.
    a. How is the mass in grams of the element converted to amount in moles?
    b. How is the mass in grams of the element converted to number of atoms? (3-3)

PROBLEMS

The Mole and Molar Mass

17. What is the mass in grams of each of the following? (Hint: See Sample Problems 3-2 and 3-5.)
    a. 1.00 mol Li
    b. 1.00 mol Al
    c. 1.00 molar mass Ca
    d. 1.00 molar mass Fe
    e. 6.022×10^{23} atoms C
    f. 6.022×10^{23} atoms Ag
18. How many moles of atoms are there in each of the following? (Hint: See Sample Problems 3-3 and 3-4.)
    a. 6.022×10^{23} atoms Ne
    b. 3.011×10^{23} atoms Mg
    c. 3.25×10^{5} g Pb
    d. 4.50×10^{-12} g O

Relative Atomic Mass

19. Three isotopes of argon occur in nature—${}^{36}_{18}Ar$, ${}^{38}_{18}Ar$, and ${}^{40}_{18}Ar$. Calculate the average atomic mass of argon to two decimal places, given the following relative atomic masses and abundances of each of the isotopes: argon-36 (35.97 amu; 0.337%), argon-38 (37.96 amu; 0.063%), and argon-40 (39.96 amu; 99.600%).

20. Naturally occurring boron is 80.20% boron-11 (atomic mass = 11.01 amu) and 19.80% of some other isotopic form of boron. What must the atomic mass of this second isotope be in order to account for the 10.81 amu average atomic mass of boron? (Write the answer to two decimal places.)

Number of Atoms in a Sample

21. How many atoms are there in each of the following?
a. 1.50 mol Na
b. 6.755 mol Pb
c. 0.250 mol Si

22. What is the mass in grams of each of the following?
a. 3.011×10^{23} atoms F
b. 1.50×10^{23} atoms Mg
c. 4.50×10^{12} atoms Cl
d. 8.42×10^{18} atoms Br
e. 25 atoms W
f. 1 atom Au

23. Determine the number of atoms in each of the following:
a. 5.40 g B
b. 8.02 g S
c. 1.50 g K
d. 0.025 50 g Pt
e. 1.00×10^{-10} g Au

MIXED REVIEW

24. Determine the mass in grams of each of the following:
a. 3.00 mol Al
b. 4.25 mol Li
c. 1.38 mol N
d. 8.075 mol Au
e. 6.50 mol Cu
f. 2.57×10^{8} mol S
g. 1.75×10^{-6} mol Hg

25. Copy and complete the following table concerning the properties of subatomic particles:

Particle	Symbol	Mass number	Actual mass	Relative charge
Electron				
Proton				
Neutron				

26. a. How is an atomic mass unit (amu) related to the mass of a carbon-12 atom?
b. What is the relative atomic mass of an atom?

27. a. What is the nucleus of an atom?
b. Who is credited with the discovery of the atomic nucleus?
c. Identify the two kinds of particles contained in the nucleus.

28. How many moles of atoms are there in each of the following?
a. 40.1 g Ca
b. 11.5 g Na
c. 5.87 g Ni
d. 150 g S
e. 2.65 g Fe
f. 0.007 50 g Ag
g. 2.25×10^{25} atoms Zn
h. 50.0 atoms Ba

29. State the law of multiple proportions, and give an example of two compounds that illustrate the law.

30. What is the approximate atomic mass of an atom if its mass is
a. 12 times that of carbon-12?
b. $\frac{1}{2}$ that of carbon-12?

31. What is an electron?

CRITICAL THINKING

32. Organizing Ideas Using two chemical compounds as an example, describe the difference between the law of definite proportions and the law of multiple proportions.

33. Constructing Models As described on pages 70 to 74, the structure of the atom was determined from observations made in painstaking experimental research. Suppose a series of experiments revealed that when an electric current is passed through gas at low pressure, the surface of the cathode-ray tube opposite the anode glows. In addition, a paddle wheel placed in the tube rolls from the anode toward the cathode when the current is on.
a. In which direction do particles pass through the gas?
b. What charge do the particles possess?

34. Inferring Relationships How much mass is converted into energy during the creation of the nucleus of a $^{235}_{92}U$ nuclide from 92 protons, 143 neutrons, and 92 electrons? (Hint: See Section 22-1.)

TECHNOLOGY & LEARNING

35. Graphing Calculator Calculate Numbers of Protons, Electrons, and Neutrons

The graphing calculator can be programmed to calculate the numbers of protons, electrons, and neutrons given the atomic and mass numbers for an atom. Given a calcium-40 atom, you will calculate the number of protons, electrons, and neutrons in the atom. Begin by programming the calculator to carry out the calculation. The program will then be used to make calculations.

A. Program the calculation.

Keystrokes for naming the program: PRGM ▶ ▶ ENTER

Name the program: N U M B E R ENTER

Program keystrokes: PRGM ▶ 2 ALPHA A 2nd : PRGM ▶ 2 ALPHA M 2nd : ALPHA M – ALPHA A STO► ALPHA N 2nd : PRGM ▶ 8 2nd : PRGM ▶ 3 2nd ALPHA " P R O T O N S 2nd TEST 1 ALPHA " 2nd : PRGM ▶ 3 ALPHA A 2nd : PRGM ▶ 3 2nd ALPHA " E L E C T R O N S 2nd TEST 1 ALPHA " 2nd : PRGM ▶ 3 ALPHA A 2nd : PRGM ▶ 3 2nd ALPHA " N E U T R O N S 2nd TEST 1 ALPHA " 2nd : PRGM ▶ 3 ALPHA N

B. Check the display.

Screen display: Prompt A: Prompt M: M–A->N: ClrHome: Disp "PROTONS=": Disp A: Disp "ELECTRONS=": Disp A: Disp "NEUTRONS=": Disp N

Press 2nd QUIT to exit the program editor.

C. Run the program.

a. Press PRGM. Select NUMBER and press ENTER ENTER. Now enter the atomic number 20 and press ENTER. Enter the mass number 40 and press ENTER. The calculator will provide the numbers of protons, electrons, and neutrons.

b. Use the NUMBER program with manganese-55.

c. Use the NUMBER program with tungsten-184.

HANDBOOK SEARCH

36. Group 14 of the *Elements Handbook* describes the reactions that produce CO and CO_2. Review this section to answer the following:

a. When a fuel burns, what determines whether CO or CO_2 will be produced?

b. What happens in the body if hemoglobin picks up CO instead of CO_2 or O_2?

c. Why is CO poisoning most likely to occur in homes that are well sealed during cold winter months?

RESEARCH & WRITING

37. Prepare a report on the series of experiments conducted by Sir James Chadwick that led to the discovery of the neutron.

38. Write a report on the contributions of Amedeo Avogadro that led to the determination of the value of Avogadro's number.

39. Trace the development of the electron microscope, and cite some of its many uses.

40. The study of atomic structure and the nucleus produced a new field of medicine called nuclear medicine. Describe the use of radioactive tracers to detect and treat diseases.

ALTERNATIVE ASSESSMENT

41. Observe a cathode-ray tube in operation, and write a description of your observations.

42. Performance Assessment Using colored clay, build a model of the nucleus of each of carbon's three naturally occurring isotopes: carbon-12, carbon-13, and carbon-14. Specify the number of electrons that would surround each nucleus.

CHAPTER 4

Arrangement of Electrons in Atoms

The emission of light is fundamentally related to the behavior of electrons.

The Development of a New Atomic Model

SECTION 4-1

OBJECTIVES

- Explain the mathematical relationship between the speed, wavelength, and frequency of electromagnetic radiation.
- Discuss the dual wave-particle nature of light.
- Discuss the significance of the photoelectric effect and the line-emission spectrum of hydrogen to the development of the atomic model.
- Describe the Bohr model of the hydrogen atom.

The Rutherford model of the atom was an improvement over previous models, but it was incomplete. It did not explain how the atom's negatively charged electrons occupy the space surrounding its positively charged nucleus. After all, it was well known that oppositely charged particles attract each other. So what prevented the negative electrons from being drawn into the positive nucleus?

In the early twentieth century, a new atomic model evolved as a result of investigations into the absorption and emission of light by matter. The studies revealed an intimate relationship between light and an atom's electrons. This new understanding led directly to a revolutionary view of the nature of energy, matter, and atomic structure.

Properties of Light

Before 1900, scientists thought light behaved solely as a wave. This belief changed when it was later discovered that light also has particle-like characteristics. Still, many of light's properties can be described in terms of waves. A quick review of these wavelike properties will help you understand the basic theory of light as it existed at the beginning of the twentieth century.

Visible light is a kind of **electromagnetic radiation,** *which is a form of energy that exhibits wavelike behavior as it travels through space.* Other kinds of electromagnetic radiation include X rays, ultraviolet and infrared light, microwaves, and radio waves. *Together, all the forms of electromagnetic radiation form the* **electromagnetic spectrum.** The electromagnetic spectrum is represented in Figure 4-1 on page 92. All forms of electromagnetic radiation move at a constant speed of about 3.0×10^8 meters per second (m/s) through a vacuum and at slightly slower speeds through matter. Because air is mostly space, the value of 3.0×10^8 m/s is usually considered light's approximate speed through air.

The significant feature of wave motion is its repetitive nature, which can be characterized by the measurable properties of wavelength and frequency. **Wavelength** (λ) *is the distance between corresponding points on adjacent waves.* Depending on the particular form of electromagnetic radiation, the unit for wavelength is the meter, centimeter, or nanometer (1 nm = 1×10^{-9} m), as shown in Figure 4-1. **Frequency** (ν) *is defined*

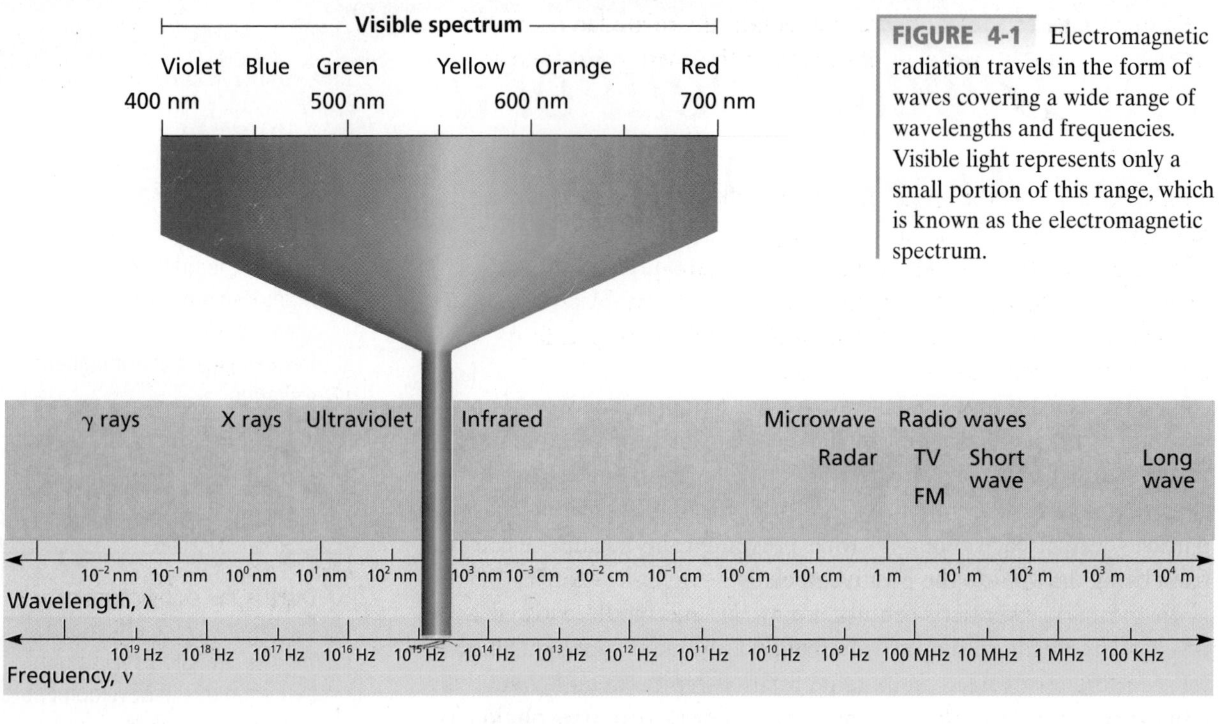

FIGURE 4-1 Electromagnetic radiation travels in the form of waves covering a wide range of wavelengths and frequencies. Visible light represents only a small portion of this range, which is known as the electromagnetic spectrum.

as the number of waves that pass a given point in a specific time, usually one second. Frequency is expressed in waves/second. One wave/second is called a hertz (Hz), named for Heinrich Hertz, who was a pioneer in the study of electromagnetic radiation. Figure 4-2 illustrates the properties of wavelength and frequency for a familiar kind of wave, a wave on the surface of water. The wavelength in Figure 4-2(a) has a longer wavelength and a lower frequency than the wave in Figure 4-2(b).

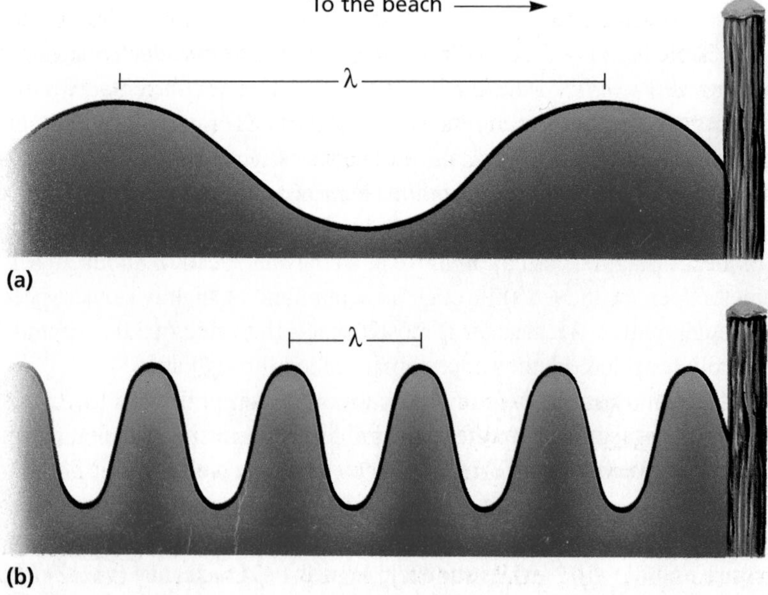

FIGURE 4-2 The distance between any two corresponding points on one of these water waves, such as from crest to crest, is the wave's wavelength, λ. We can measure the wave's frequency, ν, by observing how often the water level rises and falls at a given point, such as at the post.

Frequency and wavelength are mathematically related to each other. For electromagnetic radiation, this relationship is written as follows.

$$c = \lambda\nu$$

In the equation, c is the speed of light, λ is the wavelength of the electromagnetic wave, and ν is the frequency of the electromagnetic wave. Because c is the same for all electromagnetic radiation, the product $\lambda\nu$ is a constant. Consequently, we know that λ is inversely proportional to ν. In other words, as the wavelength of light decreases, its frequency increases, and vice versa.

The Photoelectric Effect

In the early 1900s, scientists conducted two experiments involving interactions of light and matter that could not be explained by the wave theory of light. One experiment involved a phenomenon known as the photoelectric effect. *The* **photoelectric effect** *refers to the emission of electrons from a metal when light shines on the metal,* as illustrated in Figure 4-3.

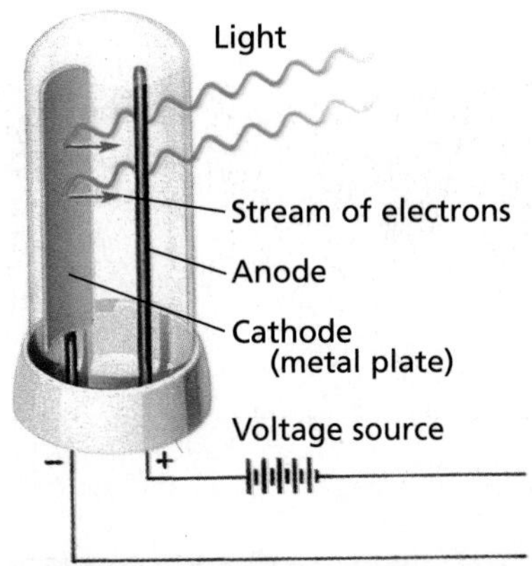

FIGURE 4-3 The photoelectric effect: electromagnetic radiation strikes the surface of the metal, ejecting electrons from the metal and creating an electric current.

The mystery of the photoelectric effect involved the frequency of the light striking the metal. For a given metal, no electrons were emitted if the light's frequency was below a certain minimum—regardless of how long the light was shone. Light was known to be a form of energy, capable of knocking loose an electron from a metal. But the wave theory of light predicted that light of any frequency could supply enough energy to eject an electron. Scientists couldn't explain why the light had to be of a minimum frequency in order for the photoelectric effect to occur.

Light as Particles

The explanation of the photoelectric effect dates back to 1900, when German physicist Max Planck was studying the emission of light by hot objects. He proposed that a hot object does not emit electromagnetic energy continuously, as would be expected if the energy emitted were in the form of waves. Instead, Planck suggested that the object emits energy in small, specific amounts called quanta. *A* **quantum** *is the minimum quantity of energy that can be lost or gained by an atom.* Planck proposed the following relationship between a quantum of energy and the frequency of radiation.

$$E = h\nu$$

In the equation, E is the energy, in joules, of a quantum of radiation, ν is the frequency of the radiation emitted, and h is a fundamental physical constant now known as Planck's constant; $h = 6.626 \times 10^{-34}$ J s.

In 1905, Albert Einstein expanded on Planck's theory by introducing the radical idea that electromagnetic radiation has a dual wave-particle nature. While light exhibits many wavelike particles, it can also be thought of as a stream of particles. Each particle of light carries a quantum of energy. Einstein called these particles photons. *A* **photon** *is a particle of electromagnetic radiation having zero rest mass and carrying a quantum of energy.* The energy of a particular photon depends on the frequency of the radiation.

$$E_{photon} = h\nu$$

Einstein explained the photoelectric effect by proposing that electromagnetic radiation is absorbed by matter only in whole numbers of photons. In order for an electron to be ejected from a metal surface, the electron must be struck by a single photon possessing at least the minimum energy required to knock the electron loose. According to the equation $E_{photon} = h\nu$, this minimum energy corresponds to a minimum frequency. If a photon's frequency is below the minimum, then the electron remains bound to the metal surface. Electrons in different metals are bound more or less tightly, so different metals require different minimum frequencies to exhibit the photoelectric effect.

The Hydrogen-Atom Line-Emission Spectrum

FIGURE 4-4 Excited neon atoms emit light when falling back to the ground state or to a lower-energy excited state.

When current is passed through a gas at low pressure, the potential energy of some of the gas atoms increases. *The lowest energy state of an atom is its* **ground state.** *A state in which an atom has a higher potential energy than it has in its ground state is an* **excited state.** When an excited atom returns to its ground state, it gives off the energy it gained in the form of electromagnetic radiation. The production of colored light in neon signs is a familiar example of this process.

When investigators passed electric current through a vacuum tube containing hydrogen gas at low pressure, they observed the emission of a characteristic pinkish glow. When a narrow beam of the emitted light was shone through a prism, it was separated into a series of specific frequencies (and therefore specific wavelengths, $\lambda = c/\nu$) of visible light. The bands of light were part of what is known as hydrogen's *line-emission spectrum.* The production of hydrogen's line-emission spectrum is illustrated in Figure 4-5.

Classical theory predicted that the hydrogen atoms would be excited by whatever amount of energy was added to them. Scientists had thus expected to observe the emission of a continuous series of electromagnetic radiation. Why had the hydrogen atoms given off only specific frequencies of light? Attempts to explain this observation led to an entirely new theory of the atom called *quantum theory.*

Whenever an excited hydrogen atom falls back from an excited state to its ground state or to a lower-energy excited state, it emits a photon of radiation. The energy of this photon ($E_{photon} = h\nu$) is equal to the difference in energy between the atom's initial state and its final state,

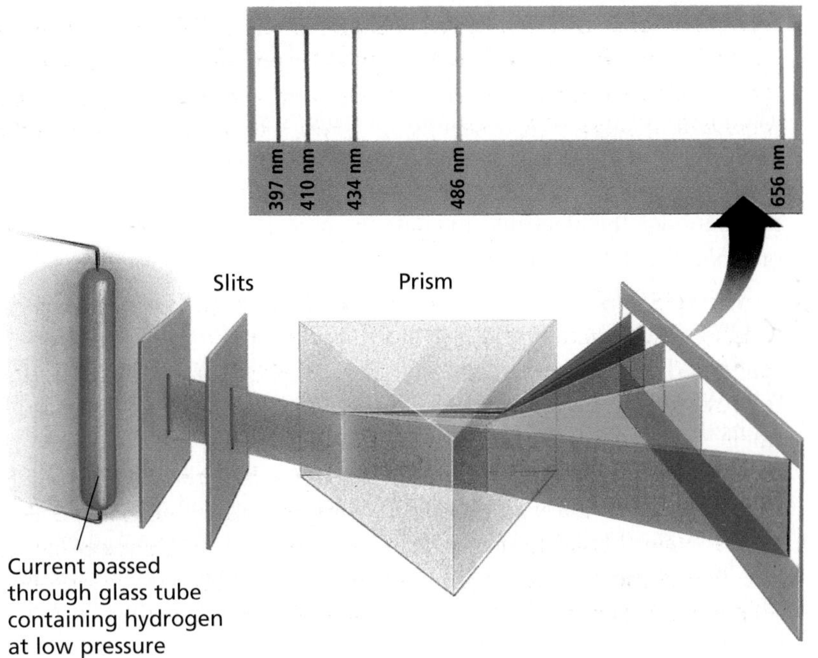

FIGURE 4-5 Excited hydrogen atoms emit a pinkish glow. When the visible portion of the emitted light is passed through a prism, it is separated into specific wavelengths that are part of hydrogen's line-emission spectrum.

as illustrated in Figure 4-6. The fact that hydrogen atoms emit only specific frequencies of light indicated that the energy differences between the atoms' energy states were fixed. This suggested that the electron of a hydrogen atom exists only in very specific energy states.

Additional series of lines were discovered in the ultraviolet and infrared regions of hydrogen's line-emission spectrum. The wavelengths of some of the spectral series are shown in Figure 4-7. They are known as the Lyman, Balmer, and Paschen series, after their discoverers. In the late nineteenth century, a mathematical relationship that related the various wavelengths of hydrogen's line-emission spectrum was developed. The challenge facing scientists was to provide a model of the hydrogen atom that accounted for this relationship.

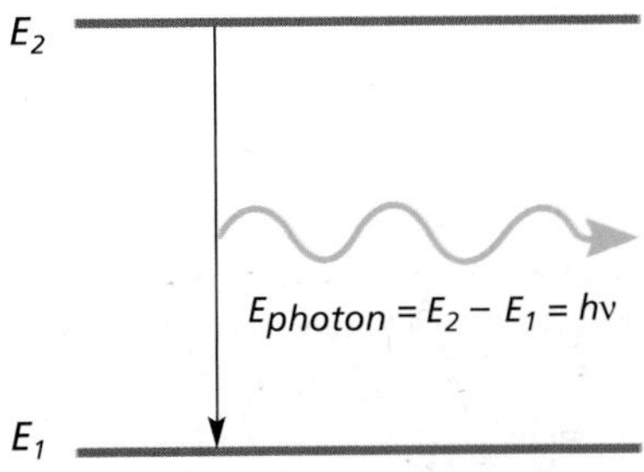

FIGURE 4-6 When an excited atom with energy E_2 falls back to energy E_1, it releases a photon that has energy $E_2 - E_1 = E_{photon} = h\nu$.

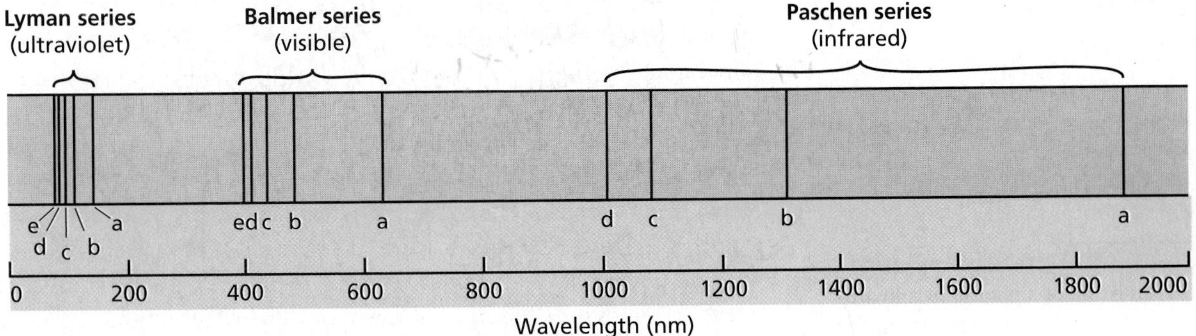

FIGURE 4-7 A series of specific wavelengths of emitted light makes up hydrogen's line-emission spectrum. The letters below the lines label hydrogen's various energy-level transitions. Niels Bohr's model of the hydrogen atom provided an explanation for these transitions.

Bohr Model of the Hydrogen Atom

The puzzle of the hydrogen-atom spectrum was solved in 1913 by the Danish physicist Niels Bohr. He proposed a model of the hydrogen atom that linked the atom's electron with photon emission. According to the model, the electron can circle the nucleus only in allowed paths, or *orbits.* When the electron is in one of these orbits, the atom has a definite, fixed energy. The electron, and therefore the hydrogen atom, is in its lowest energy state when it is in the orbit closest to the nucleus. This orbit is separated from the nucleus by a large empty space where the electron cannot exist. The energy of the electron is higher when it is in orbits that are successively farther from the nucleus.

The electron orbits or atomic energy levels in Bohr's model can be compared to the rungs of a ladder. When you are standing on a ladder, your feet are on one rung or another. The amount of potential energy that you possess corresponds to standing on the first rung, the second rung, and so forth. Your energy cannot correspond to standing between two rungs because you cannot stand in midair. In the same way, an electron can be in one orbit or another, but not in between.

How does Bohr's model of the hydrogen atom explain the observed spectral lines? While in an orbit, the electron can neither gain nor lose energy. It can, however, move to a higher energy orbit by gaining an amount of energy equal to the difference in energy between the higher-energy orbit and the initial lower-energy orbit. When a hydrogen atom is in an excited state, its electron is in a higher-energy orbit. When the atom falls back from the excited state, the electron drops down to a lower-energy orbit. In the process, a photon is emitted that has an energy equal to the energy difference between the initial higher-energy orbit and the final lower-energy orbit. Absorption and emission of radiation according to the Bohr model of the hydrogen atom are illustrated in Figure 4-8. The energy of each emitted photon corresponds to a particular frequency of emitted radiation, $E_{photon} = h\nu$.

Based on the wavelengths of hydrogen's line-emission spectrum, Bohr calculated the energies that an electron would have in the allowed energy levels for the hydrogen atom. He then used these values to show

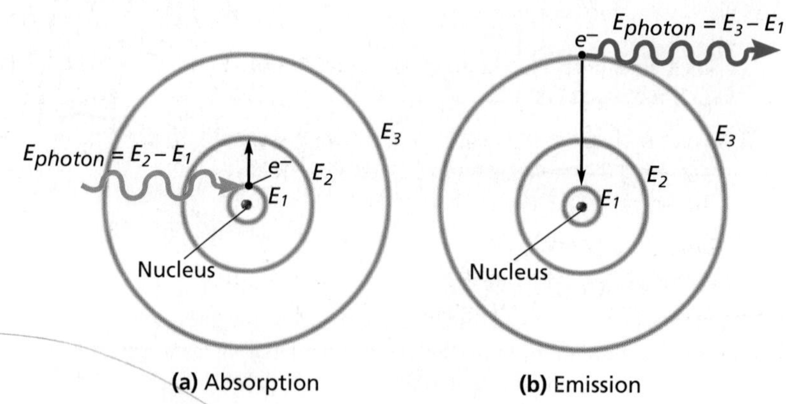

FIGURE 4-8 (a) Absorption and (b) emission of a photon by a hydrogen atom according to Bohr's model. The frequencies of light that can be absorbed and emitted are restricted because the electron can only be in orbits corresponding to the energies E_1, E_2, E_3, and so forth.

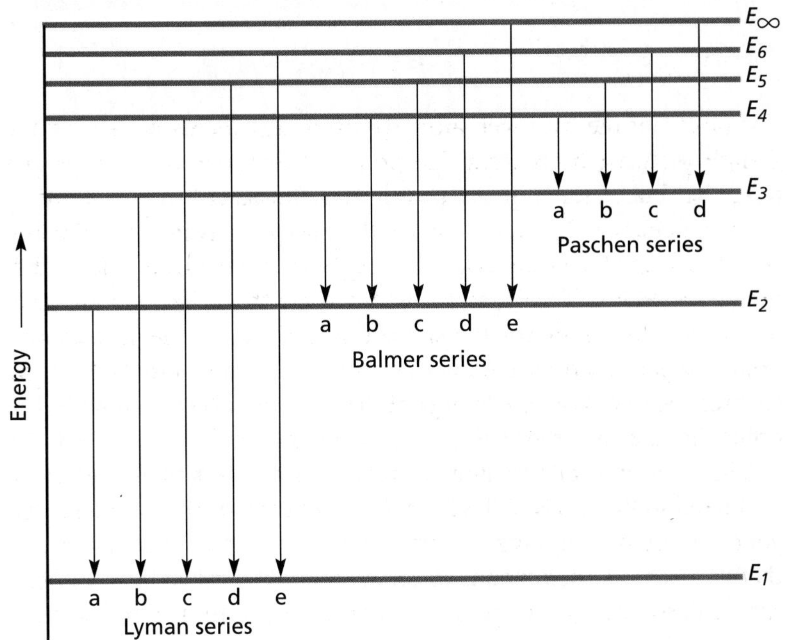

FIGURE 4-9 This electron energy-level diagram for hydrogen shows the energy transitions for the Lyman, Balmer, and Paschen spectral series. Bohr's model of the atom accounted mathematically for the energy of each of the transitions shown.

mathematically how the various spectral series of hydrogen were produced. The Lyman spectral series, for example, was shown to be the result of electrons dropping from various higher-energy levels to the ground-state energy level.

Bohr's calculated values agreed with the experimentally observed values for the lines in each series. The origins of several of the series of lines in hydrogen's line-emission spectrum are shown in Figure 4-9.

The success of Bohr's model of the hydrogen atom in explaining observed spectral lines led many scientists to conclude that a similar model could be applied to all atoms. It was soon recognized, however, that Bohr's approach did not explain the spectra of atoms with more than one electron. Nor did Bohr's theory fully explain the chemical behavior of atoms.

SECTION REVIEW

1. What was the major shortcoming of Rutherford's model of the atom?
2. Write and label the equation that relates the speed, wavelength, and frequency of electromagnetic radiation.
3. Define the following:
   a. electromagnetic radiation b. wavelength
   c. frequency d. quantum e. photon
4. What is meant by the dual wave-particle nature of light?
5. Describe the Bohr model of the hydrogen atom.

SECTION 4-2

The Quantum Model of the Atom

OBJECTIVES

- Discuss Louis de Broglie's role in the development of the quantum model of the atom.
- Compare and contrast the Bohr model and the quantum model of the atom.
- Explain how the Heisenberg uncertainty principle and the Schrödinger wave equation led to the idea of atomic orbitals.
- List the four quantum numbers, and describe their significance.
- Relate the number of sublevels corresponding to each of an atom's main energy levels, the number of orbitals per sublevel, and the number of orbitals per main energy level.

To the scientists of the early twentieth century, Bohr's model of the hydrogen atom contradicted common sense. Why did hydrogen's electron exist around the nucleus only in certain allowed orbits with definite energies? Why couldn't the electron exist in limitless orbits with slightly different energies? To explain why atomic energy states are quantized, scientists had to change the way they viewed the nature of the electron.

Electrons as Waves

The investigations into the photoelectric effect and hydrogen atomic emission revealed that light could behave as both a wave and a particle. Could electrons have a dual wave-particle nature as well? In 1924, the French scientist Louis de Broglie asked himself this very question. And the answer that he proposed led to a revolution in our basic understanding of matter.

De Broglie pointed out that in many ways the behavior of Bohr's quantized electron orbits was similar to the known behavior of waves. For example, scientists at the time knew that any wave confined to a space can have only certain frequencies. DeBroglie suggested that electrons be considered waves confined to the space around an atomic nucleus. It followed that the electron waves could exist only at specific frequencies. And according to the relationship $E = h\nu$, these frequencies corresponded to specific energies—the quantized energies of Bohr's orbits.

Other aspects of De Broglie's hypothesis that electrons have wave-like properties were soon confirmed by experiments. Investigators demonstrated that electrons, like light waves, can be bent, or diffracted. *Diffraction* refers to the bending of a wave as it passes by the edge of an object, such as the edge of an atom in a crystal. Diffraction experiments and other investigations also showed that electron beams, like waves, can interfere with each other. *Interference* occurs when waves overlap (see the Desktop Investigation on page 100). This overlapping results in a reduction of energy in some areas and an increase of energy in others. The effects of diffraction and interference can be seen in Figure 4-10.

(a)

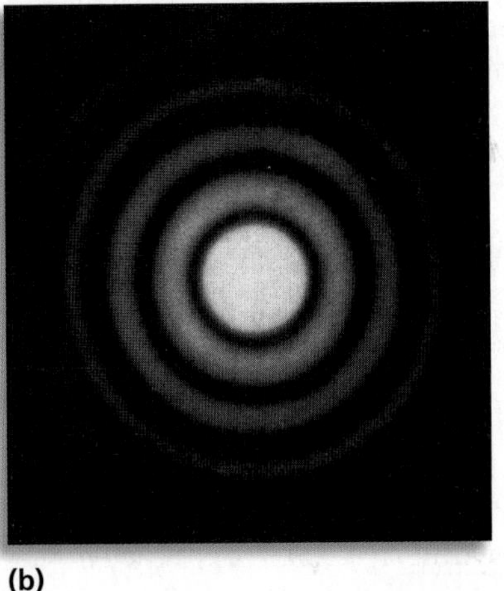

(b)

FIGURE 4-10 Diffraction patterns produced by (a) a beam of electrons passed through a crystal and (b) a beam of visible light passed through a tiny aperture. Each pattern shows the results of bent waves that have interfered with each other. The bright areas correspond to areas of increased energy, while the dark areas correspond to areas of decreased energy.

The Heisenberg Uncertainty Principle

The idea of electrons having a dual wave-particle nature troubled scientists. If electrons are both particles and waves, then where are they in the atom? To answer this question, it is important to consider a proposal first made in 1927 by the German theoretical physicist Werner Heisenberg.

Heisenberg's idea involved the detection of electrons. Electrons are detected by their interaction with photons. Because photons have about the same energy as electrons, any attempt to locate a specific electron with a photon knocks the electron off its course. As a result, there is always a basic uncertainty in trying to locate an electron (or any other particle). *The* **Heisenberg uncertainty principle** *states that it is impossible to determine simultaneously both the position and velocity of an electron or any other particle.* Although it was difficult for scientists to accept this fact at the time, it has proven to be one of the fundamental principles of our present understanding of light and matter.

The Schrödinger Wave Equation

In 1926, the Austrian physicist Erwin Schrödinger used the hypothesis that electrons have a dual wave-particle nature to develop an equation that treated electrons in atoms as waves. Unlike Bohr's theory, which assumed quantization as a fact, quantization of electron energies was a natural outcome of Schrödinger's equation. Only waves of specific energies, and therefore frequencies, provided solutions to the equation. Together with the Heisenberg uncertainty principle, the Schrödinger wave equation laid the foundation for modern quantum theory. **Quantum theory** *describes mathematically the wave properties of electrons and other very small particles.*

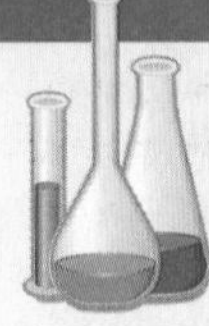

DESKTOP INVESTIGATION

The Wave Nature of Light: Interference

MATERIALS

- **scissors**
- **manila folders**
- **thumbtack**
- **masking tape**
- **aluminum foil**
- **white poster board or cardboard**
- **flashlight**

QUESTION

Does light show the wave property of interference when a beam of light is projected through a pinhole onto a screen?

PROCEDURE

Record all your observations.

1. To make the pinhole screen, cut a 20 cm × 20 cm square from a manila folder. In the center of the square, cut a 2 cm square hole. Cut a 7 cm × 7 cm square of aluminum foil. Using a thumbtack, make a pinhole in the center of the foil square. Tape the aluminum foil over the 2 cm square hole, making sure the pinhole is centered as shown in the diagram.
2. Use white poster board to make a projection screen 35 cm × 35 cm.
3. In a dark room, center the light beam from a flashlight on the pinhole. Hold the flashlight about 1 cm from the pinhole. The pinhole screen should be about 50 cm from the projection screen, as shown in the diagram. Adjust the distance to form a sharp image on the projection screen.

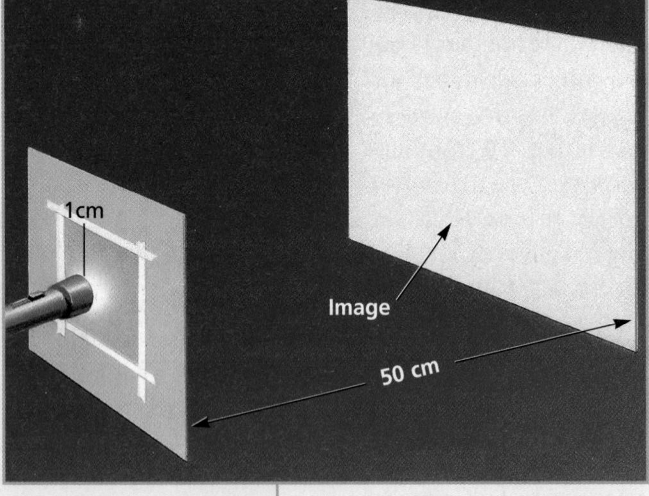

DISCUSSION

1. Did you observe interference patterns on the screen?
2. As a result of your observations, what do you conclude about the nature of light?

Solutions to the Schrödinger wave equation are known as wave functions. Based on the Heisenberg uncertainty principle, the early developers of quantum theory determined that wave functions give only the *probability* of finding an electron at a given place around the nucleus. Thus, electrons do not travel around the nucleus in neat orbits, as Bohr had postulated. Instead, they exist in certain regions called orbitals. *An* **orbital** *is a three-dimensional region around the nucleus that indicates the probable location of an electron.*

Figure 4-11 illustrates two ways of picturing one type of atomic orbital. As you will see later in this section, atomic orbitals have different shapes and sizes.

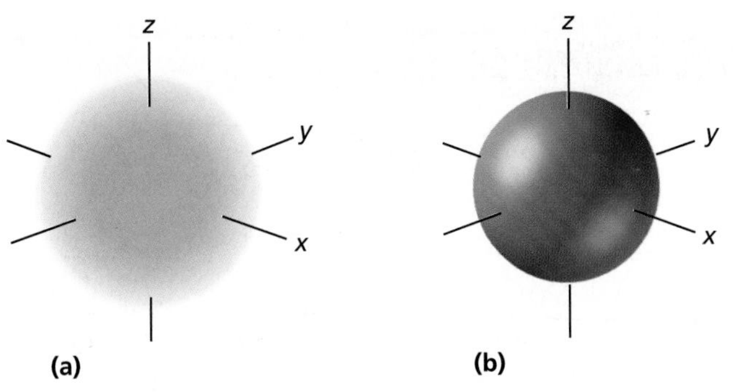

FIGURE 4-11 Two ways of showing a simple atomic orbital are presented. In (a) the probability of finding the electron is proportional to the density of the cloud. Shown in (b) is a surface within which the electron can be found a certain percentage of the time, in this case 90%.

Atomic Orbitals and Quantum Numbers

In the Bohr atomic model, electrons of increasing energy occupy orbits farther and farther from the nucleus. According to the Schrödinger equation, electrons in atomic orbitals also have quantized energies. An electron's energy level is not the only characteristic of an orbital that is indicated by solving the Schrödinger equation.

In order to completely describe orbitals, scientists use quantum numbers. **Quantum numbers** *specify the properties of atomic orbitals and the properties of electrons in orbitals.* The first three quantum numbers result from solutions to the Schrödinger equation. They indicate the main energy level, the shape, and the orientation of an orbital. The fourth, the spin quantum number, describes a fundamental state of the electron that occupies the orbital. As you read the following descriptions of the quantum numbers, refer to the appropriate columns in Table 4-2 on page 104.

Principal Quantum Number

The **principal quantum number,** *symbolized by n, indicates the main energy level occupied by the electron.* Values of n are positive integers only—1, 2, 3, and so on. As n increases, the electron's energy and its average distance from the nucleus increase (see Figure 4-12). For example, an electron for which $n = 1$ occupies the first, or lowest, main energy level and is located closest to the nucleus. As you will see, more than one electron can have the same n value. These electrons are sometimes said to be in the same electron *shell.* The total number of orbitals that exist in a given shell, or main energy level, is equal to n^2.

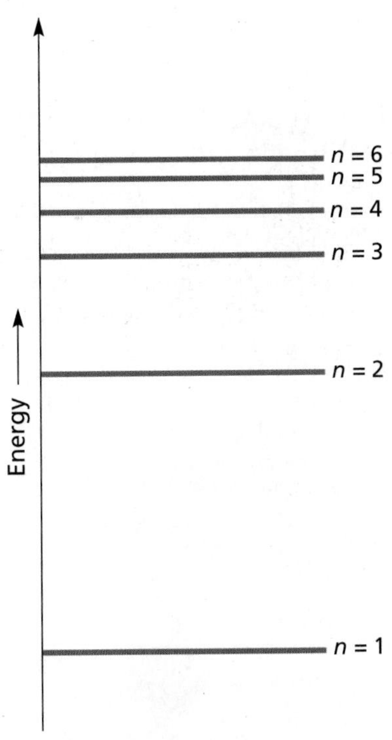

FIGURE 4-12 The main energy levels of an atom are represented by the principal quantum number, n.

Angular Momentum Quantum Number

Except at the first main energy level, orbitals of different shapes—known as *sublevels*—exist for a given value of n. *The* **angular momentum quantum number,** *symbolized by l, indicates the shape of the orbital.* For a specific main energy level, the number of orbital shapes possible is equal to n. The values of l allowed are zero and all positive integers less

TABLE 4-1 Orbital Letter Designations According to Values of *l*

l	Letter
0	*s*
1	*p*
2	*d*
3	*f*

than or equal to $n - 1$. For example, orbitals for which $n = 2$ can have one of two shapes corresponding to $l = 0$ and $l = 1$. Depending on its value of l, an orbital is assigned a letter, as shown in Table 4-1.

As shown in Figure 4-13, *s* orbitals are spherical, *p* orbitals have dumbbell shapes, and *d* orbitals are more complex. (The *f* orbital shapes are too complex to discuss here.) In the first energy level, $n = 1$, there is only one sublevel possible—an *s* orbital. As mentioned, the second energy level, $n = 2$, has two sublevels—the *s* and *p* orbitals. The third energy level, $n = 3$, has three sublevels—the *s*, *p*, and *d* orbitals. The fourth energy level, $n = 4$, has four sublevels—the *s*, *p*, *d*, and *f* orbitals. In an *n*th main energy level, there are *n* sublevels.

Each atomic orbital is designated by the principal quantum number followed by the letter of the sublevel. For example, the 1*s* sublevel is the *s* orbital in the first main energy level, while the 2*p* sublevel is the set of *p* orbitals in the second main energy level. On the other hand, a 4*d* orbital is part of the *d* sublevel in the fourth main energy level. How would you designate the *p* sublevel in the third main energy level? How many other sublevels are in the same main energy level with this one?

Magnetic Quantum Number

Atomic orbitals can have the same shape but different orientations around the nucleus. *The* **magnetic quantum number,** *symbolized by m, indicates the orientation of an orbital around the nucleus.* Here we describe the orbital orientations that correspond to various values of *m*. Because an *s* orbital is spherical and is centered around the nucleus, it has only one possible orientation. This orientation corresponds to a

FIGURE 4-13 The orbitals *s*, *p*, and *d* have different shapes. Each of the orbitals shown occupies a different region of space around the nucleus.

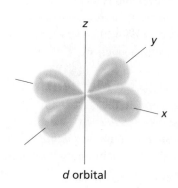

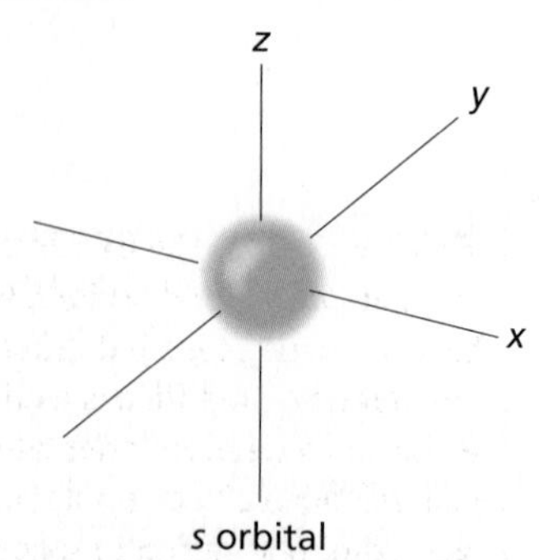

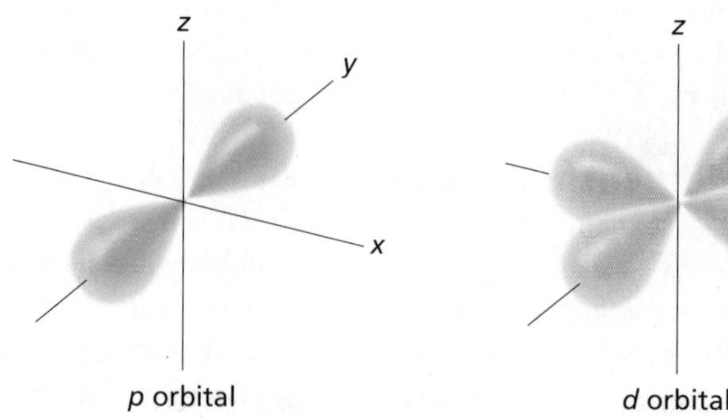

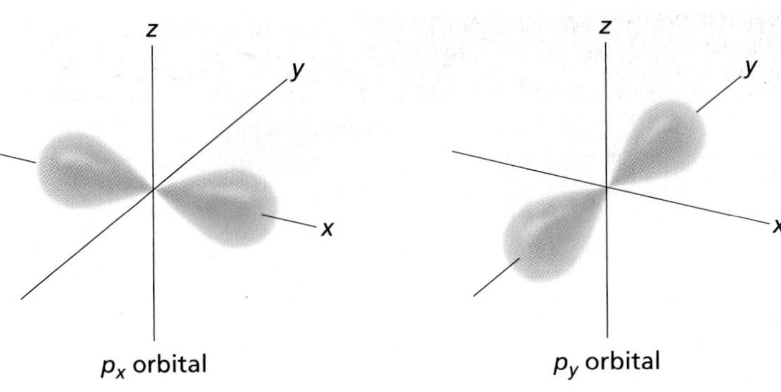

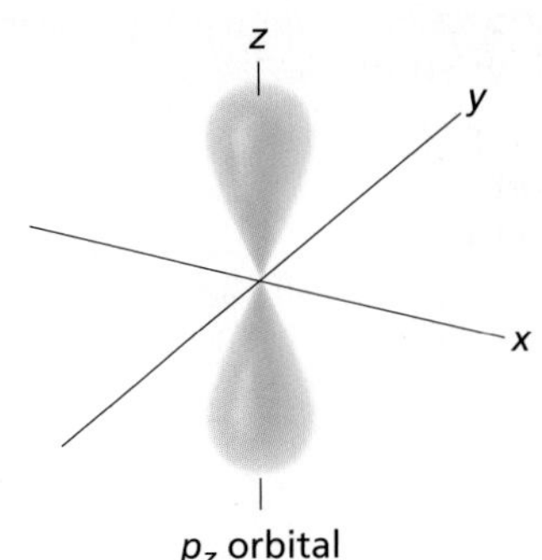

FIGURE 4-14 The subscripts *x*, *y*, and *z* indicate the three different orientations of *p* orbitals. The intersection of the *x*, *y*, and *z* axes indicates the location of the center of the nucleus.

magnetic quantum number of $m = 0$. There is therefore only one *s* orbital in each *s* sublevel. As shown in Figure 4-14, the lobes of a *p* orbital can extend along the *x*, *y*, or *z* axis of a three-dimensional coordinate system. There are therefore three *p* orbitals in each *p* sublevel, which are designated as p_x, p_y, and p_z orbitals. The three *p* orbitals occupy different regions of space and correspond, in no particular order, to values of $m = -1$, $m = 0$, and $m = +1$.

There are five different *d* orbitals in each *d* sublevel (see Figure 4-15). The five different orientations correspond to values of $m = -2$, $m = -1$, $m = 0$, $m = +1$, and $m = +2$. There are seven different *f* orbitals in each *f* sublevel.

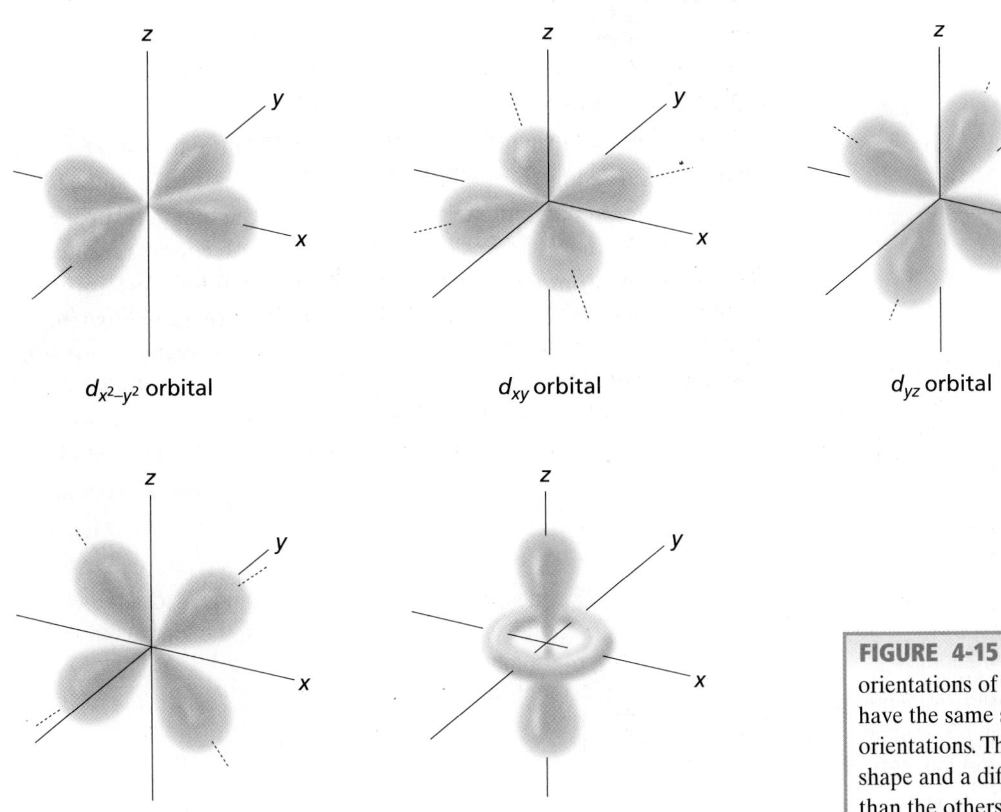

FIGURE 4-15 The five different orientations of the *d* orbitals. Four have the same shape but different orientations. The fifth has a different shape and a different orientation than the others. Each orbital occupies a different region of space.

TABLE 4-2 *Quantum Number Relationships in Atomic Structure*

Principal quantum number: main energy level (n)	Sublevels in main energy level (n sublevels)	Number of orbitals per sublevel	Number of orbitals per main energy level (n^2)	Number of electrons per sublevel	Number of electrons per main energy level ($2n^2$)
1	*s*	1	1	2	2
2	*s* *p*	1 3	4	2 6	8
3	*s* *p* *d*	1 3 5	9	2 6 10	18
4	*s* *p* *d* *f*	1 3 5 7	16	2 6 10 14	32

As you can see in Table 4-2, the total number of orbitals at a main energy level increases with the value of n. In fact, the number of orbitals at each main energy level equals the square of the principal quantum number, n^2. What is the total number of orbitals in the third energy level? Specify each of the sublevels using the three quantum numbers you've learned so far.

Spin Quantum Number

Like Earth, an electron in an orbital can be thought of as spinning on an internal axis. It spins in one of two possible directions, or states. As it spins, it creates a magnetic field. To account for the magnetic properties of the electron, theoreticians of the early twentieth century created the spin quantum number. *The* **spin quantum number** *has only two possible values—(+½, −½)—which indicate the two fundamental spin states of an electron in an orbital.* A single orbital can hold a maximum of two electrons, which must have opposite spins.

SECTION REVIEW

1. Define the following:
   a. main energy levels
   b. quantum numbers
2. a. List the four quantum numbers.
   b. What general information about atomic orbitals is provided by the four sets of quantum numbers?
3. Identify and explain the meanings associated with the possible values for each of the four quantum numbers in a set.

Electron Configurations

SECTION 4-3

OBJECTIVES

- List the total number of electrons needed to fully occupy each main energy level.
- State the Aufbau principle, the Pauli exclusion principle, and Hund's rule.
- Describe the electron configurations for the atoms of any element using orbital notation, electron-configuration notation, and, when appropriate, noble-gas notation.

The quantum model of the atom improves on the Bohr model because it describes the arrangements of electrons in atoms other than hydrogen. *The arrangement of electrons in an atom is known as the atom's* **electron configuration.** Because atoms of different elements have different numbers of electrons, a distinct electron configuration exists for the atoms of each element. Like all systems in nature, electrons in atoms tend to assume arrangements that have the lowest possible energies. The lowest-energy arrangement of the electrons for each element is called the element's *ground-state electron configuration.* A few simple rules, combined with the quantum number relationships discussed in Section 4-2, allow us to determine these ground-state electron configurations.

Rules Governing Electron Configurations

To build up electron configurations for the ground state of any particular atom, first the energy levels of the orbitals are determined. Then electrons are added to the orbitals one by one according to three basic rules. (Remember that real atoms are not built up by adding protons and electrons one at a time.)

The first rule shows the order in which electrons occupy orbitals. According to the **Aufbau principle,** *an electron occupies the lowest-energy orbital that can receive it.* Figure 4-16 shows the atomic orbitals in order of increasing energy. The orbital with the lowest energy is the 1*s* orbital. In a ground-state hydrogen atom, the electron is in this orbital. The 2*s* orbital is the next highest in energy, then the 2*p* orbitals. Beginning with the third main energy level, $n = 3$, the energies of the sublevels in different main energy levels begin to overlap.

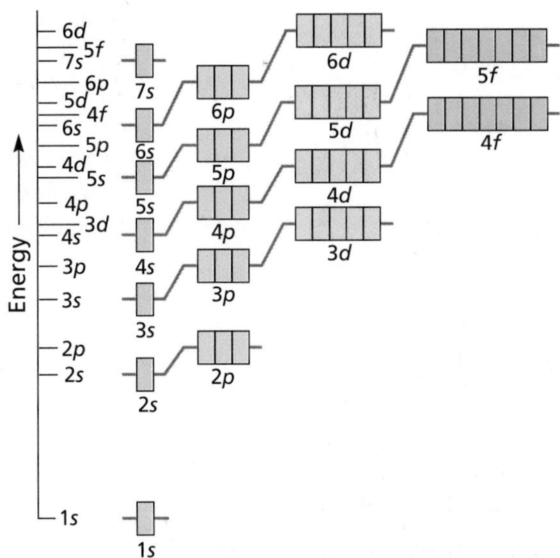

FIGURE 4-16 The order of increasing energy for atomic sublevels is shown on the vertical axis. Each individual box represents an orbital.

Note in the figure, for example, that the 4*s* sublevel is lower in energy than the 3*d* sublevel. Therefore, the 4*s* orbital is filled before any electrons enter the 3*d* orbitals. (Less energy is required for two electrons to pair up in the 4*s* orbital than for a single electron to

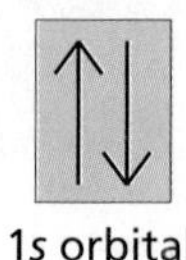

FIGURE 4-17 According to the Pauli exclusion principle, an orbital can hold two electrons of opposite spin. In this electron configuration of a helium atom, each arrow represents one of the atom's two electrons. The direction of the arrow indicates the electron's spin.

occupy a 3*d* orbital.) Once the 3*d* orbitals are fully occupied, which sublevel will be occupied next?

The second rule reflects the importance of the spin quantum number. According to the **Pauli exclusion principle,** *no two electrons in the same atom can have the same set of four quantum numbers.* The principal, angular momentum, and magnetic quantum numbers specify the energy, shape, and orientation of an orbital. The two values of the spin quantum number allow two electrons of opposite spins to occupy the orbital (see Figure 4-17).

The third rule requires placing as many unpaired electrons as possible in separate orbitals in the same sublevel. In this way, electron-electron repulsion is minimized so that the electron arrangements have the lowest energy possible. According to **Hund's rule,** *orbitals of equal energy are each occupied by one electron before any orbital is occupied by a second electron, and all electrons in singly occupied orbitals must have the same spin.* Applying this rule shows, for example, that one electron will enter each of the three *p* orbitals in a main energy level before a second electron enters any of them. This is illustrated in Figure 4-18. What is the maximum number of unpaired electrons in a *d* sublevel?

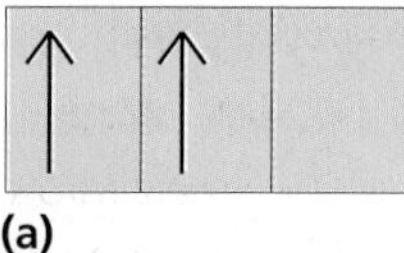

(a)

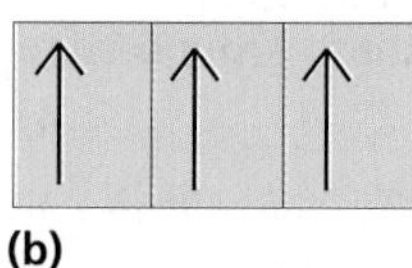

(b)

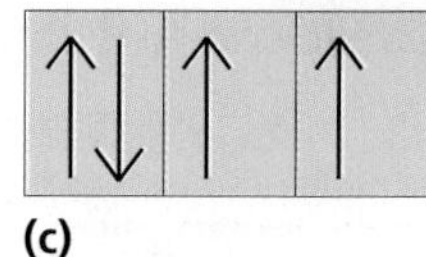

(c)

FIGURE 4-18 The figure shows how (a) two, (b) three, and (c) four electrons fill the *p* sublevel of a given main energy level according to Hund's rule.

Representing Electron Configurations

Three methods, or notations, are used to indicate electron configurations. Two of these notations will be discussed in the next two sections for the first-period elements, hydrogen and helium. The third notation applies mostly to elements of the third period and higher. It will be discussed in the section on third-period elements.

In a ground-state hydrogen atom, the single electron is in the lowest-energy orbital, the 1*s* orbital. The electron can be in either one of its two spin states. Helium has two electrons, which are paired in the 1*s* orbital.

Orbital Notation

In orbital notation, an unoccupied orbital is represented by a line, ____, with the orbital's name written underneath the line. An orbital containing one electron is represented as $\underline{\uparrow}$. An orbital containing two electrons is represented as $\underline{\uparrow\downarrow}$, showing the electrons paired and with opposite spins. The lines are labeled with the principal quantum number and sublevel letter. For example, the orbital notations for hydrogen and helium are written as follows.

$$\text{H}\ \underset{1s}{\underline{\uparrow}} \qquad \text{He}\ \underset{1s}{\underline{\uparrow\downarrow}}$$

Electron-Configuration Notation

Electron-configuration notation eliminates the lines and arrows of orbital notation. Instead, the number of electrons in a sublevel is shown by adding a superscript to the sublevel designation. The hydrogen configuration is represented by $1s^1$. The superscript indicates that one electron is present in hydrogen's 1*s* orbital. The helium configuration is represented by $1s^2$. Here the superscript indicates that there are two electrons in helium's 1*s* orbital.

SAMPLE PROBLEM 4-1

The electron configuration of boron is $1s^22s^22p^1$. How many electrons are present in an atom of boron? What is the atomic number for boron? Write the orbital notation for boron.

SOLUTION The number of electrons in a boron atom is equal to the sum of the superscripts in its electron-configuration notation: 2 + 2 + 1 = 5 electrons. The number of protons equals the number of electrons in a neutral atom. So we know that boron has 5 protons and thus has an atomic number of 5. To write the orbital notation, first draw the lines representing orbitals.

$$\underset{1s}{\underline{\quad}}\ \underset{2s}{\underline{\quad}}\ \underset{2p_x}{\underline{\quad}}\underset{2p_y}{\underline{\quad}}\underset{2p_z}{\underline{\quad}}$$

Next, add arrows showing the electron locations. The first two electrons occupy $n = 1$ energy level and fill the 1*s* orbital.

$$\underset{1s}{\underline{\uparrow\downarrow}}\ \underset{2s}{\underline{\quad}}\ \underset{2p_x}{\underline{\quad}}\underset{2p_y}{\underline{\quad}}\underset{2p_z}{\underline{\quad}}$$

The next three electrons occupy the $n = 2$ main energy level. According to the Aufbau principle, two of these occupy the lower-energy 2*s* orbital. The third occupies a higher-energy *p* orbital.

$$\underset{1s}{\underline{\uparrow\downarrow}}\ \underset{2s}{\underline{\uparrow\downarrow}}\ \underset{2p_x}{\underline{\uparrow}}\underset{2p_y}{\underline{\quad}}\underset{2p_z}{\underline{\quad}}$$

PRACTICE

1. The electron configuration of nitrogen is $1s^22s^22p^3$. How many electrons are present in a nitrogen atom? What is the atomic number of nitrogen? Write the orbital notation for nitrogen.

   Answer
   7, 7, $\underset{1s}{\underline{\uparrow\downarrow}}\ \underset{2s}{\underline{\uparrow\downarrow}}\ \underset{2p_x}{\underline{\uparrow}}\underset{2p_y}{\underline{\uparrow}}\underset{2p_z}{\underline{\uparrow}}$

2. The electron configuration of fluorine is $1s^22s^22p^5$. What is the atomic number of fluorine? How many of its *p* orbitals are filled? How many unpaired electrons does a fluorine atom contain?

   Answer
   9, 2, 1

The Noble Decade

HISTORICAL PERSPECTIVE

By the late nineteenth century, the science of chemistry had begun to be organized. The first international congress of chemistry, in 1860, established the field's first standards. And Dmitri Mendeleev's periodic table of elements gave chemists across the globe a systematic understanding of matter's building blocks. But many important findings were yet to come, including the discovery of a family of rare, unreactive gases that were unlike any substances known at the time.

Cross-Disciplinary Correspondence

In 1888, the British physicist Lord Rayleigh encountered a small but significant discrepancy in the results of one of his experiments. In an effort to redetermine the atomic mass of nitrogen, he measured the densities of several samples of nitrogen gas. Each sample had been prepared by a different method. The samples that had been isolated from chemical reactions all exhibited similar densities. But they were about one-tenth of a percent lighter than the nitrogen isolated from air, which at the time was believed to be a mixture of nitrogen, oxygen, water vapor, and carbon dioxide.

Rayleigh was at a loss to explain his discovery. Finally, in 1892, he published a letter in *Nature* magazine appealing to his colleagues for an explanation. A month later he received a reply from a Scottish chemist named William Ramsay. Ramsay related that he too had been stumped by the density difference between chemical and atmospheric nitrogen. Rayleigh decided to report his findings to the Royal Society of Chemistry, adding:

> *Until the questions arising out of these observations are thoroughly cleared up, the above number [density] for nitrogen must be received with a certain reserve.*

SEPTEMBER 29, 1892 — *NATURE*

LETTERS TO THE EDITOR.

[*The Editor does not hold himself responsible for opinions expressed by his correspondents. Neither can he undertake to return, or to correspond with the writers of, rejected manuscripts intended for this or any other part of* NATURE. *No notice is taken of anonymous communications.*]

Density of Nitrogen.

I AM much puzzled by some recent results as to the density of *nitrogen*, and shall be obliged if any of your chemical readers can offer suggestions as to the cause. According to two methods of preparation I obtain quite distinct values. The relative difference, amounting to about $\frac{1}{1000}$ part, is small in itself; but it lies entirely outside the errors of experiment, and can only be attributed to a variation in the character of the gas...

Is it possible that the difference is independent of impurity, the nitrogen itself being to some extent in a different (dissociated) state?...

RAYLEIGH.

Terling Place, Witham, September 24.

An excerpt of Lord Rayleigh's letter as originally published in Nature *magazine in 1892*

A Chemist's Approach

With Rayleigh's permission, Ramsay attempted to remove all the known components from a sample of air and analyze what, if anything, remained. First he heated magnesium in the air to remove all of the nitrogen in the form of magnesium nitride. Then he removed the oxygen, water vapor, and carbon dioxide factions. What remained was a minuscule portion of a mysterious gas.

Ramsay tried to cause the gas to react with chemically active substances, such as hydrogen, sodium, caustic soda, and platinum black. But in each case the gas remained unaltered. He decided to name this new atmospheric component argon (Greek for "inert" or "idle"):

> *The gas deserves the name "argon," for it is a most astonishingly indifferent body in as much as it is unattacked by elements of very opposite character varying from sodium and*

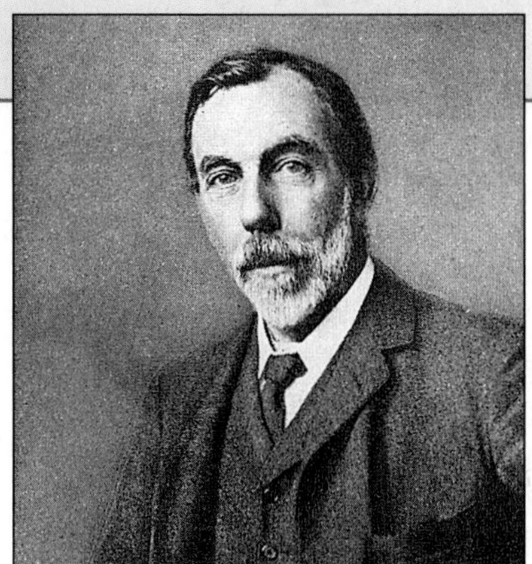

In 1893, Scottish chemist William Ramsay isolated a previously unknown component of the atmosphere.

> *magnesium on the one hand, oxygen, chlorine and sulphur [sic] on the other.*

Periodic Problems

Rayleigh and Ramsay were sure that they had discovered a new element. But this created a problem. Their calculations indicated that argon had an atomic mass of about 40. However, there was no space in the periodic table as it appeared in 1894 for such an element. The elements with atomic masses closest to that of argon were chlorine and potassium. Unfortunately, the chemical properties of the families of each of these elements were completely dissimilar to those of the strange gas.

Ramsay contemplated argon's lack of reactivity. He knew that Mendeleev had created the periodic table on the basis of valence, or the number of atomic partners an element bonds with in forming a compound. As Ramsay could not cause argon to form any compounds, he assigned it a valence of zero. And because the valence of the elements in the families of both chlorine and potassium was one, perhaps argon fit in between them. In May 1894, Ramsay wrote to Rayleigh:

> *Has it occurred to you that there is room for gaseous elements at the end of the first column of the periodic table?…Such elements should have the density 20 or thereabouts, and 0.8 pc [percent] (1/120th about) of the nitrogen of the air could so raise the density of nitrogen that it would stand to pure [chemical] nitrogen in the ratio of 230:231.*

Ramsay's insight that argon merited a new spot in the periodic table, between the chlorine family and the potassium family, was correct. And, as he would soon confirm, his newly discovered gas was indeed one of a previously unknown family of elements.

New Neighbors

In 1895, Ramsay isolated a light, inert gas from a mineral called cleveite. Physical analysis revealed that the gas was the same as one that had been identified in the sun in 1868—helium. Helium was the second zero-valent element found on Earth, and its discovery made chemists aware that the periodic table had been missing a whole column of elements.

Over the next three years, Ramsay and his assistant, Morris Travers, identified three more inert gases present in the atmosphere: neon (Greek for "new"), krypton ("hidden"), and xenon ("stranger"). Finally in 1900, German chemist Friedrich Ernst Dorn discovered radon, the last of the new family of elements known today as the noble gases. For his discovery, Ramsay received the Nobel Prize in 1904.

Groups / Periods	III b	IV b	V b	VI b	VII b	VIII b	I b	II b	III a	IV a	V a	VI a	VII a	0	I a	II a
1															H	He
2															Li	Be
3									B	C	N	O	F	Ne	Na	Mg
4									Al	Si	P	S	Cl	Ar	K	Ca
5	Sc	Ti	V	Cr	Mn	Fe Co Ni	Cu	Zn	Ga	Ge	As	Se	Br	Kr	Rb	Sr
6	Y	Zr	Nb	Mo	Tc	Ru Rh Pd	Ag	Cd	In	Sn	Sb	Te	I	Xe	Cs	Ba
7	La	Hf	Ta	W	Re	Os Ir Pt	Au	Hg	Tl	Pb	Bi	Po	At	Rn	Fr	Ra
8	Ac															
	Transition elements								Main-group elements							

One version of the periodic table as it appeared after the discovery of the noble gases. The placement of the Group 1 and 2 elements at the far right of the table shows clearly how the noble gases fit in between the chlorine family and the potassium family of elements. The "0" above the noble-gas family indicates the zero valency of the gases.

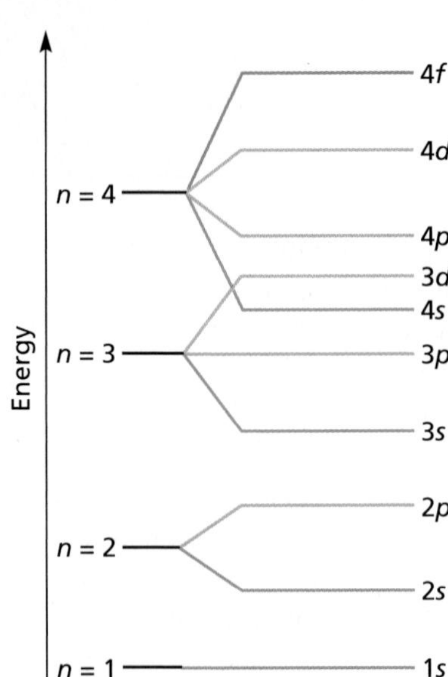

FIGURE 4-19 The direction of the arrow indicates the order in which atomic orbitals are filled according to the Aufbau principle.

Elements of the Second Period

In the first-period elements, hydrogen and helium, electrons occupy the first main energy level. The ground-state configurations in Table 4-3 illustrate how the Aufbau principle, the Pauli exclusion principle, and Hund's rule are applied to atoms of elements in the second period. Figure 4-19 shows the first four main energy levels in order of increasing energy for applying the Aufbau principle.

According to the Aufbau principle, after the 1*s* orbital is filled, the next electron occupies the *s* sublevel in the second main energy level. Thus, lithium, Li, has a configuration of $1s^22s^1$. The electron occupying the 2*s* level of a lithium atom is in the atom's highest, or outermost, occupied level. *The* **highest occupied level** *is the electron-containing main energy level with the highest principal quantum number.* The two electrons in the 1*s* sublevel of lithium are no longer in the outermost main energy level. They have become **inner-shell electrons,** *which are electrons that are not in the highest occupied energy level.*

The fourth electron in an atom of beryllium, Be, must complete the pair in the 2*s* sublevel because this sublevel is of lower energy than the 2*p* sublevel. With the 2*s* sublevel filled, the 2*p* sublevel, which has three vacant orbitals of equal energy, can be occupied. Hund's rule applies here, as is shown in the orbital notations in Table 4-3. One of the three *p* orbitals is occupied by a single electron in an atom of boron, B. Two of the three *p* orbitals are occupied by unpaired electrons in an atom of carbon, C. And all three *p* orbitals are occupied by unpaired electrons in an atom of nitrogen, N.

According to the Aufbau principle, the next electron must pair with another electron in one of the 2*p* orbitals rather than enter the third main energy level. The Pauli exclusion principle allows the electron to pair with

TABLE 4-3 *Electron Configurations of Atoms of Second-Period Elements Showing Two Notations*

Name	Symbol	Orbital notation: 1s	2s	2p			Electron-configuration notation
Lithium	Li	↑↓	↑	___	___	___	$1s^22s^1$
Beryllium	Be	↑↓	↑↓	___	___	___	$1s^22s^2$
Boron	B	↑↓	↑↓	↑	___	___	$1s^22s^22p^1$
Carbon	C	↑↓	↑↓	↑	↑	___	$1s^22s^22p^2$
Nitrogen	N	↑↓	↑↓	↑	↑	↑	$1s^22s^22p^3$
Oxygen	O	↑↓	↑↓	↑↓	↑	↑	$1s^22s^22p^4$
Fluorine	F	↑↓	↑↓	↑↓	↑↓	↑	$1s^22s^22p^5$
Neon	Ne	↑↓	↑↓	↑↓	↑↓	↑↓	$1s^22s^22p^6$

one of the electrons occupying the $2p$ orbitals as long as the spins of the paired electrons are opposite. Thus, atoms of oxygen, O, have the configuration $1s^22s^22p^4$. Oxygen's orbital notation is shown in Table 4-3.

Two $2p$ orbitals are filled in fluorine, F, and all three are filled in neon, Ne. Atoms such as those of neon, which have the s and p sublevels of their highest occupied level filled with eight electrons, are said to have an *octet* of electrons. Examine the periodic table inside the back cover of the text. Notice that neon is the last element in the second period.

Elements of the Third Period

After the outer octet is filled in neon, the next electron enters the s sublevel in the $n = 3$ main energy level. Thus, atoms of sodium, Na, have the configuration $1s^22s^22p^63s^1$. Compare the configuration of a sodium atom with that of an atom of neon in Table 4-3. Notice that the first 10 electrons in a sodium atom have the same configuration as a neon atom, $1s^22s^22p^6$. In fact, the first 10 electrons in an atom of each of the third-period elements have the same configuration as neon. This similarity allows us to use a shorthand notation for the electron configurations of the third-period elements.

Noble-Gas Notation

Neon is a member of the Group 18 elements. *The Group 18 elements (helium, neon, argon, krypton, xenon, and radon) are called the* **noble gases.** To simplify sodium's notation, the symbol for neon, enclosed in square brackets, is used to represent the complete neon configuration: $[Ne] = 1s^22s^22p^6$. This allows us to write sodium's electron configuration as $[Ne]3s^1$, which is called sodium's *noble-gas notation.* Table 4-4 shows the noble-gas notation of the electron configuration of each of the third-period elements.

TABLE 4-4 *Electron Configurations of Atoms of Third-Period Elements*

Name	Symbol	Atomic number	Number of electrons in sublevels					Noble-gas notation
			$1s$	$2s$	$2p$	$3s$	$3p$	
Sodium	Na	11	2	2	6	1		$[Ne]3s^1$
Magnesium	Mg	12	2	2	6	2		$[Ne]3s^2$
Aluminum	Al	13	2	2	6	2	1	$[Ne]3s^23p^1$
Silicon	Si	14	2	2	6	2	2	$[Ne]3s^23p^2$
Phosphorus	P	15	2	2	6	2	3	$[Ne]3s^23p^3$
Sulfur	S	16	2	2	6	2	4	$[Ne]3s^23p^4$
Chlorine	Cl	17	2	2	6	2	5	$[Ne]3s^23p^5$
Argon	Ar	18	2	2	6	2	6	$[Ne]3s^23p^6$

The last element in the third period is argon, Ar, which is a noble gas. As in neon, the highest-occupied energy level of argon has an octet of electrons, [Ne]$3s^23p^6$. In fact, each noble gas other than He has an electron octet in its highest energy level. *A* **noble-gas configuration** *is an outer main energy level fully occupied, in most cases, by eight electrons.*

Elements of the Fourth Period

The electron configurations of atoms in the fourth-period elements are shown in Table 4-5. The period begins by filling the 4*s* orbital, the empty orbital of lowest energy. Thus, the first element in the fourth period is potassium, K, which has the electron configuration [Ar]$4s^1$. The next element is calcium, Ca, which has the electron configuration [Ar]$4s^2$.

With the 4*s* sublevel filled, the 4*p* and 3*d* sublevels are the next available vacant orbitals. Figure 4-19 on page 110 shows that the 3*d* sublevel

TABLE 4-5 *Electron Configuration of Atoms of Elements in the Fourth Period*

Name	Symbol	Atomic number	Number of electrons in sublevels above 2p: 3s	3p	3d	4s	4p	Noble-gas notation
Potassium	K	19	2	6		1		*[Ar]$4s^1$
Calcium	Ca	20	2	6		2		[Ar]$4s^2$
Scandium	Sc	21	2	6	1	2		[Ar]$3d^14s^2$
Titanium	Ti	22	2	6	2	2		[Ar]$3d^24s^2$
Vanadium	V	23	2	6	3	2		[Ar]$3d^34s^2$
Chromium	Cr	24	2	6	5	1		[Ar]$3d^54s^1$
Manganese	Mn	25	2	6	5	2		[Ar]$3d^54s^2$
Iron	Fe	26	2	6	6	2		[Ar]$3d^64s^2$
Cobalt	Co	27	2	6	7	2		[Ar]$3d^74s^2$
Nickel	Ni	28	2	6	8	2		[Ar]$3d^84s^2$
Copper	Cu	29	2	6	10	1		[Ar]$3d^{10}4s^1$
Zinc	Zn	30	2	6	10	2		[Ar]$3d^{10}4s^2$
Gallium	Ga	31	2	6	10	2	1	[Ar]$3d^{10}4s^24p^1$
Germanium	Ge	32	2	6	10	2	2	[Ar]$3d^{10}4s^24p^2$
Arsenic	As	33	2	6	10	2	3	[Ar]$3d^{10}4s^24p^3$
Selenium	Se	34	2	6	10	2	4	[Ar]$3d^{10}4s^24p^4$
Bromine	Br	35	2	6	10	2	5	[Ar]$3d^{10}4s^24p^5$
Krypton	Kr	36	2	6	10	2	6	[Ar]$3d^{10}4s^24p^6$

*[Ar] = $1s^22s^22p^63s^23p^6$

is lower in energy than the $4p$ sublevel. Therefore, the five $3d$ orbitals are next to be filled. A total of 10 electrons can occupy the $3d$ orbitals. These are filled successively in the 10 elements from scandium (atomic number 21) to zinc (atomic number 30).

Scandium, Sc, has the electron configuration $[Ar]3d^1 4s^2$. Titanium, Ti, has the configuration $[Ar]3d^2 4s^2$. And vanadium, V, has the configuration $[Ar]3d^3 4s^2$. Up to this point, three electrons with the same spin have been added to three separate *d* orbitals, as required by Hund's rule.

Surprisingly, chromium, Cr, has the electron configuration $[Ar]3d^5 4s^1$. Not only did the added electron go into the fourth $3d$ orbital, but an electron also moved from the $4s$ orbital into the fifth $3d$ orbital, leaving the $4s$ orbital with a single electron. Chromium's electron configuration is contrary to what is expected according to the Aufbau principle. However, in reality the $[Ar]3d^5 4s^1$ configuration is of lower energy than a $[Ar]3d^4 4s^2$ configuration. For chromium, having six outer orbitals with unpaired electrons is a more stable arrangement than having four unpaired electrons in the $3d$ orbitals and forcing two electrons to pair up in the $4s$ orbital. On the other hand, for tungsten, W, which is in the same group as chromium, having four electrons in the $5d$ orbitals and two electrons paired in the $6s$ orbital is the most stable arrrangement. Unfortunately, there is no simple explanation for such deviations from the ideal order given in Figure 4-19.

Manganese, Mn, has the electron configuration $[Ar]3d^5 4s^2$. The added electron goes to the $4s$ orbital, completely filling this orbital while leaving the $3d$ orbitals still half-filled. Beginning with the next element, electrons continue to pair in the *d* orbitals. Thus, iron, Fe, has the configuration $[Ar]3d^6 4s^2$; cobalt, Co, has the configuration $[Ar]3d^7 4s^2$; and nickel, Ni, has the configuration $[Ar]3d^8 4s^2$. Next is copper, Cu, in which an electron moves from the $4s$ orbital to pair with the electron in the fifth $3d$ orbital. The result is an electron configuration of $[Ar]3d^{10} 4s^1$—the lowest-energy configuration for Cu.

In atoms of zinc, Zn, the $4s$ sublevel is filled to give the electron configuration $[Ar]3d^{10} 4s^2$. In atoms of the next six elements, electrons add one by one to the three $4p$ orbitals. According to Hund's rule, one electron is added to each of the three $4p$ orbitals before electrons are paired in any $4p$ orbital.

Elements of the Fifth Period

In the 18 elements of the fifth period, sublevels fill in a similar manner as in elements of the fourth period. However, they start at the $5s$ orbital instead of the $4s$. Successive electrons are added first to the $5s$ orbital, then to the $4d$ orbitals, and finally to the $5p$ orbitals. This can be seen in Table 4-6 on page 114. There are occasional deviations from the predicted configurations here also. The deviations differ from those for fourth-period elements, but in each case the preferred configuration has the lowest possible energy.

TABLE 4-6 *Electron Configurations of Atoms of Elements in the Fifth Period*

Name	Symbol	Atomic number	Number of electrons in sublevels above 3*d*					Noble-gas notation
			4*s*	4*p*	4*d*	5*s*	5*p*	
Rubidium	Rb	37	2	6		1		*$[Kr]5s^1$
Strontium	Sr	38	2	6		2		$[Kr]5s^2$
Yttrium	Y	39	2	6	1	2		$[Kr]4d^15s^2$
Zirconium	Zr	40	2	6	2	2		$[Kr]4d^25s^2$
Niobium	Nb	41	2	6	4	1		$[Kr]4d^45s^1$
Molybdenum	Mo	42	2	6	5	1		$[Kr]4d^55s^1$
Technetium	Tc	43	2	6	6	1		$[Kr]4d^55s^1$
Ruthenium	Ru	44	2	6	7	1		$[Kr]4d^75s^1$
Rhodium	Rh	45	2	6	8	1		$[Kr]4d^85s^1$
Palladium	Pd	46	2	6	10			$[Kr]4d^{10}$
Silver	Ag	47	2	6	10	1		$[Kr]4d^{10}5s^1$
Cadmium	Cd	48	2	6	10	2		$[Kr]4d^{10}5s^2$
Indium	In	49	2	6	10	2	1	$[Kr]4d^{10}5s^25p^1$
Tin	Sn	50	2	6	10	2	2	$[Kr]4d^{10}5s^25p^2$
Antimony	Sb	51	2	6	10	2	3	$[Kr]4d^{10}5s^25p^3$
Tellurium	Te	52	2	6	10	2	4	$[Kr]4d^{10}5s^25p^4$
Iodine	I	53	2	6	10	2	5	$[Kr]4d^{10}5s^25p^5$
Xenon	Xe	54	2	6	10	2	6	$[Kr]4d^{10}5s^25p^6$

*$[Kr] = 1s^22s^22p^63s^23p^63d^{10}4s^24p^6$

SAMPLE PROBLEM 4-2

a. Write both the complete electron-configuration notation and the noble-gas notation for iron, Fe.

b. How many electron-containing orbitals are in an atom of iron? How many of these orbitals are completely filled? How many unpaired electrons are there in an atom of iron? In which sublevel are the unpaired electrons located?

SOLUTION

a. The complete electron-configuration notation of iron is $1s^22s^22p^63s^23p^63d^64s^2$. The periodic table inside the back cover of the text reveals that $1s^22s^22p^63s^23p^6$ is the electron configuration of the noble gas argon, Ar. Therefore, as shown in Table 4-5 on page 112, iron's noble-gas notation is $[Ar]3d^64s^2$.

b. An iron atom has 15 orbitals that contain electrons. They consist of one 1s orbital, one 2*s* orbital, three 2*p* orbitals, one 3*s* orbital, three 3*p* orbitals, five 3*d* orbitals, and one 4*s* orbital. Eleven of these orbitals are filled. There are four unpaired electrons. They are located in the 3*d* sublevel.

PRACTICE

1. a. Write both the complete electron-configuration notation and the noble-gas notation for iodine, I. How many inner-shell electrons does an iodine atom contain?

   b. How many electron-containing orbitals are in an atom of iodine? How many of these orbitals are filled? How many unpaired electrons are there in an atom of iodine?

2. a. Write the noble-gas notation for tin, Sn. How many unpaired electrons are there in an atom of tin?

   b. How many electron-containing *d* orbitals are there in an atom of tin? Name the element in the fourth period whose atoms have the same number of highest-energy-level electrons as tin.

3. a. Without consulting the periodic table or a table in this chapter, write the complete electron configuration for the element with atomic number 25.

   b. Identify the element described in item 3a.

Answer

1. a. $1s^22s^22p^63s^23p^63d^{10}4s^24p^64d^{10}5s^25p^5$, $[Kr]4d^{10}5s^25p^5$, 46

   b. 27, 26, 1

2. a. $[Kr]4d^{10}5s^25p^2$, 2

   b. 10, germanium

3. a. $1s^22s^22p^63s^23p^63d^54s^2$

   b. manganese

Elements of the Sixth and Seventh Periods

The sixth period consists of 32 elements. It is much longer than the periods that precede it in the periodic table. To build up electron configurations for elements of this period, electrons are added first to the 6*s* orbital in cesium, Cs, and barium, Ba. Then, in lanthanum, La, an electron is added to the 5*d* orbital.

With the next element, cerium, Ce, the 4*f* orbitals begin to fill, giving cerium atoms a configuration of $[Xe]4f^15d^16s^2$. In the next 13 elements, the 4*f* orbitals are filled. Next the 5*d* orbitals are filled and the period is completed by filling the 6*p* orbitals. Because the 4*f* and the 5*d* orbitals are very close in energy, numerous deviations from the simple rules occur as these orbitals are filled. The electron configurations of the sixth-period elements can be found in the periodic table inside the back cover of the text.

The seventh period is incomplete and consists largely of synthetic elements, which will be discussed in Chapter 22.

SAMPLE PROBLEM 4-3

a. Write both the complete electron-configuration notation and the noble-gas notation for a rubidium atom.

b. Identify the elements in the second, third, and fourth periods that have the same number of highest-energy-level electrons as rubidium.

SOLUTION

a. $1s^22s^22p^63s^23p^63d^{10}4s^24p^65s^1$, $[Kr]5s^1$

b. Rubidium has one electron in its highest energy level (the fifth). The elements with the same outermost configuration are, in the second period, lithium, Li; in the third period, sodium, Na; and in the fourth period, potassium, K.

PRACTICE

1. a. Write both the complete electron-configuration notation and the noble-gas notation for a barium atom.

   b. Identify the elements in the second, third, fourth, and fifth periods that have the same number of highest-energy-level electrons as barium.

2. a. Write the noble-gas notation for a gold atom.

   b. Identify the elements in the sixth period that have one unpaired electron in their 6*s* sublevel.

Answer

1. a. $1s^22s^22p^63s^23p^63d^{10}4s^24p^64d^{10}5s^2$ $5p^66s^2$, $[Xe]6s^2$

   b. Be, Mg, Ca, Sr

2. a. $[Xe]4f^{14}5d^{10}6s^1$

   b. Au, Cs, Pt

SECTION REVIEW

1. a. What is an atom's electron configuration?
   b. What three principles guide the electron configuration of an atom?
2. What three methods are used to represent the arrangement of electrons in atoms?
3. What is an octet of electrons? Which elements contain an octet of electrons?
4. Write the complete electron-configuration notation, the noble-gas notation, and the orbital notation for the following elements:
   a. carbon b. neon c. sulfur
5. Identify the elements having the following electron configurations:
   a. $1s^22s^22p^63s^23p^3$
   b. $[Ar]4s^1$
   c. contains four electrons in its third and outer main energy level
   d. contains one paired and three unpaired electrons in its fourth and outer main energy level

CHAPTER 4 REVIEW

CHAPTER SUMMARY

4-1

- In the early twentieth century, light was determined to have a dual wave-particle nature.
- Quantum theory was developed to explain such observations as the photoelectric effect and the line-emission spectrum of hydrogen.
- Quantum theory states that electrons can exist in atoms only at specific energy levels.
- When an electron moves from one main energy level to a main energy level of lower energy, a photon is emitted. The photon's energy equals the energy difference between the two levels.
- An electron in an atom can move from one main energy level to a higher main energy level only by absorbing an amount of energy exactly equal to the difference between the two levels.

Vocabulary

electromagnetic radiation (91)
electromagnetic spectrum (91)
excited state (94)
frequency (91)
ground state (94)
photoelectric effect (93)
photon (94)
quantum (93)
wavelength (91)

4-2

- In the early twentieth century, electrons were determined to have a dual wave-particle nature.
- The Heisenberg uncertainty principle states that it is impossible to determine simultaneously both the position and velocity of an electron or any other particle.
- Quantization of electron energies is a natural outcome of the Schrödinger wave equation, which treats an atom's electrons as waves surrounding the nucleus.
- An orbital is a three-dimensional region around the nucleus that shows the region of probable electron locations.
- The four quantum numbers that describe the properties of electrons in atomic orbitals are the principal quantum number, the angular momentum quantum number, the magnetic quantum number, and the spin quantum number.

Vocabulary

angular momentum quantum number (101)
Heisenberg uncertainty principle (99)
magnetic quantum number (102)
orbital (100)
principal quantum number (101)
quantum number (101)
quantum theory (99)
spin quantum number (104)

4-3

- Electrons occupy atomic orbitals in the ground state of an atom according to the Aufbau principle, the Pauli exclusion principle, and Hund's rule.
- Electron configurations can be depicted using different types of notation. In this book, three types of notation are used: orbital notation, electron-configuration notation, and noble-gas notation.
- Electron configurations of some atoms, such as chromium, do not strictly follow the Aufbau principle, but the ground-state configuration that results is the configuration with the minimum possible energy.

Vocabulary

Aufbau principle (105)
electron configuration (105)
highest occupied energy level (110)
Hund's rule (106)
inner-shell electron (110)
noble gas (111)
noble-gas configuration (112)
Pauli exclusion principle (106)

REVIEWING CONCEPTS

1. a. List five examples of electromagnetic radiation.
   b. What is the speed of all forms of electromagnetic radiation in a vacuum?
   c. Relate the frequency and wavelength of any form of electromagnetic radiation. (4-1)
2. Prepare a two-column table. List the properties of light that can best be explained by the wave theory in one column. List those best explained by the particle theory in the second column. You may want to consult a physics textbook for reference. (4-1)
3. What are the frequency and wavelength ranges of visible light? (4-1)
4. List the colors of light in the visible spectrum in order of increasing frequency. (4-1)
5. In the early twentieth century, what two experiments involving light and matter could not be explained by the wave theory of light? (4-1)
6. a. How are the wavelength and frequency of electromagnetic radiation related?
   b. How are the energy and frequency of electromagnetic radiation related?
   c. How are the energy and wavelength of electromagnetic radiation related? (4-1)
7. Which theory of light, the wave or particle theory, best explains the following phenomena?
   a. the interference of light
   b. the photoelectric effect
   c. the emission of electromagnetic radiation by an excited atom (4-1)
8. Distinguish between the ground state and an excited state of an atom. (4-1)
9. According to the Bohr model of the hydrogen atom, how is hydrogen's emission spectrum produced? (4-1)
10. Describe two major shortcomings of the Bohr model of the atom. (4-2)
11. a. What is the principal quantum number?
    b. How is it symbolized?
    c. What are shells?
    d. How does n relate to the number of orbitals per main energy level and the number of electrons allowed per main energy level? (4-2)
12. a. What information is given by the angular momentum quantum number?
    b. What are sublevels, or subshells? (4-2)
13. For each of the following values of n, indicate the numbers and types of sublevels possible for that main energy level. (Hint: See Table 4-2.)
    a. $n = 1$
    b. $n = 2$
    c. $n = 3$
    d. $n = 4$
    e. $n = 7$ (number only) (4-2)
14. a. What information is given by the magnetic quantum number?
    b. How many orbital orientations are possible in each of the *s, p, d,* and *f* sublevels?
    c. Explain and illustrate the notation for distinguishing among the different *p* orbitals in a sublevel. (4-2)
15. a. What is the relationship between n and the total number of orbitals in a main energy level?
    b. How many total orbitals are contained in the third main energy level? in the fifth? (4-2)
16. a. What information is given by the spin quantum number?
    b. What are the possible values for this quantum number? (4-2)
17. How many electrons could be contained in the following main energy levels with n equal to:
    a. 1
    b. 3 18
    c. 4 32
    d. 6 72
    e. 7 98 (4-2)
18. a. In your own words, state the Aufbau principle.
    b. Explain the meaning of this principle in terms of an atom with many electrons. (4-3)
19. a. In your own words, state Hund's rule.
    b. What is the basis for this rule? (4-3)
20. a. In your own words, state the Pauli exclusion principle.
    b. What is the significance of the spin quantum number?

c. Compare the values of the spin quantum number for two electrons in the same orbital. (4-3)

21. a. What is meant by the highest occupied energy level in an atom?
b. What are inner-shell electrons? (4-3)

22. Determine the highest occupied energy level in the following elements:
a. He
b. Be
c. Al
d. Ca
e. Sn (4-3)

23. Write the orbital notation for the following elements. (Hint: See Sample Problem 4-1.)
a. P
b. B
c. Na
d. C (4-3)

24. Write the electron-configuration notation for an unidentified element that contains the following number of electrons:
a. 3
b. 6
c. 8
d. 13 (4-3)

25. Given that the electron configuration for oxygen is $1s^22s^22p^4$, answer the following questions:
a. How many electrons are in each atom?
b. What is the atomic number of this element?
c. Write the orbital notation for oxygen's electron configuration.
d. How many unpaired electrons does oxygen have?
e. What is the highest occupied energy level?
f. How many inner-shell electrons does the atom contain?
g. In which orbital(s) are these inner-shell electrons located? (4-3)

26. a. What are the noble gases?
b. What is a noble-gas configuration?
c. How does noble-gas notation simplify writing an atom's electron configuration? (4-3)

27. Write the noble-gas notation for the electron configuration of each of the elements that follow. (Hint: See Sample Problem 4-2.)
a. Cl
b. Ca
c. Se (4-3)

28. a. What information is given by the noble-gas notation $[Ne]3s^2$?
b. What element does this represent? (4-3)

29. Write both the complete electron-configuration notation and the noble-gas notation for each of the following elements. (Hint: See Sample Problem 4-3.)
a. Na
b. Sr
c. P (4-3)

30. Identify each of the following atoms on the basis of its electron configuration:
a. $1s^22s^22p^1$
b. $1s^22s^22p^5$
c. $[Ne]3s^2$
d. $[Ne]3s^23p^2$
e. $[Ne]3s^23p^5$
f. $[Ar]4s^1$
g. $[Ar]3d^64s^2$ (4-3)

PROBLEMS

Photons and Electromagnetic Radiation

31. Determine the frequency of light with a wavelength of 4.257×10^{-7} cm.

32. Determine the energy in joules of a photon whose frequency is 3.55×10^{17} Hz.

33. Using the two equations $E = h\nu$ and $c = \lambda\nu$, derive an equation expressing E in terms of h, c, and λ.

34. How long would it take a radio wave with a frequency of 7.25×10^5 Hz to travel from Mars to Earth if the distance between the two planets is approximately 8.00×10^7 km?

35. Cobalt-60 ($^{60}_{27}Co$) is an artificial radioisotope that is produced in a nuclear reactor for use as a gamma-ray source in the treatment of certain types of cancer. If the wavelength of the gamma radiation from a cobalt-60 source is 1.00×10^{-3} nm, calculate the energy of a photon of this radiation.

Orbitals and Electron Configuration

36. List the order in which orbitals generally fill, from the 1*s* to the 7*p* orbital.

37. Write the noble-gas notation for the electron configurations of each of the following elements:
a. As
b. Pb
c. Lr
d. Hg
e. Sn
f. Xe
g. La

38. How do the electron configurations of chromium and copper contradict the Aufbau principle?

MIXED REVIEW

39. a. Which has a longer wavelength, green or yellow light?
b. Which has a higher frequency, an X ray or a microwave?
c. Which travels at a greater speed, ultraviolet or infrared light?

40. Write both the complete electron-configuration and noble-gas notation for each of the following:
a. Ar
b. Br
c. Al

41. Given the speed of light as 3.0×10^8 m/s, calculate the wavelength of the electromagnetic radiation whose frequency is 7.500×10^{12} Hz.

42. a. What is the electromagnetic spectrum?
b. What units are used to express wavelength?
c. What unit is used to express frequencies of electromagnetic waves?

43. Given that the electron configuration for phosphorus is $1s^22s^22p^63s^23p^3$, answer the following questions:
a. How many electrons are in each atom?
b. What is the atomic number of this element?
c. Write its orbital notation.
d. How many unpaired electrons does an atom of phosphorus have?
e. What is its highest occupied energy level?
f. How many inner-shell electrons does the atom contain?
g. In which orbital(s) are these inner-shell electrons located?

44. What is the frequency of a radio wave with an energy of 1.55×10^{-24} J/photon?

45. Write the noble-gas notation for the electron configurations of each of the following elements:
a. Hf
b. Sc
c. Fe
d. At
e. Ac
f. Zn

46. Describe the major similarities and differences between Schrödinger's model of the atom and the model proposed by Bohr.

47. When sodium is heated, a yellow spectral line whose energy is 3.37×10^{-19} J/photon is produced.
a. What is the frequency of this light?
b. What is its wavelength?

48. a. What is an orbital?
b. Describe an orbital in terms of an electron cloud.

CRITICAL THINKING

49. Inferring Relationships In the emission spectrum of hydrogen shown in Figure 4-5 on page 95, each colored line is produced by the emission of photons with specific energies. Substances also produce absorption spectra when electromagnetic radiation passes through them. Certain wavelengths are absorbed. Using the diagram below, predict what the wavelengths of the absorption lines will be when white light (all the colors of the visible spectrum) is passed through hydrogen gas.

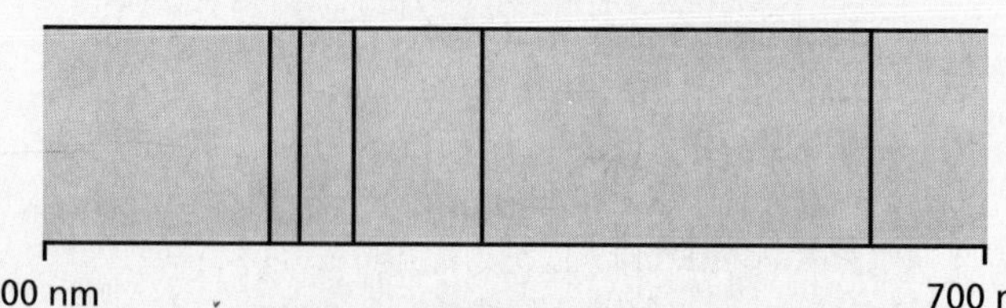

Hydrogen absorption spectrum

50. Applying Models When discussing the photoelectric effect, the minimum energy needed to remove an electron from the metal is called the *work function* and is a characteristic of the metal. For example, chromium, Cr, will emit electrons when the wavelength of the radiation is 284 nm or less. Calculate the work function for chromium. (Hint: You will need to use the two equations that describe the relationships among wavelength, frequency, speed of light, and Planck's constant.)

TECHNOLOGY & LEARNING

51. Graphing Calculator Calculating Quantum Number Relationships

The graphing calculator can be programmed to calculate relationships in atomic structure given the quantum number. Given the principal quantum number 3, you will calculate the number of orbitals, orbital shapes, and electrons. Begin by programming the calculator to carry out the calculation. The program will then be used to make calculations.

A. Program the calculation.

Keystrokes for naming the program: [PRGM] [▶] [▶] [ENTER]

Name the program: [Q] [U] [A] [N] [T] [U] [M] [ENTER]

Program keystrokes: [PRGM] [▶] [2] [ALPHA] [N]
[2nd] [:] [PRGM] [▶] [8] [2nd] [:] [PRGM] [▶] [3]
[2nd] [ALPHA] ["] [O] [R] [B] [I] [T] [A]
[L] [S] [2nd] [TEST] [1] [ALPHA] ["] [2nd] [:]
[PRGM] [▶] [3] [ALPHA] [N] [x^2] [2nd] [:] [PRGM] [▶]
[3] [2nd] [ALPHA] ["] [S] [H] [A] [P] [E]
[S] [2nd] [TEST] [1] [ALPHA] ["] [2nd] [:] [PRGM]
[▶] [3] [ALPHA] [N] [2nd] [:] [PRGM] [▶] [3] [2nd]
[ALPHA] ["] [E] [L] [E] [C] [T] [R] [O]
[N] [S] [2nd] [TEST] [1] [ALPHA] ["] [2nd] [:]
[PRGM] [▶] [3] [2] [ALPHA] [N] [x^2]

B. Check the display.

Screen display: Prompt N: ClrHome: Disp "ORBITALS=": Disp N^2: Disp "SHAPES=": Disp N: Disp "ELECTRONS=": Disp $2N^2$

Press [2nd] [QUIT] to exit the program editor.

C. Run the program.

a. Press [PRGM]. Select QUANTUM and press [ENTER] [ENTER]. Now enter the principal quantum number 3 and press [ENTER]. The calculator will provide the number of orbitals in the main energy level, the number of possible orbital shapes, and the number of electrons per main energy level.

b. Use the QUANTUM program for the principal quantum number 2.

c. Use the QUANTUM program for the principal quantum number 4.

HANDBOOK SEARCH

52. Handbook Sections 1 and 2 contain information on an analytical test and a technological application for Group 1 and 2 elements that are based on the emission of light from atoms. Review these sections to answer the following:

a. What analytical technique utilizes the emission of light from excited atoms?

b. What elements in Groups 1 and 2 can be identified by this technique?

c. What types of compounds are used to provide color in fireworks?

d. What wavelengths within the visible spectrum would most likely contain emission lines for barium?

RESEARCH & WRITING

53. Neon signs do not always contain neon gas. The various colored lights produced by the signs are due to the emission of a variety of low-pressure gases in different tubes. Research other kinds of gases used in neon signs and list the colors that they emit.

54. Prepare a report about the photoelectric effect and cite some of its practical uses. Explain the basic operation of each device or technique mentioned.

ALTERNATIVE ASSESSMENT

55. Performance Assessment A spectroscope is a device used to produce and analyze spectra. Construct a simple spectroscope and determine the absorption spectra of several elemental gases. (Your teacher will provide you with the elemental samples.)

CHAPTER 5

The Periodic Law

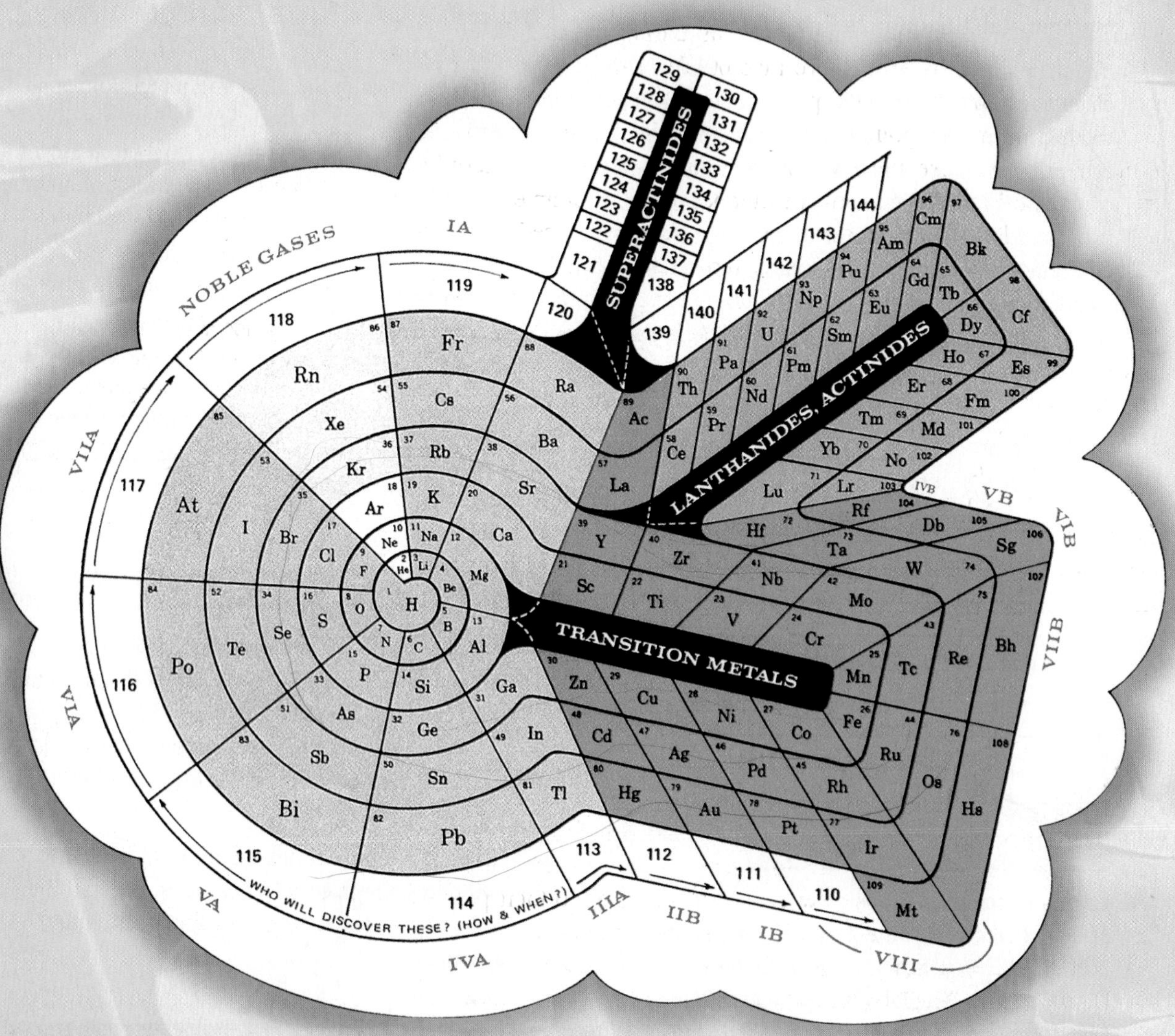

The physical and chemical properties of the elements are periodic functions of their atomic numbers.

History of the Periodic Table

SECTION 5-1

Objectives

- Explain the roles of Mendeleev and Moseley in the development of the periodic table.
- Describe the modern periodic table.
- Explain how the periodic law can be used to predict the physical and chemical properties of elements.
- Describe how the elements belonging to a group of the periodic table are interrelated in terms of atomic number.

Imagine the confusion among chemists during the middle of the nineteenth century. By 1860, more than 60 elements had been discovered. Chemists had to learn the properties of these elements as well as those of the many compounds that they formed—a difficult task. And to make matters worse, there was no method for accurately determining an element's atomic mass or the number of atoms of an element in a particular chemical compound. Different chemists used different atomic masses for the same elements, resulting in different compositions being proposed for the same compounds. This made it nearly impossible for one chemist to understand the results of another.

In September 1860, a group of chemists assembled at the First International Congress of Chemists in Karlsruhe, Germany, to settle the issue of atomic mass as well as some other matters that were making communication difficult. At the Congress, Italian chemist Stanislao Cannizzaro presented a convincing method for accurately measuring the relative mass of an atom. Cannizzaro's method enabled chemists to agree on standard values for atomic mass and initiated a search for relationships between atomic mass and other properties of the elements.

Mendeleev and Chemical Periodicity

When the Russian chemist Dmitri Mendeleev heard about the new atomic masses discussed at Karlsruhe, he decided to include the new values in a chemistry textbook that he was writing. In the book, Mendeleev hoped to organize the elements according to their properties. He went about this much as you might organize information for a research paper. He placed the name of each known element on a card, together with the atomic mass of the element and a list of its observed physical and chemical properties. He then arranged the cards according to various properties and looked for trends or patterns.

Mendeleev noticed that when the elements were arranged in order of increasing atomic mass, certain similarities in their chemical properties appeared at regular intervals. Such a repeating pattern is referred to as *periodic*. The second hand of a watch, for example, passes over any given mark at periodic, 60-second intervals. The circular waves created by a drop of water hitting a water surface are also periodic.

FIGURE 5-1 The regularly spaced water waves represent a simple periodic pattern.

FIGURE 5-2 In his first published periodic table, Mendeleev arranged the elements in vertical periods according to relative atomic mass. The atomic mass for each element is indicated by the number following the element's symbol. The unknown elements indicated by question marks at estimated atomic masses 45, 68, and 70 were later identified as scandium, Sc, gallium, Ga, and germanium, Ge.

но въ ней, мнѣ кажется, уже ясно выражается примѣнимость вы ставляемаго мною начала ко всей совокупности элементовъ, пай которыхъ извѣстенъ съ достовѣрностію. На этотъ разъ я и желалъ преимущественно найдти общую систему элементовъ. Вотъ этотъ опытъ:

			Ti=50	Zr=90	?=180.
			V=51	Nb=94	Ta=182.
			Cr=52	Mo=96	W=186.
			Mn=55	Rh=104,4	Pt=197,4
			Fe=56	Ru=104,4	Ir=198.
			Ni=Co=59	Pl=106,6,	Os=199.
H=1			Cu=63,4	Ag=108	Hg=200.
	Be=9,4	Mg=24	Zn=65,2	Cd=112	
	B=11	Al=27,4	?=68	Ur=116	Au=197?
	C=12	Si=28	?=70	Su=118	
	N=14	P=31	As=75	Sb=122	Bi=210
	O=16	S=32	Se=79,4	Te=128?	
	F=19	Cl=35,5	Br=80	I=127	
Li=7	Na=23	K=39	Rb=85,4	Cs=133	Tl=204
		Ca=40	Sr=87,6	Ba=137	Pb=207.
		?=45	Ce=92		
		?Er=56	La=94		
		?Yt=60	Di=95		
		?In=75,6	Th=118?		

а потому приходится въ разныхъ рядахъ имѣть различное измѣненіе разностей, чего нѣтъ въ главныхъ числахъ предлагаемой таблицы. Или же придется предполагать при составленіи системы очень много недостающихъ членовъ. То и другое мало выгодно. Мнѣ кажется притомъ, наиболѣе естественнымъ составить

Mendeleev created a table in which elements with similar properties were grouped together—a periodic table of the elements. His first periodic table, shown in Figure 5-2, was published in 1869. Note that Mendeleev placed iodine, I (atomic mass 127), after tellurium, Te (atomic mass 128). Although this contradicted the pattern of listing the elements in order of increasing atomic mass, it allowed Mendeleev to place tellurium in a group of elements with which it shares similar properties. Reading horizontally across Mendeleev's table, this group includes oxygen, O, sulfur, S, and selenium, Se. Iodine could also, then, be placed in the group it resembles chemically, which includes fluorine, F, chlorine, Cl, and bromine, Br.

Mendeleev's procedure left several empty spaces in his periodic table (see Figure 5-2). In 1871, the Russian chemist boldly predicted the existence and properties of the elements that would fill three of the spaces. By 1886, all three elements had been discovered. Today these elements are known as scandium, Sc, gallium, Ga, and germanium, Ge. Their properties are strikingly similar to those predicted by Mendeleev.

The success of Mendeleev's predictions persuaded most chemists to accept his periodic table and earned him credit as the discoverer of the periodic law. Two questions remained, however. (1) Why could most of the elements be arranged in the order of increasing atomic mass but a few could not? (2) What was the reason for chemical periodicity?

Moseley and the Periodic Law

The first question was not answered until more than 40 years after Mendeleev's first periodic table was published. In 1911, the English scientist Henry Moseley, who was working with Ernest Rutherford, examined the spectra of 38 different metals. When analyzing his data, Moseley discovered a previously unrecognized pattern. The elements in the periodic table were arranged in increasing order according to nuclear charge, or the number of protons in the nucleus. Moseley's work led to both the modern definition of atomic number and the recognition that atomic number, not atomic mass, is the basis for the organization of the periodic table.

Moseley's discovery was consistent with Mendeleev's ordering of the periodic table by properties rather than strictly by atomic mass. For example, according to Moseley, tellurium, with an atomic number of 52, belongs before iodine, which has an atomic number of 53. Today, Mendeleev's principle of chemical periodicity is correctly stated in what is known as the **periodic law:** *The physical and chemical properties of the elements are periodic functions of their atomic numbers.* In other words, when the elements are arranged in order of increasing atomic number, elements with similar properties appear at regular intervals.

The Modern Periodic Table

The periodic table has undergone extensive change since Mendeleev's time (see Figure 5-6 on pages 130–131). Chemists have discovered new elements and, in more recent years, synthesized new ones in the laboratory. Each of the more than 40 new elements, however, can be placed in a group of other elements with similar properties. *The* **periodic table** *is an arrangement of the elements in order of their atomic numbers so that elements with similar properties fall in the same column, or group.*

			2 He
7 N	8 O	9 F	10 Ne
15 P	16 S	17 Cl	18 Ar
33 As	34 Se	35 Br	36 Kr
51 Sb	52 Te	53 I	54 Xe
83 Bi	84 Po	85 At	86 Rn

FIGURE 5-3 The noble gases, also known as the Group 18 elements, are all rather unreactive. As you will read, the reason for this low reactivity also accounts for the special place occupied by the noble gases in the periodic table.

The Noble Gases

Perhaps the most significant addition to the periodic table came with the discovery of the noble gases. In 1894, English physicist John William Strutt (Lord Rayleigh) and Scottish chemist Sir William Ramsay discovered argon, Ar, a gas in the atmosphere that had previously escaped notice because of its total lack of chemical reactivity. Back in 1868, another noble gas, helium, He, had been discovered as a component of the sun, based on the emission spectrum of sunlight. In 1895, Ramsay showed that helium also exists on Earth.

In order to fit argon and helium into the periodic table, Ramsay proposed a new group. He placed this group between the groups now known as Group 17 (the fluorine family) and Group 1 (the lithium family). In 1898, Ramsay discovered two more noble gases to place in his new group, krypton, Kr, and xenon, Xe. The final noble gas, radon, Rn, was discovered in 1900 by the German scientist Friedrich Ernst Dorn.

The Lanthanides

The next step in the development of the periodic table was completed in the early 1900s. It was then that the puzzling chemistry of the lanthanides was finally understood. *The* **lanthanides** *are the 14 elements with atomic numbers from 58 (cerium, Ce) to 71 (lutetium, Lu).* Because these elements are so similar in chemical and physical properties, the process of separating and identifying them was a tedious task that required the effort of many chemists.

The Actinides

Another major step in the development of the periodic table was the discovery of the actinides. *The* **actinides** *are the 14 elements with atomic numbers from 90 (thorium, Th) to 103 (lawrencium, Lr).* The lanthanides and actinides belong in Periods 6 and 7, respectively, of the periodic table, between the elements of Groups 3 and 4. To save space, the lanthanides and actinides are usually set off below the main portion of the periodic table, as shown in Figure 5-6 on pages 130–131.

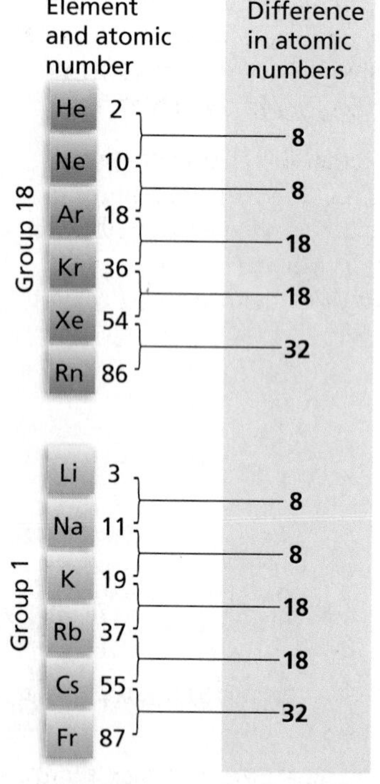

FIGURE 5-4 In each of Groups 1 and 18, the differences between the atomic numbers of successive elements are 8, 8, 18, 18, and 32, respectively. Groups 2 and 13–17 follow a similar pattern.

Periodicity

Periodicity with respect to atomic number can be observed in any group of elements in the periodic table. Consider the noble gases of Group 18. The first noble gas is helium, He. It has an atomic number of 2. The elements following helium in atomic number have completely different properties until the next noble gas, neon, Ne, which has an atomic number of 10, is reached. The remaining noble gases in order of increasing atomic number are argon (Ar, atomic number 18), krypton (Kr, atomic number 36), xenon (Xe, atomic number 54), and radon (Rn, atomic number 86). The differences in atomic number between successive noble gases are shown in Figure 5-4. Also shown in Figure 5-4 are atomic-number differences between the elements of Group 1, which are all solid, silvery metals. As you can see, the differences in atomic number between the Group 1 metals follow the same pattern as the differences in atomic number between the noble gases.

Starting with the first member of Groups 13–17, a similar periodic pattern is repeated. The atomic number of each successive element is 8, 18, 18, and 32 higher than the atomic number of the element above it. In Section 5-2, you will see that the second mystery presented by Mendeleev's periodic table—the reason for periodicity—is explained by the arrangement of the electrons around the nucleus.

DESKTOP INVESTIGATION

Designing Your Own Periodic Table

MATERIALS

- **index cards**

QUESTION

Can you design your own periodic table using information similar to that available to Mendeleev?

PROCEDURE

1. Write down the information available for each element on separate index cards. The following information is appropriate: a letter of the alphabet (A, B, C, etc.) to identify each element; atomic mass; state; density; melting point; boiling point; and any other readily observable physical properties. Do not write the name of the element on the index card, but keep a separate list indicating the letters you have assigned to each element.
2. Organize the cards for the elements in a logical pattern as you think Mendeleev might have done.

DISCUSSION

1. Keeping in mind that the information you have is similar to that available to Mendeleev in 1869, answer the following questions.
   a. Why are atomic masses given instead of atomic numbers?
   b. Can you identify each element by name?
2. How many groups of elements, or families, are in your periodic table? How many periods, or series, are in the table?
3. Predict the characteristics of any missing elements. When you have finished, check your work using your separate list of elements and a periodic table.

SECTION REVIEW

1. a. Who is credited with developing a method that led to the determination of standard relative atomic masses?
   b. Who discovered the periodic law?
   c. Who established atomic numbers as the basis for organizing the periodic table?
2. State the periodic law.
3. Name three sets of elements added to the periodic table after Mendeleev's time.
4. How do the atomic numbers of the elements within each of Groups 1, 2, and 13–18 of the periodic table vary?

SECTION 5-2

Objectives

- Describe the relationship between electrons in sublevels and the length of each period of the periodic table.
- Locate and name the four blocks of the periodic table. Explain the reasons for these names.
- Discuss the relationship between group configurations and group numbers.
- Describe the locations in the periodic table and the general properties of the alkali metals, the alkaline-earth metals, the halogens, and the noble gases.

Electron Configuration and the Periodic Table

The Group 18 elements of the periodic table (the noble gases) undergo few chemical reactions. This stability results from the gases' special electron configurations. Helium's highest occupied level, the 1*s* orbital, is completely filled with electrons. And the highest occupied levels of the other noble gases contain stable octets. Generally the electron configuration of an atom's highest occupied energy level governs the atom's chemical properties.

Periods and Blocks of the Periodic Table

While the elements are arranged vertically in the periodic table in groups that share similar chemical properties, they are also organized horizontally in rows, or *periods.* (As shown in Figure 5-6, there are a total of seven periods of elements in the modern periodic table.) As can be seen in Table 5-1, the length of each period is determined by the number of electrons that can occupy the sublevels being filled in that period.

TABLE 5-1 *Relationship Between Period Length and Sublevels Being Filled in the Periodic Table*

Period number	Number of elements in period	Sublevels in order of filling
1	2	1*s*
2	8	2*s* 2*p*
3	8	3*s* 3*p*
4	18	4*s* 3*d* 4*p*
5	18	5*s* 4*d* 5*p*
6	32	6*s* 4*f* 5*d* 6*p*
7	23 (to date)	7*s* 5*f* 6*d*, etc.

In the first period, the 1*s* sublevel is being filled. The 1*s* sublevel can hold a total of two electrons. Therefore, the first period consists of two elements—hydrogen and helium. In the second period, the 2*s* sublevel, which can hold two electrons, and the 2*p* sublevel, which can hold six electrons, are being filled. Consequently, the second period totals eight elements. Similarly, filling of the 3*s* and 3*p* sublevels accounts for the eight elements of the third period. Filling 3*d* and 4*d* sublevels in addition to the *s* and *p* sublevels adds 10 elements to both the fourth and fifth periods. Therefore, each of these periods totals 18 elements. Filling 4*f* sublevels in addition to *s, p,* and *d* sublevels adds 14 elements to the sixth period, which totals 32 elements. And as new elements are created, the 23 known elements in Period 7 could, in theory, be extended to 32.

The period of an element can be determined from the element's electron configuration. For example, arsenic, As, has the electron configuration $[Ar]3d^{10}4s^24p^3$. The 4 in $4p^3$ indicates that arsenic's highest occupied energy level is the fourth energy level. Arsenic is therefore in the fourth period in the periodic table. The period and electron configuration for each element can be found in the periodic table on pages 130–131.

Based on the electron configurations of the elements, the periodic table can be divided into four blocks, the *s, p, d,* and *f* blocks. This division is illustrated in Figure 5-5. The name of each block is determined by whether an *s, p, d,* or *f* sublevel is being filled in successive elements of that block.

FIGURE 5-5 Based on the electron configurations of the elements, the periodic table can be subdivided into four sublevel blocks.

Sublevel Blocks of the Periodic Table

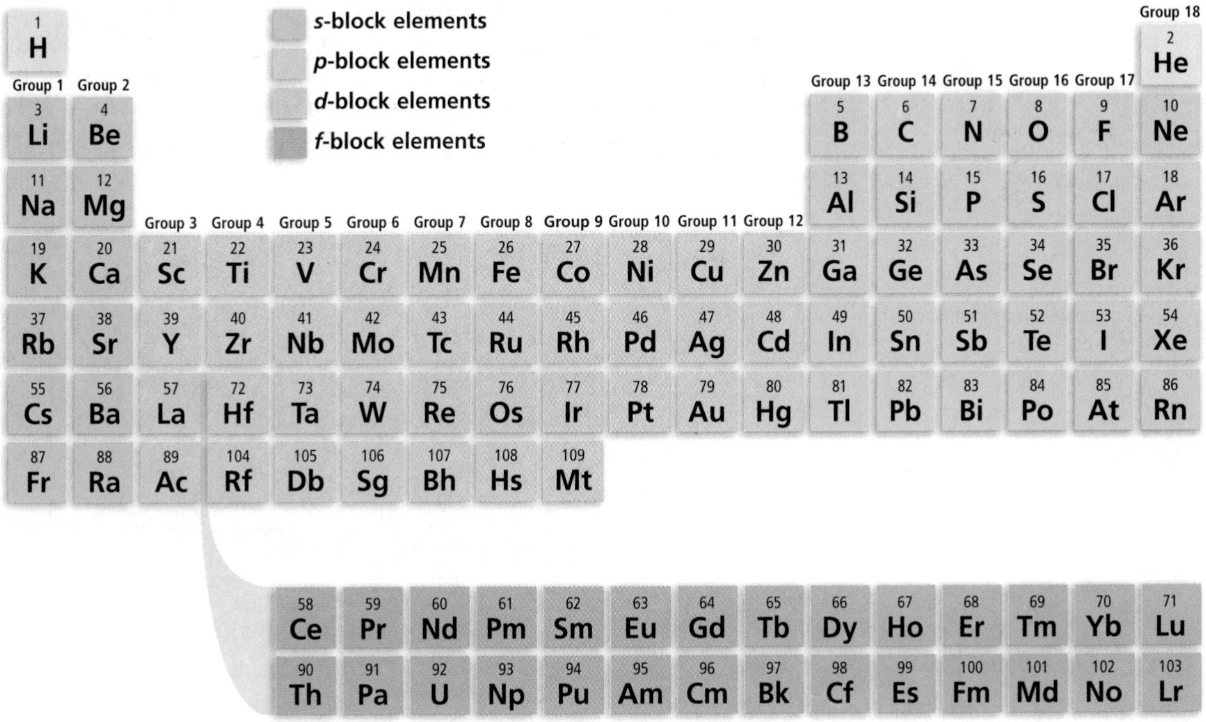

s-block elements
p-block elements
d-block elements
f-block elements

Group 1	Group 2	Group 3	Group 4	Group 5	Group 6	Group 7	Group 8	Group 9	Group 10	Group 11	Group 12	Group 13	Group 14	Group 15	Group 16	Group 17	Group 18
1 H																	2 He
3 Li	4 Be											5 B	6 C	7 N	8 O	9 F	10 Ne
11 Na	12 Mg											13 Al	14 Si	15 P	16 S	17 Cl	18 Ar
19 K	20 Ca	21 Sc	22 Ti	23 V	24 Cr	25 Mn	26 Fe	27 Co	28 Ni	29 Cu	30 Zn	31 Ga	32 Ge	33 As	34 Se	35 Br	36 Kr
37 Rb	38 Sr	39 Y	40 Zr	41 Nb	42 Mo	43 Tc	44 Ru	45 Rh	46 Pd	47 Ag	48 Cd	49 In	50 Sn	51 Sb	52 Te	53 I	54 Xe
55 Cs	56 Ba	57 La	72 Hf	73 Ta	74 W	75 Re	76 Os	77 Ir	78 Pt	79 Au	80 Hg	81 Tl	82 Pb	83 Bi	84 Po	85 At	86 Rn
87 Fr	88 Ra	89 Ac	104 Rf	105 Db	106 Sg	107 Bh	108 Hs	109 Mt									

58 Ce	59 Pr	60 Nd	61 Pm	62 Sm	63 Eu	64 Gd	65 Tb	66 Dy	67 Ho	68 Er	69 Tm	70 Yb	71 Lu
90 Th	91 Pa	92 U	93 Np	94 Pu	95 Am	96 Cm	97 Bk	98 Cf	99 Es	100 Fm	101 Md	102 No	103 Lr

Periodic Table of the Elements

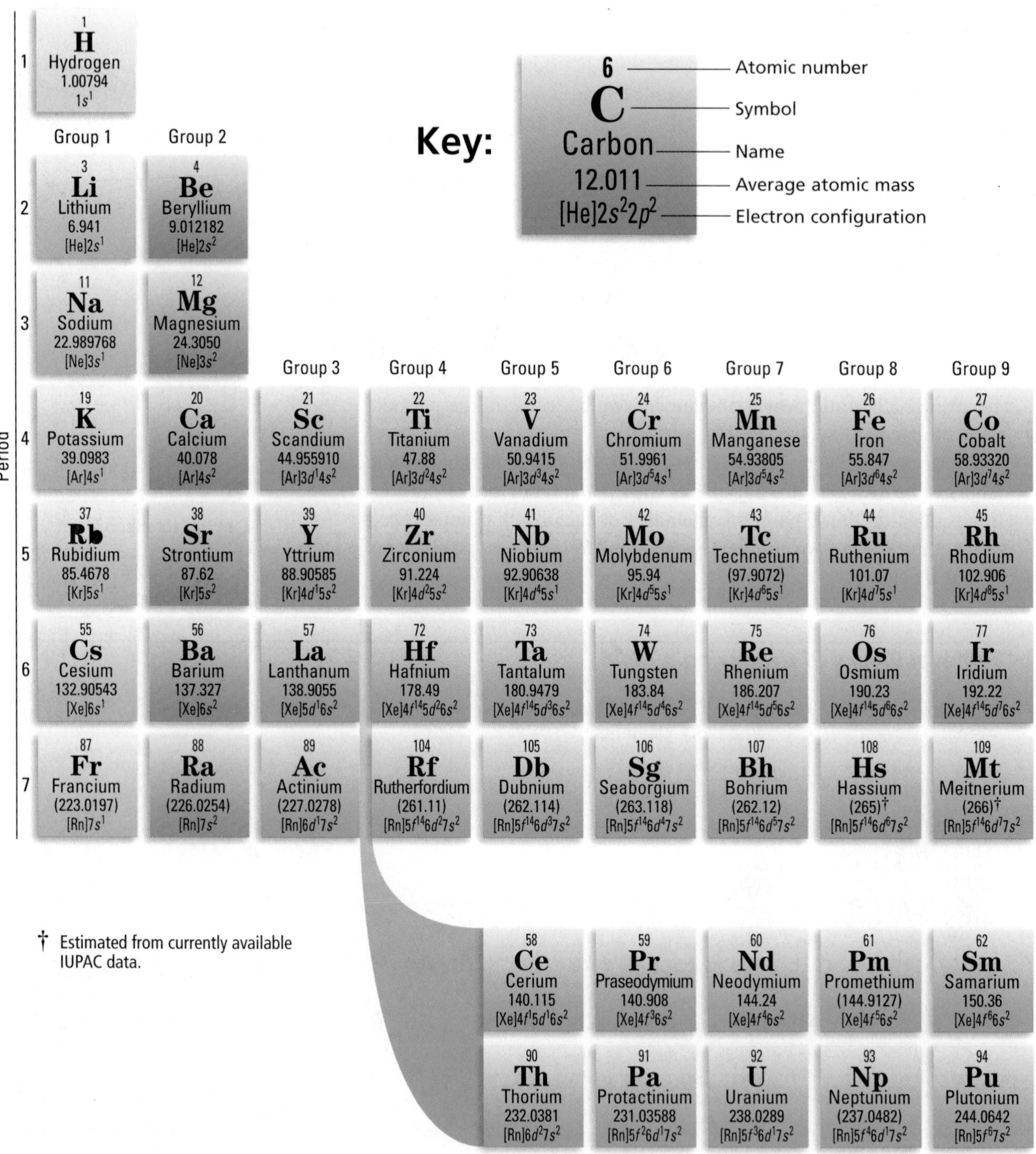

FIGURE 5-6 In the common periodic table, the elements are arranged in vertical groups and in horizontal periods.

Metals

- Alkali metals
- Alkaline-earth metals
- Transition metals
- Other metals

Metalloids

- Metalloids

Nonmetals

- Halogens
- Other nonmetals

Noble gases

- Noble gases

Group 10	Group 11	Group 12	Group 13	Group 14	Group 15	Group 16	Group 17	Group 18
								2 **He** Helium 4.002602 $1s^2$
			5 **B** Boron 10.811 [He]$2s^22p^1$	6 **C** Carbon 12.011 [He]$2s^22p^2$	7 **N** Nitrogen 14.00674 [He]$2s^22p^3$	8 **O** Oxygen 15.9994 [He]$2s^22p^4$	9 **F** Fluorine 18.9984032 [He]$2s^22p^5$	10 **Ne** Neon 20.1797 [He]$2s^22p^6$
			13 **Al** Aluminum 26.981539 [Ne]$3s^23p^1$	14 **Si** Silicon 28.0855 [Ne]$3s^23p^2$	15 **P** Phosphorus 30.9738 [Ne]$3s^23p^3$	16 **S** Sulfur 32.066 [Ne]$3s^23p^4$	17 **Cl** Chlorine 35.4527 [Ne]$3s^23p^5$	18 **Ar** Argon 39.948 [Ne]$3s^23p^6$
28 **Ni** Nickel 58.6934 [Ar]$3d^84s^2$	29 **Cu** Copper 63.546 [Ar]$3d^{10}4s^1$	30 **Zn** Zinc 65.39 [Ar]$3d^{10}4s^2$	31 **Ga** Gallium 69.723 [Ar]$3d^{10}4s^24p^1$	32 **Ge** Germanium 72.61 [Ar]$3d^{10}4s^24p^2$	33 **As** Arsenic 74.92159 [Ar]$3d^{10}4s^24p^3$	34 **Se** Selenium 78.96 [Ar]$3d^{10}4s^24p^4$	35 **Br** Bromine 79.904 [Ar]$3d^{10}4s^24p^5$	36 **Kr** Krypton 83.80 [Ar]$3d^{10}4s^24p^6$
46 **Pd** Palladium 106.42 [Kr]$4d^{10}5s^0$	47 **Ag** Silver 107.8682 [Kr]$4d^{10}5s^1$	48 **Cd** Cadmium 112.411 [Kr]$4d^{10}5s^2$	49 **In** Indium 114.818 [Kr]$4d^{10}5s^25p^1$	50 **Sn** Tin 118.710 [Kr]$4d^{10}5s^25p^2$	51 **Sb** Antimony 121.757 [Kr]$4d^{10}5s^25p^3$	52 **Te** Tellurium 127.60 [Kr]$4d^{10}5s^25p^4$	53 **I** Iodine 126.904 [Kr]$4d^{10}5s^25p^5$	54 **Xe** Xenon 131.29 [Kr]$4d^{10}5s^25p^6$
78 **Pt** Platinum 195.08 [Xe]$4f^{14}5d^96s^1$	79 **Au** Gold 196.96654 [Xe]$4f^{14}5d^{10}6s^1$	80 **Hg** Mercury 200.59 [Xe]$4f^{14}5d^{10}6s^2$	81 **Tl** Thallium 204.3833 [Xe]$4f^{14}5d^{10}6s^26p^1$	82 **Pb** Lead 207.2 [Xe]$4f^{14}5d^{10}6s^26p^2$	83 **Bi** Bismuth 208.98037 [Xe]$4f^{14}5d^{10}6s^26p^3$	84 **Po** Polonium (208.9824) [Xe]$4f^{14}5d^{10}6s^26p^4$	85 **At** Astatine (209.9871) [Xe]$4f^{14}5d^{10}6s^26p^5$	86 **Rn** Radon (222.0176) [Xe]$4f^{14}5d^{10}6s^26p^6$

63 **Eu** Europium 151.966 [Xe]$4f^76s^2$	64 **Gd** Gadolinium 157.25 [Xe]$4f^75d^16s^2$	65 **Tb** Terbium 158.92534 [Xe]$4f^96s^2$	66 **Dy** Dysprosium 162.50 [Xe]$4f^{10}6s^2$	67 **Ho** Holmium 164.930 [Xe]$4f^{11}6s^2$	68 **Er** Erbium 167.26 [Xe]$4f^{12}6s^2$	69 **Tm** Thulium 168.93421 [Xe]$4f^{13}6s^2$	70 **Yb** Ytterbium 173.04 [Xe]$4f^{14}6s^2$	71 **Lu** Lutetium 174.967 [Xe]$4f^{14}5d^16s^2$
95 **Am** Americium (243.0614) [Rn]$5f^77s^2$	96 **Cm** Curium (247.0703) [Rn]$5f^76d^17s^2$	97 **Bk** Berkelium (247.0703) [Rn]$5f^97s^2$	98 **Cf** Californium (251.0796) [Rn]$5f^{10}7s^2$	99 **Es** Einsteinium (252.083) [Rn]$5f^{11}7s^2$	100 **Fm** Fermium (257.0951) [Rn]$5f^{12}7s^2$	101 **Md** Mendelevium (258.10) [Rn]$5f^{13}7s^2$	102 **No** Nobelium (259.1009) [Rn]$5f^{14}7s^2$	103 **Lr** Lawrencium (262.11) [Rn]$5f^{14}6d^17s^2$

The atomic masses listed in this table reflect the precision of current measurements. (Values listed in parentheses are those of the element's most-stable or most-common isotope.) In calculations throughout the text, however, atomic masses have been rounded to two places to the right of the decimal.

(a)

(b)

FIGURE 5-7 (a) Like other alkali metals, potassium reacts so strongly with water that (b) it must be stored in kerosene or oil to prevent it from reacting with moisture in the air.

The *s*-Block Elements: Groups 1 and 2

The elements of the *s* block are chemically reactive *metals*. The Group 1 metals are more reactive than those of Group 2. The outermost energy level in an atom of each Group 1 element contains a single *s* electron. For example, the configurations of lithium and sodium are $[He]2s^1$ and $[Ne]3s^1$, respectively. As you will learn in Section 5-3, the ease with which the single electron is lost helps to make the Group 1 metals extremely reactive. Using *n* for the number of the highest occupied energy level, the outer, or group, configurations of the Group 1 and 2 elements are written ns^1 and ns^2, respectively.

The elements of Group 1 of the periodic table (lithium, sodium, potassium, rubidium, cesium, and francium) are known as the **alkali metals.** In their pure state, all of the alkali metals have a silvery appearance and are soft enough to cut with a knife. However, because they are so reactive, alkali metals are not found in nature as free elements. They combine vigorously with most nonmetals. And they react strongly with water to produce hydrogen gas and aqueous solutions of substances known as alkalis. Because of their extreme reactivity with air or moisture, alkali metals are usually stored in kerosene. Proceeding down the column, the elements of Group 1 melt at successively lower temperatures. With the exception of lithium, the alkali metals have melting points lower than the boiling point of water. Lithium, Li, sodium, Na, and potassium, K, are less dense than water.

The elements of Group 2 of the periodic table (beryllium, magnesium, calcium, strontium, barium, and radium) are called the **alkaline-earth metals.** Atoms of alkaline-earth metals contain a pair of electrons in their outermost *s* sublevel. Consequently, the group configuration for Group 2 is ns^2. The Group 2 metals are harder, denser, and stronger than the alkali metals. They also have higher melting points. Although they are less reactive than the alkali metals, the alkaline-earth metals are also too reactive to be found in nature as free elements.

Hydrogen and Helium

Before discussing the other blocks of the periodic table, let's consider two special cases in the classification of the elements—hydrogen and helium. Hydrogen has an electron configuration of $1s^1$, but despite the ns^1 configuration, it does not share the same properties as the elements

(a)

(b)

FIGURE 5-8 Calcium, an alkaline-earth metal, is too reactive to be found in nature in its pure state (a). Instead, it exists in compounds, such as in the minerals that make up marble (b).

of Group 1. Although it is located above the Group 1 elements in many periodic tables, hydrogen is a unique element, with properties that do not closely resemble those of any group.

Like the Group 2 elements, helium has an ns^2 group configuration. Yet it is part of Group 18. Because its highest occupied energy level is filled by two electrons, helium possesses special chemical stability, exhibiting the unreactive nature of a Group 18 element. By contrast, the Group 2 metals have no special stability; their highest occupied energy levels are not filled because each metal has an empty available *p* sublevel.

SAMPLE PROBLEM 5-1

a. Without looking at the periodic table, give the group, period, and block in which the element with the electron configuration $[Xe]6s^2$ is located.

b. Without looking at the periodic table, write the electron configuration for the Group 1 element in the third period. Is this element likely to be a more-active or less-active element than the element described in (a)?

SOLUTION

a. The element is in Group 2, as indicated by the group configuration of ns^2. It is in the sixth period, as indicated by the highest principal quantum number in its configuration, 6. The element is in the *s* block.

b. In a third-period element, the highest occupied energy level is the third main energy level, $n = 3$. The 1*s*, 2*s*, and 2*p* sublevels are completely filled (see Table 5-1). A Group 1 element has a group configuration of ns^1, which indicates a single electron in its highest *s* sublevel. Therefore, this element has the following configuration.

$$1s^2 2s^2 2p^6 3s^1 \quad \text{or} \quad [Ne]3s^1$$

Because it is in Group 1 (the alkali metals), this element is likely to be more reactive than the element described in (a), which is in Group 2 (the alkaline-earth metals).

PRACTICE

1. Without looking at the periodic table, give the group, period, and block in which the element with the electron configuration $[Kr]5s^1$ is located.
2. a. Without looking at the periodic table, write both the group configuration and the complete electron configuration for the Group 2 element in the fourth period.

   b. Refer to Figure 5-6 to identify the element described in (a). Then write the element's noble-gas notation.

   c. How does the reactivity of the element in (a) compare with the reactivity of the element in Group 1 of the same period?

Answer

1. Group 1, fifth period, *s* block
2. a. ns^2; $1s^2 2s^2 2p^6 3s^2 3p^6 4s^2$

   b. Ca, $[Ar]4s^2$

   c. The element is in Group 2, so it's probably less reactive than the Group 1 element of the same period.

FIGURE 5-9 The diagram shows the electron configuration of scandium, Sc, the Group 3 element of the third period. In general, the $(n-1)d$ sublevel in Groups 3–12 is occupied by electrons after the ns sublevel is filled.

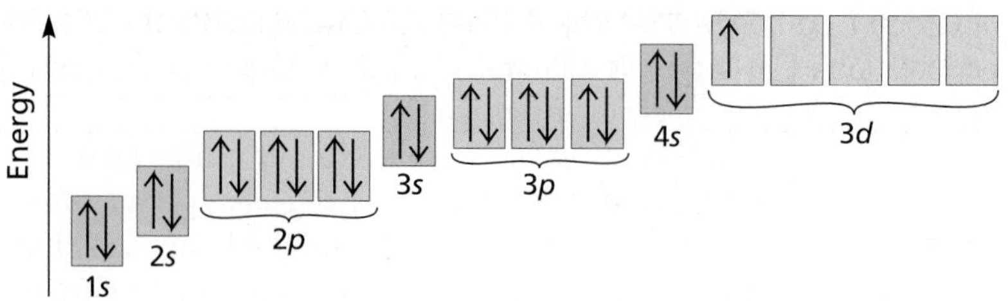

The *d*-Block Elements: Groups 3–12

In the *d*-block elements (Groups 3–12), filling of the *ns* sublevel is followed by the addition of electrons to the *d* sublevel of the preceding, or $(n-1)$, energy level (see Figure 5-9). In addition to the two *ns* electrons of Group 2, atoms of the Group 3 elements each have one electron in the *d* sublevel of the $(n-1)$ energy level. The group configuration for Group 3 is therefore $(n-1)d^1ns^2$. Nine more electrons are added to the *d* sublevel across each period. Atoms of the Group 12 elements have 10 electrons in the *d* sublevel plus two electrons in the *ns* sublevel. The group configuration for Group 12 is $(n-1)d^{10}ns^2$.

As you read in Chapter 4, some deviations from orderly *d* sublevel filling occur in Groups 4–11. As a result, elements in these *d*-block groups, unlike those in *s*-block and *p*-block groups, do not necessarily have identical outer electron configurations. For example, in Group 10, nickel, Ni, has the electron configuration $[Ar]3d^84s^2$. Palladium, Pd, has the configuration $[Kr]4d^{10}5s^0$. And platinum, Pt, has the configuration $[Xe]4f^{14}5d^96s^1$. Notice, however, that in each case the sum of the outer *s* and *d* electrons is equal to the group number. This is true for all *d*-block elements within the same group.

The d-block elements are metals with typical metallic properties and are often referred to as **transition elements.** They are good conductors of electricity and have a high luster. They are typically less reactive than the alkali metals and the alkaline-earth metals. Some are so unreactive that they do not easily form compounds, existing in nature as free elements. Palladium, platinum, and gold are among the least reactive of all the elements. Some *d*-block elements are shown in Figure 5-10.

FIGURE 5-10 Mercury, tungsten, and vanadium are transition elements. Locate them in the *d* block of the periodic table on page 129.

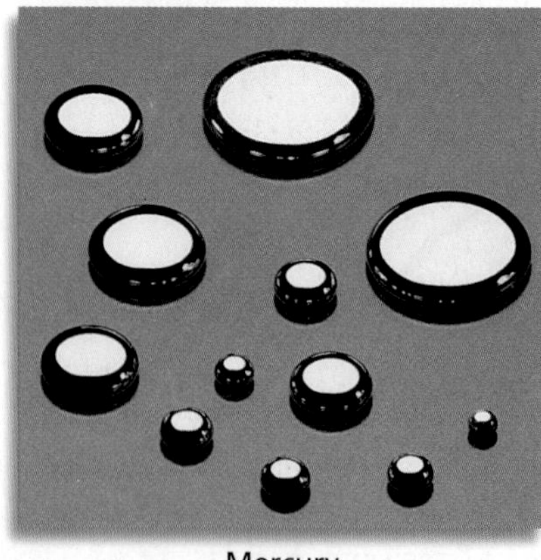

Mercury

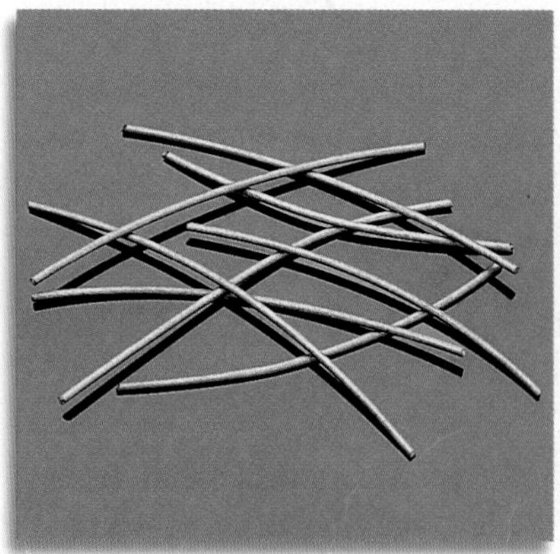

Tungsten

Vanadium

The Wild Kingdom

From *The Periodic Kingdom: A Journey Into the World of the Chemical Elements* by P. W. Atkins

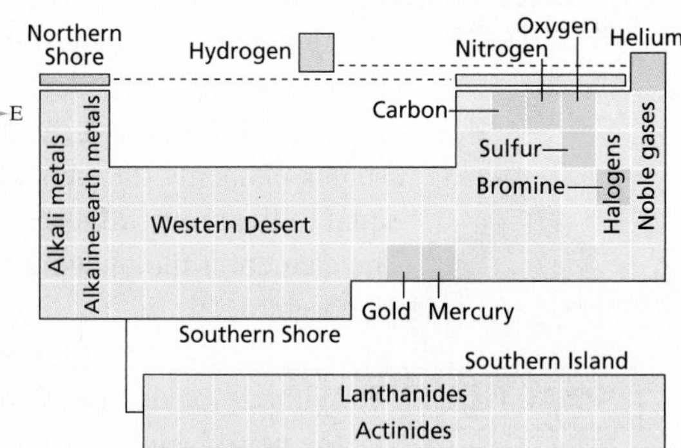

The general layout of the Periodic Kingdom.

Welcome to the Periodic Kingdom. This is a land of the imagination, but it is closer to reality than it appears to be. This is the kingdom of the chemical elements, the substances from which everything tangible is made. It is not an extensive country, for it consists of only a hundred or so regions . . . yet it accounts for everything material in our actual world. From the hundred elements that are at the center of our story, all planets, rocks, vegetation, and animals are made. These elements are the basis of the air, the oceans, and the Earth itself. We stand on the elements, we eat the elements, we *are* the elements. Because our brains are made up of elements, even our opinions are, in a sense, properties of the elements and hence inhabitants of the kingdom.

Even from . . . far above the country, we can see broad features of the landscape. There are the glittering, lustrous regions made up of metals and lying together in what we shall call the Western Desert. This desert is broadly uniform, but there is a subtlety of shades, indicating a variety of characteristics. Here and there are gentle splashes of color, such as the familiar glint of gold and the blush of copper. How remarkable it is that these desert lands make up so much of the kingdom . . . yet the kingdom supplies such luxuriance in the real world!

Generally speaking, the Western Desert was explored and exploited from east to west; that is, technology and industry made use of them in that order. Copper displaced stone to give us the Bronze Age. Then, as the explorers pressed westward, applying ever more vigorous means of discovery, they encountered iron and used it to fabricate more effective weaponry . . . The strongest states enjoyed freedom from constant aggression, and this gave them time for scholarship; thus in due course explorers were able to penetrate into more distant western regions of the desert.

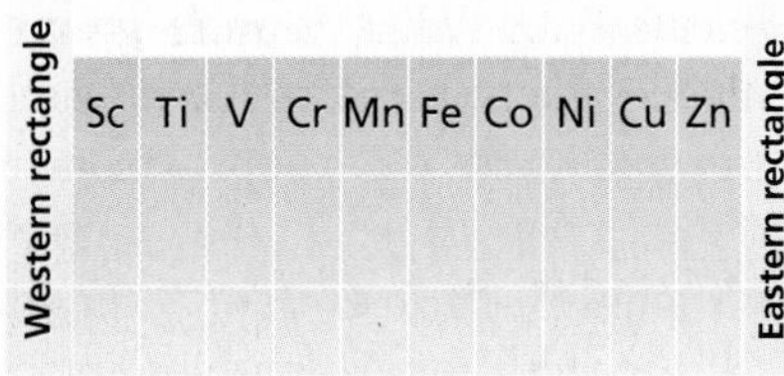

The Isthmus

Deep in the Western Desert, in an isthmuslike zone running from zinc on the east to scandium on the west, they finally stumbled upon titanium, a remarkable prize indeed. Titanium has exactly the properties that a society bent on high technology needs if it is to take to the skies: this is a metal that is tough and resistant to corrosion, yet light, and it is typical of its part of the Western Desert. Titanium and its neighbors vanadium and molybdenum, in alliance with iron, form the durable steels that enable us to chop through stone and build on a massive scale. It is a remarkable feature of the Periodic Kingdom that the Isthmus . . . has provided so many of the workhorses of our society, and that its members so readily form alliances with one another.

Reading for Meaning

What does Atkins mean when he says that metals of the Isthmus form "alliances" with one another? What does this imply about these elements?

Read Further

Atkins remarks that titanium has properties that are well-suited for a high-tech society. Find out how we obtain titanium, and list five ways that it is used in our society.

SAMPLE PROBLEM 5-2

An element has the electron configuration [Kr]$4d^5 5s^1$. Without looking at the periodic table, identify the period, block, and group in which this element is located. Then consult the periodic table to identify this element and the others in its group.

SOLUTION The number of the highest occupied energy level is five, so the element is in the fifth period. There are five electrons in the *d* sublevel. This means that the *d* sublevel is incompletely filled because it can hold 10 electrons. Therefore, the element is in the *d* block. For *d*-block elements, the sum of the electrons in the *ns* sublevel and the $(n - 1)d$ sublevel is equal to the group number, giving $5 + 1 = 6$. This is the Group 6 element in the fifth period. The element is molybdenum. The others in Group 6 are chromium, tungsten, and seaborgium.

PRACTICE

	Answer
1. Without looking at the periodic table, identify the period, block, and group in which the element with the electron configuration [Ar]$3d^8 4s^2$ is located.	1. fourth period, *d* block, Group 10
2. a. Without looking at the periodic table, write the outer electron configuration for the Group 12 element in the fifth period.	2. a. $4d^{10}5s^2$
b. Refer to the periodic table to identify the element described in (a) and to write the element's noble-gas notation.	b. Cd, [Kr]$4d^{10}5s^2$

The *p*-Block Elements: Groups 13–18

The *p*-block elements consist of all the elements of Groups 13–18 except helium. Electrons add to a *p* sublevel only after the *s* sublevel in the same energy level is filled. Therefore, atoms of all *p*-block elements contain two electrons in the *ns* sublevel. *The p-block elements together with the s-block elements are called the* **main-group elements.** For Group 13 elements, the added electron enters the *np* sublevel, giving a group configuration of ns^2np^1. Atoms of Group 14 elements contain two electrons in the *p* sublevel, giving ns^2np^2 for the group configuration. This pattern continues in Groups 15–18. In Group 18, the stable noble-gas configuration of ns^2np^6 is reached. The relationships among group numbers and electron configurations for all the groups are summarized in Table 5-2.

For atoms of *p*-block elements, the total number of electrons in the highest occupied level is equal to the group number minus 10. For example, bromine is in Group 17. It has $17 - 10 = 7$ electrons in its highest energy level. Because atoms of *p*-block elements contain two electrons in the *ns* sublevel, we know that bromine has five electrons in its outer *p* sublevel. The electron configuration of bromine is [Ar]$3d^{10}4s^24p^5$.

The properties of elements of the *p* block vary greatly. At its right-hand end, the *p* block includes all of the *nonmetals* except hydrogen and helium. All six of the *metalloids* (boron, silicon, germanium, arsenic,

TABLE 5-2 *Relationships Among Group Numbers, Blocks, and Electron Configurations*

Group number	Group configuration	Block	Comments
1, 2	$ns^{1,2}$	*s*	One or two electrons in *ns* sublevel
3–12	$(n-1)d^{1-10}ns^{0-2}$	*d*	Sum of electrons in *ns* and $(n-1)d$ levels equals group number
13–18	ns^2np^{1-6}	*p*	Number of electrons in *np* sublevel equals group number minus 12

antimony, and tellurium) are also in the *p* block. At the left-hand side and bottom of the block, there are eight *p*-block metals. The locations of the nonmetals, metalloids, and metals in the *p* block are shown with distinctive colors in Figure 5-6 and in the periodic table printed on the inside back cover of this textbook.

The elements of Group 17 (fluorine, chlorine, bromine, iodine, and astatine) are known as the **halogens.** The halogens are the most reactive of the nonmetals. They react vigorously with most metals to form the type of compounds known as salts. As you will see later, the reactivity of the halogens is based on the presence of seven electrons in their outer energy levels—one electron short of the stable noble-gas configuration. Fluorine and chlorine are gases at room temperature, bromine is a reddish liquid, and iodine is a dark purple solid. Astatine is a synthetic element prepared in only very small quantities. Most of its properties are estimated, although it is known to be a solid.

The metalloids, or semiconducting elements, fall on both sides of a line separating nonmetals and metals in the *p* block. They are mostly brittle solids with some properties of metals and some of nonmetals. The metalloid elements have electrical conductivity intermediate between that of metals, which are good conductors, and nonmetals, which are nonconductors.

The metals of the *p* block are generally harder and denser than the *s*-block alkaline-earth metals, but softer and less dense than the *d*-block metals. With the exception of bismuth, these metals are sufficiently reactive to be found in nature only in the form of compounds. Once obtained as free metals, however, they are stable in the presence of air.

FIGURE 5-11 Fluorine, chlorine, bromine, and iodine are members of Group 17 of the periodic table, also known as the halogens. Locate the halogens in the *p* block of the periodic table on page 129.

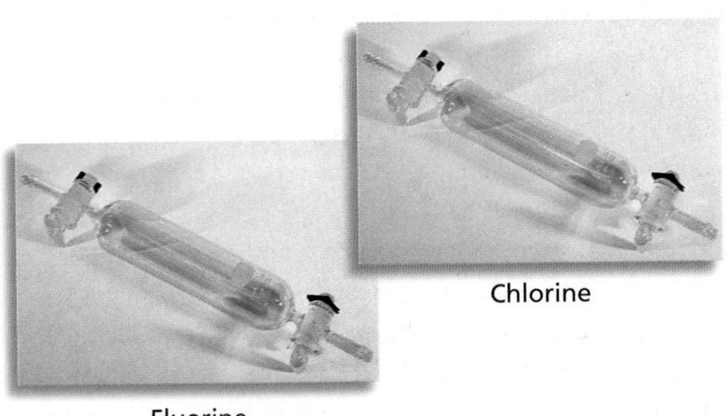
Chlorine

Fluorine

Bromine

Iodine

SAMPLE PROBLEM 5-3

Without looking at the periodic table, write the outer electron configuration for the Group 14 element in the second period. Then name the element, and identify it as a metal, nonmetal, or metalloid.

SOLUTION The group number is higher than 12, so the element is in the *p* block. The total number of electrons in the highest occupied *s* and *p* sublevels is therefore equal to the group number minus ten, $14 - 10 = 4$. With two electrons in the *s* sublevel, two electrons must also be present in the 2*p* sublevel, giving an outer electron configuration of $2s^2 2p^2$. The element is carbon, C, which is a nonmetal.

PRACTICE

1. a. Without looking at the periodic table, write the outer electron configuration for the Group 17 element in the third period.
   b. Name the element described in (a), and identify it as a metal, nonmetal, or metalloid.
2. a. Without looking at the periodic table, identify the period, block, and group of an element with the electron configuration $[Ar]3d^{10}4s^2 4p^3$.
   b. Name the element described in (a), and identify it as a metal, nonmetal, or metalloid.

Answer

1. a. $3s^2 3p^5$
   b. chlorine, nonmetal
2. a. fourth period, *p* block, Group 15
   b. arsenic, metalloid

The *f*-Block Elements: Lanthanides and Actinides

In the periodic table, the *f*-block elements are wedged between Groups 3 and 4 in the sixth and seventh periods. Their position reflects the fact that they involve the filling of the 4*f* sublevel. With seven 4*f* orbitals to be filled with two electrons each, there are a total of 14 *f*-block elements between lanthanum, La, and hafnium, Hf, in the sixth period. The lanthanides are shiny metals similar in reactivity to the Group 2 alkaline-earth metals.

There are also 14 *f*-block elements, the actinides, between actinium, Ac, and element 104, Unq, in the seventh period. In these elements the 5*f* sublevel is being filled with 14 electrons. The actinides are all radioactive. The first four actinides (thorium, Th, through neptunium, Np) have been found naturally on Earth. The remaining actinides are known only as laboratory-made elements.

SAMPLE PROBLEM 5-4

The electron configurations of atoms of four elements are written at the top of page 139. For each element, name the block and group in the periodic table in which it is located. Then name the element by consulting the periodic table on pages 130–131. Identify each

element as a metal, nonmetal, or metalloid. Finally, describe it as likely to be of high reactivity or of low reactivity.

a. **[Xe]$4f^{14}5d^{9}6s^{1}$**
b. **[Ne]$3s^{2}3p^{5}$**
c. **[Ne]$3s^{2}3p^{6}$**
d. **[Xe]$4f^{6}6s^{2}$**

SOLUTION

a. The 4*f* sublevel is filled with 14 electrons. The 5*d* sublevel is partially filled with nine electrons. Therefore, this is a *d*-block element. The element is the transition metal platinum, Pt, which is in Group 10 and has a low reactivity.

b. The incompletely filled *p* sublevel shows this to be a *p*-block element. With a total of seven electrons in the *ns* and *np* sublevels, this is a Group 17 element, a halogen. The element is chlorine, Cl, and is highly reactive.

c. This element has a noble-gas configuration and thus is in Group 18 in the *p* block. The element is argon, Ar, an unreactive nonmetal and a noble gas.

d. The incomplete 4*f* sublevel shows that it is an *f*-block element and a lanthanide. Group numbers are not assigned to the *f* block. The element is samarium, Sm. The lanthanides are all reactive metals.

PRACTICE

1. For each of the following, identify the block, period, group, group name (where appropriate), element name, element type (metal, nonmetal, or metalloid), and relative reactivity (high or low):

   a. [He]$2s^{2}2p^{5}$
   b. [Ar]$3d^{10}4s^{1}$
   c. [Kr]$5s^{1}$

Answer

1. a. *p* block, second period, Group 17, halogens, fluorine, nonmetal, high reactivity
   b. *d* block, fourth period, Group 11, transition elements, copper, metal, low reactivity
   c. *s* block, fifth period, Group 1, alkali metals, rubidium, metal, high reactivity

SECTION REVIEW

1. Into what four blocks can the periodic table be divided to illustrate the relationship between the elements' electron configurations and their placement in the periodic table?
2. What name is given to each of the following groups of elements on the periodic table:
   a. Group 1
   b. Group 2
   c. Groups 3–12
   d. Group 17
   e. Group 18
3. What is the relationship between group configuration and group number for elements in the *s, p,* and *d* blocks?
4. Without looking at the periodic table, write the outer electron configuration for the Group 15 element in the fourth period.
5. Without looking at the periodic table, identify the period, block, and group of the element with the electron configuration [Ar]$3d^{7}4s^{2}$.

SECTION 5-3

OBJECTIVES

- Define *atomic* and *ionic radii, ionization energy, electron affinity,* and *electronegativity.*
- Compare the periodic trends of atomic radii, ionization energy, and electronegativity, and state the reasons for these variations.
- Define *valence electrons,* and state how many are present in atoms of each main-group element.
- Compare the atomic radii, ionization energies, and electronegativities of the *d*-block elements with those of the main-group elements.

Electron Configuration and Periodic Properties

So far, you have learned that the elements are arranged in the periodic table according to their atomic number and that there is a rough correlation between the arrangement of the elements and their electron configurations. In this section, the relationship between the periodic law and electron configurations will be further explored.

Atomic Radii

Ideally, the size of an atom is defined by the edge of its orbital. However, this boundary is fuzzy and varies under different conditions. Therefore, to estimate the size of an atom, the conditions under which the atom exists must be specified. One way to express an atom's radius is to measure the distance between the nuclei of two identical atoms that are chemically bonded together, then divide this distance by two. As illustrated in Figure 5-12, **atomic radius** *may be defined as one-half the distance between the nuclei of identical atoms that are bonded together.*

Period Trends

Figure 5-13 gives the atomic radii of the elements and Figure 5-14 presents this information graphically. Note that there is a gradual decrease in atomic radii across the second period from lithium, Li, to neon, Ne.

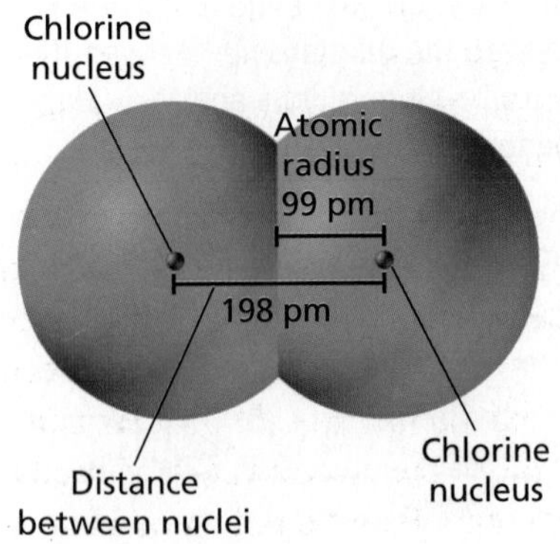

FIGURE 5-12 One method of determining atomic radius is to measure the distance between the nuclei of two identical atoms that are bonded together in an element or compound, then divide this distance by two. The atomic radius of a chlorine atom, for example, is 99 picometers (pm).

Periodic Table of Atomic Radii (pm)

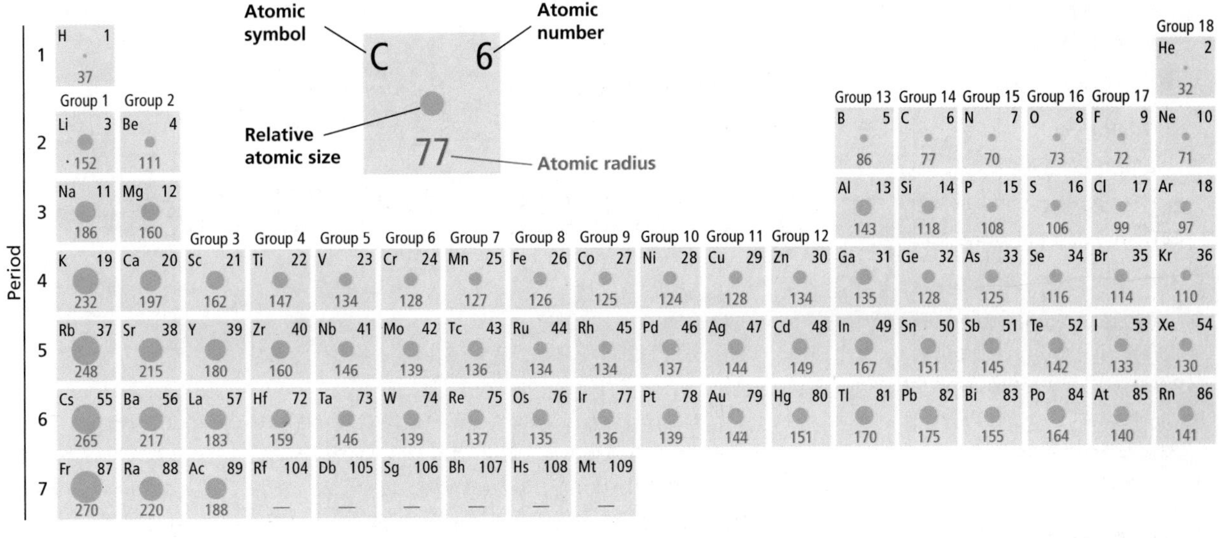

Lanthanide series

Ce 58	Pr 59	Nd 60	Pm 61	Sm 62	Eu 63	Gd 64	Tb 65	Dy 66	Ho 67	Er 68	Tm 69	Yb 70	Lu 71
182	182	181	183	180	208	180	177	178	176	176	176	—	174
Th 90	Pa 91	U 92	Np 93	Pu 94	Am 95	Cm 96	Bk 97	Cf 98	Es 99	Fm 100	Md 101	No 102	Lr 103
179	163	156	155	159	173	174	—	186	186	—	—	—	—

Actinide series

FIGURE 5-13 Atomic radii decrease from left to right across a period and increase down a group.

The trend to smaller atoms across a period is caused by the increasing positive charge of the nucleus. As electrons add to *s* and *p* sublevels in the same main energy level, they are gradually pulled closer to the more highly charged nucleus. This increased pull results in a decrease in atomic radii. The attraction of the nucleus is somewhat offset by repulsion among the increased number of electrons in the same outer energy level. As a result, the difference in radii between neighboring atoms in each period grows smaller, as shown in Figure 5-13.

Group Trends

Examine the atomic radii of the Group 1 elements in Figure 5-13. Notice that the radii of the elements increase as you read down the group. As electrons occupy sublevels in successively higher main energy levels located farther from the nucleus, the sizes of the atoms increase. *In general, the atomic radii of the main-group elements increase down a group.*

Now examine the radii of the Group 13 elements. Although gallium, Ga, follows aluminum, Al, it has a slightly smaller atomic radius than does aluminum. This is because gallium, unlike aluminum, is preceded in its period by the 10 *d*-block elements. The expected increase in gallium's radius caused by the filling of the fourth main-energy level is outweighed by a shrinking of the electron cloud caused by a nuclear charge that is considerably higher than that of aluminum.

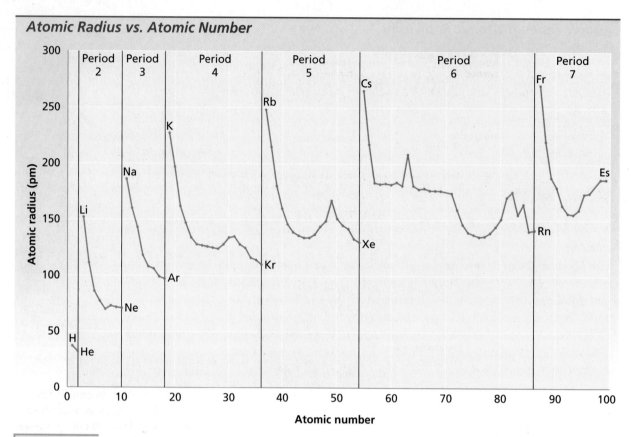

Atomic Radius vs. Atomic Number

Atomic radius (pm)

Period 2 | Period 3 | Period 4 | Period 5 | Period 6 | Period 7

H, He, Li, Ne, Na, Ar, K, Kr, Rb, Xe, Cs, Rn, Fr, Es

Atomic number

FIGURE 5-14 The plot of atomic radius versus atomic number shows period and group trends.

SAMPLE PROBLEM 5-5

a. Of the elements magnesium, Mg, chlorine, Cl, sodium, Na, and phosphorus, P, which has the largest atomic radius? Explain your answer in terms of trends in the periodic table.

b. Of the elements calcium, Ca, beryllium, Be, barium, Ba, and strontium, Sr, which has the largest atomic radius? Explain your answer in terms of trends in the periodic table.

SOLUTION **a.** All of the elements are in the third period. Of the four, sodium has the lowest atomic number and is the first element in the period. Therefore, sodium has the largest atomic radius because atomic radii decrease across a period.

b. All of the elements are in Group 2. Of the four, barium has the highest atomic number and is farthest down the group. Therefore, barium has the largest atomic radius because atomic radii increase down a group.

PRACTICE

1. Of the elements Li, O, C, and F, identify the one with the largest atomic radius and the one with the smallest atomic radius. *Answer* Li, F

2. Of the elements Br, At, F, I, and Cl, identify the one with the smallest atomic radius and the one with the largest atomic radius. *Answer* F, At

Ionization Energy

An electron can be removed from an atom if enough energy is supplied. Using A as a symbol for an atom of any element, the process can be expressed as follows.

$$A + \text{energy} \rightarrow A^+ + e^-$$

The A^+ represents an ion of element A with a single positive charge, referred to as a 1+ ion. *An* **ion** *is an atom or group of bonded atoms that has a positive or negative charge.* Sodium, for example, forms an Na^+ ion. *Any process that results in the formation of an ion is referred to as* **ionization.**

To compare the ease with which atoms of different elements give up electrons, chemists compare ionization energies. *The energy required to remove one electron from a neutral atom of an element is the* **ionization energy** *(or first ionization energy).* To avoid the influence of nearby atoms, measurements of ionization energies are made on isolated atoms in the gas phase. Figure 5-15 gives the first ionization energies for the elements in kilojoules per mole (kJ/mol). Figure 5-16 presents this information graphically.

FIGURE 5-15 In general, first ionization energies increase across a period and decrease down a group.

Periodic Table of Ionization Energies (kJ/mol)

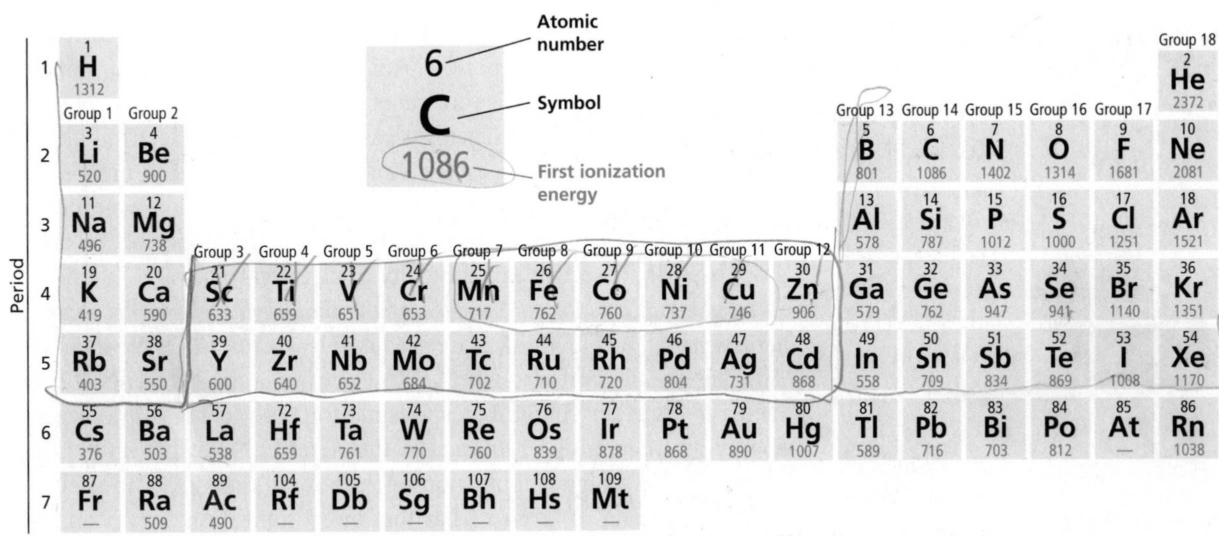

Lanthanide series

58 Ce	59 Pr	60 Nd	61 Pm	62 Sm	63 Eu	64 Gd	65 Tb	66 Dy	67 Ho	68 Er	69 Tm	70 Yb	71 Lu
534	527	533	536	545	547	592	566	573	581	589	597	603	523
90 Th	**91 Pa**	**92 U**	**93 Np**	**94 Pu**	**95 Am**	**96 Cm**	**97 Bk**	**98 Cf**	**99 Es**	**100 Fm**	**101 Md**	**102 No**	**103 Lr**
587	570	598	600	585	578	581	601	608	619	627	635	642	—

Actinide series

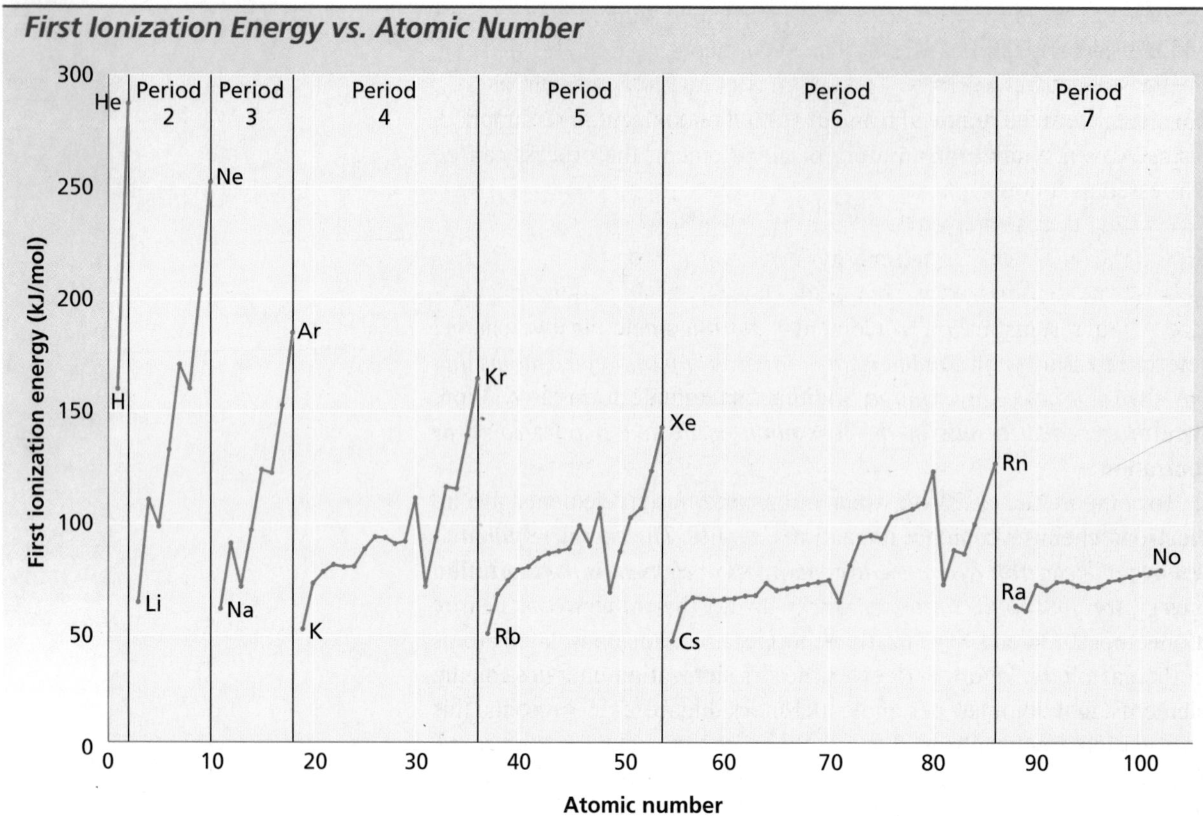

FIGURE 5-16 Plot of first ionization energy versus atomic number. As atomic number increases, both period and group trends become less pronounced.

Period Trends

In Figures 5-15 and 5-16, examine the ionization energies for the first and last elements in each period. You can see that the Group 1 metals have the lowest first ionization energies in their respective periods. Therefore, they lose electrons most easily. This ease of electron loss is a major reason for the high reactivity of the Group 1 (alkali) metals. The Group 18 elements, the noble gases, have the highest ionization energies. They do not lose electrons easily. The low reactivity of the noble gases is partly based on this difficulty of electron removal.

In general, ionization energies of the main-group elements increase across each period. This increase is caused by increasing nuclear charge. A higher charge more strongly attracts electrons in the same energy level. Increasing nuclear charge is responsible for both increasing ionization energy and decreasing radii across the periods. Note that, in general, nonmetals have higher ionization energies than metals do. In each period, the element of Group 1 has the lowest ionization energy and the element of Group 18 has the highest ionization energy.

Group Trends

Among the main-group elements, ionization energies generally decrease down the groups. Electrons removed from atoms of each succeeding element in a group are in higher energy levels, farther from the nucleus. Therefore, they are removed more easily. Also, as atomic number increases going down a group, more electrons lie between the nucleus and the electrons in the highest occupied energy levels. This partially shields the outer electrons from the effect of the nuclear charge. Together, these influences overcome the attraction of the electrons to the increasing nuclear charge.

Removing Electrons from Positive Ions

With sufficient energy, electrons can be removed from positive ions as well as from neutral atoms. The energies for removal of additional electrons from an atom are referred to as *the second ionization energy, third ionization energy,* and so on.

Table 5-3 shows the first five ionization energies for the elements of the first, second, and third periods. You can see that the second ionization energy is always higher than the first, the third is always higher than the second, and so on. This is because as electrons are removed in successive ionizations, fewer electrons remain within the atom to shield the attractive force of the nucleus. *Thus, each successive electron removed from an ion feels an increasingly stronger effective nuclear charge (the nuclear charge minus the electron shielding).*

The first ionization energies in Table 5-3 show that removing a single electron from an atom of a Group 18 element is more difficult than removing an electron from atoms of other elements in the same period. This special stability of the noble-gas configuration also applies to ions that have noble-gas configurations. Notice in Table 5-3 the large increases between the first and second ionization energies of lithium, Li, and between the second and third ionization energies of beryllium, Be. Even larger increases in ionization energy exist between the third and

TABLE 5-3 ***First Five Ionization Energies (in kJ/mol) for Elements of Periods 1–3***

	Period 1		Period 2							
	H	He	Li	Be	B	C	N	O	F	Ne
I	1312	**2372**	520	900	801	1086	1402	1314	1681	2081
II		5250	**7298**	1757	2427	2353	2856	3388	3374	3952
III			11 815	**14 849**	3660	4621	4578	5300	6050	6122
IV				21 007	**25 026**	6223	7475	7469	8408	9370
V					32 827	**37 830**	9445	10 990	11 023	12 178
			Period 3							
			Na	Mg	Al	Si	P	S	Cl	Ar
I			496	738	578	787	1012	1000	1251	1521
II			**4562**	1451	1817	1577	1903	2251	2297	2666
III			6912	**7733**	2745	3232	2912	3361	3822	3931
IV			9544	10 540	**11 578**	4356	4957	4564	5158	5771
V			13 353	13 628	14 831	**16 091**	6274	7013	6540	7238

fourth ionization energies of boron, B, and between the fourth and fifth ionization energies of carbon, C. In each case, the jump in ionization energy occurs when an ion assumes a noble-gas configuration. For example, the removal of one electron from a lithium atom ([He]$2s^1$) leaves the helium noble-gas configuration. The removal of four electrons from a carbon atom ([He]$2s^22p^2$) also leaves the helium configuration. A bigger table would show that this trend continues across the entire periodic system.

SAMPLE PROBLEM 5-6

Consider two main-group elements A and B. Element A has a first ionization energy of 419 kJ/mol. Element B has a first ionization energy of 1000 kJ/mol. For each element, decide if it is more likely to be in the *s* block or *p* block. Which element is more likely to form a positive ion?

SOLUTION Element A has a very low ionization energy, meaning that atoms of A lose electrons easily. Therefore, element A is most likely to be a metal of the *s* block because ionization energies increase across the periods. Element B has a very high ionization energy, meaning that atoms of B lose electrons with difficulty. We would expect element B to lie at the end of a period in the *p* block. Element A is more likely to form a positive ion because it has a much lower ionization energy than does element B.

PRACTICE

1. Consider the four hypothetical main-group elements Q, R, T, and X with the outer electron configurations indicated below. Then answer the questions that follow.

   Q: $3s^23p^5$ R: $3s^1$ T: $4d^{10}5s^25p^5$ X: $4d^{10}5s^25p^1$

   a. Identify the block location of each hypothetical main-group element.

   b. Which of these elements are in the same period? Which are in the same group?

   c. Which element would you expect to have the highest first ionization energy? Which would have the lowest first ionization energy?

   d. Which element would you expect to have the highest second ionization energy?

   e. Which of the elements is most likely to form a 1+ ion?

Answer

1. a. Q is in the *p* block, R is in the *s* block, T is in the *p* block, and X is in the *p* block.

   b. Q and R, and X and T are in the same period. Q and T are in the same group.

   c. Q would have the highest ionization energy, and R would have the lowest.

   d. R

   e. R

Electron Affinity

Neutral atoms can also acquire electrons. *The energy change that occurs when an electron is acquired by a neutral atom is called the atom's* **electron affinity.** Most atoms release energy when they acquire an electron.

$$A + e^- \longrightarrow A^- + \text{energy}$$

In this book, the quantity of energy released is represented by a negative number. On the other hand, some atoms must be "forced" to gain an electron by the addition of energy.

$$A + e^- + \text{energy} \longrightarrow A^-$$

The quantity of energy absorbed is represented by a positive number. An ion produced in this way will be unstable and will lose the added electron spontaneously.

Figure 5-17 shows the electron affinity in kilojoules per mole for the elements. Figure 5-18 presents these data graphically.

Period Trends

Among the elements of each period, the halogens (Group 17) gain electrons most readily. This is indicated in Figure 5-17 by the large negative values of halogens' electron affinities. The ease with which halogen atoms gain electrons is a major reason for the high reactivities of the Group 17 elements. In general, as electrons add to the same *p* sublevel of atoms with increasing nuclear charge, electron affinities become more negative across each period within the *p*-block. An exception to this trend occurs between Groups 14 and 15. Compare the electron affinities of carbon ([He]$2s^22p^2$) and nitrogen ([He]$2s^22p^3$). Adding an electron to a carbon atom gives a half-filled *p* sublevel. This occurs much more easily

FIGURE 5-17 The values listed in parentheses in this periodic table of electron affinities are approximate. Electron affinity is estimated to be –50 kJ/mol for each of the lanthanides and 0 kJ/mol for each of the actinides.

Periodic Table of Electron Affinities (kJ/mol)

Key: 6 — Atomic number; C — Symbol; –126.3 — Electron affinity

Period	Group 1	Group 2	Group 3	Group 4	Group 5	Group 6	Group 7	Group 8	Group 9	Group 10	Group 11	Group 12	Group 13	Group 14	Group 15	Group 16	Group 17	Group 18
1	1 H –75.4																	2 He (0)
2	3 Li –61.8	4 Be (0)											5 B –27.7	6 C –126.3	7 N (0)	8 O –146.1	9 F –339.9	10 Ne (0)
3	11 Na –54.8	12 Mg (0)											13 Al –44.1	14 Si –138.5	15 P –74.6	16 S –207.7	17 Cl –361.7	18 Ar (0)
4	19 K –50.1	20 Ca (0)	21 Sc –18.8	22 Ti –7.9	23 V –52.5	24 Cr –66.6	25 Mn (0)	26 Fe –16.3	27 Co –66.1	28 Ni –115.6	29 Cu –122.8	30 Zn (0)	31 Ga –30	32 Ge –135	33 As –81	34 Se –202.1	35 Br –336.5	36 Kr (0)
5	37 Rb –48.6	38 Sr (0)	39 Y –30.7	40 Zr –42.6	41 Nb –89.3	42 Mo –74.6	43 Tc –55	44 Ru –105	45 Rh –113.7	46 Pd –55.7	47 Ag –130.2	48 Cd (0)	49 In –30	50 Sn –120	51 Sb –107	52 Te –197.1	53 I –305.9	54 Xe (0)
6	55 Cs –47.2	56 Ba (0)	57 La –50	72 Hf (0)	73 Ta –32.2	74 W –81.5	75 Re –15	76 Os –110	77 Ir –156.5	78 Pt –212.8	79 Au –230.9	80 Hg (0)	81 Tl –20	82 Pb –36	83 Bi –94.6	84 Po –190	85 At –280	86 Rn (0)
7	87 Fr –47.0	88 Ra (0)	89 Ac —	104 Rf —	105 Db —	106 Sg —	107 Bh —	108 Hs —	109 Mt —									

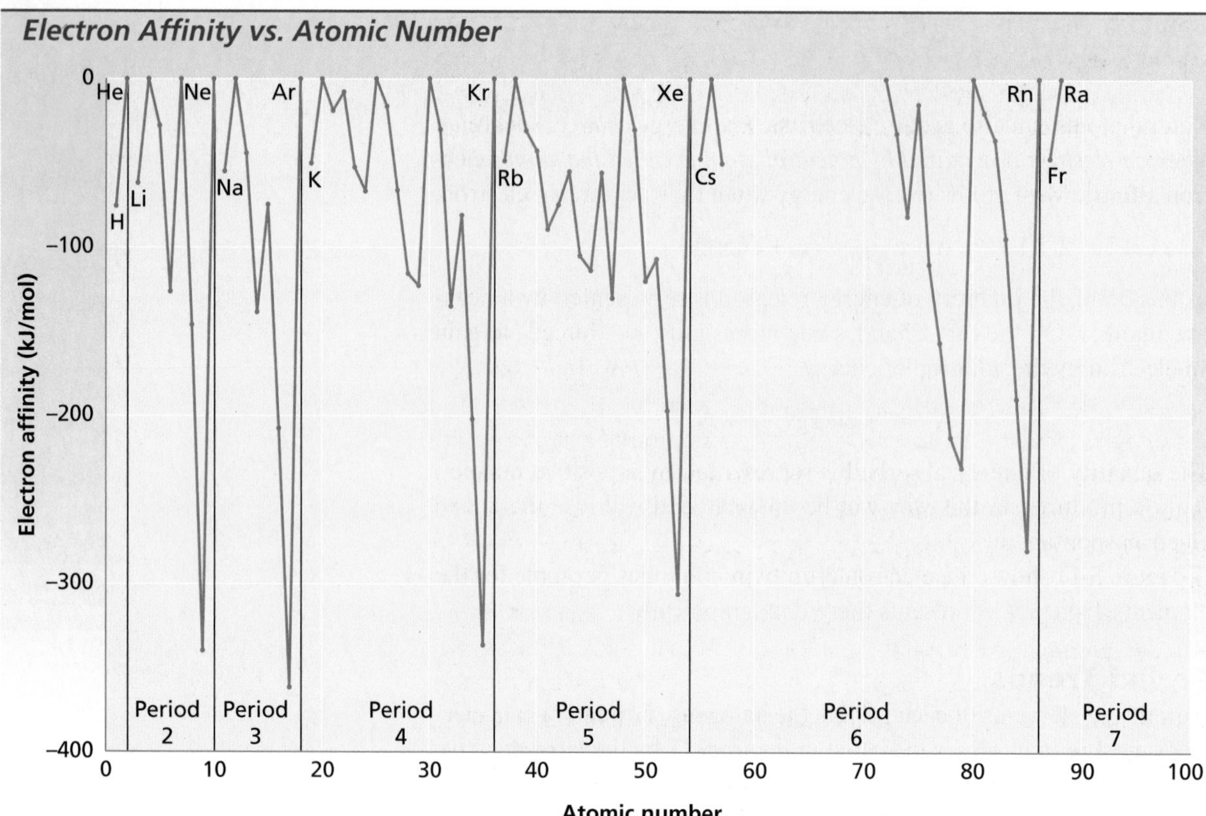

FIGURE 5-18 The plot of electron affinity versus atomic number shows that most atoms release energy when they acquire an electron, as indicated by negative values.

than forcing an electron to pair with another electron in an orbital of the already half-filled *p* sublevel of a nitrogen atom.

Group Trends

Trends for electron affinities within groups are not as regular as trends for ionization energies. As a general rule, electrons add with greater difficulty down a group. This pattern is a result of two competing factors. The first is a slight increase in effective nuclear charge down a group, which increases electron affinities. The second is an increase in atomic radius down a group, which decreases electron affinities. In general, the size effect predominates. But there are exceptions, especially among the heavy transition metals, which tend to be the same size or even decrease in radius down a group.

Adding Electrons to Negative Ions

For an isolated ion in the gas phase, it is always more difficult to add a second electron to an already negatively charged ion. Therefore, second electron affinities are all positive. Certain *p*-block nonmetals tend to form negative ions that have noble-gas configurations. The halogens do so by adding one electron. For example, chlorine has the configuration $[Ne]3s^2 3p^5$. An atom of chlorine achieves the configuration of the noble gas argon by adding an electron to form the ion Cl^- ($[Ne]3s^2 3p^6$). Adding another electron is so difficult that Cl^{2-} never occurs. Atoms of

Group 16 elements are present in many compounds as 2− ions. For example, oxygen ([He]$2s^22p^4$) achieves the configuration of the noble gas neon by adding two electrons to form the ion O^{2-}([He]$2s^22p^6$). Nitrogen achieves a neon configuration by adding three electrons to form the ion N^{3-}.

Ionic Radii

Figure 5-19 shows the radii of some of the most common ions of the elements. Positive and negative ions have specific names.

A positive ion is known as a **cation.** The formation of a cation by the loss of one or more electrons always leads to a decrease in atomic radius because the removal of the highest-energy-level electrons results in a smaller electron cloud. Also, the remaining electrons are drawn closer to the nucleus by its unbalanced positive charge.

A negative ion is known as an **anion.** The formation of an anion by the addition of one or more electrons always leads to an increase in atomic radius. This is because the total positive charge of the nucleus remains unchanged when an electron is added to an atom or an ion. So the electrons are not drawn to the nucleus as strongly as they were before the addition of the extra electron. The electron cloud also spreads out because of greater repulsion between the increased number of electrons.

Period Trends

Within each period of the periodic table, the metals at the left tend to form cations and the nonmetals at the upper right tend to form anions. Cationic radii decrease across a period because the electron cloud shrinks due to the increasing nuclear charge acting on the electrons in

FIGURE 5-19 The ionic radii of the monatomic ions most common in chemical compounds are shown. Cations are smaller and anions are larger than the atoms from which they are formed.

Periodic Table of Ionic Radii (pm)

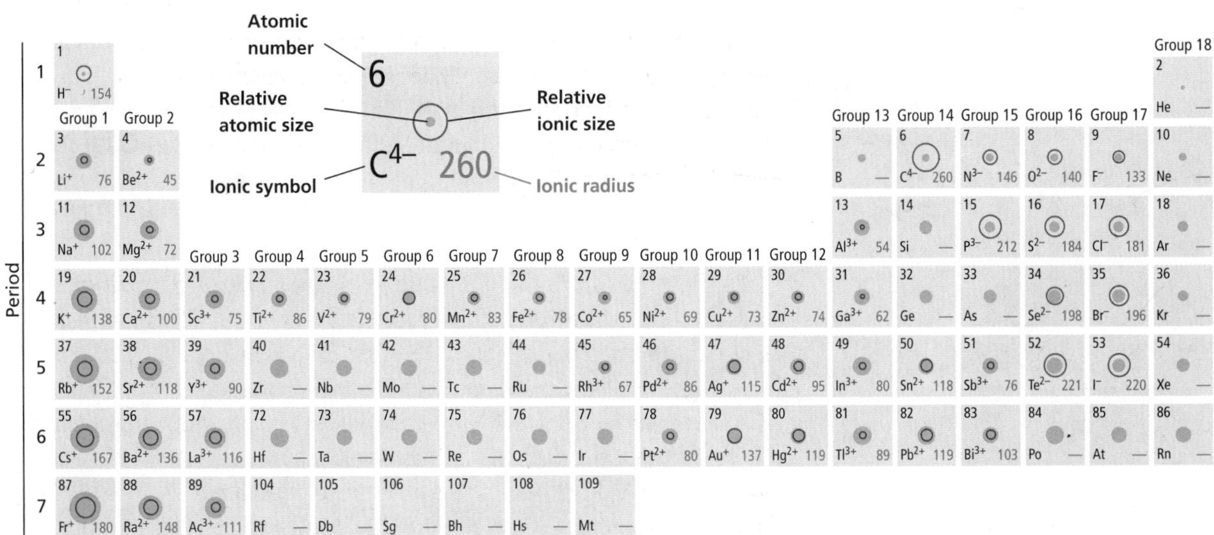

the same main energy level. Starting with Group 15, in which atoms assume stable noble-gas configurations by gaining three electrons, anions are more common than cations. Anionic radii decrease across each period for the elements in Groups 15–18. The reasons for this trend are the same as the reasons that cationic radii decrease from left to right across a period.

Group Trends

As they are in atoms, the outer electrons in both cations and anions are in higher energy levels as one reads down a group. Therefore, just as there is a gradual increase of atomic radii down a group, there is also a gradual increase of ionic radii.

Valence Electrons

Chemical compounds form because electrons are lost, gained, or shared between atoms. The electrons that interact in this manner are those in the highest energy levels. These are the electrons most subject to the influence of nearby atoms or ions. *The electrons available to be lost, gained, or shared in the formation of chemical compounds are referred to as* **valence electrons.** Valence electrons are often located in incompletely filled main-energy levels. For example, the electron lost from the 3*s* sublevel of Na to form Na^+ is a valence electron.

For main-group elements, the valence electrons are the electrons in the outermost *s* and *p* sublevels. The inner electrons are in filled energy levels and are held too tightly by the nucleus to be involved in compound formation. The Group 1 and Group 2 elements have one and two valence electrons, respectively, as shown in Table 5-4. The elements of Groups 13–18 have a number of valence electrons equal to the group number minus 10. In some cases, both the *s* and *p* sublevel valence electrons of the *p*-block elements are involved in compound formation. In other cases, only the electrons from the *p* sublevel are involved.

TABLE 5-4 Valence Electrons in Main-Group Elements

Group number	Group configuration	Number of valence electrons
1	ns^1	1
2	ns^2	2
13	ns^2p^1	3
14	ns^2p^2	4
15	ns^2p^3	5
16	ns^2p^4	6
17	ns^2p^5	7
18	ns^2p^6	8

Electronegativity

Valence electrons hold atoms together in chemical compounds. In many compounds, the negative charge of the valence electrons is concentrated closer to one atom than to another. This uneven concentration of charge has a significant effect on the chemical properties of a compound. It is therefore useful to have a measure of how strongly one atom attracts the electrons of another atom within a compound.

Linus Pauling, one of America's most famous chemists, devised a scale of numerical values reflecting the tendency of an atom to attract electrons. **Electronegativity** *is a measure of the ability of an atom in a chemical compound to attract electrons.* The most electronegative element, fluorine, is arbitrarily assigned an electronegativity value of four. Values for the other elements are then calculated in relation to this value.

Period Trends

As shown in Figure 5–20, electronegativities tend to increase across each period, although there are exceptions. The alkali and alkaline-earth metals are the least electronegative elements. In compounds, their atoms have a low attraction for electrons. Nitrogen, oxygen, and the halogens are the most electronegative elements. Their atoms attract electrons strongly in compounds. *Electronegativities tend to either decrease down a group or*

FIGURE 5-20 Shown are the electronegativities of the elements according to the Pauling scale. The most-electronegative elements are located in the upper right of the *p* block. The least-electronegative elements are located in the lower left of the *s* block.

Periodic Table of Electronegativities

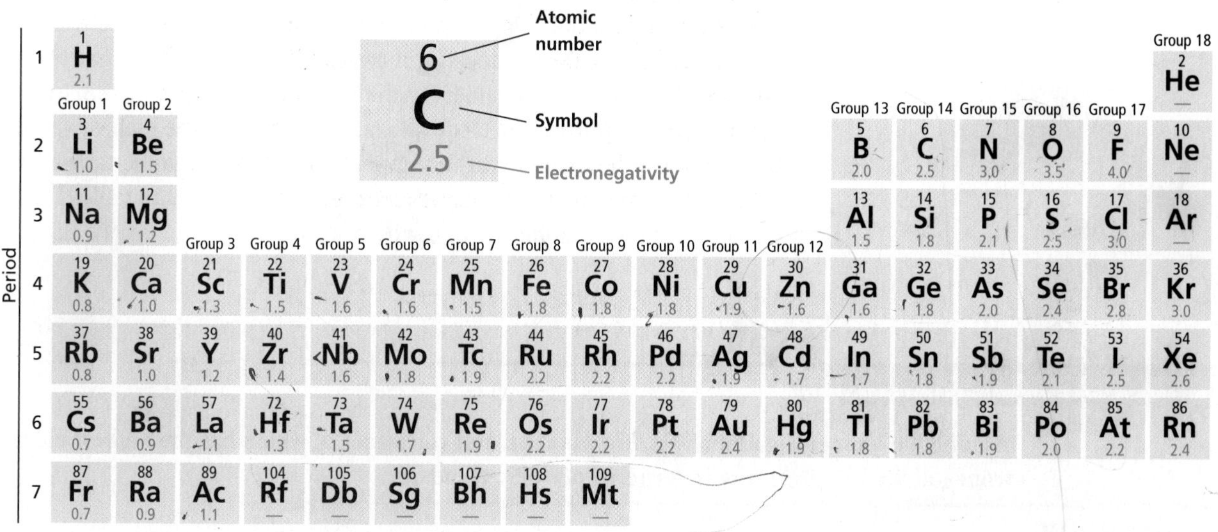

Period	Group 1	Group 2	Group 3	Group 4	Group 5	Group 6	Group 7	Group 8	Group 9	Group 10	Group 11	Group 12	Group 13	Group 14	Group 15	Group 16	Group 17	Group 18
1	1 H 2.1																	2 He —
2	3 Li 1.0	4 Be 1.5											5 B 2.0	6 C 2.5	7 N 3.0	8 O 3.5	9 F 4.0	10 Ne —
3	11 Na 0.9	12 Mg 1.2											13 Al 1.5	14 Si 1.8	15 P 2.1	16 S 2.5	17 Cl 3.0	18 Ar —
4	19 K 0.8	20 Ca 1.0	21 Sc 1.3	22 Ti 1.5	23 V 1.6	24 Cr 1.6	25 Mn 1.5	26 Fe 1.8	27 Co 1.8	28 Ni 1.8	29 Cu 1.9	30 Zn 1.6	31 Ga 1.6	32 Ge 1.8	33 As 2.0	34 Se 2.4	35 Br 2.8	36 Kr 3.0
5	37 Rb 0.8	38 Sr 1.0	39 Y 1.2	40 Zr 1.4	41 Nb 1.6	42 Mo 1.8	43 Tc 1.9	44 Ru 2.2	45 Rh 2.2	46 Pd 2.2	47 Ag 1.9	48 Cd 1.7	49 In 1.7	50 Sn 1.8	51 Sb 1.9	52 Te 2.1	53 I 2.5	54 Xe 2.6
6	55 Cs 0.7	56 Ba 0.9	57 La 1.1	72 Hf 1.3	73 Ta 1.5	74 W 1.7	75 Re 1.9	76 Os 2.2	77 Ir 2.2	78 Pt 2.2	79 Au 2.4	80 Hg 1.9	81 Tl 1.8	82 Pb 1.8	83 Bi 1.9	84 Po 2.0	85 At 2.2	86 Rn 2.4
7	87 Fr 0.7	88 Ra 0.9	89 Ac 1.1	104 Rf —	105 Db —	106 Sg —	107 Bh —	108 Hs —	109 Mt —									

Lanthanide series

58 Ce 1.1	59 Pr 1.1	60 Nd 1.1	61 Pm 1.1	62 Sm 1.2	63 Eu 1.1	64 Gd 1.2	65 Tb 1.1	66 Dy 1.2	67 Ho 1.2	68 Er 1.2	69 Tm 1.3	70 Yb 1.1	71 Lu 1.3
90 Th 1.3	91 Pa 1.5	92 U 1.4	93 Np 1.4	94 Pu 1.3	95 Am 1.3	96 Cm 1.3	97 Bk 1.3	98 Cf 1.3	99 Es 1.3	100 Fm 1.3	101 Md 1.3	102 No 1.3	103 Lr —

Actinide series

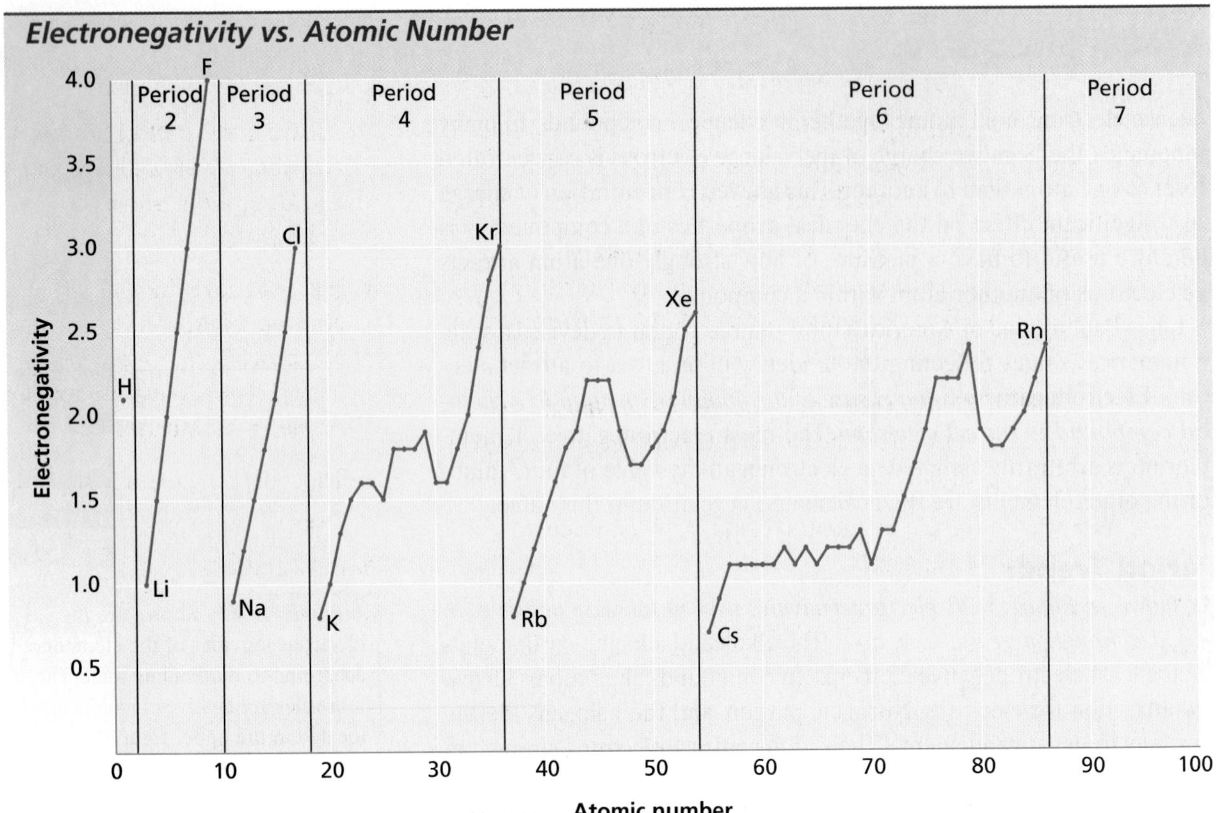

FIGURE 5-21 The plot shows electronegativity versus atomic number for Periods 1–6.

remain about the same. The noble gases are unusual in that some of them do not form compounds and therefore cannot be assigned electronegativities. When a noble gas does form a compound, its electronegativity is rather high, similar to the values for the halogens. The combination of the period and group trends in electronegativity results in the highest values belonging to the elements in the upper right of the periodic table. The lowest values belong to the elements in the lower left of the table. These trends are shown graphically in Figure 5-21.

SAMPLE PROBLEM 5-7

Among the elements gallium, Ga, bromine, Br, and calcium, Ca, which has the highest electronegativity? Explain why in terms of periodic trends.

SOLUTION

The elements are all in the fourth period. Bromine has the highest atomic number and is farthest to the right in the period. Therefore, it should have the highest electronegativity because electronegativity increases across the periods.

PRACTICE

1. Consider the five hypothetical main-group elements E, G, J, L, and M with the outer electron configurations shown at the top of page 153.

$E = 2s^22p^5$ $G = 4d^{10}5s^25p^5$ $J = 2s^22p^2$

$L = 5d^{10}6s^26p^5$ $M = 2s^22p^4$

a. Identify the block location for each element. Then determine which elements are in the same period and which are in the same group.

b. Which element would you expect to have the highest electron affinity? Which would you expect to form a 1– ion? Which should have the highest electronegativity?

c. Compare the ionic radius of the typical ion formed by the element G with the radius of its neutral atom.

d. Which element(s) contains seven valence electrons?

Answer

1. a. All are in the *p* block. E, J, and M are in the same period, and E, G, and L are in the same group.

b. E should have the highest electron affinity; E, G, and L are most likely to form 1– ions; E should have the highest electronegativity.

c. The ionic radius would be larger.

d. E, G, and L

Periodic Properties of the *d*- and *f*-Block Elements

The properties of the *d*-block elements (which are all metals) vary less and with less regularity than those of the main-group elements. This trend is indicated by the curves in Figures 5-14 and 5-16, which flatten where the *d*-block elements fall in the middle of Periods 4–6.

Recall that atoms of the *d*-block elements contain from zero to two electrons in the *s* orbital of their highest occupied energy level and one to ten electrons in the *d* sublevel of the next-lower energy level. Therefore, electrons in both the *ns* sublevel and the $(n - 1)d$ sublevel are available to interact with their surroundings. As a result, electrons in the incompletely filled *d* sublevels are responsible for many characteristic properties of the *d*-block elements.

Atomic Radii

The atomic radii of the *d*-block elements generally decrease across the periods. However, this decrease is less than that for the main-group elements because the electrons added to the $(n - 1)d$ sublevel shield the outer electrons from the nucleus. Also, note in Figure 5-14 that the radii dip to a low and then increase slightly across each of the four periods that contain *d*-block elements. As the number of electrons in the *d* sublevel increases, the radii increase because of repulsion among the electrons.

In the sixth period, the *f*-block elements fall between lanthanum (Group 3) and hafnium (Group 4). Because of the increase in atomic number that occurs from lanthanum to hafnium, the atomic radius of hafnium is actually slightly less than that of zirconium, Zr, the element immediately above it. The radii of elements following hafnium in the sixth period vary with increasing atomic number in the usual manner.

Ionization Energy

As they do for the main-group elements, ionization energies of the *d*-block and *f*-block elements generally increase across the periods. In contrast to the decrease down the main groups, however, the first ionization energies of the *d*-block elements generally increase down each group. This is because the electrons available for ionization in the outer *s* sublevels are less shielded from the increasing nuclear charge by electrons in the incomplete $(n-1)d$ sublevels.

Ion Formation and Ionic Radii

The order in which electrons are removed from all atoms of the *d*-block and *f*-block elements is *exactly the reverse of the order given by the electron-configuration notation.* In other words, electrons in the highest occupied sublevel are always removed first. For the *d*-block elements, this means that although newly added electrons occupy the *d* sublevels, the first electrons to be removed are those in the outermost *s* sublevels. For example, iron, Fe, has the electron configuration $[Ar]3d^64s^2$. It loses a 4*s* electron first to form Fe^+ ($[Ar]3d^64s^1$). Fe^+ can then lose the second 4*s* electron to form Fe^{2+} ($[Ar]3d^6$). Fe^{2+} can then lose a 3*d* electron to form Fe^{3+} ($[Ar]3d^5$).

Most *d*-block elements commonly form 2+ ions in compounds. Some, such as iron and chromium, also commonly form 3+ ions. The Group 3 elements form only ions with a 3+ charge. Copper forms 1+ and 2+ ions, and silver usually forms only 1+ ions. As expected, the cations have smaller radii than the atoms do. Comparing 2+ ions across the periods shows a decrease in size that parallels the decrease in atomic radii.

Electronegativity

The *d*-block elements all have electronegativities between 1.1 and 2.54. Only the active metals of Groups 1 and 2 have lower electronegativities. The *d*-block elements also follow the general trend for electronegativity values to increase as radii decrease, and vice versa. The *f*-block elements all have similar electronegativities, which range from 1.1 to 1.5.

SECTION REVIEW

1. State the general period and group trends among main-group elements with respect to each of the following properties:

a. atomic radii
b. first ionization energy
c. electron affinity
d. ionic radii
e. electronegativity

2. Among the main-group elements, what is the relationship between group number and the number of valence electrons among group members?

3. a. In general, how do the periodic properties of the *d*-block elements compare with those of the main-group elements?
b. Explain the comparisons made in (a).

CHAPTER 5 REVIEW

CHAPTER SUMMARY

5-1

- The periodic law states that the physical and chemical properties of the elements are periodic functions of their atomic numbers.
- The periodic table is an arrangement of the elements in order of their atomic numbers so that elements with similar properties fall in the same column.
- The columns in the periodic table are referred to as groups.

Vocabulary

actinide (126) lanthanide (126) periodic law (125) periodic table (125)

5-2

- The rows in the periodic table are called periods.
- Many chemical properties of the elements can be explained by the configurations of the elements' outermost electrons.
- The noble gases exhibit unique chemical stability because their highest occupied levels have an octet of electrons, ns^2np^6 (with the exception of helium, whose stability arises from its highest occupied level being completely filled with two electrons, $1s^2$).
- Based on the electron configurations of the elements, the periodic table can be divided into four blocks: the *s* block, the *p* block, the *d* block, and the *f* block.

Vocabulary

alkali metal (132) halogen (137) main-group element (136) transition element (134)
alkaline-earth metal (132)

5-3

- The groups and periods of the periodic table display general trends in the following properties of the elements: electron affinity, electronegativity, ionization energy, atomic radii, and ionic radii.
- The electrons in an atom that are available to be lost, gained, or shared in the formation of chemical compounds are referred to as valence electrons.
- In determining the electron configuration of an ion, the order in which electrons are removed from the atom is the reverse of the order given by the atom's electron-configuration notation.

Vocabulary

anion (149) electron affinity (147) ion (143) ionization energy (143)
atomic radius (140) electronegativity (151) ionization (143) valence electron (150)
cation (149)

REVIEWING CONCEPTS

1. Describe the contributions made by the following scientists to the development of the periodic table:
a. Stanislao Cannizzaro
b. Dmitri Mendeleev
c. Henry Moseley (5-1)

2. State the periodic law. (5-1)

3. How is the periodic law demonstrated within the groups of the periodic table? (5-1)

4. a. How do the electron configurations within the same group of elements compare?
b. Why are the noble gases relatively unreactive? (5-2)

5. What determines the length of each period in the periodic table? (5-2)

6. What is the relationship between the electron configuration of an element and the period in which that element appears in the periodic table? (5-2)

7. a. What information is provided by the specific block location of an element?
b. Identify, by number, the groups located within each of the four block areas. (5-2)

8. a. Which elements are designated as the alkali metals?
b. List four of their characteristic properties. (5-2)

9. a. Which elements are designated as the alkaline-earth metals?
b. How do their characteristic properties compare with those of the alkali metals? (5-2)

10. a. Write the usual group configuration notation for each *d*-block group.
b. How do the group numbers of those groups relate to the number of outer *s* and *d* electrons? (5-2)

11. What name is sometimes used to refer to the entire set of *d*-block elements? (5-2)

12. a. What types of elements make up the *p* block?
b. How do the properties of the *p*-block metals compare with those of the metals in the *s* and *d* blocks? (5-2)

13. a. Which elements are designated as the halogens?
b. List three of their characteristic properties. (5-2)

14. a. Which elements are metalloids?
b. Describe their characteristic properties. (5-2)

15. Which elements make up the *f* block in the periodic table? (5-2)

16. a. What are the main-group elements?
b. What trends can be observed across the various periods within the main-group elements? (5-2)

17. a. What is meant by atomic radius?
b. What trend is observed among the atomic radii of main-group elements across a period?
c. How can this trend be explained? (5-3)

18. a. What trend is observed among the atomic radii of main-group elements down a group?
b. How can this trend be explained? (5-3)

19. Define each of the following terms:
a. ion
b. ionization
c. first ionization energy
d. second ionization energy (5-3)

20. a. How do the first ionization energies of main-group elements vary across a period and down a group?
b. Explain the basis for each trend. (5-3)

21. a. What is electron affinity?
b. What signs are associated with electron affinity values, and what is the significance of each sign? (5-3)

22. a. Distinguish between a cation and an anion.
b. How does the size of each compare with the size of the neutral atom from which it is formed? (5-3)

23. a. What are valence electrons?
b. Where are such electrons located? (5-3)

24. For each of the following groups, indicate whether electrons are more likely to be lost or gained in compound formation and give the number of such electrons typically involved:
a. Group 1 d. Group 16
b. Group 2 e. Group 17
c. Group 13 f. Group 18 (5-3)

25. a. What is electronegativity?
b. Why is fluorine special in terms of electronegativity? (5-3)

26. Identify the most- and least-electronegative groups of elements in the periodic table. (5-3)

PROBLEMS

Electron Configuration and Periodic Properties

27. Write the noble-gas notation for the electron configuration of each of the following elements, and indicate the period in which each belongs:
a. Li c. Cu e. Sn
b. O d. Br

28. Without looking at the periodic table, identify the period, block, and group in which the elements with the following electron configurations are located. (Hint: See Sample Problem 5-1.)
a. $[Ne]3s^23p^4$
b. $[Kr]4d^{10}5s^25p^2$
c. $[Xe]4f^{14}5d^{10}6s^26p^5$

29. Based on the information given below, give the group, period, block, and identity of each element described. (Hint: See Sample Problem 5-2.)
a. $[He]2s^2$
b. $[Ne]3s^1$
c. $[Kr]5s^2$
d. $[Ar]4s^2$
e. $[Ar]3d^5 4s^1$

30. Without looking at the periodic table, write the expected outer electron configuration for each of the following elements. (Hint: See Sample Problem 5-3.)
a. Group 7, fourth period
b. Group 3, fifth period
c. Group 12, sixth period

31. Identify the block, period, group, group name (where appropriate), element name, element type, and relative reactivity for the elements with the following electron configurations. (Hint: See Sample Problem 5-4.)
a. $[Ne]3s^2 3p^1$
b. $[Ar]3d^{10} 4s^2 4p^6$
c. $[Kr]4d^{10} 5s^1$
d. $[Xe]4f^1 5d^1 6s^2$

Atomic Radius, Ionization, Electron Affinity, and Electronegativity

32. Of cesium, Cs, hafnium, Hf, and gold, Au, which element has the smallest atomic radius? Explain your answer in terms of trends in the periodic table. (Hint: see Sample Problem 5-5.)

33. a. Distinguish between the first, second, and third ionization energies of an atom.
b. How do the values of successive ionization energies compare?
c. Why does this occur?

34. Without looking at the electron affinity table, arrange the following elements in order of *decreasing* electron affinities: C, O, Li, Na, Rb, and F.

35. a. Without looking at the ionization energy table, arrange the following elements in order of decreasing first ionization energies: Li, O, C, K, Ne, and F.
b. Which of the elements listed in (a) would you expect to have the highest second ionization energy? Why?

36. a. Which of the following cations is least likely to form: Sr^{2+}, Al^{3+}, K^{2+}?
b. Which of the following anions is least likely to form: I^-, Cl^-, O^{2-}?

37. Which element is the most electronegative among C, N, O, Br, and S? Which group does it belong to? (Hint: See Sample Problem 5-7.)

38. The two ions K^+ and Ca^{2+} each have 18 electrons surrounding the nucleus. Which would you expect to have the smaller radius? Why?

MIXED REVIEW

39. Without looking at the periodic table, identify the period, block, and group in which each of the following elements is located:
a. $[Rn]7s^1$
b. $[Ar]3d^2 4s^2$
c. $[Kr]4d^{10} 5s^1$
d. $[Xe]4f^{14} 5d^9 6s^1$

40. a. Which elements are designated as the noble gases?
b. What is the most significant property of these elements?

41. Which of the following does not have a noble-gas configuration: Na^+, Rb^+, O^{2-}, Br^-, Ca^+, Al^{3+}, S^{2-}?

42. a. How many groups are in the periodic table?
b. How many periods are in the periodic table?
c. Which two blocks of the periodic table make up the main-group elements?

43. Write the noble-gas notation for the electron configuration of each of the following elements, and indicate the period and group in which each belongs:
a. Mg
b. P
c. Sc
d. Y

44. Use the periodic table to describe the chemical properties of the following elements:
a. fluorine, F
b. xenon, Xe
c. sodium, Na
d. gold, Au

CHAPTER 6

Chemical Bonding

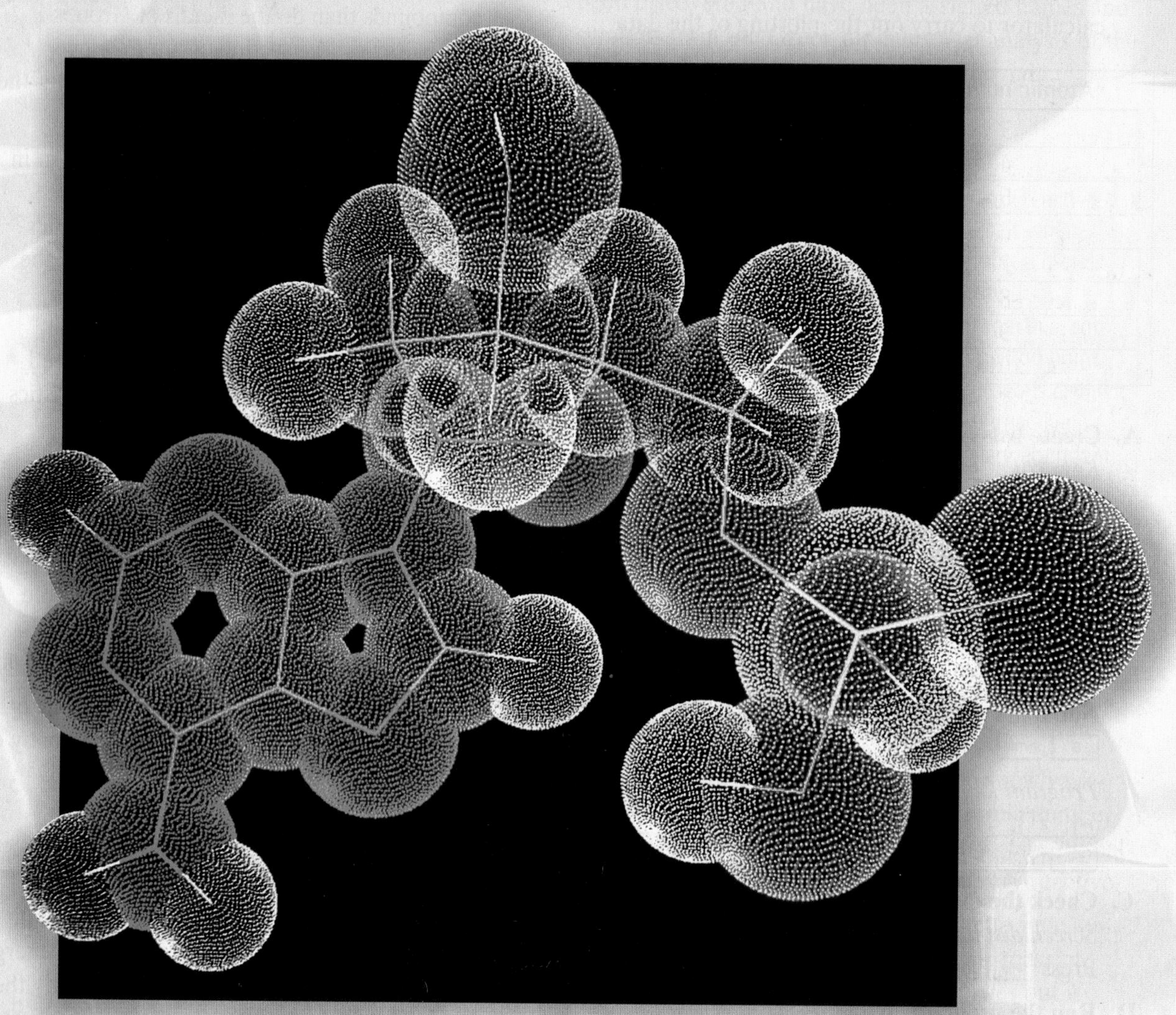

In nature, most atoms are joined to other atoms by chemical bonds.

Introduction to Chemical Bonding

SECTION 6-1

OBJECTIVES

- Define *chemical bond.*
- Explain why most atoms form chemical bonds.
- Describe ionic and covalent bonding.
- Explain why most chemical bonding is neither purely ionic nor purely covalent.
- Classify bonding type according to electronegativity differences.

Atoms seldom exist as independent particles in nature. The oxygen you breathe, the water that makes up most of your body, and virtually every other substance you can imagine are made up of combinations of atoms that are held together by chemical bonds. *A* **chemical bond** *is a mutual electrical attraction between the nuclei and valence electrons of different atoms that binds the atoms together.*

Why are most atoms chemically bonded to each other? As independent particles, they are at relatively high potential energy. Nature, however, favors arrangements in which potential energy is minimized. This means that most atoms are less stable existing by themselves than when they are combined. By bonding with each other, atoms decrease in potential energy, thereby creating more-stable arrangements of matter. The characteristics of these arrangements depend largely on the type of chemical bonding that exists between their atoms.

Types of Chemical Bonding

When atoms bond, their valence electrons are redistributed. The way in which the electrons are redistributed determines the type of bonding. In Chapter 5, you read that main-group metals tend to lose electrons to form positive ions, or cations, and nonmetals tend to gain electrons to form negative ions, or anions. *Chemical bonding that results from the electrical attraction between large numbers of cations and anions is called* **ionic bonding.** In purely ionic bonding, atoms completely give up electrons to other atoms, as illustrated in Figure 6-1 on page 162. In contrast to atoms joined by ionic bonding, atoms joined by covalent bonding share electrons. **Covalent bonding** *results from the sharing of electron pairs between two atoms* (see Figure 6-1). In a purely covalent bond, the shared electrons are "owned" equally by the two bonded atoms.

Ionic or Covalent?

Bonding between atoms of different elements is never purely ionic and is rarely purely covalent. It usually falls somewhere between these two extremes, depending on how strongly the atoms of each element attract electrons. Recall that electronegativity is a measure of an atom's ability to attract electrons. The degree to which bonding between atoms of two

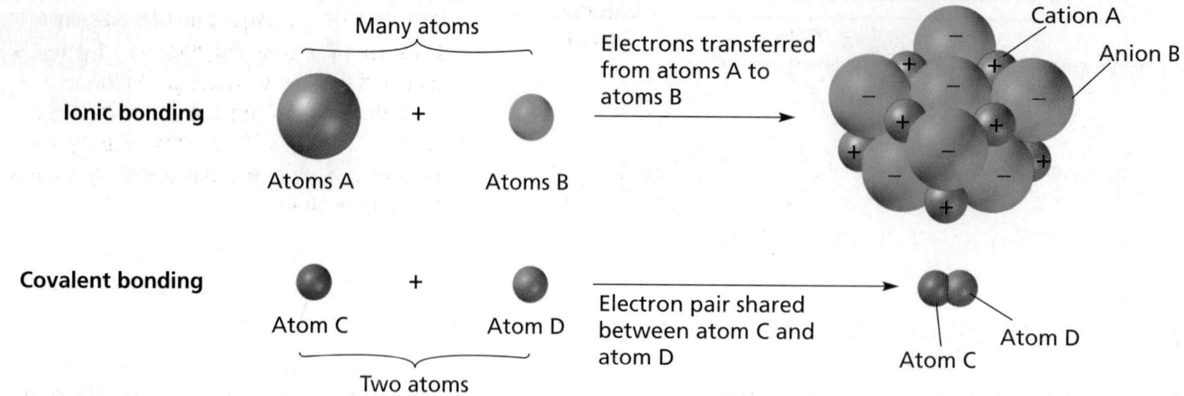

FIGURE 6-1 In ionic bonding, a large number of atoms transfer electrons. The resulting positive and negative ions combine because of mutual electrical attraction. (Shown here is a small portion of an ionic arrangement.) In covalent bonding, atoms join by sharing electron pairs to form independent molecules.

elements is ionic or covalent can be estimated by calculating the difference in the elements' electronegativities (see Figure 6-2). For example, the electronegativity difference between fluorine, F, and cesium, Cs, is 4.0 – 0.7 = 3.3. (See Figure 5-20 on page 151 for a periodic table of electronegativity values.) So, according to Figure 6-2, cesium-fluorine bonding is ionic. The valence electrons are close to the highly electronegative fluorine atoms, causing the atoms to resemble anions, and far from the barely electronegative cesium atoms, which resemble cations.

Bonding between atoms with an electronegativity difference of 1.7 or less has an ionic character of 50% or less and is classified as covalent. Bonding between two atoms of the same element is completely covalent. Hydrogen, for example, exists in nature not as isolated atoms, but as pairs of atoms held together by covalent bonds. The hydrogen-hydrogen bond has 0% ionic character. It is a **nonpolar-covalent bond,** *a covalent bond in which the bonding electrons are shared equally by the bonded atoms, resulting in a balanced distribution of electrical charge.* Bonds having 0 to 5% ionic character, corresponding to electronegativity differences of roughly 0 to 0.3, are generally considered nonpolar-covalent bonds. In bonds with significantly different electronegativities, the electrons are more strongly attracted by the more-electronegative atom. *Such bonds are* **polar,** *meaning that they have an uneven distribution of charge.* Covalent bonds having 5 to 50% ionic character, corresponding to electronegativity differences of 0.3 to 1.7, are classified as polar. *A* **polar-covalent bond** *is a covalent bond in which the bonded atoms have an unequal attraction for the shared electrons.*

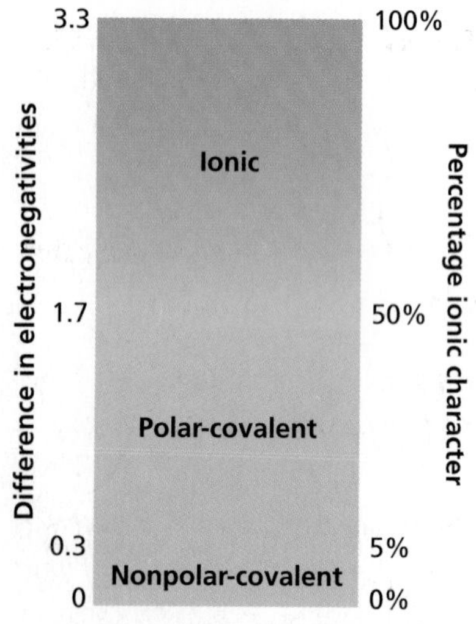

FIGURE 6-2 Difference in electronegativities is a rough measure of the character of bonding between two elements. The electronegativity of the less-electronegative element is subtracted from that of the more-electronegative element. The greater the electronegativity difference is, the more ionic the bonding is.

Nonpolar- and polar-covalent bonds are compared in Figure 6-3, which illustrates the electron density distribution in hydrogen-hydrogen and hydrogen-chlorine bonds. The electronegativity difference between chlorine and hydrogen is 3.0 – 2.1 = 0.9, indicating a polar-covalent bond. The electrons in this bond are closer to the more-electronegative chlorine atom than to the hydrogen atom, as indicated in Figure 6-3(b). Consequently, the chlorine end of the bond has a partial negative charge, indicated by the symbol $\delta-$. The hydrogen end of the bond then has an equal partial positive charge, $\delta+$.

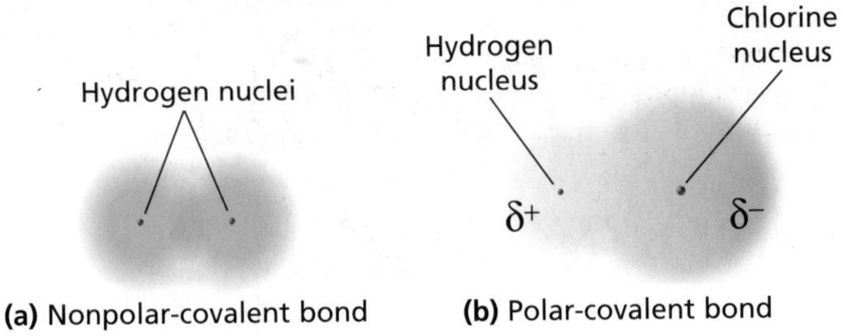

Hydrogen nuclei

Hydrogen nucleus

Chlorine nucleus

$\delta+$ $\delta-$

(a) Nonpolar-covalent bond

(b) Polar-covalent bond

FIGURE 6-3 Comparison of the electron density in (a) a nonpolar, hydrogen-hydrogen bond and (b) a polar, hydrogen-chlorine bond. Because chlorine is more electronegative than hydrogen, the electron density in the hydrogen-chlorine bond is greater around the chlorine atom.

SAMPLE PROBLEM 6-1

Use electronegativity differences and Figure 6-2 to classify bonding between sulfur, S, and the following elements: hydrogen, H; cesium, Cs; and chlorine, Cl. In each pair, which atom will be more negative?

SOLUTION From Figure 5-20 on page 151, we know that the electronegativity of sulfur is 2.5. The electronegativities of hydrogen, cesium, and chlorine are 2.1, 0.7, and 3.0, respectively. In each pair, the atom with the larger electronegativity will be the more-negative atom.

Bonding between sulfur and	Electronegativity difference	Bond type	More-negative atom
hydrogen	2.5 – 2.1 = 0.4	polar-covalent	sulfur
cesium	2.5 – 0.7 = 1.8	ionic	sulfur
chlorine	3.0 – 2.5 = 0.5	polar-covalent	chlorine

PRACTICE Use electronegativity differences and Figure 6-2 to classify bonding between chlorine, Cl, and the following elements: calcium, Ca; oxygen, O; and bromine, Br. Indicate the more-negative atom in each pair.

Answer

Bonding between chlorine and	Electronegativity difference	Bond type	More-negative atom
calcium	3.0 – 1.0 = 2.0	ionic	chlorine
oxygen	3.5 – 3.0 = 0.5	polar-covalent	oxygen
bromine	3.0 – 2.8 = 0.2	nonpolar-covalent	chlorine

SECTION REVIEW

1. Compare ionic and covalent bonding.
2. What is the role of electronegativity in determining the ionic or covalent character of the bonding between two elements?
3. What type of bonding would be expected between the following atoms?
   a. H and F
   b. Cu and S
   c. I and Br
4. List the three pairs of atoms referred to in the previous question in order of increasing ionic character of the bonding between them.

SECTION 6-2

Covalent Bonding and Molecular Compounds

OBJECTIVES

- Define *molecule* and *molecular formula.*
- Explain the relationships between potential energy, distance between approaching atoms, bond length, and bond energy.
- State the octet rule.
- List the six basic steps used in writing Lewis structures.
- Explain how to determine Lewis structures for molecules containing single bonds, multiple bonds, or both.
- Explain why scientists use resonance structures to represent some molecules.

Many chemical compounds, including most of the chemicals that are in living things and are produced by living things, are composed of molecules. *A* **molecule** *is a neutral group of atoms that are held together by covalent bonds.* A single molecule of a chemical compound is an individual unit capable of existing on its own. It may consist of two or more of the same type of atoms, as in oxygen, or of two or more different atoms, as in water or sugar (see Figure 6-4 below). *A chemical compound whose simplest units are molecules is called a* **molecular compound.**

The composition of a compound is given by its chemical formula. *A* **chemical formula** *indicates the relative numbers of atoms of each kind in a chemical compound by using atomic symbols and numerical subscripts.* The chemical formula of a molecular compound is referred to as a molecular formula. *A* **molecular formula** *shows the types and numbers of atoms combined in a single molecule of a molecular compound.* The molecular formula for water, for example, is H_2O, which reflects the fact that a single water molecule consists of one oxygen atom joined by separate covalent bonds to two hydrogen atoms. A molecule of oxygen, O_2, is an example of a diatomic molecule. *A* **diatomic molecule** *is a molecule containing only two atoms.*

(a) Water molecule

(b) Oxygen molecule

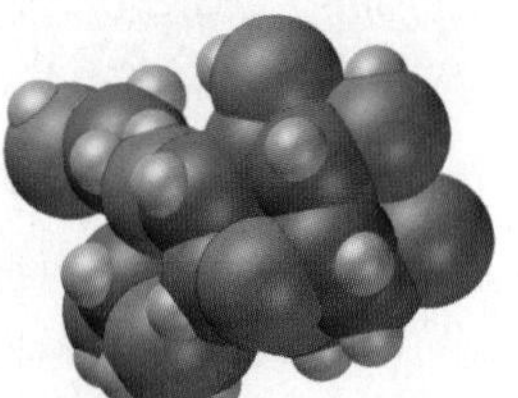

(c) Sucrose molecule

FIGURE 6-4 The models for (a) water, (b) oxygen, and (c) sucrose, or table sugar, represent a few examples of the many molecular compounds in and around us. Atoms within molecules may form one or more covalent bonds.

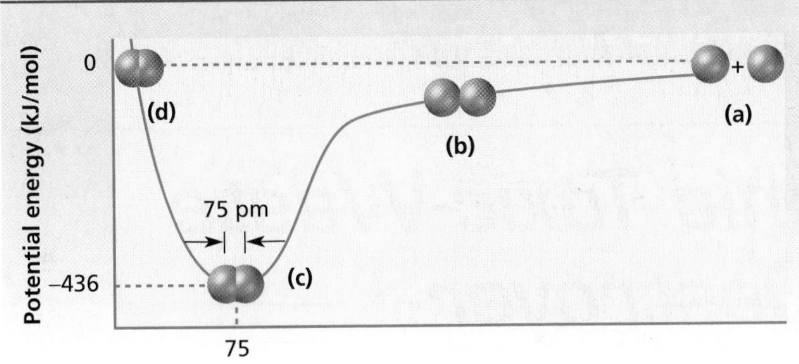

FIGURE 6-5 Potential energy changes during the formation of a hydrogen-hydrogen bond. (a) The separated hydrogen atoms do not affect each other. (b) Potential energy decreases as the atoms are drawn together by attractive forces. (c) Potential energy is at a minimum when attractive forces are balanced by repulsive forces. (d) Potential energy increases when repulsion between like charges outweighs attraction between opposite charges.

Formation of a Covalent Bond

As you read in Section 6-1, nature favors chemical bonding because most atoms are at lower potential energy when bonded to other atoms than they are at as independent particles. In the case of covalent bond formation, this idea is illustrated by a simple example, the formation of a hydrogen-hydrogen bond.

Picture two isolated hydrogen atoms separated by a distance large enough to prevent them from influencing each other. At this distance, the overall potential energy of the atoms is arbitrarily set at zero, as shown in part (a) of Figure 6-5.

Now consider what happens if the hydrogen atoms approach each other. Each atom has a nucleus containing a single positively charged proton. The nucleus of each atom is surrounded by a negatively charged electron in a spherical 1*s* orbital. As the atoms near each other, their charged particles begin to interact. As shown in Figure 6-6, the approaching nuclei and electrons are *attracted* to each other, which corresponds to a *decrease* in the total potential energy of the atoms. At the same time, the two nuclei *repel* each other and the two electrons *repel* each other, which results in an *increase* in potential energy.

The relative strength of attraction and repulsion between the charged particles depends on the distance separating the atoms. When the atoms first "see" each other, the electron-proton attraction is stronger than the electron-electron and proton-proton repulsions. Thus, the atoms are drawn to each other and their potential energy is lowered, as shown in part (b) of Figure 6-5.

The attractive force continues to dominate and the total potential energy continues to decrease until, eventually, a distance is reached at which the repulsion between the like charges equals the attraction of the opposite charges. This is shown in part (c) of Figure 6-5. At this point, which is represented by the bottom of the valley in the curve, potential energy is at a minimum and a stable hydrogen molecule forms. A closer approach of the atoms, shown in part (d) of Figure 6-5, results in a sharp rise in potential energy as repulsion becomes increasingly greater than attraction.

Both nuclei repel each other, as do both electron clouds.

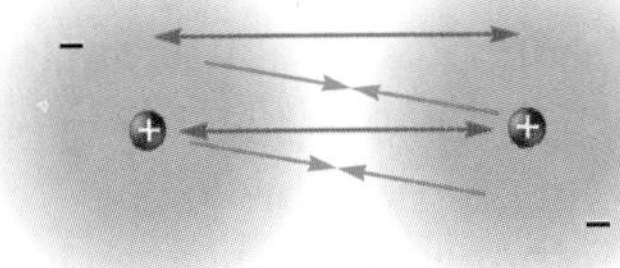

The nucleus of one atom attracts the electron cloud of the other atom, and vice versa.

FIGURE 6-6 The arrows indicate the attractive and repulsive forces between the electrons (shown as electron clouds) and nuclei of two hydrogen atoms. Attraction between particles corresponds to a decrease in potential energy of the atoms, while repulsion corresponds to an increase.

Ultrasonic Toxic-Waste Destroyer

Paints, pesticides, solvents, and sulfides—these are just a few components of the 600 million tons of toxic waste that flows out of America's factories every year. Much of this waste ends up in ground water, polluting our streams and contaminating our drinking water.

Eliminating hazardous waste is a constant challenge. Unfortunately, today's disposal methods often damage the environment as much as they help it. Incinerators burning certain chemical waste, for example, produce dioxins, one of the most dangerous class of toxins known to man.

Finding new methods to destroy toxic waste is a puzzle. Michael Hoffmann, a professor of environmental chemistry at the California Institute of Technology, thinks that part of the solution lies in sound-wave technology.

According to Hoffmann, the key to eliminating certain chemical wastes from polluted water is a phenomenon called cavitation. Cavitation occurs when the pressure in water is made to fluctuate from slightly above to slightly below normal, causing bubbles to form. The bubbles are unstable, and when they collapse they create tiny areas of extremely high pressure and heat. The pressure inside a collapsing bubble can be 1000 times greater than normal, and the temperature reaches about 5000°C—just a bit cooler than the surface of the sun. These conditions are harsh enough to combust most toxic-waste compounds in the water, breaking them down into harmless components.

Professor Hoffmann has employed a device that uses ultrasound—sound waves at frequencies just above the range of human hearing—to create cavitation in polluted water. Composed of two panels that generate ultrasound at different frequencies, the toxic-waste destroyer is simple in design and in concept. As water flows between the panels, the ultrasonic waves generated by one panel form cavitation bubbles. An instant later, the ultrasound produced by the other panel collapses the bubbles. The intense pressure and heat generated break down the complex toxic compounds into simple, innocuous substances such as carbon dioxide, chloride ions, and hydrogen ions.

"With ultrasound," says Hoffmann, "we can harness frequencies over the range of about 16 kilohertz up to 1 megahertz, and different chemical compounds are destroyed more readily at one frequency versus another frequency. So by appropriately applying a particular frequency range, we can destroy a very broad range of chemical compounds."

Some compounds take longer to destroy than others. While simple toxins can be eliminated in a few minutes, others may take several hours. Some compounds must be broken down into intermediate chemicals first and then treated again to destroy them completely. To be sure the waste is totally removed, the scientists use sophisticated tracking methods to trace what happens to every single molecule of the toxin.

The ultrasound toxic-waste destroyer treats about 10% of all types of waste, eliminating both organic and inorganic compounds, such as hydrogen cyanide, TNT, and many pesticides. While the device cannot destroy complex mixtures of compounds, such as those found in raw sewage, it does have many advantages over current technologies. Aside from having no harmful environmental side effects, ultrasonic waste destruction is much cheaper and simpler than the process of combustion. By using the destroyer preventively, manufacturers can pretreat their waste products before they dump them in the sewer, thus stopping a problem before it starts.

Characteristics of the Covalent Bond

In Figure 6-5 on page 165, the bottom of the valley in the curve represents the balance between attraction and repulsion in a stable covalent bond. At this point, the electrons of each hydrogen atom of the hydrogen molecule are shared between the nuclei. As shown below in Figure 6-7, the molecule's electrons can be pictured as occupying overlapping orbitals, moving about freely in either orbital.

The bonded atoms vibrate a bit, but as long as their potential energy remains close to the minimum, they are covalently bonded to each other. *The distance between two bonded atoms at their minimum potential energy, that is, the average distance between two bonded atoms, is the* **bond length.** The bond length of a hydrogen-hydrogen bond is 75 pm.

In forming a covalent bond, the hydrogen atoms release energy as they change from isolated individual atoms to parts of a molecule. The amount of energy released equals the difference between the potential energy at the zero level (separated atoms) and that at the bottom of the valley (bonded atoms) in Figure 6-5. The same amount of energy must be added to separate the bonded atoms. **Bond energy** *is the energy required to break a chemical bond and form neutral isolated atoms.* Scientists usually report bond energies in kilojoules per mole (kJ/mol), which indicates the energy required to break one mole of bonds in isolated molecules. For example, 436 kJ of energy is needed to break the hydrogen-hydrogen bonds in one mole of hydrogen molecules and form two moles of separated hydrogen atoms.

The energy relationships described here for the formation of a hydrogen-hydrogen bond apply generally to all covalent bonds. However, bond lengths and bond energies vary with the types of atoms that have combined. Even the energy of a bond between the same two types of atoms varies somewhat, depending on what other bonds the atoms have formed. These facts should be considered in examining the data in Table 6-1 on page 168. The first three columns in the table list bonds, bond lengths, and bond energies of atoms in specific diatomic molecules. The last three columns give average values of specified bonds in many different compounds.

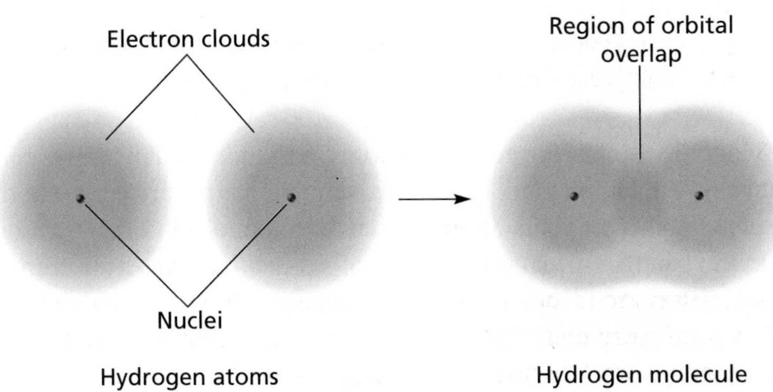

FIGURE 6-7 The orbitals of the hydrogen atoms in a hydrogen molecule overlap, allowing each electron to feel the attraction of both nuclei. The result is an increase in electron density between the nuclei.

TABLE 6-1 Bond Lengths and Bond Energies for Selected Covalent Bonds

Bond	Bond length (pm)	Bond energy (kJ/mol)	Bond	Bond length (pm)	Bond energy (kJ/mol)
H–H	74	436	C–C	154	346
F–F	141	159	C–N	147	305
Cl–Cl	199	243	C–O	143	358
Br–Br	228	193	C–H	109	418
I–I	267	151	C–Cl	177	327
H–F	92	569	C–Br	194	285
H–Cl	127	432	N–N	145	180
H–Br	141	366	N–H	101	386
H–I	161	299	O–H	96	459

All individual hydrogen atoms contain a single, unpaired electron in a 1*s* atomic orbital. When two hydrogen atoms form a molecule, they share electrons in a covalent bond. As illustrated in Figure 6-8, sharing electrons allows each atom in the hydrogen molecule to experience the effect of the stable electron configuration of helium, $1s^2$. This tendency for atoms to achieve noble-gas configurations through covalent bonding extends beyond the simple case of a hydrogen molecule.

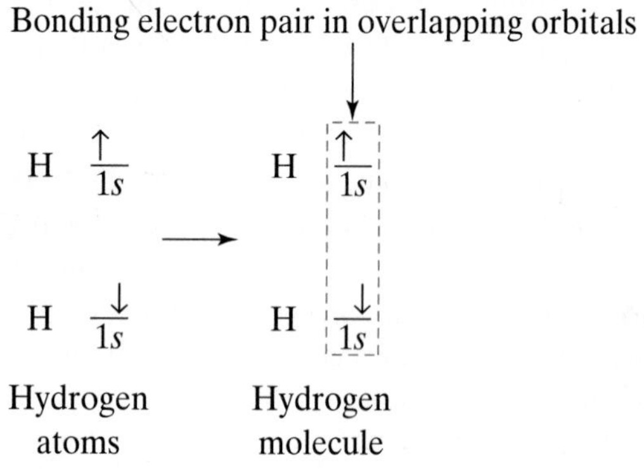

FIGURE 6-8 By sharing electrons in overlapping orbitals, each hydrogen atom in a hydrogen molecule experiences the effect of a stable $1s^2$ configuration.

Bonding electron pair in overlapping orbitals

H $\frac{\uparrow}{1s}$ H $\frac{\uparrow}{1s}$

→

H $\frac{\downarrow}{1s}$ H $\frac{\downarrow}{1s}$

Hydrogen atoms Hydrogen molecule

The Octet Rule

Unlike other atoms, the noble-gas atoms exist independently in nature. They possess a minimum of energy existing on their own because of the special stability of their electron configurations. This stability results from the fact that, with the exception of helium and its two electrons in a completely filled outer shell, the noble-gas atoms' outer *s* and *p* orbitals are completely filled by a total of eight electrons. Other main-group atoms can effectively fill their outermost *s* and *p* orbitals with electrons by sharing electrons through covalent bonding. Such bond formation follows the **octet rule:** *Chemical compounds tend to form so that each atom, by gaining, losing, or sharing electrons, has an octet of electrons in its highest occupied energy level.*

Let's examine how the bonding in a fluorine molecule illustrates the octet rule. An independent fluorine atom has seven electrons in its highest energy level ([He]$2s^22p^5$). Like hydrogen atoms, fluorine atoms bond covalently with each other to form diatomic molecules, F_2. When two fluorine atoms bond, each atom shares one of its valence electrons with its partner. The shared electron pair effectively fills each atom's outermost energy level with an octet of electrons, as illustrated in Figure 6-9(a). Figure 6-9(b) shows another example of the octet rule, in which the chlorine atom in a molecule of hydrogen chloride, HCl, achieves an outermost octet by sharing an electron pair with an atom of hydrogen.

(a)

F ↑↓ ↑↓ ↑↓ ↑↓ ↑ — 1s 2s 2p

F ↑↓ ↑↓ ↑↓ ↑↓ ↓ — 1s 2s 2p

Fluorine atoms

F ↑↓ ↑↓ ↑↓ ↑↓ ↑ — 1s 2s 2p

F ↑↓ ↑↓ ↑↓ ↑↓ ↓ — 1s 2s 2p

Fluorine molecule

Bonding electron pair in overlapping orbitals

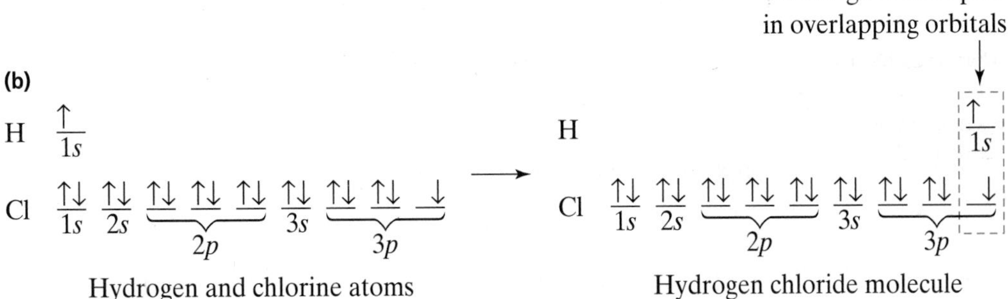

FIGURE 6-9 (a) By sharing valence electrons in overlapping orbitals, each atom in a fluorine molecule feels the effect of neon's stable configuration, [He]$2s^22p^6$. (b) In a hydrogen chloride molecule, the hydrogen atom effectively fills its 1*s* orbital with two electrons, while the chlorine atom experiences the stability of an outermost octet of electrons.

Exceptions to the Octet Rule

Most main-group elements tend to form covalent bonds according to the octet rule. However, there are exceptions. As you have seen, hydrogen forms bonds in which it is surrounded by only two electrons. Boron, B, has just three valence electrons ([He]$2s^22p^1$). Because electron pairs are shared in covalent bonds, boron tends to form bonds in which it is surrounded by six electrons. In boron trifluoride, BF_3, for example, the boron atom is surrounded by its own three valence electrons plus one from each of the three fluorine atoms bonded to it. Other elements can be surrounded by *more* than eight electrons when they combine with the highly electronegative elements fluorine, oxygen, and chlorine. In these cases of *expanded valence,* bonding involves electrons in *d* orbitals as well as in *s* and *p* orbitals.

Number of valence electrons	Electron-dot notation	Example
1	X·	Na·
2	·X·	·Mg·
3	·Ẋ·	·Ḃ·
4	·Ẋ· (dot below)	·Ċ· (dot below)
5	·Ẋ: (dot below)	·Ṅ: (dot below)
6	:Ẋ: (dot below)	:Ȯ: (dot below)
7	:Ẍ: (dot below)	:F̈: (dot below)
8	:Ẍ: (two dots below)	:N̈e: (two dots below)

FIGURE 6-10 To write an element's electron-dot notation, determine the element's number of valence electrons. Then place a corresponding number of dots around the element's symbol, as shown.

Electron-Dot Notation

Covalent bond formation usually involves only the electrons in an atom's outermost energy levels, or the atom's valence electrons. To keep track of these electrons, it is helpful to use electron-dot notation. **Electron-dot notation** *is an electron-configuration notation in which only the valence electrons of an atom of a particular element are shown, indicated by dots placed around the element's symbol.* The inner-shell electrons are not shown. For example, the electron-dot notation for a fluorine atom (electron configuration [He]$2s^2p^5$) may be written as follows.

:F̈: (dot below)

In general, an element's number of valence electrons can be determined by adding the superscripts of the element's noble-gas notation. In this book, the electron-dot notations for elements with 1–8 valence electrons are written as shown in Figure 6-10.

SAMPLE PROBLEM 6-2

a. Write the electron-dot notation for hydrogen.
b. Write the electron-dot notation for nitrogen.

SOLUTION a. A hydrogen atom has only one occupied energy level, the $n = 1$ level, which contains a single electron. Therefore, the electron-dot notation for hydrogen is written as follows.

H·

b. The group notation for nitrogen's family of elements is ns^2np^3, which indicates that nitrogen has five valence electrons. Therefore, the electron-dot notation for nitrogen is written as follows.

·Ṅ: (dot below)

Lewis Structures

Electron-dot notation can also be used to represent molecules. For example, a hydrogen molecule, H_2, is represented by combining the notations of two individual hydrogen atoms, as follows.

H:H

The pair of dots represents the shared electron pair of the hydrogen-hydrogen covalent bond. For a molecule of fluorine, F_2, the electron-dot notations of two fluorine atoms are combined.

$$:\ddot{\underset{..}{F}}:\ddot{\underset{..}{F}}:$$

Here also the pair of dots between the two symbols represents the shared pair of a covalent bond. In addition, each fluorine atom is surrounded by three pairs of electrons that are not shared in bonds. *An* **unshared pair,** *also called a* **lone pair,** *is a pair of electrons that is not involved in bonding and that belongs exclusively to one atom.*

The pair of dots representing a shared pair of electrons in a covalent bond is often replaced by a long dash. According to this convention, hydrogen and fluorine molecules are represented as follows.

$$\text{H–H} \quad :\ddot{\underset{..}{F}}\text{–}\ddot{\underset{..}{F}}:$$

These representations are all **Lewis structures,** *formulas in which atomic symbols represent nuclei and inner-shell electrons, dot-pairs or dashes between two atomic symbols represent electron pairs in covalent bonds, and dots adjacent to only one atomic symbol represent unshared electrons.* It is common to write Lewis structures that show only the electrons that are shared, using dashes to represent the bonds. *A* **structural formula** *indicates the kind, number, arrangement, and bonds but not the unshared pairs of the atoms in a molecule.* For example, F–F and H–Cl are structural formulas.

The Lewis structures (and therefore the structural formulas) for many molecules can be drawn if one knows the composition of the molecule and which atoms are bonded to each other. The following sample problem illustrates the basic steps for writing Lewis structures. The molecule described in this problem contains bonds with single shared electron pairs. A single covalent bond, or a **single bond,** *is a covalent bond produced by the sharing of one pair of electrons between two atoms.*

SAMPLE PROBLEM 6-3

Draw the Lewis structure of iodomethane, CH_3I.

SOLUTION

1. *Determine the type and number of atoms in the molecule.*
   The formula shows one carbon atom, one iodine atom, and three hydrogen atoms.

2. *Write the electron-dot notation for each type of atom in the molecule.*
   Carbon is from Group 14 and has four valence electrons. Iodine is from Group 17 and has seven valence electrons. Hydrogen has one valence electron.

$$\cdot\dot{\underset{\cdot}{C}}\cdot \quad :\ddot{\underset{\cdot}{I}}: \quad H\cdot$$

3. *Determine the total number of valence electrons in the atoms to be combined.*

$$\begin{array}{lrr} \text{C} & 1 \times 4e^- = & 4e^- \\ \text{I} & 1 \times 7e^- = & 7e^- \\ \text{H} & 3 \times 1e^- = & \underline{3e^-} \\ & & 14e^- \end{array}$$

4. *Arrange the atoms to form a skeleton structure for the molecule. If carbon is present, it is the central atom. Otherwise, the least-electronegative atom is central (except for hydrogen, which is never central). Then connect the atoms by electron-pair bonds.*

```
   H
   ..
 H:C:I
   ..
   H
```

5. *Add unshared pairs of electrons so that each hydrogen atom shares a pair of electrons and each other nonmetal is surrounded by eight electrons.*

```
   H              H
   .. ..          |  ..
 H:C:I:   or   H—C—I:
   .. ..          |  ..
   H              H
```

6. *Count the electrons in the structure to be sure that the number of valence electrons used equals the number available.*
   There are eight electrons in the four covalent bonds and six electrons in the three unshared pairs, giving the correct total of 14 valence electrons.

PRACTICE

1. Draw the Lewis structure of ammonia, NH_3.

Answer

```
  ..          ..
H:N:H  or  H—N—H
  ..          |
  H           H
```

2. Draw the Lewis structure for hydrogen sulfide, H_2S.

Answer

```
  ..          ..
H:S:H  or  H—S—H
  ..          ..
```

Multiple Covalent Bonds

Atoms of some elements, especially carbon, nitrogen, and oxygen, can share more than one electron pair. A double covalent bond, or simply a **double bond,** *is a covalent bond produced by the sharing of two pairs of electrons between two atoms.* A double bond is shown either by two side-by-side pairs of dots or by two parallel dashes. All four electrons in a double bond "belong" to both atoms. In ethene, C_2H_4, for example, two electron pairs are simultaneously shared by two carbon atoms.

```
H     H        H     H
 \   /          \   /
  C::C    or     C=C
 /    \         /   \
H      H       H     H
```

A triple covalent bond, or simply a **triple bond,** *is a covalent bond produced by the sharing of three pairs of electrons between two atoms.* For example, elemental nitrogen, N_2, like hydrogen and the halogens, normally exists as diatomic molecules. In this case, however, each nitrogen atom, which has five valence electrons, acquires three electrons to complete an octet by sharing three pairs of electrons with its partner. This is illustrated in the Lewis structure and the formula structure for N_2, as shown below.

$$:N::N: \quad \text{or} \quad N{\equiv}N$$

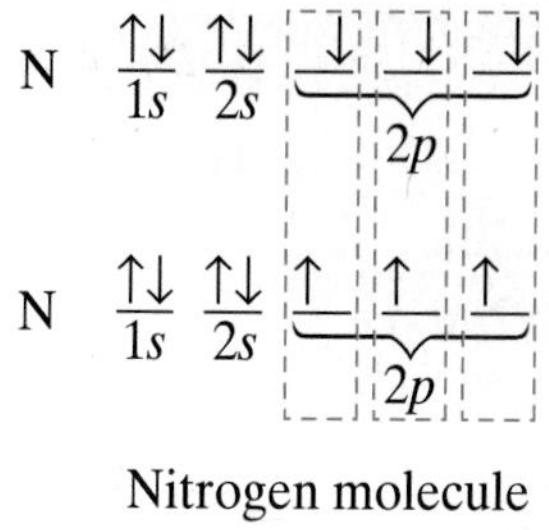

Figure 6-11 represents nitrogen's triple bond through orbital notation. Like the single bonds in hydrogen and halogen molecules, the triple bond in nitrogen molecules is nonpolar.

Carbon forms a number of compounds containing triple bonds. For example, the compound ethyne, C_2H_2, contains a carbon-carbon triple bond.

$$H:C::C:H \quad \text{or} \quad H{-}C{\equiv}C{-}H$$

Double and triple bonds are referred to as **multiple bonds,** or multiple covalent bonds. Double bonds in general have higher bond energies and are shorter than single bonds. Triple bonds are even stronger and shorter. Table 6-2 compares average bond lengths and bond energies for some single, double, and triple bonds.

In writing Lewis structures for molecules that contain carbon, nitrogen, or oxygen, one must remember that multiple bonds between pairs of these atoms are possible. (A hydrogen atom, on the other hand, has only one electron and therefore always forms a single covalent bond.) The need for a multiple bond becomes obvious if there are not enough valence electrons to complete octets by adding unshared pairs. Sample Problem 6-4 on page 174 illustrates how to deal with this situation.

N ↑↓ ↑↓ ↓ ↓ ↓
1s 2s 2p

N ↑↓ ↑↓ ↑ ↑ ↑
1s 2s 2p

Nitrogen molecule

FIGURE 6-11 In a molecule of nitrogen, N_2, each nitrogen atom is surrounded by six shared electrons plus one unshared pair of electrons. Thus, each nitrogen atom follows the octet rule in forming a triple covalent bond.

TABLE 6-2 *Bond Lengths and Bond Energies for Single and Multiple Covalent Bonds*

Bond	Bond length (pm)	Bond energy (kJ/mol)	Bond	Bond length (pm)	Bond energy (kJ/mol)
C−C	154	346	C−O	143	358
C=C	134	612	C=O	120	799
C≡C	120	835	C≡O	113	1072
C−N	147	305	N−N	145	180
C=N	132	615	N=N	125	418
C≡N	116	887	N≡N	110	942

SAMPLE PROBLEM 6-4

Draw the Lewis structure for methanal, CH_2O, which is also known as formaldehyde.

SOLUTION

1. *Determine the number of atoms of each element present in the molecule.*
The formula shows one carbon atom, two hydrogen atoms, and one oxygen atom.

2. *Write the electron-dot notation for each type of atom.*
Carbon from Group 14 has four valence electrons. Oxygen, which is in Group 16, has six valence electrons. Hydrogen has only one electron.

$$\cdot\dot{\underset{\cdot}{C}}\cdot \quad :\dot{\underset{\cdot}{O}}: \quad H\cdot$$

3. *Determine the total number of valence electrons in the atoms to be combined.*

C	**$1 \times 4e^- =$**	**$4e^-$**
O	**$1 \times 6e^- =$**	**$6e^-$**
2H	**$2 \times 1e^- =$**	**$2e^-$**
		$12e^-$

4. *Arrange the atoms to form a skeleton structure for the molecule, and connect the atoms by electron-pair bonds.*

$$\begin{array}{c} H \\ \ddot{} \\ H{:}C{:}O \end{array}$$

5. *Add unshared pairs of electrons so that each hydrogen atom shares a pair of electrons and each other nonmetal is surrounded by eight electrons.*

$$\begin{array}{c} H \\ H{:}\underset{\cdot\cdot}{\ddot{C}}{:}\underset{\cdot\cdot}{\ddot{O}}{:} \end{array}$$

6a. *Count the electrons in the Lewis structure to be sure that the number of valence electrons used equals the number available.*
The structure above has six electrons in covalent bonds and eight electrons in four lone pairs, for a total of 14 electrons. The structure has two valence electrons too many.

6b. *If too many electrons have been used, subtract one or more lone pairs until the total number of valence electrons is correct. Then move one or more lone electron pairs to existing bonds between non-hydrogen atoms until the outer shells of all atoms are completely filled.*
Subtract the lone pair of electrons from the carbon atom. Then move one lone pair of electrons from the oxygen to the bond between carbon and oxygen to form a double bond.

$$\begin{array}{c} H \\ H{:}\ddot{C}{::}\underset{\cdot\cdot}{\ddot{O}} \end{array} \quad \text{or} \quad \begin{array}{c} H \\ | \\ H{-}C{=}\underset{\cdot\cdot}{\ddot{O}} \end{array}$$

There are eight electrons in covalent bonds and four electrons in lone pairs, for a total of 12 valence electrons.

PRACTICE

1. Draw the Lewis structure for carbon dioxide, CO_2.

Answer
$\underset{\cdot\cdot}{\ddot{O}}{=}C{=}\underset{\cdot\cdot}{\ddot{O}}$

2. Draw the Lewis structure for hydrogen cyanide, which contains one hydrogen atom, one carbon atom, and one nitrogen atom.

Answer
$H{-}C{\equiv}N{:}$

Resonance Structures

Some molecules and ions cannot be represented adequately by a single Lewis structure. One such molecule is ozone, O_3, which can be represented by either of the following Lewis structures.

$$\ddot{\underset{..}{O}}{=}\ddot{O}{-}\ddot{\underset{..}{O}}\!: \quad \text{or} \quad :\!\ddot{\underset{..}{O}}{-}\ddot{O}{=}\ddot{\underset{..}{O}}$$

Notice that each structure indicates that the ozone molecule has two types of O—O bonds, one single and one double. Chemists once speculated that ozone split its time existing as one of these two structures, constantly alternating, or "resonating," from one to the other. Experiments, however, revealed that the oxygen-oxygen bonds in ozone are identical. Therefore, scientists now say that ozone has a single structure that is the average of these two structures. Together the structures are referred to as *resonance structures* or *resonance hybrids*. **Resonance** *refers to bonding in molecules or ions that cannot be correctly represented by a single Lewis structure.* To indicate resonance, a double-headed arrow is placed between a molecule's resonance structures.

$$\ddot{\underset{..}{O}}{=}\ddot{O}{-}\ddot{\underset{..}{O}}\!: \longleftrightarrow :\!\ddot{\underset{..}{O}}{-}\ddot{O}{=}\ddot{\underset{..}{O}}$$

Covalent-Network Bonding

All the covalent compounds that you have read about to this point are molecular. They consist of many identical molecules bound together by forces acting between the molecules. (You will read more about intermolecular forces in Section 6-5.) There are many covalently bonded compounds that do not contain individual molecules, but instead can be pictured as continuous, three-dimensional networks of bonded atoms. You will read more about covalently bonded networks in Chapter 11.

SECTION REVIEW

1. Define the following:
   a. bond length
   b. bond energy
2. State the octet rule.
3. How many pairs of electrons are shared in the following types of covalent bonds?
   a. a single bond
   b. a double bond
   c. a triple bond
4. Draw the Lewis structures for the following molecules:
   a. IBr
   b. CH_3Br
   c. C_2HCl
   d. $SiCl_4$
   e. F_2O

SECTION 6-3

Ionic Bonding and Ionic Compounds

OBJECTIVES

- Compare and contrast a chemical formula for a molecular compound with one for an ionic compound.
- Discuss the arrangements of ions in crystals.
- Define *lattice energy* and explain its significance.
- List and compare the distinctive properties of ionic and molecular compounds.
- Write the Lewis structure for a polyatomic ion given the identity of the atoms combined and other appropriate information.

Most of the rocks and minerals that make up the earth's crust consist of positive and negative ions held together by ionic bonding. A familiar example of an ionically bonded compound is sodium chloride, or common table salt, which is found in nature as rock salt. A sodium ion, Na^+, has a charge of 1+. A chloride ion, Cl^-, has a charge of 1–. In sodium chloride, these ions combine in a one-to-one ratio—Na^+Cl^-—so that each positive charge is balanced by a negative charge. Because chemists are aware of this balance of charge, the chemical formula for sodium chloride is usually written simply as NaCl.

An **ionic compound** *is composed of positive and negative ions that are combined so that the numbers of positive and negative charges are equal.* Most ionic compounds exist as crystalline solids (see Figure 6-12). A crystal of any ionic compound is a three-dimensional network of positive and negative ions mutually attracted to one another. As a result, in contrast to a molecular compound, an ionic compound is not composed of independent, neutral units that can be isolated and examined. The chemical formula of an ionic compound merely represents the simplest ratio of the compound's combined ions that gives electrical neutrality.

The chemical formula of an ionic compound shows the ratio of the ions present in a sample of any size. *A* **formula unit** *is the simplest collection of atoms from which an ionic compound's formula can be established.* For example, one formula unit of sodium chloride, NaCl, is one sodium cation plus one chloride anion. (In naming anions, the *-ine* ending of the element's name is dropped and replaced with *-ide*. You'll read more about naming compounds in Chapter 7.)

The ratio of ions in a formula unit depends on the charges of the ions combined. For example, to achieve electrical neutrality in the ionic compound calcium fluoride, two fluoride anions, F^-, each with a charge of 1–, must balance the 2+ charge of each calcium cation, Ca^{2+}. Therefore, the formula of calcium fluoride is CaF_2.

FIGURE 6-12 Like most ionic compounds, sodium chloride is a crystalline solid.

Formation of Ionic Compounds

Electron-dot notation can be used to demonstrate the changes that take place in ionic bonding. Ionic compounds do not ordinarily form by the combination of isolated ions, but consider for a moment a sodium atom and a chlorine atom approaching each other. The two atoms are neutral and have one and seven valence electrons, respectively.

Na·	:C̈l:
Sodium atom	Chlorine atom

We have already seen that atoms of sodium and the other alkali metals readily lose one electron to form cations. And we have seen that atoms of chlorine and the other halogens readily gain one electron to form anions. The combination of sodium and chlorine atoms to produce one formula unit of sodium chloride can thus be represented as follows.

$$\mathrm{Na}\cdot \; + \; :\ddot{\mathrm{Cl}}: \; \longrightarrow \; \mathrm{Na}^{+} \; + \; :\ddot{\mathrm{Cl}}:^{-}$$

Sodium atom   Chlorine atom   Sodium cation   Chloride anion

The transfer of an electron from the sodium atom to the chlorine atom transforms each atom into an ion with a noble-gas configuration. In the combination of calcium with fluorine, two fluorine atoms are needed to accept the two valence electrons given up by one calcium atom.

$$\cdot\mathrm{Ca}\cdot \; + \; :\ddot{\mathrm{F}}: \; + \; :\ddot{\mathrm{F}}: \; \longrightarrow \; \mathrm{Ca}^{2+} \; + \; :\ddot{\mathrm{F}}:^{-} \; + \; :\ddot{\mathrm{F}}:^{-}$$

Calcium atom   Fluorine atoms   Calcium cation   Fluoride anions

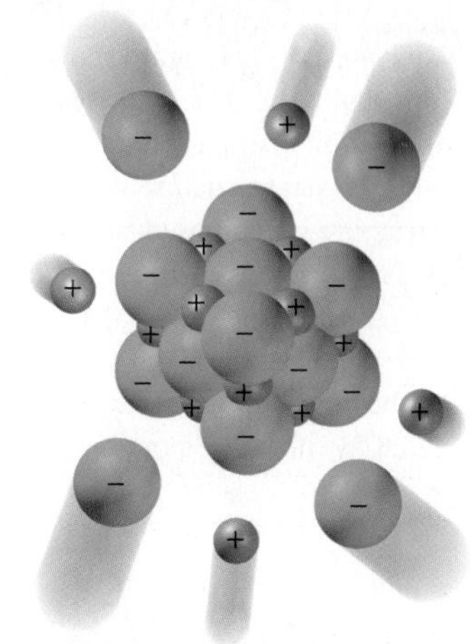

FIGURE 6-13 The ions in an ionic compound lower their potential energy by forming an orderly, three-dimensional array in which the positive and negative charges are balanced. The electrical forces of attraction between oppositely charged ions extend over long distances, causing a large decrease in potential energy.

Characteristics of Ionic Bonding

Recall that nature favors arrangements in which potential energy is minimized. In an ionic crystal, ions minimize their potential energy by combining in an orderly arrangement known as a *crystal lattice* (see Figure 6-13). The attractive forces at work within an ionic crystal include those between oppositely charged ions and those between the nuclei and electrons of adjacent ions. The repulsive forces include those between like-charged ions and those between electrons of adjacent ions. The distances between ions and their arrangement in a crystal represent a balance among all these forces. Sodium chloride's crystal structure is shown in Figure 6-14 below.

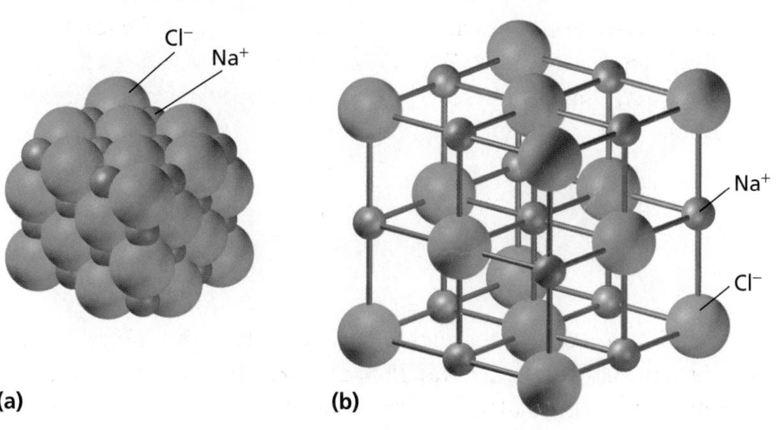

FIGURE 6-14 Two models of the crystal structure of sodium chloride are shown. (a) To illustrate the ions' actual arrangement, the sodium and chloride ions are shown with their electron clouds just touching. (b) In an expanded view, the distances between ions have been exaggerated in order to clarify the positioning of the ions in the structure.

FIGURE 6-15 The figure shows the ions that most closely surround a chloride anion and a sodium cation within the crystal structure of NaCl. The structure is composed such that (a) six Na^+ ions surround each Cl^- ion. At the same time, (b) six Cl^- ions surround each Na^+ ion (which cannot be seen but whose location is indicated by the dashed outline).

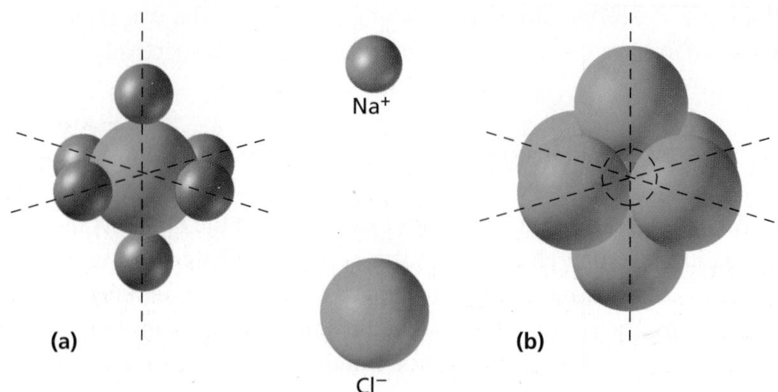

Figure 6-15 shows the crystal structure of sodium chloride in greater detail. Within the arrangement, each sodium cation is surrounded by six chloride anions. At the same time, each chloride anion is surrounded by six sodium cations. Attraction between the adjacent oppositely charged ions is much stronger than repulsion by other ions of the same charge, which are farther away.

The three-dimensional arrangements of ions and the strengths of attraction between them vary with the sizes and charges of the ions and the numbers of ions of different charges. For example, in calcium fluoride, there are two anions for each cation. Each calcium cation is surrounded by eight fluoride anions. At the same time, each fluoride ion is surrounded by four calcium cations, as shown in Figure 6-16.

To compare bond strengths in ionic compounds, chemists compare the amounts of energy released when separated ions in a gas come together to form a crystalline solid. **Lattice energy** *is the energy released when one mole of an ionic crystalline compound is formed from gaseous ions.* Lattice energy values for a few common ionic compounds are shown in Table 6-3. The negative energy values indicate that energy is *released* when the crystals are formed.

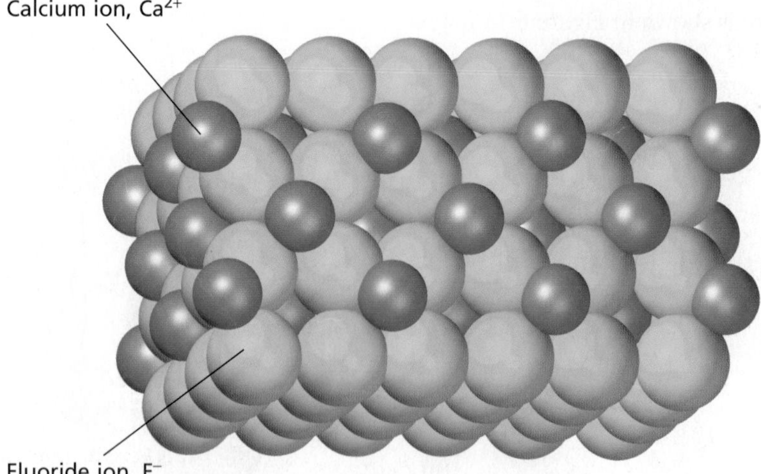

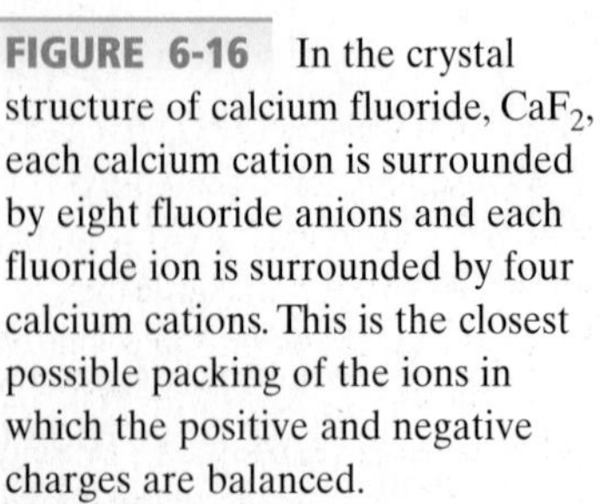

FIGURE 6-16 In the crystal structure of calcium fluoride, CaF_2, each calcium cation is surrounded by eight fluoride anions and each fluoride ion is surrounded by four calcium cations. This is the closest possible packing of the ions in which the positive and negative charges are balanced.

A Comparison of Ionic and Molecular Compounds

The force that holds ions together in ionic compounds is a very strong overall attraction between positive and negative charges. In a molecular compound, the covalent bonds of the atoms making up each molecule are also strong. But the forces of attraction *between* molecules are much weaker than the forces of ionic bonding. This difference in the strength of attraction between the basic units of molecular and ionic compounds gives rise to different properties in the two types of compounds.

The melting point, boiling point, and hardness of a compound depend on how strongly its basic units are attracted to each other. Because the forces of attraction between individual molecules are not very strong, many molecular compounds melt at low temperatures. Conversely, because of the strong forces that hold ions together, ionic compounds generally have higher melting and boiling points than do molecular compounds. Also, they do not vaporize as readily at room temperature as many molecular compounds do. In fact, many molecular compounds are already completely gaseous at room temperature.

Ionic compounds are hard but brittle. Why? In an ionic crystal, even a slight shift of one row of ions relative to another causes a large buildup of repulsive forces, as shown in Figure 6-17. These forces make it difficult for one layer to move relative to another, causing ionic compounds to be hard. If one layer is moved, however, the repulsive forces make the layers part completely, causing ionic compounds to be brittle.

In the molten state, or when dissolved in water, ionic compounds are electrical conductors because the ions can move freely to carry electrical current. In the solid state, the ions cannot move, so the compounds are not electrical conductors. Many ionic compounds are soluble in water. During dissolution, the ions separate from each other and become surrounded by water molecules. Other ionic compounds do not dissolve in water, however, because the attraction of the water molecules cannot overcome the attraction between the ions.

TABLE 6-3 ***Lattice Energies of Some Common Ionic Compounds***

Compound	Lattice energy (kJ/mol)
NaCl	−787.5
NaBr	−751.4
CaF_2	−2634.7
CaO	−3385
LiCl	−861.3
LiF	−1032
MgO	−3760
KCl	−715

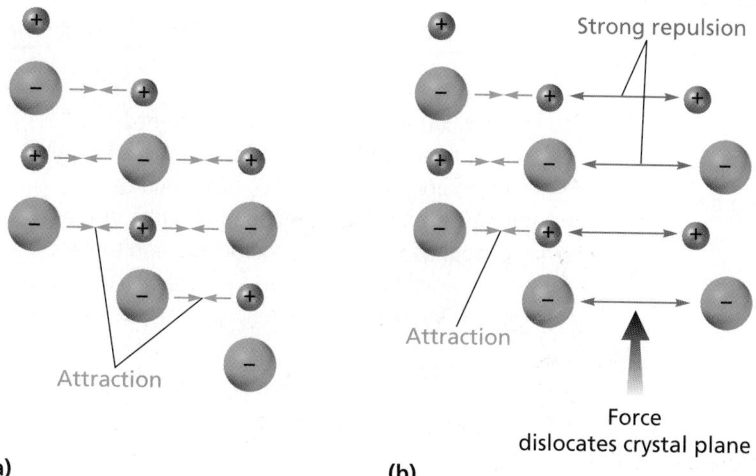

FIGURE 6-17 (a) The attraction between positive and negative ions in a crystalline ionic compound causes layers of ions to resist motion. (b) When struck with sufficient force, the layers shift so that ions of the same charge approach each other, causing repulsion. As a result, the crystal shatters along the planes.

Polyatomic Ions

Certain atoms bond covalently with each other to form a group of atoms that has both molecular and ionic characteristics. *A charged group of covalently bonded atoms is known as a* **polyatomic ion.** Polyatomic ions combine with ions of opposite charge to form ionic compounds. The charge of a polyatomic ion results from an excess of electrons (negative charge) or a shortage of electrons (positive charge). For example, an ammonium ion, the most common positively charged polyatomic ion, contains one nitrogen atom and four hydrogen atoms and has a single positive charge. Its formula is NH_4^+, sometimes written as $[NH_4]^+$ to show that the group of atoms *as a whole* has a charge of 1+. The seven protons in the nitrogen atom plus the four protons in the four hydrogen atoms give the ammonium ion a total positive charge of 11+. An independent nitrogen atom has seven electrons, and four independent hydrogen atoms have a total of four electrons. When these atoms combine to form an ammonium ion, one of their electrons is lost, giving the polyatomic ion a total negative charge of 10–.

Lewis structures for the ammonium ion and some common negative polyatomic ions—the nitrate, sulfate, and phosphate ions—are shown below. To find the Lewis structure for a polyatomic ion, follow the steps of Sample Problem 6-4 on page 174, with the following exception. If the ion is negatively charged, add to the total number of valence electrons a number of e^- corresponding to the ion's negative charge. If the ion is positively charged, subtract from the total number of valence electrons a number of e^- corresponding to the ion's positive charge.

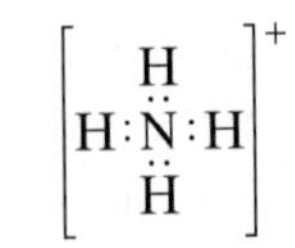

Ammonium ion 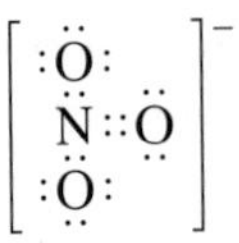 | Nitrate ion 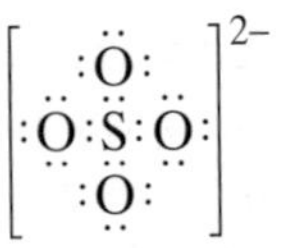 | Sulfate ion 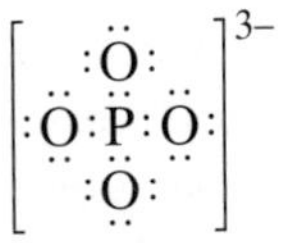 | Phosphate ion

SECTION REVIEW

1. Give two examples of an ionic compound.
2. Use electron-dot notation to demonstrate the formation of ionic compounds involving the following:
   a. Li and Cl
   b. Ca and I
3. Distinguish between ionic and molecular compounds in terms of the basic units that each is composed of.
4. Compound B has lower melting and boiling points than compound A. At the same temperature, compound B vaporizes faster and to a greater extent than compound A. If one of these compounds is ionic and the other is molecular, which would you expect to be molecular? ionic? Explain the reasoning behind your choices.

Metallic Bonding

SECTION 6-4

OBJECTIVES

- Describe the electron-sea model of metallic bonding, and explain why metals are good electrical conductors.
- Explain why metal surfaces are shiny.
- Explain why metals are malleable and ductile but ionic-crystalline compounds are not.

Chemical bonding is different in metals than it is in ionic, molecular, or covalent-network compounds. This difference is reflected in the unique properties of metals. They are excellent electrical conductors in the solid state—much better conductors than even molten ionic compounds. This property is due to the highly mobile valence electrons of the atoms that make up a metal. Such mobility is not possible in molecular compounds, in which valence electrons are localized in electron-pair bonds between neutral atoms. Nor is it possible in solid ionic compounds, in which electrons are bound to individual ions that are held in place in crystal structures.

The Metallic-Bond Model

The highest energy levels of most metal atoms are occupied by very few electrons. In *s*-block metals, for example, one or two valence electrons occupy the outermost orbital, whereas all three outermost *p* orbitals, which can hold a total of six electrons, are vacant. In addition to completely vacant outer *p* orbitals, *d*-block metals also possess many vacant *d* orbitals in the energy level just below their highest energy level.

Within a metal, the vacant orbitals in the atoms' outer energy levels overlap. This overlapping of orbitals allows the outer electrons of the atoms to roam freely throughout the entire metal. The electrons are *delocalized,* which means that they do not belong to any one atom but move freely about the metal's network of empty atomic orbitals. These mobile electrons form a *sea of electrons* around the metal atoms, which are packed together in a crystal lattice (see Figure 6-18). *The chemical bonding that results from the attraction between metal atoms and the surrounding sea of electrons is called* **metallic bonding.**

FIGURE 6-18 The model shows a portion of the crystal structure of solid sodium. The atoms are arranged so that each sodium atom is surrounded by eight other sodium atoms. The atoms are relatively fixed in position, while the electrons are free to move throughout the crystal, forming an electron sea.

Metallic Properties

The freedom of motion of electrons in a network of metal atoms accounts for the high electrical and thermal conductivity characteristic of all metals. In addition, because they contain many orbitals separated by extremely small energy differences, metals can absorb a wide range of light frequencies. This absorption of light results in the excitation of the metal atoms' electrons to higher energy levels. The electrons immediately fall back down to lower levels, emitting energy in the form of light. This de-excitation is responsible for the shiny appearance of metal surfaces.

FIGURE 6-19 Unlike ionic crystalline compounds, most metals are malleable. This property allows iron, for example, to be shaped into useful tools.

Most metals are also easy to form into desired shapes. Two important properties related to this characteristic are malleability and ductility. **Malleability** *is the ability of a substance to be hammered or beaten into thin sheets.* **Ductility** *is the ability of a substance to be drawn, pulled, or extruded through a small opening to produce a wire.* The malleability and ductility of metals are possible because metallic bonding is the same in all directions throughout the solid. One plane of atoms in a metal can slide past another without encountering any resistance or breaking any bonds. By contrast, recall from Section 6-3 that shifting the layers of an ionic crystal causes the bonds to break and the crystal to shatter.

Metallic Bond Strength

Metallic bond strength varies with the nuclear charge of the metal atoms and the number of electrons in the metal's electron sea. Both of these factors are reflected in a metal's *heat of vaporization*. When a metal is vaporized, the bonded atoms in the normal (usually solid) state are converted to individual metal atoms in the gaseous state. The amount of heat required to vaporize the metal is a measure of the strength of the bonds that hold the metal together. Some heats of vaporization for metals are given in Table 6-4.

TABLE 6-4 ***Heats of Vaporization of Some Metals (kJ/mol)***

Period	Element		
Second	Li 147	Be 297	
Third	Na 97	Mg 128	Al 294
Fourth	K 77	Ca 155	Sc 333
Fifth	Rb 76	Sr 137	Y 365
Sixth	Cs 64	Ba 140	La 402

SECTION REVIEW

1. Describe the electron-sea model of metallic bonding.
2. What is the relationship between metallic bond strength and heat of vaporization?
3. Explain why most metals are malleable and ductile but ionic crystals are not.

Molecular Geometry

SECTION 6-5

Objectives

- Explain VSEPR theory.
- Predict the shapes of molecules or polyatomic ions using VSEPR theory.
- Explain how the shapes of molecules are accounted for by hybridization theory.
- Describe dipole-dipole forces, hydrogen bonding, induced dipoles, and London dispersion forces.
- Explain what determines molecular polarity.

The properties of molecules depend not only on the bonding of atoms but also on molecular geometry—the three-dimensional arrangement of a molecule's atoms in space. The polarity of each bond, along with the geometry of the molecule, determines **molecular polarity,** *or the uneven distribution of molecular charge.* As you will read, molecular polarity strongly influences the forces that act *between* molecules in liquids and solids.

A chemical formula reveals little information about a molecule's geometry. After performing many tests designed to reveal the shapes of various molecules, chemists developed two different, equally successful theories to explain certain aspects of their findings. One theory accounts for molecular bond angles. The other is used to describe the orbitals that contain the valence electrons of a molecule's atoms.

VSEPR Theory

As shown in Figure 6-20, diatomic molecules, like those of hydrogen, H_2, and hydrogen chloride, HCl, must be linear because they consist of only two atoms. To predict the geometries of more-complicated molecules, one must consider the locations of all electron pairs surrounding the bonded atoms. This is the basis of VSEPR theory.

The abbreviation VSEPR stands for "valence-shell, electron-pair repulsion," referring to the repulsion between pairs of valence electrons of the atoms in a molecule. **VSEPR theory** *states that repulsion between the sets of valence-level electrons surrounding an atom causes these sets to be oriented as far apart as possible.* How does the assumption that electrons in molecules repel each other account for molecular shapes? For now let us consider only molecules with no unshared valence electron pairs on the central atom.

Let's examine the simple molecule BeF_2. Recall that beryllium does not follow the octet rule. The beryllium atom forms a covalent bond with each fluorine atom. It is surrounded by only the two electron pairs that it shares with the fluorine atoms.

$$:\ddot{\underset{..}{F}}:Be:\ddot{\underset{..}{F}}:$$

According to VSEPR, the shared pairs are oriented as far away from each other as possible. As shown in Figure 6-21(a) on page 184, the distance between electron pairs is maximized if the bonds to fluorine are

(a) Hydrogen, H_2

(b) Hydrogen chloride, HCl

FIGURE 6-20 Ball-and-stick models illustrate the linearity of diatomic molecules. (a) A hydrogen molecule is represented by two identical balls (the hydrogen atoms) joined by a solid bar (the covalent bond). (b) A hydrogen chloride molecule is composed of dissimilar atoms, but it is still linear.

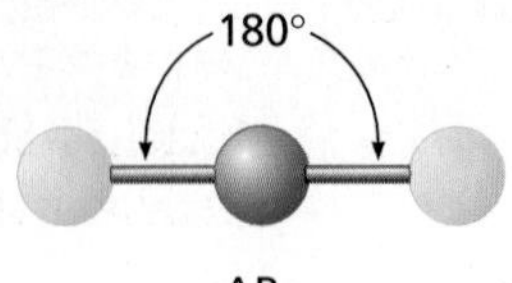

(a) Beryllium fluoride, BeF_2

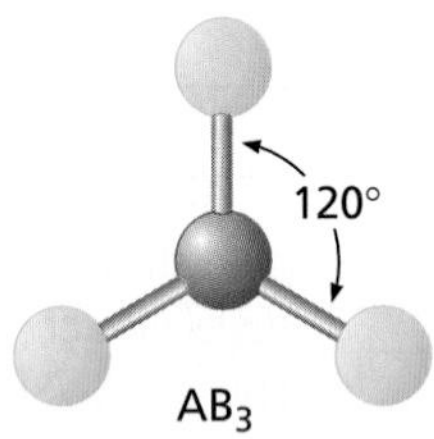

(b) Boron trifluoride, BF_3

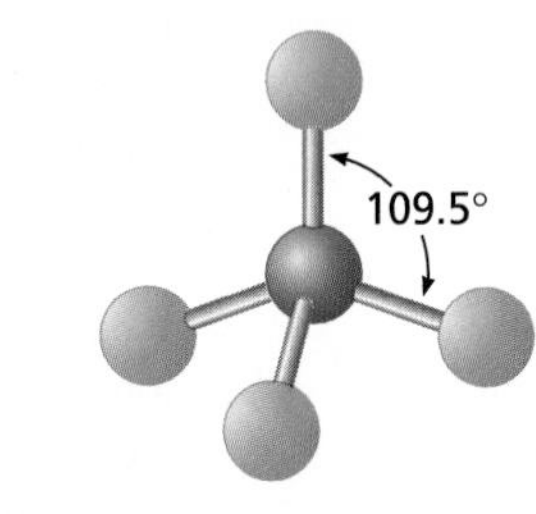

(c) Methane, CH_4

FIGURE 6-21 Ball-and-stick models show the shapes of (a) AB_2, (b) AB_3, and (c) AB_4 molecules according to VSEPR.

on opposite sides of the beryllium atom, 180° apart. Thus, all three atoms lie on a straight line. The molecule is linear.

If we represent the central atom in a molecule by the letter *A* and we represent the atoms bonded to the central atom by the letter *B*, then according to VSEPR, BeF_2 is an example of an AB_2 molecule, which is linear. Can you determine what an AB_3 molecule looks like? The three A—B bonds stay farthest apart by pointing to the corners of an equilateral triangle, giving 120° angles between the bonds. This trigonal-planar geometry is shown in Figure 6-21(b) for the AB_3 molecule boron trifluoride, BF_3.

Unlike AB_2 and AB_3 molecules, the central atoms in AB_4 molecules follow the octet rule by sharing four electron pairs with B atoms. The distance between electron pairs is maximized if each A—B bond points to one of four corners of a tetrahedron. This geometry is shown in Figure 6-21(c) for the AB_4 molecule methane, CH_4. The same figure shows that in a tetrahedral molecule, each of the bond angles formed by the A atom and any two of the B atoms is equal to 109.5°.

The shapes of various molecules are summarized in Table 6-5 on page 186. B can represent a single type of atom, a group of identical atoms, or a group of different atoms on the same molecule. The shape of the molecule will still be based on the forms given in the table. However, different sizes of B groups distort the bond angles, making some bond angles larger or smaller than those given in the table.

SAMPLE PROBLEM 6-5

Use VSEPR theory to predict the molecular geometry of aluminum trichloride, $AlCl_3$.

SOLUTION First write the Lewis structure for $AlCl_3$. Aluminum is in Group 13 and has three valence electrons.

$$\cdot\dot{\mathrm{Al}}\cdot$$

Chlorine is in Group 17 and has seven valence electrons.

$$:\ddot{\underset{\cdot}{\mathrm{Cl}}}:$$

The total number of available valence electrons is therefore $24e^-$ ($3e^-$ from aluminum and $21e^-$ from chlorine). The following Lewis structure uses all $24e^-$.

$$\begin{array}{c} :\ddot{\mathrm{Cl}}: \\ :\underset{\cdot\cdot}{\ddot{\mathrm{Cl}}}:\ddot{\mathrm{Al}}:\underset{\cdot\cdot}{\ddot{\mathrm{Cl}}}: \end{array}$$

This molecule is an exception to the octet rule because in this case Al forms only three bonds. Aluminum trichloride is an AB_3 type of molecule. Therefore, according to VSEPR theory, it should have trigonal-planar geometry.

PRACTICE

1. Use VSEPR theory to predict the molecular geometry of the following molecules:
   a. HI
   b. CBr_4
   c. $AlBr_3$
   d. SF_6
   e. CH_2Cl_2

Answer
a. linear
b. tetrahedral
c. trigonal-planar
d. octahedral
e. tetrahedral

VSEPR and Unshared Electron Pairs

Ammonia, NH_3, and water, H_2O, are examples of molecules in which the central atom has both shared and unshared electron pairs (see Table 6-5 for their Lewis structures). How does VSEPR theory account for the geometries of these molecules?

The Lewis structure of ammonia shows that in addition to the three electron pairs it shares with the three hydrogen atoms, the central nitrogen atom has one unshared pair of electrons.

$$\mathrm{H{:}\ddot{N}{:}H} \atop \mathrm{\ddot{H}}$$

VSEPR theory postulates that the lone pair occupies space around the nitrogen atom just as the bonding pairs do. Thus, as in an AB_4 molecule, the electron pairs maximize their separation by assuming the four corners of a tetrahedron. Lone pairs do occupy space, but our description of the observed shape of a molecule refers to the *positions of atoms only*. Consequently, as shown in Figure 6-22(a), the molecular geometry of an ammonia molecule is that of a pyramid with a triangular base. The general VSEPR formula for molecules such as ammonia is AB_3E, where E represents the unshared electron pair.

A water molecule has two unshared electron pairs. It is an AB_2E_2 molecule. Here, the oxygen atom is at the center of a tetrahedron, with two corners occupied by hydrogen atoms and two by the unshared pairs (Figure 6-22(b)). Again, VSEPR theory states that the lone pairs occupy space around the central atom but that the actual shape of the molecule is determined by the positions of the atoms only. In the case of water, this results in a "bent," or angular, molecule.

FIGURE 6-22 The locations of bonds and unshared electrons are shown for molecules of (a) ammonia and (b) water. Although unshared electrons occupy space around the central atoms, the shapes of the molecules depend only on the position of the molecules' atoms, as clearly shown by the ball-and-stick models.

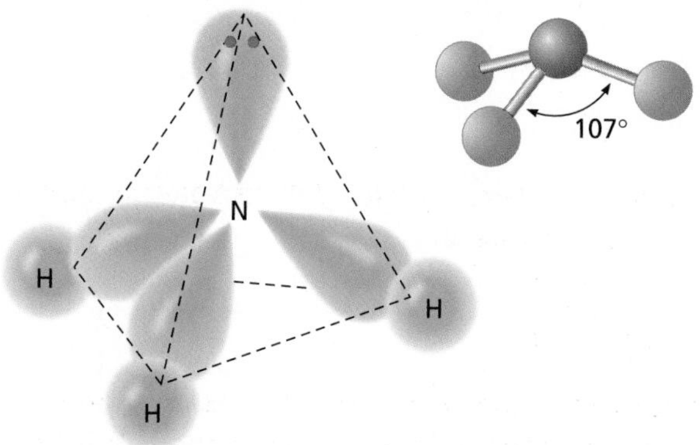

(a) Ammonia, NH_3

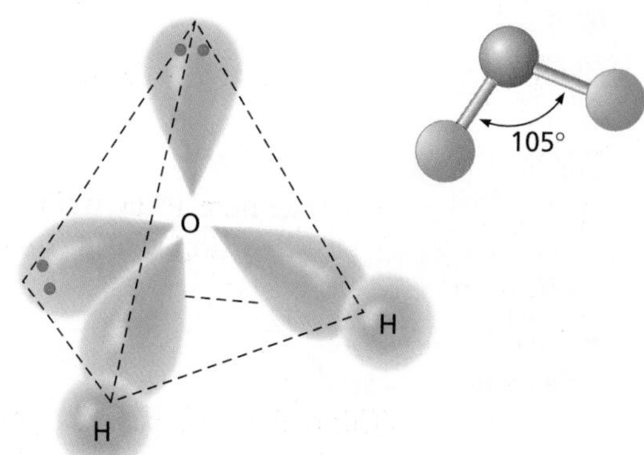

(b) Water, H_2O

In Figure 6-22(b), note that the bond angles in ammonia and water are somewhat less than the 109.5° bond angles of a perfectly tetrahedral molecule. These angles are smaller because the unshared electron pairs repel electrons more strongly than do bonding electron pairs.

Table 6-5 also includes an example of an AB_2E type molecule. This type of molecule results when a central atom forms two bonds and retains one unshared electron pair.

Finally, in VSEPR theory, double and triple bonds are treated in the same way as single bonds. And polyatomic ions are treated similarly to molecules. (Remember to consider *all* of the electron pairs present in any ion or molecule.) Thus, Lewis structures and Table 6-5 can be used together to predict the shapes of polyatomic ions as well as molecules with double or triple bonds.

TABLE 6-5 *VSEPR and Molecular Geometry*

	Molecular shape	Atoms bonded to central atom	Lone pairs of electrons	Type of molecule	Formula example	Lewis structure
Linear	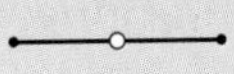	2	0	AB_2	BeF_2	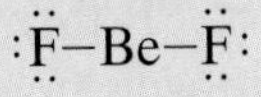 $:\ddot{F}-Be-\ddot{F}:$
Bent	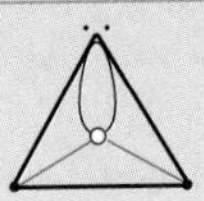	2	1	AB_2E	$SnCl_2$	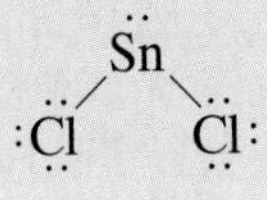 $\ddot{Sn}$ bonded to $:\ddot{Cl}$ and $\ddot{Cl}:$
Trigonal-planar	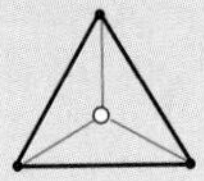	3	0	AB_3	BF_3	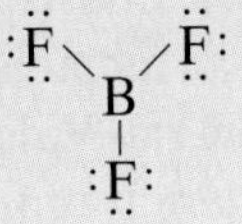 B bonded to $:\ddot{F}$, $\ddot{F}:$, $:\ddot{F}:$
Tetrahedral	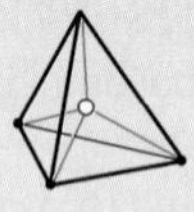	4	0	AB_4	CH_4	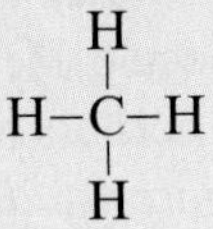 C bonded to H, H, H, H
Trigonal-pyramidal	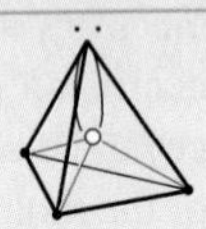	3	1	AB_3E	NH_3	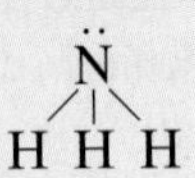 $\ddot{N}$ bonded to H, H, H
Bent	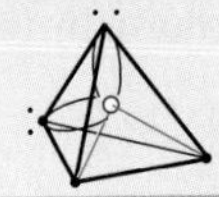	2	2	AB_2E_2	H_2O	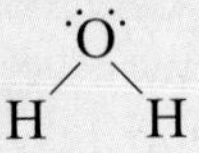 $\ddot{O}$ bonded to H, H
Trigonal-bipyramidal	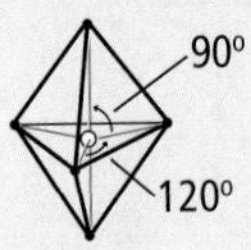 90°, 120°	5	0	AB_5	PCl_5	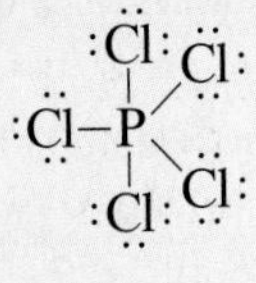 P bonded to five $:\ddot{Cl}:$
Octahedral	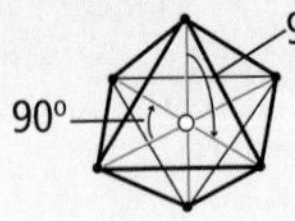 90°, 90°	6	0	AB_6	SF_6	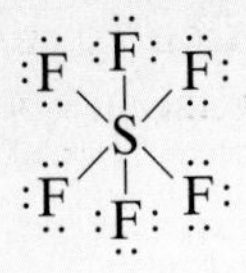 S bonded to six $:\ddot{F}:$

SAMPLE PROBLEM 6-6

a. Use VSEPR theory to predict the shape of a molecule of carbon dioxide, CO_2.
b. Use VSEPR theory to predict the shape of a chlorate ion, ClO_3^-.

SOLUTION **a.** The Lewis structure of carbon dioxide shows two carbon-oxygen double bonds and no unshared electron pairs on the carbon atom. To simplify the molecule's Lewis structure, we represent the covalent bonds with lines instead of dots.

$$\ddot{\underset{\cdot\cdot}{O}}=C=\ddot{\underset{\cdot\cdot}{O}}$$

This is an AB_2 molecule, which is linear.

b. The Lewis structure of a chlorate ion shows three oxygen atoms and an unshared pair of electrons surrounding a central chlorine atom. Again, lines are used to represent the covalent bonds.

$$\left[\begin{array}{ccc} & \ddot{Cl} & \\ :\ddot{\underset{\cdot\cdot}{O}} & :\underset{\cdot\cdot}{O}: & \ddot{\underset{\cdot\cdot}{O}}: \end{array}\right]^-$$

The chlorate ion is an AB_3E type. It has trigonal-pyramidal geometry, with the three oxygen atoms at the base of the pyramid and the chlorine atom at the top.

PRACTICE

1. Use VSEPR theory to predict the molecular geometries of the molecules whose Lewis structures are given below.

   a. $:\ddot{\underset{\cdot\cdot}{F}}-\ddot{\underset{\cdot\cdot}{S}}-\ddot{\underset{\cdot\cdot}{F}}:$ b. Cl—P—Cl (with Cl below P)

Answer
a. bent b. trigonal-pyramidal

Hybridization

VSEPR theory is useful for explaining the shapes of molecules. However, it does not reveal the relationship between a molecule's geometry and the orbitals occupied by its bonding electrons. To explain how the orbitals of an atom become rearranged when the atom forms covalent bonds, a different model is used. This model is called **hybridization,** *which is the mixing of two or more atomic orbitals of similar energies on the same atom to produce new orbitals of equal energies.*

Methane, CH_4, provides a good example of how hybridization is used to explain the geometry of molecular orbitals. The orbital notation for a carbon atom shows that it has four valence electrons, two in the 2*s* orbital and two in 2*p* orbitals.

C ↑↓ (1*s*) ↑↓ (2*s*) ↑ ↑ __ (2*p*)

We know from experiments that a methane molecule has tetrahedral geometry. How does carbon form four equivalent, tetrahedrally arranged covalent bonds by orbital overlap with four other atoms?

Two of carbon's valence electrons occupy the 2*s* orbital, and two occupy the 2*p* orbitals. Recall that the 2*s* orbital and the 2*p* orbitals have different shapes. To achieve four equivalent bonds, carbon's 2*s* and three 2*p* orbitals *hybridize* to form four new, identical orbitals called sp^3 orbitals. The superscript 3 indicates that three *p* orbitals were included in the hybridization; the superscript 1 on the *s* is understood. The sp^3 orbitals all have the same energy, which is greater than that of the 2*s* orbital but less than that of the 2*p* orbitals, as shown in Figure 6-23.

FIGURE 6-23 The sp^3 hybridization of carbon's outer orbitals combines one *s* and three *p* orbitals to form four sp^3 hybrid orbitals. Whenever hybridization occurs, the resulting hybrid orbitals are at an energy level between the levels of the orbitals that have combined.

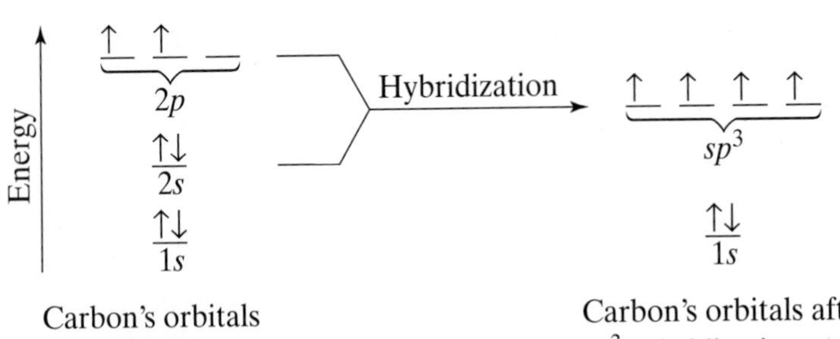

Hybrid orbitals *are orbitals of equal energy produced by the combination of two or more orbitals on the same atom.* The number of hybrid orbitals produced equals the number of orbitals that have combined. Bonding with carbon sp^3 orbitals is illustrated in Figure 6-24(a) for a molecule of methane.

Hybridization also explains the bonding and geometry of many molecules formed by Group 15 and 16 elements. The sp^3 hybridization of a nitrogen atom ([He]$2s^22p^3$) yields four hybrid orbitals—one orbital containing a pair of electrons and three orbitals that each contain an unpaired electron. Each unpaired electron is capable of forming a single bond, as shown for ammonia in Figure 6-24(b). Similarly, two of the four sp^3 hybrid orbitals on an oxygen atom ([He]$2s^22p^4$) are occupied by two electron pairs and two are occupied by unpaired electrons. Each unpaired electron can form a single bond, as shown for water in Figure 6-24(c).

FIGURE 6-24 Bonds formed by the overlap of the 1*s* orbitals of hydrogen atoms and the sp^3 orbitals of (a) carbon, (b) nitrogen, and (c) oxygen. For the sake of clarity, only the hybrid orbitals of the central atoms are shown.

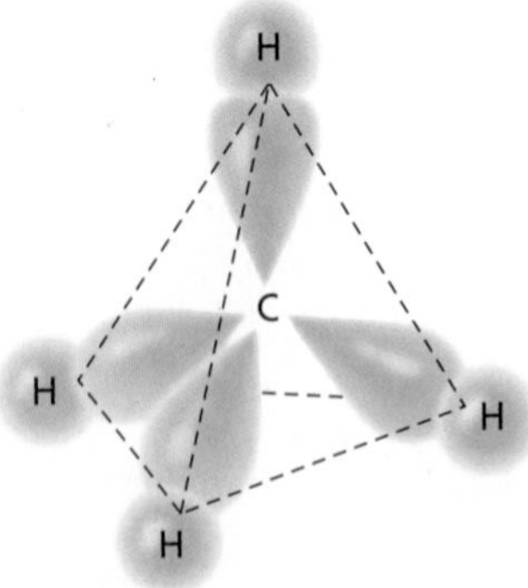

(a) Methane, CH_4

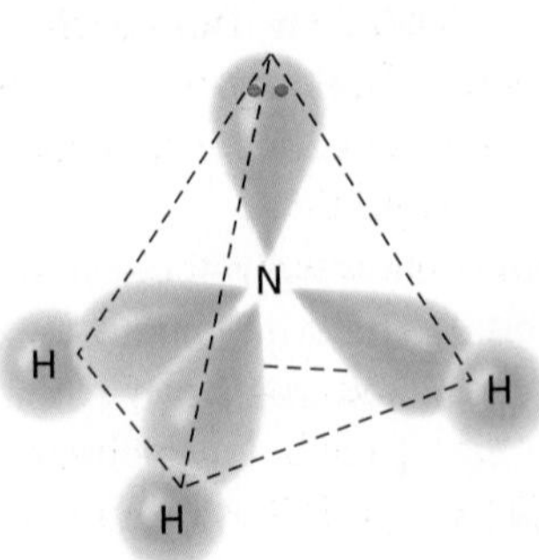

(b) Ammonia, NH_3

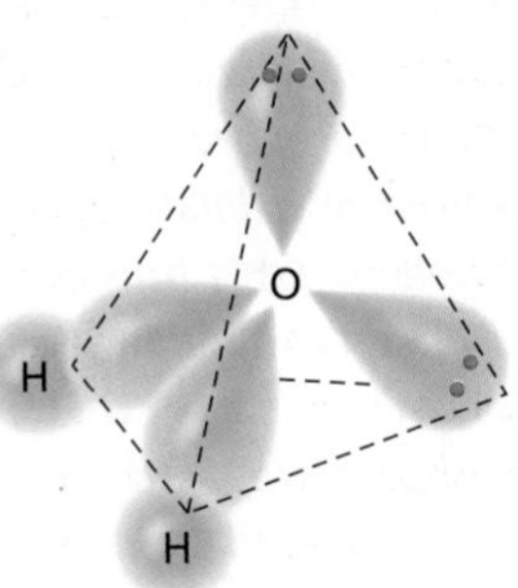

(c) Water, H_2O

TABLE 6-6 Geometry of Hybrid Orbitals

Atomic orbitals	Type of hybridization	Number of hybrid orbitals	Geometry
s, p	sp	2	180° Linear
s, p, p	sp^2	3	120° Trigonal-planar
s, p, p, p	sp^3	4	109.5° Tetrahedral

The linear geometry of molecules such as beryllium fluoride, BeF_2, (see Table 6-5) is made possible by hybridization involving the *s* orbital and one available empty *p* orbital to yield *sp* hybrid orbitals. The trigonal-planar geometry of molecules such as boron fluoride, BF_3, is made possible by hybridization involving the *s* orbital, one singly occupied *p* orbital, and one empty *p* orbital to yield sp^2 hybrid orbitals. The geometries of *sp*, sp^2, and sp^3 hybrid orbitals are summarized in Table 6-6.

Intermolecular Forces

As a liquid is heated, the kinetic energy of its particles increases. At the boiling point, the energy is sufficient to overcome the force of attraction between the liquid's particles. The particles pull away from each other and enter the gas phase. Boiling point is therefore a good measure of the force of attraction between particles of a substance. The higher the boiling point, the stronger the forces between particles.

The forces of attraction between molecules are known as **intermolecular forces.** Intermolecular forces vary in strength but are generally weaker than bonds that join atoms in molecules, ions in ionic compounds, or metal atoms in solid metals. Compare the boiling points of the metals and ionic compounds in Table 6-7 on page 190 with those of the molecular substances listed. Note that the values for ionic compounds and metals are much higher than those for molecular substances.

TABLE 6-7 Boiling Points and Bonding Types

Bonding type	Substance	bp (1 atm, °C)
Nonpolar-covalent (molecular)	H_2	−253
	O_2	−183
	Cl_2	−34
	Br_2	59
	CH_4	−164
	CCl_4	77
	C_6H_6	80
Polar-covalent (molecular)	PH_3	−88
	NH_3	−33
	H_2S	−61
	H_2O	100
	HF	20
	HCl	−85
	ICl	97
Ionic	NaCl	1413
	MgF_2	2239
Metallic	Cu	2567
	Fe	2750
	W	5660

Molecular Polarity and Dipole-Dipole Forces

The strongest intermolecular forces exist between polar molecules. Polar molecules act as tiny dipoles because of their uneven charge distribution. *A* **dipole** *is created by equal but opposite charges that are separated by a short distance.* The direction of a dipole is from the dipole's positive pole to its negative pole. A dipole is represented by an arrow with a head pointing toward the negative pole and a crossed tail situated at the positive pole. The dipole created by a hydrogen chloride molecule, which has its negative end at the more electronegative chlorine atom, is indicated as follows.

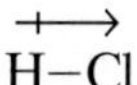

The negative region in one polar molecule attracts the positive region in adjacent molecules, and so on throughout a liquid or solid. *The forces of attraction between polar molecules are known as* **dipole-dipole forces.** These forces are short-range forces, acting only between nearby molecules. The effect of dipole-dipole forces is reflected, for example, by the significant difference between the boiling points of bromine fluoride, Br—F, and fluorine, F—F. The boiling point of polar bromine fluoride is −20°C, whereas that of nonpolar fluorine is only −188°C. The dipole-dipole forces responsible for the relatively high boiling point of BrF are illustrated schematically in Figure 6-25.

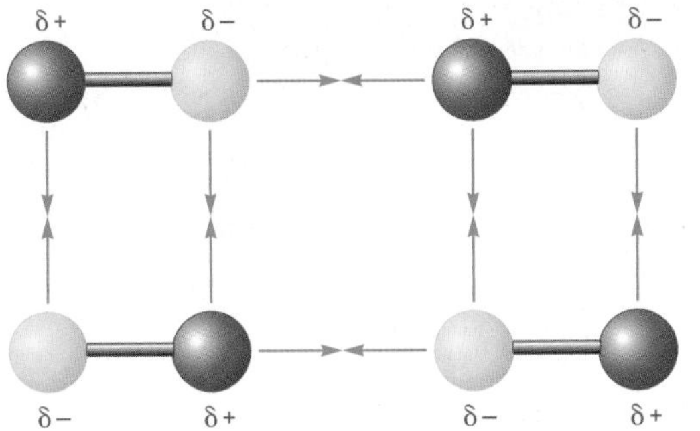

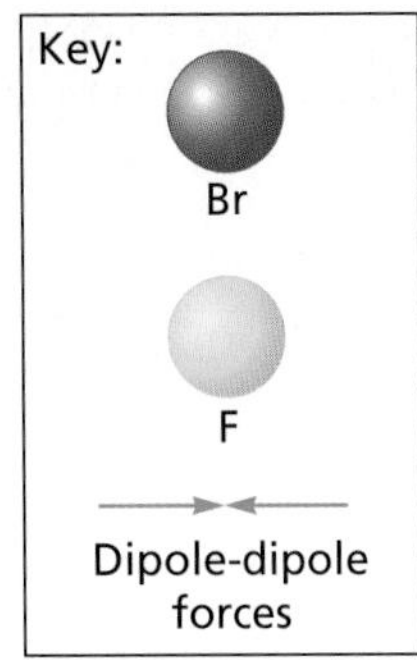

FIGURE 6-25 Ball-and-stick models illustrate the dipole-dipole forces between molecules of bromine fluoride, BrF. In each molecule, the highly electronegative fluorine atom has a partial negative charge, leaving each bromine atom with a partial positive charge. Consequently, the negative and positive ends of neighboring molecules attract each other.

The polarity of diatomic molecules such as BrF is determined by just one bond. For molecules containing more than two atoms, molecular polarity depends on both the polarity and the orientation of each bond. A molecule of water, for example, has two hydrogen-oxygen bonds in which the more-electronegative oxygen atom is the negative pole of each bond. Because the molecule is bent, the polarities of these two bonds combine to make the molecule highly polar, as shown in Figure 6-26. An ammonia molecule is also highly polar because the dipoles of the three nitrogen-hydrogen bonds are additive, combining to create a net molecular dipole. In some molecules, individual bond dipoles cancel one another, causing the resulting molecular polarity to be zero. Carbon dioxide and carbon tetrachloride are molecules of this type.

A polar molecule can *induce* a dipole in a nonpolar molecule by temporarily attracting its electrons. The result is a short-range intermolecular force that is somewhat weaker than the dipole-dipole force. The force of an induced dipole accounts for the solubility of nonpolar O_2 in water. The positive pole of a water molecule attracts the outer electrons

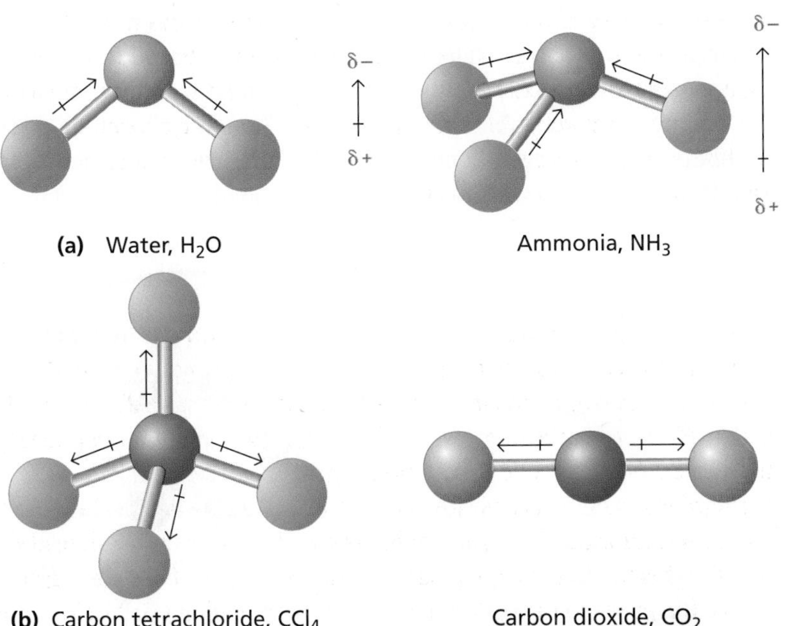

FIGURE 6-26 (a) The bond polarities in a water or an ammonia molecule are additive, causing the molecule as a whole to be polar. (b) In molecules of carbon tetrachloride and carbon dioxide, the bond polarities extend equally and symmetrically in different directions, canceling each other's effect and causing each molecule as a whole to be nonpolar.

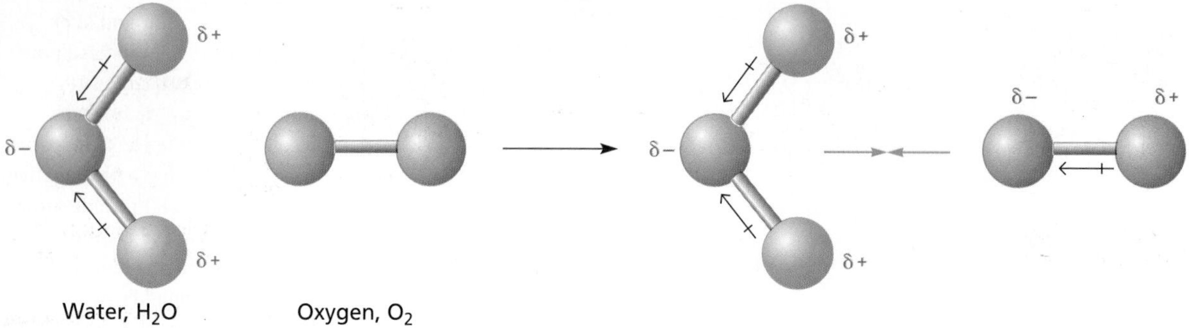

FIGURE 6-27 Dipole-induced dipole interaction. The positive pole of a water molecule causes a temporary change in the electron distribution of an oxygen molecule. The negative pole induced in the oxygen molecule is then attracted to the positive pole of the water molecule.

of an adjacent oxygen molecule. The oxygen molecule, then, has an induced negative pole on the side toward the water molecule and an induced positive pole on the opposite side. The result is an attraction to the water molecule, as shown in Figure 6-27.

Hydrogen Bonding

A particularly strong type of dipole-dipole force explains the unusually high boiling points of some hydrogen-containing compounds, such as hydrogen fluoride, HF, water, H_2O, and ammonia, NH_3. In such compounds, the large electronegativity differences between hydrogen atoms and fluorine, oxygen, or nitrogen atoms make the bonds connecting them highly polar. This gives the hydrogen atom a positive charge that is almost half as large as that of a proton. Moreover, the small size of the hydrogen atom allows the atom to come very close to an unshared pair of electrons on an adjacent molecule. *The intermolecular force in which a hydrogen atom that is bonded to a highly electronegative atom is attracted to an unshared pair of electrons of an electronegative atom in a nearby molecule is known as* **hydrogen bonding.**

Hydrogen bonds are usually represented by dotted lines connecting the hydrogen-bonded hydrogen to the unshared electron pair of the electronegative atom to which it is attracted, as illustrated for water in Figure 6-28. The effect of hydrogen bonding can be seen by comparing the boiling points in Table 6-7 on page 190. Look at phosphine, PH_3, compared with hydrogen-bonded ammonia, NH_3. How does hydrogen sulfide, H_2S, compare with strongly hydrogen-bonded water, H_2O?

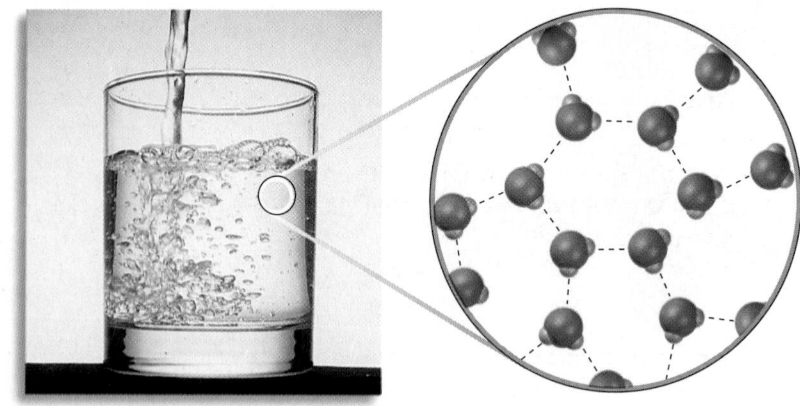

FIGURE 6-28 Space-filling models illustrate hydrogen bonding between water molecules. The dotted lines indicate the attraction between electronegative oxygen atoms and electropositive hydrogen atoms of neighboring molecules.

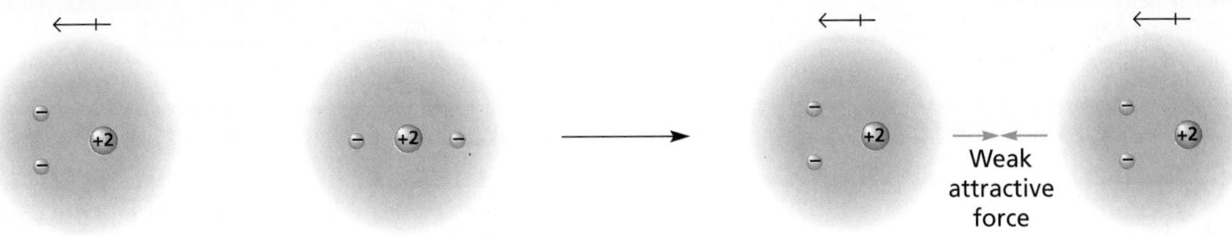

FIGURE 6-29 When an instantaneous, temporary dipole develops in a helium atom, it induces a dipole in a neighboring atom.

London Dispersion Forces

Even noble-gas atoms and molecules that are nonpolar experience a weak intermolecular attraction. In any atom or molecule—polar or nonpolar—the electrons are in continuous motion. As a result, at any instant the electron distribution may be slightly uneven. The momentary, uneven charge creates a positive pole in one part of the atom or molecule and a negative pole in another. This temporary dipole can then induce a dipole in an adjacent atom or molecule. The two are held together for an instant by the weak attraction between the temporary dipoles, as illustrated in Figure 6-29. *The intermolecular attractions resulting from the constant motion of electrons and the creation of instantaneous dipoles are called* **London dispersion forces,** after Fritz London, who first proposed their existence in 1930.

London forces act between all atoms and molecules. But they are the *only* intermolecular forces acting among noble-gas atoms, nonpolar molecules, and slightly polar molecules. This fact is reflected in the low boiling points of the noble gases and nonpolar molecular compounds listed in Table 6-7. Because London forces are dependent on the motion of electrons, their strength increases with the number of electrons in the interacting atoms or molecules. In other words, London forces increase with increasing atomic or molar mass. This trend can be seen by comparing the boiling points of the gases helium, He, and argon, Ar; hydrogen, H_2, and oxygen, O_2; and chlorine, Cl_2, and bromine, Br_2.

SECTION REVIEW

1. What two theories can be used to predict molecular geometry?
2. Draw the Lewis structure, and use the VSEPR theory to predict the molecular geometry of the following molecules:
   a. SO_2 b. CI_4 c. BCl_3
3. What are some factors that affect the geometry of a molecule?
4. Explain what is meant by sp^3 hybridization.
5. What type of intermolecular force contributes to the high boiling point of water? Explain.

CHAPTER 6 REVIEW

CHAPTER SUMMARY

6-1

- Most atoms are chemically bonded to other atoms.
- The three major types of chemical bonding are ionic, covalent, and metallic.
- In general, atoms of metals bond ionically with atoms of nonmetals, atoms of metals bond metallically with each other, and atoms of nonmetals bond covalently with each other.

Vocabulary

chemical bond (161)
covalent bonding (161)
ionic bonding (161)
nonpolar-covalent bond (162)
polar (162)
polar-covalent bond (162)

6-2

- Atoms in molecules are joined by covalent bonds.
- The bond length between two atoms in a molecule is the distance at which the potential energy of the bonded atoms is minimized.
- The octet rule states that many chemical compounds tend to form bonds so that each atom, by gaining, losing, or sharing electrons, shares or has eight electrons in its highest occupied energy level.
- A single bond is a covalent bond in which a pair of electrons is shared between two atoms. Covalent bonds in which more than one pair of electrons are shared are called multiple bonds.
- Bonding within many molecules and ions can be indicated by a Lewis structure. Molecules or ions that cannot be correctly represented by a single Lewis structure are represented by resonance structures.

Vocabulary

bond energy (167)
bond length (167)
chemical formula (164)
diatomic molecule (164)
double bond (172)
electron-dot notation (170)
Lewis structure (171)
lone pair (171)
molecular compound (164)
molecular formula (164)
molecule (164)
multiple bond (173)
octet rule (169)
resonance (175)
single bond (171)
structural formula (171)
triple bond (173)
unshared pair (171)

6-3

- An ionic compound is a three-dimensional network of positive and negative ions mutually attracted to one another.
- Because of the strong attraction between positive and negative ions, ionic compounds tend to be harder and more brittle and to have higher boiling points than materials containing only covalently bonded atoms.
- Polyatomic ions are charged groups of atoms held together by covalent bonds.

Vocabulary

formula unit (176)
ionic compound (176)
lattice energy (178)
polyatomic ion (180)

6-4

- Metallic bonding is a type of chemical bonding that results from the attraction between metal atoms and a surrounding sea of mobile electrons.
- The electron sea formed in metallic bonding gives metals their properties of high electrical and thermal conductivity, malleability, ductility, and luster.

Vocabulary

ductility (182)
malleability (182)
metallic bonding (181)

CHAPTER SUMMARY (continued)

6-5
- VSEPR theory is used to predict the shapes of molecules based on the fact that electron pairs strongly repel each other and tend to be oriented as far apart as possible.
- Hybridization theory is used to predict the shapes of molecules based on the fact that orbitals within an atom can mix to form orbitals of equal energy.
- Intermolecular forces, such as dipole-dipole forces and London dispersion forces, exist between certain types of molecules. Hydrogen bonding is a special case of dipole-dipole forces.

Vocabulary

dipole (190)
dipole-dipole force (190)
hybrid orbital (188)
hybridization (187)
hydrogen bonding (192)
intermolecular force (189)
London dispersion force (193)
molecular polarity (183)
VSEPR theory (183)

REVIEWING CONCEPTS

1. What is a chemical bond? (6-1)

2. Identify and define the three major types of chemical bonding. (6-1)

3. What is the relationship between electronegativity and the ionic character of a chemical bond? (6-1)

4. a. What is the meaning of the term *polar*, as applied to chemical bonding?
b. Distinguish between polar-covalent and nonpolar-covalent bonds. (6-1)

5. In general, what determines whether atoms will form chemical bonds? (6-1)

6. What is a molecule? (6-2)

7. a. What determines bond length?
b. In general, how are bond energies and bond lengths related? (6-2)

8. Describe the general location of the electrons in a covalent bond. (6-2)

9. As applied to covalent bonding, what is meant by an unshared or lone pair of electrons? (6-2)

10. Describe the octet rule in terms of noble-gas configurations and potential energy. (6-2)

11. Determine the number of valence electrons in an atom of each of the following elements:
a. H b. F
c. Mg d. O
e. Al f. N
g. C (6-2)

12. When drawing Lewis structures, which atom is usually the central atom? (6-2)

13. Distinguish between single, double, and triple covalent bonds by defining each and providing an illustration of each type. (6-2)

14. In writing Lewis structures, how is the need for multiple bonds generally determined? (6-2)

15. a. What is an ionic compound?
b. In what form do most ionic compounds occur? (6-3)

16. a. What is a formula unit?
b. What are the components of one formula unit of CaF_2? (6-3)

17. a. What is lattice energy?
b. In general, what is the relationship between lattice energy and the strength of ionic bonding? (6-3)

18. a. In general, how do ionic and molecular compounds compare in terms of melting points, boiling points, and ease of vaporization?
b. What accounts for the observed differences in the properties of ionic and molecular compounds?
c. Cite three physical properties of ionic compounds. (6-3)

19. a. What is a polyatomic ion?
b. Give two examples of polyatomic ions.

c. In what form do such ions often occur in nature? (6-3)

20. a. How do the properties of metals differ from those of both ionic and molecular compounds?
b. What specific property of metals accounts for their unusual electrical conductivity? (6-4)

21. What properties of metals contribute to their tendency to form metallic bonds? (6-4)

22. a. What is metallic bonding?
b. How can the strength of metallic bonding be measured? (6-4)

23. a. How is the VSEPR theory used to classify molecules?
b. What molecular geometry would be expected for F_2 and HF? (6-5)

24. According to the VSEPR theory, what molecular geometries are associated with the following types of molecules?
a. AB_2
b. AB_3
c. AB_4
d. AB_5
e. AB_6 (6-5)

25. Describe the role of each of the following in predicting molecular geometries:
a. unshared electron pairs
b. double bonds (6-5)

26. a. What are hybrid orbitals?
b. What determines the number of hybrid orbitals produced by an atom? (6-5)

27. a. What are intermolecular forces?
b. In general, how do these forces compare in strength with those in ionic and metallic bonding?
c. Where are the strongest intermolecular forces found? (6-5)

28. What is the relationship between electronegativity and the polarity of a chemical bond? (6-5)

29. a. What are dipole-dipole forces?
b. What determines the polarity of a molecule? (6-5)

30. a. What is meant by an induced dipole?
b. What is the everyday importance of this type of intermolecular force? (6-5)

31. a. What is hydrogen bonding?
b. What accounts for its extraordinary strength? (6-5)

32. What are London dispersion forces? (6-5)

PROBLEMS

Chemical Bond Character

33. Determine the electronegativity difference, the probable bond type, and the more-electronegative atom with respect to bonds formed between the following pairs of atoms. (Hint: See Sample Problem 6-1.)
a. H and I
b. S and O
c. K and Br
d. Si and Cl
e. H and F
f. Se and S
g. C and H

34. List the bonding pairs described in item 33 in order of increasing covalent character.

35. Use orbital notation to illustrate the bonding in each of the following molecules:
a. chlorine, Cl_2
b. oxygen, O_2
c. hydrogen fluoride, HF

36. The lattice energy of sodium chloride, NaCl, is −787.5 kJ/mol. The lattice energy of potassium chloride, KCl, is −715 kJ/mol. In which compound is the bonding between ions stronger? Why?

Electron-Dot Notation and Lewis Structures

37. Use electron-dot notation to illustrate the number of valence electrons present in one atom of each of the following elements. (Hint: See Sample Problem 6-2.)
a. Li
b. Ca
c. Cl
d. O

e. C
f. P
g. Al
h. S

38. Use electron-dot structures to demonstrate the formation of ionic compounds involving the following elements:
a. Na and S
b. Ca and O
c. Al and S

39. Draw Lewis structures for each of the following molecules. (Hint: See Sample Problem 6-4.)
a. contains one C and four F atoms
b. contains two H and one Se atom
c. contains one N and three I atoms
d. contains one Si and four Br atoms
e. contains one C, one Cl, and three H atoms

40. Determine the type of hybrid orbitals formed by the boron atom in a molecule of boron fluoride, BF_3. (Hint: See Sample Problems 6-5 and 6-6.)

41. Draw Lewis structures for each of the following molecules. Show resonance structures, if they exist.
a. O_2
b. N_2
c. CO
d. SO_2

42. Draw Lewis structures for each of the following polyatomic ions. Show resonance structures, if they exist.
a. OH^-
b. $H_3C_2O_2^-$
c. BrO_3^-

VSEPR Theory and Molecular Geometry

43. According to the VSEPR theory, what molecular geometries are associated with the following types of molecules?
a. AB_3E
b. AB_2E_2
c. AB_2E

44. Use hybridization to explain the bonding in methane, CH_4.

45. For each of the following polar molecules, indicate the direction of the resulting dipole:
a. H–F
b. H–Cl
c. H–Br
d. H–I

46. Determine whether each of the following bonds would be polar or nonpolar:
a. H–H
b. H–O
c. H–F
d. Br–Br
e. H–Cl
f. H–N

47. On the basis of individual bond polarity and orientation, determine whether each of the following molecules would be polar or nonpolar:
a. H_2O
b. I_2
c. CF_4
d. NH_3
e. CO_2

48. Draw a Lewis structure for each of the following molecules, and then use the VSEPR theory to predict the molecular geometry of each:
a. SCl_2
b. PI_3
c. Cl_2O
d. NH_2Cl
e. $SiCl_3Br$
f. ONCl

49. Draw a Lewis structure for each of the following polyatomic ions, and then use VSEPR theory to determine the geometry of each:
a. NO_3^-
b. NH_4^+
c. SO_4^{2-}
d. ClO_2^-

MIXED REVIEW

50. Arrange the following pairs from strongest to weakest attraction:
a. polar molecule and polar molecule
b. nonpolar molecule and nonpolar molecule
c. polar molecule and ion
d. ion and ion

51. Determine the geometry of the following molecules:
a. CCl_4
b. $BeCl_2$
c. PH_3

52. What types of atoms tend to form the following types of bonding?
a. ionic
b. covalent
c. metallic

53. What happens to the energy level and stability of two bonded atoms when they are separated and become individual atoms?

54. Draw the three resonance structures for sulfur trioxide, SO_3.

55. a. How do ionic and covalent bonding differ?
b. How does an ionic compound differ from a molecular compound?
c. How does an ionic compound differ from a metal?

56. Write the electron-dot notation for each of the following elements:
a. He
b. Cl
c. O
d. P
e. B

57. Write the structural formula for methanol, CH_3OH.

58. How many K^+ and S^{2-} ions would be in one formula unit of the ionic compound formed by these ions?

59. Explain metallic bonding in terms of the sparsely populated outermost orbitals of metal atoms.

60. Explain the role of molecular geometry in determining molecular polarity.

61. How does the energy level of a hybrid orbital compare with the energy levels of the orbitals it was formed from?

62. Aluminum's heat of vaporization is 284 kJ/mol. Beryllium's heat of vaporization is 224 kJ/mol. In which element is the bonding stronger between atoms?

63. Determine the electronegativity difference, the probable bonding type, and the more-electronegative atom for each of the following pairs of atoms:
a. Zn and O
b. Br and I
c. S and Cl

64. Draw the Lewis structure for each of the following molecules:
a. PCl_3
b. CCl_2F_2
c. CH_3NH_2

65. Write the Lewis structure for $BeCl_2$. (Hint: Beryllium atoms do not follow the octet rule.)

66. Draw a Lewis structure for each of the following polyatomic ions and determine their geometries:
a. NO_2^-
b. NO_3^-
c. NH_4^+

67. Why are most atoms chemically bonded to other atoms in nature?

CRITICAL THINKING

68. Inferring Relationships The length of a bond varies depending on the type of bond formed. Predict and compare the lengths of the carbon-carbon bonds in the following molecules. Explain your answer. (Hint: See Table 6-2.)

```
 H H          H H
 | |          | |
H-C-C-H     H-C=C-H     H-C≡C-H
 | |
 H H
```

C_2H_6	C_2H_4	C_2H_2

TECHNOLOGY & LEARNING

69. Graphing Calculator Classify Bonding Type According to Difference in Electronegativity

The graphing calculator can be programmed to classify bonding between atoms according to the difference between the atoms' electro-negativities. Begin by programming the calculator to determine the electronegativity difference between the bonded atoms. Then use the program to classify the bonding type.

A. Program the classification.

Keystrokes for naming the program: [PRGM] [▶] [▶] [ENTER]

Name the program: [B] [O] [N] [D] [T] [Y] [P] [E] [ENTER]

Program keystrokes: [PRGM] [▶] [3] [ALPHA] ["]
[ALPHA] [E] [1] [ALPHA] ["] [2nd] [:] [PRGM] [▶]
[1] [ALPHA] [A] [2nd] [:] [PRGM] [▶] [3] [ALPHA]
["] [ALPHA] [E] [2] [ALPHA] ["] [2nd] [:] [PRGM]
[▶] [1] [ALPHA] [B] [2nd] [:] [2nd] [ABS] [(]
[ALPHA] [A] [−] [ALPHA] [B] [)] [STO▸] [ALPHA] [C]
[2nd] [:] [PRGM] [1] [ALPHA] [C] [2nd] [TEST] [3]
[1] [.] [7] [2nd] [:] [PRGM] [▶] [3] [2nd]
[ALPHA] ["] [I] [O] [N] [I] [C] ["] [2nd]
[:] [PRGM] [1] [1] [.] [7] [2nd] [TEST] [4]
[ALPHA] [C] [2nd] [TEST] [▶] [1] [ALPHA] [C] [2nd] [TEST]
[4] [.] [3] [2nd] [:] [PRGM] [▶] [3] [2nd]
[ALPHA] ["] [P] [O] [L] [A] [R] ["] [2nd]
[:] [PRGM] [1] [ALPHA] [C] [2nd] [TEST] [5] [.]
[3] [2nd] [:] [PRGM] [▶] [3] [2nd] [ALPHA] ["]
[N] [O] [N] [P] [O] [L] [A] [R] ["]

B. Check the display.

Screen display: Disp " E1": Input A: Disp "E2": Input B: abs (A–B)->C: If C>1.7: Disp "IONIC": If 1.7≥C and C≥.3: Disp "POLAR": If C<.3: Disp "NONPOLAR"

Press [2nd] [QUIT] to exit the program editor.

C. Run the program.

a. Press [PRGM]. Select BONDTYPE and press [ENTER] [ENTER]. Now enter the electronegativity of the first element, 3, and press [ENTER]. Enter the electronegativity of the second element, 1, and press [ENTER]. The calculator will classify the bond type.

b. Use the BONDTYPE program to classify the bond between carbon (electronegativity of 2.5) and hydrogen (electronegativity of 2.1).

c. Use the BONDTYPE program to classify the bonds between several elements within the periodic table.

HANDBOOK SEARCH

70. Figure 6-18 on page 181 shows a model for a body-centered cubic crystal. Review the Properties tables for all of the metals in the *Elements Handbook*. What metals exist in body-centered cubic structures?

71. Group 14 of the *Elements Handbook* contains a discussion of the band theory of metals. How does this model explain the electrical conductivity of metals?

RESEARCH & WRITING

72. Prepare a report on the work of Linus Pauling.

a. Discuss his work on the nature of the chemical bond.

b. Linus Pauling was an advocate of the use of vitamin C as a preventative for colds. Evaluate Pauling's claims. Determine if there is any scientific evidence that indicates whether vitamin C helps prevent colds.

73. Covalently bonded solids, such as silicon, an element used in computer components, are harder than pure metals. Research theories that explain the hardness of covalently bonded solids and their usefulness in the computer industry. Present your findings to the class.

ALTERNATIVE ASSESSMENT

74. Devise a set of criteria that will allow you to classify the following substances as ionic or nonionic: $CaCO_3$, Cu, H_2O, NaBr, and C (graphite). Show your criteria to your instructor.

75. Performance Assessment Identify 10 common substances in and around your home, and indicate whether you would expect these substances to contain ionic, covalent, or metallic bonds.

UNIT 3

Language of Chemistry

CHAPTERS

JOHN JOSEPH GRIFFIN, FROM HIS RATIONALE FOR CHEMICAL STUDY IN *CHEMICAL RECREATIONS*

The hearing of lectures, and the reading of books, will never benefit him who attends to nothing else; for Chemistry can only be studied to advantage practically. One experiment, well-conducted, and carefully observed by the student, from first to last, will afford more knowledge than the mere perusal of a whole volume. It may be added to this, that chemical operations are, in general, the most interesting that could possibly be devised—Reader! what is more requisite to induce you to MAKE EXPERIMENTS?

(From *The Norton History of Chemistry*)

CHAPTER 7

Chemical Formulas and Chemical Compounds

Chemists use chemical names and formulas to describe the atomic composition of compounds.

Chemical Names and Formulas

SECTION 7-1

OBJECTIVES

- Explain the significance of a chemical formula.
- Determine the formula of an ionic compound formed between two given ions.
- Name an ionic compound given its formula.
- Using prefixes, name a binary molecular compound from its formula.
- Write the formula of a binary molecular compound given its name.

The total number of natural and synthetic chemical compounds runs in the millions. For some of these substances, certain common names remain in everyday use. For example, calcium carbonate is better known as limestone, and sodium chloride is usually referred to simply as table salt. And most people recognize hydrogen oxide by its popular name, water.

Unfortunately, common names usually give no information about chemical composition. To describe the atomic makeup of compounds, chemists use systematic methods for naming compounds and for writing chemical formulas. In this chapter, you will be introduced to some of the rules used to identify simple chemical compounds.

Significance of a Chemical Formula

Recall that a chemical formula indicates the relative number of atoms of each kind in a chemical compound. For a molecular compound, the chemical formula reveals the number of atoms of each element contained in a single molecule of the compound, as shown below for the hydrocarbon octane. (*Hydrocarbons* are molecular compounds composed solely of carbon and hydrogen.)

$$C_8H_{18}$$

Subscript indicates that there are 8 carbon atoms in a molecule of octane.

Subscript indicates that there are 18 hydrogen atoms in a molecule of octane.

Unlike a molecular compound, an ionic compound consists of a network of positive and negative ions held together by mutual attraction. The chemical formula for an ionic compound represents one formula unit—the simplest ratio of the compound's positive ions (cations) and its negative ions (anions). The chemical formula for aluminum sulfate, an ionic compound consisting of aluminum cations and polyatomic sulfate anions, is written as shown on page 204.

$Al_2(SO_4)_3$

Subscript 2 refers to 2 aluminum atoms.

Subscript 4 refers to 4 oxygen atoms in sulfate ion.

Subscript 3 refers to everything inside parentheses giving 3 sulfate ions, with a total of 3 sulfur atoms and 12 oxygen atoms.

Note how the parentheses are used. They surround the polyatomic anion to identify it as a unit. The subscript 3 refers to the entire unit. Notice also that there is no subscript written next to the symbol for sulfur. When there is no subscript written next to an atom's symbol, the value of the subscript is understood to be 1.

Monatomic Ions

By gaining or losing electrons, many main-group elements form ions with noble-gas configurations. For example, Group 1 metals lose one electron to give 1+ cations, such as Na^+. Group 2 metals lose two electrons to give 2+ cations, such as Mg^{2+}. *Ions formed from a single atom are known as* **monatomic ions.** The nonmetals of Groups 15, 16, and 17 gain electrons to form anions. For example, in ionic compounds nitrogen forms the 3– anion, N^{3-}. The three added electrons plus the five outermost electrons in nitrogen atoms give a completed outermost octet. Similarly, the Group 16 elements oxygen and sulfur form 2– anions, and the Group 17 halogens form 1– anions.

Not all main-group elements readily form ions, however. Rather than gain or lose electrons, atoms of carbon and silicon form covalent bonds in which they share electrons with other atoms. Other elements tend to form ions that do not have noble-gas configurations. For instance, it is difficult for the Group 14 metals tin and lead to lose four electrons to achieve a noble-gas configuration. Instead, they tend to lose the two electrons in their outer *p* orbitals but retain the two electrons in their outer *s* orbitals to form 2+ cations. (Tin and lead can also form molecular compounds in which all four valence electrons are involved in covalent bonding.)

Elements from the *d*-block form 2+, 3+, or, in a few cases, 1+ or 4+ cations. Many *d*-block elements form two ions of different charges. For example, copper forms 1+ cations and 2+ cations. Iron and chromium, on the other hand, each form 2+ cations as well as 3+ cations. And vanadium and lead commonly form 2+, 3+, and 4+ cations.

Naming cations

K^+

Potassium cation

Mg^{2+}

Magnesium cation

Naming Monatomic Ions

Monatomic cations are identified simply by the element's name, as illustrated by the examples at left. Naming monatomic anions is slightly more complicated. First, the ending of the element's name is dropped. Then the ending *-ide* is added to the root name, as illustrated by the examples at right.

The names and symbols of the common monatomic cations and anions are organized according to their charges in Table 7-1. The names of many of the ions in the table include Roman numerals. These numerals are part of the *Stock system* of naming chemical ions and elements. You will read more about the stock system and other systems of naming chemicals later in this chapter.

Naming anions

F	F^-
Fluor*ine*	Fluor*ide* anion
N	N^{3-}
Nitr*ogen*	Nitr*ide* anion

TABLE 7-1 *Some Common Monatomic Ions*

Main-group elements

1+		2+		3+	
lithium	Li^+				
sodium	Na^+	magnesium	Mg^{2+}	aluminum	Al^{3+}
potassium	K^+	calcium	Ca^{2+}		
rubidium	Rb^+	strontium	Sr^{2+}		
cesium	Cs^+	barium	Ba^{2+}		

1–		2–		3–	
fluoride	F^-	oxide	O^{2-}	nitride	N^{3-}
chloride	Cl^-	sulfide	S^{2-}		
bromide	Br^-				
iodide	I^-				

***d*-Block elements**

1+		2+		3+		4+	
copper(I)	Cu^+	cadmium	Cd^{2+}	chromium(III)	Cr^{3+}	lead(IV)	Pb^{4+}
silver	Ag^+	chromium(II)	Cr^{2+}	iron(III)	Fe^{3+}	vanadium(IV)	V^{4+}
		cobalt(II)	Co^{2+}	lead(III)	Pb^{3+}	tin(IV)	Sn^{4+}
		copper(II)	Cu^{2+}	vanadium(III)	V^{3+}		
		iron(II)	Fe^{2+}				
		lead(II)	Pb^{2+}				
		manganese(II)	Mn^{2+}				
		mercury(II)	Hg^{2+}				
		nickel(II)	Ni^{2+}				
		tin(II)	Sn^{2+}				
		vanadium(II)	V^{2+}				
		zinc	Zn^{2+}				

Binary Ionic Compounds

Compounds composed of two different elements are known as **binary compounds.** In a binary ionic compound, the total numbers of positive charges and negative charges must be equal. Therefore, the formula for such a compound can be written given the identities of the compound's ions. For example, magnesium and bromine combine to form the ionic compound magnesium bromide. Magnesium, a Group 2 metal, forms the Mg^{2+} cation. Bromine, a halogen, forms the Br^- anion when combined with a metal. In each formula unit of magnesium bromide, two Br^- anions are required to balance the 2+ charge of the Mg^{2+} cation. The compound's formula must therefore indicate one Mg^{2+} cation and two Br^- anions. The symbol for the cation is written first.

Ions combined: Mg^{2+}, Br^-, Br^- *Chemical formula:* $MgBr_2$

Note that the charges of the ions are not included in the formula. This is usually the case when writing formulas for binary ionic compounds.

As an aid to determining subscripts in formulas for ionic compounds, the positive and negative charges can be "crossed over." Crossing over is a method of balancing the charges between ions in an ionic compound. For example, the formula for the compound formed by the aluminum ion, Al^{3+}, and the oxide ion, O^{2-}, is determined as follows.

1. *Write the symbols for the ions side by side. Write the cation first.*

$$Al^{3+} \quad O^{2-}$$

2. *Cross over the charges by using the absolute value of each ion's charge as the subscript for the other ion.*

$$Al_2^{3+} \quad O_3^{2-}$$

3. *Check the subscripts and divide them by their largest common factor to give the smallest possible whole-number ratio of ions. Then write the formula.*
   Multiplying the charge by the subscript shows that the charge on two Al^{3+} cations ($2 \times 3+ = 6+$) equals the charge on three O^{2-} anions ($3 \times 2- = 6-$). The largest common factor of the subscripts is 1. The correct formula is therefore written as follows.

$$Al_2O_3$$

Naming Binary Ionic Compounds

The **nomenclature,** *or naming system,* of binary ionic compounds involves combining the names of the compound's positive and negative ions. The name of the cation is given first, followed by the name of the anion. For most simple ionic compounds, the ratio of the ions is not indicated in the compound's name because it is understood based on the relative charges of the compound's ions. The naming of a simple binary ionic compound is illustrated below.

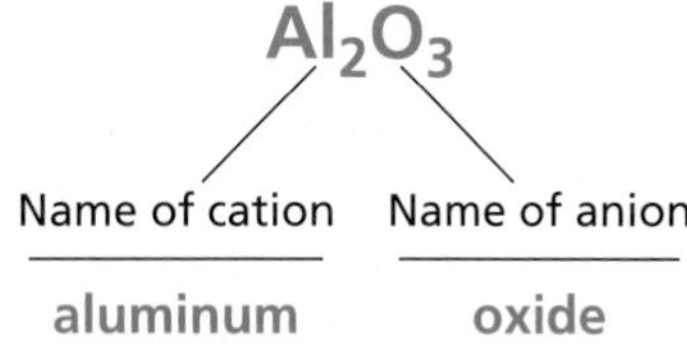

SAMPLE PROBLEM 7-1

Write the formulas for the binary ionic compounds formed between the following elements:
a. zinc and iodine **b. zinc and sulfur**

SOLUTION *Write the symbols for the ions side by side. Write the cation first.*

a. $Zn^{2+}\ I^{-}$
b. $Zn^{2+}\ S^{2-}$

Cross over the charges to give subscripts.

a. $Zn_{1}^{2+}\ I_{2}^{-}$
b. $Zn_{2}^{2+}\ S_{2}^{2-}$

Check the subscripts and divide them by their largest common factor to give the smallest possible whole-number ratio of ions. Then write the formula.

a. The subscripts are mathematically correct because they give equal total charges of $1 \times 2+ = 2+$ and $2 \times 1- = 2-$. The largest common factor of the subscripts is 1. The smallest possible whole-number ratio of ions in the compound is therefore 1:2. The subscript 1 is not written, so the formula is ZnI_2.

b. The subscripts are mathematically correct because they give equal total charges of $2 \times 2+ = 4+$ and $2 \times 2- = 4-$. The largest common factor of the subscripts is 2. The smallest whole-number ratio of ions in the compound is therefore 1:1. The correct formula is ZnS.

PRACTICE

1. Write formulas for the binary ionic compounds formed between the following elements:
a. potassium and iodine
b. magnesium and chlorine
c. sodium and sulfur
d. aluminum and sulfur
e. aluminum and nitrogen

Answer
1. a. KI
b. $MgCl_2$
c. Na_2S
d. Al_2S_3
e. AlN

2. Name the binary ionic compounds indicated by the following formulas:
a. AgCl
b. ZnO
c. $CaBr_2$
d. SrF_2
e. BaO
f. $CaCl_2$

2. a. silver chloride
b. zinc oxide
c. calcium bromide
d. strontium fluoride
e. barium oxide
f. calcium chloride

The Stock System of Nomenclature

Some elements, such as iron, form two or more cations with different charges. To distinguish the ions formed by such elements, the Stock system of nomenclature is used. This system uses a Roman numeral to indicate an ion's charge. The numeral is enclosed in parentheses and placed *immediately* after the metal name.

Fe^{2+}	Fe^{3+}
Iron(II)	Iron(III)

Names of metals that commonly form only one cation *do not* include a Roman numeral.

Na^{+}	Ba^{2+}	Al^{3+}
Sodium	Barium	Aluminum

There is no element that commonly forms more than one monatomic anion.

Naming a binary ionic compound according to the Stock system is illustrated below.

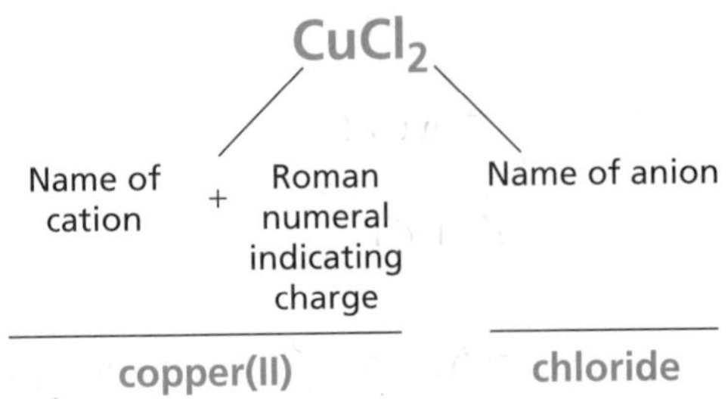

FIGURE 7-1 Different cations of the same metal form different compounds even when they combine with the same anion. Compare (a) lead(IV) oxide, PbO_2, with (b) lead(II) oxide, PbO.

SAMPLE PROBLEM 7-2

Write the formula and give the name for the compound formed by the ions Cr^{3+} and F^-.

SOLUTION *Write the symbols for the ions side by side. Write the cation first.*

$$Cr^{3+} \quad F^-$$

Cross over the charges to give subscripts.

$$Cr_1^{3+} \quad F_3^-$$

Check the subscripts and write the formula.

The subscripts are correct because they give charges of $1 \times 3+ = 3+$ and $3 \times 1- = 3-$. The largest common factor of the subscripts is 1, so the smallest whole-number ratio of the ions is 1:3. The formula is therefore CrF_3. As Table 7-1 on page 205 shows, chromium forms more than one ion. Therefore, the name of the 3+ chromium ion must be followed by a Roman numeral indicating its charge. The compound's name is chromium(III) fluoride.

PRACTICE

1. Write the formula and give the name for the compounds formed between the following ions:
   a. Cu^{2+} and Br^-
   b. Fe^{2+} and O^{2-}
   c. Pb^{2+} and Cl^-
   d. Hg^{2+} and S^{2-}
   e. Sn^{2+} and F^-
   f. Fe^{3+} and O^{2-}

2. Give the names for the following compounds:
   a. CuO
   b. CoF_3
   c. SnI_4
   d. FeS

Answer

1. a. $CuBr_2$, copper(II) bromide
   b. FeO, iron(II) oxide
   c. $PbCl_2$, lead(II) chloride
   d. HgS, mercury(II) sulfide
   e. SnF_2, tin(II) fluoride
   f. Fe_2O_3, iron(III) oxide

2. a. copper(II) oxide
   b. cobalt(III) fluoride
   c. tin(IV) iodide
   d. iron(II) sulfide

Compounds Containing Polyatomic Ions

Table 7-2 on page 210 lists some common polyatomic ions. All but the ammonium ion are negatively charged and most are **oxyanions**—*polyatomic ions that contain oxygen.* In several cases, two different oxyanions are formed by the same two elements. Nitrogen and oxygen, for example, are combined in both NO_3^- and NO_2^-. When naming compounds containing such oxyanions, the most common ion is given the ending *-ate*. The ion with one less oxygen atom is given the ending *-ite*.

NO_2^-	NO_3^-
Nitrite	Nitrate

Sometimes two elements form more than two different oxyanions. In this case, an anion with one less oxygen than the *-ite* anion is given the

TABLE 7-2 *Some Polyatomic Ions*

1+		2+			
ammonium	NH_4^+	dimercury*	Hg_2^{2+}		
1–		**2–**		**3–**	
acetate	CH_3COO^-	carbonate	CO_3^{2-}	phosphate	PO_4^{3-}
bromate	BrO_3^-	chromate	CrO_4^{2-}	arsenate	AsO_4^{3-}
chlorate	ClO_3^-	dichromate	$Cr_2O_7^{2-}$		
chlorite	ClO_2^-	hydrogen phosphate	HPO_4^{2-}		
cyanide	CN^-	oxalate	$C_2O_4^{2-}$		
dihydrogen phosphate	$H_2PO_4^-$	peroxide	O_2^{2-}		
hydrogen carbonate (bicarbonate)	HCO_3^-	sulfate	SO_4^{2-}		
hydrogen sulfate	HSO_4^-	sulfite	SO_3^{2-}		
hydroxide	OH^-				
hypochlorite	ClO^-				
nitrate	NO_3^-				
nitrite	NO_2^-				
perchlorate	ClO_4^-				
permanganate	MnO_4^-				

*The mercury(I) cation exists as two Hg^+ ions joined together by a covalent bond and is written as Hg_2^{2+}.

prefix *hypo-*. An anion with one more oxygen than the *-ate* anion is given the prefix *per-*. This nomenclature is illustrated by the four oxyanions formed between chlorine and oxygen.

ClO^-	ClO_2^-	ClO_3^-	ClO_4^-
Hypochlorite	Chlorite	Chlorate	Perchlorate

Compounds containing polyatomic ions are named in the same manner as binary ionic compounds. The name of the cation is given first, followed by the name of the anion. For example, the two compounds formed with silver by the nitrate and nitrite anions are named silver nitrate, $AgNO_3$, and silver nitrite, $AgNO_2$, respectively. When more than one polyatomic ion is present in a compound, the formula for the entire ion is surrounded by parentheses. This is illustrated on page 204 for aluminum sulfate, $Al_2(SO_4)_3$. The formula indicates that an aluminum sulfate formula unit has 2 aluminum cations and 3 sulfate anions.

SAMPLE PROBLEM 7-3

Write the formula for tin(IV) sulfate.

SOLUTION *Write the symbols for the ions side by side. Write the cation first.*

$$Sn^{4+}\ SO_4^{2-}$$

Cross over the charges to give subscripts. Add parentheses around the polyatomic ion if necessary.

$$Sn_2^{4+}\ (SO_4)_4^{2-}$$

Check the subscripts and write the formula.
The total positive charge is $2 \times 4+ = 8+$. The total negative charge is $4 \times 2- = 8-$. The charges are equal. The largest common factor of the subscripts is 2, so the smallest whole-number ratio of ions in the compound is 1:2. The correct formula is therefore $Sn(SO_4)_2$.

PRACTICE

1. Write formulas for the following ionic compounds:
   a. sodium iodide
   b. calcium chloride
   c. potassium sulfide
   d. lithium nitrate
   e. copper(II) sulfate
   f. sodium carbonate
   g. calcium nitrite
   h. potassium perchlorate

2. Give the names for the following compounds:
   a. Ag_2O
   b. $Ca(OH)_2$
   c. $KClO_3$
   d. NH_4OH
   e. $FeCrO_4$
   f. $KClO$

Answer

1. a. NaI
   b. $CaCl_2$
   c. K_2S
   d. $LiNO_3$
   e. $CuSO_4$
   f. Na_2CO_3
   g. $Ca(NO_2)_2$
   h. $KClO_4$

2. a. silver oxide
   b. calcium hydroxide
   c. potassium chlorate
   d. ammonium hydroxide
   e. iron(II) chromate
   f. potassium hypochlorite

Naming Binary Molecular Compounds

Unlike ionic compounds, molecular compounds are composed of individual covalently bonded units, or molecules. Chemists use two nomenclature systems to name binary molecules. The newer system is the Stock system for naming molecular compounds, which requires an understanding of oxidation numbers. This system will be discussed in Section 7-2.

The old system of naming molecular compounds is based on the use of prefixes. For example, the molecular compound CCl_4 is named carbon *tetra*chloride. The prefix *tetra-* indicates that four chloride atoms are present in a single molecule of the compound. The two oxides of carbon, CO and CO_2, are named carbon *mono*xide and carbon *di*oxide, respectively.

TABLE 7-3 Numerical Prefixes

Number	Prefix
1	mono-
2	di-
3	tri-
4	tetra-
5	penta-
6	hexa-
7	hepta-
8	octa-
9	nona-
10	deca-

In these names the prefix *mon-* indicates one oxygen atom and the prefix *di-* indicates two oxygen atoms. The prefixes used to specify the number of atoms, or sometimes the number of *groups* of atoms, in a molecule are listed in Table 7-3.

The rules for the prefix system of nomenclature of binary molecular compounds are as follows.

1. The less-electronegative element is given first. It is given a prefix only if it contributes more than one atom to a molecule of the compound.
2. The second element is named by combining (a) a prefix indicating the number of atoms contributed by the element, (b) the root of the name of the second element, and (c) the ending *-ide*. With few exceptions, the ending *-ide* indicates that a compound contains only two elements.
3. The *o* or *a* at the end of a prefix is usually dropped when the word following the prefix begins with another vowel, e.g., monoxide or pentoxide.

The prefix system is illustrated below.

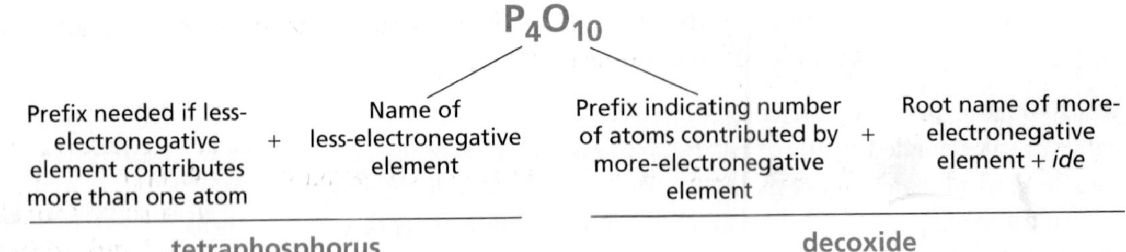

Because the less-electronegative element is written first, oxygen and the halogens are usually given second, as in carbon tetrachloride and the carbon oxides. In general, the order of nonmetals in binary compound names and formulas is C, P, N, H, S, I, Br, Cl, O, F.

TABLE 7-4 *Binary Compounds of Nitrogen and Oxygen*

Formula	Prefix-system name
N_2O	dinitrogen monoxide
NO	nitrogen monoxide
NO_2	nitrogen dioxide
N_2O_3	dinitrogen trioxide
N_2O_4	dinitrogen tetroxide
N_2O_5	dinitrogen pentoxide

The prefix system is illustrated further in Table 7-4, which lists the names of the six oxides of nitrogen. Note the application of rule 1, for example, in the name *nitrogen dioxide* for NO_2. No prefix is needed with *nitrogen* because only one atom of nitrogen, the less-electronegative element, is present in a molecule of NO_2. On the other hand, the prefix *di-* in *dioxide* is needed according to rule 2 to indicate the presence of two atoms of the more-electronegative element, oxygen. Take a moment to review the prefixes in the other names in Table 7-4.

SAMPLE PROBLEM 7-4

a. Give the name for As_2O_5.
b. Write the formula for oxygen difluoride.

SOLUTION

a. A molecule of the compound contains two arsenic atoms, so the first word in the name is "*di*arsenic." The five oxygen atoms are indicated by adding the prefix *pent-* to the word "oxide." The complete name is diarsenic pentoxide.

b. The first symbol in the formula is that for oxygen. Oxygen is first in the name because it is less electronegative than fluorine. Since there is no prefix, there must be only one oxygen atom. The prefix *di-* in *difluoride* shows that there are two fluorine atoms in the molecule. The formula is OF_2.

PRACTICE

1. Name the following binary molecular compounds:
a. SO_3
b. ICl_3
c. PBr_5

2. Write formulas for the following compounds:
a. carbon tetraiodide
b. phosphorus trichloride
c. dinitrogen trioxide

Answer

1. a. sulfur trioxide
b. iodine trichloride
c. phosphorus pentabromide

2. a. CI_4
b. PCl_3
c. N_2O_3

Covalent-Network Compounds

As you read in Chapter 6, some covalent compounds do not consist of individual molecules. Instead, each atom is joined to all its neighbors in a covalently bonded, three-dimensional network. There are no distinct units in these compounds, just as there are no such units in ionic compounds. The subscripts in a formula for a covalent-network compound indicate the smallest whole-number ratio of the atoms in the compound. Naming such compounds is similar to naming molecular compounds. Some common examples are given below.

SiC	SiO_2	Si_3N_4
Silicon carbide	Silicon dioxide	Trisilicon tetranitride

Acids and Salts

An *acid* is a distinct type of molecular compound about which you will read much more in Chapter 15. Most acids used in the laboratory can be classified as either binary acids or oxyacids. *Binary acids* are acids that consist of two elements, usually hydrogen and one of the halogens—fluorine, chlorine, bromine, iodine. *Oxyacids* are acids that contain hydrogen, oxygen, and a third element (usually a nonmetal).

Acids were first recognized as a specific class of compounds based on their properties in solutions of water. Consequently, in chemical nomenclature, the term *acid* usually refers to a solution in water of one of these special compounds rather than to the compound itself. For example, *hydrochloric acid* refers to a water solution of the molecular compound hydrogen chloride, HCl. Some common binary and oxyacids are listed in Table 7-5.

Many polyatomic ions are produced by the loss of hydrogen ions from oxyacids. A few examples of the relationship between oxyacids and oxyanions are shown below.

Sulfuric acid	H_2SO_4	Sulfate	SO_4^{2-}
Nitric acid	HNO_3	Nitrate	NO_3^-
Phosphoric acid	H_3PO_4	Phosphate	PO_4^{3-}

TABLE 7-5 *Common Binary Acids and Oxyacids*

HF	hydrofluoric acid	HNO_2	nitrous acid	HClO	hypochlorous acid
HCl	hydrochloric acid	HNO_3	nitric acid	$HClO_2$	chlorous acid
HBr	hydrobromic acid	H_2SO_3	sulfurous acid	$HClO_3$	chloric acid
HI	hydriodic acid	H_2SO_4	sulfuric acid	$HClO_4$	perchloric acid
H_3PO_4	phosphoric acid	CH_3COOH	acetic acid	H_2CO_3	carbonic acid

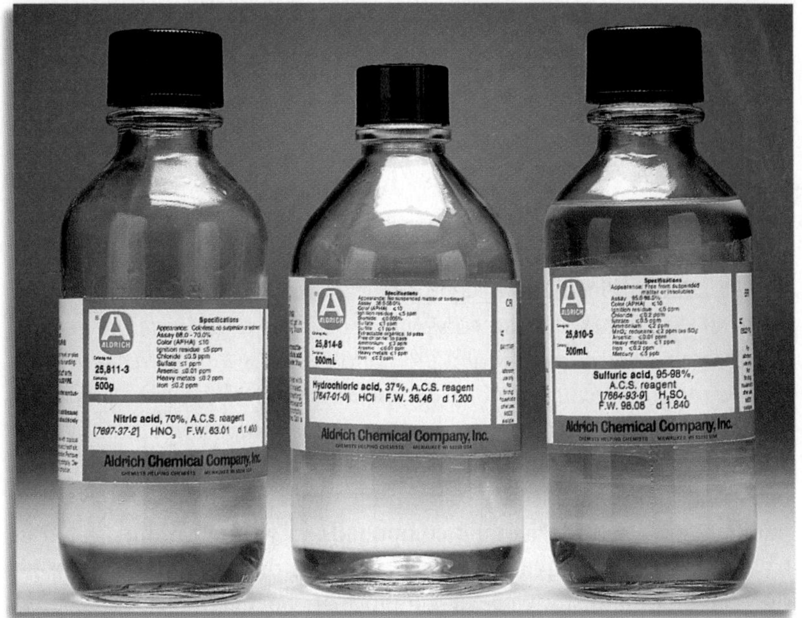

FIGURE 7-2 Some common laboratory acids. Acids should always be handled with care and according to instructions. They can burn the skin, and they burn holes in clothing.

An ionic compound composed of a cation and the anion from an acid is often referred to as a **salt.** Table salt, NaCl, contains the anion from hydrochloric acid. Calcium sulfate, $CaSO_4$, is a salt containing an anion from sulfuric acid. Some salts contain anions in which one or more hydrogen atoms from the acid are retained. Such anions are named by adding the word *hydrogen* or the prefix *bi-* to the anion name. The best known such anion comes from carbonic acid, H_2CO_3.

$$HCO_3^-$$

Hydrogen carbonate ion
Bicarbonate ion

SECTION REVIEW

1. What is the significance of a chemical formula?
2. Write formulas for the compounds formed between the following:
   a. aluminum and bromine
   b. sodium and oxygen
   c. magnesium and iodine
   d. Pb^{2+} and O^{2-}
   e. Sn^{2+} and I^-
   f. Fe^{3+} and S^{2-}
   g. Cu^{2+} and NO_3^-
   h. NH_4^+ and SO_4^{2-}
3. Name the following compounds using the Stock system:
   a. NaI
   b. MgS
   c. CaO
   d. K_2S
   e. CuBr
   f. $FeCl_2$
4. Write formulas for each of the following compounds:
   a. barium sulfide
   b. sodium hydroxide
   c. lead(II) nitrate
   d. potassium permanganate
   e. iron(II) sulfate
   f. diphosphorus trioxide
   g. disulfur dichloride
   h. carbon diselenide
   i. acetic acid
   j. chloric acid
   k. sulfurous acid
   l. phosphoric acid

SECTION 7-2

Oxidation Numbers

Objectives

- List the rules for assigning oxidation numbers.
- Give the oxidation number for each element in the formula of a chemical compound.
- Name binary molecular compounds using oxidation numbers and the Stock system.

The charges on the ions composing an ionic compound reflect the electron distribution of the compound. *In order to indicate the general distribution of electrons among the bonded atoms in a molecular compound or a molecular ion,* **oxidation numbers,** *also called* **oxidation states,** *are assigned to the atoms composing the compound or ion.* Unlike ionic charges, oxidation numbers do not have an exact physical meaning. In fact, in some cases they are quite arbitrary. However, oxidation numbers are useful in naming compounds, in writing formulas, and in balancing chemical equations. And, as will be discussed in Chapter 19, they are helpful in studying certain types of chemical reactions.

Assigning Oxidation Numbers

As a general rule in assigning oxidation numbers, shared electrons are assumed to belong to the more-electronegative atom in each bond. More specific rules for determining oxidation numbers are provided by the following guidelines.

1. The atoms in a pure element have an oxidation number of zero. For example, the atoms in pure sodium, Na, oxygen, O_2, phosphorus, P_4, and sulfur, S_8, all have oxidation numbers of zero.
2. The more-electronegative element in a binary molecular compound is assigned the number equal to the negative charge it would have as an anion. The less-electronegative atom is assigned the number equal to the positive charge it would have as a cation.
3. Fluorine has an oxidation number of –1 in all of its compounds because it is the most electronegative element.
4. Oxygen has an oxidation number of –2 in almost all compounds. Exceptions include when it is in peroxides, such as H_2O_2, in which its oxidation number is –1, and when it is in compounds with halogens, such as OF_2, in which its oxidation number is +2.
5. Hydrogen has an oxidation number of +1 in all compounds containing elements that are more-electronegative than it; it has an oxidation number of –1 in compounds with metals.
6. The algebraic sum of the oxidation numbers of all atoms in a neutral compound is equal to zero.
7. The algebraic sum of the oxidation numbers of all atoms in a polyatomic ion is equal to the charge of the ion.
8. Although rules 1 through 7 apply to covalently bonded atoms, oxidation numbers can also be assigned to atoms in ionic compounds.

A monatomic ion has an oxidation number equal to the charge of the ion. For example, the ions Na^+, Ca^{2+}, and Cl^- have oxidation numbers of +1, +2, and −1, respectively.

Let's examine the assignment of oxidation numbers to the atoms in two molecular compounds, hydrogen fluoride, HF, and water, H_2O. In HF the bond is polar, with a partial negative charge on the fluorine atom and a partial positive charge on the hydrogen atom. If HF were an ionic compound in which an electron was fully transferred to the fluorine atom, H would have a 1+ charge and F would have a 1− charge. Thus, the oxidation numbers of H and F in hydrogen fluoride are +1 and −1, respectively. In a water molecule, the oxygen atom is more electronegative than the hydrogen atoms. If H_2O were an ionic compound, the oxygen atom would have a charge of 2− and the hydrogen atoms would each have a charge of 1+. The oxidation numbers of H and O in water are therefore 1+ and 2−, respectively.

Because the sum of the oxidation numbers of the atoms in a compound must satisfy rule 6 or 7 of the guidelines on page 216, it is often possible to assign oxidation numbers when they are not known. This is illustrated in Sample Problem 7-5.

SAMPLE PROBLEM 7-5

Assign oxidation numbers to each atom in the following compounds or ions:

a. UF_6

b. H_2SO_4

c. ClO_3^-

SOLUTION a. Start by placing known oxidation numbers above the appropriate elements. From the guidelines, we know that fluorine always has an oxidation number of −1.

$$\overset{\quad -1}{UF_6}$$

Multiply known oxidation numbers by the appropriate number of atoms and place the totals underneath the corresponding elements. There are six fluorine atoms, 6 × −1 = −6.

$$\underset{\quad -6}{\overset{\quad -1}{UF_6}}$$

The compound UF_6 is molecular. According to the guidelines, the sum of the oxidation numbers must equal zero. The total of positive oxidation numbers is therefore +6.

$$\underset{+6\ -6}{\overset{\quad -1}{UF_6}}$$

Divide the total calculated oxidation number by the appropriate number of atoms. There is only one uranium atom in the molecule, so each must have an oxidation number of +6.

$$\overset{+6\ -1}{UF_6}$$
$$+6\ -6$$

b. Oxygen and sulfur are each more electronegative than hydrogen, so hydrogen has an oxidation number of +1. Oxygen is not combined with a halogen, nor is H_2SO_4 a peroxide. Therefore, the oxidation number of oxygen is –2. Place these known oxidation numbers above the appropriate symbols. Place the total of the oxidation numbers underneath.

$$\overset{+1\quad -2}{H_2SO_4}$$
$$+2\quad -8$$

The sum of the oxidation numbers must equal zero, and there is only one sulfur atom in each molecule of H_2SO_4. Each sulfur atom therefore must have an oxidation number of (+2) + (–8) = +6.

c. To assign oxidation numbers to the elements in ClO_3^-, proceed as in parts (a) and (b). Remember, however, that the total of the oxidation numbers should equal the overall charge of the anion, 1–. The oxidation number of a single oxygen atom in the ion is –2. The total oxidation number due to the three oxygen atoms is –6. For the chlorate ion to have a 1– charge, chlorine must be assigned an oxidation number of +5.

$$\overset{+5\ -2}{ClO_3^-}$$
$$+5\ -6$$

PRACTICE

1. Assign oxidation numbers to each atom in the following compounds or ions:
   a. HCl e. HNO_3 h. $HClO_3$
   b. CF_4 f. KH i. N_2O_5
   c. PCl_3 g. P_4O_{10} j. $GeCl_2$
   d. SO_2

Answer
a. +1, –1 e. +1, +5, –2 h. +1, +5, –2
b. +4, –1 f. +1, –1 i. +5, –2
c. +3, –1 g. +5, –2 j. +2, –1
d. +4, –2

Using Oxidation Numbers for Formulas and Names

As shown in Table 7-6, many nonmetals can have more than one oxidation number. (A more extensive list of oxidation numbers is given in Appendix Table A-15.) These numbers can sometimes be used in the same manner as ionic charges to determine formulas. Suppose, for example, you want to know the formula of a binary compound formed between sulfur and oxygen. From the common +4 and +6 oxidation states of sulfur, you could expect that sulfur might form SO_2 or SO_3. Both are known compounds. As with formulas for ionic compounds, of course, a formula must represent facts. Oxidation numbers alone cannot be used to predict the existence of a compound.

TABLE 7-6 Common Oxidation Numbers of Some Nonmetals That Have Variable Oxidation States*

Group 14	carbon	–4, +2, +4
Group 15	nitrogen	–3, +3, +5
	phosphorus	–3, +3, +5
Group 16	sulfur	–2, +4, +6
Group 17	chlorine	–1, +1, +3, +5, +7
	bromine	–1, +1, +3, +5, +7
	iodine	–1, +1, +3, +5, +7

*In addition to the values shown, atoms of each element in its pure state are assigned an oxidation number of zero.

In Section 7-1 we introduced the use of Roman numerals to denote ionic charges in the Stock system of naming ionic compounds. The Stock system is actually based on oxidation numbers, and it can be used as an alternative to the prefix system for naming binary molecular compounds. In the prefix system, for example, SO_2 and SO_3 are named sulfur dioxide and sulfur trioxide, respectively. Their names according to the Stock system are sulfur(IV) oxide and sulfur(VI) oxide. The international body that governs nomenclature has endorsed the Stock system, which is more practical for complicated compounds. Prefix-based names and Stock-system names are still used interchangeably for many simple compounds, however. A few additional examples of names in both systems are given below.

	Prefix system	Stock system
PCl_3	phosphorus trichloride	phosphorus(III) chloride
PCl_5	phosphorus pentachloride	phosphorus(V) chloride
N_2O	dinitrogen monoxide	nitrogen(I) oxide
NO	nitrogen monoxide	nitrogen(II) oxide
PbO_2	lead dioxide	lead(IV) oxide
Mo_2O_3	dimolybdenum trioxide	molybdenum(III) oxide

SECTION REVIEW

1. Assign oxidation numbers to each atom in the following compounds or ions:

a. HF
b. Cl_4
c. H_2O
d. PI_3
e. CS_2
f. Na_2O_2
g. H_2CO_3
h. NO_2^-
i. SO_4^{2-}
j. ClO_2^-
k. IO_3^-

2. Name each of the following binary molecular compounds according to the Stock system:

a. Cl_4
b. SO_3
c. As_2S_3
d. NCl_3

Chemistry and Art

From "Chemistry and the Plastic and Graphic Arts" by Jonathan E. Ericson in *The Central Science, Essays on the Uses of Chemistry*

The preservation of art is dependent on the control of the environment of the piece. Modern museums are air conditioned, maintaining their temperature between 68°–72°F and a 50–65% relative humidity. These controls provide the conditions under which most works of art are stable. Occasionally, the relative humidity will have to be increased or decreased depending on the stability of an individual piece. Here, art conservation has borrowed techniques from physical chemistry to determine proper conditions experimentally.

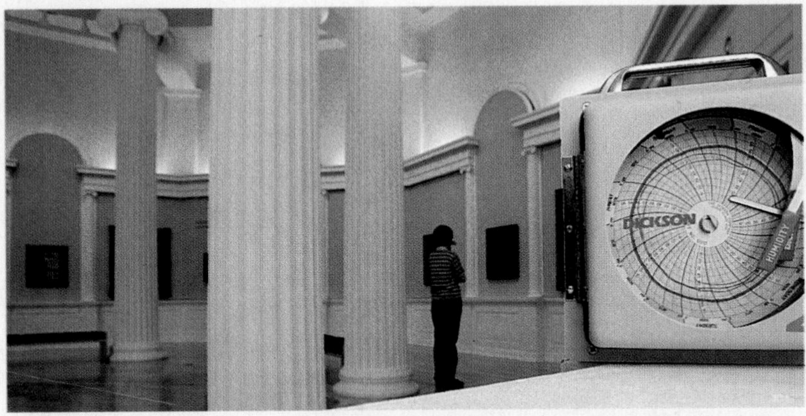

In many museums, the temperature and relative humidity of the surroundings are recorded on the same graph by a hygrothermograph, *as shown here.*

The lighting of a work of art is also a critical part of its environment. Fluorescent light and sunlight contain a lot of ultraviolet. The exposure of art to this light can cause it to fade tremendously. Papers, textiles, and organic dyes are the most sensitive to ultraviolet fading. Here, the polymer research chemist has developed special acrylic plastics like Plexiglas® UF-3, a Rohm and Haas product, which filters out the ultraviolet light.

The relative humidity, temperature and lighting are rather easy to control in a large museum. However, the agents of attack which deteriorate the appearance of the art piece are now not always so easily identified. For paintings, the accumulation of dirt and grime, discoloring of the protective varnish layer, or the tension and distortion of canvas or wood supports are destructive to the piece and distorting to us when viewing it. In the past, when cleaning a painting, people either used stronger cleaning methods like sandpaper, or, if smart, gave up. Modern chemistry has developed ways and means of using whole families of safer cleaning agents like acetone, alcohol and other organic solvents. After careful examination and evaluation, the painting conservator can remove the old varnish, reline or bond a new supporting canvas with natural wax resin or synthetic resin, spray on an isolating layer of synthetic resin, paint in pigment losses, and seal the painting with a final protective layer. All these steps use products which have been developed in the chemical laboratory.

Finally, the agents of attack are not always so subtle. Sometimes works of art fall and break, or are torn, cut, or burned. For each case and for each object, a particular conservation method is used. Almost always, the products used or the treatment itself are direct contributions of a chemist. Without chemistry, art conservation, as we know it, would be very primitive, indeed.

Reading for Meaning

What kind of molecules are acrylic plastics composed of?

Read Further

When light interacts with paper, it can cause a *photochemical reaction.* Research photochemical reactions, and explain why a newspaper left in the sunlight for a long period of time turns yellow.

Using Chemical Formulas

SECTION 7-3

OBJECTIVES

- Calculate the formula mass or molar mass of any given compound.
- Use molar mass to convert between mass in grams and amount in moles of a chemical compound.
- Calculate the number of molecules, formula units, or ions in a given molar amount of a chemical compound.
- Calculate the percentage composition of a given chemical compound.

As you have seen, a chemical formula indicates the elements as well as the relative number of atoms or ions of each element present in a compound. Chemical formulas also allow chemists to calculate a number of characteristic values for a given compound. In this section, you will learn how to use chemical formulas to calculate the *formula mass,* the *molar mass,* and the *percentage composition* by mass of a compound.

Formula Masses

In Chapter 3 we saw that hydrogen atoms have an average atomic mass of 1.007 94 amu and that oxygen atoms have an average atomic mass of 15.9994 amu. Like individual atoms, a molecule, a formula unit, or an ion has a characteristic average mass. For example, we know from the chemical formula H_2O that a single water molecule is composed of exactly two hydrogen atoms and one oxygen atom. The mass of a water molecule is found by summing the masses of the three atoms in the molecule. (In the calculation, the average atomic masses have been rounded to two decimal places.)

average atomic mass of H: 1.01 amu
average atomic mass of O: 16.00 amu

$$2 \text{ H atoms} \times \frac{1.01 \text{ amu}}{\text{H atom}} = 2.02 \text{ amu}$$

$$1 \text{ O atom} \times \frac{16.00 \text{ amu}}{\text{O atom}} = 16.00 \text{ amu}$$

average mass of H_2O molecule = 18.02 amu

The mass of a water molecule can be correctly referred to as a *molecular mass*. The mass of one NaCl formula unit, on the other hand, is not a molecular mass because NaCl is an ionic compound. The mass of *any* unit represented by a chemical formula, whether the unit is a molecule, a formula unit, or an ion, is known as the formula mass. *The* **formula mass** *of any molecule, formula unit, or ion is the sum of the average atomic masses of all the atoms represented in its formula.*

The procedure illustrated for calculating the formula mass of a water molecule can be used to calculate the mass of any unit represented by a chemical formula. In each of the problems that follow, the atomic masses from the periodic table on pages 130–131 have been rounded to two decimal places.

SAMPLE PROBLEM 7-6

Find the formula mass of potassium chlorate, $KClO_3$.

SOLUTION The mass of a formula unit of $KClO_3$ is found by summing the masses of one K atom, one Cl atom, and three O atoms. The required atomic masses can be found in the periodic table on pages 130–131. In the calculation, each atomic mass has been rounded to two decimal places.

$$1\ \cancel{\text{K atom}} \times \frac{39.10 \text{ amu}}{\cancel{\text{K atom}}} = 39.10 \text{ amu}$$

$$1\ \cancel{\text{Cl atom}} \times \frac{35.45 \text{ amu}}{\cancel{\text{Cl atom}}} = 35.45 \text{ amu}$$

$$3\ \cancel{\text{O atoms}} \times \frac{16.00 \text{ amu}}{\cancel{\text{O atom}}} = 48.00 \text{ amu}$$

$$\text{formula mass of } KClO_3 = 122.55 \text{ amu}$$

PRACTICE

1. Find the formula mass of each of the following:
   a. H_2SO_4
   b. $Ca(NO_3)_2$
   c. PO_4^{3-}
   d. $MgCl_2$

Answer
a. 98.08 amu
b. 164.10 amu
c. 94.97 amu
d. 95.21 amu

Molar Masses

In Chapter 3 you learned that the molar mass of a substance is equal to the mass in grams of one mole, or approximately 6.022×10^{23} particles, of the substance. For example, the molar mass of pure calcium, Ca, is 40.08 g/mol because one mole of calcium atoms has a mass of 40.08 g.

The molar mass of a compound is calculated by summing the masses of the elements present in a mole of the molecules or formula units that make up the compound. For example, one mole of water molecules contains exactly two moles of H atoms and one mole of O atoms. Rounded to two decimal places, a mole of hydrogen atoms has a mass of 1.01 g, and a mole of oxygen atoms has a mass of 16.00 g. The molar mass of water is calculated as follows.

$$2 \cancel{\text{mol H}} \times \frac{1.01 \text{ g H}}{\cancel{\text{mol H}}} = 2.02 \text{ g H}$$

$$1 \cancel{\text{mol O}} \times \frac{16.00 \text{ g O}}{\cancel{\text{mol O}}} = 16.00 \text{ g O}$$

$$\text{molar mass of } H_2O = 18.02 \text{ g/mol}$$

FIGURE 7-3 Every compound has a characteristic molar mass. Shown here are one mole each of nitrogen (in balloon), water (in graduated cylinder), cadmium sulfide, CdS (yellow substance), and sodium chloride, NaCl (white substance).

Figure 7-3 shows a mole of water as well as a mole of several other substances.

You may have noticed that *a compound's molar mass is numerically equal to its formula mass*. For instance, in Sample Problem 7-6 the formula mass of $KClO_3$ was found to be 122.55 amu. Therefore, because molar mass is numerically equal to formula mass, we know that the molar mass of $KClO_3$ is 122.55 g/mol.

SAMPLE PROBLEM 7-7

What is the molar mass of barium nitrate, $Ba(NO_3)_2$?

SOLUTION One mole of barium nitrate contains exactly one mole of Ba^{2+} ions and two moles of NO_3^- ions. The two moles of NO_3^- ions contain two moles of N atoms and six moles of O atoms. Therefore, the molar mass of $Ba(NO_3)_2$ is calculated as follows.

$$1 \cancel{\text{mol Ba}} \times \frac{137.33 \text{ g Ba}}{\cancel{\text{mol Ba}}} = 137.33 \text{ g Ba}$$

$$2 \cancel{\text{mol N}} \times \frac{14.01 \text{ g N}}{\cancel{\text{mol N}}} = 28.02 \text{ g N}$$

$$6 \cancel{\text{mol O}} \times \frac{16.00 \text{ g O}}{\cancel{\text{mol O}}} = 96.00 \text{ g O}$$

$$\text{molar mass of } Ba(NO_3)_2 = 261.35 \text{ g/mol}$$

PRACTICE

1. How many moles of atoms of each element are there in one mole of the following compounds?
   a. Al_2S_3
   b. $NaNO_3$
   c. $Ba(OH)_2$

Answer

1. a. 2 mol Al, 3 mol S
   b. 1 mol Na, 1 mol N, 3 mol O
   c. 1 mol Ba, 2 mol O, 2 mol H

2. Find the molar mass of each of the compounds listed in item 1.

2. a. 150.17 g/mol
   b. 85.00 g/mol
   c. 171.35 g/mol

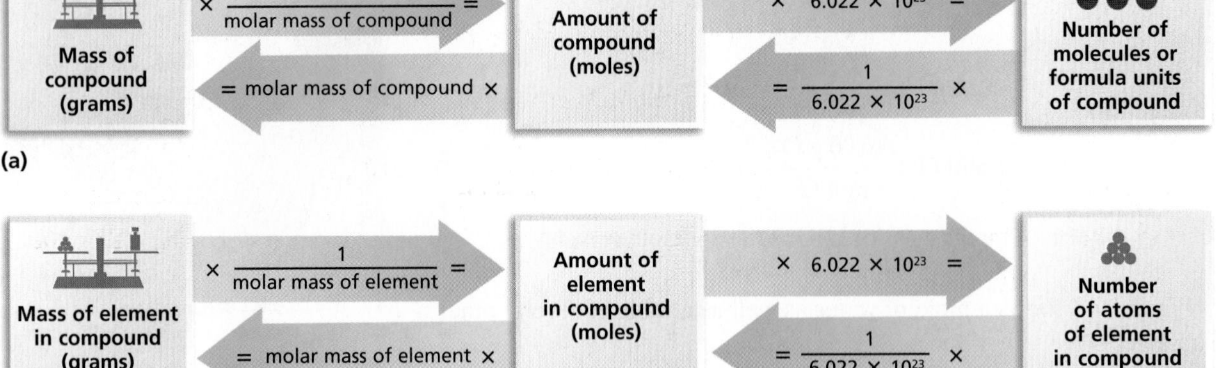

FIGURE 7-4 (a) The diagram shows the relationships between mass in grams, amount in moles, and number of molecules or atoms for a given compound. (b) Similar relationships exist for an element within a compound.

Molar Mass as a Conversion Factor

The molar mass of a compound can be used as a conversion factor to relate an amount in moles to a mass in grams for a given substance. Recall that molar mass usually has the units of grams per mole. To convert a known amount of a compound in moles to a mass in grams, multiply the amount in moles by the molar mass.

$$\text{amount in moles} \times \text{molar mass (g/mol)} = \text{mass in grams}$$

Conversions of this type for elements and compounds are summarized above in Figure 7-4.

SAMPLE PROBLEM 7-8

What is the mass in grams of 2.50 mol of oxygen gas?

SOLUTION

1 *ANALYZE*

Given: 2.50 mol O_2

Unknown: mass of O_2 in grams

2 *PLAN*

moles $O_2 \longrightarrow$ grams O_2

To convert amount of O_2 in moles to mass of O_2 in grams, multiply by the molar mass of O_2.

$$\text{amount of } O_2 \text{ (mol)} \times \text{molar mass of } O_2 \text{ (g/mol)} = \text{mass of } O_2 \text{ (g)}$$

3 COMPUTE First the molar mass of O_2 must be calculated.

$$2\ \cancel{\text{mol O}} \times \frac{16.00\ \text{g O}}{\cancel{\text{mol O}}} = 32.00\ \text{g (mass of one mole of } O_2)$$

The molar mass of O_2 is therefore 32.00 g/mol. Now do the calculation shown in step 2.

$$2.50\ \cancel{\text{mol } O_2} \times \frac{32.00\ \text{g } O_2}{\cancel{\text{mol } O_2}} = 80.0\ \text{g } O_2$$

4 EVALUATE The answer is correctly given to three significant figures and is close to an estimated value of 75 g (2.50 mol × 30 g/mol).

To convert a known mass of a compound in grams to an amount in moles, the mass must be divided by the molar mass. Or you can invert the molar mass and multiply so that units are easily canceled.

$$\text{mass in grams} \times \frac{1}{\text{molar mass (g/mol)}} = \text{amount in moles}$$

SAMPLE PROBLEM 7-9

Ibuprofen, $C_{13}H_{18}O_2$, is the active ingredient in many nonprescription pain relievers. Its molar mass is 206.29 g/mol.

a. If the tablets in a bottle contain a total of 33 g of ibuprofen, how many moles of ibuprofen are in the bottle?

b. How many molecules of ibuprofen are in the bottle?

c. What is the total mass in grams of carbon in 33 g of ibuprofen?

SOLUTION

1 ANALYZE **Given:** 33 g of $C_{13}H_{18}O_2$, molar mass 206.29 g/mol

Unknown: a. moles $C_{13}H_{18}O_2$

b. molecules $C_{13}H_{18}O_2$

c. total mass of C

2 PLAN a. grams ⟶ moles

To convert mass of ibuprofen in grams to amount of ibuprofen in moles, multiply by the inverted molar mass of $C_{13}H_{18}O_2$.

$$\text{g } C_{13}H_{18}O_2 \times \frac{1\ \text{mol } C_{13}H_{18}O_2}{206.29\ \text{g } C_{13}H_{18}O_2} = \text{mol } C_{13}H_{18}O_2$$

b. moles ⟶ molecules

To find the number of molecules of ibuprofen, multiply amount of $C_{13}H_{18}O_2$ in moles by Avogadro's number.

$$\text{mol } C_{13}H_{18}O_2 \times \frac{6.022 \times 10^{23}\ \text{molecules}}{\text{mol}} = \text{molecules } C_{13}H_{18}O_2$$

c. moles $C_{13}H_{18}O_2 \longrightarrow$ moles C $\longrightarrow$ grams C
To find the mass of carbon present in the ibuprofen, the two conversion factors needed are the amount of carbon in moles per mole of $C_{13}H_{18}O_2$ and the molar mass of carbon.

$$\text{mol } C_{13}H_{18}O_2 \times \frac{13 \text{ mol C}}{\text{mol } C_{13}H_{18}O_2} \times \frac{12.01 \text{ g C}}{\text{mol C}} = \text{g C}$$

3 COMPUTE

a. $33 \text{ g } C_{13}H_{18}O_2 \times \dfrac{1 \text{ mol } C_{13}H_{18}O_2}{206.29 \text{ g } C_{13}H_{18}O_2} = 0.16 \text{ mol } C_{13}H_{18}O_2$

b. $0.16 \text{ mol } C_{13}H_{18}O_2 \times \dfrac{6.022 \times 10^{23} \text{ molecules}}{\text{mol}} = 9.6 \times 10^{22} \text{ molecules } C_{13}H_{18}O_2$

c. $0.16 \text{ mol } C_{13}H_{18}O_2 \times \dfrac{13 \text{ mol C}}{\text{mol } C_{13}H_{18}O_2} \times \dfrac{12.01 \text{ g C}}{\text{mol C}} = 25 \text{ g C}$

The bottle contains 0.16 mol of ibuprofen, which is 9.6×10^{22} molecules of ibuprofen. The sample of ibuprofen contains 25 g of carbon.

4 EVALUATE Checking each step shows that the arithmetic is correct, significant figures have been used correctly, and units have canceled as desired.

PRACTICE

1. How many moles of compound are there in the following?
   a. 6.60 g $(NH_4)_2SO_4$
   b. 4.5 kg $Ca(OH)_2$
2. How many molecules are there in the following?
   a. 25.0 g H_2SO_4
   b. 125 g of sugar, $C_{12}H_{22}O_{11}$
3. What is the mass in grams of 6.25 mol of copper(II) nitrate?

Answer

1. a. 0.0500 mol
   b. 61 mol
2. a. 1.53×10^{23} molecules
   b. 2.20×10^{23} molecules
3. 1170 g

Percentage Composition

It is often useful to know the percentage by mass of a particular element in a chemical compound. For example, suppose the compound potassium chlorate, $KClO_3$, were to be used as a source of oxygen. It would be helpful to know the percentage of oxygen in the compound. To find the mass percentage of an element in a compound, one can divide the mass of the element in a sample of the compound by the total mass of the sample, then multiply this value by 100.

$$\frac{\text{mass of element in sample of compound}}{\text{mass of sample of compound}} \times 100 = \begin{matrix}\text{\% element in} \\ \text{compound}\end{matrix}$$

The mass percentage of an element in a compound is the same regardless of the sample's size. Therefore, a simpler way to calculate the percentage of an element in a compound is to determine how many grams of the element are present in one mole of the compound. Then divide this value by the molar mass of the compound and multiply by 100.

$$\frac{\text{mass of element in 1 mol of compound}}{\text{molar mass of compound}} \times 100 = \text{\% element in compound}$$

The percentage by mass of each element in a compound is known as the **percentage composition** *of the compound.*

SAMPLE PROBLEM 7-10

Find the percentage composition of copper(I) sulfide, Cu_2S.

SOLUTION

1 *ANALYZE* **Given:** formula, Cu_2S
Unknown: percentage composition of Cu_2S

2 *PLAN* formula ⟶ molar mass ⟶ mass percentage of each element

The molar mass of the compound must be found. Then the mass of each element present in one mole of the compound is used to calculate the mass percentage of each element.

3 *COMPUTE*

$$2\ \cancel{\text{mol Cu}} \times \frac{63.55\ \text{g Cu}}{\cancel{\text{mol Cu}}} = 127.1\ \text{g Cu}$$

$$1\ \cancel{\text{mol S}} \times \frac{32.07\ \text{g S}}{\cancel{\text{mol S}}} = 32.07\ \text{g S}$$

$$\text{molar mass of } Cu_2S = 159.2\ \text{g}$$

$$\frac{127.1\ \text{g Cu}}{159.2\ \text{g } Cu_2S} \times 100 = 79.84\%\ \text{Cu}$$

$$\frac{32.07\ \text{g S}}{159.2\ \text{g } Cu_2S} \times 100 = 20.14\%\ \text{S}$$

4 *EVALUATE* A good check is to see if the results add up to about 100%. (Because of rounding or experimental error, the total may not always be exactly 100%.)

SAMPLE PROBLEM 7-11

As some salts crystallize from a water solution, they bind water molecules in their crystal structure. Sodium carbonate forms such a *hydrate,* in which 10 water molecules are present for every formula unit of sodium carbonate. Find the mass percentage of water in sodium carbonate decahydrate, $Na_2CO_3 \cdot 10H_2O$, which has a molar mass of 286.14 g/mol.

SOLUTION

1 *ANALYZE* **Given:** chemical formula, $Na_2CO_3{\cdot}10H_2O$
molar mass of $Na_2CO_3{\cdot}10H_2O$
Unknown: mass percentage of H_2O

2 *PLAN* chemical formula $\longrightarrow$ mass H_2O per mole of $Na_2CO_3{\cdot}10H_2O$ $\longrightarrow$ % water

The mass of water per mole of sodium carbonate decahydrate must first be found. This value is then divided by the mass of one mole of $Na_2CO_3{\cdot}10H_2O$.

3 *COMPUTE* One mole of $Na_2CO_3{\cdot}10H_2O$ contains 10 mol of H_2O. Recall from page 223 that the molar mass of H_2O is 18.02 g/mol. The mass of 10 mol of H_2O is calculated as follows.

$$10\ \cancel{\text{mol}\ H_2O} \times \frac{18.02\ \text{g}\ H_2O}{\cancel{\text{mol}\ H_2O}} = 180.2\ \text{g}\ H_2O$$

$$\text{mass of } H_2O \text{ per mole of } Na_2CO_3{\cdot}10H_2O = 180.2\ \text{g}$$

The molar mass of $Na_2CO_3{\cdot}H_2O$ is 286.14 g/mol, so we know that 1 mol of the hydrate has a mass of 286.14g. The mass percentage of 10 mol of H_2O in 1 mol of $Na_2CO_3{\cdot}H_2O$ can now be calculated.

$$\text{mass percentage of } H_2O \text{ in } Na_2CO_3{\cdot}10H_2O = \frac{180.2\ \text{g}\ H_2O}{286.14\ \text{g}\ Na_2CO_3{\cdot}10H_2O} \times 100 = 62.98\%\ H_2O$$

4 *EVALUATE* Checking shows that the arithmetic is correct and that units cancel as desired.

PRACTICE

1. Find the percentage compositions of the following:
   a. $PbCl_2$ b. $Ba(NO_3)_2$
2. Find the mass percentage of water in $ZnSO_4{\cdot}7H_2O$.
3. Magnesium hydroxide is 54.87% oxygen by mass. How many grams of oxygen are in 175 g of the compound? How many moles of oxygen is this?

Answer

1. a. 74.51% Pb, 25.49% Cl
   b. 52.546% Ba, 10.72% N, 36.73% O
2. 43.86% H_2O
3. 96.0 g O; 6.00 mol O

SECTION REVIEW

1. Determine both the formula mass and molar mass of ammonium carbonate, $(NH_4)_2CO_3$.
2. How many moles of atoms of each element are there in one mole of $(NH_4)_2CO_3$?
3. What is the mass in grams of 3.25 mol $Fe_2(SO_4)_3$?
4. How many moles of molecules are there in 250 g of hydrogen nitrate, HNO_3?
5. How many molecules of aspirin, $C_9H_8O_4$, are there in a 100.0 mg tablet of aspirin?
6. Calculate the percentage composition of $(NH_4)_2CO_3$.

Determining Chemical Formulas

SECTION 7-4

OBJECTIVES

- Define *empirical formula,* and explain how the term applies to ionic and molecular compounds.
- Determine an empirical formula from either a percentage or a mass composition.
- Explain the relationship between the empirical formula and the molecular formula of a given compound.
- Determine a molecular formula from an empirical formula.

When a new substance is synthesized or is discovered, it is analyzed quantitatively to reveal its percentage composition. From this data, the empirical formula is then determined. *An* **empirical formula** *consists of the symbols for the elements combined in a compound, with subscripts showing the smallest whole-number mole ratio of the different atoms in the compound.* For an ionic compound, the formula unit is usually the compound's empirical formula. For a molecular compound, however, the empirical formula does not necessarily indicate the actual numbers of atoms present in each molecule. For example, the empirical formula of the gas diborane is BH_3, but the molecular formula is B_2H_6. In this case, the atomic ratio given by the molecular formula is equal to the empirical ratio multiplied by two.

Calculation of Empirical Formulas

To determine a compound's empirical formula from its percentage composition, begin by converting percentage composition to a mass composition. Assume that you have a 100.0 g sample of the compound. Then calculate the amount of each element in the sample. For example, the percentage composition of diborane is 78.1% B and 21.9% H. Therefore, 100.0 g of diborane contains 78.1 g of B and 21.9 g of H.

Next, the mass composition of each element is converted to a composition in moles by dividing by the appropriate molar mass.

$$78.1 \cancel{\text{g B}} \times \frac{1 \text{ mol B}}{10.81 \cancel{\text{g B}}} = 7.22 \text{ mol B}$$

$$21.9 \cancel{\text{g H}} \times \frac{1 \text{ mol H}}{1.01 \cancel{\text{g H}}} = 21.7 \text{ mol H}$$

These values give a mole ratio of 7.22 mol B to 21.7 mol H. However, this is not a ratio of smallest whole numbers. To find such a ratio, divide each number of moles by the smallest number in the existing ratio.

$$\frac{7.22 \text{ mol B}}{7.22} : \frac{21.7 \text{ mol H}}{7.22} = 1 \text{ mol B} : 3.01 \text{ mol H}$$

Because of rounding or experimental error, a compound's mole ratio sometimes consists of numbers close to whole numbers instead of exact whole numbers. In this case, the differences from whole numbers may be ignored and the nearest whole number taken. Thus, diborane contains atoms in the ratio 1 B:3 H. The compound's empirical formula is BH_3.

Sometimes mass composition is known instead of percentage composition. To determine the empirical formula in this case, convert mass composition to composition in moles. Then calculate the smallest whole-number mole ratio of atoms. This process is shown in Sample Problem 7-13.

SAMPLE PROBLEM 7-12

Quantitative analysis shows that a compound contains 32.38% sodium, 22.65% sulfur, and 44.99% oxygen. Find the empirical formula of this compound.

SOLUTION

1 *ANALYZE* **Given:** percentage composition: 32.38% Na, 22.65% S, and 44.99% O

Unknown: empirical formula

2 *PLAN* percentage composition ⟶ mass composition ⟶ composition in moles ⟶ smallest whole-number mole ratio of atoms

3 *COMPUTE* Mass composition (mass of each element in 100.0 g sample): 32.38 g Na, 22.65 g S, 44.99 g O

Composition in moles: $32.38 \text{ g Na} \times \frac{1 \text{ mol Na}}{22.99 \text{ g Na}} = 1.408 \text{ mol Na}$

$$22.65 \text{ g S} \times \frac{1 \text{ mol S}}{32.07 \text{ g S}} = 0.7063 \text{ mol S}$$

$$44.99 \text{ g O} \times \frac{1 \text{ mol O}}{16.00 \text{ g O}} = 2.812 \text{ mol O}$$

Smallest whole-number mole ratio of atoms:
The compound contains atoms in the ratio 1.408 mol Na:0.7063 mol S:2.812 mol O. To find the smallest whole-number mole ratio, divide each value by the smallest number in the ratio.

$$\frac{1.408 \text{ mol Na}}{0.7063}:\frac{0.7063 \text{ mol S}}{0.7063}:\frac{2.812 \text{ mol O}}{0.7063} = 1.993 \text{ mol Na:1 mol S:3.981 mol O}$$

Rounding each number in the ratio to the nearest whole number yields a mole ratio of 2 mol Na:1 mol S:4 mol O. The empirical formula of the compound is Na_2SO_4.

4 *EVALUATE* Calculating the percentage composition of the compound based on the empirical formula determined in the problem reveals a percentage composition of 32.37% Na, 22.58% S, and 45.05% O. These values agree reasonably well with the given percentage composition.

SAMPLE PROBLEM 7-13

Analysis of a 10.150 g sample of a compound known to contain only phosphorus and oxygen indicates a phosphorus content of 4.433 g. What is the empirical formula of this compound?

SOLUTION

1 ANALYZE

Given: sample mass = 10.150 g
phosphorus mass = 4.433 g
Unknown: empirical formula

2 PLAN

Mass composition ⟶ composition in moles ⟶ smallest whole-number ratio of atoms

3 COMPUTE

The mass of oxygen is found by subtracting the phosphorus mass from the sample mass.

$$\text{sample mass} - \text{phosphorus mass} = 10.150\text{ g} - 4.433\text{ g} = 5.717\text{ g}$$

Mass composition: 4.433 g P, 5.717 g O

Composition in moles:

$$4.433\ \cancel{\text{g P}} \times \frac{1\text{ mol P}}{30.97\ \cancel{\text{g P}}} = 0.1431\text{ mol P}$$

$$5.717\ \cancel{\text{g O}} \times \frac{1\text{ mol O}}{16.00\ \cancel{\text{g O}}} = 0.3573\text{ mol O}$$

Smallest whole-number mole ratio of atoms:

$$\frac{0.1431\text{ mol P}}{0.1431} : \frac{0.3573\text{ mol O}}{0.1431}$$

$$1\text{ mol P} : 2.497\text{ mol O}$$

The number of O atoms is not close to a whole number. But if we multiply each number in the ratio by 2, then the number of O atoms becomes 4.994 mol, which is close to 5 mol. The simplest whole-number mole ratio of P atoms to O atoms is 2:5. The compound's empirical formula is P_2O_5.

4 EVALUATE

The arithmetic is correct, significant figures have been used correctly, and units cancel as desired. The formula is reasonable because +5 is a common oxidation state of phosphorus.

PRACTICE

1. A compound is found to contain 63.52% iron and 36.48% sulfur. Find its empirical formula. *Answer* FeS

2. Find the empirical formula of a compound found to contain 26.56% potassium, 35.41% chromium, and the remainder oxygen. *Answer* $K_2Cr_2O_7$

3. Analysis of 20.0 g of a compound containing only calcium and bromine indicates that 4.00 g of calcium are present. What is the empirical formula of the compound formed? *Answer* $CaBr_2$

Calculation of Molecular Formulas

An empirical formula may or may not be a correct molecular formula. For example, diborane's empirical formula is BH_3. Any multiple of BH_3, such as B_2H_6, B_3H_9, B_4H_{12}, and so on, represents the same ratio of B atoms to H atoms. The molecular compounds ethene, C_2H_4, and cyclopropane, C_3H_6, also share an identical atomic ratio (2 H:1 C), yet they are very different substances. How is the correct formula of a molecular compound found from an empirical formula?

The relationship between a compound's empirical formula and its molecular formula can be written as follows.

$$x(\text{empirical formula}) = \text{molecular formula}$$

The number represented by x is a whole-number multiple indicating the factor by which the subscripts in the empirical formula must be multiplied to obtain the molecular formula. (The value of x is sometimes 1.) The formula masses have a similar relationship.

$$x(\text{empirical formula mass}) = \text{molecular formula mass}$$

To determine the molecular formula of a compound, you must know the compound's formula mass. For example, experimentation shows the formula mass of diborane to be 27.67 amu. The formula mass for the empirical formula, BH_3, is 13.84 amu. Dividing the experimental formula mass by the empirical formula mass gives the value of x for diborane.

$$x = \frac{27.67 \text{ amu}}{13.84 \text{ amu}} = 2.000$$

The molecular formula of diborane is therefore B_2H_6.

$$(BH_3)_2 = B_2H_6$$

Recall that a compound's molecular formula mass is numerically equal to its molar mass, so a compound's molecular formula can also be found given the compound's empirical formula and its molar mass.

SAMPLE PROBLEM 7-14

In Sample Problem 7-13, the empirical formula of a compound of phosphorus and oxygen was found to be P_2O_5. Experimentation shows that the molar mass of this compound is 283.89 g/mol. What is the compound's molecular formula?

SOLUTION

1 *ANALYZE* **Given:** empirical formula
Unknown: molecular formula

2 PLAN

$$x(\text{empirical formula}) = \text{molecular formula}$$

$$x = \frac{\text{molecular formula mass}}{\text{empirical formula mass}}$$

3 COMPUTE Molecular formula mass is numerically equal to molar mass. Thus, changing the g/mol unit of the compound's molar mass to amu yields the compound's molecular formula mass.

$$\text{molecular molar mass} = 283.89 \text{ g/mol}$$
$$\text{molecular formula mass} = 283.89 \text{ amu}$$

The empirical formula mass is found by adding the masses of each of the atoms indicated in the empirical formula.

$$\text{mass of phosphorus atom} = 30.97 \text{ amu}$$
$$\text{mass of oxygen atom} = 16.00 \text{ amu}$$
$$\text{empirical formula mass of } P_2O_5 = 2 \times 30.97 \text{ amu} + 5 \times 16.00 \text{ amu} = 141.94 \text{ amu}$$

Dividing the experimental formula mass by the empirical formula mass gives the value of x. The formula mass is numerically equal to the molar mass.

$$x = \frac{283.89 \cancel{\text{amu}}}{141.94 \cancel{\text{amu}}} = 2.0001$$

The compound's molecular formula is therefore P_4O_{10}.

$$(P_2O_5)_2 = P_4O_{10}$$

4 EVALUATE Checking the arithmetic shows that it is correct.

PRACTICE

1. Determine the molecular formula of the compound with an empirical formula of CH and a formula mass of 78.110 amu. *Answer* C_6H_6
2. A sample of a compound with a formula mass of 34.00 amu is found to consist of 0.44 g H and 6.92 g O. Find its molecular formula. *Answer* H_2O_2

SECTION REVIEW

1. A compound is found to contain 36.48% Na, 25.41% S, and 38.11% O. Find its empirical formula.
2. Find the empirical formula of a compound that contains 53.70% iron and 46.30% sulfur.
3. Analysis of a compound indicates that it contains 1.04 g K, 0.70 g Cr, and 0.86 g O. Find its empirical formula.
4. If 4.04 g of N combine with 11.46 g O to produce a compound with a formula mass of 108.0 amu, what is the molecular formula of this compound?
5. The molar mass of a compound is 92 g/mol. Analysis of a sample of the compound indicates that it contains 0.606 g N and 1.390 g O. Find its molecular formula.

CHAPTER 7 REVIEW

CHAPTER SUMMARY

7-1

- A positive monatomic ion is identified simply by the name of the appropriate element. A negative monatomic ion is named by dropping parts of the ending of the element's name and adding *-ide* to the root.
- The charge of each ion in an ionic compound may be used to determine the simplest chemical formula for the compound.
- Compounds composed of two different elements are known as binary compounds.
- Binary ionic compounds are named by combining the names of the positive and negative ions. Compounds containing polyatomic ions are named in the same manner.
- The old system of naming binary molecular compounds uses prefixes. The new system, known as the Stock system, uses oxidation numbers.

Vocabulary

binary compound (206) nomenclature (206) oxyanion (209) salt (215)
monatomic ion (204)

7-2

- Oxidation numbers, or oxidation states, are assigned to atoms in compounds according to a set of specific rules. Oxidation numbers are useful in naming compounds, in writing formulas, and in balancing chemical equations.
- Compounds containing elements that have more than one oxidation state are named using the Stock system of nomenclature.
- Stock-system names and prefix-system names are used interchangeably for many molecular compounds.
- In many molecular compounds, oxidation numbers of each element in the compound may be used to determine the compound's simplest chemical formula.
- By knowing oxidation numbers, we can name compounds without knowing whether they are ionic or molecular.

Vocabulary

oxidation number (216) oxidation state (216)

7-3

- Formula mass, molar mass, and percentage composition can be calculated from the chemical formula for a compound.
- The percentage composition of a compound is the percentage by mass of each element in the compound.
- Molar mass can be used as a conversion factor between an amount in moles and a mass in grams of a given compound or element.

Vocabulary

formula mass (221) percentage composition (227)

7-4

- An empirical formula shows the simplest whole-number ratio of atoms in a given compound.
- Each molecule of a molecular compound contains a whole-number multiple of the atoms in the empirical formula. In some cases this whole-number multiple is equal to 1.
- Empirical formulas indicate how many atoms of each element are combined in the simplest unit of a chemical compound.

Vocabulary

empirical formula (229)

REVIEWING CONCEPTS

1. a. What are monatomic ions?
b. Give three examples of monatomic ions. (7-1)

2. What is the difference between the nitrite ion and the nitrate ion? (7-1)

3. Using only the periodic table, write the symbol of the ion most typically formed by each of the following elements:
a. K
b. Ca
c. S
d. Cl
e. Ba
f. Br (7-1)

4. Write the formulas and indicate the charges for each of the following ions:
a. sodium ion
b. aluminum ion
c. chloride ion
d. nitride ion
e. iron(II) ion
f. iron(III) ion (7-1)

5. Name each of the following monatomic ions:
a. K^+
b. Mg^{2+}
c. Al^{3+}
d. Cl^-
e. O^{2-}
f. Ca^{2+} (7-1)

6. Write formulas for the binary ionic compounds formed between the following elements. (Hint: See Sample Problem 7-1.)
a. sodium and iodine
b. calcium and sulfur
c. zinc and chlorine
d. barium and fluorine
e. lithium and oxygen (7-1)

7. Give the name of each of the following binary ionic compounds. (Hint: See Sample Problem 7-2.)
a. KCl
b. $CaBr_2$
c. Li_2O
d. $MgCl_2$ (7-1)

8. Write the formulas and give the names of the compounds formed by the following ions:
a. Cr^{2+} and F^-
b. Ni^{2+} and O^{2-}
c. Fe^{3+} and O^{2-} (7-1)

9. In naming and writing formulas for binary molecular compounds, what determines the order in which the component elements are written? (7-1)

10. Name the following binary molecular compounds according to the prefix system. (Hint: See Sample Problem 7-4.)
a. CO_2
b. CCl_4
c. PCl_5
d. SeF_6
e. As_2O_5 (7-1)

11. Write formulas for each of the following binary molecular compounds. (Hint: See Sample Problem 7-4.)
a. carbon tetrabromide
b. silicon dioxide
c. tetraphosphorus decoxide
d. diarsenic trisulfide (7-1)

12. Distinguish between binary acids and oxyacids, and give two examples of each. (7-1)

13. a. What is a salt?
b. Give two examples of salts. (7-1)

14. Name each of the following acids:
a. HF
b. HBr
c. HNO_3
d. H_2SO_4
e. H_3PO_4 (7-1)

15. Name each of the following ions according to the Stock system:
a. Fe^{2+}
b. Fe^{3+}
c. Pb^{2+}
d. Pb^{4+}
e. Sn^{2+}
f. Sn^{4+} (7-2)

16. Name each of the binary molecular compounds in item 11 using the Stock system. (7-2)

17. Write formulas for each of the following compounds:
a. phosphorus(III) iodide
b. sulfur(II) chloride
c. carbon(IV) sulfide
d. nitrogen(V) oxide (7-2)

18. a. What are oxidation numbers?
b. What useful functions do oxidation numbers serve? (7-2)

19. a. Define *formula mass.*
b. In what unit is formula mass expressed? (7-3)

20. What is meant by the molar mass of a compound? (7-3)

21. What three types of information are needed in order to find an empirical formula from percentage composition data? (7-4)

22. What is the relationship between the empirical formula and the molecular formula of a compound? (7-4)

PROBLEMS

Nomenclature and Chemical Formulas

23. Write the formula and charge for each of the following ions:
a. ammonium ion
b. acetate ion
c. hydroxide ion
d. carbonate ion
e. sulfate ion
f. phosphate ion
g. copper(II) ion
h. tin(II) ion
i. iron(III) ion
j. copper(I) ion
k. mercury(I) ion
l. mercury(II) ion

24. Name each of the following ions:
a. NH_4^+
b. ClO_3^-
c. OH^-
d. SO_4^{2-}
e. NO_3^-
f. CO_3^{2-}
g. PO_4^{3-}
h. CH_3COO^-
i. HCO_3^-
j. CrO_4^{2-}

25. Write formulas for each of the following compounds:
a. sodium fluoride
b. calcium oxide
c. potassium sulfide
d. magnesium chloride
e. aluminum bromide
f. lithium nitride
g. iron(II) oxide

Oxidation Number and Stock System

26. Name each of the following ionic compounds using the Stock system:
a. NaCl
b. KF
c. CaS
d. $Co(NO_3)_2$
e. $FePO_4$
f. Hg_2SO_4
g. $Hg_3(PO_4)_2$

27. Assign oxidation numbers to each atom in the following compounds. (Hint: See Sample Problem 7-5.)
a. HI
b. PBr_3
c. GeS_2
d. KH
e. As_2O_5
f. H_3PO_4

28. Assign oxidation numbers to each atom in the following ions. (Hint: See Sample Problem 7-5.)
a. NO_3^-
b. ClO_4^-
c. PO_4^{3-}
d. $Cr_2O_7^{2-}$
e. CO_3^{2-}

Mole Relationships and Percentage Composition

29. Determine the formula mass of each of the following compounds or ions. (Hint: See Sample Problem 7-6.)
a. glucose, $C_6H_{12}O_6$
b. calcium acetate, $Ca(CH_3COO)_2$
c. the ammonium ion, NH_4^+
d. the chlorate ion, ClO_3^-

30. Determine the number of moles of each type of monatomic or polyatomic ion in one mole of the following compounds. For each polyatomic ion, determine the number of moles of each atom present in one mole of the ion.
a. KNO_3
b. Na_2SO_4
c. $Ca(OH)_2$
d. $(NH_4)_2SO_3$
e. $Ca_3(PO_4)_2$
f. $Al_2(CrO_4)_3$

31. Determine the molar mass of each compound listed in item 30. (Hint: See Sample Problem 7-7.)

32. Determine the number of moles of compound in each of the following samples. (Hint: See Sample Problem 7-9.)
a. 4.50 g H_2O
b. 471.6 g $Ba(OH)_2$
c. 129.68 g $Fe_3(PO_4)_2$

33. Determine the percentage composition of each of the following compounds. (Hint: See Sample Problem 7-10.)
a. $NaCl$
b. $AgNO_3$
c. $Mg(OH)_2$

34. Determine the percentage by mass of water in the hydrate $CuSO_4 \cdot 5H_2O$. (Hint: See Sample Problem 7-11.)

35. Determine the empirical formula of a compound containing 63.50% silver, 8.25% nitrogen, and the remainder oxygen. (Hint: See Sample Problem 7-12.)

36. Determine the empirical formula of a compound found to contain 52.11% carbon, 13.14% hydrogen, and 34.75% oxygen.

37. A 1.344 g sample of a compound contains 0.365 g Na, 0.221 g N, and 0.758 g O. What is its empirical formula? (Hint: See Sample Problem 7-13.)

MIXED REVIEW

38. Chemical analysis of citric acid shows that it contains 37.51% C, 4.20% H, and 58.29% O. What is its empirical formula?

39. Name each of the following compounds using the Stock system:
a. $LiBr$
b. $Sn(NO_3)_2$
c. $FeCl_2$
d. MgO
e. KOH
f. Fe_2O_3
g. $AgNO_3$
h. $Fe(OH)_2$
i. CrF_2

40. What is the mass in grams of each of the following samples?
a. 1.000 mol $NaCl$
b. 2.000 mol H_2O
c. 3.500 mol $Ca(OH)_2$
d. 0.625 mol $Ba(NO_3)_2$

41. Determine the formula mass and molar mass of each of the following compounds:
a. XeF_4
b. $C_{12}H_{24}O_6$
c. Hg_2I_2
d. $CuCN$

42. Write the chemical formulas for the following compounds:
a. aluminum fluoride
b. magnesium oxide
c. vanadium(V) oxide
d. cobalt(II) sulfide
e. strontium bromide
f. sulfur trioxide

43. How many atoms of each element are contained in a single formula unit of iron(III) formate, $Fe(CHO_2)_3 \cdot H_2O$? What percentage by mass of the compound is water?

44. Name each of the following acids, and assign oxidation numbers to the atoms in each:
a. HNO_2
b. H_2SO_3
c. H_2CO_3
d. HI

45. What is the molecular formula of the molecule that has an empirical formula of CH_2O and a molar mass of 120.12 g/mol?

46. Determine the percentage composition of the following compounds:
a. $NaClO$
b. H_2SO_3
c. C_2H_5COOH
d. $BeCl_2$

47. Name each of the following binary compounds:
a. MgI_2
b. NaF
c. CS_2
d. N_2O_4
e. SO_2
f. PBr_3
g. $CaCl_2$
h. AgI

48. Assign oxidation numbers to each atom in the following molecules and ions:
a. CO_2
b. NH_4^+
c. MnO_4^-
d. $S_2O_3^{2-}$
e. H_2O_2
f. P_4O_{10}
g. OF_2

49. A 175.0 g sample of a compound contains 56.15 g C, 9.43 g H, 74.81 g O, 13.11 g N, and 21.49 g Na. What is its empirical formula?

CRITICAL THINKING

50. **Analyzing Information** Sulfur trioxide is produced in the atmosphere through a reaction of sulfur dioxide and oxygen. Sulfur dioxide is a primary air pollutant. List all of the chemical information you can by analyzing the formula for sulfur trioxide.

51. **Analyzing Data** In the laboratory, a sample of pure nickel was placed in a clean, dry, weighed crucible. The crucible was heated so that the nickel would react with the oxygen in the air. After the reaction appeared complete, the crucible was allowed to cool and the mass was determined. The crucible was reheated and allowed to cool. Its mass was then determined again to be certain that the reaction was complete. The following data were collected:

Mass of crucible	= 30.02 g
Mass of nickel and crucible	= 31.07 g
Mass of nickel oxide and crucible	= 31.36 g

Determine the following information based on the data given above:

Mass of nickel	=
Mass of nickel oxide	=
Mass of oxygen	=

Based on your calculations, what is the empirical formula for the nickel oxide?

TECHNOLOGY & LEARNING

52. **Graphing Calculator** Calculate the Molar Mass of a Compound

The graphing calculator can be programmed to calculate the molar mass of a compound given the chemical formula for the compound. Begin by programming the calculator to prompt for the number of elements in the formula, the number of atoms of each element in the formula, and the atomic mass of each element in the formula. Then use the program to find the molar masses of various compounds.

A. Program the calculation.

Keystrokes for naming the program:
PRGM ▶ ▶ ENTER

Name the program: M O L M A S S ENTER

Program keystrokes: 0 STO► ALPHA T 2nd : PRGM ▶ 3 2nd ALPHA " N U M ␣ E L E M " 2nd : PRGM ▶ 1 ALPHA E 2nd : PRGM 4 ALPHA I , 1 , ALPHA E , 1) 2nd : PRGM ▶ 2 ALPHA N 2nd : PRGM ▶ 3 2nd ALPHA " E L E M ␣ M A S S " 2nd : PRGM ▶ 1 ALPHA M 2nd : ALPHA M × ALPHA N + ALPHA T STO► ALPHA T 2nd : PRGM 7 2nd : PRGM ▶ 8 2nd : PRGM ▶ 3 2nd ALPHA " M O L ␣ M A S S 2nd TEST 1 ALPHA " 2nd : PRGM ▶ 3 ALPHA T

B. Check the display.

Screen display: 0->T: Disp "NUM ELEM": Input E: For(I, 1, E, 1): Prompt N: Disp "ELEM MASS": Input M: M*N+T->T: End: ClrHome: Disp "MOL MASS=": Disp T

Press 2nd QUIT to exit the program editor.

C. Run the program.

a. Press PRGM. Select MOLMASS and press ENTER ENTER. Enter the number of elements (3) present in the compound $C_{13}H_{18}O_2$ and press ENTER. Next, enter the number of atoms of carbon (13), and press ENTER. Enter the atomic mass of carbon, rounded to two places to the right of the decimal point(12.01), and press ENTER. Repeat for the 18 atoms of hydrogen (atomic mass 1.01) and 2 atoms of oxygen (atomic mass 16.00). The calculator will calculate the molar mass for the compound.

b. Use the MOLMASS program to calculate the molar mass of Na_2SO_4.

c. Use the MOLMASS program to calculate the molar mass of $Ba(NO_3)_2$.

HANDBOOK SEARCH

53. Review the common reactions of Group 1 metals in the *Elements Handbook* and answer the following:

a. Write the formulas for Group 1 metals that form superoxides.

b. What is the charge on each cation for the formulas you wrote in (a)?

c. How does the charge on the oxide anion vary for oxides, peroxides, and superoxides?

54. Review the common reactions of Group 2 metals in the *Elements Handbook* and answer the following:

a. Write formulas for those elements that form oxides.

b. Write formulas for those elements that form peroxides.

c. Write formulas for those elements that form nitrides.

d. Most Group 2 elements form hydrides. What is hydrogen's oxidation state in these compounds?

55. Review the analytical tests for transition metals in the *Elements Handbook* and answer the following:

a. Determine the oxidation state of each metal in the precipitates shown for cadmium, zinc, and lead.

b. Determine the oxidation state of each metal in the complex ions shown for iron, manganese, and cobalt.

c. The copper compound shown is called a coordination compound. The ammonia shown in the formula exists as molecules with no charge. Determine copper's oxidation state in this compound.

56. Review the common reactions of Group 15 elements in the *Elements Handbook* and answer the following:

a. Write formulas for each of the oxides listed for the Group 15 elements.

b. Determine nitrogen's oxidation state in the oxides listed in (a).

RESEARCH & WRITING

57. Nomenclature Biologists who name newly discovered organisms use a system that is structured very much like the one used by chemists in naming compounds. The system used by biologists is called the Linnaeus system, after its creator, Carolus Linnaeus. Research this system in a biology textbook, and then note similarities and differences between the Linnaeus system and chemical nomenclature.

58. Common Chemicals Find out the systematic chemical name and write the chemical formula for each of the following common compounds:

a. baking soda
b. milk of magnesia
c. Epsom salts
d. limestone
e. lye
f. wood alcohol

ALTERNATIVE ASSESSMENT

59. Performance Assessment Your teacher will supply you with a note card with one of the following formulas on it: $NaCH_3COO{\cdot}3H_2O$, $MgCl_2{\cdot}6H_2O$, $LiC_2H_3O_2{\cdot}2H_2O$, or $MgSO_4{\cdot}7H_2O$. Design an experiment to determine the percentage of water by mass in the hydrated salt assigned to you. Be sure to explain what steps you will take to ensure that the salt is completely dry. If your teacher approves your design, obtain the salt and perform the experiment. What percentage of water does the salt contain?

60. Both ammonia, NH_3, and ammonium nitrate, NH_4NO_3, are used in fertilizers as a source of nitrogen. Which compound has the higher percentage of nitrogen? Research the physical properties of both compounds, and find out how each is manufactured and used. Explain why each compound has its own particular application. (Consider factors such as the cost of raw ingredients, the ease of manufacture, shipping fees, and so forth.)

CHAPTER 8

Chemical Equations and Reactions

The evolution of light and heat is an indication that a chemical reaction is taking place.

Describing Chemical Reactions

SECTION 8-1

Objectives

- List three observations that suggest that a chemical reaction has taken place.
- List three requirements for a correctly written chemical equation.
- Write a word equation and a formula equation for a given chemical reaction.
- Balance a formula equation by inspection.

A *chemical reaction* is the process by which one or more substances are changed into one or more different substances. In any chemical reaction, the original substances are known as the *reactants* and the resulting substances are known as the *products*. According to the law of conservation of mass, the total mass of reactants must equal the total mass of products for any given chemical reaction.

Chemical reactions are described by chemical equations. *A* **chemical equation** *represents, with symbols and formulas, the identities and relative amounts of the reactants and products in a chemical reaction.* For example, the following chemical equation shows that the reactant ammonium dichromate yields the products nitrogen, chromium(III) oxide, and water.

$$(NH_4)_2Cr_2O_7(s) \longrightarrow N_2(g) + Cr_2O_3(s) + 4H_2O(g)$$

This strongly exothermic reaction is shown in Figure 8-1.

Indications of a Chemical Reaction

To know for certain that a chemical reaction has taken place requires evidence that one or more substances have undergone a change in identity. Absolute proof of such a change can be provided only by chemical analysis of the products. However, certain easily observed changes usually indicate that a chemical reaction has occurred.

1. *Evolution of heat and light.* A change in matter that releases energy as both heat and light is strong evidence that a chemical reaction has taken place. For example, you can see in Figure 8-1 that the decomposition of ammonium dichromate is accompanied by the evolution of much heat and light. And you can see evidence that a chemical reaction occurs between natural gas and oxygen if you burn gas for cooking in your house. Some reactions release only heat or only light. But the evolution of heat or light by itself is not necessarily a sign of chemical change because many physical changes also release either heat or light.

FIGURE 8-1 The decomposition of ammonium dichromate proceeds rapidly, releasing energy in the form of light and heat.

FIGURE 8-2 (a) The reaction of vinegar and baking soda is evidenced by the production of bubbles of carbon dioxide gas. (b) When water solutions of ammonium sulfide and cadmium nitrate are combined, the yellow precipitate cadmium sulfide forms.

(a)

(b)

2. *Production of a gas.* The evolution of gas bubbles when two substances are mixed is often evidence of a chemical reaction. For example, bubbles of carbon dioxide gas form immediately when baking soda is mixed with vinegar, as shown in Figure 8-2(a).
3. *Formation of a precipitate.* Many chemical reactions take place between substances that are dissolved in liquids. If a solid appears after two solutions are mixed, a reaction has likely occurred. *A solid that is produced as a result of a chemical reaction in solution and that separates from the solution is known as a* **precipitate.** A precipitate-forming reaction is shown in Figure 8-2(b).

Characteristics of Chemical Equations

A properly written chemical equation can summarize any chemical change. The following requirements will aid you in writing and reading chemical equations correctly.

1. *The equation must represent known facts.* All reactants and products must be identified, either through chemical analysis in the laboratory or from sources that give the results of experiments.
2. *The equation must contain the correct formulas for the reactants and products.* Remember what you learned in Chapter 7 about symbols and formulas. Knowledge of the common oxidation states of the elements and of methods of writing formulas will enable you to supply formulas for reactants and products if they are not available. Recall that the elements listed in Table 8-1 exist primarily as diatomic molecules, such as H_2 and O_2. Each of these elements is represented in an equation by its molecular formula. Other elements in the elemental state are usually represented simply by their atomic symbols. For example, iron is represented as Fe and carbon is represented as C. The symbols are not given any subscripts because the elements do not form definite molecular structures. Two exceptions to this rule are

TABLE 8-1 Elements That Normally Exist as Diatomic Molecules

Element	Symbol	Molecular formula	Physical state at room temperature
Hydrogen	H	H_2	gas
Nitrogen	N	N_2	gas
Oxygen	O	O_2	gas
Fluorine	F	F_2	gas
Chlorine	Cl	Cl_2	gas
Bromine	Br	Br_2	liquid
Iodine	I	I_2	solid

sulfur, which is usually written S_8, and phosphorus, which is usually written P_4. In these cases, the formulas reflect each element's unique atomic arrangement in its natural state.

3. *The law of conservation of mass must be satisfied.* Atoms are neither created nor destroyed in ordinary chemical reactions. Therefore, the same number of atoms of each element must appear on each side of a correct chemical equation. To equalize numbers of atoms, coefficients are added where necessary. *A* **coefficient** *is a small whole number that appears in front of a formula in a chemical equation.* Placing a coefficient in front of a formula specifies the relative number of moles of the substance; if no coefficient is written, the coefficient is assumed to be 1. For example, the coefficient 4 in the equation on page 241 indicates that 4 mol of water are produced for each mole of nitrogen and chromium(III) oxide that is produced.

Word and Formula Equations

The first step in writing a chemical equation is to identify the facts to be represented. It is often helpful to write a **word equation,** *an equation in which the reactants and products in a chemical reaction are represented by words.* A word equation has only qualitative (descriptive) meaning. It does not give the whole story because it does not give the quantities of reactants used or products formed.

Consider the reaction of methane, the principal component of natural gas, with oxygen. When methane burns in air, it combines with oxygen to produce carbon dioxide and water vapor. In the reaction, methane and oxygen are the reactants, and carbon dioxide and water are the products. The word equation for the reaction of methane and oxygen is written as follows.

$$\text{methane} + \text{oxygen} \longrightarrow \text{carbon dioxide} + \text{water}$$

The arrow, $\longrightarrow$, is read as *react to yield* or *yield* (also *produce* or *form*). So the equation above is read, "methane and oxygen react to yield carbon dioxide and water," or simply, "methane and oxygen yield carbon dioxide and water."

The next step in writing a correct chemical equation is to replace the names of the reactants and products with appropriate symbols and formulas. Methane is a molecular compound composed of one carbon atom and four hydrogen atoms. Its chemical formula is CH_4. Recall that oxygen exists in nature as diatomic molecules; it is therefore represented as O_2. The correct formulas for carbon dioxide and water are CO_2 and H_2O, respectively.

A **formula equation** *represents the reactants and products of a chemical reaction by their symbols or formulas.* The formula equation for the reaction of methane and oxygen is written as follows.

$$CH_4(g) + O_2(g) \longrightarrow CO_2(g) + H_2O(g) \quad \text{(not balanced)}$$

The *g* in parentheses after each formula indicates that the corresponding substance is in the gaseous state. Like a word equation, a formula equation is a qualitative statement. It gives no information about the amounts of reactants or products.

A formula equation meets two of the three requirements for a correct chemical equation. It represents the facts and shows the correct symbols and formulas for the reactants and products. To complete the process of writing a correct equation, the law of conservation of mass must be taken into account. The relative amounts of reactants and products represented in the equation must be adjusted so that the numbers and types of atoms are the same on both sides of the equation. This process is called *balancing an equation* and is carried out by inserting coefficients. Once it is balanced, a formula equation is a correctly written chemical equation.

Look again at the formula equation for the reaction of methane and oxygen.

$$CH_4(g) + O_2(g) \longrightarrow CO_2(g) + H_2O(g) \quad \text{(not balanced)}$$

To balance the equation, begin by counting atoms of elements that are combined with atoms of other elements and that appear only once on each side of the equation. In this case, we could begin by counting either carbon or hydrogen atoms. Usually, the elements hydrogen and oxygen are balanced only after balancing all other elements in an equation. (You will read more about the rules of balancing equations later in the chapter.) Thus, we begin by counting carbon atoms.

Inspecting the formula equation reveals that there is one carbon atom on each side of the arrow. Therefore, carbon is already balanced in the equation. Counting hydrogen atoms reveals that there are four hydrogen atoms in the reactants but only two in the products. Two additional hydrogen atoms are needed on the right side of the equation. They can be added by placing the coefficient 2 in front of the chemical formula H_2O.

$$CH_4(g) + O_2(g) \longrightarrow CO_2(g) + 2H_2O(g) \quad \text{(partially balanced)}$$

A coefficient multiplies the number of atoms of each element indicated in a chemical formula. Thus, $2H_2O$ represents *four* H atoms and *two* O atoms. To add two more hydrogen atoms to the right side of the equation, you may be tempted to change the subscript in the formula of water so that H_2O becomes H_4O. However, this would be a mistake because changing the subscripts of a chemical formula changes the *identity* of the compound. H_4O is not a product in the combustion of methane. In fact, there is no such compound. One must use only coefficients to change the relative number of atoms in a chemical equation because coefficients change the numbers of atoms without changing the identities of the reactants or products.

Now consider the number of oxygen atoms. There are four oxygen atoms on the right side of the arrow in the partially balanced equation. Yet there are only two oxygen atoms on the left side of the arrow. We can increase the number of oxygen atoms on the left side to four by placing the coefficient 2 in front of the molecular formula for oxygen. This results in a correct chemical equation, or *balanced formula equation,* for the burning of methane in oxygen.

$$CH_4(g) + 2O_2(g) \longrightarrow CO_2(g) + 2H_2O(g)$$

This reaction is further illustrated in Figure 8-3.

Additional Symbols Used in Chemical Equations

Table 8-2 on page 246 summarizes the symbols commonly used in chemical equations. Sometimes a gaseous product is indicated by an arrow pointing upward, ↑, instead of (g), as shown in the table. A downward arrow, ↓, is often used to show the formation of a precipitate during a reaction in solution.

The conditions under which a reaction takes place are often indicated by placing information above or below the reaction arrow. The word *heat,*

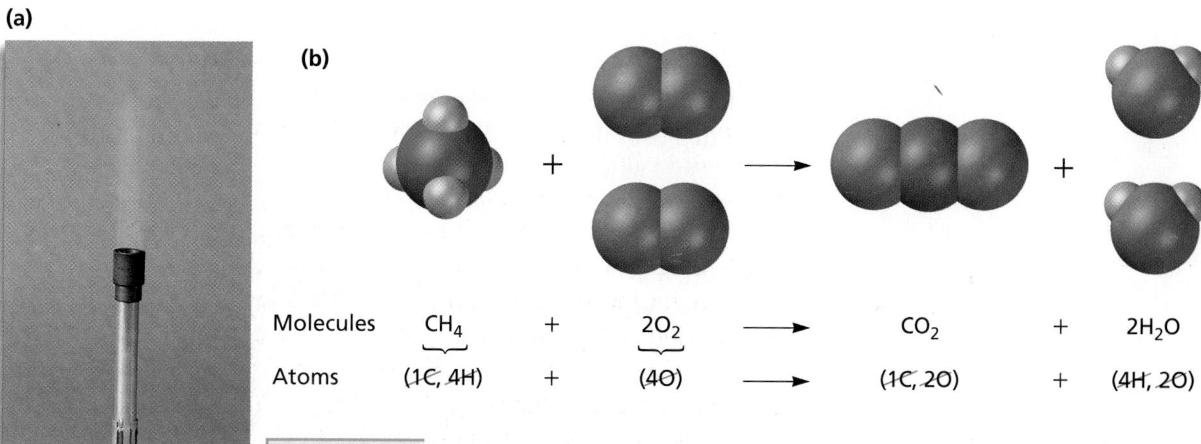

FIGURE 8-3 (a) In a Bunsen burner, methane combines with oxygen in the air to form carbon dioxide and water vapor. (b) The reaction is represented by both a molecular model and a balanced equation. Each shows that the number of atoms of each element in the reactants equals the number of atoms of each element in the products.

TABLE 8-2 *Symbols Used in Chemical Equations*

Symbol	Explanation
$\longrightarrow$	"Yields"; indicates result of reaction
$\rightleftarrows$	Used in place of a single arrow to indicate a reversible reaction
(s)	A reactant or product in the solid state; also used to indicate a precipitate
$\downarrow$	Alternative to (s), but used only to indicate a precipitate
(l)	A reactant or product in the liquid state
(aq)	A reactant or product in an aqueous solution (dissolved in water)
(g)	A reactant or product in the gaseous state
$\uparrow$	Alternative to (g), but used only to indicate a gaseous product
$\xrightarrow{\Delta}$ or $\xrightarrow{\text{heat}}$	Reactants are heated
$\xrightarrow{\text{2 atm}}$	Pressure at which reaction is carried out, in this case 2 atm
$\xrightarrow{\text{pressure}}$	Pressure at which reaction is carried out exceeds normal atmospheric pressure
$\xrightarrow{0^\circ\text{C}}$	Temperature at which reaction is carried out, in this case 0°C
$\xrightarrow{MnO_2}$	Formula of catalyst, in this case manganese dioxide, used to alter the rate of the reaction

symbolized by a Greek capital delta, Δ, indicates that the reactants must be heated. The specific temperature at which a reaction occurs may also be written over the arrow. For some reactions, it is important to specify the pressure at which the reaction occurs or to specify that the pressure must be above normal. Many reactions are speeded up and can take place at lower temperatures in the presence of a *catalyst.* A catalyst is a substance that changes the rate of a chemical reaction but can be recovered unchanged. To show that a catalyst is present, the formula for the catalyst or the word *catalyst* is written over the reaction arrow.

In many reactions, as soon as the products begin to form, they immediately begin to react with each other and re-form the reactants. In other words, the reverse reaction also occurs. The reverse reaction may occur to a greater or lesser degree than the original reaction, depending on the specific reaction and the conditions. *A* **reversible reaction** *is a chemical reaction in which the products re-form the original* *reactants.* The reversibility of a reaction is indicated by writing two arrows pointing in opposite directions. For example, the reversible reaction between iron and water vapor is written as follows.

$$3Fe(s) + 4H_2O(g) \rightleftharpoons Fe_3O_4(s) + 4H_2(g)$$

With an understanding of all the symbols and formulas used, it is possible to translate a chemical equation into a sentence. Consider the following equation.

$$2HgO(s) \xrightarrow{\Delta} 2Hg(l) + O_2(g)$$

Translated into a sentence, this equation reads, "When heated, solid mercury(II) oxide yields liquid mercury and gaseous oxygen."

It is also possible to write a chemical equation from a sentence describing a reaction. Consider the sentence, "Under pressure and in the presence of a platinum catalyst, gaseous ethene and hydrogen form gaseous ethane." This sentence can be translated into the following equation.

$$C_2H_4(g) + H_2(g) \xrightarrow[\text{Pt}]{\text{pressure}} C_2H_6(g)$$

Throughout this chapter we will often include the symbols for physical states (*s, l, g,* and *aq*) in balanced formula equations. You should be able to interpret these symbols when they are used and to supply them when the necessary information is available.

SAMPLE PROBLEM 8-1

Write word and formula equations for the chemical reaction that occurs when solid sodium oxide is added to water at room temperature and forms sodium hydroxide (dissolved in the water). Include symbols for physical states in the formula equation. Then balance the formula equation to give a balanced chemical equation.

SOLUTION The word equation must show the reactants, sodium oxide and water, to the left of the arrow. The product, sodium hydroxide, must appear to the right of the arrow.

$$\text{sodium oxide} + \text{water} \longrightarrow \text{sodium hydroxide}$$

The word equation is converted to a formula equation by replacing the name of each compound with the appropriate chemical formula. To do this requires knowing that sodium has an oxidation state of +1, that oxygen usually has an oxidation state of –2, and that a hydroxide ion has a charge of 1–.

$$Na_2O + H_2O \longrightarrow NaOH \quad \text{(not balanced)}$$

Adding symbols for the physical states of the reactants and products and the coefficient 2 in front of NaOH produces a balanced chemical equation.

$$Na_2O(s) + H_2O(l) \longrightarrow 2NaOH(aq)$$

SAMPLE PROBLEM 8-2

Translate the following chemical equation into a sentence:

$$PbCl_2(aq) + Na_2CrO_4(aq) \longrightarrow PbCrO_4(s) + 2NaCl(aq)$$

SOLUTION Each reactant is an ionic compound and is named according to the rules for such compounds. Both reactants are in aqueous solution. One product is a precipitate and the other remains in solution. The equation is translated as follows: Aqueous solutions of lead(II) chloride and sodium chromate react to produce a precipitate of lead(II) chromate plus sodium chloride in aqueous solution.

PRACTICE

1. Write word and balanced chemical equations for the following reactions. Include symbols for physical states when indicated.
   a. Solid calcium reacts with solid sulfur to produce solid calcium sulfide.
   b. Hydrogen gas reacts with fluorine gas to produce hydrogen fluoride gas. (Hint: See Table 8-1.)
   c. Solid aluminum metal reacts with aqueous zinc chloride to produce solid zinc metal and aqueous aluminum chloride.

Answer

1. a. calcium + sulfur $\longrightarrow$ calcium sulfide; $8Ca(s) + S_8(s) \longrightarrow 8CaS(s)$
   b. hydrogen + fluorine $\longrightarrow$ hydrogen fluoride; $H_2(g) + F_2(g) \longrightarrow 2HF(g)$
   c. aluminum + zinc chloride $\longrightarrow$ zinc + aluminum chloride; $2Al(s) + 3ZnCl_2(aq) \longrightarrow 3Zn(s) + 2AlCl_3(aq)$

2. Translate the following chemical equations into sentences:
   a. $CS_2(l) + 3O_2(g) \longrightarrow CO_2(g) + 2SO_2(g)$
   b. $NaCl(aq) + AgNO_3(aq) \longrightarrow NaNO_3(aq) + AgCl(s)$

2. a. Liquid carbon disulfide reacts with oxygen gas to produce carbon dioxide gas and sulfur dioxide gas.
   b. Aqueous solutions of sodium chloride and silver nitrate react to produce aqueous sodium nitrate and a precipitate of silver chloride.

Significance of a Chemical Equation

Chemical equations are very useful in doing quantitative chemical work. The arrow in a balanced chemical equation is like an equal sign. And the chemical equation as a whole is similar to an algebraic equation in that it expresses an equality. Let's examine some of the quantitative information revealed by a chemical equation.

1. *The coefficients of a chemical reaction indicate relative, not absolute, amounts of reactants and products.* A chemical equation usually shows the smallest numbers of atoms, molecules, or ions that will satisfy the law of conservation of mass in a given chemical reaction.

Consider the equation for the formation of hydrogen chloride from hydrogen and chlorine.

$$H_2(g) + Cl_2(g) \longrightarrow 2HCl(g)$$

The equation indicates that 1 molecule of hydrogen reacts with 1 molecule of chlorine to produce 2 molecules of hydrogen chloride, giving the following molecular ratio of reactants and products.

$$1 \text{ molecule } H_2 : 1 \text{ molecule } Cl_2 : 2 \text{ molecules HCl}$$

This ratio shows the smallest possible relative amounts of the reaction's reactants and products. To obtain larger relative amounts, we simply multiply each coefficient by the same number. Thus, 20 molecules of hydrogen would react with 20 molecules of chlorine to yield 40 molecules of hydrogen chloride. The reaction can also be considered in terms of amounts in moles: 1 mol of hydrogen molecules reacts with 1 mol of chlorine molecules to yield 2 mol of hydrogen chloride molecules.

2. *The relative masses of the reactants and products of a chemical reaction can be determined from the reaction's coefficients.* Recall from Figure 7-4 on page 224 that an amount of an element or compound in moles can be converted to a mass in grams by multiplying by the appropriate molar mass. We know that 1 mol of hydrogen reacts with 1 mol of chlorine to yield 2 mol of hydrogen chloride. The relative masses of the reactants and products are calculated as follows.

$$\cancel{1 \text{ mol } H_2} \times \frac{2.02 \text{ g } H_2}{\cancel{\text{mol } H_2}} = 2.02 \text{ g } H_2$$

$$\cancel{1 \text{ mol } Cl_2} \times \frac{70.90 \text{ g } Cl_2}{\cancel{\text{mol } Cl_2}} = 70.90 \text{ g } Cl_2$$

$$\cancel{2 \text{ mol HCl}} \times \frac{36.46 \text{ g HCl}}{\cancel{\text{mol HCl}}} = 72.92 \text{ g HCl}$$

The chemical equation shows that 2.02 g of hydrogen will react with 70.90 g of chlorine to yield 72.92 g of hydrogen chloride.

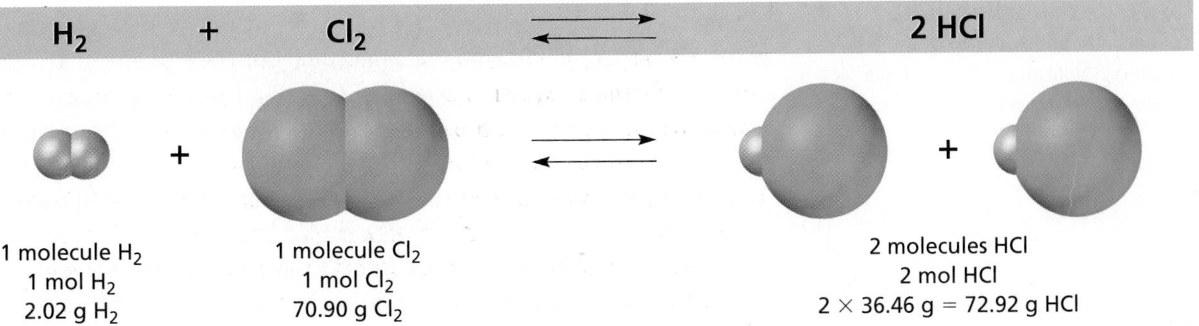

FIGURE 8-4 This representation of the reaction of hydrogen and chlorine to yield hydrogen chloride shows several ways to interpret the quantitative information of a chemical reaction.

3. *The reverse reaction for a chemical equation has the same relative amounts of substances as the forward reaction.* Because a chemical equation is like an algebraic equation, the equality can be read in either direction. Reading the hydrogen chloride formation equation on page 249 from right to left, we can see that 2 molecules of hydrogen chloride break down to form 1 molecule of hydrogen plus 1 molecule of chlorine. Similarly, 2 mol (72.92 g) of hydrogen chloride yield 1 mol (2.02 g) of hydrogen and 1 mol (70.90 g) of chlorine.

We have seen that a chemical equation provides useful quantitative information about a chemical reaction. However, there is also important information that is *not* provided by a chemical equation. For instance, an equation gives no indication of whether a reaction will actually occur. A chemical equation can be written for a reaction that may not even take place. Some guidelines about the types of simple reactions that can be expected to occur are given in Sections 8-2 and 8-3. And later chapters provide additional guidelines for other types of reactions. In all these guidelines, it is important to remember that experimentation forms the basis for predicting whether a particular chemical reaction will occur.

In addition, chemical equations give no information about the speed at which reactions occur or about how the bonding between atoms or ions changes during the reaction. These aspects of chemical reactions are discussed in Chapter 17.

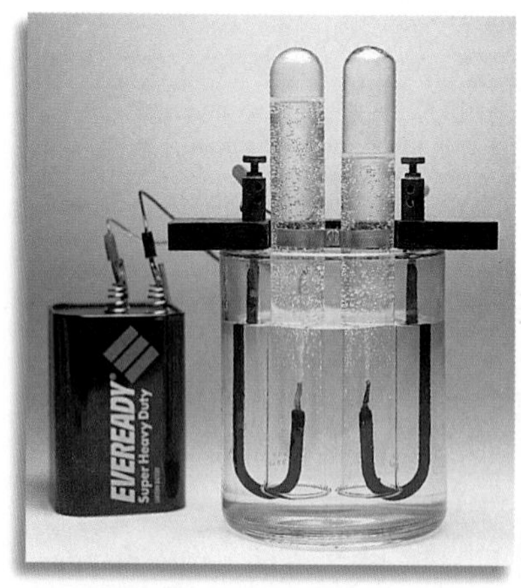

FIGURE 8-5 When an electric current is passed through water that has been made slightly conductive, the water molecules break down to yield hydrogen (in tube at right) and oxygen (in tube at left). Bubbles of each gas are evidence of the reaction. Note that twice as much hydrogen as oxygen is produced.

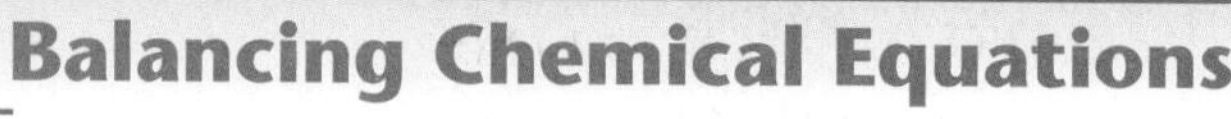

Balancing Chemical Equations

Most of the equations in the remainder of this chapter can be balanced by inspection. The following procedure demonstrates how to master balancing equations by inspection using a step-by-step approach. The equation for the decomposition of water (see Figure 8-5) will be used as an example.

1. *Identify the names of the reactants and the products, and write a word equation.* The word equation for the reaction shown in Figure 8-5 is written as follows.

$$\text{water} \longrightarrow \text{hydrogen} + \text{oxygen}$$

2. *Write a formula equation by substituting correct formulas for the names of the reactants and the products.* We know that the formula for water is H_2O. And recall that both hydrogen and oxygen exist as diatomic molecules. Therefore, their correct formulas are H_2 and O_2, respectively.

$$H_2O(l) \longrightarrow H_2(g) + O_2(g) \quad \text{(not balanced)}$$

3. *Balance the formula equation according to the law of conservation of mass.* This last step is done by trial and error. Coefficients are changed and the numbers of atoms are counted on both sides of the equation. When the numbers of each type of atom are the same for both the products and the reactants, the equation is balanced. The trial-and-error method of balancing equations is made easier by the use of the following guidelines.

- Balance the different types of atoms one at a time.
- First balance the atoms of elements that are combined and that appear only once on each side of the equation.
- Balance polyatomic ions that appear on both sides of the equation as single units.
- Balance H atoms and O atoms after atoms of all other elements have been balanced.

The formula equation in our example shows that there are two oxygen atoms on the right and only one on the left. To balance oxygen atoms, the number of H_2O molecules must be increased. Placing the coefficient 2 before H_2O gives the necessary two oxygen atoms on the left.

$$2H_2O(l) \longrightarrow H_2(g) + O_2(g) \quad \text{(partially balanced)}$$

The coefficient 2 in front of H_2O has upset the balance of hydrogen atoms. Placing the coefficient 2 in front of hydrogen, H_2, on the right, gives an equal number of hydrogen atoms (4) on both sides of the equation.

$$2H_2O(l) \longrightarrow 2H_2(g) + O_2(g)$$

4. *Count atoms to be sure that the equation is balanced.* Make sure that equal numbers of atoms of each element appear on both sides of the arrow.

$$\begin{array}{ccc} 2H_2O(l) & \longrightarrow & 2H_2(g) + O_2(g) \\ (4H + 2O) & = & (4H) \; + \; (2O) \end{array}$$

Occasionally at this point, the coefficients do not represent the smallest possible whole-number ratio of reactants and products. When this happens, the coefficients should be divided by their greatest common factor in order to obtain the smallest possible whole-number coefficients.

Balancing chemical equations by inspection becomes easier as you gain experience. Learn to avoid the most common mistakes: (1) writing incorrect chemical formulas for reactants or products and (2) trying to balance an equation by changing subscripts. Remember that subscripts cannot be added, deleted, or changed. Eventually, you will probably be able to skip writing the word equation and each separate step. However, *do not* leave out the final step of counting atoms to be sure the equation is balanced.

SAMPLE PROBLEM 8-3

The reaction of zinc with aqueous hydrochloric acid produces a solution of zinc chloride and hydrogen gas. This reaction is shown at right in Figure 8-6. Write a balanced chemical equation for the reaction.

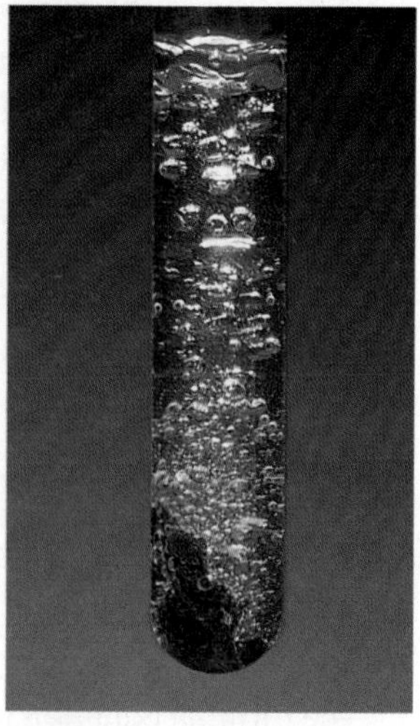

FIGURE 8-6 Solid zinc reacts with hydrochloric acid to form aqueous zinc chloride and hydrogen gas.

SOLUTION

1 ANALYZE *Write the word equation.*

$$\text{zinc} + \text{hydrochloric acid} \longrightarrow \text{zinc chloride} + \text{hydrogen}$$

2 PLAN *Write the formula equation.*

$$Zn(s) + HCl(aq) \longrightarrow ZnCl_2(aq) + H_2(g) \quad \text{(not balanced)}$$

3 COMPUTE *Adjust the coefficients.* Note that chlorine and hydrogen each appear only once on each side of the equation. We balance chlorine first because it is combined on both sides of the equation. Also, recall from the guidelines on page 251 that hydrogen and oxygen are balanced only after all other elements in the reaction are balanced. To balance chlorine, we place the coefficient 2 before HCl. Two molecules of hydrogen chloride also yield the required two hydrogen atoms on the right. Finally, note that there is one zinc atom on each side in the formula equation. Therefore, no further coefficients are needed.

$$Zn(s) + 2HCl(aq) \longrightarrow ZnCl_2(aq) + H_2(g)$$

4 EVALUATE *Count atoms to check balance.*

$$Zn(s) + 2HCl(aq) \longrightarrow ZnCl_2(aq) + H_2(g)$$
$$(1Zn) + (2H + 2Cl) = (1Zn + 2Cl) + (2H)$$

The equation is balanced.

PRACTICE

1. Write word, formula, and balanced chemical equations for each of the following reactions:
   a. Magnesium and hydrochloric acid react to produce magnesium chloride and hydrogen.
   b. Aqueous nitric acid reacts with solid magnesium hydroxide to produce aqueous magnesium nitrate and water.

Answer

1. a. Word: magnesium + hydrochloric acid ⟶ magnesium chloride + hydrogen
   Formula: $Mg + HCl \longrightarrow MgCl_2 + H_2$
   Balanced: $Mg + 2HCl \longrightarrow MgCl_2 + H_2$
   b. Word: nitric acid + magnesium hydroxide ⟶ magnesium nitrate + water
   Formula: $HNO_3(aq) + Mg(OH)_2(s) \longrightarrow Mg(NO_3)_2(aq) + H_2O(l)$
   Balanced: $2HNO_3(aq) + Mg(OH)_2(s) \longrightarrow Mg(NO_3)_2(aq) + 2H_2O(l)$

SAMPLE PROBLEM 8-4

Solid aluminum carbide, Al_4C_3, reacts with water to produce methane gas and solid aluminum hydroxide. Write a balanced chemical equation for this reaction.

SOLUTION The reactants are aluminum carbide and water. The products are methane and aluminum hydroxide. The formula equation is written as follows.

$$Al_4C_3(s) + H_2O(l) \longrightarrow CH_4(g) + Al(OH)_3(s) \quad \text{(not balanced)}$$

Begin balancing the formula equation by counting either aluminum atoms or carbon atoms. (Remember that hydrogen and oxygen atoms are balanced last.) There are four Al atoms on the left. To balance Al atoms, place the coefficient 4 before $Al(OH)_3$ on the right.

$$Al_4C_3(s) + H_2O(l) \longrightarrow CH_4(g) + 4Al(OH)_3(s) \quad \text{(partially balanced)}$$

Now balance the carbon atoms. With three C atoms on the left, the coefficient 3 must be placed before CH_4 on the right.

$$Al_4C_3(s) + H_2O(l) \longrightarrow 3CH_4(g) + 4Al(OH)_3(s) \quad \text{(partially balanced)}$$

Balance oxygen atoms next because oxygen, unlike hydrogen, appears only once on each side of the equation. There is one O atom on the left and 12 O atoms in the four $Al(OH)_3$ formula units on the right. Placing the coefficient 12 before H_2O balances the O atoms.

$$Al_4C_3(s) + 12H_2O(l) \longrightarrow 3CH_4(g) + 4Al(OH)_3(s)$$

This leaves the hydrogen atoms to be balanced. There are 24 H atoms on the left. On the right, there are 12 H atoms in the methane molecules and 12 in the aluminum hydroxide formula units, totaling 24 H atoms. The H atoms are balanced.

$$\begin{array}{ccccccc} Al_4C_3(s) & + & 12H_2O(l) & \longrightarrow & 3CH_4(g) & + & 4Al(OH)_3(s) \\ (4\cancel{Al} + 3\cancel{C}) & + & (24\cancel{H} + 12\cancel{O}) & = & (3\cancel{C} + 12\cancel{H}) & + & (4\cancel{Al} + 12\cancel{H} + 12\cancel{O}) \end{array}$$

The equation is balanced.

SAMPLE PROBLEM 8-5

Aluminum sulfate and calcium hydroxide are used in a water-purification process. When added to water, they dissolve and react to produce two insoluble products, aluminum hydroxide and calcium sulfate. These products settle out, taking suspended solid impurities with them. Write a balanced chemical equation for the reaction.

SOLUTION Each of the reactants and products is an ionic compound. Recall from Chapter 7 that the formulas of ionic compounds are determined by the charges of the ions composing each compound. The formula reaction is thus written as follows.

$$Al_2(SO_4)_3 + Ca(OH)_2 \longrightarrow Al(OH)_3 + CaSO_4 \quad \text{(not balanced)}$$

There is one Ca atom on each side of the equation, so the calcium atoms are already balanced. There are two Al atoms on the left and one Al atom on the right. Placing the coefficient 2 in front of $Al(OH)_3$ produces the same number of Al atoms on each side of the equation.

$$Al_2(SO_4)_3 + Ca(OH)_2 \longrightarrow 2Al(OH)_3 + CaSO_4 \text{ (partially balanced)}$$

Next, checking SO_4^{2-} ions shows that there are three SO_4^{2-} ions on the left side of the equation and only one on the right side. Placing the coefficient 3 before $CaSO_4$ gives an equal number of SO_4^{2-} ions on each side.

$$Al_2(SO_4)_3 + Ca(OH)_2 \longrightarrow 2Al(OH)_3 + 3CaSO_4 \text{ (partially balanced)}$$

There are now three Ca atoms on the right, however. By placing the coefficient 3 in front of $Ca(OH)_2$, we once again have an equal number of Ca atoms on each side. This last step also gives six OH^- ions on both sides of the equation.

$$Al_2(SO_4)_3(aq) + 3Ca(OH)_2(aq) \longrightarrow 2Al(OH)_3(s) + 3CaSO_4(s)$$
$$(\cancel{2Al} + \cancel{3SO_4^{2-}}) + (\cancel{3Ca} + \cancel{6OH^-}) = (\cancel{2Al} + \cancel{6OH^-}) + (\cancel{3Ca} + \cancel{3SO_4^{2-}})$$

The equation is balanced.

PRACTICE

1. Write balanced chemical equations for each of the following reactions:
   a. Solid sodium combines with chlorine gas to produce solid sodium chloride.
   b. When solid copper reacts with aqueous silver nitrate, the products are aqueous copper(II) nitrate and solid silver.
   c. In a blast furnace, the reaction between solid iron(III) oxide and carbon monoxide gas produces solid iron and carbon dioxide gas.

Answer

1. a. $2Na(s) + Cl_2(g) \longrightarrow 2NaCl(s)$
   b. $Cu(s) + 2AgNO_3(aq) \longrightarrow Cu(NO_3)_2(aq) + 2Ag(s)$
   c. $Fe_2O_3(s) + 3CO(g) \longrightarrow 2Fe(s) + 3CO_2(g)$

SECTION REVIEW

1. Describe the differences between word equations, formula equations, and chemical equations.
2. Write word and formula equations for the reaction in which aqueous solutions of sulfuric acid and sodium hydroxide react to form aqueous sodium sulfate and water.
3. Translate the following chemical equations into sentences:
   a. $2K(s) + 2H_2O(l) \longrightarrow 2KOH(aq) + H_2(g)$
   b. $2Fe(s) + 3Cl_2(g) \longrightarrow 2FeCl_3(s)$
4. Write the word, formula, and chemical equations for the reaction between hydrogen sulfide gas and oxygen gas that produces sulfur dioxide gas and water vapor.
5. Write the chemical equation for each of the following reactions:
   a. ammonium chloride + calcium hydroxide $\longrightarrow$ calcium chloride + ammonia + water
   b. hexane, C_6H_{14}, + oxygen $\longrightarrow$ carbon dioxide + water

A Chemical Mystery

From "The Chemical Adventures of Sherlock Holmes: The Hound of Henry Armitage" by Thomas G. Waddell and Thomas R. Rybolt in *The Journal of Chemical Education*

"I knew it," the old man snapped. "He was poisoned, wasn't he? . . ."

. . . But Holmes was not listening. He had picked up the dog's bowl, now empty, and was vigorously sniffing, not unlike the hound itself, at the crusted remains of the last meal . . .

An hour later I was in my chair at 221B Baker Street. Holmes was in his laboratory and I could hear him humming. In the background was the usual clattering and clanking of laboratory equipment . . . Suddenly, Holmes called to me.

"Watson, come here. I need you." . . . He calmly scribbled an equation on a slip of paper and handed it to me. "If you can balance this equation, Watson, you can solve this mystery." I looked at the page as best I could and saw the following equation with the formula of a reactant clearly missing.

$$C_6H_5\text{—}NH_2 + 3KOH + \_\_\_\_ \Rightarrow$$
$$C_6H_5\text{—}NC + 3KCl + 3H_2O$$

Holmes paced back and forth with his hands clasped behind his back. "One part aniline, three parts potassium hydroxide, and one part unknown poison yields one part phenylisocyanide, three parts potassium chloride, and three parts water. The missing reactant can be identified by balancing the equation with respect to all the atoms involved. The product phenylisocyanide . . . is derived by this reaction from that missing chemical which was the poison deliberately placed in the hound's food."

"I can follow you part of the way," I submitted. "You undoubtedly detected a foreign substance in the dog food due to a characteristic aroma."

"Correct, Watson," Holmes replied. "And as a chemist I knew immediately that the poison was *volatile* . . . We observed the compound to be a liquid at room temperature, immiscible with water, and having a density greater than 1.00! The unpleasant sweetness of it was also very helpful. The possibilities were quite limited at that point, Watson . . .

I formed a working hypothesis and performed a known chemical test for such a poisonous liquid meeting all these criteria. Did you balance the equation, Watson? The equation confirms it!"

"I can do it, Holmes. I remember that much chemistry. Let me see . . . the missing reactant must have chlorine . . . 3 units to balance Cl in the product!"

"Very good, Watson. Go on with it."

"It gets more complex, now, but look, there is one extra carbon atom in the products! Is CCl_3 the compound?"

"Carbon makes *four* bonds, Watson, not three," said Holmes with a frown.

"I have it! $CHCl_3$ balances the equation! That's *chloroform,* Holmes! Of course. It all is consistent."

Reading for Meaning

Can you infer the meaning of the word *volatile* from the story? Write down your definition. Then compare your definition with one from a chemical or technical dictionary.

Read Further

Chloroform is the common name for the compound that poisoned the hound in the story. The chemical formula for chloroform is $CHCl_3$. Write chloroform's chemical name using the prefix system of naming molecular compounds. (Hint: See Chapters 7 and 20.)

SECTION 8-2

Types of Chemical Reactions

OBJECTIVES

- Define and give general equations for *synthesis, decomposition, single-replacement,* and *double-replacement* reactions.
- Classify a reaction as synthesis, decomposition, single-replacement, double-replacement, or combustion.
- List three types of synthesis reactions and six types of decomposition reactions.
- List four types of single-replacement reactions and three types of double-replacement reactions.
- Predict the products of simple reactions given the reactants.

Thousands of known chemical reactions occur in living systems, in industrial processes, and in chemical laboratories. Often it is necessary to predict the products formed in one of these reactions. Memorizing the equations for so many chemical reactions would be a difficult task. It is therefore more useful and realistic to classify reactions according to various similarities and regularities. This general information about reaction types can then be used to predict the products of specific reactions.

There are several different ways to classify chemical reactions, and none is entirely satisfactory. The classification scheme described in this section provides an introduction to five basic types of reactions: synthesis, decomposition, single-replacement, double-replacement, and combustion. In later chapters you will be introduced to categories that are useful in classifying other types of chemical reactions.

Synthesis Reactions

In a **synthesis reaction,** *also known as a* **composition reaction,** *two or more substances combine to form a new compound.* This type of reaction is represented by the following general equation.

$$A + X \longrightarrow AX$$

A and X can be elements or compounds. AX is a compound. The following examples illustrate several kinds of synthesis reactions.

Reactions of Elements with Oxygen and Sulfur

One simple type of synthesis reaction is the combination of an element with oxygen to produce an *oxide* of the element. Almost all metals react with oxygen to form oxides. For example, when a thin strip of magnesium metal is placed in an open flame, it burns with bright white light. When the metal strip is completely burned, only a fine white powder of magnesium oxide is left. This chemical reaction, shown in Figure 8-7, is represented by the following equation.

$$2Mg(s) + O_2(g) \longrightarrow 2MgO(s)$$

The other Group 2 elements react in a similar manner, forming oxides with the formula MO, where M represents the metal. The Group 1 metals form oxides with the formula M_2O, for example, Li_2O. The Group 1 and Group 2 elements react similarly with sulfur, forming *sulfides* with the formulas M_2S and MS, respectively. Examples of these types of synthesis reactions are shown below.

$$16Rb(s) + S_8(s) \longrightarrow 8Rb_2S(s)$$
$$8Ba(s) + S_8(s) \longrightarrow 8BaS(s)$$

Some metals, such as iron, combine with oxygen to produce two different oxides.

$$2Fe(s) + O_2(g) \longrightarrow 2FeO(s)$$
$$4Fe(s) + 3O_2(g) \longrightarrow 2Fe_2O_3(s)$$

In the product of the first reaction, iron is in an oxidation state of +2. In the product of the second reaction, iron is in an oxidation state of +3. The particular oxide formed depends on the conditions surrounding the reactants. Both oxides are shown below in Figure 8-8.

Nonmetals also undergo synthesis reactions with oxygen to form oxides. Sulfur, for example, reacts with oxygen to form sulfur dioxide. And when carbon is burned in air, carbon dioxide is produced.

$$S_8(s) + 8O_2(g) \longrightarrow 8SO_2(g)$$
$$C(s) + O_2(g) \longrightarrow CO_2(g)$$

In a limited supply of oxygen, carbon monoxide is formed.

$$2C(s) + O_2(g) \longrightarrow 2CO(g)$$

Hydrogen reacts with oxygen to form hydrogen oxide, better known as water.

$$2H_2(g) + O_2(g) \longrightarrow 2H_2O(g)$$

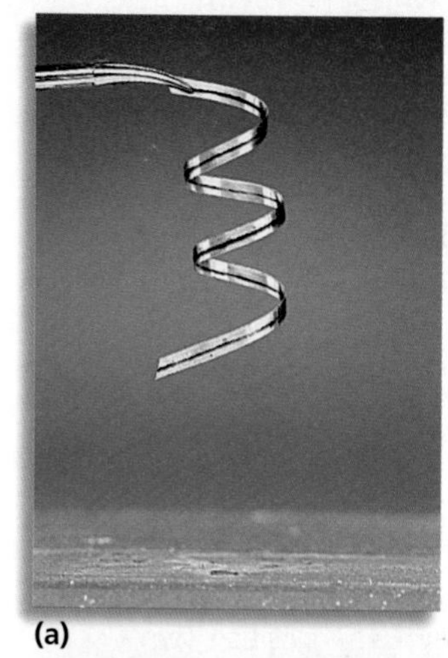

(a)

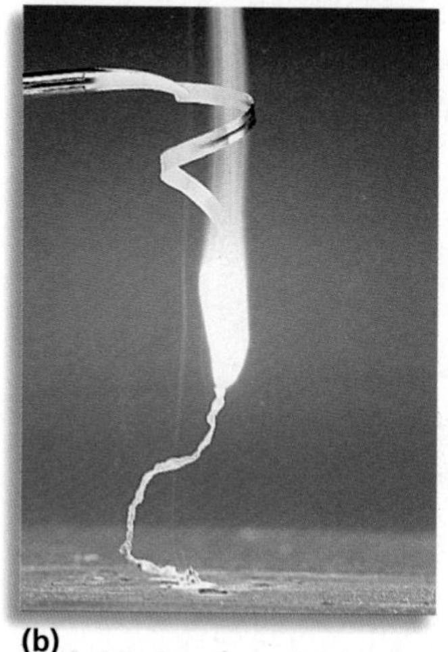

(b)

FIGURE 8-7 Magnesium, Mg, pictured in (a), undergoes a synthesis reaction with oxygen, O_2, in the air to produce magnesium oxide, MgO, as shown in (b).

(a) (b)

FIGURE 8-8 Iron, Fe, and oxygen, O_2, combine to form two different oxides: (a) iron(II) oxide, FeO, and (b) iron(III) oxide, Fe_2O_3.

Reactions of Metals with Halogens

Most metals react with the Group 17 elements, the halogens, to form either ionic or covalent compounds. For example, Group 1 metals react with halogens to form ionic compounds with the formula MX, where M is the metal and X is the halogen. Examples of this type of synthesis reaction include the reactions of sodium with chlorine and potassium with iodine.

$$2Na(s) + Cl_2(g) \longrightarrow 2NaCl(s)$$
$$2K(s) + I_2(g) \longrightarrow 2KI(s)$$

Group 2 metals react with the halogens to form ionic compounds with the formula MX_2.

$$Mg(s) + F_2(g) \longrightarrow MgF_2(s)$$
$$Sr(s) + Br_2(l) \longrightarrow SrBr_2(s)$$

The halogens undergo synthesis reactions with many different metals. Fluorine in particular is so reactive that it combines with almost all metals. For example, fluorine reacts with sodium to produce sodium fluoride. Similarly, it reacts with cobalt to form cobalt(III) fluoride and with uranium to form uranium(VI) fluoride.

$$2Na(s) + F_2(g) \longrightarrow 2NaF(s)$$
$$2Co(s) + 3F_2(g) \longrightarrow 2CoF_3(s)$$
$$U(s) + 3F_2(g) \longrightarrow UF_6(g)$$

Sodium fluoride, NaF, is added to municipal water supplies in trace amounts to provide fluoride ions, which help to prevent tooth decay in the people who drink the water. Cobalt(III) fluoride, CoF_3, is a strong fluorinating agent. And natural uranium is converted to uranium(VI) fluoride, UF_6, as the first step in the production of uranium for use in nuclear power plants.

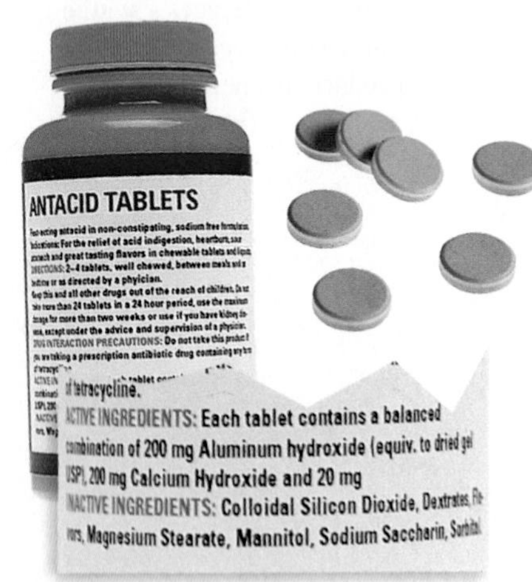

FIGURE 8-9 Calcium hydroxide, a base, can be used to *neutralize* hydrochloric acid in your stomach. You will read more about acids, bases, and neutralization in Chapter 15.

Synthesis Reactions with Oxides

Active metals are highly reactive metals. Oxides of active metals react with water to produce metal hydroxides. For example, calcium oxide reacts with water to form calcium hydroxide, an ingredient in some stomach antacids.

$$CaO(s) + H_2O(l) \longrightarrow Ca(OH)_2(s)$$

Calcium oxide, CaO, also known as lime or quicklime, is manufactured in large quantities. The addition of water to lime to produce $Ca(OH)_2$, which is also known as slaked lime, is a crucial step in the setting of cement.

Many oxides of nonmetals in the upper right portion of the periodic table react with water to produce oxyacids. For example, sulfur dioxide, SO_2, reacts with water to produce sulfurous acid.

$$SO_2(g) + H_2O(l) \longrightarrow H_2SO_3(aq)$$

In air polluted with SO_2, sulfurous acid further reacts with oxygen to form sulfuric acid, one of the main ingredients in *acid rain*.

$$2H_2SO_3(aq) + O_2(g) \longrightarrow 2H_2SO_4(aq)$$

Certain metal oxides and nonmetal oxides react with each other in synthesis reactions to form salts. For example, calcium sulfite is formed by the reaction of calcium oxide and sulfur dioxide.

$$CaO(s) + SO_2(g) \longrightarrow CaSO_3(s)$$

Decomposition Reactions

In a **decomposition reaction,** *a single compound undergoes a reaction that produces two or more simpler substances.* Decomposition reactions are the opposite of synthesis reactions and are represented by the following general equation.

$$AX \longrightarrow A + X$$

AX is a compound. A and X can be elements or compounds.

Most decomposition reactions take place only when energy in the form of electricity or heat is added. Examples of several types of decomposition reactions are given in the following sections.

Decomposition of Binary Compounds

The simplest kind of decomposition reaction is the decomposition of a binary compound into its elements. We have already examined one example of a decomposition reaction. Figure 8-5 on page 250 shows that passing an electric current through water will decompose the water into its constituent elements, hydrogen and oxygen.

$$2H_2O(l) \xrightarrow{\text{electricity}} 2H_2(g) + O_2(g)$$

The decomposition of a substance by an electric current is called **electrolysis.**

Oxides of the less-active metals, which are located in the lower center of the periodic table, decompose into their elements when heated. Joseph Priestley discovered oxygen through such a decomposition reaction in 1774, when he heated mercury(II) oxide to produce mercury and oxygen.

$$2HgO(s) \xrightarrow{\Delta} 2Hg(l) + O_2(g)$$

This reaction is shown in Figure 8-10 on page 260.

FIGURE 8-10 When mercury(II) oxide (the red-orange substance in the bottom of the test tube) is heated, it decomposes into oxygen and metallic mercury, which can be seen as droplets on the inside wall of the test tube.

Decomposition of Metal Carbonates

When a metal carbonate is heated, it breaks down to produce a metal oxide and carbon dioxide gas. For example, calcium carbonate decomposes to produce calcium oxide and carbon dioxide.

$$CaCO_3(s) \xrightarrow{\Delta} CaO(s) + CO_2(g)$$

Decomposition of Metal Hydroxides

All metal hydroxides except those containing Group 1 metals decompose when heated to yield metal oxides and water. For example, calcium hydroxide decomposes to produce calcium oxide and water.

$$Ca(OH)_2(s) \xrightarrow{\Delta} CaO(s) + H_2O(g)$$

Decomposition of Metal Chlorates

When a metal chlorate is heated, it decomposes to produce a metal chloride and oxygen. For example, potassium chlorate, $KClO_3$, decomposes in the presence of the catalyst $MnO_2(s)$ to produce potassium chloride and oxygen.

$$2KClO_3(s) \xrightarrow[MnO_2(s)]{\Delta} 2KCl(s) + 3O_2(g)$$

Decomposition of Acids

Certain acids decompose into nonmetal oxides and water. Carbonic acid is unstable and decomposes readily at room temperature to produce carbon dioxide and water.

$$H_2CO_3(aq) \longrightarrow CO_2(g) + H_2O(l)$$

When heated, sulfuric acid decomposes into sulfur trioxide and water.

$$H_2SO_4(aq) \xrightarrow{\Delta} SO_3(g) + H_2O(l)$$

Sulfurous acid, H_2SO_3, decomposes similarly.

Single-Replacement Reactions

In a **single-replacement reaction,** *also known as a* **displacement reaction,** *one element replaces a similar element in a compound.* Many single-replacement reactions take place in aqueous solution. The amount of energy involved in this type of reaction is usually smaller than the amount involved in synthesis or decomposition reactions. Single-replacement reactions can be represented by the following general equations.

$$A + BX \longrightarrow AX + B$$

or

$$Y + BX \longrightarrow BY + X$$

A, B, X, and Y are elements. AX, BX, and BY are compounds.

Replacement of a Metal in a Compound by Another Metal

Aluminum is more active than lead. When solid aluminum is placed in aqueous lead(II) nitrate, $Pb(NO_3)_2(aq)$, the aluminum replaces the lead. Solid lead and aqueous aluminum nitrate are formed.

$$2Al(s) + 3Pb(NO_3)_2(aq) \longrightarrow 3Pb(s) + 2Al(NO_3)_3(aq)$$

Replacement of Hydrogen in Water by a Metal

The most-active metals, such as those in Group 1, react vigorously with water to produce metal hydroxides and hydrogen. For example, sodium reacts with water to form sodium hydroxide and hydrogen gas.

$$2Na(s) + 2H_2O(l) \longrightarrow 2NaOH(aq) + H_2(g)$$

Less-active metals, such as iron, react with steam to form a metal oxide and hydrogen gas.

$$3Fe(s) + 4H_2O(g) \longrightarrow Fe_3O_4(s) + 4H_2(g)$$

Replacement of Hydrogen in an Acid by a Metal

The more-active metals react with certain acidic solutions, such as hydrochloric acid and dilute sulfuric acid, replacing the hydrogen in the acid. The reaction products are a metal compound (a salt) and hydrogen gas. For example, when solid magnesium reacts with hydrochloric acid, as shown in Figure 8-11, the reaction products are hydrogen gas and aqueous magnesium chloride.

$$Mg(s) + 2HCl(aq) \longrightarrow H_2(g) + MgCl_2(aq)$$

FIGURE 8-11 In this single-replacement reaction, the hydrogen in hydrochloric acid, HCl, is replaced by magnesium, Mg.

Replacement of Halogens

In another type of single-replacement reaction, one halogen replaces another halogen in a compound. Fluorine is the most-active halogen. As such, it can replace any of the other halogens in their compounds. Each halogen is less active than the one above it in the periodic table. Therefore, in Group 17 each element can replace any element below it, but not any element above it. For example, while chlorine can replace bromine in potassium bromide, it cannot replace fluorine in potassium fluoride. The reaction of chlorine with potassium bromide produces bromine and potassium chloride, whereas the combination of fluorine and sodium chloride produces sodium fluoride and solid chlorine.

$$Cl_2(g) + 2KBr(aq) \longrightarrow 2KCl(aq) + Br_2(l)$$
$$F_2(g) + 2NaCl(aq) \longrightarrow 2NaF(aq) + Cl_2(s)$$

Double-Replacement Reactions

In **double-replacement reactions,** *the ions of two compounds exchange places in an aqueous solution to form two new compounds.* One of the compounds formed is usually a precipitate, an insoluble gas that bubbles out of the solution, or a molecular compound, usually water. The other compound is often soluble and remains dissolved in solution. A double-replacement reaction is represented by the following general equation.

$$AX + BY \longrightarrow AY + BX$$

A, X, B, and Y in the reactants represent ions. AY and BX represent ionic or molecular compounds.

FIGURE 8-12 The double-replacement reaction between aqueous lead(II) nitrate, $Pb(NO_3)_2(aq)$, and aqueous potassium iodide, $KI(aq)$, yields the precipitate lead(II) iodide, $PbI_2(s)$.

Formation of a Precipitate

The formation of a precipitate occurs when the cations of one reactant combine with the anions of another reactant to form an insoluble or slightly soluble compound. For example, when an aqueous solution of potassium iodide is added to an aqueous solution of lead(II) nitrate, the yellow precipitate lead(II) iodide forms. This is shown in Figure 8-12.

$$2KI(aq) + Pb(NO_3)_2(aq) \longrightarrow PbI_2(s) + 2KNO_3(aq)$$

The precipitate forms as a result of the very strong attractive forces between the Pb^{2+} cations and the I^- anions. The other product is the water-soluble salt potassium nitrate, KNO_3. The potassium and nitrate ions do not take part in the reaction. They remain in solution as aqueous ions. The guidelines that help identify which ions form a precipitate and which ions remain in solution are developed in Chapter 14.

Formation of a Gas

In some double-replacement reactions, one of the products is an insoluble gas that bubbles out of the mixture. For example, iron(II) sulfide reacts with hydrochloric acid to form hydrogen sulfide gas and iron(II) chloride.

$$FeS(s) + 2HCl(aq) \longrightarrow H_2S(g) + FeCl_2(aq)$$

Formation of Water

In some double-replacement reactions, a very stable molecular compound, such as water, is one of the products. For example, hydrochloric acid reacts with an aqueous solution of sodium hydroxide to yield aqueous sodium chloride and water.

$$HCl(aq) + NaOH(aq) \longrightarrow NaCl(aq) + H_2O(l)$$

Combustion Reactions

In a **combustion reaction,** *a substance combines with oxygen, releasing a large amount of energy in the form of light and heat.* The combustion of hydrogen is shown below in Figure 8-13. The reaction's product is water vapor.

$$2H_2(g) + O_2(g) \longrightarrow 2H_2O(g)$$

The burning of natural gas, propane, gasoline, and wood are also examples of combustion reactions. For example, the burning of propane, C_3H_8, results in the production of carbon dioxide and water vapor.

$$C_3H_8(g) + 5O_2(g) \longrightarrow 3CO_2(g) + 4H_2O(g)$$

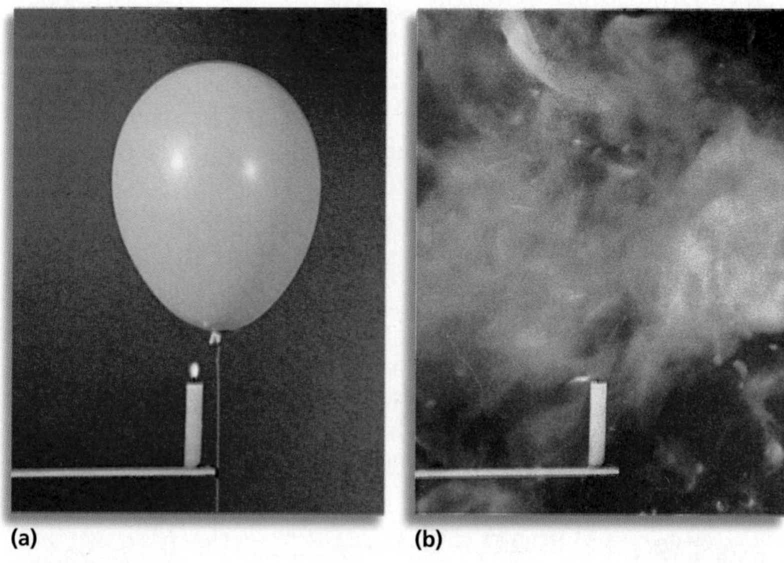

(a) (b)

FIGURE 8-13 (a) The candle supplies heat to the hydrogen and oxygen in the balloon, triggering the explosive combustion reaction shown in (b).

DESKTOP INVESTIGATION

Balancing Equations Using Models

MATERIALS

- **large and small gumdrops in at least four different colors**
- **toothpicks**

QUESTION

How can molecular models and formula-unit ionic models be used to balance chemical equations and classify chemical reactions?

PROCEDURE

Examine the partial equations in Groups A–E. Using different-colored gumdrops to represent atoms of different elements, make models of the reactions by connecting the appropriate "atoms" with toothpicks. Use your models to (1) balance equations (a) and (b) in each group, (2) determine the products for reaction (c) in each group, and (3) complete and balance each equation (c). Finally, (4) classify each group of reactions by type.

Group A

a. $H_2 + Cl_2 \longrightarrow HCl$

b. $Mg + O_2 \longrightarrow MgO$

c. $BaO + H_2O \longrightarrow$ _____

Group B

a. $H_2CO_3 \longrightarrow CO_2 + H_2O$

b. $KClO_3 \longrightarrow KCl + O_2$

c. $H_2O \xrightarrow{\text{electricity}}$ _____

Group C

a. $Ca + H_2O \longrightarrow Ca(OH)_2 + H_2$

b. $KI + Br_2 \longrightarrow KBr + I_2$

c. $Zn + HCl \longrightarrow$ _____

Group D

a. $AgNO_3 + NaCl \longrightarrow AgCl + NaNO_3$

b. $FeS + HCl \longrightarrow FeCl_2 + H_2S$

c. $H_2SO_4 + KOH \longrightarrow$ _____

Group E

a. $CH_4 + O_2 \longrightarrow CO_2 + H_2O$

b. $CO + O_2 \longrightarrow CO_2$

c. $C_3H_8 + O_2 \longrightarrow$ _____

SECTION REVIEW

1. List five types of chemical reactions.
2. Complete and balance each of the following reactions identified by type:
   a. synthesis: _____ $\longrightarrow Li_2O$
   b. decomposition: $Mg(ClO_3)_2 \longrightarrow$ _____
   c. single-replacement: $Na + H_2O \longrightarrow$ _____
   d. double-replacement: $HNO_3 + Ca(OH)_2 \longrightarrow$ _____
   e. combustion: $C_5H_{12} + O_2 \longrightarrow$ _____
3. Classify each of the following reactions as synthesis, decomposition, single-replacement, double-replacement, or combustion:
   a. $N_2(g) + 3H_2(g) \longrightarrow 2NH_3(g)$
   b. $2Li(s) + 2H_2O(l) \longrightarrow 2LiOH(aq) + H_2(g)$
   c. $2NaNO_3(s) \longrightarrow 2NaNO_2(s) + O_2(g)$
   d. $2C_6H_{14}(l) + 19O_2(g) \longrightarrow 12CO_2(g) + 14H_2O(l)$
   e. $NH_4Cl(s) \longrightarrow NH_3(g) + HCl(g)$
   f. $BaO(s) + H_2O(l) \longrightarrow Ba(OH)_2(aq)$
   g. $AgNO_3(aq) + NaCl(aq) \longrightarrow AgCl(s) + NaNO_3(aq)$
4. Complete and balance each of the following equations, and identify each by type:
   a. $Br_2 + KI \longrightarrow$ _____
   b. $Zn + HCl \longrightarrow$ _____
   c. $Ca + Cl_2 \longrightarrow$ _____
   d. $NaClO_3 \xrightarrow{\Delta}$ _____
   e. $C_7H_{14} + O_2 \longrightarrow$ _____
   f. $CuCl_2 + Na_2S \longrightarrow$ _____

Activity Series of the Elements

SECTION 8-3

OBJECTIVES

- Explain the significance of an activity series.
- Use an activity series to predict whether a given reaction will occur and what the products will be.

The ability of an element to react is referred to as the element's *activity*. The more readily an element reacts with other substances, the greater its activity is. *An* **activity series** *is a list of elements organized according to the ease with which the elements undergo certain chemical reactions.*

The order in which the elements are listed is usually determined by single-replacement reactions. The most-active element, placed at the top in the series, can replace each of the elements below it from a compound in a single-replacement reaction. An element farther down can replace any element below it but not any above it. For example, in the discussion of single-replacement reactions in Section 8-2, it was noted that each halogen will react to replace any halogen listed below it in the periodic table. Therefore, an activity series for the Group 17 elements lists them in the same order, from top to bottom, as they appear in the periodic table. This is shown in Table 8-3 on page 266.

As mentioned in Section 8-1, the fact that a chemical equation can be written does not necessarily mean that the reaction it represents will actually take place. Activity series are used to help predict whether certain chemical reactions will occur. For example, according to the activity series for metals in Table 8-3, aluminum replaces zinc. Therefore, we could predict that the following reaction does occur.

$$2Al(s) + 3ZnCl_2(aq) \longrightarrow 3Zn(s) + 2AlCl_3(aq)$$

Cobalt, however, cannot replace sodium. Therefore, we write the following.

$$Co(s) + 2NaCl(aq) \longrightarrow \text{no reaction}$$

It is important to remember that like many other aids used to predict the products of chemical reactions, activity series are based on experiment. The information that they contain is used as a general guide for assessing relative reactivities of substances and for predicting reaction outcomes. For example, the activity series reflects the fact that some metals (potassium, for example) react vigorously with water and acids, replacing hydrogen to form new compounds. Other metals, such as iron or zinc, replace hydrogen in acids such as hydrochloric acid but react with water only when the water is hot enough to become steam. Nickel, on the other hand, will replace hydrogen in acids but will not react with steam at all. And gold will not react with acid or water, either as a liquid or as steam. Such experimental observations are the basis for the activity series shown in Table 8-3.

TABLE 8-3 *Activity Series of the Elements*

Activity of metals		Activity of halogen nonmetals
Li Rb K Ba Sr Ca Na	React with cold H_2O and acids, replacing hydrogen. React with oxygen, forming oxides.	F_2 Cl_2 Br_2 I_2
Mg Al Mn Zn Cr Fe Cd	React with steam (but not cold water) and acids, replacing hydrogen. React with oxygen, forming oxides.	
Co Ni Sn Pb	Do not react with water. React with acids, replacing hydrogen. React with oxygen, forming oxides.	
H_2 Sb Bi Cu Hg	React with oxygen, forming oxides.	
Ag Pt Au	Fairly unreactive, forming oxides only indirectly.	

SAMPLE PROBLEM 8-6

Using the activity series shown in Table 8-3, explain whether each of the possible reactions listed below will occur. For those reactions that will occur, predict what the products will be.

a. $Zn(s) + H_2O(l) \xrightarrow{50°C}$ _____

b. $Sn(s) + O_2(g) \longrightarrow$ _____

c. $Cd(s) + Pb(NO_3)_2(aq) \longrightarrow$ _____

d. $Cu(s) + HCl(aq) \longrightarrow$ _____

SOLUTION

a. This is a reaction between a metal and water at 50°C. Zinc reacts with water only when it is hot enough to be steam. Therefore, no reaction will occur.

b. Any metal more active than silver will react with oxygen to form an oxide. Tin is above silver in the activity series. Therefore, a reaction will occur, and the product will be a tin oxide, either SnO or SnO_2.

c. An element will replace any element below it in the activity series from a compound in aqueous solution. Cadmium is above lead, and therefore a reaction will occur to produce lead, Pb, and cadmium nitrate, $Cd(NO_3)_2$.

d. Any metal more active than hydrogen will replace hydrogen from an acid. Copper is not above hydrogen in the series. Therefore, no reaction will occur.

PRACTICE

1. Using the activity series shown in Table 8-3, predict whether each of the possible reactions listed below will occur. For the reactions that will occur, write the products and balance the equation.

a. $Cr(s) + H_2O(l) \longrightarrow$ _____

b. $Pt(s) + O_2(g) \longrightarrow$ _____

c. $Cd(s) + 2HBr(aq) \longrightarrow$ _____

d. $Mg(s) + \text{steam} \longrightarrow$ _____

Answer

1. a. no

b. no

c. yes; $2Cd(s) + 2HBr(aq) \longrightarrow 2CdBr(aq) + H_2(g)$

d. yes; $Mg(s) + 2H_2O(g) \longrightarrow Mg(OH)_2(aq) + H_2(g)$

2. Identify the element that replaces hydrogen from acids but cannot replace tin from its compounds.

Answer 2. Pb

3. According to Table 8-3, what is the most-active transition metal?

Answer 3. Mn

SECTION REVIEW

1. How is the activity series useful in predicting chemical behavior?

2. Based on the activity series, predict whether each of the following possible reactions listed will occur:

a. $Ni(s) + H_2O(l) \longrightarrow$ _____

b. $Br_2(l) + KI(aq) \longrightarrow$ _____

c. $Au(s) + HCl(aq) \longrightarrow$ _____

d. $Cd(s) + HCl(aq) \longrightarrow$ _____

e. $Mg(s) + Co(NO_3)_2(aq) \longrightarrow$ _____

3. Write the balanced chemical equation for each of the reactions that occur in item 2.

Acid Water—A Hidden Menace

When purchasing a home with its own well, it is common practice to have the water in the well tested. Usually, the purpose of the tests is to indicate the presence of disease-causing microorganisms. Rarely is the water's acidity measured.

Many people are unaware of their water's pH value (see Chapter 16) until they are confronted with such phenomena as a blue ring materializing around a porcelain sink drain, a water heater suddenly giving out, or tropical fish that keep mysteriously dying. Each of these events could be traced to acidic water, which can also be a cause of lead poisoning.

Although much has been written about the dangers of lead paint, the possibility of lead poisoning from home water supplies has gone largely unreported. Many older homes still have lead pipes in their plumbing, while most modern homes use copper piping. Regardless of the age of the home, however, all pipe joints are sealed with lead solder. Highly acidic water can leach out both the lead from the solder joints and copper from the pipes themselves. (The presence of copper ions turns the sink drain blue.) In addition, people who are in the habit of filling their kettles in the morning without letting the tap run awhile first could be adding a number of unwanted chemicals to their tea or coffee.

Lead poisoning is of particular concern when young children are present. The absorption rate of lead in the intestinal tract of a child is much higher than that of an adult, and lead poisoning can permanently impair a child's rapidly growing nervous system. The good news is that lead poisoning and other effects of acidic water in the home can be easily prevented. Here's what you can do about it:

1. Monitor the pH of your water on a regular basis, especially if you have well water. (In the northeastern United States, acidic ground water is a common occurrence.) This can easily be done with pH test kits (see photograph) that are sold in hardware or pet stores—many tropical fish are intolerant of water with a pH that is either too high (basic) or too low (acidic). The pH of most municipal water supplies should already be regulated, but it doesn't hurt to check.
2. In the morning, let your water tap run for about half a minute before you fill your kettle or drink the water. If the water is acidic, the first flush of water will have the highest concentration of lead and copper ions.
3. Installing an alkali-injection pump is a low-cost, low-maintenance solution that can save your plumbing and lessen the risk of lead poisoning from your own water supply. The pump injects a small amount of an alkali (usually potassium carbonate or sodium carbonate) in your water-holding tank each time you activate your well's pump. This effectively neutralizes the acidity of your water. The reaction below shows the neutralizing effect of potassium carbonate on well water that has been made acidic by acid rain.

$$K_2CO_3(aq) + H_2SO_4(aq) \longrightarrow K_2SO_4(aq) + CO_2(g) + H_2O(l)$$

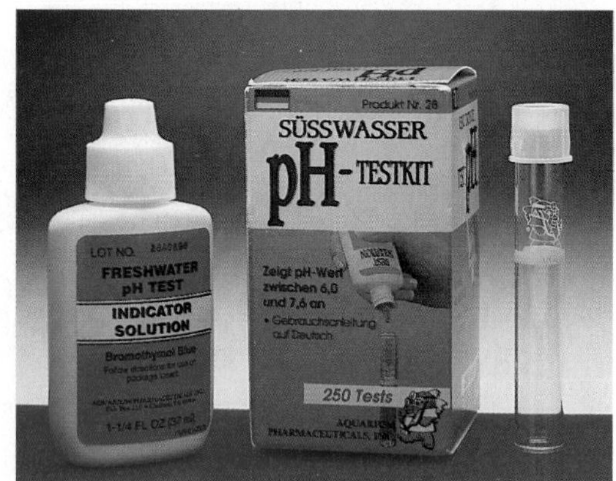

The pH of your home's water supply can be easily monitored using a test kit like the one shown here.

CHAPTER 8 REVIEW

CHAPTER SUMMARY

8-1

- Three observations that suggest a chemical reaction is taking place are the evolution of heat and light, the production of gas, and the formation of a precipitate.
- A balanced chemical equation represents, with symbols and formulas, the identities and relative amounts of reactants and products in a chemical reaction.

Vocabulary

chemical equation (241)
coefficient (243)
formula equation (244)
precipitate (242)
reversible reaction (246)
word equation (243)

8-2

- Synthesis reactions are represented by the general equation $A + X \longrightarrow AX$.
- Decomposition reactions are represented by the general equation $AX \longrightarrow A + X$.
- Single-replacement reactions are represented by the general equations $A + BX \longrightarrow AX + B$ and $Y + BX \longrightarrow BY + X$.
- Double-replacement reactions are represented by the general equation $AX + BY \longrightarrow AY + BX$.

Vocabulary

combustion reaction (263)
composition reaction (256)
decomposition reaction (259)
displacement reaction (261)
double-replacement reaction (262)
electrolysis (259)
single-replacement reaction (261)
synthesis reaction (256)

8-3

- Activity series list the elements in order of their chemical reactivity and are useful in predicting whether a chemical reaction will occur.
- Chemists determine activity series through experiments.

Vocabulary

activity series (265)

REVIEWING CONCEPTS

1. List three observations that indicate that a chemical reaction may be taking place. (8-1)
2. List the three requirements for a correctly written chemical equation. (8-1)
3. a. What is meant by the term *coefficient* in relation to a chemical equation?
   b. How does the presence of a coefficient affect the number of atoms of each type in the formula that it precedes? (8-1)
4. Give an example of a word equation, a formula equation, and a chemical equation. (8-1)
5. List some of the quantitative information revealed by a chemical equation. (8-1)
6. What limitations are associated with the use of both word and formula equations? (8-1)
7. Define each of the following:
   a. aqueous solution
   b. catalyst
   c. reversible reaction (8-1)
8. Write formulas for each of the following compounds:
   a. potassium hydroxide
   b. calcium nitrate
   c. sodium carbonate
   d. carbon tetrachloride
   e. magnesium bromide
   f. sulfur dioxide
   g. ammonium sulfate (8-1)

9. What four guidelines are useful in balancing an equation? (8-1)

10. How many atoms of each type are represented in each of the following?
a. $3N_2$
b. $2H_2O$
c. $4HNO_3$
d. $2Ca(OH)_2$
e. $3Ba(ClO_3)_2$
f. $5Fe(NO_3)_2$
g. $4Mg_3(PO_4)_2$
h. $2(NH_4)_2SO_4$
i. $6Al_2(SeO_4)_3$
j. $4C_3H_8$ (8-1)

11. Define and give general equations for the five basic types of chemical reactions introduced in Chapter 8. (8-2)

12. How are most decomposition reactions initiated? (8-2)

13. What is electrolysis? (8-2)

14. a. In what environment do many single-replacement reactions commonly occur?
b. In general, how do single-replacement reactions compare with synthesis and decomposition reactions in terms of the amount of energy involved? (8-2)

15. What is meant by the *activity* of an element? (8-3)

16. a. What is an activity series of elements?
b. What is the basis for the ordering of the elements in the activity series? (8-3)

17. a. What is the chemical principle upon which the activity series of metals is based?
b. What is the significance of the distance between two metals in the activity series? (8-3)

PROBLEMS

Chemical Equations

18. Write the chemical equation that relates to each of the following word equations. Include symbols for physical states in the equation. (Hint: See Sample Problem 8-1.)
a. solid zinc sulfide + oxygen gas ⟶ solid zinc oxide + sulfur dioxide gas
b. hydrochloric acid + aqueous magnesium hydroxide ⟶ aqueous magnesium chloride + water
c. nitric acid + aqueous calcium hydroxide ⟶ aqueous calcium nitrate + water

19. Translate each of the following chemical equations into a sentence. (Hint: See Sample Problem 8-2.)
a. $2ZnS(s) + 3O_2(g) \longrightarrow 2ZnO(s) + 2SO_2(g)$
b. $CaH_2(s) + 2H_2O(l) \longrightarrow Ca(OH)_2(aq) + 2H_2(g)$
c. $AgNO_3(aq) + KI(aq) \longrightarrow AgI(s) + KNO_3(aq)$

20. Balance each of the following:
a. $H_2 + Cl_2 \longrightarrow HCl$
b. $Al + Fe_2O_3 \longrightarrow Al_2O_3 + Fe$
c. $Pb(CH_3COO)_2 + H_2S \longrightarrow PbS + CH_3COOH$

21. The following equations are incorrect in some way. Identify and correct each error, and then balance each equation.
a. $Li + O_2 \longrightarrow LiO_2$
b. $H_2 + Cl_2 \longrightarrow H_2Cl_2$
c. $MgCO_3 \longrightarrow MgO_2 + CO_2$
d. $NaI + Cl_2 \longrightarrow NaCl + I$

22. Write chemical equations for each of the following sentences:
a. Aluminum reacts with oxygen to produce aluminum oxide.
b. Phosphoric acid, H_3PO_4, is produced through the reaction between tetraphosphorus decoxide and water.
c. Iron(III) oxide reacts with carbon monoxide to produce iron and carbon dioxide.

23. Carbon tetrachloride is used as an intermediate chemical in the manufacture of other chemicals. It is prepared in liquid form by reacting chlorine gas with methane gas. Hydrogen chloride gas is also formed in this reaction. Write the balanced chemical equation for the production of carbon tetrachloride. (Hint: See Sample Problems 8-3 and 8-4.)

24. For each of the following synthesis reactions, identify the missing reactant(s) or product(s), and then balance the resulting equation:
a. Mg + _____ ⟶ MgO
b. _____ + $O_2 \longrightarrow Fe_2O_3$
c. $Li + Cl_2 \longrightarrow$ _____
d. Ca + _____ $\longrightarrow CaI_2$

Types of Chemical Reactions

25. Complete the following synthesis reactions by writing both word and chemical equations for each:
a. sodium + oxygen $\longrightarrow$ ____
b. magnesium + fluorine $\longrightarrow$ ____

26. Complete and balance the equation for each of the following decomposition reactions:
a. $HgO \xrightarrow{\Delta}$
b. $H_2O(l) \xrightarrow{\text{electricity}}$
c. $Ag_2O \xrightarrow{\Delta}$
d. $CuCl_2 \xrightarrow{\text{electricity}}$

27. Complete and balance the equations for each of the following single-replacement reactions:
a. $Zn + Pb(NO_3)_2 \longrightarrow$ ____
b. $Al + Hg(CH_3COO)_2 \longrightarrow$ ____
c. $Al + NiSO_4 \longrightarrow$ ____
d. $Na + H_2O \longrightarrow$ ____

28. Complete and balance the equations for the following double-replacement reactions:
a. $AgNO_3(aq) + NaCl(aq) \longrightarrow$ ____
b. $Mg(NO_3)_2(aq) + KOH(aq) \longrightarrow$ ____
c. $LiOH(aq) + Fe(NO_3)_3(aq) \longrightarrow$ ____

29. Complete and balance the equation for each of the following combustion reactions:
a. $CH_4 + O_2 \longrightarrow$ ____
b. $C_3H_6 + O_2 \longrightarrow$ ____
c. $C_5H_{12} + O_2 \longrightarrow$ ____

30. Write and balance each of the following equations, and then identify each by type:
a. hydrogen + iodine $\longrightarrow$ hydrogen iodide
b. lithium + hydrochloric acid $\longrightarrow$ lithium chloride + hydrogen
c. sodium carbonate $\longrightarrow$ sodium oxide + carbon dioxide
d. mercury(II) oxide $\longrightarrow$ mercury + oxygen
e. magnesium hydroxide $\longrightarrow$ magnesium oxide + water

31. Identify the compound that could undergo decomposition to produce the following products, and then balance the final equation:
a. magnesium oxide and water
b. lead(II) oxide and water
c. lithium chloride and oxygen
d. barium chloride and oxygen
e. nickel chloride and oxygen

32. In each of the following combustion reactions, identify the missing reactant(s), product(s), or both, and then balance the resulting equation:
a. $C_3H_8 +$ ____ $\longrightarrow$ ____ $+ H_2O$
b. ____ $+ 8O_2 \longrightarrow 5CO_2 + 6H_2O$
c. $C_2H_5OH +$ ____ $\longrightarrow$ ____ + ____

33. Complete and balance each of the following reactions observed to occur, and then identify each by type:
a. zinc + sulfur $\longrightarrow$ ____
b. calcium + sodium nitrate $\longrightarrow$ ____
c. silver nitrate + potassium iodide $\longrightarrow$ ____
d. sodium iodide $\xrightarrow{\Delta}$ ____
e. toluene, C_7H_8 + oxygen $\longrightarrow$ ____
f. nonane, C_9H_{20} + oxygen $\longrightarrow$ ____

Activity Series

34. Using the activity series in Table 8-3 on page 266, predict whether each of the possible reactions listed below will occur. For the reactions that will occur, write the products and balance the equation.
a. $Ni(s) + CuCl_2(aq) \longrightarrow$ ____
b. $Zn(s) + Pb(NO_3)_2(aq) \longrightarrow$ ____
c. $Cl_2(g) + KI(aq) \longrightarrow$ ____
d. $Cu(s) + FeSO_4(aq) \longrightarrow$ ____
e. $Ba(s) + H_2O(l) \longrightarrow$ ____

35. Use the activity series to predict whether each of the following synthesis reactions will occur, and write the chemical equations for those predicted to occur:
a. $Ca(s) + O_2(g) \longrightarrow$ ____
b. $Ni(s) + O_2(g) \longrightarrow$ ____
c. $Au(s) + O_2(g) \longrightarrow$ ____

36. Based on the activity series of metals and halogens, which element within each pair is more likely to replace the other in a compound?
a. K and Na
b. Al and Ni
c. Bi and Cr
d. Cl and F
e. Au and Ag
f. Cl and I
g. Fe and Sr
h. I and F

MIXED REVIEW

37. Ammonia reacts with oxygen to yield nitrogen and water.

$$4NH_3(g) + 3O_2(g) \longrightarrow 2N_2(g) + 6H_2O(l)$$

Given this chemical equation, as well as the number of moles of the reactant or product indicated below, determine the number of moles of all remaining reactants and products.

a. 3.0 mol O_2
b. 8.0 mol NH_3
c. 1.0 mol N_2
d. 0.40 mol H_2O

38. Complete the following synthesis reactions by writing both the word and chemical equation for each:

a. potassium + chlorine $\longrightarrow$ _____
b. hydrogen + iodine $\longrightarrow$ _____
c. magnesium + oxygen $\longrightarrow$ _____

39. Use the activity series to predict which metal, Sn, Mn, or Pt, would be the best choice as a container for an acid.

40. Aqueous sodium hydroxide is produced commercially by the electrolysis of aqueous sodium chloride. Hydrogen and chlorine gases are also produced. Write the balanced chemical equation for the production of sodium hydroxide. Include the physical states of the reactants and products.

41. Balance each of the following:

a. $Ca(OH)_2 + (NH_4)_2SO_4 \longrightarrow CaSO_4 + NH_3 + H_2O$
b. $C_2H_6 + O_2 \longrightarrow CO_2 + H_2O$
c. $Cu_2S + O_2 \longrightarrow Cu_2O + SO_2$
d. $Al + H_2SO_4 \longrightarrow Al_2(SO_4)_3 + H_2$

42. Use the activity series to predict whether each of the following reactions will occur, and write the balanced chemical equations for those predicted to occur:

a. $Al(s) + O_2(g) \longrightarrow$ _____
b. $Pb(s) + ZnCl_2(s) \longrightarrow$ _____
c. $Rb(s) + Zn(NO_3)_2(aq) \longrightarrow$ _____

43. Complete and balance the equations for the following reactions, and identify the type of reaction each represents:

a. $(NH_4)_2S(aq) + ZnCl_2(aq) \longrightarrow$ _____ $+ ZnS(s)$
b. $Al(s) + Pb(NO_3)_2(aq) \longrightarrow$ _____
c. $Ba(s) + H_2O(l) \longrightarrow$ _____
d. $Cl_2(g) + KBr(aq) \longrightarrow$
e. $NH_3(g) + O_2(g) \xrightarrow{Pt} NO(g) + H_2O(l)$
f. $H_2O(l) \longrightarrow H_2(g) + O_2(g)$

44. Write and balance each of the following equations, and then identify each by type:

a. copper + chlorine $\longrightarrow$ copper(II) chloride
b. calcium chlorate $\longrightarrow$ calcium chloride + oxygen
c. lithium + water $\longrightarrow$ lithium hydroxide + hydrogen
d. lead(II) carbonate $\longrightarrow$ lead(II) oxide + carbon dioxide

45. How many moles of HCl can be made from 6.15 mol of H_2 and an excess of Cl_2?

46. What product is missing in the following equation?

$$MgO + 2HCl \longrightarrow MgCl_2 + \_\_\_\_\_$$

47. Balance the following equations:

a. $Pb(NO_3)_2(aq) + NaOH(aq) \longrightarrow Pb(OH)_2(s) + NaNO_3(aq)$
b. $C_{12}H_{22}O_{11}(l) + O_2(g) \longrightarrow CO_2(g) + H_2O(l)$
c. $Al(OH)_3(s) + H_2SO_4(aq) \longrightarrow Al_2(SO_4)_3(aq) + H_2O(l)$

48. Translate the following word equations into balanced chemical equations:

a. silver nitrate + potassium iodide $\longrightarrow$ silver iodide + potassium nitrate
b. nitrogen dioxide + water $\longrightarrow$ nitric acid + nitrogen monoxide
c. silicon tetrachloride + water $\longrightarrow$ silicon dioxide + hydrochloric acid

CRITICAL THINKING

49. Inferring Relationships Activity series are prepared by comparing single-displacement reactions between metals. Based on observations, the metals can be ranked by their ability to react. However, reactivity can be explained by the ease with which atoms of metals lose electrons. Using information from the activity series, identify the locations in the periodic table of the most-reactive metals and the least-reactive metals. Based on your knowledge of electron configurations and periodic trends, infer possible explanations for their reactivity and position in the periodic table.

50. **Analyzing Results** Formulate an activity series for the hypothetical elements A, J, Q, and Z using the reaction information provided.

$$A + ZX \longrightarrow AX + Z$$
$$J + ZX \longrightarrow \text{no reaction}$$
$$Q + AX \longrightarrow QX + A$$

HANDBOOK SEARCH

51. Find the common-reactions section for Group 1 metals in the *Elements Handbook*. Use this information to answer the following:
    a. Write a balanced chemical equation for the formation of rubidium hydroxide from rubidum oxide.
    b. Write a balanced chemical equation for the formation of cesium iodide.
    c. Classify the reactions you wrote in (a) and (b).
    d. Write word equations for the reactions you wrote in (a) and (b).
52. Find the common-reactions section for Group 13 in the *Elements Handbook*. Use this information to answer the following:
    a. Write a balanced chemical equation for the formation of gallium bromide prepared from hydrobromic acid.
    b. Write a balanced chemical equation for the formation of gallium oxide.
    c. Classify the reactions you wrote in (a) and (b).
    d. Write word equations for the reactions you wrote in (a) and (b).
53. Find the common-reactions section for Group 16 in the *Elements Handbook*. Use this information to answer the following:
    a. Write a balanced chemical equation for the formation of selenium trioxide.
    b. Write a balanced chemical equation for the formation of tellurium iodide.
    c. Classify the reactions you wrote in (a) and (b).
    d. Write word equations for the reactions you wrote in (a) and (b).

RESEARCH & WRITING

54. Trace the evolution of municipal water fluoridation. What advantages and disadvantages are associated with this practice?
55. Research how a soda-acid fire extinguisher works, and write the chemical equation for the reaction. Check your house and other structures for different types of fire extinguishers, and ask your local fire department to verify the effectiveness of each type of extinguisher.

ALTERNATIVE ASSESSMENT

56. **Performance Assessment** For one day, record situations that show evidence of a chemical change. Identify the reactants and the products, and determine whether there is proof of a chemical reaction. Classify each of the chemical reactions according to the common reaction types discussed in the chapter.
57. Design a set of experiments that will enable you to create an activity series for the elements composing the following metals and solutions:
    a. aluminum and aluminum chloride
    b. chromium and chromium(III) chloride
    c. iron and iron(II) chloride
    d. magnesium and magnesium chloride

CHAPTER 9

Stoichiometry

Stoichiometry *comes from the Greek words* stoicheion, *meaning "element,"* *and* metron, *meaning "measure."*

Introduction to Stoichiometry

SECTION 9-1

Objectives

- Define *stoichiometry.*
- Describe the importance of the *mole ratio* in stoichiometric calculations.
- Write a mole ratio relating two substances in a chemical equation.

Much of our knowledge of chemistry is based on the careful quantitative analysis of substances involved in chemical reactions. **Composition stoichiometry** (which you studied in Chapter 3) *deals with the mass relationships of elements in compounds.* **Reaction stoichiometry** *involves the mass relationships between reactants and products in a chemical reaction.* Reaction stoichiometry is the subject of this chapter and it is based on chemical equations and the law of conservation of matter. All reaction-stoichiometry calculations start with a balanced chemical equation. This equation gives the relative number of moles of reactants and products.

Reaction-Stoichiometry Problems

The reaction-stoichiometry problems in this chapter can be classified according to the information *given* in the problem and the information you are expected to find, the *unknown*. The masses are generally expressed in grams, but you will encounter both large-scale and microscale problems with other mass units, such as kg or mg. Stoichiometric problems are solved by using ratios from the balanced equation to convert the given quantity using the methods described here.

Problem Type 1: *Given* and *unknown* quantities are amounts in moles.
When you are given the amount of a substance in moles and asked to calculate the amount in moles of another substance in the chemical reaction, the general plan is

$$\begin{matrix}\text{amount of} \\ \textit{given}\text{ substance (in mol)}\end{matrix} \longrightarrow \begin{matrix}\text{amount of} \\ \textit{unknown}\text{ substance (in mol)}\end{matrix}$$

Problem Type 2: *Given* is an amount in moles and the *unknown* is a mass that is often expressed in grams.
When you are given the amount in moles of one substance and asked to calculate the mass of another substance in the chemical reaction, the general plan is

$$\begin{matrix}\text{amount of} \\ \textit{given}\text{ substance} \\ \text{(in mol)}\end{matrix} \longrightarrow \begin{matrix}\text{amount of} \\ \textit{unknown}\text{ substance} \\ \text{(in mol)}\end{matrix} \longrightarrow \begin{matrix}\text{mass of} \\ \textit{unknown}\text{ substance} \\ \text{(in g)}\end{matrix}$$

Problem Type 3: *Given* is a mass in grams and the *unknown* is an amount in moles.

When you are given the mass of one substance and asked to calculate the amount in moles of another substance in the chemical reaction, the general plan is

$$\begin{matrix}\text{mass of} & & \text{amount of} & & \text{amount of} \\ \textit{given}\text{ substance} & \longrightarrow & \textit{given}\text{ substance} & \longrightarrow & \textit{unknown}\text{ substance} \\ \text{(in g)} & & \text{(in mol)} & & \text{(in mol)}\end{matrix}$$

Problem Type 4: *Given* is a mass in grams and the *unknown* is a mass in grams.

When you are given the mass of one substance and asked to calculate the mass of another substance in the chemical reaction, the general plan is

$$\begin{matrix}\text{mass of} & & \text{amount of} & & \text{amount of} & & \text{mass of} \\ \textit{given}\text{ substance} & \longrightarrow & \textit{given}\text{ substance} & \longrightarrow & \textit{unknown}\text{ substance} & \longrightarrow & \textit{unknown}\text{ substance} \\ \text{(in g)} & & \text{(in mol)} & & \text{(in mol)} & & \text{(in g)}\end{matrix}$$

Mole Ratio

Solving any reaction-stoichiometry problem requires the use of a mole ratio to convert from moles or grams of one substance in a reaction to moles or grams of another substance. *A* **mole ratio** *is a conversion factor that relates the amounts in moles of any two substances involved in a chemical reaction.* This information is obtained directly from the balanced chemical equation. Consider, for example, the chemical equation for the electrolysis of aluminum oxide to produce aluminum and oxygen.

$$2Al_2O_3(l) \longrightarrow 4Al(s) + 3O_2(g)$$

Recall from Chapter 8 that the coefficients in a chemical equation satisfy the law of conservation of matter and represent the relative amounts in moles of reactants and products. Therefore, 2 mol of aluminum oxide decompose to produce 4 mol of aluminum and 3 mol of oxygen gas. These relationships can be expressed in the following mole ratios.

$$\frac{2 \text{ mol } Al_2O_3}{4 \text{ mol Al}} \text{ or } \frac{4 \text{ mol Al}}{2 \text{ mol } Al_2O_3}$$

$$\frac{2 \text{ mol } Al_2O_3}{3 \text{ mol } O_2} \text{ or } \frac{3 \text{ mol } O_2}{2 \text{ mol } Al_2O_3}$$

$$\frac{4 \text{ mol Al}}{3 \text{ mol } O_2} \text{ or } \frac{3 \text{ mol } O_2}{4 \text{ mol Al}}$$

For the decomposition of aluminum oxide, the appropriate mole ratio would be used as a conversion factor to convert a given amount in moles of one substance to the corresponding amount in moles of another substance. To determine the amount in moles of aluminum that can be produced from 13.0 mol of aluminum oxide, the mole ratio needed is that of Al to Al_2O_3.

$$13.0 \text{ } \cancel{\text{mol Al}_2\text{O}_3} \times \frac{4 \text{ mol Al}}{2 \text{ } \cancel{\text{mol Al}_2\text{O}_3}} = 26.0 \text{ mol Al}$$

Mole ratios are exact, so they do not limit the number of significant figures in a calculation. The number of significant figures in the answer is therefore determined only by the number of significant figures of any measured quantities in a particular problem.

Molar Mass

Recall from Chapter 7 that the molar mass is the mass, in grams, of one mole of a substance. The molar mass is the conversion factor that relates the mass of a substance to the amount in moles of that substance. To solve reaction-stoichiometry problems, you will need to determine molar masses using the periodic table.

Returning to the previous example, the decomposition of aluminum oxide, the rounded masses from the periodic table are the following.

$$Al_2O_3 = 101.96 \text{ g/mol} \qquad O_2 = 32.00 \text{ g/mol} \qquad Al = 26.98 \text{ g/mol}$$

These molar masses can be expressed by the following conversion factors.

$$\frac{101.96 \text{ g Al}_2\text{O}_3}{\text{mol Al}_2\text{O}_3} \text{ or } \frac{1 \text{ mol Al}_2\text{O}_3}{101.96 \text{ g Al}_2\text{O}_3}$$

$$\frac{26.98 \text{ g Al}}{\text{mol Al}} \text{ or } \frac{1 \text{ mol Al}}{26.98 \text{ g Al}}$$

$$\frac{32.00 \text{ g O}_2}{\text{mol O}_2} \text{ or } \frac{1 \text{ mol O}_2}{32.00 \text{ g O}_2}$$

To find the number of grams of aluminum equivalent to 26.0 mol of aluminum, the calculation would be as follows.

$$26.0 \text{ } \cancel{\text{mol Al}} \times \frac{26.98 \text{ g Al}}{\cancel{\text{mol Al}}} = 701 \text{ g Al}$$

SECTION REVIEW

1. What is stoichiometry?
2. How is a mole ratio from a reaction used in stoichiometric problems?
3. For each of the following chemical equations, write all possible mole ratios.
   a. $2HgO(s) \longrightarrow 2Hg(l) + O_2(g)$
   b. $4NH_3(g) + 6NO(g) \longrightarrow 5N_2(g) + 6H_2O(l)$
   c. $2Al(s) + 3H_2SO_4(aq) \longrightarrow Al_2(SO_4)_3(aq) + 3H_2(g)$

GREAT DISCOVERIES

The Case of Combustion

HISTORICAL PERSPECTIVE

People throughout history have transformed substances by burning them in the air. Yet at the dawn of the scientific revolution, very little was known about the process of combustion. In attempting to explain this common phenomenon, chemists of the eighteenth century developed one of the first universally accepted theories in their field. But, as one man would show, scientific theories do not always stand the test of time.

Changing Attitudes

Shunning the long-standing Greek approach of logical argument based on untested premises, investigators of the seventeenth century began to understand the laws of nature by observing, measuring, and performing experiments on the world around them. Unfortunately, this scientific method was incorporated into chemistry slowly. Although early chemists experimented extensively, most of them disregarded the importance of measurement, an oversight that would set chemistry on the wrong path for nearly a century.

Antoine-Laurent Lavoisier and his wife, Marie-Anne Pierrette Lavoisier, who assisted him. One of her important roles was to translate the papers of important scientists for her husband.

A Flawed Theory

Around 1700, combustion was assumed to be the decomposition of a material into simpler substances. People could see burning substances emitting heat, smoke, and light. To account for this, a theory was proposed that combustion depended on the emission of a substance called phlogiston, which appeared as a combination of heat and light while the material was burning but which couldn't be detected beforehand.

The phlogiston theory was used to explain many chemical observations of the day. For example, a lit candle under a glass jar burned until the surrounding air became saturated with phlogiston, at which time the flame died because the air inside could not absorb more phlogiston.

A New Phase of Study

By the 1770s, the phlogiston theory had gained universal acceptance. At that time, chemists also began to experiment with air, which was generally believed to be an element.

In 1772, when Daniel Rutherford found that a mouse kept in a closed container soon died, he explained the results based on the phlogiston theory. Like a burning candle, the mouse emitted phlogiston; when the air could hold no more phlogiston, the mouse died. Thus, Rutherford figured he had obtained "phlogisticated air."

A couple of years later, Joseph Priestley found that when he heated mercury in air, he obtained a reddish powder, which he assumed to be mercury devoid of phlogiston. But when he decided to heat the powder, he recorded an unexpected result:

> *I endeavored to extract air from [the powder by heating it]; and I presently found that . . . air was expelled from it readily. Having got about three or four times as much as the bulk of my materials, I admitted water to it, and found that water was not imbibed by it. But what surprised me more than I can well express, was, that a candle in this air burned with a remarkably vigorous flame . . .*

Appealing to the phlogiston theory, Priestley believed the highly flammable gas to be "dephlogisticated air."

Nice Try, But . . .

Antoine Laurent Lavoisier was a meticulous scientist. He realized that Rutherford and Priestley had carefully observed and described their experiments but had not bothered to weigh anything. Unlike his colleagues, Lavoisier knew the importance of using a balance:

> *The whole art of making experiments in chemistry is founded on this principle: we must always suppose an exact equality or equation between the principles [masses] of the body examined and those of the products of its analysis.*

Applying this rule, which would become known as the law of conservation of mass, Lavoisier endeavored to explain the results of Rutherford and Priestley.

He put some tin in a closed vessel and weighed the entire system. He then burned the tin. When he opened the vessel, air rushed into it, as if something had been *removed* from the air during combustion. He then weighed the burnt metal and observed a weight increase relative to the original tin. Curiously, this increase equaled the weight of the air that had rushed into the vessel. To Lavoisier, this did not support the idea of phlogiston escaping the burning material. Instead, it indicated that during combustion a portion of air was depleted.

After obtaining similar results using a variety of substances, Lavoisier concluded that air was not an element at all but a mixture composed principally of two gases, Priestley's "dephlogisticated air" (which Lavoisier renamed oxygen) and Rutherford's "phlogisticated air" (which was mostly nitrogen, with traces of other nonflammable atmospheric gases). When a substance burned, it chemically combined with oxygen, resulting in a product Lavoisier named an "oxide." Lavoisier's theory of combustion persists today. He used the name *oxygen* because he thought that all acids contained oxygen. *Oxygen* means "acid former."

The Father of Chemistry

By emphasizing the importance of quantitative analysis, Lavoisier helped establish chemistry as a science. His work on combustion laid to rest the theory of phlogiston as well as the ancient notion that air was an element. His theory also explained why hydrogen burned in oxygen to form water, or hydrogen oxide. He would later publish one of the first chemistry textbooks, which established a common naming system of compounds and elements and helped unify chemistry worldwide, earning him the reputation as the father of chemistry.

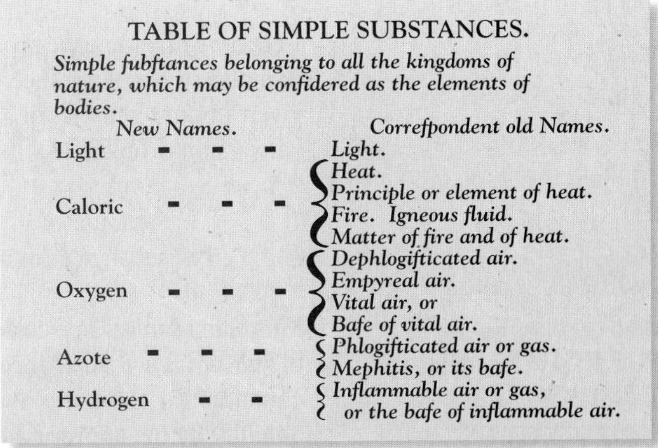

TABLE OF SIMPLE SUBSTANCES.

Simple fubftances belonging to all the kingdoms of nature, which may be confidered as the elements of bodies.

New Names.	Correfpondent old Names.
Light	Light.
Caloric	Heat. Principle or element of heat. Fire. Igneous fluid. Matter of fire and of heat.
Oxygen	Dephlogifticated air. Empyreal air. Vital air, or Bafe of vital air.
Azote	Phlogifticated air or gas. Mephitis, or its bafe.
Hydrogen	Inflammable air or gas, or the bafe of inflammable air.

Lavoisier's concept of simple substances as published in his book Elements of Chemistry *in 1789.*

SECTION 9-2

OBJECTIVES

- Calculate the amount in moles of a reactant or product from the amount in moles of a different reactant or product.
- Calculate the mass of a reactant or product from the amount in moles of a different reactant or product.
- Calculate the amount in moles of a reactant or product from the mass of a different reactant or product.
- Calculate the mass of a reactant or product from the mass of a different reactant or product.

Ideal Stoichiometric Calculations

The chemical equation plays a very important part in all stoichiometric calculations because the mole ratio is obtained directly from it. Solving any reaction-stoichiometry problem must begin with a balanced equation.

Chemical equations help us make predictions about chemical reactions without having to run the reactions in the laboratory. The reaction-stoichiometry calculations described in this chapter are theoretical. They tell us the amounts of reactants and products for a given chemical reaction under *ideal conditions,* in which all reactants are completely converted into products. However, ideal conditions are rarely met in the laboratory or in industry. Yet, theoretical stoichiometric calculations serve the very important function of showing the maximum amount of product that coud be obtained before a reaction is run in the laboratory.

Solving stoichiometric problems requires practice. These problems are extensions of the composition-stoichiometry problems you solved in Chapters 3 and 7. Practice by working the sample problems in the rest of this chapter. Using a logical, systematic approach will help you successfully solve these problems.

Conversions of Quantities in Moles

In these stoichiometric problems, you are asked to calculate the amount in moles of one substance that will react with or be produced from the given amount in moles of another substance. The plan for a simple mole conversion problem is

$$\begin{matrix}\text{amount of} \\ \textit{given}\text{ substance (in mol)}\end{matrix} \longrightarrow \begin{matrix}\text{amount of} \\ \textit{unknown}\text{ substance (in mol)}\end{matrix}$$

This plan requires one conversion factor—the stoichiometric mole ratio of the *unknown* substance to the *given* substance from the balanced equation. To solve this type of problem, simply multiply the *known* quantity by the appropriate conversion factor.

$$\textit{given}\text{ quantity} \times \text{conversion factor} = \textit{unknown}\text{ quantity}$$

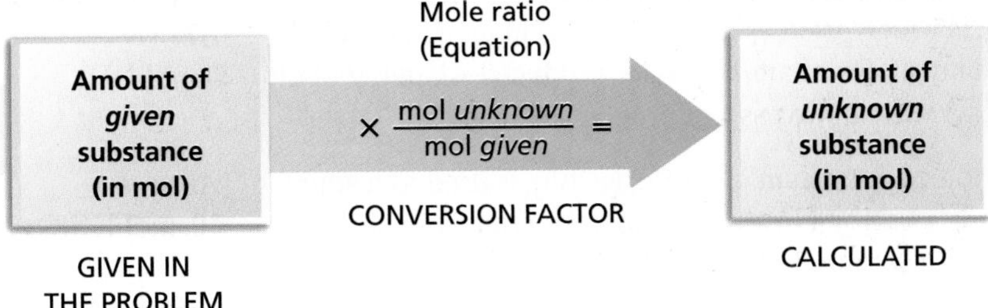

FIGURE 9-1 This is a solution plan for problems in which the given and unknown quantities are expressed in moles.

SAMPLE PROBLEM 9-1

In a spacecraft, the carbon dioxide exhaled by astronauts can be removed by its reaction with lithium hydroxide, LiOH, according to the following chemical equation.

$$CO_2(g) + 2LiOH(s) \longrightarrow Li_2CO_3(s) + H_2O(l)$$

How many moles of lithium hydroxide are required to react with 20 mol of CO_2, the average amount exhaled by a person each day?

SOLUTION

1 ANALYZE **Given:** amount of CO_2 = 20 mol
Unknown: amount of LiOH in moles

2 PLAN amount of CO_2 (in mol) $\longrightarrow$ amount of LiOH (in mol)

This problem requires one conversion factor—the mole ratio of LiOH to CO_2. The mole ratio is obtained from the balanced chemical equation. Because you are given moles of CO_2, select a mole ratio that will give you mol LiOH in your final answer. The correct ratio is the following.

$$\frac{\text{mol LiOH}}{\text{mol } CO_2}$$

This ratio gives the units mol LiOH in the answer.

$$\text{mol } CO_2 \times \overset{\text{mol ratio}}{\frac{\text{mol LiOH}}{\text{mol } CO_2}} = \text{mol LiOH}$$

3 COMPUTE Substitute the values in the equation in step 2, and compute the answer.

$$20 \cancel{\text{mol } CO_2} \times \frac{2 \text{ mol LiOH}}{1 \cancel{\text{mol } CO_2}} = 40 \text{ mol LiOH}$$

4 EVALUATE The answer is rounded correctly to one significant figure to match that in the factor 20 mol CO_2, and the units cancel to leave mol LiOH, which is the unknown. The equation shows that twice the amount in moles of LiOH react with CO_2. Therefore, the answer should be greater than 20.

PRACTICE

1. Ammonia, NH_3, is widely used as a fertilizer and in many household cleaners. How many moles of ammonia are produced when 6 mol of hydrogen gas react with an excess of nitrogen gas?

   Answer 4 mol NH_3

2. The decomposition of potassium chlorate, $KClO_3$, is used as a source of oxygen in the laboratory. How many moles of potassium chlorate are needed to produce 15 mol of oxygen?

   Answer 10. mol $KClO_3$

Conversions of Amounts in Moles to Mass

In these stoichiometric calculations, you are asked to calculate the mass (usually in grams) of a substance that will react with or be produced from a given amount in moles of a second substance. The plan for these mole to gram conversions is

$$\begin{array}{c}\text{amount of}\\ \textit{given}\text{ substance}\\ \text{(in mol)}\end{array} \longrightarrow \begin{array}{c}\text{amount of}\\ \textit{unknown}\text{ substance}\\ \text{(in mol)}\end{array} \longrightarrow \begin{array}{c}\text{mass of}\\ \textit{unknown}\text{ substance}\\ \text{(in g)}\end{array}$$

This plan requires two conversion factors—the mole ratio of the *unknown* substance to the *given* substance and the molar mass of the *unknown* substance for the mass conversion. To solve this kind of problem, you simply multiply the known quantity, which is the amount in moles, by the appropriate conversion factors.

FIGURE 9-2 This is a solution plan for problems in which the given quantity is expressed in moles and the unknown quantity is expressed in grams.

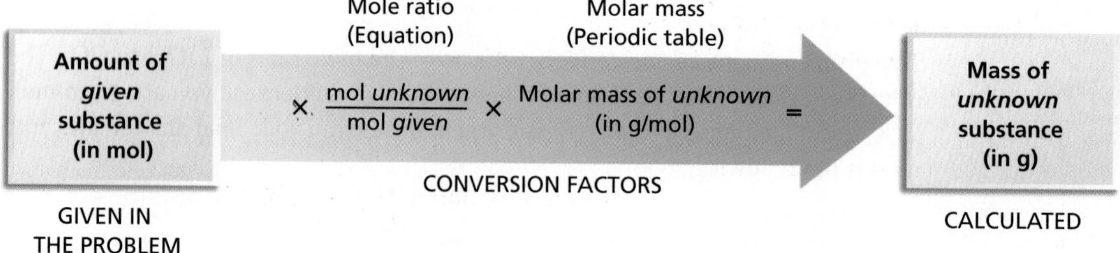

SAMPLE PROBLEM 9-2

In photosynthesis, plants use energy from the sun to produce glucose, $C_6H_{12}O_6$, and oxygen from the reaction of carbon dioxide and water. What mass, in grams, of glucose is produced when 3.00 mol of water react with carbon dioxide?

SOLUTION

1 *ANALYZE* **Given:** amount of $H_2O = 3.00$ mol

Unknown: mass of $C_6H_{12}O_6$ produced (in g)

2 PLAN You must start with a balanced equation.

$$6CO_2(g) + 6H_2O(l) \longrightarrow C_6H_{12}O_6(s) + 6O_2(g)$$

Given the amount in mol of H_2O, you need to get the mass of $C_6H_{12}O_6$ in grams. Two conversion factors are needed—the mole ratio of $C_6H_{12}O_6$ to H_2O and the molar mass of $C_6H_{12}O_6$.

$$\text{mol } H_2O \times \overset{\text{mol ratio}}{\frac{\text{mol } C_6H_{12}O_6}{\text{mol } H_2O}} \times \overset{\text{molar mass}}{\frac{\text{g } C_6H_{12}O_6}{\text{mol } C_6H_{12}O_6}} = \text{g } C_6H_{12}O_6$$

3 COMPUTE Use the periodic table to compute the molar mass of $C_6H_{12}O_6$.

$$C_6H_{12}O_6 = 180.18 \text{ g/mol}$$

$$3.00 \cancel{\text{mol } H_2O} \times \frac{1 \cancel{\text{mol } C_6H_{12}O_6}}{6 \cancel{\text{mol } H_2O}} \times \frac{180.18 \text{ g } C_6H_{12}O_6}{1 \cancel{\text{mol } C_6H_{12}O_6}} = 90.1 \text{ g } C_6H_{12}O_6$$

4 EVALUATE The answer is correctly rounded to three significant figures, to match those in 3.00 mol H_2O. The units cancel in the problem, leaving g $C_6H_{12}O_6$ as the units for the answer, which matches the unknown. The answer is reasonable because it is one-half of 180.

SAMPLE PROBLEM 9-3

What mass of carbon dioxide, in grams, is needed to react with 3.00 mol of H_2O in the photosynthetic reaction described in Sample Problem 9-2?

SOLUTION

1 ANALYZE **Given:** amount of H_2O = 3.00 mol

Unknown: mass of CO_2 in grams

2 PLAN The chemical equation from Sample Problem 9-2 is

$$6CO_2(g) + 6H_2O(l) \longrightarrow C_6H_{12}O_6(s) + 6O_2(g).$$

Two conversion factors are needed—the mole ratio of CO_2 to H_2O and the molar mass of CO_2.

$$\text{mol } H_2O \times \overset{\text{mol ratio}}{\frac{\text{mol } CO_2}{\text{mol } H_2O}} \times \overset{\text{molar mass}}{\frac{\text{g } CO_2}{\text{mol } CO_2}} = \text{g } CO_2$$

3 COMPUTE Use the periodic table to compute the molar mass of CO_2.

$$CO_2 = 44.01 \text{ g/mol}$$

$$3.00 \cancel{\text{mol } H_2O} \times \frac{6 \cancel{\text{mol } CO_2}}{6 \cancel{\text{mol } H_2O}} \times \frac{44.01 \text{ g } CO_2}{\cancel{\text{mol } CO_2}} = 132 \text{ g } CO_2$$

4 EVALUATE The answer is rounded correctly to three significant figures to match those in 3.00 mol H_2O. The units cancel to leave g CO_2, which is the unknown. The answer is close to an estimate of 120, which is 3 × 40.

PRACTICE

1. When magnesium burns in air, it combines with oxygen to form magnesium oxide according to the following equation.

   $$2Mg(s) + O_2(g) \longrightarrow 2MgO(s)$$

   What mass in grams of magnesium oxide is produced from 2.00 mol of magnesium?

   Answer
   80.6 g MgO

2. What mass in grams of oxygen combines with 2.00 mol of magnesium in this same reaction?

   Answer
   32.0 g O_2

3. What mass of glucose can be produced from a photosynthesis reaction that occurs using 10 mol CO_2?

   $$6CO_2(g) + 6H_2O(l) \longrightarrow C_6H_{12}O_6(aq) + 6O_2(g)$$

   Answer
   300 g $C_6H_{12}O_6$

Conversions of Mass to Amounts in Moles

In these stoichiometric calculations, you are asked to calculate the amount in moles of one substance that will react with or be produced from a given mass of another substance. In this type of problem you are starting with a mass (probably in grams) of some substance. The plan for this conversion is

mass of *given* substance (in g) ⟶ amount of *given* substance (in mol) ⟶ amount of *unknown* substance (in mol)

This route also requires two additional pieces of data: the molar mass of the *given* substance and the mole ratio. The molar mass is determined using masses from the periodic table. To convert the mass of a substance to moles we are using a factor which we will call the inverted molar mass. It is simply one over the molar mass. To solve this type of problem, simply multiply or divide the known quantity by the appropriate conversion factors as follows.

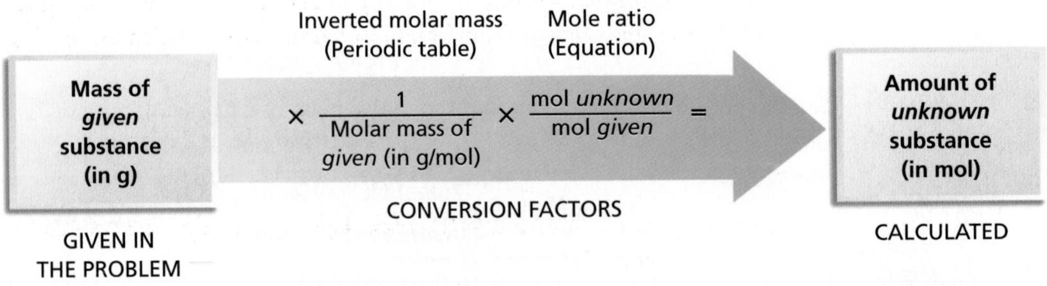

FIGURE 9-3 This is a solution plan for problems in which the given quantity is expressed in grams and the unknown quantity is expressed in moles.

SAMPLE PROBLEM 9-4

The first step in the industrial manufacture of nitric acid is the catalytic oxidation of ammonia.

$$NH_3(g) + O_2(g) \longrightarrow NO(g) + H_2O(g) \quad \text{(unbalanced)}$$

The reaction is run using 824 g of NH_3 and excess oxygen.
a. How many moles of NO are formed?
b. How many moles of H_2O are formed?

SOLUTION

1 *ANALYZE* **Given:** mass of NH_3 = 824 g
Unknown: **a.** amount of NO produced (in mol)
b. amount of H_2O produced (in mol)

2 *PLAN* First, write the balanced chemical equation.

$$4NH_3(g) + 5O_2(g) \longrightarrow 4NO(g) + 6H_2O(g)$$

Two conversion factors are needed to solve part (a)—the molar mass of NH_3 and the mole ratio of NO to NH_3. Part (b) starts with the same conversion factor as part (a), but then the mole ratio of H_2O to NH_3 is used to convert to the amount in moles of H_2O. The first conversion factor in each part is the inverted molar mass of NH_3.

$$\textbf{a. } \text{g } NH_3 \times \underbrace{\frac{1 \text{ mol } NH_3}{\text{g } NH_3}}_{\text{inverted molar mass}} \times \underbrace{\frac{\text{mol NO}}{\text{mol } NH_3}}_{\text{mol ratio}} = \text{mol NO}$$

$$\textbf{b. } \text{g } NH_3 \times \underbrace{\frac{1 \text{ mol } NH_3}{\text{g } NH_3}}_{\text{inverted molar mass}} \times \underbrace{\frac{\text{mol } H_2O}{\text{mol } NH_3}}_{\text{mol ratio}} = \text{mol } H_2O$$

3 *COMPUTE* Use the periodic table to compute the molar mass of NH_3.

$$NH_3 = 17.04 \text{ g/mol}$$

$$\textbf{a. } 824 \cancel{\text{g } NH_3} \times \frac{1 \cancel{\text{mol } NH_3}}{17.04 \cancel{\text{g } NH_3}} \times \frac{4 \text{ mol NO}}{4 \cancel{\text{mol } NH_3}} = 48.4 \text{ mol NO}$$

$$\textbf{b. } 824 \cancel{\text{g } NH_3} \times \frac{1 \cancel{\text{mol } NH_3}}{17.04 \cancel{\text{g } NH_3}} \times \frac{6 \text{ mol } H_2O}{4 \cancel{\text{mol } NH_3}} = 72.6 \text{ mol } H_2O$$

4 *EVALUATE* The answers are correctly given to three significant figures. The units cancel in the two problems to leave mol NO and mol H_2O, respectively, which are the unknowns.

PRACTICE

Oxygen was discovered by Joseph Priestley in 1774 when he heated mercury(II) oxide to decompose it to form its constituent elements.

1. How many moles of mercury(II) oxide, HgO, are needed to produce 125 g of oxygen, O_2?
   Answer 7.81 mol HgO
2. How many moles of mercury are produced?
   Answer 7.81 mol Hg

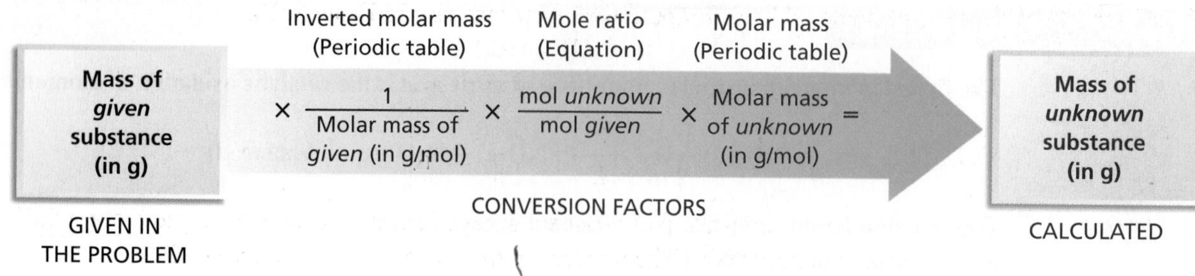

FIGURE 9-4 This is a solution plan for problems in which the given quantity is expressed in grams and the unknown quantity is also expressed in grams.

Mass-Mass Calculations

Mass-mass calculations are more practical than other mole calculations you have studied. You can never measure moles directly. You are generally required to calculate the amount in moles of a substance from its mass, which you can measure in the lab. Mass-mass problems can be viewed as the combination of the other types of problems. The plan for solving mass-mass problems is

$$\begin{array}{c}\text{mass of}\\ \textit{given}\text{ substance}\\ \text{(in g)}\end{array} \longrightarrow \begin{array}{c}\text{amount of}\\ \textit{given}\text{ substance}\\ \text{(in mol)}\end{array} \longrightarrow \begin{array}{c}\text{amount of}\\ \textit{unknown}\text{ substance}\\ \text{(in mol)}\end{array} \longrightarrow \begin{array}{c}\text{mass of}\\ \textit{unknown}\text{ substance}\\ \text{(in g)}\end{array}$$

Three additional pieces of data are needed to solve mass-mass problems: the molar mass of the *given* substance, the mole ratio, and the molar mass of the *unknown* substance.

SAMPLE PROBLEM 9-5

Tin(II) fluoride, SnF_2, is used in some toothpastes. It is made by the reaction of tin with hydrogen fluoride according to the following equation.

$$\mathbf{Sn(\mathit{s}) + 2HF(\mathit{g}) \longrightarrow SnF_2(\mathit{s}) + H_2(\mathit{g})}$$

How many grams of SnF_2 are produced from the reaction of 30.00 g of HF with Sn?

SOLUTION

1 *ANALYZE* **Given:** amount of HF = 30.00 g

Unknown: mass of SnF_2 produced in grams

2 *PLAN* The conversion factors needed are the molar masses of HF and SnF_2 and the mole ratio of SnF_2 to HF.

$$\text{g HF} \times \overset{\text{inverted molar mass}}{\frac{1 \text{ mol HF}}{\text{g HF}}} \times \overset{\text{mol ratio}}{\frac{\text{mol SnF}_2}{\text{mol HF}}} \times \overset{\text{molar mass}}{\frac{\text{g SnF}_2}{\text{mol SnF}_2}} = \text{g SnF}_2$$

3 COMPUTE Use the periodic table to compute the molar masses of HF and SnF_2.

$$HF = 20.01 \text{ g/mol}$$
$$SnF_2 = 156.71 \text{ g/mol}$$

$$30.00 \text{ g HF} \times \frac{1 \text{ mol HF}}{20.01 \text{ g HF}} \times \frac{1 \text{ mol SnF}_2}{2 \text{ mol HF}} \times \frac{156.71 \text{ g SnF}_2}{1 \text{ mol SnF}_2} = 117.5 \text{ g SnF}_2$$

4 EVALUATE The answer is correctly rounded to four significant figures. The units cancel to leave g SnF_2, which matches the unknown. The answer is close to an estimated value of 120.

PRACTICE

1. Laughing gas (nitrous oxide, N_2O) is sometimes used as an anesthetic in dentistry. It is produced when ammonium nitrate is decomposed according to the following reaction.

$$NH_4NO_3(s) \longrightarrow N_2O(g) + 2H_2O(l)$$

a. How many grams of NH_4NO_3 are required to produce 33.0 g of N_2O?

b. How many grams of water are produced in this reaction?

2. When copper metal is added to silver nitrate in solution, silver metal and copper(II) nitrate are produced. What mass of silver is produced from 100. g of Cu?

3. What mass of aluminum is produced by the decomposition of 5.0 kg of Al_2O_3?

Answer

1. a. 60.0 g NH_4NO_3

b. 27.0 g H_2O

2. 339 g

3. 2.6 kg

SECTION REVIEW

1. Balance the following equation. Then, based on the amount in moles of each reactant or product given, determine the corresponding amount in moles of all the remaining reactants and products involved.

$$NH_3 + O_2 \longrightarrow N_2 + H_2O$$

a. 4 mol NH_3
b. 4 mol N_2
c. 4.5 mol O_2

2. Hydrogen gas can be produced through the following unbalanced reaction.

$$Mg(s) + HCl(aq) \longrightarrow MgCl_2(aq) + H_2(g)$$

a. What mass of HCl is consumed by the reaction of 2.50 mol of magnesium?
b. What mass of each product is produced in part (a)?

3. Acetylene gas (C_2H_2) is produced as a result of the following reaction.

$$CaC_2(s) + 2H_2O(l) \longrightarrow C_2H_2(g) + Ca(OH)_2(aq)$$

a. If 32.0 g of CaC_2 are consumed in this reaction, how many moles of H_2O are needed?
b. How many moles of each product would be formed?

4. When sodium chloride reacts with silver nitrate, silver chloride precipitates. What mass of AgCl is produced from 75.0 g of $AgNO_3$?

5. Acetylene gas, C_2H_2, used in welding, produces an extremely hot flame when it burns in pure oxygen according to the following reaction.

$$2C_2H_2(g) + 5O_2(g) \longrightarrow 4CO_2(g) + 2H_2O(g)$$

How many grams of each product are produced when 2.50×10^4 g of C_2H_2 burns completely?

SECTION 9-3

Limiting Reactants and Percent Yield

OBJECTIVES

- Describe a method for determining which of two reactants is a limiting reactant.
- Calculate the amount in moles or mass in grams of a product, given the amounts in moles or masses in grams of two reactants, one of which is in excess.
- Distinguish between theoretical yield, actual yield, and percent yield.
- Calculate percent yield, given the actual yield and quantity of a reactant.

In the laboratory a reaction is rarely carried out with exactly the required amounts of each of the reactants. In most cases, one or more reactants is present in excess; that is, there is more than the exact amount required to react.

Once one of the reactants is used up, no more product can be formed. The substance that is completely used up first in a reaction is called the limiting reactant. *The* **limiting reactant** *is the reactant that limits the amounts of the other reactants that can combine and the amount of product that can form in a chemical reaction. The substance that is not used up completely in a reaction is sometimes called the* **excess reactant.** A limiting reactant may also be referred to as a limiting reagent.

The concept of the limiting reactant is analogous to the relationship between the number of people who want to take a certain airplane flight and the number of seats available in the airplane. If 400 people want to travel on the flight and only 350 seats are available, then only 350 people can go on the flight. The number of seats on the airplane limits the number of people who can travel. There are 50 people in excess.

The same reasoning can be applied to chemical reactions. Consider the reaction between carbon and oxygen to form carbon dioxide.

$$C(s) + O_2(g) \longrightarrow CO_2(g)$$

According to the equation, one mole of carbon reacts with one mole of oxygen to form one mole of carbon dioxide. Suppose you could mix 5 mol of C with 10 mol of O_2 and allow the reaction to take place. Figure 9-5 shows that there is more oxygen than is needed to react with the carbon. Carbon is the limiting reactant in this situation, and it limits the amount of CO_2 that is formed. Oxygen is the excess reactant, and 5 mol of O_2 will be left over at the end of the reaction.

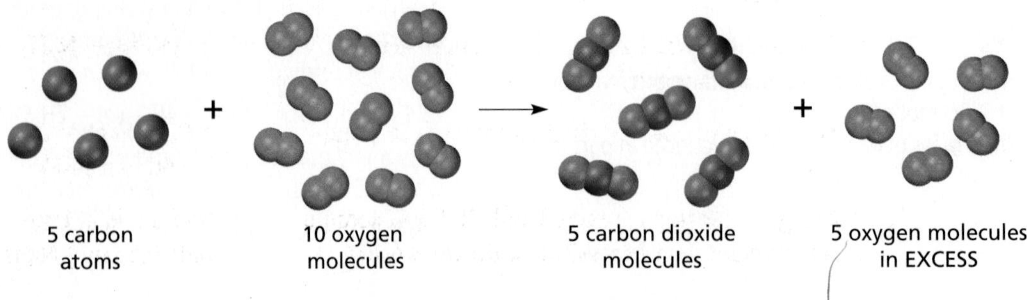

FIGURE 9-5 If you think of a mole as a multiple of molecules and atoms, you can see why the amount of O_2 is in excess.

SAMPLE PROBLEM 9-6

Silicon dioxide (quartz) is usually quite unreactive but reacts readily with hydrogen fluoride according to the following equation.

$$SiO_2(s) + 4HF(g) \longrightarrow SiF_4(g) + 2H_2O(l)$$

If 2.0 mol of HF are exposed to 4.5 mol of SiO_2, which is the limiting reactant?

SOLUTION

1 *ANALYZE*

Given: amount of HF = 2.0 mol

amount of SiO_2 = 4.5 mol

Unknown: limiting reactant

2 *PLAN*

The given amount of either reactant is used to calculate the required amount of the other reactant. The calculated amount is then compared with the amount actually available, and the limiting reactant can be identified. We will choose to calculate the moles of SiO_2 required by the given amount of HF.

$$\text{mol HF} \times \frac{\text{mol SiO}_2}{\text{mol HF}} = \text{mol SiO}_2 \text{ required}$$

3 *COMPUTE*

$$2.0 \cancel{\text{mol HF}} \times \frac{1 \text{ mol SiO}_2}{4 \cancel{\text{mol HF}}} = 0.50 \text{ mol SiO}_2 \text{ required}$$

Under ideal conditions, the 2.0 mol of HF will require 0.50 mol of SiO_2 for complete reaction. Because the amount of SiO_2 available (4.5 mol) is more than the amount required (0.50 mol), the limiting reactant is HF.

4 *EVALUATE*

The calculated amount of SiO_2 is correctly given to two significant figures. Because each mole of SiO_2 requires 4 mol of HF, it is reasonable that HF is the limiting reactant because the molar amount of HF available is less than half that of SiO_2.

PRACTICE

1. Some rocket engines use a mixture of hydrazine, N_2H_4, and hydrogen peroxide, H_2O_2, as the propellant. The reaction is given by the following equation.

$$N_2H_4(l) + 2H_2O_2(l) \longrightarrow N_2(g) + 4H_2O(g)$$

a. Which is the limiting reactant in this reaction when 0.750 mol of N_2H_4 is mixed with 0.500 mol of H_2O_2?

b. How much of the excess reactant, in moles, remains unchanged?

c. How much of each product, in moles, is formed?

2. If 20.5 g of chlorine is reacted with 20.5 g of sodium, which reactant is in excess? How do you know?

Answer

1. a. H_2O_2

b. 0.500 mol N_2H_4

c. 0.250 mol N_2, 1.00 mol H_2O

2. Sodium is in excess because only 0.58 mol Na is needed.

SAMPLE PROBLEM 9-7

The black oxide of iron, Fe_3O_4, occurs in nature as the mineral magnetite. This substance can also be made in the laboratory by the reaction between red-hot iron and steam according to the following equation.

$$3Fe(s) + 4H_2O(g) \longrightarrow Fe_3O_4(s) + 4H_2(g)$$

a. When 36.0 g of H_2O is mixed with 167 g of Fe, which is the limiting reactant?
b. What mass in grams of black iron oxide is produced?
c. What mass in grams of excess reactant remains when the reaction is completed?

SOLUTION

1 ANALYZE

Given: mass of H_2O = 36.0 g
mass of Fe = 167 g

Unknown: limiting reactant
mass of Fe_3O_4, in grams
mass of excess reactant remaining

2 PLAN

a. First convert both given masses in grams to amounts in moles. Choose one reactant and calculate the needed amount of the other to determine which is the limiting reactant. We have chosen Fe. The mole ratio from the balanced equation is 3 mol Fe for every 4 mol H_2O.

$$\text{g Fe} \times \overset{\text{inverted molar mass}}{\frac{\text{1 mol Fe}}{\text{g Fe}}} = \text{mol Fe available}$$

$$\text{g } H_2O \times \overset{\text{inverted molar mass}}{\frac{\text{1 mol } H_2O}{\text{g } H_2O}} = \text{mol } H_2O \text{ available}$$

$$\text{mol Fe} \times \overset{\text{mol ratio}}{\frac{\text{mol } H_2O}{\text{mol Fe}}} = \text{mol } H_2O \text{ required}$$

b. To find the maximum amount of Fe_3O_4 that can be produced, the given amount in moles of the limiting reactant must be used in a simple stoichiometric problem.

$$\text{limiting reactant (in mol)} \times \overset{\text{mol ratio}}{\frac{\text{mol } Fe_3O_4}{\text{mol limiting reactant}}} \times \overset{\text{molar mass}}{\frac{\text{g } Fe_3O_4}{\text{mol } Fe_3O_4}} = \text{g } Fe_3O_4 \text{ produced}$$

c. To find the amount of excess reactant remaining, the amount of the excess reactant that is consumed must first be determined. The given amount in moles of the limiting reactant must be used in a simple stoichiometric problem.

$$\text{limiting reactant (in mol)} \times \frac{\text{mol excess reactant}}{\text{mol limiting reactant}} \times \frac{\text{g excess reactant}}{\text{mol excess reactant}} = \text{g excess reactant consumed}$$

The amount of excess reactant remaining can then be found by subtracting the amount consumed from the amount originally present.

original g excess reactant − g excess reactant consumed = g excess reactant remaining

3 COMPUTE Use the periodic table to determine the molar masses of H_2O, Fe, and Fe_3O_4.

H_2O = 18.02 g/mol
Fe = 55.85 g/mol
Fe_3O_4 = 231.55 g/mol

$$36.0 \text{ g } H_2O \times \frac{1 \text{ mol } H_2O}{18.02 \text{ g } H_2O} = 2.00 \text{ mol } H_2O$$

$$167 \text{ g Fe} \times \frac{1 \text{ mol Fe}}{55.85 \text{ g Fe}} = 2.99 \text{ mol Fe}$$

$$2.99 \cancel{\text{mol Fe}} \times \frac{4 \text{ mol } H_2O}{3 \cancel{\text{mol Fe}}} = 3.99 \text{ mol } H_2O \text{ required}$$

a. The required 3.99 mol of H_2O is more than the 2.00 mol of H_2O available, so H_2O is the limiting reactant.

b. $2.00 \text{ mol } H_2O \times \frac{1 \text{ mol } Fe_3O_4}{4 \text{ mol } H_2O} \times \frac{231.55 \text{ g } Fe_3O_4}{\text{mol } Fe_3O_4} = 116 \text{ g } Fe_3O_4$

c. $2.00 \text{ mol } H_2O \times \frac{3 \text{ mol Fe}}{4 \text{ mol } H_2O} \times \frac{55.85 \text{ g Fe}}{\text{mol Fe}} = 83.8 \text{ g Fe consumed}$

167 g Fe originally present – 83.8 g Fe consumed = 83.2 g Fe remaining

4 EVALUATE Three significant digits are carried through each calculation. The result of the final subtraction is rounded to match the significance of the least accurately known number, that is, the units digit for the original mass of Fe. The mass of Fe_3O_4 is close to an estimated answer of 115, which is one-half of 230. The amount of the limiting reactant, H_2O, is about one-half the amount needed to use all of the Fe, so about one-half the Fe remains unreacted.

PRACTICE

1. Zinc and sulfur react to form zinc sulfide according to the following equation.

$$8Zn(s) + S_8(s) \longrightarrow 8ZnS(s)$$

130.78 256.8

a. If 2.00 mol of Zn are heated with 1.00 mol of S_8, identify the limiting reactant. — *Answer* 1. a. Zn

b. How many moles of excess reactant remain. — b. 0.750 mol S_8 remains

c. How many moles of the product are formed? — c. 2.00 mol ZnS

2. Carbon reacts with steam, H_2O, at high temperatures to produce hydrogen and carbon monoxide.

a. If 2.40 mol of carbon are exposed to 3.10 mol of steam, identify the limiting reactant. — 2. a. carbon

b. How many moles of each product are formed? — b. 2.40 mol H_2 and 2.40 mol CO

c. What mass of each product is formed? — c. 4.85 g H_2 and 67.2 g CO

KITCHEN INVESTIGATION

Limiting Reactants in a Recipe

MATERIALS

- 1/2 cup sugar
- 1/2 cup brown sugar
- 1 1/3 stick margarine (at room temperature)
- 1 egg
- 1/2 tsp. of salt
- 1 tsp. vanilla
- 1/2 tsp. baking soda
- 1 1/2 cup flour
- 1 1/3 cup chocolate chips
- mixing bowl
- mixing spoon
- measuring spoons and cups
- cookie sheet
- oven preheated to 350°F

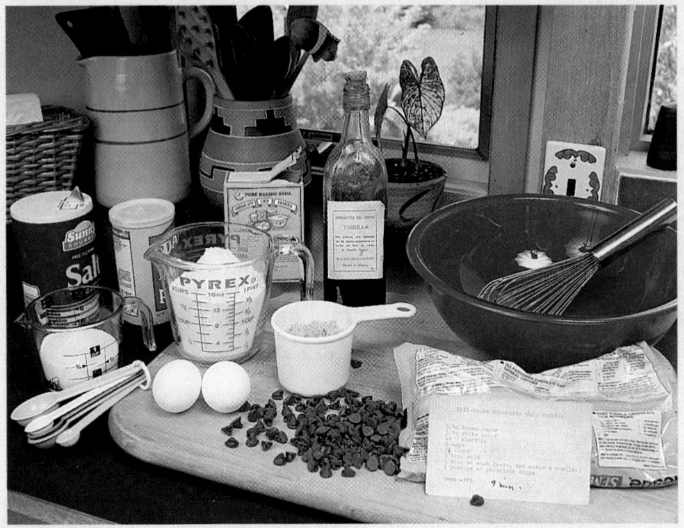

PROCEDURE

1. In the mixing bowl, combine the sugars and margarine together until smooth. (An electric mixer will make this process go much faster.)
2. Add the egg, salt, and vanilla. Mix well.
3. Stir in the baking soda, flour, and chocolate chips. Chill the dough for an hour in the refrigerator for best results.
4. Divide the dough into 24 small balls about 3 cm in diameter. Place the balls on an ungreased cookie sheet.
5. Bake at 350°F for about 10 minutes, or until the cookies are light brown.

Yield: 24 cookies

DISCUSSION

1. Suppose you are given the following amounts of ingredients:
   1 dozen eggs
   24 tsp. of vanilla
   1 lb. (82 tsp.) of salt
   1 lb. (84 tsp.) of baking soda
   3 cups of chocolate chips
   5 lb. (11 cups) of sugar
   2 lb. (4 cups) of brown sugar
   1 lb. (4 sticks) of margarine
   a. For each ingredient, calculate how many cookies could be prepared if all of that ingredient were consumed. (For example, the recipe shows that using 1 egg—with the right amounts of the other ingredients—yields 24 cookies. How many cookies can you make if the recipe is increased proportionately for 12 eggs?)
   b. To determine the limiting reactant for the new ingredients list, identify which ingredients will result in the fewest number of cookies.
   c. What is the maximum number of cookies that can be produced from the new amounts of ingredients?

Percent Yield

The amounts of products calculated in the stoichiometric problems in this chapter so far represent theoretical yields. *The* **theoretical yield** *is the maximum amount of product that can be produced from a given amount of reactant.* In most chemical reactions, the amount of product obtained is less than the theoretical yield. *The measured amount of a product obtained from a reaction is called the* **actual yield** *of that product.*

There are many reasons why the actual yield of a reaction may be less than the theoretical yield. Some of the reactant may be used in competing side reactions that reduce the amount of the desired product. Also, once a product is formed, it often is usually collected in impure form, and some of the product is often lost during the purification process. Chemists are usually interested in the efficiency of a reaction. The efficiency is expressed by comparing the actual and theoretical yields. *The* **percent yield** *is the ratio of the actual yield to the theoretical yield, multiplied by 100.*

$$\text{percent yield} = \frac{\text{actual yield}}{\text{theoretical yield}} \times 100$$

SAMPLE PROBLEM 9-8

Chlorobenzene, C_6H_5Cl, is used in the production of many important chemicals, such as aspirin, dyes, and disinfectants. One industrial method of preparing chlorobenzene is to react benzene, C_6H_6, with chlorine, which is represented by the following equation.

$$C_6H_6(l) + Cl_2(g) \longrightarrow C_6H_5Cl(s) + HCl(g)$$

When 36.8 g of C_6H_6 react with an excess of Cl_2, the actual yield of C_6H_5Cl is 38.8 g. What is the percent yield of C_6H_5Cl?

SOLUTION

1 *ANALYZE*

Given: mass of C_6H_6 = 36.8 g
mass of Cl_2 = excess
actual yield of C_6H_5Cl = 38.8 g
Unknown: percent yield of C_6H_5Cl

2 *PLAN*

First do a mass-mass calculation to find the theoretical yield of C_6H_5Cl.

$$\text{g } C_6H_6 \times \underbrace{\frac{\text{mol } C_6H_6}{\text{g } C_6H_6}}_{\text{inverted molar mass}} \times \underbrace{\frac{\text{mol } C_6H_5Cl}{\text{mol } C_6H_6}}_{\text{mol ratio}} \times \underbrace{\frac{\text{g } C_6H_5Cl}{\text{mol } C_6H_5Cl}}_{\text{molar mass}} = \text{g } C_6H_5Cl \text{ (theoretical yield)}$$

Then the percent yield can be found.

$$\text{percent yield } C_6H_5Cl = \frac{\text{actual yield}}{\text{theoretical yield}} \times 100$$

3 COMPUTE Use the periodic table to determine the molar masses of C_6H_6 and C_6H_5Cl.

$$C_6H_6 = 78.12 \text{ g/mol}$$
$$C_6H_5Cl = 112.56 \text{ g/mol}$$

$$36.8 \text{ g } \cancel{C_6H_6} \times \frac{1 \text{ mol } C_6H_6}{78.12 \text{ g } C_6H_6} \times \frac{1 \text{ mol } C_6H_5Cl}{1 \text{ mol } C_6H_6} \times \frac{112.56 \text{ g } C_6H_5Cl}{\text{mol } C_6H_5Cl} = 53.0 \text{ g } C_6H_5Cl \text{ (theoretical yield)}$$

$$\text{percent yield} = \frac{38.8 \text{ g}}{53.0 \text{ g}} \times 100 = 73.2\%$$

4 EVALUATE The answer is correctly rounded to three significant figures to match those in 36.8 g C_6H_6. The units have canceled correctly. The theoretical yield is close to an estimated value of 50 g, (one-half of 100 g). The percent yield is close to an estimated value of 70%, (70/100 × 100).

PRACTICE

1. Methanol can be produced through the reaction of CO and H_2 in the presence of a catalyst.

   Answer 79.8%

$$CO(g) + 2H_2(g) \xrightarrow{\text{catalyst}} CH_3OH(l)$$

   If 75.0 g of CO reacts to produce 68.4 g CH_3OH, what is the percent yield of CH_3OH?

2. Aluminum reacts with excess copper(II) sulfate according to the reaction given below. If 1.85 g of Al react and the percent yield of Cu is 56.6%, what mass of Cu is produced?

   Answer 3.70 g

$$Al(s) + CuSO_4(aq) \longrightarrow Al_2(SO_4)_3(aq) + Cu(s) \text{ (unbalanced)}$$

SECTION REVIEW

1. Carbon disulfide burns in oxygen to yield carbon dioxide and sulfur dioxide according to the following chemical equation.

$$CS_2(l) + 3O_2(g) \longrightarrow CO_2(g) + 2SO_2(g)$$

   a. If 1.00 mol of CS_2 is combined with 1.00 mol of O_2, identify the limiting reactant.
   b. How many moles of excess reactant remain?
   c. How many moles of each product are formed?

2. Metallic magnesium reacts with steam to produce magnesium hydroxide and hydrogen gas.
   a. If 16.2 g of Mg are heated with 12.0 g of H_2O, what is the limiting reactant?
   b. How many moles of the excess reactant are left?
   c. How many grams of each product are formed?

3. a. What is the limiting reactant when 19.9 g of CuO are exposed to 2.02 g of H_2 according to the following equation?

$$CuO(s) + H_2(g) \longrightarrow Cu(s) + H_2O(g)$$

   b. How many grams of Cu are produced?

4. Quicklime, CaO, can be prepared by roasting limestone, $CaCO_3$, according to the following reaction.

$$CaCO_3(s) \xrightarrow{\Delta} CaO(s) + CO_2(g).$$

   When 2.00×10^3 g of $CaCO_3$ are heated, the actual yield of CaO is 1.05×10^3 g. What is the percent yield?

CHAPTER 9 REVIEW

CHAPTER SUMMARY

9-1

- Reaction stoichiometry involves the mass relationships between reactants and products in a chemical reaction.
- A *mole ratio* is the conversion factor that relates the amount in moles of any two substances in a chemical reaction. The mole ratio is derived from the balanced equation.
- Amount of a substance is expressed in moles, and mass of a substance is expressed using mass units such as grams, kilograms, and milligrams.
- Mass and amount of substance are quantities, whereas moles and grams are units.
- A balanced chemical equation is necessary to solve any stoichiometric problem.

Vocabulary

composition stoichiometry (275) mole ratio (276) reaction stoichiometry (275)

9-2

- For a given chemical equation, all of a reactant may not be converted to product. Ideal stoichiometric calculations give amounts and masses of substances that may differ from those obtained when carrying out the actual reaction.

9-3

- In actual reactions, the reactants are usually combined in proportions different from the precise proportions required for complete reaction.
- The limiting reactant controls the maximum possible amount of product formed.
- Given certain quantities of reactants, the quantity of the product is always less than the maximum possible. Percent yield shows the relationship between the theoretical yield and actual yield for the product of a reaction.

$$\text{percent yield} = \frac{\text{actual yield}}{\text{theoretical yield}} \times 100$$

Vocabulary

actual yield (293) limiting reactant (288) percent yield (293) theoretical yield (293)
excess reactant (288)

REVIEWING CONCEPTS

1. a. Explain the concept of *mole ratio* as used in reaction-stoichiometry problems.
b. What is the source of this value? (9-1)

2. For each of the following chemical equations, write all possible mole ratios:
a. $2Ca + O_2 \longrightarrow 2CaO$
b. $Mg + 2HF \longrightarrow MgF_2 + H_2$ (9-1)

3. a. What is molar mass?
b. What is its role in reaction stoichiometry? (9-2)

4. Distinguish between ideal and real stoichiometric calculations. (9-3)

5. Distinguish between the limiting reactant and the excess reactant in a chemical reaction. (9-3)

6. a. Distinguish between the theoretical and actual yields in stoichiometric calculations.
b. How do the values of the theoretical and actual yields generally compare? (9-3)

7. What is the percent yield of a reaction? (9-3)

8. Why are actual yields generally less than those calculated theoretically? (9-3)

PROBLEMS

General Stoichiometry

Do not assume that the equations given are balanced.

9. Given the chemical equation $Na_2CO_3(aq)$ + $Ca(OH)_2(s) \longrightarrow 2NaOH(aq) + CaCO_3(s)$, determine to two decimal places the molar masses of all substances involved, and then write them as conversion factors.

10. Hydrogen and oxygen react under a specific set of conditions to produce water according to the following: $2H_2(g) + O_2(g) \longrightarrow 2H_2O(g)$.
a. How many moles of hydrogen would be required to produce 5.0 mol of water?
b. How many moles of oxygen would be required? (Hint: See Sample Problem 9-1.)

11. a. If 4.50 mol of ethane, C_2H_6, undergo combustion according to the unbalanced equation $C_2H_6 + O_2 \longrightarrow CO_2 + H_2O$, how many moles of oxygen are required?
b. How many moles of each product are formed?

12. Sulfuric acid reacts with sodium hydroxide according to the following:

$$H_2SO_4 + NaOH \longrightarrow Na_2SO_4 + H_2O.$$

a. Balance the equation for this reaction.
b. What mass of H_2SO_4 would be required to react with 0.75 mol of NaOH?
c. What mass of each product is formed by this reaction? (Hint: See Sample Problem 9-2.)

13. Sodium chloride is produced from its elements through a synthesis reaction. What mass of each reactant would be required to produce 25.0 mol of sodium chloride?

14. Copper reacts with silver nitrate through single replacement.
a. If 2.25 g of silver are produced from the reaction, how many moles of copper(II) nitrate are also produced?
b. How many moles of each reactant are required in this reaction? (Hint: See Sample Problem 9-4.)

15. Iron is generally produced from iron ore through the following reaction in a blast furnace: $Fe_2O_3(s) + CO(g) \longrightarrow Fe(s) + CO_2(g)$.
a. If 4.00 kg of Fe_2O_3 are available to react, how many moles of CO are needed?
b. How many moles of each product are formed?

16. Methanol, CH_3OH, is an important industrial compound that is produced from the following reaction: $CO(g) + H_2(g) \longrightarrow CH_3OH(g)$. What mass of each reactant would be needed to produce 100.0 kg of methanol? (Hint: See Sample Problem 9-5.)

17. Nitrogen combines with oxygen in the atmosphere during lightning flashes to form nitrogen monoxide, NO, which then reacts further with O_2 to produce nitrogen dioxide, NO_2.
a. What mass of NO_2 is formed when NO reacts with 384 g of O_2?
b. How many grams of NO are required to react with this amount of O_2?

18. As early as 1938, the use of NaOH was suggested as a means of removing CO_2 from the cabin of a spacecraft according to the following reaction: $NaOH + CO_2 \longrightarrow Na_2CO_3 + H_2O$.
a. If the average human body discharges 925.0 g of CO_2 per day, how many moles of NaOH are needed each day for each person in the spacecraft?
b. How many moles of each product are formed?

19. The double-replacement reaction between silver nitrate and sodium bromide produces silver bromide, a component of photographic film.
a. If 4.50 mol of silver nitrate reacts, what mass of sodium bromide is required?
b. What mass of silver bromide is formed?

20. In a soda-acid fire extinguisher, concentrated sulfuric acid reacts with sodium hydrogen carbonate to produce carbon dioxide, sodium sulfate, and water.
a. How many moles of sodium hydrogen carbonate would be needed to react with 150.0 g of sulfuric acid?
b. How many moles of each product would be formed?

21. Aspirin, $C_9H_8O_4$, is produced through the following reaction of salicylic acid, $C_7H_6O_3$, and acetic anhydride, $C_4H_6O_3$: $C_7H_6O_3(s)$ + $C_4H_6O_3(l) \longrightarrow C_9H_8O_4(s) + HC_2H_3O_2(l)$.
a. What mass of aspirin (in kg) could be produced from 75.0 mol of salicylic acid?

b. What mass of acetic anhydride (in kg) would be required?

c. At 20°C, how many liters of acetic acid, $HC_2H_3O_2$, would be formed? The density of $HC_2H_3O_2$ is 1.05 g/cm^3.

Limiting Reactant

22. Given the reactant amounts specified in each chemical equation, determine the limiting reactant in each case:

a. $HCl + NaOH \longrightarrow NaCl + H_2O$
   2.0 mol   2.5 mol

b. $Zn + 2HCl \longrightarrow ZnCl_2 + H_2$
   2.5 mol   6.0 mol

c. $2Fe(OH)_3 + 3H_2SO_4 \longrightarrow Fe_2(SO_4)_3 + 6H_2O$
   4.0 mol   6.5 mol

(Hint: See Sample Problem 9-6.)

23. For each reaction specified in Problem 22, determine the amount in moles of excess reactant that remains. (Hint: See Sample Problem 9-7.)

24. For each reaction specified in Problem 22, calculate the amount in moles of each product formed.

25. a. If 2.50 mol of copper and 5.50 mol of silver nitrate are available to react by single replacement, identify the limiting reactant.

b. Determine the amount in moles of excess reactant remaining.

c. Determine the amount in moles of each product formed.

d. Determine the mass of each product formed.

26. Sulfuric acid reacts with aluminum hydroxide by double replacement.

a. If 30.0 g of sulfuric acid react with 25.0 g of aluminum hydroxide, identify the limiting reactant.

b. Determine the mass of excess reactant remaining.

c. Determine the mass of each product formed. Assume 100% yield.

27. The energy used to power one of the Apollo lunar missions was supplied by the following overall reaction: $2N_2H_4 + (CH_3)_2N_2H_2 + 3N_2O_4 \longrightarrow 6N_2 + 2CO_2 + 8H_2O$. For the phase of the mission when the lunar module ascended from the surface of the moon, a total of 1200. kg of N_2H_4 were available to react with 1000. kg of $(CH_3)_2N_2H_2$ and 4500. kg of N_2O_4.

a. For this portion of the flight, which of the allocated components was used up first?

b. How much water, in kilograms, was put into the lunar atmosphere through this reaction?

Percent Yield

28. From theoretical and actual yields of the various chemical reactions given below, calculate the percent yield for each:

a. theoretical yield = 20.0 g, actual yield = 15.0 g

b. theoretical yield = 24.0 g, actual yield = 22.5 g

c. theoretical yield = 5.00 g, actual yield = 4.75 g

(Hint: See Sample Problem 9-8.)

29. From the theoretical and percentage yields given below, determine the actual yields:

a. theoretical yield = 12.0 g, percent yield = 90.0%

b. theoretical yield = 8.50 g, percent yield = 70.0%

c. theoretical yield = 3.45 g, percent yield = 48.0%

30. The Ostwald Process for producing nitric acid from ammonia consists of the following steps:

$$4NH_3(g) + 5O_2(g) \longrightarrow 4NO(g) + 6H_2O(g)$$
$$2NO(g) + O_2(g) \longrightarrow 2NO_2(g)$$
$$3NO_2(g) + H_2O(g) \longrightarrow 2HNO_3(aq) + NO(g)$$

If the yield in each step is 94.0%, how many grams of nitric acid can be produced from 5.00 kg of ammonia?

MIXED REVIEW

31. Magnesium is obtained from sea water. $Ca(OH)_2$ is added to sea water to precipitate $Mg(OH)_2$. The precipitate is filtered and reacted with HCl to produce $MgCl_2$. The $MgCl_2$ is electrolyzed to produce Mg and Cl_2. If 185.0 g of magnesium are recovered from 1000. g of $MgCl_2$, what is the percent yield for this reaction?

32. Phosphate baking powder is a mixture of starch, sodium hydrogen carbonate, and calcium dihydrogen phosphate. When mixed with water, phosphate baking powder releases carbon dioxide gas, causing a dough or batter to bubble and rise.

$2NaHCO_3(aq) + Ca(H_2PO_4)_2(aq) \longrightarrow$
$2Na_2HPO_4(aq) + CaHPO_4(aq) + 2CO_2(g) +$
$2H_2O(l)$

If 0.750 L of CO_2 is needed for a cake and each kilogram of baking powder contains 168 g of $NaHCO_3$, how many grams of baking powder must be used to generate this amount of CO_2? The density of CO_2 at baking temperature is about 1.20 g/L.

33. Coal gasification is a process that converts coal into methane gas. If this reaction has a percent yield of 85.0%, how much methane can be obtained from 1250 g of carbon.

$$2C(s) + 2H_2O(l) \longrightarrow CH_4(g) + CO_2(g)$$

34. If the percent yield for the coal gasification process is increased to 95%, how much methane can be obtained from 2750 g of carbon?

35. Builders and dentists must store plaster of Paris, $CaSO_4 \bullet \frac{1}{2}H_2O$, in airtight containers to prevent it from absorbing water vapor from the air and changing to gypsum, $CaSO_4 \bullet 2H_2O$. How many liters of water evolve when 5.00 L of gypsum are heated at 110°C to produce plaster of Paris? At 110°C, the density of $CaSO_4 \bullet 2H_2O$ is 2.32 g/mL, and the density of water vapor is 0.581 g/mL.

36. Gold can be recovered from sea water by reacting the water with zinc, which is refined from zinc oxide. The zinc displaces the gold in the water. What mass of gold can be recovered if 2.00 g of ZnO and an excess of sea water are available?

$2ZnO(s) + C(s) \longrightarrow 2Zn(s) + CO_2(g)$
$2Au^{3+}(aq) + 3Zn(s) \longrightarrow 3Zn^{2+}(aq) + 2Au(s)$

CRITICAL THINKING

37. Relating Ideas The chemical equation is a good source of information concerning a reaction. Explain the relationship that exists between the actual yield of a reaction product and the chemical equation of the product.

38. Analyzing Results Very seldom are chemists able to achieve a 100% yield of a product from a chemical reaction. However, the yield of a reaction is usually important because of the expense involved in producing less product. For example, when magnesium metal is heated in a crucible at high temperatures, the product magnesium oxide, MgO, is formed. Based on your analysis of the reaction, describe some of the actions you would take to increase your percent yield. The reaction is as follows:

$$2\,Mg(s) + O_2(g) \longrightarrow 2MgO(s)$$

39. Analyzing Results In the lab, you run an experiment that appears to have a percent yield of 115%. Propose reasons for this result. Can an actual yield ever exceed a theoretical yield? Explain your answer.

40. Relating Ideas Explain the stoichiometry of blowing air on a smoldering campfire to keep the coals burning.

TECHNOLOGY & LEARNING

41. Graphing Calculator Calculating Percent Yield of a Chemical Reaction

The graphing calculator can be programmed to calculate the percent yield of a chemical reaction when you enter the actual yield and the theoretical yield. Using an example in which the actual yield is 38.8 g and the theoretical yield is 53.2 g, you will calculate the percent yield. First, you will program the calculator to carry out the calculation. Then, the program will be used to make other calculations.

A. Program the calculation.

Keystrokes for naming the program: PRGM ▶ ▶ ENTER

Name the program: Y I E L D ENTER

Program keystrokes: PRGM ▶ 2 ALPHA A 2nd : PRGM ▶ 2 ALPHA T 2nd : ALPHA A ÷ ALPHA T × 1 0 0 STO▶ ALPHA P 2nd : PRGM ▶ 8 2nd : PRGM ▶ 3 2nd ALPHA " P R C N T Y L D 2nd TEST 1 ALPHA " 2nd : PRGM ▶ 3 ALPHA P

B. Check the display.

Screen display: Prompt A:Prompt T:A/T*100 ->P:ClrHome:Disp"PRCNTYLD=":Disp P

Press 2nd QUIT to exit the program editor.

C. Run the program.

a. Press PRGM. Select YIELD and press ENTER ENTER. Now enter the actual yield 38.8 and press ENTER. Enter the theoretical yield 53.2 and press ENTER. The calculator will display the percent yield.

b. Use the YIELD program with an actual yield of 68.4 g and a theoretical yield of 85.7 g.

c. Use the YIELD program with an actual yield of 27.3 g and a theoretical yield of 44.6 g.

HANDBOOK SEARCH

42. The steel-making process described in the Transition Metal section of the *Elements Handbook* shows the equation for the formation of iron carbide. Use this equation to answer the following.

a. If 3.65×10^3 kg of iron is used in a steelmaking process, what is the minimum mass of carbon needed to react with all of the iron?

b. What is the theoretical mass of iron carbide formed?

43. The reaction of aluminum with oxygen to produce a protective coating for the metal's surface is described in the discussion of aluminum in Group 13 of the *Elements Handbook*. Use this equation to answer the following.

a. What mass of aluminum oxide would theoretically be formed if a 30.0 g piece of aluminum foil reacted with excess oxygen?

b. Why would you expect the actual yield from this reaction to be far less than the mass you calculated in item (a)?

44. The reactions of oxide compounds to produce carbonates, phosphates, and sulfates are described in the section on oxides in Group 16 of the *Elements Handbook*. Use those equations to answer the following.

a. What mass of CO_2 is needed to react with 154.6 g of MgO?

b. What mass of magnesium carbonate is produced?

c. 45.7 g of P_4O_{10} is reacted with an excess of calcium oxide. What mass of calcium phosphate is produced?

RESEARCH & WRITING

45. Research the history of the Haber process for the production of ammonia. What was the significance of this process in history? How is this process related to the discussion of reaction yields in this chapter?

ALTERNATIVE ASSESSMENT

46. Performance Just as reactants combine in certain proportions to form a product, colors can be combined to create other colors. Artists do this all the time to find just the right color for their paintings. Using poster paint, determine the proportions of primary pigments used to create the following colors. Your proportions should be such that anyone could mix the color perfectly. (Hint: Don't forget to record the amount of the primary pigment and water used when you mix them.)

47. Performance Write two of your own sample problems that are descriptions of how to solve a mass-mass problem. Assume that your sample problems will be used by other students to learn how to solve mass-mass problems.

UNIT 4

Phases of Matter

CHAPTERS

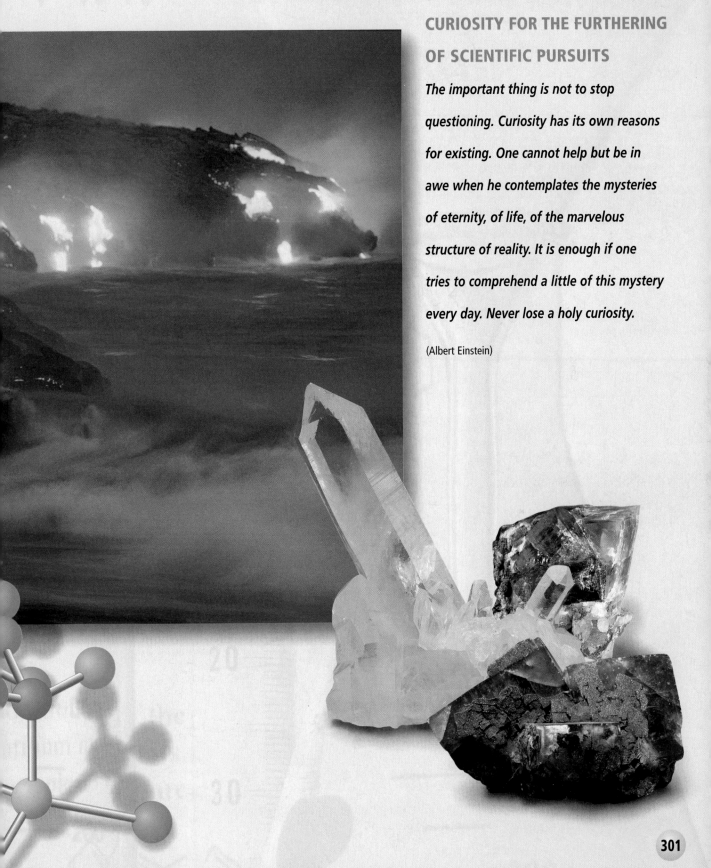

ALBERT EINSTEIN, ON THE NECESSITY OF HUMAN CURIOSITY FOR THE FURTHERING OF SCIENTIFIC PURSUITS

The important thing is not to stop questioning. Curiosity has its own reasons for existing. One cannot help but be in awe when he contemplates the mysteries of eternity, of life, of the marvelous structure of reality. It is enough if one tries to comprehend a little of this mystery every day. Never lose a holy curiosity.

(Albert Einstein)

CHAPTER 10

Physical Characteristics of Gases

The density of a gas decreases as its temperature increases.

The Kinetic-Molecular Theory of Matter

SECTION 10-1

OBJECTIVES

- State the kinetic-molecular theory of matter, and describe how it explains certain properties of matter.
- List the five assumptions of the kinetic-molecular theory of gases. Define the terms *ideal gas* and *real gas.*
- Describe each of the following characteristic properties of gases: expansion, density, fluidity, compressibility, diffusion, and effusion.
- Describe the conditions under which a real gas deviates from "ideal" behavior.

In Chapter 1, you read that matter exists on Earth in the forms of solids, liquids, and gases. Although it is impossible to observe individual particles directly, scientists have studied large groups of these particles as they occur in solids, liquids, and gases.

In the late nineteenth century, scientists developed the kinetic-molecular theory of matter to account for the behavior of the atoms and molecules that make up matter. *The* **kinetic-molecular theory** *is based on the idea that particles of matter are always in motion.* The theory can be used to explain the properties of solids, liquids, and gases in terms of the energy of particles and the forces that act between them. In this section, you will study the theory as it applies to gas molecules. In that form, it is called the kinetic-molecular theory of gases.

The Kinetic-Molecular Theory of Gases

The kinetic-molecular theory can help you understand the behavior of gas molecules and the physical properties of gases. The theory provides a model of what is called an ideal gas. *An* **ideal gas** *is an imaginary gas that perfectly fits all the assumptions of the kinetic-molecular theory.*

The kinetic-molecular theory of gases is based on the following five assumptions:

1. *Gases consist of large numbers of tiny particles that are far apart relative to their size.* These particles, usually molecules or atoms, typically occupy a volume about 1000 times greater than the volume occupied by particles in the liquid or solid state. Thus, molecules of gases are much farther apart than those of liquids or solids. Most of the volume occupied by a gas is empty space. This accounts for the lower density of gases compared with that of liquids and solids. It also explains the fact that gases are easily compressed.
2. *Collisions between gas particles and between particles and container walls are elastic collisions. An* **elastic collision** *is one in which there is no net loss of kinetic energy.* Kinetic energy is transferred between two particles during collisions. However, the total kinetic energy of the two particles remains the same as long as temperature is constant.

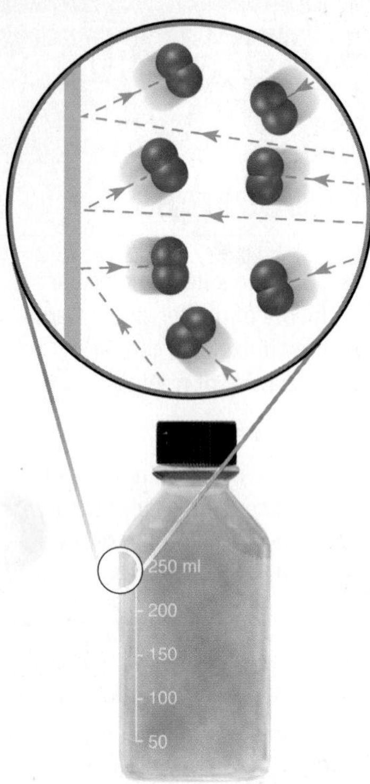

FIGURE 10-1 Gas particles travel in a straight-line motion until they collide with each other or the walls of their container.

3. *Gas particles are in constant, rapid, random motion. They therefore possess kinetic energy, which is energy of motion.* Gas particles move in all directions, as shown in Figure 10-1. The kinetic energy of the particles overcomes the attractive forces between them, except near the temperature at which the gas condenses and becomes a liquid.
4. *There are no forces of attraction or repulsion between gas particles.* You can think of ideal gas molecules as behaving like small billiard balls. When they collide, they do not stick together but immediately bounce apart.
5. *The average kinetic energy of gas particles depends on the temperature of the gas.* The kinetic energy of any moving object, including a particle, is given by the following equation.

$$KE = \frac{1}{2}mv^2$$

In the equation, m is the mass of the particle and v is its speed. Because all the particles of a specific gas have the same mass, their kinetic energies depend only on their speeds. The average speeds and kinetic energies of gas particles increase with an increase in temperature and decrease with a decrease in temperature.

All gases at the same temperature have the same average kinetic energy. Therefore, at the same temperature, lighter gas particles, such as hydrogen molecules, have higher average speeds than do heavier gas particles, such as oxygen molecules.

The Kinetic-Molecular Theory and the Nature of Gases

The kinetic-molecular theory applies only to ideal gases. Although ideal gases do not actually exist, many gases behave nearly ideally if pressure is not very high or temperature is not very low. In the following sections, you will see how the kinetic-molecular theory accounts for the physical properties of gases.

Expansion

Gases do not have a definite shape or a definite volume. They completely fill any container in which they are enclosed, and they take its shape. A gas transferred from a one-liter vessel to a two-liter vessel will quickly expand to fill the entire two-liter volume. The kinetic-molecular theory explains these facts. According to the theory, gas particles move rapidly in all directions (assumption 3) without significant attraction or repulsion between them (assumption 4).

Fluidity

Because the attractive forces between gas particles are insignificant (assumption 4), gas particles glide easily past one another. This ability to flow causes gases to behave similarly to liquids. *Because liquids and gases flow, they are both referred to as* **fluids.**

Low Density

The density of a substance in the gaseous state is about 1/1000 the density of the same substance in the liquid or solid state. That is because the particles are so much farther apart in the gaseous state (assumption 1).

Compressibility

During compression, the gas particles, which are initially very far apart (assumption 1), are crowded closer together. The volume of a given sample of a gas can be greatly decreased. Steel cylinders containing gases under pressure are widely used in industry. When they are full, such cylinders may contain 100 times as many particles of gas as would be contained in nonpressurized containers of the same size.

Diffusion and Effusion

Gases spread out and mix with one another, even without being stirred. If the stopper is removed from a container of ammonia in a room, ammonia gas will mix uniformly with the air and spread throughout the room. The random and continuous motion of the ammonia molecules (assumption 3) carries them throughout the available space. *Such spontaneous mixing of the particles of two substances caused by their random motion is called* **diffusion.**

The rate of diffusion of one gas through another depends on three properties of the gas particles: their speeds, their diameters, and the attractive forces between them. In Figure 10-2, hydrogen gas diffuses rapidly into other gases at the same temperature because its molecules are lighter and move faster than the molecules of the other gases.

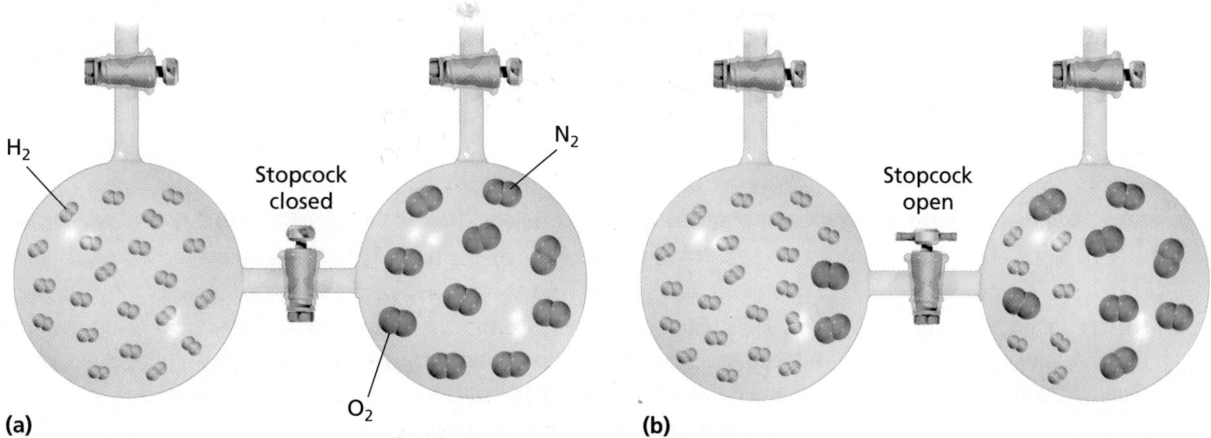

FIGURE 10-2 When hydrogen gas in a flask is allowed to mix with air at the same pressure in another flask, the low-mass molecules of hydrogen diffuse rapidly into the flask with the air. The heavier molecules of nitrogen and oxygen in the air diffuse more slowly into the flask with the hydrogen.

Diffusion is a process by which particles of a gas spread out spontaneously and mix with other gases. In contrast, **effusion** *is a process by which gas particles under pressure pass through a tiny opening.* The rates of effusion of different gases are directly proportional to the velocities of their particles. Consequently, molecules of low mass effuse faster than molecules of high mass.

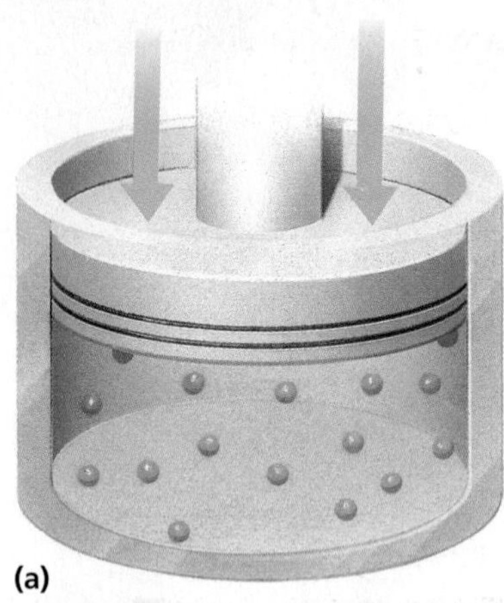

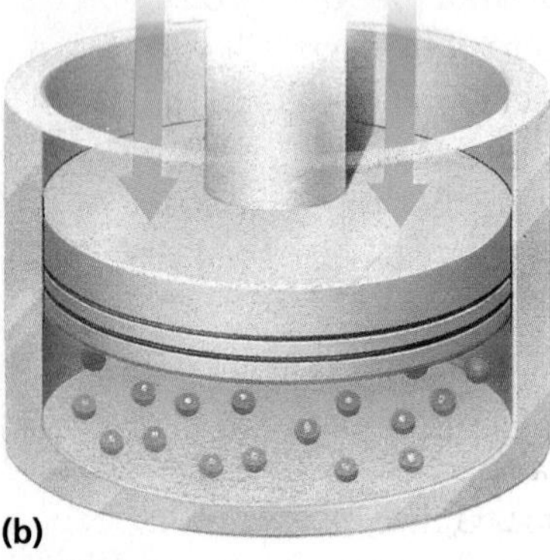

FIGURE 10-3 (a) Gas molecules in a car engine cylinder expand to fill the cylinder. (b) As pressure is exerted on them, the gas molecules move closer together, reducing their volume.

Deviations of Real Gases from Ideal Behavior

When their particles are far enough apart and have enough kinetic energy, most gases behave ideally. However, all real gases deviate to some degree from ideal-gas behavior. *A* **real gas** *is a gas that does not behave completely according to the assumptions of the kinetic-molecular theory.* In 1873, Johannes van der Waals accounted for this deviation from ideal behavior by pointing out that particles of real gases occupy space and exert attractive forces on each other. At very high pressures and low temperatures, the deviation may be considerable. Under such conditions, the particles will be closer together and their kinetic energy will be insufficient to completely overcome the attractive forces. This is illustrated in Figure 10-3.

The kinetic-molecular theory is more likely to hold true for gases whose particles have little attraction for each other. The noble gases, such as helium, He, and neon, Ne, show essentially ideal gas behavior over a wide range of temperatures and pressures. The particles of these gases are monatomic and thus nonpolar. The particles of gases, such as nitrogen, N_2, and hydrogen, H_2, are nonpolar diatomic molecules. The behavior of these gases most closely approximates that of the ideal gas under certain conditions. The more polar a gas's molecules are, the greater the attractive forces between them and the more the gas will deviate from ideal gas behavior. For example, highly polar gases, such as ammonia, NH_3, and water vapor, deviate from ideal behavior to a larger degree than nonpolar gases.

SECTION REVIEW

1. Use the kinetic-molecular theory to explain each of the following properties of gases: expansion, fluidity, low density, compressibility, and diffusion.
2. Describe the conditions under which a real gas is most likely to behave ideally.
3. State the two factors that van der Waals proposed to explain why real gases deviate from ideal behavior.
4. Which of the following gases would you expect to deviate significantly from ideal behavior: He, O_2, H_2, H_2O, N_2, HCl, or NH_3?

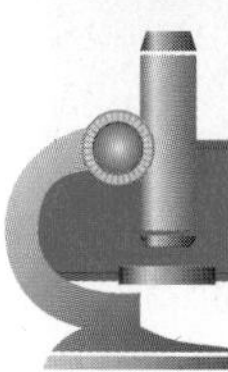

RESEARCH NOTES

Carbon Monoxide Catalyst—Stopping the Silent Killer

Colorless, odorless, and deadly—carbon monoxide, "the silent killer," causes the deaths of hundreds of Americans every year. When fuel does not burn completely in a combustion process, carbon monoxide is produced. Often this occurs in a malfunctioning heater, furnace, or fireplace. When the carbon monoxide is inhaled, it clings to the hemoglobin in the blood, leaving the body oxygen starved. Before people realize a combustion device is malfunctioning, it's often too late.

$$O_2Hb + CO \longrightarrow COHb + O_2$$

Carbon monoxide, CO, has almost 200 times the affinity to bind with the hemoglobin, Hb, in the blood as oxygen. This means if the body has a choice, it will bind to carbon monoxide over oxygen. If enough carbon monoxide is present in the blood, it can be fatal.

Carbon monoxide poisoning can be prevented by installing filters that absorb the gas. After a time, however, filters become saturated, and then carbon monoxide can pass freely into the air. The best way to prevent carbon monoxide poisoning is not just to filter out the gas, but to eliminate it completely.

The solution came to research chemists at NASA who were working on a problem with a space-based laser. In order to operate properly, NASA's space-based carbon dioxide laser needed to be fed a continuous supply of CO_2. This was necessary because as a byproduct of its operation, the laser degraded some of the CO_2 into carbon monoxide and oxygen. To address this problem, NASA scientists developed a catalyst made of tin oxide and platinum that oxidized the waste carbon monoxide back into carbon dioxide. The NASA scientists then realized that this catalyst had the potential to be used in many applications here on Earth, including removing carbon monoxide from houses and other buildings.

Typically, a malfunctioning heater circulates the carbon monoxide it produces through its air intake system back into a dwelling space. By installing the catalyst in the air intake, any carbon monoxide would be eliminated by oxidation to non-toxic carbon dioxide before it reentered the room.

"The form of our catalyst is a very thin coating on some sort of a support, or substrate as we call it," says NASA chemist David Schryer. "And that support, or substrate, can be any one of a number of things. The great thing about a catalyst is that the only thing that matters about it is its surface. So a catalyst can be incredibly thin and still be very effective."

The catalyst could also be used as a liner for airplane oxygen masks. In aircraft fires on the ground, most people are killed not by flames but by carbon monoxide poisoning. In such cases, NASA's catalyst could serve as a life-saving device.

The idea of using catalysts to oxidize gases is not a new one. Catalytic converters in cars oxidize carbon monoxide and unburned hydrocarbons to minimize pollution. Many substances are oxidized into new materials for manufacturing purposes. But both of these types of catalytic reactions occur at very high temperatures. NASA's catalyst is special, because it's able to eliminate carbon monoxide at room temperature. Also, it has the ability to oxidize formaldehyde, a noxious chemical often found in building materials, curtains, and carpets.

According to David Schryer, low-temperature catalysts constitute a whole new class of catalysts with abundant applications for the future.

SECTION 10-2

Pressure

OBJECTIVES

- Define *pressure,* and relate it to force.
- Describe how pressure is measured.
- Convert units of pressure.
- State the standard conditions of temperature and pressure.

Suppose you have a one-liter bottle of air. How much air do you actually have? The expression *a liter of air* means little unless the conditions at which the volume is measured are known. A liter of air can be compressed to a few milliliters. It can also be allowed to expand to fill an auditorium.

To describe a gas fully, you need to state four measurable quantities: volume, temperature, number of molecules, and pressure. You already know what is meant by volume, temperature, and number of molecules. In this section, you will learn about pressure and its measurement. Then, in Section 10-3, you will examine the mathematical relationships between volume, temperature, number of gas molecules, and pressure.

Pressure and Force

If you blow air into a rubber balloon, the balloon will increase in size. The volume increase is caused by the collisions of molecules of air with the inside walls of the balloon. The collisions cause an outward push, or force, against the inside walls. **Pressure** *(P) is defined as the force per unit area on a surface.* The equation defining pressure follows.

FIGURE 10-4 The pressure the ballet dancer exerts against the floor depends on the area of contact. The smaller the area of contact, the greater the pressure.

(a) Area of contact = 325 cm²

Pressure = $\frac{\text{force}}{\text{area}}$

$= \frac{500\text{ N}}{325\text{ cm}^2} = 1.5\text{ N/cm}^2$

(b) Area of contact = 13 cm²

Pressure = $\frac{\text{force}}{\text{area}}$

$= \frac{500\text{ N}}{13\text{ cm}^2} = 38.5\text{ N/cm}^2$

(c) Area of contact = 6.5 cm²

Pressure = $\frac{\text{force}}{\text{area}}$

$= \frac{500\text{ N}}{6.5\text{ cm}^2} = 77\text{ N/cm}^2$

$$pressure = \frac{force}{area}$$

The SI unit for force is the **newton,** abbreviated N. *It is the force that will increase the speed of a one kilogram mass by one meter per second each second it is applied.* At Earth's surface, each kilogram of mass exerts 9.8 N of force, due to gravity. Consider a ballet dancer with a mass of 51 kg, as shown in Figure 10-4. A mass of 51 kg exerts a force of 500 N (51×9.8) on Earth's surface. No matter how the dancer stands, she exerts that much force against the floor. However, the pressure she exerts against the floor depends on the area of contact. When she rests her weight on the soles of both feet, as shown in Figure 10-4(a), the area of contact with the floor is about 325 cm^2. The pressure, or force per unit area, when she stands in this manner is 500 N/325 cm^2. That equals roughly 1.5 N/cm^2. When she stands on her toes, as in Figure 10-4(b), the total area of contact with the floor is only 13 cm^2. The pressure exerted is then equal to 500 N/13 cm^2—roughly 38.5 N/cm^2. And when she stands on one toe, as in Figure 10-4(c), the pressure she exerts is twice that, or about 77 N/cm^2. Thus, the same force applied to a smaller area results in a greater pressure.

Gas molecules exert pressure on any surface with which they collide. The pressure exerted by a gas depends on volume, temperature, and the number of molecules present.

The atmosphere—the blanket of air surrounding Earth—exerts pressure. Figure 10-5 shows that atmospheric pressure at sea level is about equal to the weight of a 1.03 kg mass per square centimeter of surface, or 10.1 N/cm^2. The pressure of the atmosphere can be thought of as caused by the weight of the gases that compose the atmosphere. The atmosphere contains about 78% nitrogen, 21% oxygen, and 1% other gases, including

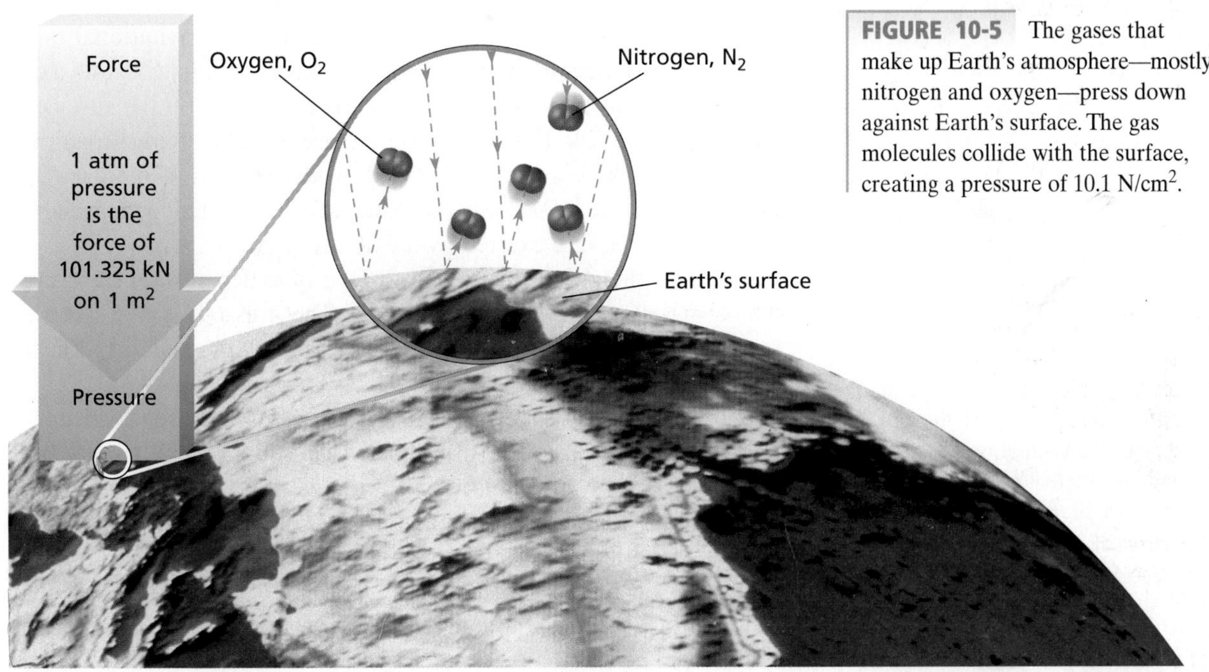

FIGURE 10-5 The gases that make up Earth's atmosphere—mostly nitrogen and oxygen—press down against Earth's surface. The gas molecules collide with the surface, creating a pressure of 10.1 N/cm^2.

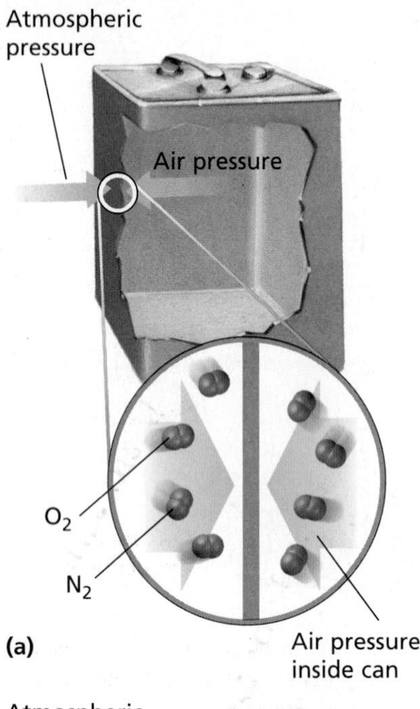

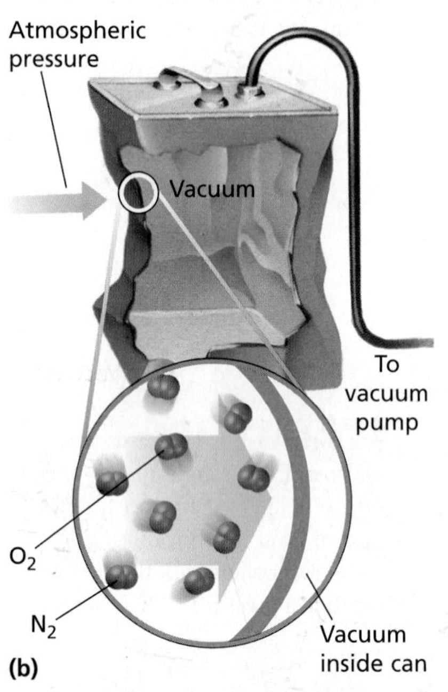

FIGURE 10-6 (a) The "empty" can has the mixture of gasses in air inside that push outward and balance the atmospheric pressure that pushes inward. (b) When the air inside the can is removed by a vacuum pump, there is insufficient force to balance the atmospheric pressure. As a result, the can collapses.

argon and carbon dioxide. Atmospheric pressure is the sum of the individual pressures of the various gases in the atmosphere.

To understand the concept of gas pressure and its magnitude, consider the model of an "empty" can shown in Figure 10-6(a). The can does contain a small amount of air. The atmosphere exerts a pressure of 10.1 N/cm^2 against the outside of the can. If the can measures 15 cm × 10 cm × 28 cm, it has a total area of 1700 cm^2. The resulting inward force on the can is greater than 1.0 metric ton of weight. The air inside the can pushes outward and balances the atmosphere's inward-pushing force. If a vacuum pump is used to remove the air from the can, as shown in Figure 10-6(b), the balancing outward force is removed. As a result, the unbalanced force due to atmospheric pressure immediately crushes the can.

Measuring Pressure

A **barometer** *is a device used to measure atmospheric pressure.* The first type of barometer, illustrated in Figure 10-7, was introduced by Evangelista Torricelli during the early 1600s. Torricelli wondered why water pumps could raise water to a maximum height of only about 34 feet. He thought that the height must depend somehow on the weight of water compared with the weight of air. He reasoned that liquid mercury, which is about 14 times as dense as water, could be raised only 1/14 as high as water. To test this idea, Torricelli sealed a long glass tube at one end and filled it with mercury. Holding the open end with his thumb, he inverted the tube into a dish of mercury without allowing any air to enter the tube. When he removed his thumb, the mercury column in the tube dropped to a height of about 30 in. (760 mm) above the surface of the mercury in the dish. He repeated the experiment with tubes of different diameters and lengths longer than 760 mm. In every case, the mercury dropped to a height of about 760 mm.

The space above the mercury in such a tube is nearly a vacuum. The mercury in the tube pushes downward because of gravitational force. The column of mercury in the tube is stopped from falling beyond a certain point because the atmosphere exerts a pressure on the surface of the mercury outside the tube. This pressure is transmitted through the fluid mercury and is exerted upward on the column of mercury. The mercury in the tube falls only until the pressure exerted by its weight is equal to the pressure exerted by the atmosphere.

The exact height of the mercury in the tube depends on the atmospheric pressure, or force per unit area. The pressure is measured directly in terms of the height of the mercury column supported in the barometer tube.

From experiments like Torricelli's, it is known that at sea level at 0°C, the average pressure of the atmosphere can support a 760 mm column of mercury. At any given place on Earth, the specific atmospheric pressure depends on the elevation and the weather conditions at the time. If the atmospheric pressure is greater than the average at sea level, the height of the mercury column in a barometer will be greater than 760 mm. If the atmospheric pressure is less, the height of the mercury column will be less than 760 mm.

All gases, not only those in the atmosphere, exert pressure. A device called a manometer can be used to measure the pressure of an enclosed gas sample, as shown in Figure 10-8. The difference in the height of mercury in the two arms of the U-tube is a measure of the oxygen gas pressure in the container.

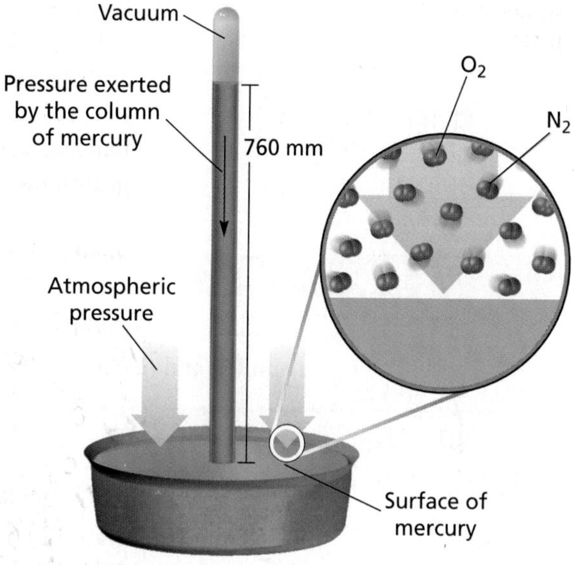

FIGURE 10-7 Torricelli discovered that the pressure of the atmosphere supports a column of mercury about 760 mm above the surface of the mercury in the dish.

Units of Pressure

A number of different units are used to measure pressure. Because atmospheric pressure is often measured by a mercury barometer, pressure can be expressed in terms of the height of a mercury column. *Thus, a common unit of pressure is* **millimeters of mercury,** *symbolized mm Hg.* The average atmospheric pressure at sea level at 0°C is 760 mm Hg.

Pressures are often measured in units of atmospheres. *One* **atmosphere of pressure** *(atm) is defined as being exactly equivalent to 760 mm Hg.*

In SI, pressure is expressed in derived units called pascals. The unit is named for Blaise Pascal, a French mathematician and philosopher who studied pressure during the seventeenth century. *One* **pascal** *(Pa) is defined as the pressure exerted by a force of one newton (1 N) acting on an area of one square meter.*

In many cases, it is more convenient to express pressure in kilopascals (kPa). The standard atmosphere (1 atm) is equal to $1.013\,25 \times 10^5$ Pa, or 101.325 kPa. The pressure units used in this book are summarized in Table 10-1.

TABLE 10-1 ***Units of Pressure***

Unit	Symbol	Definition/relationship
Pascal	Pa	SI pressure unit $1 \text{ Pa} = \dfrac{1 \text{ N}}{\text{m}^2}$
Millimeter of mercury	mm Hg	pressure that supports a 1 mm mercury column in a barometer
Atmosphere	atm	average atmospheric pressure at sea level and 0°C 1 atm = 760 mm Hg $= 1.013\,25 \times 10^5$ Pa = 101.325 kPa
Torr*	torr	1 torr = 1 mm Hg

* To honor Torricelli for his invention of the barometer, this pressure unit is called the torr.

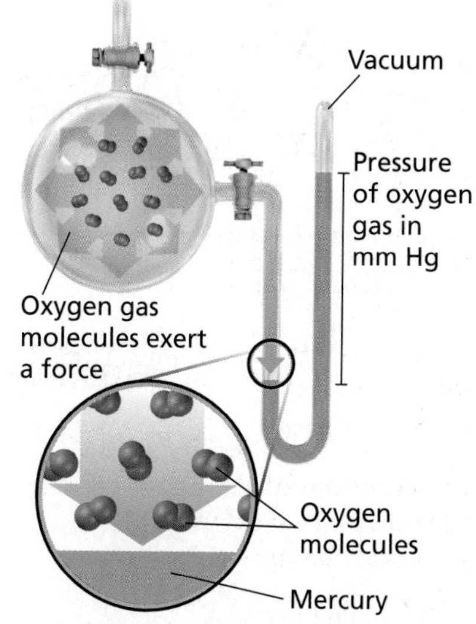

FIGURE 10-8 In the manometer above, the pressure of the oxygen gas in the flask pushes on the mercury column. The difference in the height of the mercury in the two arms of the U-tube indicates the oxygen gas pressure.

Standard Temperature and Pressure

To compare volumes of gases, it is necessary to know the temperature and pressure at which the volumes are measured. *For purposes of comparison, scientists have agreed on standard conditions of exactly 1 atm pressure and 0°C. These conditions are called* **standard temperature and pressure** *and are commonly abbreviated STP.*

SAMPLE PROBLEM 10-1

The average atmospheric pressure in Denver, Colorado, is 0.830 atm. Express this pressure (a) in mm Hg and (b) in kPa.

SOLUTION

1 *ANALYZE* **Given:** P of atmosphere = 0.830 atm

760 mm Hg = 1 atm (definition); 101.325 kPa = 1 atm (definition)

Unknown: a. P of atmosphere in mm Hg; **b.** P of atmosphere in kPa

2 *PLAN*

a. $\text{atm} \longrightarrow \text{mm Hg}; \quad \text{atm} \times \dfrac{\text{mm Hg}}{\text{atm}} = \text{mm Hg}$

b. $\text{atm} \longrightarrow \text{kPa}; \quad \text{atm} \times \dfrac{\text{kPa}}{\text{atm}} = \text{kPa}$

3 *COMPUTE*

a. $0.830\ \text{atm} \times \dfrac{760\ \text{mm Hg}}{\text{atm}} = 631\ \text{mm Hg}$

b. $0.830\ \text{atm} \times \dfrac{101.325\ \text{kPa}}{\text{atm}} = 84.1\ \text{kPa}$

4 *EVALUATE* Units have canceled to give the desired units, and answers are properly expressed to the correct number of significant digits. The known pressure is roughly 80% of atmospheric pressure. The results are therefore reasonable because each is roughly 80% of the pressure as expressed in the new units.

PRACTICE

1. Convert a pressure of 1.75 atm to kPa and to mm Hg. *Answer* 177 kPa, 1330 mm Hg
2. Convert a pressure of 570. torr to atmospheres and to kPa. *Answer* 0.750 atm, 76.0 kPa

SECTION REVIEW

1. Define *pressure.*
2. What units are used to express pressure measurements?
3. What are standard conditions for gas measurements?
4. Convert the following pressures to pressures in standard atmospheres:
   a. 151.98 kPa
   b. 456 torr
   c. 912 mm Hg

The Gas Laws

SECTION **10-3**

OBJECTIVES

- Use the kinetic-molecular theory to explain the relationships between gas volume, temperature, and pressure.
- Use Boyle's law to calculate volume-pressure changes at constant temperature.
- Use Charles's law to calculate volume-temperature changes at constant pressure.
- Use Gay-Lussac's law to calculate pressure-temperature changes at constant volume.
- Use the combined gas law to calculate volume-temperature-pressure changes.
- Use Dalton's law of partial pressures to calculate partial pressures and total pressures.

Scientists have been studying physical properties of gases for hundreds of years. In 1662, Robert Boyle discovered that gas pressure and volume are related mathematically. The observations of Boyle and others led to the development of the gas laws. *The* **gas laws** *are simple mathematical relationships between the volume, temperature, pressure, and quantity of a gas.*

Boyle's Law: Pressure-Volume Relationship

Robert Boyle discovered that doubling the pressure on a sample of gas at constant temperature reduces its volume by one-half. Tripling the gas pressure reduces its volume to one-third of the original. Reducing the pressure on a gas by one-half allows the volume of the gas to double. As one variable increases, the other decreases. Figure 10-9 shows that as the volume of gas in the syringe decreases, the pressure of the gas increases.

You can use the kinetic-molecular theory to understand why this pressure-volume relationship holds. The pressure of a gas is caused by

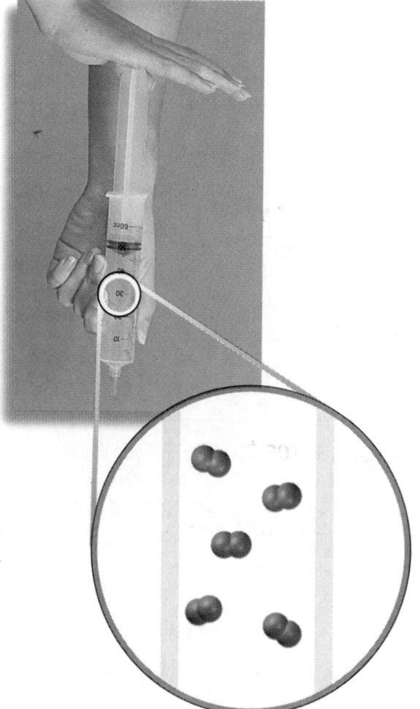

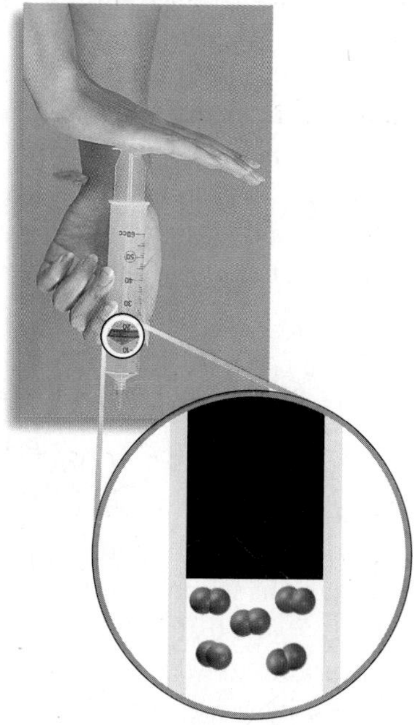

FIGURE 10-9 The volume of gas in the syringe shown in the photo is reduced when the plunger is pushed down. The gas pressure increases as the volume is reduced because the molecules collide more frequently with the walls of the container in a smaller volume.

TABLE 10-2 Volume-Pressure Data for a Gas Sample (at Constant Mass and Temperature)

Volume (mL)	Pressure (atm)	$P \times V$
1200	0.5	600
600	1.0	600
300	2.0	600
200	3.0	600
150	4.0	600
120	5.0	600
100	6.0	600

moving molecules hitting the container walls. Suppose the volume of a container is decreased but the same number of gas molecules is present at the same temperature. There will be more molecules per unit volume. The number of collisions with a given unit of wall area will increase as a result. Therefore, pressure will also increase.

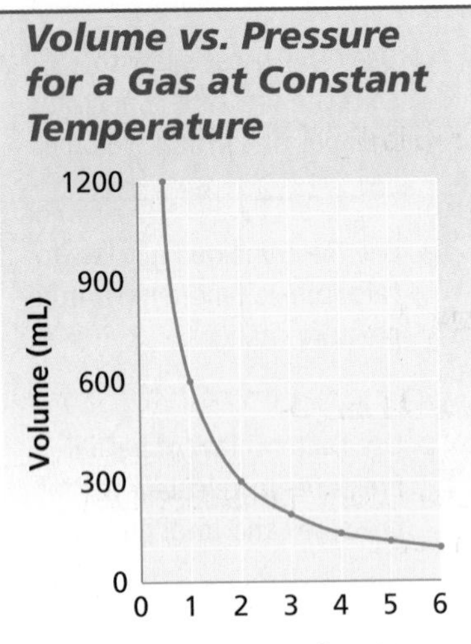

FIGURE 10-10 This graph shows that there is an inverse relationship between volume and pressure. As the pressure drops by half, the volume doubles.

Table 10-2 shows pressure and volume data for a constant mass of gas at constant temperature. Plotting the values of volume versus pressure gives a curve like that in Figure 10-10. The general volume-pressure relationship that is illustrated is called Boyle's law. **Boyle's law** *states that the volume of a fixed mass of gas varies inversely with the pressure at constant temperature.*

Mathematically, Boyle's law is expressed as follows.

$$V = k\frac{1}{P} \quad \text{or} \quad PV = k$$

The value of k is constant for a given sample of gas and depends only on the mass of gas and the temperature. (Note that for the data in Table 10-2, $k = 600$ mL•atm.) If the pressure of a given gas sample at constant temperature changes, the volume will change. However, the quantity *pressure times volume* will remain equal to the same value of k.

Boyle's law can be used to compare changing conditions for a gas. Using P_1 and V_1 to stand for initial conditions and P_2 and V_2 to stand for new conditions results in the following equations.

$$P_1V_1 = k \qquad P_2V_2 = k$$

Two quantities that are equal to the same thing are equal to each other.

$$P_1V_1 = P_2V_2$$

Given three of the four values P_1, V_1, P_2, and V_2, you can use this equation to calculate the fourth value for a system at constant temperature. For example, suppose that 1.0 L of gas is initially at 1.0 atm pressure ($V_1 = 1.0$ L, $P_1 = 1.0$ atm). The gas is allowed to expand fivefold at constant temperature to 5.0 L ($V_2 = 5.0$ L). You can then calculate the new pressure, P_2, by rearranging the equation as follows.

$$P_2 = \frac{P_1V_1}{V_2}$$

$$P_2 = \frac{(1.0 \text{ atm})(1.0 \text{ L})}{5.0 \text{ L}} = 0.20 \text{ atm}$$

The pressure has decreased to one-fifth the original pressure, while the volume increased fivefold.

SAMPLE PROBLEM 10-2

A sample of oxygen gas has a volume of 150. mL when its pressure is 0.947 atm. What will the volume of the gas be at a pressure of 0.987 atm if the temperature remains constant?

SOLUTION

1 ANALYZE

Given: V_1 of O_2 = 150. mL

P_1 of O_2 = 0.947 atm; P_2 of O_2 = 0.987 atm

Unknown: V_2 of O_2 in mL

2 PLAN

$$P_1, V_1, P_2 \longrightarrow V_2$$

Rearrange the equation for Boyle's law ($P_1V_1 = P_2V_2$) to obtain V_2.

$$V_2 = \frac{P_1V_1}{P_2}$$

3 COMPUTE

Substitute values for P_1, V_1, and P_2 to obtain the new volume, V_2.

$$V_2 = \frac{P_1V_1}{P_2} = \frac{(0.947 \text{ atm})(150. \text{ mL } O_2)}{0.987 \text{ atm}} = 144 \text{ mL } O_2$$

4 EVALUATE

When the pressure is increased slightly at constant temperature, the volume decreases slightly, as expected. Units cancel to give milliliters, a volume unit.

PRACTICE

1. A balloon filled with helium gas has a volume of 500 mL at a pressure of 1 atm. The balloon is released and reaches an altitude of 6.5 km, where the pressure is 0.5 atm. Assuming that the temperature has remained the same, what volume does the gas occupy at this height?

   Answer 1000 mL He

2. A gas has a pressure of 1.26 atm and occupies a volume of 7.40 L. If the gas is compressed to a volume of 2.93 L, what will its pressure be, assuming constant temperature?

   Answer 3.18 atm

3. Divers know that the pressure exerted by the water increases about 100 kPa with every 10.2 m of depth. This means that at 10.2 m below the surface, the pressure is 201 kPa; at 20.4 m, the pressure is 301 kPa; and so forth. Given that the volume of a balloon is 3.5 L at STP and that the temperature of the water remains the same, what is the volume 51 m below the water's surface?

   Answer 0.59 L

Charles's Law: Volume-Temperature Relationship

Balloonists, such as those in the photo at the beginning of this chapter, are making use of a physical property of gases: if pressure is constant, gases expand when heated. When the temperature increases, the volume of a fixed number of gas molecules must increase if the pressure is to stay constant. At the higher temperature, the gas molecules move faster. They collide with the walls of the container more frequently and with more force. The increased pressure causes the volume of the container to increase; then the molecules must travel farther before reaching the walls. The number of collisions per second against each unit of wall area decreases. This lower collision frequency offsets the greater collision force at the higher temperature. The pressure thus stays constant.

The quantitative relationship between volume and temperature was discovered by the French scientist Jacques Charles in 1787. Charles's experiments showed that all gases expand to the same extent when heated through the same temperature interval. Charles found that the volume changes by 1/273 of the original volume for each Celsius degree, at constant pressure and an initial temperature of 0°C. For example, raising the temperature to 1°C causes the gas volume to increase by 1/273 of the volume it had at 0°C. A 10°C temperature increase causes the volume to expand by 10/273 of the original volume at 0°C. If the temperature is increased by 273°C, the volume increases by 273/273 of the original, that is, the volume doubles.

The same regularity of volume change occurs if a gas is cooled at constant pressure, as the balloons in Figure 10-11 show. At 0°C, a 1°C decrease in temperature decreases the original volume by 1/273. At this rate of volume decrease, a gas cooled from 0°C to –273°C would be decreased by 273/273. In other words, it would have zero volume, which

FIGURE 10-11 As air-filled balloons are exposed to liquid nitrogen, they shrink greatly in volume. When they are removed from the liquid nitrogen and the air inside them is warmed to room temperature, the balloons expand to their original volume.

TABLE 10-3 ***Volume-Temperature Data for a Gas Sample (at Constant Mass and Pressure)***

Temperature (°C)	Volume (mL)
273	1092
100	746
10	566
1	548
0	546
−1	544
−73	400
−173	200
−223	100

is not actually possible. In fact, real gases cannot be cooled to −273°C. Before they reach that temperature, intermolecular forces exceed the kinetic energy of the molecules, and the gases condense to form liquids or solids.

The data in Table 10-3 illustrate the temperature-volume relationship at constant pressure for a gas sample with a volume of 546 mL at 0°C. When the gas is warmed by 1°C, it expands by 1/273 its original volume. In this case, each 1°C temperature change from 0°C causes a volume change of 2 mL, or 1/273 of 546 mL. Raising the temperature to 100°C from 0°C increases the volume by 200 mL, or 100/273 of 546 mL.

Note that in Table 10-3, the volume does not increase in direct proportion to the Celsius temperature. For example, notice what happens when the temperature is increased tenfold from 10°C to 100°C. The volume does not increase tenfold but increases only from 566 mL to 746 mL.

The Kelvin temperature scale is a scale that starts at a temperature corresponding to −273.15°C. That temperature is the lowest one possible. *The temperature −273.15°C is referred to as* **absolute zero** *and is given a value of zero in the Kelvin scale.* This fact gives the following relationship between the two temperature scales.

$$K = 273.15 + °C$$

For theoretical calculations in this book, 273.15 is rounded off to 273.

The average kinetic energy of gas molecules is more closely related to the Kelvin temperature. Gas volume and Kelvin temperature are directly proportional to each other. For example, quadrupling the Kelvin temperature causes the volume of a gas to quadruple, and reducing the Kelvin temperature by half causes the volume of a gas to decrease by half.

The relationship between Kelvin temperature and gas volume is known as Charles's law. **Charles's law** *states that the volume of a fixed mass of gas at constant pressure varies directly with the Kelvin temperature.*

TABLE 10-4 ***Volume-Temperature Data for a Gas Sample (at Constant Mass and Pressure)***

Volume (mL)	Kelvin temperature (K)	*V/T* or *k* (mL/K)
1092	546	2
746	373	2
566	283	2
548	274	2
546	273	2
544	272	2
400	200	2
100	50	2

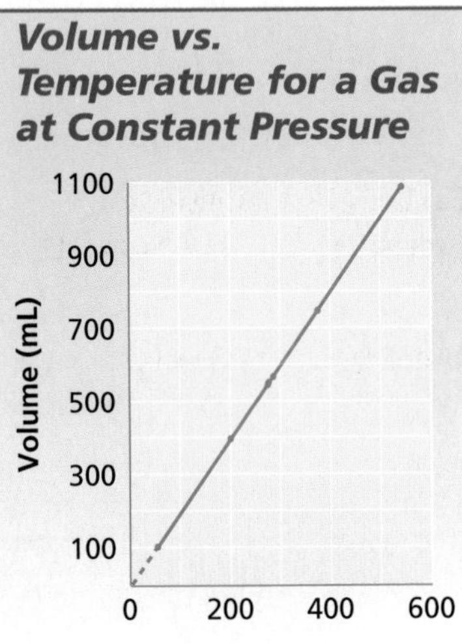

FIGURE 10-12 This graph shows the plot of the volume versus the Kelvin temperature data from Table 10-4. It gives a straight line that, when extended, indicates the volume will become 0 at –273°C. Such a plot is characteristic of directly proportional variables.

Figure 10-12 illustrates the relationship between gas volume and Kelvin temperature by plotting the data from Table 10-4. Charles's law may be expressed as follows.

$$V = kT \quad \text{or} \quad \frac{V}{T} = k$$

The value of T is the Kelvin temperature, and k is a constant. The value of k depends only on the quantity of gas and the pressure. The ratio V/T for any set of volume-temperature values always equals the same k. The form of Charles's law that can be applied directly to most volume-temperature problems involving gases is as follows.

$$\frac{V_1}{T_1} = \frac{V_2}{T_2}$$

V_1 and T_1 represent initial conditions. V_2 and T_2 represent a new set of conditions. When three of the four values V_1, T_1, V_2, and T_2 are known, this equation can be used to calculate the fourth value.

SAMPLE PROBLEM 10-3

A sample of neon gas occupies a volume of 752 mL at 25°C. What volume will the gas occupy at 50°C if the pressure remains constant?

SOLUTION

1 *ANALYZE* **Given:** V_1 of Ne = 752 mL

T_1 of Ne = 25°C + 273 = 298 K; T_2 of Ne = 50°C + 273 = 323 K

Note that Celsius temperatures have been converted to kelvins. This is a *very important* step for working the problems in this chapter.

Unknown: V_2 of Ne in mL

2 PLAN

Because the gas remains at constant pressure, an increase in temperature will cause an increase in volume. To obtain V_2, rearrange the equation for Charles's law.

$$V_2 = \frac{V_1 T_2}{T_1}$$

3 COMPUTE

Substitute values for V_1, T_1, and T_2 to obtain the new volume, V_2.

$$V_2 = \frac{V_1 T_2}{T_1} = \frac{(752 \text{ mL Ne})(323 \cancel{\text{K}})}{298 \cancel{\text{K}}} = 815 \text{ mL Ne}$$

4 EVALUATE

As expected, the volume of the gas increases as the temperature increases. Units cancel to yield milliliters, as desired. The answer contains the appropriate number of significant digits. It is also reasonably close to an estimated value of 812, calculated as $(750 \times 325)/300$.

PRACTICE

1. A helium-filled balloon has a volume of 2.75 L at 20.°C. The volume of the balloon decreases to 2.46 L after it is placed outside on a cold day. What is the outside temperature?

   Answer 262 K, or −11°C

2. A gas at 65°C occupies 4.22 L. At what Celsius temperature will the volume be 3.87 L, assuming the same pressure?

   Answer 37°C

Gay-Lussac's Law: Pressure-Temperature Relationship

You have just learned about the quantitative relationship between volume and temperature at constant pressure. What would you predict about the relationship between pressure and temperature at constant volume? You have seen that pressure is the result of collisions of molecules with container walls. The energy and frequency of collisions depend on the average kinetic energy of molecules, which depends on temperature. For a fixed quantity of gas at constant volume, the pressure should be directly proportional to the Kelvin temperature, which depends directly on average kinetic energy.

That prediction turns out to be correct. For every kelvin of temperature change, the pressure of a confined gas changes by 1/273 of the pressure at 0°C. Joseph Gay-Lussac is given credit for recognizing this in 1802. The data plotted in Figure 10-13 illustrate **Gay-Lussac's law:** *The pressure of a fixed mass of gas at constant volume varies directly with the Kelvin temperature.* Mathematically, Gay-Lussac's law is expressed as follows.

$$P = kT \quad \text{or} \quad \frac{P}{T} = k$$

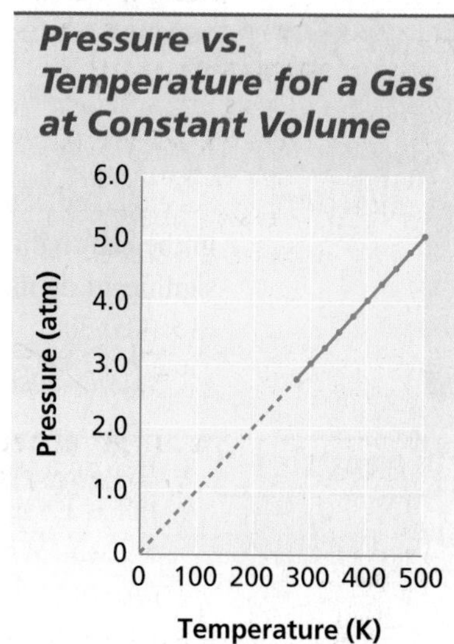

FIGURE 10-13 This graph shows that gas pressure varies directly with Kelvin temperature at constant volume.

The value of T is the temperature in kelvins, and k is a constant that depends on the quantity of gas and the volume. For a given mass of gas at constant volume, the ratio P/T is the same for any set of pressure-temperature values. Unknown values can be found using this form of Gay-Lussac's law.

$$\frac{P_1}{T_1} = \frac{P_2}{T_2}$$

When values are known for three of the four quantities, the fourth value can be calculated.

SAMPLE PROBLEM 10-4

The gas in an aerosol can is at a pressure of 3.00 atm at 25°C. Directions on the can warn the user not to keep the can in a place where the temperature exceeds 52°C. What would the gas pressure in the can be at 52°C?

SOLUTION

1 *ANALYZE* **Given:** P_1 of gas = 3.00 atm
T_1 of gas = 25°C + 273 = 298 K; T_2 of gas = 52°C + 273 = 325 K
Unknown: P_2 of gas in atm

2 *PLAN* Because the gaseous contents remain at the constant volume of the can, an increase in temperature will cause an increase in pressure. Rearrange Gay-Lussac's law to obtain P_2.

$$P_2 = \frac{P_1 T_2}{T_1}$$

3 *COMPUTE* Substitute values for P_1, T_2, and T_1 to obtain the new pressure, P_2.

$$P_2 = \frac{(3.00 \text{ atm})(325 \not{\text{K}})}{298 \not{\text{K}}} = 3.27 \text{ atm}$$

4 *EVALUATE* As expected, a temperature increase at constant volume causes the pressure of the contents in the can to increase. Units cancel correctly. The answer contains the proper number of significant digits. It also is reasonably close to an estimated value of 3.25, calculated as $(3 \times 325)/300$.

PRACTICE

1. Before a trip from New York to Boston, the pressure in an automobile tire is 1.8 atm at 20.°C. At the end of the trip, the pressure gauge reads 1.9 atm. What is the new Celsius temperature of the air inside the tire? (Assume tires with constant volume.)
   Answer 36°C

2. At 120.°C, the pressure of a sample of nitrogen is 1.07 atm. What will the pressure be at 205°C, assuming constant volume?
   Answer 1.30 atm

3. A sample of helium gas has a pressure of 1.20 atm at 22°C. At what Celsius temperature will the helium reach a pressure of 2.00 atm?
   Answer 219°C

The Combined Gas Law

A gas sample often undergoes changes in temperature, pressure, and volume all at the same time. When this happens, three variables must be dealt with at once. Boyle's law, Charles's law, and Gay-Lussac's law can be combined into a single expression that is useful in such situations. *The* **combined gas law** *expresses the relationship between pressure, volume, and temperature of a fixed amount of gas.* The combined gas law can be expressed as follows.

$$\frac{PV}{T} = k$$

In the equation, k is constant and depends on the amount of gas. The combined gas law can also be written as follows.

$$\frac{P_1V_1}{T_1} = \frac{P_2V_2}{T_2}$$

The subscripts in the equation above indicate two different sets of conditions, and T represents Kelvin temperature. From this expression, any value can be calculated if the other five are known. Note that each of the individual gas laws can be obtained from the combined gas law when the proper variable is constant. Thus, when temperature is constant, T can be canceled out of both sides of the general equation because it represents the same value ($T_1 = T_2$). This leaves Boyle's law.

$$P_1V_1 = P_2V_2$$

If the pressure is held constant, P will cancel out of both sides of the general equation, since $P_1 = P_2$. Charles's law is obtained.

$$\frac{V_1}{T_1} = \frac{V_2}{T_2}$$

Keeping the volume constant means V can be canceled out of both sides of the general equation because $V_1 = V_2$. This gives Gay-Lussac's law.

$$\frac{P_1}{T_1} = \frac{P_2}{T_2}$$

SAMPLE PROBLEM 10-5

A helium-filled balloon has a volume of 50.0 L at 25°C and 1.08 atm. What volume will it have at 0.855 atm and 10.°C?

SOLUTION

1 *ANALYZE*

Given: V_1 of He = 50.0 L
T_1 of He = 25°C + 273 = 298 K; T_2 of He = 10°C + 273 = 283 K
P_1 of He = 1.08 atm; P_2 of He = 0.855 atm

Unknown: V_2 of He in L

2 PLAN Because the gas changes in both temperature and pressure, the combined gas law is needed. Rearrange the combined gas law to solve for the final volume, V_2.

$$\frac{P_1V_1}{T_1} = \frac{P_2V_2}{T_2} \longrightarrow V_2 = \frac{P_1V_1T_2}{P_2T_1}$$

3 COMPUTE Substitute the known values into the equation to obtain a value for V_2.

$$V_2 = \frac{(1.08\ \text{atm})(50.0\ \text{L He})(283\ \text{K})}{(0.855\ \text{atm})(298\ \text{K})} = 60.0\ \text{L He}$$

4 EVALUATE Here the pressure decreases much more than the temperature decreases. As expected, the net result of the two changes gives an increase in the volume, from 50.0 L to 60.0 L. Units cancel appropriately. The answer is correctly expressed to three significant digits. It is also reasonably close to an estimated value of 50, calculated as $(50 \times 300)/300$.

PRACTICE

1. The volume of a gas is 27.5 mL at 22.0°C and 0.974 atm. What will the volume be at 15.0°C and 0.993 atm?

   Answer 26.3 mL

2. A 700. mL gas sample at STP is compressed to a volume of 200. mL, and the temperature is increased to 30.0°C. What is the new pressure of the gas in Pa?

   Answer 3.94×10^5 Pa, or 394 kPa

Dalton's Law of Partial Pressures

John Dalton, the English chemist who proposed the atomic theory, also studied gas mixtures. He found that *in the absence of a chemical reaction*, the pressure of a gas mixture is the sum of the individual pressures of each gas alone. Figure 10-14 shows a 1.0 L container filled with oxygen gas at a pressure of 0.12 atm at 0°C. In another 1.0 L container, an equal number of molecules of nitrogen gas exert a pressure of 0.12 atm at 0°C. The gas samples are then combined in a 1.0 L container. (At 0°C, oxygen gas and nitrogen gas are unreactive.) The total pressure of the mixture is found to be 0.24 atm at 0°C. The pressure that each gas exerts in the mixture is independent of that exerted by other gases present. *The pressure of each gas in a mixture is called the* **partial pressure** *of that gas.* **Dalton's law of partial pressures** *states that the total pressure of a mixture of gases is equal to the sum of the partial pressures of the component gases.* The law is true regardless of the number of different gases that are present. Dalton's law may be expressed as follows.

$$P_T = P_1 + P_2 + P_3 + \ldots$$

P_T is the total pressure of the mixture. $P_1, P_2, P_3, \ldots$ are the partial pressures of component gases 1, 2, 3, and so on.

You can understand Dalton's law in terms of the kinetic-molecular theory. The rapidly moving particles of each gas in a mixture have an equal chance to collide with the container walls. Therefore, each gas exerts a pressure independent of that exerted by the other gases present. The total pressure is the result of the total number of collisions per unit of wall area in a given time. (Note that because gas particles move independently, the other gas laws, as well as Dalton's law, can be applied to unreacting gas mixtures.)

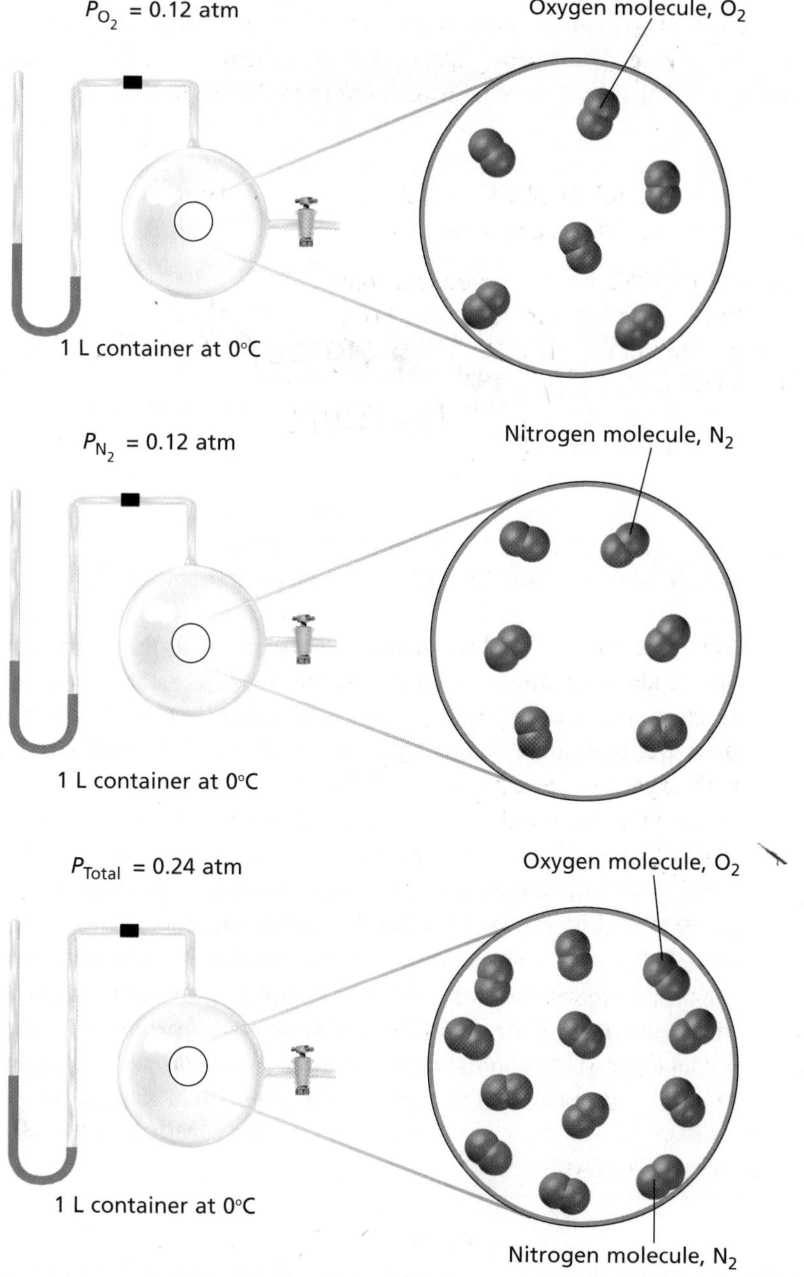

FIGURE 10-14 Samples of oxygen gas and nitrogen gas are mixed. The total pressure of the mixture is the sum of the pressures of the gases.

FIGURE 10-15
Hydrogen can be collected by water displacement by reacting zinc with sulfuric acid. The hydrogen gas produced displaces the water in the gas collecting bottle. It now contains some water vapor.

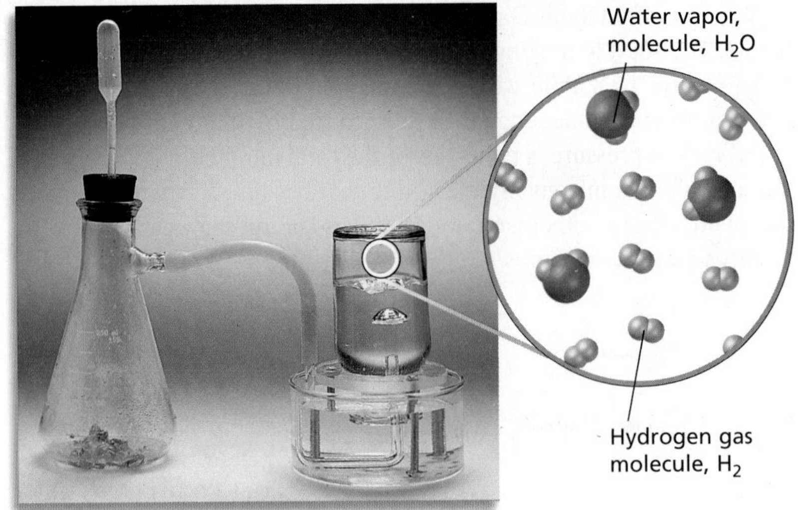

Gases Collected by Water Displacement

Gases produced in the laboratory are often collected over water, as shown in Figure 10-15. The gas produced by the reaction displaces the water, which is more dense, in the collection bottle. You can apply Dalton's law of partial pressures in calculating the pressures of gases collected in this way. A gas collected by water displacement is not pure but is always mixed with water vapor. That is because water molecules at the liquid surface evaporate and mix with the gas molecules. Water vapor, like other gases, exerts a pressure, known as *water-vapor pressure*.

Suppose you wished to determine the total pressure of the gas and water vapor inside a collection bottle. You would raise the bottle until the water levels inside and outside the bottle were the same. At that point, the total pressure inside the bottle would be the same as the atmospheric pressure, P_{atm}. According to Dalton's law of partial pressures, the following is true.

$$P_{atm} = P_{gas} + P_{H_2O}$$

Suppose you then needed to calculate the partial pressure of the dry gas collected. You would read the atmospheric pressure, P_{atm}, from a barometer in the laboratory. To make the calculation, subtract the vapor pressure of the water at the given temperature from the total pressure. The vapor pressure of water varies with temperature. You need to look up the value of P_{H_2O} at the temperature of the experiment in a standard reference table like that in Table A-8 of this book.

SAMPLE PROBLEM 10-6

Oxygen gas from the decomposition of potassium chlorate, $KClO_3$, was collected by water displacement. The barometric pressure and the temperature during the experiment were 731.0 torr and 20.0°C, respectively. What was the partial pressure of the oxygen collected?

SOLUTION

1 ANALYZE

Given: $P_T = P_{atm} = 731.0$ torr
$P_{H_2O} = 17.5$ torr (vapor pressure of water at 20.0°C, from Table A-8)
$P_{atm} = P_{O_2} + P_{H_2O}$
Unknown: P_{O_2} in torr

2 PLAN

The partial pressure of the collected oxygen is found by subtracting the partial pressure of water vapor from the atmospheric pressure, according to Dalton's law of partial pressures.

$$P_{O_2} = P_{atm} - P_{H_2O}$$

3 COMPUTE

Substituting values for P_{atm} and P_{H_2O} gives P_{O_2}.

$$P_{O_2} = 731.0 \text{ torr} - 17.5 \text{ torr} = 713.5 \text{ torr}$$

4 EVALUATE

As expected, the oxygen partial pressure is less than atmospheric pressure. It is also much larger than the partial pressure of water vapor at this temperature. The answer has the appropriate number of places. It is reasonably close to an estimated value of 713, calculated as 730 – 17.

PRACTICE

1. Some hydrogen gas is collected over water at 20.0°C. The levels of water inside and outside the gas-collection bottle are the same. The partial pressure of hydrogen is determined to be 742.5 torr. What is the barometric pressure at the time the gas is collected?
   Answer 760.0 torr

2. Helium gas is collected over water at 25°C. What is the partial pressure of the helium, given that the barometric pressure is 750.0 mm Hg?
   Answer 726.2 mm Hg

SECTION REVIEW

1. State Boyle's law, Charles's law, and the combined gas law in mathematical terms.
2. A sample of helium gas has a volume of 200.0 mL at 0.960 atm. What pressure, in atm, is needed to reduce the volume at constant temperature to 50.0 mL?
3. A certain quantity of gas has a volume of 0.750 L at 298 K. At what temperature, in degrees Celsius, would this quantity of gas be reduced to 0.500 L, assuming constant pressure?
4. An aerosol can contains gases under a pressure of 4.50 atm at 20.0°C. If the can is left on a hot, sandy beach, the pressure of the gases increases to 4.80 atm. What is the Celsius temperature on the beach?
5. Discuss the significance of the absolute-zero temperature.
6. A certain mass of oxygen was collected over water when potassium chlorate was decomposed by heating. The volume of the oxygen sample collected was 720. mL at 25.0°C and a barometric pressure of 755 torr. What would the volume of the oxygen be at STP? (Hint: First calculate the partial pressure of the oxygen, using Appendix Table A-8. Then use the combined gas law.)

CHAPTER 10 REVIEW

CHAPTER SUMMARY

10-1

- The kinetic-molecular theory of matter can be used to explain the properties of gases, liquids, and solids.
- The kinetic-molecular theory of gases describes a model of an ideal gas. The behavior of most gases is close to ideal except at very high pressures and low temperatures.
- Gases consist of large numbers of tiny, fast-moving particles that are far apart relative to their size. The average kinetic energy of the particles depends on the temperature of the gas.
- Gases exhibit expansion, fluidity, low density, compressibility, diffusion, and effusion.

Vocabulary

diffusion (305)
effusion (306)
elastic collision (303)
fluid (305)
ideal gas (303)
kinetic-molecular theory (303)
real gas (306)

10-2

- Conditions of standard temperature and pressure (STP) allow comparison of volumes of different gases.
- Pressure, volume, temperature, and number of molecules are the four measureable quantities needed to fully describe a gas.
- The gas molecules that make up the atmosphere exert pressure against Earth's surface, varying with weather conditions and elevation.
- A barometer measures the pressure of the atmosphere. The pressure of a gas in a closed container can be measured by a manometer.

Vocabulary

atmosphere of pressure (311)
barometer (310)
millimeter of mercury (311)
newton (309)
pascal (311)
pressure (308)
standard temperature and pressure (312)

10-3

- Boyle's law shows the inverse relationship between the volume and the pressure of a gas.

$$PV = k$$

- Charles's law illustrates the direct relationship between the volume of a gas and its temperature in kelvins.

$$V = kT$$

- Gay-Lussac's law represents the direct relationship between the pressure of a gas and its temperature in kelvins.

$$P = kT$$

- The combined gas law, as its name implies, combines the previous relationships into the following mathematical expression.

$$\frac{PV}{T} = k$$

- A gas exerts pressure on the walls of its container. In a mixture of unreacting gases, the total pressure equals the sum of the partial pressures of each gas.

Vocabulary

absolute zero (317)
Boyle's law (314)
Charles's law (317)
combined gas law (321)
Dalton's law of partial pressures (322)
gas laws (313)
Gay-Lussac's law (319)
partial pressure (322)

REVIEWING CONCEPTS

1. What idea is the kinetic-molecular theory based on? (10-1)
2. What is an ideal gas? (10-1)
3. State the five basic assumptions of the kinetic-molecular theory. (10-1)
4. How do gases compare with liquids and solids in terms of the distance between their molecules? (10-1)
5. What is an elastic collision? (10-1)
6. a. Write and label the equation that relates the average kinetic energy and speed of gas particles.
   b. What is the relationship between the temperature, speed, and kinetic energy of gas molecules? (10-1)
7. a. What is diffusion?
   b. What factors affect the rate of diffusion of one gas through another?
   c. What is the relationship between the mass of a gas particle and the rate at which it diffuses through another gas?
   d. What is effusion? (10-1)
8. a. Why does a gas in a closed container exert pressure?
   b. What is the relationship between the area a force is applied to and the resulting pressure? (10-2)
9. a. What is atmospheric pressure?
   b. Why does the atmosphere exert pressure?
   c. What is the value of atmospheric pressure at sea level, in newtons per square centimeter? (10-2)
10. a. Why does a column of mercury in a tube that is inverted in a dish of mercury have a height of about 760 mm at sea level?
    b. What height would be maintained by a column of water inverted in a dish of water at sea level?
    c. What accounts for the difference in the heights of the mercury and water columns? (10-2)
11. a. Identify three sets of units typically used to express pressure.
    b. Convert one atmosphere to torr.
    c. What is a pascal?
    d. What is the SI equivalent of one standard atmosphere of pressure? (10-2)
12. a. At constant pressure, how does temperature relate to the volume of a given quantity of gas?
    b. How does this explain the danger of throwing an aerosol can into a fire? (10-3)
13. a. What is the Celsius equivalent of absolute zero?
    b. What is the significance of this temperature?
    c. What is the relationship between Kelvin temperature and the average kinetic energy of gas molecules? (10-3)
14. a. Explain what is meant by the partial pressure of each gas within a mixture of gases.
    b. How do the partial pressures of gases in a mixture affect each other? (10-3)

PROBLEMS

Pressure and Temperature Conversions

15. If the atmosphere can support a column of mercury 760 mm high at sea level, what height (in mm) of each of the following could be supported, given the relative density values cited?
    a. water, whose density is approximately 1/14 that of mercury
    b. a substance with a density 1.40 times that of mercury
16. Convert each of the following into a pressure reading expressed in millimeters of mercury. (Hint: See Sample Problem 10-1.)
    a. 1.25 atm
    b. 2.48×10^{-3} atm
    c. 4.75×10^{4} atm
    d. 7.60×10^{6} atm
17. Convert each of the following into the unit specified.
    a. 125 mm Hg into atm
    b. 3.20 atm into Pa
    c. 5.38 kPa into torr
18. Convert each of the following Celsius temperatures to Kelvin temperatures.
    a. 0.°C
    b. 27°C
    c. −50.°C
    d. −273°C
19. Convert each of the following Kelvin temperatures to Celsius temperatures.
    a. 273 K
    b. 350. K
    c. 100. K
    d. 20. K

Boyle's Law

20. Use Boyle's law to solve for the missing value in each of the following. (Hint: See Sample Problem 10-2.)

a. $P_1 = 350.$ torr, $V_1 = 200.$ mL, $P_2 = 700.$ torr, $V_2 = ?$

b. $P_1 = 0.75$ atm, $V_2 = 435$ mL, $P_2 = 0.48$ atm, $V_1 = ?$

c. $V_1 = 2.4 \times 10^5$ L, $P_2 = 180$ mm Hg, $V_2 = 1.8 \times 10^3$ L, $P_1 = ?$

21. The pressure exerted on a 240. mL sample of hydrogen gas at constant temperature is increased from 0.428 atm to 0.724 atm. What will the final volume of the sample be?

22. A flask containing 155 cm^3 of hydrogen was collected under a pressure of 22.5 kPa. What pressure would have been required for the volume of the gas to have been 90.0 cm^3, assuming the same temperature?

23. A gas has a volume of 450.0 mL. If the temperature is held constant, what volume would the gas occupy if the pressure were

a. doubled? (Hint: Express P_2 in terms of P_1.)

b. reduced to one-fourth of its original value?

24. A sample of oxygen that occupies 1.00×10^6 mL at 575 mm Hg is subjected to a pressure of 1.25 atm. What will the final volume of the sample be if the temperature is held constant?

Charles's Law

25. Use Charles's law to solve for the missing value in each of the following. (Hint: See Sample Problem 10-3.)

a. $V_1 = 80.0$ mL, $T_1 = 27°C$, $T_2 = 77°C$, $V_2 = ?$

b. $V_1 = 125$ L, $V_2 = 85.0$ L, $T_2 = 127°C$, $T_1 = ?$

c. $T_1 = -33°C$, $V_2 = 54.0$ mL, $T_2 = 160.°C$, $V_1 = ?$

26. A sample of air has a volume of 140.0 mL at 67°C. At what temperature will its volume be 50.0 mL at constant pressure?

27. At standard temperature, a gas has a volume of 275 mL. The temperature is then increased to 130.°C, and the pressure is held constant. What is the new volume?

Gay-Lussac's Law

28. A sample of hydrogen at 47°C exerts a pressure of 0.329 atm. The gas is heated to 77°C at constant volume. What will its new pressure be? (Hint: See Sample Problem 10-4.)

29. To what temperature must a sample of nitrogen at 27°C and 0.625 atm be taken so that its pressure becomes 1.125 atm at constant volume?

30. The pressure on a gas at –73°C is doubled, but its volume is held constant. What will the final temperature be in degrees Celsius?

Combined Gas Law

31. A sample of gas at 47°C and 1.03 atm occupies a volume of 2.20 L. What volume would this gas occupy at 107°C and 0.789 atm? (Hint: See Sample Problem 10-5.)

32. A 350. mL air sample collected at 35°C has a pressure of 550. torr. What pressure will the air exert if it is allowed to expand to 425 mL at 57°C?

33. A gas measures 1.75 L at –23°C and 150. kPa. At what temperature would the gas occupy 1.30 L at 210. kPa?

34. A sample of oxygen at 40.°C occupies 820. mL. If this sample later occupies 1250 mL at 60.°C and 1.40 atm, what was its original pressure?

35. A gas at 7.75×10^4 Pa and 17°C occupies a volume of 850. cm^3. At what temperature, in degrees Celsius, would the gas occupy 720. cm^3 at 8.10×10^4 Pa?

36. A meteorological balloon contains 250. L of He at 22°C and 740. mm Hg. If the volume of the balloon can vary according to external conditions, what volume would it occupy at an altitude at which the temperature is –52°C and the pressure is 0.750 atm?

37. The balloon in the previous problem will burst if its volume reaches 400. L. Given the initial conditions specified in that problem, at what temperature, in degrees Celsius, will the balloon burst if its pressure at that bursting point is 0.475 atm?

38. The normal respiratory rate for a human being is 15.0 breaths per minute. The average volume of air for each breath is 505 cm^3 at 20.°C and 9.95×10^4 Pa. What is the volume of air at STP that an individual breathes in one day? Give your answer in cubic meters.

Dalton's Law of Partial Pressures

39. Three of the primary components of air are carbon dioxide, nitrogen, and oxygen. In a sample containing a mixture of only these gases at exactly one atmosphere pressure, the partial pressures of carbon dioxide and nitrogen are given as $P_{CO_2} = 0.285$ torr and $P_{N_2} =$ 593.525 torr. What is the partial pressure of oxygen? (Hint: See Sample Problem 10-6.)

40. Determine the partial pressure of oxygen collected by water displacement if the water temperature is 20.0°C and the total pressure of the gases in the collection bottle is 730.0 mm Hg.

41. A sample of gas is collected over water at a temperature of 35.0°C when the barometric pressure reading is 742.0 mm Hg. What is the partial pressure of the dry gas?

42. A sample of oxygen is collected in a 175 mL container over water at 15°C, and the barometer reads 752.0 torr. What volume would the dry gas occupy at 770.0 torr and 15°C?

43. Suppose that 120. mL of argon is collected over water at 25°C and 780.0 torr. Compute the volume of the dry argon at STP.

MIXED REVIEW

44. A sample of argon gas occupies a volume of 295 mL at 36°C. What volume will the gas occupy at 55°C, assuming constant pressure?

45. A gas has a pressure of 4.62 atm when its volume is 2.33 L. What will the pressure be when the volume is changed to 1.03 L, assuming constant temperature? Express the final pressure in torr.

46. A sample of carbon dioxide gas occupies 638 mL at 0.893 atm and 12°C. What will the pressure be at a volume of 881 mL and a temperature of 18°C?

47. At 84°C, a gas in a container exerts a pressure of 0.503 atm. Assuming the size of the container has not changed, at what Celsius temperature would the pressure be 1.20 atm?

48. A mixture of three gases, A, B, and C, is at a total pressure of 6.11 atm. The partial pressure of gas A is 1.68 atm; that of gas B is 3.89 atm. What is the partial pressure of gas C?

49. A weather balloon at Earth's surface has a volume of 4.00 L at 304 K and 755 mm Hg. If the balloon is released and the volume reaches 4.08 L at 728 mm Hg, what is the temperature?

50. A child receives a balloon filled with 2.30 L of helium from a vendor at an amusement park. The temperature outside is 311 K. What will the volume of the balloon be when the child brings it home to an air-conditioned house at 295 K? Assume that the pressure stays the same.

51. At a deep-sea station 200. m below the surface of the Pacific Ocean, workers live in a highly pressurized environment. How many liters of gas at STP must be compressed on the surface to fill the underwater environment with 2.00×10^7 L of gas at 20.0 atm? Assume that temperature remains constant.

CRITICAL THINKING

52. **Applying Models**
    a. Why do we say the graph in Figure 10-10 illustrates an inverse relationship?
    b. Why does the data plotted in Figure 10-12 show a direct relationship?

53. **Relating Ideas** Explain how different gases in a mixture can have the same average kinetic energy value, even though the masses of their individual particles differ.

54. **Inferring Conclusions** If all gases behaved as ideal gases under all conditions of temperature and pressure, there would be no solid or liquid forms of these substances. Explain.

55. **Relating Ideas** Pressure is defined as force per unit area. Yet Torricelli found that the diameter of the barometer dish and the surface area of contact between the mercury in the tube and

in the dish did not affect the height of mercury that was supported. Explain this seemingly inconsistent observation in view of the relationship between pressure and surface areas.

56. Interpreting Graphics Examine Boyle's J-tube apparatus shown here. The tube is open to the atmosphere at the top. The other end is closed and contains a gas with a volume labeled V_{gas}. If $h = 60$ mm Hg, what is the pressure exerted by the enclosed gas?

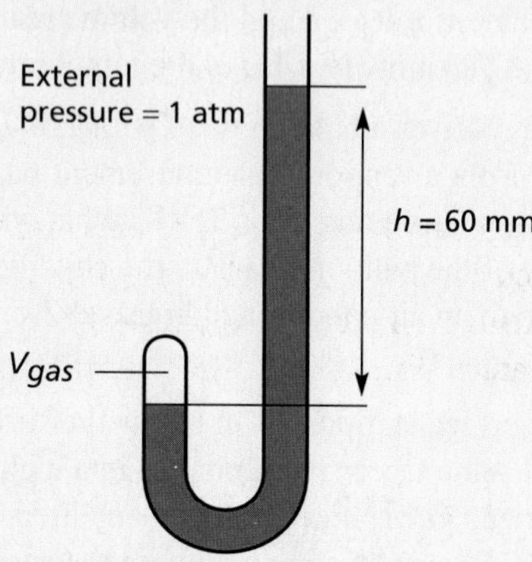

57. Velocity distribution curves are shown in the following graph for the same gas under two different conditions, A and B. Compare the behavior of the gas under conditions A and B in relation to each of the following:

a. temperature
b. average kinetic energy
c. average molecular velocity
d. gas volume
e. gas pressure

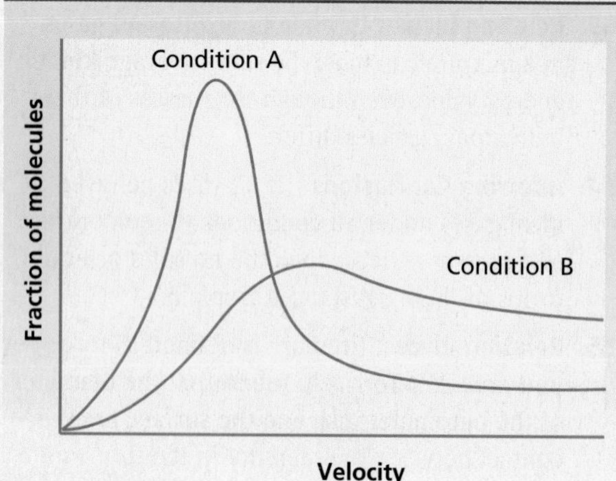

TECHNOLOGY & LEARNING

58. Graphing Calculator Deriving the Boyle's Law Equation

The graphing calculator can be programmed to derive the equation for a curve, given data such as volume versus pressure. Begin by creating a table of data. Then program the calculator to plot the data.

Pressure (atm)	Volume (mL)
0.5	1200
1.0	600
2.0	300
3.0	200
4.0	150
5.0	120
6.0	100

A. Create lists L1 and L2.

Keystrokes for creating lists: STAT 4 2nd L1 , 2nd L2 ENTER STAT 1

Enter the pressure data in L1 and the volume data in L2.

B. Program the graph.

Keystrokes for naming the program: PRGM ▶ ▶ ENTER

Name the program B O Y L E S ENTER

Program keystrokes: STAT ▶ ALPHA B 2nd L1 , 2nd L2 2nd : PRGM ▶ 8 2nd : PRGM ▶ 3 ALPHA " ALPHA V 2nd TEST 1 ALPHA " 2nd : PRGM ▶ 3 VARS 5 ▶ ▶ 1 2nd : PRGM ▶ 3 ALPHA " ÷ ALPHA P ALPHA "

C. Check the display.

Screen display: PwrReg L1, L2:ClrHome:Disp "V=":Disp a:Disp " /P"

Press 2nd QUIT to exit the program editor.

D. Run the program.

a. Press PRGM. Select BOYLES and press ENTER ENTER. The calculator will provide the Boyle's law equation and the value of the constant, *k*.

b. Use the BOYLES program with the data set shown below. Refer back to creating L1 and

L_2. Use L_1 for pressure data and L_2 for volume data. Use the program to determine the Boyle's law equation and the value of k for this new data set.

Pressure (atm)	Volume (mL)
0.5	800
1.0	400
2.0	200
3.0	133
4.0	100
5.0	80

HANDBOOK SEARCH

59. Review the melting point data in the properties tables for each group of the *Elements Handbook*. What elements on the periodic table exist as gases at room temperature?

60. Review the listing found in the *Elements Handbook* of the top 10 chemicals produced in the United States. Which of the top 10 chemicals are gases?

61. Though mercury is an ideal liquid to use in barometers, it is highly toxic. Review the transition metals section of the *Elements Handbook*, and answer the following questions:
a. What is the density of mercury? How does the density of Hg compare with that of other transition metals?
b. Why is mercury so toxic?
c. What mercury compound is becoming more common as a pollutant in lakes and streams? Why is this pollutant so dangerous to freshwater ecosystems?
d. What are the symptoms of mercury poisoning?

RESEARCH & WRITING

62. Prepare a report on the development of the modern submarine. Include a discussion of the technology that enables the submarine to withstand the tremendous pressures at great ocean depths. Also report on the equipment used to ensure a sufficient supply of oxygen for submarine crew members.

63. Design and conduct a meteorological study to examine the interrelationships among barometric pressure, temperature, humidity, and other weather variables. Prepare a report explaining your results.

64. Conduct library research on attempts made to approach absolute zero and on the interesting properties that materials exhibit near that temperature. Write a report on your findings.

ALTERNATIVE ASSESSMENT

65. The air pressure of car tires should be checked regularly for safety reasons and to prevent uneven tire wear. Find out the units of measurement on a typical tire gauge, and determine how gauge pressure relates to atmospheric pressure.

CHAPTER 11

Molecular Composition of Gases

The study of gases led to the formulation of the laws and principles that are the foundations of modern chemistry.

SECTION 11-1

Volume-Mass Relationships of Gases

Objectives

- State the law of combining volumes.
- State Avogadro's law and explain its significance.
- Define *standard molar volume of a gas,* and use it to calculate gas masses and volumes.
- Use standard molar volume to calculate the molar mass of a gas.

In this section, you will study the relationships between the volumes of gases that react with each other. You will also learn how volume, density, and molar mass are related.

Measuring and Comparing the Volumes of Reacting Gases

In the early 1800s, French chemist Joseph Gay-Lussac studied gas volume relationships involving a chemical reaction between hydrogen and oxygen. He observed that 2 L of hydrogen react exactly with 1 L of oxygen to form 2 L of water vapor, assuming the same temperature and pressure.

hydrogen gas +	oxygen gas ⟶	water vapor
2 L	1 L	2 L
2 volumes	1 volume	2 volumes

In other words, this reaction shows a simple and definite 2:1:2 relationship between the volumes of the reactants and the product. Two volumes of hydrogen react with 1 volume of oxygen to produce 2 volumes of water vapor. The 2:1:2 relationship for this reaction applies to any proportions for volume—for example, 2 mL, 1 mL, and 2 mL; 600 L, 300 L, and 600 L; or 400 cm^3, 200 cm^3, and 400 cm^3.

Gay-Lussac also noticed simple and definite proportions by volume in other reactions of gases, such as in the reaction between hydrogen gas and chlorine gas.

hydrogen gas +	chlorine gas ⟶	hydrogen chloride gas
1 L	1 L	2 L
1 volume	1 volume	2 volumes

In 1808, Gay-Lussac summarized the results of his experiments in a statement known today as **Gay-Lussac's law of combining volumes of gases.** The law states that *at constant temperature and pressure, the volumes of gaseous reactants and products can be expressed as ratios of small whole numbers.* This simple observation, combined with the findings of Avogadro, provided more insight into how gases react and combine with each other.

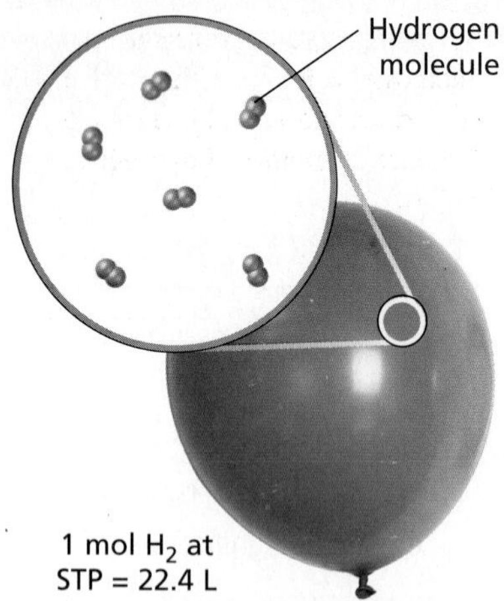

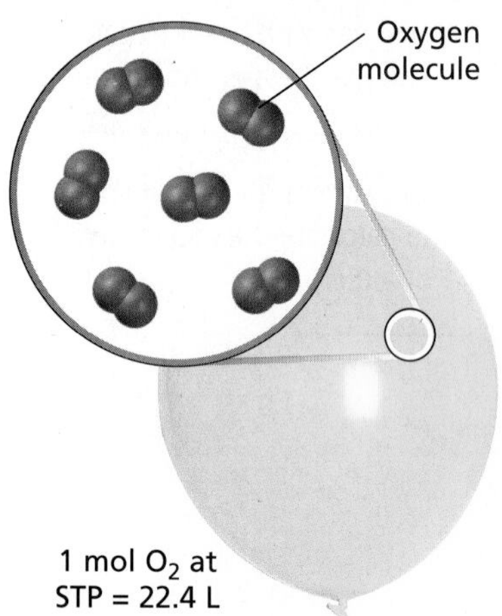

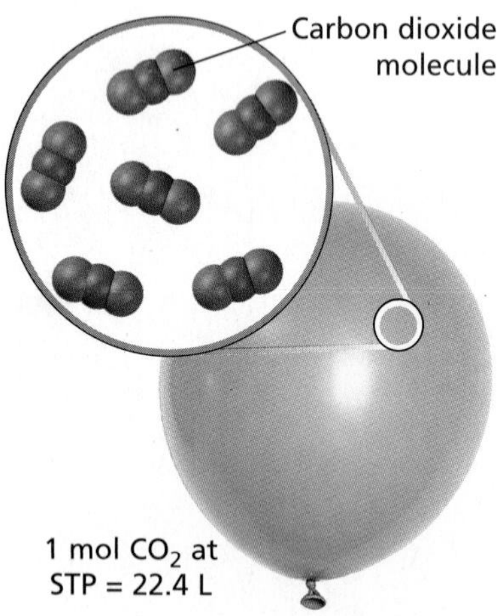

FIGURE 11-1 At the same temperature and pressure, balloons of equal volume have equal numbers of molecules, regardless of which gas they contain.

Avogadro's Law

Recall an important point of Dalton's atomic theory: atoms are indivisible. Dalton also thought that the particles of gaseous elements exist in the form of isolated single atoms. He believed that one atom of one element always combines with one atom of another element to form a single particle of the product. Accounting for some of the volume relationships observed by Gay-Lussac presented a problem for Dalton's theory. For example, in reactions such as the formation of water vapor, mentioned on the preceding page, it would seem that the oxygen involved would have to divide into two parts.

In 1811, Avogadro found a way to explain Gay-Lussac's simple ratios of combining volumes without violating Dalton's idea of indivisible atoms. He did this by rejecting Dalton's idea that reactant elements are always in monatomic form when they combine to form products. He reasoned that these molecules could contain more than one atom. Avogadro also put forth an idea known today as **Avogadro's law.** The law states that *equal volumes of gases at the same temperature and pressure contain equal numbers of molecules.* Figure 11-1 illustrates Avogadro's law. It follows that at the same temperature and pressure, the volume of any given gas varies directly with the number of molecules.

Consider the reaction of hydrogen and chlorine to produce hydrogen chloride, illustrated in Figure 11-2. According to Avogadro's law, equal volumes of hydrogen and chlorine contain the same number of molecules. Avogadro accepted Dalton's idea that atoms of hydrogen and chlorine are indivisible. However, he rejected Dalton's belief that these elements are monatomic. He concluded that the hydrogen and chlorine components must each consist of two or more atoms joined together. The simplest assumption was that hydrogen and chlorine molecules are composed of two atoms each. That assumption leads to the following balanced equation for the reaction of hydrogen with chlorine.

$$\begin{array}{ccccc} H_2(g) & + & Cl_2(g) & \longrightarrow & 2HCl(g) \\ \text{1 volume} & & \text{1 volume} & & \text{2 volumes} \\ \text{1 molecule} & & \text{1 molecule} & & \text{2 molecules} \end{array}$$

The simplest hypothetical formula for hydrogen chloride, HCl, indicates that the molecule contains one hydrogen atom and one chlorine atom. Given the ratios of the combined volumes, the simplest formulas for hydrogen and chlorine must be H_2 and Cl_2, respectively.

Avogadro's reasoning applied equally well to the combining volumes for the reaction of hydrogen and oxygen to form water vapor. The simplest hypothetical formula for oxygen indicated two oxygen atoms, which turns out to be correct. The simplest possible molecule of water indicated two hydrogen atoms and one oxygen atom per molecule, which is also correct. Experiments eventually showed that all elements that are gases near room temperature, except the noble gases, normally exist as diatomic molecules.

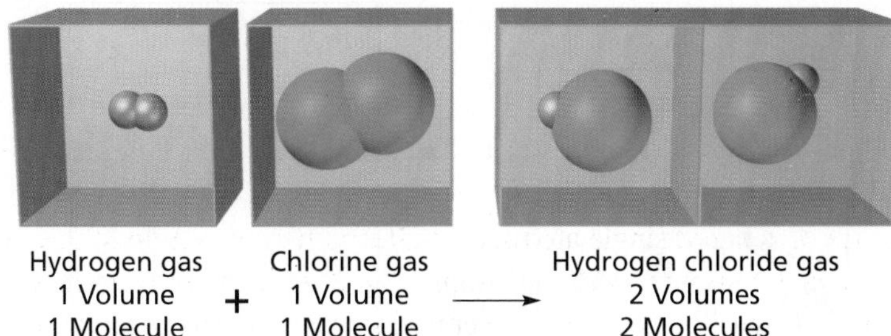

FIGURE 11-2 Hydrogen molecules combine with chlorine molecules in a 1:1 volume ratio to produce 2 volumes of hydrogen chloride.

Avogadro's law also indicates that gas volume is directly proportional to the number of moles of gas, at constant temperature and pressure. Note the equation for this relationship.

$$V = kn$$

Here, n is the number of moles of gas and k is a constant. As shown below, the coefficients in a chemical reaction involving gases indicate the relative numbers of molecules, the relative numbers of moles, and the relative volumes.

$2H_2(g)$ +	$O_2(g)$ $\longrightarrow$	$2H_2O(g)$
2 molecules	1 molecule	2 molecules
2 mol	1 mol	2 mol
2 volumes	1 volume	2 volumes

Molar Volume of Gases

Recall that one mole of a molecular substance contains Avogadro's number of molecules (6.022×10^{23}). One mole of oxygen, O_2, contains 6.022×10^{23} diatomic oxygen molecules and has a mass of 31.9988 g. One mole of hydrogen contains the same number of diatomic hydrogen molecules but has a mass of only 2.015 88 g. One mole of helium, a monatomic gas, contains Avogadro's number of helium atoms and has a mass of 4.002 602 g.

According to Avogadro's law, one mole of any gas will occupy the same volume as one mole of any other gas at the same temperature and pressure, despite mass differences. *The volume occupied by one mole of a gas at STP is known as the* **standard molar volume of a gas.** *It has been found to be 22.414 10 L.* For calculations in this book, we use 22.4 L as the standard molar volume.

Knowing the volume of a gas, you can use 1 mol/22.4 L as a conversion factor to find the number of moles, and therefore the mass, of a given volume of a known gas at STP. You can also use the molar volume of a gas to find the volume, at STP, of a known number of moles or a known mass of a gas. These types of problems can also be solved using the ideal gas law, as you will see in Section 11-2.

FIGURE 11-3 One-mole quantities of two different gases each occupy 22.4 L at STP and have equal numbers of molecules. However, their masses are different.

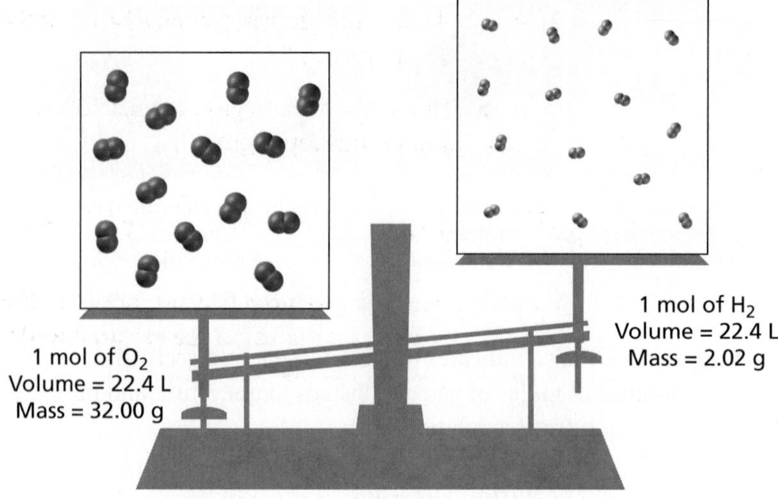

Figure 11-3 shows that 22.4 L of each gas contains the same number of molecules, but the mass of this volume is different for different gases. The mass of each is equal to the molar mass of the gas—the mass of one mole of molecules.

SAMPLE PROBLEM 11-1

A chemical reaction produces 0.0680 mol of oxygen gas. What volume in liters is occupied by this gas sample at STP?

SOLUTION

1 ANALYZE

Given: molar mass of O_2 = 0.0680 mol
Unknown: volume of O_2 in liters at STP

2 PLAN

$$\text{moles of } O_2 \longrightarrow \text{liters of } O_2 \text{ at STP}$$

The standard molar volume can be used to find the volume of a known molar amount of a gas at STP.

$$\text{mol} \times \frac{22.4 \text{ L}}{\text{mol}} = \text{volume of } O_2 \text{ in L}$$

3 COMPUTE

$$0.0680 \cancel{\text{mol}}\ O_2 \times \frac{22.4 \text{ L}}{\cancel{\text{mol}}} = 1.52 \text{ L } O_2$$

4 EVALUATE

The answer is close to an estimated value of 1.4, computed as 0.07 × 20. Units have canceled to yield liters. The calculated result is correctly expressed to three significant figures.

PRACTICE

1. At STP, what is the volume of 7.08 mol of nitrogen gas? *Answer* 159 L N_2

2. A sample of hydrogen gas occupies 14.1 L at STP. How many moles of the gas are present?

Answer
0.629 mol H_2

3. At STP, a sample of neon gas occupies 550. cm^3. How many moles of neon gas does this represent?

Answer
0.0246 mol Ne

SAMPLE PROBLEM 11-2

A chemical reaction produced 98.0 mL of sulfur dioxide gas, SO_2, at STP. What was the mass (in grams) of the gas produced?

SOLUTION

1 *ANALYZE* **Given:** volume of SO_2 at STP = 98.0 mL
Unknown: mass of SO_2 in grams

2 *PLAN*

liters of SO_2 at STP $\longrightarrow$ moles of SO_2 $\longrightarrow$ grams of SO_2

$$\text{mL} \times \frac{1\ \text{L}}{1000\ \text{mL}} \times \frac{1\ \text{mol}\ SO_2}{22.4\ \text{L}} \times \frac{\text{g}\ SO_2}{\text{mol}\ SO_2} = \text{g}\ SO_2$$

3 *COMPUTE*

$$98.0\ \cancel{\text{mL}} \times \frac{1\ \cancel{\text{L}}}{1000\ \cancel{\text{mL}}} \times \frac{1\ \cancel{\text{mol}\ SO_2}}{22.4\ \cancel{\text{L}}} \times \frac{64.07\ \text{g}\ SO_2}{\cancel{\text{mol}\ SO_2}} = 0.280\ \text{g}\ SO_2$$

4 *EVALUATE* The result is correctly expressed to three significant figures. Units cancel correctly to give the answer in grams. The known volume is roughly 1/200 of the molar volume (22 400 mL/200 = 112 mL). The answer is reasonable: the mass should also be roughly 1/200 of the molar mass (64 g/200 = 0.32 g).

PRACTICE

1. What is the mass of 1.33×10^4 mL of oxygen gas at STP?

Answer
19.0 g O_2

2. What is the volume of 77.0 g of nitrogen dioxide gas at STP?

Answer
37.5 L NO_2

3. At STP, 3 L of chlorine is produced during a chemical reaction. What is the mass of this gas?

Answer
9 g Cl_2

SECTION REVIEW

1. Explain Gay-Lussac's law of combining volumes.
2. State Avogadro's law, and explain its significance.
3. Define *molar volume.*
4. How many moles of oxygen gas are there in 135 L of oxygen at STP?
5. What volume (in mL) at STP will be occupied by 0.0035 mol of methane, CH_4?

GREAT DISCOVERIES

Chemistry's First Law

HISTORICAL PERSPECTIVE

The notion that nature abhors a vacuum was proposed by the Greek philosopher Aristotle, and his word went unchallenged for nearly 2000 years. Then in the mid-1600s, a new breed of thinkers known as experimental philosophers—later to be called scientists—began testing the long-held assumption that space must contain matter. These investigations represent some of the earliest experiments with gases, and they led to the discovery of the first empirical principle of chemistry, Boyle's law.

Overturning an Ancient Assumption

The first scientist to demonstrate the existence of a vacuum was Evangelista Torricelli. In 1643, he showed that when a glass tube 3 ft long and about an inch in diameter was sealed at one end, filled with mercury, and inverted in a container full of mercury, the mercury in the tube fell to a height of about 30 in. above the level of mercury in the container. Although some thinkers remained skeptical, it was generally accepted that the space between the mercury and the sealed end of the tube was indeed a vacuum.

Torricelli then turned his attention to how the mercury in the glass tube of his apparatus was supported. The known observation that liquids exerted a pressure on objects immersed in them inspired him to hypothesize that a "sea of air" surrounded the Earth. He further hypothesized that the air exerted pressure on the mercury in the container and thus supported the mercury in the column.

Evangelista Torricelli invented the mercury barometer.

Support for the New Theory

Although the idea of an atmosphere that has weight and exerts a pressure on the objects within it seems obvious today, it was a radical theory at the time. One of the period's great scientists, Robert Boyle, summed up the resistance encountered by him and the other experimenters investigating air:

> *The generality of men are so accustomed to judge of things by their senses that because the air is invisible they ascribe but little to it, and think of it but one remove from nothing.*

To test the effects of the atmosphere, Boyle had his talented assistant, Robert Hooke, create a piece of equipment that would revolutionize the study of air. The apparatus was an improved version of a pump designed by the famous German experimenter Otto von Guericke; the pump had a large receptacle in which a partial vacuum could be created.

Boyle placed Torricelli's setup, known today as a barometer, in the receptacle of the pump and observed the mercury column as he reduced the pressure around it. He noted that the height of the mercury decreased as the pressure surrounding the mercury in the container was lowered, strongly supporting Torricelli's atmospheric theory.

Using Hooke's pump, Boyle performed additional studies that verified the idea that air exerted pressure and had weight. His experiments also led him to the important conclusion that air was elastic, that is, it could expand and contract. It was during an investigation into air's elasticity that Boyle discovered the fundamental law that bears his name.

We can detect the difference between the atmospheric pressure on a mountaintop and the atmospheric pressure at sea level thanks to the work of Torricelli and Boyle.

An Ingenious Experiment

In response to a criticism of his findings, Boyle performed an experiment to show that air could be compressed to a pressure greater than that of the atmosphere. First he prepared a glass J tube with the short end sealed off and the long end left open. Then he poured mercury into the tube, making sure that the levels in each end were the same and letting air travel freely between the ends, to ensure that each column was at atmospheric pressure.

Boyle then poured more mercury into the long end of the tube until it was about 30 in. above the level of mercury in the short end, making the trapped air exposed to about twice as much atmospheric pressure. He observed that the volume of the trapped air was halved. He continued to add mercury until the total pressure on the trapped air was about four times that of the atmosphere. Noting that the air had been compressed to about one-quarter of its original value, Boyle discovered the inverse relationship between air's pressure and its volume:

> *It is evident, that as common air, when reduced to half its wonted extent [volume], obtained near about twice as forcible a spring [pressure] as it had before; so this thus comprest air being further thrust into half this narrow room, obtained thereby a spring about . . . four times as strong as that of common air.*

A Long-Standing Contribution

Boyle went on to show that the relationship between air pressure and volume, $P \propto 1/V$ (at constant temperature), held not only when the gas was compressed but also when it was allowed to expand. It would be up to future investigators to show that the law was a principle applying to gases in general. Together with the findings of other researchers, such as Jacques Charles, Joseph Gay-Lussac, and Amadeo Avogadro, Boyle's discovery led chemists to the famous ideal gas law, $PV = nRT$, which serves as a starting point in the study of chemistry today.

SECTION 11-2

The Ideal Gas Law

Objectives

- State the ideal gas law.
- Derive the ideal gas constant and discuss its units.
- Using the ideal gas law, calculate pressure, volume, temperature, or amount of gas when the other three quantities are known.
- Using the ideal gas law, calculate the molar mass or density of a gas.
- Reduce the ideal gas law to Boyle's law, Charles's law, and Avogadro's law. Describe the conditions under which each applies.

In Section 10-3, you learned about three quantities—pressure, volume, and temperature—needed to describe a gas sample. A gas sample can be further characterized using a fourth quantity—the number of moles. The number of molecules or moles present can affect the other three quantities. The collision rate of molecules per unit area of container wall depends on the number of molecules present. If the number of molecules is increased for a sample at constant volume and temperature, the collision rate increases. Therefore, the pressure increases, as shown by the model in Figure 11-4(a). Consider what would happen if the pressure and temperature were kept constant while the number of molecules increased. According to Avogadro's law, the volume would increase. As Figure 11-4(b) shows, an increase in volume keeps the pressure constant at constant temperature. Increasing the volume keeps the collision rate per unit of wall area constant.

You can see that gas pressure, volume, temperature, and the number of moles are all interrelated. There is a mathematical relationship that describes the behavior of a gas sample for any combination of these conditions. *The* **ideal gas law** *is the mathematical relationship of pressure, volume, temperature, and the number of moles of a gas.* It is considered an equation of state for a gas, because the particular state of a gas can be defined by its pressure, volume, temperature, and number of moles.

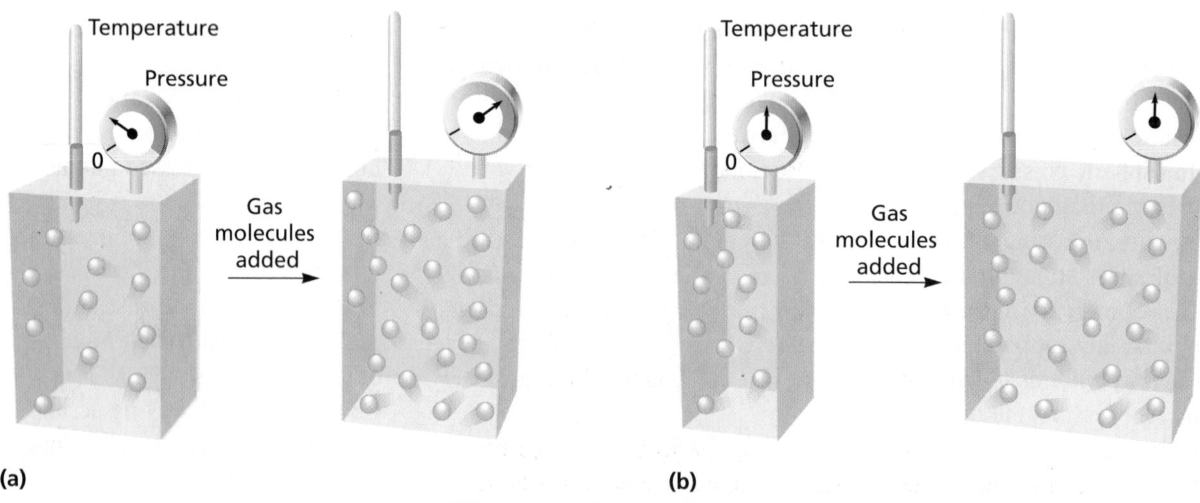

FIGURE 11-4 (a) When volume and temperature are constant, gas pressure increases as the number of molecules increases. (b) When pressure and temperature are constant, gas volume increases as the number of molecules increases.

Derivation of the Ideal Gas Law

The general equation that can be used to calculate unknown information about gas samples can be derived by combining Boyle's law, Charles's law, and Avogadro's law. First, consider each of those laws and the principle again.

Boyle's law: At constant temperature, the volume of a given mass of gas is inversely proportional to the pressure.

$$V \propto \frac{1}{P}$$

Charles's law: At constant pressure, the volume of a given mass of gas is directly proportional to the Kelvin temperature.

$$V \propto T$$

Avogadro's law: At constant pressure and temperature, the volume of a given mass of a gas is directly proportional to the number of moles.

$$V \propto n$$

A quantity—in this case, volume—that is proportional to each of several quantities is also proportional to their product. Therefore, combining the three relationships above gives the following.

$$V \propto \frac{1}{P} \times T \times n$$

Mathematically, you can change a proportionality to an equality by introducing a constant. In this case, the symbol R is used for the constant.

$$V = R \times \frac{1}{P} \times T \times n$$

The constant, R, has the same value for all gases whose behavior approaches that of an ideal gas. Rearranging the variables, the equation for the ideal gas law is derived as follows.

$$V = \frac{nRT}{P} \text{ or } PV = nRT$$

This equation states that the volume of a gas varies directly with the number of moles (or molecules) of a gas and its Kelvin temperature. The volume also varies inversely with the pressure. Under ordinary conditions, such as pressures around 1 atm, most gases exhibit behavior that is nearly ideal. The equation can then be applied with reasonable accuracy.

The ideal gas law reduces to Boyle's law, Charles's law, Gay-Lussac's law, or Avogadro's law when the right variables are held constant. For example, if n and T are constant, the product nRT is constant because R is also constant. In this case, the ideal gas law reduces to PV = a constant, which is Boyle's law.

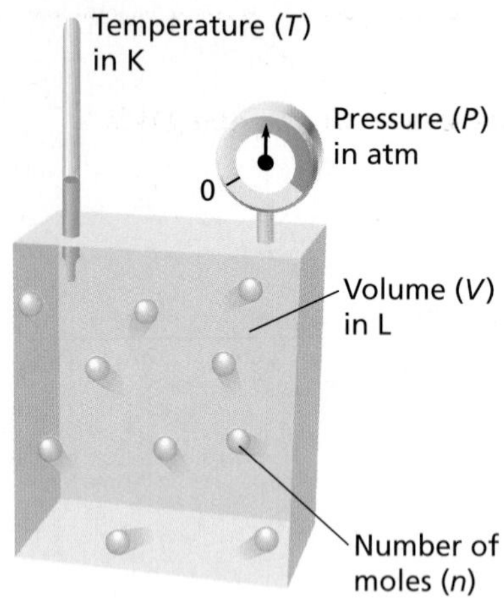

FIGURE 11-5 The ideal gas law relates the pressure, volume, number of moles, and temperature of an ideal gas. Note the value of the gas constant, R, when the other quantities are in the units shown.

The Ideal Gas Constant

In the equation representing the ideal gas law, *the constant R is known as the* **ideal gas constant.** Its value depends on the units chosen for pressure, volume, and temperature. Figure 11-5 shows that measured values of P, V, T, and n for a gas at near-ideal conditions can be used to calculate R. Recall from Section 11-1 that the volume of one mole of an ideal gas at STP (1 atm and 273.15 K) is 22.414 10 L. Substituting these values and solving the ideal gas law equation for R gives the following.

$$R = \frac{PV}{nT} = \frac{(1\ \text{atm})(22.414\ 10\ \text{L})}{(1\ \text{mol})(273.15\ \text{K})} = 0.082\ 057\ 84\ \frac{\text{L}\bullet\text{atm}}{\text{mol}\bullet\text{K}}$$

This calculated value of R is usually rounded to 0.0821 L•atm/(mol•K). Use this value in ideal gas law calculations when the volume is in liters, the pressure is in atmospheres, and the temperature is in kelvins. See Table 11-1 for the value of R when other units for n, P, V, and T are used.

Finding *P, V, T,* or *n* from the Ideal Gas Law

The ideal gas law can be applied to determine the existing conditions of a gas sample when three of the four variables, P, V, T, and n, are known. It can also be used to calculate the molar mass or density of a gas sample.

Be sure to match the units of the known quantities and the units of R. In this book, you will be using R = 0.0821 L•atm/(mol•K). Your first step in solving any ideal gas law problem should be to check the known values to be sure you are working with the correct units. If necessary, you must convert volumes to liters, pressures to atmospheres, temperatures to kelvins, and masses to numbers of moles before using the ideal gas law.

TABLE 11-1 *Numerical Values of the Gas Constant, R*

Units of *R*	Numerical value of *R*	Units of *P*	Units of *V*	Units of *T*	Units of *n*
$\frac{\text{L}\bullet\text{mm Hg}}{\text{mol}\bullet\text{K}}$	62.4	mm Hg	L	K	mol
$\frac{\text{L}\bullet\text{atm}}{\text{mol}\bullet\text{K}}$	0.0821	atm	L	K	mol
$\frac{\text{J}}{\text{mol}\bullet\text{K}}$	8.314*	atm	L	K	mol
$\frac{\text{L}\bullet\text{kPa}}{\text{mol}\bullet\text{K}}$	8.314**	kPa	L	K	mol

* 1 L•atm = 101.325 J ** SI units

SAMPLE PROBLEM 11-3

What is the pressure in atmospheres exerted by a 0.500 mol sample of nitrogen gas in a 10.0 L container at 298 K?

SOLUTION

1 *ANALYZE*

Given: V of N_2 = 10.0 L
n of N_2 = 0.500 mol
T of N_2 = 298 K
Unknown: P of N_2 in atm

2 *PLAN*

$$n, V, T \longrightarrow P$$

The gas sample undergoes no change in conditions. Therefore, the ideal gas law can be rearranged and used to find the pressure as follows.

$$P = \frac{nRT}{V}$$

3 *COMPUTE*

$$P = \frac{(0.500\ \cancel{\text{mol}})\left(\dfrac{0.0821\ \cancel{\text{L}} \cdot \text{atm}}{\cancel{\text{mol}} \cdot \cancel{\text{K}}}\right)(298\ \cancel{\text{K}})}{10.0\ \cancel{\text{L}}} = 1.22\ \text{atm}$$

4 *EVALUATE*

All units cancel correctly to give the result in atmospheres. The answer is properly limited to three significant figures. It is also close to an estimated value of 1.5, computed as $(0.5 \times 0.1 \times 300)/10$.

PRACTICE

1. What pressure, in atmospheres, is exerted by 0.325 mol of hydrogen gas in a 4.08 L container at 35°C?
   Answer 2.01 atm
2. A gas sample occupies 8.77 L at 20°C. What is the pressure, in atmospheres, given that there are 1.45 mol of gas in the sample?
   Answer 3.98 atm

SAMPLE PROBLEM 11-4

What is the volume, in liters, of 0.250 mol of oxygen gas at 20.0°C and 0.974 atm pressure?

SOLUTION

1 *ANALYZE*

Given: P of O_2 = 0.974 atm
n of O_2 = 0.250 mol
To use the value 0.0821 L•atm/(mol•K) for R, the temperature (°C) must be converted to kelvins.
T of O_2 = 20.0°C + 273 = 293 K
Unknown: V of O_2 in L

2 *PLAN*

$$P, n, T \longrightarrow V$$

The ideal gas law can be rearranged to solve for V for this sample, which undergoes no change in conditions.

$$V = \frac{nRT}{P}$$

3 *COMPUTE*

$$V = \frac{(0.250 \text{ mol } O_2)\left(\frac{0.0821 \text{ L} \cdot \text{atm}}{\text{mol} \cdot \text{K}}\right)(293 \text{ K})}{0.974 \text{ atm}} = 6.17 \text{ L } O_2$$

4 *EVALUATE* Units cancel to give liters, as desired. The answer is correctly limited to three significant digits. It is also close to an estimated value of 6, calculated as $0.2 \times 0.1 \times 300$.

PRACTICE

1. A sample that contains 4.38 mol of a gas at 250 K has a pressure of 0.857 atm. What is the volume? *Answer* 105 L
2. How many liters are occupied by 0.909 mol of nitrogen at 125°C and 0.901 atm pressure? *Answer* 33.0 L N_2

SAMPLE PROBLEM 11-5

What mass of chlorine gas, Cl_2, in grams, is contained in a 10.0 L tank at 27°C and 3.50 atm of pressure?

SOLUTION

1 *ANALYZE*

Given: P of Cl_2 = 3.50 atm
V of Cl_2 = 10.0 L
T of Cl_2 = 27°C + 273 = 300. K
Unknown: mass of Cl_2 in g

2 *PLAN* The ideal gas law can be rearranged to solve for n after temperature is converted to kelvins.

$$n = \frac{PV}{RT}$$

The number of moles is then converted to grams.

$$m \text{ (in g)} = n \times \frac{\text{g}}{\text{mol}}$$

3 *COMPUTE*

$$n = \frac{(3.50 \text{ atm})(10.0 \text{ L } Cl_2)}{\left(\frac{0.0821 \text{ L} \cdot \text{atm}}{\text{mol} \cdot \text{K}}\right)(300. \text{ K})} = 1.42 \text{ mol } Cl_2$$

$$\text{mass of } Cl_2 = 1.42 \text{ mol} \times \frac{70.90 \text{ g } Cl_2}{\text{mol}} = 101 \text{ g } Cl_2$$

4 *EVALUATE* Units cancel to leave the desired unit. The result is correctly given to three significant figures. The answer is close to the estimated value.

PRACTICE

1. How many grams of carbon dioxide gas are there in a 45.1 L container at 34°C and 1.04 atm? *Answer* 81.9 g CO_2

2. What is the mass, in grams, of oxygen gas in a 12.5 L container at 45°C and 7.22 atm? *Answer* 111 g O_2

3. A sample of carbon dioxide with a mass of 0.30 g was placed in a 250 mL container at 400. K. What is the pressure exerted by the gas? *Answer* 0.90 atm

Finding Molar Mass or Density from the Ideal Gas Law

Suppose that the pressure, volume, temperature, and mass are known for a gas sample. You can calculate the number of moles (n) in the sample, using the ideal gas law. Then you can calculate the molar mass (grams per mole) by dividing the known mass by the number of moles.

An equation showing the relationship between density, pressure, temperature, and molar mass can be derived from the ideal gas law. The number of moles (n) is equal to mass (m) divided by molar mass (M). Substituting m/M for n into $PV = nRT$ gives the following.

$$PV = \frac{mRT}{M} \text{ or } M = \frac{mRT}{PV}$$

Density (D) is mass (m) per unit volume (V). Writing this definition in the form of an equation gives $D = m/V$. You can see that m/V appears in the right-hand equation above. Introducing density into that equation gives the following.

$$M = \frac{mRT}{PV} = \frac{DRT}{P}$$

Solving for density gives this equation.

$$D = \frac{MP}{RT}$$

You can see that the density of a gas varies directly with molar mass and pressure and inversely with Kelvin temperature.

SAMPLE PROBLEM 11-6

At 28°C and 0.974 atm, 1.00 L of gas has a mass of 5.16 g. What is the molar mass of this gas?

SOLUTION

1 *ANALYZE*

Given: P of gas = 0.974 atm
V of gas = 1.00 L
T of gas = 28°C + 273 = 301 K
m of gas = 5.16 g

Unknown: M of gas in g/mol

2 PLAN

$$P, V, T, m \longrightarrow M$$

You can use the rearranged ideal gas law provided earlier in this section to find the answer.

$$M = \frac{mRT}{PV}$$

3 COMPUTE

$$M = \frac{(5.16\ \text{g})\left(\frac{0.0821\ \text{L} \cdot \text{atm}}{\text{mol} \cdot \text{K}}\right)(301\ \text{K})}{(0.974\ \text{atm})(1.00\ \text{L})} = 131\ \text{g/mol}$$

4 EVALUATE Units cancel as needed. The answer is correctly given to three significant digits. It is also close to an estimated value of 150, calculated as $(5 \times 0.1 \times 300)/1$.

PRACTICE

1. What is the molar mass of a gas if 0.427 g of the gas occupies a volume of 125 mL at 20.0°C and 0.980 atm? *Answer* 83.8 g/mol
2. What is the density of a sample of ammonia gas, NH_3, if the pressure is 0.928 atm and the temperature is 63.0°C? *Answer* 0.572 g/L NH_3
3. The density of a gas was found to be 2.0 g/L at 1.50 atm and 27°C. What is the molar mass of the gas? *Answer* 33 g/mol
4. What is the density of argon gas, Ar, at a pressure of 551 torr and a temperature of 25°C? *Answer* 1.18 g/L Ar

SECTION REVIEW

1. How does the ideal gas law reduce to the following:
   a. Boyle's law
   b. Charles's law
   c. Gay-Lussac's law
   d. Avogadro's law
2. What is the volume, in liters, of 0.100 g of $C_2H_2F_4$ vapor at 0.0928 atm and 22.3°C?
3. Why must the units P, V, T, and n match up to those of the ideal gas constant in solving problems? What would be the units for R if P is in pascals, T is in kelvins, V is in liters, and n is in moles?
4. What is the molar mass of a 1.25 g sample of gas that occupies a volume of 1.00 L at a pressure of 0.961 atm and a temperature of 27.0°C?
5. What other quantities besides pressure, mass, volume, and number of moles can be derived using the ideal gas law?

Stoichiometry of Gases

SECTION 11-3

Objectives

- Explain how Gay-Lussac's law and Avogadro's law apply to the volumes of gases in chemical reactions.
- Use a chemical equation to specify volume ratios for gaseous reactants or products, or both.
- Use volume ratios and the gas laws to calculate volumes, masses, or molar amounts of gaseous reactants or products.

You can apply the discoveries of Gay-Lussac and Avogadro to calculate the stoichiometry of reactions involving gases. For gaseous reactants or products, the coefficients in chemical equations not only indicate molar amounts and mole ratios but also reveal volume ratios. For example, consider the reaction of carbon monoxide with oxygen to give carbon dioxide.

$2CO(g)$	+ $O_2(g)$	$\longrightarrow 2CO_2(g)$
2 molecules	1 molecule	2 molecules
2 mol	1 mol	2 mol
2 volumes	1 volume	2 volumes

The possible volume ratios can be expressed in the following ways.

a. $\frac{\text{2 volumes CO}}{\text{1 volume } O_2}$ or $\frac{\text{1 volume } O_2}{\text{2 volumes CO}}$

b. $\frac{\text{2 volumes CO}}{\text{2 volumes } CO_2}$ or $\frac{\text{2 volumes } CO_2}{\text{2 volumes CO}}$

c. $\frac{\text{1 volume } O_2}{\text{2 volumes } CO_2}$ or $\frac{\text{2 volumes } CO_2}{\text{1 volume } O_2}$

Volumes can be compared in this way only if all are measured at the same temperature and pressure.

Volume-Volume Calculations

Suppose the volume of a gas involved in a reaction is known and you need to find the volume of another gaseous reactant or product, assuming both reactant and product exist under the same conditions. Use volume ratios like those given above in exactly the same way you would use mole ratios

SAMPLE PROBLEM 11-7

Propane, C_3H_8, is a gas that is sometimes used for cooking and heating. The complete combustion of propane occurs according to the following equation.

$$C_3H_8(g) + 5O_2(g) \longrightarrow 3CO_2(g) + 4H_2O(g)$$

(a) What will be the volume, in liters, of oxygen required for the complete combustion of 0.350 L of propane? (b) What will be the volume of carbon dioxide produced in the reaction? Assume that all volume measurements are made at the same temperature and pressure.

SOLUTION

1 ANALYZE **Given:** balanced chemical equation
V of propane = 0.350 L
Unknown: **a.** V of O_2 in L; **b.** V of CO_2 in L

2 PLAN **a.** V of $C_3H_8 \longrightarrow V$ of O_2; **b.** V of $C_3H_8 \longrightarrow V$ of CO_2

All volumes are to be compared at the same temperature and pressure. Therefore, volume ratios can be used like mole ratios to find the unknowns.

3 COMPUTE

$$\textbf{a. } 0.350\ \cancel{\text{L } C_3H_8} \times \frac{5 \text{ L } O_2}{1\ \cancel{\text{L } C_3H_8}} = 1.75 \text{ L } O_2$$

$$\textbf{b. } 0.350\ \cancel{\text{L } C_3H_8} \times \frac{3 \text{ L } CO_2}{1\ \cancel{\text{L } C_3H_8}} = 1.05 \text{ L } CO_2$$

4 EVALUATE Each result is correctly given to three significant figures. The answers are reasonably close to estimated values of 2, calculated as 0.4×5, and 1.2, calculated as 0.4×3, respectively.

PRACTICE

1. Assuming all volume measurements are made at the same temperature and pressure, what volume of hydrogen gas is needed to react completely with 4.55 L of oxygen gas to produce water vapor? *Answer* 9.10 L H_2

2. What volume of oxygen gas is needed to react completely with 0.626 L of carbon monoxide gas, CO, to form gaseous carbon dioxide? Assume all volume measurements are made at the same temperature and pressure. *Answer* 0.313 L O_2

Volume-Mass and Mass-Volume Calculations

Stoichiometric calculations may involve both masses and gas volumes. Sometimes the volume of a reactant or product is given and the mass of a second gaseous substance is unknown. In other cases, a mass amount may be known and a volume may be the unknown. The calculations require routes such as the following.

gas volume A $\longrightarrow$ moles A $\longrightarrow$ moles B $\longrightarrow$ mass B

or

mass A $\longrightarrow$ moles A $\longrightarrow$ moles B $\longrightarrow$ gas volume B

To find the unknown in cases like these, you must know the conditions under which both the known and unknown gas volumes have been measured. The ideal gas law is useful for calculating values at standard and nonstandard conditions.

SAMPLE PROBLEM 11-8

Calcium carbonate, $CaCO_3$, also known as limestone, can be heated to produce calcium oxide (lime), an industrial chemical with a wide variety of uses. The balanced equation for the reaction follows.

$$CaCO_3(s) \xrightarrow{\Delta} CaO(s) + CO_2(g)$$

How many grams of calcium carbonate must be decomposed to produce 5.00 L of carbon dioxide gas at STP?

SOLUTION

1 ANALYZE

Given: balanced chemical equation
desired volume of CO_2 produced at STP = 5.00 L

Unknown: mass of $CaCO_3$ in grams

2 PLAN

The known volume is given at STP. This tells us the pressure and temperature. The ideal gas law can be used to find the moles of CO_2. The mole ratios from the balanced equation can then be used to calculate the moles of $CaCO_3$ needed. (Note that volume ratios do not apply here because calcium carbonate is a solid.)

3 COMPUTE

$$n = \frac{PV}{RT} = \frac{(1\ \cancel{\text{atm}})(5.00\ \cancel{\text{L}}\ CO_2)}{\left(\frac{0.0821\ \cancel{\text{L}} \cdot \cancel{\text{atm}}}{\text{mol} \cdot \cancel{\text{K}}}\right)(273\ \cancel{\text{K}})} = 0.223\ \text{mol}\ CO_2$$

$$0.223\ \cancel{\text{mol}\ CO_2} \times \frac{1\ \cancel{\text{mol}\ CaCO_3}}{1\ \cancel{\text{mol}\ CO_2}} \times \frac{100.09\ \text{g}\ CaCO_3}{1\ \cancel{\text{mol}\ CaCO_3}} = 22.3\ \text{g}\ CaCO_3$$

4 EVALUATE

Units all cancel correctly. The answer is properly given to three significant figures. It is close to an estimated value of 20, computed as $(0.2 \times 100)/25$.

PRACTICE

1. What mass of sulfur must be used to produce 12.61 L of gaseous sulfur dioxide at STP according to the following equation? *Answer* 18.0 g S

$$S(s) + O_2(g) \longrightarrow SO_2(g)$$

2. How many grams of water can be produced from the complete reaction of 3.44 L of oxygen gas, at STP, with hydrogen gas? *Answer* 5.53 g H_2O

SAMPLE PROBLEM 11-9

Tungsten, W, a metal used in light-bulb filaments, is produced industrially by the reaction of tungsten oxide with hydrogen.

$$WO_3(s) + 3H_2(g) \longrightarrow W(s) + 3H_2O(l)$$

How many liters of hydrogen gas at 35°C and 0.980 atm are needed to react completely with 875 g of tungsten oxide?

SOLUTION

1 ANALYZE **Given:** balanced chemical equation
reactant mass of WO_3 = 875 g
P of H_2 = 0.980 atm
T of H_2 = 35°C + 273 = 308 K

Unknown: V of H_2 in L at known nonstandard conditions

2 PLAN Moles of H_2 are found by converting the mass of WO_3 to moles and then using the mole ratio. The ideal gas law is used to find the volume from the calculated number of moles of H_2.

3 COMPUTE

$$875 \text{ g } WO_3 \times \frac{1 \text{ mol } WO_3}{231.84 \text{ g } WO_3} \times \frac{3 \text{ mol } H_2}{1 \text{ mol } WO_3} = 11.3 \text{ mol } H_2$$

$$V = \frac{nRT}{P} = \frac{(11.3 \text{ mol } H_2)\left(\frac{0.0821 \text{ L} \cdot \text{atm}}{\text{mol} \cdot \text{K}}\right)(308 \text{ K})}{0.980 \text{ atm}} = 292 \text{ L } H_2$$

4 EVALUATE Unit cancellations are correct, as is the use of three significant figures for each result. The answer is reasonably close to an estimated value of 330, computed as (11 × 0.1 × 300)/1.

PRACTICE

1. What volume of chlorine gas at 38°C and 1.63 atm is needed to react completely with 10.4 g of sodium to form NaCl? *Answer* 3.54 L Cl_2
2. How many liters of gaseous carbon monoxide at 27°C and 0.247 atm can be produced from the burning of 65.5 g of carbon according to the following equation? *Answer* 544 L CO

$$2C(s) + O_2(g) \longrightarrow 2CO(g)$$

SECTION REVIEW

1. How many liters of ammonia gas can be formed from the reaction of 150. L of hydrogen gas? Assume that there is complete reaction of hydrogen with excess nitrogen gas and that all measurements are made at the same temperature and pressure.
2. How many liters of H_2 gas at STP can be produced by the reaction of 4.60 g of Na and excess water, according to the following equation?

$$2Na(s) + H_2O(l) \longrightarrow H_2(g) + Na_2O(aq)$$

3. How many grams of Na are needed to react with H_2O to liberate 4.00×10^2 mL of H_2 gas at STP?
4. What volume of oxygen gas in liters can be collected at 0.987 atm pressure and 25.0°C when 30.6 g of $KClO_3$ decompose by heating, according to the following equation?

$$2KClO_3(s) \xrightarrow[MnO_2]{\Delta} 2KCl(s) + 3O_2(g)$$

Effusion and Diffusion

SECTION 11-4

OBJECTIVES

- State Graham's law of effusion.
- Determine the relative rates of effusion of two gases of known molar masses.
- State the relationship between the molecular velocities of two gases and their molar masses.

The constant motion of gas molecules causes them to spread out to fill any container in which they are placed. As you learned in Chapter 10, the gradual mixing of two gases due to their spontaneous, random motion is known as *diffusion,* illustrated in Figure 11-6. *Effusion* is the process whereby the molecules of a gas confined in a container randomly pass through a tiny opening in the container. In this section, you will learn how effusion can be used to estimate the molar mass of a gas.

Graham's Law of Effusion

The rates of effusion and diffusion depend on the relative velocities of gas molecules. The velocity of a gas varies inversely with its mass. Lighter molecules move faster than heavier molecules at the same temperature.

Recall that the average kinetic energy of the molecules in any gas depends only on the temperature and equals $\frac{1}{2}mv^2$. For two different gases, A and B, at the same temperature, the following relationship is true.

$$\frac{1}{2}M_Av_A{}^2 = \frac{1}{2}M_Bv_B{}^2$$

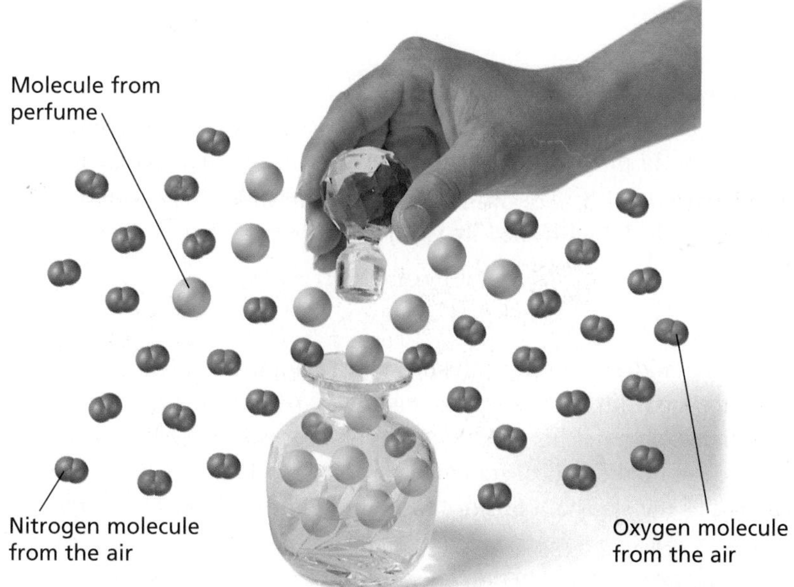

FIGURE 11-6 When a bottle of perfume is opened, some of its molecules diffuse into the air and mix with the molecules in the air. At the same time, molecules from the air, such as nitrogen and oxygen, diffuse into the bottle and mix with the gaseous scent molecules.

FIGURE 11-7 Molecules of nitrogen and oxygen from the air inside the bicycle tire effuse out through a small nail hole.

M_A and M_B represent the molar masses of gases A and B, and v_A and v_B represent their molecular velocities. Multiplying the equation by 2 gives the following.

$$M_A v_A^2 = M_B v_B^2$$

Suppose you wanted to compare the velocities of the two gases. You would first rearrange the equation above to give the velocities as a ratio.

$$\frac{v_A^2}{v_B^2} = \frac{M_B}{M_A}$$

The square root of each side of the equation is then taken.

$$\frac{v_A}{v_B} = \frac{\sqrt{M_B}}{\sqrt{M_A}}$$

This equation shows that the molecular velocities of two different gases are inversely proportional to the square roots of their molar masses. Because the rates of effusion are directly proportional to molecular velocities, the equation can be written as follows.

$$\frac{\text{rate of effusion of A}}{\text{rate of effusion of B}} = \frac{\sqrt{M_B}}{\sqrt{M_A}}$$

In the mid-1800s, the Scottish chemist Thomas Graham studied the effusion and diffusion of gases. Figure 11-7 illustrates the process of effusion. Compare this with the diffusion process. The equation derived above is a mathematical statement of some of Graham's discoveries. It describes the rates of effusion. **Graham's law of effusion** *states that the rates of effusion of gases at the same temperature and pressure are inversely proportional to the square roots of their molar masses.*

DESKTOP INVESTIGATION

Diffusion

MATERIALS

- **household ammonia**
- **perfume or cologne**
- **two 250 mL beakers**
- **two watch glasses**
- **10 mL graduated cylinder**
- **clock or watch with second hand**

QUESTION

How rapidly do gases of different molecular mass diffuse?

PROCEDURE

Record all of your results in a data table.

1. Outdoors or in a room separate from the one in which you will carry out the rest of the investigation, pour approximately 10 mL of the household ammonia into one of the 250 mL beakers, and cover it with a watch glass. Pour roughly the same amount of perfume or cologne into the second beaker. Cover it with a watch glass also.
2. Take the two samples you just prepared into a large, draft-free room. Place the samples about 12 to 15 feet apart and at the same height. Position someone as the observer midway between the two beakers. Remove both watch-glass covers at the same time.
3. Note whether the observer smells the ammonia or the perfume first. Record how long this takes. Also, record how long it takes the vapor of the other substance to reach them. Air the room after you have finished.

DISCUSSION

1. What do the times that the two vapors took to reach the observer show about their diffusion rates?
2. The molar mass of ammonia is about one-tenth the molar mass of most of the fragrant substances in perfume or cologne. What does this tell you about the relationship between molar mass and diffusion rate?

Applications of Graham's Law

Graham's experiments dealt with the densities of gases. The density of a gas varies directly with its molar mass. Therefore, the square roots of the molar masses in the equation on page 352 can be replaced by the square roots of the gas densities, giving the following relationship.

$$\frac{\text{rate of effusion of A}}{\text{rate of effusion of B}} = \frac{\sqrt{M_B}}{\sqrt{M_A}} = \frac{\sqrt{\text{density}_B}}{\sqrt{\text{density}_A}}$$

The experiment illustrated in Figure 11-8 demonstrates the relationship between distance traveled and molar mass. In the experiment, ammonia gas molecules, NH_3, and hydrogen chloride gas molecules, HCl, diffuse toward each other from opposite ends of a glass tube. A white ring forms at the point where the two gases meet and combine chemically. The white ring is solid ammonium chloride, NH_4Cl. Notice that it forms closer to the HCl end of the tube than to the NH_3 end. The lighter NH_3 molecules (molar mass 17.04 g) diffuse faster than the heavier HCl molecules (molar mass 36.46 g). This is shown by the longer distance traveled by the NH_3 molecules before the two gases meet.

Although this experiment shows that the lighter molecules diffuse faster than the heavier ones, the diffusion of the two gases in the tube does take longer than expected, based on the known velocities of NH_3 and HCl. This is because the experiment does not take place in a vacuum. The oxygen and nitrogen molecules present in the air collide with the diffusing ammonia and hydrogen chloride molecules, slowing them down. To describe all these random collisions taking place in this system would be very complicated. This is why we use these equations to represent the *relative* rates of effusion based on the *relative* velocities of the molecules.

Graham's law also provides a method for determining molar masses. The rates of effusion of gases of known and unknown molar mass can be compared at the same temperature and pressure. The unknown molar mass can then be calculated using Graham's law. One application of Graham's law was used in order to separate the heavier ${}^{238}_{92}U$ isotope from the lighter ${}^{235}_{92}U$ isotope. The uranium was converted to a gas and passed through porous membranes, where the isotopes diffused at different rates due to their different densities and were thereby separated.

FIGURE 11-8 Cotton plugs moistened with ammonia and hydrogen chloride were placed at opposite ends of the glass tube several minutes before this photograph was taken. Why does the white ring of ammonium chloride form closer to the right end than the left end?

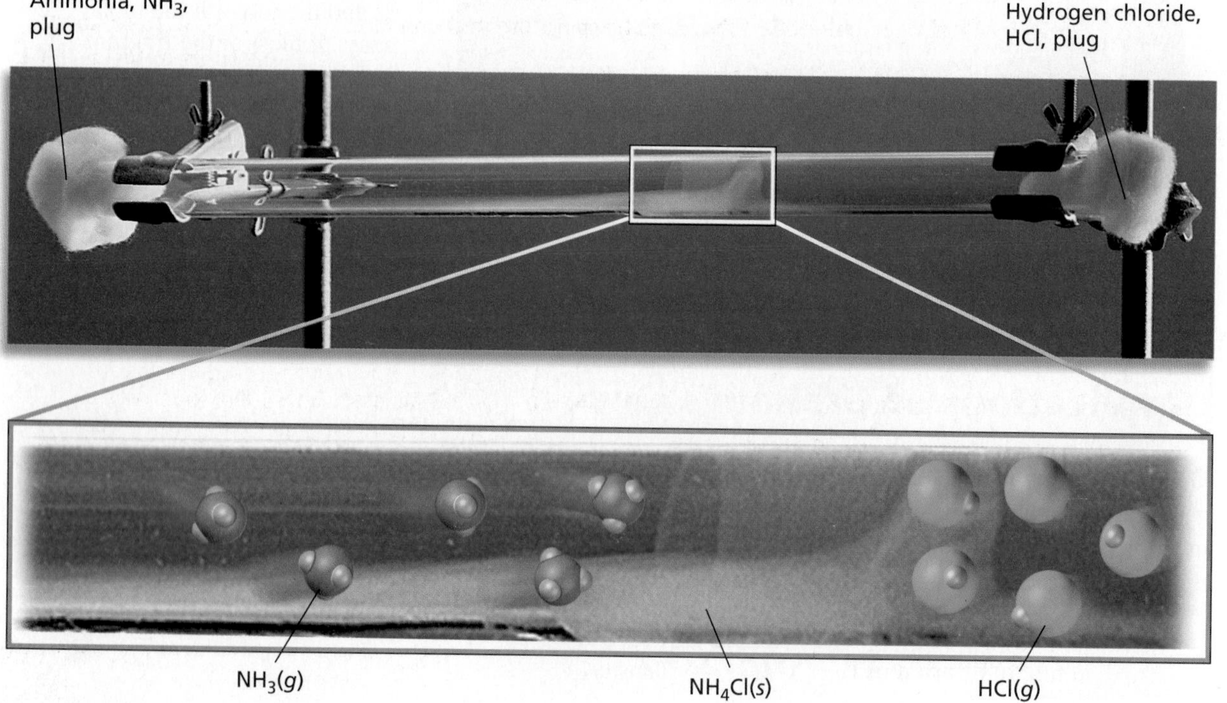

SAMPLE PROBLEM 11-10

Compare the rates of effusion of hydrogen and oxygen at the same temperature and pressure.

SOLUTION

1 *ANALYZE* **Given:** identities of two gases, H_2 and O_2

Unknown: relative rates of effusion

2 *PLAN*

$$\text{molar mass ratio} \longrightarrow \text{ratio of rates of effusion}$$

The ratio of the rates of effusion of two gases at the same temperature and pressure can be found from Graham's law.

$$\frac{\text{rate of effusion of A}}{\text{rate of effusion of B}} = \frac{\sqrt{M_B}}{\sqrt{M_A}}$$

3 *COMPUTE*

$$\frac{\text{rate of effusion of } H_2}{\text{rate of effusion of } O_2} = \frac{\sqrt{M_{O_2}}}{\sqrt{M_{H_2}}} = \frac{\sqrt{32}}{\sqrt{2}} = \sqrt{\frac{32}{2}} = \sqrt{\frac{16}{1}} = 4$$

Hydrogen effuses 4 times faster than oxygen.

4 *EVALUATE* The result is correctly limited to one significant figure. It is also equivalent to an estimated value of 4, calculated as $\sqrt{32} / \sqrt{2}$.

PRACTICE

1. A sample of hydrogen effuses through a porous container about 9 times faster than an unknown gas. Estimate the molar mass of the unknown gas.

   Answer
   160 g/mol

2. Compare the rate of effusion of carbon dioxide with that of hydrogen chloride at the same temperature and pressure.

   Answer
   CO_2 will effuse about 0.9 times as fast as HCl.

3. If neon gas travels at 400 m/s at a given temperature, estimate the rate of diffusion of butane gas, C_4H_{10}, at the same temperature.

   Answer
   Butane will diffuse at about 235 m/s.

SECTION REVIEW

1. Compare diffusion with effusion.
2. State Graham's law of effusion, and explain its meaning. Why should the equations derived from this law be used to find *relative* rates of effusion and molecular velocities?
3. Estimate the molar mass of a gas that effuses at 1.6 times the effusion rate of carbon dioxide.
4. Do all of the molecules in a 1 mol sample of oxygen gas have the same energy at 273 K?
5. List the following gases in order of increasing average molecular velocity at 25°C: H_2O, He, HCl, BrF, and NO_2.

CHAPTER 11 REVIEW

CHAPTER SUMMARY

11-1

- Gay-Lussac's law of combining volumes of gases states that the volumes of reacting gases and their products at the same temperature and pressure can be expressed as ratios of small whole numbers.
- Avogadro's law states that equal volumes of gases at the same temperature and pressure contain equal numbers of molecules. At constant temperature and pressure, gas volume is thus directly proportional to the number of moles.
- Gay-Lussac's law and Avogadro's law can be used to show that molecules of the reactive elemental gases are diatomic.
- The volume occupied by one mole of any gas at STP is called the standard molar volume. The standard molar volume of gases is 22.414 10 L at standard temperature and pressure.

Vocabulary

Avogadro's law (334) Gay-Lussac's law of combining volumes of gases (333) standard molar volume of a gas (335)

11-2

- Charles's law, Boyle's law, and Avogadro's law can be combined to create the ideal gas law. The ideal gas law is stated mathematically as follows.

$$PV = nRT$$

- The value and units of the ideal gas constant depend on the units of the variables used in the ideal gas law.
- The ideal gas law can be used to calculate gas pressure, volume, temperature, or number of moles when three of these four variables are known and the gas sample does not undergo a change in conditions.
- The ideal gas law can also be used to calculate the density or molar mass of a gas sample.

Vocabulary

ideal gas constant (342) ideal gas law (340)

11-3

- Given the volume of a reacting gas, the volume of another gaseous product or reactant can be calculated using their volume ratios, as long as reactants and products exist under the same conditions.
- Given the volume of a reactant or product gas, the mass of a second reactant or product can be calculated using the ideal gas law and mole-to-mass conversion factors.
- If the mass of a substance is known, the ideal gas law and the proper mass-to-mole conversion factors can be used to calculate the volume of a gaseous product or reactant.

11-4

- Graham's law of effusion states that the relative rates of effusion of gases at the same temperature and pressure are inversely proportional to the square roots of their molar masses.
- Graham's law reflects the fact that less massive molecules effuse faster than do more massive ones.
- Graham's law can be used to compare the rates of effusion of gases at the same temperature and pressure.
- Given the relative rates of effusion of two gases and the identity of one of them, Graham's law can be used to estimate the molar mass of the other.

Vocabulary

Graham's law of effusion (352)

REVIEWING CONCEPTS

1. a. What restrictions are there on the use of Gay-Lussac's law of combining volumes?
   b. At the same temperature and pressure, what is the relationship between the volume of a gas and the number of molecules present? (11-1)
2. According to Avogadro,
   a. what is the relationship between gas volume and number of moles at constant temperature and pressure?
   b. what is the mathematical expression denoting this relationship? (11-1)
3. What is the relationship between the number of molecules and the mass of 22.4 L of different gases at STP? (11-1)
4. Why must the temperature and pressure be specified when stating gas density values? (11-1)
5. a. Write the equation for the ideal gas law.
   b. What relationship is expressed in the ideal gas law? (11-2)
6. a. In what situation does the ideal gas law apply?
   b. Why do you have to pay particular attention to units when using this law? (11-2)
7. a. In a balanced chemical equation, what is the relationship between the molar ratios and the volume ratios of gaseous reactants and products?
   b. What restriction applies to the use of the volume ratios in solving stoichiometry problems? (11-3)
8. a. Distinguish between diffusion and effusion.
   b. At a given temperature, what factor determines the rates at which different molecules undergo these processes? (11-4)

PROBLEMS

Molar Volume and Gas Density

9. Suppose a 5.00 L sample of O_2 at a given temperature and pressure contains 1.08×10^{23} molecules. How many molecules would be contained in each of the following at the same temperature and pressure?
   a. 5.00 L H_2
   b. 5.00 L CO_2
   c. 10.00 L NH_3
10. How many molecules are contained in each of the following?
    a. 1.00 mol O_2
    b. 2.50 mol He
    c. 0.0650 mol NH_3
    d. 11.5 g NO_2
11. Find the mass of each of the following.
    a. 2.25 mol Cl_2
    b. 3.01×10^{23} molecules H_2S
    c. 25.0 molecules SO_2
12. What is the volume, in liters, of each of the following at STP? (Hint: See Sample Problem 11-1.)
    a. 1.00 mol O_2
    b. 3.50 mol F_2
    c. 0.0400 mol CO_2
    d. 1.20×10^{-6} mol He
13. How many moles are contained in each of the following at STP?
    a. 22.4 L N_2
    b. 5.60 L Cl_2
    c. 0.125 L Ne
    d. 70.0 mL NH_3
14. Find the mass, in grams, of each of the following at STP. (Hint: See Sample Problem 11-2.)
    a. 11.2 L H_2
    b. 2.80 L CO_2
    c. 15.0 mL SO_2
    d. 3.40 cm^3 F_2
15. Find the volume, in liters, of each of the following at STP.
    a. 8.00 g O_2
    b. 3.50 g CO
    c. 0.0170 g H_2S
    d. 2.25×10^5 kg NH_3

Ideal Gas Law

16. Calculate the pressure, in atmospheres, exerted by each of the following. (Hint: See Sample Problem 11-3.)
    a. 2.50 L of HF containing 1.35 mol at 320. K
    b. 4.75 L of NO_2 containing 0.86 mol at 300. K
    c. 7.50×10^2 mL of CO_2 containing 2.15 mol at 57°C

17. Calculate the volume, in liters, occupied by each of the following. (Hint: See Sample Problem 11-4.)
    a. 2.00 mol of H_2 at 300. K and 1.25 atm
    b. 0.425 mol of NH_3 at 37°C and 0.724 atm
    c. 4.00 g of O_2 at 57°C and 0.888 atm

18. Determine the number of moles of gas contained in each of the following.
    a. 1.25 L at 250. K and 1.06 atm
    b. 0.80 L at 27°C and 0.925 atm
    c. 7.50×10^2 mL at –50.°C and 0.921 atm

19. Find the mass of each of the following. (Hint: See Sample Problem 11-5.)
    a. 5.60 L of O_2 at 1.75 atm and 250. K
    b. 3.50 L of NH_3 at 0.921 atm and 27°C
    c. 125 mL of SO_2 at 0.822 atm and –53°C

20. Find the molar mass of each gas measured at the specified conditions. (Hint: See Sample Problem 11-6.)
    a. 0.650 g occupying 1.12 L at 280. K and 1.14 atm
    b. 1.05 g occupying 2.35 L at 37°C and 0.840 atm
    c. 0.432 g occupying 7.50×10^2 mL at –23°C and 1.03 atm

21. If the density of an unknown gas is 3.20 g/L at –18°C and 2.17 atm, what is the molar mass of this gas?

22. One method of estimating the temperature of the center of the sun is based on the assumption that the center consists of gases that have an average molar mass of 2.00 g/mol. If the density of the center of the sun is 1.40 g/cm^3 at a pressure of 1.30×10^9 atm, calculate the temperature in degrees Celsius.

Gas Stoichiometry

23. Carbon monoxide reacts with oxygen to produce carbon dioxide. If 1.0 L of carbon monoxide reacts with oxygen,
    a. how many liters of oxygen are required? (Hint: See Sample Problem 11-7.)
    b. how many liters of carbon dioxide are produced?

24. Acetylene gas, C_2H_2, undergoes combustion to produce carbon dioxide and water vapor. If 75.0 L of CO_2 are produced,
    a. how many liters of C_2H_2 are required?
    b. what volume of H_2O vapor is produced?
    c. what volume of O_2 is required?

25. If liquid carbon disulfide reacts with 4.50×10^2 mL of oxygen to produce the gases carbon dioxide and sulfur dioxide, what volume of each product is produced?

26. Balance the following chemical equation.
    $Mg(s) + O_2(g) \longrightarrow MgO(s)$
    Then, based on the quantity of reactant or product given, determine the corresponding quantities of the specified reactants or products, assuming that the system is at STP.
    a. 22.4 L O_2 = ___ mol $O_2 \longrightarrow$ ___ mol MgO
    b. 11.2 L O_2 = ___ mol $O_2 \longrightarrow$ ___ mol MgO
    c. 1.40 L O_2 = ___ mol $O_2 \longrightarrow$ ___ mol MgO

27. Assume that 5.60 L of H_2 at STP react with CuO according to the following equation:
    $CuO(s) + H_2(g) \longrightarrow Cu(s) + H_2O(g)$
    Make sure the equation is balanced before beginning your calculations.
    a. How many moles of H_2 react? (Hint: See Sample Problem 11-8.)
    b. How many moles of Cu are produced?
    c. How many grams of Cu are produced?

28. Assume that 8.50 L of I_2 are produced using the following reaction that takes place at STP:
    $KI(aq) + Cl_2(g) \longrightarrow KCl(aq) + I_2(g)$
    Balance the equation before beginning your calculations.
    a. How many moles of I_2 are produced?
    b. How many moles of KI were used?
    c. How many grams of KI were used?

29. Solid iron(III) hydroxide decomposes to produce iron(III) oxide and water vapor. If 0.75 L of water vapor is produced at STP,
    a. how many grams of iron(III) hydroxide were used?
    b. how many grams of iron(III) oxide are produced?

30. Suppose that 6.50×10^2 mL of hydrogen gas are produced through a replacement reaction involving solid iron and sulfuric acid, H_2SO_4, at STP. How many grams of iron(II) sulfate are also produced?

31. If 29.0 L of methane, CH_4, undergoes complete combustion at 0.961 atm and 20°C, how many liters of each product are formed?

32. If air is 20.9% oxygen by volume,
a. how many liters of air are needed for complete combustion of 25.0 L of octane vapor, C_8H_{18}?
b. what volume of each product is produced?

33. Methanol, CH_3OH, is made by causing carbon monoxide and hydrogen gases to react at high temperature and pressure. If 4.50×10^2 mL of CO and 825 mL of H_2 are mixed,
a. which reactant is present in excess?
b. how much of that reactant remains after the reaction?
c. what volume of CH_3OH is produced, assuming the same pressure?

34. Assume that 13.5 g of Al react with HCl according to the following equation, at STP:
$Al(s) + HCl(aq) \longrightarrow AlCl_3(aq) + H_2(g)$
Remember to balance the equation first.
a. How many moles of Al react?
b. How many moles of H_2 are produced?
c. How many liters of H_2 at STP are produced? (Hint: See Sample Problem 11-9.)

35. A modified Haber process for making ammonia is conducted at 550.°C and 2.50×10^2 atm. If 10.0 kg of nitrogen is used, what volume of ammonia is produced?

36. When liquid nitroglycerin, $C_3H_5(NO_3)_3$, explodes, the products are carbon dioxide, nitrogen, oxygen, and water vapor. If 5.00×10^2 g of nitroglycerin explode at STP, what is the total volume, at STP, for all gases produced?

37. The principal source of sulfur on Earth is deposits of free sulfur occurring mainly in volcanically active regions. The sulfur was initially formed by the reaction between the two volcanic vapors SO_2 and H_2S to form $H_2O(l)$ and $S(s)$. What volume of each gas, at 0.961 atm and 22°C, was needed to form a sulfur deposit of 4.50×10^5 kg on the slopes of a volcano in Hawaii?

38. A 3.25 g sample of solid calcium carbide, CaC_2, reacted with water to produce acetylene gas, C_2H_2, and aqueous calcium hydroxide. If the acetylene was collected over water at 17°C and 0.974 atm, how many milliliters of acetylene were produced?

Effusion and Diffusion

39. Quantitatively compare the rates of effusion for the following pairs of gases at the same temperature and pressure.
a. hydrogen and nitrogen (Hint: See Sample Problem 11-10.)
b. fluorine and chlorine

40. What is the ratio of the velocity of hydrogen molecules to that of neon atoms at the same temperature and pressure?

41. At a certain temperature and pressure, chlorine molecules have a velocity of 0.0380 m/s. What is the velocity of sulfur dioxide molecules under the same conditions?

42. A sample of helium effuses through a porous container 6.50 times faster than does unknown gas X. What is the molar mass of the unknown gas?

MIXED REVIEW

43. Use the ideal gas law, $PV = nRT$, to derive Boyle's law and Charles's law.

44. A container holds 265 mL of chlorine gas, Cl_2. Assuming that the gas sample is at STP, what is its mass?

45. Suppose that 3.11 mol of carbon dioxide is at a pressure of 0.820 atm and a temperature of 39°C. What is the volume of the sample, in liters?

46. Compare the rates of diffusion of carbon monoxide, CO, and sulfur trioxide, SO_3.

47. A gas sample that has a mass of 0.993 g occupies 0.570 L. Given that the temperature is 281 K and the pressure is 1.44 atm, what is the molar mass of the gas?

48. The density of a gas is 3.07 g/L at STP. Calculate the gas's molar mass.

49. How many moles of helium gas would it take to fill a gas balloon with a volume of 1000. cm^3 when the temperature is 32°C and the atmospheric pressure is 752 mm Hg?

50. A gas sample is collected at 16°C and 0.982 atm. If the sample has a mass of 7.40 g and a volume of 3.96 L, find the volume of the gas at STP and the molar mass.

51. An unknown gas effuses at 0.850 times the effusion rate of nitrogen dioxide, NO_2. Estimate the molar mass of the unknown gas.

CRITICAL THINKING

52. Evaluating Methods In solving a problem, what types of conditions involving temperature, pressure, volume, or number of moles would allow you to use
a. the combined gas law?
b. the ideal gas law?

53. Relating Ideas Write expressions relating the rates of effusion, molar masses, and densities of two different gases, A and B.

54. Evaluating Ideas Gay-Lussac's law of combining volumes holds true for relative volumes at any proportionate size. Use Avogadro's law to explain why this proportionality exists.

55. Designing Experiments Design an experiment to prove that the proportionality described in item 54 exists.

56. Interpreting Concepts The diagrams that follow represent equal volumes of four different gases.

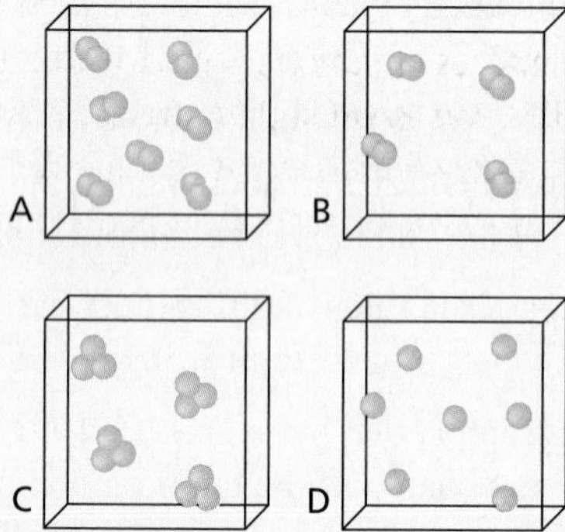

Use the diagrams to answer the following questions:
a. Are these gases at the same temperature and pressure? How do you know?
b. If the molar mass of gas B is 38 g/mol and that of gas C is 46 g/mol, which gas sample is more dense?
c. To make the densities of gas samples B and C equal, which gas should expand in volume?
d. If the densities of gas samples A and C are equal, what is the relationship between their molecular masses?

TECHNOLOGY & LEARNING

57. Graphing Calculator Calculating Pressure Using the Ideal Gas Law.

The graphing calculator can be programmed to calculate the pressure in atmospheres, given the number of moles of a gas (n), volume (V), and temperature (T). Given a 0.50 mol gas sample with a volume of 10. L at 298 K, you can calculate the pressure according to the ideal gas law. Begin by programming the calculator to carry out the calculation. Next, use the program to make calculations.

A. Program the calculation.
Keystrokes for naming the program: [PRGM] [▶] [▶] [ENTER]
Name the program: [I] [D] [E] [A] [L] [ENTER]
Program keystrokes: [PRGM] [▶] [2] [ALPHA] [N] [2nd] [:] [PRGM] [▶] [2] [ALPHA] [V] [2nd] [:] [PRGM] [▶] [2] [ALPHA] [T] [2nd] [:] [ALPHA] [N] [×] [ALPHA] [T] [×] [.] [0] [8] [2] [1] [÷] [ALPHA] [V] [STO▶] [ALPHA] [P] [2nd] [:] [PRGM] [▶] [8] [2nd] [:] [PRGM] [▶] [3] [2nd] [ALPHA] ["] [P] [R] [E] [S] [S] [U] [R] [E] [2nd] [TEST] [1] [ALPHA] ["] [2nd] [:] [PRGM] [▶] [3] [ALPHA] [P]

B. Check the display.
Screen display: Prompt N:Prompt V:Prompt T: N*T*.0821/V->P:ClrHome:Disp "PRESSURE=":Disp P
Press [2nd] [QUIT] to exit the program editor.

C. Run the program.
a. Press [PRGM]. Select IDEAL and press [ENTER] [ENTER]. Enter the mass of 0.50 mol and press [ENTER]. Enter the volume of 10. L and press [ENTER]. Enter the temperature of 298 K and press [ENTER]. The calculator will provide the pressure according to the ideal gas law.
b. Use the IDEAL program to find the pressure in atmospheres of a gas sample that has a mass of 0.25 mol and a volume of 6.0 L at 293 K.
c. Use the IDEAL program to find the pressure in atmospheres of a gas sample that has a mass of 1.3 mol and a volume of 8.0 L at 300. K.

HANDBOOK SEARCH

58. Most elements from Groups 1, 2, and 13 will react with water, acids, or bases to produce hydrogen gas. Review the common reactions information in the *Elements Handbook* and answer the following:

a. Write the equation for the reaction of barium with water.

b. Write the equation for the reaction between cesium and hydrochloric acid.

c. Write the equation for the reaction of gallium with hydrofluoric acid.

d. What mass of barium would be needed to react with excess water to produce 10.1 L of H_2 at STP?

e. What masses of cesium and hydrochloric acid would be required to produce 10.1 L of H_2 at STP?

59. Group 1 metals react with oxygen to produce oxides, peroxides, or superoxides. Review the equations for these common reactions in the *Elements Handbook* and answer the following:

a. How do oxides, peroxides, and superoxides differ?

b. What mass of product will be formed from a reaction of 5.00 L of oxygen with excess sodium? The reaction occurs at 27°C and 1 atm.

60. Most metals react with chlorine from Group 17 to produce salts. Review these common reactions in the *Elements Handbook* and answer the following:

a. Write the equation for the reaction of iron and chlorine gas.

b. What mass of iron salt is produced if an excess of iron reacts with 450. mL of chlorine gas at 27°C and 1 atm?

RESEARCH & WRITING

61. How do scuba divers use the laws and principles that describe the behavior of gases to their advantage? What precautions do they take to prevent the bends?

62. Explain the processes involved in the liquefaction of gases. What substances that are gases under normal room conditions are typically used in the liquid form? Why?

63. Research the relationship between explosives and the establishment of Nobel Prizes. Prepare a report that describes your findings.

64. Write a summary describing how Gay-Lussac's work on combining volumes relates to Avogadro's study of gases. Explain how certain conclusions about gases followed logically from consideration of the work of both scientists.

ALTERNATIVE ASSESSMENT

65. During a typical day, record every instance in which you encounter the diffusion or effusion of gases (for example, smelling perfume).

66. **Performance** Qualitatively compare the molecular masses of various gases by noting how long it takes you to smell them from a fixed distance. Work only with materials that are not dangerous, such as flavor extracts, fruit peels, and onions.

CHAPTER 12

Liquids and Solids

Courtesy of the artist and Galerie Lelong, New York

The total three-dimensional arrangement of particles of a crystal is its crystal structure.

Liquids

SECTION 12-1

Objectives

- Describe the motion of particles in liquids and the properties of liquids according to the kinetic-molecular theory.
- Discuss the process by which liquids can change into a gas. Define *vaporization.*
- Discuss the process by which liquids can change into a solid. Define *freezing.*

The water in the waves crashing on a beach and the molten lava rushing down the sides of a volcano are examples of matter in the liquid state. When you think of Earth's oceans, lakes, and rivers and the many liquids you use every day, it is hard to believe that liquids are the *least* common state of matter in the universe. Liquids are less common than solids, gases, and plasmas because a substance in the liquid state can exist only within a relatively narrow range of temperatures and pressures.

In this section, you will examine the properties of the liquid state. You will also compare them with those of the solid state and the gas state. These properties will be discussed in terms of the kinetic-molecular theory.

Properties of Liquids and the Kinetic-Molecular Theory

A liquid can be described as a form of matter that has a definite volume and takes the shape of its container. The properties of liquids can be understood by applying the kinetic-molecular theory, considering the motion and arrangement of molecules and the attractive forces between them.

As in a gas, particles in a liquid are in constant motion. However, the particles in a liquid are closer together and lower in kinetic energy than those in a gas. The attractive forces between particles in a liquid are more effective than those between particles in a gas. This attraction between liquid particles is caused by the intermolecular forces discussed in Chapter 6: dipole-dipole forces, London dispersion forces, and hydrogen bonding.

Liquids are more ordered than gases because of the stronger intermolecular forces and the lower mobility of the liquid particles. According to the kinetic-molecular theory of liquids, the particles are not bound together in fixed positions. Instead, they move about constantly. This particle mobility explains why liquids and gases are referred to as fluids. *A* **fluid** *is a substance that can flow and therefore take the shape of its container.* Most liquids naturally flow downhill because of gravity. However, some liquids can flow in other directions as well. For example, liquid helium near absolute zero has the unusual property of being able to flow uphill.

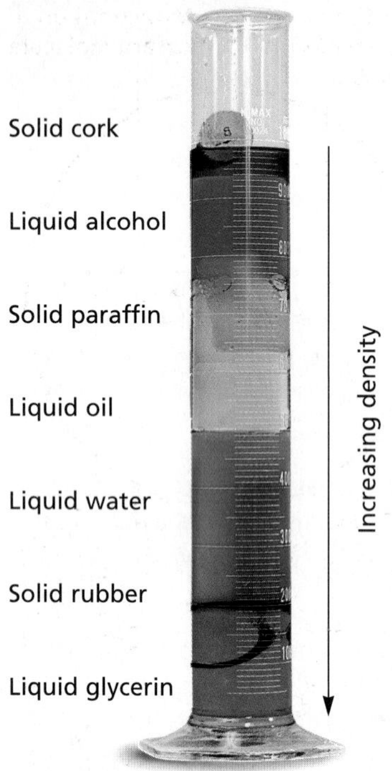

FIGURE 12-1 Solids and liquids of different densities are shown. The densest materials are at the bottom. The least dense are at the top. (Dyes have been added to the liquids to make the layers more visible.)

Relatively High Density

At normal atmospheric pressure, most substances are thousands of times denser as liquids than as gases. This higher density is a result of the close arrangement of liquid particles. Most substances are only slightly less dense (about 10%) as liquids than as solids, however. Water is one of the few substances that becomes less dense when it solidifies, as will be discussed further in Section 12-4.

At the same temperature and pressure, different liquids can differ greatly in density. Figure 12-1 shows some liquids and solids with different densities. The densities differ to such an extent that the liquids form layers.

Relative Incompressibility

When liquid water at 20°C is compressed by a pressure of 1000 atm, its volume decreases by only 4%. Such behavior is typical of all liquids and is similar to the behavior of solids. In contrast, a gas under a pressure of 1000 atm would have only about 1/1000 of its volume at normal atmospheric pressure. Liquids are much less compressible than gases because liquid particles are more closely packed together. In addition, liquids can transmit pressure equally in all directions.

Ability to Diffuse

As described in Chapter 10, gases diffuse and mix with other gas particles. Liquids also diffuse and mix with other liquids, as shown in Figure 12-2. Any liquid gradually diffuses throughout any other liquid in which it can dissolve. As in gases, diffusion in liquids occurs because of the constant, random motion of particles. Yet diffusion is much slower in liquids than in gases because liquid particles are closer together. Also, the attractive forces between the particles of a liquid slow their movement. As the temperature of a liquid is increased, diffusion occurs more rapidly. That is because the average kinetic energy, and therefore the average speed of the particles, is increased.

FIGURE 12-2 Like gases, the two liquids in this beaker diffuse over time. The green liquid food coloring from the drop will eventually form a uniform solution with the water.

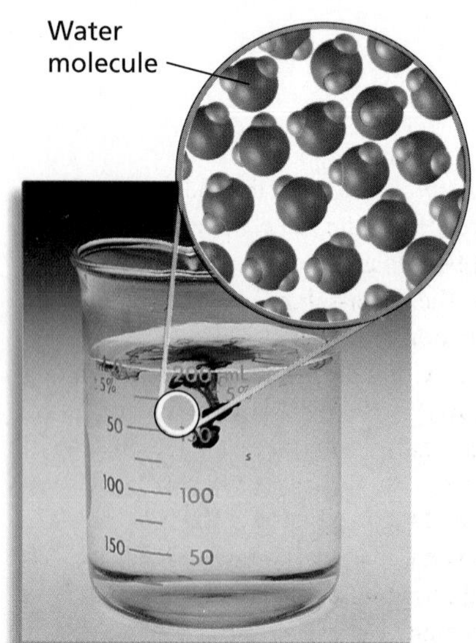

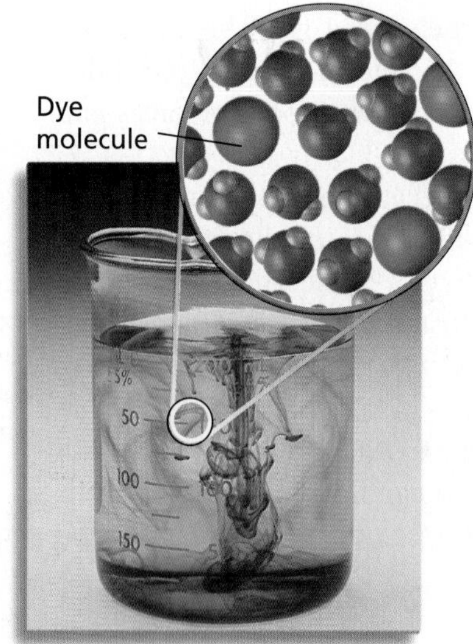

Surface Tension

A property common to all liquids, **surface tension,** *is a force that tends to pull adjacent parts of a liquid's surface together, thereby decreasing surface area to the smallest possible size.* Surface tension results from the attractive forces between particles of a liquid. The higher the force of attraction, the higher the surface tension. Water has a higher surface tension than most liquids. This is due to the hydrogen bonds water molecules can form with each other. The molecules at the surface of the water are a special case. They can form hydrogen bonds with the other water molecules beneath them and beside them, but not with the molecules in the air above them. As a result, the surface water molecules are drawn together and toward the body of the liquid, creating a high surface tension. Surface tension causes liquid droplets to take on a spherical shape because a sphere has the smallest possible surface area for a given volume. An example of this phenomenon is shown in Figure 12-3.

Capillary action, *the attraction of the surface of a liquid to the surface of a solid,* is a property closely related to surface tension. A liquid will rise quite high in a very narrow tube if a strong attraction exists between the liquid molecules and the molecules that make up the surface of the tube. This attraction tends to pull the liquid molecules upward along the surface against the pull of gravity. This process continues until the weight of the liquid balances the gravitational force. Capillary action can occur between water molecules and paper fibers, as shown in Figure 12-4. Capillary action is at least partly responsible for the transportation of water from the roots of a plant to its leaves. The same process is responsible for the concave liquid surface, called a *meniscus,* that forms in a test tube or graduated cylinder.

FIGURE 12-3 As a result of surface tension, liquids form roughly spherical drops. The net attractive forces between the particles pull the molecules on the surface inward and closer together than those farther down in the liquid, minimizing the surface area.

Evaporation and Boiling

The process by which a liquid or solid changes to a gas is **vaporization.** Evaporation is a form of vaporization. **Evaporation** *is the process by which particles escape from the surface of a nonboiling liquid and enter the gas state.*

(a)

(b)

FIGURE 12-4 The attraction between polar water molecules and polar cellulose molecules in paper fibers causes the water to move up in the paper. The water-soluble ink placed near the bottom of the paper in (a) rises up the paper along with the water, as seen in (b). As the ink moves up the paper, it is separated into its various components, producing the different bands of color. This separation occurs because the water and the paper attract the molecules of the ink components differently. These phenomena are used in the separation process of paper chromatography seen here.

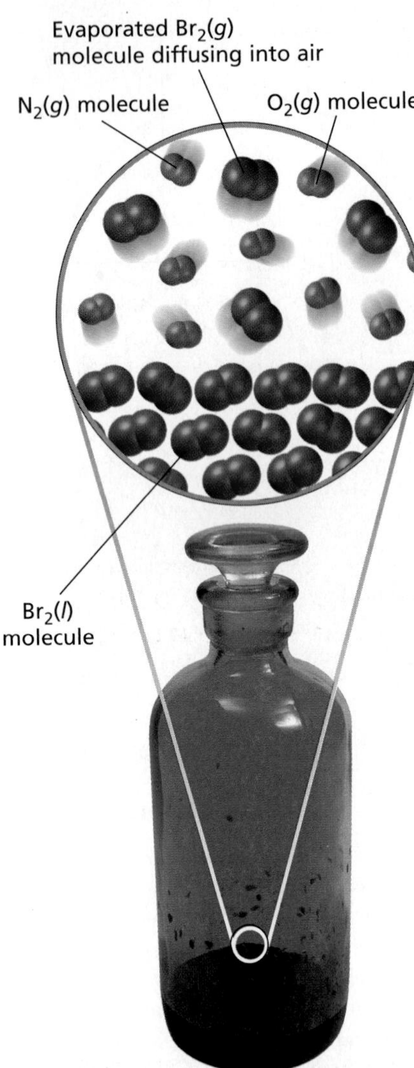

FIGURE 12-5 Liquid bromine, Br_2, evaporates near room temperature. The resulting brownish red gas diffuses into the air above the surface of the liquid.

A small amount of liquid bromine was added to the bottle shown in Figure 12-5. Within a few minutes, the air above the liquid bromine turned brownish-red. That is because some bromine molecules escaped from the surface of the liquid. They have changed into the gas state, becoming bromine vapor, which mixed with the air. A similar phenomenon occurs if you apply perfume to your wrist. Within seconds, you become aware of the perfume's fragrance. Scent molecules evaporate from your skin and diffuse through the air, to be detected by your nose.

Evaporation occurs because the particles of a liquid have different kinetic energies. Particles with higher-than-average energies move faster. Some surface particles with higher-than-average energies can overcome the intermolecular forces that bind them to the liquid. They can then escape into the gas state.

Evaporation is a crucial process in nature. Evaporation removes fresh water from the surface of the ocean, leaving behind a higher concentration of salts. In subtropical areas, evaporation occurs at a higher rate, causing the surface water to be saltier. All water that falls to Earth in the form of rain and snow previously evaporated from oceans, lakes, and rivers. Evaporation of perspiration plays an important role in keeping you cool. Perspiration, which is mostly water, cools you by absorbing body heat when it evaporates. Heat energy is absorbed from the skin, causing the cooling effect.

Boiling is the change of a liquid to bubbles of vapor that appear throughout the liquid. Boiling differs from evaporation, as you will see in Section 12-3.

Formation of Solids

When a liquid is cooled, the average energy of its particles decreases. If the energy is low enough, attractive forces pull the particles into an even more orderly arrangement. The substance then becomes a solid. *The physical change of a liquid to a solid by removal of heat is called* **freezing** *or solidification.* Perhaps the best-known example of freezing is the change of liquid water to solid water, or ice, at 0°C. Another familiar example is the solidification of paraffin at room temperature. All liquids freeze, although not necessarily at temperatures you normally encounter. Ethanol, for example, freezes near –115°C.

SECTION REVIEW

1. Describe the liquid state according to the kinetic-molecular theory.
2. List the properties of liquids.
3. How does the kinetic-molecular theory explain the following properties of liquids? (a) their relatively high density, (b) their ability to diffuse, and (c) their ability to evaporate
4. Explain why liquids in a test tube form a meniscus.
5. Compare vaporization and evaporation.

Solids

SECTION 12-2

Objectives

- Describe the motion of particles in solids and the properties of solids according to the kinetic-molecular theory.
- Distinguish between the two types of solids.
- Describe the different types of crystal symmetry. Define *crystal structure* and *unit cell.*

"Solid as a rock" is a common expression that suggests something that is hard or unyielding and that has a definite shape and volume. In this section, you will examine the properties of solids and compare them with those of liquids and gases. As with the other states of matter, the properties of solids are explained in terms of the kinetic-molecular theory.

Properties of Solids and the Kinetic-Molecular Theory

The particles of a solid are more closely packed than those of a liquid or gas. Intermolecular forces between particles are therefore much more effective in solids. Dipole-dipole attractions, London dispersion forces, and hydrogen bonding exert stronger effects in solids than in the corresponding liquids or gases. These forces tend to hold the particles of a solid in relatively fixed positions, with only vibrational movement around fixed points. Because the motions of the particles are restricted in this way, solids are more ordered than liquids and are much more ordered than gases. The importance of order and disorder in physical and chemical changes will be discussed in Chapter 17. Compare the physical appearance and molecular arrangement of the element in Figure 12-6 in solid, liquid, and gas form.

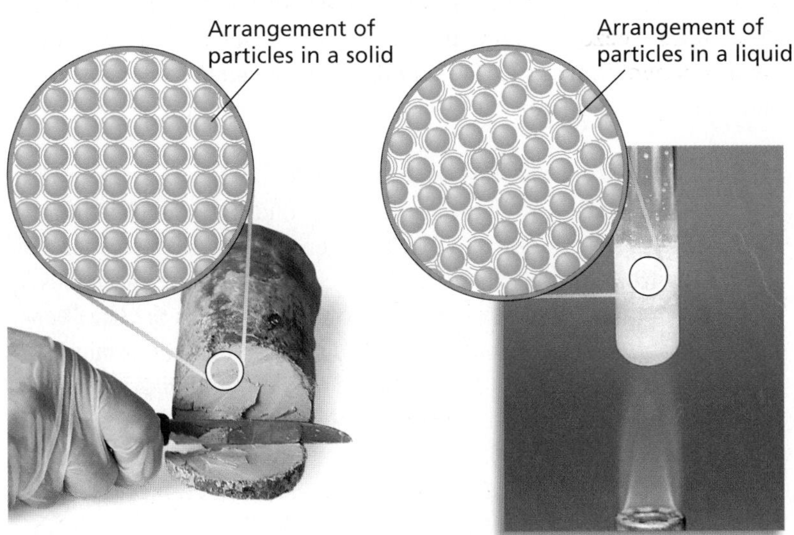

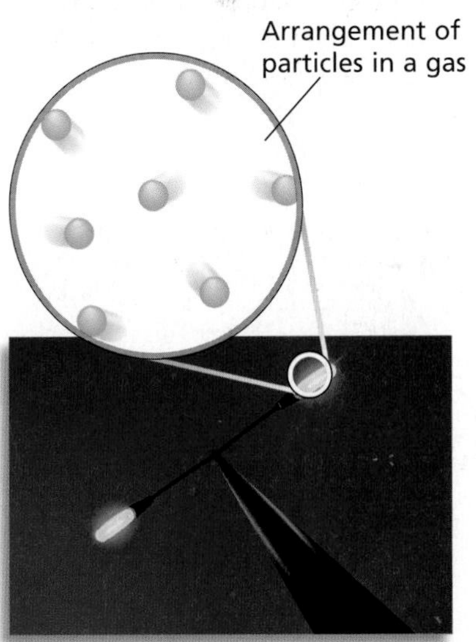

FIGURE 12-6 Particles of sodium metal in three different states are shown. Sodium exists in a gaseous state in a sodium-vapor lamp.

There are two types of solids: crystalline solids and amorphous solids. *Most solids are* **crystalline solids***—they consist of crystals. A* **crystal** *is a substance in which the particles are arranged in an orderly, geometric, repeating pattern.* Noncrystalline solids, including glass and plastics, are called amorphous solids. *An* **amorphous solid** *is one in which the particles are arranged randomly.* The two types of solids will be discussed in more detail later in this section.

(a)

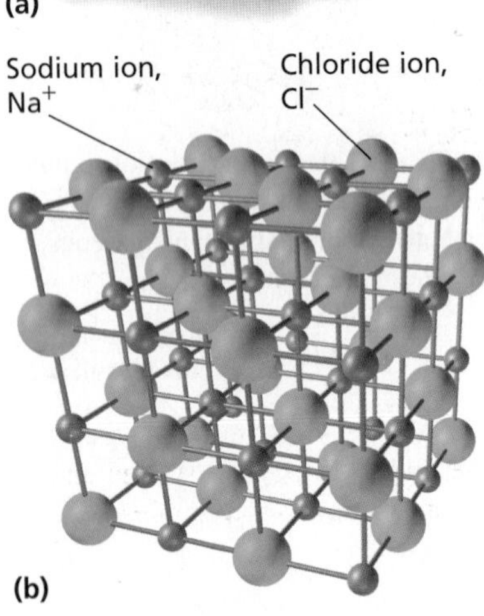

(b)

FIGURE 12-7 (a) This is a scanning electron micrograph (SEM) of a sodium chloride crystal. A sodium chloride crystal can be represented by its crystal structure (b), which is made up of individual unit cells represented regularly in three dimensions. Here, one unit cell is outlined in red.

Definite Shape and Volume

Unlike liquids and gases, solids can maintain a definite shape without a container. In addition, crystalline solids are geometrically regular. Even the fragments of a shattered crystalline solid have distinct geometric shapes that reflect their internal structure. Amorphous solids maintain a definite shape, but they do not have the distinct geometric shapes of crystalline solids. For example, glass can be molded into any shape. If it is shattered, glass fragments can have a wide variety of irregular shapes.

The volume of a solid changes only slightly with a change in temperature or pressure. Solids have definite volume because their particles are packed closely together. There is very little empty space into which the particles can be compressed. Crystalline solids generally do not flow because their particles are held in relatively fixed positions.

Definite Melting Point

Melting *is the physical change of a solid to a liquid by the addition of heat. The temperature at which a solid becomes a liquid is its* **melting point.** At this temperature, the kinetic energies of the particles within the solid overcome the attractive forces holding them together. The particles can then break out of their positions in crystalline solids, which have definite melting points. In contrast, amorphous solids, such as glass and plastics, have no definite melting point. They have the ability to flow over a range of temperatures. Therefore, amorphous solids are sometimes classified as **supercooled liquids,** *which are substances that retain certain liquid properties even at temperatures at which they appear to be solid.* These properties exist because the particles in amorphous solids are arranged randomly, much like the particles in a liquid. Unlike the particles in a true liquid, however, the particles in amorphous solids are not constantly changing their positions.

High Density and Incompressibility

In general, substances are most dense in the solid state. Solids tend to be slightly denser than liquids and much denser than gases. The higher density results from the fact that the particles of a solid are more closely packed than those of a liquid or a gas. Solid hydrogen is the least dense solid; it has a density of about 1/320 of the densest element, osmium, Os.

Solids are generally less compressible than liquids. For practical purposes, solids can be considered incompressible. Some solids, such as wood and cork, may *seem* compressible, but they are not. They contain pores that are filled with air. When subjected to intense pressure, the pores are compressed, not the wood or cork itself.

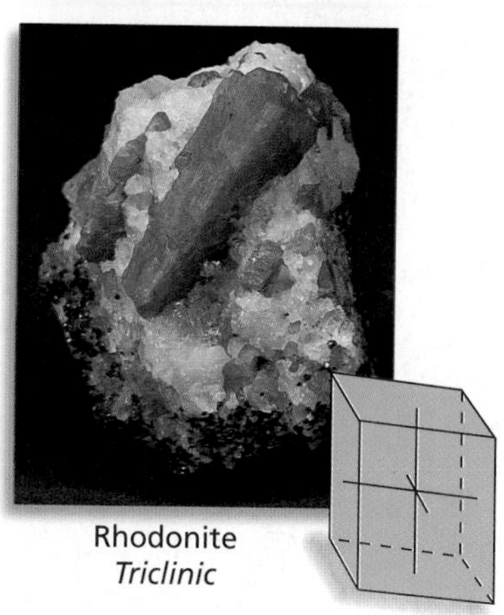

FIGURE 12-8 Shown are the seven basic crystalline systems and representative minerals of each.

Low Rate of Diffusion

If a zinc plate and a copper plate are clamped together for a long time, a few atoms of each metal will diffuse into the other. This observation shows that diffusion does occur in solids. The rate of diffusion is millions of times slower in solids than in liquids, however.

Crystalline Solids

Crystalline solids exist either as single crystals or as groups of crystals fused together. *The total three-dimensional arrangement of particles of a crystal is called a* **crystal structure.** The arrangement of particles in the crystal can be represented by a coordinate system called a lattice. *The smallest portion of a crystal lattice that shows the three-dimensional pattern of the entire lattice is called a* **unit cell.** Each crystal lattice contains many unit cells packed together. Figure 12-7 shows the relationship between a crystal lattice and its unit cell. A crystal and its unit cells can have any one of seven types of symmetry. This fact enables scientists to classify crystals by their shape. Diagrams and examples of each type of crystal symmetry are shown in Figure 12-8.

Binding Forces in Crystals

Crystal structures can also be described in terms of the types of particles in them and the types of chemical bonding between the particles.

TABLE 12-1 ***Melting and Boiling Points of Representative Crystalline Solids***

Type of substance	Formula	Melting point (°C)	Boiling point at 1 atm (°C)
Ionic	NaCl	801	1413
	MgF_2	1266	2239
Covalent network	$(SiO_2)_x$	1610	2230
	C_x (diamond)	3500	3930
Metallic	Hg	−39	357
	Cu	1083	2567
	Fe	1535	2750
	W	3410	5660
Covalent molecular (nonpolar)	H_2	−259	−253
	O_2	−218	−183
	CH_4	−182	−164
	CCl_4	−23	77
	C_6H_6	6	80
Covalent molecular (polar)	NH_3	−78	−33
	H_2O	0.	100.

According to this method of classification, there are four types of crystals. These types are listed in Table 12-1. Refer to this table as you read the following discussion.

1. *Ionic crystals.* As discussed in Chapter 6, the ionic crystal structure consists of positive and negative ions arranged in a regular pattern. The ions can be monatomic or polyatomic. Generally, ionic crystals form when Group-1 or Group-2 metals combine with Group-16 or Group-17 nonmetals or nonmetallic polyatomic ions. The strong binding forces between the positive and negative ions in the crystal structure give the ionic crystals certain properties. For example, these crystals are hard and brittle, have high melting points, and are good insulators.
2. *Covalent network crystals.* In covalent network crystals, the sites contain single atoms. Each atom is covalently bonded to its nearest neighboring atoms. The covalent bonding extends throughout a network that includes a very large number of atoms. Three-dimensional covalent network solids include diamond, C_x, quartz, $(SiO_2)_x$—shown in Figure 12-9—silicon carbide, $(SiC)_x$, and many oxides of transition metals. Such solids are essentially giant molecules. The subscript x in these formulas indicates that the component within the parentheses extends indefinitely. The network solids are nearly always very hard and brittle. They have rather high melting points and are usually nonconductors or semiconductors.
3. *Metallic crystals.* As discussed in Chapter 6, the metallic crystal structure consists of metal atoms surrounded by a sea of valence electrons. The electrons are donated by the metal atoms and belong to

the crystal as a whole. The freedom of the outer-structure electrons to move throughout the crystal explains the high electric conductivity of metals. As you can see from Table 12-1, the melting points of different metallic crystals vary greatly.

4. *Covalent molecular crystals.* The crystal structure of a covalent molecular substance consists of covalently bonded molecules held together by intermolecular forces. If the molecules are nonpolar—for example, hydrogen, H_2, methane, CH_4, and benzene, C_6H_6—then there are only weak London dispersion forces between molecules. In a polar covalent molecular crystal—for example, water, H_2O, and ammonia, NH_3—molecules are held together by dispersion forces, by somewhat stronger dipole-dipole forces, and sometimes by even stronger hydrogen bonding. The forces that hold polar or nonpolar molecules together in the structure are much weaker than the covalent chemical bonds between the atoms within each molecule. Covalent molecular crystals thus have low melting points. They are easily vaporized, are relatively soft, and are good insulators. Ice crystals, the most familiar molecular crystals, are discussed in Section 12-4.

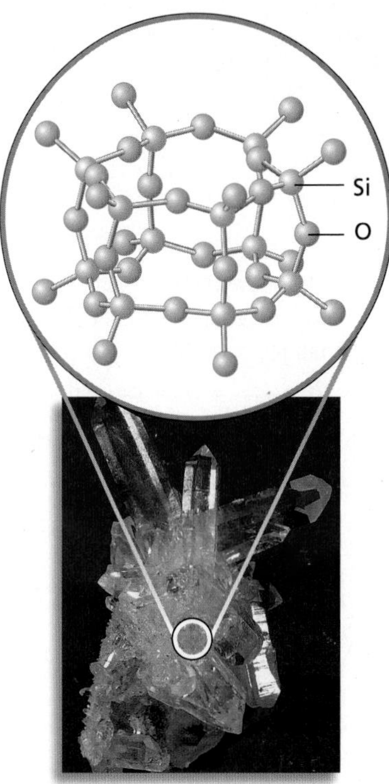

FIGURE 12-9 Covalent network crystals include three-dimensional network solids, such as this quartz, $(SiO_2)_x$, shown here with its three-dimensional atomic structure.

Amorphous Solids

The word *amorphous* comes from the Greek for "without shape." Unlike crystals, amorphous solids do not have a regular, natural shape.

Amorphous solids tend to hold their shape for a long time. However, some amorphous solids do flow, although usually very slowly. Some samples of very old window glass are thicker at the bottom than at the top suggesting that the glass has flowed downward over the years. Glasses make up a distinctive group of amorphous solids. They are made by cooling certain molten materials in such a way that they do not crystallize but remain amorphous.

There are hundreds of types of plastic and glass, all of which have thousands of important applications. Glass is used for everything from fiberglass automobile bodies to optical fibers that use light to transmit telephone conversations.

SECTION REVIEW

1. Describe the solid state according to the kinetic-molecular theory.
2. What is the difference between an amorphous solid and a crystalline solid?
3. Account for each of the following properties of solids: (a) the definite volume, (b) the relatively high density of solids, (c) the extremely low rate of diffusion.
4. Compare and contrast the four types of crystals.

SECTION 12-3

Changes of State

OBJECTIVES

- Explain the relationship between equilibrium and changes of state.
- Predict changes in equilibrium using Le Châtelier's principle.
- Explain what is meant by equilibrium vapor pressure.
- Describe the processes of boiling, freezing, melting, and sublimation.
- Interpret phase diagrams.

Matter on Earth can exist in any of these states—gas, liquid, or solid—and can change from one state to another. Table 12-2 lists the possible changes of state. In this section, you will examine these changes of state and the factors that determine them.

Equilibrium

Equilibrium *is a dynamic condition in which two opposing changes occur at equal rates in a closed system.* In a closed system, matter cannot enter or leave, but energy can. Both matter and energy can escape or enter an open system. For example, sunlight, heat from a burner, or cooling by ice can cause energy to enter or leave a system.

For an analogy of equilibrium, think of a freeway during morning rush hour. As people travel to work and school, more and more vehicles enter the freeway from side roads. This causes slow-moving traffic and can eventually cause traffic jams if large numbers of vehicles enter. The system at this point is not in equilibrium because more vehicles are entering the freeway than leaving it. As the morning rush passes, the traffic will begin to move faster, with fewer delays, because an increasing number of vehicles are leaving the freeway. When about the same number of vehicles are leaving the freeway as are entering, a fairly constant number of vehicles will be traveling on the freeway, and the system will be in equilibrium, with traffic flowing smoothly.

Equilibrium is a very important chemical concept. Here you will learn about it in relation to changes of state. In Chapter 18, you will study equilibrium in terms of chemical reactions.

TABLE 12-2 ***Possible Changes of State***

Change of state	Process	Example
Solid ⟶ liquid	melting	ice ⟶ water
Solid ⟶ gas	sublimation	dry ice ⟶ CO_2 gas
Liquid ⟶ solid	freezing	water ⟶ ice
Liquid ⟶ gas	vaporization	liquid bromine ⟶ bromine vapor
Gas ⟶ liquid	condensation	water vapor ⟶ water
Gas ⟶ solid	deposition	water vapor ⟶ ice

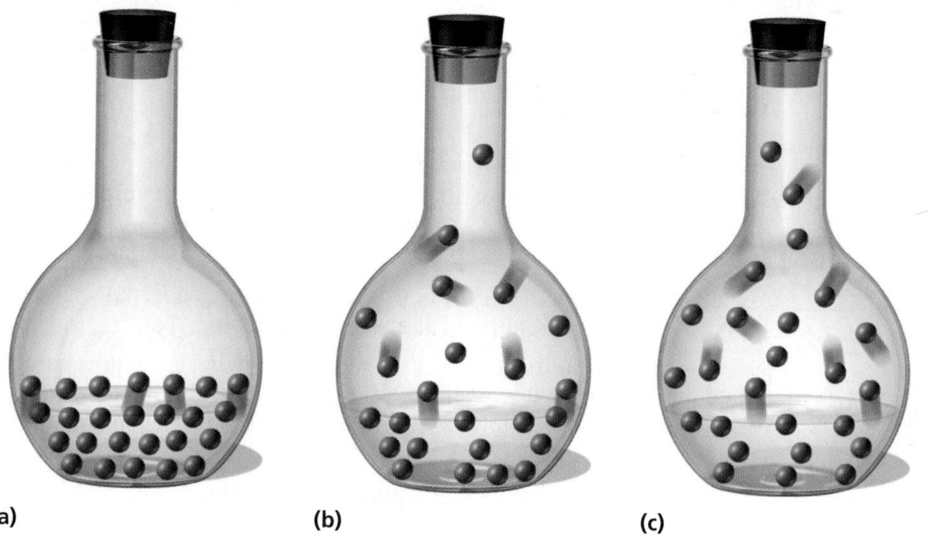

FIGURE 12-10 A liquid-vapor equilibrium develops in a closed system. (a) At first there is only liquid present, but molecules are beginning to evaporate. (b) Evaporation continues at a constant rate. Some vapor molecules are beginning to condense to liquid. (c) Equilibrium has been reached between the rate of condensation and the rate of evaporation.

Equilibrium and Changes of State

Consider the evaporation of water in a closed container in which there is initially a vacuum over the liquid, as illustrated in Figure 12-10. Assume that the contents of the container maintain the same temperature as the surroundings, approximately 25°C. The water and the container are the system. Initially, there is a single phase inside this system—the liquid phase. *A* **phase** *is any part of a system that has uniform composition and properties.*

If the energy of a water molecule at the upper surface of the liquid is high enough, the molecule can overcome the attraction of neighboring molecules, leave the liquid phase, and evaporate. Water molecules that have entered the vapor phase behave as typical gas molecules. Some of these vapor molecules move down toward the liquid surface and condense. That is, they reenter the liquid phase. **Condensation** *is the process by which a gas changes to a liquid.*

If the temperature and surface area of the liquid remain constant, the rate at which its molecules enter the vapor phase remains constant. The rate at which the water molecules pass from the vapor phase to the liquid phase depends on the concentration of molecules in the vapor phase. The concentration of vapor molecules, and therefore the rate of condensation, is initially zero, as shown in Figure 12-10(a). As time passes and evaporation continues, the concentration of vapor molecules increases. This increase in concentration results in an increase in the condensation rate. However, the rate of condensation is still lower than the rate of evaporation, as shown in Figure 12-10(b). Eventually, the concentration of vapor increases to the point at which the rate of condensation becomes equal to the rate of evaporation, as shown in Figure 12-10(c). In other words, equilibrium is reached and the amounts of liquid and of vapor now remain constant.

An Equilibrium Equation

Whenever a liquid changes to a vapor, it absorbs heat energy from its surroundings. Evaporation can therefore be represented by the following equation.

$$\text{liquid} + \text{heat energy} \longrightarrow \text{vapor}$$

Whenever a vapor condenses, it gives off heat energy to its surroundings. Condensation can therefore be represented by the following equation.

$$\text{vapor} \longrightarrow \text{liquid} + \text{heat energy}$$

The liquid-vapor equilibrium can be represented by the following equation.

$$\text{liquid} + \text{heat energy} \rightleftarrows \text{vapor}$$

The "double-yields" sign in the equation above represents a reversible reaction. It means that the reaction can proceed in either direction. The forward reaction is represented when the equation is read from left to right (liquid + heat energy $\longrightarrow$ vapor). The reverse reaction is represented when the equation is read from right to left (vapor $\longrightarrow$ liquid + heat energy).

Le Châtelier's Principle

A system will remain at equilibrium until something occurs to change this condition. It is important to understand the factors that can be used to control the equilibrium of a system. In 1888, the French chemist Henri Louis Le Châtelier developed a principle that enables us to predict how a change in conditions will affect a system at equilibrium. **Le Châtelier's principle** can be stated as follows: *When a system at equilibrium is disturbed by application of a stress, it attains a new equilibrium position that minimizes the stress.* A stress is typically a change in concentration, pressure, or temperature.

Equilibrium and Temperature

You can use Le Châtelier's principle to predict how the liquid-vapor equilibrium changes when a stress is applied in the closed container of liquid discussed earlier. For example, suppose the temperature of the system is increased from 25°C to 50°C. The equilibrium of the system can be represented by the following reversible reaction.

$$\text{liquid} + \text{heat energy} \rightleftarrows \text{vapor}$$

According to Le Châtelier's principle, the system will respond to an increase in temperature. In this case, the forward reaction is endothermic;

that is, it absorbs heat energy. The forward reaction is favored to counteract a rise in temperature and to minimize the stress applied to it (in this case, the change in temperature). The forward reaction proceeds at a higher rate than the reverse reaction until a new equilibrium is reached. At 50°C, the concentration of vapor is higher than it was at 25°C. However, at equilibrium, the reverse reaction, condensation, also occurs at a higher rate than it did at the lower temperature.

Suppose, on the other hand, that the temperature of the system in equilibrium at 25°C is lowered to 5°C. According to Le Châtelier's principle, the system will adjust to counteract the temperature decrease. Now the reverse reaction is favored because it is exothermic; that is, it releases heat energy. Equilibrium is shifted to the left and reestablished at 5°C. The vapor concentration is now lower than it originally was at 25°C.

Equilibrium and Concentration

Suppose the mass and temperature of this equilibrium system remain the same, but the volume suddenly increases. What happens to the equilibrium? Initially, the concentration of molecules in the vapor phase decreases because the same number of vapor-phase molecules now occupies a larger volume. Because the concentration of vapor molecules decreases, fewer vapor molecules strike the liquid surface per second and return to the liquid phase. Therefore, the rate of condensation decreases. The rate of evaporation, however, has remained the same and is therefore higher now than the rate of condensation. As a result, an additional net evaporation of liquid occurs. When a new equilibrium is finally reestablished, the concentration of molecules in the vapor phase is the same as in the lower-volume system. Because there was a net movement of molecules from the liquid phase to the vapor phase, the number of liquid molecules has decreased. The equilibrium has therefore shifted to the right. Table 12-3 summarizes the effects on the equilibrium of this liquid-vapor system caused by certain changes.

TABLE 12-3 ***Shifts in the Equilibrium for the Reaction Liquid + Heat Energy $\rightleftharpoons$ Vapor***

Change	Shift
Addition of liquid	right
Removal of liquid	left
Addition of vapor	left
Removal of vapor	right
Decrease in container volume	left
Increase in container volume	right
Decrease in temperature	left
Increase in temperature	right

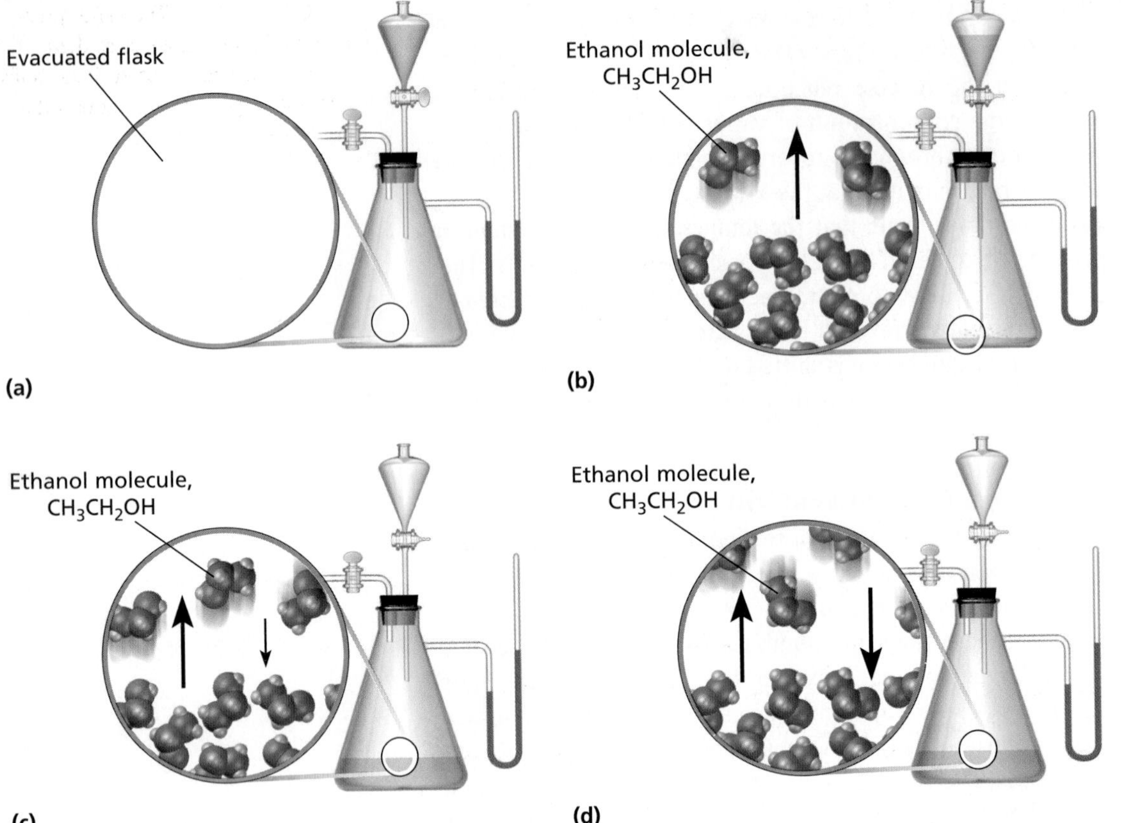

FIGURE 12-11 (a) The vapor pressure of ethanol, CH_3CH_2OH, can be measured by dispensing liquid ethanol into an evacuated flask that is part of a closed system. (b) Ethanol molecules leave the liquid surface to form vapor. (c) Molecules continue to vaporize and condense until equilibrium is reached. (d) At equilibrium, the pressure exerted by the vapor is recorded by noting the mercury levels in the side arm.

Equilibrium Vapor Pressure of a Liquid

Vapor molecules in equilibrium with a liquid in a closed system exert a pressure proportional to the vapor concentration. *The pressure exerted by a vapor in equilibrium with its corresponding liquid at a given temperature is called the* **equilibrium vapor pressure** *of the liquid.* Figure 12-11 shows the apparatus and method used to measure the equilibrium vapor pressure of a liquid.

Figure 12-12 is a graph of the equilibrium vapor pressures of water, diethyl ether, and ethanol. The curve shows that at any given temperature, vapor in equilibrium with liquid exerts a specific vapor pressure. The equilibrium vapor pressure of the liquid increases as temperature increases (although not in direct proportion).

Equilibrium Vapor Pressure and the Kinetic-Molecular Theory

The increase in equilibrium vapor pressure with increasing temperature can be explained in terms of the kinetic-molecular theory for the liquid and gaseous states. Increasing the temperature of a liquid

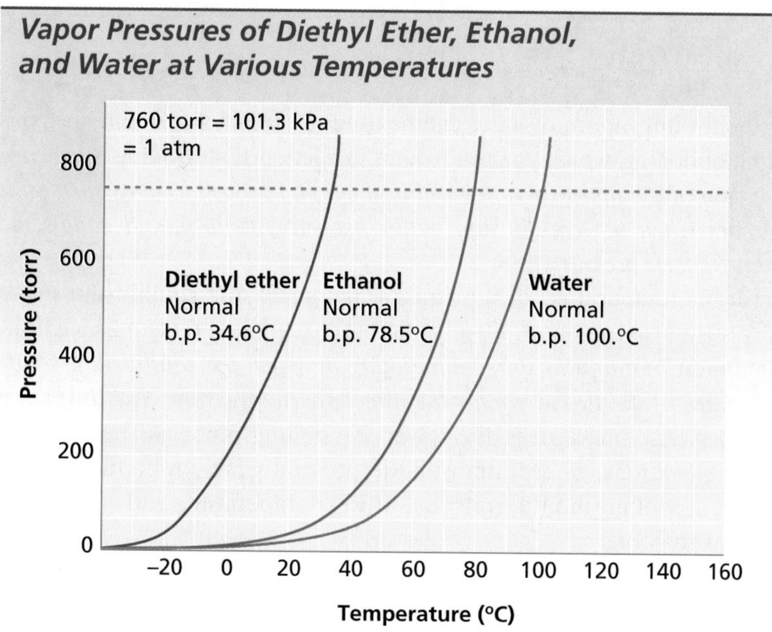

FIGURE 12-12 The vapor pressure of liquids increases as their temperature increases. A liquid boils when its vapor pressure equals the pressure of the atmosphere.

increases its average kinetic energy. That, in turn, increases the number of molecules that have enough energy to escape from the liquid phase into the vapor phase. The resulting increased evaporation rate increases the concentration of molecules in the vapor phase, which in turn increases the equilibrium vapor pressure. The liquid-vapor equilibrium is thus disturbed. However, the increase in concentration of vapor molecules then increases the rate at which molecules condense to liquid. Equilibrium is soon reestablished, but at a higher equilibrium vapor pressure.

Volatile and Nonvolatile Liquids

Because all liquids have characteristic forces of attraction between their particles, every liquid has a specific equilibrium vapor pressure at a given temperature. The stronger these attractive forces are, the smaller is the percentage of liquid particles that can evaporate at any given temperature. A low percentage of evaporation results in a low equilibrium vapor pressure. **Volatile liquids,** *which are liquids that evaporate readily,* have relatively weak forces of attraction between particles. Ether is a typical volatile liquid. Nonvolatile liquids, which evaporate slowly, have relatively strong attractive forces between particles. Molten ionic compounds are examples of nonvolatile liquids.

It is important not to confuse the equilibrium vapor pressure of a liquid with the pressure of a vapor not in equilibrium with its corresponding liquid. The equilibrium vapor pressure of a liquid depends only on temperature. The pressure of a vapor not in equilibrium with the corresponding liquid follows the gas laws, like any other gas. Its pressure is inversely proportional to its volume at constant temperature, according to Boyle's law.

Boiling

Equilibrium vapor pressures can be used to explain and define the concept of boiling, which you read about in Section 12-1. **Boiling** *is the conversion of a liquid to a vapor within the liquid as well as at its surface. It occurs when the equilibrium vapor pressure of the liquid equals the atmospheric pressure.*

If the temperature of the liquid is increased, the equilibrium vapor pressure also increases. Finally, the boiling point is reached. *The* **boiling point** *of a liquid is the temperature at which the equilibrium vapor pressure of the liquid equals the atmospheric pressure.* The lower the atmospheric pressure is, the lower the boiling point is. Therefore, at high elevations, where atmospheric pressures are lower than at sea level, a cooking liquid boils at a lower temperature and foods take longer to cook.

At the boiling point, all of the heat absorbed goes to evaporate the liquid, and the temperature remains constant as long as the pressure does not change. If the pressure above the liquid being heated is increased, the temperature of the liquid will rise until the vapor pressure equals the new pressure and the liquid boils once again. This is the principle behind the operation of a pressure cooker. The cooker is sealed so that steam pressure builds up over the surface of the boiling water inside. The pressure increases the boiling temperature of the water, resulting in shorter cooking times. Conversely, a device called a vacuum evaporator causes boiling at lower-than-normal temperatures. Vacuum evaporators are used to remove water from milk and sugar solutions. Under reduced pressure, the water boils away at a temperature low enough to avoid scorching the milk or sugar. This process is used to prepare evaporated milk and sweetened condensed milk.

At normal atmospheric pressure (1 atm, 760. torr, or 101.3 kPa), the boiling point of water is exactly 100.°C. This temperature is known as the *normal* boiling point of water. The normal boiling points of water and other liquids are shown in Figure 12-12. Note that the normal boiling point of each liquid is the temperature at which the liquid's equilibrium vapor pressure equals 760 torr.

Energy and Boiling

As you know from experience, heat must be added continuously in order to keep a liquid boiling. A pot of boiling water stops boiling almost immediately after it is removed from the stove. Suppose you were to carefully measure the temperature of a boiling liquid and its vapor. You might be surprised to find that they are at the same constant temperature. The temperature, or average kinetic energy of the particles, at the boiling point remains constant despite the continuous addition of heat. Where does the added heat energy go? It is used to overcome the attractive forces between molecules of the liquid during the liquid-to-gas change. The energy is stored in the vapor as potential energy.

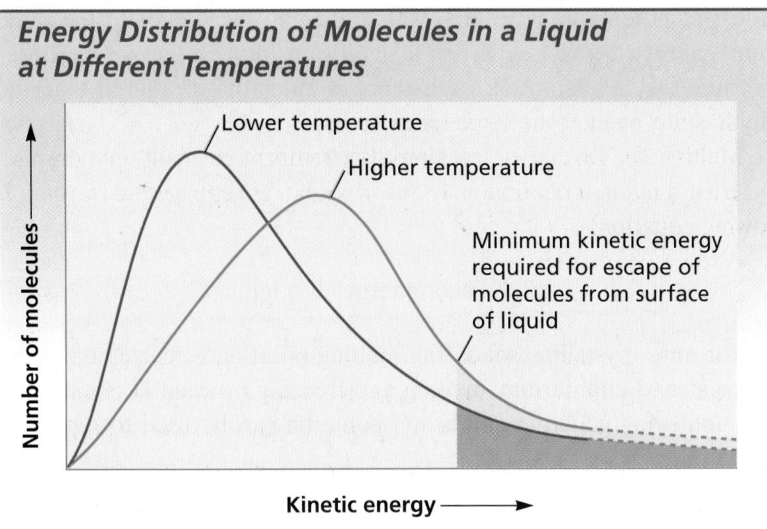

FIGURE 12-13 The number of molecules in a liquid with various kinetic energies is represented at two different temperatures. Notice the shaded area, which shows the minimum amount of kinetic energy required for evaporation to take place.

Molar Heat of Vaporization

The amount of heat energy needed to vaporize one mole of liquid at its boiling point is called the liquid's **molar heat of vaporization.** The magnitude of the molar heat of vaporization is a measure of the attraction between particles of the liquid. The stronger this attraction is, the more energy that is required to overcome it, which results in a higher molar heat of vaporization. Each liquid has a characteristic molar heat of vaporization. Compared with other liquids, water has an unusually high molar heat of vaporization due to the extensive hydrogen bonding in liquid water. This property makes water a very effective cooling agent. When water evaporates from your skin, the escaping molecules carry a great deal of heat away with them. Figure 12-13 shows the distribution of the kinetic energies of molecules in a liquid at two different temperatures. You can see that at the higher temperature a greater portion of the molecules have the kinetic energy required to escape from the liquid surface and become vapor.

Freezing and Melting

As you learned in Section 12-1, the physical change of a liquid to a solid is called freezing. Freezing involves a loss of heat energy by the liquid and can be represented by the following reaction.

$$\text{liquid} \longrightarrow \text{solid} + \text{heat energy}$$

In the case of a pure crystalline substance, this change occurs at constant temperature. *The normal* **freezing point** *is the temperature at which the solid and liquid are in equilibrium at 1 atm (760. torr, or 101.3 kPa) pressure.* At the freezing point, particles of the liquid and the solid have the same average kinetic energy. Therefore, the energy loss during freezing is a loss of potential energy that was present in the liquid. At the same time energy decreases, there is a significant increase in particle order because the solid state of a substance is much more ordered than the liquid state, even at the same temperature.

Melting, the reverse of freezing, also occurs at constant temperature. As a solid melts, it continuously absorbs heat, as represented by the following equation.

$$\text{solid} + \text{heat energy} \longrightarrow \text{liquid}$$

For pure crystalline solids, the melting point and freezing point are the same. At equilibrium, melting and freezing proceed at equal rates. The following general equilibrium equation can be used to represent these states.

$$\text{solid} + \text{heat energy} \rightleftharpoons \text{liquid}$$

At normal atmospheric pressure, the temperature of a system containing ice and liquid water will remain at 0.°C as long as both ice and water are present. That temperature will persist no matter what the surrounding temperature. As predicted by Le Châtelier's principle, adding heat to such a system shifts the equilibrium to the right. That shift increases the proportion of liquid water and decreases that of ice. Only after all the ice has melted will the addition of heat increase the temperature of the system.

Molar Heat of Fusion

The amount of heat energy required to melt one mole of solid at its melting point is its **molar heat of fusion.** The heat absorbed increases the potential energy of the solid as its particles are pulled apart, overcoming the attractive forces holding them together. At the same time, there is a significant decrease in particle order as the substance makes the transformation from solid to liquid. Similar to the molar heat of vaporization, the magnitude of the molar heat of fusion depends on the attraction between the solid particles.

Sublimation and Deposition

At sufficiently low temperature and pressure conditions, a liquid cannot exist. Under such conditions, a solid substance exists in equilibrium with its vapor instead of its liquid, as represented by the following equation.

$$\text{solid} + \text{heat energy} \rightleftharpoons \text{vapor}$$

The change of state from a solid directly to a gas is known as **sublimation.** The reverse process is called **deposition,** *the change of state from a gas directly to a solid.* Among the common substances that sublime at ordinary temperatures are dry ice (solid CO_2) and iodine. Ordinary ice sublimes slowly at temperatures lower than its melting point (0.°C). This explains how a thin layer of snow can eventually disappear, even if the temperature remains below 0.°C. Sublimation occurs in frost-free refrigerators when the temperature in the freezer compartment is periodically raised to cause any ice that has formed to sublime. A blower then removes the water vapor that has formed. The formation of frost on a cold surface is a familiar example of deposition.

Phase Diagrams

A **phase diagram** *is a graph of pressure versus temperature that shows the conditions under which the phases of a substance exist.* A phase diagram also reveals how the states of a system change with changing temperature or pressure.

Figure 12-14 shows the phase diagram for water over a range of temperatures and pressures. Note the three curves, AB, AC, and AD. Curve AB indicates the temperature and pressure conditions at which ice and water vapor can coexist at equilibrium. Curve AC indicates the temperature and pressure conditions at which liquid water and water vapor coexist at equilibrium. Similarly, curve AD indicates the temperature and pressure conditions at which ice and liquid water coexist at equilibrium. Because ice is less dense than liquid water, an increase in pressure lowers the melting point. (Most substances have a positive slope for this curve.) Point A is the triple point of water. *The* **triple point** *of a substance indicates the temperature and pressure conditions at which the solid, liquid, and vapor of the substance can coexist at equilibrium.* Point C is the critical point of water. *The* **critical point** *of a substance indicates the critical temperature and critical pressure. The* **critical temperature** *(t_c) is the temperature above which the substance cannot exist in the liquid state.* The critical temperature of water is 373.99°C. Above this temperature, water cannot be liquefied, no matter how much pressure is

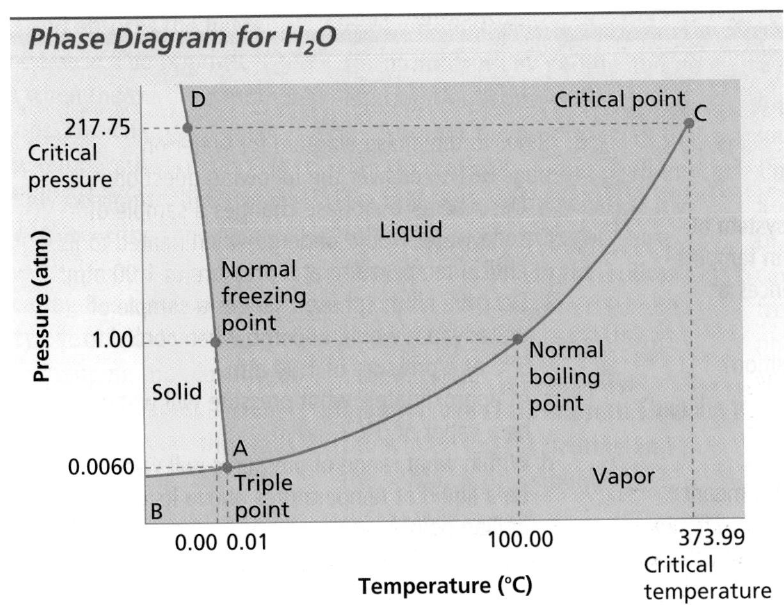

FIGURE 12-14 This phase diagram shows the relationships between the physical states of water and its pressure and temperature.

SECTION 12-4

Water

OBJECTIVES

- Describe the structure of a water molecule.
- Discuss the physical properties of water. Explain how they are determined by the structure of water.
- Calculate the amount of heat energy absorbed or released when a quantity of water changes state.

Water is a familiar substance in all three physical states: solid, liquid, and gas. On Earth, water is by far the most abundant liquid. Oceans, rivers, and lakes cover about 75% of Earth's surface. Significant quantities of water are also frozen in glaciers. Water is an essential component of all organisms; 70% to 90% of the mass of living things is water. The chemical reactions of most life processes take place in water, and water is frequently a reactant or product in such reactions. In order to better understand the importance of water, let us take a closer look at its structure and its properties.

Structure of Water

As discussed in Chapter 6, water molecules consist of two atoms of hydrogen and one atom of oxygen united by polar-covalent bonds. Research shows that a water molecule is bent. The structure can be represented as follows.

$$\underset{105°}{\overset{O}{H \quad H}}$$

The angle between the two hydrogen-oxygen bonds is about 105°. This is close to the angle expected for sp^3 hybridization of the oxygen-atom orbitals.

The molecules in solid or liquid water are linked by hydrogen bonding. The number of linked molecules decreases with increasing temperature because increases in kinetic energy make hydrogen bond formation difficult. Nevertheless, there are usually from four to eight molecules per group in liquid water, as shown in Figure 12-16. If it were not for these molecular groups, water would be a gas at room temperature. Nonpolar molecules, such as methane, CH_4, that are similar in size and mass to water molecules, do not undergo hydrogen bonding. Such substances are gases at room temperature.

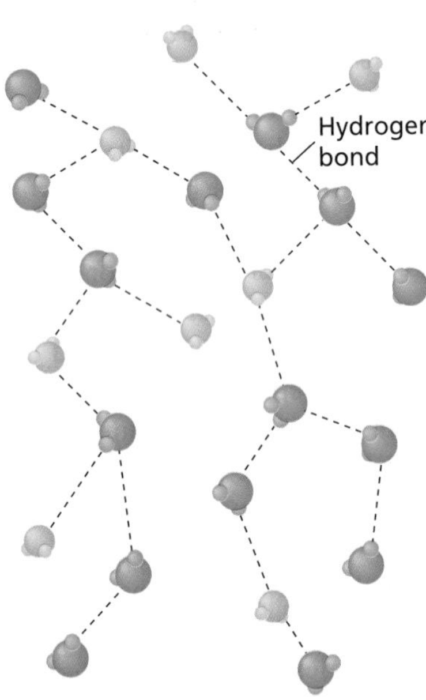

FIGURE 12-16 The structure of liquid water shows that within the water molecule, oxygen and hydrogen are covalently bonded to each other, while the molecules are held together in groups by hydrogen bonds.

Ice consists of water molecules in the hexagonal arrangement shown in Figure 12-17. The empty spaces between molecules in this pattern account for the relatively low density of ice. As ice is heated, the increased energy of the molecules causes them to move and vibrate more vigorously. When the melting point is reached, the energy of the molecules is so great that the rigid open structure of the ice crystals breaks down, and ice turns into liquid water.

Figures 12-16 and 12-17 also show that the hydrogen bonds, between molecules of liquid water at 0.°C are fewer and more disordered than those between molecules of ice at the same temperature. Because the rigid open structure of ice has broken down, water molecules can crowd closer together. Thus, liquid water is denser than ice.

As the liquid water is warmed from 0.°C, the water molecules crowd still closer together. Water molecules are as tightly packed as possible at 3.98°C. At temperatures above 3.98°C, the increasing kinetic energy of the water molecules causes them to overcome molecular attractions. The molecules move farther apart as the temperature continues to rise. As the temperature approaches the boiling point, groups of liquid water molecules absorb enough energy to break up into separate molecules. Because of hydrogen bonding between water molecules, a high kinetic energy is needed, causing water's boiling point to be relatively high (100.°C).

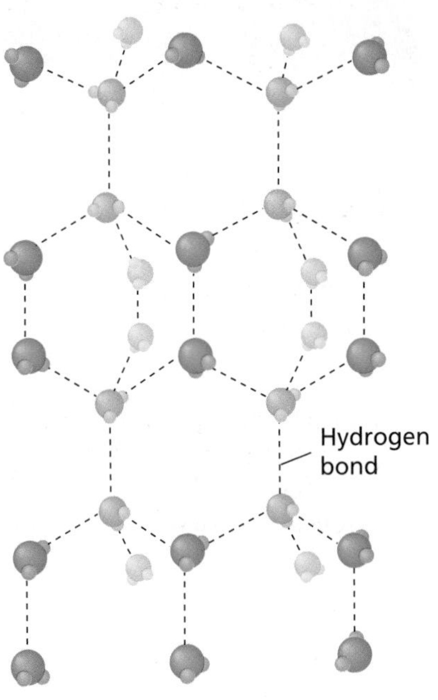

FIGURE 12-17 Ice contains the same types of bonding as liquid water. However, the structure of the hydrogen bonding is much more rigid and open than it is in liquid water.

Physical Properties of Water

At room temperature, pure liquid water is transparent, odorless, tasteless, and almost colorless. Any observable odor or taste is caused by impurities such as dissolved minerals, liquids, or gases.

As shown by its phase diagram in Figure 12-14 on page 381, water freezes at a pressure of 1 atm (101.3 kPa) and melts at 0.°C. The molar heat of fusion of ice is 6.009 kJ/mol. That value is relatively large compared with the molar heats of fusion of other solids. As you have read, water has the unusual property of expanding in volume as it freezes because its molecules form an open rigid structure. As a result, ice at 0.°C has a density of only about 0.917 g/cm^3, compared with a density of 0.999 84 g/cm^3 for liquid water at 0.°C.

This lower density explains why ice floats in liquid water. The insulating effect of floating ice is particularly important in the case of large bodies of water. If ice did not float, the water in lakes and ponds of temperate climates would freeze solid during the winter, killing nearly all the living things in it.

Under a pressure of 1 atm (101.3 kPa), water boils at 100.°C. At this temperature, water's molar heat of vaporization is 40.79 kJ/mol. Both the boiling point and the molar heat of vaporization of water are quite high compared with those of nonpolar substances of comparable molecular mass, such as methane. The values are high because of the strong hydrogen bonding that must be overcome for boiling to occur. The high molar heat of vaporization makes water useful for household steam-heating systems. The steam (vaporized water) stores a great deal of heat. When the steam condenses in radiators, great quantities of heat are released.

SAMPLE PROBLEM 12-1

How much heat energy is absorbed when 47.0 g of ice melts at STP? How much heat energy is absorbed when this same mass of liquid water boils?

SOLUTION

1 *ANALYZE*

Given: mass of $H_2O(s)$ = 47.0 g
mass of $H_2O(l)$ = 47.0 g
molar heat of fusion of ice = 6.009 kJ/mol (given on page 385)
molar heat of vaporization = 40.79 kJ/mol (given on page 385)

Unknown: heat energy absorbed when ice melts
heat energy absorbed when liquid water boils

2 *PLAN*

First, convert the mass of water in grams to moles.

$$\text{g } H_2O \times \frac{1 \text{ mol } H_2O}{\text{g } H_2O} = \text{mol } H_2O$$

Then, use the molar heat of fusion of a solid to calculate the amount of heat absorbed when the solid melts. Multiply the number of moles by the amount of energy needed to melt one mole of ice at its melting point (the molar heat of fusion of ice). Using the same method, calculate the amount of heat absorbed when water boils by using the molar heat of vaporization.

amount of substance (mol) × molar heat of fusion or vaporization (kJ/mol) = heat energy (kJ)

3 *COMPUTE*

$$47.0 \text{ g } H_2O \times \frac{1 \text{ mol } H_2O}{18.02 \text{ g } H_2O} = 2.61 \text{ mol } H_2O$$

$$2.61 \text{ mol} \times 6.009 \text{ kJ/mol} = 15.7 \text{ kJ (on melting)}$$

$$2.61 \text{ mol} \times 40.79 \text{ kJ/mol} = 106 \text{ kJ (on vaporizing or boiling)}$$

4 *EVALUATE*

Units have canceled correctly. The answers have the proper number of significant digits and are reasonably close to estimated values of 18 (3 × 6) and 120 (3 × 40), respectively.

PRACTICE

1. What quantity of heat energy is released when 506 g of liquid water freezes? *Answer* 169 kJ
2. What mass of steam is required to release 4.97×10^5 kJ of heat energy on condensation? *Answer* 2.19×10^5 g

SECTION REVIEW

1. Why is a water molecule polar?
2. How is the structure of water responsible for some of water's unique characteristics?
3. Describe the arrangement of molecules in liquid water and in ice.
4. Why does ice float? Why is this phenomenon important?

CHAPTER 12 REVIEW

CHAPTER SUMMARY

12-1

- The particles of a liquid are closer together and more ordered than those of a gas and less ordered than those of a solid.
- Liquids have a definite volume and a fairly high density, and they are relatively incompressible. Like gases, liquids can flow and are thus considered to be fluids. Liquids can dissolve other substances.
- Liquids have the ability to diffuse. They exhibit surface tension, and they can evaporate or boil. A liquid is said to freeze when it becomes a solid.

Vocabulary

capillary action (365)
evaporation (365)
fluid (363)
freezing (366)
surface tension (365)
vaporization (365)

12-2

- The particles of a solid are not nearly so free to move about as are those of a liquid or a gas. They can only vibrate about definite, fixed positions.
- Solids have a definite shape and may be crystalline or amorphous. They have a definite volume and are generally nonfluid. Additional characteristics of solids are high density, incompressibility, extremely low rates of diffusion, and definite melting points (for crystals).
- A crystal structure is the total three-dimensional array of points that describes the arrangement of the particles of a crystal. A crystal can be classified into one of seven crystalline systems on the basis of the three-dimensional shape of its unit cell. A crystal can also be described as one of four types, based on the kind of particles it contains and the type of chemical bonding between its particles.
- Amorphous solids do not have a regular shape. Instead, they take on whatever shape is imposed on them.

Vocabulary

amorphous solid (368)
crystal (368)
crystal structure (369)
crystalline solid (368)
melting (368)
melting point (368)
supercooled liquid (368)
unit cell (369)

12-3

- A liquid in a closed system will gradually reach a liquid-vapor equilibrium as the rate at which molecules condense equals the rate at which they evaporate.
- When two opposing changes occur at equal rates in the same closed system, the system is said to be in dynamic equilibrium. Le Châtelier's principle states that if any factor determining an equilibrium is changed, the system will adjust itself in a way that tends to minimize that change.
- The pressure exerted by a vapor in equilibrium with its corresponding liquid at a given temperature is the liquid's equilibrium vapor pressure. A liquid boils when the equilibrium vapor pressure of the liquid is equal to the atmospheric pressure. The amount of heat energy required to vaporize one mole of liquid at its boiling point is called the molar heat of vaporization.
- Freezing involves a loss of energy, in the form of heat, by a liquid. Melting is the physical change of a solid to a liquid by the addition of heat. The amount of heat required to melt one mole of solid at its melting point is its molar heat of fusion.

Vocabulary

boiling (378)
boiling point (378)
condensation (373)
critical point (381)
critical pressure (382)
critical temperature (381)
deposition (380)
equilibrium (372)
equilibrium vapor pressure (376)
freezing point (379)
Le Châtelier's principle (374)
molar heat of fusion (380)
molar heat of vaporization (379)
phase (373)
phase diagram (381)
sublimation (380)
triple point (381)
volatile liquid (377)

UNIT 5

Solutions and Their Behavior

CHAPTERS

G.N. LEWIS, ON THE NATURE OF SCIENTIFIC THEORIES IN THE 1900s

Perfection is rare in the science of chemistry. Our scientific theories do not spring full-armed from the brow of the creator. They are subject to slow and gradual growth, and we must candidly admit that the ionic theory in its growth has reached the 'awkward age'. Instead, however, of judging it according to the standard of perfection, let us simply ask what it has accomplished, and what it may accomplish in scientific service.

(From *The Norton History of Chemistry*)

CHAPTER 13

Solutions

Solutions are homogeneous mixtures of two or more substances in a single phase.

Types of Mixtures

SECTION 13-1

OBJECTIVES

- Distinguish between heterogeneous and homogeneous mixtures.
- List three different solute-solvent combinations.
- Compare the properties of suspensions, colloids, and solutions.
- Distinguish between electrolytes and nonelectrolytes.

It is easy to determine that some materials are mixtures because you can see their component parts. For example, soil is a mixture of various substances, including small rocks and decomposed animal and plant matter. You can see this by picking up some soil in your hand and looking at it closely. Milk, on the other hand, does not appear to be a mixture, but in fact it is. Milk is composed principally of fats, proteins, milk sugar, and water. If you look at milk under a microscope, it will look something like Figure 13-1(a). You can see round lipid droplets that measure from 1 to 10 μm in diameter. Irregularly shaped protein (casein) particles that are about 0.2 μm wide can also be seen. Both milk and soil are examples of heterogeneous mixtures because their composition is not uniform.

Salt (sodium chloride) and water form a homogeneous mixture. The sodium and chloride ions are interspersed among the water molecules, and the mixture appears uniform throughout. A model for a homogeneous mixture like salt water is shown in Figure 13-1(b).

Solutions

Suppose a sugar cube is dropped into a glass of water. You know from experience that the sugar will dissolve. Sugar is described as "soluble in water." *By* **soluble** *we mean capable of being dissolved.*

What happens as sugar dissolves? The lump gradually disappears as sugar molecules leave the surface of their crystals and mix with water molecules. Eventually all the sugar molecules become uniformly distributed among the water molecules, as indicated by the equally sweet taste of any part of the mixture. All visible traces of the solid sugar are

FIGURE 13-1 (a) Milk consists of visible particles in a nonuniform arrangement. (b) Salt water is an example of a homogeneous mixture. Ions and water molecules are in a uniform arrangement.

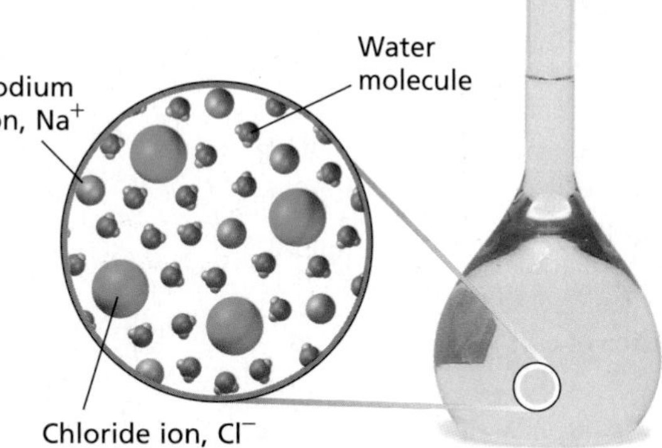

(a) Heterogeneous mixture—milk

(b) Homogeneous mixture—saltwater solution

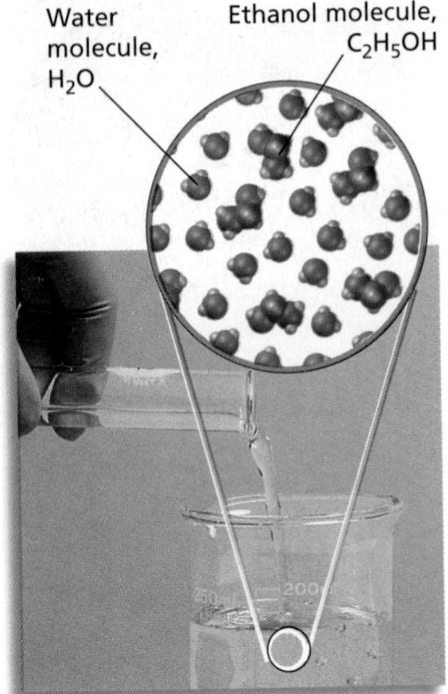

(a)

(b)

FIGURE 13-2 The solute in a solution can be a solid, liquid, or gas. (a) The ethanol-water solution is made from a liquid solute in a liquid solvent. (b) The copper(II) chloride–water solution is made from a solid solute in a liquid solvent. Note that the composition of each solution is uniform.

gone. Such a mixture is called a solution. *A* **solution** *is a homogeneous mixture of two or more substances in a single phase.* In a solution, atoms, molecules, or ions are thoroughly mixed, resulting in a mixture that has the same composition and properties throughout.

Components of Solutions

In the simplest type of solution, such as a sugar-water solution, the particles of one substance are randomly mixed with the particles of another substance. *The dissolving medium in a solution is called the* **solvent,** *and the substance dissolved in a solution is called the* **solute.** The solute is generally designated as that component of a solution that is of lesser quantity. In the ethanol-water solution shown in Figure 13-2, ethanol is the solute and water is the solvent. Occasionally, these terms have little meaning. For example, in a 50%-50% solution of ethanol and water, it would be difficult, and in fact unnecessary, to say which is the solvent and which is the solute.

In a solution, the dissolved solute particles are so small that they cannot be seen. They remain mixed with the solvent indefinitely, so long as the existing conditions remain unchanged. If the solutions in Figure 13-2 are poured through filter paper, both the solute and the solvent will pass through the paper. The solute-particle dimensions are those of atoms, molecules, and ions—which range from about 0.01 to 1 nm in diameter.

Types of Solutions

Solutions may exist as gases, liquids, or solids. Some possible solute-solvent combinations of gases, liquids, and solids in solutions are summarized in Table 13-1. In each example, one component is designated as the solvent and one as the solute.

Many alloys, such as brass (made from zinc and copper) and sterling silver (made from silver and copper), are solid solutions in which the atoms of two of more metals are uniformly mixed. By properly choosing the proportions of each metal in the alloy, many desirable properties

TABLE 13-1 ***Some Solute-Solvent Combinations for Solutions***

Solute state	Solvent state	Example
Gas	gas	oxygen in nitrogen
Gas	liquid	carbon dioxide in water
Liquid	gas	water in air
Liquid	liquid	alcohol in water
Liquid	solid	mercury in silver and tin (dental amalgam)
Solid	liquid	sugar in water
Solid	solid	copper in nickel (Monel™ alloy)

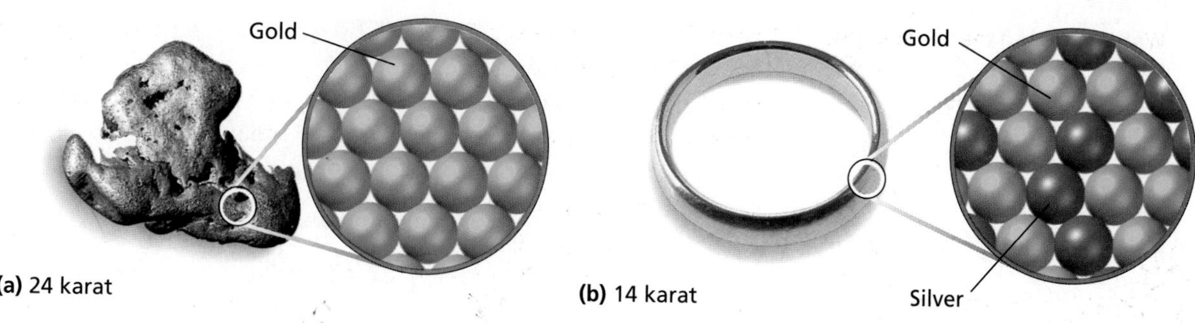

FIGURE 13-3 (a) 24-karat gold is pure gold. (b) 14-karat gold is an alloy of gold and silver. 14-karat gold is 14/24, or 58.3%, gold.

can be obtained. For example, alloys can have higher strength and greater resistance to corrosion than the pure metals. Pure gold (24K), for instance, is too soft to use in jewelry. Alloying it with silver greatly increases its strength and hardness while retaining its appearance and corrosion resistance. Figure 13-3 shows a model for comparing pure gold with a gold alloy.

Suspensions

If the particles in a solvent are so large that they settle out unless the mixture is constantly stirred or agitated, the mixture is called a **suspension.** Think of a jar of muddy water. If left undisturbed, particles of soil collect on the bottom of the jar. The soil particles are much larger and denser than water molecules. Gravity pulls them to the bottom of the container. Particles over 1000 nm in diameter—1000 times as large as atoms, molecules, or ions—form suspensions. The particles in suspension can be separated from the heterogeneous mixtures by passing the mixture through a filter.

Colloids

Particles that are intermediate in size between those in solutions and suspensions form mixtures known as colloidal dispersions, or simply **colloids.** Particles between 1 nm and 1000 nm in diameter may form colloids. After large soil particles settle out of muddy water, the water is often still cloudy because colloidal particles remain dispersed in the water. If the cloudy mixture is poured through a filter, the colloidal particles will pass through, and the mixture will remain cloudy. The particles in a colloid are small enough to be suspended throughout the solvent by the constant movement of the surrounding molecules. The colloidal particles make up the *dispersed phase,* and water is the *dispersing medium*. Examples of the various types of colloids are given in Table 13-2. Note that some familiar terms, such as *emulsion* and *foam,* refer to specific types of colloids. For example, mayonnaise is an emulsion

TABLE 13-2 *Classes of Colloids*

Class of colloid	Phases
Sol	solid dispersed in liquid
Gel	solid network extending throughout liquid
Liquid emulsion	liquid dispersed in a liquid
Foam	gas dispersed in liquid
Aerosol	solid dispersed in gas
smoke	solid dispersed in gas
fog	liquid dispersed in gas
smog	solid and liquid dispersed in gas
Solid emulsion	liquid dispersed in solid

of oil droplets in water; the egg yolk in it acts as an emulsifying agent, which helps to keep the oil droplets dispersed.

FIGURE 13-4 A beam of light distinguishes a colloid from a solution. The particles in a colloid will scatter light, making the beam visible. The mixture of gelatin and water in the jar on the right is a colloid. The mixture of water and sodium chloride in the jar on the left is a true solution.

Tyndall Effect

Many colloids appear homogeneous because the individual particles cannot be seen. The particles are, however, large enough to scatter light. You have probably noticed that a headlight beam is visible on a foggy night. This effect, known as the Tyndall effect, occurs when light is scattered by colloidal particles dispersed in a transparent medium. The Tyndall effect is a property that can be used to distinguish between a solution and a colloid, as demonstrated in Figure 13-4.

The distinctive properties of solutions, colloids, and suspensions are summarized in Table 13-3. The individual particles of a colloid can be detected under a microscope if a bright light is cast on the specimen at a right angle. The particles, which appear as tiny specks of light, are seen to move rapidly in a random motion. This motion is due to collisions of rapidly moving molecules and is called Brownian motion, after its discoverer, Robert Brown.

TABLE 13-3 *Properties of Solutions, Colloids, and Suspensions*

Solutions	Colloids	Suspensions
Homogeneous	Heterogeneous	Heterogeneous
Particle size: 0.01–1 nm; can be atoms, ions, molecules	Particle size: 1–1000 nm, dispersed; can be aggregates or large molecules	Particle size: over 1000 nm, suspended; can be large particles or aggregates
Do not separate on standing	Do not separate on standing	Particles settle out
Cannot be separated by filtration	Cannot be separated by filtration	Can be separated by filtration
Do not scatter light	Scatter light (Tyndall effect)	May scatter light, but are not transparent

DESKTOP INVESTIGATION

Observing Solutions, Suspensions, and Colloids

MATERIALS

- balance
- 7 beakers, 400 mL
- clay
- cooking oil
- flashlight
- gelatin, plain
- hot plate (to boil H_2O)
- red food coloring
- sodium borate ($Na_2B_4O_7•10H_2O$)
- soluble starch
- stirring rod
- sucrose
- test-tube rack
- water

Wear safety goggles and an apron.

PROCEDURE

1. Prepare seven mixtures, each containing 250 mL of water and one of the following substances.
   a. 12 g of sucrose
   b. 3 g of soluble starch
   c. 5 g of clay
   d. 2 mL of food coloring
   e. 2 g of sodium borate
   f. 50 mL of cooking oil
   g. 3 g of gelatin

Making the gelatin mixture:
Soften the gelatin in 65 mL of cold water, and then add 185 mL of boiling water.

2. Observe the seven mixtures and their characteristics. Record the appearance of each mixture after stirring.
3. Transfer to individual test tubes 10 mL of each mixture that does not separate after stirring. Shine a flashlight on each mixture in a dark room. Make note of the mixtures in which the path of the light beam is visible.

DISCUSSION

1. Using your observations, classify each mixture as a solution, suspension, or colloid.
2. What characteristics did you use to classify each mixture?

Solutes: Electrolytes vs. Nonelectrolytes

Substances that dissolve in water are classified according to whether they yield molecules or ions in solution. When an ionic compound dissolves, the positive and negative ions separate from each other and are surrounded by water molecules. These solute ions are free to move, making it possible for an electric current to pass through the solution. *A substance that dissolves in water to give a solution that conducts electric current is called an* **electrolyte.** Sodium chloride, NaCl, is an electrolyte, as is any soluble ionic compound. Certain highly polar molecular compounds, such as hydrogen chloride, HCl, are also electrolytes because HCl molecules form the ions H_3O^+ and Cl^- when dissolved in water.

By contrast, a solution containing neutral solute molecules does not conduct electric current because it does not contain mobile charged

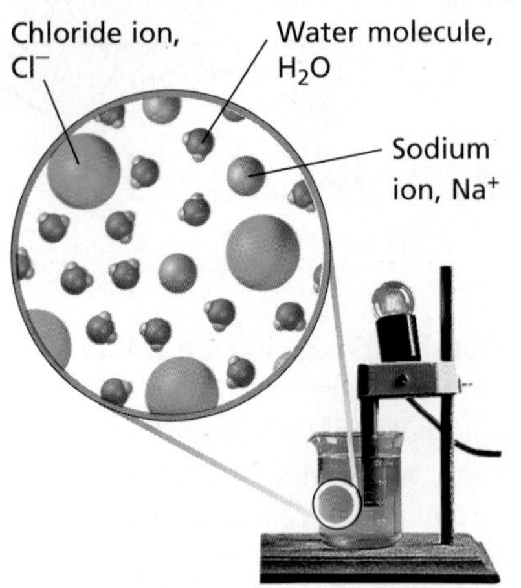

(a) Salt solution—electrolyte solute

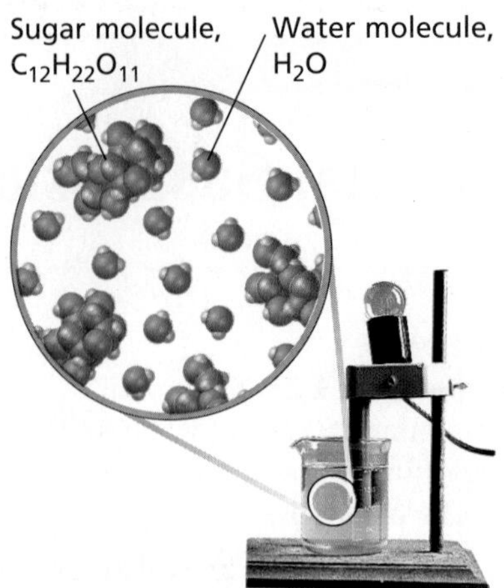

(b) Sugar solution—nonelectrolyte solute

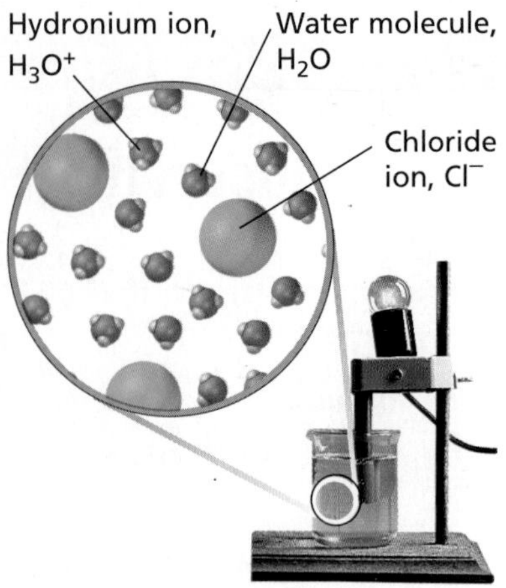

(c) Hydrochloric acid solution—electrolyte solute

FIGURE 13-5 (a) Sodium chloride dissolves in water to produce a salt solution that conducts electric current. NaCl is an electrolyte. (b) Sucrose dissolves in water to produce a sugar solution that does not conduct electricity. Sucrose is a nonelectrolyte. (c) Hydrogen chloride dissolves in water to produce a solution that conducts current. HCl is an electrolyte.

particles. *A substance that dissolves in water to give a solution that does not conduct an electric current is called a* **nonelectrolyte.** Sugar is a nonelectrolyte. Figure 13-5 shows an apparatus for testing the conductivity of solutions. The electrodes are conductors that are attached to a power supply and that make electric contact with the test solution. For a current to pass through the light-bulb filament, the test solution must provide a conducting path between the two electrodes. A nonconducting solution is like an open switch between the electrodes, and there is no current in the circuit.

The light bulb glows brightly if a solution that is a good conductor is tested. Such solutions contain solutes that are electrolytes. For a moderately conductive solution, however, the light bulb is dim. If a solution is a poor conductor, the light bulb does not glow at all. Such solutions contain solutes that are nonelectrolytes. You will learn more about the strengths and behavior of electrolytes in Chapter 14.

SECTION REVIEW

1. Classify the following as either a heterogeneous or homogeneous mixture, and explain your answers.
   a. orange juice
   b. tap water
2. What are substances called whose water solutions conduct electricity? Why does a saltwater solution conduct, while a sugar-water solution does not?
3. Make a drawing of the particles in an NaCl solution to show why this solution conducts electricity. Make a drawing of the particles in an NaCl crystal to show why pure salt does not conduct.
4. Describe one way to prove that a mixture of sugar and water is a solution and that a mixture of sand and water is not a solution.
5. Label the solute and solvent in each of the following:
   a. 14-karat gold
   b. water vapor in air
   c. carbonated, or sparkling, water
   d. hot tea

The Solution Process

SECTION 13-2

Objectives

- List and explain three factors that affect the rate at which a solid solute dissolves in a liquid solvent.
- Explain solution equilibrium, and distinguish among saturated, unsaturated, and supersaturated solutions.
- Explain the meaning of "like dissolves like" in terms of polar and nonpolar substances.
- List the three interactions that contribute to the heat of solution, and explain what causes dissolution to be exothermic or endothermic.
- Compare the effects of temperature and pressure on solubility.

Factors Affecting the Rate of Dissolution

If you have ever tried to dissolve sugar in iced tea, you know that temperature has something to do with how quickly a solute dissolves. What other factors affect how quickly you can dissolve sugar in iced tea?

Increasing the Surface Area of the Solute

Sugar dissolves as sugar molecules leave the crystal surface and mix with water molecules. The same is true for any solid solute in a liquid solvent: molecules or ions of the solute are attracted by the solvent.

Because the dissolution process occurs at the surface of the solute, it can be speeded up if the surface area of the solute is increased. Crushing sugar that is in cubes or large crystals increases the surface area. In general, the more finely divided a substance is, the greater the surface area per unit mass and the more quickly it dissolves. Figure 13-6 shows a model of solutions that are made from the same solute but have a different amount of surface area exposed to the solvent.

Agitating a Solution

The concentration of dissolved solute is very high close to the solute surface. Stirring or shaking helps to disperse the solute particles and

FIGURE 13-6 The rate at which a solid solute dissolves can be increased by increasing the surface area. A powdered solute has a greater surface area exposed to solvent particles and therefore dissolves faster than a solute in large crystals.

bring fresh solvent into contact with the solute surface. Thus, the effect of stirring is similar to that of crushing a solid—contact between the solvent and the solute surface is increased.

Heating a Solvent

You have probably noticed that sugar and many other materials dissolve more quickly in warm water than in cold water. As the temperature of the solvent increases, solvent molecules move faster, and their average kinetic energy increases. Therefore, at higher temperatures, collisions between the solvent molecules and the solute are more frequent and are of higher energy than at lower temperatures. This helps to separate solute molecules from one another and to disperse them among the solvent molecules.

Solubility

If you add spoonful after spoonful of sugar to tea, a point will be reached at which no more sugar will dissolve. For every combination of solvent with a solid solute at a given temperature, there is a limit to the amount of solute that can be dissolved. The following model describes why there is a limit.

When solid sugar is first dropped into the water, sugar molecules leave the solid surface and move about at random in the solvent. Some of these dissolved molecules may collide with the crystal and remain there (recrystallize). As more of the solid dissolves and the concentration of dissolved molecules increases, these collisions become more frequent. Eventually, molecules are returning to the crystal at the same rate at which they are going into solution, and a dynamic equilibrium is established between dissolution and crystallization, as represented by the model in Figure 13-7.

Solution equilibrium *is the physical state in which the opposing processes of dissolution and crystallization of a solute occur at equal rates.* The point at which equilibrium is reached for any solute-solvent combination is difficult to predict precisely and depends on the nature of the solute, the nature of the solvent, and the temperature.

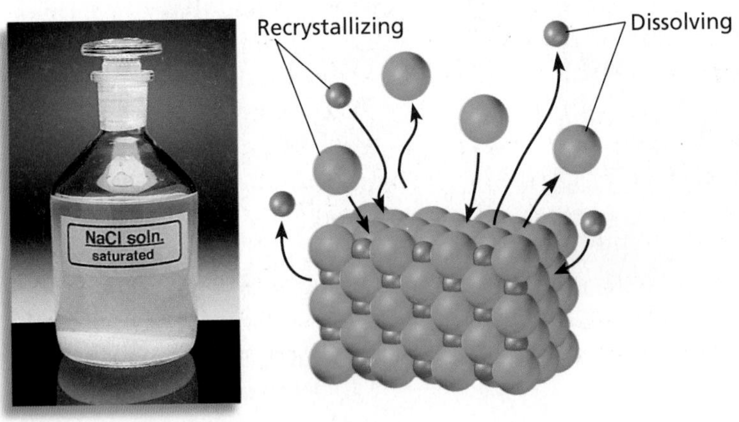

FIGURE 13-7 A saturated solution in a closed system is at equilibrium. The solute is recrystallizing at the same rate that it is dissolving, even though it appears that there is no activity in the system.

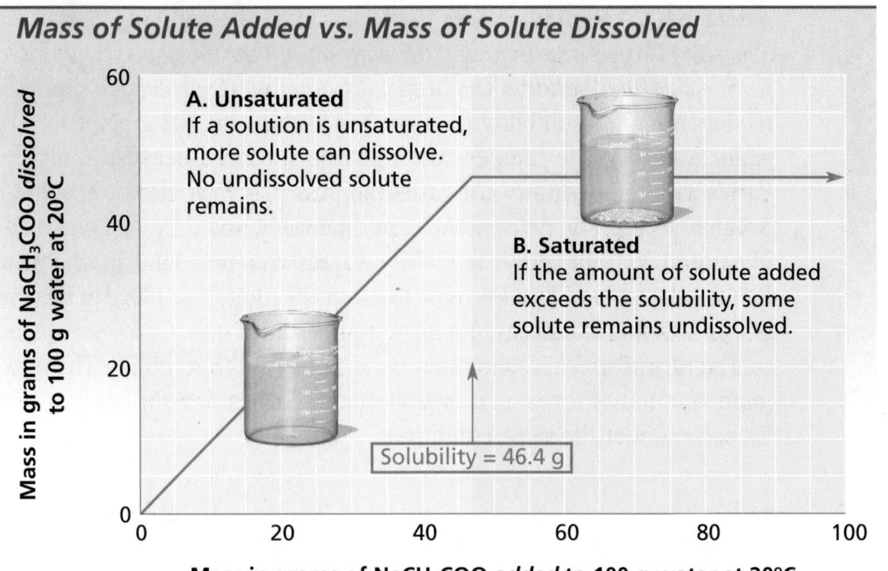

FIGURE 13-8 The graph shows the range of solute masses that will produce an unsaturated solution. Once the saturation point is exceeded, the system will contain undissolved solute.

Saturated vs. Unsaturated Solutions

A solution that contains the maximum amount of dissolved solute is described as a **saturated solution.** How can you tell that the NaCl solution pictured in Figure 13-8 is saturated? If more sodium chloride is added to the solution, it falls to the bottom and does not dissolve because an equilibrium has been established between molecules leaving and entering the solid phase. If more water is added to the saturated solution, then more sodium chloride will dissolve in it. At 20°C, 35.9 g of NaCl is the maximum amount that will dissolve in 100. g of water. *A solution that contains less solute than a saturated solution under the existing conditions is an* **unsaturated solution.**

Supersaturated Solutions

When a saturated solution of a solute whose solubility increases with temperature is cooled, the excess solute usually comes out of solution, leaving the solution saturated at the lower temperature. But sometimes, if the solution is left to cool undisturbed, the excess solute does not separate and a supersaturated solution is produced. *A* **supersaturated solution** *is a solution that contains more dissolved solute than a saturated solution contains under the same conditions.* A supersaturated solution may remain unchanged for a long time if it is not disturbed, but once crystals begin to form, the process continues until equilibrium is reestablished at the lower temperature. An example of a supersaturated solution is one prepared from a saturated solution of sodium thiosulfate, $Na_2S_2O_3$, or sodium acetate, $NaCH_3COO$. Solute is added to hot water until the solution is saturated, and the hot solution is filtered. The filtrate is left to stand undisturbed as it cools. Dropping a small crystal of the solute into the supersaturated solution ("seeding") or disturbing the solution causes a rapid formation of crystals by the excess solute.

Solubility Values

The **solubility** *of a substance is the amount of that substance required to form a saturated solution with a specific amount of solvent at a specified temperature.* The solubility of sugar, for example, is 204 g per 100. g of water at 20.°C. The temperature must be specified because solubility varies with temperature. For gases, the pressure must also be specified. Solubilities must be determined experimentally, and they vary widely, as illustrated in Table 13-4. Solubility values can be found in chemical handbooks and are usually given as grams of solute per 100. g of solvent or per 100. mL of solvent at a given temperature.

The rate at which a solid dissolves is unrelated to solubility. The maximum amount of solute that dissolves and reaches equilibrium is always the same under the same conditions.

Solute-Solvent Interactions

Lithium chloride is highly soluble in water, but gasoline is not. On the other hand, gasoline mixes readily with benzene, C_6H_6, but lithium chloride does not. Why are there such differences in solubility?

"Like dissolves like" is a rough but useful rule for predicting whether one substance will dissolve in another. What makes substances similar depends on the type of bonding, the polarity or nonpolarity of molecules, and the intermolecular forces between the solute and solvent.

TABLE 13-4 ***Solubility of Solutes as a Function of Temperature (in g solute/100. g H_2O)***

	Temperature (°C)					
Substance	**0**	**20**	**40**	**60**	**80**	**100**
$AgNO_3$	122	216	311	440	585	733
$Ba(OH)_2$	1.67	3.89	8.22	20.94	101.4	—
$C_{12}H_{22}O_{11}$	179	204	238	287	362	487
$Ca(OH)_2$	0.189	0.173	0.141	0.121	—	0.07
$Ce_2(SO_4)_3$	20.8	10.1	—	3.87	—	—
KCl	28.0	34.2	40.1	45.8	51.3	56.3
KI	128	144	162	176	192	206
KNO_3	13.9	31.6	61.3	106	167	245
LiCl	69.2	83.5	89.8	98.4	112	128
Li_2CO_3	1.54	1.33	1.17	1.01	0.85	0.72
NaCl	35.7	35.9	36.4	37.1	38.0	39.2
$NaNO_3$	73	87.6	102	122	148	180
CO_2 (gas at SP)	0.335	0.169	0.0973	0.058	—	—
O_2 (gas at SP)	0.00694	0.00537	0.00308	0.00227	0.00138	0.00

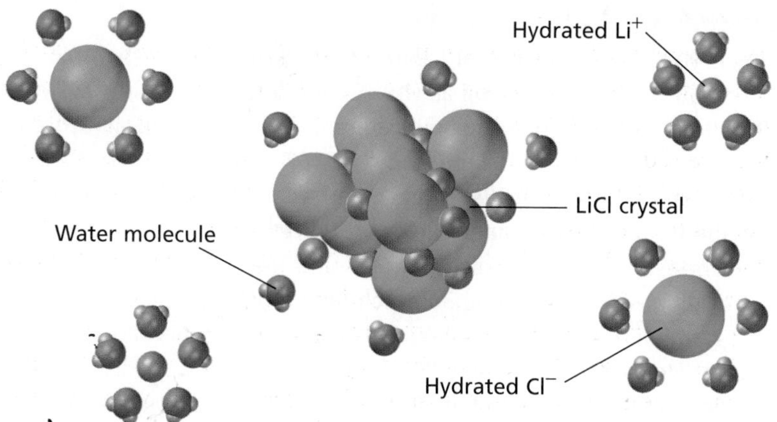

FIGURE 13-9 When LiCl dissolves, the ions are hydrated. The attraction between ions and water molecules is strong enough that each ion in solution is surrounded by water molecules.

Dissolving Ionic Compounds in Aqueous Solution

The polarity of water molecules plays an important role in the formation of solutions of ionic compounds in water. The charged ends of water molecules attract the ions in the ionic compounds and surround them to keep them separated from the other ions in the solution. Suppose we drop a few crystals of lithium chloride into a beaker of water. At the crystal surfaces, water molecules come into contact with Li^+ and Cl^- ions. The positive ends of the water molecules are attracted to Cl^- ions, while the negative ends are attracted to Li^+ ions. The attraction between water molecules and the ions is strong enough to draw the ions away from the crystal surface and into solution, as illustrated in Figure 13-9. *This solution process with water as the solvent is referred to as* **hydration.** The ions are said to be *hydrated.* As hydrated ions diffuse into the solution, other ions are exposed and are drawn away from the crystal surface by the solvent. The entire crystal gradually dissolves, and hydrated ions become uniformly distributed in the solution.

When crystallized from aqueous solutions, some ionic substances form crystals that incorporate water molecules. These crystalline compounds, known as *hydrates,* retain specific ratios of water molecules and are represented by formulas such as $CuSO_4 \bullet 5H_2O$. Heating the crystals of a hydrate can drive off the water of hydration and leave the anhydrous salt. When a crystalline hydrate dissolves in water, the water of hydration returns to the solvent. The behavior of a solute in its hydrated form is no different from the behavior of the anhydrous form. Dissolving either form results in a system containing hydrated ions and water.

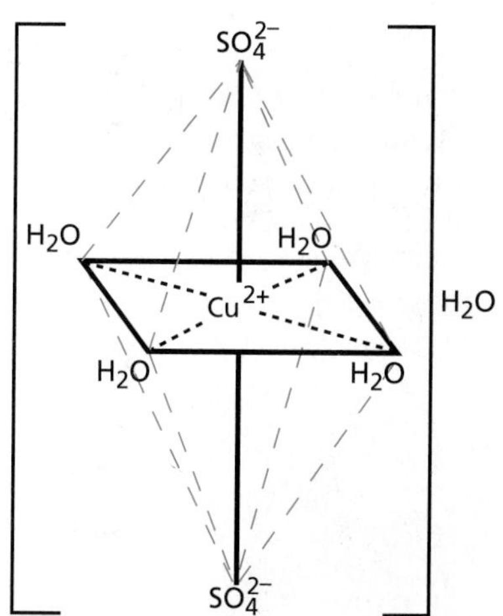

FIGURE 13-10 Hydrated copper(II) sulfate has water trapped in the crystal structure. Heating releases the water and produces the anhydrous form of the substance, which has the formula $CuSO_4$.

Nonpolar Solvents

Ionic compounds are generally not soluble in nonpolar solvents such as carbon tetrachloride, CCl_4, and toluene, $C_6H_5CH_3$. The nonpolar solvent molecules do not attract the ions of the crystal strongly enough to overcome the forces holding the crystal together.

Would you expect lithium chloride to dissolve in toluene? No, LiCl is not soluble in toluene. LiCl and $C_6H_5CH_3$ differ widely in bonding, polarity, and intermolecular forces.

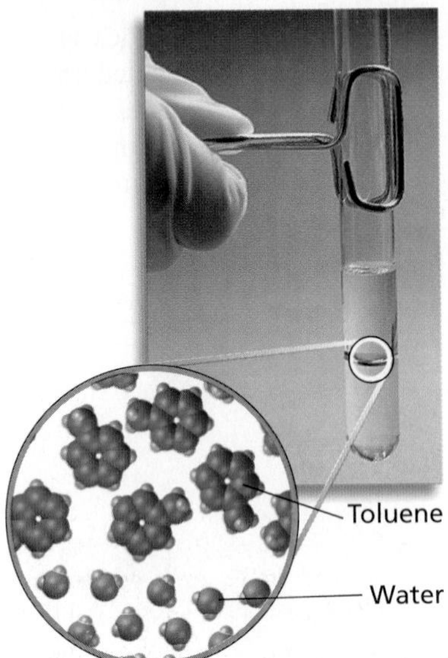

Insoluble and immiscible

FIGURE 13-11 Toluene and water are immiscible. The components of this system exist in two distinct phases.

Liquid Solutes and Solvents

When you shake a bottle of salad dressing, oil droplets become dispersed in the water. As soon as you stop shaking the bottle, the strong attraction between the water molecules squeezes out the oil droplets, forming separate layers. *Liquid solutes and solvents that are not soluble in each other are* **immiscible.** Toluene and water, shown in Figure 13-11, are another example of immiscible substances.

Nonpolar substances, such as fats, oils, and greases, are generally quite soluble in nonpolar liquids, such as carbon tetrachloride, toluene, and gasoline. The only attractions between the nonpolar molecules are relatively weak London forces. The intermolecular forces existing in the solution are therefore very similar to those in pure substances. Thus, the molecules can mix freely with one another.

Liquids that dissolve freely in one another in any proportion are said to be completely **miscible.** Benzene and carbon tetrachloride are completely miscible. The nonpolar molecules of these substances exert no strong forces of attraction or repulsion, and the molecules mix freely. Ethanol and water, shown in Figure 13-12, also mix freely, but for a different reason. The −OH group on an ethanol molecule is somewhat polar. This group can form hydrogen bonds with water as well as with other ethanol molecules. The intermolecular forces in the mixture are so similar to those in the pure liquids that the liquids are mutually soluble in all proportions.

```
   H  H
   |  |
H–C–C–OH
   |  |
   H  H
```

Gasoline contains mainly nonpolar hydrocarbons and is also an excellent solvent for fats, oils, and greases. The major intermolecular forces acting between the nonpolar molecules are relatively weak London forces.

Ethanol is intermediate in polarity between water and carbon tetrachloride. It is not as good a solvent for polar or ionic substances as water is. Sodium chloride is only slightly soluble in ethanol. On the other hand, ethanol is a better solvent than water is for less-polar substances because the molecule has a nonpolar region.

FIGURE 13-12 (a) Water and ethanol are miscible. The components of this system exist in a single phase with a uniform arrangement. (b) Hydrogen bonding between the solute and solvent enhances the solubility of ethanol in water.

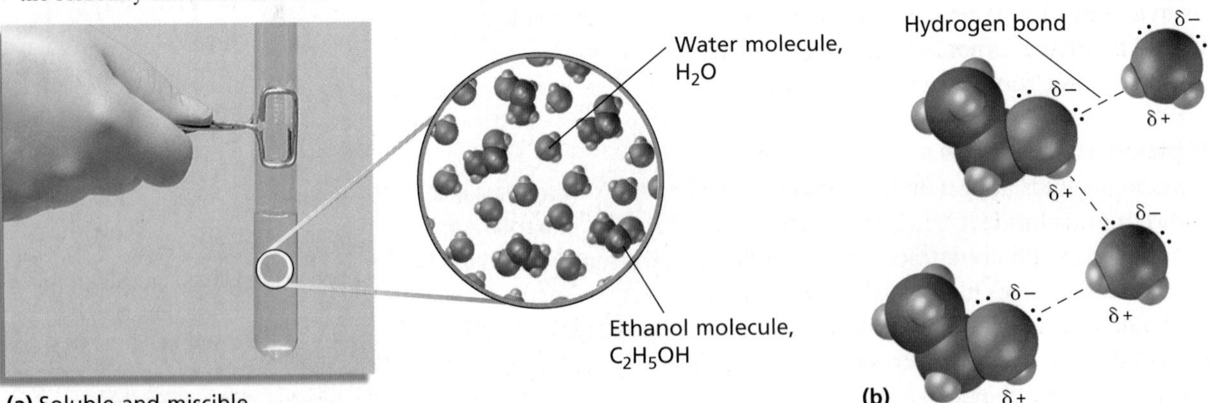

(a) Soluble and miscible

(b)

Effects of Pressure on Solubility

Changes in pressure have very little effect on the solubilities of liquids or solids in liquid solvents. However, increases in pressure increase gas solubilities in liquids.

When a gas is in contact with the surface of a liquid, gas molecules can enter the liquid. As the amount of dissolved gas increases, some molecules begin to escape and reenter the gas phase. An equilibrium is eventually established between the rates at which gas molecules enter and leave the liquid phase. As long as this equilibrium is undisturbed, the solubility of the gas in the liquid is unchanged at a given pressure.

$$\text{gas} + \text{solvent} \rightleftharpoons \text{solution}$$

Increasing the pressure of the solute gas above the solution puts stress on the equilibrium. Molecules collide with the liquid surface more often. The increase in pressure is partially offset by an increase in the rate of gas molecules entering the solution. In turn, the increase in the amount of dissolved gas causes an increase in the rate at which molecules escape from the liquid surface and become vapor. Eventually, equilibrium is restored at a higher gas solubility. As expected from Le Châtelier's principle, an increase in gas pressure causes the equilibrium to shift so that fewer molecules are in the gas phase.

FIGURE 13-13 (a) There are no gas bubbles in the unopened bottle of soda because the pressure of CO_2 applied during bottling keeps the carbon dioxide gas dissolved in the liquid. (b) When the cap on the bottle is removed, the pressure of CO_2 on the liquid is reduced, and CO_2 can escape from the liquid. The soda effervesces when the bottle is opened and the pressure is reduced.

Henry's Law

The solubility of a gas in a liquid is directly proportional to the partial pressure of that gas on the surface of the liquid. This is a statement of **Henry's law,** named after the English chemist William Henry. Henry's law applies to gas-liquid solutions at constant temperature.

Recall that when a mixture of ideal gases is confined in a constant volume at a constant temperature, each gas exerts the same pressure it would exert if it occupied the space alone. Assuming that the gases do not react in any way, each gas dissolves to the extent it would dissolve if no other gases were present.

In carbonated beverages, the solubility of CO_2 is increased by increasing the pressure. At the bottling plant, carbon dioxide gas is forced into the solution of flavored water at a pressure of 5–10 atm. The gas-in-liquid solution is then sealed in bottles or cans. When the cap is removed, the pressure is reduced to 1 atm, and some of the carbon dioxide escapes as gas bubbles. *The rapid escape of a gas from a liquid in which it is dissolved is known as* **effervescence** and is shown in Figure 13-13.

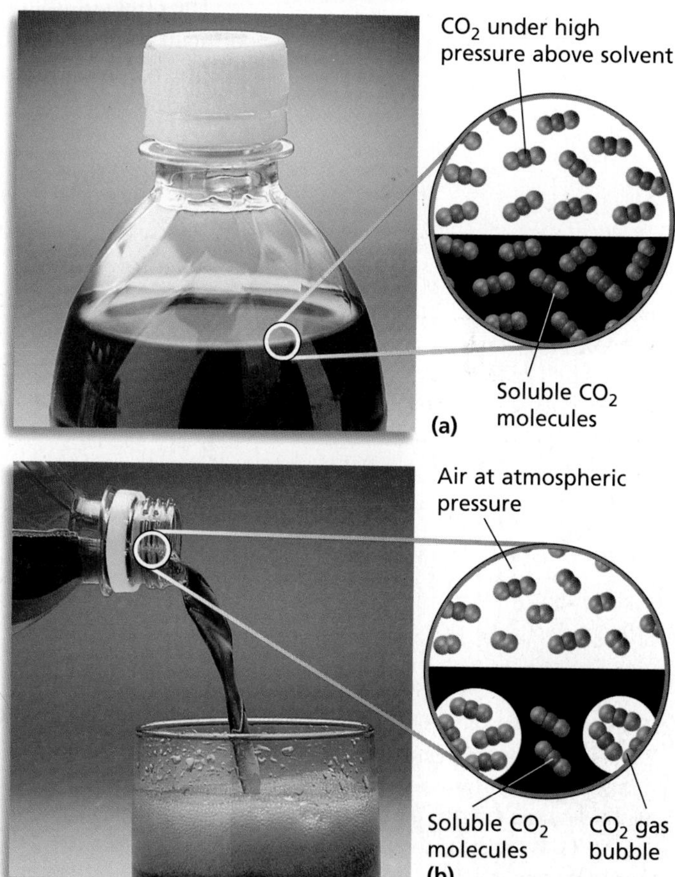

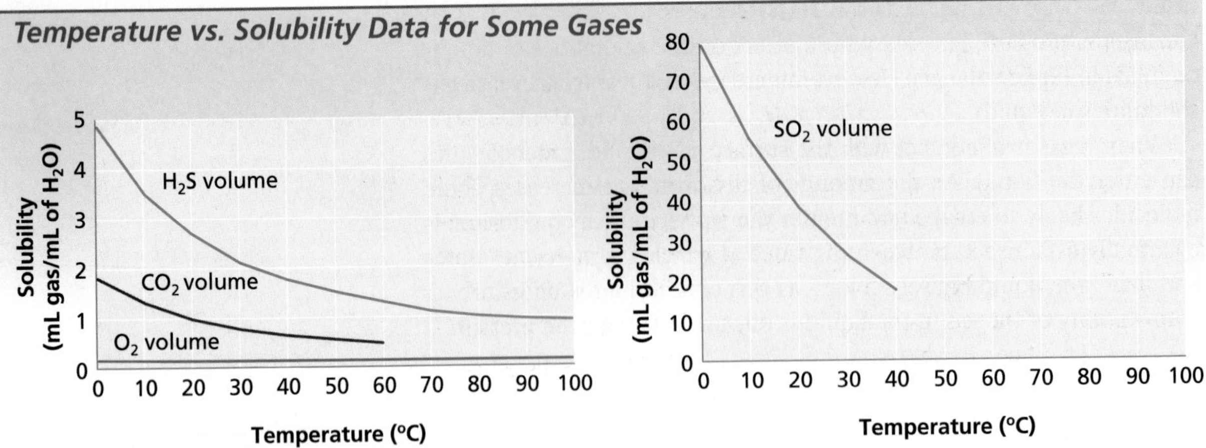

FIGURE 13-14 The solubility of gases in water decreases with increasing temperature. Which gas has the greater solubility at 30°C—CO_2 or SO_2?

Effects of Temperature on Solubility

First let's consider gas solubility. Increasing the temperature usually decreases gas solubility. As the temperature increases, the average kinetic energy of the molecules in solution increases. A greater number of solute molecules are able to escape from the attraction of solvent molecules and return to the gas phase. At higher temperatures, therefore, equilibrium is reached with fewer gas molecules in solution and gases are generally less soluble, as shown in Figure 13-14.

The effect of temperature on the solubility of solids in liquids is more difficult to predict. Often, increasing the temperature increases the solubility of solids. However, an equivalent temperature increase can

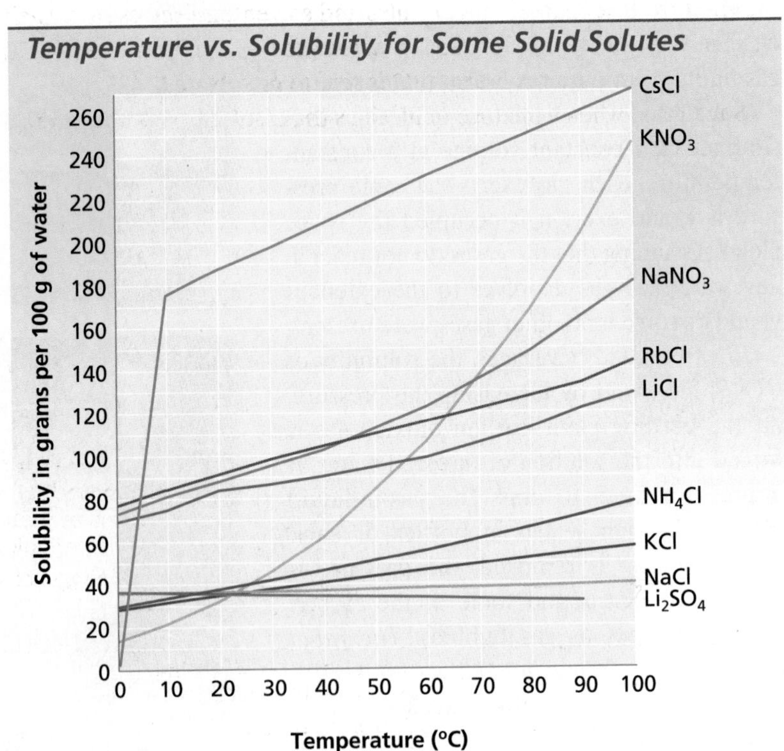

FIGURE 13-15 Solubility curves for various solid solutes generally show increasing solubility with increases in temperature. From the graph, you can see that the solubility of $NaNO_3$ is affected more by temperature than is NaCl.

result in a large increase in solubility in one case and only a slight increase in another.

In Table 13-4 and Figure 13-15, compare the effect of temperature on the solubilities of potassium nitrate, KNO_3, and sodium chloride, NaCl. About 14 g of potassium nitrate will dissolve in 100. g of water at 0.°C. The solubility of potassium nitrate increases by more than 150 g KNO_3 per 100. g H_2O when the temperature is raised to 80.°C. Under similar circumstances, the solubility of sodium chloride increases by only about 2 g NaCl per 100. g H_2O. In some cases, solubility of a solid *decreases* with an increase in temperature. For example, between 0.°C and 60.°C the solubility of cerium sulfate, $Ce_2(SO_4)_3$, decreases by about 17 g.

Heats of Solution

The formation of a solution is accompanied by an energy change. If you dissolve some potassium iodide, KI, in water, you will find that the outside of the container feels cold to the touch. But if you dissolve some lithium chloride, LiCl, in the same way, the outside of the container feels hot. The formation of a solid-liquid solution can apparently either absorb heat (KI in water) or release heat (LiCl in water).

During the formation of a solution, solvent and solute particles experience changes in the forces attracting them to other particles. Before dissolving begins, solvent molecules are held together by intermolecular forces (solvent-solvent attraction). In the solute, molecules are held together by intermolecular forces (solute-solute attraction). Energy is required to separate solute molecules and solvent molecules from their neighbors. *A solute particle that is surrounded by solvent molecules,* as shown by the model in Figure 13-9, *is said to be* **solvated.**

Solution formation can be pictured as the result of the three interactions summarized in Figure 13-16.

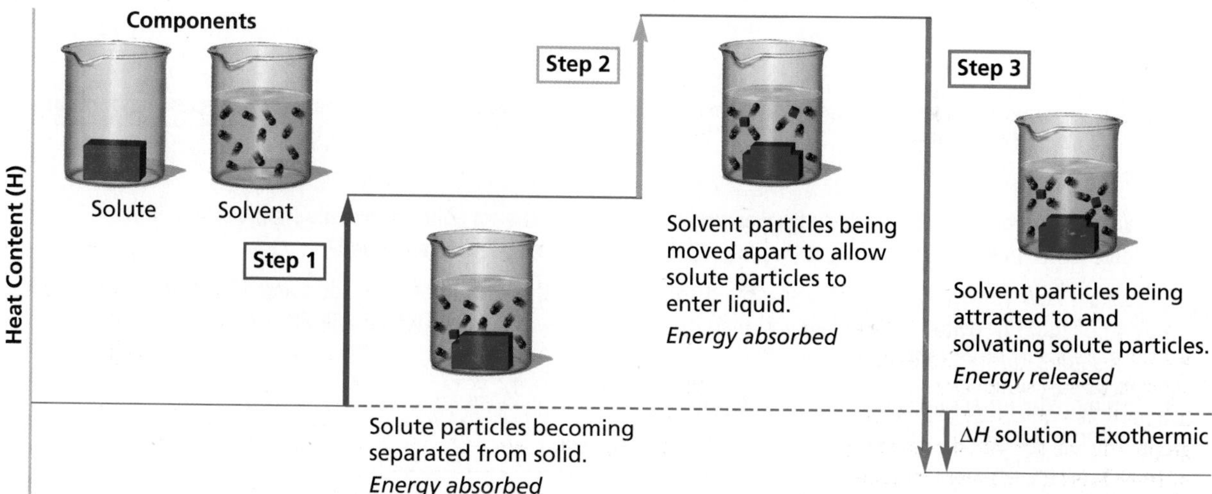
Components
Solute
Solvent
Step 1
Step 2
Step 3
Heat Content (H)
Solute particles becoming separated from solid.
Energy absorbed
Solvent particles being moved apart to allow solute particles to enter liquid.
Energy absorbed
Solvent particles being attracted to and solvating solute particles.
Energy released
ΔH solution Exothermic

FIGURE 13-16 The graph shows the changes in the heat content that occur during the formation of a solution. How would the graph differ for a system with an endothermic heat of solution?

TABLE 13-5 ***Heats of Solution (kJ/mol solute at 25°C)***

Substance	Heat of solution	Substance	Heat of solution
$AgNO_3(s)$	+22.59	$KOH(s)$	–57.61
$CH_3COOH(l)$	–1.51	$LiCl(s)$	–37.03
$HCl(g)$	–74.84	$MgSO_4(s)$	+15.9
$HI(g)$	–81.67	$NaCl(s)$	+3.88
$KCl(s)$	+17.22	$NaNO_3(s)$	+20.50
$KClO_3(s)$	+41.38	$NaOH(s)$	–44.51
$KI(s)$	+20.33	$NH_3(g)$	–30.50
$KNO_3(s)$	+34.89	$NH_4Cl(s)$	+14.78
		$NH_4NO_3(s)$	+25.69

The amount of heat energy absorbed or released when a specific amount of solute dissolves in a solvent is the **heat of solution.** From the model in Figure 13-16, you can see that the heat of solution is negative (heat is released) when the sum of attractions from Steps 1 and 2 is less than Step 3. The heat of solution is positive (heat is absorbed) when the sum of attractions from Steps 1 and 2 is greater than Step 3.

You know that heating decreases the solubility of a gas, so dissolution of gases is exothermic. How do the values for the heats of solution in Table 13-5 support this idea of exothermic solution processes for gaseous solutes?

In the gaseous state, molecules are so far apart that there are virtually no intermolecular forces of attraction between them. Therefore, the solute-solute interaction has little effect on the heat of a solution of a gas. Energy is released when a gas dissolves in a liquid because attraction between solute gas and solvent molecules outweighs the energy needed to separate solvent molecules.

SECTION REVIEW

1. Why would you expect a packet of sugar to dissolve faster in hot tea than in iced tea?
2. Explain how you would prepare a saturated solution of sugar in water. How would you then make it a supersaturated solution?
3. Explain why ethanol will dissolve in water and carbon tetrachloride will not.
4. When a solute molecule is solvated, is heat released or absorbed?
5. If a warm bottle of soda and a cold bottle of soda are opened, which will effervesce more and why?

Artificial Blood

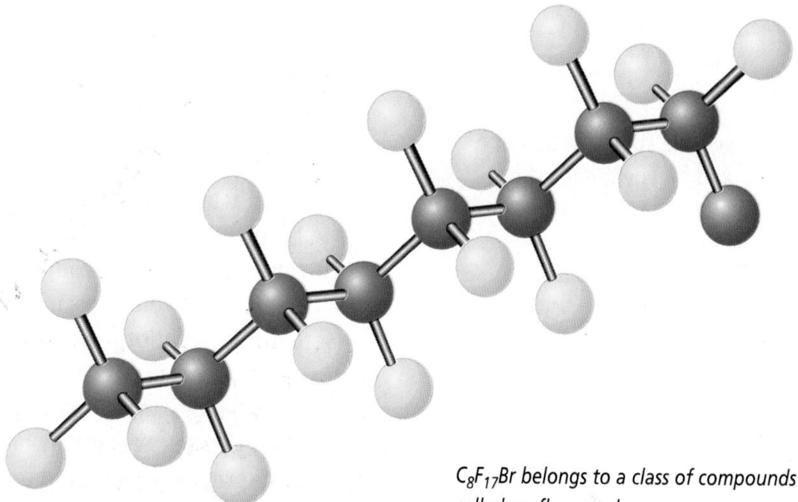

$C_8F_{17}Br$ belongs to a class of compounds called perfluorocarbons.

A patient lies bleeding on a stretcher. The doctor leans over to check the patient's wounds and barks an order to a nearby nurse: "Get him a unit of artificial blood, stat!" According to Dr. Peter Keipert, Program Director of Oxygen Carriers Development at Alliance Pharmaceutical Corp., this scenario may soon be commonplace thanks to a synthetic solution that can perform one of the main functions of human blood—transporting oxygen.

The hemoglobin inside red blood cells collects oxygen in our lungs, transports the oxygen to all the tissues of the body, and then takes carbon dioxide back to the lungs. Dr. Keipert's blood substitute accomplishes the same task, but it uses oily chemicals called perfluorocarbons instead of hemoglobin to transport the oxygen. The perfluorocarbons are carried in a water-based saline solution, but because oil and water do not mix, a bonding chemical called a surfactant is added to hold the mixture together. The perfluorocarbons are sheared into tiny droplets and then coated with the bonding molecules. One end of these molecules attaches to the perfluorocarbon, and the other end attaches to the water, creating a milky solution. The process is similar to shaking a bottle of salad dressing to form a creamy mixture of the oily portion and the liquid portion. The blood-substitute solution, called Oxygent™, is administered to a patient in the same way regular blood is. The perfluorocarbons are eventually exhaled through the lungs.

Dr. Keipert is quick to point out that Oxygent is not a true artificial blood. The solution only functions to carry gases to and from tissues; it cannot clot or perform any of the immune-system functions that blood does. Still, the substitute has several advantages over real blood. Oxygent has a shelf life of more than a year. Oxygent also eliminates many of the risks associated with blood transfusions. There is no need to match blood types, and the blood substitute prevents the possibility of spreading viruses such as HIV. Because the substitute can dissolve larger amounts of oxygen than real blood can, smaller amounts of the solution are needed.

Oxygent is currently being tested in surgical patients.

"Once this product is approved and has been demonstrated to be safe and effective in elective surgery, I think you will see its use spread into the emergency critical-care arena," says Dr. Keipert. "A patient who has lost a lot of blood and who is currently being resuscitated with normal fluids like saline solutions would be given Oxygent as an additional oxygen-delivery agent in the emergency room."

Another place where artificial blood might prove to be a lifesaver is on the battlefield, where blood is not readily available for transfusion.

SECTION 13-3

OBJECTIVES

- Given the mass of solute and volume of solvent, calculate the concentration of a solution.
- Given the concentration of a solution, determine the amount of solute in a given amount of solution.
- Given the concentration of a solution, determine the amount of solution that contains a given amount of solute.

Concentration of Solutions

The **concentration** *of a solution is a measure of the amount of solute in a given amount of solvent or solution.* Some medications are solutions of drugs—a one-teaspoon dose at the correct concentration might cure the patient, while the same dose in the wrong concentration might kill the patient.

In this section, we introduce two different ways of expressing the concentrations of solutions: molarity and molality.

Sometimes solutions are referred to as "dilute" or "concentrated." These are relative terms. "Dilute" means that there is a relatively small amount of solute in a solvent. "Concentrated," on the other hand, means that there is a relatively large amount of solute in a solvent. Note that these terms are unrelated to the degree to which a solution is saturated. A saturated solution of a substance that is not very soluble might be very dilute.

Molarity

Molarity *is the number of moles of solute in one liter of solution.* To find the molarity of a solution, you must know the molar mass of the solute. For example, a "one-molar" solution of sodium hydroxide, NaOH, contains one mole of NaOH in every liter of solution. The symbol for molarity is M, and the concentration of a one-molar solution of sodium hydroxide is written as 1 M NaOH.

One mole of NaOH has a mass of 40.0 g. If this quantity of NaOH is dissolved in enough water to make exactly 1.00 L of solution, the solution is a 1 M solution. If 20.0 g of NaOH, which is 0.500 mol, is dissolved in enough water to make 1.00 L of solution, a 0.500 M NaOH solution is produced. This relationship between molarity, moles, and volume may be expressed in the following ways.

$$\text{molarity (M)} = \frac{\text{amount of solute (mol)}}{\text{volume of solution (L)}}$$

$$= \frac{0.500 \text{ mol NaOH}}{1.00 \text{ L}}$$

$$= 0.500 \text{ M NaOH}$$

If twice the molar mass of NaOH, 80.0 g, is dissolved in enough water to make 1 L of solution, a 2 M solution is produced. The molarity of any solution can be calculated by dividing the number of moles of solute by the number of liters of solution.

Note that a 1 M solution is *not* made by adding 1 mol of solute to 1 L of *solvent*. In such a case, the final total volume of the solution would not be 1 L. Instead, 1 mol of solute is first dissolved in less than 1 L of solvent. Then the resulting solution is carefully diluted with more solvent to bring the *total volume* to 1 L, as shown in Figure 13-17. The following sample problem will show you how molarity is often used.

FIGURE 13-17 The preparation of a 0.5000 M solution of $CuSO_4 \cdot 5H_2O$ starts with calculating the mass of solute needed.

Start by calculating the mass of $CuSO_4 \cdot 5H_2O$ needed. Making a liter of this solution requires 0.5000 mol of solute. Convert the moles to mass by multiplying by the molar mass of $CuSO_4 \cdot 5H_2O$. This mass is calculated to be 124.8 g.

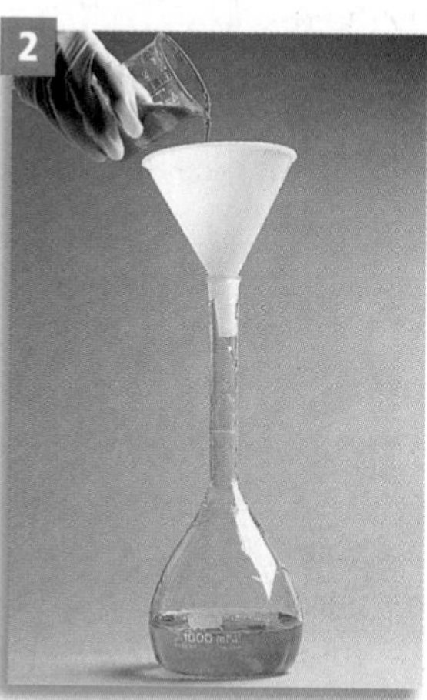

Add some solvent to the solute to dissolve it, then pour it into a 1.0 L volumetric flask.

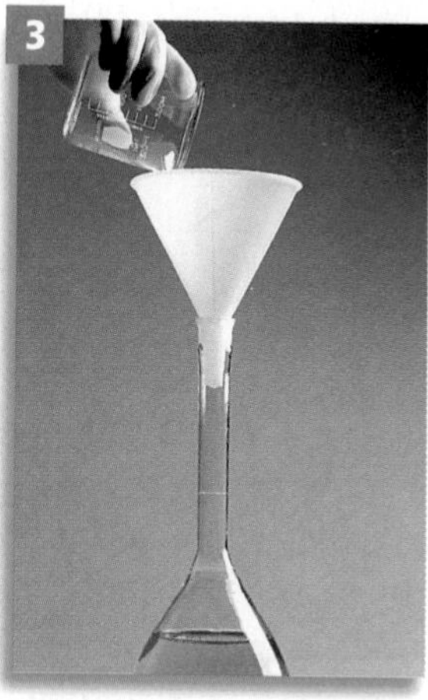

Rinse the weighing beaker with more solvent to remove all the solute, and pour the rinse into the flask. Add water until the volume of the solution nears the neck of the flask.

Put the stopper in the flask, and swirl the solution thoroughly.

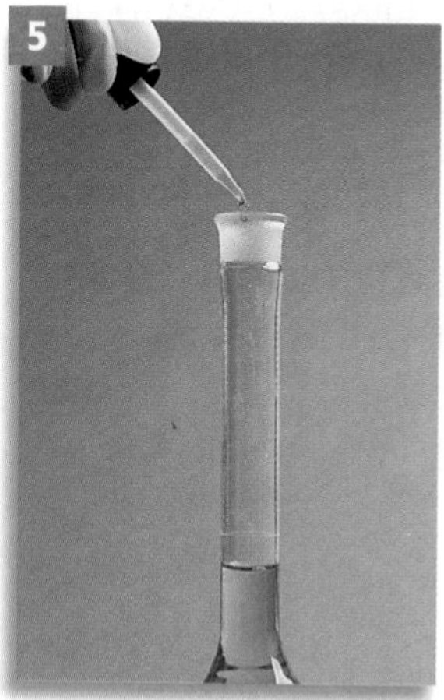

Carefully fill the flask to the 1.0 L mark with water.

Restopper the flask and invert it at least 10 times to ensure complete mixing.

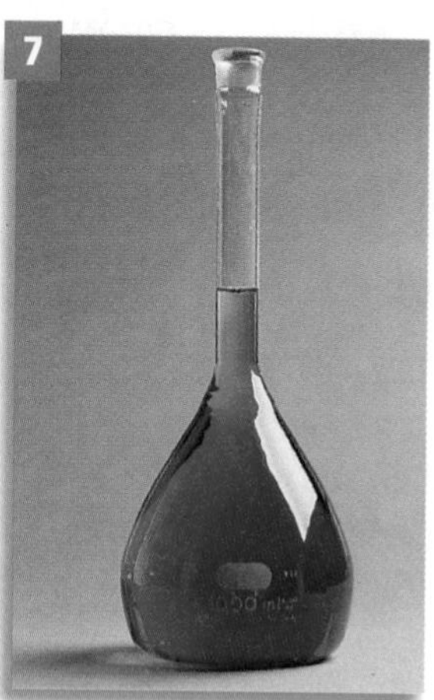

The resulting solution has 0.5000 mol of solute dissolved in 1.000 L of solution, which is a 0.5000 M concentration.

SAMPLE PROBLEM 13-1

You have 3.50 L of solution that contains 90.0 g of sodium chloride, NaCl. What is the molarity of that solution?

SOLUTION

1 *ANALYZE*

Given: solute mass = 90.0 g NaCl
solution volume = 3.50 L
Unknown: molarity of NaCl solution

2 *PLAN*

Molarity is the number of moles of solute per liter of solution. The solute is described in the problem by mass, not the amount in moles. You need one conversion (grams to moles of solute) using the inverted molar mass of NaCl to arrive at your answer.

$$\text{grams of solute} \longrightarrow \text{number of moles of solute} \longrightarrow \text{molarity}$$

$$\text{g NaCl} \times \frac{1 \text{ mol NaCl}}{\text{g NaCl}} = \text{mol NaCl}$$

$$\frac{\text{amount of solute (mol)}}{V \text{ solution (L)}} = \text{molarity of solution (M)}$$

3 *COMPUTE*

You will need the molar mass of NaCl.
NaCl = 58.44 g/mol

$$90.0 \cancel{\text{ g NaCl}} \times \frac{1 \text{ mol NaCl}}{58.44 \cancel{\text{ g NaCl}}} = 1.54 \text{ mol NaCl}$$

$$\frac{1.54 \text{ mol NaCl}}{3.50 \text{ L of solution}} = 0.440 \text{ M NaCl}$$

4 *EVALUATE*

Because each factor involved is limited to three significant digits, the answer should have three significant digits, which it does. The units cancel correctly to give the desired moles of solute per liter of solution, which is molarity.

SAMPLE PROBLEM 13-2

You have 0.8 L of a 0.5 M HCl solution. How many moles of HCl does this solution contain?

SOLUTION

1 *ANALYZE*

Given: volume of solution = 0.8 L
concentration of solution = 0.5 M HCl
Unknown: moles HCl in a given volume

2 *PLAN*

The molarity indicates the moles of solute that are in one liter of solution. Given the volume of the solution, the number of moles of solute can then be found.

$$\text{concentration (mol of HCl/L of solution)} \times \text{volume (L of solution)} = \text{mol of HCl}$$

3 COMPUTE

$$\frac{0.5 \text{ mol HCl}}{1.0 \text{ L of solution}} \times 0.8 \text{ L of solution} = 0.4 \text{ mol HCl}$$

4 EVALUATE The answer is correctly given to one significant digit. The units cancel correctly to give the desired unit, mol. There should be less than 0.5 mol HCl, because less than 1 L of solution was used.

SAMPLE PROBLEM 13-3

To produce 40.0 g of silver chromate, you will need at least 23.4 g of potassium chromate in solution as a reactant. All you have on hand in the stock room is 5 L of a 6.0 M K_2CrO_4 solution. What volume of the solution is needed to give you the 23.4 g K_2CrO_4 needed for the reaction?

SOLUTION

1 ANALYZE

Given: volume of solution = 5 L
concentration of solution = 6.0 M K_2CrO_4
mass of solute = 23.4 g K_2CrO_4
mass of product = 40.0 g Ag_2CrO_4
Unknown: volume K_2CrO_4 in L

2 PLAN The molarity indicates the moles of solute that are in 1 L of solution. Given the mass of solute needed, the amount in moles of solute can then be found. Use the molarity and the amount in moles of K_2CrO_4 to determine the volume of K_2CrO_4 that will provide 23.4 g.

$$\text{grams of solute} \longrightarrow \text{moles solute}$$
$$\text{moles solute and molarity} \longrightarrow \text{liters of solution needed}$$

3 COMPUTE To get the moles of solute, you'll need to calculate the molar mass of K_2CrO_4.

$$1 \text{ mol } K_2CrO_4 = 194.2 \text{ g } K_2CrO_4$$

$$23.4 \text{ g } K_2CrO_4 = \frac{1.0 \text{ mol } K_2CrO_4}{194.2 \text{ g } K_2CrO_4} = 0.120 \text{ mol } K_2CrO_4$$

$$6.0 \text{ M } K_2CrO_4 = \frac{0.120 \text{ mol } K_2CrO_4}{x \text{ L } K_2CrO_4 \text{ soln}}$$

$$x = 0.020 \text{ L } K_2CrO_4 \text{ soln}$$

4 EVALUATE The answer is correctly given to two significant digits. The units cancel correctly to give the desired unit, liters of solution.

PRACTICE

1. What is the molarity of a solution composed of 5.85 g of potassium iodide, KI, dissolved in enough water to make 0.125 L of solution? *Answer* 0.282 M KI
2. How many moles of H_2SO_4 are present in 0.500 L of a 0.150 M H_2SO_4 solution? *Answer* 0.0750 mol
3. What volume of 3.00 M NaCl is needed for a reaction that requires 146.3 g of NaCl? *Answer* 0.834 L

Molality

Molality *is the concentration of a solution expressed in moles of solute per kilogram of solvent.* A solution that contains 1 mol of solute, sodium hydroxide, NaOH, for example, dissolved in exactly 1 kg of solvent is a "one-molal" solution. The symbol for molality is *m*, and the concentration of this solution is written as 1 *m* NaOH.

One mole of NaOH has a molar mass of 40.0 g, so 40.0 g of NaOH dissolved in 1 kg of water results in a one-molal NaOH solution. If 20.0 g of NaOH, which is 0.500 mol of NaOH, is dissolved in exactly 1 kg of water, the concentration of the solution is 0.500 *m* NaOH.

$$\text{molality} = \frac{\text{moles solute}}{\text{mass of solvent (kg)}}$$

$$\frac{0.500 \text{ mol NaOH}}{1 \text{ kg } H_2O} = 0.500\ m \text{ NaOH}$$

If 80.0 g of sodium hydroxide, which is 2 mol, is dissolved in 1 kg of water, a 2.00 *m* solution of NaOH is produced. The molality of any solution can be found by dividing the number of moles of solute by the mass in kilograms of the solvent in which it is dissolved. Note that if the amount of solvent is expressed in grams, the mass of solvent must be converted to kilograms by multiplying by the following conversion factor.

$$1 \text{ kg}/1000 \text{ g}$$

Figure 13-18 shows how a 0.5000 *m* solution of $CuSO_4•5H_2O$ is prepared, in contrast with the 0.5000 M solution in Figure 13-17.

FIGURE 13-18 The preparation of a 0.5000 *m* solution of $CuSO_4•5H_2O$ also starts with the calculation of the mass of solute needed.

1

Calculate the mass of $CuSO_4 \cdot 5H_2O$ needed. To make this solution, each kilogram of solvent (1000 g) will require 0.5000 mol of $CuSO_4 \cdot 5H_2O$. This mass is calculated to be 124.8 g.

2

Add exactly 1 kg of solvent to the solute in the beaker. Because the solvent is water, 1 kg will equal 1000 mL.

3

Mix thoroughly.

4

The resulting solution has 0.5000 mol of solute dissolved in 1 kg of solvent.

Concentrations are expressed as molalities when studying properties of solutions related to vapor pressure and temperature changes. Molality is used because it does not change with changes in temperature. Below is a comparison of the equations for molarity and molality.

$$\text{molarity, M} = \frac{\text{amount of A (mol)}}{\text{volume of solution (L)}}$$

$$\text{molality, } m = \frac{\text{amount of A (mol)}}{\text{mass of solvent (kg)}}$$

SAMPLE PROBLEM 13-4

A solution was prepared by dissolving 17.1 g of sucrose (table sugar, $C_{12}H_{22}O_{11}$) in 125 g of water. Find the molal concentration of this solution.

SOLUTION

1 *ANALYZE*

Given: solute mass = 17.1 g $C_{12}H_{22}O_{11}$
solvent mass = 125 g H_2O

Unknown: molal concentration

2 *PLAN*

To find molality, you need moles of solute and kilograms of solvent. The given grams of sucrose must be converted to moles. The mass in grams of solvent must be converted to kilograms.

$$\text{mol } C_{12}H_{22}O_{11} = \frac{\text{g } C_{12}H_{22}O_{11}}{\text{molar mass } C_{12}H_{22}O_{11}}$$

$$\text{kg } H_2O = \text{g } H_2O \times \frac{1 \text{ kg}}{1000 \text{ g}}$$

$$\text{molality } C_{12}H_{22}O_{11} = \frac{\text{mol } C_{12}H_{22}O_{11}}{\text{kg } H_2O}$$

3 *COMPUTE*

Use the periodic table to compute the molar mass of $C_{12}H_{22}O_{11}$.
$C_{12}H_{22}O_{11} = 342.34 \text{ g/mol}$

$$17.1 \cancel{\text{g } C_{12}H_{22}O_{11}} \times \frac{1 \text{ mol } C_{12}H_{22}O_{11}}{342.34 \cancel{\text{g } C_{12}H_{22}O_{11}}} = 0.0500 \text{ mol } C_{12}H_{22}O_{11}$$

$$\frac{125 \cancel{\text{g}} \text{ } H_2O}{1000 \cancel{\text{g}}\text{/kg}} = 0.125 \text{ kg } H_2O$$

$$\frac{0.0500 \text{ mol } C_{12}H_{22}O_{11}}{0.125 \text{ kg } H_2O} = 0.400 \, m \, C_{12}H_{22}O_{11}$$

4 *EVALUATE*

The answer is correctly given to three significant digits. The unit mol solute/kg solvent is correct for molality.

SAMPLE PROBLEM 13-5

A solution of iodine, I_2, in carbon tetrachloride, CCl_4, is used when iodine is needed for certain chemical tests. How much iodine must be added to prepare a 0.480 *m* solution of iodine in CCl_4 if 100.0 g of CCl_4 is used?

SOLUTION

1 ANALYZE

Given: molality of solution = 0.480 m I_2
mass of solvent = 100.0 g CCl_4

Unknown: mass of solute

2 PLAN

Your first step should be to convert the grams of solvent to kilograms. The molality gives you the moles of solute, which can be converted to the grams of solute using the molar mass of I_2.

3 COMPUTE

Use the periodic table to compute the molar mass of I_2.
$I_2 = 253.8$ g/mol

$$100.0\ \cancel{\text{g CCl}_4} \times \frac{1\ \text{kg}}{1000\ \cancel{\text{g CCl}_4}} = 0.100\ \text{kg CCl}_4$$

$$0.480\ m = \frac{x\ \text{mol I}_2}{0.1\ \text{kg H}_2\text{O}} = 0.0480\ \text{mol I}_2$$

$$0.0480\ \cancel{\text{mol I}_2} \times \frac{253.8\ \text{g I}_2}{\cancel{\text{mol I}_2}} = 12.2\ \text{g I}_2$$

4 EVALUATE

The answer has three significant digits and the units for mass of I_2.

PRACTICE

1. What is the molality of a solution composed of 255 g of acetone, $(CH_3)_2CO$, dissolved in 200. g of water? *Answer* 22 m acetone
2. What quantity, in grams, of methanol, CH_3OH, is required to prepare a 0.244 m solution in 400. g of water? *Answer* 3.12 g CH_3OH
3. How many grams of $AgNO_3$ are needed to prepare a 0.125 m solution in 250. mL of water? *Answer* 5.31 g $AgNO_3$
4. What is the molality of a solution containing 18.2 g HCl and 250. g of water? *Answer* 1.99 m

SECTION REVIEW

1. What quantity represents the ratio of the number of moles of solute for a given volume of solution?
2. Five grams of sugar, $C_{12}H_{22}O_{11}$, are dissolved in water to make 1 L of solution. What is the concentration of this solution expressed as a molarity?

CHAPTER 13 REVIEW

CHAPTER SUMMARY

13-1

- Solutions are homogeneous mixtures.
- Mixtures are classified as solutions, suspensions, or colloids, depending on the size of the solute particles in the mixture.
- The dissolved substance is the solute. Solutions that have water as a solvent are aqueous solutions.
- Solutions can consist of solutes and solvents that are solids, liquids, or gases.
- Suspensions settle out upon standing. Colloids do not settle out, and they scatter light that is shined through them.
- Most ionic solutes and some molecular solutes form aqueous solutions that conduct an electric current. These solutes are called electrolytes.
- Nonelectrolytes are solutes that dissolve in water to form solutions that do not conduct.

Vocabulary

colloid (397)
electrolyte (399)
nonelectrolyte (400)
soluble (395)
solute (396)
solution (396)
solvent (396)
suspension (397)

13-2

- A solute dissolves at a rate that depends on the surface area of the solute, how vigorously the solution is mixed, and the temperature of the solvent.
- The solubility of a substance indicates how much of that substance will dissolve in a specified amount of solvent under certain conditions.
- The solubility of a substance depends on the temperature.
- The solubility of gases in liquids increases with increases in pressure.
- The solubility of gases in liquids decreases with increases in temperature.
- The overall energy change per mole during solution formation is called the heat of solution.

Vocabulary

effervescence (407)
heat of solution (410)
Henry's law (407)
hydration (405)
immiscible (406)
miscible (406)
saturated solution (403)
solubility (404)
solvated (409)
solution equilibrium (402)
supersaturated solution (403)
unsaturated solution (403)

13-3

- Two useful expressions of concentration are molarity and molality.
- The molar concentration of a solution represents the ratio of moles of solute to liters of solution.
- The molal concentration of a solution represents the ratio of moles of solute to kilograms of solvent.

Vocabulary

concentration (412)
molality (416)
molarity (412)

REVIEWING CONCEPTS

1. a. What is the Tyndall effect?
   b. Identify one example of this effect. (13-1)
2. Given an unknown mixture consisting of two or more substances, explain one technique that could be used to determine whether that mixture is a true solution, a colloid, or a suspension. (13-1)
3. a. What is solution equilibrium?
   b. What factors determine the point at which a given solute-solvent combination reaches equilibrium? (13-2)
4. a. What is a saturated solution?
   b. What visible evidence indicates that a solution is saturated?
   c. What is an unsaturated solution? (13-2)
5. a. What is meant by the solubility of a substance?
   b. What condition(s) must be specified when expressing the solubility of a substance? (13-2)
6. a. What rule of thumb is useful for predicting whether one substance will dissolve in another?
   b. Describe what the rule means in terms of various combinations of polar and nonpolar solutes and solvents. (13-2)
7. a. How does pressure affect the solubility of a gas in a liquid?
   b. What law is a statement of this relationship? (13-1)
   c. If the pressure of a gas above a liquid is increased, what happens to the amount of the gas that will dissolve in the liquid, if all other conditions remain constant?
   d. Two bottles of soda are opened. One is a cold bottle and the other is partially frozen. Which system will show more effervescence and why? (13-2)
8. Based on Figure 13-15, determine the solubility of each of the following in grams of solute per 100. g H_2O.
   a. $NaNO_3$ at 10°C
   b. KNO_3 at 60°C
   c. NaCl at 50°C (13-2)
9. Based on Figure 13-15, at what temperature would each of the following solubility levels be observed?
   a. 40 g KCl in 100 g H_2O
   b. 100 g $NaNO_3$ in 100 g H_2O
   c. 50 g KNO_3 in 100 g H_2O (13-2)
10. The heat of solution for $AgNO_3$ is +22.8 kJ/mol.
    a. Write the equation that represents the dissolution of $AgNO_3$ in water.
    b. Is the dissolution process endothermic or exothermic? Is the crystallization process endothermic or exothermic?
    c. As $AgNO_3$ dissolves, what change occurs in the temperature of the solution?
    d. When the system is at equilibrium, how do the rates of dissolution and crystallization compare?
    e. If the solution is then heated, how will the rates of dissolution and crystallization be affected? Why?
    f. How will the increased temperature affect the amount of solute that can be dissolved?
    g. If the solution is allowed to reach equilibrium and is then cooled, how will the system be affected? (13-2)
11. Under what circumstances might we prefer to express solution concentrations in terms of
    a. molarity?
    b. molality? (13-3)
12. What opposing forces are at equilibrium in the sodium chloride system shown in Figure 13-7? (13-2)

PROBLEMS

Solubility

13. Plot a solubility graph for $AgNO_3$ from the following data, with grams of solute (by increments of 50) per 100 grams of H_2O on the vertical axis and with temperature in °C on the horizontal axis.

Grams solute per 100 g H_2O	Temperature (°C)
122	0
216	30
311	40
440	60
585	80
733	100

a. How does the solubility of $AgNO_3$ vary with the temperature of the water?
b. Estimate the solubility of $AgNO_3$ at 35°C, 55°C, and 75°C.
c. At what temperature would the solubility of $AgNO_3$ be 275 g per 100 g of H_2O?
d. If 100 g of $AgNO_3$ were added to 100 g of H_2O at 10°C, would the resulting solution be saturated or unsaturated? What would occur if 325 g of $AgNO_3$ were added to 100 g of H_2O at 35°C?

14. If a saturated solution of KNO_3 in 100. g of H_2O at 60°C is cooled to 20°C, approximately how many grams of the solute will precipitate out of the solution? (Use Table 13-4.)

Molarity

15. Determine the molarity of each of the following solutions:
a. 20.0 g NaOH in enough H_2O to make 2.00 L of solution
b. 14.0 g NH_4Br in enough H_2O to make 150. mL of solution
c. 32.7 g H_3PO_4 in enough H_2O to make 500. mL of solution (Hint: See Sample Problem 13-1.)

16. Determine the number of grams of solute needed to make solutions of the following volumes and concentrations:
a. 1.00 L of a 3.50 M solution of H_2SO_4
b. 2.50 L of a 1.75 M solution of $Ba(NO_3)_2$
c. 500. mL of a 0.250 M solution of KOH

17. a. How many moles of NaOH are contained in 65.0 mL of a 2.20 M solution of NaOH in H_2O? (Hint: See Sample Problem 13-2.)
b. How many grams of NaOH does this represent?

18. What is the molality of a solution made by dissolving 26.42 g of $(NH_4)_2SO_4$ in enough H_2O to make 50.00 mL of solution?

19. If 100. mL of a 12.0 M HCl solution is diluted to 2.00 L, what is the molarity of the final solution?

20. How many milliliters of 16.0 M HNO_3 would be required to prepare 750. mL of a 0.500 M solution? (Hint: See Sample Problem 13-3.)

21. How many milliliters of 0.54 M $AgNO_3$ would contain 0.34 g of pure $AgNO_3$?

22. What mass of each product results if 750. mL of 6.00 M H_3PO_4 reacts according to the following equation?
$2H_3PO_4 + 3Ca(OH)_2 \longrightarrow Ca_3(PO_4)_2 + 6H_2O$

23. How many milliliters of 18.0 M H_2SO_4 are required to react with 250. mL of 2.50 M $Al(OH)_3$ if the products are aluminum sulfate and water?

24. 75.0 g of an $AgNO_3$ solution reacts with enough Cu to produce 0.250 g of Ag by single replacement. What is the molarity of the initial $AgNO_3$ solution if $Cu(NO_3)_2$ is the other product?

Molality

25. Determine the molality of each of the following solutions:
a. 294.3 g H_2SO_4 in 1.000 kg H_2O
b. 63.0 g HNO_3 in 0.250 kg H_2O
c. 10.0 g NaOH in 300. g H_2O (Hint: See Sample Problem 13-4.)

26. Determine the number of grams of solute needed to make each of the following solutions:
a. a 4.50 *m* solution of H_2SO_4 in 1.00 kg H_2O
b. a 1.00 *m* solution of HNO_3 in 2.00 kg H_2O
c. a 3.50 *m* solution of $MgCl_2$ in 0.450 kg H_2O (Hint: See Sample Problem 13-5.)

27. A solution is prepared by dissolving 17.1 g of sucrose, $C_{12}H_{22}O_{11}$, in 275 g of H_2O. What is the molality of that solution?

28. How many kilograms of H_2O must be added to 75.5 g of $Ca(NO_3)_2$ to form a 0.500 *m* solution?

29. How many grams of glucose, $C_6H_{12}O_6$, must be added to 750. g of H_2O to make a 1.25 *m* solution?

30. A solution made from ethanol, C_2H_5OH, and water is 1.75 *m*. How many grams of C_2H_5OH are contained per 250. g of water?

31. How many liters of water should be used to dissolve 65.0 g of NaCl to make a 0.450 *m* solution? Assume the density of H_2O is 1.00 g/mL.

32. If a 3.00 *m* solution of HNO_3 containing 2.25 kg H_2O is diluted so that the resulting solution contains 5.75 kg H_2O, what is the molality of that final solution?

MIXED REVIEW

33. How many moles of Na_2SO_4 are dissolved in 450 mL of a 0.250 M solution?

34. Citric acid is one component of some soft drinks. What is the molarity of citric acid in a 2 L solution made from 150 mg of citric acid, $C_6H_8O_7$?

35. How many grams KCl would be left if 350 mL of a 6.0 M KCl solution were evaporated to dryness?

36. Sodium metal reacts violently with water to form NaOH and release hydrogen gas. If 10.0 g of Na react completely with 1.00 L of water, what is the molarity of the NaOH solution formed by the reaction? Assume the final volume of the system is 1 L.

37. In cars, ethylene glycol, $C_2H_6O_2$, is used as a coolant and antifreeze. If a mechanic fills a radiator with 6.5 kg of ethylene glycol and 1.5 kg of water, what is the molality of the water?

38. Calculate the molality of a solution that contains 110.0 g toluene, $C_6H_5CH_3$, in 500. mL of ethanol, C_2H_5OH. The density of ethanol is 0.7894 g/mL.

CRITICAL THINKING

39. Predicting Outcomes You have been investigating the nature of suspensions, colloids, and solutions and have collected the following observational data on four unknown samples. From the data, infer whether each sample is a solution, suspension, or colloid.

DATA TABLE 1—SAMPLES

Sample	Color	Clarity (clear or cloudy)	Settle out	Tyndall effect
1	green	clear	no	no
2	blue	cloudy	yes	no
3	colorless	clear	no	yes
4	white	clear	no	yes

DATA TABLE 2—FILTRATE OF SAMPLES

Sample	Color	Clarity (clear or cloudy)	On filter paper	Tyndall effect
1	green	clear	nothing	no
2	blue	cloudy	gray solid	yes
3	colorless	cloudy	none	yes
4	white	clear	white solid	no

Based on your inferences in Data Table 1, you decide to conduct one more test of the particles. You filter the samples and then reexamine the filtrate. You obtain the data found in Data Table 2. Infer the classifications based on the data in Table 2.

TECHNOLOGY & LEARNING

40. Graphing Calculator Predicting Solubility from Tabular Data

The graphing calculator can be programmed to estimate data such as solubility at a given temperature. Given solubility measurements for KCl, you will use the data to predict its solubility at 50°C. Begin by creating a table of data. Then program the calculator to carry out an extrapolation. The last step will involve the solubility predictions.

A. Create lists L1 and L2.

Keystrokes for creating lists: STAT 4 2nd L1 , 2nd L2 ENTER STAT 1

Now enter the temperature data below in L1, and enter the solubility data in L2.

B. Program the extrapolation.

Keystrokes for naming the program: PRGM ▶ ▶ ENTER

DATA TABLE	
Temperature (°C)	Solubility (g/100. g H_2O)
0.0	28
20.0	34.2
40.0	40.1
60.0	45.8

Name the program: S O L U B I L ENTER

Program keystrokes: STAT ▶ 5 2nd L1 , 2nd L2 2nd : ALPHA " VARS 5 ▶ ▶ 1 X,T,θ + VARS 5 ▶ ▶ 2 ALPHA " STO▶ 2nd Y-VARS 1 1 2nd : PRGM ▶ 2 ALPHA T 2nd : ALPHA T STO▶ X,T,θ 2nd : PRGM ▶ 8 2nd : PRGM ▶ 3 2nd ALPHA " S O L U B I L 2nd TEST 1 ALPHA " 2nd : PRGM ▶ 3 2nd Y-VARS 1 1

C. Check the display.

Screen display: LinReg(ax+b) L1,L2: "ax+b" ->Y1:Prompt T: T ->X: ClrHome: Disp "SOLUB=":Disp Y1

Press 2nd QUIT to exit the program editor.

D. Run the program.

a. Press PRGM. Select SOLUBILITY and press ENTER ENTER. Now enter the temperature 50°C and press ENTER. The calculator will provide the solubility at that temperature in g/100 g H_2O.

b. Using the SOLUBILITY program with the same date set, extrapolate the solubility of KCl at 80°C.

c. Use the SOLUBILITY program with the data set shown below. Refer back to creating L1 and L2. Use L1 for temperature data and L2 for solubility data. Use the program and these data to extrapolate the solubility of KI at 50°C and 80°C.

Temperature (°C)	Solubility (g/100. g H_2O)
0.0	128
20.0	144
40.0	162
60.0	176

HANDBOOK SEARCH

41. Review the information on alloys in the *Elements Handbook.*

a. Why is aluminum such an important component of alloys?

b. What metals make up bronze?

c. What metals make up brass?

d. What is steel?

e. What is the composition of the mixture called cast iron?

42. Table 5B of the *Elements Handbook* contains carbon monoxide concentration data expressed as parts per million (ppm). The OSHA (Occupational Safety and Health Administration) limit for worker exposure to CO is 200 ppm for an eight-hour period.

a. At what concentration do harmful effects occur in less than one hour?

b. By what factor does the concentration in item (b) exceed the maximum limit set by OSHA?

RESEARCH & WRITING

43. Find out about the chemistry of emulsifying agents. How do these substances affect the dissolution of immiscible substances such as oil and water? As part of your research on this topic, find out why eggs are an emulsifying agent for baking mixtures.

ALTERNATIVE ASSESSMENT

44. Make a comparison of the electrolyte concentration in various brands of sports drinks. Using the labeling information for sugar, calculate the molarity of sugar in each product or brand. Construct a poster to show the results of your analysis of the product labels.

45. Write a set of instructions on how to prepare a solution that is 1 M $CuSO_4$ using $CuSO_4 \cdot 5H_2O$ as the solute. How do the instructions differ if the solute is anhydrous $CuSO_4$? Your instructions should include a list of all materials needed.

CHAPTER 14

Ions in Aqueous Solutions and Colligative Properties

These formations were made by the precipitation of ionic compounds from an aqueous solution.

Compounds in Aqueous Solutions

SECTION 14-1

Objectives

- Write equations for the dissolution of soluble ionic compounds in water.
- Predict whether a precipitate will form when solutions of soluble ionic compounds are combined, and write net ionic equations for precipitation reactions.
- Compare dissociation of ionic compounds with ionization of molecular compounds.
- Draw the structure of the hydronium ion, and explain why it is used to represent the hydrogen ion in solution.
- Distinguish between strong electrolytes and weak electrolytes.

As you have learned, solid compounds can be ionic or molecular. In an ionic solid, a crystal structure is made up of charged particles held together by ionic attractions. In a molecular solid, molecules are composed of covalently bonded atoms held together by intermolecular forces. When they dissolve in water, ionic compounds and molecular compounds behave differently.

Dissociation

When a compound that is made of ions dissolves in water, the ions separate from one another, as shown in Figure 14-1. *This separation of ions that occurs when an ionic compound dissolves is called* **dissociation.** For example, dissociation of sodium chloride and calcium chloride in water can be represented by the following equations. (As usual, (*s*) indicates a solid species and (*aq*) indicates a species in an aqueous solution. Note that the equation is balanced for charge as well as for atoms.)

$$NaCl(s) \xrightarrow{H_2O} Na^+(aq) + Cl^-(aq)$$

$$CaCl_2(s) \xrightarrow{H_2O} Ca^{2+}(aq) + 2Cl^-(aq)$$

Notice the number of ions produced per formula unit in the equations above. One formula unit of sodium chloride gives two ions in solution, whereas one formula unit of calcium chloride gives three ions in solution.

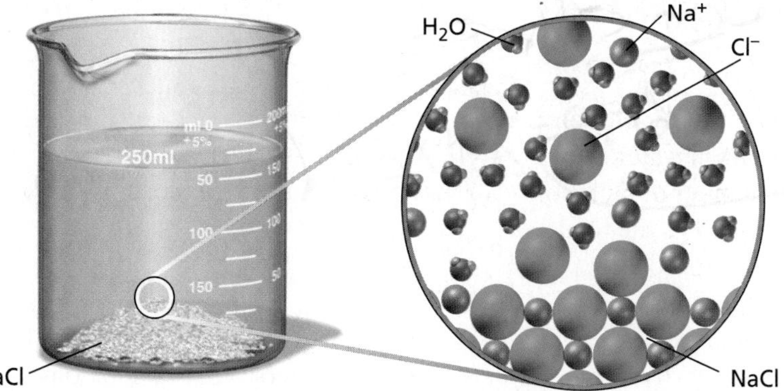

FIGURE 14-1 When NaCl dissolves in water, the ions separate as they leave the crystal.

Assuming 100% dissociation, a solution that contains 1 mol of sodium chloride contains 1 mol of Na^+ ions and 1 mol of Cl^- ions. In this book, you can assume 100% dissociation for all ionic compounds. The dissociation of NaCl can be represented as follows.

$$\underset{1\text{ mol}}{NaCl(s)} \xrightarrow{H_2O} \underset{1\text{ mol}}{Na^+(aq)} + \underset{1\text{ mol}}{Cl^-(aq)}$$

A solution that contains 1 mol of calcium chloride contains 1 mol of Ca^{2+} ions and 2 mol of Cl^- ions—a total of 3 mol of ions.

$$\underset{1\text{ mol}}{CaCl_2(s)} \xrightarrow{H_2O} \underset{1\text{ mol}}{Ca^{2+}(aq)} + \underset{2\text{ mol}}{2Cl^-(aq)}$$

SAMPLE PROBLEM 14-1

Write the equation for the dissolution of aluminum sulfate, $Al_2(SO_4)_3$, in water. How many moles of aluminum ions and sulfate ions are produced by dissolving 1 mol of aluminum sulfate? What is the total number of moles of ions produced by dissolving 1 mol of aluminum sulfate?

SOLUTION

1 ANALYZE **Given:** amount of solute = 1 mol $Al_2(SO_4)_3$
solvent identity = water
Unknown: **a.** moles of aluminum ions and sulfate ions
b. total number of moles of solute ions produced

2 PLAN The coefficients in the balanced dissociation equation will reveal the mole relationships, so you can use the equation to determine the number of moles of solute ions produced.

$$Al_2(SO_4)_3(s) \xrightarrow{H_2O} 2Al^{3+}(aq) + 3SO_4^{2-}(aq)$$

3 COMPUTE **a.** 1 mol $Al_2(SO_4)_3 \rightarrow$ 2 mol Al^{3+} + 3 mol SO_4^{2-}
b. 2 mol Al^{3+} + 3 mol SO_4^{2-} = 5 mol of solute ions

4 EVALUATE The equation is correctly balanced. Because one formula unit of $Al_2(SO_4)_3$ produces 5 ions, 1 mol of $Al_2(SO_4)_3$ produces 5 mol of ions.

PRACTICE

1. Write the equation for the dissolution of each of the following in water, and then determine the number of moles of each ion produced as well as the total number of moles of ions produced.
   a. 1 mol ammonium chloride
   b. 1 mol sodium sulfide
   c. 0.5 mol barium nitrate

Answer
a. $NH_4Cl(s) \xrightarrow{H_2O} NH_4^+(aq) + Cl^-(aq)$; 1 mol NH_4^+, 1 mol Cl^-, 2 mol ions
b. $Na_2S(s) \xrightarrow{H_2O} 2Na^+(aq) + S^{2-}(aq)$; 2 mol Na^+, 1 mol S^{2-}, 3 mol ions
c. $Ba(NO_3)_2(s) \xrightarrow{H_2O} Ba^{2+}(aq) + 2NO_3^-(aq)$; 0.5 mol Ba^{2+}, 1 mol NO_3^-, 1.5 mol ions

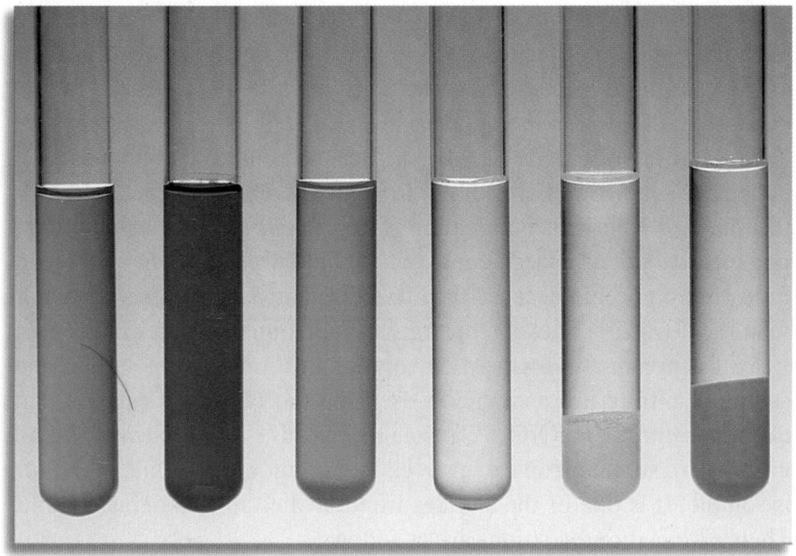

FIGURE 14-2 Ionic compounds can be soluble or insoluble in water. $NiCl_2$, $KMnO_4$, $CuSO_4$, and $Pb(NO_3)_2$ are soluble in water. AgCl and CdS are insoluble in water.

Precipitation Reactions

Although no compound is completely insoluble, compounds of very low solubility can be considered insoluble for most practical purposes. Some examples of ionic compounds that are soluble and insoluble in water are shown in Figure 14-2. It is difficult to write solubility rules that cover all possible conditions. However, we can write some general guidelines to help predict whether a compound made of a certain combination of ions is soluble. These general solubility guidelines are given in Table 14-1.

By looking at the table you can tell that most sodium compounds are soluble. Sodium carbonate, Na_2CO_3, is soluble because it contains sodium. Its dissociation equation is as follows.

$$Na_2CO_3(s) \xrightarrow{H_2O} 2Na^+(aq) + CO_3^{2-}(aq)$$

TABLE 14-1 ***General Solubility Guidelines***

1. Most sodium, potassium, and ammonium compounds are soluble in water.
2. Most nitrates, acetates, and chlorates are soluble.
3. Most chlorides are soluble, except those of silver, mercury(I), and lead. Lead(II) chloride is soluble in hot water.
4. Most sulfates are soluble, except those of barium, strontium, and lead.
5. Most carbonates, phosphates, and silicates are insoluble, except those of sodium, potassium, and ammonium.
6. Most sulfides are insoluble, except those of calcium, strontium, sodium, potassium, and ammonium.

Is calcium phosphate, $Ca_3(PO_4)_2$, soluble or insoluble? According to Table 14-1, most phosphates are insoluble. Calcium phosphate is not one of the exceptions listed, so it is insoluble. Dissociation equations cannot be written for insoluble compounds.

The information in Table 14-1 is also useful in predicting what will happen if solutions of two different soluble compounds are mixed. If the mixing results in a combination of ions that forms an insoluble compound, a double-replacement reaction and precipitation will occur. Precipitation occurs because the attraction between the ions is greater than the attraction between the ions and surrounding water molecules.

Will a precipitate form when solutions of ammonium sulfide and cadmium nitrate are combined? By using the table you can tell that cadmium nitrate, $Cd(NO_3)_2$, is soluble because it is a nitrate and all nitrates are soluble. You can also tell that ammonium sulfide, $(NH_4)_2S$, is soluble. It is one of the sulfides listed in the table as being soluble. Their dissociation equations are as follows.

$$(NH_4)_2S(s) \xrightarrow{H_2O} 2NH_4^+(aq) + S^{2-}(aq)$$

$$Cd(NO_3)_2(s) \xrightarrow{H_2O} Cd^{2+}(aq) + 2NO_3^-(aq)$$

FIGURE 14-3 Ammonium sulfide is a soluble compound that dissociates in water to form NH_4^+ and S^{2-} ions. Cadmium nitrate is a soluble compound that dissociates in water to form NO_3^- and Cd^{2+} ions. Precipitation of cadmium sulfide occurs when the two solutions are mixed.

The two possible products of a double-replacement reaction between $(NH_4)_2S$ and $Cd(NO_3)_2$ are ammonium nitrate, NH_4NO_3, and cadmium sulfide, CdS. (The question marks indicate that the states are unknown.)

$$(NH_4)_2S(aq) + Cd(NO_3)_2(aq) \longrightarrow NH_4NO_3(?) + CdS(?)$$

To decide whether a precipitate can form, you must know the solubilities of these two compounds. Consulting Table 14-1, you can see that NH_4NO_3 is soluble in water. However, CdS is insoluble. You can therefore predict that when solutions of ammonium sulfide and cadmium nitrate are combined, ammonium nitrate will not precipitate and cadmium sulfide will. As illustrated in Figure 14-3, crystals of CdS form when the solutions are mixed. In the following equation, the designations (*aq*) and (*s*) show that ammonium nitrate remains in solution and cadmium sulfide precipitates.

$$(NH_4)_2S(aq) + Cd(NO_3)_2(aq) \longrightarrow NH_4NO_3(aq) + CdS(s)$$

Net Ionic Equations

Reactions of ions in aqueous solution are usually represented by net ionic equations rather than formula equations. *A* **net ionic equation** *includes only those compounds and ions that undergo a chemical change in a reaction in an aqueous solution.* To write a net ionic equation, you first write an overall ionic equation. All soluble compounds are shown as dissociated ions in solution. The precipitates are shown as solids. The precipitation of cadmium sulfide described previously can be shown by the following overall ionic equation.

$$Cd^{2+}(aq) + 2NO_3^-(aq) + 2NH_4^+(aq) + S^{2-}(aq) \longrightarrow CdS(s) + 2NO_3^-(aq) + 2NH_4^+(aq)$$

Notice that the ammonium ion, NH_4^+, and the nitrate ion, NO_3^-, appear on both sides of this equation. Therefore, they have not undergone any chemical change and are still present in their original form. *Ions that do not take part in a chemical reaction and are found in solution both before and after the reaction are* **spectator ions.**

To convert an ionic equation into a net ionic equation, the spectator ions are canceled on both sides of the equation. Eliminating the NH_4^+ and NO_3^- ions from the overall ionic equation above gives the following net ionic equation.

$$Cd^{2+}(aq) + S^{2-}(aq) \longrightarrow CdS(s)$$

This net ionic equation applies not only to the reaction between $(NH_4)_2S$ and $Cd(NO_3)_2$ but also to *any* reaction in which a precipitate of cadmium sulfide forms when the ions are combined in solution. For example, it is also the net ionic equation for the precipitation of CdS when $CdSO_4$ and H_2S react.

SAMPLE PROBLEM 14-2

Identify the precipitate that forms when aqueous solutions of zinc nitrate and ammonium sulfide are combined. Write the equation for the possible double-replacement reaction. Then write the formula equation, overall ionic equation, and net ionic equation for the reaction.

SOLUTION

1 ANALYZE

Given: identity of reactants: zinc nitrate and ammonium sulfide
reaction medium: aqueous solution

Unknown: **a.** equation for the possible double-replacement reaction **b.** identity of the precipitate **c.** formula equation **d.** overall ionic equation **e.** net ionic equation

2 PLAN

Write the possible double-replacement reaction between $Zn(NO_3)_2$ and $(NH_4)_2S$. Use Table 14-1 to determine if any of the products are insoluble and will precipitate. Write a formula equation and an overall ionic equation, then cancel the spectator ions to produce a net ionic equation.

3 COMPUTE

a. The equation for the possible double-replacement reaction is as follows.

$$Zn(NO_3)_2(aq) + (NH_4)_2S(aq) \longrightarrow ZnS(?) + 2NH_4NO_3(?)$$

b. Table 14-1 reveals that zinc sulfide is not a soluble sulfide and is therefore a precipitate. Ammonium nitrate is soluble according to the table.

c. The formula equation is as follows.

$$Zn(NO_3)_2(aq) + (NH_4)_2S(aq) \longrightarrow ZnS(s) + 2NH_4NO_3(aq)$$

d. The overall ionic equation is as follows.

$$Zn^{2+}(aq) + 2NO_3^-(aq) + 2NH_4^+(aq) + S^{2-}(aq) \longrightarrow ZnS(s) + 2NH_4^+(aq) + 2NO_3^-(aq)$$

e. The ammonium and nitrate ions appear on both sides of the equation as spectator ions. The net ionic equation is as follows.

$$Zn^{2+}(aq) + S^{2-}(aq) \longrightarrow ZnS(s)$$

PRACTICE

1. Will a precipitate form if solutions of potassium sulfate and barium nitrate are combined? If so, write the net ionic equation for the reaction.

   Answer
   Yes;
   $Ba^{2+}(aq) + SO_4^{2-}(aq) \longrightarrow BaSO_4(s)$

2. Will a precipitate form if solutions of potassium nitrate and magnesium sulfate are combined? If so, write the net ionic equation for the reaction.

   Answer
   No

3. Will a precipitate form if solutions of barium chloride and sodium sulfate are combined? If so, identify the spectator ions and write the net ionic equation.

   Answer
   Yes; Na^+ and Cl^-;
   $Ba^{2+}(aq) + SO_4^{2-}(aq) \longrightarrow BaSO_4(s)$

4. Write the net ionic equation for the precipitation of nickel(II) sulfide.

   Answer
   $Ni^{2+}(aq) + S^{2-}(aq) \longrightarrow NiS(s)$

Ionization

Some molecular compounds can also form ions in solution. Usually such compounds are polar. *Ions are formed from solute molecules by the action of the solvent in a process called* **ionization.** The more-general meaning of this term is the creation of ions where there were none. Note that *ionization* is different from *dissociation*. When an ionic compound dissolves, the ions that were already present separate from one another. When a molecular compound dissolves and ionizes in a polar solvent, ions are formed where none existed in the undissolved compound. Like all ions in aqueous solution, the ions formed by such a molecular solute are hydrated. The heat released during the hydration of the ions provides the energy needed to break the covalent bonds.

In general, the extent to which a solute ionizes in solution depends on the strength of the bonds within the molecules of the solute and the strength of attraction between the solute and solvent molecules. If the strength of a bond within the solute molecule is weaker than the attractive forces of the solvent molecules, then the covalent bond of the solute breaks and the molecule is separated into ions. Hydrogen chloride, HCl, is a molecular compound that ionizes in aqueous solution. It contains a highly polar bond. The attraction between a polar HCl molecule and the polar water molecules is strong enough to break the HCl bond, forming hydrogen ions and chloride ions.

$$HCl \xrightarrow{H_2O} H^+(aq) + Cl^-(aq)$$

The Hydronium Ion

The H^+ ion attracts other molecules or ions so strongly that it does not normally exist alone. The ionization of hydrogen chloride in water is better described as a chemical reaction in which a proton is transferred directly from HCl to a water molecule, where it becomes covalently bonded to oxygen and forms H_3O^+.

$$H_2O(l) + HCl(g) \longrightarrow H_3O^+(aq) + Cl^-(aq)$$

This process is represented in Figure 14-4. *The H_3O^+ ion is known as the* **hydronium ion.**

The hydration of the H^+ ion to form the hydronium ion is a very favorable reaction. The energy released makes a large contribution to the energy needed to ionize a molecular solute. Many molecular compounds that ionize in an aqueous solution contain hydrogen and form H_3O^+.

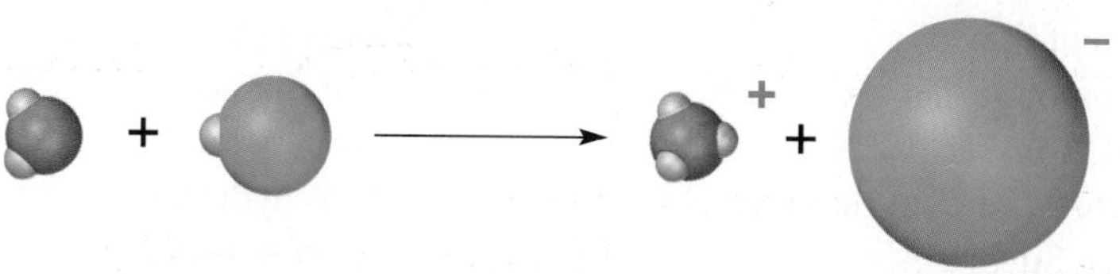

FIGURE 14-4 When hydrogen chloride gas dissolves in water, it ionizes to form an H^+ ion and a Cl^- ion. The H^+ ion immediately bonds to a water molecule, forming a hydronium ion. The aqueous solution of hydrogen chloride is called hydrochloric acid.

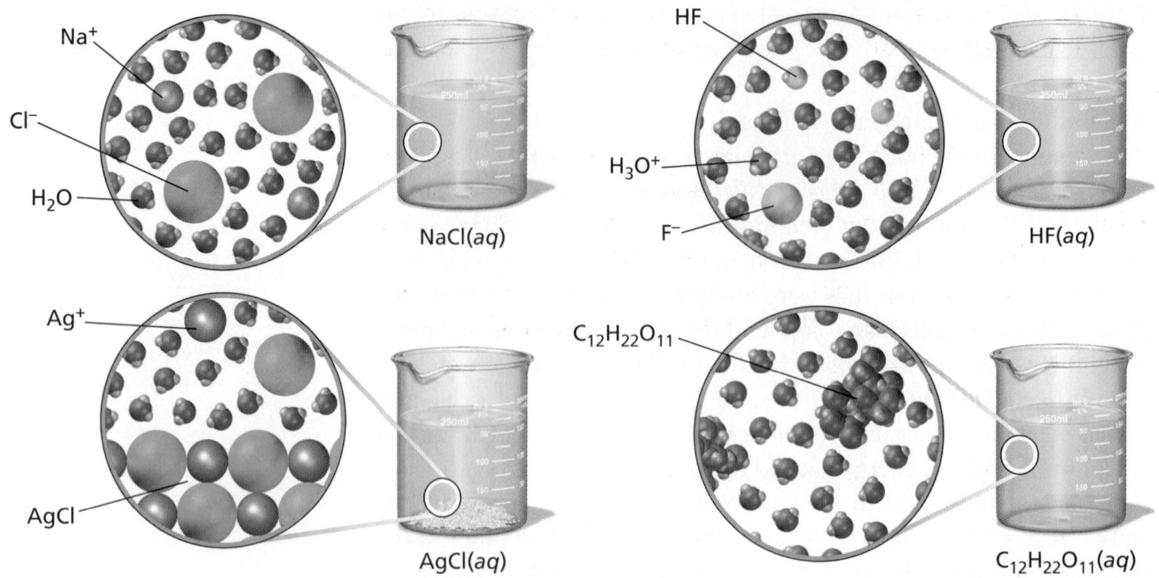

FIGURE 14-5 Strong electrolytes, such as NaCl and AgCl, yield only ions when they dissolve in aqueous solution. Weak electrolytes, such as HF, exist as both ions and unionized molecules in aqueous solution. Nonelectrolytes, such as sucrose, $C_{12}H_{22}O_{11}$, do not form any ions in aqueous solution.

Strong and Weak Electrolytes

As discussed in Chapter 13, substances that yield ions and conduct an electric current in solution are electrolytes. Substances that do not yield ions and do not conduct an electric current in solution are non-electrolytes. Hydrogen chloride is one of a series of compounds composed of hydrogen and the members of Group 17 (known as the halogens). The hydrogen halides are all molecular compounds with single polar-covalent bonds. All are gases, all are very soluble in water, and all are electrolytes. Hydrogen chloride, hydrogen bromide, and hydrogen iodide strongly conduct an electric current in an aqueous solution. However, hydrogen fluoride only weakly conducts an electric current at the same concentration. The strength with which substances conduct an electric current is related to their ability to form ions in solution, as shown in Figure 14-5.

Strong Electrolytes

Hydrogen chloride, hydrogen bromide, and hydrogen iodide are 100% ionized in dilute aqueous solution. *Any compound of which all or almost all of the dissolved compound exists as ions in an aqueous solution is a* **strong electrolyte.** Hydrogen chloride, hydrogen bromide, and hydrogen iodide are all acids in aqueous solution. These acids, several other acids, and all ionic compounds are strong electrolytes.

The distinguishing feature of strong electrolytes is that, to whatever extent they dissolve in water, they yield only ions. For example, an ionic compound may be highly soluble in water and dissociate into ions in solution, such as NaCl. Other ionic compounds may not dissolve much, but the amount that does dissolve exists solely as hydrated ions in solution. Silver chloride, AgCl, is almost insoluble, with a solubility of only

0.000 089 g/100. g of water. However, it is considered a strong electrolyte because the small amount that does dissolve exists as dissociated ions.

Weak Electrolytes

An aqueous solution of some molecular compounds contains not only dissolved ions but also some dissolved molecules that are not ionized. Hydrogen fluoride, HF, dissolves in water to give an acid solution known as hydrofluoric acid. However, the hydrogen-fluorine bond is much stronger than the bonds between hydrogen and the other halogens. When hydrogen fluoride dissolves, some molecules ionize. But the reverse reaction—the transfer of H^+ ions back to F^- ions to form hydrogen fluoride molecules—also takes place.

$$HF(aq) + H_2O(l) \rightleftharpoons H_3O^+(aq) + F^-(aq)$$

Thus, the concentration of dissolved unionized HF molecules remains high and the concentration of H_3O^+ and F^- ions remains low.

Hydrogen fluoride is an example of a *weak electrolyte. A* **weak electrolyte** *is a compound of which a relatively small amount of the dissolved compound exists as ions in an aqueous solution.* This is in contrast to a nonelectrolyte, of which none of the dissolved compound exists as ions. Another example of a weak electrolyte is acetic acid, CH_3COOH. Only a small percentage of the acetic acid molecules ionize in aqueous solution.

$$CH_3COOH(aq) + H_2O(l) \rightleftharpoons CH_3COO^-(aq) + H_3O^+(aq)$$

The description of an electrolyte as strong or weak must not be confused with the description of a solution as concentrated or dilute. Strong and weak electrolytes differ in the degree of ionization or dissociation. Concentrated and dilute solutions differ in the amount of solute dissolved in a given quantity of a solvent. Hydrochloric acid is always a strong electrolyte. This is true even in a solution that is 0.000 01 M—a very dilute solution. By contrast, acetic acid is always considered a weak electrolyte, even in a 10 M solution—a fairly concentrated solution.

SECTION REVIEW

1. Write the equation for the dissolution of $Sr(NO_3)_2$ in water. How many moles of strontium ions and nitrate ions are produced by dissolving 0.5 mol of strontium nitrate?
2. Will a precipitate form if solutions of magnesium acetate and strontium chloride are combined?
3. What determines whether a molecular compound will be ionized in a polar solvent?
4. How is a hydronium ion formed when a compound like HCl dissolves in water? What is its net charge?
5. Explain why HCl is a strong electrolyte and HF is a weak electrolyte.

GREAT DISCOVERIES

The Riddle of Electrolysis

HISTORICAL PERSPECTIVE

When Michael Faraday performed his electrochemical experiments, little was known about the relationship between matter and electricity. Chemists were still debating the existence of atoms, and the discovery of the electron was more than 50 years away. Combining his talents in electrical and chemical investigation, Faraday pointed researchers to the intimate connection between chemical reactions and electricity while setting the stage for the development of a new branch of chemistry.

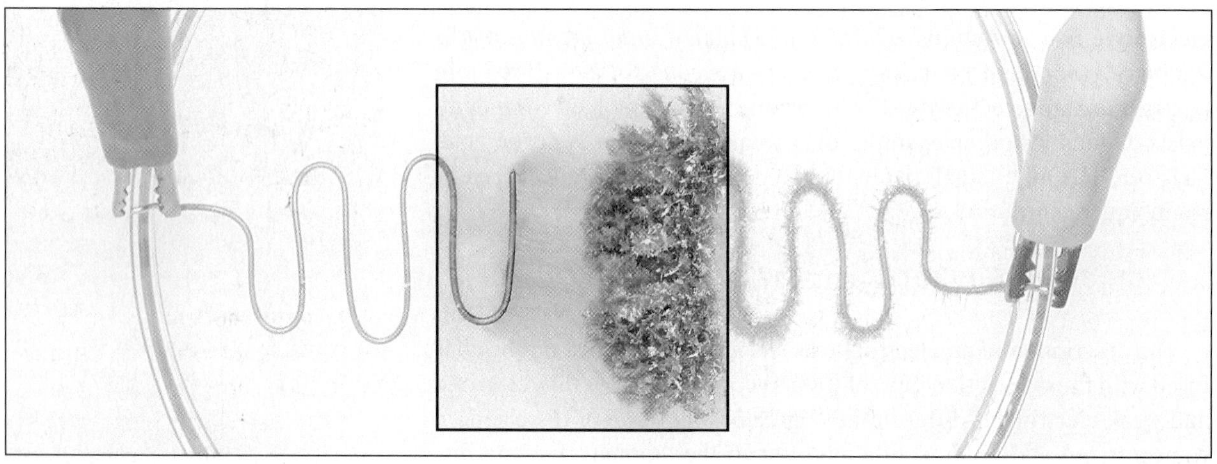

Chlorine being produced by electrolysis.

Electrifying Experiments

In 1800, the Italian physicist Alessandro Volta introduced his "voltaic pile," better known as the battery. The stack of alternating zinc and silver disks provided scientists with a source of electric current for the first time, and, in the words of the famous English chemist Humphry Davy, it acted as "an alarm-bell to experimenters in every part of Europe."

That same year, chemists discovered a new phenomenon using Volta's device. They immersed the two poles of a battery at different locations in a container of water. As current flowed, the water decomposed into its elemental components, with hydrogen evolving at the positive pole of the battery and oxygen evolving at the negative pole. Similar experiments using solutions of certain solids dissolved in water resulted in the decomposition of the solids, with the two products of their breakdown also evolving at opposite poles of the battery. This electrochemical decomposition was later named electrolysis.

The Roots of Electrolytic Theory

The discovery of electrolysis led two pioneering chemists to ponder the connection between chemical forces and electricity. One of them was Davy:

> *Is not what has been called chemical affinity merely the union . . . of particles in naturally opposite states? And are not chemical attractions of particles and electrical attractions of masses owing to one property and governed by one simple law?*

The Swedish chemist Jöns Jacob Berzelius took Davy's idea a step further. He postulated that matter consisted of combinations of "electropositive" and "electronegative" substances, classifying the parts by the pole at which they accumulated during electrolysis.

These ideas inspired two early electrolytic theories, each of which ultimately proved incorrect but which contributed to our present understanding of the phenomenon. The "contact theory"—the more widely accepted at the time—proposed that electrolytic current was due merely to the contact of the battery's metals with the electrolytic solution. According to this theory, the solution was composed of Berzelius's positive and negative substances, which were attracted by and drawn to the appropriate poles from a distance—in the same way a magnet attracts iron filings. The "chemical theory," on the other hand, attributed the flow of current to undefined changes in the solution's components.

Faraday Provides a Spark

Although Michael Faraday is best remembered for his work in electromagnetism, he began his career as Humphry Davy's laboratory assistant at the Royal Institution, in London, and went on to be the professor of chemistry there for over 30 years. In the 1830s, Faraday devised several ingenious experiments to determine whether the flow of current in an electrolytic solution is dependent solely on the contact of the battery's poles with the solution. In a typical setup, one of the poles was separated from the solution and electricity was permitted to enter it by way of a spark. In all cases, Faraday observed current to flow in the electrolytic cell despite one or both of the poles not being in direct contact with the electrolytic solution. In 1833 he wrote the following:

> *I conceive the effects [of electrolysis] to arise from forces which are* internal, *relative to the matter under decomposition, and not* external, *as they might be considered, if directly dependent on the poles. I suppose that the effects are due to a modification, by the electric current, of the chemical affinity of the particles through or by which the current is passing.*

Although the battery's poles were, in fact, later shown to play a part in the flow of the current, Faraday had established the active role of the electrolytic solution in electrolysis. And in realizing that electricity affected the chemical nature of the solution, he anticipated the ideas of oxidation and reduction—despite the fact that the concepts of electrons and ions were half a century away.

Faraday's Legacy

Faraday continued to study the role of the electrolytic solution, or *electrolyte* as he named it, in electrolysis. (He also coined most of the other modern terms of electrolysis, for example, *electrode, ion, anode, cathode, anion,* and *cation.*) These investigations culminated in the discovery of his basic laws of electrolysis.

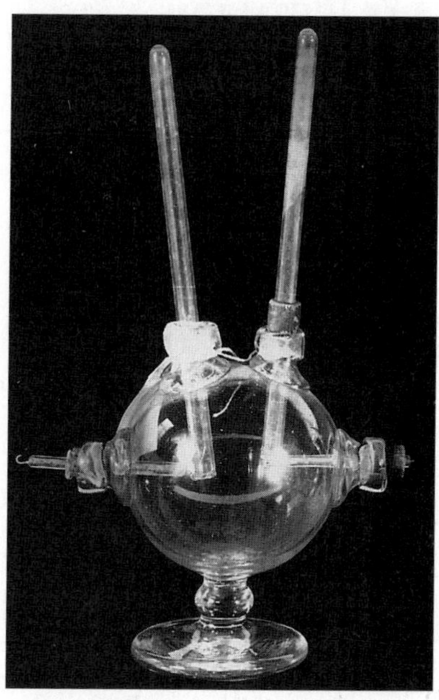

Instrument used by Michael Faraday in his studies of electrolysis.

Still valid today, these principles not only put electrolysis on a quantitative footing, leading to our current understanding of the phenomenon, but also bolstered the atomic theory, which was still seriously contested by many chemists at the time. And perhaps most important, Faraday's experiments inspired his successors to further clarify the chemical nature of solutions. This ultimately led to Svante Arrhenius's theory of electrolytic dissociation and the evolution of a new division in the chemical field, known today as physical chemistry.

SECTION 14-2

Colligative Properties of Solutions

Objectives

- List four colligative properties, and explain why they are classified as colligative properties.
- Calculate freezing-point depression, boiling-point elevation, and solution molality of nonelectrolytic solutions.
- Calculate the expected changes in freezing point and boiling point of an electrolytic solution.
- Discuss causes of the differences between expected and experimentally observed colligative properties of electrolytic solutions.

The presence of solutes affects the properties of the solutions. Some of these properties are not dependent on the nature of the dissolved substance but only on how many dissolved particles are present. *Properties that depend on the concentration of solute particles but not on their identity are called* **colligative properties.** In calculations involving some colligative properties, the concentration is given in terms of molality, *m*.

Vapor-Pressure Lowering

The boiling point and freezing point of a solution differ from those of the pure solvent. The graph in Figure 14-6 shows that a nonvolatile solute raises the boiling point and lowers the freezing point. *A* **nonvolatile substance** *is one that has little tendency to become a gas under existing conditions.*

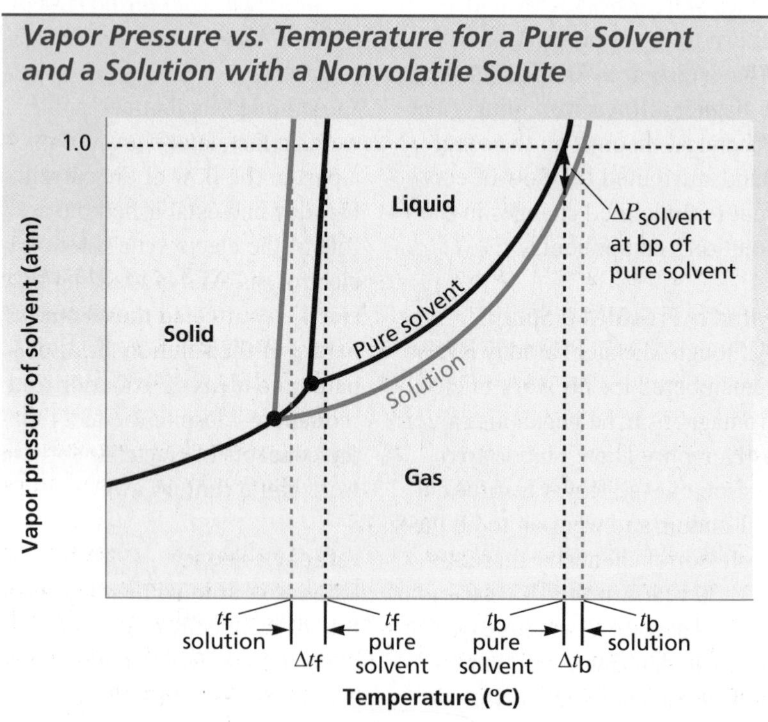

FIGURE 14-6 Vapor pressure as a function of temperature is shown for a pure solvent, a solution of that solvent, and a nonvolatile solute. The vapor pressure of the solution is lower than the vapor pressure of the pure solvent. (This can be seen by noting the decrease in pressure between the pure solvent and the solution at the temperature that is the boiling point of the pure solvent.) The solute thus reduces the freezing point and elevates the boiling point.

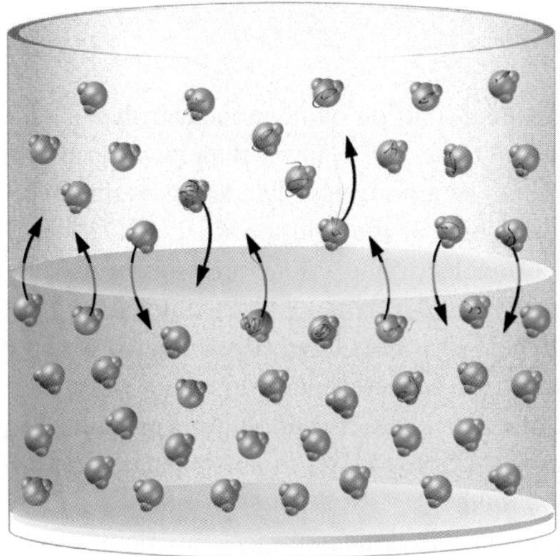

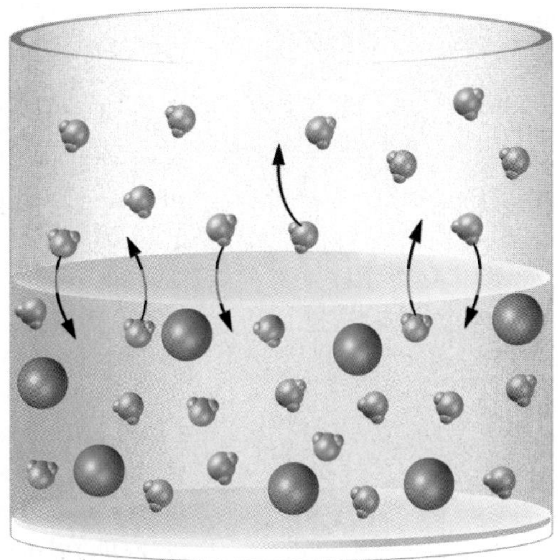

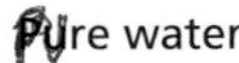

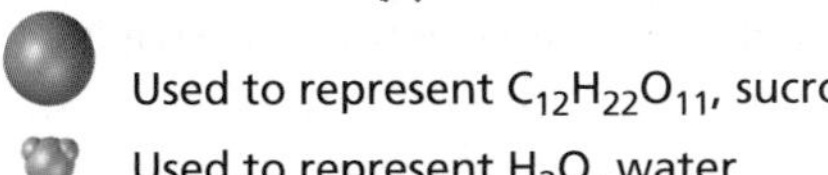

FIGURE 14-7 The vapor pressure of water over pure water is greater than the vapor pressure of water over an aqueous solution containing a nonvolatile solute, such as sucrose.

To understand why a nonvolatile solute changes the boiling point and freezing point, you must consider equilibrium vapor pressure, which was discussed in Chapter 12. Vapor pressure is the pressure caused by molecules that have escaped from the liquid phase to the gaseous phase. Experiments show that the vapor pressure of a solvent containing a nonvolatile solute is lower than the vapor pressure of the pure solvent at the same temperature, as shown in Figure 14-7. Notice the liquid-vapor boundaries. Vapor pressure can be thought of as a measure of the tendency of molecules to escape from a liquid. Addition of sucrose, a nonvolatile solute, lowers the concentration of water molecules at the surface of the liquid. This lowers the tendency of water molecules to leave the solution and enter the vapor phase. Thus, the vapor pressure of the solution is lower than the vapor pressure of pure water.

Nonelectrolyte solutions of the same molality have the same concentration of particles. Dilute solutions of the same solvent and equal molality of any nonelectrolyte solute lower vapor pressure equally. For example, a 1 m aqueous solution of the nonelectrolyte glucose, $C_6H_{12}O_6$, lowers the vapor pressure of water 5.5×10^{-4} atm at 25°C. A 1 m solution of sucrose, $C_{12}H_{22}O_{11}$, another nonelectrolyte, also lowers the vapor pressure 5.5×10^{-4} atm. Because vapor-pressure lowering depends on the concentration of a nonelectrolyte solute and is independent of solute identity, it is a colligative property.

Refer to the graph in Figure 14-6. Because the vapor pressure has been lowered, the solution remains liquid over a larger temperature range. This lowers the freezing point and raises the boiling point. It follows that changes in boiling point and freezing point also depend on the concentration of solute and are therefore colligative properties.

Freezing-Point Depression

The freezing point of a 1-molal solution of any nonelectrolyte solute in water is found by experiment to be 1.86°C lower than the freezing point of water. That is, when 1 mol of a nonelectrolyte solute is dissolved in 1 kg of water, the freezing point of the solution is –1.86°C instead of 0.00°C. When 2 mol of a nonelectrolyte solute is dissolved in 1 kg of water, the freezing point of the solution is –3.72°C. This is 2 × –1.86°C. In fact, for any concentration of a nonelectrolyte solute in water, the decrease in freezing point can be determined by using the value of –1.86°C/*m*. This value, called the **molal freezing-point constant (K_f),** *is the freezing-point depression of the solvent in a 1-molal solution of a nonvolatile, nonelectrolyte solute.*

Each solvent has its own characteristic molal freezing-point constant. The values of K_f for some common solvents are given in Table 14-2. These values are most accurate for dilute solutions at 1 atmosphere of pressure. Some variations are introduced in the value of K_f at other pressures and with more-concentrated solutions. The table also shows the values of a related quantity called K_b, which you will study next.

As stated earlier, the freezing point of a solution containing 1 mol of a nonelectrolyte solute in water is 1.86°C lower than the normal freezing point of water. *The* **freezing-point depression,** Δt_f, *is the difference between the freezing points of the pure solvent and a solution of a nonelectrolyte in that solvent, and it is directly proportional to the molal concentration of the solution.* As shown by the previous example, if the molal concentration is doubled, the freezing-point depression is doubled. Freezing-point depression can be calculated by the following equation.

$$\Delta t_f = K_f m$$

K_f is expressed as °C/*m*, *m* is expressed in mol solute/kg solvent (molality), and Δt_f is expressed in °C. Sample Problems 14-3 and 14-4 show how this relationship can be used to determine the freezing-point depression and molal concentration of a solution.

TABLE 14-2 ***Molal Freezing-Point and Boiling-Point Constants***

Solvent	Normal f.p. (°C)	Molal f.p. constant, K_f (°C/*m*)	Normal b.p. (°C)	Molal b.p. constant, K_b (°C/*m*)
Acetic acid	16.6	–3.90	117.9	3.07
Camphor	178.8	–39.7	207.4	5.61
Ether	–116.3	–1.79	34.6	2.02
Naphthalene	80.2	–6.94	217.7	5.80
Phenol	40.9	–7.40	181.8	3.60
Water	0.00	–1.86	100.0	0.51

SAMPLE PROBLEM 14-3

What is the freezing-point depression of water in a solution of 17.1 g of sucrose, $C_{12}H_{22}O_{11}$, and 200. g of water? What is the actual freezing point of the solution?

SOLUTION

1 *ANALYZE* **Given:** solute mass and chemical formula = 17.1 g $C_{12}H_{22}O_{11}$

solvent mass and identity = 200.0 g water

Unknown: **a.** freezing-point depression

b. freezing point of the solution

2 *PLAN* Find the molal freezing-point constant for water in Table 14-2. To use the equation for freezing-point depression, $\Delta t_f = K_f m$, you need to determine the molality of the solution.

$$\text{mass of solute (g)} \times \frac{1 \text{ mol solute}}{\text{molar mass of solute (g)}} = \text{amount of solute (mol)}$$

$$\frac{\text{amount of solute (mol)}}{\text{mass of solvent (g)}} \times \frac{1000 \text{ g water}}{1 \text{ kg water}} = \text{molality}$$

$$\Delta t_f = K_f m$$

$$\text{f.p. solution} = \text{f.p. solvent} + \Delta t_f$$

3 *COMPUTE*

$$17.1 \cancel{\text{g } C_{12}H_{22}O_{11}} \times \frac{1 \text{ mol } C_{12}H_{22}O_{11}}{342.34 \cancel{\text{g } C_{12}H_{22}O_{11}}} = 0.0500 \text{ mol } C_{12}H_{22}O_{11}$$

$$\frac{0.0500 \text{ mol } C_{12}H_{22}O_{11}}{200. \cancel{\text{g water}}} \times \frac{1000 \cancel{\text{g water}}}{\text{kg water}} = \frac{0.250 \text{ mol } C_{12}H_{22}O_{11}}{\text{kg water}} = 0.250\ m$$

a. $\Delta t_f = 0.250 \cancel{m} \times (-1.86°\text{C}/\cancel{m}) = -0.465°\text{C}$

b. f.p. solution = 0.000°C + (−0.465°C) = −0.465°C

SAMPLE PROBLEM 14-4

A water solution containing an unknown quantity of a nonelectrolyte solute is found to have a freezing point of −0.23°C. What is the molal concentration of the solution?

SOLUTION

1 *ANALYZE* **Given:** freezing point of solution = −0.23°C

Unknown: molality of the solution

2 *PLAN* Water is the solvent, so you will need the value of K_f, the molal-freezing-point constant for water, from Table 14-2. The Δt_f for this solution is the difference between the f.p. of water and the f.p. of the solution. Use the equation for freezing-point depression to calculate molality.

$$\Delta t_f = \text{f.p. of solution} - \text{f.p. of pure solvent}$$

$$\Delta t_f = K_f m \qquad \text{Solve for molality, } m.$$

$$m = \frac{\Delta t_f}{K_f}$$

3 COMPUTE

$$\Delta t_f = -0.23°C - 0.00°C = -0.23°C$$

$$m = \frac{-0.23°\cancel{C}}{-1.86°\cancel{C}/m} = 0.12\ m$$

4 EVALUATE As shown by the unit cancellation, the answer gives the molality, as desired. The answer is properly limited to two significant digits.

PRACTICE

1. A solution consists of 10.3 g of the nonelectrolyte glucose, $C_6H_{12}O_6$, dissolved in 250. g of water. What is the freezing-point depression of the solution? *Answer* −0.426°C

2. In a laboratory experiment, the freezing point of an aqueous solution of glucose is found to be −0.325°C. What is the molal concentration of this solution? *Answer* 0.175 *m*

3. If 0.500 mol of a nonelectrolyte solute are dissolved in 500.0 g of ether, what is the freezing point of the solution? *Answer* −118.1°C

4. The freezing point of an aqueous solution that contains a non-electrolyte is −9.0°C.
   a. What is the freezing-point depression of the solution?
   b. What is the molal concentration of the solution?

   Answer a. −9.0°C b. 4.8 *m*

Boiling-Point Elevation

As you learned in Chapter 12, the boiling point of a liquid is the temperature at which the vapor pressure of the liquid is equal to the prevailing atmospheric pressure. Therefore, a change in the vapor pressure of the liquid will cause a corresponding change in the boiling point. As stated earlier, the vapor pressure of a solution containing a nonvolatile solute is lower than the vapor pressure of the pure solvent. This means that more heat will be required to raise the vapor pressure of the solution to equal the atmospheric pressure. Thus, the boiling point of a solution is higher than the boiling point of the pure solvent.

The **molal boiling-point constant** (K_b) *is the boiling-point elevation of the solvent in a 1-molal solution of a nonvolatile, nonelectrolyte solute.* The boiling-point elevation of a 1-molal solution of any nonelectrolyte solute in water has been found by experiment to be 0.51°C. Thus, the molal boiling-point constant for water is 0.51°C/*m*.

For different solvents, the boiling-point elevations of 1-molal solutions have different values. Some other values for K_b are included in Table 14-2. Like the freezing-point constants, these values are most accurate for dilute solutions.

The **boiling-point elevation,** Δt_b, *is the difference between the boiling points of the pure solvent and a nonelectrolyte solution of that solvent,* *and it is directly proportional to the molal concentration of the solution.* Boiling-point elevation can be calculated by the following equation.

$$\Delta t_b = K_b m$$

When Δt_b is expressed in °C/*m* and *m* is expressed in mol of solute/kg of solvent, Δt_b is the boiling-point elevation in °C.

SAMPLE PROBLEM 14-5

What is the boiling-point elevation of a solution made from 20.0 g of a nonelectrolyte solute and 400.0 g of water? The molar mass of the solute is 62.0 g.

SOLUTION

1 ANALYZE

Given: solute mass = 20.0 g
solute molar mass = 62.0 g
solvent mass and identity = 400.0 g of water
Unknown: boiling-point elevation

2 PLAN

Find the molal boiling-point constant for water in Table 14-2. To use the equation for boiling-point elevation, $\Delta t_b = K_b m$, you need to determine the molality of the solution.

$$\text{mass of solute (g)} \times \frac{1 \text{ mol solute}}{\text{molar mass of solute (g)}} = \text{amount of solute (mol)}$$

$$\frac{\text{amount of solute (mol)}}{\text{mass of solvent (g)}} \times \frac{1000 \text{ g water}}{1 \text{ kg water}} = \text{molality}$$

$$\Delta t_b = K_b m$$

3 COMPUTE

$$20.0 \cancel{\text{ g of solute}} \times \frac{1 \text{ mol solute}}{62.0 \cancel{\text{ g of solute}}} = 0.323 \text{ mol of solute}$$

$$\frac{0.323 \text{ mol of solute}}{400.0 \cancel{\text{ g water}}} \times \frac{1000 \cancel{\text{ g water}}}{1 \text{ kg water}} = 0.808 \frac{\text{mol solute}}{\text{kg water}} = 0.808\ m$$

$$\Delta t_b = 0.51\text{°C}/\cancel{m} \times 0.808\ \cancel{m} = 0.41\text{°C}$$

PRACTICE

1. A solution contains 50.0 g of sucrose, $C_{12}H_{22}O_{11}$, a nonelectrolyte, dissolved in 500.0 g of water. What is the boiling-point elevation? — *Answer* 0.15°C

2. A solution contains 450.0 g of sucrose, $C_{12}H_{22}O_{11}$, a nonelectrolyte, dissolved in 250 g of water. What is the boiling point of the solution? — *Answer* 102.7°C

3. If the boiling point elevation of an aqueous solution containing a nonvolatile electrolyte is 1.02°C, what is the molality of the solution? — *Answer* 2.0 *m*

4. The boiling point of an aqueous solution containing a nonvolatile electrolyte is 100.75°C.
   a. What is the boiling-point elevation?
   b. What is the molality of the solution?

   Answer a. 0.75°C b. 1.5 *m*

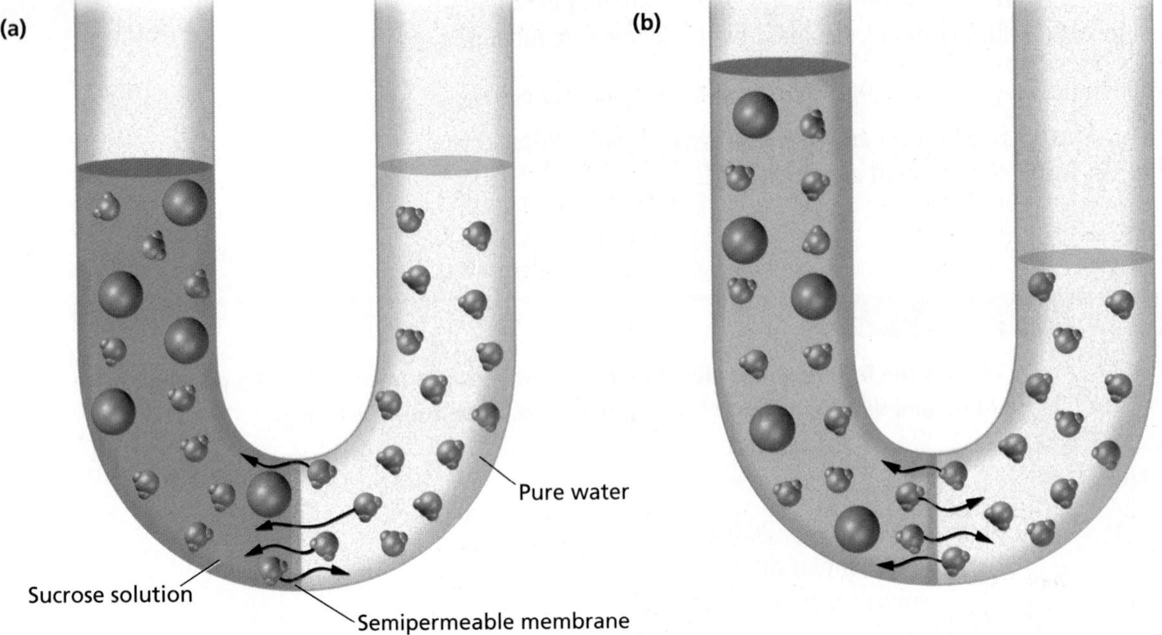

FIGURE 14-8 (a) When pure water and an aqueous sucrose solution are separated by a semipermeable membrane, the net movement of water molecules through the membrane is from the pure water side into the aqueous solution. (b) The level of the solution rises until pressure exerted by the height of the solution equals the osmotic pressure, at which point no net movement of water molecules occurs.

Osmotic Pressure

Figure 14-8 illustrates another colligative property. In the figure, an aqueous sucrose solution is separated from pure water by a semipermeable membrane. **Semipermeable membranes** *allow the movement of some particles while blocking the movement of others.* The level of the sucrose solution will rise until a certain height is reached. What causes the level of the solution to rise?

The semipermeable membrane allows water molecules, but not sucrose molecules, to pass through. The sucrose molecules on the solution side allow fewer water molecules to strike the membrane than strike on the pure water side in the same amount of time. Thus, the rate at which water molecules leave the pure water side is greater than the rate at which they leave the solution. This causes the level of the solution to rise. The level rises until the pressure exerted by the height of the solution is large enough to force water molecules back through the membrane from the solution at a rate equal to that at which they enter from the pure water side.

The movement of solvent through a semipermeable membrane from the side of lower solute concentration to the side of higher solute concentration is **osmosis.** Osmosis occurs whenever two solutions of different concentrations are separated by a semipermeable membrane. **Osmotic pressure** *is the external pressure that must be applied to stop osmosis.* In the example given above, osmosis caused the level of the solution to rise until the height of the solution provided the pressure necessary to stop osmosis. Because osmotic pressure is dependent on the concentration of solute particles and not on the type of solute particles, it is a colligative property. The greater the concentration of a solution, the greater the osmotic pressure of the solution.

Regulation of osmosis is vital to the life of a cell because cell membranes are semipermeable. Cells lose water and shrink when placed in a solution of higher concentration. They gain water and swell when placed in a solution of lower concentration. In vertebrates, cells are protected from swelling and shrinking by blood and lymph that surround the cells. Blood and lymph are equal in concentration to the concentration inside the cell.

Electrolytes and Colligative Properties

Early investigators were puzzled by experiments in which certain substances depressed the freezing point or elevated the boiling point of a solvent more than expected. For example, a 0.1 *m* solution of sodium chloride, NaCl, lowers the freezing point of the solvent nearly twice as much as a 0.1 *m* solution of sucrose. A 0.1 *m* solution of calcium chloride, $CaCl_2$, lowers the freezing point of the solvent nearly three times as much as a 0.1 *m* solution of sucrose. The effect on boiling points is similar.

To understand why this is so, contrast the behavior of sucrose with that of sodium chloride in aqueous solutions. Each sucrose molecule dissolves to produce only one particle in solution. So 1 mol of sucrose dissolves to produce only 1 mol of particles in solution. Sugar is a nonelectrolyte. NaCl, however, is a strong electrolyte. Each mole of NaCl dissolves to produce 2 mol of particles in solution: 1 mol of sodium ions and 1 mol of chloride ions. Figure 14-9 compares the production of particles in solution for three different solutes. As you can see, electrolytes produce more than 1 mol of solute particles for each mole of compound dissolved.

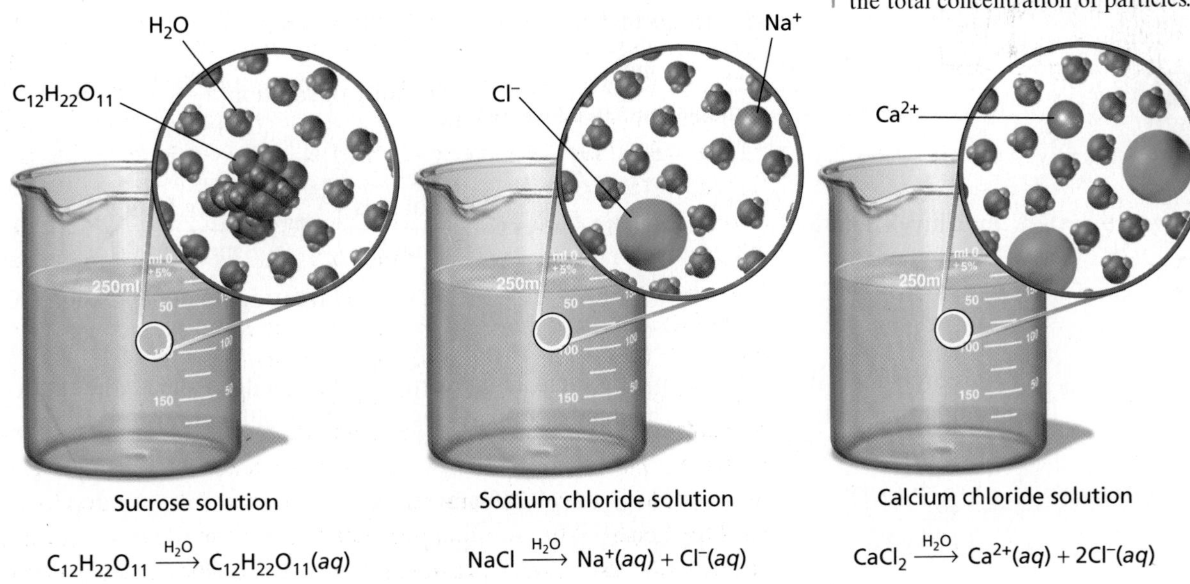

FIGURE 14-9 Compare the number of particles produced per formula unit for these three solutes. Colligative properties depend on the total concentration of particles.

Calculated Values for Electrolyte Solutions

Remember that colligative properties depend on the total concentration of solute particles regardless of their identity. So electrolytes cause changes in colligative properties proportional to the total molality in terms of all dissolved particles instead of formula units. For the same molal concentrations of sucrose and sodium chloride, you would expect the effect on colligative properties to be twice as large for sodium chloride as for sucrose. What about barium nitrate, $Ba(NO_3)_2$? Each mole of barium nitrate yields 3 mol of ions in solution.

$$Ba(NO_3)_2(s) \longrightarrow Ba^{2+}(aq) + 2NO_3^-(aq)$$

You would expect a $Ba(NO_3)_2$ solution of a given molality to lower the freezing point of its solvent three times as much as a nonelectrolytic solution of the same molality.

SAMPLE PROBLEM 14-6

What is the expected change in the freezing point of water in a solution of 62.5 g of barium nitrate, $Ba(NO_3)_2$, in 1.00 kg of water?

SOLUTION

1 *ANALYZE*

Given: solute mass and formula = 62.5 g $Ba(NO_3)_2$
solvent mass and identity = 1.00 kg water
$\Delta t_f = K_f m$
Unknown: expected freezing-point depression

2 *PLAN*

The molality can be calculated by converting the solute mass to moles and then dividing by the number of kilograms of solvent. That molality is in terms of formula units of $Ba(NO_3)_2$ and must be converted to molality in terms of dissociated ions in solution. It must be multiplied by the number of moles of ions produced per mole of formula unit. This adjusted molality can then be used to calculate the freezing-point depression.

$$\frac{\text{mass of solute (g)}}{\text{mass of solvent (kg)}} \times \frac{\text{1 mol solute}}{\text{molar mass of solute (g)}} = \text{molality of solution}\left(\frac{\text{mol}}{\text{kg}}\right)$$

$$\text{molality of solution}\left(\frac{\text{mol}}{\text{kg}}\right) \times \text{molality conversion}\left(\frac{\text{mol ions}}{\text{mol}}\right) \times K_f\left(\frac{°\text{C} \cdot \text{kg H}_2\text{O}}{\text{mol ions}}\right)$$
$$= \text{expected freezing-point depression (°C)}$$

This problem is similar to Sample Problem 14-5, except that the solute is ionic rather than a nonionizing molecular solute. The number of particles in solution will therefore equal the number of ions of the solute.

3 *COMPUTE*

$$\frac{62.5 \text{ g } \cancel{Ba(NO_3)_2}}{1.00 \text{ kg } H_2O} \times \frac{\text{mol } Ba(NO_3)_2}{261.35 \text{ g } \cancel{Ba(NO_3)_2}} = \frac{0.239 \text{ mol } Ba(NO_3)_2}{\text{kg } H_2O}$$

$Ba(NO_3)_2(s) \longrightarrow Ba^{2+}(aq) + 2NO_3^-(aq)$

Each formula unit of barium nitrate yields three ions in solution.

$$\frac{0.239 \cancel{\text{mol Ba(NO}_3)_2}}{\cancel{\text{kg H}_2\text{O}}} \times \frac{3 \cancel{\text{mol ions}}}{\cancel{\text{mol Ba(NO}_3)_2}} \times \frac{-1.86°\text{C} \cdot \cancel{\text{kg H}_2\text{O}}}{\cancel{\text{mol ions}}} = -1.33°\text{C}$$

4 EVALUATE The units cancel properly to give the desired answer in °C. The answer is correctly given to three significant digits. The mass of the solute is approximately one-fourth its molar mass and would give 0.75 mol of ions in the 1 kg of solvent, so the estimated answer of $0.75 \times -1.86°C = -1.4°C$ supports our computation.

PRACTICE

1. What is the expected freezing-point depression for a solution that contains 2.0 mol of magnesium sulfate dissolved in 1.0 kg of water? *Answer* −7.4°C
2. What is the expected boiling-point elevation of water for a solution that contains 150 g of sodium chloride dissolved in 1.0 kg of water? *Answer* 2.7°C
3. The freezing point of an aqueous sodium chloride solution is −20.0°C. What is the molality of the solution? *Answer* 5.4 *m* NaCl

Actual Values for Electrolyte Solutions

It is important to remember that the values just calculated are only *expected* values. As stated earlier, a 0.1 *m* solution of sodium chloride lowers the freezing point *nearly* twice as much as a 0.1 *m* solution of sucrose. The actual values of the colligative properties for all strong electrolytes are *almost* what would be expected based on the number of particles they produce in solution. Some specific examples are given in Table 14-3. The freezing-point depression of a compound that produces two ions per formula unit is almost twice that of a nonelectrolytic solution. The freezing-point depression of a compound that produces three ions per formula unit is almost three times that of a nonelectrolytic solution.

TABLE 14-3 *Molal Freezing-Point Depressions for Aqueous Solutions of Ionic Solutes*

Solute	Concentration (*m*)	Δt_{fp}, observed (°C)	Δt_{fp}, nonelectrolyte solution (°C)	Δt_{fp}, observed / Δt_{fp}, nonelectrolyte solution
KCl	0.1	−0.345	−0.186	1.85
	0.01	−0.0361	−0.0186	1.94
	0.001	−0.00366	−0.00186	1.97
$MgSO_4$	0.1	−0.225	−0.186	1.21
	0.01	−0.0285	−0.0186	1.53
	0.001	−0.00338	−0.00186	1.82
$BaCl_2$	0.1	−0.470	−0.186	2.53
	0.01	−0.0503	−0.0186	2.70
	0.001	−0.00530	−0.00186	2.84

FIGURE 14-10 The salts applied to icy roads are electrolytes. They lower the freezing point of water and melt the ice.

Look at the values given for KCl solutions in Table 14-3. The freezing-point depression of a 0.1 *m* KCl solution is only 1.85 times greater than that of a nonelectrolyte solution. However, as the concentration decreases, the freezing-point depression comes closer to the value that is twice that of a nonelectrolytic solution.

The differences between the expected and calculated values are caused by the attractive forces that exist between dissociated ions in aqueous solution. The attraction between the hydrated ions in the solution is small compared with those in the crystalline solid. However, forces of attraction do interfere with the movements of the aqueous ions. The more concentrated a solution is, the closer together the ions are, and the greater the attraction between ions is. Only in very dilute solutions is the average distance between the ions large enough and the attraction between ions small enough for the solute ions to move about almost completely freely.

Peter Debye and Erich Hückel introduced a theory in 1923 to account for this attraction between ions in dilute aqueous solutions. According to this theory, the attraction between dissociated ions of ionic solids in dilute aqueous solutions is caused by an ionic atmosphere that surrounds each ion. This means that each ion is, on average, surrounded by more ions of opposite charge than of like charge. This clustering effect hinders the movements of solute ions. A cluster of hydrated ions can act as a single unit rather than as individual ions. Thus, the effective total concentration is less than expected, based on the number of ions known to be present.

Ions of higher charge attract other ions very strongly. They therefore cluster more and have lower effective concentrations than ions with smaller charge. For example, ions formed by $MgSO_4$ have charges of 2+ and 2−. Ions formed by KCl have charges of 1+ and 1−. Note in Table 14-3 that $MgSO_4$ in a solution does not depress the freezing point as much as the same concentration of KCl.

SECTION REVIEW

1. What colligative properties are displayed by each of the following situations?
   a. Antifreeze is added to a car's cooling system to prevent freezing when the air temperature is below 0°C.
   b. Ice melts on sidewalks after salt has been spread on them.
2. Two moles of a nonelectrolytic solute are dissolved in 1 kg of an unknown solvent. The solution freezes at 7.8°C *below* its normal freezing point. What is the molal freezing-point constant of the unknown solvent? What is your prediction for the identity of the solvent?
3. If two solutions of equal amounts in a U-tube are separated by a semipermeable membrane, will the level of the more-concentrated solution or the less-concentrated solution rise?
4. a. Calculate the expected freezing-point depression of a 0.2 *m* KNO_3 solution.
   b. Will the value you calculated match the actual freezing-point depression for this solution? Why or why not?

CHAPTER 14 REVIEW

CHAPTER SUMMARY

14-1

- The separation of ions that occurs when an ionic solid dissolves is called dissociation.
- When two different ionic solutions are mixed, a precipitate may form if ions from the two solutions react to form an insoluble compound.
- A net ionic equation for a reaction in aqueous solution includes only compounds and ions that change chemically in the reaction. Spectator ions are ions that do not take part in such a reaction.
- Formation of ions from solute molecules is called ionization. A molecular compound may ionize in a water solution if the attraction of the polar water molecules is strong enough to break the polar-covalent bonds of the solute molecules.
- An H_3O^+ ion is called a hydronium ion.
- All, or almost all, of a dissolved strong electrolyte exists as ions in an aqueous solution, whereas a relatively small amount of a dissolved weak electrolyte exists as ions in an aqueous solution.

Vocabulary

dissociation (425)
hydronium ion (431)
ionization (431)
net ionic equation (429)
spectator ions (429)
strong electrolyte (432)
weak electrolyte (433)

14-2

- Colligative properties of solutions depend only on the total number of solute particles present. Boiling-point elevation, freezing-point depression, vapor-pressure lowering, and osmotic pressure are all colligative properties.
- The molal boiling-point and freezing-point constants are used to calculate boiling-point elevations and freezing-point depressions of solvents containing nonvolatile solutes.
- Electrolytes have a greater effect on the freezing and boiling points of solvents than do nonelectrolytes.
- Except in very dilute solutions, the values of colligative properties of electrolytic solutions are less than expected because of the attraction between ions in solution.

Vocabulary

boiling-point elevation, Δt_b (440)
colligative properties (436)
freezing-point depression, Δt_f (438)
molal boiling-point constant, K_b (440)
molal freezing-point constant, K_f (438)
nonvolatile substance (436)
osmosis (442)
osmotic pressure (442)
semipermeable membrane (442)

REVIEWING CONCEPTS

1. How many moles of ions are contained in 1 L of a 1 M solution of KCl? of $Mg(NO_3)_2$? (14-1)

2. Use Table 14-1 to predict whether each of the following compounds is considered soluble or insoluble:

a. KCl
b. $NaNO_3$
c. AgCl
d. $BaSO_4$
e. $Ca_3(PO_4)_2$
f. $Pb(ClO_3)_2$
g. $(NH_4)_2S$
h. $PbCl_2$ (in cold water)
i. FeS
j. $Al_2(SO_4)_3$ (14-1)

3. What is a net ionic equation? (14-1)

4. a. What is ionization?
b. Distinguish between ionization and dissociation. (14-1)

5. a. Define and distinguish between strong electrolytes and weak electrolytes.
   b. Give two examples of each type. (14-1)

6. What determines the strength with which a solute acts as an electrolyte? (14-1)

7. Distinguish between the use of the terms *strong* and *weak* and the use of the terms *dilute* and *concentrated* when used to describe electrolyte solutions. (14-1)

8. How does the presence of a nonvolatile solute affect each of the following properties of the solvent into which the solute is dissolved?
   a. vapor pressure
   b. freezing point
   c. boiling point
   d. osmotic pressure (14-2)

9. Using Figure 14-6 as a guide, make a sketch of a vapor pressure-versus-temperature curve that shows the comparison of pure water, a solution with x amount of solute, and a solution with $2x$ the amount of solute. What is the relationship between Δt_f for the x curve and Δt_f for the $2x$ curve? (14-2)

10. a. Why does the level of the more-concentrated solution rise when two solutions of different concentrations are separated by a semipermeable membrane?
    b. When does the level of the solution stop rising?
    c. When the level stops rising, what is the net movement of water molecules across the membrane? (14-2)

11. a. Compare the effects of nonvolatile electrolytes with the effects of nonvolatile nonelectrolytes on the freezing and boiling points of solvents in which they are dissolved.
    b. Why are such differences observed? (14-2)

12. Why does the actual freezing-point depression of an electrolytic solution differ from the freezing-point depression calculated on the basis of the concentration of particles? (14-2)

PROBLEMS

Dissociation

13. Write the equation for the dissolution of each of the following ionic compounds in water. (Hint: See Sample Problem 14-1.)
    a. KI
    b. $NaNO_3$
    c. $MgCl_2$
    d. Na_2SO_4

14. For the compounds listed in the previous problem, determine the number of moles of each ion produced as well as the total number of moles of ions produced when 1 mol of each compound dissolves in water.

15. Write the equation for the dissolution of each of the following in water, and then indicate the total number of moles of solute ions formed.
    a. 0.50 mol strontium nitrate
    b. 0.50 mol sodium phosphate

Precipitation Reactions

16. Using Table 14-1, write the balanced chemical equation, write the overall ionic equation, identify the spectator ions and possible precipitates, and write the net ionic equation for each of the following reactions. (Hint: See Sample Problem 14-2.)
    a. mercury(II) chloride (aq) + potassium sulfide (aq) $\longrightarrow$
    b. sodium carbonate (aq) + calcium chloride (aq) $\longrightarrow$
    c. copper(II) chloride (aq) + ammonium phosphate (aq) $\longrightarrow$

17. Copper(II) chloride and lead(II) nitrate react in aqueous solutions by double replacement. Write the balanced chemical equation, the overall ionic equation, and the net ionic equation for this reaction. If 13.45 g of copper(II) chloride react, what is the maximum amount of precipitate that could be formed?

18. Identify the spectator ions in the reaction between KCl and $AgNO_3$ in an aqueous solution.

Freezing-Point Depression of Nonelectrolytes

19. Determine the freezing-point depression of H_2O in each of the following solutions. (Hint: See Sample Problem 14-3.)
a. 1.50 m solution of $C_{12}H_{22}O_{11}$ (sucrose) in H_2O
b. 171 g of $C_{12}H_{22}O_{11}$ in 1.00 kg H_2O
c. 77.0 g of $C_{12}H_{22}O_{11}$ in 400. g H_2O

20. Determine the molality of each solution of an unknown nonelectrolyte in water, given the following freezing-point depressions. (Hint: See Sample Problem 14-4.)
a. −0.930°C
b. −3.72°C
c. −8.37°C

21. A solution contains 20.0 g of $C_6H_{12}O_6$ (glucose) in 250. g of water.
a. What is the freezing-point depression of the solvent?
b. What is the freezing point of the solution?

22. How many grams of antifreeze, $C_2H_4(OH)_2$, would be required per 500. g of water to prevent the water from freezing at a temperature of −20.0°C?

23. Pure benzene, C_6H_6, freezes at 5.45°C. A solution containing 7.24 g of $C_2Cl_4H_2$ in 115 g of benzene (specific gravity = 0.879) freezes at 3.55°C. Based on these data, what is the molal freezing-point constant for benzene?

24. If 1.500 g of a solute having a molar mass of 125.0 g were dissolved in 35.00 g of camphor, what would be the resulting freezing point of the solution?

Boiling-Point Elevation of Nonelectrolytes

25. Determine the boiling-point elevation of H_2O in each of the following solutions. (Hint: See Sample Problem 14-5.)
a. 2.5 m solution of $C_6H_{12}O_6$ (glucose) in H_2O
b. 3.20 g $C_6H_{12}O_6$ in 1.00 kg H_2O
c. 20.0 g $C_{12}H_{22}O_{11}$ (sucrose) in 500. g H_2O

26. Determine the molality of each water solution given the following boiling points:
a. 100.25°C
b. 101.53°C
c. 102.805°C

Colligative Properties of Electrolytes

27. Given 1.00 m aqueous solutions of each of the following electrolytic substances, what is the expected change in the freezing point of the solvent? (Hint: See Sample Problem 14-6.)
a. KI
b. $CaCl_2$
c. $Ba(NO_3)_2$

28. What is the expected change in the freezing point of water for an aqueous solution that is 0.015 m $AlCl_3$?

29. What is the expected freezing point of a solution containing 85.0 g of NaCl dissolved in 450. g of water?

30. Determine the expected boiling point of a solution made by dissolving 25.0 g of barium chloride in 0.150 kg of water.

31. The change in the boiling point of water for an aqueous solution of potassium iodide is 0.65°C. Determine the apparent molal concentration of potassium iodide.

32. The freezing point of an aqueous solution of barium nitrate is −2.65°C. Determine the apparent molal concentration of barium nitrate.

33. Calculate the expected freezing point of a solution containing 1.00 kg of H_2O and 0.250 mol of NaCl.

34. Experimental data for a 1.00 m MgI_2 aqueous solution indicate an actual change in the freezing point of water of −4.78°C. Determine the expected change in the freezing point of water. Suggest a possible reason for discrepancies between the experimental and the expected values.

MIXED REVIEW

35. Given 0.01 m aqueous solutions of each of the following, arrange the solutions in order of increasing change in the freezing point of the solution.
a. NaI
b. $CaCl_2$
c. K_3PO_4
d. $C_6H_{12}O_6$ (glucose)

36. What is the molal concentration of an aqueous calcium chloride solution that freezes at –2.43°C?

37. a. Write the balanced formula equation that shows the possible products of a double replacement reaction between calcium nitrate and sodium chloride.
b. Using Table 14-1, determine if there is a precipitate.

38. Write a balanced equation to show what occurs when hydrogen bromide dissolves and reacts with water. Include a hydronium ion in the equation.

39. Write the equation for the dissolution of each of the following in water, and then indicate the total number of moles of solute ions formed.
a. 0.275 mol potassium sulfide
b. 0.15 mol aluminum sulfate

40. Calculate the expected change in the boiling point of water in a solution made up of 131.2 g of silver nitrate, $AgNO_3$, in 2.00 kg of water.

41. Nitrous acid, HNO_2, is a weak electrolyte. Nitric acid, HNO_3, is a strong electrolyte. Write equations to represent the ionization of each in water. Include the hydronium ion, and show the appropriate kind of arrow in each case.

42. Find the boiling point of an aqueous solution containing a nonelectrolyte that freezes at –6.51°C.

43. Write a balanced equation for the dissolution of sodium carbonate, Na_2CO_3, in water. Find the number of moles of each ion produced when 0.20 mol of sodium carbonate dissolves. Then find the total number of moles of ions.

44. Given the reaction below and the information in Table 14-1, write the net ionic equation for the reaction.

potassium phosphate (*aq*) + lead (II) nitrate (*aq*)

45. Find the expected freezing point of a water solution that contains 268 g of aluminum nitrate, $Al(NO_3)_3$, in 8.50 kg of water.

CRITICAL THINKING

46. Applying Models
a. You are conducting a freezing-point determination in the laboratory using an aqueous solution of KNO_3. The observed freezing point of the solution is –1.15°C. Using a pure water sample, you recorded the freezing point of the pure solvent on the same thermometer as 0.25°C. Determine the molal concentration of KNO_3. Assume that there are no forces of attraction between ions.
b. You are not satisfied with the result in part (a) because you suspect that you should not ignore the effect of ion interaction. You take a 10.00 mL sample of the solution. After carefully evaporating the water from the solution, you obtain a mass of 0.415 g KNO_3. Determine the actual molal concentration of KNO_3 and the percentage difference between the predicted concentration and the actual concentration of KNO_3. Assume that the solution's density is 1.00 g/mL.

47. Analyzing Information The observed freezing-point depression for electrolyte solutions is sometimes less than the calculated value. Why does this occur? Is the difference greater for concentrated solutions or dilute solutions?

48. Analyzing Information The osmotic pressure of a dilute solution can be calculated as follows.

$\pi = MRT$
π = osmotic pressure
M = concentration in moles per liter
R = ideal gas constant
T = absolute temperature of the solution

How does the osmotic-pressure equation compare with the ideal gas law?

HANDBOOK SEARCH

49. Common reactions for Group 13 elements are found in the *Elements Handbook*. Review this material and answer the following.
   a. Write net ionic equations for each of the example reactions shown on page 751.
   b. Which reactions did not change when written in net ionic form? Why?

50. Common reactions for Group 14 elements are found in the *Elements Handbook*. Review this material and answer the following.
   a. Write net ionic equations for each of the example reactions shown on page 755.
   b. Which reactions did not change when written in net ionic form? Why?

RESEARCH & WRITING

51. Find out how much salt a large northern city, such as New York City or Chicago, uses on its streets in a typical winter. What environmental problems result from this use of salt? What substitutes for salt are being used to melt ice and snow?

52. Research the role of electrolytes and electrolytic solutions in your body. Find out how electrolytes work in the functioning of nerves and muscles. What are some of the health problems that can arise from an imbalance of electrolytes in body fluids?

ALTERNATIVE ASSESSMENT

53. Performance Determine the freezing point of four different mixtures of water and ethylene glycol (use commercial antifreeze). What mixture shows the lowest freezing point?

54. Performance Find the optimum mixture of salt and ice for reducing the temperature of the chilling bath for an ice-cream freezer. Use your data to write a set of instructions on how to prepare the chilling bath for making ice cream.

55. Performance Using a low-voltage dry cell, assemble a conductivity apparatus. Secure several unknown aqueous solutions of equal molality from your instructor, and use the apparatus to distinguish the electrolytes from the nonelectrolytes. Among those identified as electrolytes, rank their relative strengths as conductors from good to poor.

56. Performance Using equal volumes of the unknown solutions from the preceding activity, explain how you could use the freezing-point depression concept to distinguish the electrolytes from the nonelectrolytes. Explain how you could determine the number of ions contained per molecule among the solutes identified as electrolytes. Design and conduct an experiment to test your theories.

CHAPTER 15

Acids and Bases

Acids and bases change the color of compounds called indicators.

Properties of Acids and Bases

SECTION 15-1

OBJECTIVES

- List five general properties of aqueous acids and bases.
- Name common binary acids and oxyacids, given their chemical formulas.
- List five acids commonly used in industry and the laboratory, and give two properties of each.
- Define *acid* and *base* according to Arrhenius's theory of ionization.
- Explain the differences between strong and weak acids and bases.

How many foods can you think of that are sour? Chances are that almost all the foods you thought of, like those in Figure 15-1(a), owe their sour taste to an acid. Sour milk contains *lactic acid.* Vinegar, which can be produced by fermenting juices, contains *acetic acid. Phosphoric acid* gives a tart flavor to many carbonated beverages. Most fruits contain some kind of acid. Lemons, oranges, grapefruits, and other citrus fruits contain *citric acid.* Apples contain *malic acid,* and grape juice contains *tartaric acid*.

Many substances known as bases are commonly found in household products, such as those in Figure 15-1(b). Household ammonia is an ammonia-water solution that is useful for all types of general cleaning. Sodium hydroxide, NaOH, known by the common name *lye,* is present in some commercial drain and oven cleaners. Milk of magnesia is a suspension in water of magnesium hydroxide, $Mg(OH)_2$, which is not very water-soluble. It is used as an antacid to relieve discomfort caused by excess hydrochloric acid in the stomach. Aluminum hydroxide, $Al(OH)_3$, and sodium hydrogen carbonate, $NaHCO_3$, are also bases commonly found in antacids.

(a)

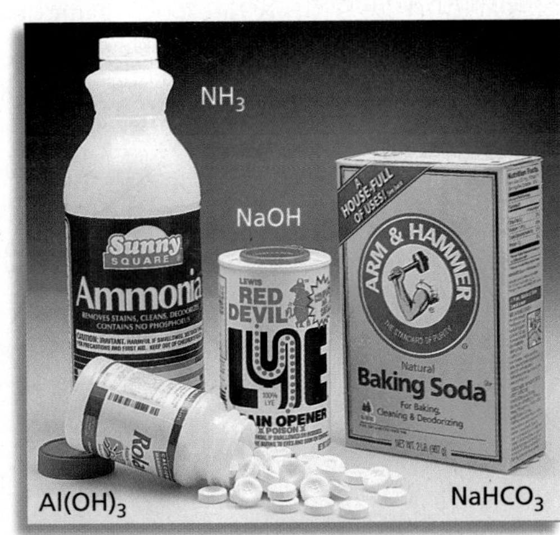

(b)

FIGURE 15-1 (a) Fruits and fruit juices contain acids such as citric acid and ascorbic acid. Carbonated beverages contain benzoic acid, phosphoric acid, and carbonic acid. (b) Many household cleaners contain bases such as ammonia and sodium hydroxide.

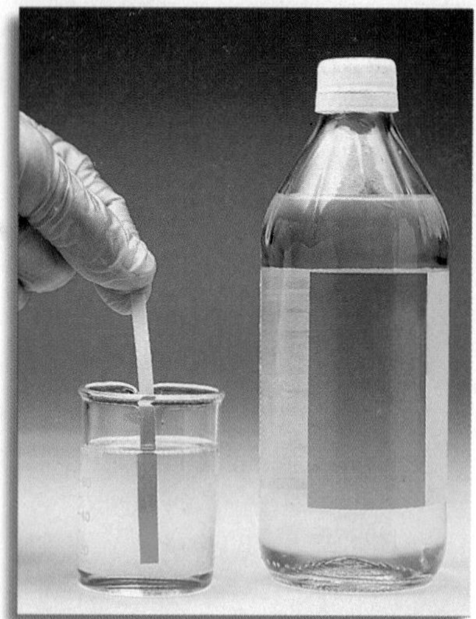

FIGURE 15-2 A strip of pH paper dipped into vinegar turns red, showing that vinegar is an acid.

Acids

Acids were first recognized as a distinct class of compounds because of the common properties of their aqueous solutions. These properties are listed below.

1. *Aqueous solutions of acids have a sour taste.* Taste, however, should NEVER be used as a test to evaluate any chemical substance. Many acids, especially in concentrated solutions, are corrosive; that is, they destroy body tissue and clothing. Many are also poisons.
2. *Acids change the color of acid-base indicators.* When pH paper is used as an indicator, the paper turns certain colors in acidic solution. This reaction is demonstrated in Figure 15-2.
3. *Some acids react with active metals to release hydrogen gas, H_2.* Recall that metals can be ordered in terms of an activity series. Metals above hydrogen in the series undergo single-replacement reactions with certain acids. Hydrogen gas is formed as a product, as shown by the reaction of barium with sulfuric acid.

$$Ba(s) + H_2SO_4(aq) \longrightarrow BaSO_4(aq) + H_2(g)$$

4. *Acids react with bases to produce salts and water.* When chemically equivalent amounts of acids and bases react, the three properties just described disappear because the acid is "neutralized." The reaction products are water and an ionic compound called a *salt.*
5. *Acids conduct electric current.* Because acids form ions in aqueous solutions, acids are electrolytes.

Acid Nomenclature

A **binary acid** *is an acid that contains only two different elements: hydrogen and one of the more-electronegative elements.* Many common inorganic acids are binary acids. The hydrogen halides—HF, HCl, HBr, and HI—are all binary acids.

The procedure used to name binary acids is illustrated by the examples given in Table 15-1. In pure form, each acid listed in the table is a gas. Aqueous solutions of these compounds are known by the acid names. From the table you can see that naming binary compounds can be summarized as follows.

TABLE 15-1 ***Names of Binary Acids***

Formula	Acid name
HF	hydrofluoric acid
HCl	hydrochloric acid
HBr	hydrobromic acid
HI	hydriodic acid
H_2S	hydrosulfuric acid

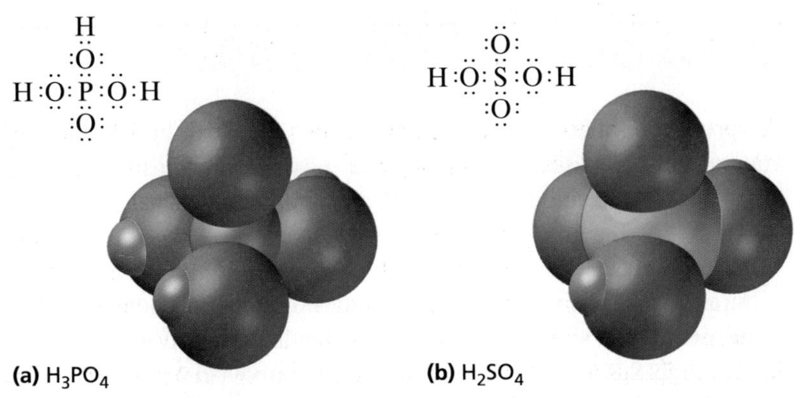

FIGURE 15-3 Structures of (a) phosphoric acid and (b) sulfuric acid

Binary Acid Nomenclature

1. The name of a binary acid begins with the prefix *hydro-*.
2. The root of the name of the second element follows this prefix.
3. The name then ends with the suffix *-ic.*

An **oxyacid** *is an acid that is a compound of hydrogen, oxygen, and a third element, usually a nonmetal.* Nitric acid, HNO_3, is an oxyacid. The structures of two other oxyacids are shown in Figure 15-3. Oxyacids are one class of ternary acids, which are acids that contain three different elements. Usually, the elements in an oxyacid formula are written as one or more hydrogen atoms followed by a polyatomic anion. The name of an oxyacid is based on this anion. Some common oxyacids and their anions are given in Table 15-2.

TABLE 15-2 ***Names of Common Oxyacids and Oxyanions***

Formula	Acid name	Anion
CH_3COOH	acetic acid	CH_3COO^-, acetate
H_2CO_3	carbonic acid	CO_3^{2-}, carbonate
HIO_3	iodic acid	IO_3^-, iodate
$HClO$	hypochlorous acid	ClO^-, hypochlorite
$HClO_2$	chlorous acid	ClO_2^-, chlorite
$HClO_3$	chloric acid	ClO_3^-, chlorate
$HClO_4$	perchloric acid	ClO_4^-, perchlorate
HNO_2	nitrous acid	NO_2^-, nitrite
HNO_3	nitric acid	NO_3^-, nitrate
H_3PO_3	phosphorous acid	PO_3^{3-}, phosphite
H_3PO_4	phosphoric acid	PO_4^{3-}, phosphate
H_2SO_3	sulfurous acid	SO_3^{2-}, sulfite
H_2SO_4	sulfuric acid	SO_4^{2-}, sulfate

Some Common Industrial Acids

The properties of acids make them important chemicals both in the laboratory and in industry. Sulfuric acid, nitric acid, phosphoric acid, hydrochloric acid, and acetic acid are all common industrial acids.

Sulfuric Acid

Sulfuric acid is the most commonly produced industrial chemical in the world. More than 47 million tons of it are made each year in the United States alone. Sulfuric acid is used in large quantities in petroleum refining and metallurgy as well as in the manufacture of fertilizer. It is also essential to a vast number of industrial processes, including the production of metals, paper, paint, dyes, detergents, and many chemical raw materials. Sulfuric acid is the acid used in automobile batteries.

Because it can form hydrates with water, concentrated sulfuric acid is an effective dehydration (water-removing) agent. It can be used to remove water from gases with which it does not react. Sugar and certain other organic compounds are also dehydrated by sulfuric acid. Skin contains organic compounds that are attacked by concentrated sulfuric acid, which can cause serious burns.

Nitric Acid

Pure nitric acid is a volatile, unstable liquid rarely used in industry or laboratories. Dissolving the acid in water provides stability. Nitric acid stains proteins yellow. The feather in Figure 15-4 was stained by nitric acid. The acid has a suffocating odor, stains skin, and can cause serious burns. It is used in making explosives, many of which are nitrogen-containing compounds. It is also used to make rubber, plastics, dyes, and

FIGURE 15-4 Concentrated nitric acid stains a feather yellow.

pharmaceuticals. Initially, nitric acid solutions are colorless; however, upon standing, they gradually become yellow because of slight decomposition to brown nitrogen dioxide gas.

Phosphoric Acid

Phosphorus, along with nitrogen and potassium, is an essential element for plants and animals. The bulk of phosphoric acid produced each year is used directly for manufacturing fertilizers and animal feed. Dilute phosphoric acid has a pleasant but sour taste and is not toxic. It is used as a flavoring agent in beverages and as a cleaning agent for dairy equipment. Phosphoric acid is also important in the manufacture of detergents and ceramics.

Hydrochloric Acid

The stomach produces HCl to aid in digestion. Industrially, hydrochloric acid is important for "pickling" iron and steel. Pickling is the immersion of metals in acid solutions to remove surface impurities. This acid is also used in industry as a general cleaning agent, in food processing, in the activation of oil wells, in the recovery of magnesium from sea water, and in the production of other chemicals.

A dilute solution of hydrochloric acid, commonly referred to as muriatic acid, may be found in hardware stores. It is used to maintain the correct acidity in swimming pools and for general cleaning of masonry.

Acetic Acid

Concentrated acetic acid is a clear, colorless, pungent-smelling liquid known as glacial acetic acid. This name derives from the fact that pure acetic acid has a freezing point of only 17°C. It can form crystals in a cold room. The fermentation of certain plants produces vinegars containing acetic acid. White vinegar contains 4–8% acetic acid.

Acetic acid is important industrially in synthesizing chemicals used in the manufacture of plastics. It is a raw material in the production of food supplements—for example, lysine, an essential amino acid. Acetic acid is also used as a fungicide.

Bases

How do bases differ from acids? You can answer this question by comparing the following properties of bases with those of acids.

1. *Aqueous solutions of bases taste bitter.* You may have noticed this fact if you have ever gotten soap, a basic substance, in your mouth. As with acids, taste should NEVER be used to test a substance to see if it is a base. Many bases are caustic; they attack the skin and tissues, causing severe burns.
2. *Bases change the color of acid-base indicators.* As Figure 15-5 shows, an indicator will be a different color in a basic solution than it would be in an acidic solution.

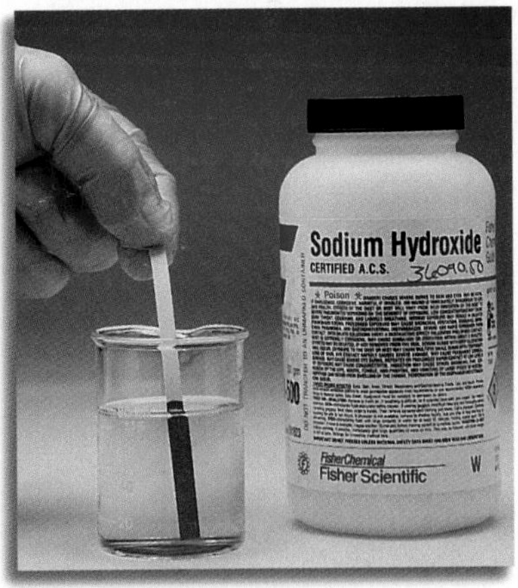

FIGURE 15-5 pH paper turns blue in the presence of this solution of sodium hydroxide.

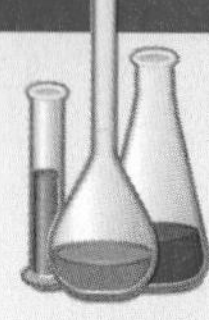

DESKTOP INVESTIGATION

Household Acids and Bases

MATERIALS

- **dishwashing liquid, dishwasher detergent, laundry detergent, laundry stain remover, fabric softener, and bleach**
- **fresh red cabbage**
- **hot plate**
- **beaker, 500 mL or larger**
- **beakers, 50 mL**
- **mayonnaise, baking powder, baking soda, white vinegar, cider vinegar, lemon juice, soft drinks, mineral water, and milk**
- **spatula**
- **tap water**
- **tongs**

Wear safety goggles and an apron.

QUESTION

Which of the household substances are acids, and which are bases?

PROCEDURE

Record all your results in a data table.

Red cabbage can be made into an acid-base indicator.

1. To make an acid-base indicator, extract juice from red cabbage. First, cut up some red cabbage and place it in a large beaker. Add enough water so that the beaker is half full. Then bring the mixture to a boil. Let it cool, and pour off the cabbage juice. Save the solution.
2. Assemble foods, beverages, and cleaning products to be tested.
3. If the substance being tested is a liquid, pour about 5 mL into a small beaker. If it is a solid, place a small amount into a beaker, and moisten it with about 5 mL of water.
4. Add a drop or two of the red-cabbage juice to the solution being tested, and note the color. The solution will turn red if it is acidic and green if it is basic.

DISCUSSION

1. Are the cleaning products acids, bases, or neutral?
2. What are acid/base characteristics of foods and beverages?
3. Did you find consumer warning labels on basic or acidic products?

3. *Dilute aqueous solutions of bases feel slippery.* You encounter this property of aqueous bases whenever you wash with soap.
4. *Bases react with acids to produce salts and water.* The properties of an acid disappear with the addition of an equivalent amount of a base. It could also be said that "neutralization" of the base occurs when these two substances react to produce salts and water.
5. *Bases conduct electric current.* Like acids, bases form ions in aqueous solutions and are thus electrolytes.

Arrhenius Acids and Bases

Svante Arrhenius, a Swedish chemist who lived from 1859 to 1927, understood that aqueous solutions of acids and bases conducted electric current. Arrhenius therefore theorized that acids and bases must produce ions in solution. *An* **Arrhenius acid** *is a chemical compound that increases the concentration of hydrogen ions, H^+, in aqueous solution.* In other words, an acid will ionize in solution, increasing the number of hydrogen ions present. *An* **Arrhenius base** *is a substance that increases the concentration of hydroxide ions, OH^-, in aqueous solution.* Some bases are ionic hydroxides. These bases dissociate in solution to release hydroxide ions into the solution. Other bases are substances that react with water to remove a hydrogen ion, leaving hydroxide ions in the solution.

Aqueous Solutions of Acids

The acids described by Arrhenius are molecular compounds with ionizable hydrogen atoms. Their water solutions are known as *aqueous acids*. All pure aqueous acids are electrolytes.

Acid molecules are sufficiently polar so that one or more hydrogen ions are attracted by water molecules. Negatively charged anions are left behind. As explained in Chapter 14, the hydrogen ion in aqueous solution is best represented as H_3O^+, the hydronium ion. The ionization of an HNO_3 molecule is shown by the following equation. Figure 15-6 shows how the hydrogen atoms combine with water to form a hydronium ion when nitric acid is diluted.

$$HNO_3(l) + H_2O(l) \longrightarrow H_3O^+(aq) + NO_3^-(aq)$$

Similarly, ionization of a hydrogen chloride molecule in hydrochloric acid can be represented in the following way.

$$HCl(g) + H_2O(l) \longrightarrow H_3O^+(aq) + Cl^-(aq)$$

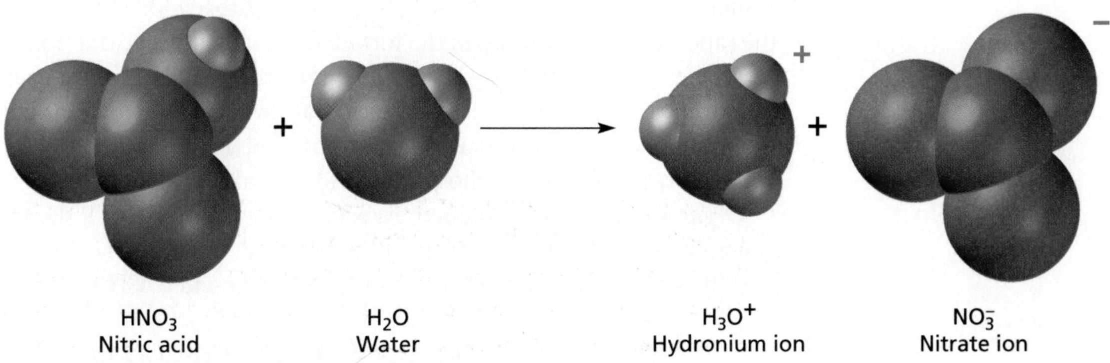

FIGURE 15-6 Arrhenius's observations form the basis of a definition of acids. Arrhenius acids, such as the nitric acid shown here, produce hydronium ions in aqueous solution.

TABLE 15-3 *Common Aqueous Acids*

Strong acids	Weak acids
$H_2SO_4 \longrightarrow H^+ + HSO_4^-$	$HSO_4^- \rightleftarrows H^+ + SO_4^{2-}$
$HClO_4 \longrightarrow H^+ + ClO_4^-$	$H_3PO_4 \rightleftarrows H^+ + H_2PO_4^-$
$HCl \longrightarrow H^+ + Cl^-$	$HF \rightleftarrows H^+ + F^-$
$HNO_3 \longrightarrow H^+ + NO_3^-$	$CH_3COOH \rightleftarrows H^+ + CH_3COO^-$
$HBr \longrightarrow H^+ + Br^-$	$H_2CO_3 \rightleftarrows H^+ + HCO_3^-$
$HI \longrightarrow H^+ + I^-$	$H_2S \rightleftarrows H^+ + HS^-$
	$HCN \rightleftarrows H^+ + CN^-$
	$HCO_3^- \rightleftarrows H^+ + CO_3^{2-}$

Strength of Acids

A **strong acid** *is one that ionizes completely in aqueous solution.* A strong acid is a strong electrolyte. Perchloric acid, $HClO_4$, hydrochloric acid, HCl, and nitric acid, HNO_3, are examples of strong acids. The strength of an acid depends on the polarity of the bond between hydrogen and the element to which it is bonded and the ease with which that bond can be broken. Acid strength increases with increasing polarity and decreasing bond energy.

Acids that are weak electrolytes are known as **weak acids.** The aqueous solution of a weak acid contains hydronium ions, anions, and dissolved acid molecules. Hydrocyanic acid is an example of a weak electrolyte. In aqueous solution, both the ionization of HCN and the reverse reaction occur simultaneously. Although hydronium and cyanide ions are present in solution, the reverse reaction is favored. Most of the solution is composed of hydrogen cyanide and water.

$$HCN(aq) + H_2O(l) \rightleftarrows H_3O^+(aq) + CN^-(aq)$$

Common aqueous acids are listed in Table 15-3. Each strong acid is assumed to ionize completely to give up one hydrogen ion. Notice that the ratio of hydrogen atoms in the formula does not affect acid strength. Molecules with multiple hydrogen ions may not readily give up each hydrogen. The fact that phosphoric acid has three hydrogen atoms per molecule does not mean that it is a strong acid. None of these ionize completely in solution, so phosphoric acid is weak.

Organic acids, which contain the acidic carboxyl group —COOH, are generally weak acids. For example, acetic acid, CH_3COOH, ionizes slightly in water to give hydronium ions and acetate ions, CH_3COO^-.

$$CH_3COOH(aq) + H_2O(l) \rightleftarrows H_3O^+(aq) + CH_3COO^-(aq)$$

A molecule of acetic acid contains four hydrogen atoms. However, only one of the hydrogen atoms is ionizable. The hydrogen atom in the carboxyl group in acetic acid is the one that is "acidic" and forms the hydronium ion. This acidic hydrogen can be seen in the structural diagram in Figure 15-7.

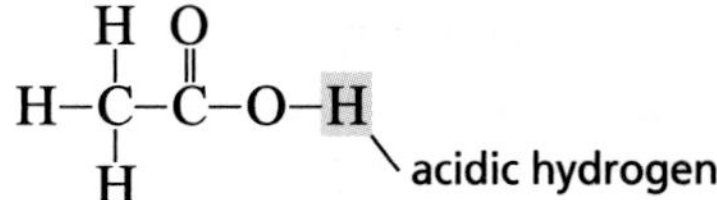

FIGURE 15-7 Acetic acid contains four hydrogen atoms, but only one of them is "acidic" or forms the hydronium ion in solution.

Aqueous Solutions of Bases

Most bases are ionic compounds containing metal cations and the hydroxide anion, OH^-. Because these bases are ionic, they dissociate to some extent when placed in solution. *When a base completely dissociates in water to yield aqueous OH^- ions, the solution is referred to as* **alkaline.** Sodium hydroxide, NaOH, is a common laboratory base. It is water-soluble and dissociates as shown by the equation below.

$$NaOH(s) \xrightarrow{H_2O} Na^+(aq) + OH^-(aq)$$

You will remember from Chapter 5 that sodium is one of the alkali metals. This group gets its name from the fact that the hydroxides of Li, Na, K, Rb, and Cs all form alkaline solutions.

Not all bases are ionic compounds. A base commonly used in household cleaners is ammonia, NH_3, which is molecular. Ammonia is a base because it produces hydroxide ions when it reacts with water molecules, as shown in the equation below.

$$NH_3(g) + H_2O(l) \rightleftarrows NH_4^+(aq) + OH^-(aq)$$

Strength of Bases

As with acids, the strength of a base also depends on the extent to which the base dissociates, or adds hydroxide ions to the solution. For example, potassium hydroxide, KOH, is a strong base because it completely dissociates into its ions in dilute aqueous solutions.

$$KOH(s) \xrightarrow{H_2O} K^+(aq) + OH^-(aq)$$

Strong bases are strong electrolytes, just as strong acids are strong electrolytes. Table 15-4 lists some strong bases.

TABLE 15-4 ***Common Aqueous Bases***

Strong bases	Weak bases
$Ca(OH)_2 \longrightarrow Ca^{2+} + 2OH^-$	$NH_3 + H_2O \rightleftarrows NH_4^+ + OH^-$
$Sr(OH)_2 \longrightarrow Sr^{2+} + 2OH^-$	$C_6H_5NH_2 + H_2O \rightleftarrows C_6H_5NH_3^+ + OH^-$
$Ba(OH)_2 \longrightarrow Ba^{2+} + 2OH^-$	
$NaOH \longrightarrow Na^+ + OH^-$	
$KOH \longrightarrow K^+ + OH^-$	
$RbOH \longrightarrow Rb^+ + OH^-$	
$CsOH \longrightarrow Cs^+ + OH^-$	

FIGURE 15-8 The hydroxides of most *d*-block metals are nearly insoluble in water, as is shown by the gelatinous precipitate, copper(II) hydroxide, $Cu(OH)_2$, in the beaker on the right.

Bases that are not very soluble do not produce a large number of hydroxide ions when added to water. Some metal hydroxides, such as $Cu(OH)_2$, are not very soluble in water, as seen in Figure 15-8. They cannot produce strongly alkaline solutions. The alkalinity of aqueous solutions depends on the concentration of OH^- ions in solution. It is unrelated to the number of hydroxide ions in the undissolved compound.

Now consider ammonia, which is highly soluble but is a weak electrolyte. The concentration of OH^- ions in an ammonia solution is relatively low. Ammonia is therefore a *weak base*. Many organic compounds that contain nitrogen atoms are also weak bases. For example, aniline, a substance used to make dyes, is a weak base.

$$C_6H_5NH_2(aq) + H_2O(l) \rightleftharpoons C_6H_5NH_3^+(aq) + OH^-(aq)$$

SECTION REVIEW

1. a. What are five general properties of aqueous acids?
   b. Name some common substances that have one or more of these properties.
2. Name the following acids: a. HBrO b. $HBrO_3$.
3. a. What are five general properties of aqueous bases?
   b. Name some common substances that have one or more of these properties.
4. a. Why are strong acids also strong electrolytes?
   b. Is every strong electrolyte also a strong acid?

Logic in the Laboratory

From "Acid and Water: A Socratic Dialogue," by David Todd, in *The Journal of Chemical Education.*

Tutor: . . . tell me, how does one set about diluting an acid with water?
Student: The rule is: pour the acid into the water.
Tutor: Why so? . . .
Student: I believe much heat is given off if you do it the wrong way, and the mixture can boil up in your face.
Tutor: Indeed, that is so . . . But is there not heat also developed if you pour the acid into the water?
Student: (thoughtfully) I suppose so. But then I can only assume since there is a rule, that the heat developed is a lot less if you do it that way.
Tutor: Let us reason together. Have you ever heard of Hess's Law?
Student: . . . Doesn't it have something to do with A going to B and the heat change involved?
Tutor: Very good . . . Now suppose I start with 100 g of water in one container at 25°C, and 100 g of concentrated sulfuric acid at 25°C in another container. We can call these two items your A . . . Now let us assume that the two have been mixed—regardless of the mode of mixing—do we not obtain a diluted acid that is 100 g of acid and 100 g of water, and we can call this B?
Student: (cautiously) Well—almost. But . . . concentrated sulfuric acid is 96% by weight H_2SO_4 and 4% water. This means that B is 96 g of pure sulfuric acid and 104 g of water.
Tutor: . . . Oh, excellent! You have had good teachers . . . But the main point is that B has the same composition regardless of the route by which it is obtained. Agreed?
Student: That most assuredly must be so.
Tutor: Now let us return to Hess's Law. It states that the heat change involved in going from A to B is the same regardless of the path taken.
Student: . . . Yes—I remember it now. You mean this law says that the same heat is evolved (or absorbed) if I pour the acid into the water, or vice versa?
Tutor: Yes, indeed.
Student: (now bewildered) You mean that the rule is nonsense, and therefore useless?
Tutor: Oh no, not at all! . . .
Student: . . . Hmm. Then in that case there must be some other factor involved.
Tutor: Indeed there is . . . May I drop a hint? . . . Which is the more dense—concentrated sulfuric acid or water?
Student: The acid . . .
Tutor: Good. Now if I put water on the acid, does it float or sink?
Student: Of course it will stay on the top.
Tutor: Right. And it will begin to mix . . .

Always dilute by pouring acid into water.

Student: I get it—it reacts on the surface, generates a lot of heat, and some of the diluted acid can boil up in my face.
Tutor: Exactly. And if I pour the acid into . . .
Student: (interrupting) Yes, yes—of course. The acid falls down through the water generating the heat in the entire body of the liquid—not just on the surface. So it won't form steam and boil up in my face . . . now I see the reason for the rule.

Reading for Meaning

In your own words, sum up the reason behind the rule for diluting acid with water.

Read Further

Hydrofluoric acid is a fairly weak acid. It does not burn the skin the way sulfuric acid and other strong acids do. Find an explanation of why this acid can greatly damage body tissue when it comes into contact with the skin.

SECTION 15-2

Brønsted-Lowry Acids and Bases

OBJECTIVES

- Define and recognize *Brønsted-Lowry acids* and *bases.*
- Define and relate *conjugate acid* and *conjugate base.*
- Determine the formula for the conjugate acid of a given base and for the conjugate base of a given acid.
- Explain why the conjugate base of a strong acid is a weak base and why the conjugate acid of a strong base is a weak acid.
- Explain what determines whether an amphoteric compound acts as an acid or a base.

For most uses, scientists found the Arrhenius definition of acids and bases to be adequate. However, as scientists further investigated acid-base behavior, they found that some substances acted as acids or bases when they were not in a water solution. Because the Arrhenius definition requires that the substances be aqueous, the definitions of acids and bases had to be revised.

In 1923, the Danish chemist J. N. Brønsted and the English chemist T. M. Lowry independently expanded the Arrhenius acid definition. *A* **Brønsted-Lowry acid** *is a molecule or ion that is a proton donor.* Because H^+ is a proton, all acids as defined by Arrhenius donate protons to water and are Brønsted-Lowry acids as well. Substances other than molecules, such as certain ions, can also donate protons. Such substances are not Arrhenius acids but are included in the category of Brønsted-Lowry acids.

Hydrogen chloride acts as a Brønsted-Lowry acid when it is dissolved in ammonia. It transfers protons to the solvent much as it does in water.

$$HCl + NH_3 \longrightarrow NH_4^+ + Cl^-$$

A proton is transferred from the hydrogen chloride molecule, HCl, to the ammonia molecule, NH_3. The ammonium ion, NH_4^+, is formed. Electron-dot formulas show the similarity of this reaction to the reaction of HCl with water.

$$\mathrm{H{:}\ddot{\underset{..}{Cl}}{:}} + \mathrm{H{:}\underset{\textstyle H}{\ddot{O}}{:}} \longrightarrow \left[\mathrm{H{:}\underset{\textstyle H}{\ddot{O}}{:}H}\right]^+ + \mathrm{{:}\ddot{\underset{..}{Cl}}{:}}^-$$

$$\mathrm{H{:}\ddot{\underset{..}{Cl}}{:}} + \mathrm{H{:}\underset{\textstyle H}{\ddot{N}}{:}H} \longrightarrow \left[\mathrm{H{:}\overset{\textstyle H}{\underset{\textstyle H}{\ddot{\underset{..}{N}}}}{:}H}\right]^+ + \mathrm{{:}\ddot{\underset{..}{Cl}}{:}}^-$$

In both reactions, hydrogen chloride is a Brønsted-Lowry acid.

Water can also act as a Brønsted-Lowry acid. Consider, for example, the following reaction, in which the water molecule donates a proton to the ammonia molecule.

$$H_2O(l) + NH_3(g) \rightleftharpoons NH_4^+(aq) + OH^-(aq)$$

$$\mathrm{H{:}\underset{\textstyle H}{\ddot{O}}{:}} + \mathrm{H{:}\underset{\textstyle H}{\ddot{N}}{:}H} \rightleftharpoons \left[\mathrm{H{:}\overset{\textstyle H}{\underset{\textstyle H}{\ddot{\underset{..}{N}}}}{:}H}\right]^+ + \left[\mathrm{{:}\underset{\textstyle H}{\ddot{O}}{:}}\right]^-$$

FIGURE 15-9 Hydrogen chloride gas escapes from a hydrochloric acid solution and combines with ammonia gas that has escaped from an aqueous ammonia solution. The resulting cloud is solid ammonium chloride.

A **Brønsted-Lowry base** *is a molecule or ion that is a proton acceptor.* In the reaction between hydrochloric acid and ammonia, ammonia accepts a proton from the hydrochloric acid. It is a Brønsted-Lowry base. The Arrhenius hydroxide bases, such as NaOH, are not, strictly speaking, Brønsted-Lowry bases. That is because as compounds they are not proton acceptors. The OH^- ion produced in solution is the Brønsted-Lowry base. It is the species that can accept a proton.

In a **Brønsted-Lowry acid-base reaction,** *protons are transferred from one reactant (the acid) to another (the base).* Figure 15-9 shows the reaction between the Brønsted-Lowry acid HCl and the Brønsted-Lowry base, NH_3.

Monoprotic and Polyprotic Acids

An acid that can donate only one proton (hydrogen ion) per molecule is known as a **monoprotic acid.** Perchloric acid, $HClO_4$, hydrochloric acid, HCl, and nitric acid, HNO_3, are all monoprotic. The following equation shows how a molecule of the monoprotic acid HCl donates a proton to a water molecule. HCl is completely ionized; it has no more hydrogen atoms to lose.

$$HCl(g) + H_2O(l) \longrightarrow H_3O^+(aq) + Cl^-(aq)$$

A **polyprotic acid** *is an acid that can donate more than one proton per molecule.* Sulfuric acid, H_2SO_4, and phosphoric acid, H_3PO_4, are examples of polyprotic acids. The ionization of a polyprotic acid occurs in stages. The acid loses its hydrogen atoms one at a time. Sulfuric acid

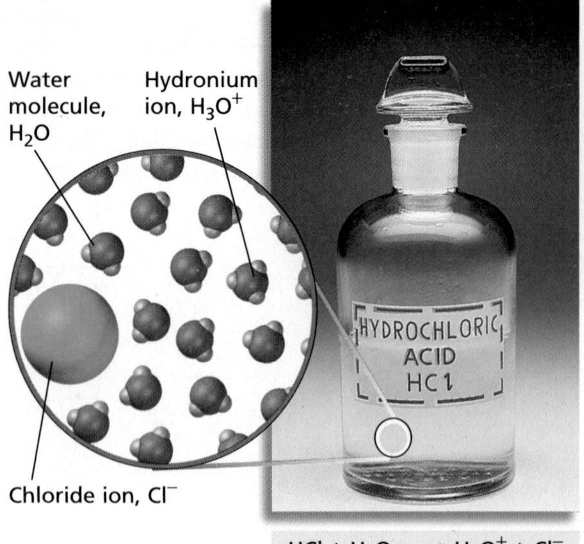

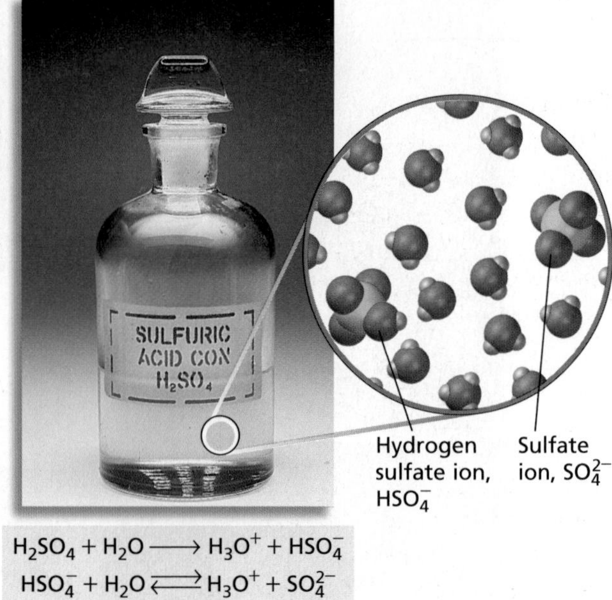

FIGURE 15-10 Hydrochloric acid, HCl, is a strong monoprotic acid. A dilute HCl solution contains hydronium ions and chloride ions. Sulfuric acid, H_2SO_4, is a strong diprotic acid. A dilute H_2SO_4 solution contains hydrogen sulfate ions from the first ionization, sulfate ions from the second ionization, and hydronium ions from both ionizations.

ionizes in two stages. In its first ionization, sulfuric acid is a strong acid. It is completely converted to hydrogen sulfate ions, HSO_4^-.

$$H_2SO_4(l) + H_2O(l) \longrightarrow H_3O^+(aq) + HSO_4^-(aq)$$

The hydrogen sulfate ion is itself a weak acid. It establishes the following equilibrium in solution.

$$HSO_4^-(aq) + H_2O(l) \rightleftharpoons H_3O^+(aq) + SO_4^{2-}(aq)$$

All stages of ionization of a polyprotic acid occur in the same solution. Sulfuric acid solutions therefore contain H_3O^+, HSO_4^-, and SO_4^{2-} ions.

Sulfuric acid is the type of polyprotic acid that *can donate two protons per molecule, and it is therefore known as a* **diprotic acid.** Ionizations of a monoprotic acid and a diprotic acid are shown in Figure 15-10.

Phosphoric acid is the type of polyprotic acid known as a **triprotic acid**—*an acid able to donate three protons per molecule.* The equations for these reactions are shown below.

$$H_3PO_4(aq) + H_2O(l) \rightleftharpoons H_3O^+(aq) + H_2PO_4^-(aq)$$
$$H_2PO_4^-(aq) + H_2O(l) \rightleftharpoons H_3O^+(aq) + HPO_4^{2-}(aq)$$
$$HPO_4^{2-}(aq) + H_2O(l) \rightleftharpoons H_3O^+(aq) + PO_4^{3-}(aq)$$

A solution of phosphoric acid contains H_3O^+, H_3PO_4, $H_2PO_4^-$, HPO_4^{2-}, and PO_4^{3-}. As with most polyprotic acids, the concentration of ions formed in the first ionization is the greatest. There are lesser concentrations of the respective ions during each succeeding ionization. Phosphoric acid is a weak acid in each step of its ionization.

Conjugate Acids and Bases

The Brønsted-Lowry definitions of acids and bases provide a basis for studying proton-transfer reactions. Suppose that a Brønsted-Lowry acid gives up a proton; the remaining ion or molecule can re-accept that proton and can act as a base. Such a base is known as a conjugate base. Thus, *the species that remains after a Brønsted-Lowry acid has given up a proton is the* **conjugate base** *of that acid.* For example, the fluoride ion is the conjugate base of hydrogen fluoride.

$$\underset{\text{acid}}{HF(aq)} + H_2O(l) \rightleftarrows \underset{\text{conjugate base}}{F^-(aq)} + H_3O^+(aq)$$

In this reaction, the water molecule is a Brønsted-Lowry base. It accepts a proton to form H_3O^+, which is an acid. The hydronium ion is the conjugate acid of water. *The species that is formed when a Brønsted-Lowry base gains a proton is the* **conjugate acid** *of that base.*

$$HF(aq) + \underset{\text{base}}{H_2O(l)} \rightleftarrows F^-(aq) + \underset{\text{conjugate acid}}{H_3O^+(aq)}$$

In general, Brønsted-Lowry acid-base reactions are equilibrium systems meaning that both the forward and reverse reactions occur. They involve two acid-base pairs, known as conjugate acid-base pairs.

$$\underset{\text{acid}_1}{HF(aq)} + \underset{\text{base}_2}{H_2O(l)} \rightleftarrows \underset{\text{base}_1}{F^-(aq)} + \underset{\text{acid}_2}{H_3O^+(aq)}$$

The subscripts designate the two conjugate acid-base pairs: (1) HF and F^- and (2) H_3O^+ and H_2O. In every conjugate acid-base pair, the acid has one more proton than its conjugate base.

Strength of Conjugate Acids and Bases

The extent of the reaction between a Brønsted-Lowry acid and base depends on the relative strengths of the acids and bases involved. Consider the following example. Hydrochloric is a strong acid. It gives up protons readily. It follows that the Cl^- ion has little tendency to attract and retain a proton. Consequently, the Cl^- ion is an extremely weak base.

$$\underset{\text{strong acid}}{HCl(g)} + \underset{\text{base}}{H_2O(l)} \longrightarrow \underset{\text{acid}}{H_3O^+(aq)} + \underset{\text{weak base}}{Cl^-(aq)}$$

This observation leads to an important conclusion: *the stronger an acid is, the weaker its conjugate base; the stronger a base is, the weaker its conjugate acid.*

This concept allows strengths of different acids and bases to be compared to predict the outcome of a reaction. As an example, consider the reaction of perchloric acid, $HClO_4$, and water.

$$HClO_4(aq) + H_2O(l) \longrightarrow H_3O^+(aq) + ClO_4^-(aq)$$

stronger acid — stronger base — weaker acid — weaker base

The hydronium ion is too weak an acid to compete successfully with perchloric acid in donating a proton; $HClO_4$ is the stronger acid. In this reaction, the chlorate ion, ClO_4^-, and H_2O are both bases. Because $HClO_4$ is a very strong acid, ClO_4^- is an extremely weak base. Therefore, H_2O competes more strongly than ClO_4^- to acquire a proton. The reaction proceeds such that the stronger acid reacts with the stronger base to produce the weaker acid and base.

Now consider a comparable reaction between water and acetic acid.

$$CH_3COOH(aq) + H_2O(l) \longleftarrow H_3O^+(aq) + CH_3COO^-(aq)$$

weaker acid — weaker base — stronger acid — stronger base

TABLE 15-5 ***Relative Strengths of Acids and Bases***

	Conjugate acid	Formula	Conjugate base	Formula	
Strong acids	chloric acid	$HClO_3$	chlorate ion	ClO_3^-	Very weak bases
	hydrobromic acid	HBr	bromide ion	Br^-	
	hydrochloric acid	HCl	chloride ion	Cl^-	
	hydriodic acid	HI	iodide ion	I^-	
	nitric acid	HNO_3	nitrate ion	NO_3^-	
	perchloric acid	$HClO_4$	perchlorate ion	ClO_4^-	
	sulfuric acid	H_2SO_4	hydrogen sulfate ion	HSO_4^-	
Increasing acid strength ↑	hydronium ion	H_3O^+	water	H_2O	Increasing base strength ↓
	chlorous acid	$HClO_2$	chlorite ion	ClO_2^-	
	hydrogen sulfate ion	HSO_4^-	sulfate ion	SO_4^{2-}	
	phosphoric acid	H_3PO_4	dihydrogen phosphate ion	$H_2PO_4^-$	
	hydrofluoric acid	HF	fluoride ion	F^-	
	acetic acid	CH_3COOH	acetate ion	CH_3COO^-	
	carbonic acid	H_2CO_3	hydrogen carbonate ion	HCO_3^-	
	hydrosulfuric acid	H_2S	hydrosulfide ion	HS^-	
	dihydrogen phosphate ion	$H_2PO_4^-$	hydrogen phosphate ion	HPO_4^{2-}	
	hypochlorous acid	$HClO$	hypochlorite ion	ClO^-	
	ammonium ion	NH_4^+	ammonia	NH_3	
	hydrogen carbonate ion	HCO_3^-	carbonate ion	CO_3^{2-}	
	hydrogen phosphate ion	HPO_4^{2-}	phosphate ion	PO_4^{3-}	
	water	H_2O	hydroxide ion	OH^-	
	ammonia	NH_3	amide ion	NH_2^-	
	hydrogen	H_2	hydride ion	H^-	

$$H^-(aq) + H_2O(l) \longrightarrow OH^-(aq) + H_2(g)$$

FIGURE 15-11 Because H^- is an extremely strong base, calcium hydride, CaH_2, reacts vigorously with water to produce hydrogen gas. The hydride ion accepts a proton from water, which acts as an acid in this reaction.

The H_3O^+ ion concentration in this solution is much lower than it was in the $HClO_4$ solution because acetic acid is a weak acid. The CH_3COOH molecule does not compete successfully with the H_3O^+ ion in donating protons to a base. The acetate ion, CH_3COO^-, is a stronger base than H_2O. Therefore, the H_2O molecule does not compete successfully with the CH_3COO^- ion in accepting a proton. The H_3O^+ ion is the stronger acid, and the CH_3COO^- ion is the stronger base. Thus, the reaction to the left is more favorable.

Note that in the reactions for both perchloric acid and acetic acid, the favored direction is toward the weaker acid and the weaker base. This observation leads to a second important general conclusion: *proton-transfer reactions favor the production of the weaker acid and the weaker base.* For a reaction to approach completion, the reactants must be much stronger as an acid and as a base than the products.

By comparing many different acids and bases, a table of relative strengths, such as Table 15-5, can be assembled. Note that a very strong acid, such as $HClO_4$, has a very weak conjugate base, ClO_4^-. The strongest base listed in the table, the hydride ion, H^-, has the weakest conjugate acid, H_2. A violent proton-transfer reaction could result from bringing together a very strong acid and a very strong base in certain proportions because the reaction has almost no tendency to go in the reverse direction. Such a reaction would give off a great deal of heat and would be dangerous. In fact, even the reaction between hydride ions and water—a much weaker acid than perchloric acid—is quite vigorous. The reaction is illustrated in Figure 15-11.

Amphoteric Compounds

You have probably noticed that water can be either an acid or a base. *Any species that can react as either an acid or a base is described as*

amphoteric. For example, consider the first ionization of sulfuric acid, in which water acts as a base.

$$\underset{\text{acid}}{H_2SO_4(aq)} + \underset{\text{base}}{H_2O(l)} \longrightarrow \underset{\text{acid}}{H_3O^+(aq)} + \underset{\text{base}}{HSO_4^-(aq)}$$

However, water acts as an acid in the following reaction.

$$\underset{\text{base}}{NH_3(g)} + \underset{\text{acid}}{H_2O(l)} \rightleftharpoons \underset{\text{acid}}{NH_4^+(aq)} + \underset{\text{base}}{OH^-(aq)}$$

Thus, water can act as either an acid or a base and is amphoteric. Such a substance acts as either an acid or a base, depending on the strength of the acid or base with which they are reacting. For example, if water reacts with a compound that is a stronger acid than itself, water acts as a base. If water reacts with a weaker acid, water acts as an acid.

–OH in a Molecule

Molecular compounds containing –OH groups can be acidic or amphoteric. For the compound to be acidic, a water molecule must be able to attract a hydrogen atom from a hydroxyl group. This occurs more easily when the O–H bond is very polar. Any feature of a molecule that increases the polarity of the O–H bond increases the acidity of a molecular compound. The covalently bonded –OH group in an acid is referred to as a *hydroxyl group*. The small, more-electronegative atoms of nonmetals at the upper right in the periodic table form compounds with acidic hydroxyl groups. All oxyacids are molecular electrolytes that contain one or more of these O–H bonds. Such compounds include chloric and perchloric acids.

Figure 15-12 shows the electron-dot formulas of the four oxyacids of chlorine. Notice that all of the oxygen atoms are bonded to the chlorine atom. Each hydrogen atom is bonded to an oxygen atom. Aqueous solutions of these molecules are acids because the O–H bonds are broken as the hydrogen is attracted away by water molecules.

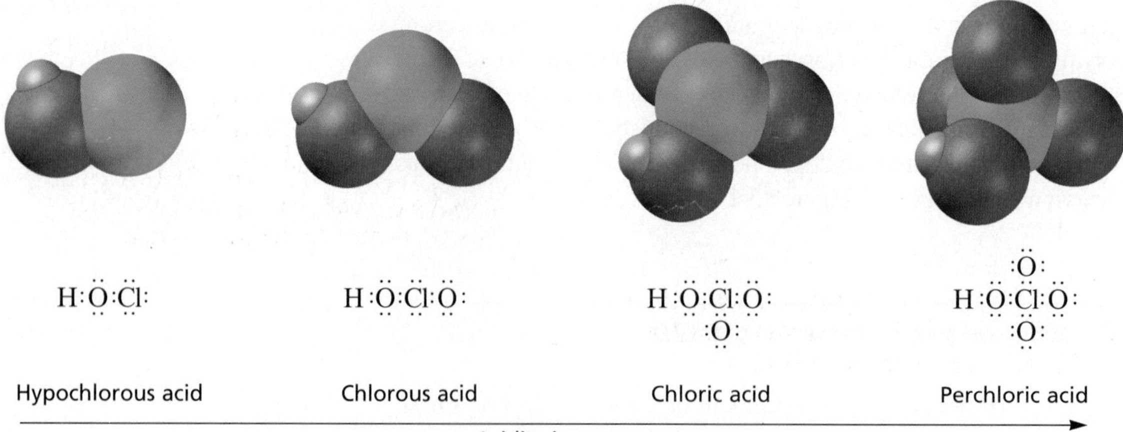

FIGURE 15-12 Each oxyacid of chlorine contains one chlorine atom and one hydrogen atom. They differ in the number of oxygen atoms they contain. The effect of the changing O–H bond polarity can be seen in the increasing acid strength from hypochlorous acid to perchloric acid.

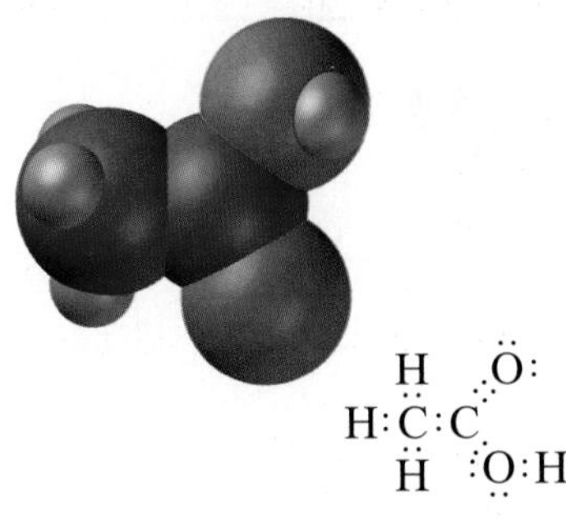

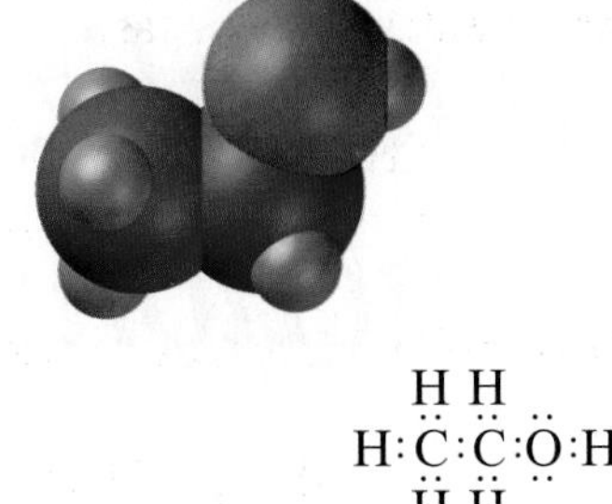

(a) CH_3COOH
Acetic acid

(b) C_2H_5OH
Ethanol

FIGURE 15-13 (a) Acetic acid is acidic. The second oxygen atom on the carbon draws electron density away from the –OH group, making the O–H bond more polar.
(b) Ethanol is essentially neutral. It has no second oxygen atom.

The behavior of a compound is affected by the number of oxygen atoms bonded to the atom connected to the – OH group. The larger the number of such oxygen atoms is, the more acidic the compound is likely to be. The electronegative oxygen atoms draw electron density away from the O–H bond and make it more polar. For example, chromium forms three different compounds containing – OH groups, as shown below.

basic	*amphoteric*	*acidic*
$Cr(OH)_2$	$Cr(OH)_3$	H_2CrO_4
chromium(II) hydroxide	chromium(III) hydroxide	chromic acid

Notice that as the number of oxygen atoms increases, so does the acidity of the compound.

Consider also the compounds shown in Figure 15-13. In acetic acid, but not in ethanol, a second oxygen atom is bonded to the carbon atom connected to the –OH group. That explains why acetic acid is acidic but ethanol is not, even though the same elements form each compound.

SECTION REVIEW

1. a. Label each reactant and each product in the reaction below as a proton donor or a proton acceptor and as acidic or basic.
   b. Label the conjugate acid-base pairs in the reaction.

$$H_2CO_3 + H_2O \rightleftharpoons HCO_3^- + H_3O^+$$

2. a. Refer to Table 15-5 to compare the strengths of the two acids in the preceding equation. Do the same for the two bases.
   b. Determine which direction—forward or reverse—is favored in the reaction.
3. Explain how the presence of several oxygen atoms in a compound containing an –OH group can make the compound acidic.

SECTION 15-3

Neutralization Reactions

OBJECTIVES

- Explain the process of neutralization.
- Explain how acid rain damages marble structures.
- Define a *Lewis acid* and a *Lewis base.*
- Name a compound that is an acid under the Lewis definition but is not an acid under the Brønsted-Lowry definition.

There are many common examples of acidic compounds reacting with basic compounds, each neutralizing the other. Sodium bicarbonate, $NaHCO_3$, and tartaric acid, $C_4H_6O_6$, are two components in baking powder. When allowed to react in solution, the two compounds produce carbon dioxide. The escaping carbon dioxide causes foods, such as biscuits, to rise. An antacid soothes an overly acidic stomach by neutralizing the stomach acid.

Strong Acid-Strong Base Neutralization

An acid-base reaction occurs in aqueous solution between hydrochloric acid, a strong acid that completely dissociates to produce H_3O^+, and sodium hydroxide, a strong base that completely dissociates to produce OH^-. The formula equation for this reaction is written as follows.

$$HCl(aq) + NaOH(aq) \longrightarrow NaCl(aq) + H_2O(l)$$

In an aqueous solution containing 1 mol of sodium hydroxide, NaOH dissociates as represented by the following equation.

$$NaOH(aq) \longrightarrow Na^+(aq) + OH^-(aq)$$

A solution containing 1 mol of hydrochloric acid ionizes as represented by the following equation.

$$HCl(aq) + H_2O(l) \longrightarrow H_3O^+(aq) + Cl^-(aq)$$

If the two solutions are mixed, as in Figure 15-14, a reaction occurs between the aqueous ions. Notice that sodium chloride, NaCl, and water are produced. The overall ionic equation is shown below.

$$H_3O^+(aq) + Cl^-(aq) + Na^+(aq) + OH^-(aq) \longrightarrow Na^+(aq) + Cl^-(aq) + 2H_2O(l)$$

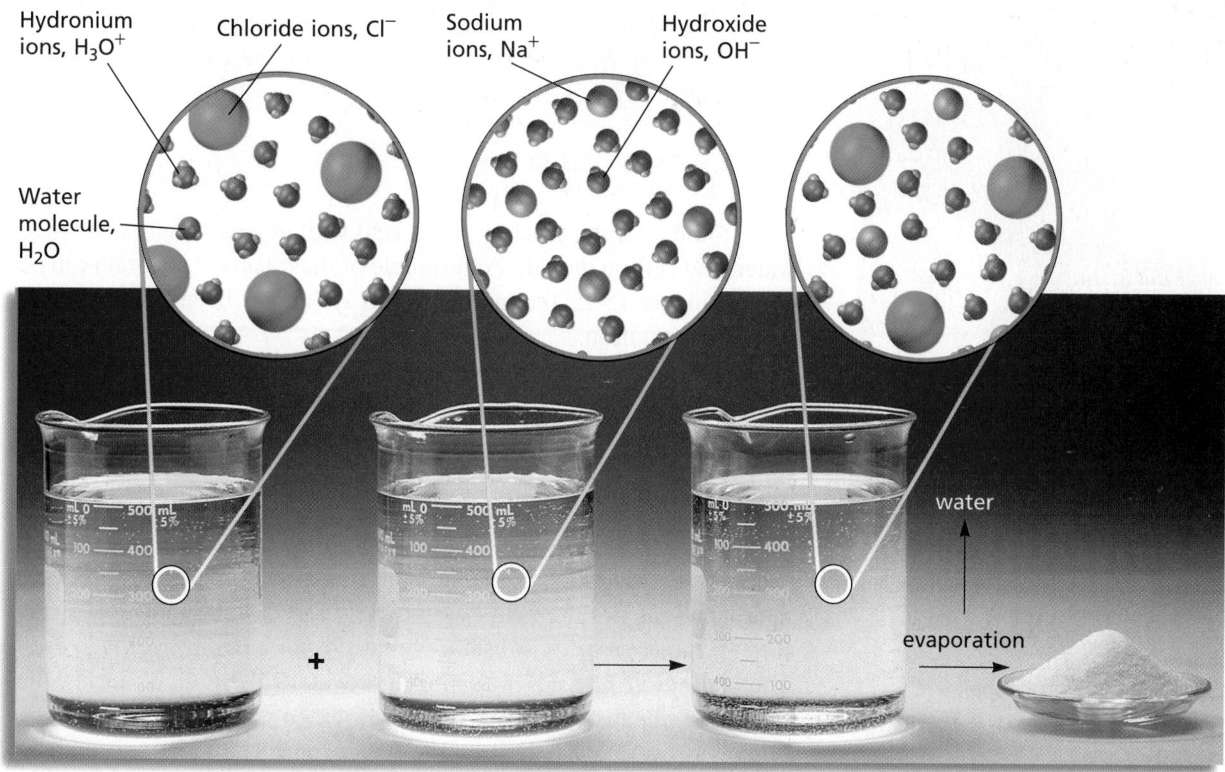

FIGURE 15-14 When aqueous hydrochloric acid, HCl, reacts with aqueous sodium hydroxide, NaOH, the reaction produces aqueous sodium chloride, NaCl. Ions that are present in each solution are represented by the models.

Because they appear on both sides of the overall ionic equation, Na^+ and Cl^- are spectator ions. The only participants in the reaction are the hydronium ion and the hydroxide ion, as shown in the following net ionic equation.

$$H_3O^+(aq) + OH^-(aq) \longrightarrow 2H_2O(l)$$

There are equal numbers of H_3O^+ and OH^- ions in this reaction, and they are fully converted to water. In aqueous solutions, **neutralization** *is the reaction of hydronium ions and hydroxide ions to form water molecules.*

Notice that water is not the only product of a neutralization. A salt is also produced. *A* **salt** *is an ionic compound composed of a cation from a base and an anion from an acid.*

Acid Rain

Many industrial processes produce gases such as NO, NO_2, CO_2, SO_2, and SO_3. These compounds can dissolve in atmospheric water to produce acidic solutions that fall to the ground in the form of rain or snow. The following reaction shows how sulfur trioxide, SO_3, dissolves in water to produce sulfuric acid.

$$SO_3(g) + H_2O(l) \longrightarrow H_2SO_4(aq)$$

FIGURE 15-15 Acid precipitation causes extensive damage to buildings and other structures.

Marble found in many buildings and statues is composed of calcium carbonate, $CaCO_3$. When acid snow or rain falls on these structures, the following reaction takes place.

$$CaCO_3(s) + 2H_3O^+(aq) \longrightarrow Ca^{2+}(aq) + CO_2(g) + 3H_2O(l)$$

The products of the reaction of an acid with any carbonate are a salt, water, and carbon dioxide. An example of the effects of acid rain can be seen in Figure 15-15; the marble in this statue has been eroded by the acidic precipitation.

Lewis Acids and Bases

The Arrhenius and Brønsted-Lowry definitions describe most acids and bases. Both definitions assume that the acid contains or produces hydrogen ions. A third acid classification, based on bonding and structure, includes, as acids, substances that do not contain hydrogen ions. This definition of acids was introduced in 1923 by G. N. Lewis, the American chemist whose name was given to electron-dot structures. Lewis's definition emphasizes the role of electron pairs in acid-base reactions. *A* **Lewis acid** *is an atom, ion, or molecule that accepts an electron pair to form a covalent bond.*

The Lewis definition is the broadest of the three acid definitions you have read about so far. It applies to any species that can accept an electron pair to form a covalent bond with another species. A bare proton (hydrogen ion) is a Lewis acid in reactions in which it forms a covalent bond, as shown below.

$$H^+(aq) + :NH_3(aq) \longrightarrow [H{-}NH_3]^+(aq) \text{ or } [NH_4]^+(aq)$$

The formula for a Lewis acid need not include hydrogen. Even a silver ion can be a Lewis acid, accepting electron pairs from ammonia to form covalent bonds.

$$Ag^+(aq) + 2:NH_3(aq) \longrightarrow [H_3N{-}Ag{-}NH_3]^+(aq) \text{ or } [Ag(NH_3)_2]^+$$

Any compound in which the central atom has three valence electrons and forms three covalent bonds can react as a Lewis acid. It does so by accepting a pair of electrons to form a fourth covalent bond, completing an electron octet. Boron trifluoride, for example, is an excellent Lewis acid. It forms a fourth covalent bond with many molecules and ions. Its reaction with a fluoride ion is shown below.

$$\begin{array}{c}:\ddot{F}:\\:\ddot{F}:\ddot{B}\\:\ddot{F}:\end{array} + :\ddot{F}:^- \longrightarrow \left[\begin{array}{c}:\ddot{F}:\\:\ddot{F}:\ddot{B}:\ddot{F}:\\:\ddot{F}:\end{array}\right]^-$$

$$BF_3(aq) + F^-(aq) \longrightarrow BF_4^-(aq)$$

TABLE 15-6 *Acid-Base Systems*

Type	Acid	Base
Arrhenius	H^+ or H_3O^+ producer	OH^- producer
Brønsted-Lowry	proton (H^+) donor	proton (H^+) acceptor
Lewis	electron-pair acceptor	electron-pair donor

The Lewis definition of acids can apply to species in any phase. For example, boron trifluoride is a Lewis acid in the gas-phase combination with ammonia.

$$\begin{array}{c} :\ddot{F}: \\ :\ddot{F}:B \\ :\underset{..}{F}: \end{array} + \begin{array}{c} H \\ :\ddot{N}:H \\ \ddot{H} \end{array} \longrightarrow \begin{array}{c} :\ddot{F}:H \\ :\ddot{F}:\ddot{B}:\ddot{N}:H \\ :\underset{..}{\ddot{F}}:H \end{array}$$

A **Lewis base** *is an atom, ion, or molecule that donates an electron pair to form a covalent bond.* An anion is a Lewis base in a reaction in which it forms a covalent bond by donating an electron pair. In the example of boron trifluoride reacting with the fluoride anion, F^- donates an electron pair to boron trifluoride. F^- acts as a Lewis base.

$$BF_3(aq) + :\ddot{\underset{..}{F}}:^-(aq) \longrightarrow BF_4^-(aq)$$

A **Lewis acid-base reaction** *is the formation of one or more covalent bonds between an electron-pair donor and an electron-pair acceptor.*

Note that although the three acid-base definitions differ, many compounds may be categorized as acids or bases according to all three descriptions. For example, ammonia is an Arrhenius base because OH^- ions are created when ammonia is in solution, it is a Brønsted-Lowry base because it accepts a proton in an acid-base reaction, and it is a Lewis base in all reactions in which it donates its lone pair to form a covalent bond. A comparison of the three acid-base definitions is given in Table 15-6.

SECTION REVIEW

1. Consider the following three reactions. Identify the Arrhenius bases, Brønsted-Lowry bases, and Lewis bases in these reactions. Explain your answers.
   a. $NaOH(s) \longrightarrow Na^+(aq) + OH^-(aq)$
   b. $HF(aq) + H_2O(l) \longrightarrow F^-(aq) + H_3O^+(aq)$
   c. $H^+(aq) + NH_3(aq) \longrightarrow NH_4^+(aq)$

2. Complete and balance the equations for the following acid-base reactions:
   a. $H_2CO_3 + Sr(OH)_2 \longrightarrow$
   b. $HClO_4 + NaOH \longrightarrow$
   c. $HBr + Ba(OH)_2 \longrightarrow$
   d. $NaHCO_3 + H_2SO_4 \longrightarrow$

CHAPTER 15 REVIEW

CHAPTER SUMMARY

15-1

- Acids have a sour taste and react with active metals. They change the colors of acid-base indicators, react with bases to produce salts and water, and conduct electricity.
- Bases have a bitter taste, feel slippery to the skin in dilute aqueous solutions, change the colors of acid-base indicators, react with acids to produce salts and water, and conduct electricity.
- An Arrhenius acid contains hydrogen and ionizes in aqueous solution to form hydrogen ions. An Arrhenius base produces hydroxide ions in aqueous solution.
- The strength of an Arrhenius acid or base is determined by the extent to which it ionizes or dissociates in aqueous solutions.

Vocabulary

alkaline (461)
Arrhenius acid (459)
Arrhenius base (459)
binary acid (454)
oxyacid (455)
strong acid (460)
weak acid (460)

15-2

- A Brønsted-Lowry acid is a proton donor. A Brønsted-Lowry base is a proton acceptor.
- Acids are described as monoprotic, diprotic, or triprotic, depending on whether they can donate one, two, or three protons per molecule, respectively, in aqueous solutions.
- In every Brønsted-Lowry acid-base reaction, there are two conjugate acid-base pairs.
- A strong acid has a weak conjugate base; a strong base has a weak conjugate acid.
- Proton-transfer reactions favor the production of weaker acids and bases.
- The acidic or basic behavior of a molecule containing $-OH$ groups may depend on the electronegativity of other atoms in the molecule and the number of oxygen atoms bonded to the atom connected to the $-OH$ group.

Vocabulary

amphoteric (470)
Brønsted-Lowry acid (464)
Brønsted-Lowry acid-base reaction (465)
Brønsted-Lowry base (465)
conjugate acid (467)
conjugate base (467)
diprotic acid (466)
monoprotic acid (465)
polyprotic acid (465)
triprotic acid (466)

15-3

- A neutralization reaction produces water and an ionic compound called a salt.
- Acid rain neutralizes the calcium carbonate in marble structures, causing them to deteriorate.
- A Lewis acid is an electron-pair acceptor. A Lewis base is an electron-pair donor.

Vocabulary

Lewis acid (474)
Lewis acid-base reaction (475)
Lewis base (475)
neutralization (473)
salt (473)

REVIEWING CONCEPTS

1. Compare and contrast the general properties of acids and bases. (15-1)

2. a. Distinguish between binary acids and oxyacids in terms of their component elements and the systems used in naming them.
b. Give three examples of each. (15-1)

3. Identify and describe the characteristic properties of five common acids used in industry. Give some examples of the typical uses of each. (15-1)

4. Although $HCl(aq)$ exhibits Arrhenius acidic properties, pure HCl gas and HCl dissolved in a nonpolar solvent exhibit no acidic properties in the Arrhenius sense. Explain why. (15-1)

5. a. What distinguishes strong acids from weak acids?
b. Give two examples of each. (15-1)

6. H_3PO_4, which contains three hydrogen atoms per molecule, is a weak acid, whereas HCl, which contains only one hydrogen atom per molecule, is a strong acid. Explain why. (15-1)

7. a. What determines the strength of an Arrhenius base?
b. Give one example each of solutions that are strongly and weakly basic. (15-1)

8. Distinguish among a monoprotic, a diprotic, and a triprotic acid. Give an example of each. (15-2)

9. Define and give an equation to illustrate each of the following:
a. a conjugate base
b. a conjugate acid (15-2)

10. a. What is the relationship between the strength of an acid and that of its conjugate base?
b. What is the relationship between the strength of a base and its conjugate acid? (15-2)

11. a. What trend is there in the favored direction of proton-transfer reactions?
b. What determines the extent to which a proton-transfer reaction occurs? (15-2)

12. a. What is meant by the term *amphoteric*?
b. Give an example of a substance or ion with amphoteric characteristics. (15-2)

13. For each reaction listed, identify the proton donor or acid and the proton acceptor or base. Label each conjugate acid-base pair.
a. $CH_3COOH(aq) + H_2O(l) \rightleftharpoons H_3O^+(aq) + CH_3COO^-(aq)$
b. $HCO_3^-(aq) + H_2O(l) \rightleftharpoons H_2CO_3(aq) + OH^-(aq)$
c. $HNO_3 + SO_4^{2-} \longrightarrow HSO_4^- + NO_3^-$ (15-2)

14. Based on the information given in Table 15-5, determine the following relative to HF, H_2S, HNO_3, and CH_3COOH:
a. strongest acid
b. weakest acid
c. strongest conjugate base among the four produced by the acids listed
d. weakest conjugate base among the four produced (15-2)

15. Explain why the conjugate base of a strong acid is a weak base and the conjugate acid of a strong base is a weak acid. (15-2)

16. Which of the three acid definitions is the broadest? Explain. (15-3)

PROBLEMS

Acid Nomenclature

17. Name each of the following binary acids:
a. HCl
b. H_2S

18. Name each of the following oxyacids:
a. HNO_3
b. H_2SO_3
c. $HClO_3$
d. HNO_2

19. Write formulas for each of the following binary acids:
a. hydrofluoric acid
b. hydriodic acid

20. Write formulas for each of the following oxyacids:
a. perbromic acid
b. chlorous acid
c. phosphoric acid
d. hypochlorous acid

Acid-Base Theory

21. a. Write the balanced equations that describe the two-stage ionization of sulfuric acid in a dilute aqueous solution.
b. How do the degrees of ionization in the two steps compare?

22. Dilute HCl(*aq*) and KOH(*aq*) are mixed in chemically equivalent quantities. Write the following:
a. formula equation for the reaction
b. overall ionic equation
c. net ionic equation

23. Repeat item 22 with $H_3PO_4(aq)$ and NaOH(*aq*).

24. Write the formula equation and net ionic equation for each of the following reactions:
a. $Zn(s) + HCl(aq) \longrightarrow$
b. $Al(s) + H_2SO_4(aq) \longrightarrow$

25. Write the formula equation and net ionic equation for the reaction between Ca(*s*) and HCl(*aq*).

Neutralization Reactions

26. Complete the following neutralization reactions. Balance each reaction, and then write the overall ionic and net ionic equation for each.
a. $HCl(aq) + NaOH(aq) \longrightarrow$
b. $HNO_3(aq) + KOH(aq) \longrightarrow$
c. $Ca(OH)_2(aq) + HNO_3(aq) \longrightarrow$
d. $Mg(OH)_2(aq) + HCl(aq) \longrightarrow$

27. Write the formula equation, the overall ionic equation, and the net ionic equation for the neutralization reaction involving aqueous solutions of H_3PO_4 and $Mg(OH)_2$. Assume that the solutions are sufficiently dilute so that no precipitates form.

28. Write the balanced chemical equation for each of the following reactions between an acid and a carbonate:
a. $BaCO_3(s) + HCl(aq) \longrightarrow$
b. $MgCO_3(s) + HNO_3(aq) \longrightarrow$
c. $Na_2CO_3(s) + H_2SO_4(aq) \longrightarrow$
d. $CaCO_3(s) + H_3PO_4(aq) \longrightarrow$

Stoichiometry

29. *Acid precipitation* is the term generally used to describe rain or snow that is more acidic than normal. One cause of acid precipitation is the formation of sulfuric and nitric acids from various sulfur and nitrogen oxides produced in volcanic eruptions, forest fires, and thunderstorms. In a typical volcanic eruption, for example, 3.50×10^8 kg of SO_2 may be produced. If this amount of SO_2 were converted to H_2SO_4 according to the two-step process given below, how many kilograms of H_2SO_4 would be produced from such an eruption?

$$SO_2 + \frac{1}{2}O_2 \longrightarrow SO_3$$
$$SO_3 + H_2O \longrightarrow H_2SO_4$$

30. Zinc reacts with 100. mL of 6.00 M cold, aqueous sulfuric acid through single replacement.
a. How many grams of zinc sulfate are produced?
b. How many liters of hydrogen gas would be released at STP?

31. A seashell, composed largely of calcium carbonate, is placed in a solution of HCl. As a result, 1500 mL of dry CO_2 gas at STP is produced. The other products are $CaCl_2$ and H_2O.
a. Based on this information, how many grams of $CaCO_3$ are consumed in the reaction?
b. What volume of 2.00 M HCl solution is used in this reaction?

32. A 211 g sample of barium carbonate, $BaCO_3$, is placed in a solution of nitric acid. Assuming that the acid is present in excess, what mass and volume of dry carbon dioxide gas at STP will be produced?

MIXED REVIEW

33. Suppose that dilute $HNO_3(aq)$ and LiOH(*aq*) are mixed in chemically equivalent quantities. Write the following for the reaction:
a. formula equation
b. overall ionic equation
c. net ionic equation

34. Write the balanced chemical equation for the reaction between hydrochloric acid and magnesium metal.

35. Write equations for the three-step ionization of phosphoric acid, H_3PO_4. Compare the degree of ionization for the three steps.

36. Name or give the molecular formula for each of the following acids:
a. HF
b. acetic acid
c. phosphorous acid
d. $HClO_4$
e. H_3PO_4
f. hydrobromic acid
g. HClO
h. H_2CO_3
i. sulfuric acid

CRITICAL THINKING

37. Analyzing Conclusions In the eighteenth century, Antoine Lavoisier experimented with oxides such as CO_2 and SO_2. He observed that they formed acidic solutions. His observations led him to infer that for a substance to exhibit acidic behavior, it must contain oxygen. However, today that is known to be incorrect. Provide evidence to refute Lavoisier's conclusion.

HANDBOOK SEARCH

38. Group 16 of the *Elements Handbook* contains a section covering the acid-base chemistry of oxides. Review this material and answer the following:
a. What type of compounds form acidic oxides?
b. What is an acidic anhydride?
c. List three examples of compounds that are classified as acidic anhydrides.
d. What type of compounds form basic oxides? Why are they basic oxides?

39. a. Look at Table 7A in the *Elements Handbook*. What periodic trends do you notice regarding acid-base character of oxides?
b. How is the nature of the product affected by the concentration of NaOH in a reaction with CO_2?

RESEARCH & WRITING

40. Explain how sulfuric acid production serves as a measure of a country's economy. Write a report on your findings.

41. Performance Conduct library research to find out about the buffering of solutions. Include information on why buffering is typically carried out and on the kinds of materials used. Write a brief report on your findings.

42. Obtain some pH paper from your teacher. Determine whether the soil around your house is acidic or basic. Find one type of plant that would grow well in that type of soil and one that would not.

ALTERNATIVE ASSESSMENT

43. Antacids are designed to neutralize excess hydrochloric acid secreted by the stomach during digestion. Carbonates, bicarbonates, and hydroxides are the active ingredients in bringing about the neutralization reactions in the most widely used antacids. Examine the labels of several common antacids, and identify the active ingredients.

44. Performance Design an experiment that compares three brands of antacids in terms of reaction speed and amount of acid neutralized.

CHAPTER 16

Acid-Base Titration and pH

The pH of solutions is important to the chemistry of life.

Aqueous Solutions and the Concept of pH

SECTION 16-1

OBJECTIVES

- Describe the self-ionization of water.
- Define *pH,* and give the pH of a neutral solution at 25°C.
- Explain and use the pH scale.
- Given $[H_3O^+]$ or $[OH^-]$, find pH.
- Given pH, find $[H_3O^+]$ or $[OH^-]$.

Hydronium Ions and Hydroxide Ions

You have already seen that acids and bases form hydronium ions and hydroxide ions in aqueous solutions. However, ions formed from the solute are not the only ions present in an aqueous solution. Hydronium ions and hydroxide ions are also provided by the solvent, water.

Self-Ionization of Water

Careful electrical-conductivity experiments have shown that pure water is an extremely weak electrolyte. This is evidence that water undergoes self-ionization, as shown in the model in Figure 16-1. *In the* **self-ionization of water,** *two water molecules produce a hydronium ion and a hydroxide ion by transfer of a proton.* The following self-ionization equilibrium takes place.

$$H_2O(l) + H_2O(l) \rightleftharpoons H_3O^+(aq) + OH^-(aq)$$

Conductivity measurements show that concentrations of H_3O^+ and OH^- in pure water are each only 1.0×10^{-7} mol/L of water at 25°C.

There is a standard notation to represent concentration in moles per liter. The formula of the particular ion or molecule is enclosed in brackets, [ ]. For example, the symbol $[H_3O^+]$ means "hydronium ion concentration in moles per liter," or "molar hydronium ion concentration." In water at 25°C, $[H_3O^+] = 1.0 \times 10^{-7}$ M and $[OH^-] = 1.0 \times 10^{-7}$ M.

The mathematical product of $[H_3O^+]$ and $[OH^-]$ remains constant in water and dilute aqueous solutions at constant temperature. This

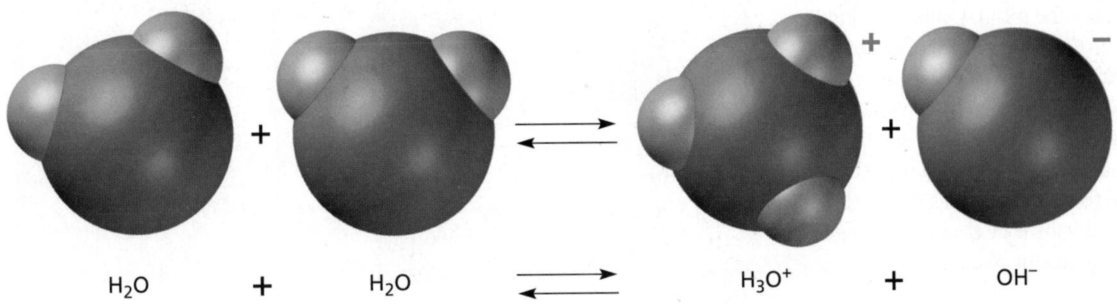

FIGURE 16-1 Water undergoes self-ionization to a slight extent. A proton is transferred from one water molecule to another. A hydronium ion, H_3O^+, and a hydroxide ion, OH^-, are produced.

TABLE 16-1 K_w at Selected Temperatures

Temperature (°C)	K_w (M^2)
0	1.2×10^{-15}
10	3.0×10^{-15}
25	1.0×10^{-14}
50	5.3×10^{-14}

constant mathematical product is called the *ionization constant of water*, K_w, and is expressed by the following equation.

$$K_w = [H_3O^+][OH^-]$$

For example, in water and dilute aqueous solutions at 25°C, the following relationship is valid.

$$K_w = [H_3O^+][OH^-] = (1.0 \times 10^{-7}\ M)(1.0 \times 10^{-7}\ M) = 1.0 \times 10^{-14}\ M^2$$

The ionization of water increases as temperature increases. Therefore, the ion product, K_w, also increases as temperature increases, as shown in Table 16-1. However, at any given temperature K_w is always a constant value. The value $1.0 \times 10^{-14}\ M^2$ is assumed to be constant within the ordinary range of room temperatures. In this chapter, you can assume that these conditions are present unless otherwise stated.

Neutral, Acidic, and Basic Solutions

Because the hydronium ion and hydroxide ion concentrations are the same in pure water, it is *neutral.* Any solution in which $[H_3O^+] = [OH^-]$ is also neutral. Recall from Chapter 15 that acids increase the concentration of H_3O^+ in aqueous solutions, as shown in Figure 16-2(a). Solutions in which the $[H_3O^+]$ is greater than the $[OH^-]$ are *acidic.* Bases increase the concentration of OH^- in aqueous solutions, as shown in Figure 16-2(b). In *basic* solutions, the $[OH^-]$ is greater than the $[H_3O^+]$.

As stated earlier, the $[H_3O^+]$ and the $[OH^-]$ of a neutral solution at 25°C both equal 1.0×10^{-7} M. Therefore, if the $[H_3O^+]$ is increased to greater than 1.0×10^{-7} M, the solution is acidic. A solution containing 1.0×10^{-5} mol H_3O^+ ion/L at 25°C is acidic because 1.0×10^{-5} is greater than 1.0×10^{-7}. If the $[OH^-]$ is increased to greater than 1.0×10^{-7} M, the solution is basic. A solution containing 1.0×10^{-4} mol OH^- ions/L at 25°C is basic because 1.0×10^{-4} is greater than 1.0×10^{-7}.

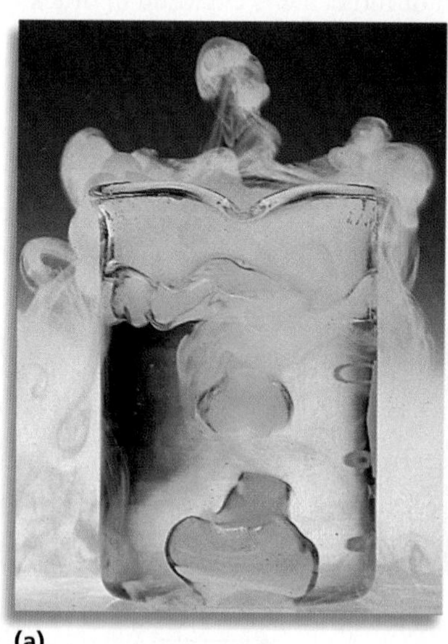
(a)

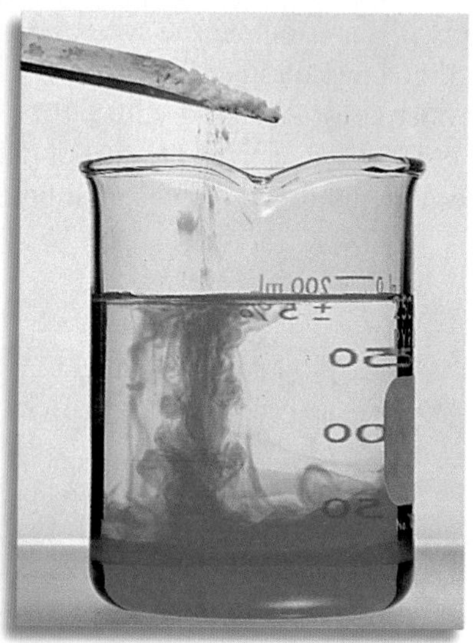
(b)

FIGURE 16-2 (a) Addition of dry ice, carbon dioxide, to water increases the $[H_3O^+]$, which is shown by the color change of the indicator bromthymol blue to yellow. The white mist is formed by condensation of water droplets because the dry ice is cold. (b) Addition of sodium peroxide to water increases the $[OH^-]$, which is shown by the color change of the indicator phenolphthalein to pink.

Calculating $[H_3O^+]$ and $[OH^-]$

Recall that strong acids and bases are considered completely ionized or dissociated in weak aqueous solutions. A review of strong acids and bases is given in Table 16-2. Notice that 1 mol of NaOH will yield 1 mol of OH^- in an aqueous solution. Therefore, a 1.0×10^{-2} M NaOH solution has an $[OH^-]$ of 1.0×10^{-2} M, as shown by the following.

$$\underset{1\ \text{mol}}{\text{NaOH}(s)} \xrightarrow{H_2O} \underset{1\ \text{mol}}{\text{Na}^+(aq)} + \underset{1\ \text{mol}}{\text{OH}^-(aq)}$$

$$\frac{1.0 \times 10^{-2}\ \cancel{\text{mol NaOH}}}{\text{L solution}} \times \frac{1\ \text{mol OH}^-}{1\ \cancel{\text{mol NaOH}}} = \frac{1.0 \times 10^{-2}\ \text{mol OH}^-}{\text{L solution}}$$

$$= 1.0 \times 10^{-2}\ \text{M OH}^-$$

Notice that the $[OH^-]$ is greater than 1.0×10^{-7} M. This solution is basic.

Because the K_w of an aqueous solution is a relatively constant $1.0 \times 10^{-14}\ M^2$ at ordinary room temperatures, the concentration of either ion can be determined if the concentration of the other ion is known. The $[H_3O^+]$ of this solution is calculated as follows.

$$K_w = [H_3O^+][OH^-] = 1.0 \times 10^{-14}\ M^2$$

$$[H_3O^+] = \frac{1.0 \times 10^{-14}\ M^2}{[OH^-]} = \frac{1.0 \times 10^{-14}\ M^2}{1.0 \times 10^{-2}\ M} = 1.0 \times 10^{-12}\ M$$

The $[OH^-]$, 1.0×10^{-2} M, is greater than the $[H_3O^+]$, 1.0×10^{-12} M, as is true for all basic solutions.

Now consider a 1.0×10^{-4} M H_2SO_4 solution. Because H_2SO_4 is a diprotic acid, the $[H_3O^+]$ is 2.0×10^{-4} M, as shown by the following.

$$\underset{1\ \text{mol}}{H_2SO_4(l)} + \underset{2\ \text{mol}}{2H_2O(l)} \longrightarrow \underset{2\ \text{mol}}{2H_3O^+(aq)} + \underset{1\ \text{mol}}{SO_4^{2-}(aq)}$$

$$\frac{1.0 \times 10^{-4}\ \cancel{\text{mol } H_2SO_4}}{\text{L solution}} \times \frac{2\ \text{mol } H_3O^+}{1\ \cancel{\text{mol } H_2SO_4}} = \frac{2.0 \times 10^{-4}\ \text{mol } H_3O^+}{\text{L solution}}$$

$$= 2.0 \times 10^{-4}\ \text{M } H_3O^+$$

Notice that the $[H_3O^+]$ is greater than 1.0×10^{-7} M. This solution is acidic. The $[OH^-]$ of this solution is calculated as follows.

$$K_w = [H_3O^+][OH^-] = 1.0 \times 10^{-14}\ M^2$$

$$[OH^-] = \frac{1.0 \times 10^{-14}\ M^2}{[H_3O^+]} = \frac{1.0 \times 10^{-14}\ M^2}{2.0 \times 10^{-4}\ M} = 5.0 \times 10^{-10}\ M$$

As is true for all acidic solutions, the $[H_3O^+]$ is greater than the $[OH^-]$.

You may have realized that in order for K_w to remain constant, an increase in either the $[H_3O^+]$ or the $[OH^-]$ in an aqueous solution causes a decrease in the concentration of the other ion. Another example of the calculation of the $[H_3O^+]$ and $[OH^-]$ of an acidic solution is shown in Sample Problem 16-1.

TABLE 16-2 *Common Strong Acids and Bases*

Strong Acids	Strong Bases
HCl	NaOH
HBr	KOH
HI	RbOH
$HClO_4$	CsOH
HNO_3	$Ca(OH)_2$
H_2SO_4	$Sr(OH)_2$
	$Ba(OH)_2$

SAMPLE PROBLEM 16-1

A 1.0×10^{-4} M solution of HNO_3 has been prepared for a laboratory experiment.
a. Calculate the $[H_3O^+]$ of this solution. b. Calculate the $[OH^-]$.

SOLUTION

1 ANALYZE

Given: Concentration of the solution = 1.0×10^{-4} M HNO_3
Unknown: a. $[H_3O^+]$ b. $[OH^-]$

2 PLAN

HNO_3 is a strong acid, which means that it is essentially 100% ionized in dilute solutions. One molecule of acid produces one hydronium ion. The concentration of the hydronium ions thus equals the concentration of the acid. Because the ion product, $[H_3O^+][OH^-]$, is a constant, $[OH^-]$ can easily be determined by using the value for $[H_3O^+]$.

a. $$\underset{1\text{ mol}}{HNO_3(l)} + \underset{1\text{ mol}}{H_2O(l)} \longrightarrow \underset{1\text{ mol}}{H_3O^+(aq)} + \underset{1\text{ mol}}{NO_3^-(aq)} \quad \text{(assuming 100\% ionization)}$$

$$\text{molarity of } HNO_3 = \frac{\text{mol } HNO_3}{\text{L solution}}$$

$$\frac{\text{mol } HNO_3}{\text{L solution}} \times \frac{1\text{ mol } H_3O^+}{1\text{ mol } HNO_3} = \frac{\text{mol } H_3O^+}{\text{L solution}} = \text{molarity of } H_3O^+$$

b. $$[H_3O^+][OH^-] = 1.0 \times 10^{-14}\ M^2$$

$$[OH^-] = \frac{1.0 \times 10^{-14}\ M^2}{[H_3O^+]}$$

3 COMPUTE

a. $$\frac{1.0 \times 10^{-4}\ \cancel{\text{mol } HNO_3}}{\text{L solution}} \times \frac{1\text{ mol } H_3O^+}{1\ \cancel{\text{mol } HNO_3}} = \frac{1.0 \times 10^{-4}\text{ mol } H_3O^+}{\text{L solution}} = 1.0 \times 10^{-4}\ M\ H_3O^+$$

b. $$[OH^-] = \frac{1.0 \times 10^{-14}\ M^2}{[H_3O^+]} = \frac{1.0 \times 10^{-14}\ M^2}{1.0 \times 10^{-4}\ \cancel{M}} = 1.0 \times 10^{-10}\ M$$

4 EVALUATE

Because the $[H_3O^+]$, 1.0×10^{-4}, is greater than 1.0×10^{-7}, the $[OH^-]$ must be less than 1.0×10^{-7}. The answers are correctly expressed to two significant digits.

PRACTICE

1. Determine the hydronium and hydroxide ion concentrations in a solution that is 1×10^{-4} M HCl.
   Answer $[H_3O^+] = 1 \times 10^{-4}$ M; $[OH^-] = 1 \times 10^{-10}$ M

2. Determine the hydronium and hydroxide ion concentrations in a solution that is 1.0×10^{-3} M HNO_3.
   Answer $[H_3O^+] = 1.0 \times 10^{-3}$ M; $[OH^-] = 1.0 \times 10^{-11}$ M

3. Determine the hydronium and hydroxide ion concentrations in a solution that is 3.0×10^{-2} M NaOH.
   Answer $[H_3O^+] = 3.3 \times 10^{-13}$ M; $[OH^-] = 3.0 \times 10^{-2}$ M

4. Determine the hydronium and hydroxide ion concentrations in a solution that is 1.0×10^{-4} M $Ca(OH)_2$.
   Answer $[H_3O^+] = 5.0 \times 10^{-11}$ M; $[OH^-] = 2.0 \times 10^{-4}$ M

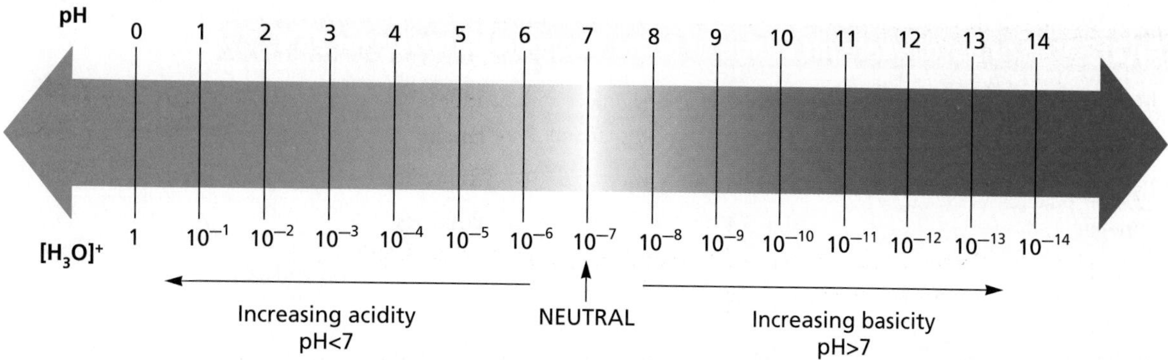

FIGURE 16-3 As the concentration of hydronium ions increases, the solution becomes more acidic and the pH decreases. As the concentration of hydronium ions decreases, the solution becomes more basic and the pH increases.

The pH Scale

Expressing acidity or basicity in terms of the concentration of H_3O^+ or OH^- can be cumbersome because the values tend to be very small. A more convenient quantity, called pH, also indicates the hydronium ion concentration of a solution. The letters *pH* stand for the French words *pouvoir hydrogène,* meaning "hydrogen power." *The* **pH** *of a solution is defined as the negative of the common logarithm of the hydronium ion concentration,* $[H_3O^+]$. The pH is expressed by the following equation.

$$pH = -\log [H_3O^+]$$

The common logarithm of a number is the power to which 10 must be raised to equal the number. A neutral solution at 25°C has a $[H_3O^+]$ of 1×10^{-7} M. The logarithm of 1×10^{-7} is −7.0. The pH is determined as follows.

$$pH = -\log [H_3O^+] = -\log (1 \times 10^{-7}) = -(-7.0) = 7.0$$

The relationship between the pH and $[H_3O^+]$ is shown on the scale in Figure 16-3.

Likewise, the **pOH** *of a solution is defined as the negative of the common logarithm of the hydroxide ion concentration,* $[OH^-]$.

$$pOH = -\log[OH^-]$$

A neutral solution at 25°C has a $[OH^-]$ of 1×10^{-7} M. Therefore, the pOH is 7.0.

Remember that the values of $[H_3O^+]$ and $[OH^-]$ are related by K_w. The negative logarithm of K_w at 25°C, 1×10^{-14}, is 14.0. You may have noticed that the sum of the pH and the pOH of a neutral solution at 25°C is also equal to 14.0. The following relationship is true at 25°C.

$$pH + pOH = 14.0$$

At 25°C the range of pH values of aqueous solutions generally falls between 0 and 14, as shown in Table 16-3.

TABLE 16-3 ***Approximate pH Range of Some Common Materials (at 25°C)***

Material	pH	Material	pH
Gastric juice	1.0–3.0	Bread	5.0–6.0
Lemons	2.2–2.4	Rainwater	5.4–5.8
Vinegar	2.4–3.4	Potatoes	5.6–6.0
Soft drinks	2.0–4.0	Milk	6.3–6.6
Apples	2.9–3.3	Saliva	6.5–7.5
Grapefruit	3.0–3.3	Pure water	7.0
Oranges	3.0–4.0	Blood	7.3–7.5
Cherries	3.2–4.0	Eggs	7.6–8.0
Tomatoes	4.0–4.4	Sea water	8.0–8.5
Bananas	4.5–5.7	Milk of magnesia	10.5

Suppose the $[H_3O^+]$ in a solution is greater than the $[OH^-]$, as is true for acidic solutions. For example, the pH of an acidic solution at 25°C with a $[H_3O^+]$ of 1×10^{-6} M is 6.0.

$$pH = -\log [H_3O^+] = -\log (1 \times 10^{-6}) = -(-6.0) = 6.0$$

The pH of this solution is less than 7. This is the case for all acidic solutions at 25°C. The following calculation shows that the pOH is greater than 7.0, as is true for all acidic solutions at 25°C.

$$pOH = 14.0 - pH = 14.0 - 6.0 = 8.0$$

Similar calculations show that the pH of a basic solution at 25°C is more than 7.0 and the pOH is less than 7.0. These and other relationships are listed in Table 16-4. Remember that as the temperature changes, the exact values will change because the value of K_w changes. However, the relationships between the values will remain the same.

TABLE 16-4 ***$[H_3O^+]$, $[OH^-]$, pH, and pOH of Solutions***

Solution	General condition	At 25°C
Neutral	$[H_3O^+] = [OH^-]$ $pH = pOH$	$[H_3O^+] = [OH^-] = 1 \times 10^{-7}$ M $pH = pOH = 7.0$
Acidic	$[H_3O^+] > [OH^-]$ $pH < pOH$	$[H_3O^+] > 1 \times 10^{-7}$ M $[OH^-] < 1 \times 10^{-7}$ M $pH < 7.0$ $pOH > 7.0$
Basic	$[H_3O^+] < [OH^-]$ $pH > pOH$	$[H_3O^+] < 1 \times 10^{-7}$ M $[OH^-] > 1 \times 10^{-7}$ M $pH > 7.0$ $pOH < 7.0$

Calculations Involving pH

If either the $[H_3O^+]$ or pH of a solution is known, the other can be calculated. Significant figures involving pH must be handled carefully. Because pH represents a logarithm, the number to the *left of the decimal* only locates the decimal point. It is not included when counting significant figures. So there must be as many significant figures to the *right of the decimal* as there are in the number whose logarithm was found. For example, a $[H_3O^+]$ value of 1×10^{-7} has *one* significant figure. Therefore, the pH, or –log, of this value must have one digit to the right of the decimal. Thus, pH = 7.0 has the correct number of significant figures.

Calculating pH from $[H_3O^+]$

You have already seen the simplest pH problems. In these problems, the $[H_3O^+]$ of the solution is an integral power of 10, such as 1 M or 0.01 M. The pH of this type of solution is the exponent of the hydronium ion concentration with the sign changed. For example, the pH of a solution in which $[H_3O^+]$ is 1×10^{-5} M is 5.0.

SAMPLE PROBLEM 16-2

What is the pH of a 1.0×10^{-3} M NaOH solution?

SOLUTION

1 ANALYZE

Given: Identity and concentration of solution = 1.0×10^{-3} M NaOH
Unknown: pH of solution

2 PLAN

concentration of base $\longrightarrow$ concentration of OH^- $\longrightarrow$ concentration of H_3O^+ $\longrightarrow$ pH

NaOH is completely dissociated when it is dissolved in water. A 1.0×10^{-3} M NaOH solution therefore produces a $[OH^-]$ equal to 1.0×10^{-3} M. The ion product of $[H_3O^+]$ and $[OH^-]$ is a constant, 1.0×10^{-14} M^2. By substitution, the $[H_3O^+]$ can be determined. The pH can then be calculated.

3 COMPUTE

$$[H_3O^+][OH^-] = 1.0 \times 10^{-14}\ M^2$$

$$[H_3O^+] = \frac{1.0 \times 10^{-14}\ M^2}{[OH^-]} = \frac{1.0 \times 10^{-14}\ M^{\not 2}}{1.0 \times 10^{-3}\ \not M} = 1.0 \times 10^{-11}\ M$$

$$pH = -\log [H_3O^+] = -\log (1.0 \times 10^{-11}) = 11.00$$

4 EVALUATE

The answer correctly indicates that NaOH forms a solution with pH > 7, which is basic.

PRACTICE

1. Determine the pH of the following solutions:
   a. 1×10^{-3} M HCl
   b. 1×10^{-5} M HNO_3
   c. 1×10^{-4} M NaOH
   d. 1.0×10^{-2} M KOH

Answer
a. pH = 3.0
b. pH = 5.0
c. pH = 10.0
d. pH = 12.00

Using a Calculator to Calculate pH from $[H_3O^+]$

Some problems involve hydronium ion concentrations that are not equal to integral powers of 10. These problems require a calculator. Most scientific calculators have a "log" key. Consult the instructions for your particular calculator.

An estimate of pH can be used to check your calculations. For example, suppose the $[H_3O^+]$ of a solution is 3.4×10^{-5} M. Because 3.4×10^{-5} lies between 10^{-4} and 10^{-5}, the pH of the solution must be between 4 and 5. Sample Problem 16-3 continues the actual calculation of the pH value for a solution with $[H_3O^+] = 3.4 \times 10^{-5}$ M.

SAMPLE PROBLEM 16-3

What is the pH of a solution if the $[H_3O^+]$ is 3.4×10^{-5} M?

SOLUTION

1 *ANALYZE*

Given: $[H_3O^+] = 3.4 \times 10^{-5}$ M
Unknown: pH of solution

2 *PLAN*

$$[H_3O^+] \longrightarrow pH$$

The only difference between this problem and previous pH problems is that you will determine the logarithm of 3.4×10^{-5} using your calculator. You can convert numbers to logarithms on most calculators by using the "log" key.

3 *COMPUTE*

$$\begin{aligned} pH &= -\log [H_3O^+] \\ &= -\log (3.4 \times 10^{-5}) \\ &= 4.47 \end{aligned}$$

On most calculators, this problem is entered in the following steps.

[3] [.] [4] [EE] [5] [+/−] [LOG] [+/−]

4 *EVALUATE*

The pH of a 1×10^{-5} M H_3O^+ solution is 5.0. Therefore, it follows that a solution with a greater concentration of hydronium ions would be more acidic and have a pH less than 5.

PRACTICE

1. What is the pH of a solution if the $[H_3O^+]$ is 6.7×10^{-4} M?
   Answer pH = 3.17
2. What is the pH of a solution with a hydronium ion concentration of 2.5×10^{-2} M?
   Answer pH = 1.60
3. Determine the pH of a 2.5×10^{-6} M HNO_3 solution.
   Answer pH = 5.60
4. Determine the pH of a 2.0×10^{-2} M $Sr(OH)_2$ solution.
   Answer pH = 12.60

Calculating $[H_3O^+]$ and $[OH^-]$ from pH

You have now learned to calculate the pH of a solution, given its $[H_3O^+]$. Suppose that you are given the pH of a solution instead. How can you determine its hydronium ion concentration?

You already know the following equation.

$$pH = -\log [H_3O^+]$$

Remember that the base of common logarithms is 10. Therefore, the antilog of a common logarithm is 10 raised to that number.

$$\log [H_3O^+] = -pH$$
$$[H_3O^+] = \text{antilog } (-pH)$$
$$[H_3O^+] = 10^{-pH}$$

The simplest cases are those in which pH values are integers. The exponent of 10 that gives the $[H_3O^+]$ is the negative of the pH. For an aqueous solution that has a pH of 2, for example, the $[H_3O^+]$ is equal to 10^{-2} M. Likewise, when the pH is 0, the $[H_3O^+]$ is 1 M because $10^0 = 1$. Sample Problem 16-4 shows how to convert a pH value that is a positive integer. Sample Problem 16-5 shows how to use a calculator to convert a pH that is not an integral number.

SAMPLE PROBLEM 16-4

Determine the hydronium ion concentration of an aqueous solution that has a pH of 4.0.

SOLUTION

1 *ANALYZE*

Given: pH = 4.0
Unknown: $[H_3O^+]$

2 *PLAN*

$$pH \longrightarrow [H_3O^+]$$

This problem requires that you rearrange the pH equation and solve for the $[H_3O^+]$. Because 4.0 has one digit to the right of the decimal, the answer must have one significant figure.

$$pH = -\log [H_3O^+]$$
$$\log [H_3O^+] = -pH$$
$$[H_3O^+] = \text{antilog } (-pH)$$
$$[H_3O^+] = 1 \times 10^{-pH}$$

3 *COMPUTE*

$$[H_3O^+] = 1 \times 10^{-pH}$$
$$[H_3O^+] = 1 \times 10^{-4} \text{ M}$$

4 *EVALUATE*

A solution with a pH of 4.0 is acidic. The answer, 1×10^{-4} M, is greater than 1.0×10^{-7} M, which is correct for an acidic solution.

SAMPLE PROBLEM 16-5

The pH of a solution is measured and determined to be 7.52.
a. What is the hydronium ion concentration? **c. Is the solution acidic or basic?**
b. What is the hydroxide ion concentration?

SOLUTION

1 *ANALYZE* **Given:** pH of the solution = 7.52
Unknown: **a.** $[H_3O^+]$ **b.** $[OH^-]$ **c.** Is the solution acidic or basic?

2 *PLAN*

$$pH \longrightarrow [H_3O^+] \longrightarrow [OH^-]$$

This problem is very similar to previous pH problems. You will need to substitute values into the $pH = -\log [H_3O^+]$ equation and use a calculator. Once the $[H_3O^+]$ is determined, the ion-product constant $[H_3O^+][OH^-] = 1.0 \times 10^{-14}$ may be used to calculate $[OH^-]$.

3 *COMPUTE* **a.**

$$pH = -\log [H_3O^+]$$
$$\log [H_3O^+] = -pH$$
$$[H_3O^+] = \text{antilog}(-pH) = \text{antilog}(-7.52) = 1.0 \times 10^{-7.52} = 3.0 \times 10^{-8} \text{ M } H_3O^+$$

On most calculators, this is entered in one of the following two ways.

[7] [.] [5] [2] [+/−] [2nd] [10^x] *or* [7] [.] [5] [2] [+/−] [INV] [LOG]

b. $[H_3O^+][OH^-] = 1.0 \times 10^{-14} \text{ M}^2$

$$[OH^-] = \frac{1.0 \times 10^{-14} \text{ M}^2}{[H_3O^+]}$$

$$= \frac{1.0 \times 10^{-14} \text{ M}^{\cancel{2}}}{3.0 \times 10^{-8} \cancel{\text{M}}} = 3.3 \times 10^{-7} \text{ M } OH^-$$

c. A pH of 7.52 is slightly greater than a pH of 7. This means that the solution is slightly basic.

4 *EVALUATE* Because the solution is slightly basic, a hydroxide ion concentration slightly larger than 10^{-7} M is predicted. A hydronium ion concentration slightly less than 10^{-7} M is also predicted. The answers agree with these predictions.

PRACTICE

1. The pH of a solution is determined to be 5.0. What is the hydronium ion concentration of this solution?
   Answer $[H_3O^+] = 1 \times 10^{-5}$ M
2. The pH of a solution is determined to be 12.0. What is the hydronium ion concentration of this solution?
   Answer $[H_3O^+] = 1 \times 10^{-12}$ M
3. The pH of an aqueous solution is measured as 1.50. Calculate the $[H_3O^+]$and the $[OH^-]$.
   Answer $[H_3O^+] = 3.2 \times 10^{-2}$ M; $[OH^-] = 3.2 \times 10^{-13}$ M
4. The pH of an aqueous solution is 3.67. Determine $[H_3O^+]$.
   Answer $[H_3O^+] = 2.1 \times 10^{-4}$ M

TABLE 16-5 ***Relationship of $[H_3O^+]$ to $[OH^-]$ and pH (at 25°C)***

Solution	$[H_3O^+]$	$[OH^-]$	pH
1.0×10^{-2} M KOH	1.0×10^{-12}	1.0×10^{-2}	12.00
1.0×10^{-2} M NH_3	2.4×10^{-11}	4.2×10^{-4}	10.62
Pure H_2O	1.0×10^{-7}	1.0×10^{-7}	7.00
1.0×10^{-3} M HCl	1.0×10^{-3}	1.0×10^{-11}	3.00
1.0×10^{-1} M CH_3COOH	1.3×10^{-3}	7.7×10^{-12}	2.88

pH Calculations and the Strength of Acids and Bases

So far, we have discussed the pH of solutions that contain only strong acids or strong bases. We must also consider weak acids and weak bases. Table 16-5 lists the $[H_3O^+]$, the $[OH^-]$, and the pH for several solutions.

KOH, the solute in the first solution listed, is a soluble ionic compound and a strong base. The molarity of a KOH solution directly indicates the $[OH^-]$, and the $[H_3O^+]$ can be calculated. Once the $[H_3O^+]$ is known, the pH can be calculated as in Sample Problem 16-3. If the pH of this solution is measured experimentally, it will be the same as this calculated value. Methods for experimentally determining the pH of solutions will be presented in Section 16-2. Hydrochloric acid, HCl, is a strong acid, and similar calculations can be made for solutions that contain HCl.

Solutions of weak acids, such as acetic acid, CH_3COOH, present a different problem. The $[H_3O^+]$ cannot be calculated directly from the molar concentration because not all of the acetic acid molecules are ionized. The same problem occurs for weak bases such as ammonia, NH_3. The pH of these solutions must be measured experimentally. The $[H_3O^+]$ and $[OH^-]$ can then be calculated from the measured pH values.

SECTION REVIEW

1. What is the concentration of hydronium and hydroxide ions in pure water at 25°C?

2. Why does the pH scale generally range from 0 to 14 in aqueous solutions?

3. Why does a pH of 7 represent a neutral solution at 25°C?

4. Identify each of the following as being true of acidic or basic solutions at 25°C:

a. $[H_3O^+] = 1 \times 10^{-3}$ M   c. pH = 5.0
b. $[OH^-] = 1 \times 10^{-4}$ M   d. pH = 8.0

5. A solution contains 4.5×10^{-3} HCl. Determine the following for the solution:
a. $[H_3O^+]$ b. $[OH^-]$ c. pH

6. A $Ca(OH)_2$ solution has a pH of 8.0. Determine the following for the solution:
a. $[H_3O^+]$ b. $[OH^-]$ c. $[Ca(OH)_2]$

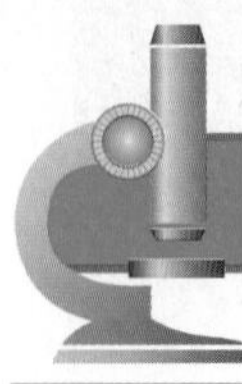

Liming Streams

In 1987, Dr. Ken Simmons tested some rainbow trout in the waters of north-central Massachusetts' Whetstone Brook. He placed the trout in cages in the brook so that their behavior and survival could be monitored. Three days later, they were all dead. Acid rain had lowered the pH level of the water to a point at which the trout simply could not survive.

Biologists studied trout to determine the effectiveness of liming Whetstone Brook to raise the pH.

Acid rain begins with the fossil fuels we burn to power our cars and factories. The fumes released by those fuels contain sulfur dioxides and nitrous oxides that combine with the water vapor in the atmosphere and turn it acidic. While normal rainwater has a pH level around 5.7, acid rain's pH can be less than 4.2.

Usually soil has enough natural buffers in it to counteract acidic rain, but beneath streams like the Whetstone, the soil is very sandy and lacks the buffering agents that neutralize the acid. The rain lowers the pH level of the brook and significantly affects most of the organisms living in it. Some fish, like the rainbow trout, simply die. Other species refuse to spawn in acidic waters, as the Whetstone's brown trout did in 1987. And if aluminum in nearby soil runs into the stream, it reacts with the acidic water, making it poisonous to the fish.

The year that the brown trout refused to spawn, the pH level of Whetstone Brook averaged 5.97. The population of all the trout dropped dangerously low, and in 1989, Dr. Simmons and other researchers instituted an experiment to decrease the acidity of the stream. They created a system to constantly add calcium carbonate, or limestone, in measured amounts to part of the brook. The limestone, ground into a powder, dissolved instantly and acted as a buffer against the acid, raising the pH level of the water.

The experiment lasted three years and managed to raise the average pH level of the stream from 5.97 to 6.54, meeting the scientists' goal. At the same time, the amount of toxic aluminum in the limed area decreased, while it increased in untreated parts of the brook. The lime also increased the ability of the water in the stream to buffer acidic pulses from acid rain and snowmelt.

The success of the project was most convincingly demonstrated by the stream's residents. The population of brook trout increased, the mortality rate of brown trout decreased, and for the first time in years, fish actually began to move into the stream from its source, the Millers River. In 1991, Dr. Simmons again tested rainbow trout in the waters of the Whetstone. This time, they all survived.

"We clearly don't view it as a solution," says Dr. Simmons. "It's a band-aid approach, but we need data to make intelligent management decisions as to how useful or harmful liming could be. And I think that is the key thing this study has shown. It has provided us with information that we can use."

Determining pH and Titrations

SECTION 16-2

OBJECTIVES

- Describe how an acid-base indicator functions.
- Explain how to carry out an acid-base titration.
- Calculate the molarity of a solution from titration data.

Indicators and pH Meters

An approximate value for the pH of a solution can be obtained using acid-base indicators. **Acid-base indicators** *are compounds whose colors are sensitive to pH.* In other words, the color of an indicator changes as the pH of a solution changes.

Indicators change colors because they are either weak acids or weak bases. In solution, the equilibrium of an indicator that is a weak acid can be represented by the equation below, which is modeled in Figure 16-4.

$$\mathrm{H}In \rightleftarrows \mathrm{H}^+ + In^-$$

(In^- is the symbol of the anion part of the indicator.) The colors that an indicator displays result from the fact that HIn and In^- are different colors.

In acidic solutions, any In^- ions that are present act as Brønsted bases and accept protons from the acid. The indicator is then present in largely nonionized form, HIn. The indicator has its acid-indicating color, as shown for litmus in Figure 16-4.

In basic solutions, the OH^- ions from the base combine with the H^+ ions produced by the indicator. The indicator molecules further ionize to offset the loss of H^+ ions. The indicator is thus present largely in the form of its anion, In^-. The solution now displays the base-indicating color, which for litmus is blue.

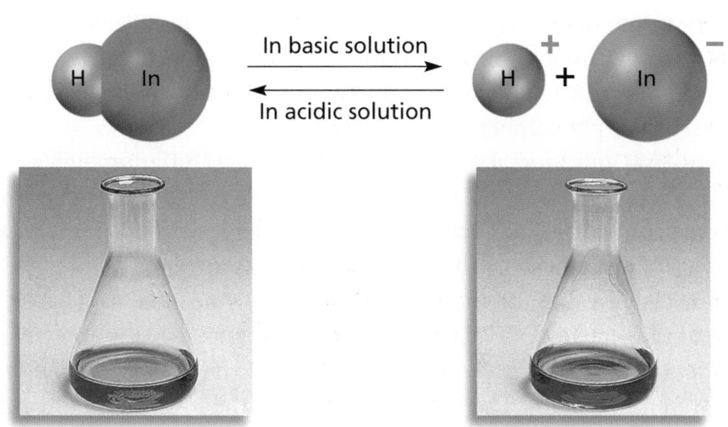

FIGURE 16-4 Basic solutions shift the equilibrium of litmus to the right. The ionized form, In^-, then predominates, and the litmus turns blue. Acidic solutions shift the equilibrium of the indicator litmus to the left. The nonionized form, HIn, predominates, and the litmus turns red.

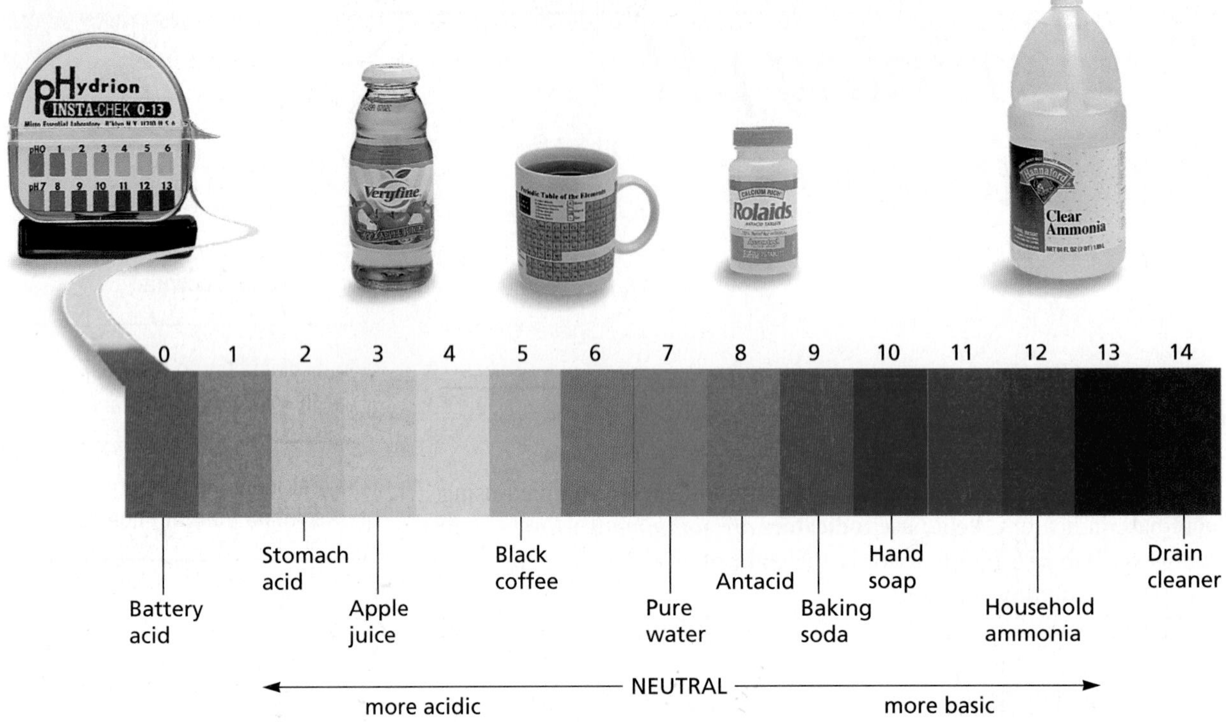

FIGURE 16-5 The pH of a solution can be determined by comparing the color it turns pH paper with the scale of the paper. The colors of pH paper at various pH values are shown, as are the pH values for some common materials.

Indicators come in many different colors. The exact pH range over which an indicator changes color also varies. *The pH range over which an indicator changes color is called its* **transition interval.** Table 16-6 gives the color changes and transition intervals for a number of common acid-base indicators.

Indicators that change color at pH lower than 7, such as methyl orange, are simply stronger acids than the other types of indicators. They tend to ionize more completely than the others. The In^- anions that these indicators produce are weaker Brønsted bases and have less tendency to accept protons from any acid being tested. These indicators therefore do not shift to their nonionized (HIn) form unless the concentration of H^+ is fairly high. The color transition of these indicators occurs at rather low pH. In contrast, indicators that undergo transition in the higher pH range, such as phenolphthalein, are weaker acids.

Universal indicators are made by mixing several different indicators. Paper soaked in universal indicator solution is called pH paper. This paper can turn almost any color of the rainbow and provides a fairly accurate way of distinguishing the pH of solutions, as shown in Figure 16-5.

If an exact value for the pH of a solution is needed, a pH meter, shown in Figure 16-6, should be used. *A* **pH meter** *determines the pH of a solution by measuring the voltage between the two electrodes that are placed in the solution.* The voltage changes as the hydronium ion concentration in the solution changes.

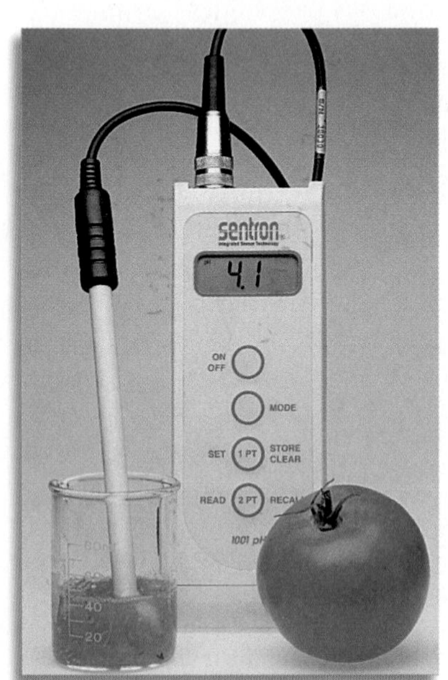

FIGURE 16-6 A pH meter measures the exact pH of a solution.

TABLE 16-6 Color Ranges of Various Indicators Used in Titrations

Titration type	Indicator	Acid color	Transition color	Base color
Strong-acid/ strong-base	litmus			
	bromthymol blue			
Strong-acid/ weak-base	methyl orange			
	bromphenol blue			
Weak-acid/ strong-base	phenolphthalein			
	phenol red			

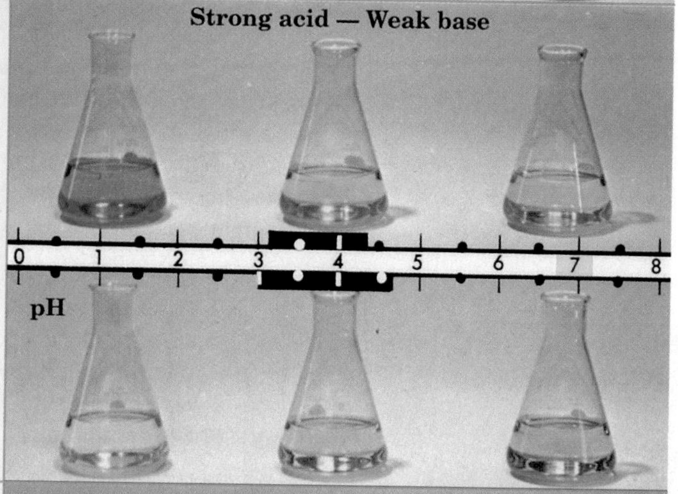

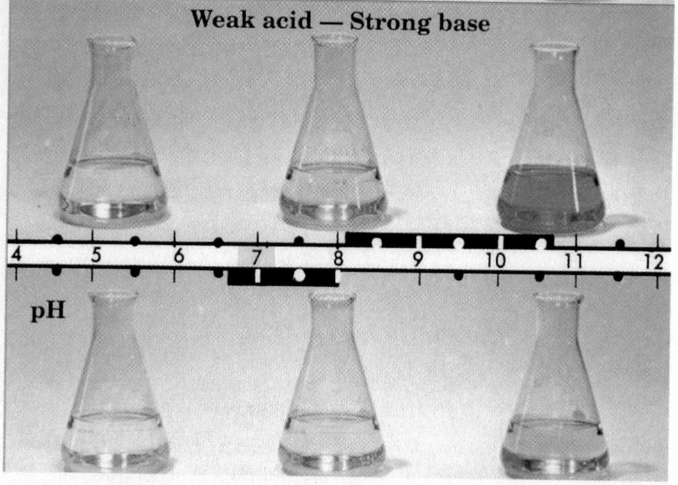

DESKTOP INVESTIGATION

Testing the pH of Rainwater

MATERIALS

- **rainwater**
- **distilled water**
- **500 mL jars**
- **thin, transparent metric ruler (± 0.1 cm)**
- **pH test paper: narrow range, ± 0.2–0.3, or pH meter**

QUESTION

Do you have acid precipitation in your area?

PROCEDURE

Record all your results in a data table.

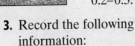

1. Each time it rains, set out five clean jars to collect the rainwater. If the rain continues for more than 24 hours, put out new containers at the end of each 24-hour period until the rain stops. (The same procedure can be used with snow if the snow is allowed to melt before measurements are taken. You may need to use larger containers if a heavy snowfall is expected.)
2. After the rain stops or at the end of each 24-hour period, measure the depth of the water to the nearest 0.1 cm with a thin plastic ruler. Test the water with the pH paper to determine its pH to the nearest 0.2–0.3.

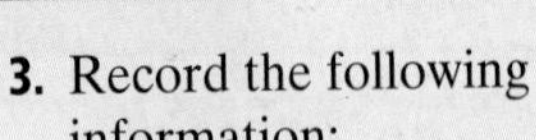

3. Record the following information:
   a. the date and time the collection was started
   b. the date and time the collection was ended
   c. the location where the collection was made (town and state)
   d. the amount of rainfall in centimeters
   e. the pH of the rainwater
4. Find the average pH of each collection that you have made for each rainfall, and record it in the data table.
5. Collect samples on at least five different days. The more samples you collect, the more reliable your data will be.
6. For comparison, determine the pH of pure water by testing five samples of distilled water with pH paper. Record your results in a separate data table, and then calculate an average pH for distilled water.

DISCUSSION

1. What is the pH of distilled water?
2. What is the pH of normal rainwater? How do you explain any differences between the pH readings?
3. What are the drawbacks of using a ruler to measure the depth of collected water? How could you increase the precision of your measurement?
4. Does the amount of rainfall or the time of day the sample is taken have an effect on its pH? Explain any variability among samples.
5. What conclusion can you draw from this investigation? Explain how your data support your conclusion.

Titration

As you know, neutralization reactions occur between acids and bases. The OH^- ion acquires a proton from the H_3O^+ ion, forming two molecules of water. The following equation summarizes this reaction.

$$H_3O^+(aq) + OH^-(aq) \longrightarrow 2H_2O(l)$$

This equation shows that one mol of hydronium ions, 19.0 g, and one mol of hydroxide ions, 17.0 g, are chemically equivalent masses. They combine in a one-to-one ratio. Neutralization occurs when hydronium ions and hydroxide ions are supplied in equal numbers by reactants, as shown in Figure 16-7.

One liter of a 0.10 M HCl solution contains 0.10 mol of hydronium ions. Now suppose that 0.10 mol of solid NaOH is added to 1 L of 0.10 M HCl solution. The NaOH dissolves and supplies 0.10 mol of hydroxide ions to the solution. HCl and NaOH are present in chemically equivalent amounts. Hydronium and hydroxide ions, which are present in equal numbers, combine until the product $[H_3O^+]$ $[OH^-]$ returns to the value of 1×10^{-14} M^2. NaCl, the salt produced in the reaction, is the product of the neutralization of a strong acid and a strong base. The resulting solution is neutral.

Because acids and bases react, the progressive addition of an acid to a base (or a base to an acid) can be used to compare the concentrations of the acid and the base. **Titration** *is the controlled addition and measurement of the amount of a solution of known concentration required to react completely with a measured amount of a solution of unknown concentration.* Titration provides a sensitive means of determining the chemically equivalent volumes of acidic and basic solutions.

FIGURE 16-7 The solution on the left turns pH paper red because it is acidic. The solution on the right turns pH paper blue because it is basic. When equal numbers of H_3O^+ and OH^- from the acidic and basic solutions react, the resulting solution is neutral. The neutral solution turns pH paper green.

Equivalence Point

The point at which the two solutions used in a titration are present in chemically equivalent amounts is the **equivalence point.** Indicators and pH meters can be used to determine the equivalence point. A pH meter will show a large voltage change occurring at the equivalence point. If an indicator is used, it must change color over a range that includes the pH of the equivalence point, as shown in Figure 16-8. *The point in a titration at which an indicator changes color is called the* **end point** *of the indicator.*

Refer to Table 16-6. Some indicators, such as litmus, change color at about pH 7. However, the color-change interval for litmus is broad, pH 5.5–8.0. This broad range makes it difficult to determine an accurate pH. Bromthymol blue is better because it has a limited transition interval, pH 6.0–7.6. Indicators that undergo transition at about pH 7 are used to determine the equivalence point of strong-acid/strong-base titrations because the neutralization of strong acids with strong bases produces a salt solution with a pH of approximately 7.

Indicators that change color at pH lower than 7 are useful in determining the equivalence point of strong-acid/weak-base titrations. Methyl orange is an example of this type. The equivalence point of a strong-acid/weak-base titration is acidic because the salt formed is itself a weak acid. Thus the salt solution has a pH lower than 7.

Indicators that change color at pH higher than 7 are useful in determining the equivalence point of weak-acid/strong-base titrations. Phenolphthalein is an example. These reactions produce salt solutions whose pH is greater than 7. This occurs because the salt formed is a weak base.

You may be wondering what type of indicator is used to determine the equivalence point of weak-acid/weak-base titrations. The surprising answer is "none at all." The pH of the equivalence point of weak acids and weak bases may be almost any value, depending on the relative strengths of the reactants. The color transition of an indicator helps

(a)

(b)

FIGURE 16-8 Indicators change color at the end point of a titration. Phenolphthalein (a) turns pink and methyl red (b) turns red at the end point of these titrations.

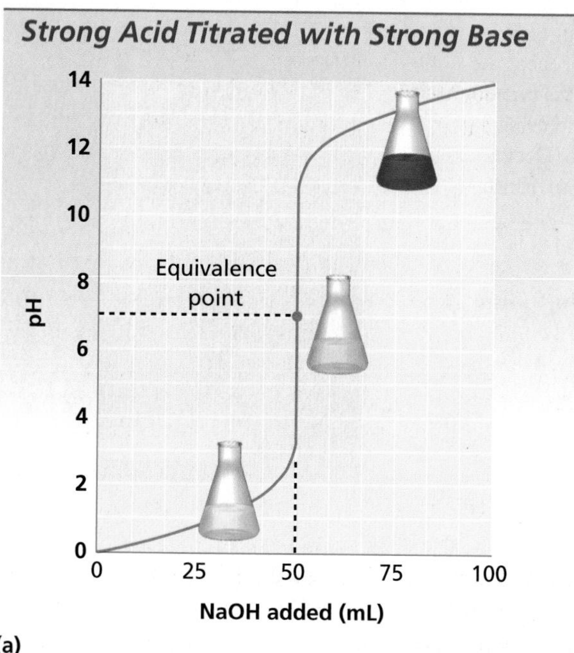

(a)

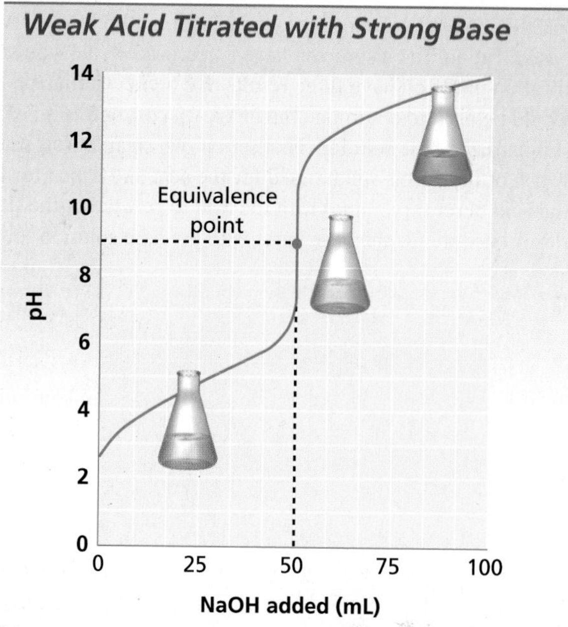

(b)

FIGURE 16-9 (a) When a strong acid, such as 50.0 mL of 1.00 M HCl, is titrated with a strong base, such as 1.00 M NaOH, the equivalence point occurs at about pH 7. (b) When a weak acid, such as 50 mL of 1.00 M CH_3COOH, is titrated with a strong base, such as 1.00 M NaOH, the equivalence point occurs at a pH above 7.

very little in determining whether reactions between such acids and bases are complete.

In a titration, successive additions of an aqueous base can be made to a measured volume of an aqueous acid. As base is added, the pH changes from a low numerical value to a high one. The change in pH occurs slowly at first, then rapidly through the equivalence point, then slowly again as the solution becomes more basic. Typical pH curves for strong-acid/strong-base and weak-acid/strong-base titrations are shown in Figure 16-9.

Molarity and Titration

Figure 16-10 on pages 500–501 shows the proper method of carrying out a titration. If the concentration of one solution is known precisely, the concentration of the other solution in a titration can be calculated from the chemically equivalent volumes. *The solution that contains the precisely known concentration of a solute is known as a* **standard solution.** It is often called simply the "known" solution.

To be certain of the concentration of the known solution, that solution must first be compared with a solution of a primary standard. *A* **primary standard** *is a highly purified solid compound used to check the concentration of the known solution in a titration.* The known solution is prepared first, and its volume is adjusted to give roughly the desired concentration. The concentration is then determined more precisely by titrating the solution with a carefully measured quantity of a solution of the primary standard.

FIGURE 16-10 Following is the proper method for carrying out an acid-base titration. To be sure you have an accurate value, you should repeat the titration until you have three results that agree within 0.05 mL. This procedure would be used to determine the unknown concentration of an acid using a standardized base solution. First set up two clean burets as shown. Decide which of the burets will be used for the acid and which for the base. Rinse the acid buret three times with the acid to be used in the titration. Repeat this procedure for the base buret with the base solution to be used.

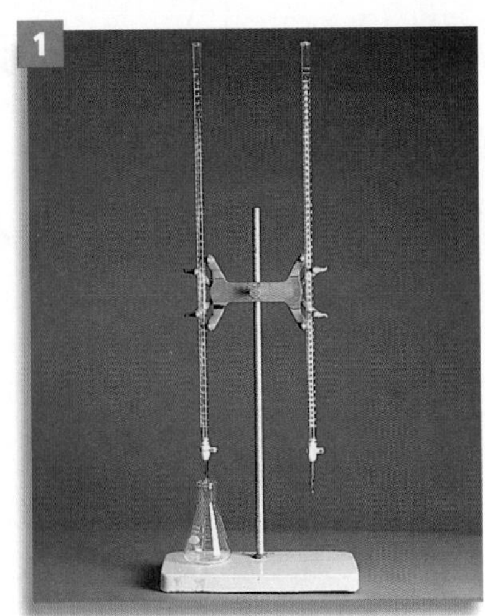

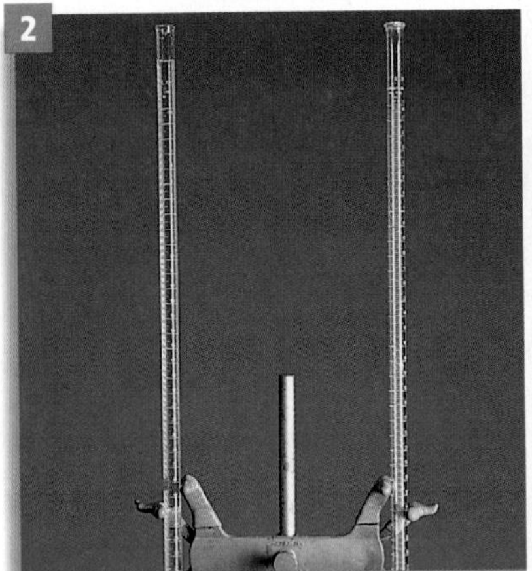

Fill the first buret to a point above the calibration mark with the acid of unknown concentration.

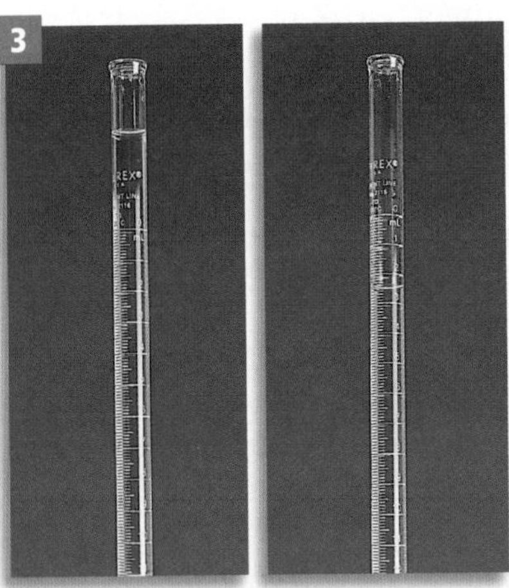

Release some acid from the buret to remove any air bubbles and to lower the volume to the calibrated portion of the buret.

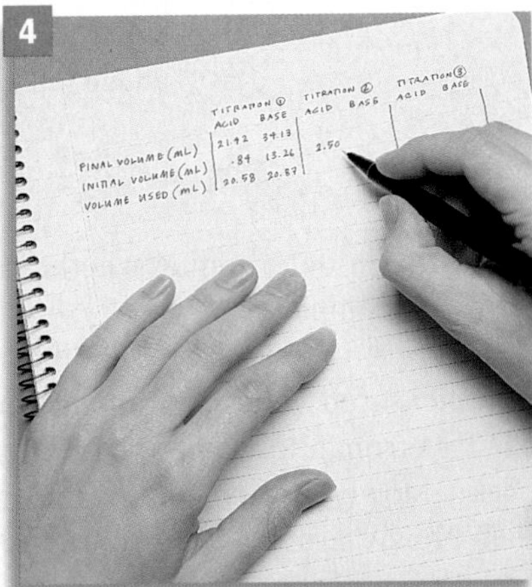

Record the reading at the top of the acid in the buret to the nearest 0.01 mL as your starting point.

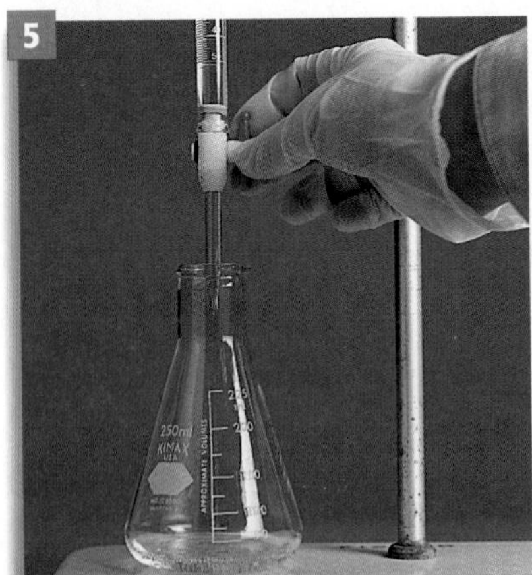

Release a predetermined volume of the acid (determined by your teacher or lab procedure) into a clean, dry Erlenmeyer flask.

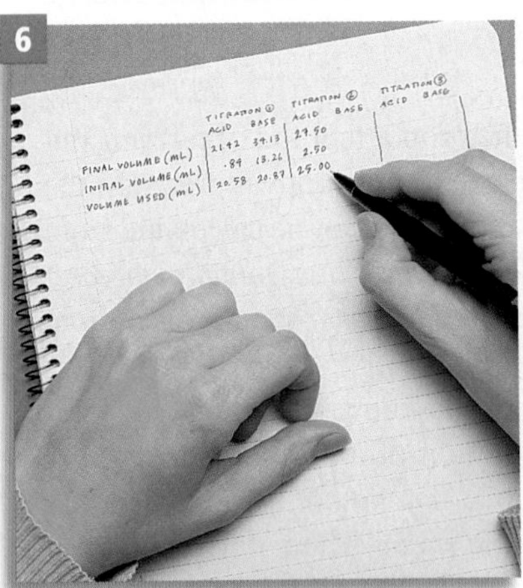

Subtract the current volume reading on the buret from the initial reading. This is the exact volume of the acid released into the flask. Record it to the nearest 0.01 mL.

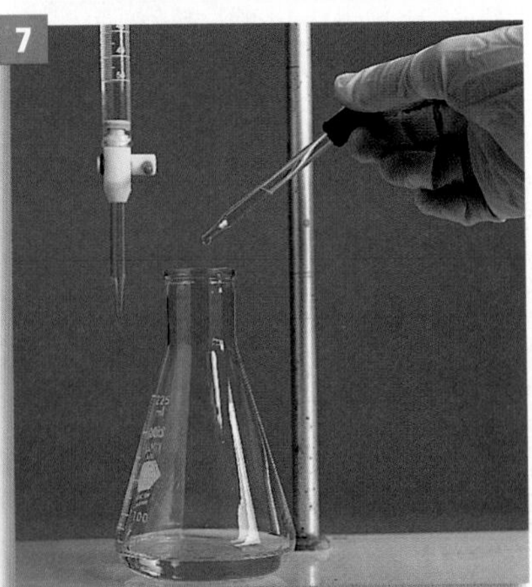

Add three drops of the appropriate indicator (in this case phenolphthalein) to the flask.

Fill the other buret with the standard base solution to a point above the calibration mark. The concentration of the standard base is known to a certain degree of precision because it was previously titrated with an exact mass of solid acid.

8

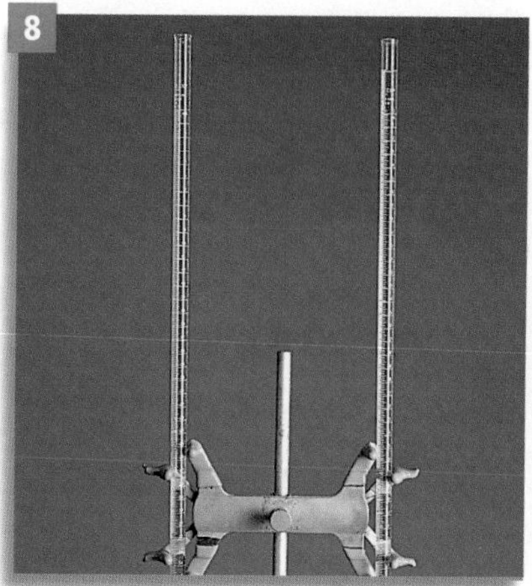

9

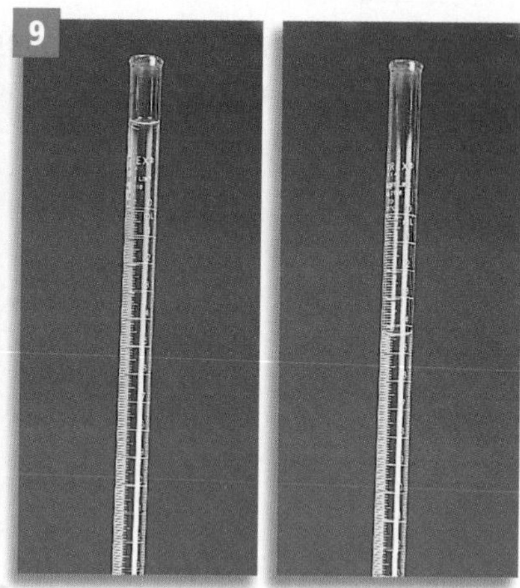

Release some base from the buret to remove any air bubbles and to lower the volume to the calibrated portion of the buret.

10

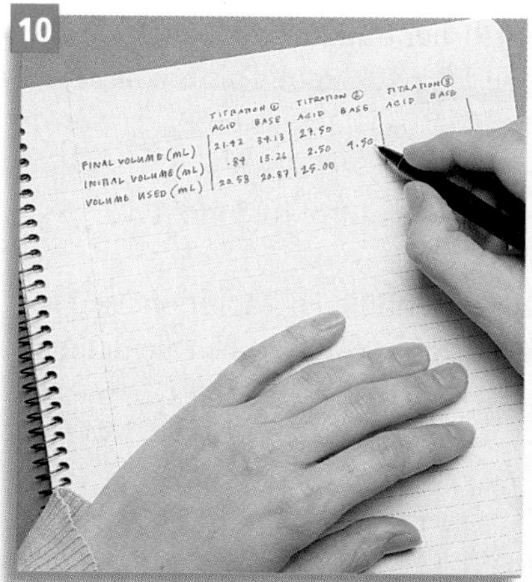

Record the reading at the top of the base to the nearest 0.01 mL as your starting point.

11

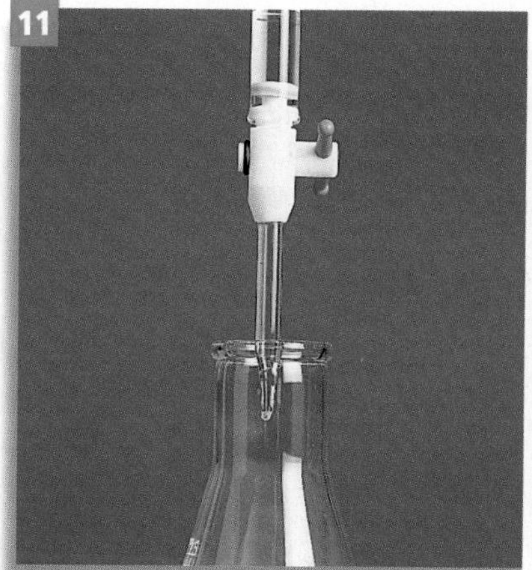

Place the Erlenmeyer flask under the base buret as shown. Notice that the tip of the buret extends into the mouth of the flask.

12

Slowly release base from the buret into the flask while constantly swirling the contents of the flask. The pink color of the indicator should fade with swirling.

13

The titration is nearing the end point when the pink color stays for longer periods of time. At this point, add base drop by drop.

14

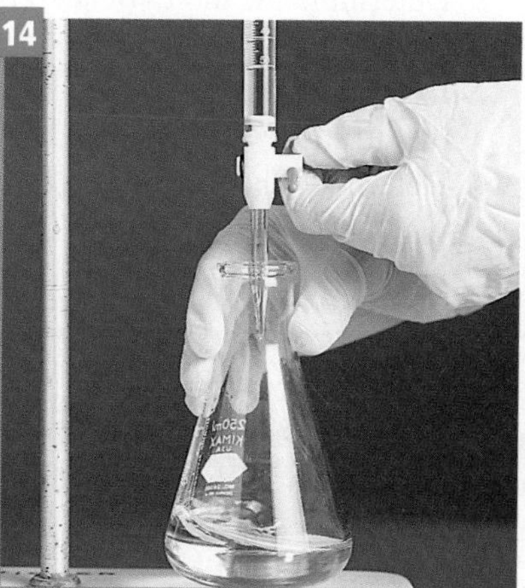

The equivalence point is reached when a very light pink color remains after 30 seconds of swirling.

15

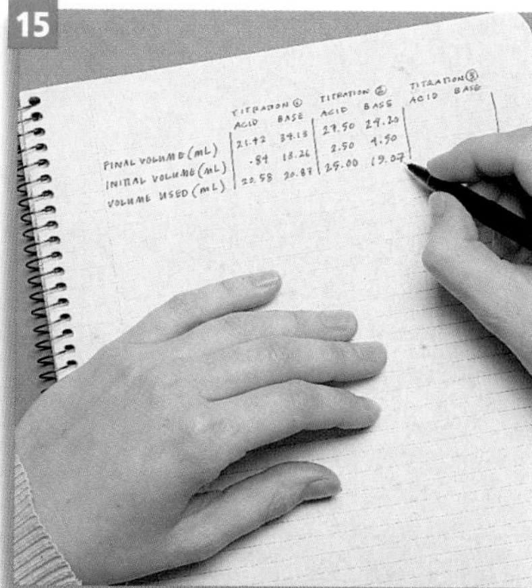

Subtract the current volume reading on the buret from the initial reading. This is the exact volume of the base released into the flask. Record it to the nearest 0.01 mL.

The known solution can be used to determine the molarity of another solution by titration. Suppose 20.0 mL of 5.0×10^{-3} M NaOH is required to reach the end point in the titration of 10.0 mL of HCl of unknown concentration. How can these titration data be used to determine the molarity of the acidic solution?

Begin with the balanced neutralization reaction equation. From the equation, determine the chemically equivalent amounts of HCl and NaOH.

$$\underset{1\text{ mol}}{HCl(aq)} + \underset{1\text{ mol}}{NaOH(aq)} \longrightarrow \underset{1\text{ mol}}{NaCl(aq)} + \underset{1\text{ mol}}{H_2O(l)}$$

Calculate the number of moles of NaOH used in the titration.

$$\frac{5.0 \times 10^{-3}\text{ mol NaOH}}{\cancel{\text{L}}} \times \frac{1\ \cancel{\text{L}}}{1000\ \cancel{\text{mL}}} \times 20.0\ \cancel{\text{mL}} = 1.0 \times 10^{-4}\text{ mol NaOH used}$$

Because 1 mol of NaOH is needed to neutralize 1 mol of HCl, the amount of HCl in the titration must be 1.0×10^{-4} mol. This is confirmed by the following equation.

$$1.0 \times 10^{-4}\ \cancel{\text{mol NaOH}} \times \frac{1\text{ mol HCl}}{1\ \cancel{\text{mol NaOH}}} = 1.0 \times 10^{-4}\text{ mol HCl}$$

This amount of acid must be in the 10.0 mL of the HCl solution used for the titration. The molarity of the HCl solution can now be calculated.

$$\frac{1 \times 10^{-4}\text{ mol HCl}}{10.0\ \cancel{\text{mL}}} \times \frac{1000\ \cancel{\text{mL}}}{1\text{ L}} = \frac{1.0 \times 10^{-2}\text{ mol HCl}}{\text{L}}$$

$$= 1.0 \times 10^{-2}\text{ M HCl}$$

Sample Problem 16-6 illustrates the following four steps.

1. Start with the balanced equation for the neutralization reaction, and determine the chemically equivalent amounts of the acid and base.
2. Determine the moles of acid (or base) from the known solution used during the titration.
3. Determine the moles of solute of the unknown solution used during the titration.
4. Determine the molarity of the unknown solution.

SAMPLE PROBLEM 16-6

In a titration, 27.4 mL of 0.0154 M $Ba(OH)_2$ is added to a 20.0 mL sample of HCl solution of unknown concentration. What is the molarity of the acid solution?

SOLUTION

1 *ANALYZE*

Given: volume and concentration of known solution = 27.4 mL of 0.0154 M $Ba(OH)_2$
volume of unknown HCl solution = 20.0 mL

Unknown: molarity of acid solution

2 PLAN

1. balanced neutralization equation ⟶ chemically equivalent amounts

$$\begin{array}{cccc} Ba(OH)_2 + & 2HCl \longrightarrow & BaCl_2 + & 2H_2O \\ 1\text{ mol} & 2\text{ mol} & 1\text{ mol} & 2\text{ mol} \end{array}$$

2. volume of known basic solution used (mL) ⟶ amount of base used (mol)

$$\frac{\text{mol } Ba(OH)_2}{\text{L}} \times \text{mL of } Ba(OH)_2 \text{ solution} \times \frac{1\text{ L}}{1000\text{ mL}} = \text{mol } Ba(OH)_2$$

3. mole ratio, moles of base used ⟶ moles of acid used from unknown solution

$$\frac{2\text{ mol HCl}}{1\text{ mol } Ba(OH)_2} \times \text{mol } Ba(OH)_2 \text{ in known solution} = \text{mol HCl in unknown solution}$$

4. volume of unknown, moles of solute in unknown ⟶ molarity of unknown

$$\frac{\text{amount of solute in unknown solution (mol)}}{\text{volume of unknown solution (mL)}} \times \frac{1000\text{ mL}}{1\text{ L}} = \text{molarity of unknown solution}$$

3 COMPUTE

1. The mole ratio from the equation is 1 mol $Ba(OH)_2$ for every 2 mol HCl.

2. $$\frac{0.0154\text{ mol } Ba(OH)_2}{\cancel{\text{L}}} \times 27.4\ \cancel{\text{mL}} \times \frac{1\ \cancel{\text{L}}}{1000\ \cancel{\text{mL}}} = 4.22 \times 10^{-4}\text{ mol } Ba(OH)_2$$

3. $$\frac{2\text{ mol HCl}}{1\ \cancel{\text{mol } Ba(OH)_2}} \times 4.22 \times 10^{-4}\ \cancel{\text{mol } Ba(OH)_2} = 8.44 \times 10^{-4}\text{ mol HCl}$$

4. $$\frac{8.44 \times 10^{-4}\text{ mol HCl}}{20.0\ \cancel{\text{mL}}} \times \frac{1000\ \cancel{\text{mL}}}{1\text{ L}} = \frac{4.22 \times 10^{-2}\text{ mol HCl}}{\text{L}} = 4.22 \times 10^{-2}\text{ M HCl}$$

PRACTICE

1. A 15.5 mL sample of 0.215 M KOH solution required 21.2 mL of aqueous acetic acid solution in a titration experiment. Calculate the molarity of the acetic acid solution.

   Answer 0.157 M CH_3COOH

2. By titration, 17.6 mL of aqueous H_2SO_4 neutralized 27.4 mL of 0.0165 M LiOH solution. What was the molarity of the aqueous acid solution?

   Answer 0.0128 M H_2SO_4

SECTION REVIEW

1. Name an appropriate indicator for titrating the following:
   a. a strong acid and a weak base
   b. a strong base and a weak acid
2. Suppose you have an NaOH solution of unknown concentration. Name three possible substances that could be used in "known" solutions to titrate the NaOH solution.
3. If 20.0 mL of 0.0100 M aqueous HCl is required to neutralize 30.0 mL of an aqueous solution of NaOH, determine the molarity of the NaOH solution.
4. Suppose that 20.0 mL of 0.10 M $Ca(OH)_2$ is required to neutralize 12.0 mL of aqueous HCl solution. What is the molarity of the HCl solution?

CHAPTER 16 REVIEW

CHAPTER SUMMARY

16-1

- Pure water undergoes self-ionization to give 1.0×10^{-7} M H_3O^+ and 1.0×10^{-7} M OH^- at 25°C.
- $pH = -\log[H_3O^+]$; $pOH = -\log[OH^-]$
  At 25°C, pH + pOH = 14.0.
- At 25°C, acids have a pH of less than 7, bases have a pH of greater than 7, and neutral solutions have a pH of 7.
- If a solution contains a strong acid or a strong base, the $[H_3O^+]$, $[OH^-]$ and pH can be calculated from the molarity of the solution. If a solution contains a weak acid or a weak base, the $[H_3O^+]$ and the $[OH^-]$ must be calculated from an experimentally measured pH.

Vocabulary

pH (485) | pOH (485) | self-ionization of water (481)

16-2

- The pH of a solution can be determined using either a pH meter or acid-base indicators.
- Titration uses a solution of known concentration to determine the concentration of a solution of unknown concentration.
- To determine the end point of a titration, indicators are chosen that change color over ranges that include the pH of the equivalence point.
- When the molarity and volume of a known solution used in a titration are known, then the molarity of a given volume of an unknown solution can be found.

Vocabulary

end point (498) | acid-base indicator (493) | primary standard (499) | titration (497)
equivalence point (498) | pH meter (494) | standard solution (499) | transition interval (494)

REVIEWING CONCEPTS

1. Why does pure water weakly conduct an electric current? (16-1)
2. What does it mean when the formula of a particular ion or molecule is enclosed in brackets? (16-1)
3. a. What is the $[H_3O^+]$ of pure water at 25°C?
   b. Is this true at all temperatures? Why or why not? (16-1)
4. a. What is always true about the $[H_3O^+]$ value of acidic solutions?
   b. What is true about the $[H_3O^+]$ value of acidic solutions at 25°C? (16-1)
5. a. Describe what is meant by the pH of a solution.
   b. Write the equation for determining pH.
   c. Explain and illustrate what is meant by the common logarithm of a number. (16-1)
6. Identify each of the following as being true of acidic, basic, or neutral solutions at 25°C:
   a. $[H_3O^+] = 1.0 \times 10^{-7}$ M
   b. $[H_3O^+] = 1.0 \times 10^{-10}$ M
   c. $[OH^-] = 1.0 \times 10^{-7}$ M
   d. $[OH^-] = 1.0 \times 10^{-11}$ M
   e. $[H_3O^+] = [OH^-]$
   f. pH = 3.0
   g. pH = 13.0 (16-1)
7. Arrange the following common substances in order of increasing pH:
   a. eggs
   b. apples
   c. tomatoes
   d. milk
   e. bananas
   f. potatoes
   g. lemons
   h. milk of magnesia
   i. sea water (16-1)
8. What is meant by the transition interval of an indicator? (16-2)
9. Explain how an indicator's equilibrium determines the color the indicator displays at a given pH. (16-2)

10. a. Other than through indicators, how can the equivalence point of a titration experiment or the pH of a solution be determined?
b. What can be observed about the rate of change in the pH of a solution near the end point of a titration? (16-2)

11. a. What is meant by the end point of a titration?
b. What is the role of an indicator in the titration process?
c. On what basis is an indicator selected for a particular titration experiment? (16-2)

12. For each of the four possible types of acid-base titration combinations, indicate the approximate pH at the end point. Also name a suitable indicator for detecting that end point. (16-2)

13. Based on Figures 14-9(a) and 14-9(b), draw a pH curve for a strong-acid/weak-base titration. (16-2)

14. An unknown solution is colorless when tested with phenolphthalein but causes the indicator phenol red to turn red. Based on this information, what is the approximate pH of this solution? (16-2)

PROBLEMS

pH Calculations

15. Calculate the $[H_3O^+]$ and $[OH^-]$ for each of the following. (Hint: See Sample Problem 16-1.)
a. 0.03 M HCl
b. 1×10^{-4} M NaOH
c. 5×10^{-3} M H_2SO_4
d. 0.01 M $Ca(OH)_2$

16. Determine the pH of each of the following solutions. (Hint: See Sample Problem 16-2.)
a. 1.0×10^{-2} M HCl
b. 1.0×10^{-3} M HNO_3
c. 1.0×10^{-5} M HI
d. 1.0×10^{-4} M HBr

17. Given the following $[OH^-]$ values, determine the pH of each solution.
a. 1.0×10^{-6} M
b. 1.0×10^{-9} M
c. 1.0×10^{-2} M
d. 1.0×10^{-7} M

18. Determine the pH of each solution.
a. 1.0×10^{-2} M NaOH
b. 1.0×10^{-3} M KOH
c. 1.0×10^{-4} M LiOH

19. Determine the pH of solutions with each of the following $[H_3O^+]$. (Hint: See Sample Problem 16-3.)
a. 2.0×10^{-5} M
b. 4.7×10^{-7} M
c. 3.8×10^{-3} M

20. Given the following pH values, determine the $[H_3O^+]$ for each solution. (Hint: See Sample Problem 16-4.)
a. 3.0
b. 7.00
c. 11.0
d. 5.0

21. Given the following pH values, determine the $[OH^-]$ for each solution.
a. 7.00
b. 11.00
c. 4.00
d. 6.00

22. Determine $[H_3O^+]$ for solutions with the following pH values. (Hint: See Sample Problem 16-5.)
a. 4.23 b. 7.65 c. 9.48

23. A nitric acid solution is found to have a pH of 2.70. Determine each of the following:
a. $[H_3O^+]$
b. $[OH^-]$
c. the number of moles of HNO_3 required to prepare 5.50 L of this solution
d. the mass of the moles of HNO_3 in the solution in part (c)
e. the milliliters of concentrated acid needed to prepare the solution in part (c) (Concentrated nitric acid is 69.5% HNO_3 by mass and has a density of 1.42 g/mL.)

Titrations

24. For each of the following acid-base titration combinations, determine the number of moles of the first substance listed that would be the chemically equivalent amount of the second substance.
a. NaOH with 1.0 mol HCl
b. HNO_3 with 0.75 mol KOH
c. $Ba(OH)_2$ with 0.20 mol HF
d. H_2SO_4 with 0.60 mol $Al(OH)_3$

25. Suppose that 15.0 mL of 2.50×10^{-2} M aqueous H_2SO_4 is required to neutralize 10.0 mL of an aqueous solution of KOH. What is the molarity of the KOH solution? (Hint: See Sample Problem 16-6.)

26. In a titration experiment, a 12.5 mL sample of 1.75×10^{-2} M $Ba(OH)_2$ just neutralized 14.5 mL of HNO_3 solution. Calculate the molarity of the HNO_3 solution.

MIXED REVIEW

27. a. What is the $[OH^-]$ of a 4.0×10^{-4} M solution of $Ca(OH)_2$?
b. What is the $[H_3O^+]$ of the solution?

28. Given the following $[H_3O^+]$ values, determine the pH of each solution.
a. 1.0×10^{-7} M
b. 1.0×10^{-3} M
c. 1.0×10^{-12} M
d. 1.0×10^{-5} M

29. In a titration, 25.9 mL of 3.4×10^{-3} M $Ba(OH)_2$ neutralized 16.6 mL of HCl solution. What is the molarity of the HCl solution?

30. What is the $[H_3O^+]$ for a solution that has a pH of 6.0?

31. Suppose that a 5.0×10^{-5} M solution of $Ba(OH)_2$ is prepared. What is the pH of the solution?

32. Find the molarity of a $Ca(OH)_2$ solution, given that 428 mL of it is neutralized in a titration by 115 mL of 6.7×10^{-3} M HNO_3.

33. a. Calculate the pH of a solution that has an $[H_3O^+]$ of 8.4×10^{-11} M.
b. Calculate the $[H_3O^+]$ of a solution that has a pH of 2.50.

34. a. What is the concentration of OH^- in a 5.4×10^{-5} M solution of magnesium hydroxide, $Mg(OH)_2$?
b. Calculate the concentration of H_3O^+ for this solution.

35. a. Calculate the molarity of H_3O^+ in a solution that has a pH of 8.90.
b. Calculate the concentration of OH^- in the solution.

36. What is the pH of a solution for which $[OH^-]$ equals 6.9×10^{-10} M?

37. Suppose that 10.1 mL of HNO_3 is neutralized by 71.4 mL of a 4.2×10^{-3} M solution of KOH in a titration. Calculate the concentration of the HNO_3 solution.

CRITICAL THINKING

38. Interpreting Graphics The following titration curve resulted from the titration of an unknown acid with 0.10 M NaOH. Analyze the curve. Make inferences related to the type of acidic solution titrated.

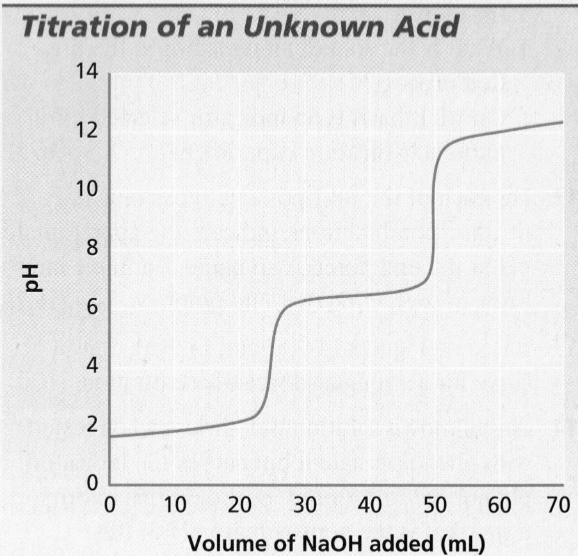

TECHNOLOGY & LEARNING

39. Graphing Calculator Graphing Titration Data

The graphing calculator can be programmed to graph data such as pH versus volume of base. Graphing the titration data will allow you to determine which combination of acid and base is represented by the shape of the graph. Begin by creating a table of data. Then program the calculator to plot the data.

Volume (mL)	pH
0	1
10	1.1
20	1.4
30	1.6
40	2.9
50	11.9
60	12.2
70	12.5

A. Create lists L1 and L2.

Keystrokes for creating lists: STAT 4 2nd L1 , 2nd L2 ENTER STAT 1

Enter the volume in L1, and enter the pH data in L2.

B. Program the graph.

Press Y= and clear the graphs.

Keystrokes for naming the program: PRGM ▶ ▶ ENTER

Name the program: T I T R A T N ENTER

Program keystrokes: 2nd STAT PLOT 1 2nd STAT PLOT ▶ 2 , 2nd L1 , 2nd L2 , 2nd STAT PLOT ▶ ▶ 3) 2nd : ZOOM 9

C. Check the display.

Screen display: Plot1(xyLine, L1, L2, .): ZoomStat

Press 2nd QUIT to exit the program editor.

D. Run the program.

a. Press PRGM. Select TITRATN and press ENTER ENTER. The calculator will provide a graph of the data. This is a titration between a strong acid and a strong base.

b. Use the TITRATN program with the data given below. Refer back to creating L1 and L2. Use L1 for volume data and L2 for pH data. Use the program to graph this data. This is a titration between a weak acid and a strong base.

Volume (mL)	pH
0	3
10	4.5
20	4.8
30	5.1
40	11.9
50	12.1
60	12.3

HANDBOOK SEARCH

40. The normal pH of blood is about 7.4. When the pH shifts above or below that level, the results are acidosis or alkalosis. Review the section on blood pH in Group 14 of the *Elements Handbook,* then answer the following.

a. What chemical species keep H_3O^+ in blood at the appropriate pH?

b. What condition results when there is an excess of CO_2 in the blood?

c. What is hyperventilation and how does it affect blood pH?

RESEARCH & WRITING

41. Examine the labels of at least five brands of shampoo. Note what is written there, if anything, regarding the pH of the shampoo. Do library research to find out why such pH ranges are chosen and why other ranges might be harmful to hair or eyes.

42. Water quality depends on the ability of treatment facilities to reprocess waste water. Most waste-water treatment facilities use living organisms to break down wastes. Because of acid rain, the acidic nature of waste water can pose problems for a waste-treatment plant. Conduct library research on this topic and write a brief report. Include an inference to how pH affects waste-water treatment.

ALTERNATIVE ASSESSMENT

43. Performance Use pH paper to determine the approximate pH of various brands of orange juice, which contains citric acid.

44. Performance Design and conduct an experiment to extract possible acid-base indicators from sources such as red cabbage, berries, and flower petals. Use known acidic, basic, and neutral solutions to test the action of each indicator that you are able to isolate.

45. Performance Design and conduct an experiment to study the pH of rain for an extended period of time. Coordinate your data-collection activities with those of your classmates so that the pH readings are made in a variety of locations. Try to determine whether any patterns emerge in terms of location, season, time of day, degree of industrialization in the area, and so on.

UNIT 6

Chemical Reactions

CHAPTERS

LOUIS PASTEUR, ON THE GLOBAL IMPORTANCE OF SCIENTIFIC PROGRESS

Science knows no country, because knowledge belongs to humanity, and is the torch which illuminates the world. Science is the highest personification of the nation because that nation will remain the first which carries the furthest the works of thought and intelligence.

(From *Pasteur and Modern Science*)

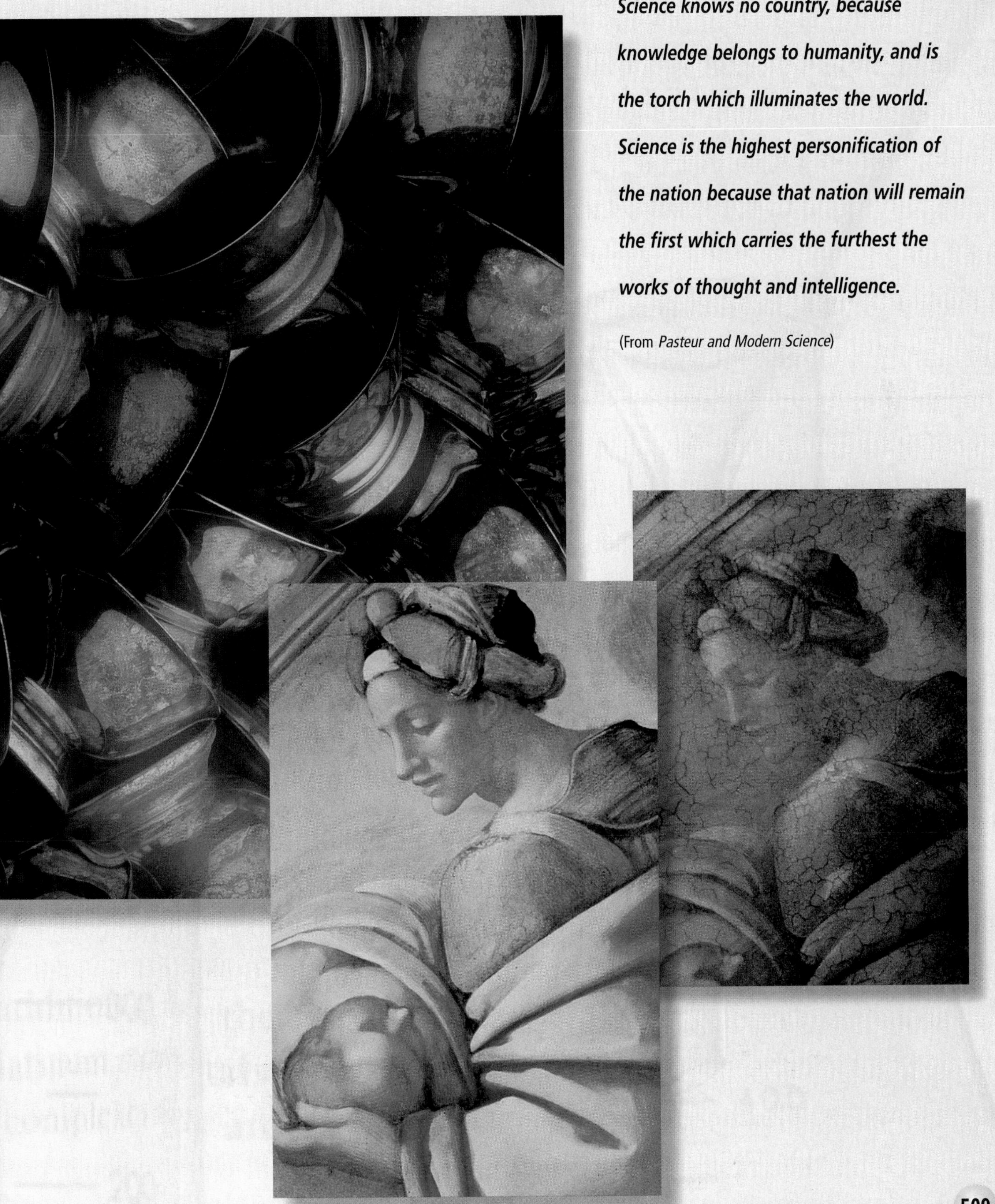

CHAPTER 17

Reaction Energy and Reaction Kinetics

Many chemical reactions give off or take in energy in the form of heat.

Thermochemistry

SECTION 17-1

OBJECTIVES

- Define *temperature,* and state the units in which it is measured.
- Define *heat* and state its units.
- Perform specific-heat calculations.
- Explain heat of reaction, heat of formation, heat of combustion, and enthalpy.
- Solve problems involving heats of reaction, heats of formation, and heats of combustion.

Virtually every chemical reaction is accompanied by a change in energy. Chemical reactions usually absorb or release energy in the form of heat. You learned in Chapter 12 that heat is also absorbed or released in physical changes, such as melting a solid or condensing a vapor. **Thermochemistry** *is the study of the changes in heat energy that accompany chemical reactions and physical changes.*

Heat and Temperature

The heat absorbed or released in a chemical or physical change is measured in a **calorimeter.** In one kind of calorimeter, known quantities of reactants are sealed in a reaction chamber, which is immersed in a known quantity of water in an insulated vessel. Therefore, the heat given off (or absorbed) during the reaction is equal to the heat absorbed (or given off) by the known quantity of water. The amount of heat is determined from the temperature change of the known mass of surrounding water. The data collected from calorimetry experiments are temperature changes because heat cannot be measured directly; but temperature, which is affected by the transfer of heat, is directly measurable. To see why this is so, let us look at the definitions of heat and temperature and at how temperature is measured.

Temperature *is a measure of the average kinetic energy of the particles in a sample of matter.* The greater the kinetic energy of the particles in a sample, the higher the temperature and the hotter it feels. To assign a numerical value to temperature, it is necessary to define a temperature scale. For calculations in thermochemistry, we use the Celsius and Kelvin scales. A Celsius temperature can be converted to Kelvin temperature by adding 273.15, although in most calculations in this book, it is acceptable to add 273.

The ability to measure temperature is thus based on heat transfer. The amount of heat transferred is usually measured in joules. *A* **joule** *is the SI unit of heat energy as well as all other forms of energy.* The joule, abbreviated J, is derived from the units for force and length.

$$N \times m = \frac{kg \times m^2}{s^2}$$

Because the joule is a rather small unit compared with other units for heat, the kilojoule, kJ, is also commonly used as a heat unit.

FIGURE 17-1 The direction of heat flow is determined by the temperature differences between the objects within a system. The heat is transferred from the hotter brass bar to the cooler water. This heat transfer will continue until the bar and the water reach the same temperature.

Heat *or* **heat energy** *can be thought of as the sum total of the kinetic energies of the particles in a sample of matter.* Heat always flows spontaneously from matter at a higher temperature to matter at a lower temperature, as shown in Figure 17-1. The temperature of the cool water in the beaker increases as heat flows into it. Likewise, the temperature of the hot brass bar decreases as heat flows away from it. When the temperature of the water equals the temperature of the brass bar, heat no longer flows within the system.

Heat Capacity and Specific Heat

The quantity of heat transferred during a temperature change depends on the nature of the material changing temperature, the mass of the material changing temperature, and the size of the temperature change. One gram of iron heated to 100.0°C and cooled to 50.0°C in a calorimeter transfers 22.5 J of heat to the surrounding water. But one gram of silver transfers 11.8 J of heat under the same conditions. The difference results from the metals' differing capacities for absorbing heat. A quantity called specific heat can be used to compare heat absorption capacities for different materials. **Specific heat** *is the amount of heat energy required to raise the temperature of one gram of substance by one Celsius degree (1°C) or one kelvin (1 K)*(because the sizes of the degree divisions on both scales are equal). Values of specific heat can be given in units of joules per gram per Celsius degree, J/(g•°C), joules per gram per kelvin, J/(g•K), or calories per gram per Celsius degree, cal/(g•°C). Table 17-1 gives the specific heats of some common substances. Notice the extremely high specific heat of water, one of the highest of most common substances.

Specific heat is usually measured under constant pressure conditions, so its symbol, c_p, contains a subscripted p as a reminder to the reader.

TABLE 17-1 Specific Heats of Some Common Substances at 298.15 K

Substance	Specific heat J/(g•K)
Water (*l*)	4.18
Water (*s*)	2.06
Water (*g*)	1.87
Ammonia (*g*)	2.09
Benzene (*l*)	1.74
Ethanol (*l*)	2.44
Ethanol (*g*)	1.42
Aluminum (*s*)	0.897
Calcium (*s*)	0.647
Carbon, graphite (*s*)	0.709
Copper (*s*)	0.385
Gold (*s*)	0.129
Iron (*s*)	0.449
Mercury (*l*)	0.140
Lead (*s*)	0.129

In the following mathematical equation, c_p is the specific heat at a given pressure, q is the heat lost or gained, m is the mass of the sample, and ΔT represents the difference between the initial and final temperatures.

$$c_p = \frac{q}{m \times \Delta T}$$

This equation can be rearranged to give an equation that can be used to find the quantity of heat gained or lost with a change in temperature.

$$q = c_p \times m \times \Delta T$$

SAMPLE PROBLEM 17-1

A 4.0 g sample of glass was heated from 274 K to 314 K, a temperature increase of 40 K, and was found to have absorbed 32 J of heat.

a. What is the specific heat of this type of glass?

b. How much heat will the same glass sample gain when it is heated from 314 K to 344 K?

SOLUTION

1 *ANALYZE* **Given:** $m = 4.0$ g

$\Delta T = 40$ K

$q = 32$ J

Unknown: c_p in J/(g•K)

2 PLAN

a. The specific heat, c_p, of the glass is calculated using the equation given for specific heat.

$$c_p = \frac{q}{m \times \Delta T}$$

b. The rearranged specific heat equation is used to find the heat gained when the glass was heated.

$$q = c_p \times m \times \Delta T$$

3 COMPUTE

a. $$\frac{32 \text{ J}}{(4.0 \text{ g})(40 \text{ K})} = 0.20 \text{ J/(g•K)}$$

b. $$\frac{0.20 \text{ J}}{(\text{g•K})}(4.0 \text{ g})(71 \text{ K} - 41 \text{ K})$$

$$\frac{0.20 \text{ J}}{(\text{g•K})}(4.0 \text{ g})(30 \text{ K}) = 24 \text{ J}$$

4 EVALUATE The units combine or cancel correctly to give the specific heat in J/(g•K) and the heat in J. Both answers are correct given to two significant digits.

PRACTICE

1. Determine the specific heat of a material if a 35 g sample absorbed 48 J as it was heated from 293 K to 313 K. *Answer* 0.069 J/(g•K)
2. If 980 kJ of energy are added to 6.2 L of water at 291 K, what will the final temperature of the water be? *Answer* 329 K

Heat of Reaction

The **heat of reaction** *is the quantity of heat released or absorbed during a chemical reaction.* You can think of heat of reaction as the difference between the stored energy of the reactants and the products.

If a mixture of hydrogen and oxygen is ignited, water will form and heat energy will release explosively. The heat that is released comes from the reactants as they form products. Because heat is released, the reaction is exothermic, and the heat content of the product, water, must be less than the heat content of the reactants before ignition. The following chemical equation for this reaction indicates that when 2 mol of hydrogen gas at room temperature are burned, 1 mol of oxygen gas is consumed and 2 mol of water vapor are formed.

$$2H_2(g) + O_2(g) \longrightarrow 2H_2O(g)$$

The equation does not tell you that heat is evolved during the reaction. Experiments have shown that 483.6 kJ of heat are evolved when 2 mol of gaseous water are formed at 298.15 K from its elements. Modifying the chemical equation to show the amount of heat produced during the reaction gives the following expression.

$$2H_2(g) + O_2(g) \longrightarrow 2H_2O(g) + 483.6 \text{ kJ}$$

This expression is an example of a **thermochemical equation,** *an equation that includes the quantity of heat released or absorbed during the reaction as written.* In any thermochemical equation, we must always interpret the coefficients as *numbers of moles* and never as *numbers of molecules.* The quantity of heat for this or any reaction depends on the amounts of reactants and products. The quantity of heat released during the formation of water from H_2 and O_2 is proportional to the quantity of water formed. Producing twice as much water vapor would require twice as many moles of reactants and would release 2×483.6 kJ of heat, as shown in the following thermochemical equation.

$$4H_2(g) + 2O_2(g) \longrightarrow 4H_2O(g) + 967.2 \text{ kJ}$$

Producing one-half as much water would require one-half as many moles of reactants and would release only one-half as much heat, or $1/2 \times 483.6$ kJ. The thermochemical equation for this reaction follows.

$$H_2(g) + \frac{1}{2}O_2(g) \longrightarrow H_2O(g) + 241.8 \text{ kJ}$$

Fractional coefficients are sometimes used in thermochemical equations.

The situation is reversed in an endothermic reaction because products have a higher heat content than reactants. The decomposition of water vapor is endothermic; it is the reverse of the reaction that forms water vapor. The amount of heat absorbed by water molecules to form hydrogen and oxygen equals the amount of heat released when the elements combine to form the water. This is to be expected because the difference between the heat content of reactants and products is unchanged. Heat now appears on the reactant side of the thermochemical equation that follows, indicating that it was absorbed during the reaction.

$$2H_2O(g) + 483.6 \text{ kJ} \longrightarrow 2H_2(g) + O_2(g)$$

The physical states of reactants and products must always be included in thermochemical equations because they influence the overall amount of heat exchanged. For example, the heat needed for the decomposition of water would be greater than 483.6 kJ if we started with ice because extra heat would be needed to melt the ice and to change the liquid into a vapor.

The heat absorbed or released during a chemical reaction at constant pressure is represented by ΔH. The H is the symbol for a quantity called **enthalpy,** *the heat content of a system at constant pressure.* Most chemical reactions are run in open vessels, so the pressure is constant and equal to atmospheric pressure. However, it is not practical to talk just about heat content or enthalpy because we have no way to directly measure the enthalpy of a system. Only *changes* in enthalpy can be measured. The

FIGURE 17-2 In an exothermic chemical reaction, heat is released from the system, meaning the enthalpy change is negative.

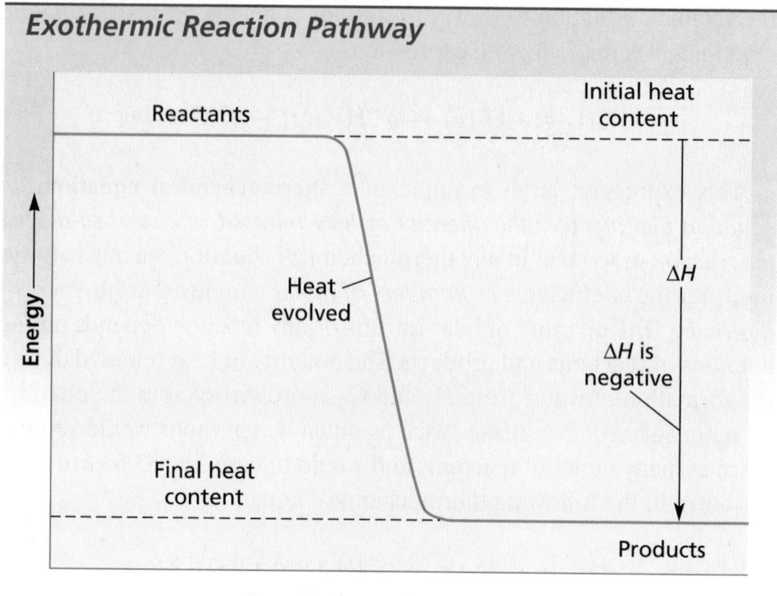

Greek letter Δ (a capital "delta") stands for "change in." Therefore, ΔH is read as "change in enthalpy." *An* **enthalpy change** *is the amount of heat absorbed or lost by a system during a process at constant pressure.* The enthalpy change is always the difference between the enthalpies of the products and the reactants. The following equation expresses an enthalpy change mathematically.

$$\Delta H = H_{products} - H_{reactants}$$

Thermochemical equations are usually written by designating the value of ΔH rather than writing the heat as a reactant or product. By convention, for an exothermic reaction, ΔH is always given a minus sign because the system loses energy. So the thermochemical equation for the exothermic formation of 2 mol of gaseous water from its elements now has the following form.

$$2H_2(g) + O_2(g) \longrightarrow 2H_2O(g) \quad \Delta H = -483.6 \text{ kJ/mol}$$

Figure 17-2 graphically shows the course of an exothermic reaction. The initial heat content of the reactants is greater than the final heat content of the products. This means heat is evolved, or given off, during the reaction; this is described as a negative enthalpy change.

For an endothermic reaction, ΔH is always given a positive value because the system gains energy. Thus, the endothermic decomposition of 2 mol of gaseous water has the following thermochemical equation.

$$2H_2O(g) \longrightarrow 2H_2(g) + O_2(g) \quad \Delta H = +483.6 \text{ kJ/mol}$$

The course of an endothermic reaction is illustrated in Figure 17-3.

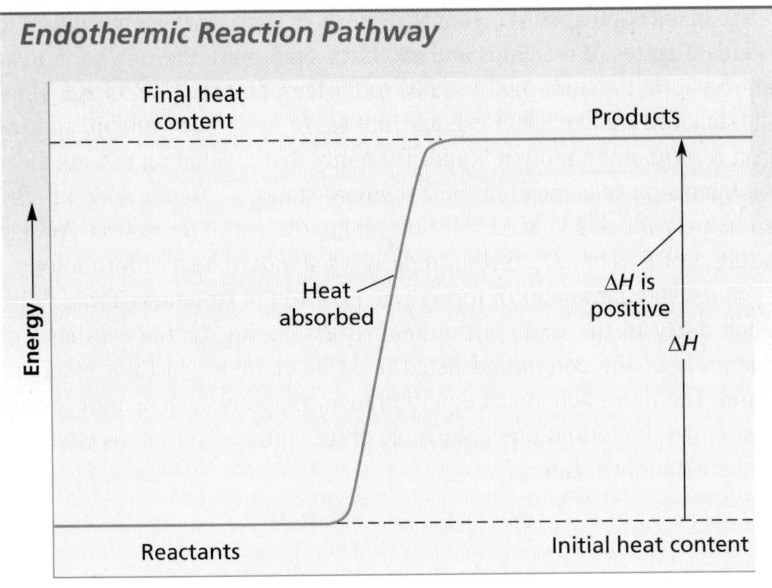

FIGURE 17-3 In an endothermic chemical reaction, the enthalpy change is positive because heat is absorbed into the system.

Heat is absorbed in this reaction because the initial heat content of the reactants is lower than the final heat content of the products. In this case, ΔH is designated as positive.

Keep in mind the following concepts when using thermochemical equations.

1. The coefficients in a balanced thermochemical equation represent the numbers of *moles* of reactants and products and never the numbers of *molecules.* This allows us to write these coefficients as fractions rather than whole numbers when necessary.
2. The physical state of the product or reactant involved in a reaction is an important factor and therefore must be included in the thermochemical equation.
3. The change in energy represented by a thermochemical equation is directly proportional to the number of moles of substances undergoing a change. For example, if 2 mol of water are decomposed, twice as much energy, 483.6 kJ, is needed than for the decomposition of 1 mol of water.
4. The value of the energy change, ΔH, is usually not significantly influenced by changing temperature.

Heat of Formation

The formation of water from hydrogen and oxygen is a composition reaction—the formation of a compound from its elements. Thermochemical data are often recorded as the heats of such composition reactions. *The* **molar heat of formation** *is the heat released or absorbed when one mole of a compound is formed by combination of its elements.*

To make comparisons meaningful, heats of formation are given for the standard states of reactants and products—these are the states found at atmospheric pressure and, usually, room temperature (298.15 K). Thus, the standard state of water is liquid, not gas or solid. The standard state of iron is solid, not a molten liquid. To signify that a value represents measurements on substances in their standard states, a ^{0} sign is added to the enthalpy symbol, giving ΔH^0 for the standard heat of a reaction. Adding a subscript *f*, as in ΔH_f^0, further indicates a standard heat of formation.

Some standard heats of formation are given in Appendix Table A-14. Each entry in the table is the heat of formation for the synthesis of *one mole* of the compound listed from its elements in their standard states. The thermochemical equation to accompany each heat of formation shows the formation of one mole of the compound from its elements in their standard states.

Stability and Heat of Formation

If a large amount of energy is released when a compound is formed, the compound has a high negative heat of formation. Such compounds are very stable. Once they start, the reactions forming them usually proceed vigorously and without outside assistance.

Elements in their standard states are *defined* as having $\Delta H_f^0 = 0$. The ΔH_f^0 of carbon dioxide is −393.5 kJ/mol of gas produced. Therefore, carbon dioxide is more stable than the elements from which it was formed. You can see in Appendix Table A-14 that the majority of the heats of formation are negative.

Compounds with relatively positive values of heats of formation, or only slightly negative values, are relatively unstable and will spontaneously decompose into their elements if the conditions are appropriate. Hydrogen iodide, HI, is a colorless gas that decomposes somewhat when stored at room temperature. It has a relatively high positive heat of formation of +26.5 kJ/mol. As it decomposes, violet iodine vapor becomes visible throughout the container of the gas.

Compounds with a high positive heat of formation are sometimes very unstable and may react or decompose violently. For example, ethyne (acetylene), C_2H_2, ($\Delta H_f^0 = +226.7$ kJ/mol) reacts violently with oxygen and must be stored in cylinders as a solution in acetone. Mercury fulminate, $HgC_2N_2O_2$, has a very high heat of formation of +270 kJ/mol. Its instability makes it useful as a detonator for explosives.

Heat of Combustion

Combustion reactions produce a considerable amount of energy in the form of light and heat when a substance is combined with oxygen. *The heat released by the complete combustion of one mole of a substance is*

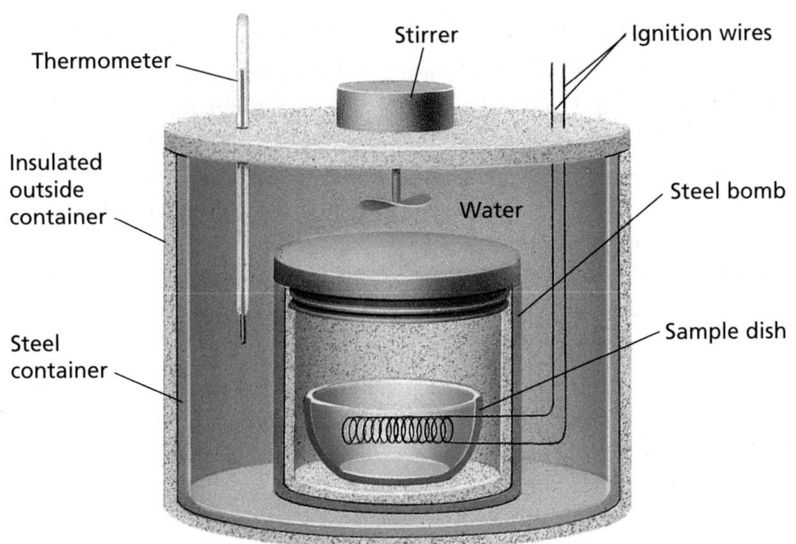

FIGURE 17-4 This is a diagram of a simple combustion calorimeter. A weighed sample is ignited by an electric spark and burned in the sample dish in an atmosphere of pure oxygen. The heat generated by the combustion reaction warms the steel bomb and the water surrounding it. The thermometer measures the initial and final temperatures of the water, and this temperature change is then used to calculate the heat evolved by the reaction.

called the **heat of combustion** *of the substance.* Heat of combustion is defined in terms of *one mole of reactant*, whereas the heat of formation is defined in terms of *one mole of product.* All substances are in their standard state. The general enthalpy notation, ΔH, applies to heats of reaction, but the addition of a subscripted *c*, ΔH_c, refers specifically to heat of combustion. A list of heats of combustion can be found in Appendix Table A-5.

Carbon dioxide and water are the products of the complete combustion of organic compounds containing only carbon and hydrogen or carbon, hydrogen, and oxygen. Knowing this, we can write the thermochemical equation for the combustion of any organic compound listed by balancing the equation for the reaction of one mole of the compound. For example, propane is a major component of the fuel used for outdoor gas grills. Its reaction with oxygen in the air, forming carbon dioxide and water as products and releasing energy in the form of heat and light, is a common example of combustion. The complete combustion of one mole of propane, C_3H_8, is described by the following thermochemical equation.

$$C_3H_8(g) + 5O_2(g) \longrightarrow 3CO_2(g) + 4H_2O(l) \quad \Delta H_c^0 = -2219.2 \text{ kJ/mol}$$

Figure 17-4 shows a combustion calorimeter, one of the instruments used to determine heats of combustion.

Calculating Heats of Reaction

Thermochemical equations can be rearranged and added to give enthalpy changes for reactions not included in the data tables. The basis for calculating heats of reaction is known as **Hess's law:** *The overall enthalpy change in a reaction is equal to the sum of enthalpy changes for the individual steps in the process.* The energy difference between reactants and products is independent of the route taken to get from one to the other. In fact, measured heats of reaction can be combined to calculate heats of reaction that are difficult or impossible to actually measure.

To demonstrate how to apply Hess's law, we will work through the calculation of the heat of formation for the formation of methane gas, CH_4, from its elements, hydrogen gas and solid carbon (graphite), at 298.15 K (25°C).

$$C(s) + 2H_2(g) \longrightarrow CH_4(g) \quad \Delta H_f^0 = ?$$

In order to calculate the change in enthalpy for this reaction, we can use the combustion reactions of the elements, carbon and hydrogen, and of methane.

$$C(s) + O_2(g) \longrightarrow CO_2(g) \qquad \Delta H_c^0 = -393.5 \text{ kJ/mol}$$

$$H_2(g) + \tfrac{1}{2}O_2(g) \longrightarrow H_2O(l) \qquad \Delta H_c^0 = -285.8 \text{ kJ/mol}$$

$$CH_4(g) + 2O_2(g) \longrightarrow CO_2(g) + 2H_2O(l) \quad \Delta H_c^0 = -890.8 \text{ kJ/mol}$$

The general principles for combining thermochemical equations follow.

1. If a reaction is reversed, the sign of ΔH is also reversed.
2. Multiply the coefficients of the known equations so that when added together they give the desired thermochemical equation.

In this case, reverse the combustion equation for methane, and remember to change the sign of ΔH from negative to positive. This will change the exothermic reaction to an endothermic one.

$$CO_2(g) + 2H_2O(l) \longrightarrow CH_4(g) + 2O_2(g) \quad \Delta H^0 = +890.8 \text{ kJ/mol}$$

Now we notice that 2 formula units of water are used as a reactant; therefore, 2 formula units of water will be needed as a product. In the combustion reaction for hydrogen as it is written, it only produces one formula unit of water. We must multiply the coefficients of this combustion reaction and the value of ΔH by 2 in order to achieve the desired quantity of water.

$$2H_2(g) + O_2(g) \longrightarrow 2H_2O(l) \quad \Delta H_c^0 = 2(-285.8 \text{ kJ/mol})$$

We are now ready to add the three equations together using Hess's law to give the heat of formation for methane and the balanced equation.

$$C(s) + \cancel{O_2(g)} \longrightarrow \cancel{CO_2(g)} \qquad \Delta H_c^0 = -393.5 \text{ kJ/mol}$$

$$2H_2(g) + \cancel{O_2(g)} \longrightarrow \cancel{2H_2O(l)} \qquad \Delta H_c^0 = 2(-285.8 \text{ kJ/mol})$$

$$\underline{\cancel{CO_2(g)} + \cancel{2H_2O(l)} \longrightarrow CH_4(g) + \cancel{2O_2(g)} \quad \Delta H^0 = +890.8 \text{ kJ/mol}}$$

$$C(s) + 2H_2(g) \longrightarrow CH_4(g) \qquad \Delta H_f^0 = -74.3 \text{ kJ/mol}$$

SAMPLE PROBLEM 17-2

Calculate the heat of reaction for the combustion of nitrogen monoxide gas, NO, to form nitrogen dioxide gas, NO_2, as given in the following thermochemical equation.

$$NO(g) + \frac{1}{2}O_2(g) \longrightarrow NO_2(g)$$

Use the heat-of-formation data in Appendix Table A-14. Solve by combining the known thermochemical equations. Verify the result by using the general equation for finding heats of reaction from heats of formation.

SOLUTION

1 ANALYZE

Given: $\frac{1}{2}N_2(g) + \frac{1}{2}O_2(g) \longrightarrow NO(g) \qquad \Delta H_f^0 = +90.29 \text{ kJ/mol}$

$\frac{1}{2}N_2(g) + O_2(g) \longrightarrow NO_2(g) \qquad \Delta H_f^0 = +33.2 \text{ kJ/mol}$

Unknown: ΔH^0 for $NO(g) + \frac{1}{2}O_2(g) \longrightarrow NO_2(g)$

2 PLAN

The ΔH requested can be found by adding the ΔHs of the component reactions as specified in Hess's law. The desired equation has $NO(g)$ and $\frac{1}{2}O_2(g)$ as reactants and $NO_2(g)$ as the product.

$$NO(g) + \frac{1}{2}O_2(g) \longrightarrow NO_2(g)$$

We need an equation with NO as a reactant. Reversing the first reaction for the formation of NO from its elements and the sign of ΔH yields the following thermochemical equation.

$$NO(g) \longrightarrow \frac{1}{2}N_2(g) + \frac{1}{2}O_2(g) \qquad \Delta H^0 = -90.29 \text{ kJ/mol}$$

The other equation should have NO_2 as a product, so we can retain the second equation for the formation of NO_2 from its elements as it stands.

$$\frac{1}{2}N_2(g) + O_2(g) \longrightarrow NO_2(g) \qquad \Delta H_f^0 = +33.2 \text{ kJ/mol}$$

3 COMPUTE

$$NO(g) \longrightarrow \cancel{\frac{1}{2}N_2(g)} + \cancel{\frac{1}{2}O_2(g)} \qquad \Delta H^0 = -90.29 \text{ kJ/mol}$$

$$\cancel{\frac{1}{2}N_2(g)} + O_2(g) \longrightarrow NO_2(g) \qquad \Delta H_f^0 = +33.2 \text{ kJ/mol}$$

$$NO(g) + \frac{1}{2}O_2(g) \longrightarrow NO_2(g) \qquad \Delta H^0 = -57.1 \text{ kJ/mol}$$

Note the cancellation of the $\frac{1}{2}N_2(g)$ and the partial cancellation of the $O_2(g)$.

4 EVALUATE

The unnecessary reactants and products cancel to give the desired equation. The general relationship between the heat of a reaction and the heats of formation of the reactants and products is described in the following word equation.

$$\Delta H^0 = \text{sum of } \Delta H_f^0 \text{ of products} - \text{sum of } \Delta H_f^0 \text{ of reactants}$$

To find the necessary sums, the ΔH_f^0 value for each reactant and each product must be multiplied by their respective coefficient in the desired equation. For the reaction of NO with O_2, applying this equation gives the following value for ΔH^0.

$$\Delta H^0 = \Delta H_f^0(NO_2) - [\Delta H_f^0(NO) + 0]$$
$$= +33.2 \text{ kJ/mol} - 90.29 \text{ kJ/mol} = -57.1 \text{ kJ/mol}$$

Note that zero is the assigned value for the heats of formation of elements in their standard states.

PRACTICE

1. Calculate the heat of reaction for the combustion of methane gas, CH_4, to form $CO_2(g) + H_2O(l)$. Write all equations involved.

   Answer
   –890.36 kJ/mol

2. Carbon occurs in two distinct forms. It can be the soft, black material found in pencils and lock lubricants, called graphite, or it can be the hard, brilliant gem we know as diamond. Calculate ΔH^0 for the conversion of graphite to diamond for the following reaction.

   Answer
   2 kJ/mol

$$C_{graphite}(s) \longrightarrow C_{diamond}(s)$$

The combustion reactions you will need follow.

$$C_{graphite}(s) + O_2(g) \longrightarrow CO_2(g) \qquad \Delta H_c^0 = -394 \text{ kJ/mol}$$
$$C_{diamond}(s) + O_2(g) \longrightarrow CO_2(g) \qquad \Delta H_c^0 = -396 \text{ kJ/mol}$$

Determining Heat of Formation

When carbon is burned in a limited supply of oxygen, carbon monoxide is produced. In this reaction, carbon is probably first oxidized to carbon dioxide. Then part of the carbon dioxide is reduced with carbon to give some carbon monoxide. Because these two reactions occur simultaneously and we get a mixture of CO and CO_2, it is not possible to directly measure the heat of formation of $CO(g)$ from $C(s)$ and $O_2(g)$.

$$C(s) + \tfrac{1}{2}O_2(g) \longrightarrow CO(g) \quad \Delta H_f^0 = ?$$

However, we do know the heat of formation of carbon dioxide and the heat of combustion of carbon monoxide.

$$C(s) + O_2(g) \longrightarrow CO_2(g) \qquad \Delta H_f^0 = -393.5 \text{ kJ/mol}$$
$$CO(g) + \tfrac{1}{2}O_2(g) \longrightarrow CO_2(g) \quad \Delta H_c^0 = -283.0 \text{ kJ/mol}$$

We reverse the second equation because we need CO as a product. Adding gives the desired heat of formation of carbon monoxide.

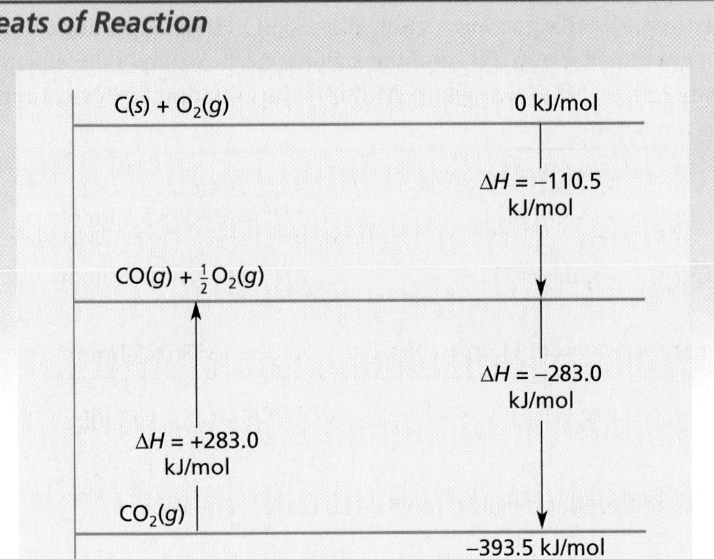

FIGURE 17-5 This diagram shows the heat of reaction for carbon dioxide, CO_2, and carbon monoxide, CO.

$$\begin{array}{ll} C(s) + O_2(g) \longrightarrow \cancel{CO_2(g)} & \Delta H_f^0 = -393.5 \text{ kJ/mol} \\ \cancel{CO_2(g)} \longrightarrow CO(g) + \cancel{\tfrac{1}{2}O_2(g)} & \Delta H_c^0 = +283.0 \text{ kJ/mol} \\ \hline C(s) + \tfrac{1}{2}O_2(g) \longrightarrow CO(g) & \Delta H^0 = -110.5 \text{ kJ/mol} \end{array}$$

Figure 17-5 is a model for the process described in this section. If we plot the reactions based on their relative heat content, you can see the relationship among the values obtained for the heat of formation of carbon monoxide. The formation of CO_2 is plotted at a level corresponding to –393.5 kJ/mol. The diagram shows the reverse of the combustion reaction (+283.0 kJ/mol) is added to that level. From the diagram, you see the difference, which represents the formation of CO. This value is –110.5 kJ/mol.

SAMPLE PROBLEM 17-3

Calculate the heat of formation of pentane, C_5H_{12}, using the information on heats of formation in Appendix Table A-14 and the information on heats of combustion in Appendix Table A-5. Solve by combining the known thermochemical equations.

SOLUTION

1 *ANALYZE*

Given: $C(s) + O_2(g) \longrightarrow CO_2(g)$ $\Delta H_f^0 = -393.5$ kJ/mol

$H_2(g) + \frac{1}{2}O_2(g) \longrightarrow H_2O(l)$ $\Delta H_f^0 = -285.8$ kJ/mol

$C_5H_{12}(g) + 8O_2(g) \longrightarrow 5CO_2(g) + 6H_2O(l)$ $\Delta H_c^0 = -3535.6$ kJ/mol

Unknown: ΔH_f^0 for $5C(s) + 6H_2(g) \longrightarrow C_5H_{12}(g)$

2 PLAN Combine the given equations according to Hess's law. We need C_5H_{12} as a product, so we reverse the equation for combustion of C_5H_{12} and the sign for ΔH_c^0. Multiply the equation for formation of CO_2 by 5 to give 5C as a reactant. Multiply the equation for formation of H_2O by 6 to give $6H_2$ as a reactant.

3 COMPUTE

$$5C(s) + \cancel{5O_2(g)} \longrightarrow \cancel{5CO_2(g)} \qquad \Delta H_f^0 = 5(-393.5 \text{ kJ/mol})$$

$$6H_2(g) + \cancel{3O_2(g)} \longrightarrow \cancel{6H_2O(l)} \qquad \Delta H_f^0 = 6(-285.8 \text{ kJ/mol})$$

$$\cancel{5CO_2(g)} + \cancel{6H_2O(l)} \longrightarrow C_5H_{12}(g) + \cancel{8O_2(g)} \qquad \Delta H^0 = +3536.6 \text{ kJ/mol}$$

$$5C(s) + 6H_2(g) \longrightarrow C_5H_{12}(g) \qquad \Delta H_f^0 = -145.7 \text{ kJ/mol}$$

4 EVALUATE The unnecessary reactants and products cancel to give the correct equation.

PRACTICE

1. Calculate the heat of formation of butane, C_4H_{10}, using the balanced chemical equation and information in Appendix Table A-5 and Table A-14. Write out the solution according to Hess's law. *Answer* −124.7 kJ/mol

2. Calculate the heat of combustion of 1 mol of nitrogen, N_2, to form NO_2 using the balanced chemical equation and Appendix Table A-14. (Hint: The heat of combustion of N_2 will be equal to the sum of the heats of formation of the combustion products of N_2 minus the heat of formation of N_2.) *Answer* + 66.36 kJ/mol

3. Calculate the heat of formation for sulfur dioxide, SO_2, from its elements, sulfur and oxygen. Use the balanced chemical equation and the following information. *Answer* −296.1 kJ/mol

$$S(s) + \tfrac{3}{2}O_2(g) \longrightarrow SO_3(g) \qquad \Delta H_c^0 = -395.2 \text{ kJ/mol}$$

$$2SO_2(g) + O_2(g) \longrightarrow 2SO_3(g) \qquad \Delta H_c^0 = -198.2 \text{ kJ/mol}$$

SECTION REVIEW

1. What is meant by enthalpy?
2. What is meant by heat of reaction? How is heat of reaction related to the enthalpy, or heat content, of reactants and products?
3. Describe the relationship between a compound's stability and its heat of formation.
4. What is the importance of Hess's law to thermodynamic calculations?
5. How much heat would be absorbed by 75 g of iron when heated from 295 K to 301 K?
6. When 1 mol of methane is burned at constant pressure, 890 kJ of heat energy is released. If a 3.2 g sample of methane is burned at constant pressure, what will be the value of ΔH? (Hint: Convert the grams of methane to moles. Also make sure your answer has the correct sign for an exothermic process.)

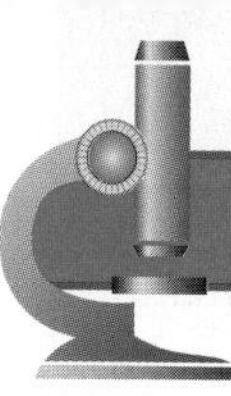

Self-Heating Meals

Who would have thought that corrosion could be useful? The HeaterMeals Company did. This company uses the properties of saltwater corrosion to heat TV-type dinners, and now it is taking packaged foods to a new level of convenience.

HeaterMeals' products, as their name implies, come with their own self-contained heat source. Each meal contains a package of food, a tray that holds a porous pouch containing magnesium and iron alloy, and a 2 oz pouch filled with salt water. When the salt water is poured into the tray with the porous pouch, it begins to vigorously corrode the metals. The sealed, precooked food package is then placed on top of the tray and returned to its box, where the temperature of the food package is raised by 100°F, heating the meal in 14 minutes.

This product uses supercorrosion to give you a hot meal.

Corrosion, the process by which a metal reacts with air or water, is usually an undesirable event, such as when iron corrodes to form rust. With HeaterMeals, however, the corrosion process is speeded up to produce an exothermic reaction—with the excess heat as the desired result.

According to Drew McLandrich, of The HeaterMeals Company, the idea for using self-heating metallic alloy powders has been around since the 1930s. "But," says McLandrich, "there really have been no significant uses of the product until the Desert Storm conflict, which led to the military's taking this heating technology and adopting it for field use so that soldiers could heat a meal-ready-to-eat."

"We've made about 80 million heaters for the military in the last 10 years. Lately we've been successfully marketing them to long-distance truck drivers. The product is in about 800 truck stops in 48 states. Additional users include hunters, campers, fishermen, boaters—anybody who would like a hot meal anytime, anyplace. That would also mean football fans at half time of a football game, picnickers, and people at a music concert. So it has a great variety of uses," McLandrich says.

The company has plans to develop other products using the controlled use of "supercorrosion." "A beverage could be heated," says McLandrich, "and we do have prototypes for a baby-bottle warmer. We're also working on making a portable hot cup of coffee or a hot cup of tea or cocoa."

So the next time you need a hot meal and there are no kitchens or restaurants around, consider the possibilities of supercorrosion and exothermic reactions.

SECTION 17-2

Driving Force of Reactions

Objectives

- Explain the relationship between enthalpy change and the tendency of a reaction to occur.
- Explain the relationship between entropy change and the tendency of a reaction to occur.
- Discuss the concept of free energy, and explain how the value of this quantity is calculated and interpreted.
- Describe the use of free energy change to determine the tendency of a reaction to occur.

The change in heat content of a reaction system is one of two factors that allow chemists to predict whether a reaction will occur spontaneously and to explain how it occurs. The randomness of the particles in a system is the second factor affecting whether a reaction will occur spontaneously.

Enthalpy and Reaction Tendency

The great majority of chemical reactions in nature are exothermic. As these reactions proceed, energy is liberated and the products have less energy than the original reactants. The products are also more resistant to change, more stable, than the original reactants. The tendency throughout nature is for a reaction to proceed in a direction that leads to a lower energy state.

We might think that endothermic reactions, in which energy is absorbed, cannot occur spontaneously because the products are at higher potential energy and are less stable than the original reactants. They would be expected to proceed only with the assistance of an outside influence, such as continued heating. However, some endothermic reactions *do* occur spontaneously. Something other than enthalpy must determine whether a reaction will occur.

Entropy and Reaction Tendency

A naturally occurring endothermic process is melting. An ice cube melts spontaneously at room temperature as heat is transferred from the warm air to the ice. The well-ordered arrangement of water molecules in the ice crystal is lost, and the less-orderly liquid phase of higher energy content is formed. A system that can go from one state to another without an enthalpy change does so by becoming more disordered.

Look at the physical states of the reactants in the chemical equation for the decomposition of ammonium nitrate.

$$2NH_4NO_3(s) \longrightarrow 2N_2(g) + 4H_2O(l) + O_2(g)$$

On the left side are 2 mol of solid ammonium nitrate. The right-hand side of the equation shows 3 mol of gaseous molecules plus 4 mol of a liquid. The arrangement of particles on the right-hand side of the equation is more random than the arrangement on the left side of the equation and hence is less ordered. Figure 17-6(a) and (b) show the reactant and products of this decomposition reaction.

These examples illustrate that there is a tendency in nature to proceed in a direction that increases the disorder of a system. A disordered system is one that lacks a regular arrangement of its parts. This factor is called entropy. **Entropy,** *S*, can be defined in a simple qualitative way as *a measure of the degree of randomness of the particles, such as molecules, in a system*. In order to understand the concept of entropy, consider solids, liquids, and gases. In a solid, the particles are fixed in position in their small regions of space, but they are vibrating back and forth. Even so, we can determine with fair precision the location of the particles. The degree of randomness is low, so the entropy is low. When the solid melts, the particles are still very close together but they can move about somewhat. The system is more random, and it is more difficult to describe the location of the particles. The entropy is higher. When the liquid evaporates, the particles are moving rapidly and are also much farther apart. Locating an individual particle is much more difficult and the system is much more random. The entropy of the gas is still higher than that of the liquid. A general but not absolute rule is that the entropy of liquids is higher than that of solids, and the entropy of gases is higher than that of liquids. But this rule must be used with caution. For example, the entropy of liquid mercury is much lower than that of some solids.

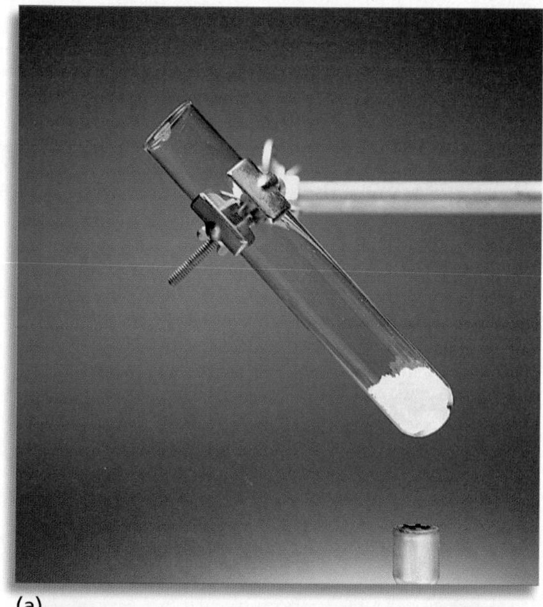

(a)

(b)

FIGURE 17-6 When ammonium nitrate, NH_4NO_3, decomposes, the entropy of the reaction system increases as (a) one solid reactant becomes (b) two gaseous products and one liquid product.

At absolute zero, random motion ceases, so the entropy of a pure crystalline solid is by definition zero at absolute zero. As heat is added, the randomness of the molecular motion increases. Measurements of heat absorbed and calculations are used to determine the absolute entropy or standard molar entropy, and values are then recorded in tables. These molar values are reported as kJ/(mol•K). Entropy change, which can also be measured, is defined as the difference between the entropy of the products and the reactants. Therefore, an increase in entropy is represented by a positive value for ΔS, and a decrease in entropy is represented by a negative value for ΔS.

The process of forming a solution almost always involves an increase in entropy because there is an increase in randomness. This is true for mixing gases, dissolving a liquid in another liquid, and dissolving a solid in a liquid.

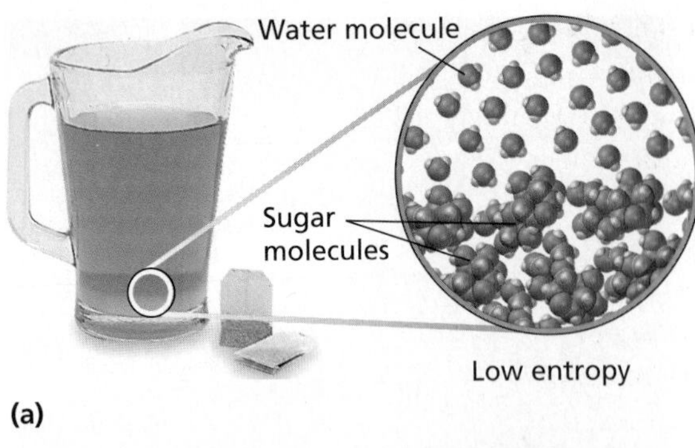

(a)

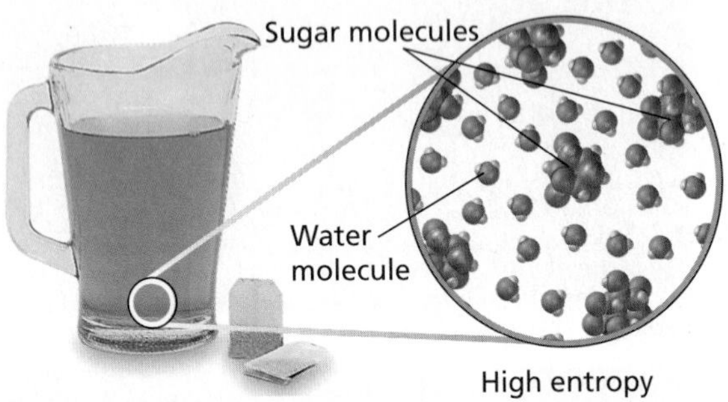

(b)

FIGURE 17-7 When a solid dissolves in a liquid, the entropy of the system increases.

Figure 17-7 illustrates the entropy change that takes place when solid sugar is dissolved in liquid tea. In the sugar-water system shown in Figure 17-7(a), the solid sugar has just been added to the tea, but most of it has not yet dissolved. The entropy is low because the majority of the sugar molecules are in one region at the bottom of the pitcher and the majority of the water molecules can be found everywhere else in the pitcher. After the sugar dissolves in the tea, shown in Figure 17-7(b), the sugar molecules are thoroughly mixed throughout the tea solution. Sugar molecules and water molecules might be found anywhere in the solution, so the entropy, the randomness, of the system increases. This would give ΔS a positive value for this solid-liquid system. You can imagine the same series of events happening for a system of gases mixing with each other or a system of liquids mixing. In each case, ΔS would have a positive value once the solution was formed.

Free Energy

Processes in nature are driven in two directions: toward lowest enthalpy and toward highest entropy. When these two oppose each other, the dominant factor determines the direction of change. To predict which factor will dominate for a given system, a function has been defined to relate the enthalpy and entropy factors at a given temperature. *This combined enthalpy-entropy function is called the* **free energy,** *G, of the system.* This function simultaneously assesses both the enthalpy-change and entropy-change tendencies. Natural processes proceed in the direction that lowers the free energy of a system.

Only the *change* in free energy can be measured. The change in free energy can be defined in terms of the changes in enthalpy and entropy. *At a constant pressure and temperature, the* **free-energy change,** *ΔG, of a system is defined as the difference between the change in enthalpy, ΔH, and the product of the Kelvin temperature and the entropy change, which is defined as $T\Delta S$.*

$$\Delta G^0 = \Delta H^0 - T\Delta S^0$$

Note that this expression is for substances in their standard states. The product $T\Delta S$ and the quantities ΔG and ΔH have the same units, usually kJ/mol. The units of ΔS for use in this equation are usually kJ/(mol•K).

Each of the variables in the free-energy equation can have positive or negative values. This leads to four possible combinations of terms.

Table 17-2 shows us that if ΔH is negative and ΔS is positive, then both terms on the right in the free energy equation are negative. Both factors contribute to the process being spontaneous. Therefore, ΔG will always be negative, and the reaction is definitely spontaneous. On the other hand, if ΔH is positive (endothermic process) and ΔS is negative (decrease in randomness), then the reaction as written will definitely not occur. When the enthalpy and entropy changes are operating in different directions, sometimes one will predominate and sometimes the other will predominate. There are cases in which the enthalpy change is negative and the entropy change is negative. The enthalpy factor leads to a spontaneous process, but the negative entropy change opposes this. This is true in the following reaction. (The entropy decreases because there is a decrease in moles of gas.)

$$C_2H_4(g) + H_2(g) \longrightarrow C_2H_6(g)$$

There is a fairly large decrease in entropy, $\Delta S^0 = -0.1207$ kJ/(mol•K). However, the reaction is strongly exothermic, with a $\Delta H^0 = -136.9$ kJ/mol. The reaction proceeds because the enthalpy term predominates.

$$\begin{aligned}\Delta G^0 = \Delta H^0 - T\Delta S^0 &= -136.9 \text{ kJ/mol} - 298 \text{ K}[-0.1207 \text{ kJ/(mol•K)}]\\ &= -101.1 \text{ kJ/mol}\end{aligned}$$

We can contrast this with the common commercial process for the manufacture of syngas, a mixture of CO and H_2. (This gas mixture is the starting point for the synthesis of a number of large-volume commercial chemicals, such as methanol, CH_3OH.)

$$CH_4(g) + H_2O(g) \longrightarrow CO(g) + 3H_2(g)$$

This reaction is endothermic, with $\Delta H^0 = +206.1$ kJ/mol and $\Delta S^0 = +0.215$ kJ/(mol•K), at standard conditions. The resulting ΔG is positive at room temperature. This tells us that the reaction will not occur at room temperature even though the entropy change is favorable.

$$\begin{aligned}\Delta G^0 = \Delta H^0 - T\Delta S^0 &= +206.1 \text{ kJ/mol} - 298 \text{ K}[+0.125 \text{ kJ/(mol•K)}]\\ &= +142.0 \text{ kJ/mol}\end{aligned}$$

TABLE 17-2 Relating Enthalpy, Entropy, and Free Energy Changes to Reaction Occurrence

ΔH	ΔS	ΔG
– value (exothermic)	+ value (disordering)	always negative
– value (exothermic)	– value (ordering)	negative at *lower* temperatures
+ value (endothermic)	+ value (disordering)	negative at *higher* temperatures
+ value (endothermic)	– value (ordering)	never negative

SAMPLE PROBLEM 17-4

For the reaction $NH_4Cl(s) \longrightarrow NH_3(g) + HCl(g)$, at 298.15 K, ΔH^0 = 176 kJ/mol and ΔS^0 = 0.285 kJ/(mol•K). Calculate ΔG^0, and tell whether this reaction can proceed in the forward direction at 298.15 K.

SOLUTION

1 *ANALYZE*

Given: $\Delta H^0 = 176$ kJ/mol at 298.15 K
$\Delta S^0 = 0.285$ kJ/(mol•K) at 298.15 K
Unknown: ΔG^0 at 298.15 K

2 *PLAN*

$$\Delta S, \Delta H, T \rightarrow \Delta G$$

The value of ΔG can be calculated according to the following equation.

$$\Delta G^0 = \Delta H^0 - T\Delta S^0$$

3 *COMPUTE*

$$\begin{aligned} \Delta G^0 &= 176 \text{ kJ/mol} - 298 \text{ K } [0.285 \text{ kJ/(mol•K)}] \\ &= 176 \text{ kJ/mol} - 84.9 \text{ kJ/mol} \\ &= 91 \text{ kJ/mol} \end{aligned}$$

4 *EVALUATE*

The answer is reasonably close to an estimated value of 110, calculated as $200 - (300 \times 0.3)$. The positive value of ΔG shows that this reaction is not naturally occurring at 298.15 K.

PRACTICE

1. For the vaporization reaction $Br_2(l) \longrightarrow Br_2(g)$, $\Delta H^0 = 31.0$ kJ/mol and $\Delta S^0 = 93.0$ J/(mol•K). At what temperature will this process be spontaneous?

   Answer above 333 K

SECTION REVIEW

1. What kind of enthalpy change favors a spontaneous reaction?
2. What is entropy, and how does it relate to spontaneity of reactions?
3. List several changes that result in an entropy increase.
4. Define *free energy*, and explain how its change is calculated.
5. Explain the relationship between free-energy change and spontaneity of reactions.
6. In the reaction in Sample Problem 17-4, why does the entropy increase?
7. How should increasing temperature affect the value of ΔG for the reaction in Sample Problem 17-4?
8. Predict the sign of ΔS^0 for each of the following reactions:
   a. the thermal decomposition of solid calcium carbonate

   $$CaCO_3(s) \longrightarrow CaO(s) + CO_2(g)$$

   b. the oxidation of SO_2 in air

   $$2SO_2(g) + O_2(g) \longrightarrow 2SO_3(g)$$

The Reaction Process

SECTION 17-3

OBJECTIVES

- Explain the concept of reaction mechanism.
- Use the collision theory to interpret chemical reactions.
- Define *activated complex.*
- Relate activation energy to heat of reaction.

The enthalpy change, entropy change, and free energy of a chemical reaction are independent of the actual route by which a reaction occurs. What happens between the initial and final states of a reaction system is described by the energy pathway that a reaction follows and the changes that take place on the molecular level when substances interact.

Reaction Mechanisms

If you mix aqueous solutions of HCl and NaOH, an extremely rapid neutralization reaction occurs.

$$H_3O^+(aq) + Cl^-(aq) + Na^+(aq) + OH^-(aq) \longrightarrow 2H_2O(l) + Na^+(aq) + Cl^-(aq)$$

The reaction is practically instantaneous; the rate is limited only by the speed with which the H_3O^+ and OH^- ions can diffuse through the water to meet each other. On the other hand, reactions between ions of the same charge and between molecular substances are not instantaneous. Negative ions repel each other, and the electron clouds of molecules repel each other strongly at very short distances. Therefore, only ions or molecules with very high kinetic energy can overcome repulsive forces and get close enough to react. In this section we will limit our discussion to reactions between molecules.

Colorless hydrogen gas consists of pairs of hydrogen atoms bonded together as diatomic molecules, H_2. Violet-colored iodine vapor is also diatomic, consisting of pairs of iodine atoms bonded together as I_2 molecules. A chemical reaction between these two gases at elevated temperatures produces hydrogen iodide, HI, a colorless gas. Hydrogen iodide molecules, in turn, tend to decompose and re-form hydrogen and iodine molecules, producing the violet gas shown in Figure 17-8. The following chemical equations describe these two reactions.

$$H_2(g) + I_2(g) \longrightarrow 2HI(g)$$
$$2HI(g) \longrightarrow H_2(g) + I_2(g)$$

Such equations indicate only which molecular species disappear as a result of the reactions and which species are produced. They do not show the **reaction mechanism,** *the step-by-step sequence of reactions by which the overall chemical change occurs.*

FIGURE 17-8 Clear hydrogen iodide gas, HI, decomposes into clear hydrogen gas and violet iodine gas.

Although only the net chemical change is directly observable for most chemical reactions, experiments can often be designed that suggest the probable sequence of steps in a reaction mechanism. Each reaction step is usually a simple process. The equation for each step represents the *actual* atoms, ions, or molecules that participate in that step. Even a reaction that appears from its balanced equation to be a simple process may actually be the result of several simple steps.

For many years, the formation of hydrogen iodide was considered a simple one-step process. It was thought to involve the interaction of two molecules, H_2 and I_2, in the forward reaction and two HI molecules in the reverse reaction. Experiments eventually showed, however, that a direct reaction between H_2 and I_2 does not take place.

Alternative mechanisms for the reaction were proposed based on the experimental results. The steps in each reaction mechanism had to add together to give the overall equation. Note that two of the species in the mechanism steps—I and H_2I—do not appear in the net equation. *Species that appear in some steps but not in the net equation are known as* **intermediates.** (Notice that they cancel each other out in the following mechanisms.) The first possible mechanism has the following two-step pathway.

Step 1: $I_2 \rightleftarrows 2\cancel{I}$

Step 2: $2\cancel{I} + H_2 \rightleftarrows 2HI$

$$I_2 + H_2 \rightleftarrows 2HI$$

The second possible mechanism has a three-step pathway.

Step 1: $I_2 \rightleftarrows 2\cancel{I}$

Step 2: $\cancel{I} + H_2 \rightleftarrows \cancel{H_2I}$

Step 3: $\cancel{H_2I} + \cancel{I} \rightleftarrows 2HI$

$$I_2 + H_2 \rightleftarrows 2HI$$

The reaction between hydrogen gas and iodine vapor to produce hydrogen iodide gas is an example of a **homogeneous reaction,** *a reaction whose reactants and products exist in a single phase*—in this case, the gas phase. This reaction system is also an example of a homogeneous chemical system because all reactants and products in all intermediate steps are in the same phase.

Collision Theory

In order for reactions to occur between substances, their particles (molecules, atoms, or ions) must collide. Furthermore, these collisions must result in interactions. *The set of assumptions regarding collisions and reactions is known as* **collision theory.** Chemists use this theory to interpret many of their observations about chemical reactions.

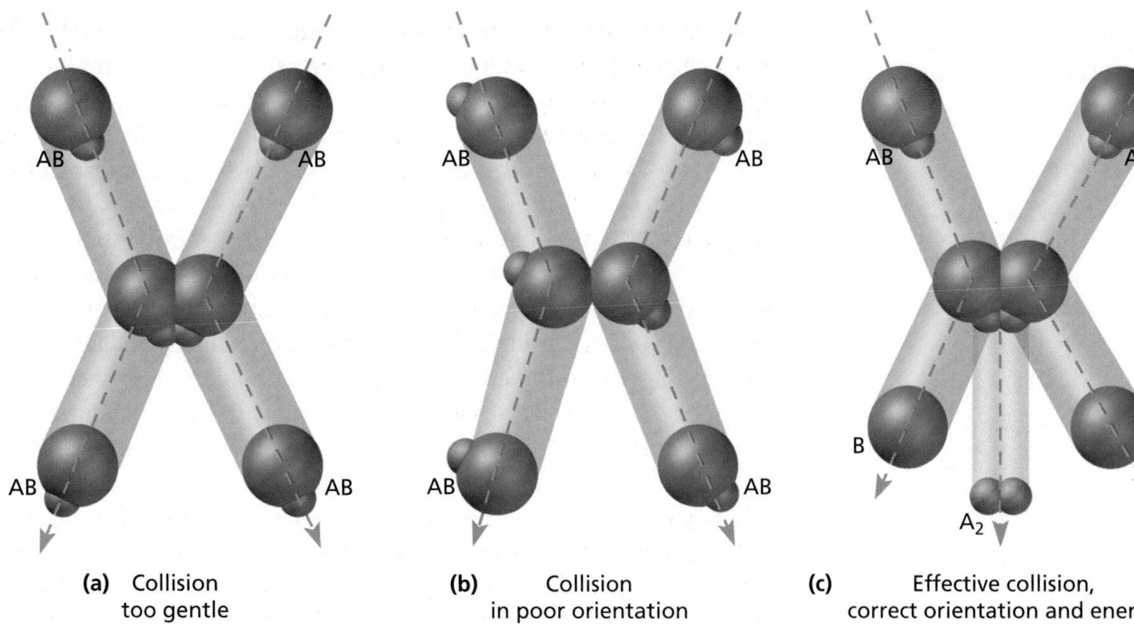

(a) Collision too gentle

(b) Collision in poor orientation

(c) Effective collision, correct orientation and energy

FIGURE 17-9 Three possible collision patterns for AB molecules are shown. Not every collision produces a chemical reaction.

Consider what might happen on a molecular scale in one step of a homogeneous reaction system. We will analyze a proposed first step in a hypothetical decomposition reaction.

$$AB + AB \rightleftharpoons A_2 + 2B$$

According to the collision theory, the two AB molecules must collide in order to react. Furthermore, they must collide while favorably oriented and with enough energy to merge the valence electrons and disrupt the bonds of the molecules. If they do so, a reshuffling of bonds leads to the formation of the products, one A_2 molecule and two B atoms. An effective collision is modeled in Figure 17-9(c).

If a collision is too gentle, the two molecules simply rebound from each other unchanged. This effect is illustrated in Figure 17-9(a). Similarly, a collision in which the reactant molecules are poorly oriented has little effect. The colliding molecules rebound without reacting. A poorly oriented collision is shown in Figure 17-9(b).

Thus, collision theory provides two reasons why a collision between reactant molecules may fail to produce a new chemical species: the collision is not energetic enough to supply the required energy, or the colliding molecules are not oriented in a way that enables them to react with each other.

Activation Energy

Consider the reaction for the formation of water from the diatomic gases oxygen and hydrogen according to the following equation.

$$2H_2(g) + O_2(g) \longrightarrow 2H_2O(l)$$

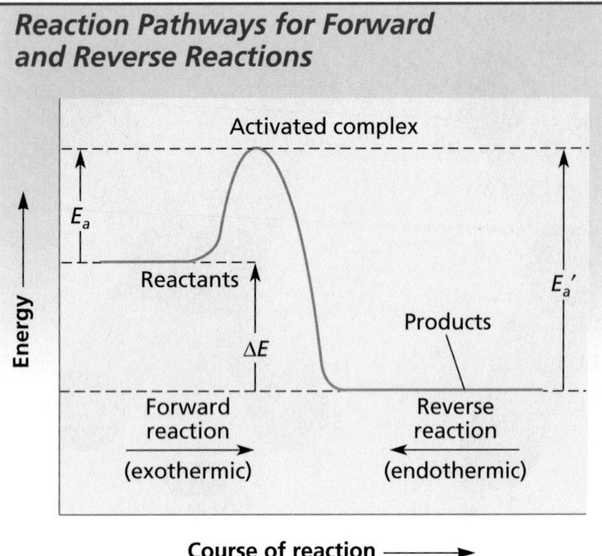

FIGURE 17-10 The difference between the activation energies for the reverse and forward reactions of a reversible reaction equals the energy change in the reaction, ΔE. The quantity for ΔE is the same for both directions, but is negative for the exothermic direction and positive for the endothermic direction.

The heat of formation is quite high: ΔH_f^0 = –285.8 kJ/mol at 298.15 K. The free-energy change is also large: ΔG^0 = –237.1 kJ/mol. Why, then, don't oxygen and hydrogen combine spontaneously and immediately to form water when they are mixed at room temperature?

Hydrogen and oxygen gases exist as diatomic molecules. When the molecules approach each other, the electron clouds repel each other and the molecules bounce off and never actually meet. In order for a reaction to occur, the colliding molecules must have enough kinetic energy to actually intermingle the valence electrons. In other words, the bonds of these molecular species must be broken in order for new bonds to be formed between oxygen and hydrogen atoms. Bond breaking is an endothermic process, and bond forming is exothermic. Even though the net process for forming water is exothermic, an initial input of energy is needed to overcome the repulsion forces that occur between reactant molecules when they are brought very close together. This initial energy input activates the reaction.

Once an exothermic reaction is started, the energy released is enough to sustain the reaction by activating other molecules. Thus, the reaction rate keeps increasing. It is limited only by the time required for reactant particles to acquire the energy and make contact. Energy from a flame or a spark, or the energy associated with high temperatures or radiations, may start exothermic reactants along the pathway of reaction. A generalized reaction pathway for an exothermic reaction is shown as the forward reaction in Figure 17-10. The minimum amount of energy needed to activate this reaction is the activation energy represented by E_a. **Activation energy** *is the minimum energy required to transform the reactants into an activated complex.*

The reverse reaction, decomposition of water molecules, is endothermic because the water molecules lie at an energy level that is lower than that of the hydrogen and oxygen molecules. The water molecules require a larger activation energy before they can decompose to re-form oxygen and hydrogen. The energy needed to activate an endothermic reaction is greater than that required for the original exothermic change and is represented by E_a' in Figure 17-10. The difference between E_a' and E_a is equal to the energy change in the reaction, ΔE. This energy change has the same numerical value for the forward reaction as it has for the reverse reaction but with the opposite sign.

The Activated Complex

When molecules collide, some of their high kinetic energy is converted into internal potential energy within the colliding molecules. If enough energy is converted, molecules with suitable orientation become activated. New bonds can then form. In this brief interval of bond breakage and bond formation, the collision complex is in a transition state. Some sort of partial bonding exists in this transitional structure. *A transitional structure that results from an effective collision and that persists while old bonds are breaking and new bonds are forming is called an* **activated complex.** The exact structure of this activated complex is not known.

Figure 17-11 graphically breaks down the reaction pathway of the formation of hydrogen iodide gas described on page 532 into three steps. Beginning with the reactants, H_2 and I_2, we see that a certain amount of activation energy, E_{a1}, is needed to form the activated complex that leads to the formation of the intermediates H_2 and 2I. During the course of the reaction, more activation energy, E_{a2}, is needed to form the activated complex leading to the intermediates H_2I and I. In order to arrive at the final product, 2HI, another increase in activation energy is necessary, as seen by the highest peak labeled E_{a3}.

An activated complex is formed when an effective collision raises the internal energies of the reactants to their minimum level for reaction, as shown in Figure 17-10. Both forward and reverse reactions go through the same activated complex. A bond that is broken in the activated complex for the forward reaction must be re-formed in the activated complex for the reverse reaction. Observe that an activated complex occurs at a high-energy position along the reaction pathway. In this sense, the activated complex defines the activation energy for the system.

From the kinetic-molecular theory presented in Chapter 10, you know that the speeds and therefore the kinetic energies of the molecules increase as the temperature increases. An increase in speed causes more collisions, which can cause an increase in reactions. However, the increase in reactions depends on more than simply the number of collisions, as Figure 17-9 illustrates. The collisions between molecules must possess sufficient energy to form an activated complex or a reaction will not take place. Raising the temperature of a reaction provides more molecules with this activation energy, causing an increase in reactions.

In its brief existence, the activated complex has partial bonding that is characteristic of both reactant and product. In this state, it may re-form the original bonds and separate back into the reactant particles, or it may form new bonds and separate into product particles. Usually, the formation of products is just as likely as the formation of reactants. Do not confuse the activated complex with the relatively stable intermediate products of different steps of a reaction mechanism. The activated complex, unlike intermediate products, is a very short-lived molecular complex in which bonds are in the process of being broken and formed.

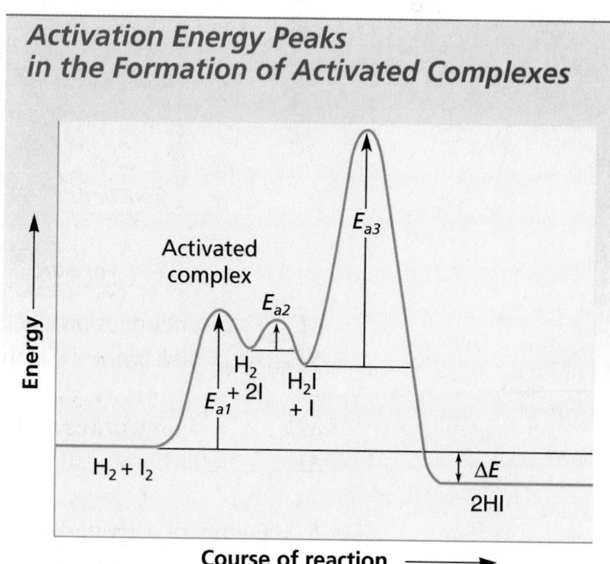

FIGURE 17-11 This energy profile graphically shows the formation of activated complexes during the gas-phase reaction $H_2 + I_2 \longrightarrow 2HI$.

SAMPLE PROBLEM 17-5

Copy the energy diagram below, and label the reactants, products, ΔE, E_a, *and* E_a'. Determine the value of $\Delta E_{forward}$, $\Delta E_{reverse}$, E_a, and E_a'.

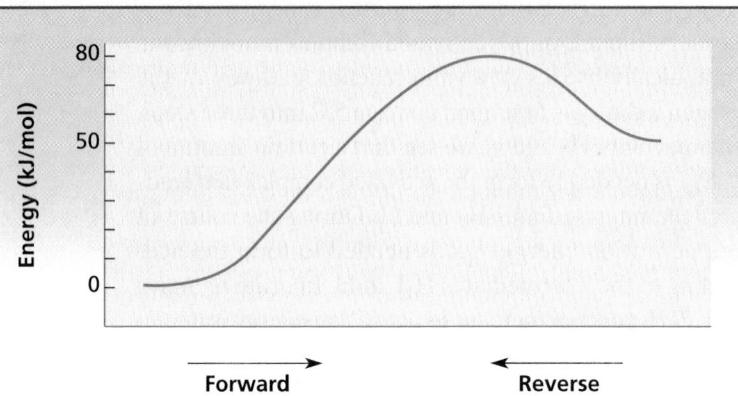

SOLUTION The energy level of reactants is always at the left-hand end of such a curve, and the energy level of products is always at the right-hand end. The energy change in the reaction, ΔE, is the difference between these two energy levels. The activation energy differs in the forward and reverse directions. As E_a, it is the difference between the reactant energy level and the peak in the curve. As E_a', it is the difference between the product energy level and the peak in the curve. It is the minimum energy needed to achieve effective reaction in either direction.

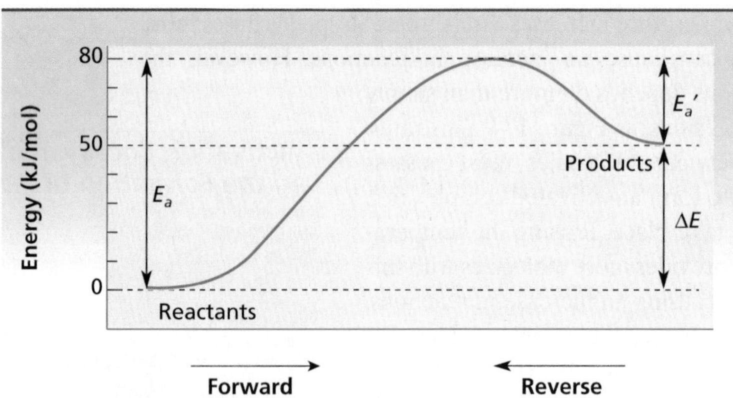

$\Delta E_{forward}$ = energy of products – energy of reactants
$\Delta E_{forward}$ = 50 kJ/mol – 0 kJ/mol = +50 kJ/mol

$\Delta E_{reverse}$ = energy of reactants – energy of products
$\Delta E_{reverse}$ = 0 kJ/mol – 50 kJ/mol = – 50 kJ/mol

E_a = energy of activated complex – energy of reactants
E_a = 80 kJ/mol – 0 kJ/mol = 80 kJ/mol

E_a' = energy of activated complex – energy of products
E_a' = 80 kJ/mol – 50 kJ/mol = 30 kJ/mol

PRACTICE

1. Use the method shown in the sample problem to redraw and label the following energy diagram. Determine the value of $\Delta E_{forward}$, $\Delta E_{reverse}$, E_a, and E_a'.

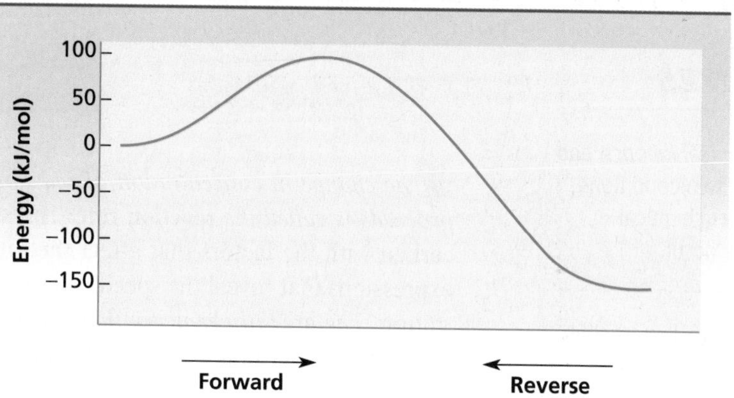

Answer

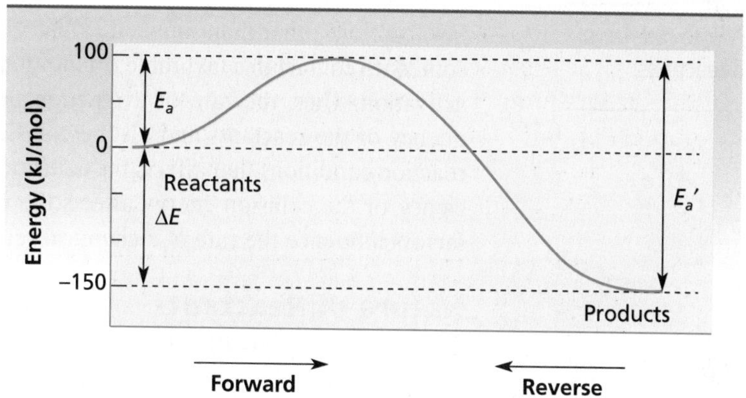

$\Delta E_{forward} = -150$ kJ/mol
$\Delta E_{reverse} = +150$ kJ/mol

$E_a = 100$ kJ/mol
$E_a' = 250$ kJ/mol

SECTION REVIEW

1. What is meant by reaction mechanism?
2. What factors determine whether a molecular collision produces a reaction?
3. What is activation energy?
4. What is an activated complex?
5. How is activation energy related to the energy of reaction?

SECTION 17-4

Reaction Rate

OBJECTIVES

- Define *chemical kinetics*, and explain the two conditions necessary for chemical reactions to occur.
- Discuss the five factors that influence reaction rate.
- Define *catalyst*, and discuss two different types.
- Explain and write rate laws for chemical reactions.

The change in concentration of reactants per unit time as a reaction proceeds is called the **reaction rate.** The study of reaction rates is concerned with the factors that affect the rate and with the mathematical expressions that reveal the specific dependencies of the rate on concentration. *The area of chemistry that is concerned with reaction rates and reaction mechanisms is called* **chemical kinetics.**

Rate-Influencing Factors

For reactions other than simple decompositions to occur, particles must come into contact in a favorable orientation and with enough energy for activation. Thus, the rate of a reaction depends on the collision frequency of the reactants and on the collision efficiency. Any change in reaction conditions that affects the collision frequency, the collision efficiency, or the collision energy affects the reaction rate. Five important factors influence the rate of a chemical reaction.

Nature of Reactants

Substances vary greatly in their tendencies to react. For example, hydrogen combines vigorously with chlorine under certain conditions. Under the same conditions, it may react only feebly with nitrogen. Sodium and oxygen combine much more rapidly than iron and oxygen under similar conditions. Bonds are broken and other bonds are formed in reactions. The rate of reaction depends on the particular reactants and bonds involved.

Surface Area

Gaseous mixtures and dissolved particles can mix and collide freely; therefore, reactions involving them can occur rapidly. In heterogeneous reactions, the reaction rate depends on the area of contact of the reaction substances. **Heterogeneous reactions** *involve reactants in two different phases.* These reactions can occur only when the two phases are in contact. Thus, the surface area of a solid reactant is an important factor in determining rate. An increase in surface area increases the rate of heterogeneous reactions.

Solid zinc reacts with aqueous hydrochloric acid to produce zinc chloride and hydrogen gas according to the following equation.

$$Zn(s) + 2HCl(aq) \longrightarrow ZnCl_2(aq) + H_2(g)$$

This reaction occurs at the surface of the zinc solid. A cube of zinc measuring 1 cm on each edge presents only 6 cm^2 of contact area. The same amount of zinc in the form of a fine powder might provide a contact area on the order of 10^4 times the original area. Consequently, the reaction rate of the powdered solid is much higher.

A lump of coal burns slowly when kindled in air. The rate of burning can be increased by breaking the lump into smaller pieces, exposing more surface area. If the piece of coal is powdered and then ignited while suspended in air, it burns explosively. This is the cause of some explosions in coal mines.

Temperature

An increase in temperature increases the average kinetic energy of the particles in a substance; this can result in a greater number of effective collisions when the substance is allowed to react with another substance. If the number of effective collisions increases, the reaction rate will increase.

To be effective, the energy of the collisions must be equal to or greater than the activation energy. At higher temperatures, more particles possess enough energy to form the activated complex when collisions occur. Thus, a rise in temperature produces an increase in collision energy as well as in collision frequency.

Decreasing the temperature of a reaction system has the opposite effect. The average kinetic energy of the particles decreases, so they collide less frequently and with less energy, producing fewer effective collisions. Beginning near room temperature, the reaction rates of many common reactions roughly double with each 10 K (10°C) rise in temperature. This rule of thumb should be used with caution, however. The actual rate increase with a given rise in temperature must be determined experimentally.

Concentration

Pure oxygen has five times the concentration of oxygen molecules that air has at the same pressure; consequently, a substance that oxidizes in air oxidizes more vigorously in pure oxygen. For example, in Figure 17-12, the light produced when the lump of charcoal is combusted in a bottle of pure oxygen is much more intense than the light produced when the charcoal lump is heated in air until combustion begins. The oxidation of charcoal is a heterogeneous reaction system in which one reactant is a gas. The reaction rate depends not only on the amount of exposed charcoal surface but also on the concentration of the reacting species, O_2.

In homogeneous reaction systems, reaction rates depend on the concentration of the reactants. Predicting the mathematical relationship between rate and concentration is difficult because most chemical reactions occur in a series of steps, and only one of these steps determines the reaction rate. If the number of effective collisions increases, the rate

(a)

(b)

FIGURE 17-12 Carbon burns faster in pure oxygen (a) than in air (b) because the concentration of the reacting species, O_2, is greater.

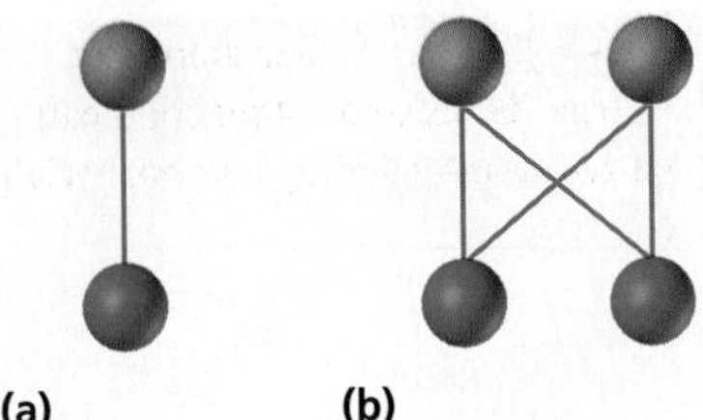
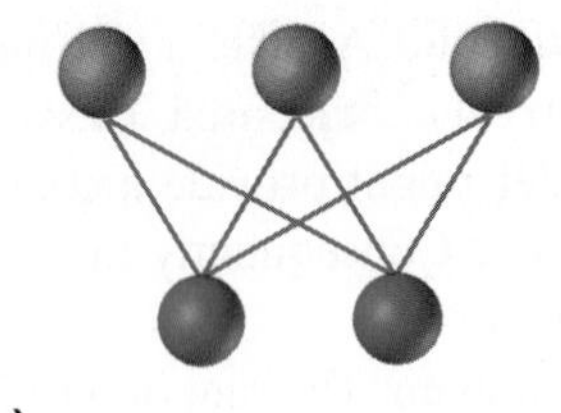
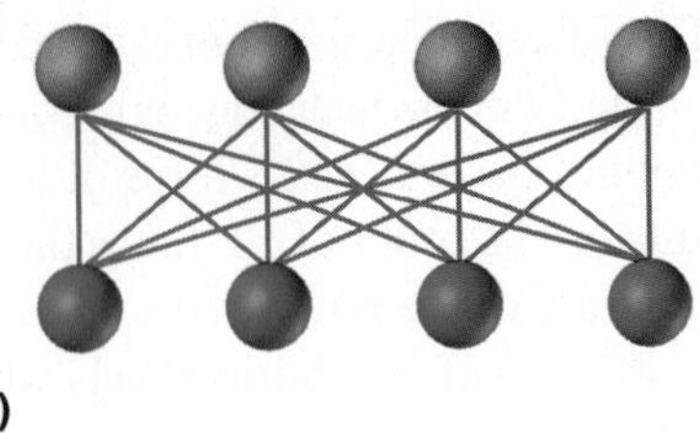

FIGURE 17-13 The concentration of reacting species affects the number of collisions and therefore the reaction rate.

increases as well. In general, an increase in rate is expected if the concentration of one or more of the reactants is increased, as depicted by the model in Figure 17-13. In the system with only two molecules, shown in Figure 17-13(a), only one collision can possibly occur. When there are four molecules in the system, as in Figure 17-13(b), there can be four possible collisions. Under constant conditions, as the number of molecules in the system increases, so does the total number of possible collisions between them. Figure 17-13(c) and (d) show a five- and eight-molecule system, allowing six and sixteen possible collisions, respectively. Lowering the concentration should have the opposite effect. The actual effect of concentration changes on reaction rate, however, must be determined experimentally.

Presence of Catalysts

Some chemical reactions proceed quite slowly. Sometimes their reaction rates can be increased dramatically by the presence of a catalyst. *A* **catalyst** *is a substance that changes the rate of a chemical reaction without itself being permanently consumed. The action of a catalyst is called* **catalysis.** The catalysis of the decomposition reaction of hydrogen peroxide by manganese dioxide is shown in Figure 17-14. A catalyst provides an alternative energy pathway or reaction mechanism in which the potential-energy barrier between reactants and products is lowered. The catalyst may be effective in forming an alternative activated complex that requires a lower activation energy—as suggested in the energy profiles of the decomposition of hydrogen peroxide, H_2O_2, shown in Figure 17-15—according to the following equation.

$$2H_2O_2(l) \longrightarrow O_2(g) + 2H_2O(l)$$

FIGURE 17-14 The reaction rate of the decomposition of hydrogen peroxide, H_2O_2, can be increased by using a catalyst. The catalyst used here is manganese dioxide, MnO_2, a black solid. A 30% H_2O_2 solution is added dropwise onto the MnO_2 in the beaker and rapidly decomposes to O_2 and H_2O. Both the oxygen and water appear as gases because the high heat of reaction causes the water to vaporize.

Catalysts do not appear among the final products of reactions they accelerate. They may participate in one step along a reaction pathway and be regenerated in a later step. In large-scale and cost-sensitive reaction systems, catalysts are recovered and reused. *A catalyst that is in the same phase as all the reactants and products in a reaction system is called a* **homogeneous catalyst.** *When its phase is different from that of the reactants, it is called a* **heterogeneous catalyst.** Metals are often used as heterogeneous catalysts. The catalysis of many reactions is promoted by adsorption of reactants on the metal surfaces, which has the effect of increasing the concentration of the reactants.

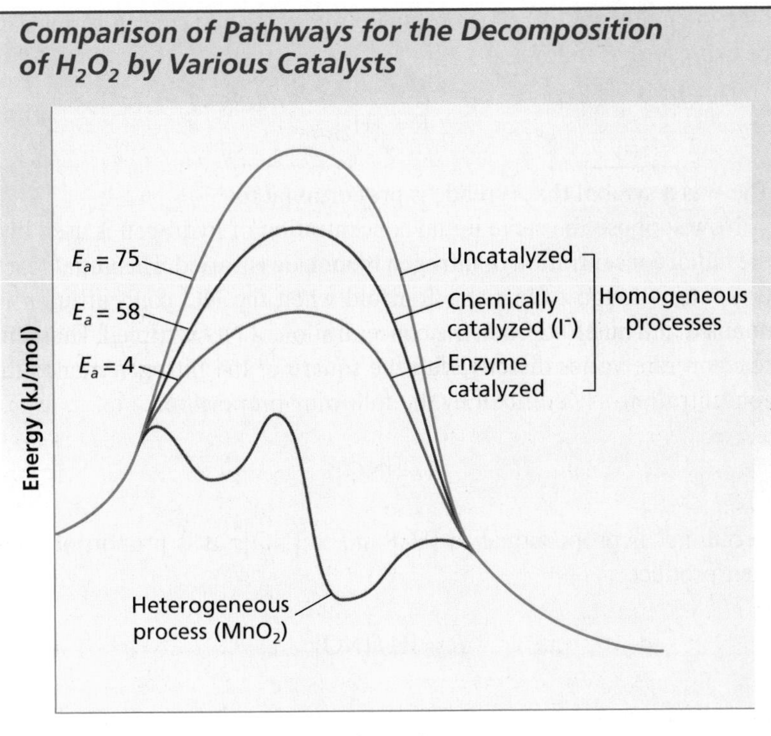

FIGURE 17-15 The activation energy for a chemical reaction can be reduced by adding an appropriate catalyst.

Rate Laws for Reactions

The relationship between the rate of a reaction and the concentration of one reactant is determined experimentally by first keeping the concentrations of other reactants and the temperature of the system constant. Then the reaction rate is measured for various concentrations of the reactant in question. A series of such experiments reveals how the concentration of each reactant affects the reaction rate.

Hydrogen gas reacts with nitrogen monoxide gas at constant volume and at an elevated constant temperature, according to the following equation.

$$2H_2(g) + 2NO(g) \longrightarrow N_2(g) + 2H_2O(g)$$

Four moles of reactant gases produce three moles of product gases; thus, the pressure of the system diminishes as the reaction proceeds. The rate of the reaction can, therefore, be determined by measuring the change of pressure in the vessel with time.

Suppose a series of experiments is conducted using the same initial concentration of nitrogen monoxide but different initial concentrations of hydrogen. The initial reaction rate is found to vary directly with the hydrogen concentration: doubling the concentration of H_2 doubles the rate, and tripling the concentration of H_2 triples the rate. If R represents the reaction rate and $[H_2]$ is the concentration of hydrogen in moles per liter, the mathematical relationship between rate and concentration can be expressed as follows.

$$R \propto [H_2]$$

The $\propto$ is a symbol that is read "is proportional to."

Now suppose the same initial concentration of hydrogen is used but the initial concentration of nitrogen monoxide is varied. The initial reaction rate is found to increase fourfold when the NO concentration is doubled and ninefold when the concentration of NO is tripled. Thus, the reaction rate varies directly with the square of the nitrogen monoxide concentration, as described by the following proportion.

$$R \propto [NO]^2$$

Because R is proportional to $[H_2]$ and to $[NO]^2$, it is proportional to their product.

$$R \propto [H_2][NO]^2$$

By introduction of an appropriate proportionality constant, k, the expression becomes an equality.

$$R = k[H_2][NO]^2$$

An equation that relates reaction rate and concentrations of reactants is called the **rate law** *for the reaction.* It is applicable for a specific reaction at a given temperature. A rise in temperature increases the reaction rates of most reactions. The value of k usually increases as the temperature increases, but the relationship between reaction rate and concentration almost always remains unchanged.

Rate Laws and Reaction Pathway

The form of the rate law depends on the reaction mechanism. For a reaction that occurs in a *single step,* the reaction rate of that step is proportional to the product of the reactant concentrations, each of which is raised to its stoichiometric coefficient. For example, suppose one molecule of gas A collides with one molecule of gas B to form two molecules of substance C, according to the following equation.

$$A + B \longrightarrow 2C$$

One particle of each reactant is involved in each collision. Thus, doubling the concentration of either reactant will double the collision frequency. It will also double the reaction rate *for this step.* Therefore, the rate for this step is directly proportional to the concentration of A and B. The rate law for this one-step reaction follows.

$$R = k[A][B]$$

Now suppose the reaction is reversible. In the reverse step, two molecules of C must decompose to form one molecule of A and one of B.

$$2C \longrightarrow A + B$$

Thus, the reaction rate for this reverse step is directly proportional to $[C] \times [C]$. The rate law for the step is as follows.

$$R = k[C]^2$$

The power to which the molar concentration of each reactant is raised in the rate laws above corresponds to the coefficient for the reactant in the balanced chemical equation. Such a relationship holds *only* if the reaction follows a simple one-step path, that is, if the reaction occurs at the molecular level exactly as written in the chemical equation.

If a chemical reaction proceeds in a sequence of steps, the rate law is determined from the slowest step because it has the lowest rate. *This slowest-rate step is called the* **rate-determining step** *for the chemical reaction.*

Consider the reaction of nitrogen dioxide and carbon monoxide.

$$NO_2(g) + CO(g) \longrightarrow NO(g) + CO_2(g)$$

The reaction is believed to be a two-step process represented by the following mechanism.

Step 1: $NO_2 + NO_2 \longrightarrow NO_3 + NO$ slow
Step 2: $NO_3 + CO \longrightarrow NO_2 + CO_2$ fast

In the first step, two molecules of NO_2 collide, forming the intermediate species NO_3. This structure then collides with one molecule of CO and reacts quickly to produce one molecule each of NO_2 and CO_2. The first step is the slower of the two and is therefore the rate-determining step. We can write the rate law from this rate-determining step.

$$R = k[NO_2]^2$$

This tells us that two molecules of NO_2 are the reactants in the slower, rate-determining step. CO does not appear in the rate law because it reacts after the rate-determining step, so the reaction rate will not depend on [CO].

The general form for the rate law is given by the following equation.

$$R = k[A]^n[B]^m \ldots$$

The reaction rate is represented by R, k is the rate constant, and [A] and [B]... represent the molar concentrations of reactants. The n and m are the respective powers to which the concentrations are raised. They must be determined from *experimental data.*

SAMPLE PROBLEM 17-6

Nitrogen dioxide and fluorine react in the gas phase according to the following equation.

$$2NO_2(g) + F_2(g) \longrightarrow 2NO_2F(g)$$

A proposed mechanism for this reaction follows.

Step 1:	$NO_2 + F_2 \longrightarrow NO_2F + F$	**slow**
Step 2:	$F + NO_2 \longrightarrow NO_2F$	**fast**

Identify the rate-determining step and write an acceptable rate law.

SOLUTION If we combine these two steps, the intermediate, F, cancels out and we are left with the original equation. The first step is the slower step, and is considered the rate-determining step. We can write the rate law from this rate-determining step.

$$R = k\,[NO_2][F_2]$$

SAMPLE PROBLEM 17-7

A reaction involving reactants X and Y was found to occur by a one-step mechanism: $X + 2Y \longrightarrow XY_2$. Write the rate law for this reaction, and then determine the effect of each of the following on the reaction rate:

a. doubling the concentration of X
b. doubling the concentration of Y
c. using one-third the concentration of Y

SOLUTION Because the equation represents a single-step mechanism, the rate law can be written from the equation (otherwise, it could not be). The rate will vary directly with the concentration of X, which has an implied coefficient of 1 in the equation, and will vary directly with the square of the concentration of Y, which has the coefficient of 2: $R = k[X][Y]^2$.

a. Doubling the concentration of X will double the rate ($R = k[2X][Y]^2$).
b. Doubling the concentration of Y will increase the rate fourfold ($R = k[X][2Y]^2$).
c. Using one-third the concentration of Y will reduce the rate to one-ninth of its original value ($R = k[X][\frac{1}{3}Y]^2$).

PRACTICE

1. The rate of a reaction involving L, M, and N is found to double if the concentration of L is doubled, to increase eightfold if the concentration of M is doubled, and to double if the concentration of N is doubled. Write the rate law for this reaction.

   Answer $R = k[L][M]^3[N]$

2. At temperatures below 498 K, the following reaction takes place.

   $$NO_2(g) + CO(g) \longrightarrow CO_2(g) + NO(g)$$

   Doubling the concentration of NO_2 quadruples the rate of CO_2 being formed if the CO concentration is held constant. However, doubling the concentration of CO has no effect on the rate of CO_2 formation. Write a rate-law expression for this reaction.

   Answer $R = k[NO_2]^2$

DESKTOP INVESTIGATION

Factors Influencing Reaction Rate

MATERIALS

- Bunsen burner
- paper ash
- copper foil strip
- graduated cylinder, 10 mL
- magnesium ribbon
- matches
- paper clip
- sandpaper
- steel wool
- 2 sugar cubes
- white vinegar
- zinc strip
- 6 test tubes, 16 × 150 mm
- tongs

Wear safety goggles and an apron.

QUESTION

How do the type of reactants, surface area of reactants, concentration of reactants, and catalysts affect the rates of chemical reactions?

PROCEDURE

Remove all combustible material from the work area. Wear safety goggles and an apron. Record all your results in a data table.

1. Add 10 mL of vinegar to each of three test tubes. To one test tube, add a 3 cm piece of magnesium ribbon; to a second, add a 3 cm zinc strip; and to a third, add a 3 cm copper strip. (All metals should be the same width.) If necessary, sandpaper the metals until they are shiny.
2. Using tongs, hold a paper clip in the hottest part of the burner flame for 30 s. Repeat with a ball of steel wool 2 cm in diameter.
3. To one test tube, add 10 mL of vinegar; to a second, add 5 mL of vinegar plus 5 mL of water; and to a third, add 2.5 mL of vinegar plus 7.5 mL of water. To each of the three test tubes, add a 3 cm piece of magnesium ribbon.
4. Using tongs, hold a sugar cube and try to ignite it with a match. Then try to ignite it in a burner flame. Rub paper ash on a second cube, and try to ignite it with a match.

DISCUSSION

1. What are the rate-influencing factors in each step of the procedure?
2. What were the results from each step of the procedure? How do you interpret each result?

SECTION REVIEW

1. What is studied in the branch of chemistry that is known as chemical kinetics?
2. List the five important factors that influence the rate of chemical reactions.
3. What is a catalyst? Explain the effect of a catalyst on the rate of chemical reactions. How does a catalyst influence the activation energy required by a particular reaction?
4. What is meant by a rate law for a chemical reaction? Explain the conditions under which a rate law can be written from a chemical equation. When can a rate law not be written from a single step?

CHAPTER 18

Chemical Equilibrium

The creation of stalactites and stalagmites is the result of a reversible chemical reaction.

The Nature of Chemical Equilibrium

SECTION 18-1

Objectives

- Define *chemical equilibrium.*
- Explain the nature of the equilibrium constant.
- Write chemical equilibrium expressions and carry out calculations involving them.

In systems that are in equilibrium, opposing processes occur at the same time and at the same rate. For example, when an excess of sugar is placed in water, sugar molecules go into solution. At equilibrium, molecules of sugar are crystallizing at the same rate that molecules from the crystal are dissolving. The rate of evaporation of a liquid in a closed vessel can eventually be equaled by the rate of condensation of its vapor. The resulting equilibrium vapor pressure is a characteristic of the liquid at the prevailing temperature. Le Châtelier's principle, which you read about in Chapter 12, can help predict the outcome of changes made to these equilibrium systems. The preceding examples are physical equilibria. The concept of equilibrium and Le Châtelier's principle also apply to chemical processes.

Reversible Reactions

Theoretically, every reaction can proceed in two directions, forward and reverse. Thus, essentially all chemical reactions are considered to be reversible under suitable conditions. *A chemical reaction in which the products can react to re-form the reactants is called a* **reversible reaction.**

Mercury(II) oxide decomposes when heated.

$$2HgO(s) \xrightarrow{\Delta} 2Hg(l) + O_2(g)$$

Mercury and oxygen combine to form mercury(II) oxide when heated gently.

$$2Hg(l) + O_2(g) \xrightarrow{\Delta} 2HgO(s)$$

FIGURE 18-1 When heated, mercury(II) oxide decomposes into the elements from which it was formed. Liquid mercury reacts with oxygen to re-form mercury(II) oxide. Together these reactions represent a reversible chemical process.

Figure 18-1 shows both of these reactions taking place. Suppose mercury(II) oxide is heated in a closed container from which neither the mercury nor the oxygen can escape. Once decomposition has begun, the mercury and oxygen released can recombine to form mercury(II) oxide again. Thus, both reactions can proceed at the same time. Under these conditions, the rate of the composition reaction will eventually equal that of the decomposition reaction. At equilibrium, mercury and oxygen will combine to form mercury(II) oxide at the same rate that mercury(II) oxide decomposes into mercury and oxygen. The amounts of mercury(II) oxide, mercury, and oxygen can then be expected to remain constant as long as these conditions persist. At this point, a state of dynamic equilibrium has been reached between the two chemical reactions. Both reactions continue, but there is no net change in the composition of the system. *A reversible chemical reaction is in* **chemical equilibrium** *when the rate of its forward reaction equals the rate of its reverse reaction and the concentrations of its products and reactants remain unchanged.* The chemical equation for the reaction at equilibrium is written using double arrows to indicate the overall reversibility of the reaction.

$$2HgO(s) \rightleftharpoons 2Hg(l) + O_2(g)$$

Equilibrium, a Dynamic State

Many chemical reactions are reversible under ordinary conditions of temperature and concentration. They will reach a state of equilibrium unless at least one of the substances involved escapes or is removed from the reaction system. In some cases, however, the forward reaction is nearly completed before the rate of the reverse reaction becomes high enough to establish equilibrium. Here, the products of the forward reaction are favored, meaning that at equilibrium there is a higher concentration of products than of reactants. The favored reaction that produces this situation is referred to as a reaction to the right because the convention for writing chemical reactions is that *left-to-right* is forward and *right-to-left* is reverse. An example of such a system is the dissociation of hydrobromic acid in aqueous solution.

$$HBr(aq) + H_2O(l) \rightleftharpoons H_3O^+(aq) + Br^-(aq)$$

Notice that the equation is written showing an inequality of the two arrow lengths. The forward reaction is represented by the longer arrow to imply that products are favored in this reaction.

In other cases, the forward reaction is barely under way when the rate of the reverse reaction becomes equal to that of the forward reaction, and equilibrium is established. In these cases, the products of the reverse reaction are favored and the original reactants are formed. That is, at equilibrium there is a higher concentration of reactants than of products. The favored reaction that produces this situation is referred to as a reaction to the left. An example of such a system is the acid-base reaction between carbonic acid and water.

$$H_2CO_3(aq) + H_2O(l) \rightleftharpoons H_3O^+(aq) + HCO_3^-(aq)$$

In still other cases, both forward and reverse reactions occur to nearly the same extent before chemical equilibrium is established. Neither reaction is favored, and considerable concentrations of both reactants and products are present at equilibrium. An example is the dissociation of sulfurous acid in water.

$$H_2SO_3(aq) + H_2O(l) \rightleftharpoons H_3O^+(aq) + HSO_3^-(aq)$$

Chemical reactions ordinarily are used to convert available reactants into more desirable products. Chemists try to convert as much of these reactants as possible into products. The extent to which reactants are converted to products can be determined from the numerical value of the equilibrium constant.

The Equilibrium Expression

Suppose two substances, A and B, react to form products C and D. In turn, C and D react to produce A and B. Under appropriate conditions, equilibrium occurs for this reversible reaction. This hypothetical equilibrium reaction is described by the following general equation.

$$n\text{A} + m\text{B} \rightleftharpoons x\text{C} + y\text{D}$$

Initially, the concentrations of C and D are zero and those of A and B are maximum. Figure 18-2 shows that over time the rate of the forward reaction decreases as A and B are used up. Meanwhile, the rate of the reverse reaction increases as C and D are formed. When these two reaction rates become equal, equilibrium is established. The individual concentrations of A, B, C, and D undergo no further change if conditions remain the same.

After equilibrium is attained, the concentrations of products and reactants remain constant, so a ratio of their concentrations should also

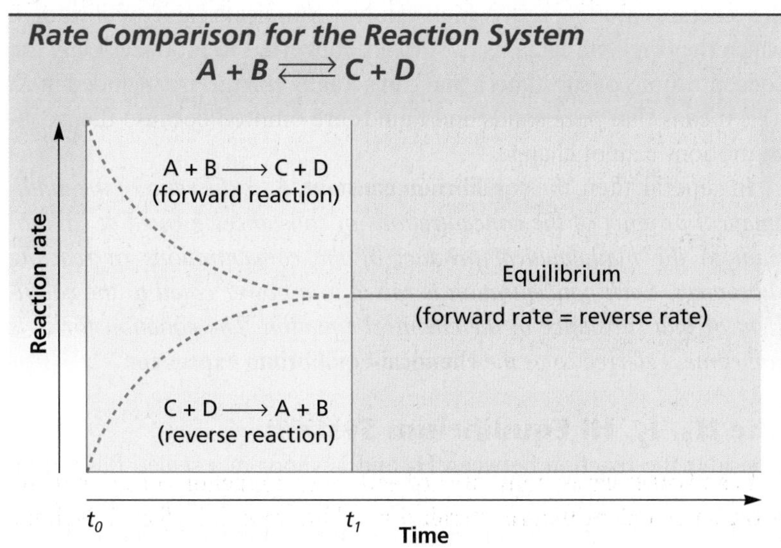

FIGURE 18-2 Shown are reaction rates for the hypothetical equilibrium reaction system $\text{A} + \text{B} \rightleftharpoons \text{C} + \text{D}$. From the time A and B are mixed together at t_0, the rate of the forward reaction declines and the rate of the reverse reaction increases until both forward and reverse reaction rates are equal at t_1, when the equilibrium condition begins.

remain constant. The ratio of the mathematical product $[C]^x \times [D]^y$ to the mathematical product $[A]^n \times [B]^m$ for this reaction has a definite value at a given temperature. It is the equilibrium constant of the reaction and is designated by the letter K. The following equation describes the equilibrium constant for the hypothetical equilibrium system. [C] indicates the concentration of C in moles per liter. The concentrations of the other substances are indicated similarly.

$$K = \frac{[C]^x[D]^y}{[A]^n[B]^m}$$

The concentrations of substances on the right side of the chemical equation appear in the numerator of the equilibrium ratio, with each concentration raised to a power equal to the coefficient of that substance in the chemical equation. These substances are the products of the forward reaction. The concentrations of substances on the left side of the chemical equation are found in the denominator of the equilibrium ratio, with each concentration raised to a power equal to the coefficient of that substance in the chemical equation. These substances are the reactants of the forward reaction. The constant K is independent of the initial concentrations. It is, however, dependent on the temperature of the system.

The Equilibrium Constant

The numerical value of K for a particular equilibrium system is obtained experimentally. The chemist must analyze the equilibrium mixture and determine the concentrations of all substances. The value of K for a given equilibrium reaction at a given temperature shows the extent to which the reactants are converted into the products of the reaction. If K is equal to 1, the products of the concentrations raised to the appropriate power in the numerator and denominator have the same value. Therefore, at equilibrium, there are roughly equal concentrations of reactants and products. If the value of K is low, the forward reaction occurs only very slightly before equilibrium is established, and the reactants are favored. A high value of K indicates an equilibrium in which the original reactants are largely converted to products. Only the concentrations of substances that can actually change are included in K. This means that *pure* solids and liquids are omitted because their concentrations cannot change.

In general, then, *the* **equilibrium constant,** *K, is the ratio of the mathematical product of the concentrations of substances formed at equilibrium to the mathematical product of the concentrations of reacting substances. Each concentration is raised to a power equal to the coefficient of that substance in the chemical equation. The equation for K is sometimes referred to as the* **chemical-equilibrium expression.**

The H_2, I_2, HI Equilibrium System

Consider the reaction between H_2 and I_2 vapor in a sealed flask at an elevated temperature. The rate of reaction can be followed by observing the rate at which the violet color of the iodine vapor diminishes, as shown in Figure 18-3. Suppose the hydrogen gas is present in excess and the reaction continues until all the iodine is used up. The color of the flask's contents, provided by the iodine gas, will gradually become less intense because the product, hydrogen iodide, HI, and the excess hydrogen are both colorless gases.

In actuality, the color fades to a constant intensity but does not disappear because the reaction is reversible. Hydrogen iodide decomposes to re-form hydrogen and iodine. The rate of this reverse reaction increases as the concentration of hydrogen iodide increases. The rate of the forward reaction decreases accordingly. The concentrations of hydrogen and iodine decrease as they are used up. As the rates of the opposing reactions become equal, an equilibrium is established. The constant color achieved indicates that equilibrium exists among hydrogen, iodine, and hydrogen iodide. The net chemical equation for the reaction system at equilibrium follows.

$$H_2(g) + I_2(g) \rightleftharpoons 2HI(g)$$

From this chemical equation, the following chemical-equilibrium expression can be written. The concentration of HI is raised to the power of 2 because the coefficient of HI in the balanced chemical equation is 2.

$$K = \frac{[HI]^2}{[H_2][I_2]}$$

Chemists have carefully measured the concentrations of H_2, I_2, and HI in equilibrium mixtures at various temperatures. In some experiments, the flasks were filled with hydrogen iodide at known pressure. The flasks were held at fixed temperatures until equilibrium was established. In other experiments, hydrogen and iodine were the original substances. Experimental data, together with the calculated values for K, are listed in Table 18-1. Experiments 1 and 2 began with hydrogen iodide. Experiments 3 and 4 began with hydrogen and iodine. Note the close agreement obtained for the numerical values of the equilibrium constant in all cases.

At 425°C, the equilibrium constant for this equilibrium reaction system has the average value of 54.34. This value for K should hold for any system of H_2, I_2, and HI at equilibrium at this temperature. If the

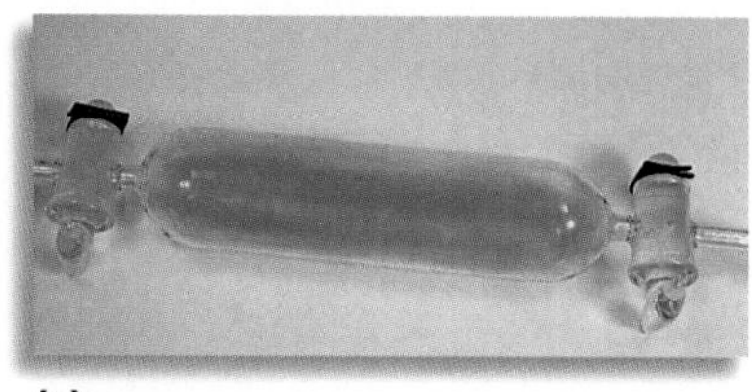

(a)

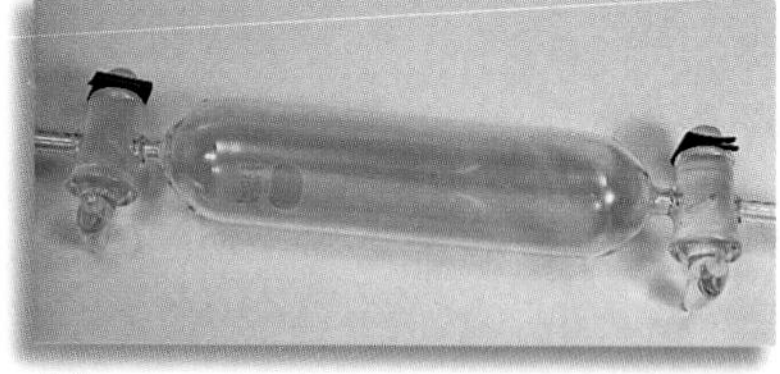

(b)

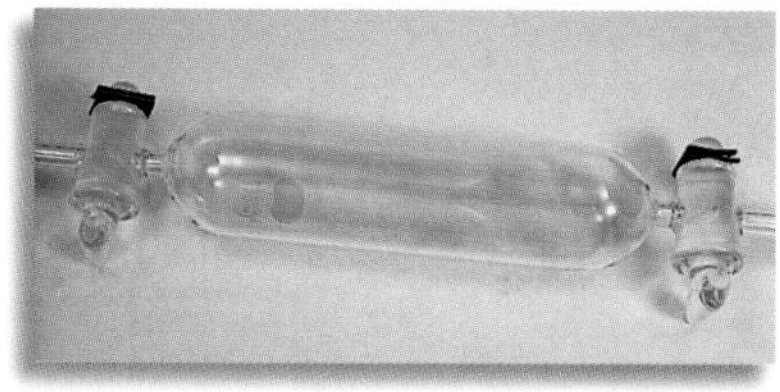

(c)

FIGURE 18-3 Hydrogen iodide gas is produced from gaseous hydrogen and iodine. The violet color of iodine gas (a) becomes fainter as the reaction consumes the iodine (b). The violet does not disappear but reaches a constant intensity when the reaction reaches equilibrium (c).

TABLE 18-1 ***Typical Equilibrium Concentrations of H_2, I_2, and HI in mol/L at 425°C***

Experiment	$[H_2]$	$[I_2]$	$[HI]$	$K = \frac{[HI]^2}{[H_2][I_2]}$
1	0.4953×10^{-3}	0.4953×10^{-3}	3.655×10^{-3}	54.46
2	1.141×10^{-3}	1.141×10^{-3}	8.410×10^{-3}	54.33
3	3.560×10^{-3}	1.250×10^{-3}	15.59×10^{-3}	54.62
4	2.252×10^{-3}	2.336×10^{-3}	16.85×10^{-3}	53.97

calculation for K yields a different result, there must be a reason. Either the H_2, I_2, and HI system has not reached equilibrium or the system is not at 425°C.

The balanced chemical equation for an equilibrium system helps establish the expression for the equilibrium constant. The data in Table 18-1 show that the validity of this expression is confirmed when the actual values of the equilibrium concentrations of reactants and products are determined experimentally. The values of K are calculated from these concentrations. No information concerning the kinetics of the reacting systems is required.

Once the value of the equilibrium constant is known, the equilibrium-constant expression can be used to calculate concentrations of reactants or products at equilibrium. Suppose an equilibrium system at 425°C is found to contain 0.015 mol/L each of H_2 and I_2. To find the concentration of HI in this system, the chemical equilibrium expression can be rearranged as shown in the two equations that follow.

$$K = \frac{[HI]^2}{[H_2][I_2]}$$

$$[HI] = \sqrt{K[H_2][I_2]}$$

Using the known K value and the given concentrations for H_2 and I_2, the equation can be solved for [HI].

$$[HI] = \sqrt{0.015 \times 0.015 \times 54.34}$$
$$= 0.11 \text{ mol/L}$$

SAMPLE PROBLEM 18-1

An equilibrium mixture of N_2, O_2, and NO gases at 1500 K is determined to consist of 6.4×10^{-3} mol/L of N_2, 1.7×10^{-3} mol/L of O_2, and 1.1×10^{-5} mol/L of NO. What is the equilibrium constant for the system at this temperature?

SOLUTION

1 ANALYZE

Given: $[N_2] = 6.4 \times 10^{-3}$ mol/L
$[O_2] = 1.7 \times 10^{-3}$ mol/L
$[NO] = 1.1 \times 10^{-5}$ mol/L

Unknown: K

2 PLAN

The balanced chemical equation is $N_2(g) + O_2(g) \rightleftharpoons 2NO(g)$.

The chemical equilibrium expression is $K = \frac{[NO]^2}{[N_2][O_2]}$.

3 COMPUTE

Substitute the given values for the concentrations into the equilibrium expression.

$$K = \frac{(1.1 \times 10^{-5} \cancel{\text{mol/L}})^2}{(6.4 \times 10^{-3} \cancel{\text{mol/L}})(1.7 \times 10^{-3} \cancel{\text{mol/L}})} = 1.1 \times 10^{-5}$$

4 **EVALUATE** The answer has the correct number of significant digits and is close to an estimated value of 8×10^{-6}, calculated as $\frac{1 \times 10^{-10}}{(6 \times 10^{-3})(2 \times 10^{-3})}$.

PRACTICE

1. At equilibrium a mixture of N_2, H_2, and NH_3 gas at 500°C is determined to consist of 0.602 mol/L of N_2, 0.420 mol/L of H_2, and 0.113 mol/L of NH_3. What is the equilibrium constant for the reaction $N_2(g) + 3H_2(g) \rightleftharpoons 2NH_3(g)$ at this temperature? *Answer* 0.286

2. The reaction $AB_2C(g) \rightleftharpoons B_2(g) + AC(g)$ reached equilibrium at 900 K in a 5.00 L vessel. At equilibrium 0.084 mol of AB_2C, 0.035 mol of B_2, and 0.059 mol of AC were detected. What is the equilibrium constant at this temperature for this system? *Answer* 4.9×10^{-3}

3. At equilibrium a 1.0 L vessel contains 20.0 mol of H_2, 18.0 mol of CO_2, 12.0 mol of H_2O, and 5.9 mol of CO at 427°C. What is the value of K at this temperature for the following reaction? *Answer* 0.20
$$CO_2(g) + H_2(g) \rightleftharpoons CO(g) + H_2O(g)$$

4. A reaction between gaseous sulfur dioxide and oxygen gas to produce gaseous sulfur trioxide takes place at 600°C. At that temperature, the concentration of SO_2 is found to be 1.50 mol/L, the concentration of O_2 is 1.25 mol/L, and the concentration of SO_3 is 3.50 mol/L. Using the balanced chemical equation, calculate the equilibrium constant for this system. *Answer* 4.36

SECTION REVIEW

1. What is meant by *chemical equilibrium*?
2. What is an equilibrium constant?
3. How does the value of an equilibrium constant relate to the relative quantities of reactants and products at equilibrium?
4. What is meant by a *chemical-equilibrium expression*?
5. Hydrochloric acid, HCl, is a strong acid that dissociates completely in water to form H_3O^+ and Cl^-. Would you expect the value of K for the reaction $HCl(aq) + H_2O(l) \rightleftharpoons H_3O^+(aq) + Cl^-(aq)$ to be 1×10^{-2}, 1×10^{-5}, or "very large"? Justify your answer.
6. Write the chemical-equilibrium expression for the reaction $4HCl(g) + O_2(g) \rightleftharpoons 2Cl_2(g) + 2H_2O(g)$.
7. At equilibrium at 2500 K, [HCl] = 0.0625 mol/L and $[H_2] = [Cl_2] = 0.0045$ mol/L for the reaction $H_2(g) + Cl_2(g) \rightleftharpoons 2HCl(g)$. Find the value of K.
8. An equilibrium mixture at 425°C is found to consist of 1.83×10^{-3} mol/L of H_2, 3.13×10^{-3} mol/L of I_2, and 1.77×10^{-2} mol/L of HI. Calculate the equilibrium constant, K, for the reaction $H_2(g) + I_2(g) \rightleftharpoons 2HI(g)$.
9. For the reaction $H_2(g) + I_2(g) \rightleftharpoons 2HI(g)$ at 425°C, calculate [HI], given $[H_2] = [I_2] = 4.79 \times 10^{-4}$ mol/L and $K = 54.3$.

Fixing the Nitrogen Problem

HISTORICAL PERSPECTIVE

Each year, the chemical industry synthesizes tons of nitrogenous fertilizer, increasing agricultural production around the globe. But prior to 1915, humans had to rely solely on natural resources for fertilizer, and the dwindling supply of these materials caused widespread fear of world starvation. A crisis was averted, however, through the discovery of an answer to the "nitrogen problem," a term used at the time to describe the shortage of useful nitrogen despite its abundance in the atmosphere.

The Malthusian Threat

In 1798, Thomas Malthus published his famous "Essay on Population," a report predicting that the world's food supplies could not keep up with the growing human population and that famine, death, and misery were inevitable. Malthus's warning seemed to be echoed in the 1840s by the great Irish potato famine. In fact, the rest of Europe likely would have suffered serious food shortages as well had crop yields per acre not been increased through the use of fertilizers containing nitrogen.

Few living things can utilize the gas that forms 78 percent of the atmosphere; they need nitrogen that has been combined with other elements, or "fixed," to survive.

But soil often lacks sufficient amounts of the organisms that fix nitrogen for plants, so fertilizers containing usable nitrogen compounds are added. In 1898, two-thirds of the world's supply of these compounds came from Chile, where beds of sodium nitrate, or Chilean saltpeter, were abundant.

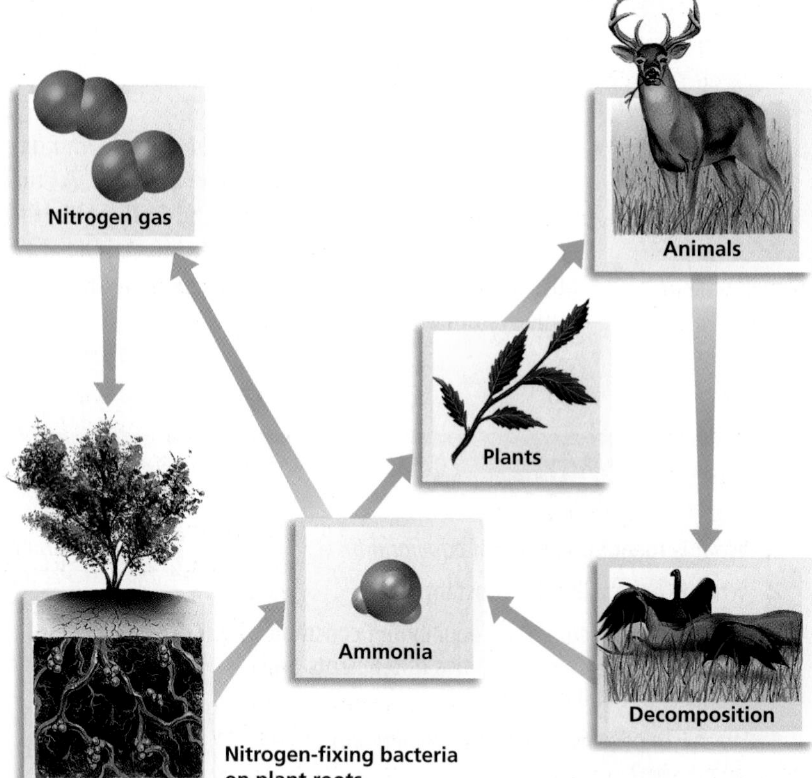

Nitrogen is released when living things die and also from animal wastes and plant material. A few kinds of bacteria are able to break the bond holding the nitrogen molecule together, freeing the nitrogen atoms to combine with hydrogen to form ammonia. Plants can absorb the nitrogen in this form from the soil. Animals then benefit from the nitrogen by eating the plants.

But, as the chemist William Crookes emphasized in his speech to the British Association that year, these reserves were limited; it was up to his colleagues to discover alternatives and prevent Malthus's dire forecast from coming true:

> *It is the chemist who must come to the rescue of the threatened communities. It is through the*

laboratory that starvation may ultimately be turned to plenty.

The Haber-Nernst Controversy

As early as the 1890s, chemists had shown that ammonia, a practical source of fixed nitrogen, could be synthesized at high temperatures and at atmospheric pressure from elemental hydrogen and nitrogen. The problem was that the end-product was present in such minute amounts that the process was not industrially practical.

In 1904, the German chemist Fritz Haber seemed to confirm this assessment. He tried reacting hydrogen and nitrogen at temperatures of up to 1020°C using pure iron as well as other metals as a catalyst. He found that the amount of ammonia was a mere 0.005–0.012% at equilibrium. Thus, he concluded:

> *From dull red heat upwards no catalyst can produce more than traces of ammonia under ordinary pressure; and even at greatly increased pressures the position of the equilibrium must remain very unfavorable.*

Haber had apparently closed the door on the synthesis of ammonia from its elements. But in 1906, Walther Nernst, using his new heat theorem, calculated the reaction's theoretical ammonia concentration at equilibria corresponding to several pressures. He found that his value at atmospheric pressure disagreed significantly with Haber's, and he publicly challenged Haber's values.

Today ammonia is produced on an industrial scale in plants like this one.

Haber was convinced that he was right. Applying Le Châtelier's principle, he ran the reaction at increased pressure to attain an amount of ammonia that could be measured more accurately.

Haber and his assistants confirmed their original findings, and Nernst later conceded a mathematical error. But more important, the new round of experiments indicated that a reasonable amount of ammonia might be attained at pressures of 200 atm (402 kPa) using a uranium or osmium catalyst.

Scaling Up

Large-scale equipment that could withstand such high pressures was unheard of at the time, and osmium and uranium were far too scarce to be cost-effective for industry. Nevertheless, in 1909, the German firm BASF bought the rights to Haber's findings and put its gifted chemical engineer Karl Bosch in charge of creating an industrial-scale system that would make the process profitable.

After nearly five years, Bosch and the company's chief chemist, Alwin Mittasch, succeeded in developing a suitable reactor that could handle the reaction's high pressures. They also discovered that a catalyst of iron containing small amounts of impurities was an effective replacement for the rare metals used by Haber. Haber was impressed:

> *It is remarkable how matter continually reveals new facets. Iron . . . which we have studied a hundred times in the pure state, now works in the impure state . . . I congratulate you.*

An Eerie Epilogue

By September of 1913, BASF was producing 20 metric tons of ammonia a day using the Haber-Bosch process. Eventually, enough ammonia was produced by the chemical industry to free Germany and the world of dependence on Chilean saltpeter for fertilizer. Chemists had thwarted the Malthusian threat. Yet, sadly, the victory proved bittersweet; the new ammonia synthesis also became the basis of the production of nitric acid, which was used to make many of the explosives employed in the wars that rocked Europe and the rest of the globe in the first half of the twentieth century.

SECTION 18-2

Shifting Equilibrium

OBJECTIVES

- Discuss the factors that disturb equilibrium.
- Discuss conditions under which reactions go to completion.
- Describe the common-ion effect.

In systems that have attained chemical equilibrium, the forward and reverse reactions are proceeding at equal rates. Any change that alters the rates of these reactions disturbs the original equilibrium. The system then seeks a new equilibrium state. By shifting an equilibrium in the desired direction, chemists can often improve the yield of the product they are seeking.

Predicting the Direction of Shift

Le Châtelier's principle provides a means of predicting the influence of stress factors on equilibrium systems. As you may recall from Chapter 12, Le Châtelier's principle states that *if a system at equilibrium is subjected to a stress, the equilibrium is shifted in the direction that tends to relieve the stress.* This principle is true for all dynamic equilibria, chemical as well as physical. Changes in pressure, concentration, and temperature illustrate the application of Le Châtelier's principle to chemical equilibrium.

Changes in Pressure

A change in pressure affects only equilibrium systems in which gases are involved. For changes in pressure to affect the system, the total number of moles of gas on the left side of the equation must be different from the total number of moles on the right side of the equation. For example, the balanced chemical equation for the decomposition of solid $CaCO_3$ given on page 564 indicates that 0 mol reactant gases produce 1 mol of product gases. The change in moles is 1. Therefore, a high pressure favors the reverse reaction because fewer CO_2 molecules are produced. A similar argument can explain the increased production of CO_2 that accompanies a low system pressure.

Next let us consider the Haber process for the synthesis of ammonia.

$$N_2(g) + 3H_2(g) \rightleftharpoons 2NH_3(g)$$

This situation is somewhat different. First consider an increase in pressure as the applied stress. Can the system shift in a way that reduces the stress? Yes. An increase in pressure causes increases in the concentrations of all species. The system can reduce the number of molecules, and hence the total pressure, by shifting the equilibrium to the right. For each four molecules of reactants there are two molecules of products. By producing more NH_3, and using up N_2 and H_2, the system can reduce the number of molecules. This leads to a decrease in pressure. However, the new equilibrium pressure is still higher than before, although not as high as the pressure caused by the initial stress.

An increase in pressure on confined gases causes an increase in the concentrations of these gases. So changes in pressure may shift the equilibrium position, but they do not affect the value of the equilibrium constant.

Ammonia produced in the Haber process is continuously removed by condensation to liquid. This condensation removes most of the product from the gas phase in which the reaction occurs. The resulting decrease in the partial pressure of NH_3 gas in the reaction vessel is the same as a decrease in product concentration and shifts the equilibrium to the right.

The introduction of an inert gas, such as helium, into the reaction vessel for the synthesis of ammonia increases the total pressure in the vessel. But it does not change the partial pressures of the reaction gases present. Therefore, increasing pressure by adding a gas that is not a reactant or a product *cannot* affect the equilibrium position of the reaction system.

(a)

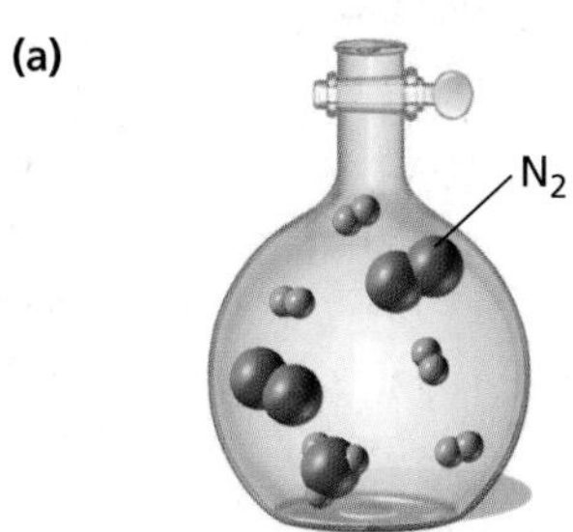

(b)

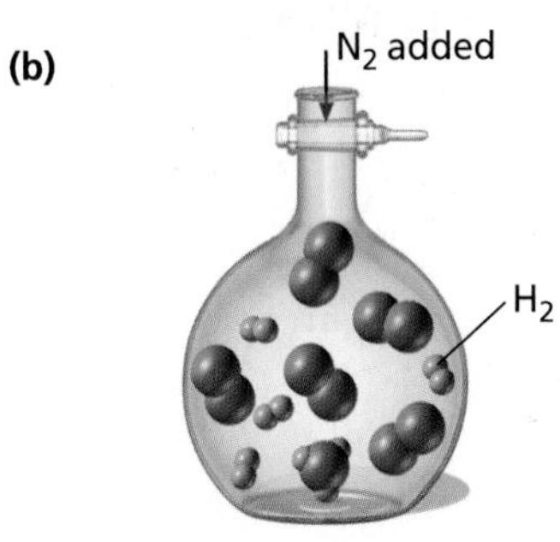

(c)

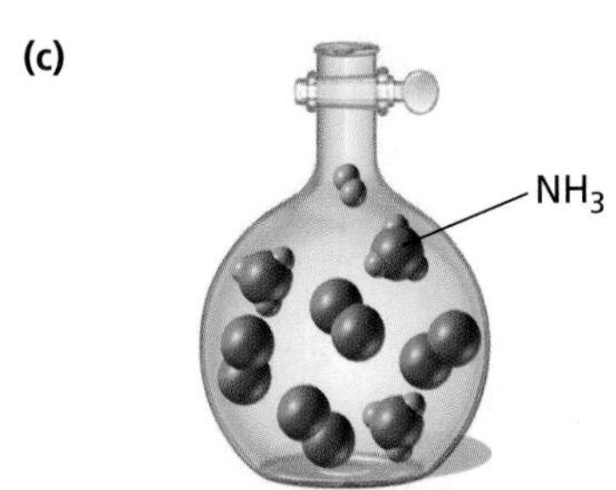

$$N_2(g) + 3H_2(g) \rightleftharpoons 2NH_3(g)$$

FIGURE 18-4 (a) H_2, N_2, and NH_3 are in equilibrium within a closed system. (b) Addition of more N_2 causes a stress on the initial equilibrium. (c) The new equilibrium position for this system. There is now a higher concentration of N_2, a lower concentration of H_2, and a higher concentration of NH_3.

Changes in Concentration

An increase in the concentration of a reactant is a stress on the equilibrium system. It causes an increase in collision frequency and, generally, an increase in reaction rate. Consider the following hypothetical reaction.

$$A + B \rightleftharpoons C + D$$

An increase in the concentration of A shifts the equilibrium to the right. Both A and B are used up faster, and more of C and D is formed. The equilibrium is reestablished with a lower concentration of B. The equilibrium has shifted to reduce the stress caused by the increase in concentration of A. Figure 18-4 illustrates the effect on a system in equilibrium produced by increasing the concentration of a reactant. Similarly, an increase in the concentration of B drives the reaction to the right. An increase in the concentration of either C or D increases the rate of the reverse reaction, and the equilibrium shifts to the left. A decrease in the concentration of C or D has the same effect on the position of the equilibrium as does an increase in the concentration of A or B; the equilibrium shifts to the right.

Changes in concentration have no effect on the value of the equilibrium constant. This is because such changes have an equal effect on the numerator and the denominator of the chemical equilibrium expression. Thus, all concentrations give the same value or numerical ratio for the equilibrium constant when equilibrium is reestablished.

Many chemical processes involve heterogeneous reactions in which the reactants and products are in different phases. The *concentrations* of pure substances in solid and liquid phases are not changed by adding or removing quantities of the substance. This is because concentration is density-dependent, and the density of these phases is constant, regardless of the total amounts present. *A pure substance in a condensed phase, solid or liquid, can be eliminated from the expression for the equilibrium constant.* The concentration of the pure solid or liquid is set equal to 1 in the equilibrium expression, signifying that the concentration is assumed to remain unchanged in the equilibrium system.

The following equation describes the equilibrium system established by the decomposition of solid calcium carbonate.

$$CaCO_3(s) \rightleftharpoons CaO(s) + CO_2(g)$$

The products are a solid and a gas, leading to the following expression for the equilibrium constant.

$$K = \frac{[CaO][CO_2]}{[CaCO_3]} = \frac{[1][CO_2]}{[1]} = [CO_2]$$

Carbon dioxide is the only substance in the system subject to changes in concentration. Because it is a gas, the system is affected by pressure changes.

Changes in Temperature

Reversible reactions are exothermic in one direction and endothermic in the other. The effect of changing the temperature of an equilibrium mixture depends on which of the opposing reactions is endothermic and which is exothermic.

According to Le Châtelier's principle, the addition of heat shifts the equilibrium so that heat is absorbed. This favors the endothermic reaction. The removal of heat favors the exothermic reaction. A rise in temperature increases the rate of any reaction. In an equilibrium system, however, the rates of the opposing reactions are raised unequally. Thus, the value of the equilibrium constant for a given system is affected by the temperature.

The synthesis of ammonia by the Haber process is exothermic, as indicated by the heat shown on the product side of the equation.

$$N_2(g) + 3H_2(g) \rightleftharpoons 2NH_3(g) + 92 \text{ kJ}$$

A high temperature is not desirable because it favors the decomposition of ammonia, the endothermic reaction. At low temperatures, however, the forward reaction is too slow to be commercially useful. The temperature used represents a compromise between kinetic and equilibrium requirements. It is high enough that equilibrium is established rapidly but low enough that the equilibrium concentration of ammonia is significant. Moderate temperature (about 500°C) and very high pressure (700–1000 atm) produce a satisfactory yield of ammonia.

The production of colorless dinitrogen tetroxide gas, N_2O_4, from dark brown NO_2 gas is also an exothermic reaction. Figure 18-5 shows how temperature affects the equilibrium of this system. In this figure, all three flasks contain the same total mass of gas, a mixture of NO_2 and N_2O_4. In Figure 18-5(a) the temperature of the system is lowered to 0°C. This causes the equilibrium of the system to shift to the right, allowing more of the colorless N_2O_4 gas to form, which produces a light brown color.

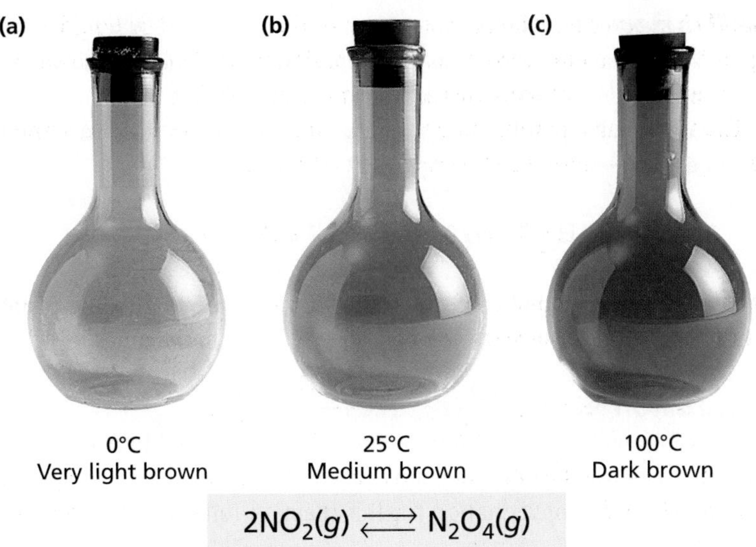

FIGURE 18-5 Different temperatures can cause an equilibrium system to shift and seek a new equilibrium position.

The system is at equilibrium at 25°C in Figure 18-5(b). The system contains an equilibrium mixture of NO_2 and N_2O_4, producing a medium brown color. As the temperature is raised to 100°C, as shown in Figure 18-5(c), the equilibrium shifts to the left, causing more of the dark brown NO_2 gas to form.

For an endothermic reaction, such as the decomposition of calcium carbonate, heat shows up on the reactant side of the equation.

$$556 \text{ kJ} + CaCO_3(s) \rightleftarrows CaO(s) + CO_2(g)$$

An increase in temperature caused by adding heat to the system causes the value of K to increase and the equilibrium to shift to the right.

The reactions of the system are also accelerated by a suitable catalyst. However, catalysts have no effect on relative equilibrium amounts. They only affect the rates at which equilibrium is reached. This is because catalysts increase the rates of forward and reverse reactions in a system by equal factors. Therefore, they do not affect K.

Reactions That Go to Completion

Some reactions involving compounds formed by the chemical interaction of ions in solutions appear to go to completion in the sense that the ions are almost completely removed from solution. The extent to which reacting ions are removed from solution depends on the solubility of the compound formed and, if the compound is soluble, on the degree of ionization. Thus, a product that escapes as a gas, precipitates as a solid, or is only slightly ionized effectively removes from solution the bulk of the reacting ions that compose it. Consider some specific examples of situations in which such ionic reactions go to completion.

Formation of a Gas

Unstable substances formed as products of ionic reactions decompose spontaneously. An example is carbonic acid, H_2CO_3, the acid in carbonated water, such as club soda, which yields a gas as a decomposition product.

$$H_2CO_3(aq) \longrightarrow H_2O(l) + CO_2(g)$$

This reaction goes practically to completion because one of the products, CO_2, escapes as a gas if the container is open to the air.

Formation of a Precipitate

When solutions of sodium chloride and silver nitrate are mixed, a white precipitate of silver chloride immediately forms, as shown in Figure 18-6. The overall ionic equation for this reaction follows.

$$Na^+(aq) + Cl^-(aq) + Ag^+(aq) + NO_3^-(aq) \longrightarrow Na^+(aq) + NO_3^-(aq) + AgCl(s)$$

If chemically equivalent amounts of the two solutes are used, only Na^+ ions and NO_3^- ions remain in solution in appreciable quantities. Almost all of the Ag^+ ions and Cl^- ions combine and separate from the solution as a precipitate of AgCl. This is because AgCl is only very sparingly soluble in water. It separates by precipitation from what turns out to be a saturated solution of its particles. The reaction thus effectively goes to completion because an essentially insoluble product is formed.

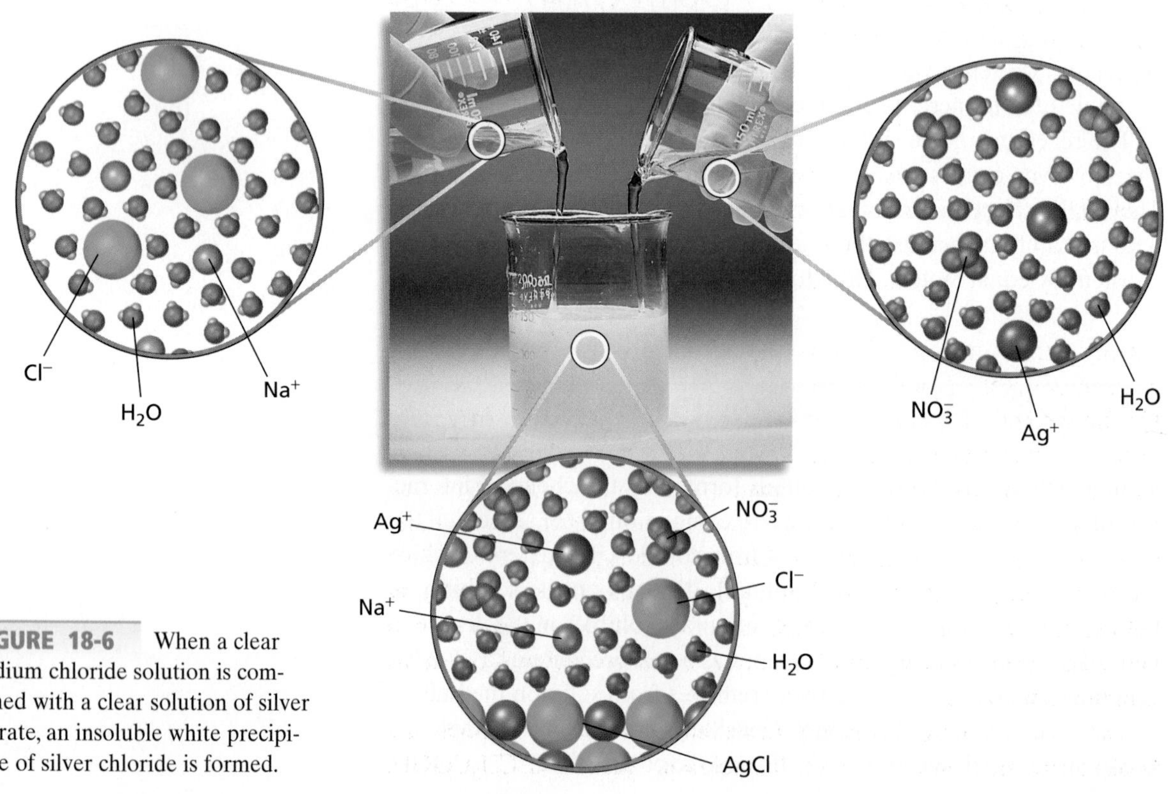

FIGURE 18-6 When a clear sodium chloride solution is combined with a clear solution of silver nitrate, an insoluble white precipitate of silver chloride is formed.

Formation of a Slightly Ionized Product

Neutralization reactions between H_3O^+ ions from aqueous acids and OH^- ions from aqueous bases result in the formation of water molecules, which are only slightly ionized. A reaction between HCl and NaOH illustrates this process. Aqueous HCl supplies H_3O^+ ions and Cl^- ions to the solution, and aqueous NaOH supplies Na^+ ions and OH^- ions, as shown in the following overall ionic equation.

$$H_3O^+(aq) + Cl^-(aq) + Na^+(aq) + OH^-(aq) \longrightarrow Na^+(aq) + Cl^-(aq) + 2H_2O(l)$$

Neglecting the spectator ions, the net ionic equation is as follows.

$$H_3O^+(aq) + OH^-(aq) \longrightarrow 2H_2O(l)$$

Because it is only slightly ionized, the water exists almost entirely as covalently bonded molecules. Thus, insofar as they are initially present in equal amounts, hydronium ions and hydroxide ions are almost entirely removed from the solution. The reaction effectively runs to completion because the product is only slightly ionized.

Common-Ion Effect

An equilibrium reaction may be driven in the desired direction by applying Le Châtelier's principle. Suppose hydrogen chloride gas is bubbled into a saturated solution of sodium chloride. Hydrogen chloride is extremely soluble in water, and it is almost completely ionized.

$$HCl(g) + H_2O(l) \longrightarrow H_3O^+(aq) + Cl^-(aq)$$

The equilibrium for a saturated solution of sodium chloride is described by the following equation.

$$NaCl(s) \rightleftharpoons Na^+(aq) + Cl^-(aq)$$

As the hydrogen chloride dissolves in sufficient quantity, it increases the concentration of Cl^- ions in the solution, which is a stress on the equilibrium system. The system can compensate, according to Le Châtelier's principle, by combining some of the added Cl^- ions with an equivalent amount of Na^+ ions. This causes some solid NaCl to precipitate out, relieving the stress of added chloride. The new equilibrium has a greater concentration of Cl^- ions but a decreased concentration of Na^+ ions. However, the product of $[Na^+]$ and $[Cl^-]$ still has the same value as before. *This phenomenon, in which the addition of an ion common to two solutes brings about precipitation or reduced ionization, is called the* **common-ion effect.**

The common-ion effect is also observed when one ion species of a weak electrolyte is added in excess to a solution. Acetic acid, CH_3COOH, is such an electrolyte. A 0.1 M CH_3COOH solution is only about 1.4% ionized to produce hydronium ions and acetate ions, CH_3COO^-. The ionic equilibrium is shown by the following equation.

$$CH_3COOH(aq) + H_2O(l) \rightleftharpoons H_3O^+(aq) + CH_3COO^-(aq)$$

Small additions of sodium acetate, $NaCH_3COO$ (an ionic salt that is completely dissociated in water), to a solution containing acetic acid greatly increase the acetate-ion concentration. The equilibrium shifts in the direction that uses up some of the acetate ions in accordance with Le Châtelier's principle. More molecules of acetic acid are formed and the concentration of hydronium ions is reduced. In general, the addition of a salt with an ion common to the solution of a weak electrolyte reduces the ionization of the electrolyte. Figure 18-7 shows a 0.25 M CH_3COOH solution on the left that has a pH of about 2.7. Mixing that with the 0.10 M $NaCH_3COO$ solution in the center produces the solution on the right, which has a pH of about 4.5, indicating lower $[H_3O^+]$ and thus lowered acetic acid ionization. (The universal indicator used turns red in acidic solutions, green in weakly basic solutions, and yellow in neutral solutions.)

FIGURE 18-7 The solution of CH_3COOH on the left is combined with the solution of $NaCH_3COO$ in the center. Both contain the common ion, CH_3COO^-. They produce the solution on the right, which is only slightly acidic due to the decreased ionization of the acid.

SECTION REVIEW

1. Name three ways the chemical equilibrium can be disturbed.
2. Describe three situations in which ionic reactions go to completion.
3. Describe the common-ion effect.
4. Identify the common ion in each of the following situations.
   a. 5 g of NaCl is added to a 2.0 M solution of HCl
   b. 50 mL of 1.0 M $NaCH_3COO$ is added to 1.0 M CH_3COOH
   c. 10 pellets of NaOH are added to 100 mL of water
5. Predict the effect that decreasing pressure would have on each of the following reaction systems at equilibrium.
   a. $H_2(g) + Cl_2(g) \rightleftharpoons 2HCl(g)$
   b. $NH_4Cl(s) \rightleftharpoons NH_3(g) + HCl(g)$
   c. $2H_2O_2(aq) \rightleftharpoons 2H_2O(l) + O_2(g)$
   d. $3O_2(g) \rightleftharpoons 2O_3(g)$
6. When solid carbon reacts with oxygen gas to form carbon dioxide, 393.51 kJ of heat are released. Does this reaction become more favorable or less favorable as the temperature decreases? Explain.

Equilibria of Acids, Bases, and Salts

SECTION 18-3

Objectives

- Explain the concept of acid-ionization constants, and write acid-ionization equilibrium expressions.
- Review the ionization constant of water.
- Explain buffering.
- Compare cation and anion hydrolysis.

Ionization Constant of a Weak Acid

About 1.4% of the solute molecules in a 0.1 M acetic acid solution are ionized at room temperature. The remaining 98.6% of the acetic acid molecules, CH_3COOH, remain nonionized. Thus, the solution contains three species of particles in equilibrium: CH_3COOH molecules, H_3O^+ ions, and acetate ions, CH_3COO^-. The equilibrium constant for this system expresses the equilibrium ratio of ions to molecules. From the equilibrium equation for the ionization of acetic acid, the equilibrium constant equation can be written.

$$CH_3COOH + H_2O \rightleftharpoons H_3O^+ + CH_3COO^-$$

$$K = \frac{[H_3O^+][CH_3COO^-]}{[CH_3COOH][H_2O]}$$

At the 0.1 M concentration, water molecules greatly exceed the number of acetic acid molecules. Without introducing a measurable error, one can assume that the molar concentration of H_2O molecules remains constant in such a solution. Thus, because both K and $[H_2O]$ are constant, the product $K[H_2O]$ is constant.

$$K[H_2O] = \frac{[H_3O^+][CH_3COO^-]}{[CH_3COOH]}$$

The left side of the equation can be simplified by setting $K[H_2O] = K_a$.

$$K_a = \frac{[H_3O^+][CH_3COO^-]}{[CH_3COOH]}$$

The term K_a is called the **acid-ionization constant.** The acid ionization constant, K_a, like the equilibrium constant, K, is constant for a specified temperature but has a new value for each new temperature.

The acid-ionization constant for a weak acid represents a small value. To determine the numerical value of the ionization constant for acetic acid at a specific temperature, the equilibrium concentrations of H_3O^+ ions, CH_3COO^- ions, and CH_3COOH molecules must be known. The ionization of a molecule of CH_3COOH in water yields one H_3O^+ ion and one CH_3COO^- ion. These concentrations can, therefore, be found experimentally by measuring the pH of the solution.

TABLE 18-2 *Ionization of Acetic Acid*

Molarity	% ionized	$[H_3O^+]$	$[CH_3COOH]$	K_a
0.100	1.33	0.00133	0.0987	1.79×10^{-5}
0.0500	1.89	0.000945	0.0491	1.82×10^{-5}
0.0100	4.17	0.000417	0.00958	1.81×10^{-5}
0.00500	5.86	0.000293	0.00471	1.82×10^{-5}
0.00100	12.6	0.000126	0.000874	1.82×10^{-5}

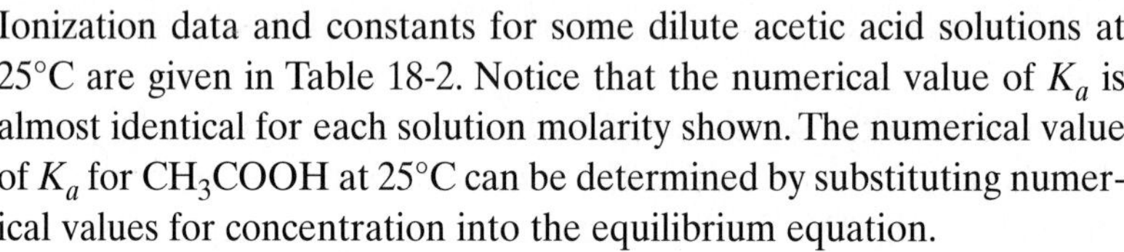

Ionization data and constants for some dilute acetic acid solutions at 25°C are given in Table 18-2. Notice that the numerical value of K_a is almost identical for each solution molarity shown. The numerical value of K_a for CH_3COOH at 25°C can be determined by substituting numerical values for concentration into the equilibrium equation.

$$K_a = \frac{[H_3O^+][CH_3COO^-]}{[CH_3COOH]}$$

At constant temperature, an increase in the concentration of CH_3COO^- ions through the addition of sodium acetate, $NaCH_3COO$, disturbs the equilibrium. This disturbance causes a decrease in $[H_3O^+]$ and an increase in $[CH_3COOH]$. Eventually, the equilibrium is reestablished with the *same* value of K_a. But there is a higher concentration of nonionized acetic acid molecules and a lower concentration of H_3O^+ ions than before the extra CH_3COO^- was added. Changes in the hydronium-ion concentration affect pH. In this example, the reduction in $[H_3O^+]$ means an increase in the pH of the solution.

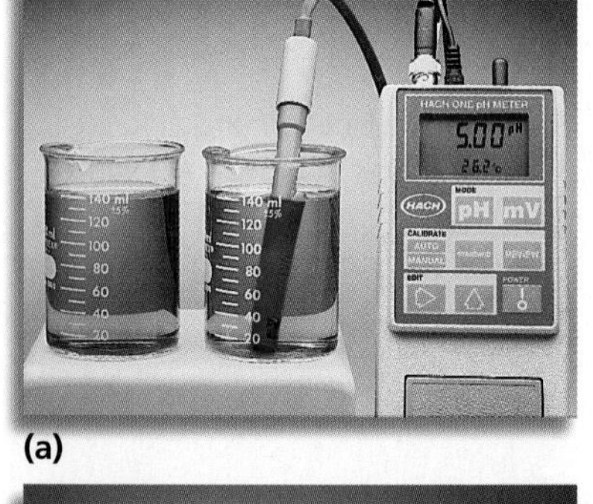

(a)

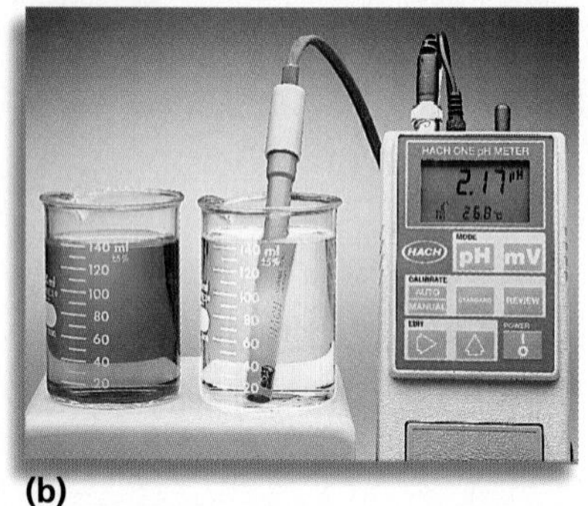

(b)

FIGURE 18-8 (a) The beaker on the left contains a buffered solution and an indicator with a pH of about 7. The beaker on the right contains mostly water with a trace amount of acid and an indicator. The pH meter shows a pH of 5.00 for this solution. (b) After 5 mL of 0.10 M HCl is added to both beakers, the beaker on the left does not change color, indicating no substantial change in its pH. However, the beaker on the right undergoes a definite color change, and the pH meter shows a pH of 2.17.

Buffers

The solution just described contains both a weak acid, CH_3COOH, and a salt of the weak acid, $NaCH_3COO$. The solution can react with either an acid or a base. When small amounts of acids or bases are added, the pH of the solution remains nearly constant. The weak acid and the common ion, CH_3COO^-, act as a "buffer" against significant changes in the pH of the solution. *Because it can resist changes in pH, this solution is a* **buffered solution.** Figure 18-8 shows how a buffered and a nonbuffered solution react to the addition of an acid.

Suppose a small amount of acid is added to the acetic acid–sodium acetate solution. Acetate ions react with most of the added hydronium ions to form nonionized acetic acid molecules.

$$CH_3COO^-(aq) + H_3O^+(aq) \longrightarrow CH_3COOH(aq) + H_2O(l)$$

The hydronium ion concentration and the pH of the solution remain practically unchanged.

Suppose a small amount of a base is added to the original solution. The OH^- ions of the base react with and remove hydronium ions to form nonionized water molecules. Acetic acid molecules then ionize and restore the equilibrium concentration of hydronium ions.

$$CH_3COOH(aq) + H_2O(l) \longrightarrow H_3O^+(aq) + CH_3COO^-(aq)$$

The pH of the solution again remains practically unchanged.

A solution of a weak base containing a salt of the base behaves in a similar manner. The hydroxide ion concentration and the pH of the solution remain essentially constant with small additions of acids or bases. Suppose a base is added to an aqueous solution of ammonia that also contains ammonium chloride. Ammonium ions donate a proton to the added hydroxide ions to form nonionized water molecules.

$$NH_4^+(aq) + OH^-(aq) \longrightarrow NH_3(aq) + H_2O(l)$$

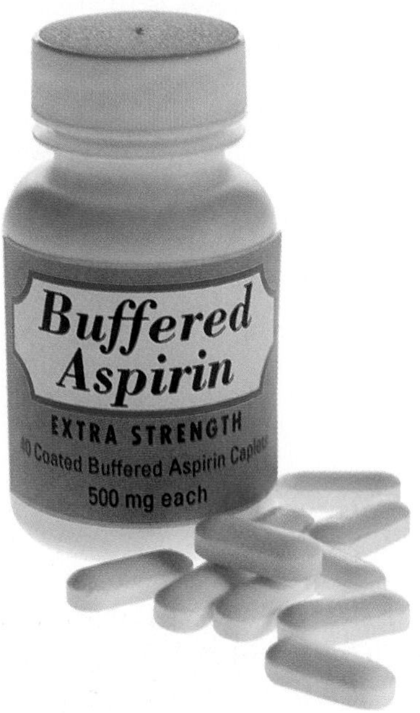

FIGURE 18-9 Many consumer products are buffered to protect the body from potentially harmful pH changes.

If a small amount of an acid is added to the solution instead, hydroxide ions from the solution accept protons from the added hydronium ions to form nonionized water molecules. Ammonia molecules in the solution then ionize and restore the equilibrium concentration of hydronium ions and the pH of the solution.

$$NH_3(aq) + H_2O(l) \longrightarrow NH_4^+(aq) + OH^-(aq)$$

Buffer action has many important applications in chemistry and physiology. Human blood is naturally buffered to maintain a pH of between 7.3 and 7.5. This is essential because large changes in pH would lead to serious disturbances of normal body functions. Figure 18-9 shows an example of one of the many medicines buffered to prevent large and potentially damaging changes in pH.

Ionization Constant of Water

Recall from Chapter 16 that the self-ionization of water is an equilibrium reaction.

$$H_2O(l) + H_2O(l) \rightleftarrows H_3O^+(aq) + OH^-(aq)$$

Equilibrium is established with a very low concentration of H_3O^+ and OH^- ions. The following expression for the equilibrium constant is derived from the balanced chemical equation.

$$K_w = [H_3O^+][OH^-] = 1.0 \times 10^{-14}$$

Hydrolysis of Salts

Salts are formed during the neutralization reaction between a Brønsted acid and a Brønsted base. When a salt dissolves in water, it produces positive ions (cations) of the base from which it was formed and negative ions (anions) of the acid from which it was formed. Therefore, the solution might be expected to be neutral. The aqueous solutions of some salts, such as NaCl and KNO_3, are neutral, having a pH of 7. However, when sodium carbonate dissolves in water, the resulting solution turns red litmus paper blue, indicating a pH greater than 7. Ammonium chloride produces an aqueous solution that turns blue litmus paper red, indicating a pH less than 7. Salts formed from the combination of strong or weak acids and bases are shown in Figure 18-10.

The variation in pH values can be accounted for by examining the ions formed when each of these salts dissociates. If the ions formed are from weak acids or bases, they react chemically with the water solvent, and the pH of the solution will have a value other than 7. *A reaction between water molecules and ions of a dissolved salt is* **hydrolysis.** If the anions react with water, the process is anion hydrolysis and results in a more basic solution. If the cations react with water molecules, the process is cation hydrolysis and results in a more acidic solution.

Anion Hydrolysis

In the Brønsted sense, the anion of the salt is the conjugate base of the acid from which it was formed. It is also a proton acceptor. If the acid is weak, its conjugate base (the anion) will be strong enough to remove protons from some water molecules, proton donors, to form OH^- ions. An

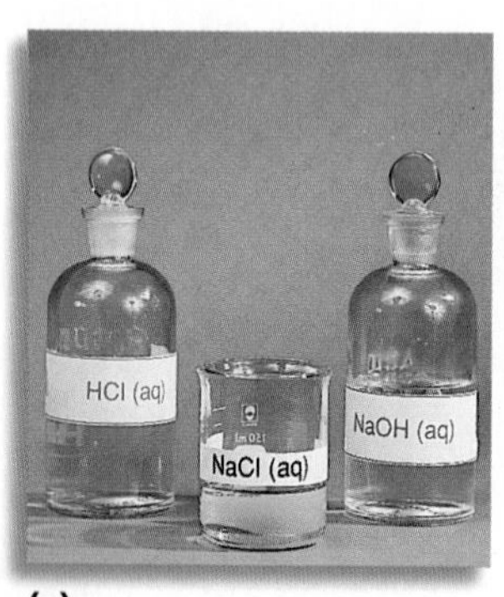

(a)

(b)

(c)

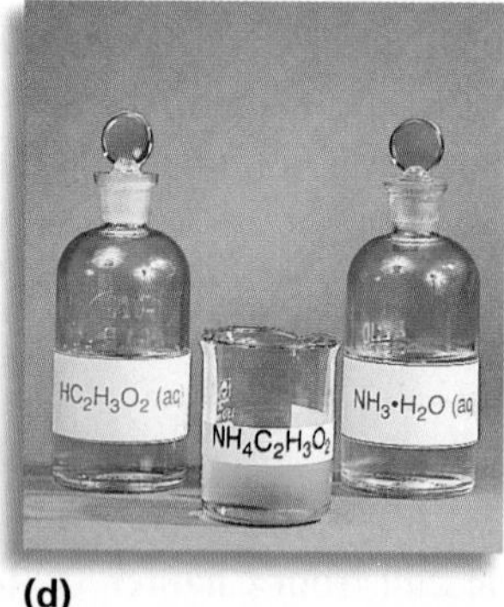

(d)

FIGURE 18-10 The universal indicator shows that the pH of salt solutions varies, depending on the strength of the acid and the base that formed the salt. (a) NaCl is formed from a strong acid and a strong base; the color of the indicator shows the pH is neutral. (b) The indicator shows the pH of the sodium acetate solution is basic. This was formed from a strong base and a weak acid. (c) The strong acid and weak base combination in ammonium chloride produces an acidic solution, as shown by the color of the indicator. (d) The weak acid and weak base that form ammonium acetate are of comparable strength. A solution of ammonium acetate is essentially neutral.

equilibrium is established in which the net effect of the anion hydrolysis is an increase in the hydroxide-ion concentration, $[OH^-]$, of the solution.

The equilibrium equation for a typical weak acid in water, HA, forming hydronium ion and an anion, A^-, is as follows.

$$HA(aq) + H_2O(l) \rightleftharpoons H_3O^+(aq) + A^-(aq)$$

From this equation, the generalized expression for K_a can be written. Recall that the concentration of water in dilute aqueous solutions is essentially constant, so it is included in the equilibrium constant instead of in the concentration ratio.

$$K_a = \frac{[H_3O^+][A^-]}{[HA]}$$

The hydrolysis reaction between water and the anion, A^-, that is produced by the dissociation of the weak acid, HA, is represented by the general equilibrium equation that follows.

$$A^-(aq) + H_2O(l) \rightleftharpoons HA(aq) + OH^-(aq)$$

In the forward reaction, the anion, A^-, acquires a proton from the water molecule to form the weak acid, HA, and hydroxide ion, OH^-. The extent of OH^- ion formation and the position of the equilibrium depends on the relative strength of the anion, A^-. The lower the K_a value of HA, the stronger the attraction for protons that A^- will have compared with OH^-, and the greater the production of OH^- ion will be. Therefore, as the relative strength of A^- increases, the equilibrium position lies farther to the right.

Aqueous solutions of sodium carbonate are strongly basic. The sodium ions, Na^+, in sodium carbonate do not undergo hydrolysis in aqueous solution, but the carbonate ions, CO_3^{2-}, react as a Brønsted base. A CO_3^{2-} anion acquires a proton from a water molecule to form the slightly ionized hydrogen carbonate ion, HCO_3^-, and the OH^- ion.

$$CO_3^{2-}(aq) + H_2O(l) \rightleftharpoons HCO_3^-(aq) + OH^-(aq)$$

The OH^- ion concentration increases until equilibrium is established. Consequently, the H_3O^+ ion concentration decreases so that the product $[H_3O^+][OH^-]$ remains equal to the ionization constant, K_w, of water at the temperature of the solution. Thus, the pH is *higher* than 7, and the solution is basic.

Cation Hydrolysis

In the Brønsted sense, the cation of the salt is the conjugate acid of the base from which it was formed. It is also a proton donor. If the base is weak, the cation is an acid strong enough to donate a proton to a water molecule, a proton acceptor, to form H_3O^+ ions. An equilibrium is established in which the net effect of the cation hydrolysis is an increase in the hydronium-ion concentration, $[H_3O^+]$, of the solution.

The following equilibrium equation for a typical weak base, B, is used to derive the generalized expression for K_b, the base dissociation constant.

$$B(aq) + H_2O(l) \rightleftharpoons BH^+(aq) + OH^-(aq)$$

$$K_b = \frac{[BH^+][OH^-]}{[B]}$$

The hydrolysis reaction between water and the cation, BH^+, produced by the dissociation of the weak base, B, is represented by the general equilibrium equation that follows.

$$BH^+(aq) + H_2O(l) \rightleftharpoons H_3O^+(aq) + B(aq)$$

In the forward reaction, the cation BH^+ donates a proton to the water molecule to form the hydronium ion and the weak base, B. The extent of H_3O^+ ion formation and the position of the equilibrium depend on the relative strength of the cation, BH^+. The lower the K_b value of B, the stronger the donation of protons that BH^+ will have compared with H_3O^+, and the greater the production of H_3O^+ ions will be. Therefore, as the relative strength of BH^+ increases, the equilibrium position lies farther to the right.

Ammonium chloride, NH_4Cl, dissociates in water to produce NH_4^+ ions, Cl^- ions, and an acidic solution. Chloride ions are the conjugate base of a strong acid, HCl, so they show no noticeable tendency to hydrolyze in aqueous solution. Ammonium ions, however, are the conjugate acid of a weak base, NH_3. Ammonium ions donate protons to water molecules. Equilibrium is established with an increased $[H_3O^+]$, so the pH is *lower* than 7.

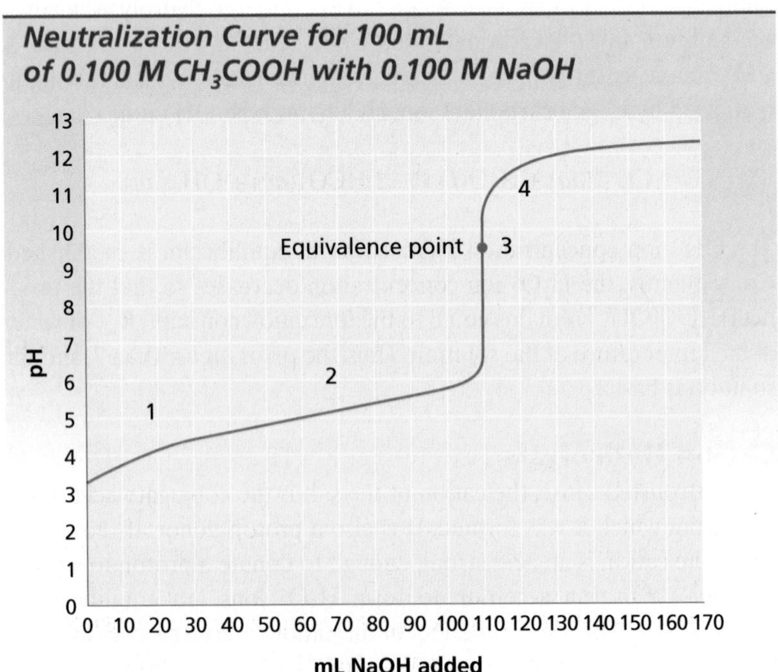

FIGURE 18-11 At point *1* on the titration curve, only acetic acid is present. The pH depends on the weak acid alone. At *2* there is a mixture of CH_3COOH and CH_3COO^-. Adding NaOH changes the pH slowly. At point *3* all acid has been converted to CH_3COO^-. This hydrolyzes to produce a slightly basic solution. At *4* the pH is due to the excess OH^- that has been added.

Hydrolysis in Acid-Base Reactions

Hydrolysis can help explain why the end point of a neutralization reaction can occur at a pH other than 7. The hydrolysis properties of salts are determined by the relative strengths of the acids and bases from which the salts were formed. Salts can be placed in four general categories, depending on their hydrolysis properties: strong acid–strong base, strong acid–weak base, weak acid–strong base, and weak acid–weak base.

Salts of strong acids and strong bases produce neutral solutions because neither the cation of a strong base nor the anion of a strong acid hydrolyze appreciably in aqueous solutions. HCl(*aq*) is a strong acid, and NaOH(*aq*) is a strong base. Neither the Na^+ cation of the strong base nor the Cl^- anion of the strong acid undergoes hydrolysis in water solutions. Therefore, aqueous solutions of NaCl are neutral. Similarly, KNO_3 is the salt of the strong acid HNO_3 and the strong base KOH. Measurements show that the pH of an aqueous KNO_3 solution is always very close to 7.

The aqueous solutions of salts formed from reactions between weak acids and strong bases are basic, as Figure 18-11 shows. Anions of the dissolved salt are hydrolyzed in the water solvent, and the pH of the solution is raised, indicating that the hydroxide-ion concentration has increased. Aqueous solutions of sodium acetate, $NaCH_3COO$, are basic. The acetate ions, CH_3COO^-, undergo hydrolysis because they are the anions of the weak acid–acetic acid. The cations of the salt are the positive ions from a strong base, NaOH, and do not hydrolyze appreciably.

Figure 18-12 shows that salts of strong acids and weak bases produce acidic aqueous solutions. Cations of the dissolved salt are hydrolyzed in

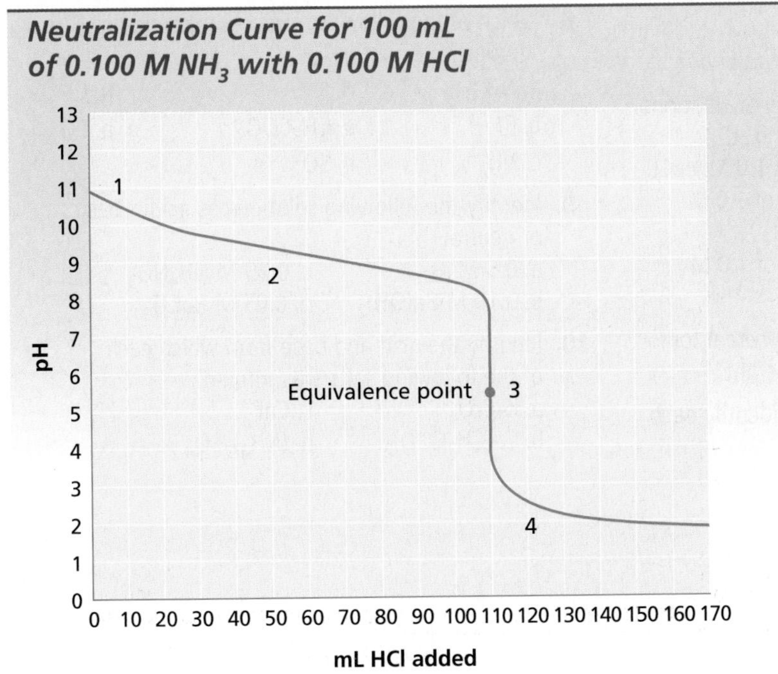

FIGURE 18-12 At point *1* on the titration curve, only aqueous ammonia is present. The pH is determined by the base alone. At *2* there is a mixture of NH_3 and NH_4^+. Adding HCl changes the pH slowly. At point *3* all aqueous ammonia has been converted to NH_4^+. At *4* the pH is determined by the excess H_3O^+ that is being added.

the water solvent, and the pH of the solution is lowered, indicating that the hydronium-ion concentration has increased. In this case, the cations of the salt undergo hydrolysis because they are the positive ions from a weak base. The anions of the salt are the negative ions from a strong acid and do not hydrolyze appreciably. Ammonium chloride, NH_4Cl, is a salt that produces an acidic solution.

Salts of weak acids and weak bases can produce either acidic, neutral, or basic aqueous solutions, depending on the salt dissolved. This is because both ions of the dissolved salt are hydrolyzed extensively. If both ions are hydrolyzed equally, the solution remains neutral. The ions in ammonium acetate, NH_4CH_3COO, hydrolyze equally, producing a neutral solution, as shown in Figure 18-10(d) on page 572.

In cases in which both the acid and the base are very weak, the salt may undergo essentially complete decomposition to the products of hydrolysis. For example, when aluminum sulfide is placed in water, both a precipitate and a gas are formed. The reaction is symbolized by the following chemical equation.

$$Al_2S_3(s) + 6H_2O(l) \longrightarrow 2Al(OH)_3(s) + 3H_2S(g)$$

Both products are sparingly soluble in water and are removed from solution.

SECTION REVIEW

1. What is meant by an *acid-ionization constant*?
2. How is an acid-ionization equilibrium expression written?
3. What is meant by the term *buffered solution*?
4. Which of the following combinations of solutions would form buffers when they are mixed?
   a. 50 mL of 1.0 M HCl and 50 mL of 1.0 M NaCl
   b. 25 mL of 0.5 M HNO_2 and 50 mL of 1.0 M $NaNO_2$
   c. 25 mL of 1.0 M HNO_2 and 25 mL of 1.0 M NaCl
5. What is meant by the *ion-product constant* for water? What is the value of this constant?
6. For each of the following reactions, identify each conjugate acid-base pair.
   a. $H_2CO_3 + H_2O \rightleftharpoons HCO_3^- + H_3O^+$
   b. $H_2O + H_2O \rightleftharpoons H_3O^+ + OH^-$
   c. $H_2S + NH_3 \rightleftharpoons HS^- + NH_4^+$
   d. $H_2PO_4^- + H_2O \rightleftharpoons H_3PO_4 + OH^-$
7. What is hydrolysis? Compare cation and anion hydrolysis.
8. Which of the following ions hydrolyze in aqueous solution?
   a. NO_3^-
   b. F^-
   c. NH_4^+
   d. K^+
   e. CH_3COO^-
   f. SO_4^{2-}
   g. CO_3^{2-}
   h. PO_4^{3-}
9. Identify the following solutions as acidic, basic, or neutral.
   a. 0.5 M KI
   b. 0.10 M $Ba(OH)_2$
   c. 0.25 M NH_4NO_3
   d. 0.05 M $BaCO_3$
10. Identify the acid and base from which each of the following salts was formed.
    a. K_2CrO_4
    b. $Ca(CH_3COO)_2$
    c. CaF_2
    d. $(NH_4)_2SO_4$

Solubility Equilibrium

SECTION 18-4

Objectives

- Explain what is meant by *solubility-product constants,* and calculate their values.
- Calculate solubilities using solubility-product constants.
- Carry out calculations to predict whether precipitates will form when solutions are combined.

Ionic solids dissolve in water until they are in equilibrium with their ions. An equilibrium expression can be written from the balanced chemical equation of the solid's dissociation. Concentrations of the ions can be determined from the balanced chemical equation and solubility data. The ion concentrations can then be used to determine the value of the equilibrium constant. The numerical value for the equilibrium constant can be used to predict whether precipitation occurs when solutions of various concentrations are combined.

Solubility Product

A saturated solution contains the maximum amount of solute possible at a given temperature in equilibrium with an undissolved excess of the substance. A saturated solution is not necessarily a concentrated solution. The concentration may be high or low, depending on the solubility of the solute.

A general rule is often used to express solubilities qualitatively. By this rule, a substance is said to be *soluble* if the solubility is *greater than* 1 g per 100 g of water. It is said to be *insoluble* if the solubility is *less than* 0.1 g per 100 g of water. Substances whose solubilities fall between these limits are described as *slightly soluble.*

The equilibrium principles developed in this chapter apply to all saturated solutions of sparingly soluble salts. Silver chloride is so sparingly soluble in water that it is sometimes described as insoluble. Its solution reaches saturation at a very low concentration of its ions. All Ag^+ and Cl^- ions in excess of this concentration eventually precipitate as AgCl.

Consider the equilibrium system in a saturated solution of silver chloride containing an excess of the solid salt. This system is represented by the following equilibrium equation and equilibrium-constant expression.

$$AgCl(s) \rightleftharpoons Ag^+(aq) + Cl^-(aq)$$

$$K = \frac{[Ag^+][Cl^-]}{[AgCl]}$$

Because the concentration of a pure substance in the solid or liquid phase remains constant, adding more solid AgCl to this equilibrium system does not change the concentration of the undissolved AgCl present. Thus, [AgCl] does not appear in the final expression. Rearranging the equilibrium expression so that both constants are on the same side of the equation gives the solubility-product constant K_{sp}. *The* **solubility-product constant** *of a substance is the product of the molar concentrations of its ions in a saturated solution, each raised to the power that is the coefficient of that ion in the chemical equation.*

$$K[\text{AgCl}] = [\text{Ag}^+][\text{Cl}^-]$$
$$K_{sp} = [\text{Ag}^+][\text{Cl}^-]$$

This equation is the solubility-equilibrium expression for the reaction. It expresses the fact that the solubility-product constant, K_{sp}, of AgCl is the product of the molar concentrations of its ions in a saturated solution.

Calcium fluoride is another sparingly soluble salt. The equilibrium in a saturated CaF_2 solution is described by the following equation.

$$\text{CaF}_2(s) \rightleftarrows \text{Ca}^{2+}(aq) + 2\text{F}^-(aq)$$

The solubility-product constant has the following form.

$$K_{sp} = [\text{Ca}^{2+}][\text{F}^-]^2$$

Notice that this constant is the product of the molar concentration of Ca^{2+} ions and the molar concentration of F^- ions squared, as required by the general chemical equilibrium expression.

The numerical value of K_{sp} can be determined from solubility data. Data listed in Appendix Table A-13 indicate that a maximum of 8.9×10^{-5} g of AgCl can dissolve in 100. g of water at 10°C. One mole of AgCl has a mass of 143.32 g. The saturation concentration, or solubility, of AgCl can therefore be expressed in moles per liter of water, which is very nearly equal to moles per liter of solution.

$$\frac{8.9 \times 10^{-5}\text{ g AgCl}}{100.\text{ g H}_2\text{O}} \times \frac{1\text{ g H}_2\text{O}}{\text{mL H}_2\text{O}} \times \frac{1000\text{ mL}}{\text{L}} \times \frac{1\text{ mol AgCl}}{143.32\text{ g AgCl}}$$
$$= 6.2 \times 10^{-6}\text{ mol/L}$$

Silver chloride dissociates in solution, contributing equal numbers of Ag^+ and Cl^- ions. The ion concentrations in the saturated solution are therefore 6.2×10^{-6} mol/L.

$$[\text{Ag}^+] = 6.2 \times 10^{-6}$$
$$[\text{Cl}^-] = 6.2 \times 10^{-6}$$

and

$$K_{sp} = [\text{Ag}^+][\text{Cl}^-]$$
$$K_{sp} = (6.2 \times 10^{-6})(6.2 \times 10^{-6})$$
$$K_{sp} = (6.2 \times 10^{-6})^2$$
$$K_{sp} = 3.8 \times 10^{-11}$$

This result is the solubility-product constant of AgCl at 10°C.

From Appendix Table A-13, the solubility of CaF_2 is 1.7×10^{-3} g/100 g of water at 26°C. Expressed in moles per liter, as before, this concentration becomes 2.2×10^{-4} mol/L. CaF_2 dissociates in solution to yield twice as many F^- ions as Ca^{2+} ions. The ion concentrations in the saturated solution are 2.2×10^{-4} for the calcium ion and $2(2.2 \times 10^{-4})$, or 4.4×10^{-4}, for the fluoride ion. Note that at equilibrium at 26°C, $[Ca^{2+}]$ equals the solubility of 2.2×10^{-4} mol/L but $[F^-]$ equals twice the solubility, or 4.4×10^{-4} mol/L. The number of moles of positive and negative ions per mole of compound must always be accounted for when using K_{sp} and solubilities.

$$K_{sp} = [Ca^{2+}][F^-]^2$$
$$K_{sp} = (2.2 \times 10^{-4})(4.4 \times 10^{-4})^2$$
$$K_{sp} = 4.3 \times 10^{-11}$$

Thus, the solubility-product constant of CaF_2 is 4.3×10^{-11} at 26°C.

It is difficult to measure very small concentrations of a solute with precision. For this reason, solubility data from different sources may report different values of K_{sp} for a substance. Thus, calculations of K_{sp} ordinarily should be limited to two significant figures. Representative values of K_{sp} at 25°C for some sparingly soluble compounds are listed in Table 18-3. Assume that all data used in K_{sp} calculations have been taken at 25°C unless otherwise specified.

At this point, you should note the difference between the solubility of a given solid and its solubility-product constant. Remember that the

TABLE 18-3 *Solubility-Product Constants, K_{sp}, at 25°C*

Salt	Ion product	K_{sp}	Salt	Ion product	K_{sp}
$AgCH_3COO$	$[Ag^+][CH_3COO^-]$	1.9×10^{-3}	$CuCl$	$[Cu^+][Cl^-]$	1.2×10^{-6}
$AgBr$	$[Ag^+][Br^-]$	5.0×10^{-13}	CuS	$[Cu^{2+}][S^{2-}]$	6.3×10^{-36}
Ag_2CO_3	$[Ag^+]^2[CO_3^{2-}]$	8.1×10^{-12}	FeS	$[Fe^{2+}][S^{2-}]$	6.3×10^{-18}
$AgCl$	$[Ag^+][Cl^-]$	1.8×10^{-10}	$Fe(OH)_2$	$[Fe^{2+}][OH^-]^2$	8.0×10^{-16}
AgI	$[Ag^+][I^-]$	8.3×10^{-17}	$Fe(OH)_3$	$[Fe^{3+}][OH^-]^3$	4×10^{-38}
Ag_2S	$[Ag^+]^2[S^{2-}]$	6.3×10^{-50}	HgS	$[Hg^{2+}][S^{2-}]$	1.6×10^{-52}
$Al(OH)_3$	$[Al^{3+}][OH^-]^3$	1.3×10^{-33}	$MgCO_3$	$[Mg^{2+}][CO_3^{2-}]$	3.5×10^{-8}
$BaCO_3$	$[Ba^{2+}][CO_3^{2-}]$	5.1×10^{-9}	$Mg(OH)_2$	$[Mg^{2+}][OH^-]^2$	1.8×10^{-11}
$BaSO_4$	$[Ba^{2+}][SO_4^{2-}]$	1.1×10^{-10}	MnS	$[Mn^{2+}][S^{2-}]$	2.5×10^{-13}
CdS	$[Cd^{2+}][S^{2-}]$	8.0×10^{-27}	$PbCl_2$	$[Pb^{2+}][Cl^-]^2$	1.6×10^{-5}
$CaCO_3$	$[Ca^{2+}][CO_3^{2-}]$	2.8×10^{-9}	$PbCrO_4$	$[Pb^{2+}][CrO_4^{2-}]$	2.8×10^{-13}
CaF_2	$[Ca^{2+}][F^-]^2$	5.3×10^{-9}	$PbSO_4$	$[Pb^{2+}][SO_4^{2-}]$	1.6×10^{-8}
$Ca(OH)_2$	$[Ca^{2+}][OH^-]^2$	5.5×10^{-6}	PbS	$[Pb^{2+}][S^{2-}]$	8.0×10^{-28}
$CaSO_4$	$[Ca^{2+}][SO_4^{2-}]$	9.1×10^{-6}	SnS	$[Sn^{2+}][S^{2-}]$	1.0×10^{-25}
$CoCO_3$	$[Co^{2+}][CO_3^{2-}]$	1.4×10^{-13}	$SrSO_4$	$[Sr^{2+}][SO_4^{2-}]$	3.2×10^{-7}
CoS	$[Co^{2+}][S^{2-}]$	4.0×10^{-21}	ZnS	$[Zn^{2+}][S^{2-}]$	1.6×10^{-24}

solubility-product constant is an equilibrium constant representing the product of the molar concentrations of its ions in a saturated solution. It has only one value for a given solid at a given temperature. The *solubility* of a solid is an equilibrium position that represents the amount of the solid required to form a saturated solution with a specific amount of solvent. It has an infinite number of possible values at a given temperature and is dependent on other conditions, such as the presence of a common ion.

SAMPLE PROBLEM 18-2

Calculate the solubility-product constant, K_{sp}, for copper(I) chloride, CuCl, given that the solubility of this compound at 25°C is 1.08×10^{-2} g/100. g H_2O.

SOLUTION

1 ANALYZE

Given: solubility of CuCl = 1.08×10^{-2} g CuCl/100. g H_2O

Unknown: K_{sp}

2 PLAN

Start by converting the solubility of CuCl in g/100. g H_2O to mol/L. You will need the molar mass of CuCl to get moles CuCl from grams CuCl. Then use the solubility of the $[Cu^+]$ and $[Cl^-]$ ions in the K_{sp} expression and solve for K_{sp}.

$$\frac{\text{g CuCl}}{\text{100. g } H_2O} \times \frac{\text{1 g } H_2O}{\text{1 mL } H_2O} \times \frac{\text{1000 mL}}{\text{1 L}} \times \frac{\text{1 mol CuCl}}{\text{g CuCl}} = \text{solubility in mol/L}$$

$$CuCl(s) \rightleftarrows Cu^+(aq) + Cl^-(aq)$$

$$K_{sp} = [Cu^+][Cl^-]$$

$$[Cu^+] = [Cl^-] = \text{solubility in mol/L}$$

3 COMPUTE

The molar mass of CuCl is 99.0 g/mol.

$$\text{solubility} = \frac{1.08 \times 10^{-2} \text{ g CuCl}}{\text{100. g } H_2O} \times \frac{\text{1 g } H_2O}{\text{1 mL}} \times \frac{\text{1000 mL}}{\text{L}} \times \frac{\text{1 mol CuCl}}{\text{99.0 g CuCl}} =$$

$$1.09 \times 10^{-3} \text{ mol/L CuCl}$$

$$[Cu^+] = [Cl^-] = 1.09 \times 10^{-3} \text{ mol/L}$$

$$K_{sp} = (1.09 \times 10^{-3})(1.09 \times 10^{-3}) = 1.19 \times 10^{-6}$$

4 EVALUATE

The answer contains the proper number of significant digits and is close to the K_{sp} value given in Table 18-3.

PRACTICE

1. Calculate the solubility-product constant, K_{sp}, of lead(II) chloride, $PbCl_2$, which has a solubility of 1.0 g/100. g H_2O at a temperature other than 25°C.
   Answer 1.9×10^{-4}

2. Five grams of Ag_2SO_4 will dissolve in 1 L of water. Calculate the solubility product constant for this salt.
   Answer 2×10^{-5}

Calculating Solubilities

Once known, the solubility-product constant can be used to determine the solubility of a sparingly soluble salt. Suppose you wish to know how much barium carbonate, $BaCO_3$, can be dissolved in 1 L of water at 25°C. From Table 18-3, K_{sp} for $BaCO_3$ has the numerical value 5.1×10^{-9}. The equilibrium equation is written as follows.

$$BaCO_3(s) \rightleftharpoons Ba^{2+}(aq) + CO_3^{2-}(aq)$$

Given the value for K_{sp}, we can write the solubility-equilibrium expression as follows.

$$K_{sp} = [Ba^{2+}][CO_3^{2-}] = 5.1 \times 10^{-9}$$

Therefore, $BaCO_3$ dissolves until the product of the molar concentrations of Ba^{2+} ions and CO_3^{2-} ions equals 5.1×10^{-9}. The solubility-equilibrium equation shows that Ba^{2+} ions and CO_3^{2-} ions enter the solution in equal numbers as the salt dissolves. Thus, they have the same concentration. Let $[Ba^{2+}] = x$. Then $[CO_3^{2-}] = x$ also.

$$[Ba^{2+}][CO_3^{2-}] = K_{sp} = 5.1 \times 10^{-9}$$

$$(x)(x) = 5.1 \times 10^{-9}$$

$$x = \sqrt{5.1 \times 10^{-9}}$$

The solubility of $BaCO_3$ is 7.14×10^{-5} mol/L.

Thus, the solution concentration is 7.14×10^{-5} M for Ba^{2+} ions and 7.14×10^{-5} M for CO_3^{2-} ions.

SAMPLE PROBLEM 18-3

Calculate the solubility of silver acetate, $AgCH_3COO$, in mol/L, given the K_{sp} value for this compound listed in Table 18-3.

SOLUTION

1 *ANALYZE*

Given: $K_{sp} = 1.9 \times 10^{-3}$

Unknown: solubility of $AgCH_3COO$

2 *PLAN*

$$AgCH_3COO \rightleftharpoons Ag^+(aq) + CH_3COO^-(aq)$$

$$K_{sp} = [Ag^+][CH_3COO^-]$$

$[Ag^+] = [CH_3COO^-]$, so let $[Ag^+] = x$ and $[CH_3COO^-] = x$

3 *COMPUTE*

$$K_{sp} = [Ag^+][CH_3COO^-]$$

$$K_{sp} = x^2$$

$$x^2 = 1.9 \times 10^{-3}$$

$$x = \sqrt{1.9 \times 10^{-3}}$$

Solubility of $AgCH_3COO = \sqrt{1.9 \times 10^{-3}} = 4.4 \times 10^{-2}$ mol/L

4 EVALUATE The answer has the proper number of significant digits and is close to an estimated value of 5.0×10^{-2} calculated as $\sqrt{2.5 \times 10^{-3}}$.

PRACTICE

1. Calculate the solubility of cadmium sulfide, CdS, in mol/L, given the K_{sp} value listed in Table 18-3. *Answer* 8.9×10^{-14} mol/L
2. Determine the concentration of strontium ions in a saturated solution of strontium sulfate, $SrSO_4$, if the K_{sp} for $SrSO_4$ is 3.2×10^{-7}. *Answer* 5.7×10^{-4} mol/L

Precipitation Calculations

In an earlier example, $BaCO_3$ served as the source of both Ba^{2+} and CO_3^{2-} ions. Because each mole of $BaCO_3$ yields one mole of Ba^{2+} ions and one mole of CO_3^{2-} ions, the concentrations of the two ions were equal. However, the equilibrium condition does not require that the two ion concentrations be equal. Equilibrium will still be established so that the ion product $[Ba^{2+}][CO_3^{2-}]$ does not exceed the value of K_{sp} for the system.

Similarly, if the ion product $[Ca^{2+}][F^-]^2$ is less than the value of K_{sp} at a particular temperature, the solution is unsaturated. If the ion product is greater than the value for K_{sp}, CaF_2 precipitates. This precipitation reduces the concentrations of Ca^{2+} and F^- ions until equilibrium is established.

Suppose that unequal quantities of $BaCl_2$ and Na_2CO_3 are dissolved in water and that the solutions are mixed. If the ion product $[Ba^{2+}][CO_3^{2-}]$ exceeds the K_{sp} of $BaCO_3$, a precipitate of $BaCO_3$ forms. Precipitation continues until the ion concentrations decrease to the point at which $[Ba^{2+}][CO_3^{2-}]$ equals the K_{sp}.

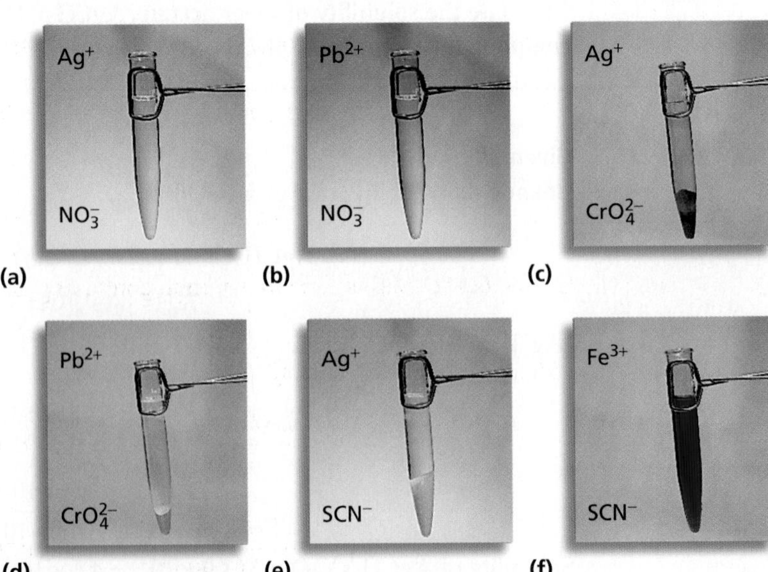

FIGURE 18-13 Nitrate salts of Ag^+ (a) and Pb^{2+} (b) are soluble. When chromate ions, CrO_4^{2-}, combine with Ag^+ (c) or Pb^{2+} (d), an insoluble salt forms. Thiocyanate ions, SCN^-, can form an insoluble salt with Ag^+ (e) or a soluble salt with Fe^{3+} (f).

Substances differ greatly in their tendencies to form precipitates when mixed in moderate concentrations. The photos in Figure 18-13 show the behavior of some negative ions in the presence of certain metallic ions. Note that some of the combinations have produced precipitates and some have not. The solubility product can be used to predict whether a precipitate forms when two solutions are mixed.

SAMPLE PROBLEM 18-4

Will a precipitate form if 20.0 mL of 0.010 M $BaCl_2$ is mixed with 20.0 mL of 0.0050 M Na_2SO_4?

SOLUTION

1 ANALYZE

Given: concentration of $BaCl_2$ = 0.010 M
volume of $BaCl_2$ = 20.0 mL
concentration of Na_2SO_4 = 0.0050 M
volume of Na_2SO_4 = 20.0 mL

Unknown: whether a precipitate forms

2 PLAN

The two possible new pairings of ions are NaCl and $BaSO_4$. Of these, $BaSO_4$ is a sparingly soluble salt. It will precipitate if the ion product $[Ba^{2+}][SO_4^{2-}]$ in the combined solution exceeds K_{sp} for $BaSO_4$. From the list of solubility products in Table 18-3, the K_{sp} is found to be 1.1×10^{-10}. The solubility-equilibrium equation follows.

$$BaSO_4(s) \rightleftharpoons Ba^{2+}(aq) + SO_4^{2-}(aq)$$

The solubility-equilibrium expression is written as follows.

$$K_{sp} = [Ba^{2+}][SO_4^{2-}] = 1.1 \times 10^{-10}$$

First $[Ba^{2+}]$ and $[SO_4^{2-}]$ in the above solution must be found. Then the ion product is calculated and compared with the K_{sp}.

3 COMPUTE

Calculate the mole quantities of Ba^{2+} and SO_4^{2-} ions.

$$0.020\ \cancel{L} \times \frac{0.010\ \text{mol}\ Ba^{2+}}{\cancel{L}} = 0.000\ 20\ \text{mol}\ Ba^{2+}$$

$$0.020\ \cancel{L} \times \frac{0.0050\ \text{mol}\ SO_4^{2-}}{\cancel{L}} = 0.000\ 10\ \text{mol}\ SO_4^{2-}$$

Calculate the total volume of solution containing Ba^{2+} and SO_4^{2-} ions.

$$0.020\ \text{L} + 0.020\ \text{L} = 0.040\ \text{L}$$

Calculate the Ba^{2+} and SO_4^{2-} ion concentrations in the combined solution.

$$\frac{0.000\ 20\ \text{mol}\ Ba^{2+}}{0.040\ \text{L}} = 5.0 \times 10^{-3}\ \text{mol/L}\ Ba^{2+}$$

$$\frac{0.000\ 10\ \text{mol}\ SO_4^{2-}}{0.040\ \text{L}} = 2.5 \times 10^{-3}\ \text{mol/L}\ SO_4^{2-}$$

Trial value of the ion product:

$$[Ba^{2+}][SO_4^{2-}] = (5.0 \times 10^{-3})(2.5 \times 10^{-3})$$
$$= 1.2 \times 10^{-5}$$

The ion product is much greater than the value of K_{sp}, so precipitation occurs.

4 EVALUATE The answer contains the appropriate number of significant digits and is close to an estimated value of 1×10^{-5}, calculated as $(5 \times 10^{-3})(2 \times 10^{-3})$; because $10^{-5} > 10^{-10}$, precipitation should occur.

PRACTICE

1. Does a precipitate form when 100. mL of 0.0025 M $AgNO_3$ and 150. mL of 0.0020 M NaBr solutions are mixed?

   Answer
   AgBr precipitates.

2. Does a precipitate form when 20. mL of 0.038 M $Pb(NO_3)_2$ and 30. mL of 0.018 M KCl solutions are mixed?

   Answer
   $PbCl_2$ does *not* precipitate.

Limitations on the Use of K_{sp}

The solubility-product principle can be very useful when applied to solutions of sparingly soluble substances. It *cannot* be applied very successfully to solutions of moderately soluble or very soluble substances. This is because the positive and negative ions attract each other, and this attraction becomes appreciable when the ions are close together. Sometimes it is necessary to consider two equilibria simultaneously. For example, if either ion hydrolyzes, the salt will be more soluble than predicted when only the solubility-product constant is used. The solubility product is also sensitive to changes in solution temperature to the extent that the solubility of the dissolved substance is affected by such changes. All of these factors limit the conditions under which the solubility-product principle can be applied.

SECTION REVIEW

1. What is a solubility-product constant? How are such constants determined?
2. How are solubility-product constants used to calculate solubilities?
3. What is an ion product?
4. How are calculations to predict possible precipitation carried out?
5. What is the value of K_{sp} for Ag_2SO_4 if 5.40 g is soluble in 1.00 L of water?
6. Determine whether a precipitate will form if 20.0 mL of 1.00×10^{-7} M $AgNO_3$ is mixed with 20.0 mL of 2.00×10^{-9} M NaCl at 25°C.

CHAPTER 18 REVIEW

CHAPTER SUMMARY

18-1

- A reaction system in which the forward and reverse reactions occur simultaneously and at the same rate is said to be in *equilibrium.* Both reactions continue, but there is no net change in the composition of the system.
- At equilibrium, the ratio of the product of the mole concentrations of substances formed to the product of the mole concentrations of reactants, each raised to the appropriate power, has a definite numerical value, *K,* which is the equilibrium constant at a given temperature. For values of *K* greater than 1, the products of the forward reaction are favored. For values of *K* less than 1, the products of the reverse reaction are favored.

Vocabulary

chemical equilibrium (554)
chemical-equilibrium expression (556)
equilibrium constant (556)
reversible reaction (553)

18-2

- Any change that alters the rate of either the forward or reverse reaction disturbs the equilibrium of the system. According to Le Châtelier's principle, the equilibrium is shifted in the direction that relieves the stress.
- Catalysts increase the rates of forward and reverse reactions equally, and they do not shift an equilibrium or change the value of *K*.
- The common-ion effect is recognized when a solution containing ions like those of a reactant in an equilibrium system is added to the system. Le Châtelier's principle explains the response of the system to the stress.

Vocabulary

common-ion effect (567)

18-3

- The equilibrium expression for the ionization constant of the weak acid HA follows.

$$K_a = \frac{[H_3O^+][A^-]}{[HA]}$$

- Salts formed from strong bases and weak acids produce aqueous solutions that are basic because of *anion hydrolysis.*
- Salts formed from strong acids and weak bases produce aqueous solutions that are acidic because of *cation hydrolysis.*
- Salts formed from strong acids and strong bases do not hydrolyze in water, and their solutions are neutral.
- Salts formed from weak acids and weak bases may produce neutral, acidic, or basic solutions, depending on the relative amounts of cation and anion hydrolysis. They may also hydrolyze completely in water solution.

Vocabulary

acid-ionization constant (569)
buffered solution (570)
hydrolysis (572)

18-4

- Ions of salts that are very sparingly soluble form saturated aqueous solutions at low concentrations. The solubility-equilibrium expression for such salts yields a useful constant—the solubility-product constant, K_{sp}. The value of K_{sp} equals the product of the molar concentrations of solute ions in the saturated solution raised to a power equal to the coefficient in the balanced equation for the solution of one mole.

Vocabulary

solubility-product constant (578)

REVIEWING CONCEPTS

1. Describe and explain how the concentrations of A, B, C, and D change from the time when A and B are first combined to the point at which equilibrium is established for the reaction $A + B \rightleftharpoons C + D$. (18-1)
2. a. Write the general expression for an equilibrium constant based on the equation $nA + mB + \ldots \rightleftharpoons xC + yD + \ldots$
   b. What information is provided by the value of K for a given equilibrium system at a specified temperature? (18-1)
3. In general, which reaction is favored (forward, reverse, or neither) if the value of K at a specified temperature is
   a. equal to 1.
   b. very small.
   c. very large. (18-1)
4. Predict whether each of the following pressure changes would favor the forward or reverse reaction.

   $$2NO(g) + O_2(g) \rightleftharpoons 2NO_2(g)$$

   a. increased pressure
   b. decreased pressure (18-2)
5. In heterogeneous reaction systems, what types of substances do not appear in the equilibrium constant expression? Why? (18-2)
6. Explain the effect of a catalyst on an equilibrium system. (18-2)
7. Predict the effect of each of the following on the indicated equilibrium system in terms of which reaction will be favored (forward, reverse, or neither).

   $$H_2(g) + Cl_2(g) \rightleftharpoons 2HCl(g) + 184 \text{ kJ}$$

   a. addition of Cl_2
   b. removal of HCl
   c. increased pressure
   d. decreased temperature
   e. removal of H_2
   f. decreased pressure
   g. addition of a catalyst
   h. increased temperature
   i. decreased system volume (18-2)
8. How would parts (a) through (i) of item 7 affect the new equilibrium concentration of HCl and the value of K at the new equilibrium? (18-2)
9. Changes in the concentrations of the reactants and products at equilibrium have no impact on the value of the equilibrium constant. Explain. (18-2)
10. What relative pressure (high or low) would result in the production of the maximum level of CO_2 according to the following? Explain. (18-2)

    $$2CO(g) + O_2(g) \rightleftharpoons 2CO_2(g)$$

11. What relative conditions (reactant concentrations, pressure, and temperature) would favor a high equilibrium concentration of the underlined substance in each of the following equilibrium systems?
    a. $2CO(g) + O_2(g) \rightleftharpoons \underline{2CO_2(g)} + 167 \text{ kJ}$
    b. $Cu^{2+}(aq) + 4NH_3(aq) \rightleftharpoons \underline{Cu(NH_3)_4^{2+}(aq)} + 42 \text{ kJ}$
    c. $2HI(g) + 12.6 \text{ kJ} \rightleftharpoons H_2(g) + \underline{I_2(g)}$
    d. $4HCl(g) + O_2(g) \rightleftharpoons 2H_2O(g) + \underline{2Cl_2(g)} + 113 \text{ kJ}$
    e. $H_2O(l) + 42 \text{ kJ} \rightleftharpoons \underline{H_2O(g)}$ (18-2)
12. A combustion reaction proceeding in air under standard pressure is transferred to an atmosphere of pure oxygen under the same pressure.
    a. What effect would you observe?
    b. How can you account for this effect? (18-2)
13. What two factors determine the extent to which reacting ions are removed from solution? (18-2)
14. Identify the three conditions under which ionic reactions can run to completion, and write an equation for each. (18-2)
15. a. Write the ion-product constant expression for water.
    b. What is the value of this constant at 25°C? (18-3)
16. List and distinguish between the four general categories of salts, based on their hydrolysis properties, and give an example of each. (18-3)

17. The pH of a solution containing both acetic acid and sodium acetate is higher than that of a solution containing the same concentration of acetic acid alone. Explain. (18-3)

18. The ionization constant, K_a, for acetic acid is 1.8×10^{-5} at 25°C. Explain the significance of this value. (18-3)

19. a. From the development of K_a described in Section 18-3, show how you would express an ionization constant, K_b, for the weak base NH_3.
b. In this case, $K_b = 1.8 \times 10^{-5}$. What is the significance of this numerical value to equilibrium? (18-3)

20. A saturated solution is not necessarily a concentrated solution. Explain. (18-4)

21. What rule of thumb is used to distinguish between soluble, insoluble, and slightly soluble substances? (18-4)

22. What is the major solubility characteristic of those types of substances typically involved in solubility-equilibrium systems? (18-4)

23. What is the relationship between K_{sp} and the product of the ion concentrations in terms of determining whether a solution of those ions is saturated? (18-4)

PROBLEMS

Equilibrium Constant

24. Determine the value of the equilibrium constant for each reaction given, assuming that the equilibrium concentrations are found to be those specified. (Concentrations are in mol/L.) (Hint: See Sample Problem 18-1.)
a. $A + B \rightleftharpoons C$; [A] = 2.0; [B] = 3.0; [C] = 4.0
b. $D + 2E \rightleftharpoons F + 3G$; [D] = 1.5; [E] = 2.0; [F] = 1.8; [G] = 1.2
c. $N_2(g) + 3H_2(g) \rightleftharpoons 2NH_3(g)$; $[N_2] = 0.45$; $[H_2] = 0.14$; $[NH_3] = 0.62$

25. An equilibrium mixture at a specific temperature is found to consist of 1.2×10^{-3} mol/L HCl, 3.8×10^{-4} mol/L O_2, 5.8×10^{-2} mol/L H_2O, and 5.8×10^{-2} mol/L Cl_2 according to the following: $4HCl(g) + O_2(g) \rightleftharpoons 2H_2O(g) + 2Cl_2(g)$. Determine the value of the equilibrium constant for this system.

26. At 450°C the value of the equilibrium constant for the following system is 6.59×10^{-3}. If $[NH_3] = 1.23 \times 10^{-4}$ and $[H_2] = 2.75 \times 10^{-3}$ at equilibrium, determine the concentration of N_2 at that point.

$$N_2(g) + 3H_2(g) \rightleftharpoons 2NH_3(g)$$

27. The value of the equilibrium constant for the reaction below is 40.0 at a specified temperature. What would be the value of that constant for the reverse reaction under the same conditions? $H_2(g) + I_2(g) \rightleftharpoons 2HI(g)$

Solubility-Product Constant

28. The ionic substance EJ dissociates to form E^{2+} and J^{2-} ions. The solubility of EJ is 8.45×10^{-6} mol/L. What is the value of the solubility-product constant? (Hint: See Sample Problem 18-2.)

29. Calculate the solubility-product constant K_{sp} for each of the following, based on the solubility information provided:
a. $BaSO_4 = 2.4 \times 10^{-4}$ g/100. g H_2O at 20°C
b. $Ca(OH)_2 = 0.173$ g/100. g H_2O at 20°C

30. Calculate the solubility of a substance MN that ionizes to form M^{2+} and N^{2-} ions, given that $K_{sp} = 8.1 \times 10^{-6}$. (Hint: See Sample Problem 18-3.)

31. Use the K_{sp} values given in Table 18-3 to evaluate the solubility of each of the following in moles per liter.
a. AgBr
b. CoS

32. Complete each of the following relative to the reaction that occurs when 25.0 mL of 0.0500 M $Pb(NO_3)_2$ is combined with 25.0 mL of 0.0400 M Na_2SO_4 if equilibrium is reached at 25°C.
a. Write the solubility-equilibrium equation at 25°C.
b. Write the solubility-equilibrium expression for the net reaction.

33. The ionic substance T_3U_2 ionizes to form T^{2+} and U^{3-} ions. The solubility of T_3U_2 is 3.77×10^{-20} mol/L. What is the value of the solubility-product constant?

34. A solution of AgI contains 2.7×10^{-10} mol/L Ag^+. What is the maximum I^- concentration that can exist in this solution?

35. Calculate whether a precipitate will form if 0.35 L of 0.0044 M $Ca(NO_3)_2$ and 0.17 L of 0.000 39 M NaOH are mixed at 25°C. (See Table 18-3 for K_{sp} values.) (Hint: See Sample Problem 18-4.)

36. Determine whether a precipitate will form if 1.70 g of solid $AgNO_3$ and 14.5 g of solid NaCl are dissolved in 200. mL of water to form a solution at 25°C.

37. If 2.50 g of solid $Fe(NO_3)_3$ is added to 100. mL of a 1.0×10^{-20} M NaOH solution, will a precipitate form?

MIXED REVIEW

38. Calcium carbonate is only slightly soluble in water.

a. Write the equilibrium equation for calcium carbonate in solution.

b. Write the equilibrium-constant expression, K, and the solubility-product constant expression, K_{sp}, for the equilibrium in a saturated solution of $CaCO_3$.

39. Calculate the concentration of Hg^{2+} ions in a saturated solution of HgS(*s*). What volume of solution contains one Hg^{2+} ion?

40. Calculate the equilibrium constant, K, for the following reaction at 900°C.

$$H_2(g) + CO_2(g) \rightleftharpoons H_2O(g) + CO(g)$$

The components were analyzed and it was found that $[H_2] = 0.61$ mol/L, $[CO_2] = 1.6$ mol/L, $[H_2O] = 1.1$ mol/L, and $[CO] = 1.4$ mol/L.

41. A solution in equilibrium with solid barium phosphate is found to have a barium ion concentration of 5×10^{-4} M and a K_{sp} of 3.4×10^{-23}. Calculate the concentration of phosphate ion.

42. At 25°C, the value of K is 1.7×10^{-13} for the following reaction.

$$N_2O(g) + \tfrac{1}{2}O_2(g) \rightleftharpoons 2NO(g)$$

It is determined that $[N_2O] = 0.0035$ mol/L and $[O_2] = 0.0027$ mol/L. Using this information, what is the concentration of NO(*g*) at equilibrium?

43. Determine if a precipitate will form when 0.96 g Na_2CO_3 is combined with 0.20 g $BaBr_2$ in a 10 L solution ($K_{sp} = 2.8 \times 10^{-9}$).

44. For the formation of ammonia, the equilibrium constant is calculated to be 5.2×10^{-5} at 25°C. After analysis, it is determined that $[N_2] = 2.00$ M and $[H_2] = 0.80$ M. How many grams of ammonia are in the 10 L reaction vessel at equilibrium? Use the following equilibrium equation.

$$N_2(g) + 3H_2(g) \rightleftharpoons 2NH_3(g)$$

45. Tooth enamel is composed of the mineral hydroxyapatite, $Ca_5(PO_4)_3OH$, which has a K_{sp} of 6.8×10^{-37}. The molar solubility of hydroxyapatite is 2.7×10^{-5} mol/L. When hydroxyapatite is reacted with fluoride, the OH^- is replaced with the F^- ion on the mineral, forming fluorapatite, $Ca_5(PO_4)_3F$. (The latter is harder and less susceptible to caries.) The K_{sp} of fluorapatite is 1×10^{-60}. Calculate the solubility of fluorapatite in water. Given your calculations, can you support the fluoridation of drinking water?

CRITICAL THINKING

46. Predicting Outcomes When gasoline burns in an automobile engine, nitric oxide is formed from oxygen and nitrogen. Nitric oxide is a major air pollutant. High temperatures like those found in a combustion engine are needed for the reaction. The reaction follows.

$$N_2(g) + O_2(g) \rightleftharpoons 2NO(g)$$

K for the reaction is 0.01 at 2000°C. If 4.0 mol of N_2, 0.1 mol of O_2, and 0.08 mol of NO are placed in a 1.0 L vessel at 2000°C, predict which reaction will be favored.

TECHNOLOGY & LEARNING

47. Graphing Calculator Calculating the Equilibrium Constant, K, for a System

The graphing calculator can be programmed to calculate K for a system, given the concentrations of the products and the concentrations of the reactants.

Given the balanced chemical equation

$$H_2(g) + I_2(g) \longrightarrow 2HI(g)$$

and the equilibrium mixture at 425°C, you can calculate the equilibrium constant for the system. Begin by programming the calculator to carry out the calculation. Next, use the program to make calculations.

A. Program the calculation.

Keystrokes for naming the program: PRGM ▶ ▶ ENTER

Name the program: C O N S T A N T ENTER

Program keystrokes: PRGM ▶ 2 ALPHA H 2nd : PRGM ▶ 2 ALPHA I 2nd : PRGM ▶ 2 ALPHA P 2nd : ALPHA P x^2 ÷ (ALPHA H × ALPHA I) STO▸ ALPHA K 2nd : PRGM ▶ 8 2nd : PRGM ▶ 3 2nd ALPHA " E Q U I L ␣ C O N S T 2nd TEST 1 ALPHA " 2nd : PRGM ▶ 3 ALPHA K

B. Check the display.

Screen display: Prompt H:Prompt I:Prompt P:P²/(H*I)->K:ClrHome:Disp "EQUIL CONST=":Disp K

Press 2nd QUIT to exit the program editor.

C. Run the program.

a. Press PRGM. Select CONSTANT and press ENTER ENTER. Now enter the equilibrium concentration of H_2 as 0.004 5647 mol/L and press ENTER. Enter the equilibrium concentration of I_2 as 0.000 7378 mol/L and press ENTER. Enter the equilibrium concentration of the product HI as 0.013 544 mol/L and press ENTER. The calculator will provide the equilibrium constant.

b. Use the CONSTANT program to calculate K with equlibrium concentrations of $[H_2]$ = 0.001 83 mol/L, $[I_2]$ = 0.003 13 mol/L, and [HI] = 0.0177 mol/L.

c. Use the CONSTANT program to calculate K with equlibrium concentrations of $[H_2]$ = 0.000 479 mol/L, $[I_2]$ = 0.000 479 mol/L, and [HI] = 0.003 53 mol/L.

HANDBOOK SEARCH

48. An equilibrium system helps maintain the pH of the blood. Review the material on the carbon dioxide–bicarbonate ion equilibrium system in Group 14 of the *Elements Handbook* and answer the following.

a. Write the equation for the equilibrium system that responds to changes in H_3O^+ concentration.

b. Use Le Châtelier's principle to explain how hyperventilation affects this system.

c. How does this system maintain pH when acid is added?

49. The reactions used to confirm the presence of transition metal ions often involve the formation of precipitates. Review the analytical tests for the transition metals in the *Elements Handbook*. Use that information and Table 18-3 to determine the minimum concentration of Zn^{2+} needed to produce a precipitate that confirms the presence of Zn. Assume enough sulfide ion reagent is added to the unknown solution in the test tube to produce a sulfide ion concentration of 1.4×10^{-20} M.

RESEARCH & WRITING

50. Find photos of several examples of stalagmites and stalactites in various caves. Investigate the equilibrium processes involved in the formation of stalagmites and stalactites.

51. Carry out library research on the use of catalysts in industrial processes. Explain what types of catalysts are used for specific processes, such as the Haber process.

ALTERNATIVE ASSESSMENT

52. Performance Fill a drinking glass with water and add sugar by the teaspoonful, stirring after each addition. Continue adding the sugar until some of the sugar remains undissolved, even after vigorous stirring. Now heat the sugar-water solution. How are you using Le Châtelier's principle to shift the equilibrium of the system?

CHAPTER 19

Oxidation-Reduction Reactions

Oxidation-reduction reactions propel rockets into space.

Oxidation and Reduction

SECTION 19-1

OBJECTIVES

- Assign oxidation numbers to reactant and product species.
- Define *oxidation* and *reduction.*
- Explain what an oxidation-reduction reaction (redox reaction) is.

Oxidation-reduction reactions involve a transfer of electrons. Oxidation involves the loss of electrons, whereas reduction involves the gain of electrons. Reduction and oxidation half-reactions must occur simultaneously. These processes can be identified through the understanding and use of oxidation numbers (oxidation states).

Oxidation States

Oxidation states were defined in Chapter 7. The oxidation number assigned to an element in a molecule is based on the distribution of electrons in that molecule. The rules by which oxidation numbers are assigned were given in Chapter 7. These rules are summarized in Table 19-1.

TABLE 19-1 ***Rules for Assigning Oxidation Numbers***

Rule	Example
1. The oxidation number of any uncombined element is 0.	The oxidation number of Na(*s*) is 0.
2. The oxidation number of a monatomic ion equals the charge on the ion.	The oxidation number of Cl^- is –1.
3. The more-electronegative element in a binary compound is assigned the number equal to the charge it would have if it were an ion.	The oxidation number of O in NO is –2.
4. The oxidation number of fluorine in a compound is always –1.	The oxidation number of F in LiF is –1.
5. Oxygen has an oxidation number of –2 unless it is combined with F, when it is +2, or it is in a peroxide, such as H_2O_2, when it is –1.	The oxidation number of O in NO_2 is –2.
6. The oxidation state of hydrogen in most of its compounds is +1 unless it is combined with a metal, in which case it is –1.	The oxidation number of H in LiH is –1.
7. In compounds, the elements of Groups 1 and 2 as well as aluminum have oxidation numbers of +1, +2, and +3, respectively.	The oxidation number of Ca in $CaCO_3$ is +2.
8. The sum of the oxidation numbers of all atoms in a neutral compound is 0.	The oxidation number of C in $CaCO_3$ is +4.
9. The sum of the oxidation numbers of all atoms in a polyatomic ion equals the charge of the ion.	The oxidation number of P in $H_2PO_4^-$ is +5.

FIGURE 19-1 The color of solutions containing chromium compounds changes with the oxidation state of chromium.

Chromium provides a very visual example of different oxidation numbers. Different oxidation states of chromium have dramatically different colors, as can be seen in Figure 19-1. The chromium(II) chloride solution is blue, chromium(III) chloride solution is green, potassium chromate solution is yellow, and potassium dichromate solution is orange.

Oxidation

Reactions in which the atoms or ions of an element experience an increase in oxidation state are **oxidation** *processes.* The combustion of metallic sodium in an atmosphere of chlorine gas is shown in Figure 19-2. The sodium ions and chloride ions produced during this strongly exothermic reaction form a cubic crystal lattice in which sodium cations are ionically bonded to chloride anions. The chemical equation for this reaction is written as follows.

$$2Na(s) + Cl_2(g) \longrightarrow 2NaCl(s)$$

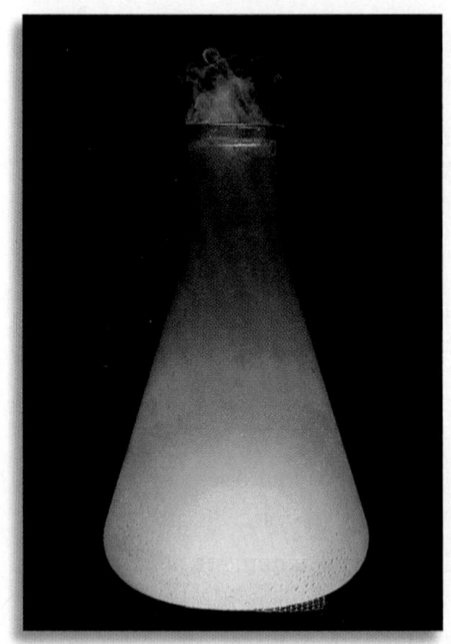

FIGURE 19-2 Sodium and chlorine react violently to form NaCl. The synthesis of NaCl from its elements illustrates the oxidation-reduction process.

The formation of sodium ions illustrates an oxidation process because each sodium atom loses an electron to become a sodium ion. The oxidation state is represented by placing an oxidation number above the symbol of the atom and the ion.

$$\overset{0}{Na} \longrightarrow \overset{+1}{Na^+} + e^-$$

The oxidation state of sodium has changed from 0, its elemental state, to the +1 state of the ion (Rules 1 and 7, Table 19-1). *A species whose oxidation number increases is* **oxidized.** The sodium atom is *oxidized* to a sodium ion.

Reduction

Reactions in which the oxidation state of an element decreases are **reduction** *processes.* Consider the behavior of chlorine in its reaction with sodium. Each chlorine atom accepts an electron and becomes a chloride ion. The oxidation state of chlorine decreases from 0 to –1 for the chloride ion (Rules 1 and 2, Table 19-1).

$$\overset{0}{Cl_2} + 2e^- \longrightarrow 2\overset{-1}{Cl^-}$$

A species that undergoes a decrease in oxidation state is **reduced.** The chlorine atom is reduced to the chloride ion.

FIGURE 19-3 Copper is oxidized and nitrogen monoxide is produced when this penny is placed in a nitric acid solution.

Oxidation and Reduction as a Process

Electrons are produced in oxidation and acquired in reduction. Therefore, for oxidation to occur during a chemical reaction, reduction must occur simultaneously. Furthermore, the number of electrons produced in oxidation must equal the number of electrons acquired in reduction. This makes sense when you recall that electrons are negatively charged and that for charge to be conserved, the number of electrons lost must equal the number of electrons gained. You learned in Chapter 8 that mass is conserved in any chemical reaction. Therefore, the masses of the elements that undergo oxidation and reduction and the electrons that are exchanged are conserved.

A transfer of electrons causes changes in the oxidation states of one or more elements. *Any chemical process in which elements undergo changes in oxidation number is an* **oxidation-reduction reaction.** This name is often shortened to **redox reaction.** An example of a redox reaction can be seen in Figure 19-3, in which copper is being oxidized and NO_3^- from nitric acid is being reduced. *The part of the reaction involving oxidation or reduction alone can be written as a* **half-reaction.** The overall equation for a redox reaction is the sum of two half-reactions. Because the number of electrons involved is the same for oxidation and reduction, they cancel each other out and do not appear in the overall chemical equation. Equations for the reaction between nitric acid and copper illustrate the relationship between half-reactions and the overall redox reaction.

$$3\overset{0}{Cu} \longrightarrow 3\overset{+2}{Cu^{2+}} + 6e^- \qquad \text{(oxidation)}$$

$$2\overset{+5}{N}\overset{-2}{O_3^-} + 6e^- + 8\overset{+1}{H^+} \longrightarrow 2\overset{+2}{N}\overset{-2}{O} + 4\overset{+1}{H_2}\overset{-2}{O} \qquad \text{(reduction)}$$

$$3\overset{0}{Cu} + 2\overset{+5}{N}O_3^- + 8H^+ \longrightarrow 3\overset{+2}{Cu^{2+}} + 2\overset{+2}{N}O + 4H_2O \qquad \text{(redox reaction)}$$

Notice that electrons lost in oxidation appear on the product side of the oxidation half-reaction. Electrons are gained in reduction and appear as reactants in the reduction half-reaction. When metallic copper reacts in nitric acid, three copper atoms are oxidized to Cu^{2+} ions as two nitrogen atoms are reduced from a +5 oxidation state to a +2 oxidation state. Atoms are conserved. This is illustrated by the balanced chemical equation for the reaction between copper and nitric acid.

If none of the atoms in a reaction change oxidation state, the reaction is *not* a redox reaction. For example, sulfur dioxide gas, SO_2, dissolves in water to form an acidic solution containing a low concentration of sulfur*ous* acid, H_2SO_3.

$$\overset{+4\,-2}{SO_2} + \overset{+1\,-2}{H_2O} \longrightarrow \overset{+1\,+4\,-2}{H_2SO_3}$$

The oxidation states of all elemental species remain unchanged in this composition reaction. Therefore, it is *not* a redox reaction.

When a solution of sodium chloride is added to a solution of silver nitrate, an ion-exchange reaction occurs and white silver chloride precipitates.

$$\overset{+1}{Na^+} + \overset{-1}{Cl^-} + \overset{+1}{Ag^+} + \overset{+5\,-2}{NO_3^-} \longrightarrow \overset{+1}{Na^+} + \overset{+5\,-2}{NO_3^-} + \overset{+1\,-1}{AgCl}$$

The oxidation state of each monatomic ion remains unchanged. Again, this reaction is not an oxidation-reduction reaction.

Redox Reactions and Covalent Bonds

Both the synthesis of NaCl from its elements and the reaction between copper and nitric acid involve ionic bonding. Substances with covalent bonds also undergo redox reactions. An oxidation number, unlike an ionic charge, has no physical meaning. That is, the oxidation number assigned to a particular atom is based on its electronegativity relative to the other atoms to which it is bonded in a given molecule; it is not based on any real charge on the atom. For example, an ionic charge of 1− results from the complete gain of one electron by an atom or other neutral species, whereas an oxidation state of −1 means an increased attraction for a bonding electron. A change in oxidation number does not require a change in actual charge.

When hydrogen burns in chlorine, a covalent bond forms from the sharing of two electrons. The two bonding electrons in the hydrogen chloride molecule are not shared equally. Rather, the pair of electrons is more strongly attracted to the chlorine atom because of its higher electronegativity.

$$\overset{0}{H_2} + \overset{0}{Cl_2} \longrightarrow 2\overset{+1\,-1}{HCl}$$

As specified by Rule 3 in Table 19-1, chlorine in HCl is assigned an oxidation number of −1. Thus, the oxidation number for the chlorine atoms changes from 0, its oxidation number in the elemental state, to −1; chlorine atoms are reduced. As specified by Rule 1, the oxidation number of each hydrogen atom in the hydrogen molecule is 0. As specified by Rule 6, the oxidation state of the hydrogen atom in the HCl molecule is +1;

the hydrogen atom is oxidized. No electrons have been totally lost or gained by either atom. Hydrogen has donated a share of its bonding electron to the chlorine; it has not completely transferred that electron. The assignment of oxidation numbers allows the determination of the partial transfer of electrons in compounds that are not ionic. Thus, increases or decreases in oxidation number can be seen in terms of complete or partial loss or gain of electrons.

Reactants and products in redox reactions are not limited to monatomic ions and uncombined elements. Elements in molecular compounds or polyatomic ions can also be oxidized and reduced if they have more than one non-zero oxidation state. An example of this is provided in the reaction between the copper penny and nitric acid when the nitrate ion, NO_3^-, is converted to nitrogen monoxide, NO. Nitrogen is reduced in this reaction. Usually we refer to the oxidation or reduction of the entire molecule or ion. Instead of saying the nitrogen atom is reduced, we say the nitrate ion is reduced to nitrogen monoxide.

$$\cdots + \overset{+5}{N}O_3^- \longrightarrow \overset{+2}{N}O + \cdots$$

SECTION REVIEW

1. How are oxidation numbers assigned?

2. Label each of the following half-reactions as either an oxidation or a reduction half-reaction:

a. $\overset{0}{Br_2} + 2e^- \longrightarrow 2\overset{-1}{Br^-}$

b. $\overset{0}{Na} \longrightarrow \overset{+1}{Na^+} + e^-$

c. $2\overset{-1}{Cl^-} \longrightarrow \overset{0}{Cl_2} + 2e^-$

d. $\overset{0}{Cl_2} + 2e^- \longrightarrow 2\overset{-1}{Cl^-}$

e. $\overset{+1}{Na^+} + e^- \longrightarrow \overset{0}{Na}$

f. $\overset{0}{Fe} \longrightarrow \overset{+2}{Fe^{2+}} + 2e^-$

g. $\overset{+2}{Cu^{2+}} + 2e^- \longrightarrow \overset{0}{Cu}$

h. $\overset{+3}{Fe^{3+}} + e^- \longrightarrow \overset{+2}{Fe^{2+}}$

3. Which of the following equations represent redox reactions?

a. $2KNO_3(s) \longrightarrow 2KNO_2(s) + O_2(g)$

b. $H_2(g) + CuO(s) \longrightarrow Cu(s) + H_2O(l)$

c. $NaOH(aq) + HCl(aq) \longrightarrow NaCl(aq) + H_2O(l)$

d. $H_2(g) + Cl_2(g) \longrightarrow 2HCl(g)$

e. $SO_3(g) + H_2O(l) \longrightarrow H_2SO_4(aq)$

4. For each redox equation identified in the previous question, determine which element is oxidized and which is reduced.

5. Use the equations below for the redox reaction between aluminum metal and sodium metal to answer the following.

$$3\overset{0}{Na} \longrightarrow 3\overset{+1}{Na^+} + 3e^- \qquad \text{(oxidation)}$$

$$\overset{+3}{Al^{3+}} + 3e^- \longrightarrow \overset{0}{Al} \qquad \text{(reduction)}$$

$$3\overset{0}{Na} + \overset{+3}{Al^{3+}} \longrightarrow 3\overset{+1}{Na} + \overset{0}{Al} \qquad \text{(redox reaction)}$$

a. Explain how this reaction illustrates that charge is conserved in a redox reaction.

b. Explain how this reaction illustrates that mass is conserved in a redox reaction.

c. Explain why electrons do not appear as reactants or products in the combined equation.

Skunk-Spray Remedy

So that pretty black cat with the white stripe down its back wasn't a cat after all? Well, hold off on dumping Fido into a tomato-juice bath. Chemistry has a much better way of conquering skunk spray.

Paul Krebaum, the inventor of a new deskunking formula, says that while working as a materials engineer, he constantly had to deal with the less-than-pleasant smell of the hydrogen sulfide gas that was released from one of his experiments. Mr. Krebaum was losing popularity with his neighbors, and venting off the gas only partially solved the problem. A better solution, he decided, would be to find a way to eliminate the smell entirely.

Mr. Krebaum rifled through his old chemistry books and found that hydrogen peroxide could oxidize these sulfur-containing compounds to much less smelly components. And by decreasing the chemical's volatility, the reaction prevents the gas from reaching our noses quite so easily. He immediately whipped up a hydrogen peroxide mixture, and it worked like a charm.

The reaction by which the hydrogen sulfide was destroyed, producing sulfate compounds that do not have the unpleasant odor, can be seen in the following equation.

$$2NaOH + 4H_2O_2 + H_2S \longrightarrow Na_2SO_4 + 6H_2O$$

Skunk spray gets its odor from chemicals called mercaptans.

"The receptors that are in your nose are sensitive to sulfur in its low oxidation state," says Mr. Krebaum. "However, they are not sensitive to sulfur in its high oxidation state."

Some time later, a friend of Mr. Krebaum's complained to him that a skunk had sprayed his pet. Because the odor in a skunk's spray also comes from compounds containing sulfur in a low oxidation state, Mr. Krebaum thought his solution might also work on this age-old problem. He mixed up a milder version to try out on the pet: 1 quart 3% hydrogen peroxide solution, 1/4 cup baking soda, and 1 teaspoon liquid soap. His friend tried it out, and the result was one wet and unhappy—but much less smelly—pet.

Mr. Krebaum says that the hydrogen peroxide in the remedy actually oxidizes the compounds, while the baking soda reduces the acidity of the mixture and the soap helps to wash out the greasy skunk spray. The reaction that occurs can be seen in the following equation. The symbol *R* represents all the other elements in the sulfur-containing compound found in skunk spray.

$$RSH + 3H_2O_2 + NaHCO_3 \longrightarrow RSO_3Na + 4H_2O + CO_2$$

The pet should be washed thoroughly with the mixture, taking care to avoid its eyes. If the mixture is left on for a few minutes—long enough for the reaction to occur—and then rinsed away with tap water, the smell will disappear.

There's no reason to fear ending up with a platinum blond pet, Mr. Krebaum says. The formula does not bleach or have any other negative side effects. He does have one warning: mix the formula just before using it because the mixture breaks down quickly. The reaction releases oxygen, so the formula should not be put into a sealed container; it will build up pressure and could eventually blow the top. For these reasons, bottles of "Krebaum's Skunkinator" will not be appearing on drug-store shelves any time soon.

SECTION 19-2

Balancing Redox Equations

Objectives

- Explain what must be conserved in redox equations.
- Balance redox equations by using the half-reaction method.

Equations for simple redox reactions can be balanced by inspection, which you learned to do in Chapter 8. Most redox equations, however, require more systematic methods. The equation-balancing process requires the use of oxidation numbers. In a balanced equation, both charge and mass are conserved. Although oxidation and reduction half-reactions occur together, their reaction equations are balanced separately, then combined to give the balanced redox-reaction equation.

Half-Reaction Method

The *half-reaction method,* or ion-electron method, for balancing redox equations consists of seven steps. Oxidation numbers are assigned to all atoms and polyatomic ions to determine which species are part of the redox process. The oxidation and reduction equations are balanced separately for mass and charge. They are then added together to produce a complete balanced equation. These seven steps are applied to balance the reaction of hydrogen sulfide and nitric acid. Sulfuric acid, nitrogen dioxide, and water are the products of the reaction.

1. ***Write the formula equation if it is not given in the problem. Then write the ionic equation.***

Formula equation: $H_2S + HNO_3 \longrightarrow H_2SO_4 + NO_2 + H_2O$

Ionic equation: $H_2S + H^+ + NO_3^- \longrightarrow 2H^+ + SO_4^{2-} + NO_2 + H_2O$

2. ***Assign oxidation numbers to each element and ion. Delete substances containing an element that does not change oxidation state.***

$$\overset{+1\,-2}{H_2S} + \overset{+1}{H^+} + \overset{+5\,-2}{NO_3^-} \longrightarrow \overset{+1}{3H^+} + \overset{+6\,-2}{SO_4^{2-}} + \overset{+4\,-2}{NO_2} + \overset{+1\,-2}{H_2O}$$

The sulfur changes oxidation state from –2 to +6. The nitrogen changes oxidation state from +5 to +4. The other substances are deleted.

$$\overset{+1\,-2}{H_2S} + \overset{+5\,-2}{NO_3^-} \longrightarrow \overset{+6\,-2}{SO_4^{2-}} + \overset{+4\,-2}{NO_2}$$

The remaining species are used in step 3.

3. *Write the half-reaction for oxidation.* In this example, the sulfur is being oxidized.

$$\overset{-2}{H_2S} \longrightarrow \overset{+6}{SO_4^{2-}}$$

- *Balance the mass.* To balance the oxygen in this half-reaction, H_2O must be added to the left side. This gives 10 extra hydrogen atoms on that side of the equation. Therefore, 10 hydrogen ions are added to the right side. In basic solution, OH^- ions and water may be used to balance mass.

$$\overset{-2}{H_2S} + 4H_2O \longrightarrow \overset{+6}{SO_4^{2-}} + 10H^+$$

- *Balance the charge.* Electrons are added to the side having the greater positive net charge. The left side of the equation has no net charge; the right side has a net charge of 8+. For the charges to balance, each side must have the same net charge. Therefore, 8 electrons are added to the product side so that it has no charge and balances with the reactant side of the equation. Notice that the oxidation of sulfur from a state of –2 to +6 indicates a loss of 8 electrons.

$$\overset{-2}{H_2S} + 4H_2O \longrightarrow \overset{+6}{SO_4^{2-}} + 10H^+ + 8e^-$$

The oxidation half-reaction is now balanced.

4. *Write the half-reaction for reduction.* In this example, nitrogen is being reduced from a +5 state to a +4 state.

$$\overset{+5}{NO_3^-} \longrightarrow \overset{+4}{NO_2}$$

- *Balance the mass.* H_2O must be added to the product side of the reaction to balance the oxygen atoms. Therefore, two hydrogen ions must be added to the reactant side to balance the hydrogen atoms.

$$\overset{+5}{NO_3^-} + 2H^+ \longrightarrow \overset{+4}{NO_2} + H_2O$$

- *Balance the charge.* Electrons are added to the side having the greater positive net charge. The left side of the equation has a net charge of 1+. Therefore, 1 electron must be added to this side to balance the charge.

$$\overset{+5}{NO_3^-} + 2H^+ + e^- \longrightarrow \overset{+4}{NO_2} + H_2O$$

The reduction half-reaction is now balanced.

5. *Conserve charge by adjusting the coefficients in front of the electrons so that the number lost in oxidation equals the number gained in reduction.* Write the ratio of the number of electrons lost to the number of electrons gained.

$$\frac{e^- \text{ lost in oxidation}}{e^- \text{ gained in reduction}} = \frac{8}{1}$$

This ratio is already in its lowest terms. If it were not, it would need to be reduced. Multiply the oxidation half-reaction by 1 (it remains unchanged) and the reduction half-reaction by 8. The number of electrons lost now equals the number of electrons gained.

$$1\left(\overset{-2}{H_2S} + 4H_2O \longrightarrow \overset{+6}{SO_4^{2-}} + 10H^+ + 8e^-\right)$$

$$8\left(\overset{+5}{NO_3^-} + 2H^+ + e^- \longrightarrow \overset{+4}{NO_2} + H_2O\right)$$

6. ***Combine the half-reactions, and cancel out anything common to both sides of the equation.***

$$\overset{-2}{H_2S} + 4H_2O \longrightarrow \overset{+6}{SO_4^{2-}} + 10H^+ + 8e^-$$

$$\overset{+5}{8NO_3^-} + 16H^+ + 8e^- \longrightarrow \overset{+4}{8NO_2} + 8H_2O$$

$$\overset{+5}{8NO_3^-} + \overset{6}{\cancel{16}}H^+ + \cancel{8e^-} + \overset{-2}{H_2S} + \cancel{4H_2O} \longrightarrow$$

$$\overset{+4}{8NO_2} + \overset{4}{\cancel{8}}H_2O + \overset{+6}{SO_4^{2-}} + \cancel{10H^+} + \cancel{8e^-}$$

Each side of the above equation has $10H^+$, $8e^-$, and $4H_2O$. These cancel each other out and do not appear in the balanced equation.

$$\overset{+5}{8NO_3^-} + \overset{-2}{H_2S} + 6H^+ \longrightarrow \overset{+4}{8NO_2} + 4H_2O + \overset{+6}{SO_4^{2-}}$$

7. ***Combine ions to form the compounds shown in the original formula equation. Check to ensure that all other ions balance.*** The NO_3^- ion appeared as nitric acid in the original equation. There are only 6 hydrogen ions to pair with the 8 nitrate ions. Therefore, 2 hydrogen ions must be added to complete this formula. If 2 hydrogen ions are added to the left side of the equation, 2 hydrogen ions must also be added to the right side of the equation.

$$8HNO_3 + H_2S \longrightarrow 8NO_2 + 4H_2O + SO_4^{2-} + 2H^+$$

The sulfate ion appeared as sulfuric acid in the original equation. The hydrogen ions added to the right side are used to complete the formula for sulfuric acid.

$$8HNO_3 + H_2S \longrightarrow 8NO_2 + 4H_2O + H_2SO_4$$

A final check must be made to ensure that all elements are correctly balanced.

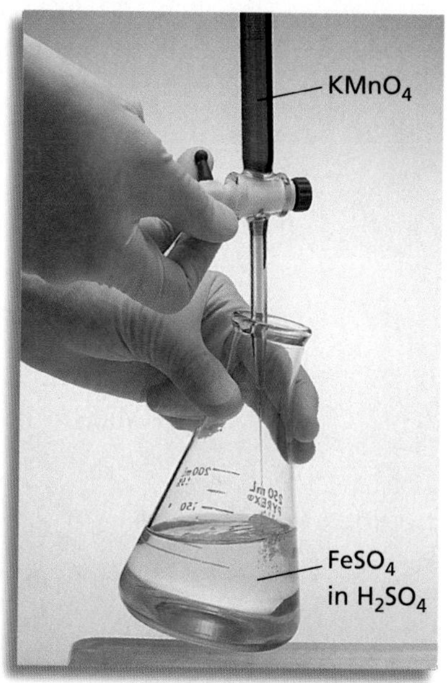

FIGURE 19-4 As a $KMnO_4$ solution is titrated into an acidic solution of $FeSO_4$, deep purple MnO_4^- ions are reduced to colorless Mn^{2+} ions. When all Fe^{2+} ions are oxidized, MnO_4^- ions are no longer reduced to colorless Mn^{2+} ions. Thus, the first faint appearance of the MnO_4^- color indicates the end point of the titration.

SAMPLE PROBLEM 19-1

Write a balanced equation for the reaction shown in Figure 19-4. A deep purple solution of potassium permanganate is titrated into a colorless solution of iron(II) sulfate and sulfuric acid. The products are iron(III) sulfate, manganese(II) sulfate, potassium sulfate, and water—all of which are colorless.

SOLUTION **1.** *Write the formula equation if it is not given in the problem. Then write the ionic equation.*

$$KMnO_4 + FeSO_4 + H_2SO_4 \longrightarrow Fe_2(SO_4)_3 + MnSO_4 + K_2SO_4 + H_2O$$

$$K^+ + MnO_4^- + Fe^{2+} + SO_4^{2-} + 2H^+ + SO_4^{2-} \longrightarrow$$
$$2Fe^{3+} + 3SO_4^{2-} + Mn^{2+} + SO_4^{2-} + 2K^+ + SO_4^{2-} + H_2O$$

2. *Assign oxidation numbers to each element and ion. Delete substances containing an element that does not change oxidation state.*

$$\overset{+1}{K^+} + \overset{+7\ -2}{MnO_4^-} + \overset{+2}{Fe^{2+}} + \overset{+6\ -2}{SO_4^{2-}} + \overset{+1}{2H^+} + \overset{+6\ -2}{SO_4^{2-}} \longrightarrow$$
$$\overset{+3}{2Fe^{3+}} + \overset{+6\ -2}{3SO_4^{2-}} + \overset{+2}{Mn^{2+}} + \overset{+6\ -2}{SO_4^{2-}} + \overset{+1}{2K^+} + \overset{+6\ -2}{SO_4^{2-}} + \overset{+1\ -2}{H_2O}$$

Only ions or molecules whose oxidation numbers change are retained.

$$\overset{+7\ -2}{MnO_4^-} + \overset{+2}{Fe^{2+}} \longrightarrow \overset{+3}{Fe^{3+}} + \overset{+2}{Mn^{2+}}$$

3. *Write the half-reaction for oxidation.* The iron shows the increase in oxidation number. Therefore, it is oxidized.

$$\overset{+2}{Fe^{2+}} \longrightarrow \overset{+3}{Fe^{3+}}$$

- *Balance the mass.* The mass is already balanced.

- *Balance the charge.*

$$\overset{+2}{Fe^{2+}} \longrightarrow \overset{+3}{Fe^{3+}} + e^-$$

4. *Write the half-reaction for reduction.* Manganese shows a change in oxidation number from +7 to +2. It is reduced.

$$\overset{+7}{MnO_4^-} \longrightarrow \overset{+2}{Mn^{2+}}$$

- *Balance the mass.* Water and hydrogen ions must be added to balance the oxygen atoms in the permanganate ion.

$$\overset{+7}{MnO_4^-} + 8H^+ \longrightarrow \overset{+2}{Mn^{2+}} + 4H_2O$$

- *Balance the charge.*

$$\overset{+7}{MnO_4^-} + 8H^+ + 5e^- \longrightarrow \overset{+2}{Mn^{2+}} + 4H_2O$$

5. *Adjust the coefficients to conserve charge.*

$$\frac{e^- \text{ lost in oxidation}}{e^- \text{ gained in reduction}} = \frac{1}{5}$$

$$5(Fe^{2+} \longrightarrow Fe^{3+} + e^-)$$
$$1(MnO_4^- + 8H^+ + 5e^- \longrightarrow Mn^{2+} + 4H_2O)$$

6. *Combine the half-reactions and cancel.*

$$
\begin{array}{l}
5Fe^{2+} \longrightarrow 5Fe^{3+} + 5e^- \\
MnO_4^- + 8H^+ + 5e^- \longrightarrow Mn^{2+} + 4H_2O \\
\hline
MnO_4^- + 5Fe^{2+} + 8H^+ + \cancel{5e^-} \longrightarrow Mn^{2+} + 5Fe^{3+} + 4H_2O + \cancel{5e^-}
\end{array}
$$

7. *Combine ions to form compounds from the original equation.* The iron(III) product appears in the original equation as $Fe_2(SO_4)_3$. Every iron(III) sulfate molecule requires two iron ions. Therefore, the entire equation must be multiplied by 2 to provide an even number of iron ions.

$$2(5Fe^{2+} + MnO_4^- + 8H^+ \longrightarrow 5Fe^{3+} + Mn^{2+} + 4H_2O)$$
$$10Fe^{2+} + 2MnO_4^- + 16H^+ \longrightarrow 10Fe^{3+} + 2Mn^{2+} + 8H_2O$$

The iron(II), iron(III), manganese(II), and 2 hydrogen ions in the original equation are paired with sulfate ions. Iron(II) sulfate requires 10 sulfate ions, and sulfuric acid requires 8 sulfate ions. To balance the equation, 18 sulfate ions must be added to each side. On the product side, 15 of these ions form iron(III) sulfate, and 2 of them form manganese(II) sulfate. That leaves 1 sulfate ion unaccounted for. The permanganate ion requires the addition of 2 potassium ions to each side. These 2 potassium ions form potassium sulfate on the product side of the reaction.

$$10FeSO_4 + 2KMnO_4 + 8H_2SO_4 \longrightarrow 5Fe_2(SO_4)_3 + 2MnSO_4 + K_2SO_4 + 8H_2O$$

Final inspection shows that atoms and charges are balanced.

PRACTICE

1. Copper reacts with hot, concentrated sulfuric acid to form copper(II) sulfate, sulfur dioxide, and water. Write and balance the equation for this reaction.
   Answer $Cu + 2H_2SO_4 \longrightarrow CuSO_4 + SO_2 + 2H_2O$
2. Write and balance the equation for the reaction between nitric acid and potassium iodide. The products are potassium nitrate, iodine, nitrogen monoxide, and water.
   Answer $8HNO_3 + 6KI \longrightarrow 6KNO_3 + 3I_2 + 2NO + 4H_2O$
3. Rust occurs when iron reacts with oxygen and water to form iron(II) hydroxide. Write and balance the equation for this reaction.
   Answer $2Fe + O_2 + 2H_2O \longrightarrow 2Fe(OH)_2$

SECTION REVIEW

1. What two quantities are conserved in redox equations?
2. Why do we add H^+ and H_2O to some half-reactions and OH^- and H_2O to others?
3. Balance the following oxidation-reduction reaction:
   $Na_2SnO_2 + Bi(OH)_3 \longrightarrow Bi + Na_2SnO_3 + H_2O$

DESKTOP INVESTIGATION

Redox Reactions

MATERIALS

- aluminum foil
- beaker, 250 mL
- 1 M copper(II) chloride solution, $CuCl_2$
- 3% hydrogen peroxide
- manganese dioxide
- metric ruler
- scissors
- test-tube clamp
- test tube, 16 × 150 mm
- wooden splint

Wear safety goggles and an apron.

PROCEDURE

Record all your results in a data table.

1. Put 10 mL of hydrogen peroxide in a test tube, and add a small amount of manganese dioxide (equal to the size of about half a pea). What is the result?
2. Insert a glowing wooden splint into the test tube (see diagram). What is the result? If oxygen is produced, a glowing wooden splint inserted into the test tube will glow brighter.
3. Fill the 250 mL beaker halfway with the copper(II) chloride solution.
4. Cut foil into 2 cm × 12 cm strips.
5. Add the aluminum strips to the copper(II) chloride solution. Use a glass rod to stir the mixture, and observe for 12 to 15 minutes. What is the result?

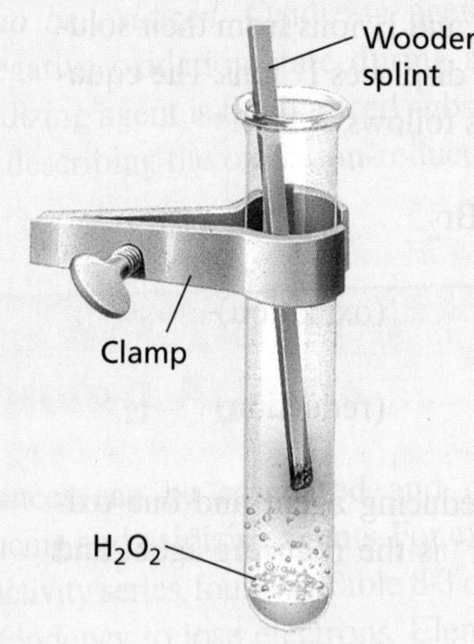

DISCUSSION

1. Write balanced equations showing what happened in each of the reactions.
2. Write a conclusion for the two experiments.

Autooxidation

Some substances can be both reduced and oxidized easily. For example, peroxide ions, O_2^{2-}, have a relatively unstable covalent bond between the two oxygen atoms. The electron-dot formula is written as follows.

$$\left[:\ddot{\underset{..}{O}}:\ddot{\underset{..}{O}}:\right]^{2-}$$

Each oxygen atom has an oxidation number of −1. The peroxide ion structure represents an intermediate oxidation state between O_2 and O^{2-}. Therefore, the peroxide ion is highly reactive.

Hydrogen peroxide, H_2O_2, contains the reactive peroxide ion. It decomposes into water and molecular oxygen, as shown in the equation below.

$$2H_2\overset{-1}{O}_2 \longrightarrow 2H_2\overset{-2}{O} + \overset{0}{O}_2$$

FIGURE 19-6 A bombardier beetle can repel large predators such as frogs with a chemical defense mechanism that uses the autooxidation of hydrogen peroxide.

Notice that in this reaction, hydrogen peroxide is both oxidized and reduced. Oxygen atoms that become part of gaseous oxygen molecules are oxidized. The oxidation number of theses oxygen atoms increases from −1 to 0. Oxygen atoms that become part of water are reduced. The oxidation number of these oxygen atoms decreases from −1 to −2. *A process in which a substance acts as both an oxidizing agent and a reducing agent is called* **autooxidation.** A substance that undergoes autooxidation is both *self-oxidizing* and *self-reducing.*

The bombardier beetle defends itself by spraying its enemies with an unpleasant hot chemical mixture as shown in Figure 19-6. The catalyzed autooxidation of hydrogen peroxide produces hot oxygen gas. This gas gives the insect an ability to eject irritating chemicals from its abdomen with explosive force.

SECTION REVIEW

1. Describe the chemical activity of the alkali metals and of the halogens on the basis of oxidizing and reducing strength.

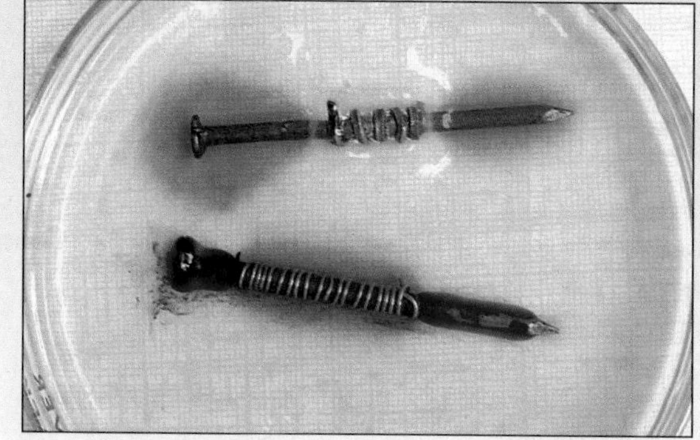

2. The photo on the left depicts two redox reactions. Use them to answer the following questions:
   a. When zinc is wrapped around an iron nail, is the iron reduced or oxidized?
   b. When copper is wrapped around an iron nail, is the iron reduced or oxidized?
3. Would Cl_2 be reduced by I^-? Explain.
4. Which is the stronger oxidizing agent in each of the following pairs?
   Cu^{2+} or Al^{3+}, H_2 or H_3O^+, Cr or Cu?
5. What is meant by *autooxidation*?

SECTION 19-4

Electrochemistry

OBJECTIVES

- Explain what is required for an electrochemical cell.
- Describe the nature of voltaic cells.
- Describe the nature of electrolytic cells.
- Explain the process of electroplating.
- Describe the chemistry of a rechargeable cell.
- Calculate cell potentials from a table of standard electrode potentials.

Oxidation-reduction reactions involve energy changes. Because these reactions involve electron transfer, the net *release* or net *absorption* of energy can occur in the form of electrical energy rather than heat energy. This property allows for a great many practical applications of redox reactions. It also makes possible quantitative predictions and comparisons of the oxidizing and reducing abilities of different substances. *The branch of chemistry that deals with electricity-related applications of oxidation-reduction reactions is called* **electrochemistry.**

Electrochemical Cells

Oxidation-reduction reactions involve a transfer of electrons from the substance oxidized to the substance reduced. If the two substances are in contact with one another, a transfer of heat energy accompanies the electron transfer. In Figure 19-7 a zinc strip is in contact with a copper(II) sulfate solution. The zinc strip loses electrons to the copper(II) ions in solution. Copper(II) ions accept the electrons and fall out of solution as copper atoms. As electrons are transferred between zinc atoms and copper(II) ions, energy is released in the form of heat, as indicated by the rise in temperature.

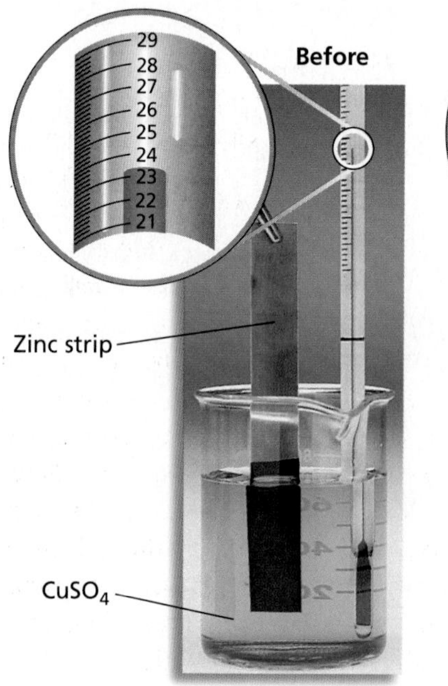

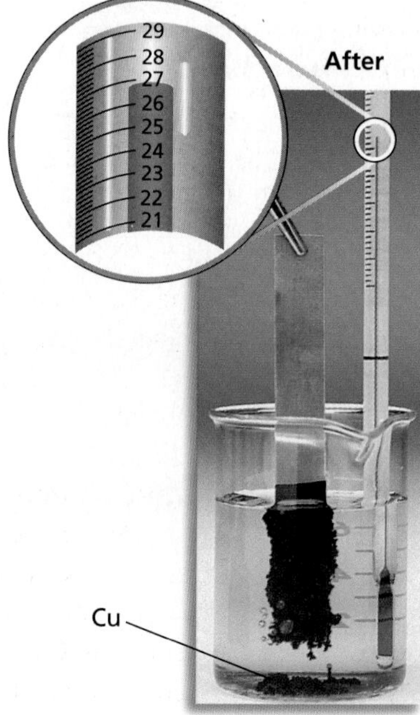

FIGURE 19-7 Heat energy given off when electrons are transferred directly from Zn atoms to Cu^{2+} ions causes the temperature of the aqueous $CuSO_4$ solution to rise.

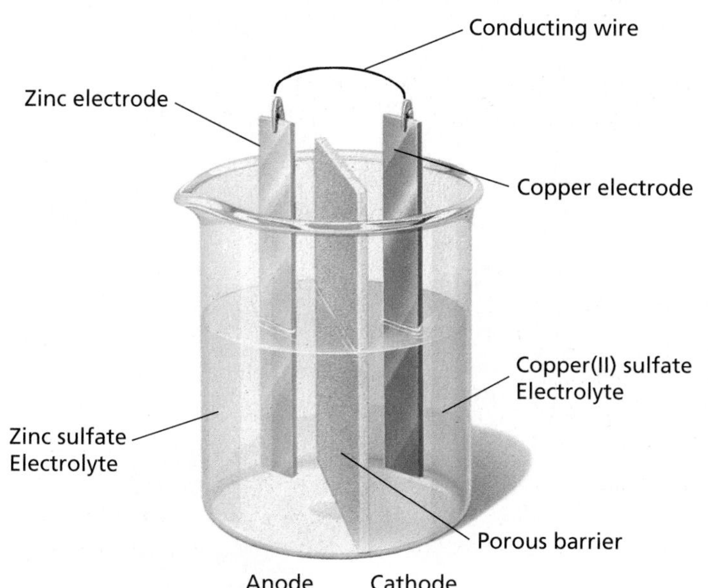

FIGURE 19-8 An electrochemical cell consists of two electrodes. Each electrode is in contact with an electrolyte; the electrode and the electrolyte make up a half-cell. The two electrodes are connected by a wire, and a porous barrier separates the two electrolytes.

If, however, the substance that is oxidized during the reaction is separated from the substance that is reduced during the reaction, the electron transfer is accompanied by a transfer of electrical energy instead of heat. One means of separating oxidation and reduction half-reactions is with a *porous barrier.* This barrier prevents the metal atoms of one half-reaction from mixing with the ions of the other half-reaction. Ions in the two solutions can move through the porous barrier. Electrons can be transferred from one side to the other through an external connecting wire. Electric current moves in a closed loop path, or *circuit,* so this movement of electrons through the wire is balanced by the movement of ions in solution.

Altering the system in Figure 19-7 so that electrical current is produced instead of heat would simply involve separating the copper and zinc. The Zn strip is in an aqueous solution of $ZnSO_4$. The Cu strip is in an aqueous solution of $CuSO_4$. Both solutions conduct electricity, so, as you learned in Chapter 14, they are classified as electrolytes. *An* **electrode** *is a conductor used to establish electrical contact with a nonmetallic part of a circuit, such as an electrolyte.* In Figure 19-8, Zn and Cu are electrodes. *A single electrode immersed in a solution of its ions is a* **half-cell.** The Zn strip in aqueous $ZnSO_4$ is an **anode,** *the electrode where oxidation takes place.* The Cu strip in $CuSO_4$ is a **cathode,** *the electrode where reduction takes place.* The copper half-cell can be written as Cu^{2+}/Cu, and the zinc half-cell can be written as Zn^{2+}/Zn. The two half-cells together make an electrochemical cell. *An* **electrochemical cell** *is a system of electrodes and electrolytes in which either chemical reactions produce electrical energy or an electric current produces chemical change.* An electrochemical cell may be represented by the following notation: cathode | anode. For the example, the cell made up of zinc and copper could be written as Cu | Zn. There are two types of electrochemical cells: voltaic (also called galvanic) and electrolytic.

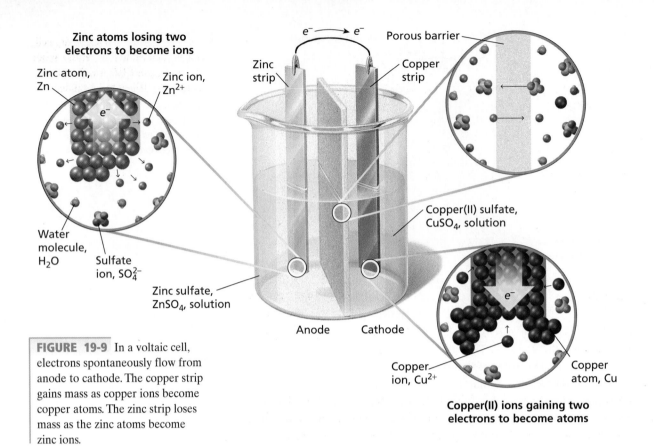

Zinc atoms losing two electrons to become ions

Zinc atom, Zn

Zinc ion, Zn^{2+}

Water molecule, H_2O

Sulfate ion, SO_4^{2-}

Zinc sulfate, $ZnSO_4$, solution

Zinc strip

Copper strip

Porous barrier

Copper(II) sulfate, $CuSO_4$, solution

Anode

Cathode

Copper ion, Cu^{2+}

Copper atom, Cu

Copper(II) ions gaining two electrons to become atoms

FIGURE 19-9 In a voltaic cell, electrons spontaneously flow from anode to cathode. The copper strip gains mass as copper ions become copper atoms. The zinc strip loses mass as the zinc atoms become zinc ions.

Voltaic Cells

If the redox reaction in an electrochemical cell occurs naturally and produces electrical energy, the cell is a **voltaic cell.** Cations in the solution are reduced when they gain electrons at the surface of the cathode to become metal atoms. This half-reaction for the voltaic cell shown in Figure 19-9 is as follows.

$$Cu^{2+}(aq) + 2e^- \longrightarrow Cu(s)$$

The half-reaction occurring at the anode is as follows.

$$Zn(s) \longrightarrow Zn^{2+}(aq) + 2e^-$$

Electrons given up at the anode pass along the external connecting wire to the cathode.

The movement of electrons through the wire must be balanced by a movement of ions in the solution. Anions move toward the anode to replace the negatively charged electrons that are moving away. Cations move toward the cathode as positive charge is lost through reduction. Thus, in Figure 19-9, sulfate ions in the $CuSO_4$ solution can move through the barrier into the $ZnSO_4$ solution. Likewise, the Zn^{2+} ions pass through the barrier into the $CuSO_4$ solution.

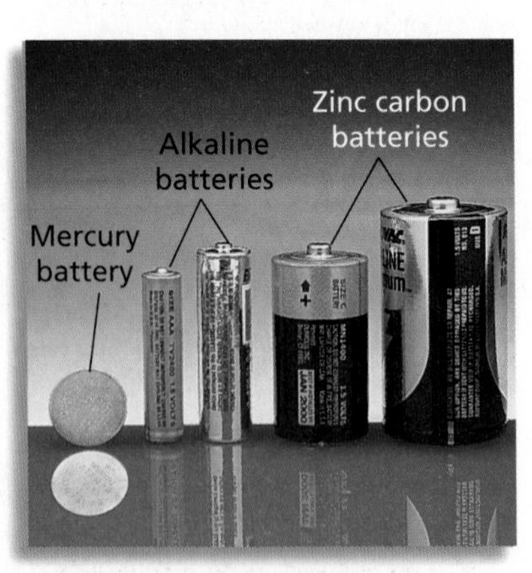

FIGURE 19-10 Many common batteries are simple voltaic dry cells.

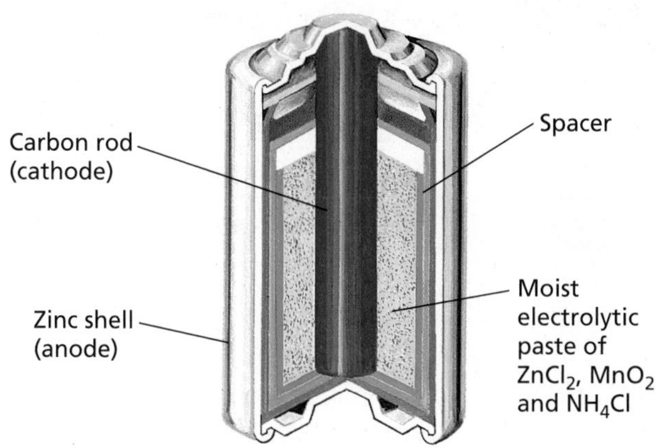

FIGURE 19-11 In a zinc dry cell, zinc is oxidized to Zn^{2+} at the anode, and manganese(IV) is reduced to manganese(III) at the cathode.

The dry cells pictured in Figure 19-10 are common sources of electrical energy. Like the wet cell previously described, dry cells are voltaic cells. The three most common types of dry cells are the zinc-carbon battery, the alkaline battery, and the mercury battery. They differ in the substances being oxidized and reduced.

Zinc-Carbon Dry Cells

Batteries such as those used in flashlights are zinc-carbon dry cells. These cells consist of a zinc container, which serves as the anode, filled with a moist paste of MnO_2, graphite, and NH_4Cl, as illustrated in Figure 19-11. When the external circuit is closed, zinc atoms are oxidized at the negative electrode, or anode.

$$\overset{0}{\mathrm{Zn}}(s) \longrightarrow \overset{+2}{\mathrm{Zn}}{}^{2+}(aq) + 2e^-$$

Electrons move across the circuit and reenter the cell through the carbon rod. The carbon rod is the cathode or positive electrode. Here MnO_2 is reduced in the presence of H_2O according to the following half-reaction.

$$2\overset{+4}{\mathrm{Mn}}\mathrm{O_2}(s) + \mathrm{H_2O}(l) + 2e^- \longrightarrow \overset{+3}{\mathrm{Mn}}{}_2\mathrm{O_3}(s) + 2\mathrm{OH}^-(aq)$$

Alkaline Batteries

The batteries found in a small, portable cassette player or other small electronic device are frequently alkaline dry cells. These cells do not have a carbon rod cathode, as in the zinc-carbon cell. The absence of the carbon rod allows them to be smaller. Figure 19-12 shows a model of an alkaline battery. This cell uses a paste of Zn metal and potassium hydroxide instead of a solid metal anode. The half-reaction at the anode is as follows.

$$\overset{0}{\mathrm{Zn}}(s) + 2\mathrm{OH}^-(aq) \longrightarrow \overset{+2}{\mathrm{Zn}}(\mathrm{OH})_2(s) + 2e^-$$

The reduction half-reaction, the reaction at the cathode, is exactly the same as that for the zinc-carbon dry cell.

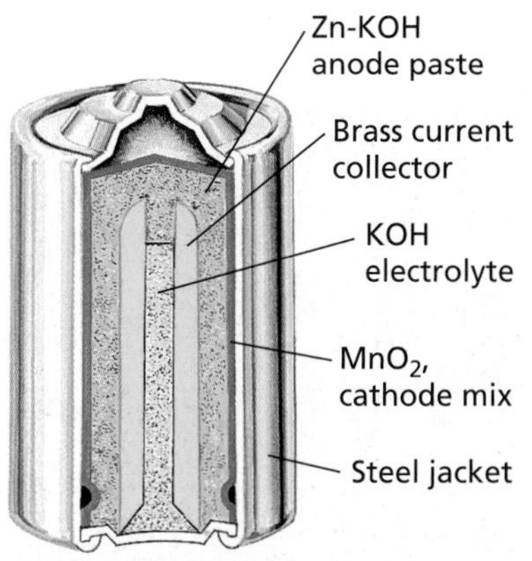

FIGURE 19-12 KOH makes the electrolyte paste in this battery basic. Thus, it is called an alkaline dry cell.

FIGURE 19-13 It is important that mercury batteries be recycled and not just discarded because mercury is a poisonous substance.

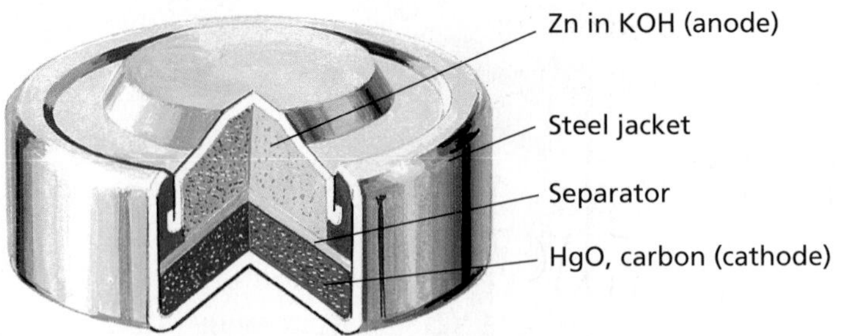

Mercury Batteries

The tiny batteries found in hearing aids, calculators, and camera flashes are mercury batteries, as shown in Figure 19-13. The anode half-reaction is identical to that found in the alkaline dry cell. However, the cathode, or reduction, half-reaction is different. The cathode half-reaction is described by the following equation.

$$\overset{+2}{\text{Hg}}\text{O}(s) + \text{H}_2\text{O}(l) + 2e^- \longrightarrow \overset{0}{\text{Hg}}(l) + 2\text{OH}^-(aq)$$

Electrolytic Cells

Some oxidation-reduction reactions do not occur spontaneously, but can be driven by electrical energy. *The process in which an electric current is used to produce an oxidation-reduction reaction is* **electrolysis.** *If electrical energy is required to produce a redox reaction and bring about a chemical change in an electrochemical cell, it is an* **electrolytic cell.** Most commercial uses of redox reactions make use of electrolytic cells.

An electrolytic cell is depicted in Figure 19-14. The electrode of the cell connected to the negative terminal of the battery acquires an excess of electrons and becomes the cathode of the electrolytic cell. The electrode of the cell connected to the positive terminal of the battery loses electrons to the battery; it is the anode of the electrolytic cell. The battery can be thought of as an electron pump simultaneously supplying electrons to the cathode and recovering electrons from the anode. This energy input from the battery drives the electrode reactions in the electrolytic cell.

A comparison of electrolytic and voltaic cells can be seen in Figure 19-14. The voltaic cell shown in Figure 19-14 has a copper cathode and a zinc anode. If a battery is connected so that the positive terminal contacts the copper electrode and the negative terminal contacts the zinc electrode, the electrons flow in the opposite direction. The battery forces the cell to reverse its reaction; the zinc electrode becomes the cathode, and the copper electrode becomes the anode. The half-reaction at the anode, in which copper metal is oxidized, can be written as follows.

$$\overset{0}{\text{Cu}} \longrightarrow \overset{+2}{\text{Cu}}{}^{2+} + 2e^-$$

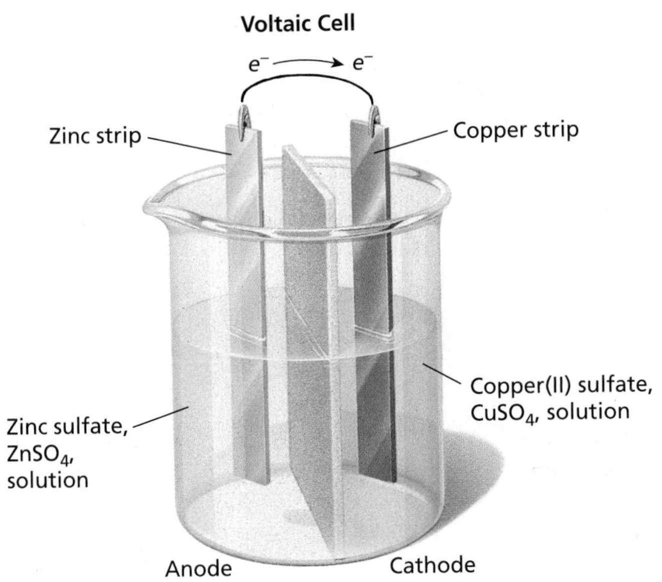

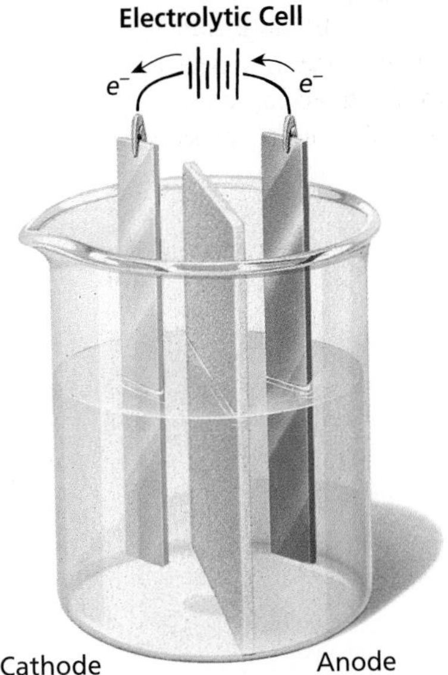

FIGURE 19-14 The direction in which the electrons move reverses if a voltaic cell is connected to a direct current source to become an electrolytic cell.

The reduction half-reaction of zinc at the cathode is written as follows.

$$\overset{+2}{Zn^{2+}} + 2e^- \longrightarrow \overset{0}{Zn}$$

There are two important differences between the voltaic cell and the electrolytic cell.

1. The anode and cathode of an electrolytic cell are connected to a battery or other direct-current source, whereas a voltaic cell serves as a source of electrical energy.
2. Electrolytic cells are those in which electrical energy from an external source causes *nonspontaneous* redox reactions to occur. Voltaic cells are those in which *spontaneous* redox reactions produce electricity. In an electrolytic cell, electrical energy is converted to chemical energy; in a voltaic cell, chemical energy is converted to electrical energy.

Electroplating

Metals such as copper, silver, and gold are difficult to oxidize. In an electrolytic cell, these inactive metals form ions at an anode that are easily reduced at a cathode. This type of cell allows solid metal from one electrode to be deposited on the other electrode. *An electrolytic process in which a metal ion is reduced and a solid metal is deposited on a surface is called* **electroplating.**

An electroplating cell contains a solution of a salt of the plating metal. It has an object to be plated (the cathode) and a piece of the plating metal (the anode). A silver-plating cell contains a solution of a soluble silver salt and a silver anode. The cathode is the object to be plated. The silver anode is connected to the positive electrode of a battery or to some other source of direct current. The object to be plated is connected to the negative electrode.

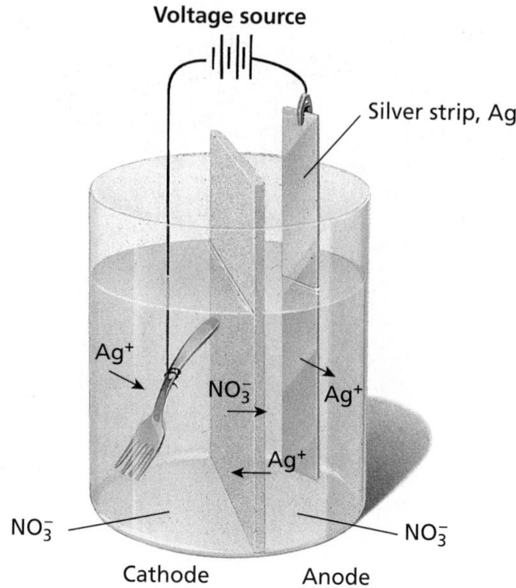

FIGURE 19-15 Electroplating is often used to avoid corrosion (a redox reaction) at the object's surface by putting a layer of an inactive metal on a more-active metal.

A cell in which silver is being electroplated onto a fork can be seen in Figure 19-15. Silver ions are reduced at the cathode according to the following equation and deposited as metallic silver when electrons flow through the circuit.

$$\overset{+1}{Ag^+} + e^- \longrightarrow \overset{0}{Ag}$$

Meanwhile, metallic silver is removed from the anode as ions. Silver atoms are oxidized at the anode according to the following half-reaction.

$$\overset{0}{Ag} \longrightarrow \overset{+1}{Ag^+} + e^-$$

This action maintains the Ag^+ ion concentration of the solution. Thus, in effect, silver is transferred from the anode to the cathode of the cell.

Rechargeable Cells

A rechargeable cell combines the oxidation-reduction chemistry of both voltaic cells and electrolytic cells. When a rechargeable cell converts chemical energy to electrical energy, it operates as a voltaic cell. But when the cell is recharged, it operates as an electrolytic cell, converting electrical energy to chemical energy.

The standard 12 V automobile battery, shown in Figure 19-16, is a set of six rechargeable cells. The anode in each cell is lead submerged in a solution of H_2SO_4. The anode half-reaction is described by the following equation.

$$Pb(s) + SO_4^{2-}(aq) \longrightarrow PbSO_4(s) + 2e^-$$

Electrons move through the circuit to the cathode, where PbO_2 is reduced according to the following equation.

$$PbO_2(s) + 4H^+(aq) + SO_4^{2-}(aq) + 2e^- \longrightarrow PbSO_4(s) + 2H_2O(l)$$

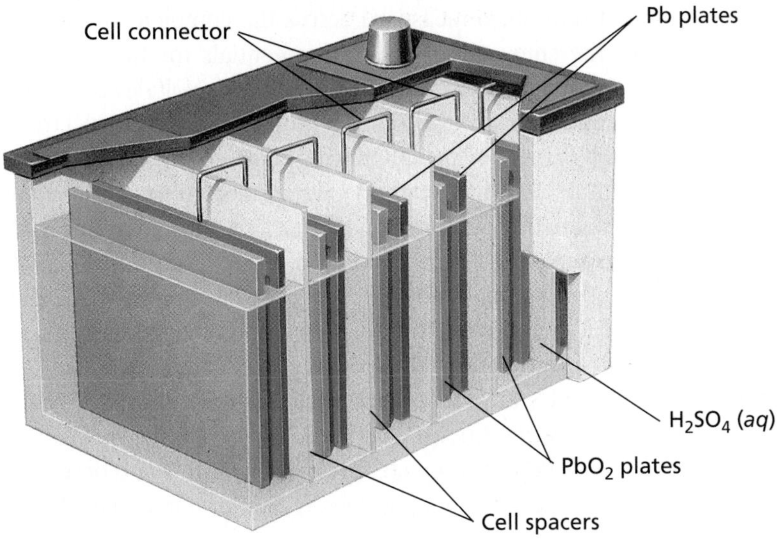

FIGURE 19-16 The rechargeable cells of a car battery produce electricity from reactions between lead(IV) oxide, lead, and sulfuric acid.

The net oxidation-reduction reaction for the discharge cycle is described by the following chemical equation.

$$Pb(s) + PbO_2(s) + 2H_2SO_4(aq) \longrightarrow 2PbSO_4(s) + 2H_2O(l)$$

A car's battery produces the electric energy needed to start its engine. Sulfuric acid, present as its ions, is consumed, and lead(II) sulfate accumulates as a white powder on the electrodes. Once the car is running, the half-reactions are reversed by a voltage produced by the alternator. The Pb, PbO_2, and H_2SO_4 are regenerated. A battery can be recharged as long as all reactants necessary for the electrolytic reaction are present, and all reactions are reversible.

Electrode Potentials

Reconsider the voltaic cell shown in Figure 19-9. There are two electrodes, Zn and Cu. According to Table 19-3, these two metals each have different tendencies for accepting electrons. *This tendency for the half-reaction of either copper or zinc to occur as a reduction half-reaction in an electrochemical cell can be quantified as a* **reduction potential.** There are two half-cells in Figure 19-9: a strip of zinc placed in a solution of $ZnSO_4$ and a strip of copper placed in a solution of $CuSO_4$. *The difference in potential between an electrode and its solution is known as its* **electrode potential.** When these two half-cells are connected and the reaction begins, a difference in potential develops between the electrodes. This potential difference, or voltage, is a measure of the energy required to move a certain electric charge between the electrodes. Potential difference is measured in volts. A voltmeter connected across the Cu | Zn voltaic cell measures a potential difference of about 1.10 V when the solution concentrations of Zn^{2+} and Cu^{2+} ions are each 1 M.

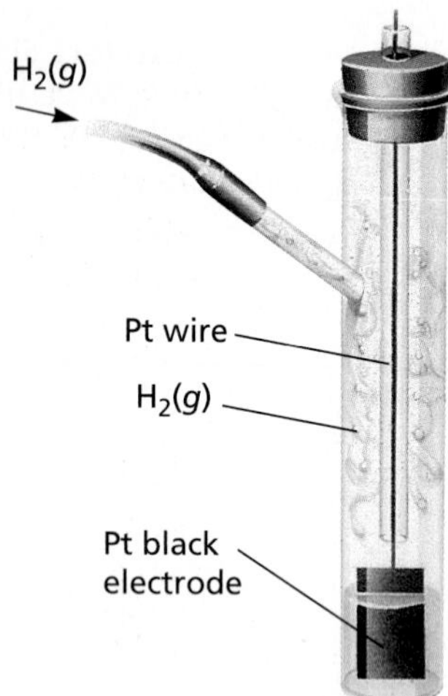

FIGURE 19-17 A hydrogen electrode is the standard reference electrode for measuring electrode potentials. The electrode surface in contact with the solution is actually a layer of hydrogen adsorbed onto the surface of the platinum.

The potential difference measured across the complete voltaic cell roughly equals the sum of the electrode potentials for the two half-reactions. But while the potential difference across a voltaic cell is easily measured, there is no way to measure an individual electrode potential directly. This is because there can be no transfer of electrons unless both the anode and the cathode are connected to form a complete circuit. A relative value for the potential of a half-reaction can be determined by connecting it to a standard half-cell as a reference. This standard half-cell, shown in Figure 19-17, is called a standard hydrogen electrode, or SHE. It consists of a platinum electrode dipped into a 1.00 M acid solution surrounded by hydrogen gas at 1 atm pressure and 25°C. Other electrodes are ranked according to their ability to reduce hydrogen under these conditions.

The anodic reaction for the standard hydrogen electrode is described by the forward half-reaction in the following equilibrium equation.

$$\overset{0}{H_2}(g) \rightleftharpoons \overset{+1}{2H^+}(aq) + 2e^-$$

The cathodic half-reaction is the reverse. An arbitrary potential of 0.00 V is assigned to both of these half-reactions. Therefore, any voltage measurement obtained is attributed to the half-cell connected to the SHE. *A half-cell potential measured relative to a potential of zero for the standard hydrogen electrode is a* **standard electrode potential,** E^0. Electrode potentials are expressed as potentials for reduction. These reduction potentials provide a reliable indication of the tendency of a substance to be reduced. Half-reactions for some common electrodes and their standard electrode potentials are listed in Table 19-4.

Positive E^0 values indicate that hydrogen is more willing to give up its electrons than the other electrode. Half-reactions with positive reduction potentials are favored. Effective oxidizing agents, such as Cu and F_2, have positive E^0 values. Half-reactions with negative reduction potentials are not favored; these half-reactions prefer oxidation over reduction. Negative E^0 values indicate that the metal or other electrode is more willing to give up electrons than hydrogen. Effective reducing agents, such as Li and Zn, have negative E^0 values. When a half-reaction

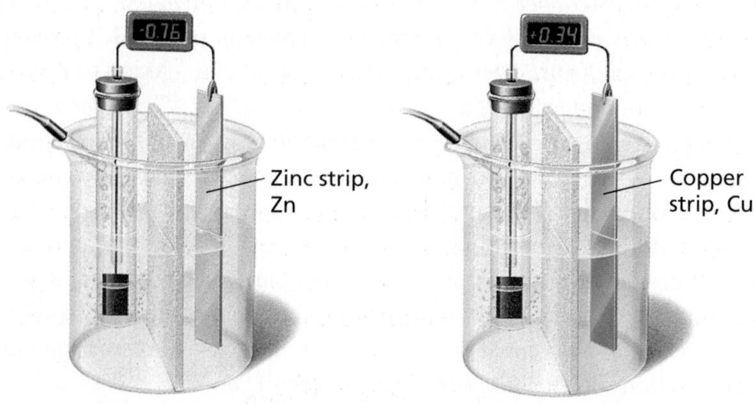

FIGURE 19-18 The electrode potentials of zinc and copper half-cells are measured by coupling them with a standard hydrogen electrode.

TABLE 19-4 *Standard Reduction Potentials*

Half-cell reaction	Standard electrode potential, E^0 (in volts)	Half-cell reaction	Standard electrode potential, E^0 (in volts)
$F_2 + 2e^- \rightleftarrows 2F^-$	+2.87	$Fe^{3+} + 3e^- \rightleftarrows Fe$	−0.04
$MnO_4^- + 8H^+ + 5e^- \rightleftarrows Mn^{2+} + 4H_2O$	+1.50	$Pb^{2+} + 2e^- \rightleftarrows Pb$	−0.13
$Au^{3+} + 3e^- \rightleftarrows Au$	+1.50	$Sn^{2+} + 2e^- \rightleftarrows Sn$	−0.14
$Cl_2 + 2e^- \rightleftarrows 2Cl^-$	+1.36	$Ni^{2+} + 2e^- \rightleftarrows Ni$	−0.26
$Cr_2O_7^{2-} + 14H^+ + 6e^- \rightleftarrows 2Cr^{3+} + 7H_2O$	+1.23	$Co^{2+} + 2e^- \rightleftarrows Co$	−0.28
$MnO_2 + 4H^+ + 2e^- \rightleftarrows Mn^{2+} + 2H_2O$	+1.22	$Cd^{2+} + 2e^- \rightleftarrows Cd$	−0.40
$Br_2 + 2e^- \rightleftarrows 2Br^-$	+1.07	$Fe^{2+} + 2e^- \rightleftarrows Fe$	−0.45
$Hg^{2+} + 2e^- \rightleftarrows Hg$	+0.85	$S + 2e^- \rightleftarrows S^{2-}$	−0.48
$Ag^+ + e^- \rightleftarrows Ag$	+0.80	$Cr^{3+} + 3e^- \rightleftarrows Cr$	−0.74
$Hg_2^{2+} + 2e^- \rightleftarrows 2Hg$	+0.80	$Zn^{2+} + 2e^- \rightleftarrows Zn$	−0.76
$Fe^{3+} + e^- \rightleftarrows Fe^{2+}$	+0.77	$Al^{3+} + 3e^- \rightleftarrows Al$	−1.66
$MnO_4^- + e^- \rightleftarrows MnO_4^{2-}$	+0.56	$Mg^{2+} + 2e^- \rightleftarrows Mg$	−2.37
$I_2 + 2e^- \rightleftarrows 2I^-$	+0.54	$Na^+ + e^- \rightleftarrows Na$	−2.71
$Cu^{2+} + 2e^- \rightleftarrows Cu$	+0.34	$Ca^{2+} + 2e^- \rightleftarrows Ca$	−2.87
$Cu^{2+} + e^- \rightleftarrows Cu^+$	+0.15	$Ba^{2+} + 2e^- \rightleftarrows Ba$	−2.91
$S + 2H^+(aq) + 2e^- \rightleftarrows H_2S(aq)$	+0.14	$K^+ + e^- \rightleftarrows K$	−2.93
$2H^+(aq) + 2e^- \rightleftarrows H_2$	0.00	$Li^+ + e^- \rightleftarrows Li$	−3.04

is written as an oxidation reaction, the sign of its electrode potential is changed as shown for the oxidation and reduction half-reactions for zinc.

$$Zn^{2+} + 2e^- \longrightarrow Zn \qquad E^0 = -0.76\text{ V}$$
$$Zn \longrightarrow Zn^{2+} + 2e^- \qquad E^0 = +0.76\text{ V}$$

To measure the reduction potential of a zinc half-cell, it is connected to a standard hydrogen electrode, as in Figure 19-18. The potential difference across the cell is −0.76 V. The negative number indicates that electrons flow through the external circuit from the zinc electrode, where zinc is oxidized, to the hydrogen electrode, where aqueous hydrogen ions are reduced.

A copper half-cell coupled with the standard hydrogen electrode gives a potential difference measurement of +0.34 V. This positive number indicates that $Cu^{2+}(aq)$ ions are more readily reduced than $H^+(aq)$ ions.

Standard electrode potentials can be used to predict if a redox reaction will occur naturally. A naturally occurring reaction will have a positive value for E^0_{cell}, which is calculated using the following equation.

$$E^0_{cell} = E^0_{cathode} - E^0_{anode}$$

When evaluating the two half-reactions in a voltaic cell, the reaction with the lower standard reduction potential is the anode. Oxidation occurs at the anode, so the half-cell reaction is the reverse of the reduction reaction found in Table 19-4. When a reaction is reversed, the actual half-cell potential is the negative of the standard reduction potential. For this reason, the total potential of a cell is calculated by subtracting the standard reduction potential for the reaction at the anode (E^0_{anode}) from the standard reduction potential for the reaction at the cathode ($E^0_{cathode}$).

Consider cells made from Fe in a solution of $Fe(NO_3)_3$ and Ag in a solution of $AgNO_3$. Table 19-4 gives the following half-reactions and E^0 values for these half-cells.

$$Fe^{3+}(aq) + 3e^- \longrightarrow Fe(s) \qquad E^0 = -0.04 \text{ V}$$
$$3Ag^+(aq) + 3e^- \longrightarrow 3Ag(s) \qquad E^0 = +0.80 \text{ V}$$

Fe in $Fe(NO_3)_3$ is the anode because it has the lower reduction potential, and Ag in $AgNO_3$ is therefore the cathode. The reduction of silver ions is multiplied by 3 so that the number of electrons lost in that reaction equals the number of electrons gained in the oxidation of iron. The standard reduction potentials for the anode and cathode are as follows.

$$E^0_{anode} = -0.04 \text{ V}$$
$$E^0_{cathode} = +0.80 \text{ V}$$

Note that when a half-reaction is multiplied by a constant, the E^0 value is not multiplied but remains the same.

Using the above equation, the potential for the cell can be calculated as follows.

$$E^0_{cell} = E^0_{cathode} - E^0_{anode}$$
$$E^0_{cell} = +0.80 \text{ V} - (-0.04 \text{ V})$$
$$E^0_{cell} = 0.84 \text{ V}$$

If the calculated value for E^0_{cell} is negative, the reaction does not occur naturally in the direction written, so it will not occur in a voltaic cell. It can be made to occur in an electrolytic cell.

SECTION REVIEW

1. What is a voltaic cell?
2. Describe an electrolytic cell.
3. Explain the process of electroplating.
4. What is a rechargeable cell?
5. What is electrode potential, and how is it used to calculate information about an electrochemical reaction?
6. Given the Na^+/Na and K^+/K half-cells, determine the overall electrochemical reaction that proceeds spontaneously and the E^0 value.

CHAPTER 19 REVIEW

CHAPTER SUMMARY

19-1

- Oxidation numbers are assigned by the set of rules listed in Table 19-1.
- Oxidation-reduction reactions consist of two half-reactions that must occur simultaneously.
- Oxidation-reduction reactions are identified by examining the reactants and products for changes in oxidation numbers of their constituent atoms.
- Oxidation involves the loss of electrons, and reduction involves the gain of electrons.

Vocabulary

half-reaction (593)
oxidation (592)
oxidation-reduction reaction (593)
oxidized (592)
redox reaction (593)
reduced (593)
reduction (593)

19-2

- Both charge and mass are conserved in a balanced redox equation.
- In the half-reaction method for balancing equations, oxidation and reduction equations are balanced separately for atoms and charge. Then they are combined to give a complete balanced equation.

19-3

- In redox reactions, the substance that is *reduced* acts as an *oxidizing agent* because it *acquires* electrons from the substance oxidized.
- The substance that is *oxidized* in a redox reaction is the *reducing agent* because it *supplies* the electrons to the substance reduced.

Vocabulary

autooxidation (605)
oxidizing agent (602)
reducing agent (602)

19-4

- Some oxidation-reduction reactions that occur naturally can be sources of electrical energy in voltaic cells. Other oxidation-reduction reactions that do not occur naturally can be driven by an external source of electrical energy in electrolytic cells; this process is called electrolysis.
- An electrode and its electrolyte in an electrochemical cell are called a half-cell.
- The potential difference between the electrode and its solution is called the electrode potential. The sum of the electrode potentials of the two half-reactions of an electrochemical cell is roughly equal to the potential difference across the cell.
- Standard electrode potentials are measured relative to a standard hydrogen electrode. They indicate the relative strengths of substances as oxidizing and reducing agents.

Vocabulary

anode (607)
cathode (607)
electrochemical cell (607)
electrochemistry (606)
electrode (607)
electrode potential (613)
electrolysis (610)
electrolytic cell (610)
electroplating (611)
half-cell (607)
reduction potential (613)
standard electrode potential (614)
voltaic cell (608)

REVIEWING CONCEPTS

1. a. Distinguish between the processes of oxidation and reduction.
b. Write an equation to illustrate each. (19-1)

2. Which of the following are redox reactions?
a. $2Na + Cl_2 \longrightarrow 2NaCl$
b. $C + O_2 \longrightarrow CO_2$
c. $2H_2O \longrightarrow 2H_2 + O_2$
d. $NaCl + AgNO_3 \longrightarrow AgCl + NaNO_3$
e. $NH_3 + HCl \longrightarrow NH_4^+ + Cl^-$
f. $2KClO_3 \longrightarrow 2KCl + 3O_2$
g. $H_2 + Cl_2 \longrightarrow 2HCl$
h. $H_2SO_4 + 2KOH \longrightarrow K_2SO_4 + 2H_2O$
i. $Zn + CuSO_4 \longrightarrow ZnSO_4 + Cu$ (19-1)

3. For each oxidation-reduction reaction in the previous question, identify the substance oxidized and the substance reduced. (19-1)

4. a. Identify the most active reducing agent among all common elements.
b. Why are all of the elements in its group in the periodic table very active reducing agents?
c. Identify the most active oxidizing agent among the common elements. (19-3)

5. Based on Table 19-3, identify the strongest and weakest reducing agents among the substances listed within each of the following groupings:
a. Ca, Ag, Sn, Cl^-
b. Fe, Hg, Al, Br^-
c. F^-, Pb, Mn^{2+}, Na
d. Cr^{3+}, Cu^{2+}, NO_3^-, K^+
e. Cl_2, S, Zn^{2+}, Ag^+
f. Li^+, F_2, Ni^{2+}, Fe^{3+} (19-3)

6. Use Table 19-3 to respond to each of the following:
a. Would Al be oxidized by Ni^{2+}?
b. Would Cu be oxidized by Ag^+?
c. Would Pb be oxidized by Na^+?
d. Would F_2 be reduced by Cl^-?
e. Would Br_2 be reduced by Cl^-? (19-3)

7. Distinguish between a voltaic cell and an electrolytic cell in terms of the nature of the reaction involved. (19-4)

8. a. What is electroplating?
b. Distinguish between the nature of the anode and cathode in such a process. (19-4)

9. a. Explain what is meant by the potential difference between the two electrodes in an electrochemical cell.
b. How, and in what units, is this potential difference measured? (19-4)

10. The standard hydrogen electrode is assigned an electrode potential of 0.00 V. Explain why this voltage is assigned. (19-4)

11. a. What information is provided by the electrode potential of a given half-cell?
b. What does the relative value of the potential of a given half-reaction indicate about its oxidation-reduction tendency? (19-4)

PROBLEMS

Redox Equations

12. Each of the following atom/ion pairs undergoes the oxidation number change indicated below. For each pair, determine whether oxidation or reduction has occurred, and then write the electronic equation indicating the corresponding number of electrons lost or gained.
a. $K \longrightarrow K^+$
b. $S \longrightarrow S^{2-}$
c. $Mg \longrightarrow Mg^{2+}$
d. $F^- \longrightarrow F_2$
e. $H_2 \longrightarrow H^+$
f. $O_2 \longrightarrow O^{2-}$
g. $Fe^{3+} \longrightarrow Fe^{2+}$
h. $Mn^{2+} \longrightarrow MnO_4^-$

13. Identify the following reactions as redox or nonredox:
a. $2NH_4Cl(aq) + Ca(OH)_2(aq) \longrightarrow 2NH_3(aq) + 2H_2O(l) + CaCl_2(aq)$
b. $2HNO_3(aq) + 3H_2S(g) \longrightarrow 2NO(g) + 4H_2O(l) + 3S(s)$
c. $[Be(H_2O)_4]^{2+}(aq) + H_2O(l) \longrightarrow H_3O^+(aq) + [Be(H_2O)_3OH]^+(aq)$

14. Arrange the following in order of increasing oxidation number of the xenon atom:
$CsXeF_8$, Xe, XeF_2, $XeOF_2$, XeO_3, XeF

15. Determine the oxidation number of each atom indicated in the following:

a. H_2
b. H_2O
c. Al
d. MgO
e. Al_2S_3
f. HNO_3
g. H_2SO_4
h. $Ca(OH)_2$
i. $Fe(NO_3)_2$
j. O_2

16. Balance the oxidation-reduction equation below by using the half-reaction method in response to each requested step. (Hint: See Sample Problem 19-1.)

$$K + H_2O \longrightarrow KOH + H_2$$

a. Write the ionic equation, and assign oxidation numbers to all atoms to determine what is oxidized and what is reduced.
b. Write the equation for the reduction, and balance it for both atoms and charge.
c. Write the equation for the oxidation, and balance it for both atoms and charge.
d. Adjust the oxidation and reduction equations by multiplying the coefficients as needed so that electrons lost equal electrons gained, and add the two resulting equations.
e. Add species as necessary to balance the overall formula equation.

17. Use the method in the previous problem to balance each of the reactions below.

a. $HI + HNO_2 \longrightarrow NO + I_2 + H_2O$
b. $FeCl_3 + H_2S \longrightarrow FeCl_2 + HCl + S$

18. Balance the equation for the reaction in which hot, concentrated sulfuric acid reacts with zinc to form zinc sulfate, hydrogen sulfide, and water.

Voltaic and Electrolytic Cells

19. For each of the following pairs of half-cells, determine the overall electrochemical reaction that proceeds spontaneously:

a. Cu^{2+}/Cu, Ag^+/Ag
b. Cd^{2+}/Cd, Co^{2+}/Co
c. Na^+/Na, Ni^{2+}/Ni
d. I_2/I^-, Br_2/Br^-

20. Determine the values of E^0 for the cells in the previous problem.

21. Suppose chemists had chosen to make the $I_2 + 2e^- \rightleftharpoons 2I^-$ half-cell the standard electrode and had assigned it a potential of zero volts.

a. What would be the E^0 value for the $Br_2 + 2e^- \rightleftharpoons 2Br^-$ half-cell?
b. What would be the E^0 value for the $Al^{3+} + 3e^- \rightleftharpoons Al$ half-cell?
c. How much change would be observed in the E^0 value for the reaction involving $Br_2 + I^-$ using the I_2 half-cell as the standard?

22. If a strip of Ni were dipped into a solution of $AgNO_3$, what would be expected to occur? Explain, using E^0 values and equations.

23. a. What would happen if an aluminum spoon were used to stir a solution of $Zn(NO_3)_2$?
b. Could a strip of Zn be used to stir a solution of $Al(NO_3)_3$? Explain, using E^0 values.

24. How do the redox reactions for each of the following types of batteries differ?

a. zinc-carbon
b. alkaline
c. mercury

25. a. Why are some standard reduction potentials positive and some negative?
b. Compare the E^0 value for a metal with the reactivity of that metal.

MIXED REVIEW

26. Balance the following redox equations:

a. $SbCl_5 + KI \longrightarrow KCl + I_2 + SbCl_3$
b. $Ca(OH)_2 + NaOH + ClO_2 + C \longrightarrow NaClO_2 + CaCO_3 + H_2O$

27. Predict whether each of the following reactions will occur spontaneously as written by determining the E^0 value for potential reaction. Write and balance the overall equation for each reaction that does occur.

a. $Mg + Sn^{2+} \longrightarrow$
b. $K + Al^{3+} \longrightarrow$
c. $Li^+ + Zn \longrightarrow$
d. $Cu + Cl_2 \longrightarrow$

28. Why is it possible for alkaline batteries to be smaller than zinc-carbon dry cells?

29. Draw a diagram of a voltaic cell whose two half-reactions consist of Ag in $AgNO_3$ and Ni in $NiSO_4$. Identify the anode and cathode, and indicate the directions in which the electrons and ions are moving.

30. Identify the following reactions as redox or nonredox:

a. $Mg(s) + ZnCl_2(aq) \longrightarrow Zn(s) + MgCl_2(aq)$

b. $H_2(g) + OF_2(g) \longrightarrow H_2O(g) + HF(g)$

c. $2KI(aq) + Pb(NO_3)_2(aq) \longrightarrow PbI_2(s) + 2KNO_3(aq)$

d. $CaO(s) + H_2O(l) \longrightarrow Ca(OH)_2(aq)$

e. $3CuCl_2(aq) + 2(NH_4)_3PO_4(aq) \longrightarrow 6NH_4Cl(aq) + Cu_3(PO_4)_2(s)$

f. $CH_4(g) + 2O_2(g) \longrightarrow CO_2(g) + 2H_2O(g)$

31. Can a solution of $Sn(NO_3)_2$ be stored in an aluminum container? Explain, using E^0 values.

32. A voltaic cell is made up of a cadmium electrode in a solution of $CdSO_4$ and a zinc electrode in a solution of $ZnSO_4$. The two half-cells are separated by a porous barrier.

a. Which is the cathode, and which is the anode?

b. In which direction are the electrons flowing?

c. Write balanced equations for the two half-reactions, and write a net equation for the combined reaction.

33. Would the following pair of electrodes make a good battery? Explain.

$$Cd \longrightarrow Cd^{2+} + 2e^-$$
$$Fe \longrightarrow Fe^{2+} + 2e^-$$

34. Arrange the following in order of decreasing oxidation number of the nitrogen atom: N_2, NH_3, N_2O_4, N_2O, N_2H_4, NO_3^-

CRITICAL THINKING

35. Applying Models Explain how the oxidation-reduction chemistry of both the voltaic cell and the electrolytic cell are combined in the chemistry of rechargeable cells.

36. Applying Ideas In lead/acid batteries, such as your car battery, the charge of the battery can be determined by measuring the density of the battery fluid. Explain how this is possible.

37. Interpreting Graphics A galvanic cell is pictured below. The temperature is 25°C. Identify the species that is oxidized if current is allowed to flow.

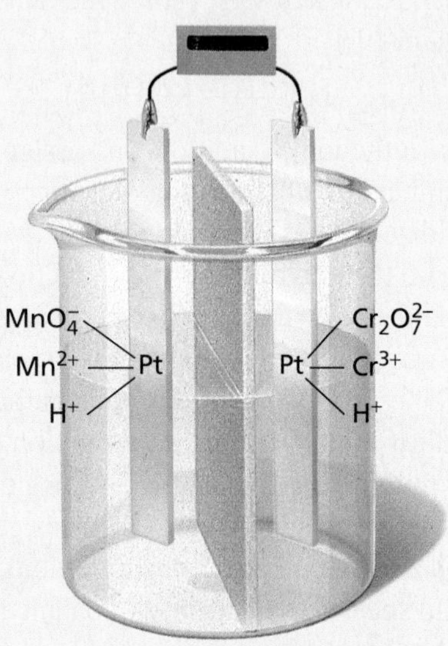

TECHNOLOGY & LEARNING

38. Graphing Calculator: Calculate the Equilibrium Constant Using the Standard Cell Voltage

The graphing calculator can be programmed to calculate the equilibrium constant for an electrochemical cell using an equation called the Nernst equation, given the standard potential and the number of electrons transferred. Given that the standard potential is 2.041 V and that two electrons are transferred, you will calculate the equilibrium constant. Begin by programming the calculator to carry out the calculation. Next, the program will be used to make calculations.

A. Program the calculation.

Keystrokes for naming the program:

PRGM ▶ ▶ ENTER.

Name the program: N E R N S T ENTER.

Program keystrokes: PRGM ► 2 ALPHA E 2nd : PRGM ► 2 ALPHA N 2nd : 1 0 ^ (ALPHA E × ALPHA N ÷ . 0 5 9 2) STO► ALPHA K 2nd : PRGM ► 8 2nd : PRGM ► 3 2nd ALPHA " E Q U I L I B R I U M ␣ C O N S T A N T 2nd TEST 1 ALPHA " 2nd : PRGM ► 3 ALPHA K

B. Check the display.

Screen display: Prompt E: Prompt N: 10^(E*N/.0592) ⟶ K: ClrHome: Disp "EQUILIBRIUM CONSTANT=": Disp K

Press 2nd QUIT to exit the program editor.

C. Run the program.

a. Press PRGM. Select NERNST and press ENTER ENTER. Now enter the standard potential of 2.041 V and press ENTER. Enter 2 for the number of electrons transferred and press ENTER. The calculator will provide the equilibrium constant.

b. Use the NERNST program to calculate the equilibrium constant for a reaction with a standard potential of 0 V and 2 electrons transferred.

HANDBOOK SEARCH

39. Several reactions of aluminum are shown in the common reactions section for Group 13 of the *Elements Handbook*. Use these reactions to answer the following.

a. Which of the five reactions shown are oxidation-reduction reactions? How do you know?

b. For the redox reactions you listed for item a, identify the substance oxidized and the substance reduced.

c. Write half-reactions for each equation you listed in item a.

40. Aluminum is described in Group 13 of the *Elements Handbook* as a self-protecting metal. This property of aluminum results from a redox reaction.

a. Write the redox equation for the oxidation of aluminum.

b. Write the half-reactions for this reaction showing the number of electrons transferred.

c. What problems are associated with the buildup of aluminum oxide on electrical wiring made of aluminum?

RESEARCH & WRITING

41. Go to the library and find out what you can about the electroplating industry in the United States. What are the top three metals used for plating, and how many metric tons of each are used in the United States each year for electroplating?

42. Investigate the types of batteries being considered for electric cars. Write a report on the advantages and disadvantages of these types of batteries.

ALTERNATIVE ASSESSMENT

43. Performance Take an inventory of the types of batteries used in your home. Find out the voltage supplied by each battery and what electrochemical reaction each uses. Suggest why that electrochemical reaction is used in each case.

44. In our portable society, batteries have become a necessary power supply. As consumers, we want to purchase batteries that will last as long as possible. Advertisements tell us that some batteries last longer than others, but do they really? Design an investigation to answer the question. Is there a difference in longevity among the major brands of AA batteries? Add a cost-effectiveness component to your design.

45. When someone with a silver filling in a tooth bites down on an aluminum gum wrapper, saliva acts as an electrolyte. The system is an electrochemical cell which produces a small jolt of pain. Explain what occurs using half-cell reactions and $E°$ values.

UNIT 7

Organic and Nuclear Chemistry

CHAPTERS

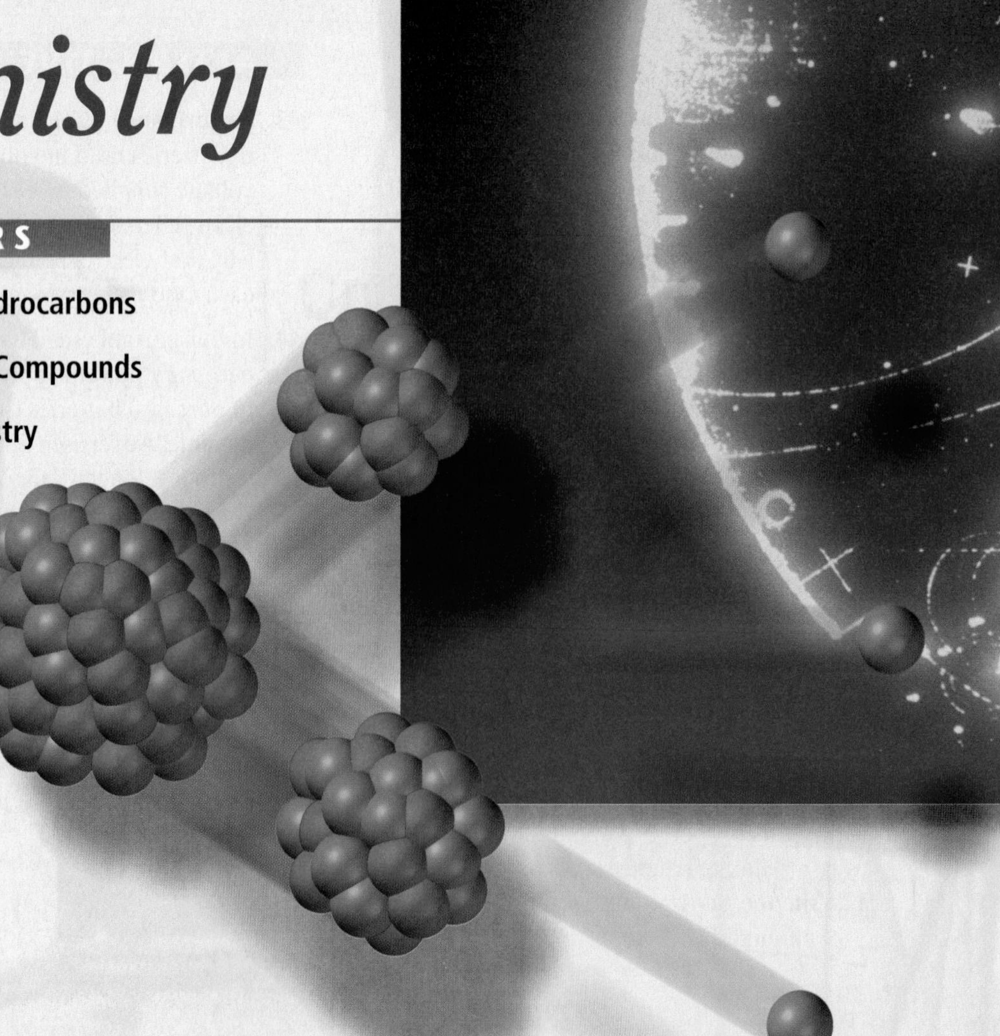

JOHN DEWEY, ON THE INSPIRATION FOR SCIENTIFIC ADVANCEMENT

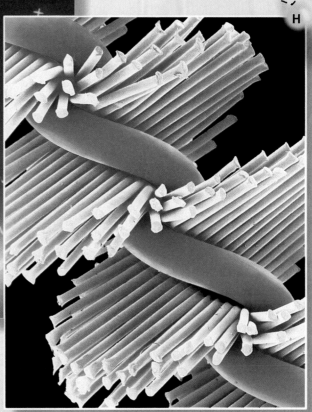

Every great advance in science has issued from a new audacity of imagination.

(John Dewey, *The Quest for Certainty*)

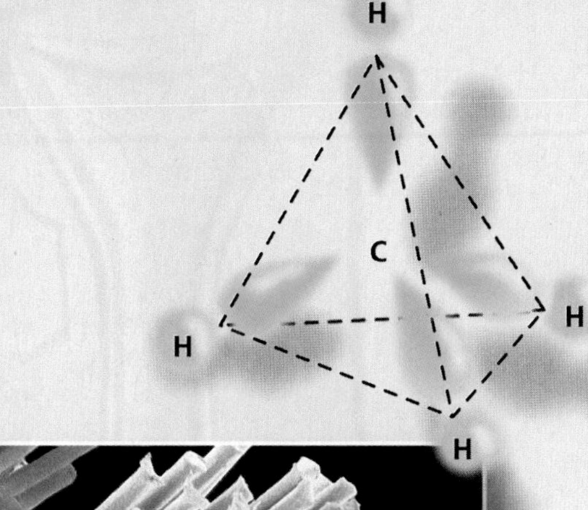

CHAPTER 20

Carbon and Hydrocarbons

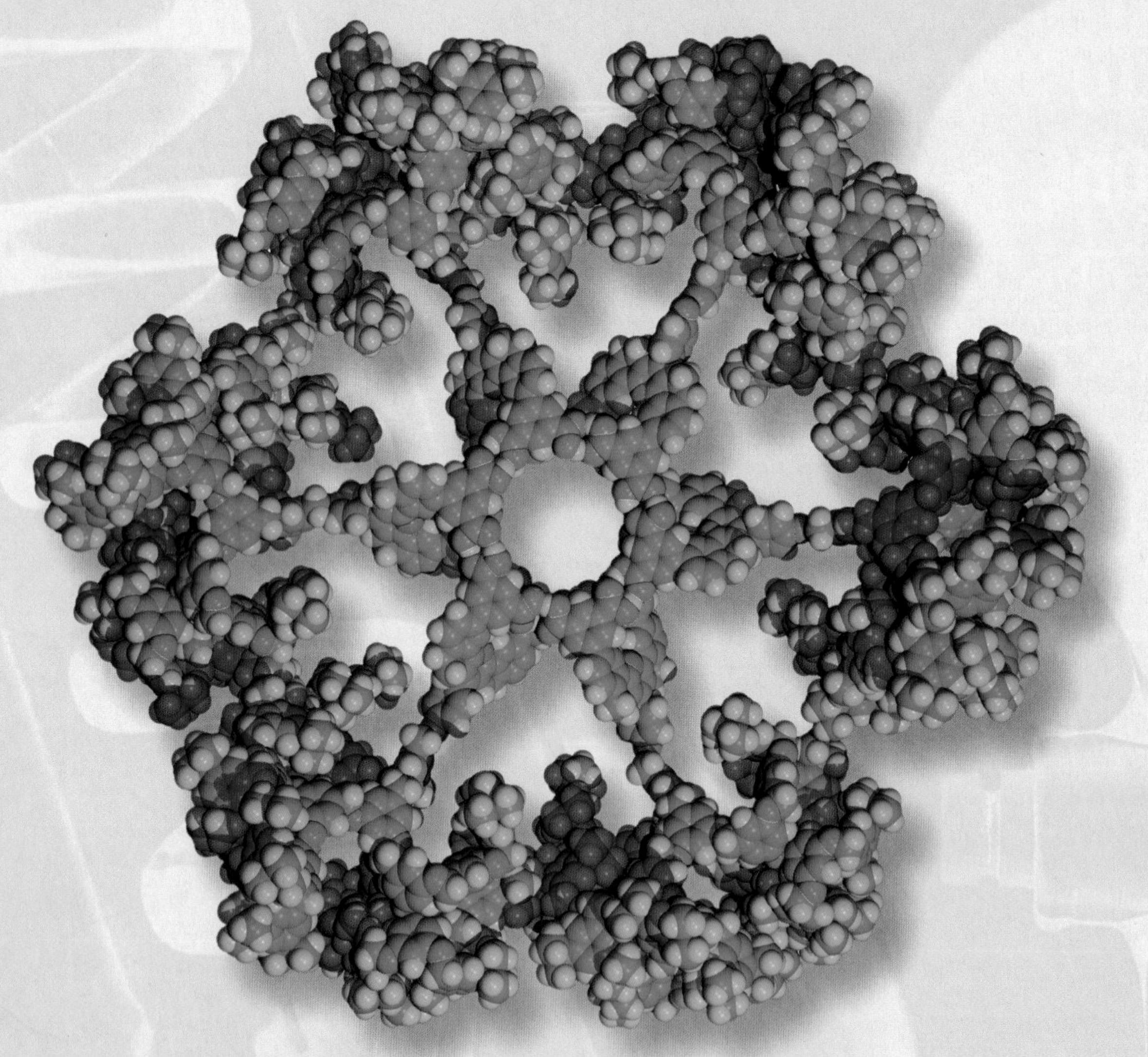

Three-dimensional models help us visualize the shape of carbon compounds.

Abundance and Importance of Carbon

SECTION 20-1

OBJECTIVES

- Relate the ability of a carbon atom to form covalent bonds to its atomic structure and hybrid orbitals.
- Identify the different allotropes of carbon and their structural differences.
- Explain how the different structures of carbon allotropes affect their properties.

Carbon is found in nature both as an element and in combined form. Although carbon ranks about 17th in abundance by mass among the elements in the Earth's crust, it is exceedingly important because it is found in all living matter. Carbon is present in body tissues and in the foods you eat. It is also found in common fuels, such as coal, petroleum, natural gas, and wood.

Structure and Bonding of Carbon

Carbon, the first member of Group 14, has mostly nonmetallic properties. In its ground state, a carbon atom has an electronic configuration of $1s^22s^22p^2$. The two $1s$ electrons are tightly bound to the nucleus. The two $2s$ electrons and the two $2p$ electrons are the valence electrons. Carbon atoms show a very strong tendency to share electrons and form covalent bonds.

As was covered in Chapter 6, hybridization can be used to explain the bonding and geometry of most carbon compounds. Carbon atoms that form four single bonds have four sp^3 orbitals. These orbitals are directed toward the four corners of a regular tetrahedron, as shown in Figure 20-1. This results in the tetrahedral shape of methane, CH_4, and the zigzag pattern of molecules with multiple single-bonded carbon atoms, such as C_4H_{10}.

FIGURE 20-1 The orbital models show how the orientation of sp^3 hybrid orbitals relates to the geometry of CH_4 and C_4H_{10}.

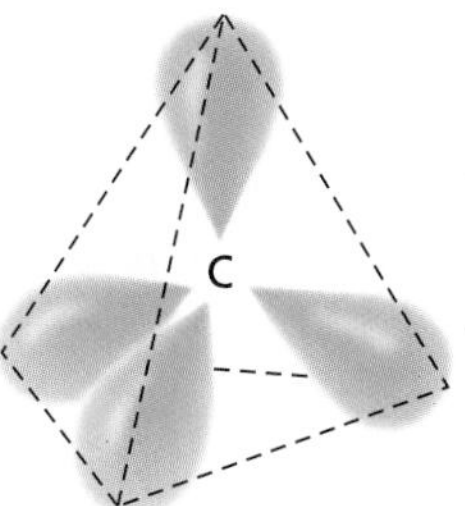

sp^3 hybrid orbitals

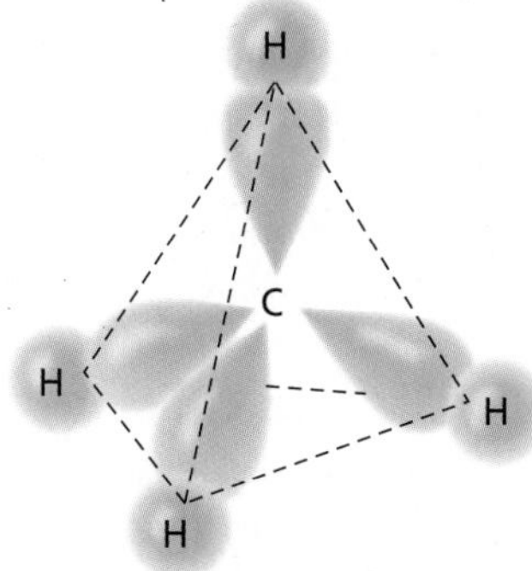

CH_4 orbital overlap

CH_4

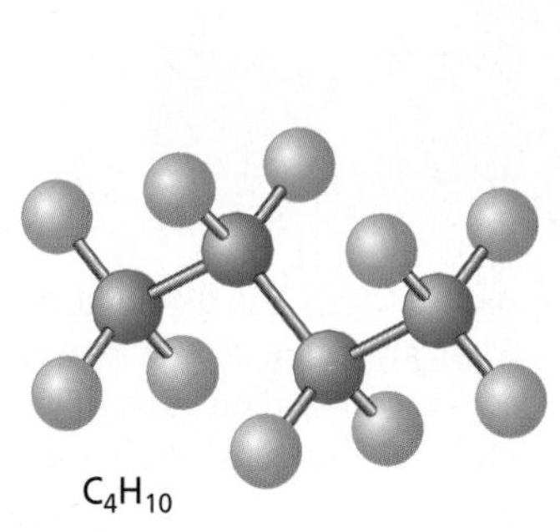

C_4H_{10}

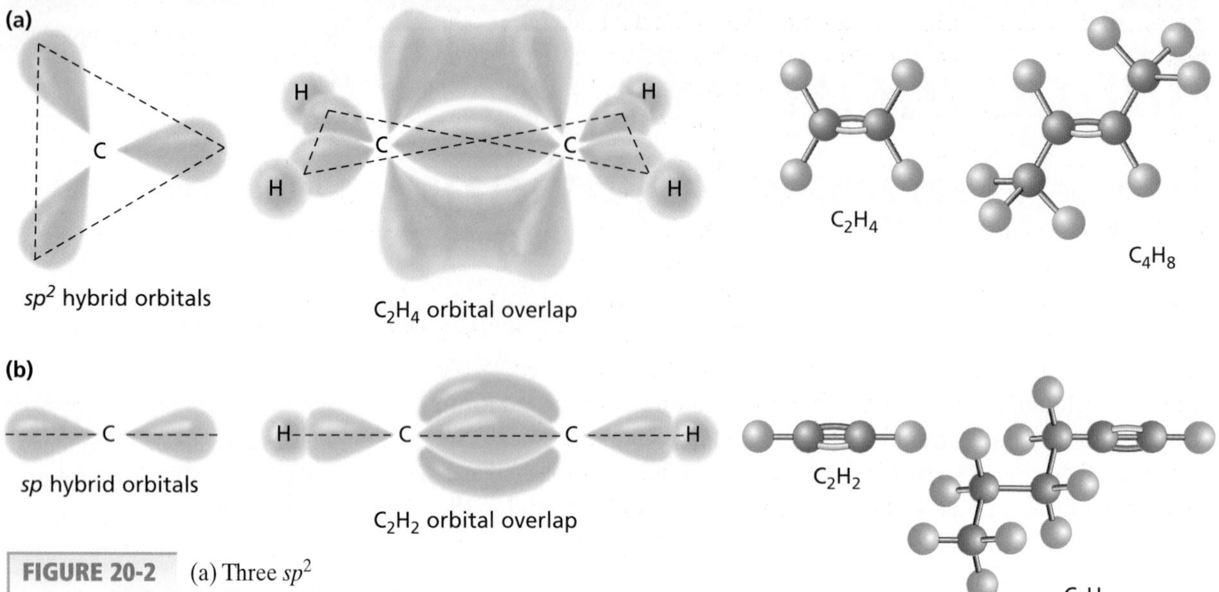

FIGURE 20-2 (a) Three sp^2 hybrid orbitals lie in the same plane. The C_2H_4 orbital overlap model shows the orientation of sp^2 hybrid orbitals in molecules that contain a double bond, such as C_2H_4 and C_4H_8. (b) The C_2H_2 orbital overlap model shows the orientation of *sp* hybrid orbitals in molecules that contain a triple bond, such as C_2H_2 and C_6H_{10}.

Carbon atoms form double bonds through sp^2 hybridization, as shown in Figure 20-2(a). When carbon atoms form double bonds, the sp^2 hybrid orbitals of both carbon atoms lie in the same plane, as shown in the orbital overlap model of ethene, C_2H_4. Because the hydrogen atoms of C_2H_4 also bond with carbon sp^2 orbitals, all six atoms lie in the same plane. The three-dimensional models of C_2H_4 and C_4H_8 show the geometry of molecules containing carbon-carbon double bonds.

Carbon triple bonds are linear due to the linear arrangement of two *sp* hybrid orbitals, as shown in Figure 20-2(b). This can be seen in the orbital overlap model for ethyne, C_2H_2. The three-dimensional models of C_2H_2 and C_6H_{10} show the geometry of molecules containing carbon-carbon triple bonds.

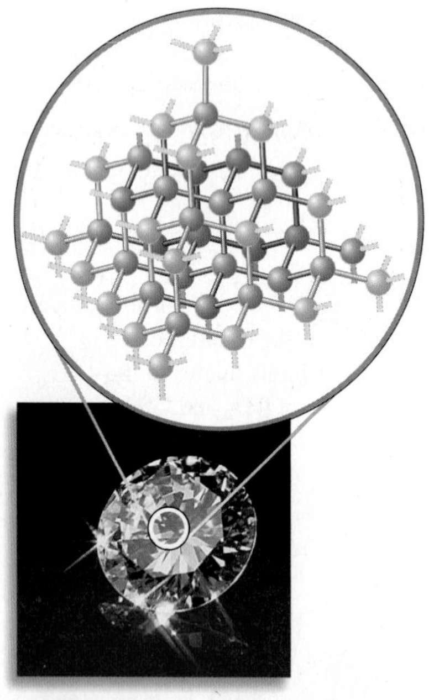

FIGURE 20-3 In diamond, the carbon atoms are densely packed because each carbon atom is bonded to four tetrahedrally oriented carbon atoms.

Allotropes of Carbon

Carbon occurs in several solid allotropic forms that have dramatically different properties. **Diamond** *is a colorless, crystalline, solid form of carbon.* **Graphite** *is a soft, black, crystalline form of carbon that is a fair conductor of electricity.* **Fullerenes** *are dark-colored solids made of spherically networked carbon-atom cages.*

Diamond

Diamond is the hardest material known. It is the most dense form of carbon—about 3.5 times more dense than water. It also has an extremely high melting point (greater than 3500°C). These properties of diamond can be explained by its structure. The model in Figure 20-3 shows carbon atoms in diamond bonded covalently in a network fashion. Each carbon atom is tetrahedrally oriented to its four nearest neighbors. The distance between the carbon-atom nuclei has been measured to be 154 pm. Because of diamond's extreme hardness and high melting point, its major industrial uses are for cutting, drilling, and grinding. Diamonds used in industry are not of gem quality.

Another property of diamond is its ability to conduct heat. A diamond crystal conducts heat more than five times more readily than silver or copper, which are the best metallic conductors. In diamond, heat is conducted by the transfer of energy of vibration from one carbon atom to the next. In a diamond crystal, this process is very efficient because the carbon atoms have a small mass. The forces holding the atoms together are strong and can easily transfer vibratory motion among the atoms. However, unlike metals, diamond does not conduct electricity. Because all the valence electrons are used in forming localized covalent bonds, none of the electrons can migrate.

Graphite

Graphite is nearly as remarkable for its softness as diamond is for its hardness. It feels greasy and crumbles easily, characteristics that are readily explained by its structure. The carbon atoms in graphite are arranged in layers that form thin hexagonal plates, as shown by the model in Figure 20-4.

The distance between the nuclei of adjacent carbon atoms within a layer has been measured to be 142 pm. This distance is less than the distance between adjacent carbon atom nuclei in diamond. However, the distance between the nuclei of atoms in adjacent layers measures 335 pm. Because the average distance between carbon atoms in graphite is greater than the average distance in diamond, graphite has a lower density.

The layers of carbon atoms in graphite are too far apart to be held together by covalent bonds. Only weak London dispersion forces hold the layers together. Because of the weak attraction, the layers can slide across one another. This property allows graphite to be used as a lubricant and in pencil "lead."

Within each layer, each carbon atom is bonded to only three other carbon atoms. These bonds are examples of resonance hybrid bonds, which were discussed in Chapter 6. The bonding electrons of resonance hybrid bonds can be thought of as delocalized. **Delocalized electrons** *are electrons shared by more than two atoms.*

Graphite is a fairly good conductor of electricity, even though it is a nonmetal, because the delocalized electrons move freely within each layer. Like diamond, graphite has a high melting point (3652°C). This is because the structure created by delocalized electrons results in a strongly bonded covalent network. Another use of graphite is in graphite fibers. Graphite fibers are stronger and stiffer than steel, but less dense. The strength of the bonds within a layer makes graphite difficult to pull apart in the direction parallel to the surface of the layers. The strength and light weight of graphite fiber have led to its use in products such as sporting goods and aircraft.

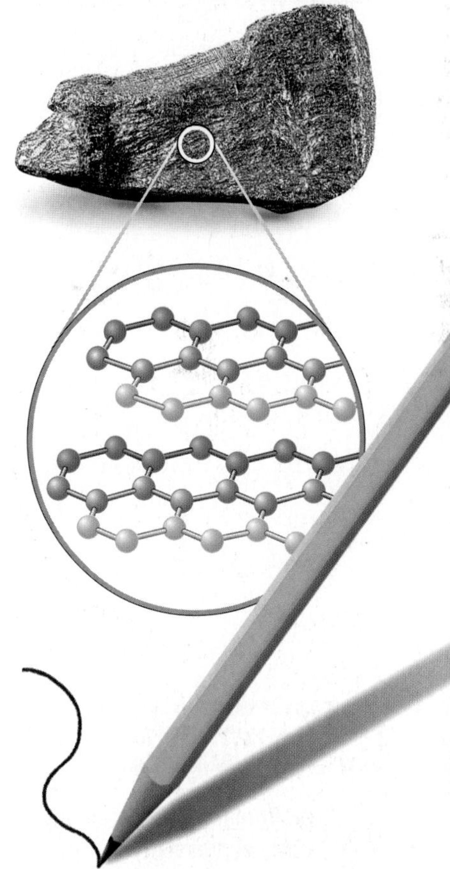

FIGURE 20-4 Notice the space between layers in the ball-and-stick model of graphite. Graphite pencils mark on paper because adjacent layers can slide past each other.

(a)

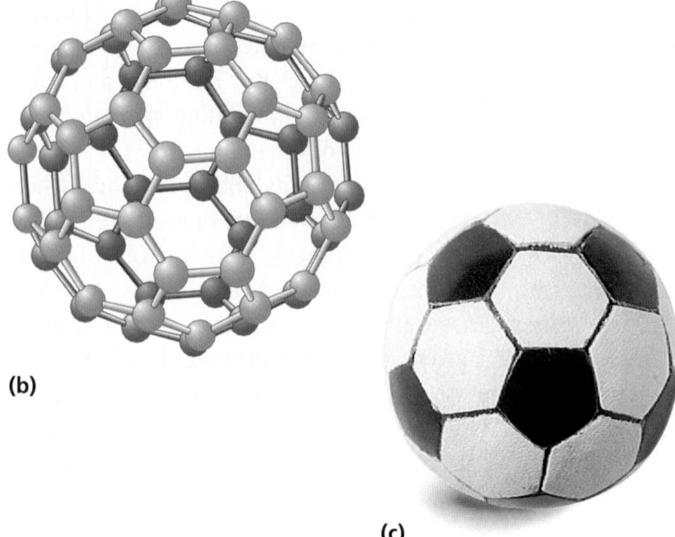

(b)

(c)

FIGURE 20-5 (a) Buckminsterfullerene was named after Buckminster Fuller, who invented geodesic domes like the one shown here. (b) The structure of buckminsterfullerene resembles the pattern of a soccer ball (c).

Fullerenes

In the mid-1980s a new allotropic form of carbon was discovered. The 1996 Nobel Prize in chemistry was awarded to Richard E. Smalley, Robert F. Curl, and Harold W. Kroto, leaders of the research teams that discovered this class of compounds, fullerenes.

Fullerenes are part of the soot that forms when carbon-containing materials are burned with limited oxygen. Their structures consist of near-spherical cages of carbon atoms. The most stable of these is C_{60}, shown in Figure 20-5. C_{60} is formed by 60 carbon atoms arranged in interconnected five- and six-membered rings.

Because of its structural resemblance to geodesic domes, Richard Smalley and his co-workers at Rice University named C_{60} "buckminsterfullerene" in honor of the geodesic-dome architect, Buckminster Fuller. The whole family of carbon-atom cages, which have a wide range in the number of carbon atoms, are therefore called fullerenes. Because the structure of C_{60} also resembles the design of a soccer ball, C_{60} is also known less formally as buckyball. Scientists are currently trying to find practical uses for these compounds.

SECTION REVIEW

1. What makes carbon an important element in the study of chemistry?
2. What type of hybrid orbital is found in carbon double bonds? In carbon triple bonds?
3. How does the structure of graphite relate to its properties and uses?
4. a. How are the structures of different fullerenes similar?
   b. How do they differ?

Organic Compounds

SECTION 20-2

OBJECTIVES

- Explain how the structure and bonding of carbon lead to the diversity and number of organic compounds.
- Explain the importance and limitations of molecular and structural formulas.
- Compare structural and geometric isomers.

All organic compounds contain carbon atoms. However, not all carbon-containing compounds are classified as organic. There are a few exceptions, such as Na_2CO_3, CO, and CO_2, that are considered inorganic. **Organic compounds,** then, can be defined as *covalently bonded compounds containing carbon, excluding carbonates and oxides*. Figure 20-6 shows a few familiar items that contain organic compounds.

Carbon Bonding and the Diversity of Organic Compounds

The diversity of organic compounds results from the uniqueness of carbon's structure and bonding. Carbon's electronic structure allows it to bind to itself to form chains and rings, to bind covalently to other elements, and to bind to itself and other elements in different arrangements.

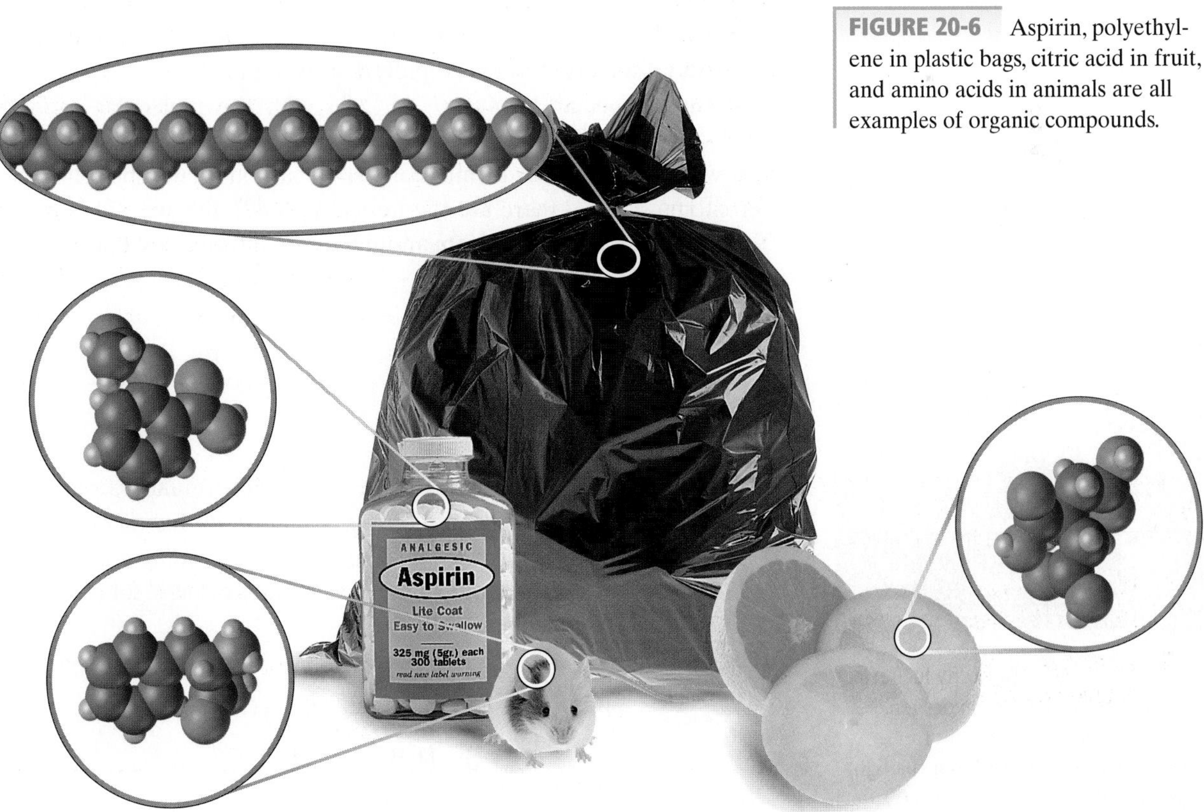

FIGURE 20-6 Aspirin, polyethylene in plastic bags, citric acid in fruit, and amino acids in animals are all examples of organic compounds.

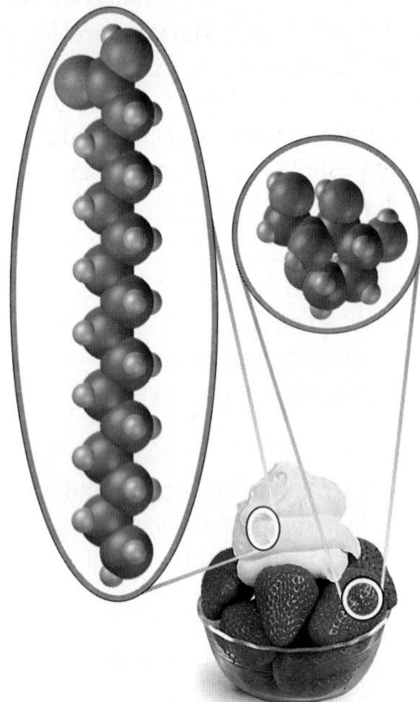

FIGURE 20-7 Compare the shape of a fatty acid found in cream with that of fructose, found in fruit. In the fatty acid, the carbon atoms are in chains. In fructose, carbon atoms form a ring.

Carbon-Carbon Bonding

Carbon atoms are unique in their ability to form long chains and rings of covalently bonded carbon atoms. This type of bonding is known as **catenation,** *the covalent binding of an element to itself to form chains or rings.* This produces a multitude of chain, branched-chain, and ring structures. In addition, carbon atoms in these structures can be linked by single, double, or triple covalent bonds. Examples of molecules containing carbon-atom rings and chains are shown in Figure 20-7.

Carbon Bonding to Other Elements

Besides binding to other carbon atoms, carbon atoms bind readily to elements with similar electronegativities. Organic compounds consist of carbon and these other elements. **Hydrocarbons** *are the simplest organic compounds, composed of only carbon and hydrogen.* Other organic compounds contain hydrocarbon backbones to which other elements, primarily O, N, S, and the halogens, are attached. An example of a molecule in which carbon atoms are bound to other elements is shown in Figure 20-8.

Arrangement of Atoms

The bonding capabilities of carbon also allow for different arrangements of atoms. This means that some compounds may contain the same atoms but have different properties because the atoms are arranged differently. For example, the molecular formula C_2H_6O represents both ethanol and dimethyl ether. *Compounds that have the same molecular formula but different structures are called* **isomers.** As the number of carbon atoms in a molecular formula increases, the number of possible isomers increases rapidly. For example, there are 18 isomers with the molecular formula C_8H_{18}, 35 with the molecular formula C_9H_{20}, and 75 with the molecular formula $C_{10}H_{22}$. Calculations show that with the molecular formula of just 40 carbon atoms and 82 hydrogen atoms, $C_{40}H_{82}$, there are theoretically 69 491 178 805 831 isomers. To distinguish one isomer from another, more information than just the molecular formula is needed.

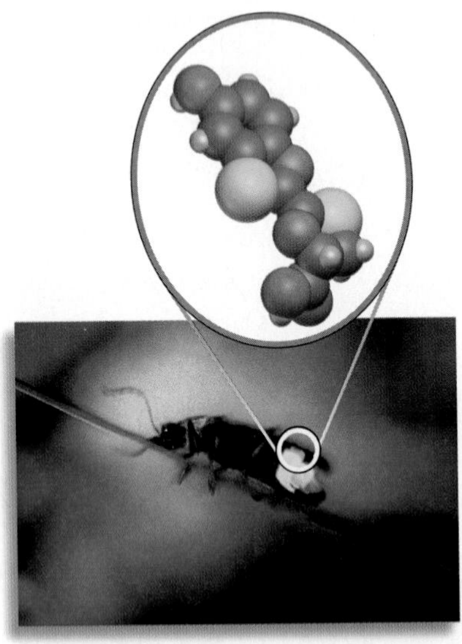

FIGURE 20-8 In firefly luciferin, carbon atoms bind to hydrogen, oxygen, nitrogen, and sulfur. Luciferin is responsible for the light emitted from the tail of a firefly.

Structural Formulas

For this reason, organic chemists use structural formulas to represent organic compounds. *A* **structural formula** *indicates the number and types of atoms present in a molecule and also shows the bonding arrangement of the atoms.* For example, one possible structural formula for an isomer of C_4H_{10} is the following.

```
  H     H     H
  |     |     |
H-C-----C-----C-H
  |     |     |
  H   H-C-H   H
        |
        H
```

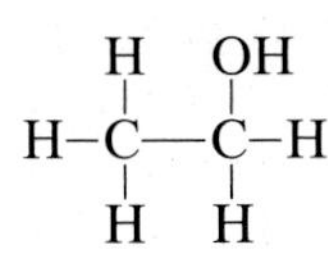

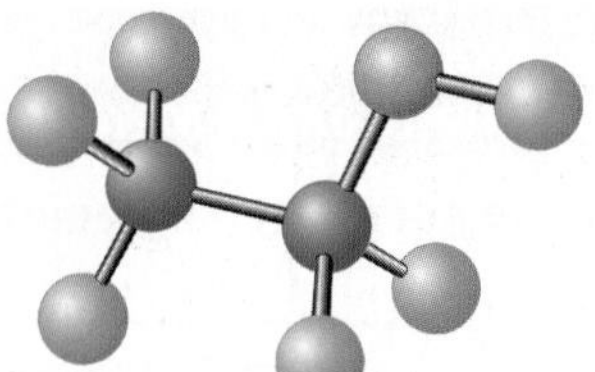
Ball-and-stick model

Space-filling model

FIGURE 20-9 The structure of ethanol can be represented in different ways. Ball-and-stick and space-filling models represent the three-dimensional shape of the molecule.

Structural formulas are sometimes condensed to make them easier to read. In one type of condensed structure, hydrogen single covalent bonds are not shown. The hydrogen atoms are understood to bind to the atom they are written beside. The following structural and condensed structural formulas represent the same molecule.

```
    H       H       H
    |       |       |
  H-C-------C-------C-H                      CH3-CH-CH3
    |       |       |     is the same as         |
    H     H-C-H     H                           CH3
            |
            H
```

It is important to remember that the structural formula does not accurately show the three-dimensional shape of the molecule. Three-dimensional shape is depicted with drawings or models, as shown for ethanol in Figure 20-9.

Isomers

You have learned that isomers are compounds that have the same molecular formula but different structural formulas. Isomers can be further classified by structure and geometry.

Structural Isomers

Structural isomers *are isomers in which the atoms are bonded together in different orders.* For example, the atoms of the molecular formula C_4H_{10} can be arranged in two different ways.

```
                             H       H       H
   H H H H                   |       |       |
   | | | |                 H-C-------C-------C-H
 H-C-C-C-C-H                 |       |       |
   | | | |                   H     H-C-H     H
   H H H H                           |
                                     H
   butane                      2-methylpropane
```

Notice that the formula for butane shows a continuous chain of four carbon atoms. The chain may be bent or twisted, but it is continuous. The formula of 2-methylpropane shows a continuous chain of three carbon atoms, with the fourth carbon atom attached to the second carbon atom of the chain.

TABLE 20-1 Physical Properties of the Structural Isomers Butane and 2-Methylpropane

	Melting point (°C)	Boiling point (°C)	Density at 20°C (g/mL)
butane	–138.4	–0.5	0.5788
2-methylpropane	–159.4	–11.633	0.549

Structural isomers can have different physical or chemical properties. For example, butane and 2-methylpropane have different melting points, boiling points, and densities, as shown in Table 20-1.

Geometric Isomers

Geometric isomers *are isomers in which the order of atom bonding is the same but the arrangement of atoms in space is different.* Consider the molecule 1,2-dichloroethene, which contains a double bond. The double bond prevents free rotation and holds groups to either side of the molecule. This means there can be two different 1,2-dichloroethene geometric isomers as shown below.

```
Cl     Cl          H      Cl
  \   /              \   /
   C=C                C=C
  /   \              /   \
H       H          Cl      H
   cis                trans
```

Because the two chlorine atoms are on the same side of the molecule in the first structure, it is called *cis*. In the second molecule, the two chlorine atoms are on opposite sides of the molecule, and so the molecule is called *trans*. Notice that in both molecules the bonding order of the atoms is the same: each carbon atom in the double bond is also bound to one chlorine atom and one hydrogen atom.

Now consider the molecule 1,2-dichloroethane. Atoms attached to the carbon atoms can rotate freely around the single carbon-carbon bond, as shown in Figure 20-10. There are no geometric isomers of 1,2-dichloroethane. *In order for geometric isomers to exist, there must be a rigid structure in the molecule to prevent free rotation around a bond.*

Now consider two apparent structures for another molecule with a double bond, chloroethene.

```
H      Cl          H      H
  \   /              \   /
   C=C                C=C
  /   \              /   \
H       H          H       Cl
```

Although these structures may appear different at first glance, they are actually the same. In *both* structures, two hydrogen atoms are on one side of the molecule, and one chlorine atom and one hydrogen atom are on the other. *A molecule can have a geometric isomer only if two carbon atoms in a rigid structure each have two different groups attached.*

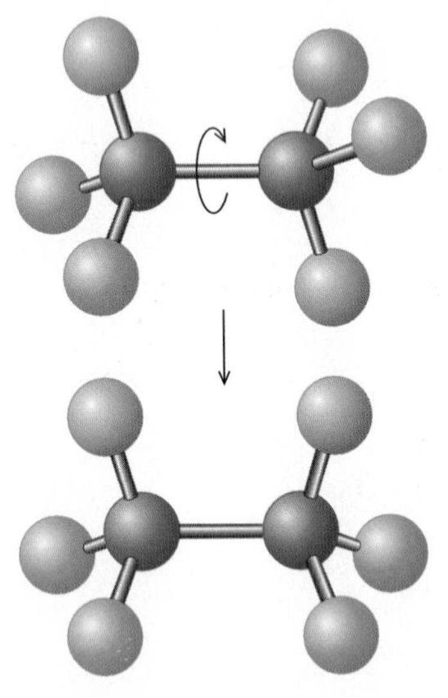

FIGURE 20-10 Unlike double bonds, single bonds allow free rotation within a molecule. Groups attached to the carbon atoms are not held to one side of the molecule, so there are no geometric isomers.

$$\begin{array}{ccc} CH_3CH_2 & & (CH_2)_9CH_2O\overset{O}{\overset{\|}{C}}CH_3 \\ & C{=}C & \\ H & & H \end{array}$$

Cis-11-tetradecenyl acetate

$$\begin{array}{ccc} H & & (CH_2)_9CH_2O\overset{O}{\overset{\|}{C}}CH_3 \\ & C{=}C & \\ CH_3CH_2 & & H \end{array}$$

Trans-11-tetradecenyl acetate

FIGURE 20-11 Males of the Iowa strain of the European corn borer respond most strongly to mixtures of the female sex attractant pheromone that are 96% *cis* isomer. But males of the New York strain respond most strongly to mixtures containing 97% *trans* isomer.

Like structural isomers, geometric isomers differ in physical and chemical properties. Some geometric isomers are known to differ in physiological behavior as well. For example, insects can communicate by chemicals called pheromones and may distinguish between the geometric isomers of pheromones. One geometric isomer of a pheromone may be physiologically active, while the other will be only slightly active or not at all. The European corn borer, shown in Figure 20-11, distinguishes between isomers of its sex-attractant pheromone. Another example of differences between geometric isomers is found in fatty acids. Natural unsaturated fatty acids are primarily *cis*-fatty acids. Hydrogenation is used to convert vegetable oil, which contains unsaturated fatty acids, into a solid fat, such as margarine or vegetable shortening. During hydrogenation *trans*-fatty acids are produced. Research has shown that there may be health risks associated with diets high in *trans*-fatty acids.

SECTION REVIEW

1. What are three characteristics of carbon that contribute to the diversity of organic compounds?
2. Define the term *isomer,* and distinguish between structural and geometric isomers.
3. Which of the following types of molecular representations can be used to show differences between isomers? Explain why each can or cannot.
   a. molecular formula
   b. structural formula
   c. three-dimensional drawing or model
4. Which of the following can represent the same molecule?

   a.
   $$\begin{array}{ccccccccccc} & & H & & H & & H & & H & & H \\ & & | & & | & & | & & | & & | \\ H & - & C & - & C & - & C & - & C & - & C & - & H \\ & & | & & | & & | & & | & & | \\ & & H & & H & & H & & H & & H \end{array}$$

   b. $CH_3{-}CH_2{-}CH_2{-}CH_3$

   c.
   $$\begin{array}{l} CH_3{-}CH_2{-}CH_2 \\ \qquad\qquad\quad | \\ \qquad\qquad\; CH_2 \\ \qquad\qquad\quad | \\ \qquad\qquad\; CH_3 \end{array}$$

   d. C_5H_{12}

SECTION 20-3

Saturated Hydrocarbons

OBJECTIVES

- Recognize the important structural feature of saturated hydrocarbons, alkanes.
- Be able to name and write structural formulas for alkanes.
- Explain how structures of alkanes relate to their properties and how those properties affect the uses of specific alkanes.

Hydrocarbons are grouped mainly by the type of bonding between carbon atoms. **Saturated hydrocarbons** *are hydrocarbons in which each carbon atom in the molecule forms four single covalent bonds with other atoms.*

Alkanes

Hydrocarbons that contain only single bonds are **alkanes.** In Table 20-2, the molecular formulas, structural formulas, and space-filling models are given for alkanes with one to four carbon atoms. If you examine the molecular formulas for successive alkanes in Table 20-2, you will see a clear pattern. Each member of the series differs from the preceding one by one carbon atom and two hydrogen atoms. For example, propane, C_3H_8, differs from ethane, C_2H_6, by one carbon atom and two hydrogen atoms, a $-CH_2-$ group.

```
   H H            H H H
   | |            | | |
 H-C-C-H        H-C-C-C-H
   | |            | | |
   H H            H H H
 ethane          propane
```

Compounds that differ in this fashion belong to a homologous series. A **homologous series** *is one in which adjacent members differ by a constant unit.* It is not necessary to remember the molecular formulas for all members of a homologous series. Instead, a general molecular formula can be used to determine the formulas. Look at the molecular formulas for ethane and propane, C_2H_6 and C_3H_8. They both fit the formula C_nH_{2n+2}. For ethane, $n = 2$, so there are two carbon atoms and $(2 \times 2) + 2 = 6$ hydrogen atoms. For propane, $n = 3$, so there are three carbon atoms and $(2 \times 3) + 2 = 8$ hydrogen atoms. Now consider a molecule for which we do not know the molecular formula. Suppose a member of this series has 30 carbon atoms in its molecules. Then $n = 30$, and there are $(2 \times 30) + 2 = 62$ hydrogen atoms. The formula is $C_{30}H_{62}$.

Notice that for alkanes with three or fewer carbon atoms, only one molecular structure is possible. However, in alkanes with more than three carbon atoms, the chains can be straight or branched. Thus,

TABLE 20-2 *Alkanes with One to Four Carbon Atoms*

Molecular formulas	Structural formulas	Space-filling models
CH_4	$H-\underset{H}{\overset{H}{C}}-H$ **methane**	
C_2H_6	$H-\underset{H}{\overset{H}{C}}-\underset{H}{\overset{H}{C}}-H$ **ethane**	
C_3H_8	$H-\underset{H}{\overset{H}{C}}-\underset{H}{\overset{H}{C}}-\underset{H}{\overset{H}{C}}-H$ **propane**	
C_4H_{10}	$H-\underset{H}{\overset{H}{C}}-\underset{H}{\overset{H}{C}}-\underset{H}{\overset{H}{C}}-\underset{H}{\overset{H}{C}}-H$ **butane** $H-\underset{H}{\overset{H}{C}}-\underset{H-\underset{H}{C}-H}{\overset{H}{C}}-\underset{H}{\overset{H}{C}}-H$ **2-methylpropane**	

alkanes with four or more carbon atoms have structural isomers. There are two possible structural isomers for alkanes with four carbon atoms, butane and 2-methylpropane.

Cycloalkanes

Cycloalkanes *are alkanes in which the carbon atoms are arranged in a ring, or cyclic, structure.* The structural formulas for cycloalkanes are often drawn in a simplified form. It is understood that there is a carbon atom at each corner and enough hydrogen atoms to complete the four bonds to each carbon atom.

$$\begin{array}{ccc} & CH_2 & \\ CH_2 & & CH_2 \\ & CH_2-CH_2 & \end{array}$$

cyclopentane

or

cyclopentane

Because there are no free ends where carbon atoms are attached to three hydrogen atoms, there are two fewer hydrogen atoms in cycloalkanes than in noncyclic alkanes.

$$\begin{array}{c} \text{H H H H} \\ \text{H–C–C–C–C–H} \\ \text{H H H H} \end{array}$$

butane
C_4H_{10}

$$\begin{array}{c} \text{H H} \\ \text{H–C–C–H} \\ \text{H–C–C–H} \\ \text{H H} \end{array}$$

cyclobutane
C_4H_8

The general structure for cycloalkanes, C_nH_{2n}, shows that they have $2 \times n$ hydrogen atoms. This is two fewer hydrogen atoms than noncyclic alkanes, C_nH_{2n+2}, which have $(2 \times n) + 2$ hydrogen atoms.

Systematic Names of Alkanes

Historically, the names of many organic compounds were derived from the sources in which they were found. As more organic compounds were discovered, a systematic naming method became necessary. The systematic method used primarily in this book was developed by the International Union of Pure and Applied Chemistry, IUPAC.

The basic part of the systematic name of an organic compound is the name of the longest carbon chain, or parent hydrocarbon, in the molecule. Table 20-3 gives the names of the prefixes for carbon-atom chains up to 10 carbon atoms long. Beginning with *pent-*, the prefixes are Greek or Latin numerical prefixes.

TABLE 20-3 *Carbon-Atom Chain Prefixes*

Number of carbon atoms	Prefix
1	meth-
2	eth-
3	prop-
4	but-
5	pent-
6	hex-
7	hept-
8	oct-
9	non-
10	dec-

Unbranched-Chain Alkane Nomenclature

To name an unbranched alkane, find the prefix in Table 20-3 that corresponds to the number of carbon atoms in the chain of the hydrocarbon. Then add the suffix *-ane* to the prefix. An example is shown below.

$$\overset{1}{C}H_3-\overset{2}{C}H_2-\overset{3}{C}H_2-\overset{4}{C}H_2-\overset{5}{C}H_2-\overset{6}{C}H_2-\overset{7}{C}H_3$$

heptane

The molecule has a chain seven carbon atoms long, so the prefix *hept-* is added to the suffix *-ane* to form *heptane*.

TABLE 20-4 ***Some Straight-Chain Alkyl Groups***

Alkane	Name	Alkyl group	Name
CH_4	methane	$-CH_3$	methyl
CH_3-CH_3	ethane	$-CH_2-CH_3$	ethyl
$CH_3-CH_2-CH_3$	propane	$-CH_2-CH_2-CH_3$	propyl
$CH_3-CH_2-CH_2-CH_3$	butane	$-CH_2-CH_2-CH_2-CH_3$	butyl
$CH_3-CH_2-CH_2-CH_2-CH_3$	pentane	$-CH_2-CH_2-CH_2-CH_2-CH_3$	pentyl

Branched-Chain Alkane Nomenclature

The naming of branched-chain alkanes also follows a systematic method. The hydrocarbon branches of alkanes are alkyl groups. **Alkyl groups** *are groups of atoms that are formed when one hydrogen atom is removed from an alkane molecule.* Alkyl groups are named by replacing the suffix *-ane* of the parent alkane with the suffix *-yl.* Some examples are shown in Table 20-4. Alkyl group names are used when naming branched-chain alkanes. We will only present the method for naming simple branched-chain alkanes with only straight-chain alkyl groups. Consider the following molecule.

$$\begin{array}{cccccccc} & & & CH_3 & CH_3 & & & \\ & & & | & | & & & \\ CH_3- & CH_2- & CH_2- & CH- & CH- & CH- & CH_2- & CH_3 \\ & & & & & | & & \\ & & & & & CH-CH_3 & & \\ & & & & & | & & \\ & & & & & CH_3 & & \end{array}$$

To name this molecule, locate the parent hydrocarbon. The parent hydrocarbon is the longest continuous chain that contains the most straight-chain branches. In this molecule, there are two chains that are eight carbon atoms long. The parent hydrocarbon is the chain that contains the most straight-chain branches. Do not be tricked by the way the molecule is drawn. The longest chain may be shown bent.

$$\begin{array}{cccccccc} & & & CH_3 & CH_3 & & & \\ & & & | & | & & & \\ CH_3- & CH_2- & CH_2- & CH- & CH- & CH- & CH_2- & CH_3 \\ & & & & & | & & \\ & & & & & CH-CH_3 & & \\ & & & & & | & & \\ & & & & & CH_3 & & \end{array}$$

NOT

$$\begin{array}{cccccccc} & & & CH_3 & CH_3 & & & \\ & & & | & | & & & \\ CH_3- & CH_2- & CH_2- & CH- & CH- & CH- & CH_2- & CH_3 \\ & & & & & | & & \\ & & & & & CH-CH_3 & & \\ & & & & & | & & \\ & & & & & CH_3 & & \end{array}$$

To name the parent hydrocarbon, add the suffix *-ane* to the prefix *oct-* (for a carbon-atom chain with eight carbon atoms) to form *octane.* Now identify and name the alkyl groups.

$$
\begin{array}{cccccccccccccccc}
 & & & & & & CH_3 & & CH_3 & & & & & & & \\
 & & & & & & | & & | & & & & & & & \\
CH_3 & - & CH_2 & - & CH_2 & - & CH & - & CH & - & CH & - & CH_2 & - & CH_3 \\
 & & & & & & & & & & | & & & & \\
 & & & & & & & & & & CH & - & CH_3 & & \\
 & & & & & & & & & & | & & & & \\
 & & & & & & & & & & CH_3 & & & &
\end{array}
$$

The three $-CH_3$ groups are methyl groups. The $-CH_2-CH_3$ group is an ethyl group. Arrange the names in alphabetical order in front of the name of the parent hydrocarbon.

ethyl methyloctane

To show that there are three methyl groups present, attach the prefix *tri-* to the name *methyl* to form *trimethyl.*

ethyl **tri**methyloctane

Now we need to show the locations of the alkyl groups on the parent hydrocarbon. Number the octane chain so that the alkyl groups have the lowest numbers possible.

$$
\begin{array}{cccccccccccccccc}
 & & & & & & CH_3 & & CH_3 & & & & & & & \\
^{8} & & ^{7} & & ^{6} & & ^{5}| & & ^{4}| & & ^{3} & & & & & \\
CH_3 & - & CH_2 & - & CH_2 & - & CH & - & CH & - & CH & - & CH_2 & - & CH_3 \\
 & & & & & & & & & & | & & & & \\
 & & & & & & & & & & {}^{2}CH & - & CH_3 & & \\
 & & & & & & & & & & | & & & & \\
 & & & & & & & & & & {}^{1}CH_3 & & & &
\end{array}
$$

NOT

$$
\begin{array}{cccccccccccccccc}
 & & & & & & CH_3 & & CH_3 & & & & & & & \\
^{1} & & ^{2} & & ^{3} & & ^{4}| & & ^{5}| & & ^{6} & & & & & \\
CH_3 & - & CH_2 & - & CH_2 & - & CH & - & CH_2 & - & CH & - & CH_2 & - & CH_3 \\
 & & & & & & & & & & | & & & & \\
 & & & & & & & & & & {}^{7}CH & - & CH_3 & & \\
 & & & & & & & & & & | & & & & \\
 & & & & & & & & & & {}^{8}CH_3 & & & &
\end{array}
$$

Place the location numbers of *each* of the alkyl groups in front of their name. Separate the numbers from the names of the alkyl groups with hyphens. The ethyl group is on carbon *3*.

3-ethyl trimethyloctane

Because there are three methyl groups, there will be three numbers in front of *trimethyl,* separated by commas.

3-ethyl-**2,4,5-**trimethyloctane

The full name is 3-ethyl-2,4,5-trimethyloctane.

The procedure for naming simple branched-chain alkanes can be summarized as follows.

Alkane Nomenclature

1. **Name the parent hydrocarbon.** Find the longest continuous chain of carbon atoms with straight-chain branches. Add the suffix *-ane* to the prefix corresponding to the number of carbon atoms in the chain.
2. **Add the names of the alkyl groups.** Add the names of the alkyl groups in front of the name of the parent hydrocarbon in alphabetical order. When there is more than one branch of the same alkyl group present, attach the appropriate numerical prefix to the name, *di* = 2, *tri* = 3, *tetra* = 4, and so on. Do this after the names have been put in alphabetical order.
3. **Number the carbon atoms in the parent hydrocarbon.** If one or more alkyl groups are present, number the carbon atoms in the continuous chain so that the alkyl groups have the lowest position numbers possible. If there are two equivalent lowest positions with two different alkyl groups, give the lowest number to the alkyl group that comes first in the name. (This will be the alkyl group that is first in alphabetical order, *before* any prefixes are attached.)
4. **Insert position numbers.** Put the position numbers of each alkyl group in front of the name of that alkyl group.
5. **Punctuate the name.** Separate the position numbers from the names with hyphens. If there is more than one number in front of a name, separate the numbers by commas.

SAMPLE PROBLEM 20-1

Name the following simple branched-chain alkane:

$$\begin{array}{l} CH_3-\underset{\displaystyle CH_3}{\underset{|}{CH}}-CH_2-\underset{\displaystyle CH_3}{\underset{|}{CH}}-\underset{\displaystyle CH_3}{\underset{|}{CH}}-CH_3 \end{array}$$

SOLUTION

1. Identify and name the parent hydrocarbon.

$$CH_3-\underset{\displaystyle CH_3}{\underset{|}{CH}}-CH_2-\underset{\displaystyle CH_3}{\underset{|}{CH}}-\underset{\displaystyle CH_3}{\underset{|}{CH}}-CH_3$$

Because the longest continuous chain contains six carbon atoms, the parent hydrocarbon is *hexane*.

2. Identify and name the alkyl groups attached to the chain.

$$CH_3-\underset{\displaystyle CH_3}{\underset{|}{CH}}-CH_2-\underset{\displaystyle CH_3}{\underset{|}{CH}}-\underset{\displaystyle CH_3}{\underset{|}{CH}}-CH_3$$

There is only one type of alkyl group, with one carbon atom. Alkyl groups with one carbon atom are methyl groups. Add the name *methyl* in front of the name of the continuous chain. Add the prefix *tri-* to show that there are three methyl groups present.

trimethylhexane

3. Number the carbon atoms in the continuous chain so that the alkyl groups have the lowest numbers possible.

```
 6     5      4      3     2     1
CH3–CH–CH2–CH–CH–CH3
       |             |     |
      CH3           CH3  CH3
```

4. The methyl groups are on the carbon atoms numbered *2*, *3*, and *5*. Put the numbers of the positions of the alkyl groups, separated by commas, in front of the name of the alkyl group. Separate the numbers from the name with a hyphen.

2,3,5-trimethylhexane

The complete name is 2,3,5-trimethylhexane.

SAMPLE PROBLEM 20-2

Draw the condensed structural formula of 3-ethyl-4-methylhexane.

SOLUTION

1. Identify the name of the parent hydrocarbon.

3-ethyl-4-methyl**hexane**

The parent hydrocarbon is hexane, so there are six carbon atoms in the chain. Draw and number the carbon atoms in the chain.

```
1    2    3    4    5    6
C––––C––––C––––C––––C––––C
```

2. Identify the alkyl groups, and determine the number of carbon atoms in the alkyl groups.

3-**ethyl**-4-**methyl**hexane

Methyl groups have one carbon atom and ethyl groups have two carbon atoms.

```
  H                 H H
  |                 | |
–C–H              –C–C–H
  |                 | |
  H                 H H
```

3. Locate the position numbers for the ethyl and methyl groups.

3-ethyl-**4**-methylhexane

Draw the alkyl groups on the parent hydrocarbon in the correct positions.

```
                 H
                 |
               H–C–H
 1    2    3    4|   5    6
 C––––C––––C––––C––––C––––C
           |
         H–C–H
           |
         H–C–H
           |
           H
```

Notice that in this molecule the methyl group and the ethyl group were in equivalent positions from the end of the chain. They are both on third carbons from the end. In such a case, the alkyl group that comes first in the name is given the lower number.

4. Add the correct number of hydrogen atoms so that each carbon atom has four single bonds. This is the complete, uncondensed, structural formula.

$$\begin{array}{ccccccccccc}
 & & & & & & \mathrm{H} & & & & \\
 & & & & & & | & & & & \\
 & \mathrm{H} & & \mathrm{H} & & \mathrm{H} & \mathrm{H{-}C{-}H} & \mathrm{H} & & \mathrm{H} & \\
 & | & & | & & | & | & | & & | & \\
\mathrm{H}{-} & \mathrm{C} & - & \mathrm{C} & - & \mathrm{C} & - \mathrm{C} - & \mathrm{C} & - & \mathrm{C} & {-}\mathrm{H} \\
 & | & & | & & | & | & | & & | & \\
 & \mathrm{H} & & \mathrm{H} & & \mathrm{H{-}C{-}H} & \mathrm{H} & \mathrm{H} & & \mathrm{H} & \\
 & & & & & | & & & & & \\
 & & & & & \mathrm{H{-}C{-}H} & & & & & \\
 & & & & & | & & & & & \\
 & & & & & \mathrm{H} & & & & &
\end{array}$$

5. To draw the condensed structural formula, show only the bonds between carbon atoms.

$$\begin{array}{cccccc}
 & & & CH_3 & & \\
 & & & | & & \\
CH_3- & CH_2- & CH- & CH- & CH_2- & CH_3 \\
 & & | & & & \\
 & & CH_2 & & & \\
 & & | & & & \\
 & & CH_3 & & &
\end{array}$$

PRACTICE

1. Name the following molecule:

$$\begin{array}{cccc}
CH_3- & CH- & CH_2- & CH_3 \\
 & | & & \\
 & CH_3 & &
\end{array}$$

Answer

2-methylbutane

2. Draw the condensed structural formula for 3,3-diethyl-2,5-dimethylnonane.

Answer

$$\begin{array}{ccccccccc}
 & & CH_3 & & & & & & \\
 & & | & & & & & & \\
 & & CH_2 & & & & & & \\
 & & | & & & & & & \\
CH_3- & CH- & C- & CH_2- & CH- & CH_2- & CH_2- & CH_2- & CH_3 \\
 & | & | & & | & & & & \\
 & CH_3 & CH_2 & & CH_3 & & & & \\
 & & | & & & & & & \\
 & & CH_3 & & & & & &
\end{array}$$

3. Draw the condensed structural formulas for the two structural isomers of methylpentane and name the isomers.

Answer

$$\begin{array}{ccccc}
CH_3- & CH- & CH_2- & CH_2- & CH_3 \\
 & | & & & \\
 & CH_3 & & &
\end{array}$$

2-methylpentane

$$\begin{array}{ccccc}
CH_3- & CH_2- & CH- & CH_2- & CH_3 \\
 & & | & & \\
 & & CH_3 & &
\end{array}$$

3-methylpentane

Cycloalkane Nomenclature

When naming simple cycloalkanes, the cycloalkane is the parent hydrocarbon. Cycloalkanes are named by adding the prefix *cyclo-* to the name of the straight-chain alkane with the same number of hydrocarbons.

$CH_3{-}CH_2{-}CH_3$

propane

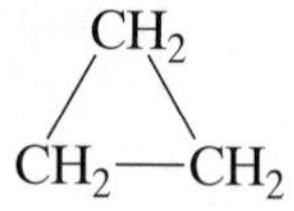

cyclopropane

When there is only one alkyl group attached to the ring, no position number is necessary. When there is more than one alkyl group attached to the ring, the carbon atoms in the ring are numbered to give the lowest numbers possible to the alkyl groups. This means that one of the alkyl groups will always be in position 1.

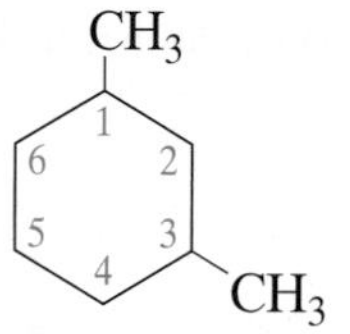

1,3-dimethylcyclohexane *NOT* ~~1,5-dimethylcyclohexane~~

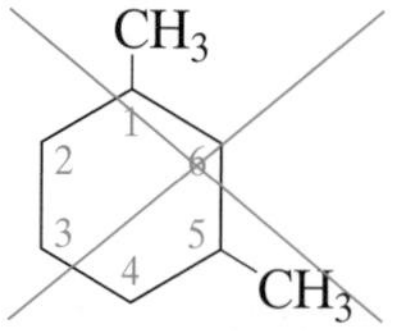

Naming cycloalkanes can be summarized as follows.

Cycloalkane Nomenclature

Use the rules for alkane nomenclature on page 639, with the following exceptions.

1. **Name the parent hydrocarbon.** Count the number of carbon atoms in the ring. Add the prefix *cyclo-* to the name of the corresponding straight-chain alkane.
2. **Add the names of the alkyl groups.**
3. **Number the carbon atoms in the parent hydrocarbon.** If there are two or more alkyl groups attached to the ring, number the carbon atoms in the ring. Assign position number one to the alkyl group that comes first in alphabetical order. Then, number in the direction that gives the rest of the alkyl groups the lowest numbers possible.
4. **Insert position numbers.**
5. **Punctuate the name.**

Two examples of correctly named cycloalkanes are given below.

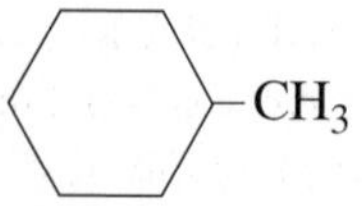

methylcyclohexane

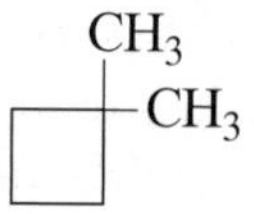

1,1-dimethylcyclobutane

TABLE 20-5 *Properties of Straight-Chain Alkanes*

Molecular formula	IUPAC name	Boiling point (°C)	State at room temperature
CH_4	methane	−164	gas
C_2H_6	ethane	−88.6	
C_3H_8	propane	−42.1	
C_4H_{10}	butane	−0.5	
C_5H_{12}	pentane	36.1	liquid
C_8H_{18}	octane	125.7	
$C_{10}H_{22}$	decane	174.1	
$C_{17}H_{36}$	heptadecane	301.8	solid
$C_{20}H_{42}$	eicosane	343	

Properties and Uses of Alkanes

Properties for some straight-chain alkanes are listed in Table 20-5. The trends in these properties can be explained by examining the structure of alkanes. The carbon-hydrogen bonds of alkanes are nonpolar. The only forces of attraction between nonpolar molecules are weak intermolecular forces, or London dispersion forces. The strength of London dispersion forces increases as the mass of a molecule increases.

Physical States

The physical states at which some alkanes exist at room temperature and atmospheric pressure are found in Table 20-5. Alkanes with the lowest molecular mass, those with one to four carbon atoms, are gases. **Natural gas** *is a fossil fuel composed primarily of alkanes containing one to four carbon atoms*. The existence of these alkanes as gases agrees with the idea that very small molecules have weak London dispersion forces between them and are not held together tightly. Larger alkanes are liquids. Gasoline and kerosene consist mostly of liquid alkanes. Stronger London dispersion forces are required to hold molecules close enough together to form liquids. Alkanes with a very high molecular mass are solids, corresponding to a greater increase in London dispersion forces. Paraffin wax contains solid alkanes. It can be used in candles, as shown in Figure 20-12.

FIGURE 20-12 Paraffin wax, used in candles, contains solid alkanes. Molecules of paraffin wax contain 26 to 30 carbon atoms.

Boiling Points

The boiling points of alkanes, also shown in Table 20-5, increase with increasing molecular mass. As London dispersion forces increase, more energy, or heat, is required to pull the molecules apart. This property is used in the separation of petroleum, a major source of alkanes. **Petroleum** *is a complex mixture of different hydrocarbons that varies greatly in composition.* The hydrocarbon molecules in petroleum contain from one to more than fifty carbon atoms. This range allows the separation of

TABLE 20-6 *Petroleum Fractions*

Fraction	Size range of molecules	Boiling-point range (°C)
Gasoline	C_4–C_{12}	up to 200
Kerosene	C_{10}–C_{14}	180–290
Middle distillate, such as heating oil, gas-turbine fuel, diesel	C_{12}–C_{20}	185–345
Wide-cut gas oil, such as lubricating oil, waxes	C_{20}–C_{36}	345–540
Asphalt	above C_{36}	residues

petroleum into different portions with different boiling point ranges, as shown in Table 20-6. In **fractional distillation,** shown in Figure 20-13, *components of a mixture are separated on the basis of boiling point, by condensation of vapor in a fractionating column*. During its fractional distillation, petroleum is heated to about 370°C. Nearly all the components of the petroleum are vaporized at this temperature. As the vapors rise in the fractionating column, or tower, they are gradually cooled.

FIGURE 20-13 (a) Fractional distillation takes place in petroleum refinery towers. (b) This is a model of a fractional distillation tower. Because fractions contain hydrocarbons of different masses, they condense and are drawn off at different levels.

(a)

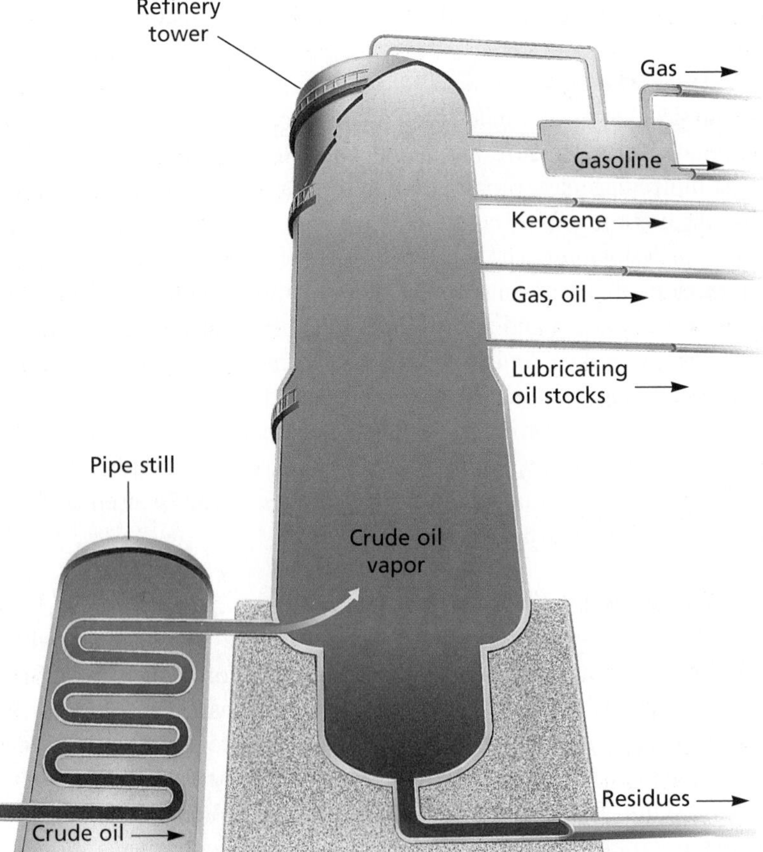

(b)

Alkanes with higher boiling points have higher condensation temperatures and condense for collection lower in the tower. For example, lubricating oils, which have higher condensation temperatures than gasoline has, are collected lower in the fractionating tower.

Combustion

Alkanes are less reactive than other hydrocarbons because of the stability of their single covalent bonds. One reaction alkanes do undergo is combustion. Because alkanes make up a large proportion of gaseous and liquid fossil fuels, combustion is their most important reaction. Complete combustion of hydrocarbons produces energy, CO_2, and H_2O. The reaction for the combustion of methane produces 890 kJ/mol.

$$CH_4 + 2O_2 \longrightarrow CO_2 + 2H_2O$$

One concern about the combustion of fossil fuels is their possible contribution to the greenhouse effect. CO_2 is one of the atmospheric molecules that absorbs infrared radiation. Increased levels of CO_2 through the combustion of fossil fuels may increase the amount of infrared energy absorbed by the atmosphere to a level that can increase the average temperature of Earth.

Engines can be powered by gasoline combustion. When fuel ignites spontaneously before it is reached by the flame front, there is a decrease in the amount of power gained, and engine knocking results. Straight-chain hydrocarbons are more likely to ignite spontaneously than branched-chain hydrocarbons. This tendency is the basis for the octane rating scale. *The* **octane rating** *of a fuel is a measure of its burning efficiency and its antiknock properties.* The octane rating scale is based on mixtures of 2,2,4-trimethylpentane, a highly branched alkane, and heptane, a straight-chain alkane. The term *octane* comes from the common name of 2,2,4-trimethylpentane, *isooctane*. Pure 2,2,4-trimethylpentane is very resistant to knocking and is assigned an octane number of 100. Pure heptane has an octane number of 0 and burns with a lot of knocking. Increasing the percentage of branched-chain alkanes in gasoline is one way to increase octane rating. The octane rating on gasoline pumps is shown in Figure 20-14.

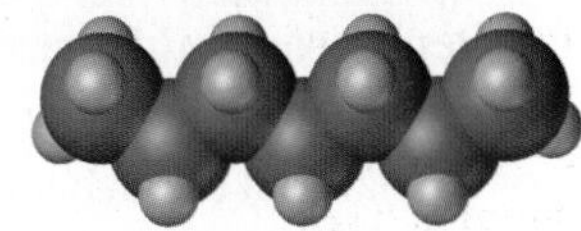
heptane

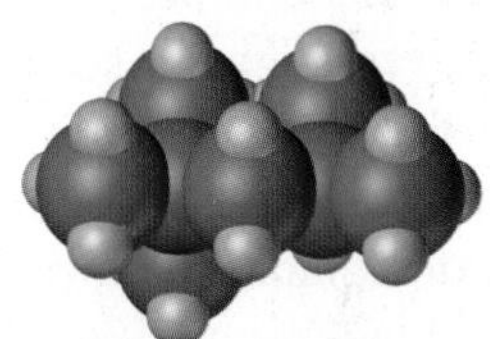
2,2,4-trimethylpentane

FIGURE 20-14 The octane rating scale is based on a rating of 100 for 2,2,4-trimethylpentane and 0 for heptane. Compare their molecular shapes.

SECTION REVIEW

1. What is the basic structural characteristic of alkanes?
2. Draw all of the condensed structural formulas that can represent C_5H_{12}.
3. Give the systematic name for each of the compounds whose formulas appear in item 2.
4. Relate the properties of some alkanes to their uses.
5. Draw the condensed structural formulas of 2-ethyl-3-methylpentane and 1-methyl-3-propylcyclopentane.

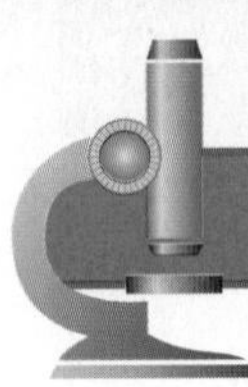

RESEARCH NOTES

Synthetic Diamonds

Diamonds made to order? Almost. A thin coating of diamond film may not be pretty to look at, but it does offer many useful properties to industry. A number of methods are being developed to produce diamond coatings cheaply and efficiently. If successful, the processes will affect the way tools, containers, computer chips, and a host of other items are manufactured.

James Adair is an associate professor of material science and engineering at the University of Florida. "Natural diamonds are made at very, very high pressures and heat," Adair says. "Basically, it's a naturally occurring process that literally took millennia to form the diamond. We make diamonds in a couple of minutes." The process involves sticking very fine diamond particles on all kinds of different surfaces. Chemical Vapor Deposition is then used to grow more diamond from these diamond "seeds."

In Chemical Vapor Deposition, the objects to be coated—in this case, the diamond seeds—are placed inside a chamber filled with methane and other gases. The gases are subjected to microwave radiation, which breaks them down into hydrogen, carbon, and its mixtures (carbon-hydrogen radicals). Diamond crystals grow when these carbon atoms coat the diamond-seed crystals.

Another method of coating with diamond, invented by metallurgist Pravin Mistry, uses lasers to scan the object to be coated. The energy of the lasers breaks down CO_2 (supplied by a gas delivery system) into carbon and oxygen atoms and vaporizes the surface of the object, forming a superheated plasma. The plasma serves as an environment for bonding the carbon atoms into a coating of crystalline diamond.

This picture, taken with an electron microscope, shows synthetic diamond formed by Chemical Vapor Deposition.

One of the biggest challenges in making synthetic diamond coatings is making sure that the carbon crystallizes correctly to form diamond and not graphite. Graphite is useful for making lubricants and pencil leads, but it is not as strong and durable as diamond. In the crystalline molecular structure of graphite, the spaces between carbon atoms are relatively far apart. The process must compress the spaces to form a compact octagonal diamond crystal.

Diamond is one of the hardest materials known to man, so diamond coatings would be particularly useful for making machine tools, work surfaces, and other applications where a durable protective covering is needed. Diamond also has the highest thermoconductivity of any material, which means that it transports heat very effectively. You wouldn't want to drink from a diamond coffee cup because the cup would warm up rapidly and you'd burn your lips. But diamond's ability to conduct heat makes it very useful as a coating on silicon computer chips.

"For microelectronics," says Adair, "dealing with the heat generated by the circuit is one of the biggest problems. If the heat builds up too much within a silicon circuit, it can literally melt the silicon. And it's not going to act as a very good computer brain for you. Diamond can pull that heat out of the silicon chip, so the circuit can run a little bit cooler."

If a computer chip is prevented from getting too hot, it can perform faster. And a faster chip can lead to a new breed of computers with enhanced capabilities.

Unsaturated Hydrocarbons

SECTION 20-4

OBJECTIVES

- Distinguish between the structures of alkenes, alkynes, and aromatic hydrocarbons.
- Be able to name and write structural formulas for unsaturated hydrocarbons.
- Explain how structures of unsaturated hydrocarbons relate to their properties and how those properties affect the uses of specific hydrocarbons.

Hydrocarbons that do not contain the maximum amount of hydrogen are referred to as unsaturated. **Unsaturated hydrocarbons** *are hydrocarbons in which not all carbon atoms have four single covalent bonds.*

Alkenes

Alkenes *are hydrocarbons that contain double covalent bonds.* Some examples of alkenes are given in Table 20-7. Notice that because alkenes have a double bond, the simplest alkene, ethene, has two carbon atoms.

Carbon atoms linked by double bonds cannot bind as many atoms as those that are linked by only single bonds. An alkene with one double bond has two fewer hydrogen atoms than the corresponding alkane.

```
  H H H            H
  | | |            |
H-C-C-C-H    H-C=C-C-H
  | | |        | | |
  H H H        H H H
   C3H8         C3H6
```

C_3H_8 $\qquad$ C_3H_6

Thus, the general formula for noncyclic alkenes with one double bond is C_nH_{2n}.

TABLE 20-7 *Structures of Alkenes*

	ethene	propene	*trans*-2-butene	*cis*-2-butene
Structural formula	$H_2C{=}CH_2$ (H, H / C=C / H, H)	H, H / C=C / CH_3, H	H, CH_3 / C=C / CH_3, H	H, H / C=C / CH_3, CH_3
Ball-and-stick model				

Because alkenes have a double bond, they can have geometric isomers, as shown in the examples below.

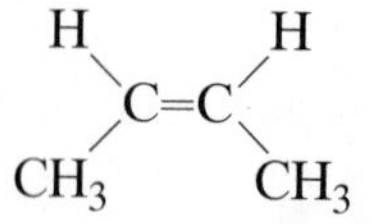

$$\begin{array}{ccc} \text{H} & & \text{H} \\ & \text{C}=\text{C} & \\ \text{CH}_3 & & \text{CH}_3 \end{array}$$

cis-2-butene

$$\begin{array}{ccc} \text{H} & & \text{CH}_3 \\ & \text{C}=\text{C} & \\ \text{CH}_3 & & \text{H} \end{array}$$

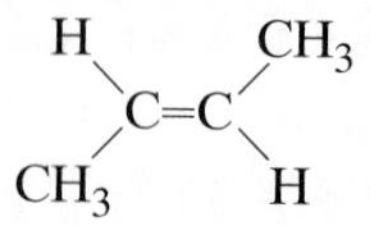

trans-2-butene

Systematic Names of Alkenes

The rules for naming a simple alkene are similar to those for naming an alkane. The parent hydrocarbon is the longest continuous chain of carbon atoms *that contains the double bond.* If there is only one double bond, the suffix *-ene* is added to the carbon-chain prefix. Here, the longest chain that contains the double bond has five carbon atoms and one double bond, so the parent hydrocarbon is pentene.

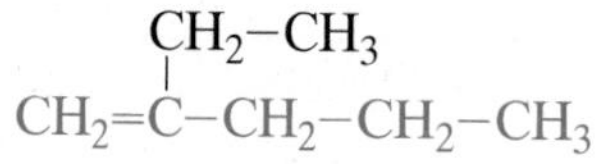

$$\begin{array}{l} \quad\;\; CH_2-CH_3 \\ \quad\;\; | \\ CH_2=C-CH_2-CH_2-CH_3 \end{array}$$

pentene

NOT

$$\begin{array}{l} \quad\;\; CH_2-CH_3 \\ \quad\;\; | \\ CH_2=C-CH_2-CH_2-CH_3 \end{array}$$

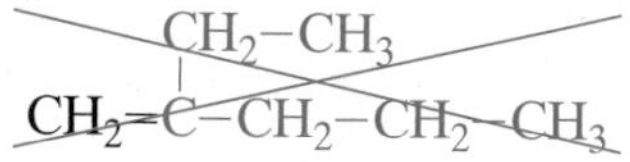

~~hexane~~

The carbon atoms in the chain are numbered so that the first carbon atom in the double bond has the lowest number. The number indicating the position of the double bond is placed before the name of the hydrocarbon chain and separated by a hyphen.

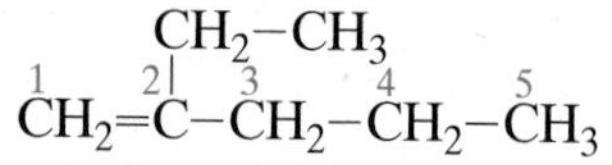

$$\begin{array}{l} \quad\;\; CH_2-CH_3 \\ \overset{1}{C}H_2=\overset{2}{C}-\overset{3}{C}H_2-\overset{4}{C}H_2-\overset{5}{C}H_3 \end{array}$$

1-pentene

The position number and name of the alkyl group are placed in front of the double-bond position number. This alkyl group has two carbon atoms, an ethyl group. It is on the second carbon atom of the parent hydrocarbon.

2-ethyl-1-pentene

The molecule is 2-ethyl-1-pentene.

If there is more than one double bond, the suffix is modified to indicate the number of double bonds: 2 = *-adiene*, 3 = *-atriene*, and so on.

$$CH_2=CH-CH_2-CH=CH_2$$

1,4-pentadiene

If numbering from both ends gives equivalent positions for the double bonds in an alkene with two double bonds, then the chain is numbered from the end nearest the first alkyl group.

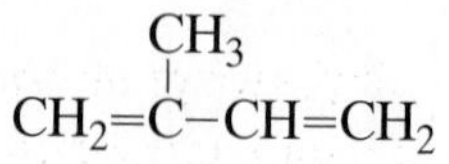

$$\begin{array}{l} \quad\;\; CH_3 \\ \quad\;\; | \\ CH_2=C-CH=CH_2 \end{array}$$

2-methyl-1,3-butadiene

The procedure for naming alkenes can be summarized as follows.

Alkene Nomenclature

Use the rules for alkane nomenclature on page 639, with the following exceptions.

1. **Name the parent hydrocarbon.** Locate the longest continuous chain that *contains the double bond(s)*. If there is only one double bond, add the suffix *-ene* to the prefix corresponding to the number of carbon atoms in this chain. If there is more than one double bond, modify the suffix to indicate the number of double bonds. For example, 2 = *-adiene,* 3 = *-atriene,* and so on.
2. **Add the names of the alkyl groups.**
3. **Number the carbon atoms in the parent hydrocarbon.** Number the carbon atoms in the chain so that the first carbon atom in the double bond nearest the end of the chain has the lowest number. If numbering from both ends gives equivalent positions for two double bonds, then number from the end nearest the first alkyl group.
4. **Insert position numbers.** Place double-bond position numbers immediately before the name of the parent hydrocarbon alkene. Place alkyl group position numbers immediately before the name of the corresponding alkyl group.
5. **Punctuate the name.**

SAMPLE PROBLEM 20-3

Name the following alkene.

$$\begin{array}{l} CH_3 \\ | \\ CH_3-CH-C=CH_2 \\ | \\ CH_2-CH_3 \end{array}$$

SOLUTION

1. Identify and name the parent hydrocarbon.

$$\begin{array}{l} CH_3 \\ | \\ CH_3-CH-C=CH_2 \\ | \\ CH_2-CH_3 \end{array}$$

The parent hydrocarbon has four carbon atoms and one double bond, so it is named *butene.*

2. Identify and name the alkyl groups.

$$\begin{array}{l} CH_3 \\ | \\ CH_3-CH-C=CH_2 \\ | \\ CH_3-CH_3 \end{array}$$

The alkyl groups are *ethyl* and *methyl.*
Place their names in front of the name of the parent hydrocarbon in alphabetical order.

ethyl methyl butene

3. Number the carbon chain to give the double bond the lowest position.

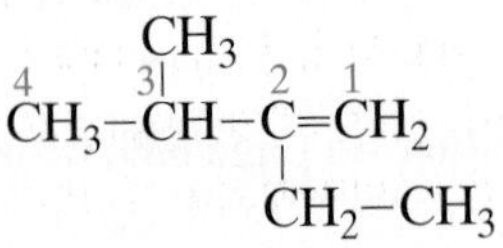

$$\begin{array}{ccccccc} & & \overset{3}{CH_3} & & & & \\ \overset{4}{CH_3} & - & \overset{|}{CH} & - & \overset{2}{C} & = & \overset{1}{CH_2} \\ & & & & | & & \\ & & & & CH_2-CH_3 & & \end{array}$$

4. Place the position number of the double bond in front of butene. Place the position numbers of the alkyl groups in front of each alkyl group. Separate the numbers from the name with hyphens.

The first carbon in the double bond is in position *1*.
The ethyl group is on carbon *2*.
The methyl group is on carbon *3*.

2-ethyl-**3**-methyl-**1**-butene

The full name is 2-ethyl-3-methyl-1-butene.

PRACTICE

1. Name the following alkene:

$$CH_3-CH_2-CH_2-CH=CH-CH_3$$

Answer
2-hexene

2. Draw the condensed structural formula for 4-methyl-1,3-pentadiene.

Answer

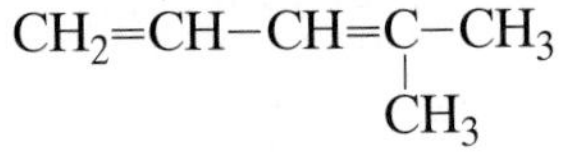

$$\begin{array}{c} CH_2=CH-CH=\underset{\displaystyle CH_3}{\underset{|}{C}}-CH_3 \end{array}$$

3. Name the following alkenes:

a. 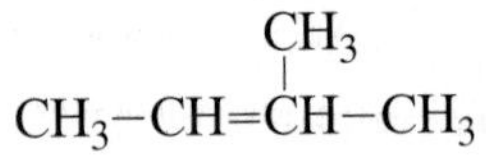
$$CH_3-CH=\overset{\displaystyle CH_3}{\overset{|}{C}}H-CH_3$$

b. 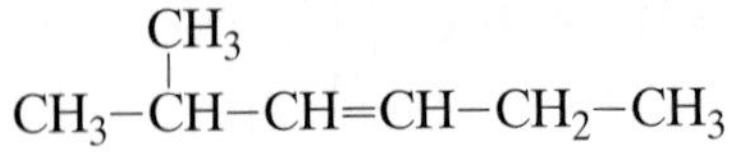
$$CH_3-\overset{\displaystyle CH_3}{\overset{|}{C}}H-CH=CH-CH_2-CH_3$$

Answer
a. 2-methyl-2-butene
b. 2-methyl-3-hexene

FIGURE 20-15 α-farnesene is a solid alkene found in the natural wax covering of apples. Can you determine the IUPAC name for this large alkene?

Properties and Uses of Alkenes

Alkenes are nonpolar and show trends in properties similar to those of alkanes in boiling points and physical states. For example, α-farnesene has 15 carbon atoms and 4 double bonds, as shown in Figure 20-15. This large alkene is a solid at room temperature and atmospheric pressure. It is found in the natural wax covering of apples. Ethene, the smallest alkene, is a gas.

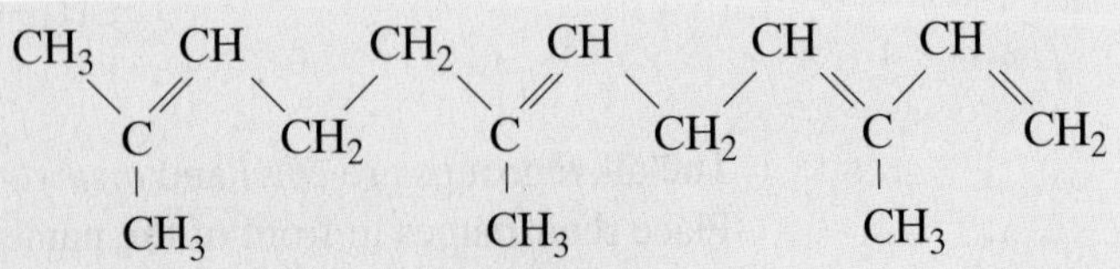

$$CH_3-\underset{\displaystyle CH_3}{\underset{|}{C}}=CH-CH_2-CH_2-CH=\underset{\displaystyle CH_3}{\underset{|}{C}}-CH_2-CH_2-CH=\underset{\displaystyle CH_3}{\underset{|}{C}}-CH=CH_2$$

α-farnesene

Ethene is the hydrocarbon commercially produced in the greatest quantity in the United States. It is used in the synthesis of many plastics and commercially important alcohols. Ethene is also an important plant hormone. Induction of flowering and fruit ripening, as shown in Figure 20-16, are effects of ethene hormone action that can be manipulated by commercial growers.

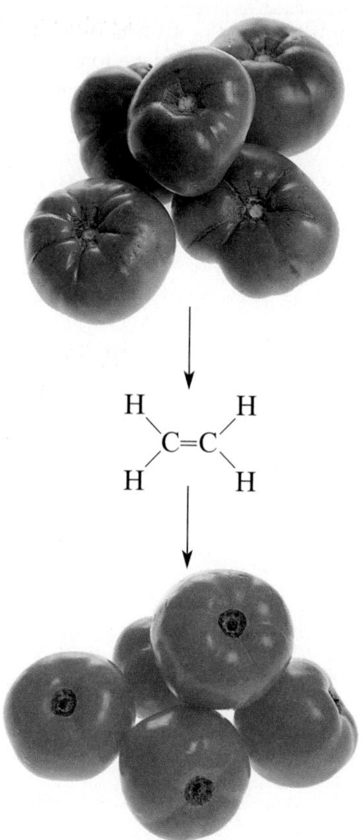

FIGURE 20-16 Ethene is a plant hormone that triggers fruit ripening. Its small size allows it to travel as a gas.

Alkynes

Hydrocarbons with triple covalent bonds are **alkynes.** Like the double bond of alkenes, the triple bond of alkynes requires that the simplest alkyne has two carbon atoms.

$$H-C\equiv C-H$$

ethyne

The general formula for the alkynes is C_nH_{2n-2}. Alkynes have four fewer hydrogen atoms than the corresponding alkanes and two fewer than the corresponding alkenes.

H_3C-CH_3 (H–C–C–H with H's above and below each C)	$H_2C=CH_2$	$H-C\equiv C-H$
C_2H_6	C_2H_4	C_2H_2

Systematic Naming of Alkynes

Alkyne nomenclature is almost the same as alkene nomenclature. The only difference is that the *-ene* suffix of the corresponding alkene is replaced with *-yne*. A complete list of rules follows.

Alkyne nomenclature

Use the rules for alkane nomenclature on page 639, with the following exceptions.

1. **Name the parent hydrocarbon.** Locate the longest continuous chain that *contains the triple bond(s)*. If there is only one triple bond, add the suffix *-yne* to the prefix corresponding to the number of carbon atoms in the chain. If there is more than one triple bond, modify the suffix to indicate the number of triple bonds. For example, 2 = *-adiyne,* 3 = *-atriyne,* and so on.
2. **Add the names of the alkyl groups.**
3. **Number the carbon atoms in the parent hydrocarbon.** Number the carbon atoms in the chain so that the first carbon atom in the triple bond nearest the end of the chain has the lowest number. If numbering from both ends gives the same positions for two triple bonds, then number from the end nearest the first alkyl group.

FIGURE 20-17 Ethyne is the fuel used in oxyacetylene torches. Oxyacetylene torches can reach temperatures of over 3000°C.

4. **Insert position numbers.** Place the position numbers of the triple bonds immediately before the name of the parent hydrocarbon alkyne. Place alkyl group position numbers immediately before the name of the corresponding alkyl group.
5. **Punctuate the name.**

Two examples of correctly named alkynes are given below.

$$CH_3{-}CH_2{-}CH_2{-}C{\equiv}CH$$

1-pentyne

$$CH{\equiv}C{-}\underset{\displaystyle CH_3}{\underset{|}{CH}}{-}CH_3$$

3-methyl-1-butyne

Properties and Uses of Alkynes

As with the other hydrocarbons, alkynes exhibit the same trends in boiling points and physical state and are nonpolar. The smallest alkyne, ethyne, is a gas. The combustion of ethyne when it is mixed with pure oxygen produces the intense heat of welding torches, as shown in Figure 20-17. The common name of ethyne is *acetylene,* and these welding torches are commonly called oxyacetylene torches.

Aromatic Hydrocarbons

Aromatic hydrocarbons *are hydrocarbons with six-membered carbon rings and delocalized electrons.* **Benzene** *is the primary aromatic hydrocarbon.* The molecular formula of benzene is C_6H_6. One possible structural formula is a six-carbon atom ring with three double bonds.

(Structural formula: benzene ring with alternating double bonds, each carbon bonded to one H)

However, benzene does not behave chemically like an alkene. All of the carbon–carbon bonds in the molecule are the same. Like graphite, benzene contains resonance hybrid bonds. The structure of the benzene ring allows the delocalized electrons to be spread over the ring. The entire molecule lies in the same plane, as shown in Figure 20-18. The following structural formulas are often used to show this spreading of electrons. In the condensed form, the hydrogen atom at each corner is understood.

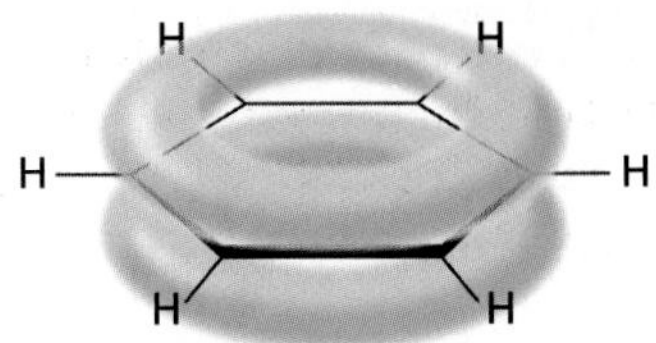

FIGURE 20-18 Electron orbitals in benzene overlap to form continuous orbitals that allow the delocalized electrons to spread uniformly over the entire ring.

H H H H H H

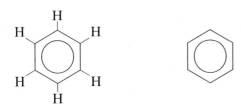

Aromatic hydrocarbons can be thought of as derivatives of benzene. The simplest have one benzene ring, as shown in the following example.

CH_3

methylbenzene

Systematic Names of Aromatic Hydrocarbons

The simplest aromatic hydrocarbons are named as alkyl-substituted benzenes. The names of the alkyl groups are added in front of the word *benzene* according to the rules for other hydrocarbons. As with cycloalkanes, the carbon atoms in the ring do not need to be numbered if there is only one alkyl group. If there is more than one alkyl group, the carbons are numbered in order to give all of the alkyl groups the lowest possible numbers. Following are some examples.

$CH_2-CH_2-CH_3$ CH_3 CH_3

propylbenzene 1,3-dimethylbenzene

The rules for naming simple aromatic hydrocarbons can be summarized as follows.

Simple Aromatic Hydrocarbon Nomenclature

Use the rules for alkane nomenclature on page 639, with the following exceptions.

1. **Name the parent hydrocarbon.** The parent hydrocarbon is the benzene ring, *benzene*.
2. **Add the names of the alkyl groups.**

3. **Number the carbon atoms in the parent hydrocarbon.** If there are two or more alkyl groups attached to the benzene ring, number the carbon atoms in the ring. Assign position number one to the alkyl group that comes first in alphabetical order. Then number in the direction that gives the rest of the alkyl groups the lowest numbers possible.
4. **Insert position numbers.**
5. **Punctuate the name.**

SAMPLE PROBLEM 20-4

Draw the condensed structural formula for 1,2-dimethylbenzene.

SOLUTION

1. Identify the parent hydrocarbon in the name.

   1,2-dimethyl**benzene**

2. Draw the benzene ring.

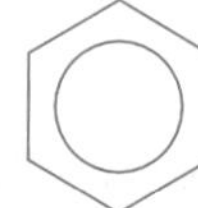

3. Number the carbon atoms in the benzene ring.

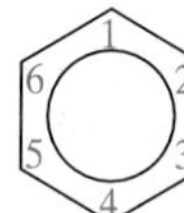

4. Identify any alkyl groups.

   1,2-di**methyl**benzene

   There are only methyl groups in this molecule. The prefix *di-* is attached to the word *methyl,* so there are two methyl groups.

5. Locate the position numbers for the methyl groups.

   1,2-dimethylbenzene

6. Attach the methyl groups to the carbon atoms numbered *1* and *2*.

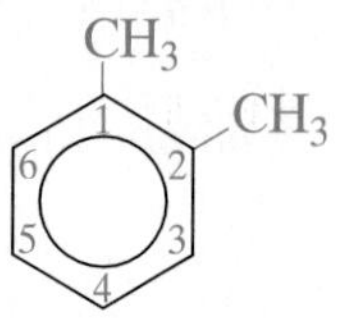

CH_3 CH_3

7. The complete structural formula for 1,2-dimethylbenzene is as follows.

CH_3 CH_3

PRACTICE

1. Name the following compound:

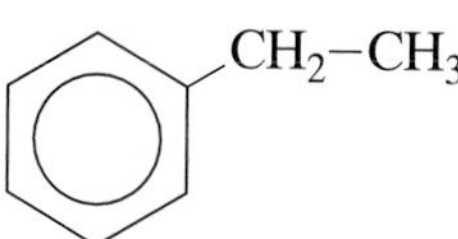

Answer
ethylbenzene

2. Draw the condensed structural formula for 1-ethyl-4-methylbenzene.

Answer

CH_3-CH_2- ⬡ $-CH_3$

Properties and Uses of Aromatic Hydrocarbons

Benzene rings are chemically very stable, a property that can be explained by the concept of delocalized electrons. Therefore, aromatic hydrocarbons are less reactive than alkenes and alkynes are. In the past, benzene was used as a nonpolar solvent because of this stability. However, benzene is both a poison and a carcinogen. Like other hydrocarbons, benzene is nonpolar and has limited solubility in water. It appears that oxidation of the benzene ring, in an attempt to solubilize it for elimination from the body, produces toxic molecules. This has led to the replacement of benzene as a solvent with methylbenzene, which is less toxic. Another aromatic hydrocarbon, 3,4-benzpyrene, is found in coal tar, tar from cigarette smoke, and soot in heavily polluted urban areas. Studies have shown this compound can cause cancer.

SECTION REVIEW

1. List the basic structural features that characterize each of the following:
   a. alkenes
   b. alkynes
   c. aromatic hydrocarbons
2. Draw three condensed structural formulas that can represent C_4H_8.
3. Give the systematic name for each compound whose formula appears in item 2.
4. Give examples of a property or use of three unsaturated hydrocarbons.
5. Draw the condensed structural formula for each of the following:
   a. 1,3-butadiene
   b. 2-pentyne
   c. 1,2-diethylbenzene

CHAPTER 20 REVIEW

CHAPTER SUMMARY

20-1

- Carbon is important because all living matter contains carbon.
- Hybridized orbitals allow carbon atoms to form single, double, and triple covalent bonds.
- Carbon occurs in several solid allotropic forms, such as diamond, graphite, and fullerenes, all of which have different structures and properties.

Vocabulary

delocalized electrons (627) diamond (626) fullerene (626) graphite (626)

20-2

- All organic compounds contain carbon, but not all carbon-containing compounds are classified as organic.
- The number of possible organic compounds is virtually unlimited because of the bonding properties of carbon. The unique catenation ability of carbon allows it to link together to form long chains and rings. The ability of carbon to bind other elements and to allow different arrangements of atoms adds to the diversity of carbon compounds.
- Isomers are compounds with the same molecular formula but different structures. Structural formulas are needed to show the bonding order and arrangement of atoms in an organic molecule to distinguish between isomers.
- Structural isomers are isomers in which the atoms are bonded together in different orders. Geometric isomers are isomers in which the order of atom bonding is the same but the atoms are arranged differently in space.

Vocabulary

catenation (630) hydrocarbon (630) organic compound (629) structural isomer (631)
geometric isomer (632) isomer (630) structural formula (630)

20-3

- In saturated hydrocarbons, each carbon atom has four single covalent bonds. Alkanes are saturated hydrocarbons.
- Organic compounds are named according to a systematic method developed by IUPAC.
- Alkanes contain only single bonds. Because alkanes consist of saturated single covalent bonds, these compounds are not very reactive. One important reaction they do undergo is combustion.
- Trends in physical states, boiling points, and combustion properties correspond to trends in alkane size and amount of branching.

Vocabulary

alkane (634) fractional distillation (644) natural gas (643) petroleum (643)
alkyl group (637) homologous series (634) octane rating (645) saturated hydrocarbon (634)
cycloalkane (635)

20-4

- Carbon atoms in unsaturated hydrocarbons do not all have four single covalent bonds. Alkenes, alkynes, and aromatic hydrocarbons are unsaturated hydrocarbons.
- Alkenes contain carbon-carbon double bonds and can have geometric isomers. The smallest alkene, ethene, is an important industrial and agricultural chemical.
- Alkynes contain carbon-carbon triple bonds.
- Benzene and derivatives of benzene are aromatic hydrocarbons. The concept of delocalized electrons helps explain the stability of the benzene ring.

Vocabulary

alkene (647) aromatic hydrocarbon (652) benzene (652) unsaturated hydrocarbon (647)
alkyne (651)

REVIEWING CONCEPTS

1. What is the orientation of the four covalent bonds and the sp^3 orbitals of a carbon atom? (20-1)
2. Name and describe the structures of three allotropic forms of carbon. (20-1)
3. What properties of diamond determine most of its industrial uses? (20-1)
4. Why does graphite conduct electricity while diamond does not? (20-1)
5. Explain why the structure of graphite makes it useful as a lubricant. (20-1)
6. Describe the structure of buckminsterfullerene. (20-1)
7. a. What is catenation?
   b. How does catenation contribute to the diversity of organic compounds? (20-2)
8. What are hydrocarbons, and what is their importance? (20-2)
9. a. What information about a compound is provided by a structural formula?
   b. How are structural formulas used in organic chemistry? (20-2)
10. Can molecules with the molecular formulas C_4H_{10} and $C_4H_{10}O$ be structural isomers of one another? Why or why not? (20-2)
11. Can molecules with only single bonds (and no rings) have geometric isomers? Why or why not? (20-2)
12. a. What do the terms *saturated* and *unsaturated* mean when applied to hydrocarbons?
    b. What other meanings do these terms have in chemistry?
    c. Classify alkenes, alkanes, alkynes, and aromatic hydrocarbons as either saturated or unsaturated. (20-3 and 20-4)
13. Classify each of the following as an alkane, alkene, alkyne, or aromatic hydrocarbon.
    a. benzene ring with two substituents: $CH_2{-}CH_3$ and $CH_2{-}CH_3$
    b. $CH_3{-}CH{=}CH_2$
    c. $CH{\equiv}C{-}CH(CH_3){-}CH_2{-}CH_3$
    d. $CH_3{-}CH(CH_3){-}CH_2{-}CH_2{-}CH_2{-}CH_2{-}CH_3$
    (20-3 and 20-4)
14. Give the general formula for the members of the following:
    a. alkane series
    b. alkene series
    c. alkyne series (20-3 and 20-4)
15. Give the molecular formula for each type of hydrocarbon if it contains seven carbon atoms.
    a. an alkane
    b. an alkene
    c. an alkyne (20-3 and 20-4)
16. a. What is a homologous series?
    b. By what method are straight-chain hydrocarbons named?
    c. Name the straight-chain alkane with the molecular formula $C_{10}H_{22}$. (20-3)
17. What are cycloalkanes? (20-3)
18. a. What trend occurs in the boiling points of alkanes?
    b. How would you explain this trend?
    c. How is the trend in alkane boiling points used in petroleum fractional distillation? (20-3)
19. How does the structure of alkanes affect the octane rating of gasoline? (20-3)
20. Write a balanced equation for the complete combustion of each of the following:
    a. methane
    b. ethyne (20-3 and 20-4)
21. Which types of isomers are possible for alkanes (with no rings), alkenes, and alkynes? Why? (20-3 and 20-4)
22. Give examples of ethene's commercial uses. (20-4)
23. a. Alkyne nomenclature is very similar to the nomenclature of what other group of hydrocarbons?
    b. How do these nomenclatures differ? (20-4)

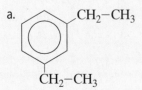

24. Give one use for ethyne. (20-4)

25. a. What are delocalized electrons?
b. What is their effect on the reactivity of aromatic hydrocarbons? (20-4)

26. What is the name of the parent hydrocarbon of simple aromatic hydrocarbons? (20-4)

27. Describe a possible cause of benzene toxicity. (20-4)

PROBLEMS

Structural Formulas

28. Draw the condensed structural formula for the following:

```
   H      H      H
   |      |      |
H–C=C─────C─────C–H
   |      |      |
   H    H–C–H    H
          |
          H
```

29. Identify each of the following pairs of formulas as representing the same or different molecules:

a. C_5H_{12} AND

```
   H H    H      H
   | |    |      |
H–C–C─────C─────C–H
   | |    |      |
   H H  H–C–H    H
          |
          H
```

b. $CH_3–CH_2–CH_3$

AND

```
   H H H H
   | | | |
H–C–C–C–C–H
   | | | |
   H H H H
```

c. C_6H_{10} AND

```
CH3–CH=C–CH3
        |
        CH3
```

d.

```
   H     H        H    H H
   |     |        |    | |
H–C──────C────────C────C–C–H
   |     |        |    | |
   H   H–C–H    H–C–H  H H
         |        |
         H        H
```

AND

```
CH3–CH–CH–CH2–CH3
     |   |
    CH3 CH3
```

Isomers

30. Identify whether each pair represents the same molecule or structural isomers.

a. $CH_3–CH_2–CH_2–CH_3$

```
CH3
|
CH2–CH2–CH3
```

b.

```
                CH3
                |
CH3–CH–CH2–CH2
     |
    CH2
     |
    CH3
```

```
               CH3
                |
CH3–CH–CH2–CH–CH3
     |
    CH3
```

c.

```
            O                        O
            ‖                        ‖
CH3–CH2–C–OH        CH3–O–CH2–CH
```

31. Draw structural formulas for the five isomers of C_6H_{14}.

32. Draw the geometric isomers of the following molecule. Label each isomer as *cis* or *trans.*

$CH_3–CH=CH–CH_2–CH_3$

33. a. Which of the following can have geometric isomers?

$CH_3–CH=CH–Cl$

```
             CH3
              |
CH3–CH=C–CH3
```

$CH_3–CH_2–CH=CH–CH_2–CH_3$

b. Draw the geometric isomers for those that can have geometric isomers.
c. Label each geometric isomer as *cis* or *trans.*

Alkane Nomenclature

34. Name the following molecules. (Hint: See Sample Problem 20-1.)

a. $CH_3–CH_2–CH_2–CH_2–CH_2–CH_2–CH_3$

b.

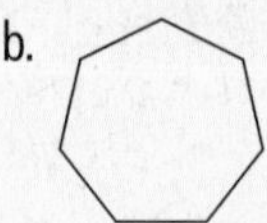

c.

```
       CH3
        |
CH3–C–CH2–CH–CH–CH3
        |          |
       CH3        CH3
```

d.
```
CH3              CH2-CH2-CH2-CH3
 |                |
CH3-C-CH2-CH2-CH-CH2-CH2-CH2-CH3
 |
CH3
```

35. Give the complete, uncondensed, structural formula for each of the following alkanes. (Hint: See Sample Problem 20-2.)
a. decane
b. 3,3-dimethylpentane

36. Give the condensed structural formula for each of the following alkanes:
a. 1,1-dimethylcyclopropane
b. 2,2,4,4-tetramethylpentane

37. For each of the following, determine whether the alkane is named correctly. If it is not, give the correct name.

a.
```
CH3-CH2-CH2
         |
        CH3
```
1-methylpropane

b.
```
CH3-CH2-CH2-CH2-CH2-CH2-CH2
                         |
                        CH2
                         |
                        CH3
```
nonane

c.
```
            CH3
             |
CH3-CH2-CH2-CH-CH2-CH3
```
4-methylhexane

d.
```
                  CH3
                   |
CH3-CH2-CH-CH2-CH-CH3
         |
        CH2-CH3
```
4-ethyl-2-methylhexane

Alkene Nomenclature

38. Name the following alkenes. (Hint: See Sample Problem 20-3.)

a. $CH_2{=}CH{-}CH_2{-}CH_2{-}CH_3$

b.
```
CH3      H
   \    /
    C=C
   /    \
CH3      CH3
```

c.
```
           CH3
            |
CH2=CH-C-CH2-CH3
            |
           CH2
            |
           CH3
```

d.
```
CH=C-CH2-CH2-CH=CH2
   |
  CH3
```

39. Draw the condensed structural formula for each of the following alkenes:
a. 2-methyl-2-hexene
b. 3-ethyl-2,2-dimethyl-3-heptene

40. Draw structural formulas for geometric isomers of each of the following:
a. $CH_3{-}CH_2{-}CH_2{-}CH{=}CH{-}CH_3$
b. 3-methyl-2-pentene

Alkyne Nomenclature

41. Name the following alkynes:

a. $CH{\equiv}C{-}CH_3$

b.
```
CH3-C≡C-CH-CH3
         |
        CH3
```

c.
```
CH3-CH-C≡C-CH-CH3
     |       |
    CH3     CH3
```

d. $CH{\equiv}C{-}CH_2{-}CH_2{-}CH_2{-}C{\equiv}CH$

42. Draw the condensed structural formula for each of the following alkynes:
a. 1-decyne
b. 6,6-dimethyl-3-heptyne

Aromatic Hydrocarbon Nomenclature

43. Name the following aromatic hydrocarbons. (Hint: See Sample Problem 20-4.)

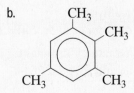

a.

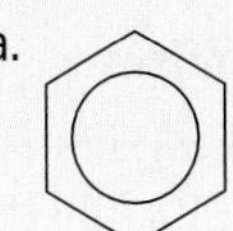

b. (benzene ring with substituents)
```
      CH3
          CH3
CH3       CH3
```

44. Draw the condensed structural formula for each of the following molecules:
a. 1,3,5-trimethylbenzene
b. 1,3-dimethylbenzene

Calculations with Carbon Compounds

45. The jewelers' mass unit for diamond is the carat. By definition, 1 carat equals exactly 200 mg. What is the volume of a 1.00 carat diamond? The density of diamond is 3.51 g/cm^3.

46. For 100.0 g of butadiene, C_4H_6, calculate the following:
a. number of moles b. number of molecules

47. An alkene has the molecular formula $C_{12}H_{24}$. Determine its percent composition.

48. Assuming that the volumes of carbon dioxide and of propane are measured under the same experimental conditions, what volume of carbon dioxide is produced by the complete combustion of 15.0 L of propane?

49. Assume a gasoline is isooctane, which has a density of 0.692 g/mL. What is the mass in kilograms of 12.0 gal of the gasoline (1 gal = 3.78 L)?

MIXED REVIEW

50. a. Draw the complete, uncondensed structural formula for 4-methyloctane.
b. Convert it into the condensed structural formula.
c. Determine the molecular formula for the molecule from both the structure you drew and the general molecular formula for alkanes. Compare the two. Are they the same?

51. Draw and name two different condensed structural formulas for molecules of each of the following types of hydrocarbons containing eight carbon atoms:
a. alkane c. alkyne
b. alkene d. aromatic hydrocarbon

52. Draw the condensed structural formulas for 4,4-dimethyl-2-pentyne and 2,2-dimethyl-4-propyloctane.

53. Draw the three structural isomers for an alkyne containing five carbon atoms and one triple bond. Name the molecules you draw.

54. Which of the following molecules have geometric isomers? Draw all possible geometric isomers. Label the molecules you draw as either *cis* or *trans*.
a. butane
b. 2-pentene
c. 2-hexyne
d. 2-methyl-1-butene

55. Identify the following pairs as the same compound, isomers, or different compounds that are not isomers:

a.
```
    H     H       H     H H
    |     |       |     | |
  H-C-----C-------C-----C-C-H
    |     |       |     | |
    H   H-C-H     H     H H
          |
          H
```
AND
```
CH3-CH-CH2-CH2-CH3
    |
    CH3
```

b. C_4H_8

AND
```
    H H H H
    | | | |
  H-C-C-C-C-H
    | | | |
    H H H H
```

c.
```
            CH3
            |
       CH3  CH2
       |    |
   CH3-C----CH2
       |
       CH3
```
AND
```
       CH3
       |
   CH3-C-CH2-CH2-CH3
       |
       CH3
```

d.
```
   CH3-C=CH-CH2-CH3
       |
       CH3
```
AND
```
   CH3-CH2-CH-CH=CH2
           |
           CH3
```

CRITICAL THINKING

56. **Inferring Conclusions** Why are organic compounds with covalent bonds usually less stable when heated than inorganic compounds with ionic bonds?

57. **Inferring Relationships** The element that appears in the greatest number of compounds is hydrogen. The element found in the second greatest number of compounds is carbon. Why are there more hydrogen compounds than carbon compounds?

58. **Relating Ideas** As the number of carbon atoms in an alkane molecule increases, does the percentage of hydrogen increase, decrease, or remain the same?

HANDBOOK SEARCH

59. The top 10 chemicals produced in the United States are listed in Table 7B of the *Elements Handbook*. Review this material, and answer the following:
    a. Which of the top ten compounds are organic?
    b. Write structural formulas for the compounds you listed in item (a).
    c. To what homologous series do each of these compounds belong?

60. The reaction of methane with oxygen produces two different oxides of carbon. Review this material in the *Elements Handbook,* and answer the following:
    a. What conditions determine whether the product of the methane reaction is CO_2 or CO?
    b. If a home heating system is fueled by natural gas, what difference does it make if the combustion produces CO_2 or CO?

61. Silicon is similar to carbon in forming long-chain compounds. Review the material on silicon in the *Elements Handbook* and answer the following.
    a. How does a long-chain silicon compound differ in composition from a long-chain carbon compound?
    b. The simplest alkane is methane. Methyl groups are found in all alkanes. What is a common subunit of a silicate? What is the geometry of that subunit?

62. Mercury in the environment poses a hazard to living things. Review the section on mercury poisoning in the *Elements Handbook.*
    a. Draw a structure formula for the organic mercury compound described in that section.
    b. What is the IUPAC name for this compound?

RESEARCH & WRITING

63. *Chemical and Engineering News* publishes a list once a year of the top 50 chemicals. Find out which chemicals on the current year's list are hydrocarbons, and report your findings to the class.

64. Consult reference materials at the library, and read about products made from hydrocarbons. Keep a list of the number of petroleum-related products you use in a single day.

ALTERNATIVE ASSESSMENT

65. **Performance** Models are often used to visualize the three-dimensional shape of molecules. Using gumdrops as atoms and toothpicks to bond them together, construct models of different hydrocarbons. Use large gumdrops for carbon and smaller gumdrops for hydrogen. Refer to Figures 20-1 and 20-2 for guidelines on the three-dimensional shapes of hydrocarbons.

66. **Performance** Using your gumdrop models, demonstrate why alkenes can have geometric isomers, while alkanes cannot.

CHAPTER 21

Other Organic Compounds

Organic compounds are used to make many of the products we use every day.

Functional Groups and Classes of Organic Compounds

SECTION 21-1

OBJECTIVES

- Define *functional group,* and explain why functional groups are important.
- Identify alcohols, alkyl halides, and ethers based on the functional group present in each.
- Classify alcohols, alkyl halides, and ethers from names and structural formulas.
- Relate properties of alcohols, alkyl halides, and ethers to their structures. Describe how these properties influence the uses of specific organic compounds.

A **functional group** *is an atom or group of atoms that is responsible for the specific properties of an organic compound;* the bonds within functional groups are often the site of chemical reactivity. A given functional group undergoes the same types of chemical reactions in every molecule in which it is found. Therefore, all compounds that contain the same functional group have similar properties and can be classified together.

Alcohols

Alcohols *are organic compounds that contain one or more hydroxyl groups.* The general formula for a class of organic compounds consists of the functional group and the letter *R,* which stands for the rest of the molecule. The general formula for alcohols is $R{-}OH$. Systematic names of organic compounds indicate which functional groups are present in a molecule. The rules for naming simple alcohols are as follows.

Alcohol Nomenclature

1. **Name the parent compound.** Locate the longest continuous chain of carbon atoms *that contains the hydroxyl group*. If there is only one hydroxyl group, add the suffix *-ol* to the prefix corresponding to the number of carbon atoms in this chain. If there is more than one hydroxyl group, use the full name of the corresponding alkane and add the suffix modified to indicate the number of hydroxyl groups. For example, *-diol* = 2, *-triol* = 3, and so on.
2. **Number the carbon atoms in the parent chain.** Number the carbon atoms in the chain so that the hydroxyl groups have the lowest numbers possible.
3. **Insert position numbers.** Place the hydroxyl position number or numbers immediately before the name of the parent alcohol.
4. **Punctuate the name.** Separate the position numbers from the name with a hyphen. If there is more than one position number, separate the position numbers with commas.

TABLE 21-1 *Boiling Points of Some Alcohols and Alkanes*

Compound	Molecular formula	Molar mass (g/mol)	Boiling point (°C)
methanol	CH_3OH	32	64.7
ethane	C_2H_6	30	–88
ethanol	C_2H_5OH	46	78.3
propane	C_3H_8	44	–42.1
1-propanol	C_3H_7OH	60	97.2
butane	C_4H_{10}	58	–0.50

Following are three examples of correctly named alcohols.

$$CH_3{-}CH_2{-}\underset{}{\overset{OH}{\overset{|}{CH_2}}} \qquad CH_3{-}CH_2{-}\overset{OH}{\overset{|}{CH}}{-}CH_3 \qquad CH_3{-}\overset{OH}{\overset{|}{\underset{OH}{\underset{|}{C}}}}{-}CH_2{-}CH_3$$

1-propanol 2-butanol 2,2-butanediol

Properties and Uses of Alcohols

As shown in Table 21-1, the boiling points of alcohols tend to be higher than those of alkanes of comparable molar mass. For example, the molar mass of ethanol, 46 g/mol, is close to that of propane, 44 g/mol. However, their boiling points are very different. The boiling point of ethanol is 78.3°C, while the boiling point of propane is –42.1°C. In addition, boiling points increase as the number of hydroxyl groups in the molecule increases. This trend can be seen in Table 21-2, which shows the boiling points for alcohols with one, two, and three hydroxyl groups.

The boiling point trends shown in Tables 21-1 and 21-2 can be explained by hydrogen bonding, which was discussed in Chapter 6. Compared to alkanes, extra energy is required to break hydrogen bonds between alcohol molecules before conversion from a liquid to a gas. When more than one hydroxyl group is present, multiple hydrogen bonds may exist. In this case, even more energy is required to break the hydrogen bonds before the liquid is converted to a gas.

TABLE 21-2 *Multiple Hydroxyl Groups and Boiling Points*

Alcohol	Number of hydroxyl groups	Boiling point (°C)
ethanol	1	78.3
1,2-ethanediol	2	197.3
1-propanol	1	97.2
1,2-propanediol	2	188
1,2,3-propanetriol	3	258–260

TABLE 21-3 *Solubility of Some Alcohols in Water*

Alcohol	Molecular formula	Solubility (g/100. g of water)
methanol	CH_3OH	∞ (completely soluble)
1-butanol	C_4H_9OH	7.4
1-pentanol	$C_5H_{11}OH$	2.7
1-octanol	$C_8H_{17}OH$	0.06

As explained in Chapter 13, alcohols are soluble in water because of hydrogen bonding. However, the solubility of alcohols in water tends to decrease with an increase in the size of the molecule. The longer the hydrocarbon chain in an alcohol, the larger the nonpolar, insoluble portion of the molecule. Table 21-3 illustrates this trend.

Hydrogen bonding in alcohols can also explain other properties and uses of alcohols. Cold creams, lipsticks, body lotions, and similar products generally include 1,2,3-propanetriol, commonly called glycerol, to keep them moist. A model for glycerol is shown in Figure 21-1. Multiple hydroxyl groups allow glycerol to form many hydrogen bonds with water molecules in the air or in the surrounding material.

Alcohols are sometimes used today as alternative fuels and as octane enhancers in fuel for automobiles. Ethanol is combined with gasoline, for example, in a one-to-nine ratio to produce gasohol. Some experts have promoted the use of gasohol as a fuel for automobiles because it burns more cleanly, helps save valuable petroleum reserves, and reduces our nation's dependence on foreign imports of petroleum. However, there are also disadvantages. The combustion of ethanol produces only 60% as much energy per gram as does the combustion of gasoline. Ethanol also encourages water absorption in the fuel.

FIGURE 21-1 Glycerol contains three hydroxyl groups. This structure allows it to form multiple hydrogen bonds with water. Glycerol is added as a moisturizer to skin products.

All simple alcohols are poisonous to some extent. When ethanol is consumed, it is broken down by the enzyme alcohol dehydrogenase. This enzyme rapidly converts ethanol to an oxidized form known as acetaldehyde, which is then converted to acetic acid. Acetic acid, a component of household vinegar, is relatively harmless to the human body. However, consuming large amounts of ethanol can be fatal. The lethal dose varies for different individuals. The amount of ethanol found in a liter of hard liquor, 400 mL, is usually fatal.

Other simple alcohols are attacked by alcohol dehydrogenase more slowly, making these alcohols more toxic than ethanol. For example, methanol, or wood alcohol, is converted to formaldehyde and formic acid, both of which are toxic. A great deal of damage can be done to cells before these chemicals are completely metabolized by the body. Methanol is about 10 times more toxic than is ethanol. Toxic effects of methanol include damage to the optic nerve, coma, and death.

Alkyl Halides

Alkyl halides *are organic compounds in which one or more halogen atoms—fluorine, chlorine, bromine, or iodine—are substituted for one or more hydrogen atoms in a hydrocarbon.* Because $-X$ is often used to represent any halogen, an alkyl halide may be represented by the general formula $R{-}X$. The rules for naming simple alkyl halides in the IUPAC system are as follows.

Alkyl Halide Nomenclature

1. **Name the parent compound.** Locate the longest continuous chain of carbon atoms *that contains the halogen.* Add the prefixes for the attached halogen atoms to the name of the alkane corresponding to the number of carbon atoms in this chain. The prefixes to use are *fluoro-* for fluorine, *chloro-* for chlorine, *bromo-* for bromine, and *iodo-* for iodine. If more than one kind of halogen atom is present, add the halogen prefixes in alphabetical order. If there is more than one atom of the same halogen, add the appropriate prefix (*di-*, *tri-*, and so on) after the prefixes are arranged in alphabetical order.
2. **Number the carbon atoms in the parent chain.** Number the carbon-atom chain so that the sum of the halogen numbers is as low as possible. If there are different halogen atoms in equivalent positions, give the lower number to the one that comes first in alphabetical order.
3. **Insert position numbers.** Place the halogen position number or numbers immediately before the halogen prefixes.
4. **Punctuate the name.** Separate the position numbers from the name with hyphens. If there is more than one position number, separate the position numbers with commas.

SAMPLE PROBLEM 21-1

Name the alkyl halide shown.

```
    H  H  H
    |  |  |
 H–C–C–C–H
    |  |  |
   Br Br  H
```

SOLUTION

1. Locate the longest continuous chain of carbon atoms that contains the halogen.

```
    H  H  H
    |  |  |
 H–C–C–C–H
    |  |  |
   Br Br  H
```

The chain has three carbon atoms, so the name of the chain is *propane.*

2. Identify and name the halogen atoms attached to the chain.

```
    H  H  H
    |  |  |
 H–C–C–C–H
    |  |  |
   Br Br  H
```

Bromine atoms are attached to the chain. Add the prefix *bromo-* in front of *propane.* Add the prefix *di-* to show that there are two bromine atoms present.

dibromopropane

3. Number the carbon-atom chain so that the sum of the halogen numbers is as low as possible.

```
    H    H    H
    |    |    |
 H–C¹–C²–C³–H
    |    |    |
   Br   Br    H
```

4. The bromine atoms are on carbons numbered *1* and *2.* Place these numbers immediately before the halogen prefix. Separate the numbers from the prefix with a hyphen. Separate the numbers with a comma.

1,2-dibromopropane

The complete name is 1,2-dibromopropane.

PRACTICE

1. Name each of the following alkyl halides:

a.
```
        Br
        |
 CH3–CH–CH3
```

b.
```
         F    F    F
         |    |    |
 CH3–CH–CH–CH2
```

2. Draw condensed structures for each of the following alkyl halides:

a. 2-iodopropane

b. 1,1,1,2-tetrabromobutane

Answer

1. a. 2-bromopropane
   b. 1,2,3-trifluorobutane

2. a.
```
        I
        |
 CH3–CH–CH3
```

b.
```
     Br Br
     |  |
 Br–C–CH–CH2–CH3
     |
     Br
```

Properties and Uses of Alkyl Halides

Alkyl halides are some of the most widely used organic chemicals. A family of alkyl halides that has received widespread attention in recent years is the chlorofluorocarbons, or CFCs. *CFCs* are alkyl halides that contain both chlorine and fluorine. The formulas for two widely used CFCs, Freon-11 and Freon-12, are shown below.

```
     Cl                    F
     |                     |
  F—C—Cl               Cl—C—F
     |                     |
     Cl                    Cl
```

trichlorofluoromethane (Freon-11)

dichlorodifluoromethane (Freon-12)

CFC-11 and CFC-12 are odorless, nontoxic, nonflammable, and very stable. They also easily change physical states. These properties make them useful in a number of commercial operations. They have been used in the manufacture of plastic foam and as liquid refrigerants in commercial refrigerators. At the height of their production, in 1985, more than 700 million kilograms of CFC-11 and CFC-12 were manufactured worldwide.

However, CFCs contribute to the destruction of ozone in the upper atmosphere, as shown in Figure 21-2. When released into the atmosphere, CFCs can break down and release free chlorine atoms.

$$CCl_2F_2 \xrightarrow{\text{solar radiation}} Cl + CClF_2$$

The released chlorine atoms attack molecules of ozone (O_3) found in the upper atmosphere. The ozone is converted to diatomic oxygen.

$$Cl + O_3 \longrightarrow ClO + O_2$$

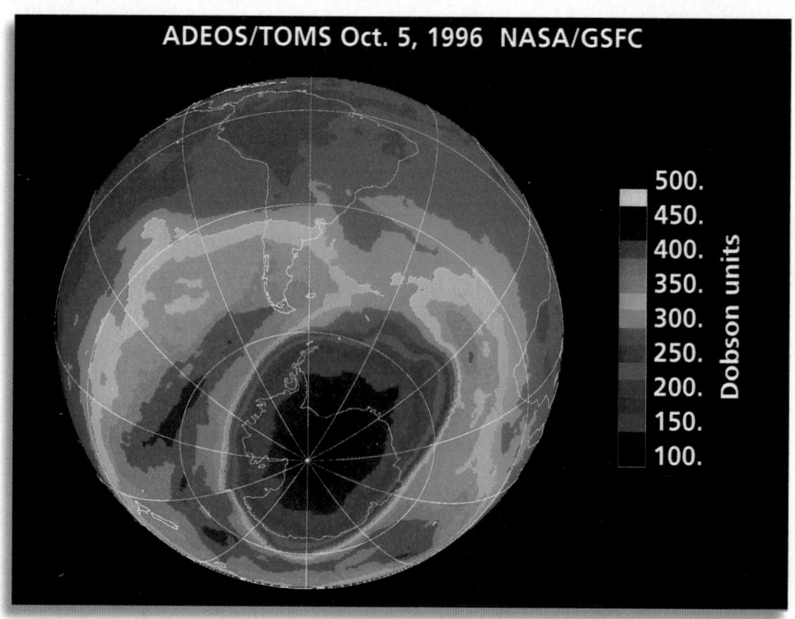

FIGURE 21-2 The depletion of ozone in the upper atmosphere above Antarctica has produced an area of very low ozone concentration, called the ozone hole. The black region over Antarctica shows the area of greatest ozone depletion.

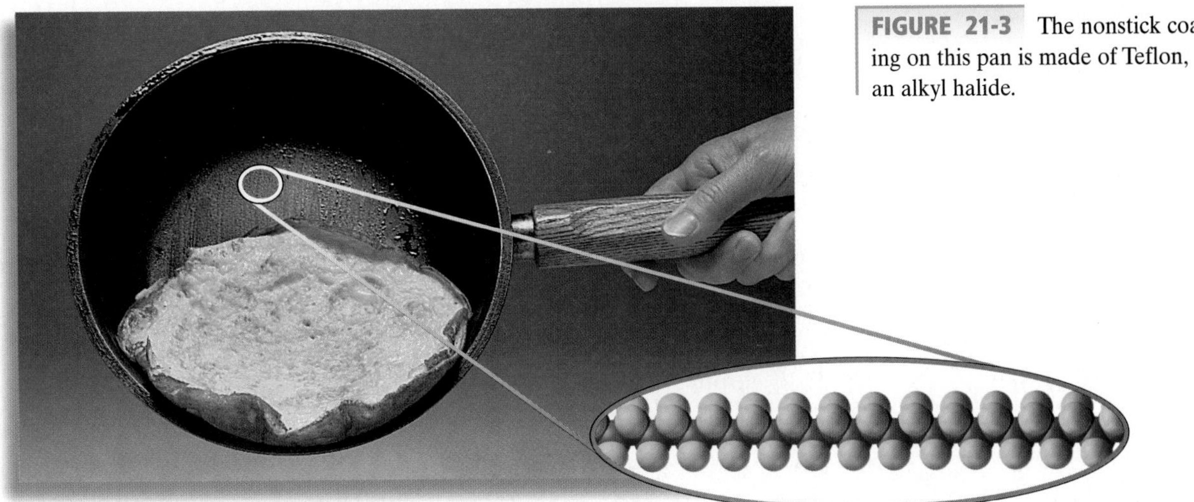

FIGURE 21-3 The nonstick coating on this pan is made of Teflon, an alkyl halide.

Chlorine atoms are eventually regenerated.

$$ClO + O \longrightarrow Cl + O_2$$

This makes it possible for a single chlorine atom to destroy thousands of ozone molecules.

The depletion of ozone in the upper atmosphere has produced an area of very low concentration, called an ozone hole, over Antarctica. Ozone molecules absorb ultraviolet radiation from sunlight, preventing most of that radiation from reaching Earth. Ultraviolet radiation is known to cause skin cancer in humans, to kill some forms of microscopic life, to damage plant tissue, and to cause other harmful effects in plants and animals. Because CFCs are a major cause of ozone depletion, more than 100 nations signed an agreement in 1987 to reduce the amount of CFCs produced.

Another alkyl halide is tetrafluoroethene, C_2F_4. It is joined in long chains to make a material with the trade name Teflon. Because of the unreactive carbon-fluorine bond, Teflon is inactive and stable to about 325°C. It also has a low coefficient of friction, which means that other objects slide smoothly over its surface. These properties enable Teflon to be used in heat-resistant machine parts that cannot be lubricated. It is also used in making utensils with "nonstick" surfaces, such as the frying pan in Figure 21-3.

Ethers

Ethers *are organic compounds in which two hydrocarbon groups are bonded to the same atom of oxygen.* They can be represented by the general formula $R{-}O{-}R'$. In this formula, R' may be the same hydrocarbon group as R or a different one. The IUPAC rules for naming simple ethers are as follows.

Ether Nomenclature

1. **Name the parent compound.** The word *ether* will come at the end of the name.
2. **Add the names of the alkyl groups.** If there are two different alkyl groups, arrange the names in alphabetical order in front of the word *ether.* If both alkyl groups are the same, the prefix *di-* is added to the name of the alkyl group in front of the word *ether.*
3. **Leave appropriate spaces in the name.** There should be spaces between the names of the alkyl groups and between the alkyl groups and the word *ether.*

SAMPLE PROBLEM 21-2

Name the ether shown below.

$$CH_3{-}CH_2{-}O{-}CH_2{-}CH_3$$

SOLUTION

1. The word *ether* will come at the end of the name.

ether

2. Identify and name the two alkyl groups joined to the oxygen atom.

$$CH_3{-}CH_2{-}O{-}CH_2{-}CH_3$$

They are both ethyl groups. Add the prefix *di-* to show that there are two ethyl groups present. Place *diethyl* in front of *ether,* separated by a space.

diethyl ether

The full name is diethyl ether. Diethyl ether is the most common ether and is also known as ethyl ether, or just ether.

PRACTICE

1. Name each of the following ethers:
   a. $CH_3{-}CH_2{-}CH_2{-}O{-}CH_3$
   b. (cyclopentyl)–O–(cyclopentyl)
   c. $CH_3{-}O{-}CH_2{-}CH_2{-}CH_2{-}CH_2{-}CH_3$

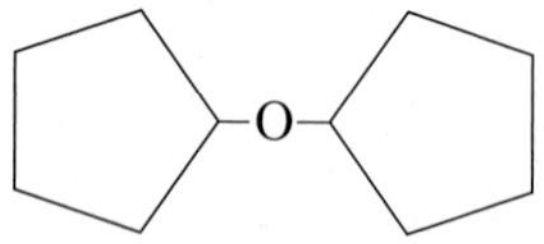

Answer
1. a. methyl propyl ether
   b. dicyclopentyl ether
   c. methyl pentyl ether

2. Draw condensed structures for the following ethers:
   a. ethyl propyl ether
   b. dicyclohexyl ether
   c. butyl methyl ether

2. a. $CH_3{-}CH_2{-}O{-}CH_2{-}CH_2{-}CH_3$
   b. (cyclohexyl)–O–(cyclohexyl)
   c. $CH_3{-}CH_2{-}CH_2{-}CH_2{-}O{-}CH_3$

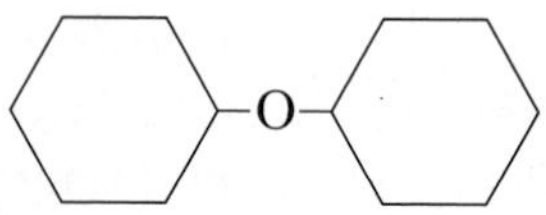

TABLE 21-4 Comparison of the Boiling Points of an Ether, Alkane, and Alcohol

Compound	Molar mass	Boiling point (°C)
diethyl ether	74	34.6
pentane	72	36.1
1-butanol	74	117.7

Properties and Uses of Ethers

The solubilities of ethers and alcohols in water are similar. For example, diethyl ether and 1-butanol have the same molar mass. They also have approximately the same solubility, 6 g/100 g of water for diethyl ether and 7.4 g/100 g of water for 1-butanol. This similarity can be explained by the fact that ethers, like alcohols, can form hydrogen bonds with water molecules.

In comparison, the boiling points of ethers are much lower than those of alcohols of similar molar mass. But, they are about the same as those of alkanes of similar molar mass. This trend is clear from the comparison in Table 21-4. This trend can also be explained by hydrogen bonding. Unlike alcohols, ethers cannot form hydrogen bonds with each other because they do not have a hydrogen atom bonded to a highly electronegative atom. Therefore, no extra energy is needed to break hydrogen bonds for ethers to boil.

Like alkanes, ethers are not very reactive compounds. This property explains their most common uses as solvents. In many organic reactions in which water cannot be used as a solvent, an ether is used instead.

Methyl-*tertiary*-butyl ether (MTBE) is the most widely used ether. It is another gasoline octane enhancer. At one time, tetraethyl lead, $(C_2H_5)_4Pb$, was widely used for this purpose. However, concerns about the release of lead into the environment have increased, so tetraethyl lead has been replaced by MTBE and other octane enhancers.

SECTION REVIEW

1. Give the general formula and class of organic compounds for each of the following:

a. $CH_3{-}OH$

b. $CH_3{-}O{-}CH_3$

c. $Br{-}CH_2{-}CH_2{-}CH_3$

2. Give the name of each of the following:

a. $CH_3{-}CH(OH){-}CH_3$ (OH attached to the middle carbon)

b. $CH_3{-}O{-}CH_3$

c. $CH_3{-}CH_2{-}CHF_2$ (both F attached to the terminal CH)

3. Compare the boiling points of alcohols, ethers, and alkanes, and explain one reason for the differences.

4. Draw condensed structures for each of the following:

a. 1,2-propanediol

b. ethyl methyl ether

c. dichloromethane

SECTION 21-2

More Classes of Organic Compounds

OBJECTIVES

- Identify aldehydes, ketones, carboxylic acids, esters, and amines based on the functional group present in each.
- Classify aldehydes, ketones, carboxylic acids, esters, and amines from names and structural formulas.
- Relate properties of aldehydes, ketones, carboxylic acids, esters, and amines to their structures. Describe how these properties influence the uses of specific organic compounds.

Aldehydes and Ketones

Aldehydes and ketones contain the *carbonyl group*, shown below.

$$-\overset{\overset{\displaystyle O}{\|}}{C}-$$

The difference between aldehydes and ketones is the location of the carbonyl group. **Aldehydes** *are organic compounds in which the carbonyl group is attached to a carbon atom at the end of a carbon-atom chain.* **Ketones** *are organic compounds in which the carbonyl group is attached to carbon atoms within the chain.* These differences can be seen in their general formulas, shown below.

$$R-\overset{\overset{\displaystyle O}{\|}}{C}-H \qquad R-\overset{\overset{\displaystyle O}{\|}}{C}-R'$$

aldehyde — ketone

The IUPAC rules for naming simple aldehydes and ketones are as follows.

Aldehyde Nomenclature

Name the parent compound. Locate the longest continuous chain *that contains the carbonyl group*. Add the suffix *-al* to the prefix corresponding to the number of carbon atoms in this chain.

Following are three examples of correctly named aldehydes.

$$H-\overset{\overset{\displaystyle O}{\|}}{C}-H \qquad CH_3-\overset{\overset{\displaystyle O}{\|}}{C}-H \qquad CH_3-CH_2-\overset{\overset{\displaystyle O}{\|}}{C}-H$$

methanal — ethanal — propanal

Ketone Nomenclature

1. **Name the parent compound.** Locate the longest continuous chain *that contains the carbonyl group.* Add the suffix *-one* to the prefix corresponding to the number of carbon atoms in this chain.

2. **Number the carbon atoms in the parent chain.** Number the carbon atoms in the chain so that the carbon atom in the carbonyl group has the lowest possible number.
3. **Insert position numbers.** Place the carbonyl position number in front of the name.
4. **Punctuate the name.** Separate the position number from the name with a hyphen.

Following are three examples of correctly named ketones.

$$CH_3-\overset{\overset{\displaystyle O}{\|}}{C}-CH_3 \qquad CH_3-\overset{\overset{\displaystyle O}{\|}}{C}-CH_2-CH_3 \qquad CH_3-CH_2-\overset{\overset{\displaystyle O}{\|}}{C}-CH_2-CH_3$$

2-propanone 2-butanone 3-pentanone

Properties and Uses of Aldehydes and Ketones

The simplest aldehyde is methanal, also known as formaldehyde. It was once commonly used in biology laboratories as a preservative for dead animals. Its most important commercial use, however, is in the production of plastics. One of the first commercial plastics, bakelite, was made by combining phenol and formaldehyde in a long chain.

The simplest ketone is 2-propanone, whose common name is acetone. Acetone is found in some nail-polish removers because it dissolves the organic substances in nail polish. However, artificial fingernails are made of plastics that are also dissolved by acetone. Today other solvents are being used more frequently as nail-polish removers.

Aldehydes and ketones are often responsible for odors and flavors. For example, cinnamaldehyde contributes to the odor and flavor of cinnamon. Figure 21-4 gives some examples of odors and flavors that come from aldehydes and ketones.

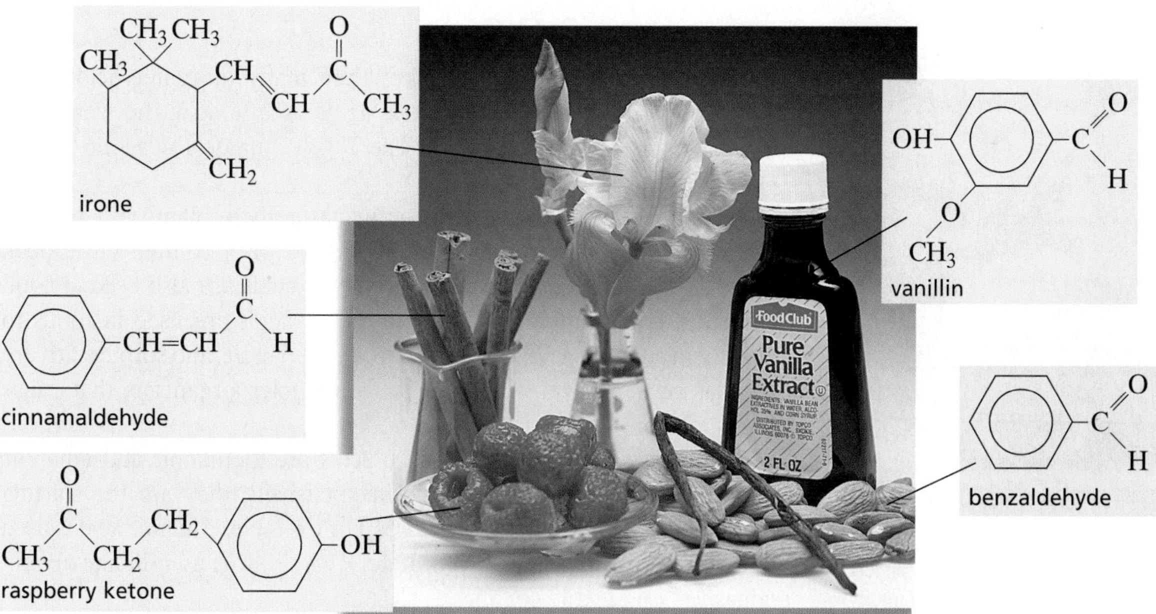

FIGURE 21-4 Many common odors and flavors come from aldehydes and ketones.

Carboxylic Acids

Carboxylic acids *are organic compounds that contain the carboxyl functional group.* The carboxyl group always comes at the end of a carbon-atom chain. A member of this class of organic compounds can be represented by the general formula shown below.

$$R-\overset{\displaystyle O}{\overset{\|}{C}}-OH$$

The rules for naming simple carboxylic acids are as follows.

Carboxylic Acid Nomenclature

Name the parent compound. Locate the longest continuous chain *that contains the carboxyl group.* If there is only one carboxyl group, add the suffix *-oic acid* to the prefix corresponding to the number of carbon atoms in this chain. If there is more than one carboxyl group, use the full name of the corresponding alkane, and add the suffix modified to indicate the number of carboxyl groups. For example, *-dioic acid* = 2, *-trioic acid* = 3, and so on.

Following are three examples of correctly named carboxylic acids.

$H-\overset{\displaystyle O}{\overset{\|}{C}}-OH$	$CH_3-CH_2-CH_2-\overset{\displaystyle O}{\overset{\|}{C}}-OH$	$OH-\overset{\displaystyle O}{\overset{\|}{C}}-\overset{\displaystyle O}{\overset{\|}{C}}-OH$
methanoic acid	butanoic acid	ethanedioic acid

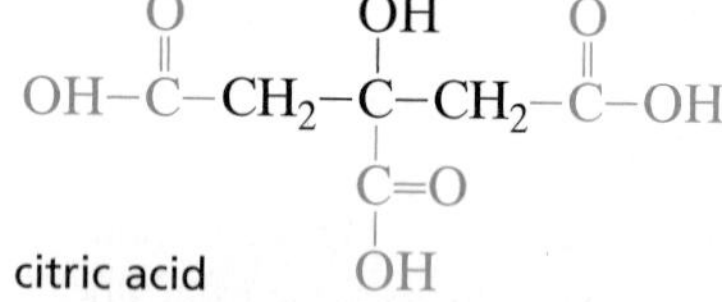

Properties and Uses of Carboxylic Acids

Carboxylic acids, like inorganic acids, react to lose a hydrogen ion and become a negatively charged ion in water.

$$R-\overset{\displaystyle O}{\overset{\|}{C}}-OH \underset{}{\overset{H_2O}{\rightleftharpoons}} R-\overset{\displaystyle O}{\overset{\|}{C}}-O^- + H^+$$

Carboxylic acids are much weaker than many inorganic acids, such as hydrochloric, sulfuric, and nitric acids. Acetic acid, the weak acid in vinegar, is a carboxylic acid. The IUPAC name for acetic acid is ethanoic acid.

A number of carboxylic acids occur naturally in plants and animals. For example, citrus fruits, shown in Figure 21-5, contain citric acid. Table 21-5 lists more examples. Carboxylic acids are also used as food additives. For example, ethanoic and citric acids are used in foods to give them a tart or acidic flavor. Benzoic, propanoic, and sorbic acids are used as preservatives. All three acids kill microorganisms that cause foods to spoil.

The most widely used carboxylic acids are methanoic and ethanoic acids. Because they can be made inexpensively, they are the starting material for many chemical processes. For example, ethanoic acid is used in the production of polyvinyl acetate, PVA. PVA is used in latex paint, adhesives, and textile coatings.

FIGURE 21-5 Citric acid, found in citrus fruits, contains three carboxylic acid groups shown in red on the structural formula.

TABLE 21-5 *Some Carboxylic Acids and Their Natural Sources*

Carboxylic acid	Structural formula	Source
methanoic acid	$H-\overset{\overset{O}{\parallel}}{C}-OH$	ants
butanoic acid	$CH_3-CH_2-CH_2-\overset{\overset{O}{\parallel}}{C}-OH$	rancid butter
hexanoic acid	$CH_3-CH_2-CH_2-CH_2-CH_2-\overset{\overset{O}{\parallel}}{C}-OH$	milk fats, coconut oil, palm oil
lactic acid	$CH_3-\overset{\overset{OH}{\vert}}{C}H-\overset{\overset{O}{\parallel}}{C}-OH$	sour milk, blood, and muscle fluid
malic acid	$OH-\overset{\overset{O}{\parallel}}{C}-CH_2-\overset{\overset{OH}{\vert}}{C}H-\overset{\overset{O}{\parallel}}{C}-OH$	apples
oxalic acid	$OH-\overset{\overset{O}{\parallel}}{C}-\overset{\overset{O}{\parallel}}{C}-OH$	rhubarb

Esters

Esters *are organic compounds with carboxylic acid groups in which the hydrogen of the hydroxyl group has been replaced by an alkyl group.* The general formula for an ester is given below.

$$R-\overset{\overset{O}{\parallel}}{C}-O-R'$$

The IUPAC system for naming simple esters is as follows.

Ester Nomenclature

1. **Name the parent compound.** Name the carboxylic acid from which the ester was formed (see page 674). Change the *-oic acid* ending in the name of this acid to *-oate*. This gives the second half of the ester's name.
2. **Add the name of the alkyl group.** Identify and name the alkyl group that has replaced the hydrogen of the hydroxyl group. Add the name of the alkyl group to the front of the name.
3. **Leave appropriate spaces in the name.** There should be a space between the name of the alkyl group and the name of the parent compound.

Following are two examples of correctly named esters.

$$CH_3-\overset{\overset{O}{\parallel}}{C}-O-CH_2-CH_3$$

ethyl ethanoate

$$CH_3-CH_2-CH_2-\overset{\overset{O}{\parallel}}{C}-O-CH_2-CH_3$$

ethyl butanoate

TABLE 21-6 *Common Flavors and Odors Produced by Esters*

Ester	Structural formula	Flavor or odor
ethyl butanoate	$CH_3{-}CH_2{-}CH_2{-}\overset{\overset{\large O}{\|}}{C}{-}O{-}CH_2{-}CH_3$	pineapple
methyl salicylate	OH (on benzene ring), $\overset{\overset{\large O}{\|}}{C}{-}O{-}CH_3$ (on adjacent ring carbon)	wintergreen oil
geraniol formate	$H{-}\overset{\overset{\large O}{\|}}{C}{-}O{-}CH_2{-}CH{=}\overset{\overset{\large CH_3}{\mid}}{C}{-}CH_2{-}CH_2{-}CH{=}\overset{\overset{\large CH_3}{\mid}}{C}{-}CH_3$	rose
methyl anthranilate	NH_2 (on benzene ring), $\overset{\overset{\large O}{\|}}{C}{-}O{-}CH_3$ (on adjacent ring carbon)	grape juice and jasmine
linalyl acetate	$CH_3{-}\overset{\overset{\large O}{\|}}{C}{-}O{-}\underset{\underset{\large CH=CH_2}{\mid}}{\overset{\overset{\large CH_3}{\mid}}{C}}{-}CH_2{-}CH_2{-}CH{=}\overset{\overset{\large CH_3}{\mid}}{C}{-}CH_3$	lavender

Properties and Uses of Esters

Esters are common in plants and are responsible for some distinctive flavors and odors. Table 21-6 lists some of these esters and the flavors and odors with which they are associated. At one time, compounds such as those listed in the table were obtained only from natural materials. But chemists have learned how to synthesize these and many other naturally occurring compounds for use as food additives. Figure 21-6 shows the structure of isoamyl acetate, which is found in bananas and is also used as an artificial flavoring.

FIGURE 21-6 The ester in bananas can be synthesized and used as a flavoring.

Amines

Amines *are organic compounds that can be considered to be derivatives of ammonia, NH_3.* They can be represented by the following general formula.

$$\begin{array}{c} R\text{—N—}R'' \\ | \\ R' \end{array}$$

Amines are often named by a common system rather than the IUPAC system. The steps in naming a simple amine by the common system are as follows.

Amine Nomenclature

1. **Name the parent compound.** The end of the name will be *-amine*.
2. **Add the names of the alkyl groups.** Arrange the names of the alkyl groups attached to the nitrogen atom in alphabetical order. Add the prefixes *di-* or *tri-* in front of the group name if two or three, respectively, of the same kind are included in the amine. Combine these names in front of *-amine* to form one word.

Following are three examples of correctly named amines.

$$\begin{array}{c} CH_3\text{—N—H} \\ | \\ H \end{array} \qquad \begin{array}{c} CH_3\text{—N—H} \\ | \\ CH_2\text{—}CH_3 \end{array} \qquad \begin{array}{c} CH_3\text{—N—}CH_3 \\ | \\ CH_3 \end{array}$$

methylamine (primary amine)	ethylmethylamine (secondary amine)	trimethylamine (tertiary amine)

Amines are categorized as primary, secondary, or tertiary, depending on the number of hydrogen atoms that have been replaced in the ammonia molecule. As shown in the structures above, *in a* **primary amine,** *one hydrogen atom of an ammonia molecule has been replaced by an alkyl group. In a* **secondary amine,** *two hydrogen atoms of an ammonia molecule have been replaced by alkyl groups. In a* **tertiary amine,** *all three hydrogen atoms of an ammonia molecule have been replaced by alkyl groups.*

Properties and Uses of Amines

The chemical properties of the amines depend largely on the electronic structure of the nitrogen atom, which has an unshared pair of electrons. This region of negative charge makes amines weak bases in aqueous solutions. The unshared pair of electrons on the amine molecule attracts a positive hydrogen atom in a water molecule. The hydrogen atom bonds with the amine, forming a positively charged ion and leaving the hydroxide ion behind.

$$\begin{array}{c} R\text{—}\ddot{N}\text{—}R'' \\ | \\ R' \end{array} + H\text{—O—H} \rightleftharpoons \begin{array}{c} H^+ \\ | \\ R\text{—N—}R'' \\ | \\ R' \end{array} + OH^-$$

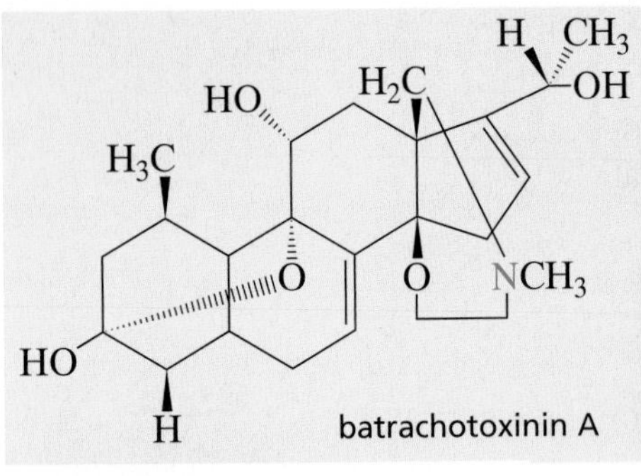

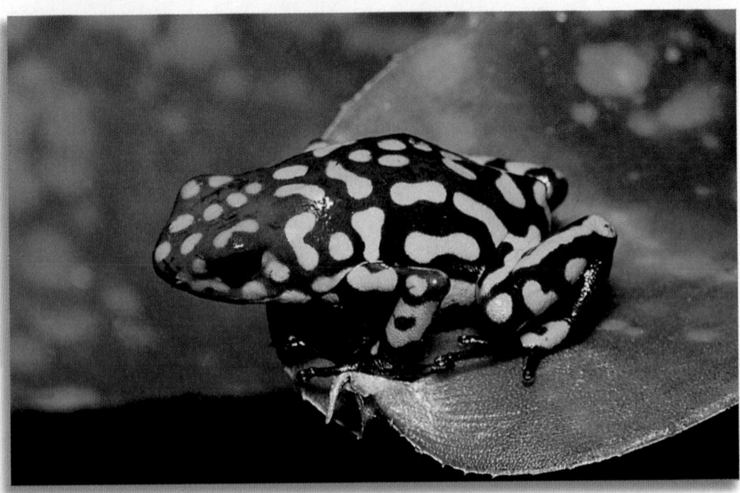

FIGURE 21-7 The poison dart frog produces toxic amines, one of which is batrachotoxinin A, that kill nerve cells. The nitrogen atom of the amine is shown in red.

An interesting example of this reaction involves two naturally occurring and toxic amines known as batrachotoxin and batrachotoxinin A. These compounds are produced by the poison dart frog, shown in Figure 21-7, which lives in the rain forests of Colombia. In water solution, both of these amines gain protons and become positively charged ions. Because these two amines carry the same charge as sodium ions found in the nervous system, they behave in much the same way. Both batrachotoxin ions can move through the openings in nerve cells, called sodium channels. These ions are much larger than sodium ions, however. They force the sodium channels to remain in the "open" position, and sodium ions are allowed to flood a nerve cell. This causes the nerve cell to continuously transmit nerve impulses, resulting in rapid death of the cell.

Amines are common in nature. They are often formed during the breakdown of proteins in animal cells. Two such amines are putrescine and cadaverine. Their names reflect their foul odors and presence in decaying bodies. Another amine, methylamine, has the unpleasant odor associated with dead fish. Yet a fourth amine, skatole, is found in feces and contributes to their odor.

The class of organic compounds known as alkaloids also consists of amines. *Alkaloids* are naturally occurring amine products of plants that have physiological effects on animals. Examples of alkaloids include caffeine, nicotine, morphine, and coniine (found in poisonous hemlock). Alkaloids tend to have complex chemical structures, as illustrated by the structure for morphine.

$N-CH_3$

OH O OH

morphine (a tertiary amine)

Table 21-7 summarizes the functional groups and the general formulas of the classes of organic compounds discussed in Sections 21-1 and 21-2.

TABLE 21-7 *Classes of Organic Compounds*

Class	Functional group	General formula
alcohol	$-OH$	$R{-}OH$
alkyl halide	$-X, X = F, Cl, Br, I$	$R{-}X$
ether	$-O-$	$R{-}O{-}R'$
aldehyde	$-\overset{\overset{O}{\|}}{C}-H$	$R-\overset{\overset{O}{\|}}{C}-H$
ketone	$-\overset{\overset{O}{\|}}{C}-$	$R-\overset{\overset{O}{\|}}{C}-R'$
carboxylic acid	$-\overset{\overset{O}{\|}}{C}-OH$	$R-\overset{\overset{O}{\|}}{C}-OH$
ester	$-\overset{\overset{O}{\|}}{C}-O-$	$R-\overset{\overset{O}{\|}}{C}-O-R'$
amine	$-\underset{\|}{N}-$	$R-\underset{\underset{R'}{\|}}{N}-R''$

SECTION REVIEW

1. Give the general formula and class of organic compounds for each of the following:
   a. $CH_3-CH_2-\overset{\overset{O}{\|}}{C}-OH$
   b. $CH_3-\overset{\overset{O}{\|}}{C}-H$
   c. $CH_3-CH_2-NH_2$
   d. $CH_3-\overset{\overset{O}{\|}}{C}-O-CH_2-CH_3$
   e. $CH_3-\overset{\overset{O}{\|}}{C}-CH_3$
2. Give the name of each of the following:
   a. $CH_3-\overset{\overset{O}{\|}}{C}-CH_2-CH_2-CH_3$
   b. $CH_3-CH_2-CH_2-\overset{\overset{O}{\|}}{C}-OH$
   c. $CH_3-NH-CH_3$
3. Draw condensed structures for each of the following:
   a. ethyl ethanoate
   b. triethylamine
   c. butanal
4. How are aldehydes and ketones alike? How do they differ from each other?
5. How do the strengths of organic acids compare with the strengths of most inorganic acids?
6. Show the reaction that occurs when amines are dissolved in water.

Unraveling the Mystery of DNA

HISTORICAL PERSPECTIVE

Today genetic engineers can identify, modify, and even transplant genes, but virtually nothing was known about the chemical mechanism of heredity at the beginning of the twentieth century. The term* gene *was coined in 1909 to describe a molecule that existed only in theory at the time. By the century's midpoint, however, scientists were poised to discover the molecular structure of the gene and explain the biochemical process that is the foundation of modern genetics.

The Chemical Nature of the Gene

By the early part of the twentieth century, scientists knew that genes were one of two types of organic macromolecules: proteins or nucleic acids. Most researchers believed genes to be the former until 1944, when it was shown that hereditary information could be transmitted from one bacterial cell to another by DNA alone. Gradually, genetic investigators switched their attention to nucleic acids.

An Important New Technique

One problem early researchers encountered was the inability to directly observe the minuscule genes. The development of a relatively new crystallographic technique proved vital to the elucidation of the structures of DNA and other biological macromolecules.

The technique was called X-ray diffraction, and it involved shining X rays onto crystallized molecular samples to take "snapshots" of their structure. Until the 1950s, most X-ray crystallography was focused on proteins. But in 1951, James D. Watson, a young American biologist, realized the potential value of X-ray diffraction in determining a model for DNA. That year, he accepted a position at Cambridge University's famous Cavendish Laboratory, where he could learn more about the procedure.

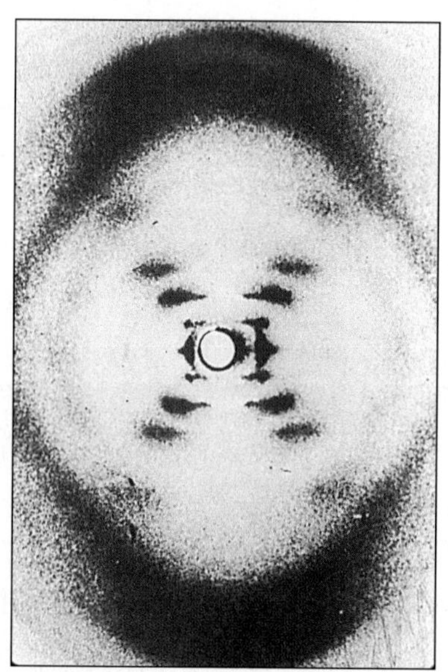

X-ray diffraction patterns of DNA taken by Rosalind Franklin were used to determine the structure of DNA.

A Fateful Union

At Cavendish, Watson befriended a theoretical biophysicist by the name of Francis Crick. Crick did not know how to perform X-ray diffraction either, but he was adept at analyzing the images that resulted from the technique. Fortunately for both men, Cavendish was also home to two experts on X-ray crystallography of DNA: Maurice Wilkins and Rosalind Franklin. With the help of these two experimentalists, Watson and Crick set out to solve the riddle of DNA.

A Big Leap

One of the team's biggest clues was the result of an investigation into proteins by Linus Pauling, the eminent chemist at the California Institute of Technology. Pauling determined the basic shape of a polypeptide chain, or protein strand, to be an α-helix, a large molecule of repeating units that twist around a central axis.

This discovery inspired Watson and Crick to look for a similar but more complex structure in DNA.

Watson wrote:

In the α-helix, a single polypeptide . . . chain folds up into a helical arrangement held together by hydrogen bonds between groups on the same chain. Maurice told Francis, however, that the diameter of the DNA molecule was thicker than would be the case if only one polynucleotide (DNA strand) were present. This made him think that the DNA molecule was a compound helix composed of several polynucleotide chains twisted about each other.

Watson and Crick began contemplating DNA configurations with two, three, and four helices. They correctly hypothesized that the sugar and phosphate groups of DNA's basic repeating units, or nucleotides, alternated to form the molecular backbone of each helix in the molecule. However, they mistakenly situated the sugar-phosphate backbones in the center of the molecule, with the bases of the nucleotides jutting outward.

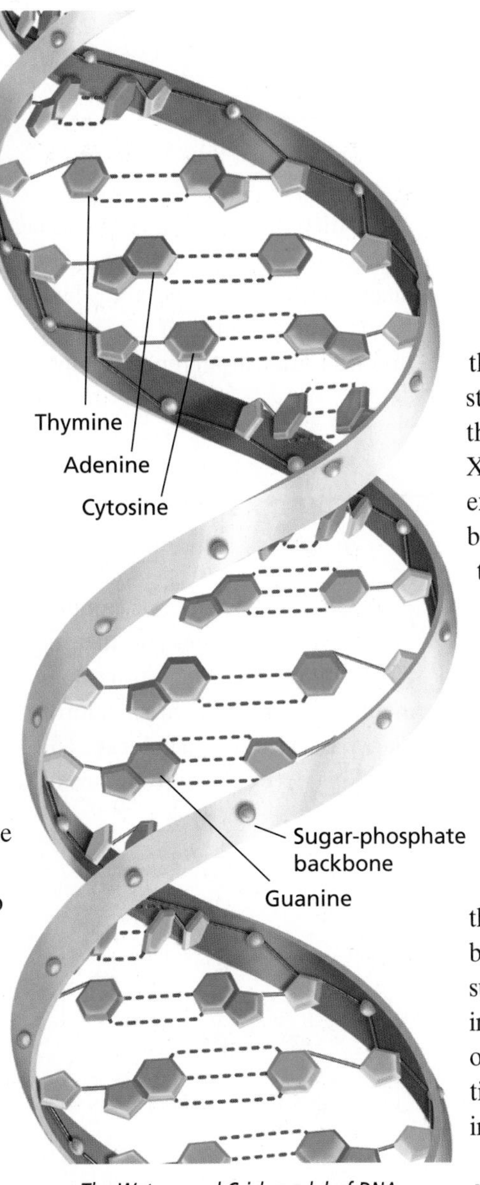

The Watson-and-Crick model of DNA contains opposing adenine-thymine and guanine-cytosine bases on a double-helical sugar-phosphate backbone.

The DNA Solution

Redirected by new X-ray diffraction data obtained by Franklin, Watson began considering models with the sugar-phosphate chains on the outside of the DNA molecule and the bases pointing inward. After reviewing a titration study of DNA indicating that many, if not all, of the molecule's bases formed hydrogen bonds with each other, he speculated:

Conceivably the crux of the matter was a rule governing hydrogen bonding between bases. . . . I thus started wondering whether each DNA molecule consisted of two chains with identical base sequences held together by hydrogen bonds between pairs of identical bases.

Crick immediately observed that this like-with-like double-stranded model did not satisfy the symmetry requirements of the X-ray data. When Watson began exploring models pairing different bases, he uncovered the final clue to the DNA mystery:

Suddenly I became aware that an adenine-thymine pair held together by two hydrogen bonds was identical in shape to a guanine-cytosine pair held together by at least two hydrogen bonds.

Fitting the opposing adenine-thymine and guanine-cytosine bases inside the double-helical sugar-phosphate backbones resulted in the first correct molecular model of DNA, for which the two scientists were awarded a Nobel Prize in 1962.

A Long-Standing Theory

For 45 years, the model of DNA discovered by James Watson and Francis Crick has served as the basis of biochemical genetics. The model has enabled scientists to explain genetic mutations, to predict or correct certain genetic disorders, and to genetically engineer organisms to have desirable traits.

SECTION 21-3

Organic Reactions

OBJECTIVES

- Describe and distinguish between the organic reactions: substitution, addition, condensation, and elimination.
- Relate some functional groups to some characteristic reactions.

Substitution Reactions

A **substitution reaction** *is one in which one or more atoms replace another atom or group of atoms in a molecule.* The reaction between an alkane, such as methane, and a halogen, such as chlorine, to form an alkyl halide is an example of a substitution reaction. Notice that in this reaction, a chlorine atom replaces a hydrogen atom on the methane molecule.

$$\underset{\text{methane}}{CH_4} + \underset{\text{chlorine}}{Cl{-}Cl} \longrightarrow \underset{\text{chloromethane}}{CH_3Cl} + \underset{\text{hydrogen chloride}}{H{-}Cl}$$

Additional compounds can be formed by replacing the other hydrogen atoms remaining in the methane molecule. The products are dichloromethane, trichloromethane, and tetrachloromethane. Trichloromethane is also known as chloroform, and tetrachloromethane is also known as carbon tetrachloride. CFCs are formed by further substitution reactions between chloroalkanes and HF.

$$CCl_4 + H{-}F \xrightarrow{SbF_5} CCl_3F + H{-}Cl$$

$$CCl_3F + H{-}F \xrightarrow{SbF_5} CCl_2F_2 + H{-}Cl$$

Addition Reactions

An **addition reaction** *is one in which an atom or molecule is added to an unsaturated molecule and increases the saturation of the molecule.* A common type of addition reaction is hydrogenation. In *hydrogenation,* one or more hydrogen atoms are added to an unsaturated molecule. Vegetable oils contain unsaturated fatty acids, long chains of carbon atoms with many double bonds. The following equation shows just one portion of an oil molecule. When hydrogen gas is blown through an oil, hydrogen atoms may add to the double bonds in the oil molecule.

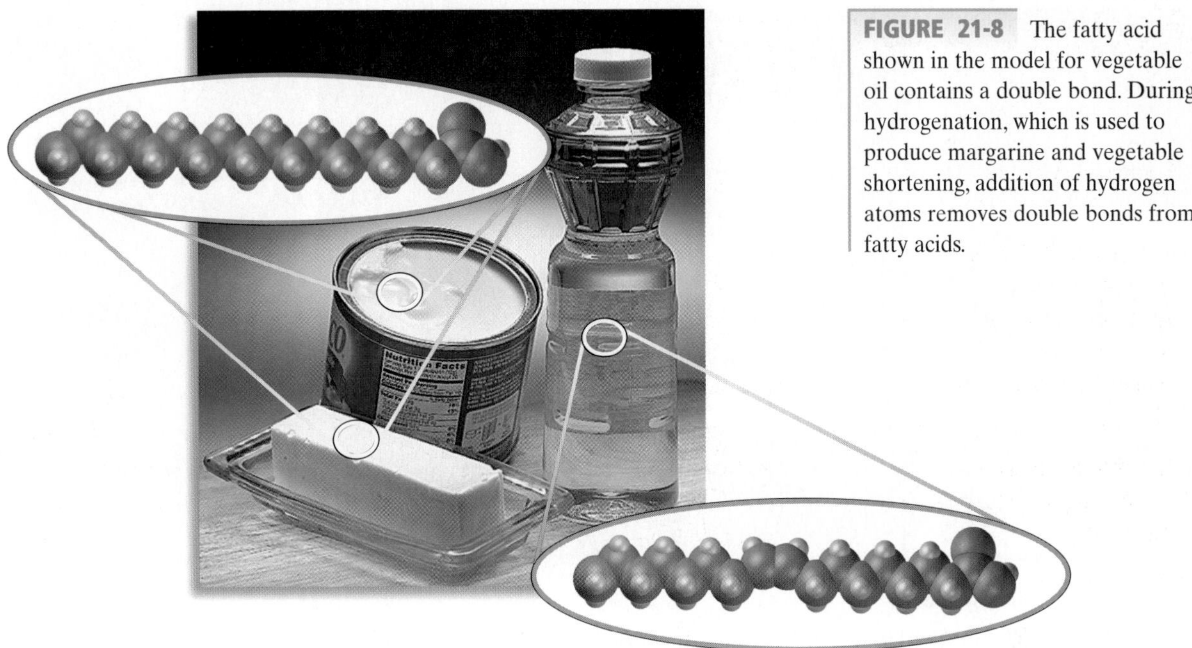

FIGURE 21-8 The fatty acid shown in the model for vegetable oil contains a double bond. During hydrogenation, which is used to produce margarine and vegetable shortening, addition of hydrogen atoms removes double bonds from fatty acids.

$$\left(\begin{array}{ccccccccccccc} H & & H & & H & & H & & H & & H & & H \\ | & & | & & | & & | & & | & & | & & | \\ C & - & C & = & C & - & C & - & C & = & C & - & C \\ | & & & & & & | & & & & & & | \\ H & & & & & & H & & & & & & H \end{array}\right) + H_2 \xrightarrow{\text{catalyst}} \left(\begin{array}{ccccccccccccc} H & & H & & H & & H & & H & & H & & H \\ | & & | & & | & & | & & | & & | & & | \\ C & - & C & = & C & - & C & - & C & - & C & - & C \\ | & & & & & & | & & | & & | & & | \\ H & & & & & & H & & H & & H & & H \end{array}\right)$$

The molecule still consists of long chains of carbon atoms, but it contains far fewer double bonds. The conversion of these double bonds to single bonds changes the material from an oil, a liquid, into a fat, a solid. When you see the word *hydrogenated* on a food product, you know that an oil has been converted to a fat by this process. Examples of an oil and hydrogenated fats are shown in Figure 21-8.

Condensation Reactions

A **condensation reaction** *is one in which two molecules or parts of the same molecule combine.* A small molecule, such as water, is usually removed during the reaction. An example is the reaction between two amino acids, which contain both amine and carboxyl groups. One hydrogen from the amine group of one amino acid combines with the hydroxyl from the carboxyl group of the other amino acid to form a molecule of water. When repeated many times, this reaction forms a protein molecule.

$$\underset{\text{amino acid}}{\begin{array}{ccccccc} & & H & & H & & O \\ & & | & & | & & \| \\ H & - & N & - & C & - & C-OH \\ & & & & | & & \\ & & & & R & & \end{array}} + \underset{\text{amino acid}}{\begin{array}{ccccccc} & & H & & H & & O \\ & & | & & | & & \| \\ H & - & N & - & C & - & C-OH \\ & & & & | & & \\ & & & & R' & & \end{array}} \longrightarrow \underset{\text{dipeptide}}{\begin{array}{ccccccccccc} & & H & & H & & O & & H & & H \quad O \\ & & | & & | & & \| & & | & & | \quad \| \\ H & - & N & - & C & - & C & - & N & - & C - C-OH \\ & & & & | & & & & & & | \qquad \\ & & & & R & & & & & & R' \qquad \end{array}} + \underset{\text{water}}{H_2O}$$

FIGURE 21-9 Sucrose is dehydrated when it reacts with concentrated sulfuric acid. Elimination of water produces a compound that is mostly carbon.

Elimination Reactions

An **elimination reaction** *is one in which a simple molecule, such as water or ammonia, is removed from adjacent carbon atoms of a larger molecule.* A simple example of an elimination reaction is the heating of ethanol in the presence of concentrated sulfuric acid. Under these conditions, a hydrogen atom from one carbon atom and a hydroxyl group from the second carbon atom are removed from the ethanol molecule. The hydrogen atom and hydroxyl group combine to form a molecule of water.

$$\underset{\text{ethanol}}{\text{H}-\underset{\text{H}}{\overset{\text{H}}{\text{C}}}-\underset{\text{H}}{\overset{\text{OH}}{\text{C}}}-\text{H}} \xrightarrow[\Delta]{H_2SO_4} \underset{\text{ethene}}{\text{H}-\underset{\text{H}}{\text{C}}=\underset{\text{H}}{\text{C}}-\text{H}} + \underset{\text{water}}{H_2O}$$

Another example of an elimination reaction is the dehydration of sucrose with concentrated sulfuric acid, shown in Figure 21-9.

SECTION REVIEW

1. Can an addition reaction occur between chlorine and ethane? Why or why not?
2. Does an addition reaction increase or decrease the saturation of a molecule?
3. What functional groups does the molecule of water come from in the condensation reaction between two amino acids?
4. Explain how elimination reactions could be considered the opposite of addition reactions.

Polymers

SECTION 21-4

Objectives

- Explain the relationship between monomers and polymers.
- Describe how the differences in the general structures of linear, branched, and cross-linked polymers contribute to their properties.
- Identify the two main types of polymers and the basic reaction mechanisms by which they are made.
- Relate the structures of specific polymers to their properties and uses.

Polymers *are large molecules made of many small units joined to each other through organic reactions. The small units are* **monomers.** A polymer can be made from identical or different monomers. *A polymer made from two different monomers is a* **copolymer.**

Polymers are all around us. The foods we eat and clothes we wear are made of polymers. Some of the most common natural polymers include starch, cellulose, and proteins. Some synthetic polymers may be familiar to you as plastics and synthetic fibers.

Polymer Thermal Properties and Structure

Polymers can be classified by the way they behave when heated. *A* **thermoplastic polymer** *melts when heated and can be reshaped many times. A* **thermosetting polymer** *does not melt when heated but keeps its original shape.* The thermal properties of polymers can be explained by whether their structure is linear, branched, or cross-linked, as shown in Figure 21-10.

The molecules of a *linear polymer* are free to move. They slide back and forth against each other easily when heated. So, linear polymers are thermoplastic. The molecules of a *branched polymer* contain side chains that prevent the molecules from sliding across each other easily. More heat is required to melt a branched polymer than a linear polymer. However, branched polymers are still likely to be thermoplastic. In *cross-linked polymers* adjacent molecules in the polymer have formed bonds with each other. Individual molecules are not able to slide past each other when heated. Cross-linked polymers retain their shape when heated and are thermosetting polymers.

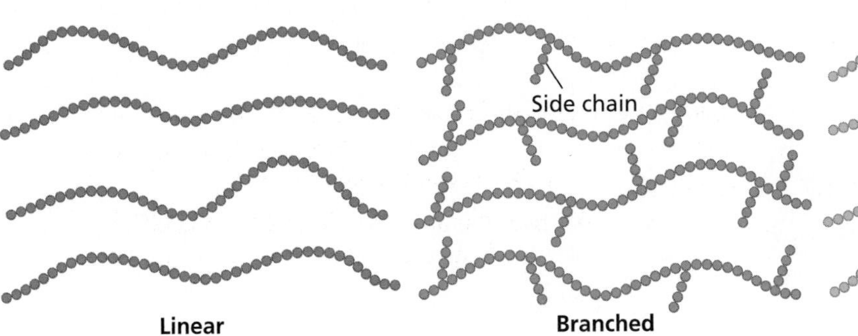

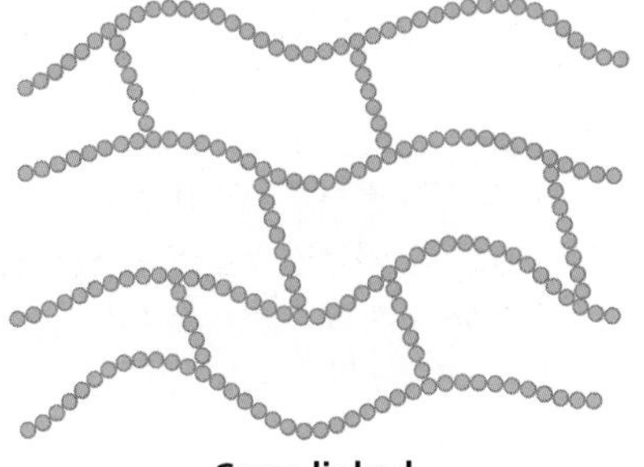

FIGURE 21-10 Compare the structures of linear, branched, and cross-linked polymers. Linear polymers are completely free to slide past one another. Branched and cross-linked polymers have inhibited movement.

Addition Polymers

An **addition polymer** *is a polymer formed by chain addition reactions between monomers that contain a double bond.* For example, molecules of ethene can polymerize with each other to form polyethene, commonly called polyethylene.

$$\underset{\text{ethene}}{n\ CH_2{=}CH_2} \xrightarrow{\text{catalyst}} \underset{\text{polyethylene}}{\left(CH_2{-}CH_2\right)_n}$$

The letter *n* shows that the addition reaction can be repeated multiple times to form a polymer *n* monomers long. In fact, this reaction can be repeated hundreds or thousands of times.

Forms of Polyethylene and Related Polymers

Various forms of polyethylene, shown in Figure 21-11, have different molecular structures. High-density polyethylene (HDPE) is a linear polymer. It has a high density because linear molecules can pack together closely. One use of HDPE is in plastic containers such as milk and juice bottles because HDPE tends to remain stiff and rigid.

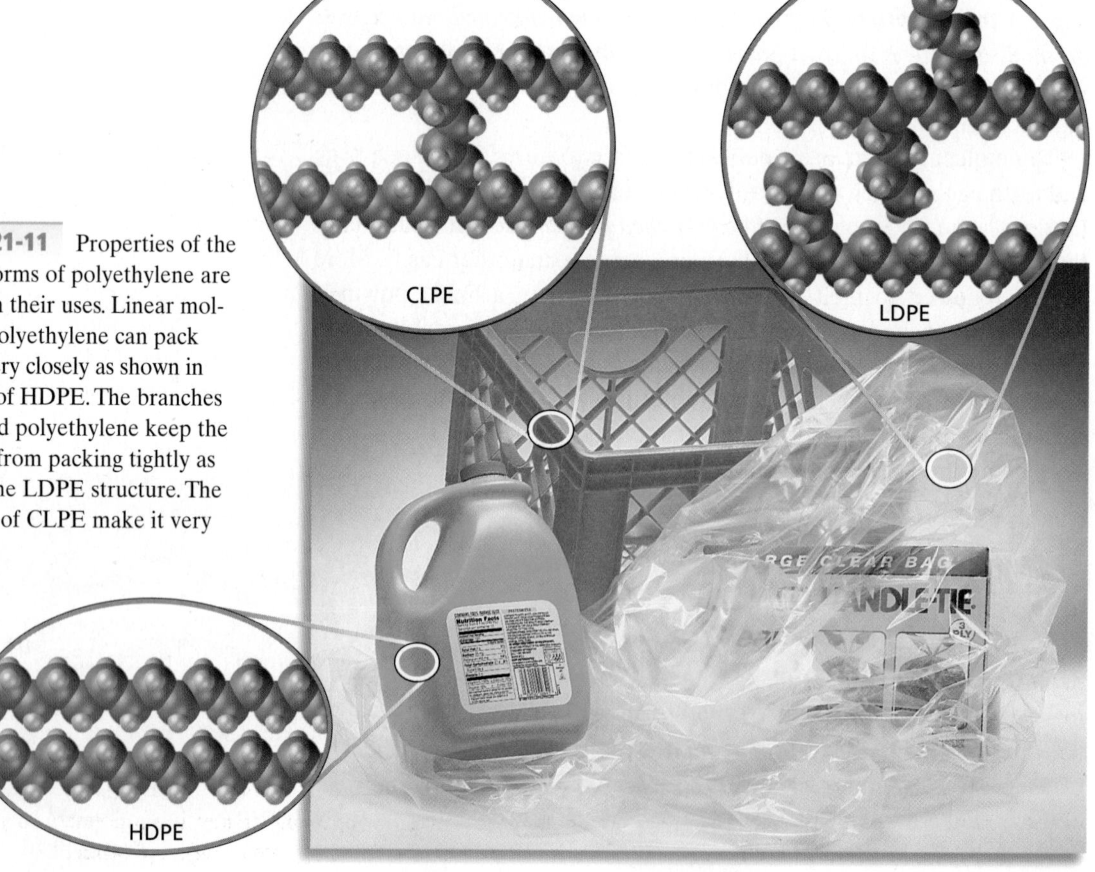

FIGURE 21-11 Properties of the different forms of polyethylene are reflected in their uses. Linear molecules of polyethylene can pack together very closely as shown in the model of HDPE. The branches of branched polyethylene keep the molecules from packing tightly as shown in the LDPE structure. The cross-links of CLPE make it very strong.

TABLE 21-8 ***Some Addition Polymers***

Monomer structure	Monomer name	Polymer name	Typical use
$CH_2{=}CH{-}CH_3$	propylene	polypropylene	plastic bottles
$CH_2{=}CH{-}Cl$	vinyl chloride	polyvinyl chloride (PVC)	piping
$CH_2{=}CH{-}CN$	acrylonitrile	polyacrylonitrile	fabrics
$CH_2{=}CH{-}C_6H_5$ (benzene ring)	styrene	polystyrene	insulation
$CH_2{=}CH{-}O{-}\overset{\overset{O}{\Vert}}{C}{-}CH_3$	vinyl acetate	polyvinyl acetate (PVA)	latex paints

If ethylene is heated to about 200°C at pressures above 200 atmospheres, random branching of the molecule chains may occur during polymerization. Branches form when hydrogen atoms are removed from the molecule and ethylene molecules add at these locations. Branched-chain molecules are not able to cluster as closely together as are linear molecules. Thus, the density of branched-chain polyethylene is less than that of linear polyethylene. The branched-chain form of polyethylene is known as low-density polyethylene (LDPE). LDPE tends to be less rigid than HDPE and is used, for example, in plastic bags.

When hydrogen atoms are removed from polyethylene molecules, two adjacent molecule chains may bond with each other. This forms a cross-link between the two molecules. Cross-linked polyethylene (CLPE) is even tougher and more rigid than HDPE. It is used for objects that need to be very strong.

Addition polymers similar to polyethylene can be made by substituting an atom or group of atoms for a hydrogen atom in ethene to form the monomer. Table 21-8 lists examples of these addition polymers.

Polystyrene, a polymer found in this table, is another example of a polymer whose properties can be altered. Pure polystyrene can be melted and converted to a clear plastic for use in cold-drink cups. Or it can be formed into tiny beads and then soaked in a low-boiling-point liquid, such as pentane. When this mixture is heated, the liquid vaporizes, expands, and forms tiny bubbles within the polystyrene. The product of this reaction is an opaque white material used to make hot-drink cups. Figure 21-12 shows the two types of polystyrene cups.

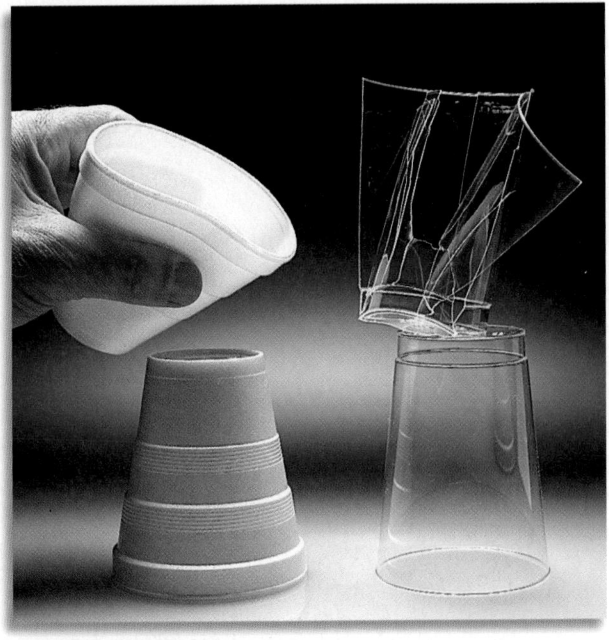

FIGURE 21-12 Both of these cups are made of polystyrene. Depending on how it is processed, polystyrene can be either flexible or rigid and brittle.

FIGURE 21-13 A rubber tree secretes latex, which is a suspension of rubber particles in water. When the rubber particles are precipitated, a gooey, sticky mass forms. An idealized model of natural rubber is shown in this figure.

Natural and Synthetic Rubber

Natural rubber is produced by the rubber tree, *Hevea brasiliensis,* shown in Figure 21-13. It is formed as the result of an addition reaction. The monomer in this reaction is 2-methyl-1,3-butadiene, commonly called isoprene.

$$2n\ \underset{\text{isoprene}}{H_2C{=}C(CH_3){-}CH{=}CH_2} \xrightarrow[\text{polymerization}]{\text{addition}} \underset{\text{polyisoprene}}{\left({-}CH_2{-}C(CH_3){=}CH{-}CH_2{-}CH_2{-}C(CH_3){=}CH{-}CH_2{-} \right)_n}$$

Natural rubber has relatively few practical applications. When warmed, individual molecules of polyisoprene slide easily back and forth past each other. The rubber gets soft and gooey, making it useless for many purposes.

A process for converting natural rubber into a useful commercial product was accidentally discovered by Charles Goodyear in 1839. Goodyear found that the addition of sulfur to molten rubber produces a material that remains very hard and tough when cooled. He called this process vulcanization. **Vulcanization** *is a cross-linking process between adjacent polyisoprene molecules that occurs when the molecules are heated with sulfur atoms.* Sulfur atoms bond to a carbon atom in one molecule and a second carbon atom in a second molecule, forming a cross-link between the two molecules. This is shown in the model in Figure 21-14. Vulcanization enabled rubber to be used in a wide variety of products, such as hoses, rainwear, and tires.

In the first year of World War II, Japan controlled large portions of Southeast Asia, where most of the world's natural rubber is obtained. The United States and other Allied nations were forced to develop

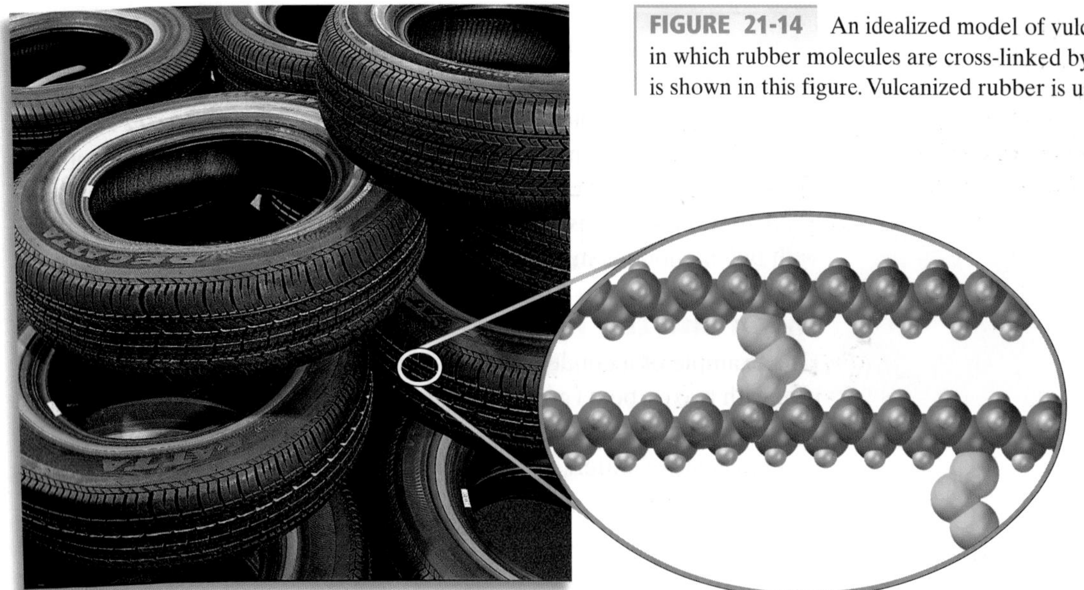

FIGURE 21-14 An idealized model of vulcanized rubber, in which rubber molecules are cross-linked by sulfur atoms, is shown in this figure. Vulcanized rubber is used in tires.

synthetic substitutes for natural rubber. Some synthetic rubbers have superior properties to natural rubber. Styrene-butadiene rubber, SBR, is a copolymer made in the reaction between styrene and butadiene, as shown below. SBR is used primarily in tires.

$$3n\ \underset{\text{1,3-butadiene}}{H_2C{=}CH{-}CH{=}CH_2} + n\ \underset{\text{styrene}}{H_2C{=}CH(C_6H_5)} \xrightarrow[\text{polymerization}]{\text{addition}}$$

$$\left(\!{-}CH_2{-}CH{=}CH{-}CH_2{-}CH_2{-}CH{=}CH{-}CH_2{-}CH_2{-}CH(C_6H_5){-}CH_2{-}CH{=}CH{-}CH_2{-}\right)_n$$

styrene-butadiene rubber (SBR)

Another substitute, neoprene, is formed during the polymerization of 2-chlorobutadiene. Notice that 2-chlorobutadiene is identical to isoprene, the monomer of natural rubber, except for the presence of a chlorine atom in place of a methyl group on the number 2 carbon atom.

$$\underset{\text{2-chlorobutadiene}}{H_2C{=}C(Cl){-}CH{=}CH_2} \qquad \underset{\text{isoprene}}{H_2C{=}C(CH_3){-}CH{=}CH_2}$$

High-Barrier Plastics

Flat-tasting soda? Technology comes to the rescue. No matter how tightly the cap is sealed on a bottle, soda loses fizz over time. Because of the nature of plastic, the carbon dioxide gas actually escapes through the bottle. Benny Freeman, an associate professor of chemical engineering at North Carolina State University, has discovered how to stop this problem with a special kind of plastic called a liquid crystalline polymer.

According to Freeman, the long molecules that make up plastics do not fit together tightly. Instead, the plastic molecules are constantly moving and opening gaps between them similar to the spaces between noodles in a bowl of cooked spaghetti. Gas molecules can pass through these gaps. In a soda bottle, the carbon dioxide in the soft drink dissolves into the plastic and escapes into the air outside the bottle. Conversely, plastic containers also allow oxygen to seep inside them, which can spoil the food they hold.

Liquid crystal polymers, or LCPs, are made up of long molecules, like other plastics are. But, says Freeman, the molecules in an LCP are much straighter, and they all point in the same general direction. Instead of looking like a bowl of cooked pasta, LCP molecules more closely resemble a box of uncooked spaghetti packed tightly together. Because it is much more difficult for gas molecules to squeeze through them, LCPs make an ideal material for manufacturing containers.

The LCP layer (white) of this power cable makes the cable waterproof.

Aside from keeping soft drinks fresh, these high-barrier plastics could also replace glass packaging for foods that easily become stale or spoiled from contact with oxygen. That would result in much lighter, unbreakable containers. Gas tanks made from LCPs would better prevent harmful fumes from leaking out and polluting the environment. The liquid crystal polymer could also be used as a protectant to line underground electrical cables, making them virtually waterproof and corrosion-free. LCPs are currently used to line the breathing system of NASA's new spacesuits.

Liquid crystal polymers have not yet made it into the packaging of everyday products because they are several times more expensive than regular plastics. Freeman thinks that this problem could be solved by creating what he calls a composite structure: a combination of regular plastic and the liquid crystal polymer. By sandwiching a thin LCP layer between two pieces of regular plastic, manufacturers could keep their costs down and still keep your soda fizzy.

"We are," says Freeman, "at the beginning of really understanding where you can start to manipulate structure and actually control properties in these materials. There's still a lot of uncharted territory in the science of these materials."

CHAPTER 21 REVIEW

CHAPTER SUMMARY

21-1

- A functional group is an atom or group of atoms responsible for the properties of the organic compound in which the functional group occurs.
- Systematic names of organic compounds indicate the type and position of functional groups present.
- Alcohols contain the hydroxyl functional group. Their tendency to form hydrogen bonds affects their properties and uses.
- In alkyl halides, one or more hydrogen atoms of an alkane have been replaced by one or more halogen atoms. One class of alkyl halides, the CFCs, has many important industrial uses but may cause serious environmental problems.
- Two alkyl groups are joined to an oxygen atom in ethers. Generally unreactive, they are widely used as solvents.
- The physical and chemical properties of organic classes often reflect the ability or inability of molecules in each class to form hydrogen bonds.

Vocabulary

alcohol (663) alkyl halide (666) ether (669) functional group (663)

21-2

- Aldehydes and ketones both contain the carbonyl group and are responsible for some odors and flavors. In aldehydes, the carbonyl group is attached to the first carbon atom of a carbon-atom chain. In ketones, it is in the middle of a carbon-atom chain.
- Carboxylic acids contain carboxyl groups. They act as weak acids in aqueous solutions.
- In esters, the hydrogen atom of a carboxylic acid group has been replaced with an alkyl group. Natural and synthetic esters give many foods their flavors and odors.
- Amines are derivatives of ammonia in which one or more hydrogen atoms are replaced by one or more alkyl groups. They behave as weak bases in aqueous solutions.

Vocabulary

aldehyde (672) carboxylic acid (674) ketone (672) secondary amine (677)
amine (677) ester (675) primary amine (677) tertiary amine (677)

21-3

- Both addition and substitution reactions add atoms to a molecule. In substitution reactions, an atom or group of atoms is replaced. In addition reactions, an atom or group of atoms is added to a double or triple bond.
- A small molecule is usually removed during both condensation reactions and elimination reactions. In a condensation reaction, two molecules or parts of the same molecule combine. In an elimination reaction, a large molecule eliminates a small molecule.

Vocabulary

addition reaction (682) condensation reaction (683) elimination reaction (684) substitution reaction (682)

CHAPTER SUMMARY continued

21-4
- Polymers are large molecules made of many repeating units called monomers. A copolymer consists of two different monomers.
- Thermosetting polymers cannot be melted once they are formed. Thermoplastic polymers can be melted more than once.
- The physical properties of polymers are strongly influenced by the presence or absence of branching and cross-linking among polymer chains.
- Addition reactions require that the monomers contain a double bond. Polyethylene and related polymers as well as natural and synthetic rubbers are addition polymers.
- Monomers of condensation polymers must contain two functional groups. Nylon 66, other polyamides, and polyesters are condensation polymers.

Vocabulary

addition polymer (686)
condensation polymer (690)
copolymer (685)
monomer (685)
polymer (685)
thermoplastic polymer (685)
thermosetting polymer (685)
vulcanization (688)

REVIEWING CONCEPTS

1. Write the general formula for each of the following:
   a. alcohol
   b. ether
   c. alkyl halide (21-1)
2. Based on the boiling points of water and methanol, in which would you expect to observe a greater degree of hydrogen bonding? Explain your answer. (21-1)
3. a. Why is glycerol used in moisturizing skin lotions?
   b. How does this relate to the chemical structure of glycerol? (21-1)
4. State two advantages and one disadvantage of using gasohol as an automotive fuel. (21-1)
5. Why are CFCs regarded as an environmental hazard? (21-1)
6. Alcohols and ethers are both organic compounds that contain oxygen. Explain how their chemical structures differ. (22-1)
7. What is the important chemical property of ethers that leads to their most common use? (21-1)
8. Write the general formula for each of the following:
   a. aldehyde
   b. ketone
   c. carboxylic acid
   d. ester
   e. amine (21-2)
9. Aldehydes and ketones both contain the same functional group. Why are they classified as separate classes of organic compounds? (21-2)
10. a. Cinnamaldehyde is responsible for what odor?
    b. The IUPAC name for cinnamaldehyde is 3-phenyl-2-propenal. Based on its names, to which class of compounds does it belong? (21-2)
11. Why can a carboxyl group not be in the middle of a carbon-atom chain? (21-2)
12. a. Show the reaction that occurs when carboxylic acids are dissolved in water.
    b. What property of carboxylic acids does this reaction illustrate? (21-2)
13. How are esters related to carboxylic acids? (21-2)
14. What element do amines contain besides carbon and hydrogen? (21-2)
15. Explain why an amine acts as a base. (21-2)
16. Show the reaction that occurs when aqueous solutions of carboxylic acids and amines are mixed. (21-2)

17. Name five classes of organic compounds that are often responsible for odors and flavors. (21-2)

18. What classes of organic compounds contain oxygen? (21-1 and 21-2)

19. What type of chemical reaction would you expect to occur between 2-octene and hydrogen bromide, HBr? (21-3)

20. How many molecules of chlorine, Cl_2, can be added to a molecule of 1-propene? a molecule of 1-propyne? (21-3)

21. Compare substitution and addition reactions. (21-3)

22. What problems would you expect to encounter in trying to hydrogenate hexane, C_6H_{14}? (21-3)

23. In a chemical reaction, two small molecules are joined and a water molecule is produced. What type of reaction took place? (21-3)

24. Do elimination reactions increase or decrease the saturation of a molecule? (21-3)

25. Name three common natural products and three synthetic products made of polymers. (21-4)

26. Some automobile engine parts are made of polymers. Do you think these polymers are thermosetting or thermoplastic? Explain your answer. (21-4)

27. Classify each of the following as thermosetting or thermoplastic:
a. linear polymer
b. branched polymer
c. cross-linked polymer (21-4)

28. What are two reactions by which polymers can be formed? (21-4)

29. What is the structural requirement for a molecule to be a monomer in an addition polymer? (21-4)

30. Explain the structural molecular differences between the following three types of polyethylene: HDPE, LDPE, and CLPE. (21-4)

31. What is the primary structural difference between polyethylene and polystyrene? (21-4)

32. Give a molecular explanation for the fact that natural rubber melts when it is heated but vulcanized rubber does not. (21-4)

33. How is the molecular structure of neoprene different from that of natural rubber? (21-4)

34. Could ethanoic acid be used as a monomer in a condensation polymer? Why or why not? (21-4)

35. Draw the structural formula for the amide group. (21-4)

36. Why is polyester wrinkle resistant? (21-4)

PROBLEMS

Organic Compound Nomenclature

37. Name the following alcohols:

a. $CH_3{-}OH$

b. 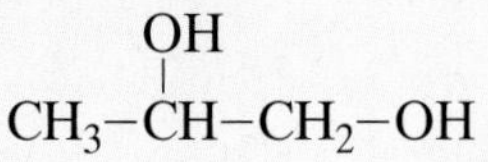
$$\begin{array}{c} \quad OH \\ \quad | \\ CH_3{-}CH{-}CH_2{-}OH \end{array}$$

c.
$$\begin{array}{c} \quad OH \\ \quad | \\ CH_3{-}CH{-}CH_2{-}CH_3 \end{array}$$

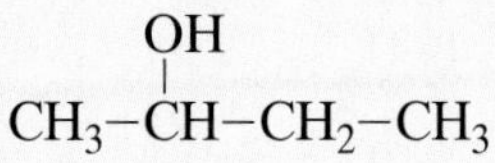

d. 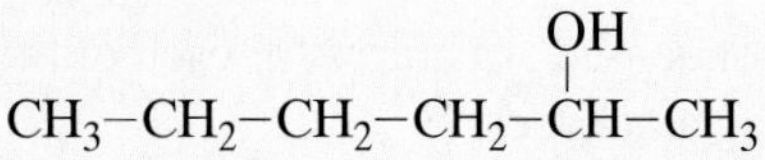
$$\begin{array}{c} \qquad\qquad\qquad\qquad OH \\ \qquad\qquad\qquad\qquad | \\ CH_3{-}CH_2{-}CH_2{-}CH_2{-}CH{-}CH_3 \end{array}$$

38. Draw condensed structures for each of the following alcohols:
a. 2,3-pentanediol
b. 1-pentanol
c. 1,2,3-propanetriol
d. ethanol

39. Name the following alkyl halides. (Hint: See Sample Problem 21-1.)

a. $CH_3{-}I$

b. $Cl{-}CH_2{-}CH_2{-}Cl$

c. 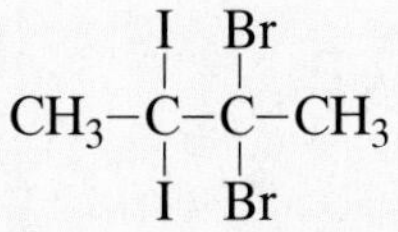
$$\begin{array}{c} \quad I \;\; Br \\ \quad | \;\;\; | \\ CH_3{-}C{-}C{-}CH_3 \\ \quad | \;\;\; | \\ \quad I \;\; Br \end{array}$$

d. 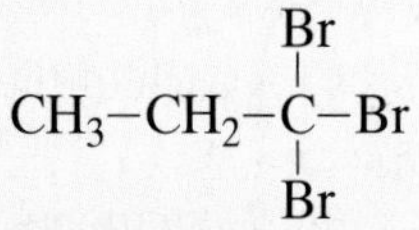
$$\begin{array}{c} \qquad\qquad Br \\ \qquad\qquad | \\ CH_3{-}CH_2{-}C{-}Br \\ \qquad\qquad | \\ \qquad\qquad Br \end{array}$$

40. Draw condensed structures for each of the following alkyl halides:
a. 2,3,4-trichloropentane
b. 1,1-diiodopropane
c. 1-fluorohexane
d. 2,2-dichloro-1,1-difluoropropane

41. Name the following ethers. (Hint: See Sample Problem 21-2.)

a. $CH_3{-}CH_2{-}CH_2{-}O{-}CH_2{-}CH_2{-}CH_3$

b. $CH_3-O-CH_2-CH_3$

c. $CH_3-CH_2-CH_2-CH_2-O-CH_2-CH_2-CH_2-CH_3$

d. $CH_3-CH_2-O-CH_2-CH_2-CH_2-CH_3$

42. Draw condensed structures for each of the following ethers:

a. dimethyl ether
b. methyl propyl ether
c. butyl propyl ether
d. ethyl heptyl ether

43. Name the following aldehydes:

a. $CH_3-\overset{\overset{\large O}{\|}}{C}-H$

b. $CH_3-CH_2-CH_2-CH_2-\overset{\overset{\large O}{\|}}{C}-H$

c. $CH_3-CH_2-\overset{\overset{\large O}{\|}}{C}-H$

d. $CH_3-CH_2-CH_2-\overset{\overset{\large O}{\|}}{C}-H$

44. Draw condensed structures for each of the following aldehydes:

a. methanal
b. hexanal
c. octanal
d. ethanal

45. Name the following ketones:

a. $CH_3-\overset{\overset{\large O}{\|}}{C}-CH_3$

b. $CH_3-CH_2-\overset{\overset{\large O}{\|}}{C}-CH_3$

c. $CH_3-CH_2-CH_2-\overset{\overset{\large O}{\|}}{C}-CH_3$

d. $CH_3-CH_2-\overset{\overset{\large O}{\|}}{C}-CH_2-CH_3$

46. Draw condensed structures for each of the following ketones:

a. 3-hexanone
b. 2-pentanone
c. 2-octanone
d. 2-hexanone

47. Name the following carboxylic acids:

a. $CH_3-CH_2-\overset{\overset{\large O}{\|}}{C}-OH$

b. $CH_3-CH_2-CH_2-CH_2-\overset{\overset{\large O}{\|}}{C}-OH$

c. $H-\overset{\overset{\large O}{\|}}{C}-OH$

d. $OH-\overset{\overset{\large O}{\|}}{C}-CH_2-CH_2-\overset{\overset{\large O}{\|}}{C}-OH$

48. Draw condensed structures for each of the following carboxylic acids:

a. butanoic acid
b. hexanoic acid
c. hexanedioic acid
d. heptanoic acid

49. Name the following esters:

a. $CH_3-\overset{\overset{\large O}{\|}}{C}-O-CH_3$

b. $H-\overset{\overset{\large O}{\|}}{C}-O-CH_3$

c. $CH_3-CH_2-\overset{\overset{\large O}{\|}}{C}-O-CH_2-CH_3$

d. $CH_3-CH_2-CH_2-CH_2-\overset{\overset{\large O}{\|}}{C}-O-CH_2-CH_3$

50. Draw condensed structures for each of the following esters:

a. butyl ethanoate
b. ethyl methanoate
c. propyl propanoate
d. methyl butanoate

51. Name the following amines:

a. CH_3-NH_2

b. $CH_3-CH_2-NH-CH_2-CH_3$

c. $CH_3-CH_2-NH-CH_3$

d. $CH_3-\underset{\underset{\large CH_3}{|}}{\ddot{N}}-CH_3$

52. Draw condensed structures for each of the following amines:

a. butylmethylamine
b. ethylamine
c. diethylmethylamine
d. ethylpropylamine

Organic Reactions

53. Which of the following reactions is a substitution reaction?

a. $CH_2{=}CH_2 + Cl_2 \longrightarrow Cl-CH_2-CH_2-Cl$

b. $CH_3-CH_2-CH_2-CH_3 + Cl_2 \longrightarrow$
$Cl-CH_2-CH_2-CH_2-CH_3 + HCl$

c. $$CH_3{-}OH + CH_3{-}\overset{O}{\overset{\|}{C}}{-}OH \longrightarrow CH_3{-}\overset{O}{\overset{\|}{C}}{-}O{-}CH_3 + H_2O$$

54. Which of the following reactions is an addition reaction?

a. $$CH_3{-}CH_2{-}CH{=}CH_2 + Br_2 \longrightarrow CH_3{-}CH_2{-}\underset{Br}{\underset{|}{CH}}{-}CH_2{-}Br$$

b. cyclohexanol (ring with OH) $\xrightarrow[\text{heat}]{85\%\ H_3PO_4}$ cyclohexene + H_2O

c. $$CH_3{-}\overset{O}{\overset{\|}{C}}{-}OH + CH_3{-}OH \longrightarrow CH_3{-}\overset{O}{\overset{\|}{C}}{-}O{-}CH_3 + H_2O$$

55. Which of the following reactions is a condensation reaction?

a. $$CH_3C{\equiv}CH + HBr \xrightarrow{\text{ether}} CH_3{-}\overset{Br}{\overset{|}{C}}{=}CH_2$$

b. cyclohexene + Br_2 $\xrightarrow{CCl_4}$ 1,2-dibromocyclohexane (ring with Br, Br)

c. $$CH_3{-}CH_2{-}OH + CH_3{-}CH_2{-}OH \xrightarrow{H_2SO_4} CH_3{-}CH_2{-}O{-}CH_2{-}CH_3 + H_2O$$

56. Which of the following reactions is an elimination reaction?

a. $$CH_2{=}CH{-}CH_2{-}CH_3 + Cl_2 \longrightarrow Cl{-}CH_2{-}\underset{Cl}{\underset{|}{CH}}{-}CH_2{-}CH_3$$

b. $$CH_3{-}\overset{OH}{\overset{|}{CH}}{-}CH_3 \xrightarrow{H_3O^+} CH_3{-}CH{=}CH_2 + H_2O$$

c. $$CH_3CH_3 + Cl_2 \xrightarrow[\text{heat}]{\text{light or}} CH_3CH_2Cl + HCl$$

Calculations with Organic Compounds

57. Calculate the molecular mass of trichlorofluoromethane.

58. A compound is found to contain 54.5% carbon, 9.1% hydrogen, and 36.4% oxygen.

a. Determine the simplest formula.

b. The molecular mass of this compound is 88.1 g. What is the molecular formula?

59. The hydronium ion concentration in 0.05 M acetic acid is 9.4×10^{-4} mol/L. What is the pH of the solution?

60. What volume of ethanol must be diluted with water to prepare 500. mL of 0.750 M solution? The density of ethanol is given as 0.789 g/mL.

61. 1,2-ethanediol, also called ethylene glycol, is commonly used as an antifreeze. The density of ethylene glycol is 1.432 g/mL.

a. Calculate the theoretical freezing point of the water in a 50% (by volume) solution of ethylene glycol.

b. The actual freezing point of such a solution is about −37°C. Account for any difference between this and the value you calculated.

MIXED REVIEW

62. Classify each of the following reactions as an elimination reaction or a condensation reaction:

a. $$CH_3{-}\overset{Br}{\overset{|}{C}}{=}CH_2 + NaNH_2 \longrightarrow CH_3C{\equiv}CH + NaBr + NH_3$$

b. $$CH_3{-}CH_2{-}\overset{OH}{\overset{|}{CH}}{-}CH_3 \xrightarrow[\text{heat}]{85\%\ H_3PO_4} CH_3{-}CH{=}CH{-}CH_3 + H_2O$$

c. $$CH_3{-}CH_2{-}OH + CH_3{-}\overset{O}{\overset{\|}{C}}{-}OH \longrightarrow CH_3{-}CH_2{-}O{-}\overset{O}{\overset{\|}{C}}{-}CH_3 + H_2O$$

d. $$C_6H_5{-}\overset{O}{\overset{\|}{C}}H + CH_3{-}\overset{O}{\overset{\|}{C}}{-}CH_3 \xrightarrow{OH^-} C_6H_5{-}CH{=}CH{-}\overset{O}{\overset{\|}{C}}{-}CH_3 + H_2O$$

CHAPTER 22

Nuclear Chemistry

Particle detectors are important tools in the study of nuclear chemistry.

The Nucleus

SECTION 22-1

OBJECTIVES

- Explain what a nuclide is, and describe the different ways nuclides can be represented.
- Define and relate the terms *mass defect* and *nuclear binding energy.*
- Explain the relationship between nucleon number and stability of nuclei.
- Explain why nuclear reactions occur, and know how to balance a nuclear equation.

Atomic nuclei are made of *protons and neutrons, which are collectively called* **nucleons.** In nuclear chemistry, *an atom is referred to as a* **nuclide** *and is identified by the number of protons and neutrons in its nucleus.* Nuclides can be represented in two ways: when a symbol such as $^{228}_{88}\text{Ra}$ is used, the superscript is the mass number and the subscript is the atomic number; the same nuclide can also be written as radium-228.

Mass Defect and Nuclear Stability

Because an atom is made of protons, neutrons, and electrons, you might expect the mass of an atom to be the same as the mass of an equal number of isolated protons, neutrons, and electrons. However, this is not the case. Let's consider a $^{4}_{2}\text{He}$ atom as an example. The combined mass of two protons, two neutrons, and two electrons is calculated below.

$$\begin{aligned}
\text{2 protons:}\quad & (2 \times 1.007\ 276\ \text{amu}) = 2.014\ 552\ \text{amu} \\
\text{2 neutrons:}\quad & (2 \times 1.008\ 665\ \text{amu}) = 2.017\ 330\ \text{amu} \\
\text{2 electrons:}\quad & (2 \times 0.000\ 5486\ \text{amu}) = \underline{0.001\ 097\ \text{amu}} \\
& \text{total combined mass: } 4.032\ 979\ \text{amu}
\end{aligned}$$

However, the atomic mass of a $^{4}_{2}\text{He}$ atom has been measured to be 4.002 60 amu. The measured mass, 4.002 60 amu, is 0.030 38 amu *less* than the calculated mass, 4.032 98 amu. *The difference between the mass of an atom and the sum of the masses of its protons, neutrons, and electrons is called the* **mass defect.**

Nuclear Binding Energy

What causes the loss in mass? According to Albert Einstein's equation $E = mc^2$, mass can be converted to energy, and energy to mass. The mass defect is caused by the conversion of mass to energy upon formation of the nucleus. The mass units of the mass defect can be converted to energy units by using Einstein's equation. First, convert 0.030 38 amu to kilograms to match the units for energy, $\text{kg}\cdot\text{m}^2/\text{s}^2$.

$$0.030\ 38\ \text{amu} \times \frac{1.6605 \times 10^{-27}\ \text{kg}}{1\ \text{amu}} = 5.0446 \times 10^{-29}\ \text{kg}$$

FIGURE 22-1 This graph shows the relationship between binding energy per nucleon and mass number. The binding energy per nucleon is a measure of the stability of a nucleus.

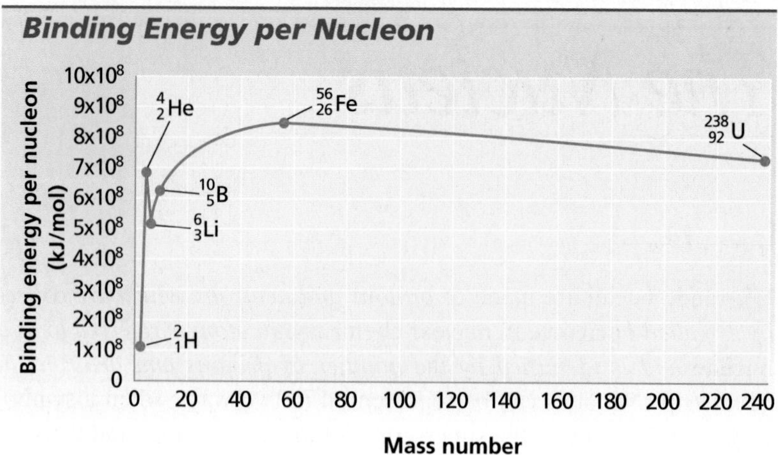

The energy equivalent can now be calculated.

$$E = mc^2$$
$$E = (5.0446 \times 10^{-29}\ \text{kg})(3.00 \times 10^8\ \text{m/s})^2$$
$$= 4.54 \times 10^{-12}\ \text{kg•m}^2/\text{s}^2 = 4.54 \times 10^{-12}\ \text{J}$$

This is the **nuclear binding energy,** *the energy released when a nucleus is formed from nucleons.* This energy can also be thought of as the amount of energy required to break apart the nucleus. Therefore, the nuclear binding energy is also a measure of the stability of a nucleus.

Binding Energy per Nucleon

The binding energy per nucleon is used to compare the stability of different nuclides, as shown in Figure 22-1. *The* **binding energy per nucleon** *is the binding energy of the nucleus divided by the number of nucleons it contains.* The higher the binding energy per nucleon, the more tightly the nucleons are held together. Elements with intermediate atomic masses have the greatest binding energies per nucleon and are therefore the most stable.

Nucleons and Nuclear Stability

Stable nuclides have certain characteristics. When the number of protons in stable nuclei is plotted against the number of neutrons, as shown in Figure 22-2, a beltlike graph is obtained. *This stable nuclei cluster over a range of neutron-proton ratios is referred to as the* **band of stability.** Among atoms having low atomic numbers, the most stable nuclei are those with a neutron-proton ratio of approximately 1:1. For example, $^{4}_{2}He$, a stable isotope of helium with two neutrons and two protons, has a neutron-proton ratio of 1:1. As the atomic number increases, the stable neutron-proton ratio increases to about 1.5:1. For example, $^{206}_{82}Pb$, with 124 neutrons and 82 protons, has a neutron-proton ratio of 1.51:1.

This trend can be explained by the relationship between the nuclear force and the electrostatic forces between protons. Protons in a nucleus repel all other protons through electrostatic repulsion, but the short

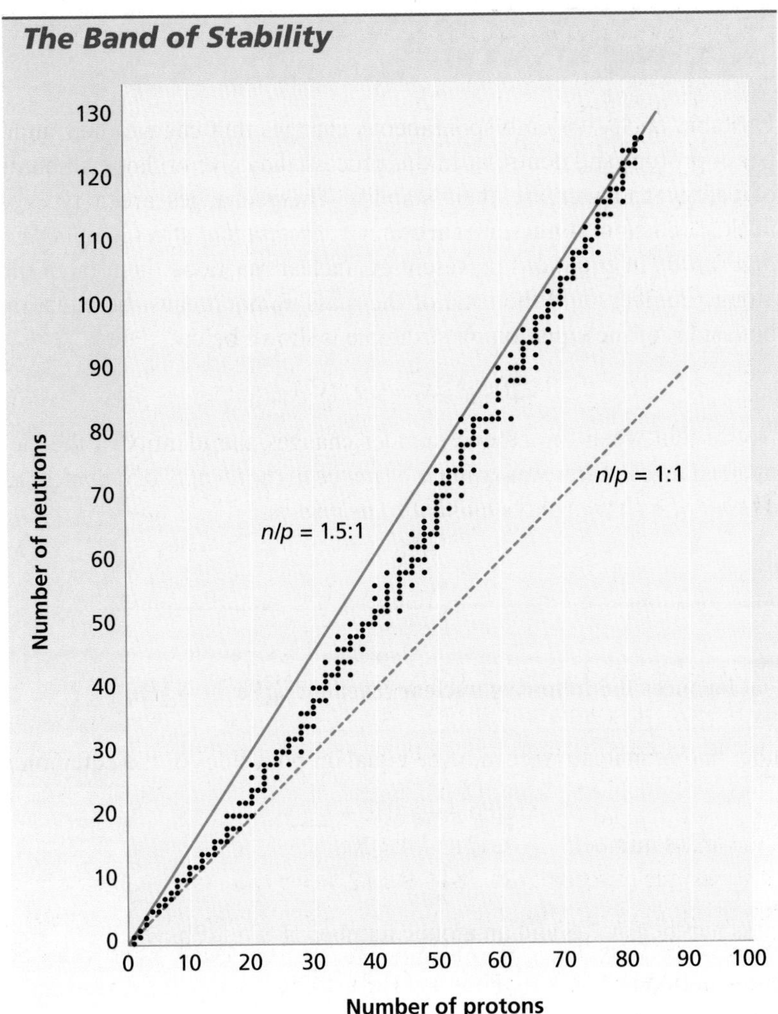

FIGURE 22-2 The neutron-proton ratios of stable nuclides cluster together in a region known as the band of stability. As the number of protons increases, the ratio increases from 1:1 to about 1.5:1.

range of the nuclear force allows them to attract only protons very close to them, as shown in Figure 22-3. So as the number of protons in a nucleus increases, the electrostatic force between protons increases faster than the nuclear force. More neutrons are required to increase the nuclear force and stabilize the nucleus. Beyond the atomic number 83, bismuth, the repulsive force of the protons is so great that no stable nuclides exist.

Stable nuclei tend to have even numbers of nucleons. Out of 265 stable nuclides, 159 have even numbers of both protons and neutrons. Only four nuclides have odd numbers of both. This indicates that stability of a nucleus is greatest when the nucleons—like electrons—are paired.

The most stable nuclides are those having 2, 8, 20, 28, 50, 82, or 126 protons, neutrons, or total nucleons. This extra stability at certain numbers supports a theory that nucleons—like electrons—exist at certain energy levels. According to the **nuclear shell model,** *nucleons exist in different energy levels, or shells, in the nucleus. The numbers of nucleons that represent completed nuclear energy levels—2, 8, 20, 28, 50, 82, and 126—are called* **magic numbers.**

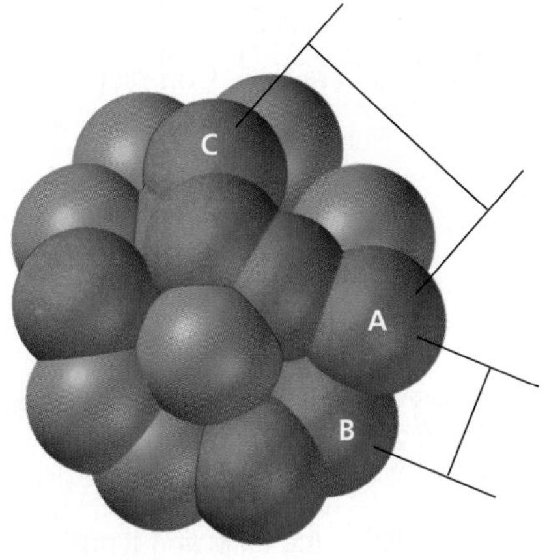

FIGURE 22-3 Proton A attracts proton B through the nuclear force but repels it through the electrostatic force. Proton A mainly repels proton C through the electrostatic force because the nuclear force reaches only a few nucleon diameters.

Nuclear Reactions

Unstable nuclei undergo spontaneous changes that change their number of protons and neutrons. In this process, they give off large amounts of energy and increase their stability. These changes are a type of nuclear reaction. *A* **nuclear reaction** *is a reaction that affects the nucleus of an atom.* In equations representing nuclear reactions, the total of the atomic numbers and the total of the mass numbers must be equal on both sides of the equation. An example is shown below.

$$^{9}_{4}\text{Be} + ^{4}_{2}\text{He} \longrightarrow ^{12}_{6}\text{C} + ^{1}_{0}n$$

Notice that when the atomic number changes, the identity of the element changes. *A* **transmutation** *is a change in the identity of a nucleus as a result of a change in the number of its protons.*

SAMPLE PROBLEM 22-1

Identify the product that balances the following nuclear reaction: $^{212}_{84}\text{Po} \longrightarrow ^{4}_{2}\text{He} + \underline{\ ?\ }$

SOLUTION

1. The total mass number and atomic number must be equal on both sides of the equation.

$$^{212}_{84}\text{Po} \longrightarrow ^{4}_{2}\text{He} + \underline{\ ?\ }$$

mass number:	$212 - 4 = 208$
atomic number:	$84 - 2 = 82$

2. The nuclide has a mass number of 208 and an atomic number of 82, $^{208}_{82}\text{Pb}$.
3. The balanced nuclear equation is $^{212}_{84}\text{Po} \longrightarrow ^{4}_{2}\text{He} + ^{208}_{82}\text{Pb}$

PRACTICE

Complete the following nuclear equations:

1. $^{218}_{84}\text{Po} \longrightarrow ^{4}_{2}\text{He} + \underline{\ ?\ }$ — *Answer* $^{218}_{84}\text{Po} \longrightarrow ^{4}_{2}\text{He} + ^{214}_{82}\text{Pb}$
2. $^{253}_{99}\text{Es} + ^{4}_{2}\text{He} \longrightarrow ^{1}_{0}n + \underline{\ ?\ }$ — *Answer* $^{253}_{99}\text{Es} + ^{4}_{2}\text{He} \longrightarrow ^{1}_{0}n + ^{256}_{101}\text{Md}$
3. $^{142}_{61}\text{Pm} + \underline{\ ?\ } \longrightarrow ^{142}_{60}\text{Nd}$ — *Answer* $^{142}_{61}\text{Pm} + ^{0}_{-1}e \longrightarrow ^{142}_{60}\text{Nd}$

SECTION REVIEW

1. Define the following terms:
   a. nuclide
   b. nucleon
   c. mass defect
   d. nuclear binding energy
2. How is nuclear stability related to the neutron-proton ratio?
3. Why do unstable nuclides undergo nuclear reactions?
4. Complete and balance the following nuclear equations:
   a. $^{187}_{75}\text{Re} + \underline{\ ?\ } \longrightarrow ^{188}_{75}\text{Re} + ^{1}_{1}\text{H}$
   b. $^{9}_{4}\text{Be} + ^{4}_{2}\text{He} \longrightarrow \underline{\ ?\ } + ^{1}_{0}n$
   c. $^{22}_{11}\text{Na} + \underline{\ ?\ } \longrightarrow ^{22}_{10}\text{Ne}$

Radioactive Decay

SECTION 22-2

OBJECTIVES

- Define and relate the terms *radioactive decay* and *nuclear radiation.*
- Describe the different types of radioactive decay and their effects on the nucleus.
- Define the term *half-life,* and explain how it relates to the stability of a nucleus.
- Define and relate the terms *decay series, parent nuclide,* and *daughter nuclide.*
- Explain how artificial radioactive nuclides are made, and discuss their significance.

In 1896, Henri Becquerel was studying the possible connection between light emission of some uranium compounds after exposure to sunlight and X-ray emission. He wrapped a photographic plate in a lightproof covering and placed a uranium compound on top of it. He then placed them in sunlight. The photographic plate was exposed even though it was protected from visible light, suggesting exposure by X rays. When he tried to repeat his experiment, cloudy weather prevented him from placing the experiment in sunlight. To his surprise, the plate was still exposed. This meant that sunlight was not needed to produce the rays that exposed the plate. The rays were produced by radioactive decay. **Radioactive decay** *is the spontaneous disintegration of a nucleus into a slightly lighter and more stable nucleus, accompanied by emission of particles, electromagnetic radiation, or both.* The radiation that exposed the plate was **nuclear radiation,** *particles or electromagnetic radiation emitted from the nucleus during radioactive decay.*

Uranium is a **radioactive nuclide,** *an unstable nucleus that undergoes radioactive decay*. Studies by Marie Curie and Pierre Curie found that of the elements known in 1896, only uranium and thorium were radioactive. In 1898, the Curies discovered two new radioactive metallic elements, polonium and radium. Since that time, many other radioactive nuclides have been identified. In fact, all of the nuclides beyond atomic number 83 are unstable and radioactive.

Types of Radioactive Decay

A nuclide's type and rate of decay depend on the nucleon content and energy level of the nucleus. Some common types of radioactive nuclide emissions are summarized in Table 22-1.

TABLE 22-1 ***Radioactive Nuclide Emissions***

Type	Symbol	Charge	Mass (amu)
Alpha particle	${}^{4}_{2}He$	2+	4.002 60
Beta particle	${}^{0}_{-1}\beta$	1−	0.000 5486
Positron	${}^{0}_{+1}\beta$	1+	0.000 5486
Gamma ray	γ	0	0

FIGURE 22-4 An alpha particle, identical to a helium nucleus, is emitted during the radioactive decay of some very heavy nuclei.

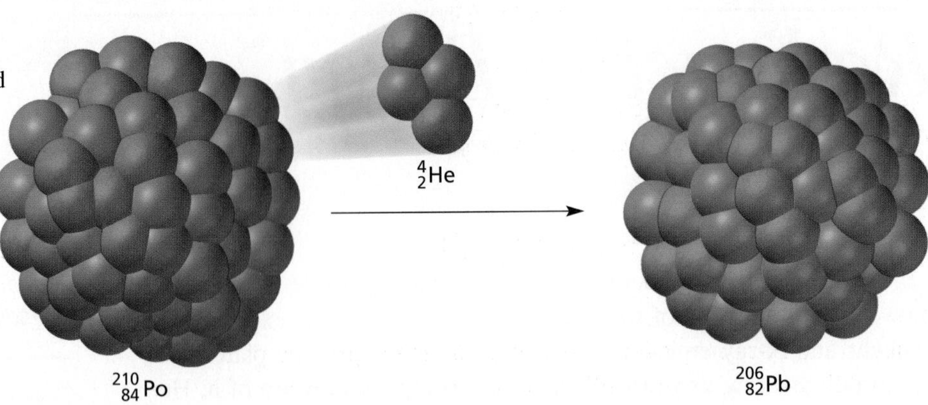

Alpha Emission

An **alpha particle** *(α) is two protons and two neutrons bound together and is emitted from the nucleus during some kinds of radioactive decay.* Alpha particles are basically helium nuclei and have a charge of 2+. They are often represented with the symbol $^{4}_{2}He$. Alpha emission is restricted almost entirely to very heavy nuclei. In these nuclei, both the number of neutrons and the number of protons need to be reduced in order to increase the stability of the nucleus. An example of alpha emission is the decay of $^{210}_{84}Po$ into $^{206}_{82}Pb$, shown in Figure 22-4. The atomic number decreases by two, and the mass number decreases by four.

$$^{210}_{84}Po \longrightarrow ^{206}_{82}Pb + ^{4}_{2}He$$

Beta Emission

Elements above the band of stability are unstable because they have too many neutrons. To decrease the number of neutrons, a neutron can be converted into a proton and an electron. The electron is emitted from the nucleus as a beta particle. *A* **beta particle** *(β) is an electron emitted from the nucleus during some kinds of radioactive decay.*

$$^{1}_{0}n \longrightarrow ^{1}_{1}p + ^{0}_{-1}\beta$$

An example of beta emission, shown in Figure 22-5, is the decay of $^{14}_{6}C$ into $^{14}_{7}N$. Notice that the atomic number increases by one and the mass number stays the same.

$$^{14}_{6}C \longrightarrow ^{14}_{7}N + ^{0}_{-1}\beta$$

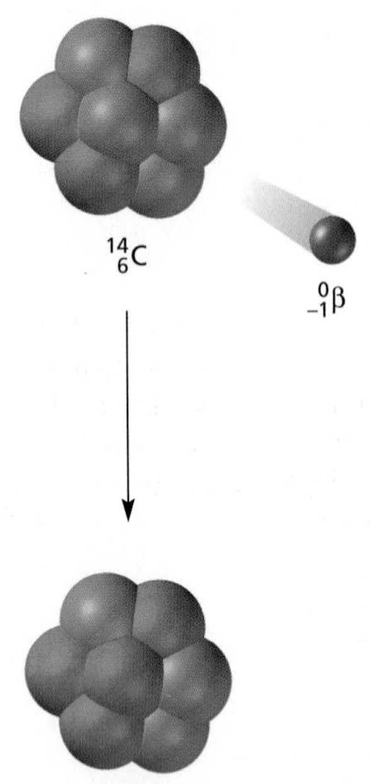

FIGURE 22-5 Beta emission causes the transmutation of $^{14}_{6}C$ into $^{14}_{7}N$. Beta emission is a type of radioactive decay in which a proton is converted to a neutron with the emission of a beta particle.

Positron Emission

Elements below the band of stability have too many protons to be stable. To decrease the number of protons, a proton can be converted into a neutron by emitting a positron. *A* **positron** *is a particle that has the same mass as an electron, but has a positive charge, and is emitted from the nucleus during some kinds of radioactive decay.*

$$^{1}_{1}p \longrightarrow {}^{1}_{0}n + {}^{0}_{+1}\beta$$

An example of positron emission is the decay of $^{38}_{19}K$ into $^{38}_{18}Ar$. Notice that the atomic number decreases by one but the mass number stays the same.

$$^{38}_{19}K \longrightarrow {}^{38}_{18}Ar + {}^{0}_{+1}\beta$$

Electron Capture

Another type of decay for nuclides with too many protons is electron capture. *In* **electron capture,** *an inner orbital electron is captured by the nucleus of its own atom.* The inner orbital electron combines with a proton, and a neutron is formed.

$$^{0}_{-1}e + {}^{1}_{1}p \longrightarrow {}^{1}_{0}n$$

An example of electron capture is the radioactive decay of $^{106}_{47}Ag$ into $^{106}_{46}Pd$. Just as in positron emission, the atomic number decreases by one but the mass number stays the same.

$$^{106}_{47}Ag + {}^{0}_{-1}e \longrightarrow {}^{106}_{46}Pd$$

Gamma Emission

Gamma rays *(γ) are high-energy electromagnetic waves emitted from a nucleus as it changes from an excited state to a ground energy state.* The position of gamma rays in the electromagnetic spectrum is shown in Figure 22-6. The emission of gamma rays is another piece of evidence supporting the nuclear shell model. According to the nuclear shell model, gamma rays are produced when nuclear particles undergo transitions in nuclear-energy levels. This is similar to the emission of light when an electron drops to a lower-energy level, which was covered in Chapter 4. Gamma emission usually occurs immediately following other types of decay, when other types of decay leave the nucleus in an excited state.

FIGURE 22-6 Gamma rays, like visible light, are a form of electromagnetic radiation, but they have a much shorter wavelength and are much higher in energy than visible light.

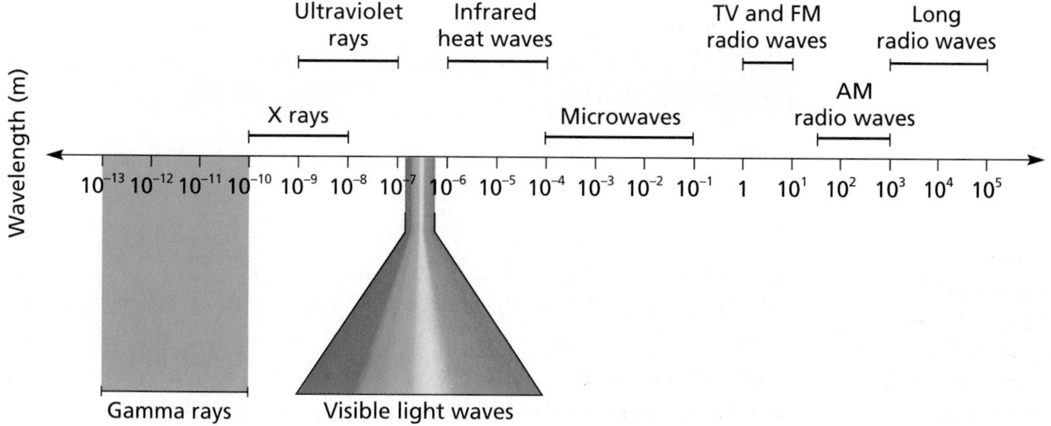

FIGURE 22-7 The half-life of radium-226 is 1599 years. Half of the remaining radium-226 decays by the end of each half-life.

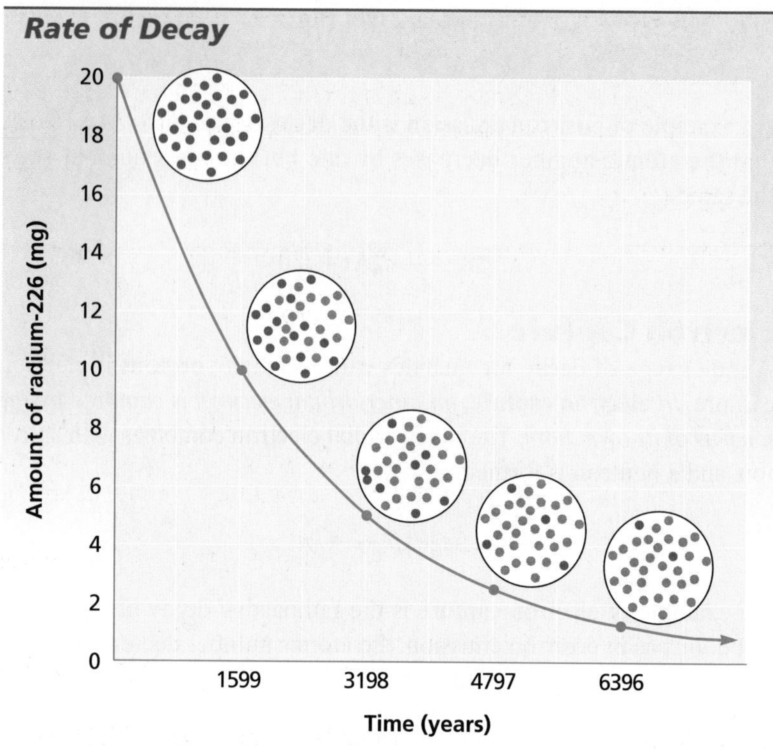

Half-Life

No two radioactive isotopes decay at the same rate. **Half-life,** $t_{1/2}$, *is the time required for half the atoms of a radioactive nuclide to decay.* Look at the graph of the decay of radium-226 in Figure 22-7. Radium-226 has a half-life of 1599 years. Half of a given amount of radium-226 decays in 1599 years. In another 1599 years, half of the remaining radium-226 decays. This cycle continues until there is a negligible amount of radium-226. Each radioactive nuclide has its own half-life. More-stable nuclides decay slowly and have longer half-lives. Less-stable nuclides decay very quickly and have shorter half-lives, sometimes just a fraction of a second. Some representative radioactive nuclides, along with their half-lives and types of decay, are given in Table 22-2.

TABLE 22-2 ***Representative Radioactive Nuclides and Their Half-Lives***

Nuclide	Half-life	Nuclide	Half-life
$^{3}_{1}H$	12.32 years	$^{214}_{84}Po$	163.7 μs
$^{14}_{6}C$	5715 years	$^{218}_{84}Po$	3.0 min
$^{32}_{15}P$	14.28 days	$^{218}_{85}At$	1.6 s
$^{40}_{19}K$	1.3×10^{9} years	$^{238}_{92}U$	4.46×10^{9} years
$^{60}_{27}Co$	10.47 min	$^{239}_{94}Pu$	2.41×10^{4} years

SAMPLE PROBLEM 22-2

Phosphorus-32 has a half-life of 14.3 days. How many milligrams of phosphorus-32 remain after 57.2 days if you start with 4.0 mg of the isotope?

SOLUTION

1 ANALYZE

Given: original mass of phosphorus-32 = 4.0 mg
half-life of phosphorus-32 = 14.3 days
time elapsed = 57.2 days

Unknown: mass of phosphorus-32 remaining after 57.2 days

2 PLAN

To determine the number of milligrams of phosphorus-32 remaining, we must first find the number of half-lives that have passed in the time elapsed. Then the amount of phosphorus-32 is determined by reducing the original amount by half for every half-life that has passed.

$$\text{number of half-lives} = \text{time elapsed (days)} \times \frac{\text{1 half-life}}{\text{14.3 days}}$$

$$\text{amount of phosphorus-32 remaining} = \text{original amount of phosphorus-32} \times \tfrac{1}{2} \text{ for each half-life}$$

3 COMPUTE

$$\text{number of half-lives} = 57.2 \cancel{\text{days}} \times \frac{\text{1 half-life}}{14.3 \cancel{\text{days}}} = \text{4 half-lives}$$

$$\text{amount of phosphorus-32 remaining} = 4.0 \text{ mg} \times \tfrac{1}{2} \times \tfrac{1}{2} \times \tfrac{1}{2} \times \tfrac{1}{2} = 0.25 \text{ mg}$$

4 EVALUATE

A period of 57.2 years is four half-lives for phosphorus-32. At the end of one half-life, 2.0 mg of phosphorus-32 remains; 1.0 mg remains at the end of two half-lives; 0.50 mg remains at the end of three half-lives; and 0.25 mg remains at the end of four half-lives.

PRACTICE

1. The half-life of polonium-210 is 138.4 days. How many milligrams of polonium-210 remain after 415.2 days if you start with 2.0 mg of the isotope? — *Answer* 0.25 mg
2. Assuming a half-life of 1599 years, how many years will be needed for the decay of $\frac{15}{16}$ of a given amount of radium-226? — *Answer* 6396 years
3. The half-life of radon-222 is 3.824 days. After what time will one-fourth of a given amount of radon remain? — *Answer* 7.648 days
4. The half-life of cobalt-60 is 10.47 min. How many milligrams of cobalt-60 remain after 104.7 min if you start with 10.0 mg? — *Answer* 0.00977 mg
5. The half-life of uranium-238 is 4.46×10^9 years. If 4.46×10^9 years ago a sample contained 4.0 mg of uranium-238, how many milligrams of uranium-238 does the sample contain today? — *Answer* 2.0 mg
6. The half-life of polonium-218 is 3.0 min. If you start with 16 mg, how long will it be before only 1.0 mg remains? — *Answer* 12 min

Decay Series

One nuclear reaction is not always enough to produce a stable nuclide. *A* **decay series** *is a series of radioactive nuclides produced by successive radioactive decay until a stable nuclide is reached. The heaviest nuclide of each decay series is called the* **parent nuclide.** *The nuclides produced by the decay of the parent nuclides are called* **daughter nuclides.** All naturally occurring nuclides with atomic numbers greater than 83 are radioactive and belong to one of three natural decay series. The parent nuclides are uranium-238, uranium-235, and thorium-232. The transmutations of the uranium-238 decay series are charted in Figure 22-8.

Locate the parent nuclide, uranium-238, on the chart. As the nucleus of uranium-238 decays, it emits an alpha particle. The mass number of the nuclide, and thus the vertical position on the graph, decreases by four. The atomic number, and thus the horizontal position, decreases by two. The daughter nuclide is an isotope of thorium.

$$^{238}_{92}U \longrightarrow\, ^{234}_{90}Th + \,^{4}_{2}He$$

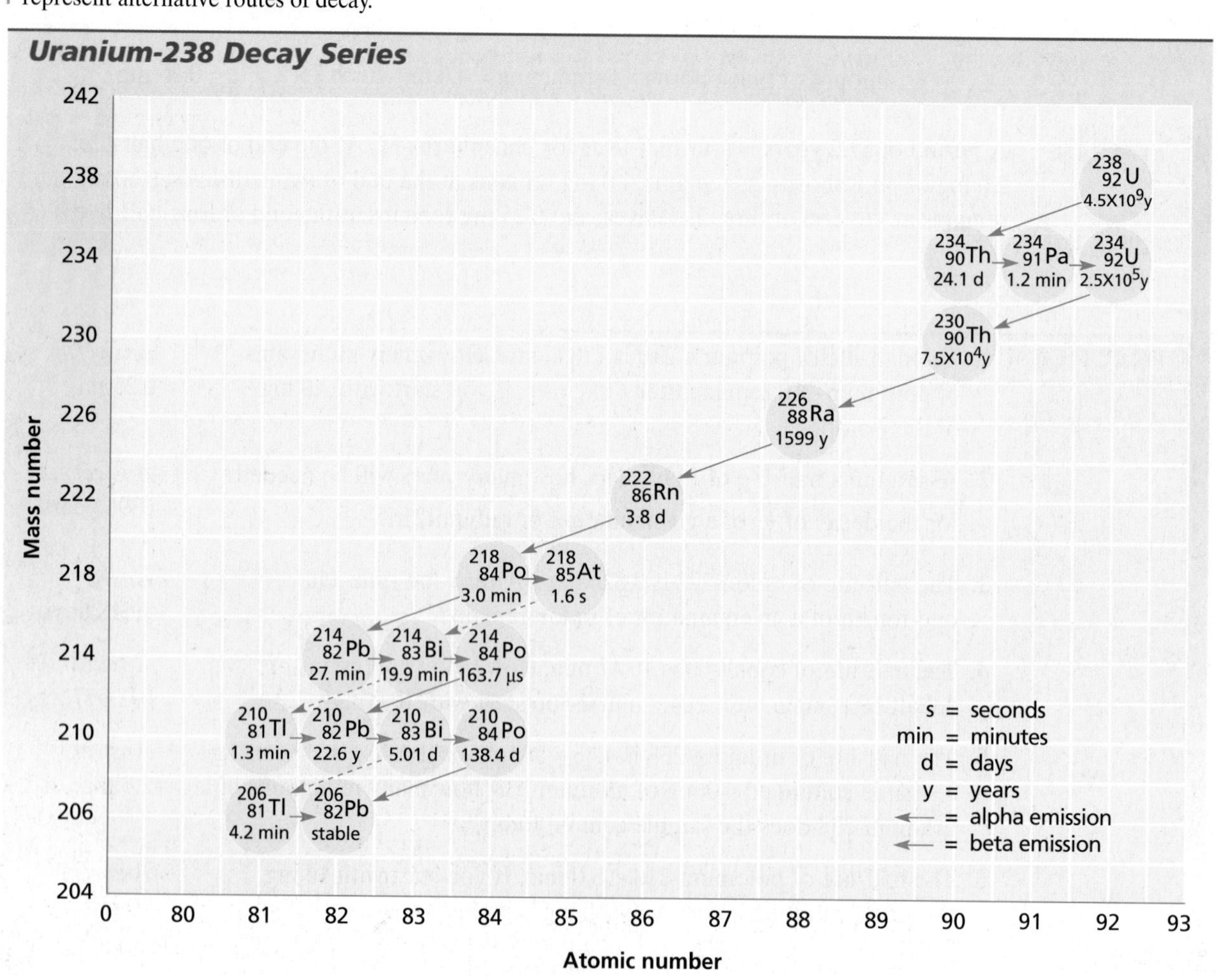

FIGURE 22-8 This chart shows the transmutations that occur as $^{238}_{92}U$ decays to the final, stable nuclide, $^{206}_{82}Pb$. Decay usually follows the solid arrows. The dotted arrows represent alternative routes of decay.

The half-life of $^{234}_{90}Th$, about 24 days, is indicated on the chart. It decays by giving off beta particles. This increases its atomic number, and thus its horizontal position, by one. The mass number, and thus its vertical position, remains the same.

$$^{234}_{90}Th \longrightarrow ^{234}_{91}Pa + ^{0}_{-1}\beta$$

The remaining atomic number and mass number changes shown on the decay chart are also explained in terms of the particles given off. In the final step, $^{210}_{84}Po$ loses an alpha particle to form $^{206}_{82}Pb$. This is a stable, nonradioactive isotope of lead. Notice that $^{206}_{82}Pb$ contains 82 protons, a magic number. It contains the extra-stable nuclear configuration of a completed nuclear shell.

Artificial Transmutations

Artificial radioactive nuclides are radioactive nuclides not found naturally on Earth. They are made by **artificial transmutations,** *bombardment of stable nuclei with charged and uncharged particles.* Because neutrons have no charge, they can easily penetrate the nucleus of an atom. However, positively charged alpha particles, protons, and other ions are repelled by the nucleus. Because of this repulsion, great quantities of energy are required to bombard nuclei with these particles. The necessary energy may be supplied by accelerating these particles in the magnetic or electrical field of a particle accelerator. An example of an accelerator is shown in Figure 22-9.

FIGURE 22-9 This is an aerial view of the Fermi International Accelerator Laboratory (Fermilab), in Illinois. The particle accelerators are underground. The Tevatron ring, the larger particle accelerator, has a circumference of 4 mi. The smaller ring in the background is a new accelerator, the Main Injector.

TABLE 22-3 Reactions for the First Preparation of Several Transuranium Elements

Atomic number	Name	Symbol	Nuclear reaction
93	neptunium	Np	$^{238}_{92}U + ^{1}_{0}n \longrightarrow ^{239}_{92}U$ $^{239}_{92}U \longrightarrow ^{239}_{93}Np + ^{0}_{-1}\beta$
94	plutonium	Pu	$^{238}_{93}Np \longrightarrow ^{238}_{94}Pu + ^{0}_{-1}\beta$
95	americium	Am	$^{239}_{94}Pu + 2^{1}_{0}n \longrightarrow ^{241}_{95}Am + ^{0}_{-1}\beta$
96	curium	Cm	$^{239}_{94}Pu + ^{4}_{2}He \longrightarrow ^{242}_{96}Cm + ^{1}_{0}n$
97	berkelium	Bk	$^{241}_{95}Am + ^{4}_{2}He \longrightarrow ^{243}_{97}Bk + 2^{1}_{0}n$
98	californium	Cf	$^{242}_{96}Cm + ^{4}_{2}He \longrightarrow ^{245}_{98}Cf + ^{1}_{0}n$
99	einsteinium	Es	$^{238}_{92}U + 15^{1}_{0}n \longrightarrow ^{253}_{99}Es + 7^{0}_{-1}\beta$
100	fermium	Fm	$^{238}_{92}U + 17^{1}_{0}n \longrightarrow ^{255}_{100}Fm + 8^{0}_{-1}\beta$
101	mendelevium	Md	$^{253}_{99}Es + ^{4}_{2}He \longrightarrow ^{256}_{101}Md + ^{1}_{0}n$
102	nobelium	No	$^{246}_{96}Cm + ^{12}_{6}C \longrightarrow ^{254}_{102}No + 4^{1}_{0}n$
103	lawrencium	Lr	$^{252}_{98}Cf + ^{10}_{5}B \longrightarrow ^{258}_{103}Lr + 4^{1}_{0}n$

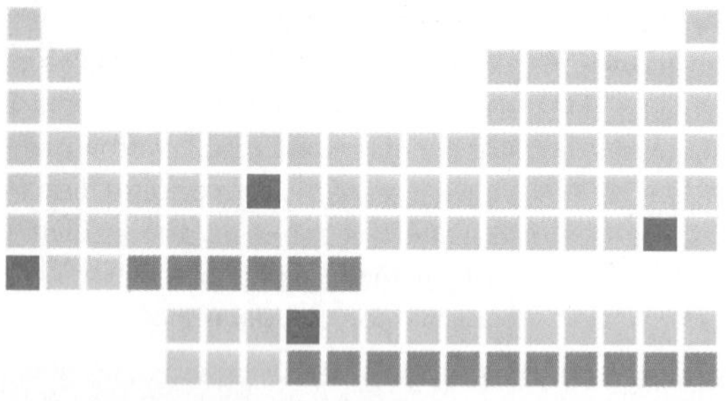

FIGURE 22-10 Artificial transmutations filled the gaps in the periodic table, shown in red, and extended the periodic table with the transuranium elements, shown in blue.

Artificial Radioactive Nuclides

Radioactive isotopes of all the natural elements have been produced by artificial transmutation. In addition, production of technetium, astatine, francium, and promethium by artificial transmutation has filled gaps in the periodic table. Their positions are shown in Figure 22-10.

Artificial transmutations are also used to produce the transuranium elements. **Transuranium elements** *are elements with more than 92 protons in their nuclei.* All of these elements are radioactive. The nuclear reactions for the synthesis of several transuranium elements are shown in Table 22-3. Currently, 17 artificially prepared transuranium elements have been reported. The positions of the transuranium elements in the periodic table are also shown in Figure 22-10.

SECTION REVIEW

1. Define *radioactive decay.*
2. a. What are the different types of radioactive decay?
   b. List the types of radioactive decay that involve conversion of particles.
3. What fraction of a given sample of a radioactive nuclide remains after four half-lives?
4. When does a decay series end?
5. Distinguish between natural and artificial radioactive nuclides.

Nuclear Radiation

SECTION 22-3

OBJECTIVES

- Compare the penetrating ability and shielding requirements of alpha particles, beta particles, and gamma rays.
- Define the terms *roentgen* and *rem,* and distinguish between them.
- Describe three devices used in radiation detection.
- Discuss applications of radioactive nuclides.

In Becquerel's experiment, nuclear radiation from the uranium compound penetrated the lightproof covering and exposed the film. Different types of nuclear radiation have different penetrating abilities. Nuclear radiation includes alpha particles, beta particles, and gamma rays.

Alpha particles have a range of only a few centimeters in air and have a low penetrating ability due to their large mass and charge. They cannot penetrate skin. However, they can cause damage if ingested or inhaled. Beta particles travel at speeds close to the speed of light and have a penetrating ability about 100 times greater than that of alpha particles. They have a range of a few meters in air. Gamma rays have the greatest penetrating ability. The penetrating abilities and shielding requirements of different types of nuclear radiation are shown in Figure 22-11.

Radiation Exposure

Nuclear radiation can transfer its energy to the electrons of atoms or molecules and cause ionization. *The* **roentgen** *is a unit used to measure nuclear radiation; it is equal to the amount of radiation that produces* 2×10^9 *ion pairs when it passes through 1* cm^3 *of dry air.* Ionization can damage living tissue. Radiation damage to human tissue is measured in rems (roentgen equivalent, man). *One* **rem** *is the quantity of ionizing radiation that does as much damage to human tissue as is done by 1 roentgen of high-voltage X rays.* Cancer and genetic effects caused by DNA mutations are long-term radiation damage to living tissue. DNA

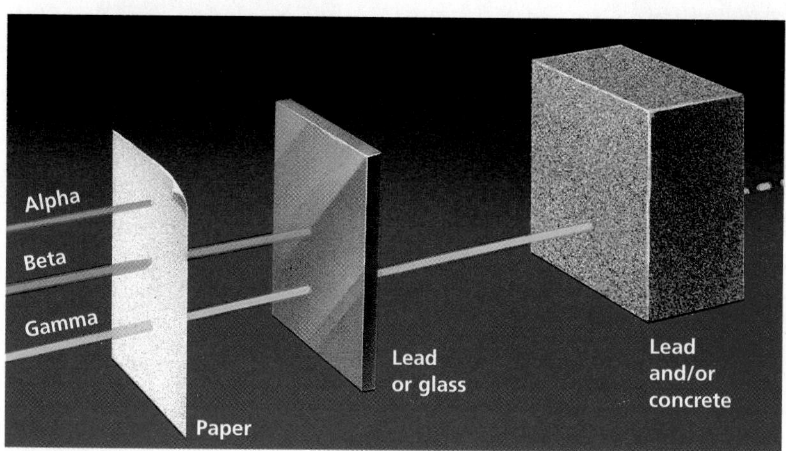

FIGURE 22-11 The different penetrating abilities of alpha particles, beta particles, and gamma rays require different levels of shielding. Alpha particles can be shielded with just a sheet of paper. Lead or glass is often used to shield beta particles. Gamma rays are the most penetrating and require shielding with thick layers of lead or concrete, or both.

can be mutated directly by interaction with radiation or indirectly by interaction with previously ionized molecules.

Everyone is exposed to environmental background radiation. Average exposure for people living in the United States is estimated to be about 0.1 rem per year. However, actual exposure varies. The maximum permissible dose of radiation exposure for a person in the general population is 0.5 rem per year. Airline crews and people who live at high altitudes have increased exposure levels because of increased cosmic-ray levels at high altitudes. Radon-222 trapped inside homes may also cause increased exposure. Because it is a gas, radon can move up from the soil into homes through cracks and holes in the foundation. Radon trapped in homes increases the risk of lung cancer among smokers.

Radiation Detection

Film badges, Geiger-Müller counters, and scintillation counters are three devices commonly used to detect and measure nuclear radiation. A film badge and a Geiger-Müller counter are shown in Figure 22-12. As previously mentioned, nuclear radiation exposes film just as visible light does. This property is used in film badges. **Film badges** *use exposure of film to measure the approximate radiation exposure of people working with radiation.* **Geiger-Müller counters** *are instruments that detect radiation by counting electric pulses carried by gas ionized by radiation.* Geiger-Müller counters are typically used to detect beta-particle radiation. Radiation can also be detected when it transfers its energy to substances that *scintillate,* or absorb ionizing radiation and emit visible light. **Scintillation counters** *are instruments that convert scintillating light to an electric signal for detecting radiation.*

FIGURE 22-12 Film badges (a) and Geiger-Müller counters (b) are both used to detect nuclear radiation.

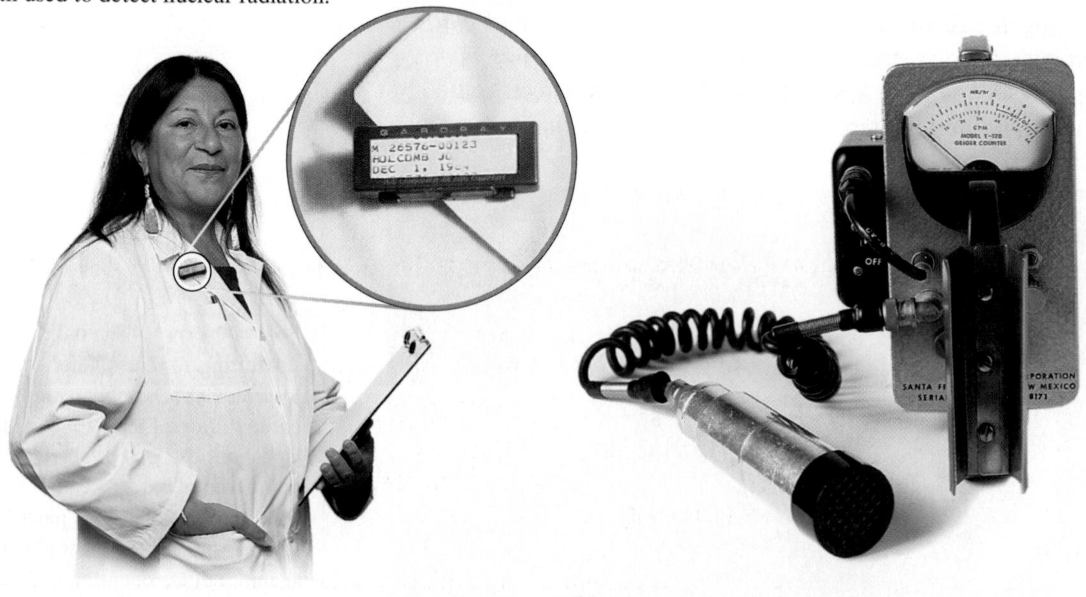

(a)

(b)

Applications of Nuclear Radiation

Many applications are based on the fact that stable and radioactive isotopes of a given element behave essentially the same in both chemical and physical processes. A few uses of radioactive nuclides are discussed below.

Radioactive Dating

Radioactive dating *is the process by which the approximate age of an object is determined based on the amount of certain radioactive nuclides present.* Such an estimate is based on the fact that radioactive substances decay with known half-lives. Age is estimated by measuring either the accumulation of a daughter nuclide or the disappearance of the parent nuclide.

Carbon-14 is radioactive and has a half-life of approximately 5715 years. It can be used to estimate the age of organic material up to about 50 000 years old. Nuclides with longer half-lives are used to estimate the age of older objects; methods using nuclides with long half-lives have been used to date minerals and lunar rocks more than 4 billion years old.

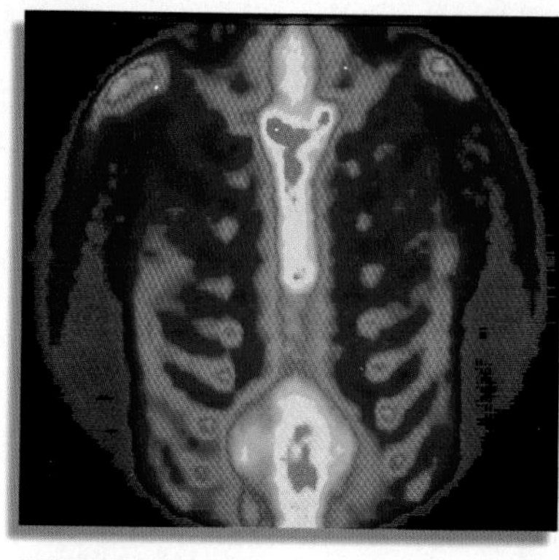

FIGURE 22-13 Radioactive nuclides, such as technetium-99, can be used to detect bone cancer. In this procedure, technetium-99 accumulates in areas of abnormal bone metabolism. Detection of the nuclear radiation then shows the location of bone cancer.

Radioactive Nuclides in Medicine

In medicine, radioactive nuclides, such as the artificial radioactive nuclide cobalt-60, are used to destroy certain types of cancer cells. Many radioactive nuclides are also used as **radioactive tracers,** *which are radioactive atoms that are incorporated into substances so that movement of the substances can be followed by radiation detectors.* Detection of radiation from radioactive tracers can be used to diagnose cancer and other diseases. See Figure 22-13.

Radioactive Nuclides in Agriculture

In agriculture, radioactive tracers in fertilizers are used to determine the effectiveness of the fertilizer. The amount of radioactive tracer absorbed by a plant indicates the amount of fertilizer absorbed. Nuclear radiation is also used to prolong the shelf life of food. For example, gamma rays from cobalt-60 can be used to kill bacteria and insects that spoil and infest food.

SECTION REVIEW

1. What is required to shield alpha particles and why?
2. a. What is the average exposure of people living in the United States to environmental background radiation?
   b. How does this relate to the maximum permissible dose?
3. What device is used to measure the radiation exposure of people working with radiation?
4. Explain why nuclear radiation can be used to preserve food.

The Dating Game

From *Science Matters: Achieving Science Literacy*, by Robert M. Hazen and James Trefil

The best way to think about the behavior of radioactive nuclei is to picture popcorn popping on your stove. Kernels don't all pop at once. A few kernels pop, then a few more, spaced out over several minutes. No theory of nuclear stability exists that will always predict all the details of radioactive decay. We don't understand, for example, why some atoms take billions of years to decay while others disappear in a matter of seconds. However, we can measure the phenomenon with great precision.

Any given collection of radioactive nuclei behaves in roughly the same way as another, with individual nuclei decaying at different times. The overall rate of decay is called the half-life, defined as the time it takes for half of the nuclei in a given sample to decay.

The fact that an atom's chemical reactions are largely independent of the nucleus leads to some very important ways of dating artifacts and rocks. The most familiar of these, the so-called carbon-14 method, is based on the fact that, in addition to normal, stable carbon-12, a certain amount of the radioactive element carbon-14 is always present in the environment. (Carbon-14 results from the collision of cosmic rays with nitrogen atoms in the upper atmosphere.) Because the chemistry of the two isotopes of carbon is identical, a certain amount of carbon-14 finds its way into all living tissues. When an organism dies, it stops taking in carbon-14, and its complement of these atoms starts to fall off as they decay. Given that the half-life of carbon-14 is 5730 years*, we can work out how long it's been since that piece of material was taking in fresh carbon-14.

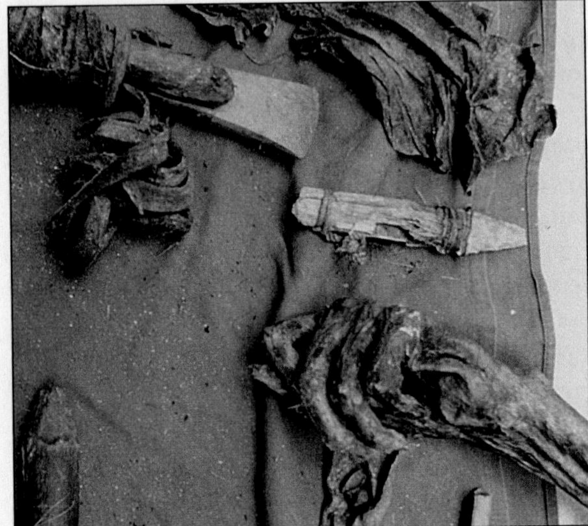

The Iceman was found frozen in the Alps in 1991. Carbon-14 dating has placed his lifetime at somewhere between 3500 and 3000 B.C.

For example, if you find that a piece of wood contains only half the amount of carbon-14 it had when it was formed, you know that the tree from which it came died about 5730 years ago. If the piece of material happens to be leather from a grave site or an elk shoulder blade used as a shovel, you have a pretty good idea of the age of the civilization that produced the artifact. This makes carbon dating an important tool in anthropology.

The same general technique can be used to date many rocks. The mineral's atomic structure tells us how much of a given isotope must have been present at the beginning, and measuring the amount left (or, equivalently, the number of decays that have occurred) will tell us how many half-lives it's been since the rock was formed.

Reading for Meaning

The authors mention that carbon-14 results from the collision of cosmic rays with nitrogen atoms. What kind of nuclear reaction would account for this transmutation?

Read Further

Research other isotopes used for radiometric dating. List the elements, their half-lives, and examples of materials dated, as well as the ages of the materials, using the isotopes.

* This is the value given by this source.

Nuclear Fission and Nuclear Fusion

SECTION 22-4

OBJECTIVES

- Define *nuclear fission, chain reaction,* and *nuclear fusion,* and distinguish between them.
- Explain how a fission reaction is used to generate power.
- Discuss the possible benefits and the current difficulty of controlling fusion reactions.

Nuclear Fission

Review Figure 22-1 on page 702, which shows that nuclei of intermediate mass are the most stable. *In* **nuclear fission,** *a very heavy nucleus splits into more-stable nuclei of intermediate mass.* This process releases enormous amounts of energy. Nuclear fission can occur spontaneously or when nuclei are bombarded by particles. When uranium-235 is bombarded with slow neutrons, a uranium nucleus may capture one of the neutrons, making it very unstable. The nucleus splits into medium-mass parts with the emission of more neutrons. The mass of the products is less than the mass of the reactants. The missing mass is converted to energy.

Nuclear Chain Reaction

When fission of an atom bombarded by neutrons produces more neutrons, a chain reaction can occur. *A* **chain reaction** *is a reaction in which the material that starts the reaction is also one of the products and can start another reaction.* As shown in Figure 22-14, two or three neutrons can be given off when uranium-235 fission occurs. These neutrons can cause the fission of other uranium-235 nuclei. Again neutrons are emitted, which

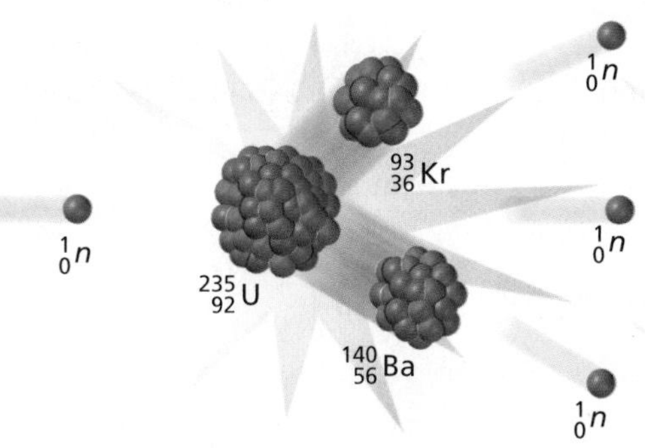

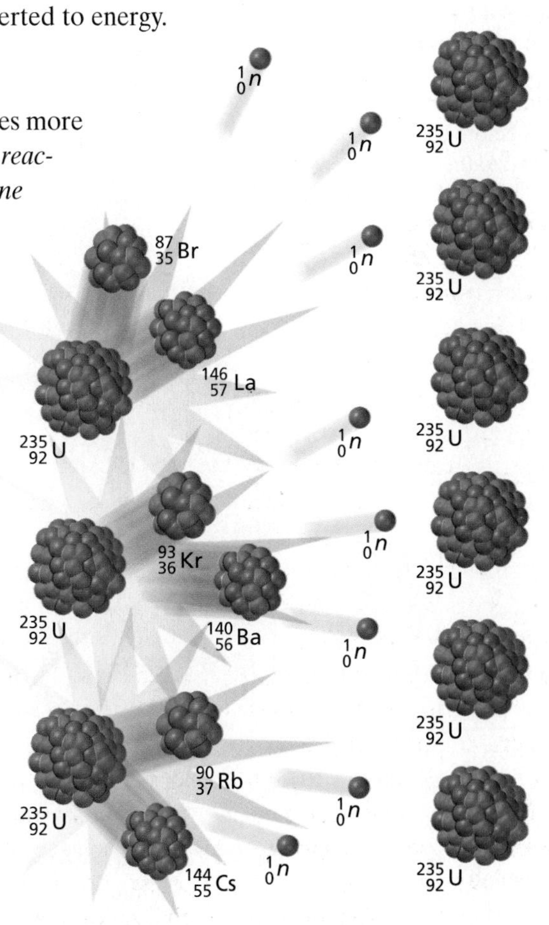

FIGURE 22-14 Fission induction of uranium-235 by bombardment with neutrons can lead to a chain reaction when a critical mass of uranium-235 is present.

can cause the fission of still other uranium-235 nuclei. This chain reaction continues until all of the uranium-235 atoms have split or until the neutrons fail to strike uranium-235 nuclei. If the mass of uranium-235 is below a certain minimum, too many neutrons will escape without striking other nuclei, and the chain reaction will stop. *The minimum amount of nuclide that provides the number of neutrons needed to sustain a chain reaction is called the* **critical mass.** Uncontrolled chain reactions provide the explosive energy of atomic bombs. **Nuclear reactors** *are devices that use controlled-fission chain reactions to produce energy or radioactive nuclides.*

Nuclear Power Plants

Nuclear power plants *use heat from nuclear reactors to produce electrical energy.* They have five main components: shielding, fuel, control rods, moderator, and coolant. The components, shown in Figure 22-15, are surrounded by shielding. **Shielding** *is radiation-absorbing material that is used to decrease radiation exposure from nuclear reactors, especially gamma rays.* Uranium-235 is typically used as the fissionable fuel to produce heat, which is absorbed by the coolant. **Control rods** *are neutron-absorbing rods that help control the reaction by limiting the number of free neutrons.* Because fission of uranium-235 is more efficiently induced by slow neutrons, a **moderator** *is used to slow down the fast neutrons produced by fission.* Current problems with nuclear power plant development include environmental requirements, safety of operation, plant construction costs, and storage of spent fuel.

FIGURE 22-15 In this model of a nuclear power plant, pressurized water is heated by fission of uranium-235. This water is circulated to a steam generator. The steam drives a turbine to produce electricity. Cool water from a lake or river is then used to condense the steam into water. The warm water from the condenser may be cooled in cooling towers before being reused or returned to the lake or river.

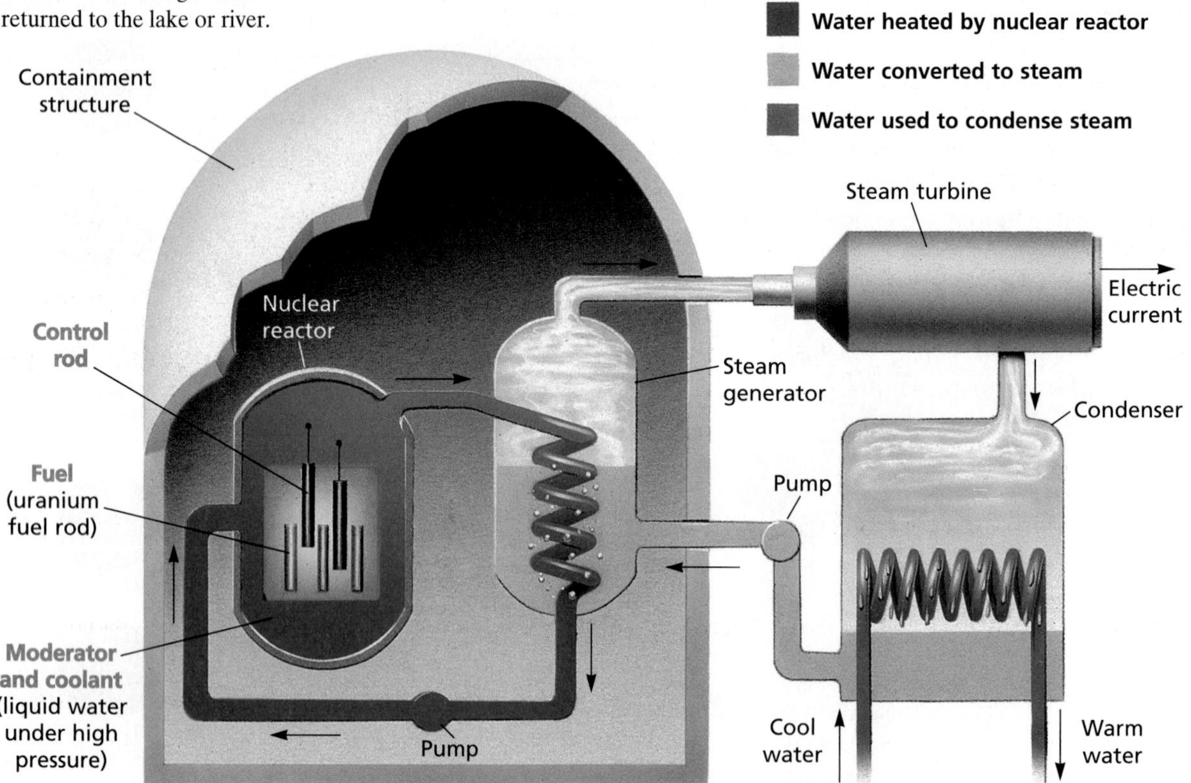

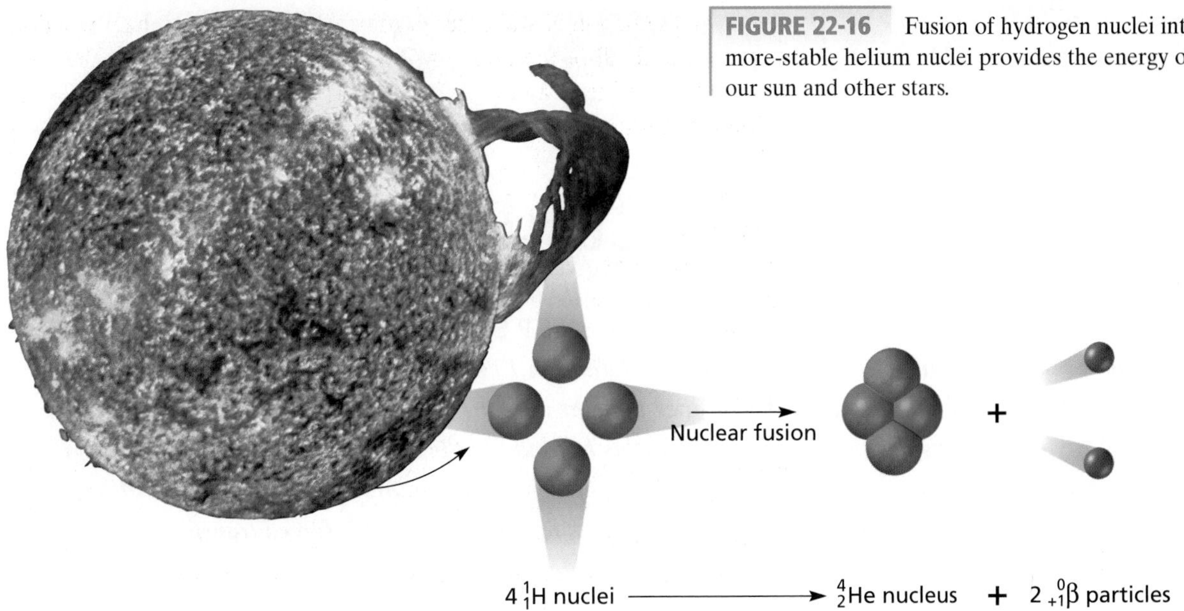

FIGURE 22-16 Fusion of hydrogen nuclei into more-stable helium nuclei provides the energy of our sun and other stars.

Nuclear Fusion

The high stability of nuclei with intermediate masses can also be used to explain nuclear fusion. *In* **nuclear fusion,** *light-mass nuclei combine to form a heavier, more stable nucleus.* Nuclear fusion releases even more energy per gram of fuel than nuclear fission. In our sun and other stars, four hydrogen nuclei combine at extremely high temperature and pressure to form a helium nucleus with a loss of mass and release of energy. This reaction is illustrated in Figure 22-16.

Uncontrolled fusion reactions of hydrogen are the source of energy for the hydrogen bomb, as shown in Figure 22-17. A fission reaction is used to provide the heat and pressure necessary to trigger the fusion of nuclei. Current research indicates that fusion reactions may be controllable if a few problems can be overcome. One of the problems is that no known material can withstand the initial temperatures, about 10^8 K, required to induce fusion at high temperatures. Methods that contain the fusion reactions within a magnetic field or induce fusion at lower temperatures are being investigated.

FIGURE 22-17 The enormous amount of energy released by fusion reactions is illustrated by this explosion of a hydrogen bomb.

SECTION REVIEW

1. Distinguish between nuclear fission and nuclear fusion.
2. Define *chain reaction.*
3. List the five main components of a nuclear power plant.
4. Explain how fusion is one of our sources of energy.

GREAT DISCOVERIES

An Unexpected Finding

HISTORICAL PERSPECTIVE

The discovery of the artificial transmutation of uranium in 1934 triggered great excitement in science. Chemists preoccupied with identifying what they thought were the final missing elements of the periodic table suddenly had to consider the existence of elements beyond atomic number 92, while physicists began to probe the stability of the nucleus more deeply. By 1939, nuclear investigators in both fields had collaborated to provide a stunning explanation for the mysterious results of uranium's forced transformation.

Neutrons in Italy

In 1934, uranium, with an atomic number of 92, had the most protons of any known element. But that year, Italian physicist Enrico Fermi believed he had synthesized elements with atomic numbers beyond the limit recognized on the periodic table. After bombarding a sample of uranium with neutrons, Fermi and his co-workers recorded measurements that seemed to indicate that some uranium nuclei had absorbed neutrons and then undergone beta decay:

$$^{238}_{92}\text{U} + ^{1}_{0}n \longrightarrow ^{239}_{92}\text{U} \longrightarrow ^{239}_{93}? + ^{0}_{-1}\beta$$

Fermi's report noted additional, subsequent beta decays, inspiring Fermi to hypothesize the existence of a whole new series of "transuranic" elements:

$$^{238}_{92}\text{U} \longrightarrow ^{238}_{93}? + ^{0}_{-1}\beta \longrightarrow ^{238}_{94}?? + ^{0}_{-1}\beta \longrightarrow ^{238}_{95}??? + ^{0}_{-1}\beta$$

Unfortunately, the Italian group could not verify the existence of the transuranes because, as Fermi put it, "We did not know enough chemistry to separate the products of uranium disintegration from one another."

This apparatus from Otto Hahn's lab was used to produce fission reactions.

Curiosity in Berlin

Fermi's experiments caught the attention of a physicist in Berlin, Lise Meitner. Knowing that she could not perform the difficult task of chemically separating radionuclides either, Meitner persuaded a colleague, radiochemist Otto Hahn, to help her explain Fermi's results. Joined by expert chemical analyst Fritz Strassman, the team began investigating neutron-induced uranium decay at the end of 1934.

From the onset, the Berlin team, along with all other scientists at the time, operated under two false assumptions. The first involved the makeup of the bombarded nuclei. In every nuclear reaction observed previously, the resulting nucleus had never differed from the original by more than a few protons or neutrons. Thus, it was assumed that the products of neutron bombardment were radioisotopes of elements at most a few places in the periodic table before or beyond the atoms being bombarded (as Fermi had presumed in hypothesizing the transuranes).

The second assumption concerned the periodicity of the transuranes. Because the elements Ac, Th, Pa, and U chemically resembled the third-row transition elements, La, Hf, Ta, and W, it was believed that elements beyond U would correspondingly resemble those following W. Thus, the transuranes were thought to be homologues of Re, Os, Ir, Pt, and so on. This belief was generally unquestioned and seemed to be confirmed. In fact, by 1937 Hahn was sure of the chemical evidence of transuranes:

> *In general, the chemical behavior of the transuranes . . . is such that their position in the periodic system is no longer in doubt. Above all, their chemical distinction from all previously known elements needs no further discussion.*

Meitner's Exile

By 1938, the political situation in Germany had become dangerous for Meitner. Because she was of Jewish descent, she was being targeted by the Nazis and fled to Sweden to escape persecution. Meanwhile in Berlin, the anti-Nazi Hahn and Strassman had to be careful of every move they made under the watchful eyes of the fascists around them.

Despite censorship, the Berlin team continued to communicate through letters. Meitner could not come up with a satisfying physical explanation for the chemical results of Hahn and Strassman, and she insisted that her partners re-examine their findings. As Strassman later recalled, Meitner

> *. . . urgently requested that [the] experiments be scrutinized very carefully and intensively one more time. . . . Fortunately L. Meitner's opinion and judgment carried so much weight with us in Berlin that the necessary control experiments were immediately undertaken.*

The politics of World War II prevented Lise Meitner from receiving the Nobel Prize in physics for explaining nuclear fission.

A Shocking Discovery

Prompted by Meitner, Hahn and Strassman realized that they had been looking in the wrong place to determine the cause of their results. In analyzing a fraction of a solution assay that they had previously ignored, they found the critical evidence they had been seeking.

The analysis indicated that barium appeared to be a result of neutron bombardment of uranium. Suspecting the spectacular truth but lacking confidence, Hahn wrote to Meitner for an explanation. After consultation with her nephew, Otto Frisch, Meitner proposed that the uranium nuclei had been broken apart into elemental fragments, one of which was Ba. On January 3, 1939, she wrote to Hahn:

> *I am quite certain that the two of you really do have a splitting to Ba, and I find that to be a truly beautiful result, for which I most heartily congratulate you and Strassman.*

Thus, the transuranes turned out to be merely radioisotopes of known elements—atomic fragments of uranium atoms that had burst apart on being struck by neutrons.

For the discovery of this unexpected phenomenon, which Meitner named nuclear fission, the talented Hahn was awarded the 1944 Nobel Prize in chemistry. Due to wartime politics, however, Lise Meitner did not receive the corresponding award in physics, and she was not popularly recognized for her role in clarifying the process that she first explained and named until well after her death in 1968.

CHAPTER 22 REVIEW

CHAPTER SUMMARY

22-1

- The difference between the sum of the masses of the nucleons and electrons in an atom and the actual mass of an atom is the mass defect, or nuclear binding energy.
- Nuclear stability tends to be greatest when nucleons are paired, when there are magic numbers of nucleons, and when there are certain neutron-proton ratios.
- Nuclear reactions, represented by nuclear equations, can involve the transmutation of nuclides.

Vocabulary

band of stability (702)
binding energy per nucleon (702)
magic numbers (703)
mass defect (701)
nuclear binding energy (702)
nuclear reaction (704)
nuclear shell model (703)
nucleon (701)
nuclide (701)
transmutation (704)

22-2

- Radioactive nuclides become more stable by radioactive decay, a type of nuclear reaction.
- Alpha emission, beta emission, positron emission, electron capture, and gamma emission are all types of radioactive decay. A nuclide's type of decay is related to its nucleon content and the energy level of the nucleus.
- The half-life of a radioactive nuclide is the length of time that it takes for half of a given number of atoms of the nuclide to decay.
- A decay series is a series of related nuclides starting with a parent nuclide and ending with a stable daughter nuclide.
- Artificial transmutations are used to produce artificial radioactive nuclides, including the transuranium elements.

Vocabulary

alpha particle (706)
artificial transmutation (711)
beta particle (706)
daughter nuclide (710)
decay series (710)
electron capture (707)
gamma ray (707)
half-life (708)
nuclear radiation (705)
parent nuclide (710)
positron (706)
radioactive decay (705)
radioactive nuclide (705)
transuranium element (712)

22-3

- Alpha particles, beta particles, and gamma rays have different penetrating abilities and therefore different shielding requirements.
- Film badges, Geiger-Müller counters, and scintillation detectors are used to detect radiation.
- Everyone is exposed to environmental background radiation. Exposure levels vary.
- Radioactive nuclides have many uses, including radioactive dating, radioactive tracers, cancer therapy and detection, and food preservation.

Vocabulary

film badge (714)
Geiger-Müller counter (714)
radioactive dating (715)
radioactive tracer (715)
rem (713)
roentgen (713)
scintillation counter (714)

22-4

- Nuclear fission and nuclear fusion are nuclear reactions in which the splitting and fusing of nuclei produce more stable nuclei and release enormous amounts of energy.
- Controlled fission reactions are used to produce energy and radioactive nuclides.
- Fusion reactions produce the sun's heat and light. If fusion reactions could be controlled, they would produce more usable energy per gram of fuel than fission reactions.

Vocabulary

chain reaction (717)
control rod (718)
critical mass (718)
moderator (718)
nuclear fission (717)
nuclear fusion (719)
nuclear power plant (718)
nuclear reactor (718)
shielding (718)

REVIEWING CONCEPTS

1. a. How does mass defect relate to nuclear binding energy?
   b. How does binding energy per nucleon vary with mass number?
   c. How does binding energy per nucleon affect the stability of a nucleus? (22-1)
2. Describe three ways in which the number of protons and the number of neutrons in a nucleus affect its stability. (22-1)
3. Where on the periodic table are most of the natural radioactive nuclides located? (22-2)
4. What changes in atomic number and mass number occur in each of the following types of radioactive decay?
   a. alpha emission
   b. beta emission
   c. positron emission
   d. electron capture (22-2)
5. Which types of radioactive decay cause the transmutation of a nuclide? (Hint: Review the definition of *transmutation*.) (22-2)
6. Explain how beta emission, positron emission, and electron capture affect the neutron-proton ratio. (22-2)
7. Write out the nuclear reactions that show particle conversion for the following types of radioactive decay:
   a. beta emission
   b. positron emission
   c. electron capture (22-2)
8. Compare and contrast electrons, beta particles, and positrons. (22-2)
9. a. What are gamma rays?
   b. How do scientists think they are produced? (22-2)
10. How does the half-life of a nuclide relate to its stability? (22-2)
11. List the three parent nuclides of the natural decay series. (22-2)
12. How are artificial radioactive isotopes produced? (22-2)
13. Neutrons are more effective than protons or alpha particles for bombarding atomic nuclei. Why? (22-2)
14. Why are all of the transuranium elements radioactive? (Hint: See Section 22-1.) (22-2)
15. Why can a radioactive material affect photographic film even though the film is well wrapped in black paper? (22-3)
16. How does the penetrating ability of gamma rays compare with that of alpha particles and beta particles? (22-3)
17. How does nuclear radiation damage biological tissue? (22-3)
18. Explain how film badges, Geiger-Müller counters, and scintillation detectors are used to detect radiation and measure radiation exposure.(22-3)
19. How is the age of an object containing a radioactive nuclide estimated? (22-3)
20. How is the fission of a uranium-235 nucleus induced? (22-4)
21. How does the fission of uranium-235 produce a chain reaction? (22-4)
22. Describe the purposes of the five major components of a nuclear power plant. (22-4)
23. Describe the reaction that produces the sun's energy. (22-4)
24. What is one problem that must be overcome before energy-producing controlled fusion reactions are a reality? (22-4)

PROBLEMS

Mass Defect

25. The mass of a $^{20}_{10}Ne$ atom is 19.992 44 amu. Calculate its mass defect.
26. The mass of a $^{7}_{3}Li$ atom is 7.016 00 amu. Calculate its mass defect.

Nuclear Binding Energy

27. Calculate the nuclear binding energy of one lithium-6 atom. The measured atomic mass of lithium-6 is 6.015 amu.
28. Calculate the binding energies of the following two nuclei, and indicate which releases more

energy when formed. You will need information from the periodic table and the text.
a. atomic mass 34.988011 amu, $^{35}_{19}K$
b. atomic mass 22.989767 amu, $^{23}_{11}Na$

29. a. What is the binding energy per nucleon for each nucleus in the previous problem?
b. Which nucleus is more stable?

30. The mass of $^{7}_{3}Li$ is 7.016 00 amu. Calculate the binding energy per nucleon for $^{7}_{3}Li$. Convert the mass in amu to binding energy in joules.

Neutron-Proton Ratio

31. Calculate the neutron-proton ratios for the following nuclides:
a. $^{12}_{6}C$
b. $^{3}_{1}H$
c. $^{206}_{82}Pb$
d. $^{134}_{50}Sn$

32. a. Locate the nuclides in problem 31 on the graph in Figure 22-2. Which ones lie on the band of stability?
b. For the stable nuclides, determine whether their neutron-proton ratio tends toward 1:1 or 1.5:1.

Nuclear Equations

33. Balance the following nuclear equations. (Hint: See Sample Problem 22-1.)
a. $^{43}_{19}K \longrightarrow {}^{43}_{20}Ca + \underline{\ ?\ }$
b. $^{233}_{92}U \longrightarrow {}^{229}_{90}Th + \underline{\ ?\ }$
c. $^{11}_{6}C + \underline{\ ?\ } \longrightarrow {}^{11}_{5}B$
d. $^{13}_{7}N \longrightarrow {}^{0}_{+1}\beta + \underline{\ ?\ }$

34. Write the nuclear equation for the release of an alpha particle by $^{210}_{84}Po$.

35. Write the nuclear equation for the release of a beta particle by $^{210}_{82}Pb$.

Half-Life

36. The half-life of plutonium-239 is 24 110 years. Of an original mass of 100.g, how much remains after 96 440 years? (Hint: See Sample Problem 22-2.)

37. The half-life of thorium-227 is 18.72 days. How many days are required for three-fourths of a given amount to decay?

38. The half-life of protactinium-234 is 6.69 hours. What fraction of a given amount remains after 26.76 hours?

39. How many milligrams remain of a 15.0 mg sample of radium-226 after 6396 years? The half-life of radium-226 is 1599 years.

MIXED REVIEW

40. Balance the following nuclear reactions;
a. $^{239}_{93}Np \longrightarrow {}^{0}_{-1}\beta + \underline{\ ?\ }$
b. $^{9}_{4}Be + {}^{4}_{2}He \longrightarrow \underline{\ ?\ }$
c. $^{32}_{15}P + \underline{\ ?\ } \longrightarrow {}^{33}_{15}P$
d. $^{236}_{92}U \longrightarrow {}^{94}_{36}Kr + \underline{\ ?\ } + 3{}^{1}_{0}n$

41. The energy released by the formation of a nucleus of $^{56}_{26}Fe$ is 7.89×10^{-11} J. Use Einstein's equation, $E = mc^2$, to determine how much mass is lost (in kilograms) in this process.

42. The parent nuclide of the thorium decay series is $^{232}_{90}Th$. The first four decays are as follows: alpha emission, beta emission, beta emission, and alpha emission. Write the nuclear equations for this series of emissions.

43. Calculate the binding energy per nucleon of $^{238}_{92}U$ in joules. The atomic mass of a $^{238}_{92}U$ nucleus is 238.050 784 amu.

44. After 4797 years, how much of an original 0.250 g of radium-226 remains? Its half-life is 1599 years.

45. Calculate the binding energy for one mole of deuterium atoms. The measured mass of deuterium is 2.0140 amu.

46. The half-life of radium-224 is 3.66 days. What was the original mass of radium-224 if 0.0500 g remains after 7.32 days?

47. Calculate the neutron-proton ratios for the following nuclides, and determine where they lie in relation to the band of stability.
a. $^{235}_{92}U$
b. $^{16}_{8}O$
c. $^{56}_{26}Fe$
d. $^{156}_{60}Nd$

CRITICAL THINKING

48. Why do we compare binding energy per nuclear particle of different nuclides instead of the total binding energy per nucleus?

49. Why is the constant rate of decay of radioactive nuclei so important in radioactive dating?

TECHNOLOGY & LEARNING

50. Graphing Calculator Calculating the Amount of Radioactive Material

The graphing calculator can be programmed to graph the relationship between the amount of radioactive material and elapsed time. Given the half-life of the radioactive material and the initial amount of the material in grams, you will graph the relationship between the amount of radioactive material and the elapsed time. Then, with an elapsed time, you will trace the graph to calculate the amount of radioactive material. Begin by programming the calculator to prompt for the half-life and initial amount of material, and then create a graph. Trace the curve to make the calculation.

A. Program the graph.

Press [Y=] and clear the graphs.

Keystrokes for naming the program: [PRGM] [▶] [▶] [ENTER]

Name the program: [R] [A] [D] [I] [O] [A] [C] [T] [ENTER]

Program keystrokes: [PRGM] [▶] [2] [ALPHA] [H] [2nd] [:] [PRGM] [▶] [2] [ALPHA] [I] [2nd] [:] [PRGM] [▶] [2] [ALPHA] [T] [2nd] [:] [ALPHA] ["] [ALPHA] [I] [×] [2nd] [e^x] [(] [(−)] [.] [6] [9] [3] [×] [X,T,θ] [÷] [ALPHA] [H] [)] [ALPHA] ["] [STO▶] [2nd] [Y-VARS] [1] [1] [2nd] [:] [0] [STO▶] [VARS] [1] [1] [2nd] [:] [ALPHA] [T] [STO▶] [VARS] [1] [2] [2nd] [:] [ALPHA] [T] [÷] [5] [STO▶] [VARS] [1] [3] [2nd] [:] [0] [STO▶] [VARS] [1] [4] [2nd] [:] [ALPHA] [I] [STO▶] [VARS] [1] [5] [2nd] [:] [ALPHA] [I] [÷] [5] [STO▶] [VARS] [1] [6] [2nd] [:] [PRGM] [▶] [4]

B. Check the display.

Screen display: Prompt H: Prompt I: Prompt T: "I*e^(-.693*X/H)" ->Y_1: 0->Xmin: T->Xmax: T/5->Xscl: 0->Ymin: I->Ymax: I/5->Yscl: DispGraph Press [2nd] [QUIT] to exit the program editor.

C. Run the program.

Turn off any statistics plots by pressing [2nd] [STAT PLOT] [4] [ENTER]. Press [PRGM]. Select RADIOACT and press [ENTER] [ENTER]. Now enter the half-life of the radioactive material (1600.) and press [ENTER]. Enter the initial amount of the material (5.00) and press [ENTER]. Enter the maximum elapsed time (8000.) and press [ENTER]. The calculator will provide a graph of the relationship.

D. Find the amount of radioactive material.

a. Press [TRACE] and [◀] or [▶] to move the tracer along the curve. Find the elapsed time of 5000. years by moving the tracer near x = 5000. The y value represents an approximate amount of radioactive material left. Press [ZOOM] [2] [ENTER] to zoom in on the point. Retrace to find a more accurate estimate for the amount left. Press [ZOOM] [▶] [1] to return to the original graphing window.

b. Use the current graph to determine the amount left after 1000. years. Now use the RADIOACT program to determine the amount of plutonium-236 left after 10.0 years, given the half-life of plutonium-236 is 2.85 years and the initial amount was 5.00 grams.

RESEARCH & WRITING

51. Investigate the history of the Manhattan Project.

52. Research the 1986 nuclear reactor accident at Chernobyl, Ukraine. What factors combined to cause the accident?

53. Find out about the various fusion-energy research projects that are being conducted in the United States and other parts of the world. How close are the researchers to finding an economical method of producing energy? What obstacles must still be overcome?

ALTERNATIVE ASSESSMENT

54. Your local grocery store may sell irradiated foods. Find out what stores in your area sell irradiated foods and determine whether you have any of these foods at home. What are the shelf lives of these foods before and after irradiation? Report your findings to the class.

Elements Handbook

100
90
80
70
60
50
40
30
200 mL
±5%
150
100
500 ml
PYREX®

GROUP 1
ALKALI METALS

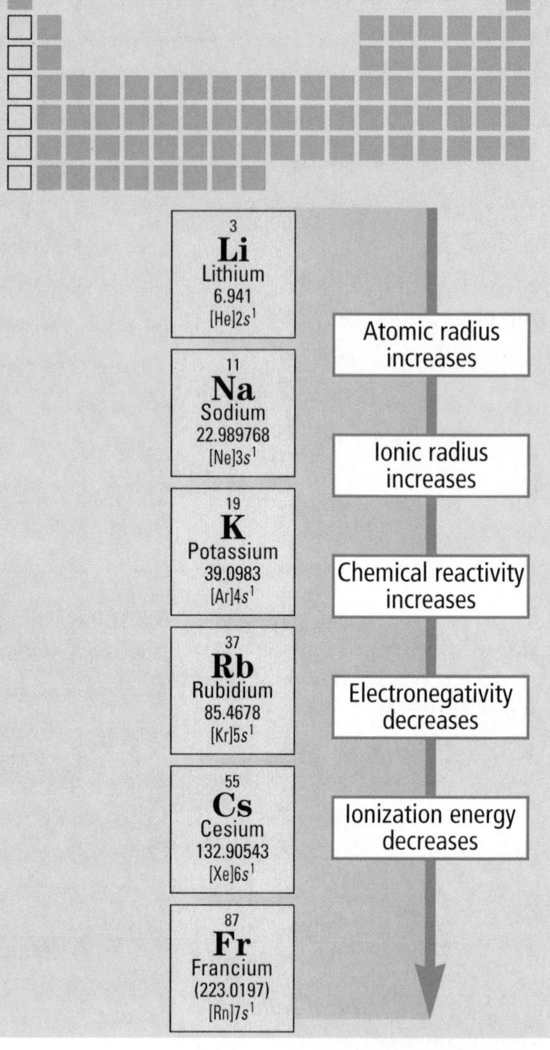

CHARACTERISTICS

- do not occur in nature as elements
- are reactive metals obtained by reducing the 1+ ions in their natural compounds
- are stored under kerosene or other hydrocarbon solvent because they react with water vapor or oxygen in air
- consist of atoms with one electron in the outermost energy level
- form colorless ions, each with a 1+ charge
- form ionic compounds
- form water-soluble bases
- are strong reducing agents
- consist of atoms that have low ionization energies
- are good conductors of electricity and heat
- are ductile, malleable, and soft enough to be cut with a knife
- have a silvery luster, low density, and low melting point

Lithium was discovered in 1817. It is found in most igneous rocks and is used in batteries as an anode because it has a very low reduction potential. Lithium is soft and is stored in oil or kerosene to prevent it from reacting with the air.

Sodium derives its name from the word soda. It was first isolated in 1807 from the electrolysis of caustic soda, NaOH. Sodium is soft enough to be cut with a knife. It is shiny until it reacts with oxygen, which causes the surface to lose its luster.

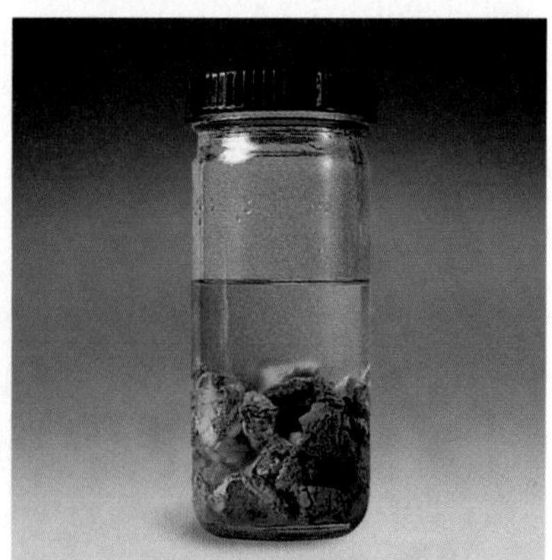

Potassium was first isolated in 1807 from the electrolysis of caustic potash, KOH.

COMMON REACTIONS

With Water and Acids to Form Bases and Hydrogen Gas

Example: $2Na(s) + 2H_2O(l) \longrightarrow 2NaOH(aq) + H_2(g)$
Li, K, Rb, and Cs also follow this pattern.

Example: $2Na(s) + 2HCl(aq) \longrightarrow 2NaCl(aq) + H_2(g)$
Li, K, Rb, and Cs also follow this pattern.

With Halogens to Form Salts

Example: $2Na(s) + F_2(g) \longrightarrow 2NaF(s)$
Li, K, Rb, and Cs also follow this pattern in reacting with F_2, Cl_2, Br_2, and I_2.

With Oxygen to Form Oxides, Peroxides, or Superoxides

Lithium forms an oxide.
$4Li(s) + O_2(g) \longrightarrow 2Li_2O(s)$

Sodium forms a peroxide.
$2Na(s) + O_2(g) \longrightarrow Na_2O_2(s)$

Alkali metals with higher molecular masses can form superoxides.
$K(s) + O_2(g) \longrightarrow KO_2(s)$
Rb and Cs also follow this pattern.

Alkali-Metal Oxides with Water to Form Bases

Oxides of Na, K, Rb, and Cs can be prepared indirectly. These basic anhydrides form hydroxides in water.

Example: $K_2O(s) + H_2O(l) \longrightarrow 2KOH(aq)$
Li, Na, Rb, and Cs also follow this pattern.

A small piece of potassium dropped into water will react explosively, releasing H_2 to form a strongly basic hydroxide solution. The heat of the reaction ignites the hydrogen gas that is produced.

Sodium reacts vigorously with chlorine to produce NaCl. Most salts of Group 1 metals are white crystalline compounds.

ANALYTICAL TEST

Alkali metals are easily detected by flame tests because each metal imparts a characteristic color to a flame.

When sodium and potassium are both present in a sample, the yellow color of the sodium masks the violet color of the potassium. The violet color can be seen only when the combined sodium-potassium flame is viewed through a cobalt-blue glass. The glass blocks the yellow flame of sodium and makes it possible to see the violet flame of potassium.

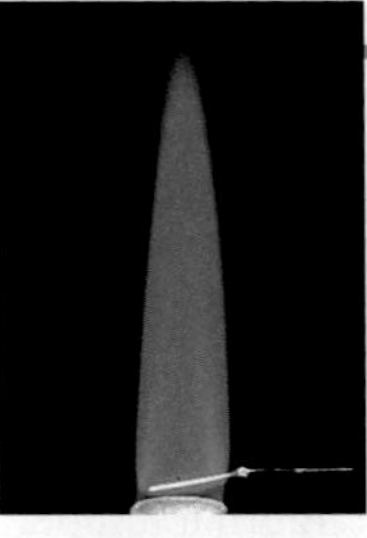

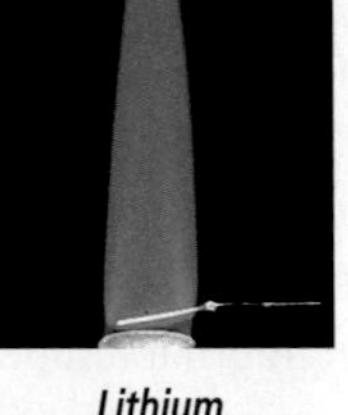

Lithium

Sodium

Potassium

Rubidium

Cesium

PROPERTIES OF THE GROUP 1 ELEMENTS

	Li	Na	K	Rb	Cs	Fr
Melting point (°C)	180.5	97.8	63.25	38.89	28.5	27
Boiling point (°C)	1342	882.9	760	691	668	—
Density (g/cm^3)	0.534	0.971	0.862	1.53	1.87	—
Ionization energy (kJ/mol)	520	496	419	403	376	—
Atomic radius (pm)	152	186	232	248	265	270
Ionic radius (pm)	76	102	138	152	167	180
Common oxidation number in compounds	+1	+1	+1	+1	+1	—
Crystal structure	bcc*	bcc	bcc	bcc	bcc	—
Hardness (Mohs' scale)	0.6	0.4	0.5	0.3	0.2	—

*body-centered cubic

APPLICATION *Technology*

Sodium Vapor Lighting

The flame test for sodium shows a bright line between 589.0 and 589.5 nm, which is the yellow range of the emission spectrum. Sodium can be vaporized at high temperatures in a sealed tube and made to give off light using two electrodes connected to a power source. Sodium vapor lighting is often used along highways and in parking lots because it provides good illumination while using less energy than other types of lighting.

Sodium vapor lighting comes in both low-pressure and high-pressure bulbs. Low-pressure lamps reach an internal temperature of 270°C to vaporize the sodium under a pressure of about 1 Pa. High-pressure lamps contain mercury and xenon in addition to sodium. These substances reach an internal temperature of 1100°C under a pressure of about 100 000 Pa. The high-pressure lamp provides a higher light intensity. The design of both types of lamps must take into account the high reactivity of sodium, which increases at high temperatures. Because ordinary glass will react with sodium at 250°C, a special sodium-resistant glass is used for low-pressure lamps. High-pressure lamps use an aluminum oxide material for the column containing the sodium, mercury, and xenon. Both types of lamps contain tungsten electrodes.

The light intensity per watt for sodium vapor lamps far exceeds that of fluorescent lamps, high-pressure mercury vapor lamps, tungsten halogen lamps, and incandescent bulbs.

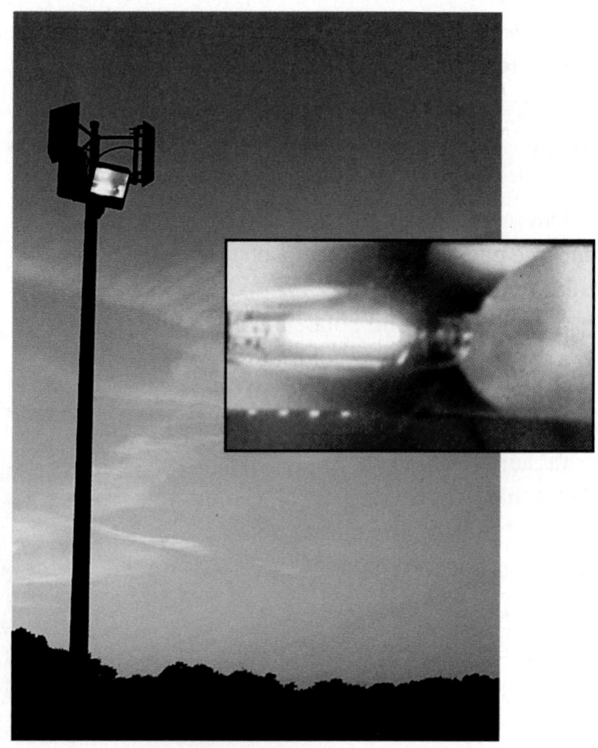

APPLICATION *Health*

Electrolyte Balance in the Body

The elements of Group 1 are important to a person's diet and body maintenance because they form ionic compounds, that are present in the body as solutions of the ions. All ions carry an electric charge, so they are electrolyte solutes. Two of the most important electrolyte solutes found in the body are K^+ and Na^+ ions. Both ions facilitate the transmission of nerve impulses and control the amount of water retained by cells.

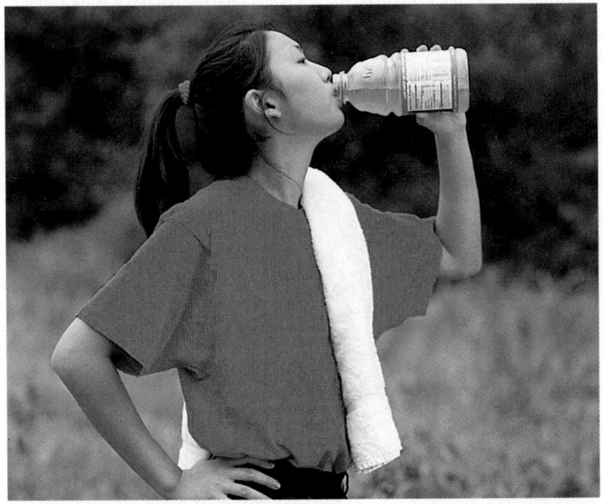

During situations where the body is losing water rapidly through intense sweating or diarrhea for a prolonged period (more than 5 hours), a sports drink can hydrate the body and restore electrolyte balance.

TABLE 1A Sodium-Potassium Composition of Body Fluids

Cation	Inside cells (mmol/L)	Outside cells or in plasma (mmol/L)
Na^+	12	145
K^+	140	4

The sodium and potassium ion concentrations of body fluids are shown in Table 1A. Sodium ions are found primarily in the fluid outside cells, while potassium ions are largely found in the fluid inside cells. Anions are present in the fluids to balance the electrical charge of the Na^+ and K^+ cations.

Abnormal electrolyte concentrations in blood serum can indicate the presence of disease. The ion concentrations that vary as a result of disease are Na^+, K^+, Cl^-, and HCO_3^-. Sodium ion concentration is a good indicator of the water balance between blood and tissue cells. Potassium ion level is an indicator of kidney failure. Chloride ion is the anion that balances the positive charge of the sodium ion in the fluid outside the cells. It also diffuses into a cell to maintain normal electrolyte balance when hydrogen carbonate ions diffuse out of the cell into the blood. Table 1B shows medical conditions associated with electrolyte imbalances.

TABLE 1B Electrolyte Imbalances

Electrolyte	Normal range (mmol/L)	Causes of imbalance: Excess	Causes of imbalance: Deficiency
Sodium, Na^+	135–145	hypernatremia (increased urine excretion; excess water loss)	hyponatremia (dehydration; diabetes-related low blood pH; vomiting; diarrhea)
Potassium, K^+	3.5–5.0	hyperkalemia (renal failure; low blood pH)	hypokalemia (gastrointestinal conditions)
Hydrogen carbonate, HCO_3^-	24–30	hypercapnia (high blood pH; hypoventilation)	hypocapnia (low blood pH; hyperventilation; dehydration)
Chloride, Cl^-	100–106	hyperchloremia (anemia; heart conditions; dehydration)	hypochloremia (acute infection; burns; hypoventilation)

Sodium-Potassium Pump in the Cell Membrane

The process of active transport allows a cell to maintain its proper electrolyte balance. To keep the ion concentrations at the proper levels shown in Table 1B, a sodium-potassium pump embedded in the cell membrane shuttles sodium ions out of the cell across the cell membrane. A model for the action of the sodium-potassium pump is shown below.

Nerve Impulses and Ion Concentration

An uneven distribution of Na^+ and K^+ ions across nerve cell membranes is essential for the normal operation of the nervous system. The uneven distribution of ions creates a voltage across nerve cell membranes. When a nerve cell is stimulated, sodium ions diffuse into the cell from the surrounding fluid, raising voltage across the nerve cell membrane from –70 mV to nearly +60 mV. Potassium ions then diffuse out of the cell into the surrounding fluid, restoring the voltage across the nerve cell membrane to –70 mV. This voltage fluctuation initiates the transmission of a nerve impulse. The amount of Na^+ inside the cell increases slightly, and the amount of K^+ outside the cell also increases. But the sodium-potassium pump will restore these ions to their proper concentrations.

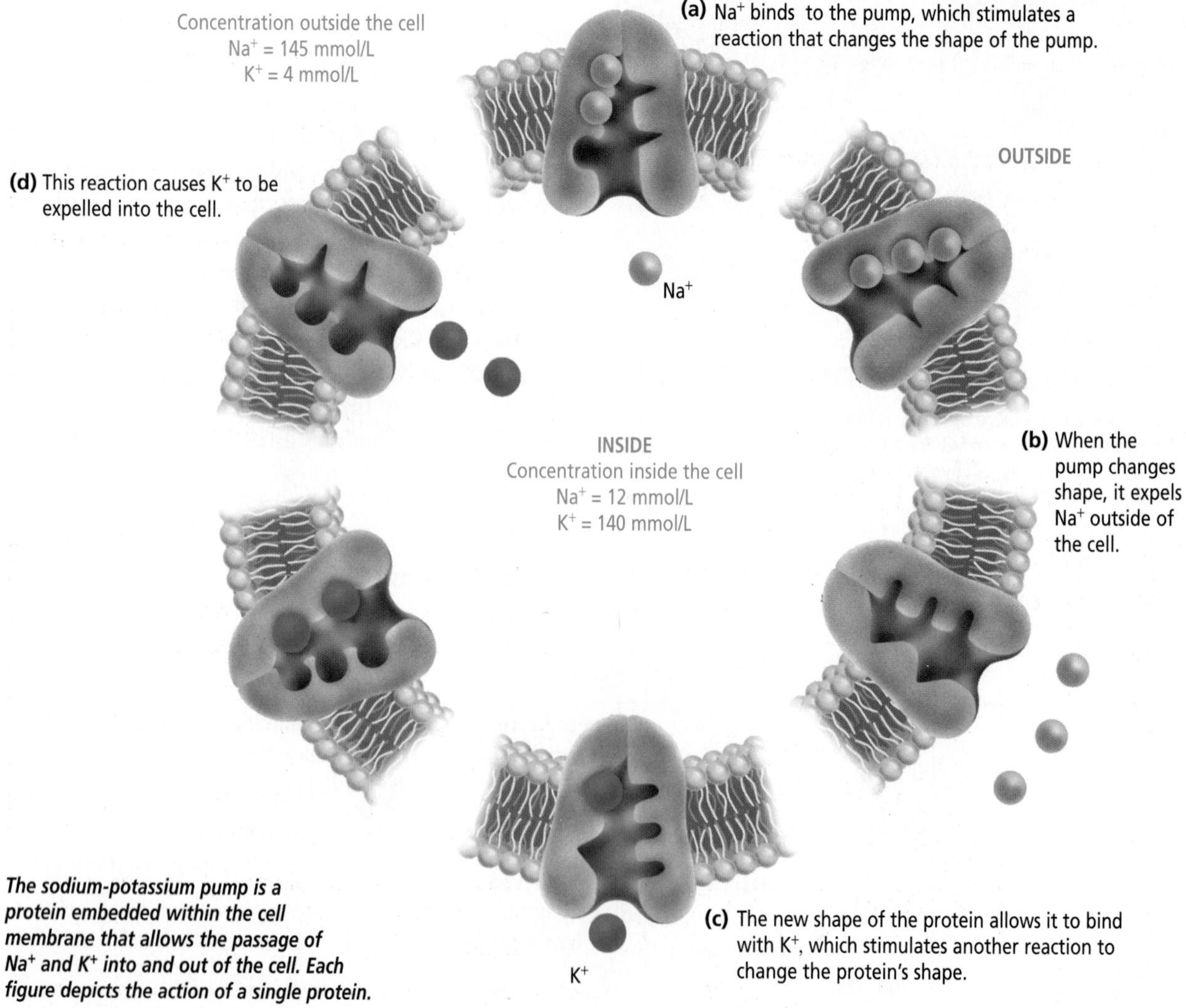

Concentration outside the cell
Na+ = 145 mmol/L
K+ = 4 mmol/L
(a) Na+ binds to the pump, which stimulates a reaction that changes the shape of the pump.
OUTSIDE
(d) This reaction causes K+ to be expelled into the cell.
Na+
INSIDE
Concentration inside the cell
Na+ = 12 mmol/L
K+ = 140 mmol/L
(b) When the pump changes shape, it expels Na+ outside of the cell.
(c) The new shape of the protein allows it to bind with K+, which stimulates another reaction to change the protein's shape.
K+

The sodium-potassium pump is a protein embedded within the cell membrane that allows the passage of Na^+ and K^+ into and out of the cell. Each figure depicts the action of a single protein.

What's your sodium IQ?

Though sodium is an important mineral in your body, a diet that is high in sodium is one of several factors linked to high blood pressure, also know as hypertension. High Na^+ levels cause water retention, which results in increased blood pressure. Sodium is not the direct cause of all hypertension, but reducing sodium levels in the diet can affect individuals with a condition known as salt-sensitive hypertension. Therefore, the Dietary Guidelines for Americans recommend consuming salt and sodium in moderation. Test your knowledge about sodium in foods with the questions below.

1. Which of the following condiments do you think has the lowest salt content?
a. mustard c. catsup e. vinegar
b. steak sauce d. pickles

2. One-fourth of a teaspoon of salt contains about _____ of sodium.
a. 10 mg c. 500 mg e. 1 kg
b. 100 g d. 500 g

Nutrition Facts
Serving Size ¾ cup (30g)
Servings Per Container About 14

Amount Per Serving	Corn Crunch	with ½ cup skim milk
Calories	120	160
Calories from Fat	15	20
	% **Daily Value****	
Total Fat 2g*	**3**%	**3**%
Saturated Fat 0g	[illegible]	**0**%
Cholesterol 0mg	**0**%	[illegible]
Sodium 160mg	**7**%	**9**%
Potassium 65mg	[illegible]	[illegible]
Total Carbohydrate 25g	**8**%	**10**%
Dietary Fiber 3g		
Sugars 3g		
Other Carbohydrate 11g		
Protein 2g		

*Amount in Cereal. A serving of cereal plus skim milk provides 2g fat, less 5mg cholesterol, 220mg sodium, 270mg potassium, 31g carbohydrate (19g sugars) and 6g protein.

**Percent Daily Values are based on a 2,000 calorie diet. Your daily values may be higher or lower depending on your calorie needs:

	Calories	2,000	2,500
Total Fat	Less than	65g	80g
Sat Fat	Less than	20g	25g
Cholesterol	Less than	300mg	300mg
Sodium	Less than	2,400mg	2,400mg
Potassium		3,500mg	3,500mg
Total Carbohydrate		300g	375g
Dietary Fiber		25g	30g

3. According to FDA regulations for food product labels, a food labeled *salt-free* must contain less than _____ mg of sodium ion per serving.
a. 100 c. 0.001 e. 0.00005
b. 5 d. 0.005

4. The Nutrition Facts label for a particular food reads "Sodium 15 mg." This is the amount of sodium ion per _____.
a. package c. serving e. RDA
b. teaspoon d. ounce

5. The recommended average daily intake of sodium ion for adults is 2400 mg. For a low-sodium diet the intake should be _____.
a. 200 mg c. 750 mg e. 150 mg
b. 2000 mg d. 500 mg

6. Each of the following ingredients can be found in the ingredients lists for some common food products. Which ones indicate that the product contains sodium?
a. trisodium phosphate d. sodium sulfate
b. sodium bicarbonate e. MSG
c. sodium benzoate f. baking soda

7. Which of the following spices is NOT a salt substitute?
a. caraway seeds d. ginger
b. dill e. onion salt
c. mace

8. Most salt in the diet comes from salting foods too heavily at the dinner table.
a. true b. false

9. Which of the following foods are high in sodium?
a. potato chips c. doughnuts e. figs
b. pizza d. banana

10. Your body requires about 200 mg of sodium ion, or 500 mg of salt per day. Why do these numbers differ?

Answers 1. e; 2. c; 3. b; 4. c; 5. c; 6. all of them; 7. e; 8. b, processed foods can contain very high levels of sodium; 9. a, b, c; 10. Salt is not pure sodium.

GROUP 2
ALKALINE EARTH METALS

CHARACTERISTICS

- do not occur naturally in their elemental state
- occur most commonly as the carbonates, phosphates, silicates, and sulfates
- occur naturally as compounds that are either insoluble or only slightly soluble in water
- consist of atoms that contain two electrons in their outermost energy level
- consist of atoms that tend to lose two electrons per atom, forming ions with a 2+ charge
- are less reactive than alkali metals
- form ionic compounds primarily
- react with water to form bases and hydrogen gas
- are good conductors of heat and electricity
- are ductile and malleable
- have a silvery luster
- include the naturally radioactive element radium

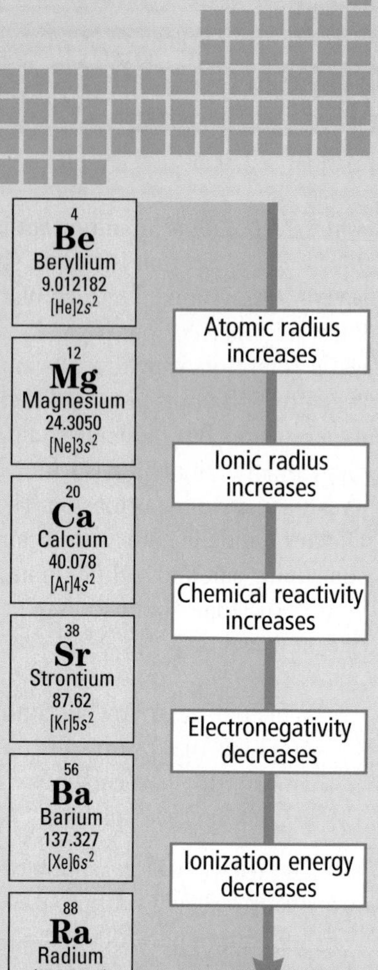

Calcium carbonate is a major component of marble.

Beryllium is found in the mineral compound beryl. Beryl crystals include the dark green emerald and the blue-green aquamarine. The colors of these gems come from other metal impurities.

The mineral dolomite, $CaCO_3{\cdot}MgCO_3$, is a natural source of both calcium and magnesium.

COMMON REACTIONS

With Water to Form Bases and Hydrogen Gas

Example: $Mg(s) + 2H_2O(l) \longrightarrow Mg(OH)_2(aq) + H_2(g)$

Ca, Sr, and Ba also follow this pattern.

Calcium reacts with water to form hydrogen gas.

With Acids to Form Salts and Hydrogen Gas

Example: $Mg(s) + 2HCl(aq) \longrightarrow MgCl_2(aq) + H_2(g)$

Be, Ca, Sr, and Ba also follow this pattern.

With Halogens to Form Salts

Example: $Mg(s) + F_2(g) \longrightarrow MgF_2(s)$

Ca, Sr, and Ba also follow this pattern in reacting with F_2, Cl_2, Br_2, and I_2.

With Oxygen to Form Oxides or Peroxides

Magnesium forms an oxide.

$2Mg(s) + O_2(g) \longrightarrow 2MgO(s)$

Be and Ca also follow this pattern.

Strontium forms a peroxide.

$Sr(s) + O_2(g) \longrightarrow SrO_2(s)$

Ba also reacts in this way.

With Hydrogen to Form Hydrides

Example: $Mg(s) + H_2(g) \longrightarrow MgH_2(s)$

Ca, Sr, and Ba also follow this pattern.

With Nitrogen to Form Nitrides

Example: $3Mg(s) + N_2(g) \longrightarrow Mg_3N_2(s)$

Be and Ca also follow this pattern.

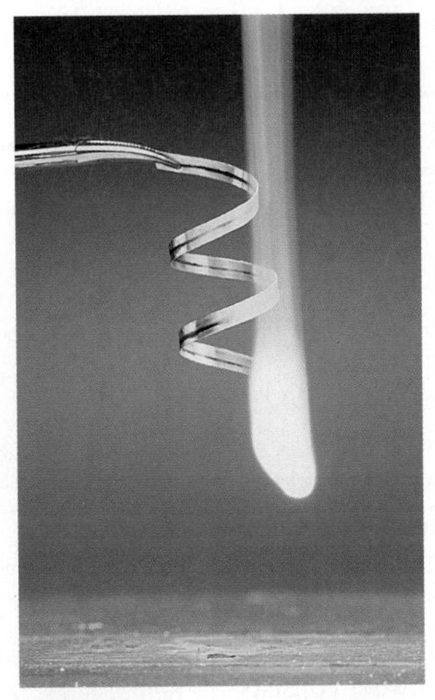

Magnesium burns in air to form MgO and Mg_3N_2.

Magnesium reacts with HCl to produce $MgCl_2(aq)$.

ANALYTICAL TEST

Flame tests can be used to identify three of the alkaline earth elements. The colors of both calcium and strontium can be masked by the presence of barium, which produces a green flame.

Calcium

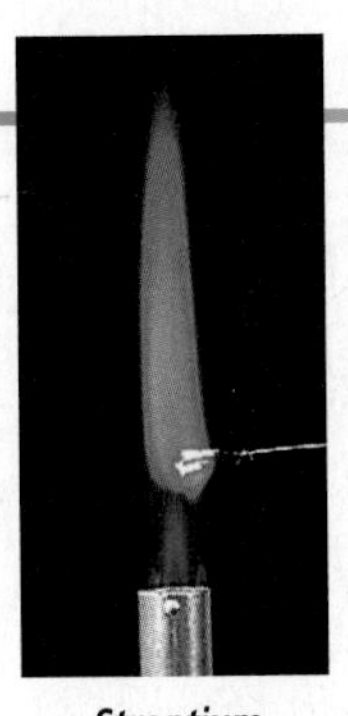

Strontium

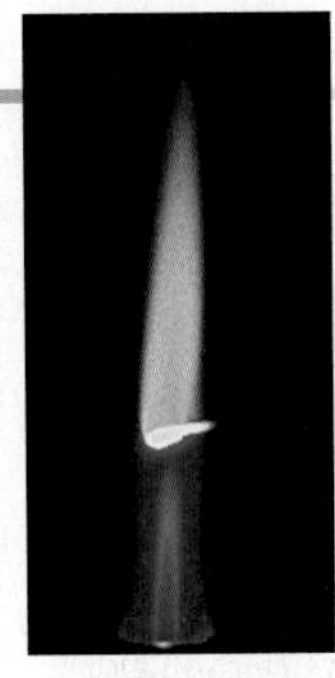

Barium

PROPERTIES OF THE GROUP 2 ELEMENTS

	Be	Mg	Ca	Sr	Ba	Ra
Melting point (°C)	1287 ± 5	649	839 ± 2	769	725	700
Boiling point (°C)	2467	1107	1484	1384	1640	1737
Density (g/cm^3)	1.85	1.74	1.54	2.6	3.51	—
Ionization energy (kJ/mol)	900	738	590	550	503	509
Atomic radius (pm)	111.3	160	197	215	217.3	220
Ionic radius (pm)	45	72	100	118	136	148
Common oxidation number in compounds	+2	+2	+2	+2	+2	+2
Crystal structure	hcp*	hcp	fcc**	fcc	bcc	bcc
Hardness (Mohs' scale)	4.0	2.0	1.5	1.8	1.5	—

*hexagonal close-packed **face-centered cubic

APPLICATION *Technology*

Fireworks

Fireworks are made from pyrotechnics—chemical substances that produce light and smoke when they are ignited. Pyrotechnics are also used in flares, smoke bombs, explosives, and matches. An aerial fireworks device is a rocket made of a cylinder, chemicals inside the cylinder, and fuses attached to the cylinder. The illustration on the right shows how the device works. The lift charge at the bottom of the cylinder consists of a small amount of black gunpowder. When the side fuse ignites the gunpowder, it explodes like a small bomb. The gunpowder consists of potassium nitrate, charcoal, and sulfur. When these three chemicals react with one another, they produce gases. In this case, the gases produced are carbon monoxide, carbon dioxide, sulfur dioxide, and nitrogen monoxide. These hot gases expand very rapidly, providing the thrust that lifts the rocket into the sky.

About the time the shell reaches its maximum altitude and minimum speed, the time fuse ignites the chemicals contained in the cylinder. The chemicals inside the cylinder determine the color of the burst.

Time-delay fuses activate the reactions in the other chambers
Ignition fuse activates the reaction in the bottom chamber
Red star bursts
Blue star bursts
Flash and sound mixture
Black powder propellant

The cylinder of a multiple-burst rocket contains separate reaction chambers connected by fuses. A common fuse ignites the propellant and the time-delay fuse in the first reaction chamber.

Chemical Composition and Color

One of the characteristics of fireworks that we enjoy most is their variety of rich colors. These colors are created in much the same way as the colors produced during a flame test. In a fireworks device, the chloride salt is heated to a high temperature, causing the excited atoms to give off a burst of light. The color of light produced depends on the metal used. The decomposition of barium chloride, $BaCl_2$, for example, produces a burst of green light, whereas strontium chloride, $SrCl_2$, releases red light.

People who design fireworks combine artistry with a technical knowledge of chemical properties. They have found ways to combine different colors within a single cylinder and to make parts of the cylinder explode at different times. Fireworks designers have a technical knowledge of projectile motion that is used to determine the height, direction, and angle at which a fireworks device will explode to produce a fan, fountain, flower, stream, comet, spider, star, or other shape.

Strontium and the Visible Spectrum

When heated, some metallic elements and their compounds emit light at specific wavelengths that are characteristic of the element or compound. Visible light includes wavelengths between about 380 and 780 nanometers. The figure below shows the emission spectrum for strontium. When heated, strontium gives off the maximum amount of visible light at about 700 nanometers, which falls in the red-light region of the visible spectrum.

The emission spectrum for strontium shows strong bands in the red region of the visible light spectrum.

4000 4500 5000 5500 6000 6500 7000 7500

Flares

Flares operate on a chemical principle that is different from that of fireworks. A typical flare consists of finely divided magnesium metal and an oxidizing agent. When the flare is ignited, the oxidizing agent reacts with the magnesium metal to produce magnesium oxide. This reaction releases so much energy that it produces a glow like that of the filaments in a light bulb. The brilliant white light produced by the flare is caused by billions of tiny particles of magnesium that glow when they react. If slightly larger particles of magnesium metal are used in the flare, the system glows for a longer period of time because the particles' reaction with the oxidizing agent is slower.

A colored flare can be thought of as a combination of a white flare and a chemical that produces colored light when burned. For example, a red flare can be made from magnesium metal, an oxidizing agent, and a compound of strontium. When the flare is ignited, the oxidizing agent and magnesium metal react, heating the magnesium to white-hot temperatures. The heat from this reaction causes the strontium compound to give off its characteristic red color.

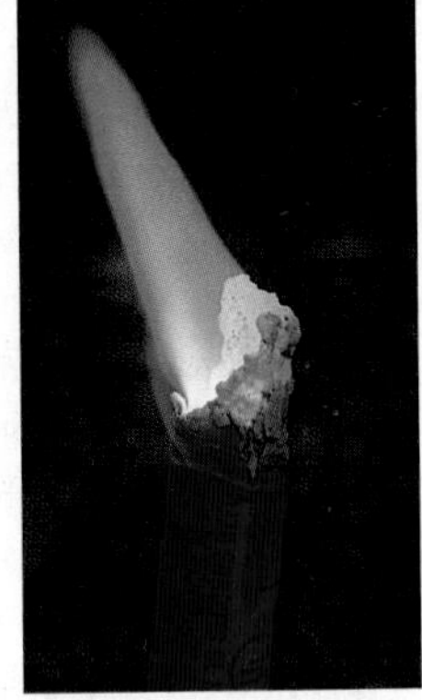

A flare is made up of billions of reacting magnesium particles.

For safety reasons, some fireworks manufacturers store their products in metal sheds separated by sand banks. Also, people who work with fireworks are advised to wear cotton clothing because cotton is less likely than other fabrics to develop a static charge, which can cause a spark and accidentally ignite fireworks.

APPLICATION *Health*

Calcium: An Essential Mineral in the Diet

Calcium is the most abundant mineral in the body. It is the mineral that makes up a good portion of the teeth and the bone mass of the body. A small percentage of calcium in the body is used in the reactions by which cells communicate and in the regulation of certain body processes. Calcium is so important to normal body functioning that if the calcium level of the blood falls far below normal, hormones signal the release of calcium from bone and signal the gastrointestinal tract to absorb more calcium during the digestion process.

A prolonged diet that is low in calcium is linked to a disease characterized by a decrease in bone mass, a condition called osteoporosis. Reduced bone mass results in brittle bones that fracture easily. Osteoporosis generally occurs later in life and is more prevalent in females. However, because you achieve peak bone mass during the late teens or early twenties, it is critical that your diet meet the recommended requirements to increase your peak bone mass. The recommended dietary intake for calcium is 1000 mg per day. Maintaining that level in the diet along with regular exercise through adulthood are thought to reduce the rate of bone loss later in life. Excess calcium in the diet (consuming more than 2500 mg daily) can interfere with the absorption of other minerals.

Dairy products are generally good sources of calcium.

Magnesium: An Essential Mineral in the Diet

Though magnesium has several functions in the body, one of the more important functions is its role in the absorption of calcium by cells. Magnesium, like sodium and potassium, is involved in the transmission of nerve impulses. Like calcium, magnesium is a component of bone.

A major source of magnesium in the diet is plants. Magnesium is the central atom in the green plant pigment chlorophyll. The structure of chlorophyll in plants is somewhat similar to the structure of heme—the oxygen-carrying molecule in animals. (See page 758 for the heme structure.)

TABLE 2A Good Sources of Calcium in the Diet

Food	Serving size	Calcium present (mg)
Broccoli	6.3 oz	82
Cheddar cheese	1 oz	204
Cheese pizza, frozen	pizza for one	375
Milk, low-fat 1%	8 oz	300
Tofu, regular	4 oz	130
Vegetable pizza, frozen	pizza for one	500
Yogurt, low-fat	8 oz	415
Yogurt, plain whole milk	8 oz	274

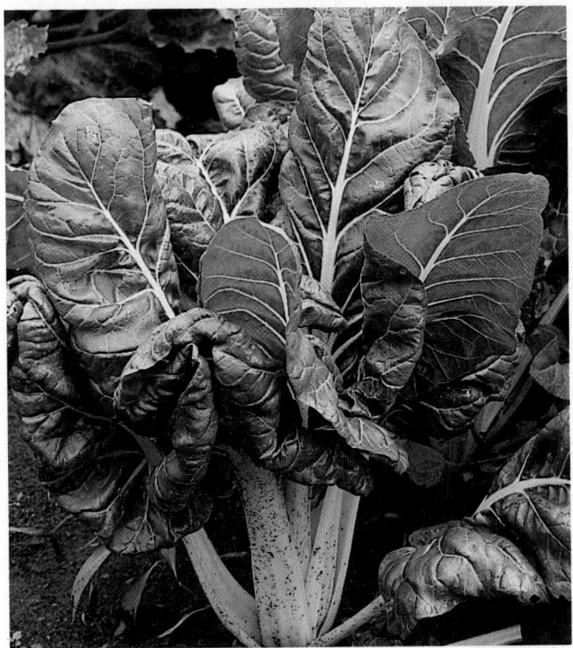

Spinach is a good source of magnesium. Magnesium is the central atom in the green plant pigment chlorophyll. The chlorophyll structure is shown on the right.

The recommended dietary intake of magnesium is 400 mg per day. This is equivalent to just 4 oz of bran cereal. Because magnesium levels are easily maintained by a normal diet, it is unusual for anyone to have a magnesium deficiency. Most magnesium deficiencies are the result of factors that decrease magnesium absorption. People with gastrointestinal disorders, alcohol abusers, and the critically ill are most likely to have these types of absorption problems.

Excess magnesium in the diet is excreted by the kidneys, so there are no cumulative toxic effects.

CH_2 CH_2 CH_3 CH_3 CH_2CH_3 N N Mg N N CH_3 CH_3 H CH_3 CH_3 CH_3 CH_3 CH_3 O C O CH_2 CH_2 H H C O CH_3O O

TABLE 2B Good Sources of Magnesium in the Diet

Food	Serving size	Magnesium present (mg)
Barley, raw	1 cup	244
Beef, broiled sirloin	4 oz	36
Cabbage, raw	1 med. head	134
Cashews, dry-roasted	1 oz	74
Chicken, roasted breast	4 oz	31
Lima beans, boiled	1/2 cup	63
Oatmeal	1 oz	39
Potato, baked	7.1 oz	115
Prunes, dried	4 oz	51
Rice bran	8 oz	648
Salmon, canned	4 oz	39
Spinach, raw	10 oz	161

GROUPS 3–12
TRANSITION METALS

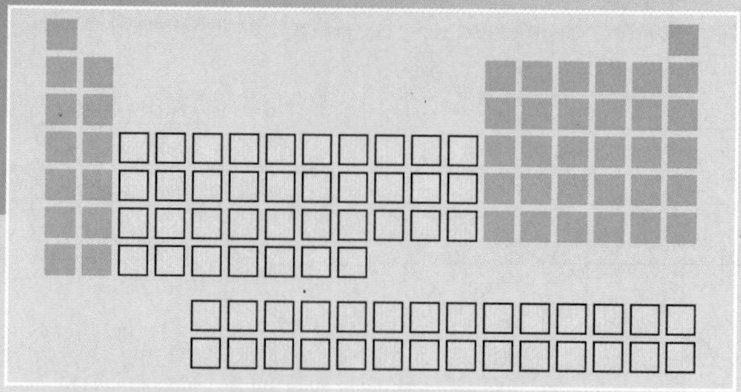

CHARACTERISTICS

- consist of metals in Groups 3 through 12
- contain one or two electrons in their outermost energy level
- are usually harder and more brittle than metals in Groups 1 and 2
- have higher melting and boiling points than metals in Groups 1 and 2
- are good conductors of heat and electricity
- are malleable and ductile
- have a silvery luster, except copper and gold
- include radioactive elements with numbers 89 through 109
- include mercury, the only liquid metal at room temperature
- have chemical properties that differ from each other
- tend to have two or more common oxidation states
- often form colored compounds
- may form complex ions

Iron ore is obtained from surface mines. Hematite, Fe_2O_3, is the most common iron ore.

Copper ores are also obtained from surface mines. Copper ore is shown here.

Gold, silver, platinum, palladium, iridium, rhodium, ruthenium, and osmium are sometimes referred to as the noble metals because they are not very reactive. These inert metals are found in coins, jewelry, and metal sculptures.

COMMON REACTIONS

Because this region of the periodic table is so large, you would expect great variety in the types of reaction characteristics of transition metals. For example, copper oxidizes in air to form the green patina you see on the Statue of Liberty. Copper reacts with concentrated HNO_3 but not with dilute HNO_3. Zinc, on the other hand, reacts readily with dilute HCl. Iron oxidizes in air to form rust, but chromium is generally unreactive in air. Some common reactions for transition elements are shown by the following.

May form two or more different ions

Example: $Fe(s) \longrightarrow Fe^{2+}(aq) + 2e^-$

Example: $Fe(s) \longrightarrow Fe^{3+}(aq) + 3e^-$

May react with oxygen to form oxides

Example: $4Cr(s) + 3O_2(g) \longrightarrow 2Cr_2O_3(s)$

Example: $2Cu(s) + O_2(g) \longrightarrow 2CuO(s)$

May react with halogens to form halides

Example: $Ni(s) + Cl_2(g) \longrightarrow NiCl_2(s)$

May form complex ions

See examples in the lower right.

Copper reacts with oxygen in air.

Copper reacts with concentrated nitric acid.

Zinc reacts with dilute hydrochloric acid.

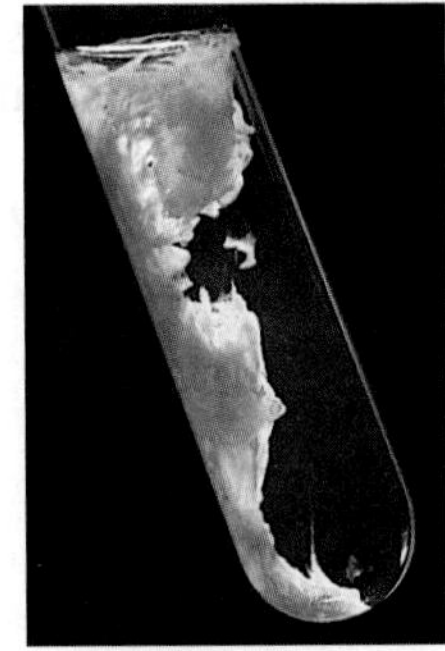

Soluble iron(III) salts form insoluble $Fe(OH)_3$ when they are reacted with a hydroxide base.

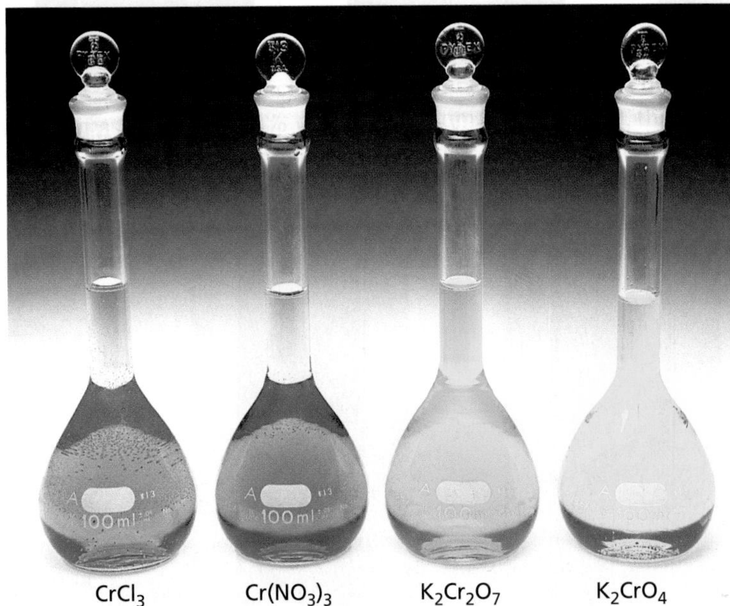

Chromium has several common oxidation states, represented here by aqueous solutions of its compounds. The violet and green solutions contain chromium in the +3 state, and the yellow and orange solutions contain chromium in the +6 oxidation state.

Complex ions belong to a class of compounds called coordination compounds. Coordination compounds show great variety in colors. Several transition-metal coordination compounds are shown.

TABLE 3A Transition Metals and Gemstone Colors

Gemstone	Color	Element
Amethyst	purple	iron
Aquamarine	blue	iron
Emerald	green	iron/titanium
Garnet	red	iron
Peridot	yellow-green	iron
Ruby	red	chromium
Sapphire	blue	iron/titanium
Spinel	colorless to red to black	varies
Turquoise	blue	copper

Verneuil's method, although somewhat modified, is still the one most widely used today for the manufacture of colored gemstones. When magnesium oxide is substituted for aluminum oxide, a colorless spinel-like product is formed. The addition of various transition metals then adds a tint to the spinel that results in the formation of synthetic emerald, aquamarine, tourmaline, or other gemstones. Synthetic gems look very much like their natural counterparts.

Synthetic sapphire

Synthetic ruby

APPLICATION *Technology*

Alloys

An alloy is a mixture of a metal and one or more other elements. In most cases, the second component of the mixture is also a metal.

Alloys are desirable because mixtures of elements usually have properties different from and often superior to the properties of individual metals. For example, many alloys that contain iron are harder, stronger, and more resistant to oxidation than iron itself.

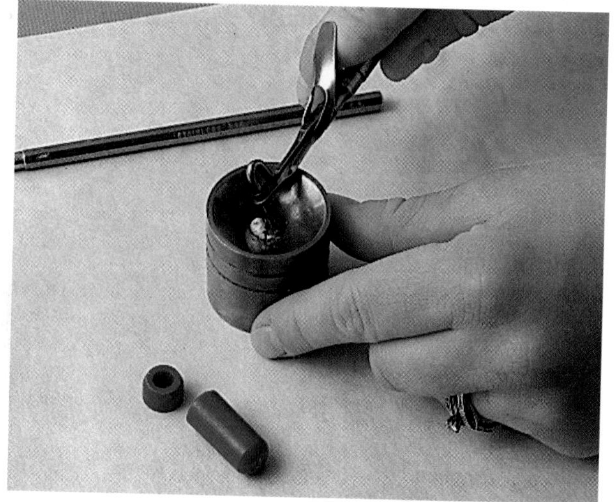

Amalgams are alloys that contain mercury. They are soft and pliable when first produced, but later become solid and hard. Dental fillings were once made of an amalgam of mercury and silver. Concerns about the possible toxicity of mercury led to the development of other filling materials.

Cast Iron and Steel

The term *steel* applies to any alloy consisting of iron and less than 1.5% carbon, and often other elements. When iron ore is treated with carbon in the form of coke to extract pure iron metal, some of the carbon also reacts with the iron to produce a form of iron carbide known as cementite. The reaction can be represented by the following equation.

$$3Fe + C \longrightarrow Fe_3C$$

Cast iron is a mixture that consists of some pure iron, known as ferrite, some cementite, and some carbon atoms trapped within the crystalline structure of the iron and cementite. The rate at which cast iron is cooled changes the proportion of these three components. If the cast iron is cooled slowly, the ferrite and cementite tend to separate from each other, forming a banded product that is tough but not very hard. However, if the cast iron is cooled quickly, the components of the original mixture cannot separate from each other, forming a product that is both tough and hard.

Stainless steel, which is hard and resists corrosion, is made of iron and chromium (12–30%). The properties of stainless steel make it a suitable alloy for making cutlery and utensils.

TABLE 3B Composition and Uses of Some Alloys

Name of alloy	Composition	Uses
Brass	copper with up to 50% zinc	inexpensive jewelry; hose nozzles and couplings; piping; stamping dies
Bronze	copper with up to 12% tin	coins and medals; heavy gears; tools; electrical hardware
Coin metal	copper: 75% nickel: 25%	United States coins
Duralumin	aluminum: 95% copper: 4% magnesium: 0.5% manganese: <1%	aircraft, boats, railroad cars, and machinery because of its high strength and resistance to corrosion
Nichrome	nickel: 80–85% chromium: 15–20%	heating elements in toasters, electric heaters, etc.
Phosphor bronze	bronze with a small amount of phosphorus	springs, electrical springs, boat propellers
Solder	lead: 50%, tin: 50% or tin: 98%, silver: 2%	joining two metals to each other joining copper pipes
Sterling silver	silver: 92.5% copper: 7.5%	jewelry, art objects, flatware
Type metal	lead: 75–95% antimony: 2–18% tin: trace	used to make type for printing because it expands as it cools

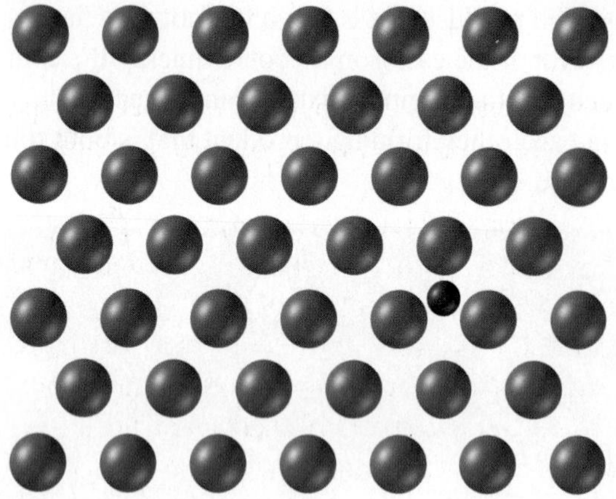

Interstitial crystal
A smaller atom or ion fits into a small space between particles in the array.

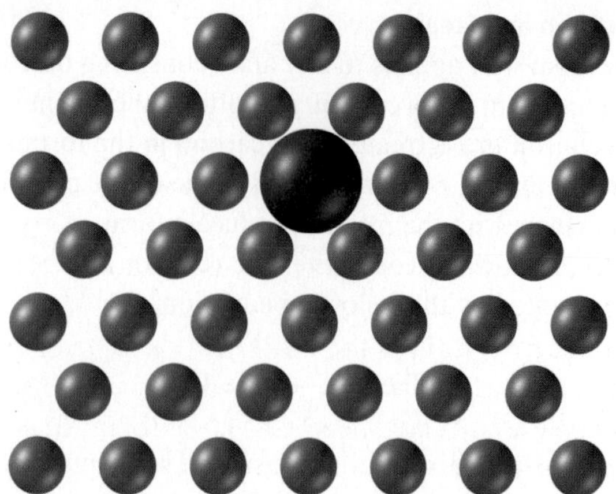

Substitutional crystal
A larger atom or ion is substituted for a particle in the array.

Structures and Preparation of Alloys

Alloys generally crystallize in one of two ways, depending on relative sizes of atoms. If the atoms of one of the metals present are small enough to fit into the spaces between the atoms of the second metal, they form an alloy with an interstitial structure (*inter* means "between," and *stitial* means "spaces"). If atoms of the two metals are of similar size or if one is larger, the atoms of one metal can substitute for the atoms of the second metal in its crystalline structure. Such alloys are substitutional alloys. Models for both types of crystals are shown above.

Techniques for making alloys depend on the metals used in the mixture. In some cases, the two metals can simply be melted together to form a mixture. The composition of the mixture often varies within a range, evidence that the final product is indeed a mixture and not a compound. In other cases, one metal may be melted first and the second dissolved in it. Brass is prepared in this way. If copper and zinc were heated together to a high temperature, zinc (bp 907°C) would evaporate before copper (mp 1084°C) melted. Therefore, the copper is melted first, and the zinc is added to it.

Brass has a high luster and resembles gold when cleaned and polished. A brass object can be coated with a varnish to prevent reactions of the alloy with air and water.

Sterling silver is more widely used than pure silver because it is stronger and more durable.

APPLICATION *The Environment*

Mercury Poisoning

Mercury is the only metal that is liquid at room temperature. It has a very high density compared with most other common transition metals and has a very high surface tension and high vapor pressure. Mercury and many of its compounds must be handled with extreme care because they are highly toxic. Mercury spills are especially hazardous because the droplets scatter easily and are often undetected during cleanup. These droplets release toxic vapors into the air.

Overexposure to mercury vapor or its compounds can occur by absorption through the skin, respiratory tract, or digestive tract. Mercury is a cumulative poison, which means that its concentration in the body increases as exposure increases.

Mercury that enters the body damages the kidneys, heart, and brain. The action of mercury on the brain affects the nervous system. Symptoms of mercury poisoning include numbness, tunnel vision, garbled speech, bleeding and inflammation of the gums, muscle spasms, anemia, and emotional disorders, such as depression, irritability, and personality changes. The saying "mad as a hatter" probably came about because of mercury poisoning. Mercury salts were once routinely used to process the felt used in hats. Hatters often displayed the nerve and mental impairments associated with overexposure to mercury.

Methylmercury in Freshwater Ecosystems

Mercury, Hg, can be found in our environment and in our food supply. Fortunately, the body has some protective mechanisms to deal with trace amounts of mercury. However, levels of mercury and of an organic mercury compound, methylmercury $(CH_3)_2Hg$, are increasing in the environment due to mercury mining operations and runoff from the application of pesticides and fungicides.

Mercury is easily converted to methylmercury by bacteria. Methylmercury is more readily absorbed by cells than mercury itself. As a result, methylmercury accumulates in the food chain as shown in the diagram below. A serious incident of methylmercury poisoning occurred in Japan in the 1950s. People living in Minamata, Japan, were exposed to high levels of methylmercury from eating shellfish.

In the United States there is concern about mercury levels in fish from some freshwater lakes. Though environmental regulations have reduced the level of lake pollutants, it takes time to see a reduction in the concentration of an accumulated poison.

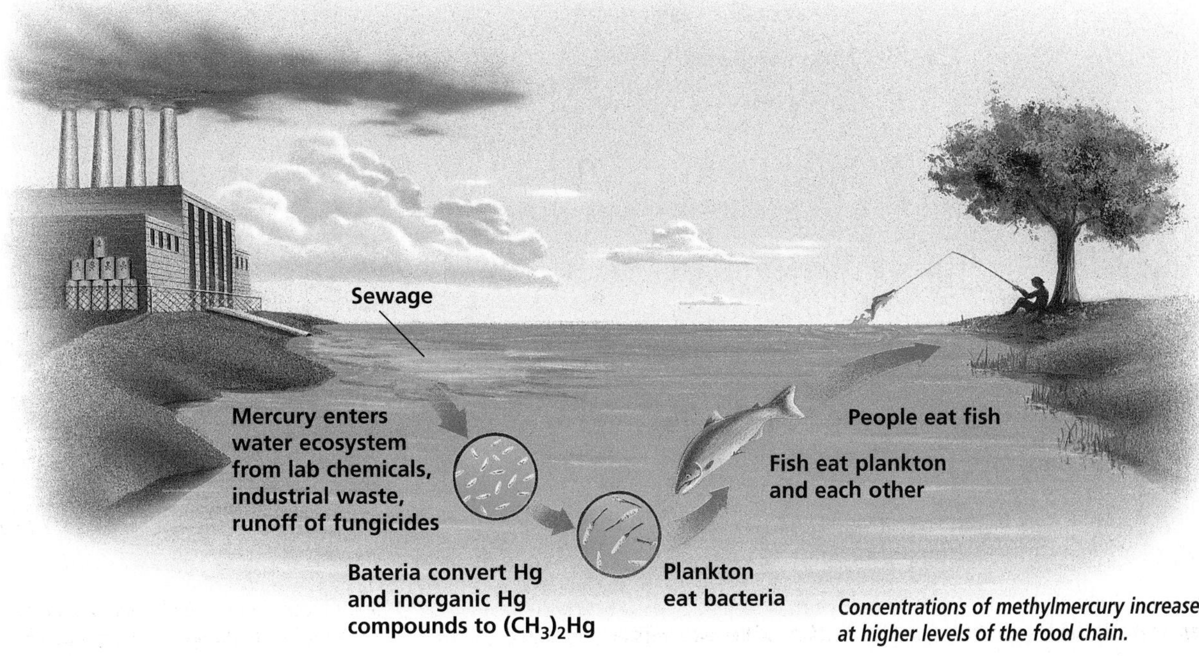
Sewage
Mercury enters water ecosystem from lab chemicals, industrial waste, runoff of fungicides
Bateria convert Hg and inorganic Hg compounds to $(CH_3)_2Hg$
Plankton eat bacteria
Fish eat plankton and each other
People eat fish

Concentrations of methylmercury increase at higher levels of the food chain.

ATION **Health**

ıts in the Body

r most abundant elements in the ›xygen, carbon, hydrogen, and nitro- 'e the major components of organic ıolecules, such as carbohydrates, pro- ıns, fats, and nucleic acids. Other elements compose a dietary category of compounds called minerals. Minerals are considered the inorganic elements of the body. Minerals fall into two categories—the major minerals and the trace minerals, or trace elements, as they are sometimes called. Notice in the periodic table below that most elements in the trace elements category of minerals are transition metals.

Trace elements are minerals with dietary daily requirements of 100 mg or less. They are found in foods derived from both plants and animals. Though these elements are present in very small quantities, they perform a variety of essential functions in the body, as shown in Table 3C on the next page.

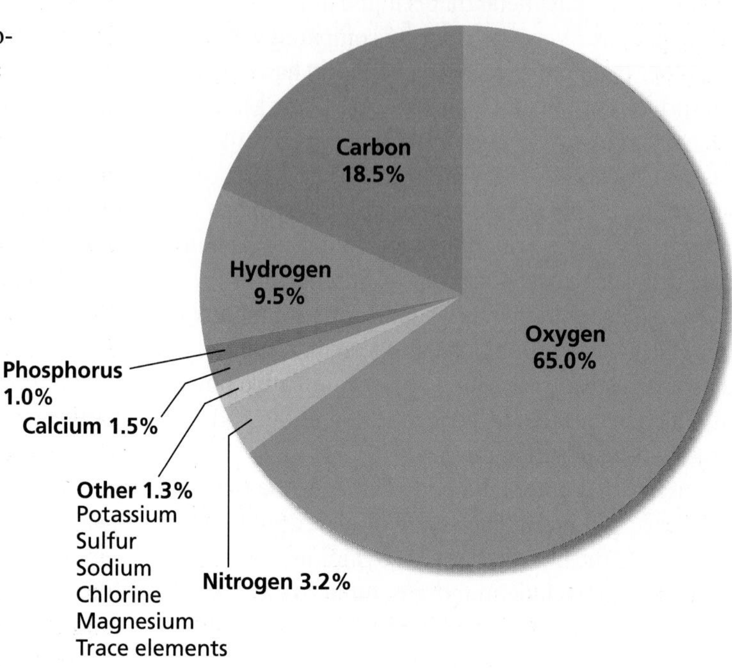

Elements in organic matter

Major minerals

Trace elements

Group 1	Group 2	Group 3	Group 4	Group 5	Group 6	Group 7	Group 8	Group 9	Group 10	Group 11	Group 12	Group 13	Group 14	Group 15	Group 16	Group 17	Group 18
1 H																	2 He
3 Li	4 Be											5 B	6 C	7 N	8 O	9 F	10 Ne
11 Na	12 Mg											13 Al	14 Si	15 P	16 S	17 Cl	18 Ar
19 K	20 Ca	21 Sc	22 Ti	23 V	24 Cr	25 Mn	26 Fe	27 Co	28 Ni	29 Cu	30 Zn	31 Ga	32 Ge	33 As	34 Se	35 Br	36 Kr
37 Rb	38 Sr	39 Y	40 Zr	41 Nb	42 Mo	43 Tc	44 Ru	45 Rh	46 Pd	47 Ag	48 Cd	49 In	50 Sn	51 Sb	52 Te	53 I	54 Xe
55 Cs	56 Ba	57 La	72 Hf	73 Ta	74 W	75 Re	76 Os	77 Ir	78 Pt	79 Au	80 Hg	81 Tl	82 Pb	83 Bi	84 Po	85 At	86 Rn
87 Fr	88 Ra	89 Ac															

TABLE 3C Transition Metal Trace Elements

Transition metal	Function
Vanadium, Cadmium	function not fully determined, but linked to a reduced growth rate and impaired reproduction
Chromium	needed for glucose transport to cells
Manganese	used in the enzyme reactions that synthesize cholesterol and metabolize carbohydrates
Iron	central atom in the heme molecule—a component of hemoglobin, which binds oxygen in the blood for transport to cells
Cobalt	a component of vitamin B_{12}
Nickel	enzyme cofactor in the metabolism of fatty acids and amino acids
Copper	a major component of an enzyme that functions to protect cells from damage
Zinc	needed for tissue growth and repair and as an enzyme cofactor
Molybdenum	enzyme cofactor in the production of uric acid

Role of Iron

Most iron in the body is in hemoglobin. Fe^{3+} is the central ion in the heme molecule, which is a component of the proteins hemoglobin and myoglobin. Hemoglobin in red blood cells transports oxygen to cells and picks up carbon dioxide as waste. Myoglobin is a protein that stores oxygen to be used in muscle contraction. Iron is also in the proteins of the electron transport system and the immune system.

Mechanisms of the body control the rate of iron absorption from food in the diet. When iron reserves are low, chemical signals stimulate cells of the intestines to absorb more iron during digestion. If the diet is low in iron, causing a deficiency, hemoglobin production stops and a condition called iron-deficiency anemia results. The blood cells produced during this state are stunted and unable to deliver adequate oxygen to cells. As a result, a person with iron-deficiency anemia feels tired and weak and has difficulty maintaining normal body temperature. The recommended daily intake of iron is 15 mg. The recommended level doubles for pregnant women. Iron supplements are for people who do not get enough iron in their daily diets. Table 3D lists some foods that are good sources of iron in the diet. Too much iron can be toxic because the body stores iron once it is absorbed. Abusing iron supplements can cause severe liver and heart damage.

TABLE 3D Sources of Iron in Foods

Food	Serving size	Iron present (mg)
Beef roast (lean cut)	4 oz	3.55
Beef, T-bone steak (lean cut)	4 oz	3.40
Beef, ground (hamburger)	4 oz	2.78
Broccoli	6.3 oz	1.50
Chicken, breast	4 oz	1.35
Chicken, giblets	4 oz	7.30
Oatmeal, instant enriched	1 pkg	8.35
Pita bread, white enriched	6 1/2 in. diameter	1.40
Pork roast	4 oz	1.15
Prunes	4 oz	2.00
Raisins	4 oz	1.88

GROUP 13
BORON FAMILY

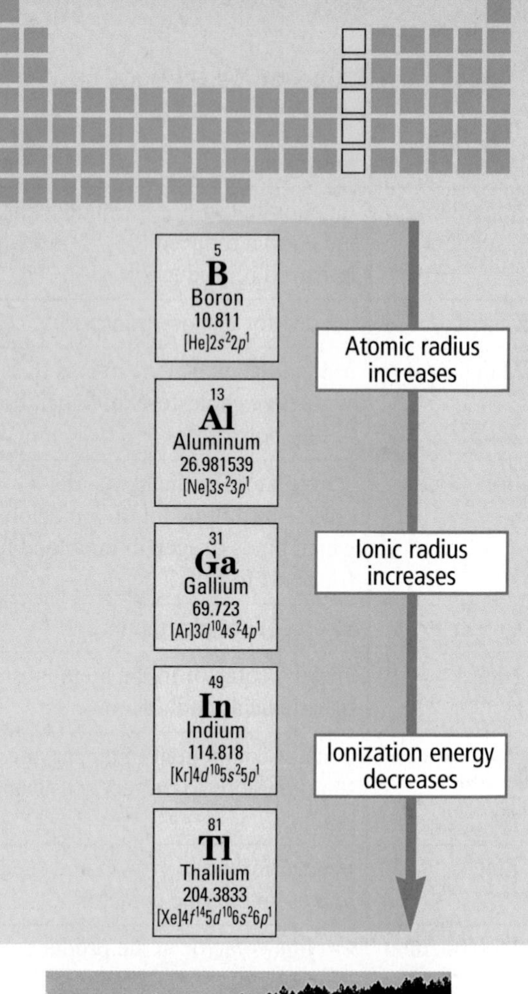

CHARACTERISTICS

- do not occur naturally in element form
- are scarce in nature (except aluminum, which is the most abundant metallic element)
- consist of atoms that have three electrons in their outer energy level
- are metallic solids (except boron, which is a solid metalloid)
- are soft and have low melting points (except boron, which is hard and has a high melting point)
- are chemically reactive at moderate temperatures (except boron)

Boron is a covalent solid. Other members of the family are metallic solids.

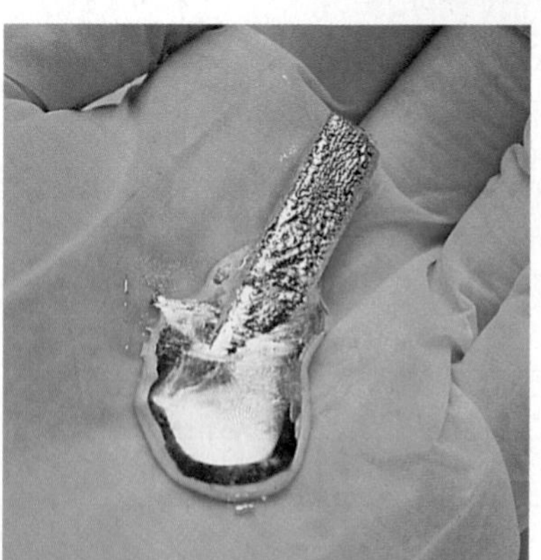

The warmth of a person's hand will melt gallium. Gallium metal has the lowest melting point (29.77°C) of any metal except mercury.

Aluminum is the most abundant metal in Earth's crust. It exists in nature as an ore called bauxite.

COMMON REACTIONS

The reaction chemistry of boron differs greatly from that of the other members of this family. Pure boron is a covalent network solid, whereas the other members of the family are metallic crystals in pure form. Boron resembles silicon more closely than it resembles the other members of its family.

With Strong Bases to Form Hydrogen Gas and a Salt

Example: $2Al(s) + 2NaOH(aq) + 2H_2O(l) \longrightarrow 2NaAlO_2(aq) + 3H_2(g)$

Ga also follows this pattern.

With Dilute Acids to Form Hydrogen Gas and a Salt

Example: $2Al(s) + 6HCl(aq) \longrightarrow 2AlCl_3(aq) + 3H_2(g)$

Ga, In, and Tl follow this pattern in reacting with dilute HF, HCl, HBr, and HI.

Example: $Al(s) + 4HNO_3(aq) \longrightarrow Al(NO_3)_3(aq) + NO(g) + 2H_2O(l)$

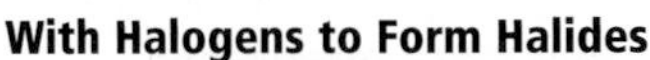

With Halogens to Form Halides

Example: $2Al(s) + 3Cl_2(g) \longrightarrow 2AlCl_3(s)$

B, Al, Ga, In, and Tl also follow this pattern in reacting with F_2, Cl_2, Br_2, and I_2 (except BF_3).

With Oxygen to Form Oxides

Example: $4Al(s) + 3O_2(g) \longrightarrow 2Al_2O_3(s)$

Ga, In, and Tl also follow this pattern.

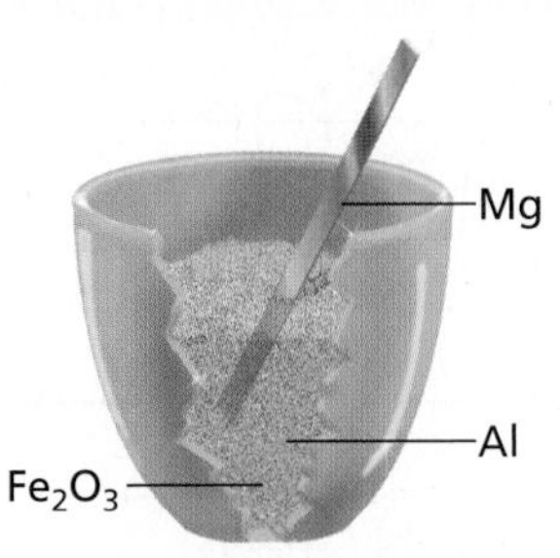

A mixture of powdered aluminum and iron(III) oxide is called thermite. Al reacts with Fe_2O_3 using Mg ribbon as a fuse to provide activation energy. The energy produced by the thermite reaction is sufficient to produce molten iron as a product.

ANALYTICAL TEST

Other than atomic absorption spectroscopy, there is no simple analytical test for all the members of the boron family.

The confirmatory test for the presence of aluminum in qualitative analysis is the formation of $Al(OH)_3$, which may be hard to detect in solution. The precipitate is made visible by the addition of a red dye called aluminum reagent.

Aluminum forms a thin layer of Al_2O_3, which protects the metal from oxidation and makes it suitable for outdoor use.

PROPERTIES OF THE GROUP 13 ELEMENTS

	B	Al	Ga	In	Tl
Melting point (°C)	2300	660.37	29.77	156.61	303.5
Boiling point (°C)	2550	2467	2203	2080	1457
Density (g/cm^3)	2.34	2.702	5.904	7.31	11.85
Ionization energy (kJ/mol)	801	578	579	558	589
Atomic radius (pm)	86	143.1	135	167	170
Ionic radius (pm)	—	53.5	62.0	80.0	88.5
Common oxidation number in compounds	+3	+3	+1, +3	+1, +3	+1, +3
Crystal structure	monoclinic	fcc	orthorhombic	fcc	hcp
Hardness (Mohs' scale)	9.3	2.75	1.5	1.2	1.2

APPLICATION *Technology*

Aluminum

Chemically, aluminum is much more active than copper, and it belongs to the category of *self-protecting metals*. These metals are oxidized when exposed to oxygen in the air and form a hard, protective metal oxide on the surface. The oxidation of aluminum is shown by the following reaction.

$$4Al(s) + 3O_2(g) \longrightarrow 2Al_2O_3(s)$$

This oxide coating protects the underlying metal from further reaction with oxygen or other substances. Self-protecting metals are valuable in themselves or when used to coat iron and steel to keep them from corroding.

Aluminum is a very good conductor of electric current. Many years ago, most high-voltage electric power lines were made of copper. Although copper is a better conductor of electricity than aluminum, copper is heavier and more expensive. Today more than 90% of high-voltage transmission lines are made of relatively pure aluminum. The aluminum wire does not have to be self-supporting because steel cable is incorporated to bear the weight of the wire in the long spans between towers.

In the 1960s, aluminum electric wiring was used in many houses and other buildings. Over time, however, because the aluminum oxidized, Al_2O_3 built up and increased electric resistance at points where wires connected to outlets, switches, and other metals. As current flowed through the extra resistance, enough heat was generated to cause a fire. Though some homes have been rewired, aluminum wiring is still prevalent in many homes.

These high-voltage transmission lines are made of aluminum supported with steel cables.

Aluminum Alloys

Because aluminum has a low density and is inexpensive, it is used to construct aircraft, boats, sports equipment, and other lightweight, high-strength objects. The pure metal is not strong, so it is mixed with small quantities of other metals—usually manganese, copper, magnesium, zinc, or silicon—to produce strong low-density alloys. Typically, 80% of a modern aircraft frame consists of aluminum alloy.

Aluminum and its alloys are good conductors of heat. An alloy of aluminum and manganese is used to make cookware. High-quality pots and pans made of stainless steel may have a plate of aluminum on the bottom to help conduct heat quickly to the interior.

Automobile radiators made of aluminum conduct heat as hot coolant from the engine enters the bottom of the radiator. The coolant is deflected into several channels. These channels are covered by thin vanes of aluminum, which conduct heat away from the coolant and transfer it to the cooler air rushing past. By the time the coolant reaches the top of the radiator, its temperature has dropped so that when it flows back into the engine it can absorb more heat. To keep the process efficient, the outside of a radiator should be kept unobstructed and free of dirt buildup.

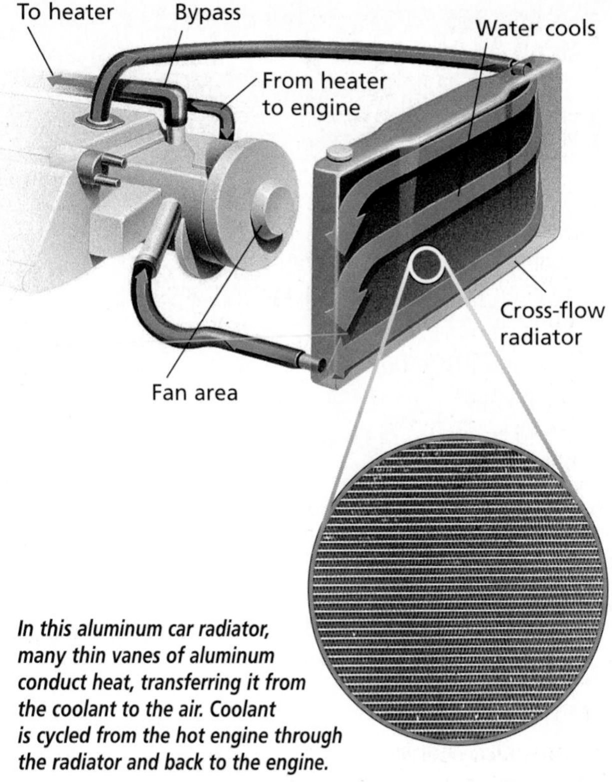

In this aluminum car radiator, many thin vanes of aluminum conduct heat, transferring it from the coolant to the air. Coolant is cycled from the hot engine through the radiator and back to the engine.

TABLE 4A Alloys of Aluminum and Their Uses

Principal alloying element(s)*	Characteristics	Application examples
Manganese	moderately strong, easily worked	cookware, roofing, storage tanks, lawn furniture
Copper	strong, easily formed	aircraft structural parts; large, thin structural panels
Magnesium	strong, resists corrosion, easy to weld	parts for boats and ships, outdoor decorative objects, tall poles
Zinc and magnesium	very strong, resists corrosion	aircraft structural parts, vehicle parts, anything that needs high strength and low weight
Silicon	expands little on heating and cooling	aluminum castings
Magnesium and silicon	resists corrosion, easily formed	exposed parts of buildings, bridges

* All these alloys have small amounts of other elements.

GROUP 14
CARBON FAMILY

CHARACTERISTICS

- include a nonmetal (carbon), two metalloids (silicon and germanium), and two metals (tin and lead)
- vary greatly in both physical and chemical properties
- occur in nature in both combined and elemental forms
- consist of atoms that contain four electrons in the outermost energy level
- are relatively unreactive
- tend to form covalent compounds (tin and lead also form ionic compounds)

Atomic number	Symbol	Name	Atomic mass	Electron configuration
6	C	Carbon	12.011	$[He]2s^22p^2$
14	Si	Silicon	28.0855	$[Ne]3s^23p^2$
32	Ge	Germanium	72.61	$[Ar]3d^{10}4s^24p^2$
50	Sn	Tin	118.710	$[Kr]4d^{10}5s^25p^2$
82	Pb	Lead	207.2	$[Xe]4f^{14}5d^{10}6s^26p^2$

Atomic radius increases

Ionization energy decreases

Lead has a low reactivity and is resistant to corrosion. It is very soft, highly ductile, and malleable. Lead is toxic and, like mercury, it is a cumulative poison.

Silicon has a luster but does not exhibit metallic properties. Most silicon in nature is a silicon oxide, which occurs in sand and quartz, which is shown here.

Tin, which is shown on the right, is a self-protecting metal like lead, but unlike lead it has a high luster. Tin occurs in nature in cassiterite ore, which is shown above.

COMMON REACTIONS

With Oxygen to Form Oxides

Example: $Sn(s) + O_2(g) \longrightarrow SnO_2(s)$

Pb follows this pattern, as do C, Si, and Ge at high temperatures.

With Acids to Form Salts and Hydrogen Gas

Only the metallic elements of this group react slowly with aqueous acids.

Example: $Sn(s) + 2HCl(aq) \longrightarrow SnCl_2(aq) + H_2(g)$

Both Sn and Pb can also react to form tin(IV) and lead(IV) salts, respectively.

With Halogens to Form Halides

Example: $Sn(s) + 2Cl_2(g) \longrightarrow SnCl_4(s)$

Si, Ge, and Pb follow this pattern, reacting with F_2, Cl_2, Br_2, and I_2.

ANALYTICAL TEST

The only way to identify all the elements of this group is by atomic absorption spectroscopy. Ionic compounds of tin and lead can be identified in aqueous solutions by adding a solution containing sulfide ions. The formation of a yellow precipitate indicates the presence of Sn^{4+}, and the formation of a black precipitate indicates the presence of Pb^{2+}.

$$Sn^{4+}(aq) + 2S^{2-}(aq) \longrightarrow SnS_2(s)$$
$$Pb^{2+}(aq) + S^{2-}(aq) \longrightarrow PbS(s)$$

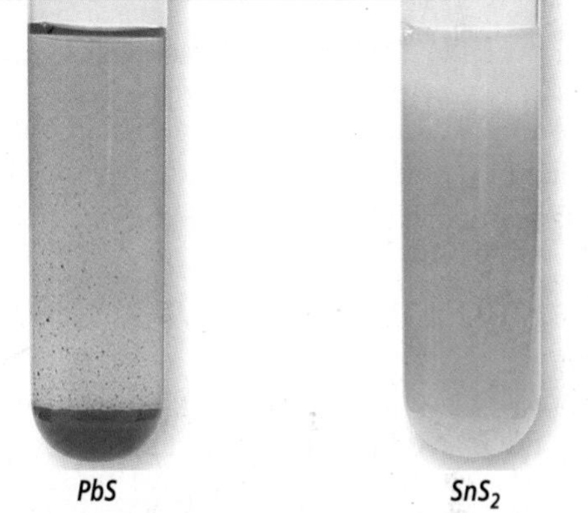

PROPERTIES OF THE GROUP 14 ELEMENTS

	C	Si	Ge	Sn	Pb
Melting point (°C)	3500/3652*	1410	937.4	231.88	327.502
Boiling point (°C)	3930	2355	2830	2260	1740
Density (g/cm^3)	3.51/2.25*	2.33 ± 0.01	5.323	7.28	11.343
Ionization energy (kJ/mol)	1086	787	762	709	716
Atomic radius (pm)	77.2	118	128	151	175
Ionic radius† (pm)	260	—	—	118	119
Common oxidation number in compounds	+4, −4	+4	+2, +4	+2, +4	+2, +4
Crystal structure	cubic/hexagonal*	cubic	cubic	tetragonal	fcc
Hardness (Mohs' scale)	10/0.5*	6.5	6.0	1.5	1.5

* The data are for two allotropic forms: diamond/graphite.
† Ionic radii are for 4+ ions.

APPLICATION *Chemical Industry*

Carbon and the Reduction of Iron Ore

Some metals, especially iron, are separated from their ores through reduction reactions in a blast furnace. The blast furnace gets its name from the fact that air or pure oxygen is blown into the furnace, where it oxidizes carbon to form carbon monoxide, CO. Carbon and its compounds are important reactants in this process.

What happens inside the blast furnace to recover the iron from its ore? The actual chemical changes that occur are complex. A simplified explanation begins with the reaction of oxygen in hot air with coke, a form of carbon. Some of the coke burns to form carbon dioxide.

$$C(s) + O_2(g) \longrightarrow CO_2(g)$$

As the concentration of oxygen is increased, the carbon dioxide comes in contact with pieces of hot coke and is reduced to carbon monoxide.

$$CO_2(g) + C(s) \longrightarrow 2CO(g)$$

The carbon monoxide now acts as a reducing agent to reduce the iron oxides in the ore to metallic iron.

$$Fe_2O_3(s) + 3CO(g) \longrightarrow 2Fe(l) + 3CO_2(g)$$

The reduction is thought to occur in steps as the temperature in the furnace increases. The following are some of the possible steps.

$$Fe_2O_3 \longrightarrow Fe_3O_4 \longrightarrow FeO \longrightarrow Fe$$

The white-hot liquid iron collects in the bottom of the furnace and is removed every four or five hours. The iron may be cast in molds or converted to steel in another process.

Limestone, present in the center of the furnace, decomposes to form calcium oxide and carbon dioxide.

$$CaCO_3(s) \longrightarrow CaO(s) + CO_2(g)$$

The calcium oxide then combines with silica, a silicon compound, to form calcium silicate slag.

The relatively high carbon content of iron produced in a blast furnace makes the metal hard but brittle. It also has other impurities, like sulfur and phosphorus, that cause the recovered iron to be brittle. The conversion of iron to steel is essentially a purification process in which impurities are removed by oxidation. This purification process is carried out in another kind of furnace at very high temperatures. All steel contains 0.02 to 1.5% carbon. In fact, steels are graded by their carbon content. Low-carbon steels typically contain 0.02 to 0.3% carbon. Medium-carbon steels typically contain 0.03 to 0.7% carbon. High-carbon steels contain 0.7 to 1.5% carbon.

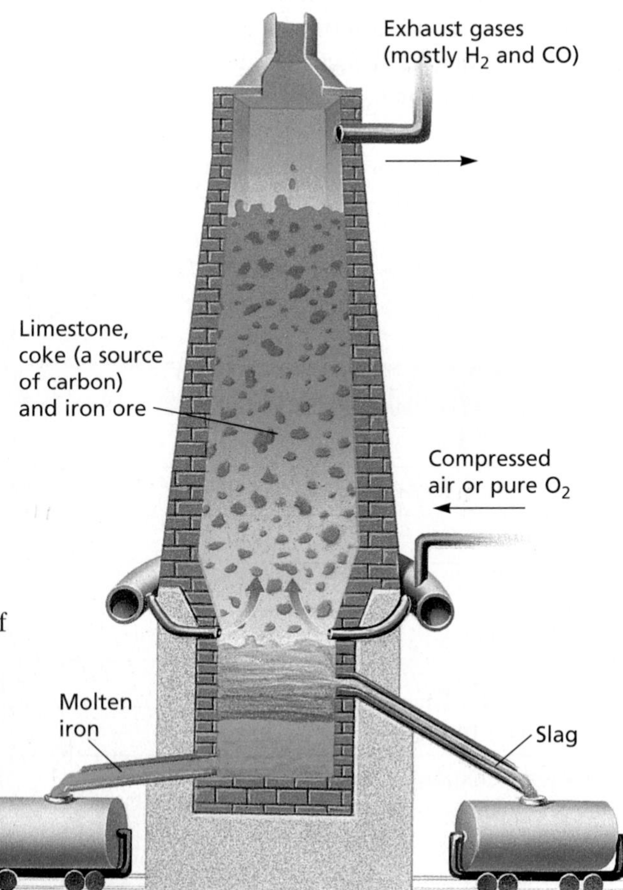

Molten iron flowing from the bottom of a blast furnace has been reduced from its ore through a series of reactions at high temperatures in different regions of the furnace.

TION **Biochemistry**

Dioxide and Respiration

nisms, including humans, carry out cellular In this process, cells break down food and release the energy used to build those during photosynthesis. Glucose, $C_6H_{12}O_6$, is substance broken down in respiration. The hemical equation expresses this process.

$$O_6 + 6O_2 \longrightarrow 6CO_2 + 6H_2O + \text{energy}$$

and most other vertebrate animals, the eded for this reaction is delivered to cells obin found in red blood cells. Oxygen binds globin as blood passes through capillaries s, as represented by the following reaction.

$$Hb + O_2 \longrightarrow HbO_2$$

ents the hemoglobin molecule, and HbO_2 oxyhemoglobin, which is hemoglobin d oxygen. When the red blood cells pass pillaries near cells that have depleted en supply through respiration, the reaction nd oxyhemoglobin gives up its oxygen.

$$HbO_2 \longrightarrow Hb + O_2$$

Carbon dioxide produced during respiration is a waste product that must be expelled from an organism. Various things happen when CO_2 enters the blood. Seven percent dissolves in the plasma, about 23% binds loosely to hemoglobin, and the remaining 70% reacts reversibly with water in plasma to form hydrogen carbonate, HCO_3^- ions. To form HCO_3^- ions, CO_2 first combines with H_2O to form carbonic acid, H_2CO_3, in a reversible reaction.

$$CO_2(aq) + H_2O(l) \rightleftharpoons H_2CO_3(aq)$$

The dissolved carbonic acid ionizes to HCO_3^- ions and aqueous H^+ ions in the form of H_3O^+.

$$H_2CO_3(aq) + H_2O \rightleftharpoons H_3O^+(aq) + HCO_3^-(aq)$$

The combined equilibrium reaction follows.

$$CO_2(aq) + 2H_2O(l) \rightleftharpoons H_3O^+(aq) + HCO_3^-(aq)$$

When the blood reaches the lungs, the reaction reverses and the blood releases CO_2, which is then exhaled to the surroundings.

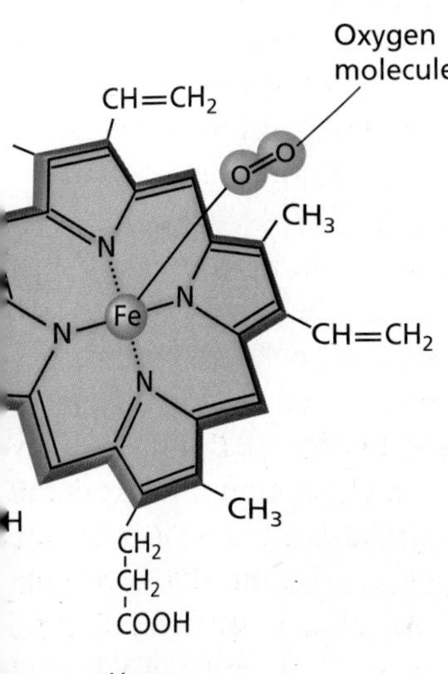

Heme

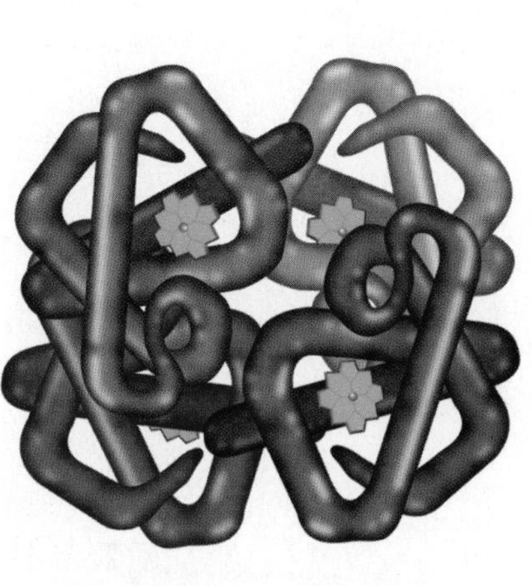
Hemoglobin (protein)

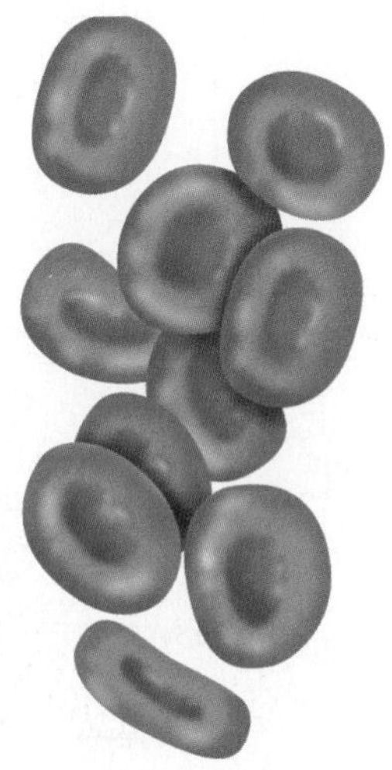
Red blood cells

*carrier molecule, heme, is a component of the more-complex protein Note that each **hemoglobin** molecule has four heme subunits. Hemoglobin ent of red blood cells.*

Carbon Dioxide

Carbon dioxide is a colorless gas with a faintly irritating odor and a slightly sour taste. The sour taste is the result of the formation of carbonic acid when CO_2 dissolves in the water in saliva. It is a stable gas that does not burn or support combustion. At temperatures lower than 31°C and at pressures higher than 72.9 atm, CO_2 condenses to the liquid form. A phase diagram for CO_2 is found in the chapter review section of Chapter 12. At normal atmospheric pressure, solid CO_2 (dry ice) sublimes at –78.5°C. The linear arrangement of carbon dioxide molecules makes them nonpolar.

CO_2 is produced by the burning of organic fuels and from respiration processes in most living things. Most CO_2 released into the atmosphere is used by plants during photosynthesis. Recall that photosynthesis is the process by which green plants and some forms of algae and bacteria make food. During photosynthesis, CO_2 reacts with H_2O, using the energy from sunlight. The relationships among the various processes on Earth that convert carbon to carbon dioxide are summarized the carbon cycle, which is pictu

Carbon Monoxide

Carbon monoxide is a poisonou naturally by decaying plants, ce volcanic eruptions, and the oxic the atmosphere.

Because CO is colorless, odc it is difficult to detect. It is sligl air and slightly soluble in wate uses are in the reduction of iro 756, and the production of org as ethanol.

$$CO(g) + 2H_2(g) —$$

Carbon monoxide is also pı incomplete combustion of org combustion of methane occur oxygen is limited.

$$2CH_4(g) + 3O_2(g) \longrightarrow$$

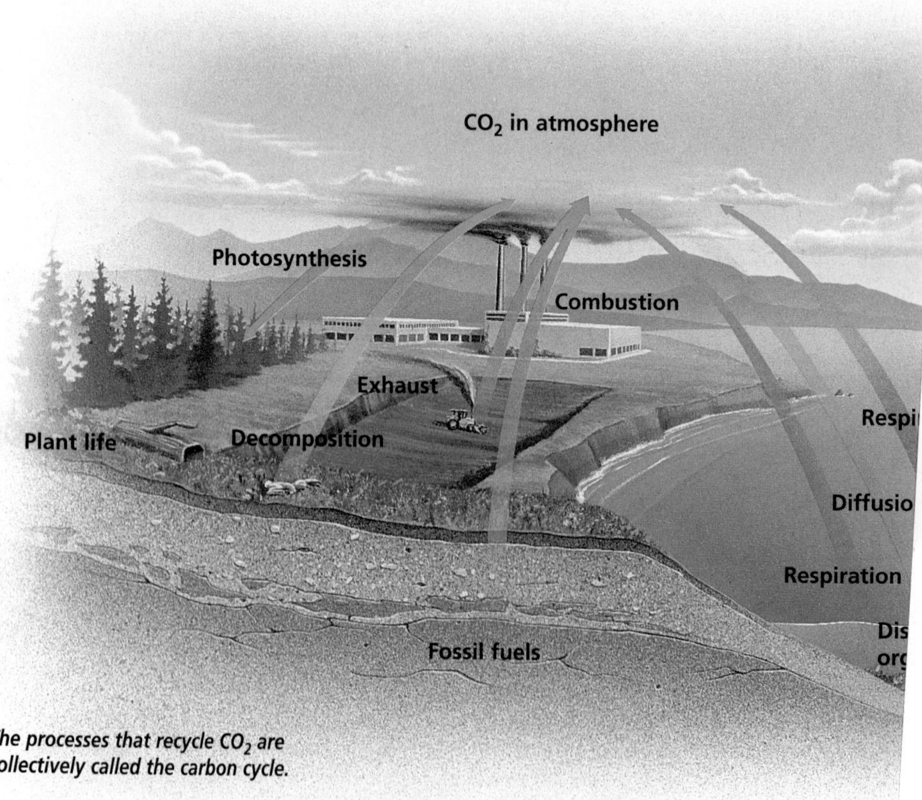

The processes that recycle CO_2 are collectively called the carbon cycle.

Exchange of CO_2 and O_2 in the Lungs

Why does CO_2 leave the blood as it passes through the lung's capillaries, and why does O_2 enter the blood? The exchange is caused by the difference in concentrations of CO_2 and O_2 in the blood and in the atmosphere. Oxygen is 21% of the atmosphere. Although the amount of CO_2 varies from place to place, it averages about 0.033% of the atmosphere. Thus, O_2 is about 640 times more concentrated in the atmosphere than is CO_2.

Substances tend to diffuse from regions of higher concentration toward regions of lower concentration. Thus, when blood reaches the capillaries of the lung, O_2 from the air diffuses into the blood, where its pressure is only 40 mm Hg, while CO_2 diffuses out of the blood, where its pressure is 45 mm Hg, and into the air. The diagram below summarizes the process.

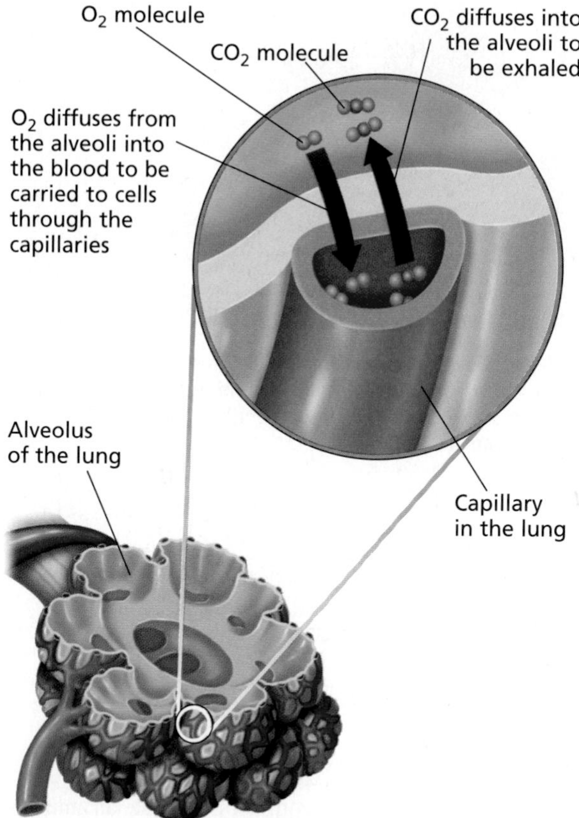
O_2 molecule
CO_2 molecule
CO_2 diffuses into the alveoli to be exhaled
O_2 diffuses from the alveoli into the blood to be carried to cells through the capillaries
Alveolus of the lung
Capillary in the lung

The pressure of O_2 in the blood entering the lung is much lower than it is in the atmosphere. As a result, O_2 diffuses into the blood. The opposite situation exists for CO_2, so it diffuses from the blood into the air. Note that blood leaving the lung still contains a significant concentration of CO_2.

Acidosis and Alkalosis

In humans, blood is maintained between pH 7.3 and 7.5. The pH of blood is dependent on the concentration of CO_2 in the blood. Look again at this equilibrium system.

$$CO_2(aq) + 2H_2O(l) \rightleftharpoons H_3O^+(aq) + HCO_3^-(aq)$$

Notice that the right side of the equation contains the H_3O^+ ion, which determines the pH of the blood. If excess H_3O^+ enters the blood from tissues, the reverse reaction is favored. Excess H_3O^+ combines with HCO_3^- to produce more CO_2 and H_2O. If the H_3O^+ concentration begins to fall, the forward reaction is favored, producing additional H_3O^+ and HCO_3^-. To keep H_3O^+ in balance, adequate amounts of both CO_2 and HCO_3^- must be present. If something occurs that changes these conditions, a person can become very ill and can even die.

Hyperventilation occurs when a person breathes too rapidly for an extended time. Too much CO_2 is eliminated, causing the reverse reaction to be favored, and H_3O^+ and HCO_3^- are used up. As a result, the person develops a condition known as alkalosis because the pH of the blood rises to an abnormal alkaline level. The person begins to feel lightheaded and faint, and, unless treatment is provided, he or she may fall into a coma. Alkalosis is treated by having the victim breathe air that is rich in CO_2. One way to accomplish this is to have the person breathe with a bag held tightly over the nose and mouth. Alkalosis is also caused by fever, infection, intoxication, hysteria, and prolonged vomiting.

The reverse of alkalosis is a condition known as acidosis. This condition is often caused by a depletion of HCO_3^- ions from the blood, which can occur as a result of kidney dysfunction. The kidney controls the excretion of HCO_3^- ions. If there are too few HCO_3^- ions in solution, the forward reaction is favored and H_3O^+ ions accumulate, which lowers the blood's pH. Acidosis can also result from the body's inability to expel CO_2, which can occur during pneumonia, emphysema, and other respiratory disorders. Perhaps the single most common cause of acidosis is uncontrolled diabetes, in which acids normally excreted in the urinary system are instead retained by the body.

APPLICATION **The Environment**

Carbon Monoxide Poisoning

Standing on a street corner in any major city exposes a person to above-normal concentrations of carbon monoxide from automobile exhaust. Carbon monoxide also reacts with hemoglobin. The following reaction takes place in the capillaries of the lung.

$$Hb + CO \longrightarrow HbCO$$

Unlike CO_2 or O_2, CO binds strongly to hemoglobin. Carboxyhemoglobin, HbCO, is 200 times more stable than oxyhemoglobin, HbO_2. So as blood circulates, more and more CO molecules bind to hemoglobin, reducing the amount of O_2 bond sites available for transport. Eventually, CO occupies so many hemoglobin binding sites that cells die from lack of oxygen. Symptoms of carbon monoxide poisoning include headache, mental confusion, dizziness, weakness, nausea, loss of muscular control, and decreased heart rate and respiratory rate. The victim loses consciousness and will die without treatment.

If the condition is caught in time, a victim of carbon monoxide poisoning can be revived by breathing pure oxygen. This treatment causes carboxyhemoglobin to be converted slowly to oxyhemoglobin according to the following chemical equation.

$$O_2 + HbCO \longrightarrow CO + HbO_2$$

Mild carbon monoxide poisoning usually does not have long-term effects. In severe cases, cells are destroyed. Damage to brain cells is irreversible.

The level of danger posed by carbon monoxide depends on two factors: the concentration of the gas in the air and the amount of time that a person is exposed to the gas. Table 5A shows the effects of increasing levels of carbon monoxide in the bloodstream. These effects vary considerably depending on a person's activity level and metabolic rate.

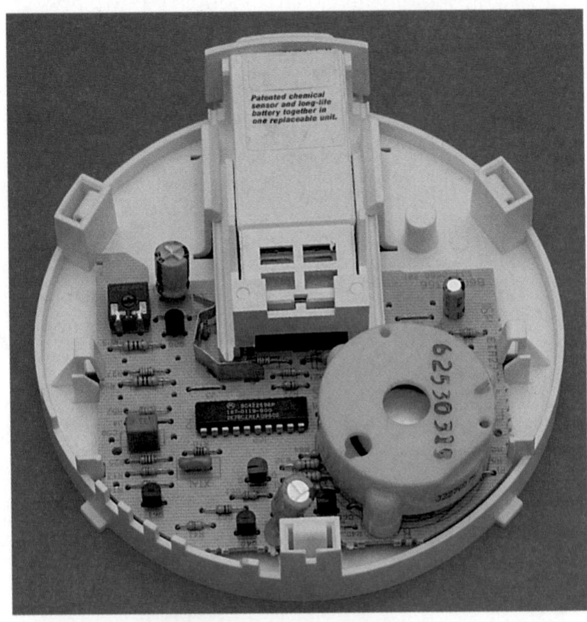

Carbon monoxide detectors are now available to reduce the risk of poisoning from defective home heating systems. The Consumer Products Safety Commission recommends that all homes have a CO detector with a UL label.

TABLE 5A Symptoms of CO Poisoning at Increasing Levels of CO Exposure and Concentration

Concentration of CO in air (ppm)*	Hemoglobin molecules as HbCO	Visible effects
100 for 1 hour or less	10% or less	no visible symptoms
500 for 1 hour or less	20%	mild to throbbing headache, some dizziness, impaired perception
500 for an extended period of time	30–50%	headache, confusion, nausea, dizziness, muscular weakness, fainting
1000 for 1 hour or less	50–80%	coma, convulsions, respiratory failure, death

* ppm is parts per million

APPLICATION *Biochemistry*

Macromolecules

Large organic polymers are called macromolecules (the prefix *macro* means "large"). Macromolecules play important roles in living systems. Most macromolecules essential to life belong to four main classes, three of which we know as nutrients in food:

1. **Proteins** Hair, tendons, ligaments, and silk are made of protein. Other proteins act as hormones, transport substances throughout the body, and fight infections. Enzymes are proteins that control the body's chemical reactions. Proteins provide energy, yielding 17 kJ/g.
2. **Carbohydrates** Sugars, starches, and cellulose are carbohydrates. Carbohydrates are sources of energy, yielding 17 kJ/g.
3. **Lipids** Fats, oils, waxes, and steroids are lipids, nonpolar substances that do not dissolve in water. Fats are sources of energy, yielding 38 kJ/g.
4. **Nucleic acids** The nucleic acids are DNA and RNA. In most organisms, DNA is used to store hereditary information and RNA helps to assemble proteins.

Proteins

Proteins are macromolecules formed by condensation reactions between amino acid monomers. Proteins contain carbon, oxygen, hydrogen, nitrogen, and usually some sulfur.

All amino acids have a carboxyl group, —COOH, and an amino group, —NH_2, attached to a central carbon atom, which is also attached to hydrogen, —H. Amino acids differ from one another at the fourth bond site of the central carbon, which is attached to a functional group (called an *R* group). *R* groups differ from one amino acid to another, as shown in the structures for several amino acids below. The proteins of all organisms contain a set of 20 common amino acids. The reaction that links amino acids is a condensation reaction, which is described in Chapter 21.

Each protein has its own unique sequence of amino acids. A complex organism has at least several thousand different proteins, each with a special structure and function. For instance, *insulin,* a hormone that helps the body regulate the level of sugar in the blood, is made up of two linked chains.

Amino acids have the same general structure. These examples show some of the variations within this class of compounds.

```
   NH2  O
   |   //
R-C-C
   |   \
   H    OH
```
General structure

```
    NH2  O
     |  //
CH3-C-C
     |  \
     H   OH
```
Alanine

```
     O        NH2  O
     ||        |  //
H2N-C-CH2-C-C
               |  \
               H   OH
```
Asparagine

```
     O              NH2  O
     ||              |  //
H2N-C-CH2-CH2-C-C
                     |  \
                     H   OH
```
Glutamine

```
         H     NH2  O
         |      |  //
CH3CH2-C----C-C
         |      |  \
        CH3     H   OH
```
Isoleucine

```
  CH3      NH2  O
  |         |  //
H-C-CH2-C-C
  |         |  \
  CH3       H   OH
```
Leucine

```
                  NH2  O
                   |  //
CH3-S-CH2-CH2-C-C
                   |  \
                   H   OH
```
Methionine

```
          NH2  O
           |  //
(C6H5)-CH2-C-C
           |  \
           H   OH
```
Phenylalanine

```
         NH2  O
          |  //
CH3-CH-C-C
     |    |  \
     OH   H   OH
```
Threonine

```
                NH2  O
                 |  //
HO-(C6H4)-CH2-C-C
                 |  \
                 H   OH
```
Tyrosine

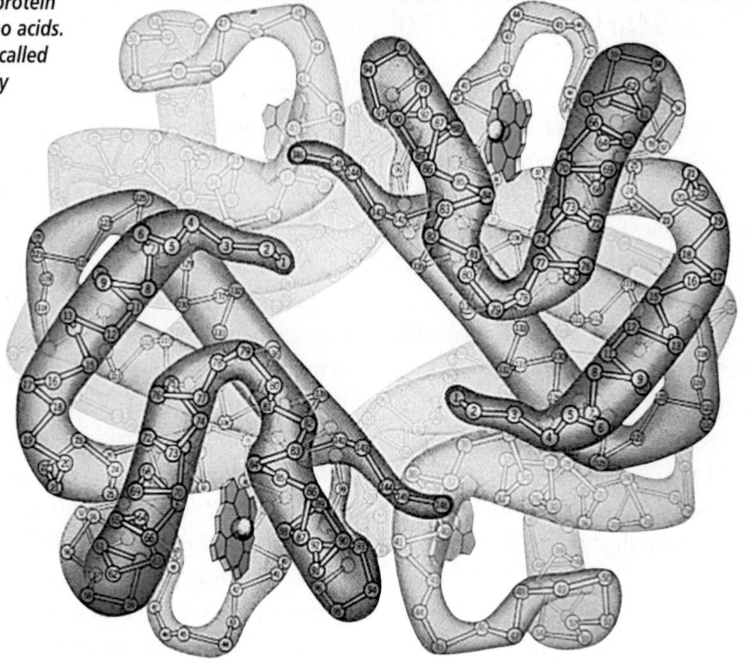

Hemoglobin is a complex protein made of hundreds of amino acids. Its 3-dimensional shape is called a tertiary structure. Tertiary structures break down when a protein is denatured.

The chains are held together by S—S bonds between sulfur atoms in two cysteine amino acids. Insulin is one of the smaller proteins, containing only 51 amino acids. In contrast, hemoglobin, which carries oxygen in the blood, is a large protein consisting of four long chains with the complicated three-dimensional structures shown above. Proteins can lose their shape with increases in temperature or changes in the chemical composition of their environment. When they are returned to normal surroundings, they may fold or coil up again and re-form their original structure.

Changing even one amino acid can change a protein's structure and function. For example, the difference between normal hemoglobin and the hemoglobin that causes sickle cell anemia is just two amino acids.

Enzymes

You learned how enzymes alter reaction rates in Chapter 17. Some enzymes cannot bind to their substrates without the help of additional molecules. These may be *minerals,* such as calcium or iron ions, or helper molecules called *coenzymes* that play accessory roles in enzyme-catalyzed reactions. Many vitamins are coenzymes or parts of coenzymes.

Vitamins are organic molecules that we cannot manufacture and hence need to eat in small amounts.

Vitamin C, $C_6H_8O_6$
Water-soluble

Vitamin A, $C_{20}H_{30}O$
Fat-soluble

You can see why we need vitamins and minerals in our diet—to enable our enzymes to work. You can also see why we need only small amounts of them. Minerals and coenzymes are not destroyed in biochemical reactions. Like enzymes, coenzymes and minerals can be used over and over again.

Temperature and pH have the most significant effects on the rates of reactions catalyzed by enzymes. Most enzymes work best in a solution of approximately neutral pH. Most body cells have a pH of 7.4. However, some enzymes function only in acidic or basic environments. For example, pepsin, the collective

The protein in fish is denatured by the low pH of lime juice. Notice that the flesh shown with the limes has turned white compared with the flesh at normal pH.

term for the digestive enzymes found in the human stomach, works best at a very acidic pH of about 1.5. Cells that line the stomach secrete hydrochloric acid to produce this low pH environment. When food travels down the digestive tract, it carries these enzymes out of the stomach into the intestine. In the intestine, stomach enzymes stop working because sodium bicarbonate in the intestine raises the pH to about 8. Digestive enzymes in the intestine are formed by the pancreas and work best at pH 8.

Most chemical reactions, including enzyme reactions, speed up with increases in temperature. However, high temperatures (above about 60°C) destroy, or denature, protein by breaking up the three-dimensional structure. For example, the protein in an egg or a piece of meat denatures when the egg or meat is cooked. Proteins in the egg white become opaque when denatured. Heating can preserve food by denaturing the enzymes of organisms that cause decay. In milk pasteurization, the milk is heated to denature enzymes that would turn it sour. Refrigeration and freezing also help preserve food by slowing the enzyme reactions that cause decay.

Carbohydrates

Carbohydrates are sugars, starches, and related compounds. The monomers of carbohydrates are monosaccharides, or simple sugars, such as fructose and glucose. A monosaccharide contains carbon, hydrogen, and oxygen in about a 1:2:1 ratio, which is an empirical formula of CH_2O.

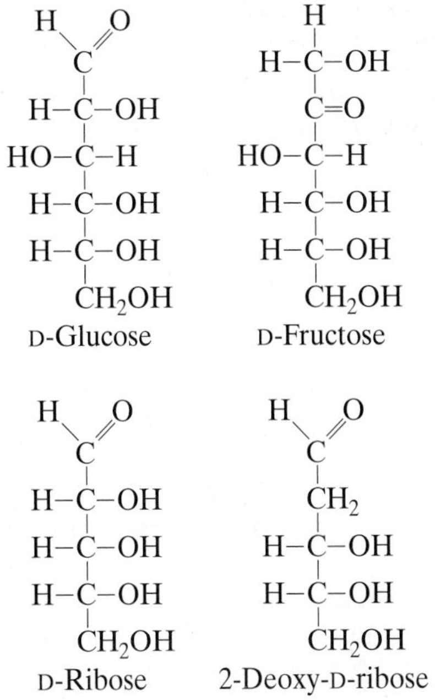

D-Glucose: H–C(=O)–, H–C–OH, HO–C–H, H–C–OH, H–C–OH, CH_2OH

D-Fructose: H–C(H)–OH, C=O, HO–C–H, H–C–OH, H–C–OH, CH_2OH

D-Ribose: H–C(=O)–, H–C–OH, H–C–OH, H–C–OH, CH_2OH

2-Deoxy-D-ribose: H–C(=O)–, CH_2, H–C–OH, H–C–OH, CH_2OH

Monosaccharides chain representation

Two monosaccharides may be joined together to form a disaccharide. Sucrose, shown below, is a disaccharide. A disaccharide can be hydrolyzed to produce the monosaccharides that formed it. By a series of condensation reactions, many monosaccharides can be joined to form a polymer called a polysaccharide (commonly known as a complex carbohydrate).

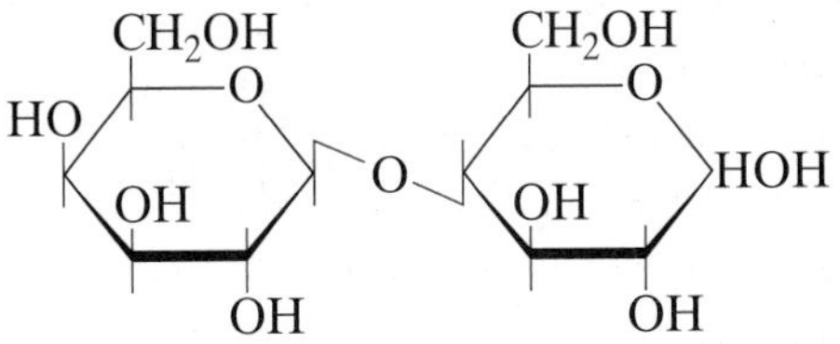

Lactose—made from glucose and galactose

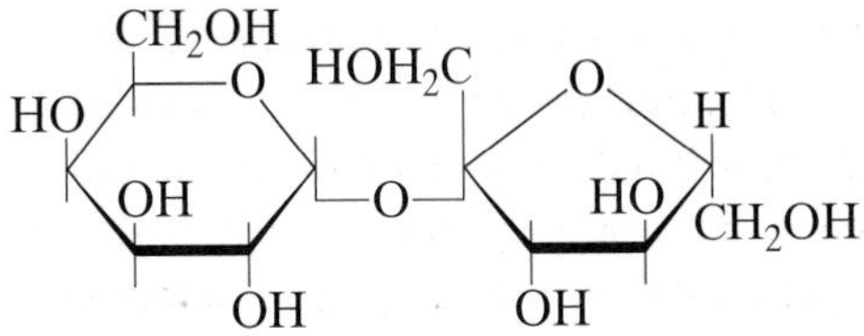

Sucrose—made from glucose and fructose

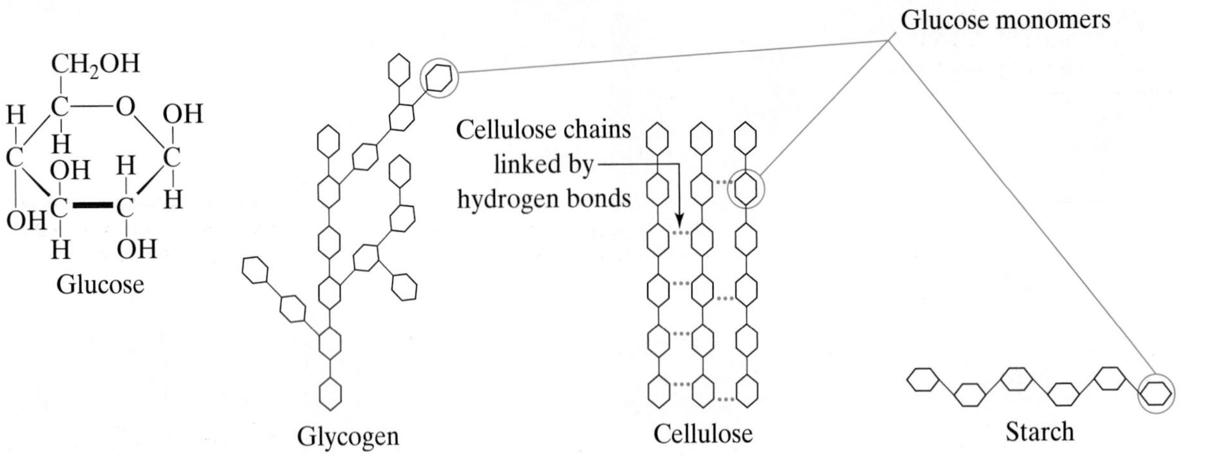

Glucose is the structural unit for glycogen, cellulose, and starch. Notice that these three polymers differ in the arrangement of glucose monomers.

Three important polysaccharides made of glucose monomers are glycogen, starch, and cellulose. Animals store energy in glycogen. The liver and muscles remove glucose from the blood and condense it into glycogen, which can later be hydrolyzed back into glucose and used to supply energy as needed.

Starch consists of two kinds of glucose polymers. It is hydrolyzed in plants to form glucose for energy and for building material to produce more cells. The structural polysaccharide cellulose is probably the most common organic compound on Earth. Glucose monomers link cellulose chains together at the hydroxyl groups to form cellulose fibers. Cotton fibers consist almost entirely of cellulose.

Lipids

Lipids are a varied group of organic compounds that share one property: they are not very soluble in water. Lipids contain a high proportion of C—H bonds, and they dissolve in nonpolar organic solvents, such as ether, chloroform, and benzene.

Fatty acids are the simplest lipids. A fatty acid consists of an unbranched chain of carbon and hydrogen atoms with a carboxyl group at one end. Bonding within the carbon chain gives both saturated and unsaturated fatty acids, just as the simple hydrocarbons (see Chapter 20) can be saturated or unsaturated.

The bonds in a carboxyl group are polar, and so the carboxyl end of a fatty acid attracts water

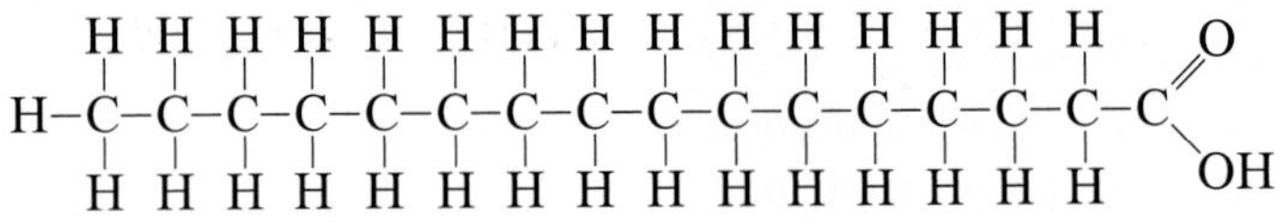

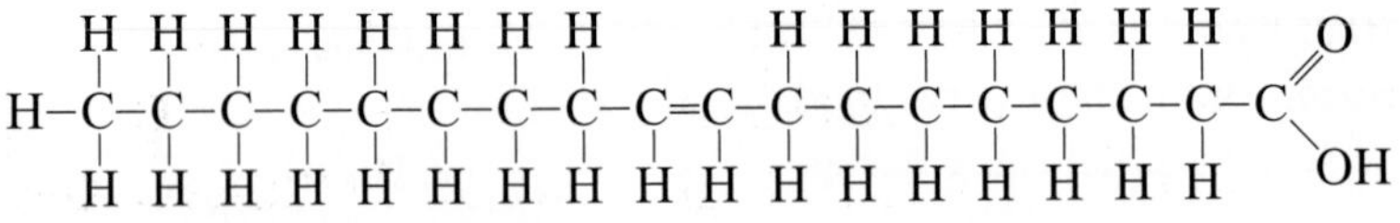

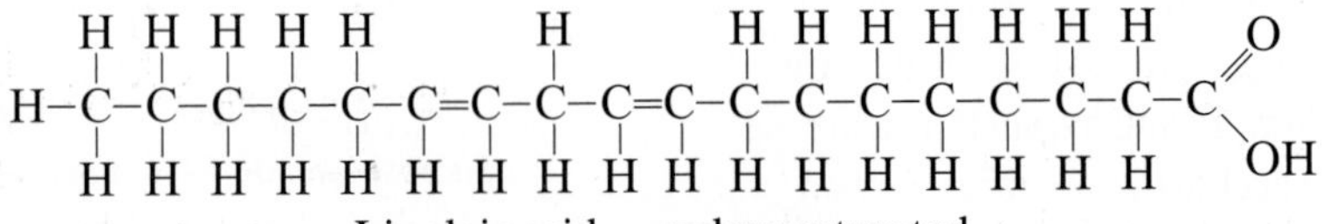

These examples of common fatty acids show the differences in saturation level.

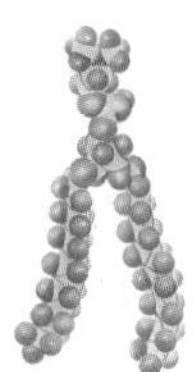

This phospholipid molecule contains two fatty-acid chains.

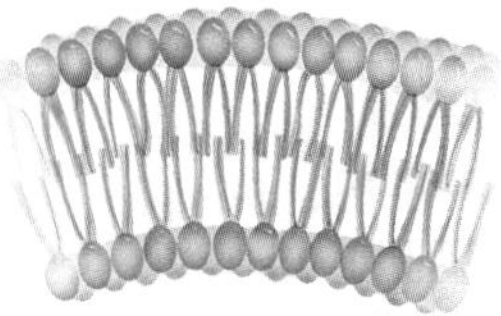

The lipid bilayer is the framework of the cell membrane.

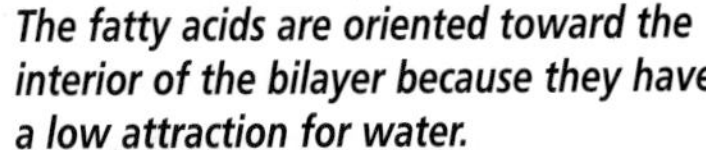

The fatty acids are oriented toward the interior of the bilayer because they have a low attraction for water.

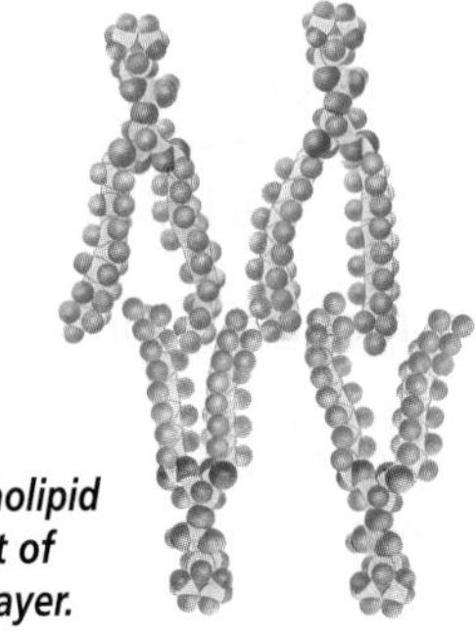

This phospholipid chain is part of the lipid bilayer.

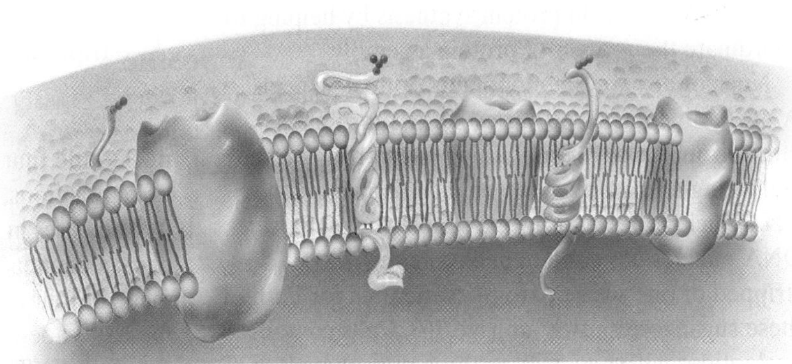

molecules. The carbon-hydrogen bonds of a lipid's hydrocarbon chain are nonpolar, however. The polar end will dissolve in water, and the other end will dissolve in nonpolar organic compounds. This behavior enables fatty acids to form membranes when they are dropped into water. It also gives soaps and detergents their cleaning power.

Lipids are the main compounds in biological membranes, such as the cell membrane. Because lipids are insoluble, the lipid bilayer of a cell membrane is adapted to keep the contents of the cell inside separated from the outer environment of the cell.

The structural component of a cell membrane is a phospholipid. The "head" of the phospholipid is polar, and the fatty acid tails are nonpolar, as shown in the model above.

Most fatty acids found in foods and soaps belong to a class of compounds called triglycerides. The fat content shown on a nutrition label for packaged food represents a mixture of the triglycerides in the food. Triglycerides have the general structure shown below.

Fatty acids are usually combined with other molecules to form classes of biomolecules called glycolipids (made from a carbohydrate and a lipid) or lipoproteins (made from a lipid and a protein). These compounds are also parts of more-complex lipids found in the body.

$$\begin{array}{ccccc}
\mathrm{H_3C{-}(C_nH_{2n}){-}COOH} & & \mathrm{HO{-}CH_2} & & \mathrm{H_3C{-}(C_nH_{2n}){-}C(=O){-}O{-}CH_2} \\
\mathrm{H_3C{-}(C_nH_{2n}){-}COOH} & + & \mathrm{HO{-}CH} & \longrightarrow & \mathrm{H_3C{-}(C_nH_{2n}){-}C(=O){-}O{-}CH} \\
\mathrm{H_3C{-}(C_nH_{2n}){-}COOH} & & \mathrm{HO{-}CH_2} & & \mathrm{H_3C{-}(C_nH_{2n}){-}C(=O){-}O{-}CH_2} \\
\text{Saturated fatty acids} & + & \text{Glycerol} & \longrightarrow & \text{Triglyceride}
\end{array}$$

Triglycerides are made from three long-chain fatty acids bonded to a glycerol backbone.

APPLICATION *Technology*

Semiconductors

When electrons can move freely through a material, the material is a conductor. The electrons in metals are loosely held and require little additional energy to move from one vacant orbital to the next. A set of overlapping orbitals is called a *conduction band.* Because electrons can easily jump to the conduction band, metals conduct electricity when only a very small voltage is applied.

Semiconductors conduct a current if the voltage applied is large enough to excite the outer-level electrons of their atoms into the higher energy levels. With semiconductors, more energy, and thus a higher voltage, is required to cause conduction. By contrast, nonmetals are insulators because they do not conduct at ordinary voltages. Too much energy is needed to raise their outer electrons into conduction bands.

Semiconductor devices include transistors; diodes, including light-emitting diodes (LEDs); some lasers; and photovoltaic cells ("solar" cells). Though silicon is the basis of most semiconductor devices in the computer industry, pure silicon has little use as a semiconductor. Instead, small amounts of impurities are added to increase its conductive properties. Adding impurities to silicon is called *doping,* and the substances added are *dopants.* The dopant is usually incorporated into just the surface layer of a silicon chip. Typical dopants include the Group 15 elements phosphorus and arsenic and the Group 13 elements boron, aluminum, gallium, and indium.

A silicon atom has four electrons in its outer energy level whereas Group 13 atoms have three and Group 15 atoms have five. Adding boron to silicon creates a mix of atoms having four valence electrons and atoms having three valence electrons. Boron atoms form only three bonds with silicon, whereas silicon forms four bonds with other silicon atoms. The unbonded spot between a silicon atom

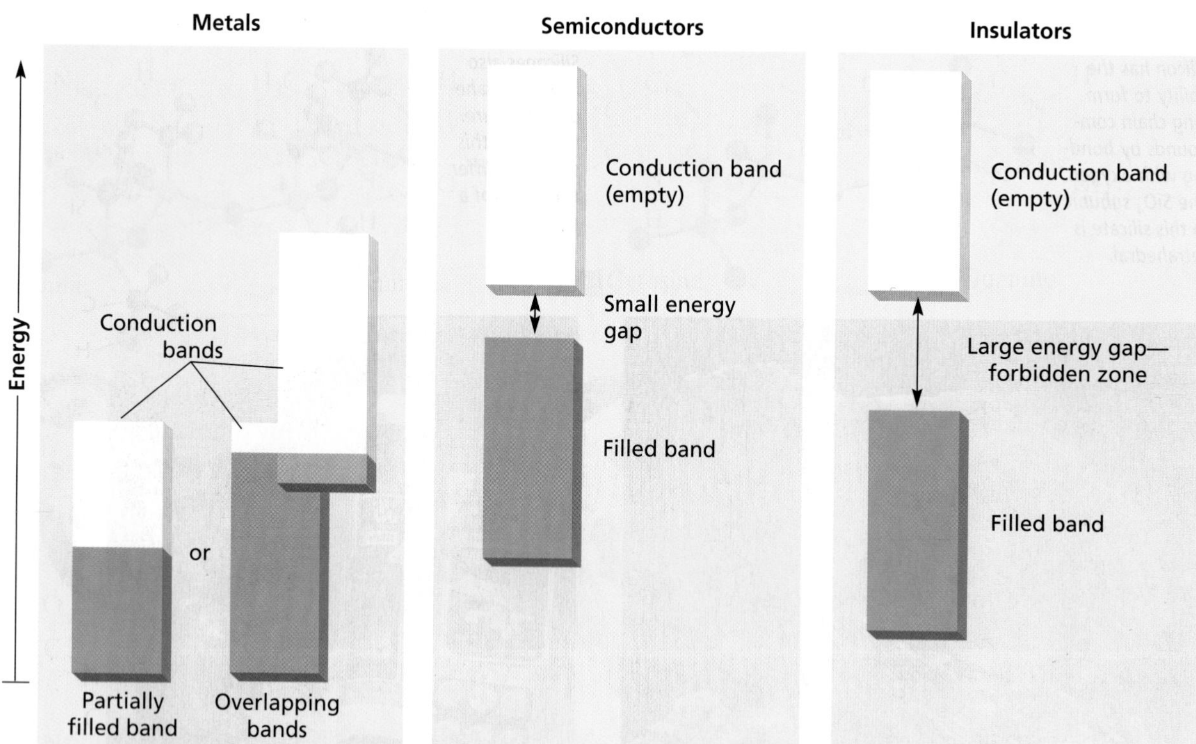

This model shows the difference in the levels of energy required to excite electrons into the conduction band in metals, semiconductors, and insulators. The forbidden zone is too great an energy gap in insulators for these elements to function as conductors. The energy gap for semiconductors is small enough that it can be crossed under certain conditions.

Dopants
Semiconductor elements
Forms semiconductor compounds

Group 1	Group 2	Group 3	Group 4	Group 5	Group 6	Group 7	Group 8	Group 9	Group 10	Group 11	Group 12	Group 13	Group 14	Group 15	Group 16	Group 17	Group 18
1 H																	2 He
3 Li	4 Be											5 B	6 C	7 N	8 O	9 F	10 Ne
11 Na	12 Mg											13 Al	14 Si	15 P	16 S	17 Cl	18 Ar
19 K	20 Ca	21 Sc	22 Ti	23 V	24 Cr	25 Mn	26 Fe	27 Co	28 Ni	29 Cu	30 Zn	31 Ga	32 Ge	33 As	34 Se	35 Br	36 Kr
37 Rb	38 Sr	39 Y	40 Zr	41 Nb	42 Mo	43 Tc	44 Ru	45 Rh	46 Pd	47 Ag	48 Cd	49 In	50 Sn	51 Sb	52 Te	53 I	54 Xe
55 Cs	56 Ba	57 La	72 Hf	73 Ta	74 W	75 Re	76 Os	77 Ir	78 Pt	79 Au	80 Hg	81 Tl	82 Pb	83 Bi	84 Po	85 At	86 Rn
87 Fr	88 Ra	89 Ac															

Semiconductor elements and dopants fall in the metalloid region of the periodic table. Semiconductor compounds often contain metals.

and a boron atom is a hole that a free electron can occupy. Because this hole "attracts" an electron, it is viewed as if it were positively charged. Semiconductors that are doped with boron, aluminum, or gallium are *p-type semiconductors,* the *p* standing for "positive." P-type semiconductors conduct electricity better than pure silicon because they provide spaces that moving electrons can occupy as they flow through the material.

Doping silicon with phosphorus or arsenic produces the opposite effect. When phosphorus is added to silicon, it forms four bonds to silicon atoms and has a nonbonding electron left over. This extra electron is free to move through the material when a voltage is applied, thus increasing its conductivity compared with pure silicon. These extra electrons have a negative charge. Therefore, the material is an *n-type semiconductor*. Compare these two types of semiconductors in the models below.

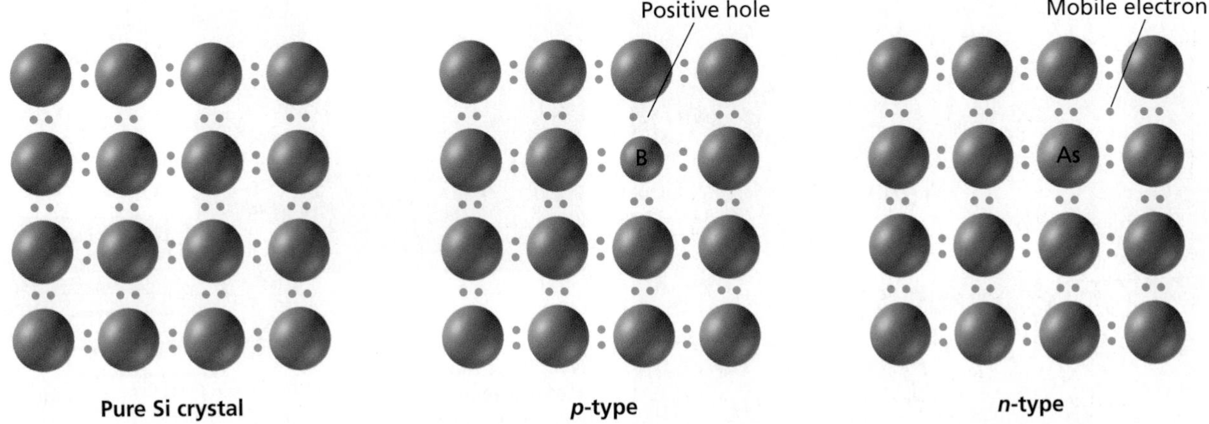

Each silicon atom in the pure crystal is surrounded by four pairs of electrons. The p-type semiconductor model contains an atom of boron with a hole that an electron can occupy. The n-type semiconductor model contains an atom of arsenic, which provides the extra electron that can move through the crystal.

GROUP 15
NITROGEN FAMILY

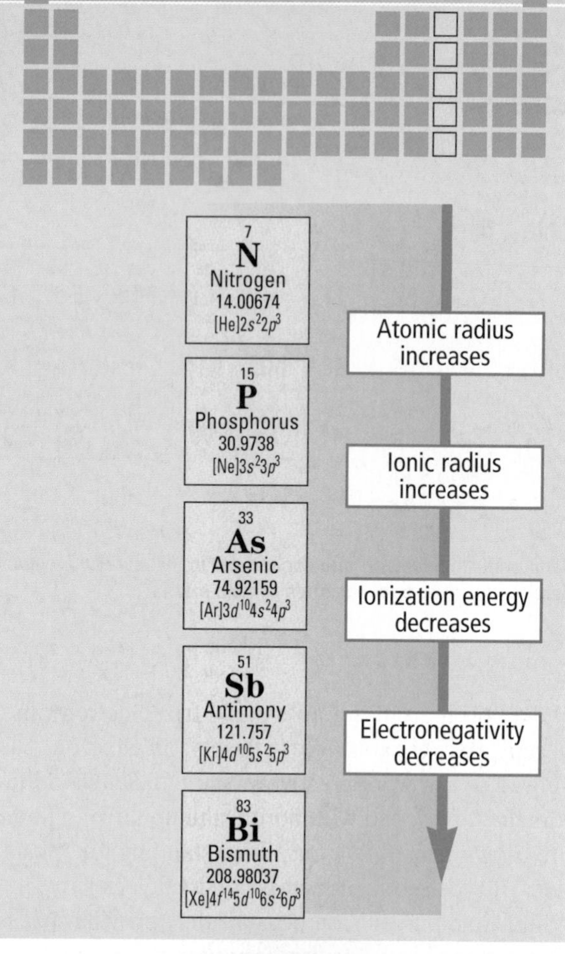

CHARACTERISTICS

- consist of two nonmetals (nitrogen and phosphorus), two metalloids (arsenic and antimony), and one metal (bismuth)
- Nitrogen is most commonly found as atmospheric N_2; phosphorus as phosphate rock; and arsenic, antimony, and bismuth as sulfides or oxides. Antimony and bismuth are also found as elements.
- range from very abundant elements (nitrogen and phosphorus) to relatively rare elements (arsenic, antimony, and bismuth)
- consist of atoms that contain five electrons in their outermost energy level
- tend to form covalent compounds, most commonly with oxidation numbers of +3 or +5
- exist in two or more allotropic forms, except nitrogen and bismuth
- are solids at room temperature, except nitrogen

You can see the contrast in physical properties among the elements of this family. Arsenic, antimony, and bismuth are shown.

Some matches contain phosphorus compounds in the match head. Safety matches contain phosphorus in the striking strip on the matchbox.

Phosphorus exists in three allotropic forms. White phosphorus must be kept underwater because it catches on fire when exposed to air. The red and black forms are stable in air.

COMMON REACTIONS

With Oxygen to Form Oxides

Example: $P_4(s) + 5O_2(g) \longrightarrow P_4O_{10}(s)$

As, Sb, and Bi follow this reaction pattern, but as monatomic elements. N reacts as N_2 to form N_2O_3 and N_2O_5.

With Metals to Form Binary Compounds

Example: $3Mg(s) + N_2(g) \longrightarrow Mg_3N_2(s)$

ANALYTICAL TEST

Other than atomic absorption spectroscopy, there are no simple analytical tests for the presence of nitrogen or phosphorus compounds in a sample. Antimony produces a pale green color in a flame test, and arsenic produces a light blue color. Arsenic, antimony, and bismuth are recognized in qualitative analyses by their characteristic sulfide colors.

Formation of sulfides is the confirmatory qualitative analysis test for the presence of bismuth, antimony, and arsenic.

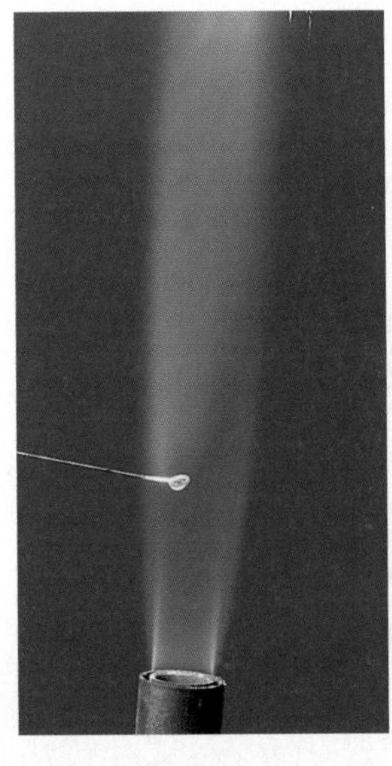

Arsenic flame test

Antimony flame test

PROPERTIES OF THE GROUP 15 ELEMENTS

	N	P*	As	Sb	Bi
Melting point (°C)	–209.86	44.1	817 (28 atm)	630.5	271.3
Boiling point (°C)	–195.8	280	613 (sublimes)	1750	1560 ± 5
Density (g/cm^3)	1.25×10^{-3}	1.82	5.727	6.684	9.80
Ionization energy (kJ/mol)	1402	1012	947	834	703
Atomic radius (pm)	70	108	124.8	145	154.7
Ionic radius (pm)	171 (N^{3-})	212 (P^{3-})	—	76 (Sb^{3+})	103 (Bi^{3+})
Common oxidation number in compounds	–3, +3, +5	–3, +3, +5	+3, +5	+3, +5	+3
Crystal structure†	cubic (as a solid)	cubic	rhombohedral	hcp	rhombohedral
Hardness (Mohs' scale)	none (gas)	—	3.5	3.0	2.25

* Data given apply to white phosphorus.

† Crystal structures are for the most common allotropes.

APPLICATION *Biology*

Plants and Nitrogen

All organisms, including plants, require certain elements to survive and grow. These elements include carbon, hydrogen, oxygen, nitrogen, phosphorus, potassium, sulfur, and several other elements needed in small amounts. An organism needs nitrogen to synthesize structural proteins, enzymes, and the nucleic acids DNA and RNA.

Carbon, hydrogen, and oxygen are available to plants from carbon dioxide in the air and from water in both the air and the soil. Although nitrogen gas, N_2, makes up 78% of air and is necessary for plants' survival, plants cannot take nitrogen out of the air and incorporate it into their cells. Plants need nitrogen in the form of a compound that they can take in and use. The strong triple covalent bond in N_2 is not easily broken. The process of using atmospheric N_2 to make NH_3 is called *nitrogen fixation*. Several kinds of nitrogen-fixing bacteria live in the soil and in the root nodules of plants called legumes. Legumes obtain the nitrogen they need through a symbiotic relationship with nitrogen-fixing bacteria. Legumes include peas, beans, clover, alfalfa, and locust trees. The bacteria convert nitrogen into ammonia, NH_3, which is then absorbed by the host plants.

Because wheat, rice, corn, and potatoes cannot perform the same feat as legumes, these plants depend on nitrogen-fixing bacteria in the soil. Soil bacteria convert NH_3 into nitrate ions, NO_3^-, the form of nitrogen that can be absorbed and used by plants. These plants also often need nitrogen fertilizers to supplement the work of the bacteria. Besides supplying nitrogen, fertilizers are manufactured to contain phosphorus, potassium, and trace minerals.

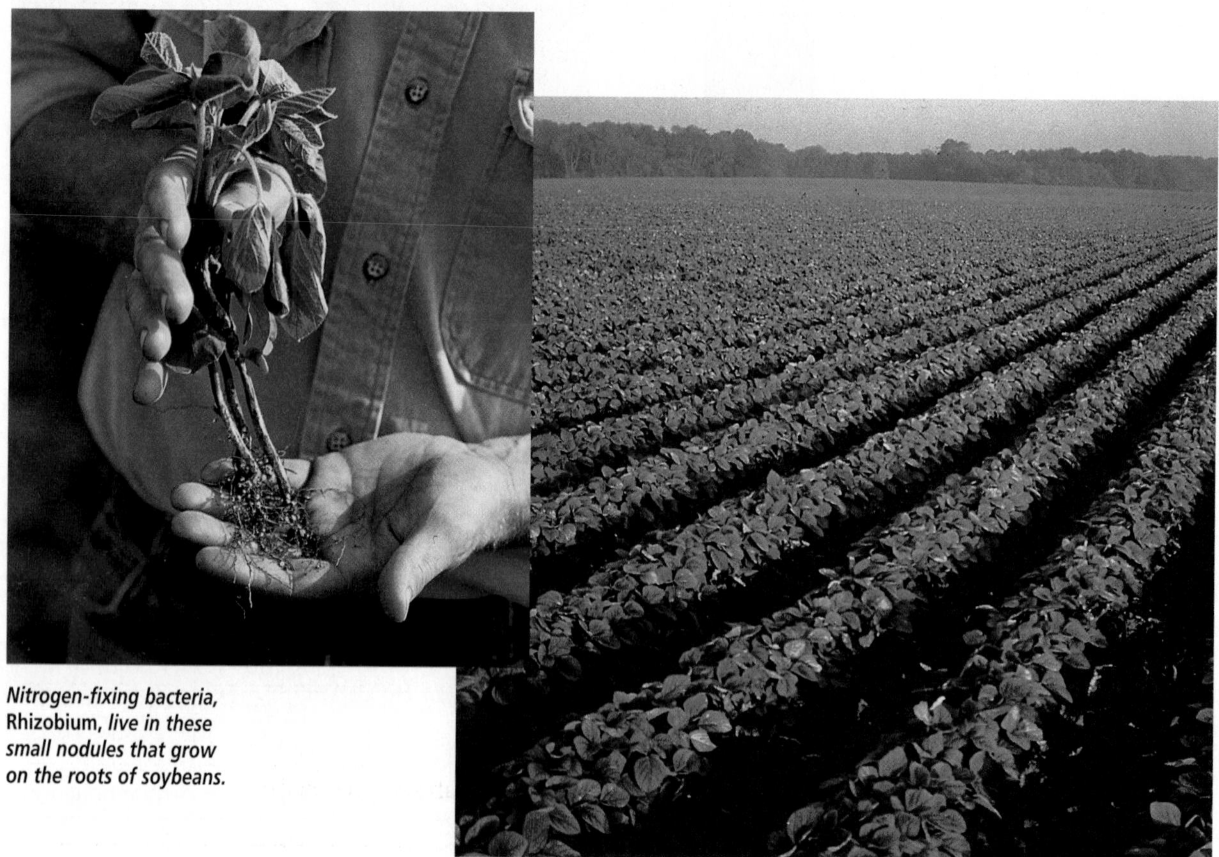

Nitrogen-fixing bacteria, Rhizobium, *live in these small nodules that grow on the roots of soybeans.*

Soybeans are legumes that live in a symbiotic relationship with nitrogen-fixing bacteria.

APPLICATION Chemical Industry

Fertilizers

Fertilizers can supply nitrogen to plants in the form of ammonium sulfate, ammonium nitrate, and urea, all of which are made from NH_3. Now you know why there is such a demand for ammonia. Though some soils contain sufficient concentrations of phosphorus and potassium, most soils need additional nitrogen for adequate plant growth. Ammonia, ammonium nitrate, or urea can fill that need.

Most fertilizers contain all three major plant nutrients N, P, and K, and are called *complete fertilizers*. A typical complete fertilizer might contain ammonium nitrate or sodium nitrate to provide nitrogen. Calcium dihydrogen phosphate, $Ca(H_2PO_4)_2$, or the anhydrous form of phosphoric acid, P_2O_5, can provide phosphorus. Potassium chloride, KCl, contains sufficient K_2O impurities to provide potassium.

The proportion of each major nutrient in a fertilizer is indicated by a set of three numbers printed on the container. These numbers are the N-P-K formula of the fertilizer and indicate the percentage of N, P, and K, respectively. A fertilizer graded as 6-12-6, for example, contains 6% nitrogen, 12% phosphorus, and 6% potassium by weight and all in the form of compounds.

Nitrogen stimulates overall plant growth. Phosphorus promotes root growth and flowering. Potassium regulates the structures in leaves that allow CO_2 to enter the leaf and O_2 and H_2O to exit. Fertilizers are available in N-P-K formulas best suited for their intended use. For example, plants that produce large amounts of carbohydrates (sugars) need more potassium than most other types of plants. Grain crops need higher concentrations of phosphorus. Lawn fertilizers applied in the spring are generally high in nitrogen to stimulate shoot growth in grasses. Lawn fertilizers applied in the fall of the year should have a higher phosphorus content to stimulate root growth during the winter.

TABLE 6A Some Commercial Fertilizers and Uses

Fertilizer composition (N-P-K)	Uses
1-2-1 ratio 10-20-10 15-30-15	early-spring application for trees and shrubs with flowers and fruit; general-purpose feedings of the following: cucumbers, peppers, tomatoes
3-1-2 ratio 12-4-8 15-5-10 21-7-4 16-4-8 20-5-10	lawns and general-purpose feedings of the following: trees, shrubs, most berries, apple trees, grapes, vines, walnut trees, broccoli, cabbage, carrots, onions
High nitrogen 33-0-0 21-0-0 40-4-4 36-6-6	pecan trees, lawns, early feedings of corn
Balanced 13-13-13	general purpose feeding of the following: broccoli, cabbage, melons, potatoes
Special purpose: acid-loving flowering shrubs 12-10-4	azaleas, rhododendrons, camellias, gardenias
Special purpose 18-24-16	roses
Special purpose: flowering 12-55-6	flowering plants and shrubs (annuals and perennials)
Special purpose: root growth 5-20-10	starter fertilizer for transplants

GROUP 16
OXYGEN FAMILY

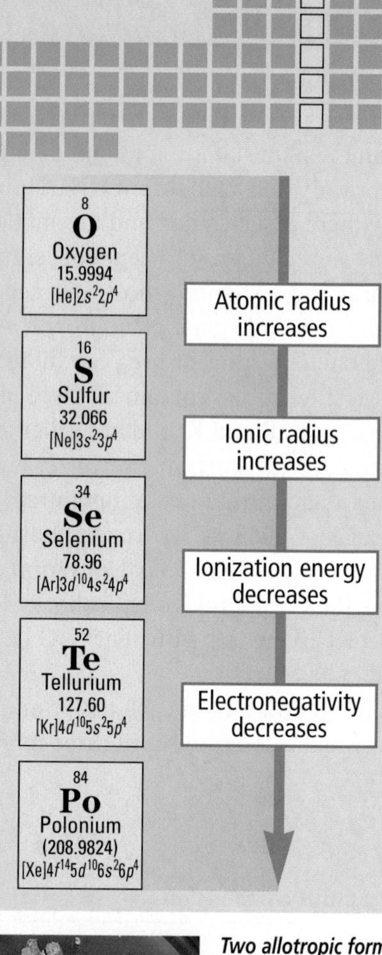

CHARACTERISTICS

- occur naturally as elements and in combined states
- consist of three nonmetals (oxygen, sulfur, and selenium), one metalloid (tellurium), and one metal (polonium)
- consist of atoms that have six electrons in their outermost energy level
- tend to form covalent compounds with other elements
- exist in several allotropic forms
- tend to exist as diatomic and polyatomic molecules, such as O_2, O_3, S_6, S_8, and Se_8
- commonly exist in compounds with the −2 oxidation state but often exhibit other oxidation states

Sulfur is found naturally in underground deposits and in the steam vents near volcanoes.

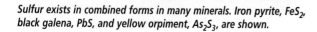

Sulfur exists in combined forms in many minerals. Iron pyrite, FeS_2, black galena, PbS, and yellow orpiment, As_2S_3, are shown.

Two allotropic forms of sulfur are orthorhombic and monoclinic. Each has a different crystal structure.

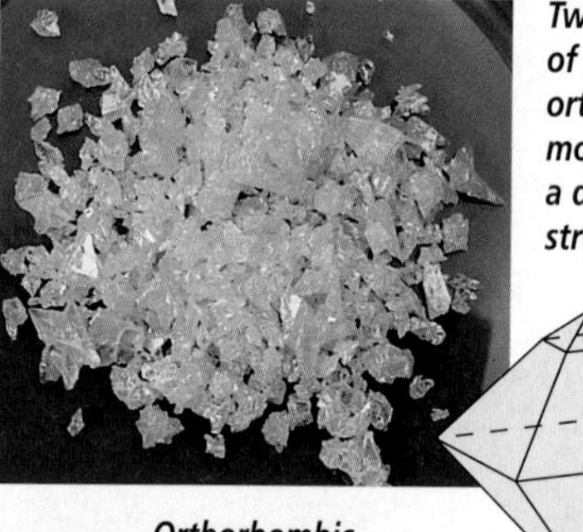

Orthorhombic

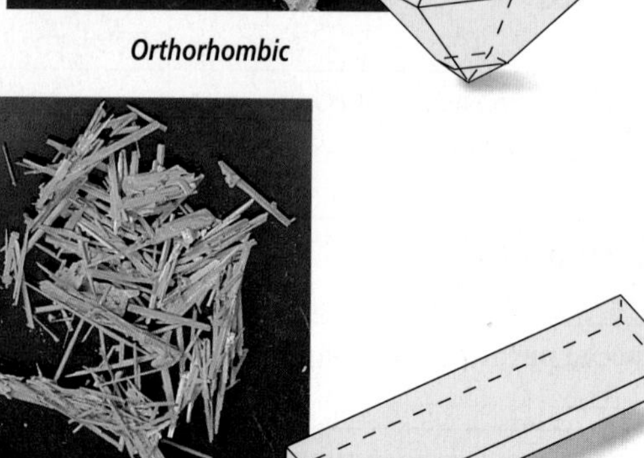

Monoclinic

COMMON REACTIONS

With Metals to Form Binary Compounds

Example: $8Mg(s) + S_8(l) \longrightarrow 8MgS(s)$

O_2, Se, and Te follow this pattern in reacting with Na, K, Ca, Mg, and Al.

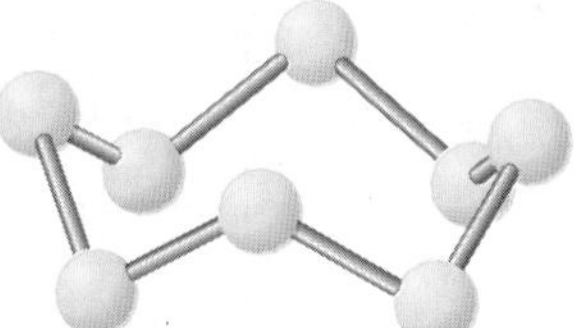

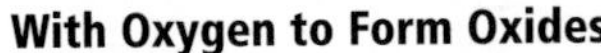

With Oxygen to Form Oxides

Example: $Se(s) + O_2(g) \longrightarrow SeO_2(s)$

S, Te, and Po follow this pattern. S, Se, and Te can form SO_3, SeO_3, and TeO_3.

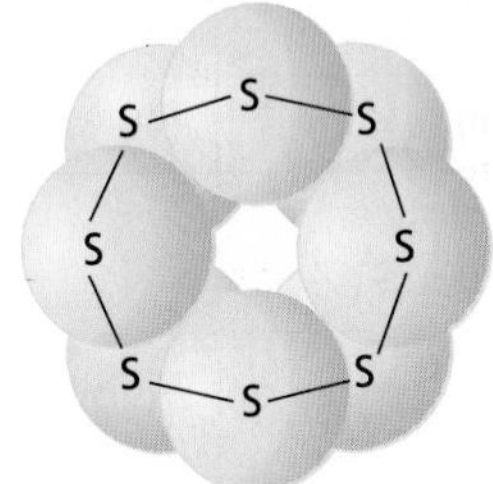

With Halogens to Form Binary Compounds

Example: $S_8(l) + 8Cl_2(g) \longrightarrow 8SCl_2(l)$

O, Se, Te, and Po follow this pattern in reacting with F_2, Cl_2, Br_2, and I_2.

With Hydrogen to Form Binary Compounds

$2H_2(g) + O_2(g) \longrightarrow 2H_2O(l)$

Sulfur exists as S_8 molecules in which the atoms are bonded in a ring, as shown by the ball-and-stick and space-filling models.

ANALYTICAL TEST

Other than atomic absorption spectroscopy, there is no simple analytical test to identify all elements of this family. Selenium and tellurium can be identified by flame tests. A light blue flame is characteristic of selenium, and a green flame is characteristic of tellurium. Oxygen can be identified by the splint test, in which a glowing splint bursts into flame when thrust into oxygen. Elemental sulfur is typically identified by its physical characteristics, especially its color and its properties when heated. It melts to form a viscous brown liquid and burns with a blue flame.

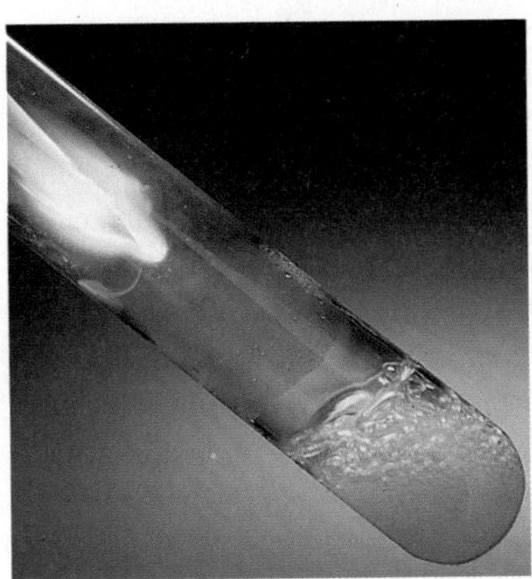

A glowing splint thrust into oxygen bursts into a bright flame.

Sulfur burns with a characteristically deep blue flame.

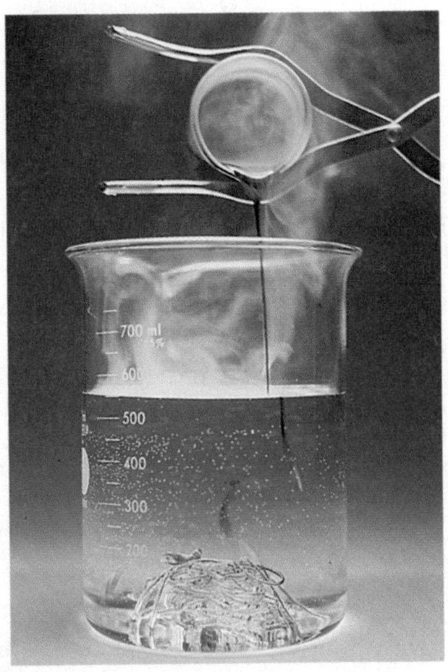

Molten sulfur returns to its orthorhombic form upon cooling.

PROPERTIES OF THE GROUP 16 ELEMENTS

	O	S	Se	Te	Po
Melting point (°C)	−218.4	119.0	217	449.8	254
Boiling point (°C)	−182.962	444.674	685	989.9	962
Density (g/cm^3)	1.429×10^{-3}	1.96	4.82	6.24	9.4
Ionization energy (kJ/mol)	1314	1000	941	869	812
Atomic radius (pm)	73	106	116	142	164
Ionic radius (pm)	140	184	198	221	—
Common oxidation number in compounds	−2	−2, +4, +6	−2, +2, +4, +6	−2, +2, +4, +6	−2, +2, +4, +6
Crystal structure*	orthorhombic, rhombohedral, cubic (when solid)	orthorhombic, monoclinic	hexagonal	hexagonal	cubic, rhombohedral
Hardness (Mohs' scale)	none (gas)	2.0	2.0	2.3	—

* Most elements of this family can have more than one crystal structure.

APPLICATION *Chemical Industry*

Oxides

Oxides of the reactive metals are ionic compounds. The oxide ion from any soluble oxide reacts immediately with water to form hydroxide ions as represented by the following equation.

$$O^{2-}(aq) + H_2O(l) \longrightarrow 2OH^-(aq)$$

The reactive metal oxides of Groups 1 and 2 react vigorously with water and release a large amount of heat. The product of the reaction is a metal hydroxide. The following equation is an example of this reaction.

$$Na_2O(s) + H_2O(l) \longrightarrow 2NaOH(aq)$$

A basic oxide can be thought of as the dehydrated form of a hydroxide base. Oxides of the less reactive metals, such as magnesium, can be prepared by using thermal decomposition to drive off the water.

$$Mg(OH)_2(s) \xrightarrow{heat} MgO(s) + H_2O(g)$$

Hydroxides of the reactive metals of Group 1 are too stable to decompose in this manner.

If a hydroxide formed by a metal oxide is water-soluble, it dissolves to form a basic solution. An oxide that reacts with water to form a basic solution is called a basic oxide or a basic anhydride. Table 7A on the next page lists oxides that form basic solutions with water.

Molecular Oxides

Nonmetals, located on the right side of the periodic table, form molecular oxides. For example, sulfur forms two gaseous oxides: sulfur dioxide, SO_2, and sulfur trioxide, SO_3. In reactions typical of nonmetal oxides, each of the sulfur oxides reacts with water to form an oxyacid.

An oxide that reacts with water to form an acid is called an acidic oxide or an acid anhydride. As with the basic anhydrides, each acid anhydride can be thought of as the dehydrated form of the appropriate oxyacid. For example, when sulfuric acid decomposes, the loss of H_2O leaves the oxide SO_3, which is an anhydride.

$$H_2SO_4(aq) \xrightarrow{heat} H_2O(g) + SO_3(g)$$

Amphoteric Oxides

Table 7A lists some common oxides of main-group elements. You can see that the active metal oxides are basic and that the nonmetal oxides are acidic. Between these lies a group of oxides, the *amphoteric oxides*. The bonding in amphoteric oxides is intermediate between ionic and covalent bonding. As a result, oxides of this type show behavior intermediate between that of acidic oxides and basic oxides, and react as both acids and bases.

Aluminum oxide, Al_2O_3, is a typical amphoteric oxide. With hydrochloric acid, aluminum oxide acts as a base. The reaction produces a salt and water.

$$Al_2O_3(s) + 6HCl(aq) \longrightarrow 2AlCl_3(aq) + 3H_2O(l)$$

With aqueous sodium hydroxide, aluminum oxide acts as an acid. The reaction forms a soluble ionic compound and water. That compound contains aluminate ions, AlO_2^-. (The AlO_2^- formula is used here rather than the more precise hydrated aluminate formula, $Al(OH)_4^-$.)

$$Al_2O_3(s) + 2NaOH(aq) \longrightarrow 2NaAlO_2(aq) + H_2O(l)$$

Reactions of Oxides

In the reaction between an acid and a metal oxide, the products are a salt and water—the same as the products in a neutralization reaction. For example, when magnesium oxide reacts with dilute sulfuric acid, magnesium sulfate and water are produced.

$$MgO(s) + H_2SO_4(dil.\ aq) \longrightarrow MgSO_4(aq) + H_2O(l)$$

The reaction between a basic metal oxide, such as MgO, and an acidic nonmetal oxide, such as CO_2, tends to produce an oxygen-containing salt. The dry oxides are mixed and heated without water. Salts such as metal carbonates, phosphates, and sulfates can be made by this synthesis reaction.

$$MgO(s) + CO_2(g) \longrightarrow MgCO_3(s)$$

$$6CaO(s) + P_4O_{10}(s) \longrightarrow 2Ca_3(PO_4)_2(s)$$

$$CaO(s) + SO_3(g) \longrightarrow CaSO_4(s)$$

Reactions of Hydroxides with Nonmetal Oxides

Nonmetal oxides tend to be acid anhydrides. The reaction of a hydroxide base with a nonmetal oxide is an acid-base reaction. The product is either a salt or a salt and water, depending on the identities and relative quantities of reactants. For example, 2 mol of the hydroxide base sodium hydroxide and 1 mol of the nonmetal oxide carbon dioxide form sodium carbonate, which is a salt, and water.

$$CO_2(g) + 2NaOH(aq) \longrightarrow Na_2CO_3(aq) + H_2O(l)$$

However, if sodium hydroxide is limited, only sodium hydrogen carbonate is produced.

$$CO_2(g) + NaOH(aq) \longrightarrow NaHCO_3(aq)$$

TABLE 7A Periodicity of Acidic and Basic Oxides of Main-Group Elements

Group Number						
1	2	13	14	15	16	17
Li_2O basic	BeO amphoteric	B_2O_3 acidic	CO_2 acidic	N_2O_5 acidic		
Na_2O basic	MgO basic	Al_2O_3 amphoteric	SiO_2 acidic	P_4O_{10} acidic	SO_3 acidic	Cl_2O acidic
K_2O basic	CaO basic	Ga_2O_3 amphoteric	GeO_2 amphoteric	As_4O_6 amphoteric	SeO_3 acidic	
Rb_2O basic	SrO basic	In_2O_3 basic	SnO_2 amphoteric	Sb_4O_6 amphoteric	TeO_3 acidic	I_2O_5 acidic
Cs_2O basic	BaO basic	Tl_2O_3 basic	PbO_2 amphoteric	Bi_2O_3 basic		

APPLICATION *The Environment*

Ozone

Ozone, O_3, is an allotrope of oxygen that is important for life on Earth. Like O_2, O_3 is a gas at room temperature. However, unlike O_2, O_3 is a poisonous bluish gas with an irritating odor at high concentrations. The triatomic ozone molecule is bent with a bond angle of about 116.5°. The O—O bonds in ozone are shorter and stronger than a single bond, but longer and weaker than a double bond. The ozone molecule is best represented by two resonance hybrid structures.

$$:\ddot{O}{=}\ddot{O}{-}\ddot{O}: \longleftrightarrow :\ddot{O}{-}\ddot{O}{=}\ddot{O}:$$

Ozone forms naturally in Earth's atmosphere more than 24 km above the Earth's surface in a layer called the stratosphere. There, O_2 molecules absorb energy from ultraviolet light and split into free oxygen atoms.

$$O_2(g) \xrightarrow{\text{ultraviolet light}} 2O$$

A free oxygen atom has an unpaired electron and is highly reactive. A chemical species that has one or more unpaired or unshared electrons is referred to as a *free radical*. A free radical is a short-lived fragment of a molecule. The oxygen free radical can react with a molecule of O_2 to produce an ozone molecule.

$$O + O_2(g) \longrightarrow O_3(g)$$

A molecule of O_3 can then absorb ultraviolet light and split to produce O_2 and a free oxygen atom.

$$O_3(g) \xrightarrow{\text{ultraviolet light}} O_2(g) + O$$

The production and breakdown of ozone in the stratosphere are examples of *photochemical* processes, in which light causes a chemical reaction.

In this way, O_3 is constantly formed and destroyed in the stratosphere, and its concentration is determined by the balance among these reactions. The breakdown of ozone absorbs the sun's intense ultraviolet light in the range of wavelengths between 290 nm and 320 nm. Light of these wavelengths damages and kills living cells, so if these wavelengths were to reach Earth's surface in large amounts, life would be impossible. Even now, the normal amount of ultraviolet light reaching Earth's surface is a major cause of skin cancer and the damage to DNA molecules that causes mutations. One life-form that is very sensitive to ultraviolet radiation is the phytoplankton in the oceans. These organisms carry out photosynthesis and are the first level of oceanic food webs.

Ozone and Air Pollution

Ozone in the lower atmosphere is a harmful pollutant. Ozone is highly reactive and can oxidize organic compounds. The products of these reactions are harmful substances that, when mixed with air, water vapor, and dust, make up *photochemical smog*. This mixture is the smog typically found in cities.

Typically, ozone is produced in a complex series of reactions involving unburned hydrocarbons and nitrogen oxides given off from engines in the form of exhaust and from fuel-burning power plants. When fuel burns explosively in the cylinder of an internal-combustion engine, some of the nitrogen in the cylinder also combines with oxygen to form NO, a very reactive nitrogen-oxide free radical.

$$N_2(g) + O_2(g) \longrightarrow 2NO$$

When the free radical reaches the air, it reacts with oxygen to produce NO_2 radicals, which react with water in the air to produce HNO_3.

$$2NO + O_2(g) \longrightarrow 2NO_2$$
$$3NO_2 + H_2O(l) \longrightarrow NO + 2HNO_3(aq)$$

In sunlight, nitrogen dioxide decomposes to give nitric oxide and an atom of oxygen. Note that the NO produced is free to undergo the previous reaction once more.

$$NO_2 \xrightarrow{\text{sunlight}} NO + O$$

Just as it is in the stratosphere, a free oxygen atom in the lower atmosphere is highly reactive and reacts with a molecule of diatomic oxygen to form ozone.

$$O + O_2(g) \longrightarrow O_3(g)$$

APPLICATION *Chemical Industry*

Sulfuric Acid

Sulfuric acid is the so-called "king of chemicals" because it is produced in the largest volume in the United States. It is produced by the contact process. This process starts with the production of SO_2 by burning sulfur or roasting iron pyrite, FeS_2. The purified sulfur dioxide is mixed with air and passed through hot iron pipes containing a catalyst. The contact between the catalyst, SO_2, and O_2 produces sulfur trioxide, SO_3, and gives the contact process its name. SO_3 is dissolved in concentrated H_2SO_4 to produce pyrosulfuric acid, $H_2S_2O_7$.

$$SO_3(g) + H_2SO_4(aq) \longrightarrow H_2S_2O_7(aq)$$

The pyrosulfuric acid is then diluted with water to produce sulfuric acid.

$$H_2S_2O_7(aq) + H_2O(l) \longrightarrow 2H_2SO_4(aq)$$

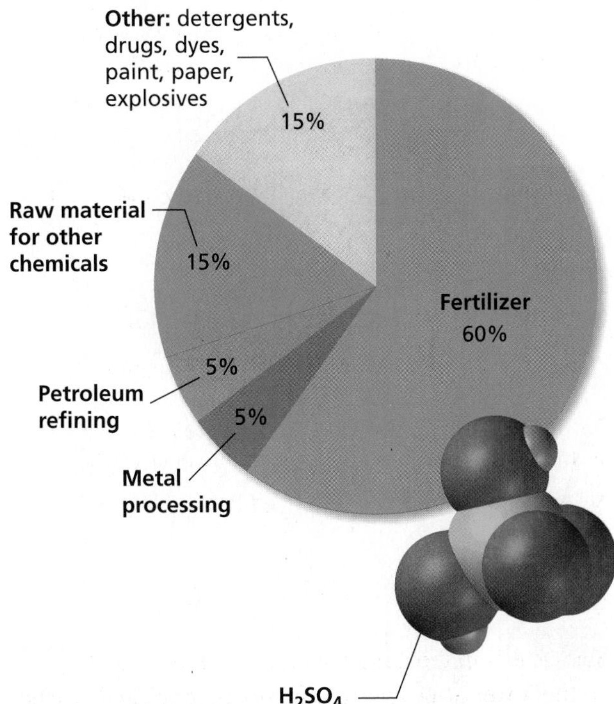

Important uses of the U.S. supply of sulfuric acid

Properties and Uses of Sulfuric Acid

Concentrated sulfuric acid is a good oxidizing agent. During the oxidation process, sulfur is reduced from +6 to +4 or −2. The change in oxidation state for a reaction depends on the concentration of the acid and on the nature of the reducing agent used in the reaction.

Sulfuric acid is also an important dehydrating agent. Gases that do not react with H_2SO_4 can be dried by being bubbled through concentrated sulfuric acid. Organic compounds, like sucrose, are dehydrated to leave carbon, as shown by the following reaction.

$$C_{12}H_{22}O_{11}(s) + 11H_2SO_4(aq) \longrightarrow 12C(s) + 11H_2SO_4{\cdot}H_2O(l)$$

The decomposition of sucrose proceeds rapidly, as shown in Figure 21-9 on page 684.

About 60% of the sulfuric acid produced in this country is used to make superphosphate, which is a mixture of phosphate compounds used in fertilizers.

TABLE 7B Top Ten Chemicals Produced in the U.S.

Rank	Chemical	Physical state	Formula
1	sulfuric acid	*l*	H_2SO_4
2	nitrogen	*g*	N_2
3	oxygen	*g*	O_2
4	ethylene	*g*	C_2H_4
5	calcium oxide (lime)	*s*	CaO
6	ammonia	*g*	NH_3
7	phosphoric acid	*l*	H_3PO_4
8	sodium hydroxide	*s*	NaOH
9	propylene	*g*	C_3H_6
10	chlorine	*g*	Cl_2

GROUP 17
HALOGEN FAMILY

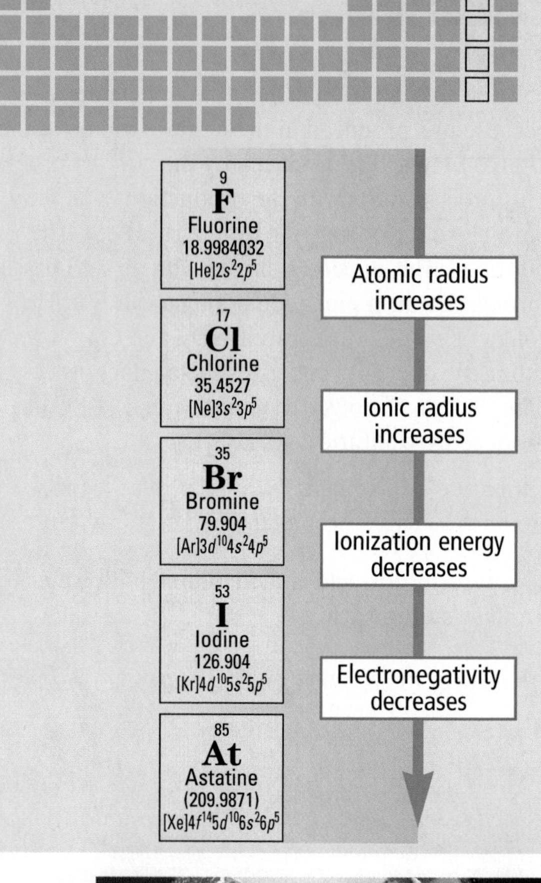

CHARACTERISTICS

- are all nonmetals and occur in combined form in nature, mainly as metal halides
- are found in the rocks of Earth's crust and dissolved in sea water
- range from fluorine, the 13th most abundant element, to astatine, which is one of the rarest elements
- exist at room temperature as a gas (F_2 and Cl_2), a liquid (Br_2), and a solid (I_2 and At)
- consist of atoms that have seven electrons in their outermost energy level
- tend to gain one electron to form a halide, X^- ion, but also share electrons and have positive oxidation states
- are reactive, with fluorine being the most reactive of all nonmetals

Halogens are the only family that contains elements representing all three states of matter at room temperature. Chlorine is a yellowish green gas; bromine is a reddish brown liquid; and iodine is a purple-black solid.

Iodine sublimes to produce a violet vapor that recrystallizes on the bottom of the evaporating dish filled with ice.

COMMON REACTIONS*

With Metals to Form Halides

Example: $Mg(s) + Cl_2(g) \longrightarrow MgCl_2(s)$
Example: $Sn(s) + 2F_2(g) \longrightarrow SnF_4(s)$
The halide formula depends on the oxidation state of the metal.

With Hydrogen to Form Hydrogen Halides

Example: $H_2(g) + F_2(g) \rightarrow 2HF(g)$
Cl_2, Br_2, and I_2 also follow this pattern.

With Nonmetals and Metalloids to Form Halides

Example: $Si(s) + 2Cl_2(g) \longrightarrow SiCl_4(s)$
Example: $N_2(g) + 3F_2(g) \longrightarrow 2NF_3(g)$
Example: $P_4(s) + 6Br_2(l) \longrightarrow 4PBr_3(s)$
The formula of the halide depends on the oxidation state of the metalloid or nonmetal.

With Other Halogens to Form Interhalogen Compounds

Example: $Br_2(l) + 3F_2(g) \longrightarrow 2BrF_3(l)$

* Chemists assume that astatine undergoes similar reactions, but few chemical tests have been made.

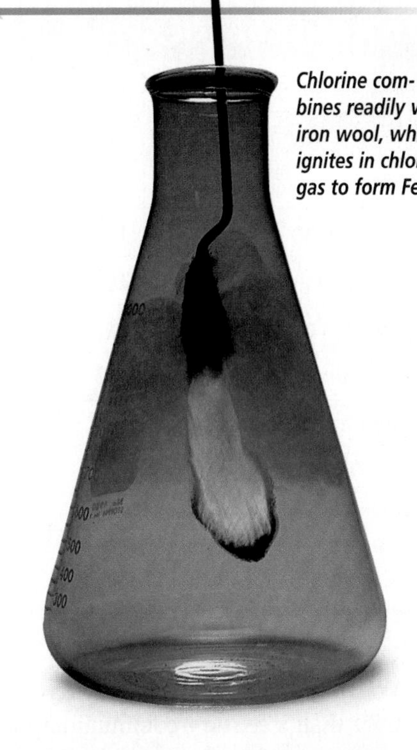

Chlorine combines readily with iron wool, which ignites in chlorine gas to form $FeCl_3$.

Hydrofluoric acid is used to etch patterns into glass.

Shown here from left to right are precipitates of AgCl, AgBr, and AgI.

ANALYTICAL TEST

As with most elements, the presence of each of the halogens can be determined by atomic absorption spectroscopy. Fluorides react with concentrated sulfuric acid, H_2SO_4, to release hydrogen fluoride gas. Three of the halide ions can be identified in solution by their reactions with silver nitrate.

$Cl^-(aq) + Ag^+(aq) \longrightarrow AgCl(s)$
$Br^-(aq) + Ag^+(aq) \longrightarrow AgBr(s)$
$I^-(aq) + Ag^+(aq) \longrightarrow AgI(s)$

PROPERTIES OF THE GROUP 17 ELEMENTS

	F	Cl	Br	I	At
Melting point (°C)	–219.62	–100.98	–7.2	113.5	575.2
Boiling point (°C)	–188.14	–34.6	58.78	184.35	610
Density (g/cm^3)	1.69×10^{-3}	3.214×10^{-3}	3.119	4.93	not known
Ionization energy (kJ/mol)	1681	1251	1140	1008	—
Atomic radius (pm)	71.7	99	114	133	140
Ionic radius (pm)	133	181	196	220	—
Common oxidation number in compounds	–1	–1, +1, +3, +5, +7	–1, +1, +3, +5, +7	–1, +1, +3, +5, +7	–1, +5
Crystal structure	cubic	orthorhombic	orthorhombic	orthorhombic	not known

APPLICATION *The Environment*

Chlorine in Water Treatment

For more than a century, communities have treated their water to reduce the amount of toxic substances in the water and to prevent disease. A treatment process widely used in the United States is chlorination. All halogens kill bacteria and other microorganisms. Chlorine, however, is the only halogen acceptable for large-scale treatment of public water supplies.

When chlorine is added to water, the following reaction produces HCl and hypochlorous acid, HOCl.

$$Cl_2(g) + H_2O(l) \longrightarrow HCl(aq) + HOCl(aq)$$

Hypochlorous acid is a weak acid that ionizes to give hydrogen ions and hypochlorite ions, OCl^-.

$$HOCl(aq) + H_2O(l) \longrightarrow H_3O^+(aq) + OCl^-(aq)$$

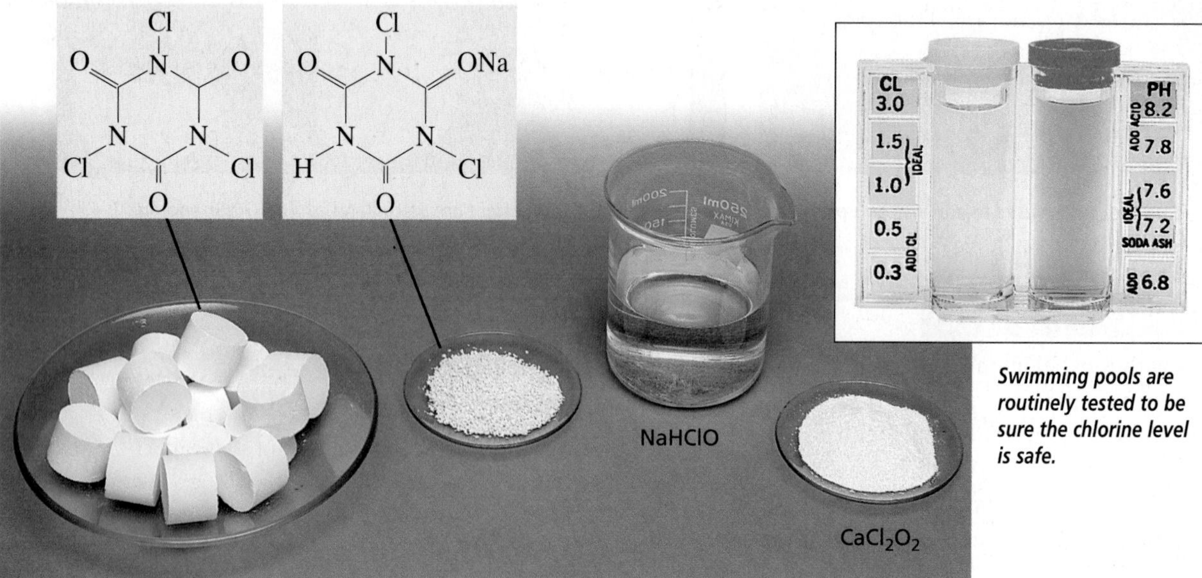

The "chlorine" used in swimming pools is really the compounds shown above and not chlorine at all.

Swimming pools are routinely tested to be sure the chlorine level is safe.

The OCl^- ions are strong oxidizing agents that can destroy microorganisms.

In some water-treatment plants, calcium hypochlorite, $Ca(ClO)_2$, a salt of hypochlorous acid, is added to water to provide OCl^- ions. Similar treatments are used in swimming pools.

Nearly a hundred cities in the United States and thousands of communities in Europe use chlorine in the form of chlorine dioxide, ClO_2, as their primary means of disinfecting water. The main drawback to the use of ClO_2 is that it is unstable and cannot be stored. Instead, ClO_2 must be prepared on location by one of the following reactions involving sodium chlorite, $NaClO_2$.

$$10NaClO_2(aq) + 5H_2SO_4(aq) \longrightarrow 8ClO_2(g) + 5Na_2SO_4(aq) + 2HCl(aq) + 4H_2O(l)$$

$$2NaClO_2(aq) + Cl_2(g) \longrightarrow 2ClO_2(g) + 2NaCl(aq)$$

The expense of using ClO_2 makes it less desirable than Cl_2 in water-treatment systems unless there are other considerations. For example, the use of ClO_2 is likely to result in purified water with less of the aftertaste and odor associated with water purified by Cl_2.

Fluoride and Tooth Decay

In the 1940s, scientists noticed that people living in communities that have natural water supplies with high concentrations of fluoride ions, F^-, have significantly lower rates of dental caries (tooth decay) than most of the population.

In June 1944, a study on the effects of water fluoridation began in two Michigan cities, Muskegon and Grand Rapids, where the natural level of fluoride in drinking water was low (about 0.05 ppm). In Grand Rapids, sodium fluoride, NaF, was added to the drinking water to raise levels to 1.0 ppm. In Muskegon, no fluoride was added. Also included in the study was Aurora, Illinois, a city that was similar to Grand Rapids and Muskegon, except that it had a natural F^- concentration of 1.2 ppm in the water supply. After 10 years, the rate of tooth decay in Grand Rapids had dropped far below that in Muskegon and was about the same as it was in Aurora.

Tooth enamel is made of a strong, rocklike material consisting mostly of calcium hydroxyphosphate, $Ca_5(PO_4)_3(OH)$, also known as apatite. Apatite is an insoluble and very hard compound—ideal for tooth enamel. Sometimes, however, saliva becomes more acidic, particularly after a person eats a high-sugar meal. Acids ionize to produce hydronium ions, which react with the hydroxide ion, OH^-, in the apatite to form water. The loss of OH^- causes the apatite to dissolve.

$$Ca_5(PO_4)_3(OH)(s) + H_3O^+(aq) \longrightarrow 5Ca^{2+}(aq) + 3PO_4^{3-}(aq) + 2H_2O(l)$$

Saliva supplies more OH^- ions, and new apatite is formed, but slowly.

If fluoride ions are present in saliva, some fluorapatite, $Ca_5(PO_4)_3F$, also forms.

$$5Ca^{2+}(aq) + 3PO_4^{3-}(aq) + F^-(aq) \longrightarrow Ca_5(PO_4)_3F(s)$$

Fluorapatite resists attack by acids, so the tooth enamel resists decay better than enamel containing no fluoride.

When the beneficial effect of fluoride had been established, public health authorities proposed that fluoride compounds be added to water supplies in low-fluoride communities. Fluoridation started in the 1950s, and by 1965, nearly every medical and dental association in the United States had endorsed fluoridation of water supplies. In the past decade, however, that trend slowed as opposition to fluoridation grew.

It is now known that high concentrations of fluorides are dangerous to human health. One anti-fluoridation organization claims that fluorides can cause or contribute to diabetes, heart disease, stroke, cancer, miscarriages, stillbirths, and disabling childhood diseases. Although little scientific evidence exists for such claims, we do know that the intake of too much fluoride can cause fluorosis, in which teeth become mottled with brown spots. Fluorosis seems to occur when a person takes in too much fluoride in addition to that in the water supply.

Laboratory Program

900 ml
± 5%
800
700
600
500
400
No. 9820
No. 9820
No. 9820
No. 9820
No. 9820
No. 9820

Safety in the Chemistry Laboratory

Any chemical can be dangerous if it is misused. Always follow the instructions for the experiment. Pay close attention to the safety notes. Do not do anything differently unless told to do so by your teacher.

Chemicals, even water, can cause harm. The challenge is to know how to use chemicals correctly. If you follow the rules stated below, pay attention to your teacher's directions, follow cautions on chemical labels and in the experiments, then you will be using chemicals correctly.

THESE SAFETY RULES ALWAYS APPLY IN THE LAB

1. **Always wear a lab apron and safety goggles.** Even if you aren't working on an experiment at the time, laboratories contain chemicals that can damage your clothing. Keep the apron strings tied.

   Some chemicals can cause eye damage, and even blindness. If your safety goggles are uncomfortable or cloud up, ask your teacher for help. Try lengthening the strap, washing the goggles with soap and warm water, or using an anti-fog spray.

2. **No contact lenses in the lab.** Even while wearing safety goggles, chemicals can get between contact lenses and your eyes and cause irreparable eye damage. If your doctor requires that you wear contact lenses instead of glasses, then you should wear eye-cup safety goggles in the lab. Ask your doctor or your teacher how to use this very important and special eye protection.

3. **NEVER work alone in the laboratory.** You should do lab work only under the supervision of your teacher.

4. **Wear the right clothing for lab work.** Necklaces, neckties, dangling jewelry, long hair. and loose clothing can knock things over or catch on fire. Tuck in neckties or take them off. Do not wear a necklace or other dangling jewelry, including hanging earrings. It might also be a good idea to remove your wristwatch so that it is not damaged by a chemical splash.

   Pull back long hair, and tie it in place. Nylon and polyester fabrics burn and melt more readily than cotton, so wear cotton clothing if you can. It's best to wear fitted garments, but if your clothing is loose or baggy, tuck it in or tie it back so that it does not get in the way or catch on fire.

   Wear shoes that will protect your feet from chemical spills—no open-toed shoes or sandals, and no shoes with woven leather straps. Shoes made of solid leather or polymer are much better than shoes made of cloth. It is also important to wear pants, not shorts or skirts.

5. **Only books and notebooks needed for the experiment should be in the lab.** Do not bring textbooks, purses, bookbags, backpacks, or other items into the lab; keep these things in your desk or locker.

6. **Read the entire experiment before entering the lab.** Memorize the safety precautions. Be familiar with the instructions for the experiment. Only materials and equipment authorized by your teacher should be used. When you do your lab work, follow the instructions and safety precautions described in the directions for the experiment.

7. **Read chemical labels.** Follow the instructions and safety precautions stated on the labels.

8. **Walk with care in the lab.** Sometimes you will have to carry chemicals *from* the supply station to your lab station. Avoid bumping into other students and spilling the chemicals. Stay at your lab station at other times.

9. **Food, beverages, chewing gum, cosmetics, and smoking are NEVER allowed in the lab.** (You should already know this.)

10. **NEVER taste chemicals or touch them with your bare hands.** Keep your hands away from your face and mouth while working, even if you are wearing gloves.

11. **Use a sparker to light a Bunsen burner.** Do not use matches. Be sure that all gas valves are turned off and that all hot plates are turned off and unplugged when you leave the lab.

12. **Be careful with hot plates, Bunsen burners, and other heat sources.** Keep your body and clothing away from flames. Do not touch a hot plate after it has just been turned off because it is probably hotter than you think. The same is true of glassware, crucibles, and other things after removing them from the flame of a Bunsen burner or from a drying oven.

13. **Do not use electrical equipment with frayed or twisted wires.**

14. **Be sure your hands are dry before using electrical equipment**. Before plugging an electrical cord into a socket, be sure the electrical equipment is turned off. When you are finished with it, turn it off. Before you leave the lab, unplug it, but be sure to turn it off FIRST.

15. **Do not let electrical cords dangle from work stations, dangling cords can cause tripping or electrical shocks.** The area under and around electrical equipment should be dry; cords should not lie in puddles of spilled liquid.

16. **Know fire drill procedures and the locations of exits.**

17. **Know the location and operation of safety showers and eyewash stations.**

18. **If your clothes catch on fire, walk to the safety shower, stand under it, and turn it on.**

19. **If you get a chemical in your eyes, walk immediately to the eyewash station, turn it on, and lower your head so your eyes are in the running water.** Hold your eyelids open with your thumbs and fingers, and roll your eyeballs around. You have to flush your eyes continuously for at least 15 minutes. Call your teacher while you are doing this.

20. **If you have a spill on the floor or lab bench, call your teacher rather than trying to clean it up by yourself.** Your teacher will tell you if it is OK for you to do the cleanup; if not, your teacher will know how the spill should be cleaned up safely.

21. **If you spill a chemical on your skin, wash it off using the sink faucet, and call your teacher.** If you spill a solid chemical on your clothing brush it off carefully without scattering it on somebody else, and call your teacher. If you get liquid on your clothing, wash it off right away using the sink faucet, and call your teacher. If the spill is on your pants or somewhere else that will not fit under the sink faucet, use the safety shower. Remove the pants or other affected clothing while under the shower, and call your teacher. (It may be temporarily embarrassing to remove pants or other clothing in front of your class, but failing to flush that chemical off your skin could cause permanent damage.)

22. **The best way to prevent an accident is to stop it before it happens.** If you have a close call, tell your teacher so that you and your teacher can find a way to prevent it from happening again. Otherwise, the next time, it could be a harmful accident instead of just a close call.

23. **All accidents should be reported to your teacher, no matter how minor.** Also, if you get a headache, feel sick to your stomach, or feel dizzy, tell your teacher immediately.

24. **For all chemicals, take only what you need.** However, if you do happen to take too much and have some left over, DO NOT put it back in the bottle. If somebody accidentally puts a chemical into the wrong bottle, the next person to use it will have a contaminated sample. Ask your teacher what to do with any leftover chemicals.

25. **NEVER take any chemicals out of the lab.** (This is another one that you should already know. You probably know the remaining rules also, but read them anyway.)

26. **Horseplay and fooling around in the lab are very dangerous.** NEVER be a clown in the laboratory

27. **Keep your work area clean and tidy.** After your work is done, clean your work area and all equipment.

28. **Always wash your hands with soap and water before you leave the lab.**

29. **Whether or not the lab instructions remind you, all of these rules apply all of the time.**

SAFETY SYMBOLS

To highlight specific types of precautions, the following symbols are used throughout the lab program. Remember that no matter what safety symbols you see in the lab instructions, all 29 of the safety rules previously described should be followed at all times.

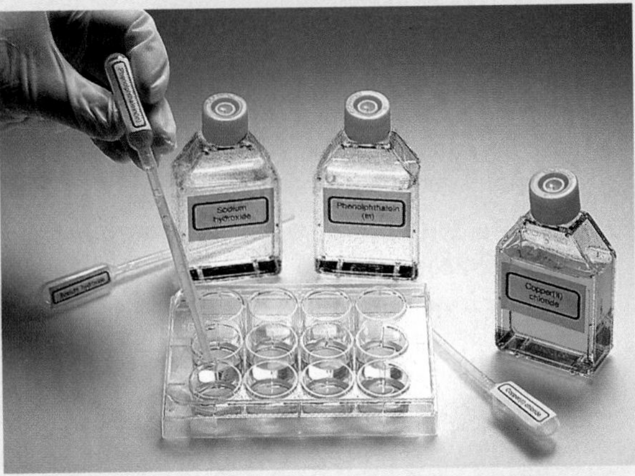

CLOTHING PROTECTION

- Wear laboratory aprons in the laboratory. Keep the apron strings tied so that they do not dangle.

EYE SAFETY

- Wear safety goggles in the laboratory at all times. Know how to use the eyewash station.

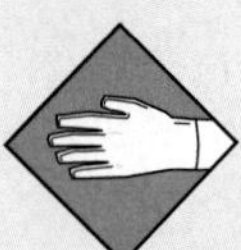

HAND SAFETY

- If a chemical gets on your skin or clothing or in your eyes, rinse it immediately, and alert your teacher.

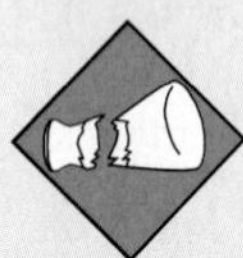

GLASSWARE SAFETY

- Never place glassware, containers of chemicals, or anything else near the edges of a lab bench or table.

CHEMICAL SAFETY

- Never return unused chemicals to the original container.
- It helps to label the beakers and test tubes containing chemicals. (This is not a new rule, just a good idea.)
- Never transfer substances by sucking on a pipet or straw; use a suction bulb.

CAUSTIC SAFTEY

- If a chemical is spilled on the floor or lab bench, tell your teacher, but do not clean it up yourself unless your teacher says it is OK to do so.

HEATING SAFETY

- When heating a chemical in a test tube, always point the open end of the test tube away from yourself and other people.

CLEAN UP

- Never taste, eat, or swallow any chemicals in the laboratory. Do not eat or drink any food from laboratory containers. Beakers are not cups, and evaporating dishes are not bowls.

WASTE DISPOSAL

- Some chemicals are harmful to our environment. You can help protect the environment by following the instructions for proper disposal.

SAFETY QUIZ

Look at the list of rules and identify whether a specific rule applies, or if the rule presented is a new rule.

1. Tie back long hair, and confine loose clothing. (Rule ? applies.)
2. Never reach across an open flame. (Rule ? applies.)
3. Use proper procedures when lighting Bunsen burners. Turn off hot plates, Bunsen burners, and other heat sources when not in use. (Rule ? applies.)
4. Heat flasks or beakers on a ringstand with wire gauze between the glass and the flame. (Rule ? applies.)
5. Use tongs when heating containers. Never hold or touch containers while heating them. Always allow heated materials to cool before handling them. (Rule ? applies.)
6. Turn off gas valves when not in use. (Rule ? applies.)

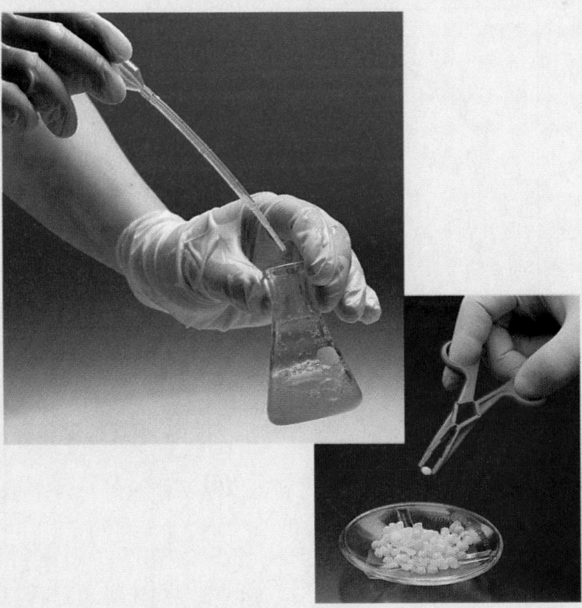

7. Use flammable liquids only in small amounts. (Rule ? applies.)
8. When working with flammable liquids, be sure that no one else is using a lit Bunsen burner or plans to use one. (Rule ? applies.)
9. Check the condition of glassware before and after using it. Inform your teacher of any broken, chipped, or cracked glassware because it should not be used. (Rule ? applies.)
10. Do not pick up broken glass with your bare hands. Place broken glass in a specially designated disposal container. (Rule ? applies.)
11. Never force glass tubing into rubber tubing, rubber stoppers, or wooden corks. To protect your hands, wear heavy cloth gloves or wrap toweling around the glass and the tubing, stopper, or cork, and gently push in the glass. (Rule ? applies.)
12. Do not inhale fumes directly. When instructed to smell a substance, use your hand to wave the fumes toward your nose, and inhale gently. (Rule ? applies.)
13. Keep your hands away from your face and mouth. (Rule ? applies.)
14. Always wash your hands before leaving the laboratory. (Rule ? applies.)

Finally, if you are wondering how to answer the questions that asked what additional rules apply to the safety symbols, here is the correct answer.

Any time you see any of the safety symbols, you should remember that all 29 of the numbered laboratory rules always apply.

Extraction and Filtration

Extraction, the separation of substances in a mixture by using a solvent, depends on solubility. For example, sand can be separated from salt by adding water to the mixture. The salt dissolves in the water, and the sand settles to the bottom of the container. The sand can be recovered by decanting the water. The salt can then be recovered by evaporating the water.

Filtration separates substances based on differences in their physical states or in the size of their particles. For example, a liquid can be separated from a solid by pouring the mixture through a paper-lined funnel or, if the solid is more dense than the liquid, the solid will settle to the bottom of the container, leaving the liquid on top. The liquid can then be decanted, leaving the solid.

SETTLING AND DECANTING

1. Fill an appropriate-sized beaker with the solid-liquid mixture provided by your teacher. Allow the beaker to sit until the bottom is covered with solid particles and the liquid is clear.
2. Grasp the beaker with one hand. With the other hand, pick up a stirring rod and hold it along the lip of the beaker. Tilt the beaker slightly so that liquid begins to pour out in a slow, steady stream, as shown in Figure A.

FIGURE A
Settling and decanting

GRAVITY FILTRATION

1. Prepare a piece of filter paper as shown in Figure B. Fold it in half and then in half again. Tear the corner of the filter paper, and open the filter paper into a cone. Place it in the funnel.

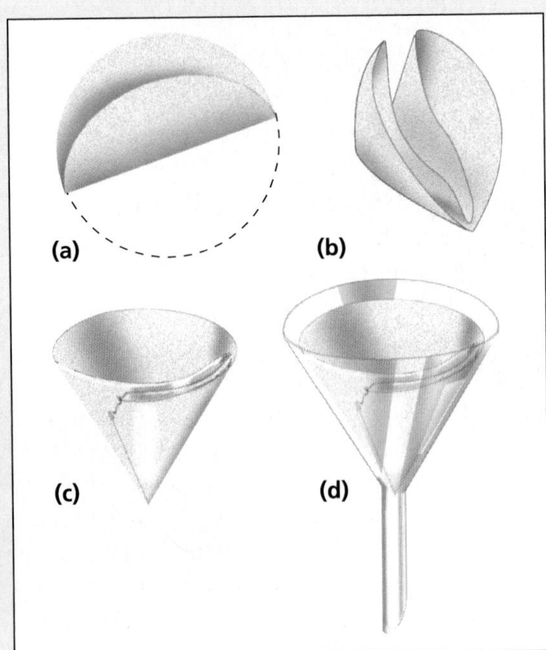

FIGURE B

FIGURE C Gravity filtration

2. Put the funnel, stem first, into a filtration flask, or suspend it over a beaker using an iron ring, as shown in Figure C.
3. Wet the filter paper with distilled water from a wash bottle. The paper should adhere to the sides of the funnel, and the torn corner should prevent air pockets from forming between the paper and the funnel.
4. Pour the mixture to be filtered down a stirring rod into the filter. The stirring rod directs the mixture into the funnel and reduces splashing.
5. Do not let the level of the mixture in the funnel rise above the edge of the filter paper.
6. Use a wash bottle to rinse all of the mixture from the beaker into the funnel.

VACUUM FILTRATION

1. Check the T attachment to the faucet. Turn on the water. Water should run without overflowing the sink or spitting while creating a vacuum. To test for a vacuum, cover the opening of the horizontal arm of the T with your thumb or index finger. If you feel your thumb being pulled inward, you have a vacuum. Note the number of turns of the knob that are needed to produce the flow of water that creates a vacuum.
2. Turn the water off. Attach the pressurized rubber tubing to the *horizontal* arm of the T. (You do not want water to run through the tubing.)
3. Attach the free end of the rubber tubing to the side arm of a filter flask. Check for a vacuum. Turn on the water so that it rushes out of the faucet (refer to step 1). Place the palm of your hand over the opening of the Erlenmeyer flask. You should feel the vacuum pull your hand inward. If you do not feel any pull or if the pull is weak, increase the flow of water. If increasing the flow of water fails to work, shut off the water and make sure your tubing connections are tight.
4. Insert the neck of a Büchner funnel into a one-hole rubber stopper until the stopper is about two-thirds to three-fourths up the neck of the funnel. Place the funnel stem into the Erlenmeyer flask so that the stopper rests in the mouth of the flask, as shown in Figure D.
5. Obtain a piece of round filter paper. Place it inside the Büchner funnel over the holes. Turn on the water as in step 1. Hold the filter flask with one hand, place the palm of your hand over the mouth of the funnel, and check for a vacuum.
6. Pour the mixture to be filtered into the funnel. Use a wash bottle to rinse all of the mixture from the beaker into the funnel.

FIGURE D Vacuum filtration

PRE-LAB • PAGE 790

EXTRACTION AND FILTRATION

Mixture Separation

OBJECTIVES

- ***Observe*** **the chemical and physical properties of a mixture.**
- ***Relate*** **knowledge of chemical and physical properties to the task of purifying the mixture.**
- ***Analyze*** **the success of methods of purifying the mixture.**

MATERIALS

- **aluminum foil**
- **cotton balls**
- **distilled water**
- **filter funnels**
- **filter paper**
- **forceps**
- **magnet**
- **paper clips**
- **paper towels**
- **Petri dish**
- **pipets**
- **plastic forks**
- **plastic spoons**
- **plastic straws**
- **rubber stoppers**
- **sample of mixture and components (sand, iron filings, salt, poppy seeds)**
- **test tubes and rack**
- **tissue paper**
- **transparent tape**
- **wooden splints**

BACKGROUND

The ability to separate and recover pure substances from mixtures is extremely important in scientific research and industry. Chemists need to work with pure substances, but naturally occurring materials are seldom pure. Often, differences in the physical properties of the components in a mixture provide the means for separating them. In this experiment, you will have an opportunity to design, develop, and implement your own procedure for separating a mixture. The mixture you will work with contains salt, sand, iron filings, and poppy seeds. All four substances are in dry, granular form.

SAFETY

Always wear safety goggles and a lab apron to protect your eyes and clothing. If you get a chemical in your eyes, immediately flush the chemical out at the eyewash station while calling to your teacher. Know the location of the emergency lab shower and eyewash station and the procedures for using them.

Call your teacher in the event of a spill. Spills should be cleaned up promptly according to your teacher's directions.

PREPARATION

1. Your task will be to plan and carry out the separation of a mixture. Before you can plan your experiment, you will need to investigate the properties of each component in the mixture. The properties will be used to design your mixture separation. Copy the data table on the following page in your lab notebook, and use it to record your observations.

DATA TABLE

Properties	Sand	Iron filings	Salt	Poppy seeds
Dissolves				
Floats				
Magnetic				
Other				

PROCEDURE

1. Obtain separate samples of each of the four mixture components from your teacher. Use the equipment you have available to make observations of the components and determine their properties. You will need to run several tests with each substance, so don't use all of your sample on the first test. Look for things like whether the substance is magnetic, whether it dissolves, or whether it floats. Record your observations in your data table.
2. Make a plan for what you will do to separate a mixture that includes the four components from step 1. Review your plan with your teacher.
3. Obtain a sample of the mixture from your teacher. Using the equipment you have available, run the procedure you have developed.

CLEANUP AND DISPOSAL

4. Clean your lab station. Clean all equipment, and return it to its proper place. Dispose of chemicals and solutions in the containers designated by your teacher. Do not pour any chemicals down the drain or throw anything in the trash unless your teacher directs you to do so. Wash your hands thoroughly after all work is finished and before you leave the lab.

ANALYSIS AND INTERPRETATION

1. **Evaluating Methods:** On a scale of 1 to 10, how successful were you in separating and recovering each of the four components: sand, salt, iron filings, and poppy seeds? Consider 1 to be the best and 10 to be the worst. Justify your ratings based on your observations.

CONCLUSIONS

1. **Evaluating Methods:** How did you decide on the order of your procedural steps? Would any order have worked?
2. **Designing Experiments:** If you could do the lab over again, what would you do differently? Be specific.
3. **Designing Experiments:** Name two materials or tools that weren't available that might have made your separation easier.
4. **Applying Ideas:** For each of the four components, describe a specific physical property that enabled you to separate the component from the rest of the mixture.

EXTENSIONS

1. **Evaluating Methods:** What methods could be used to determine the purity of each of your recovered components?
2. **Designing Experiments:** How could you separate each of the following two-part mixtures?
   a. lead filings and iron filings
   b. sand and gravel
   c. sand and finely ground polystyrene foam
   d. salt and sugar
   e. alcohol and water
   f. nitrogen and oxygen

Conservation of Mass

OBJECTIVES

- ***Observe*** **the signs of a chemical reaction.**
- ***Infer*** **that a reaction has occurred.**
- ***Compare*** **masses of reactants and products.**
- ***Resolve*** **chemical discrepancies.**
- ***Design*** **experiments.**
- ***Relate*** **observations to the law of conservation of mass.**

MATERIALS

- **2 L plastic soda bottle**
- **5% acetic acid solution (vinegar)**
- **balance**
- **clear plastic cups, 2**
- **graduated cylinder**
- **hook-insert cap for bottle**
- **microplunger**
- **sodium hydrogen carbonate (baking soda)**

BACKGROUND

The law of conservation of mass states that matter is neither created nor destroyed during a chemical reaction. Therefore, the mass of a system should remain constant during any chemical process. In this experiment, you will determine whether mass is conserved by examining a simple chemical reaction and comparing the mass of the system before the reaction with its mass after the reaction.

SAFETY

Always wear safety goggles and a lab apron to protect your eyes and clothing. If you get a chemical in your eyes, immediately flush the chemical out at the eyewash station while calling to your teacher. Know the location of the emergency lab shower and eyewash station and the procedure for using them.

Do not touch any chemicals. If you get a chemical on your skin or clothing, wash the chemical off at the sink while calling to your teacher. Make sure you carefully read the labels and follow the precautions on all containers of chemicals that you use. If no precautions are stated on the label, ask your teacher what precautions to follow. Do not taste any chemicals or items used in the laboratory. Never return leftover chemicals to their original containers; take only small amounts to avoid wasting supplies.

Call your teacher in the event of a spill. Spills should be cleaned up promptly according to your teacher's directions.

PREPARATION

1. Make two data tables in your lab notebook, one for Part I and another for Part II. In each table, create three columns labeled *Initial mass (g)*,

Final mass (g), and *Change in mass (g)*. Each table should also have space for observations of the reaction.

PROCEDURE—PART I

1. Obtain a microplunger and tap it down into a sample of baking soda until the bulb end is packed with a plug of the powder (4–5 mL of baking soda should be enough to pack the bulb).

2. Hold the microplunger over a plastic cup, and squeeze the sides of the microplunger to loosen the plug of baking soda so that it falls into the cup.

3. Use a graduated cylinder to measure 100 mL of vinegar, and pour it into a second plastic cup.

4. Place the two cups side by side on the balance pan and measure the total mass of the system (before reaction) to the nearest 0.01 g. Record the mass in your data table.

5. Add the vinegar to the baking soda a little at a time to prevent the reaction from getting out of control, as shown in Figure A. Allow the vinegar to slowly run down the inside of the cup. Observe and record your observations about the reaction.

FIGURE A Slowly add the vinegar to prevent the reaction from getting out of control.

6. When the reaction is complete, place both cups on the balance and determine the total final mass of the system to the nearest 0.01 g. Calculate any change in mass. Record both the final mass and any change in mass in your data table.

7. Examine the plastic bottle and the hook-insert cap. Try to develop a modified procedure that will test the law of conservation of mass more accurately than the procedure in Part I.

8. In your notebook, write the answers to items 1 through 3 in Analysis and Interpretation—Part I.

PROCEDURE—PART II

9. Your teacher should approve the procedure you designed in Procedure—Part 1, step 7. Implement your procedure with the same chemicals and quantities you used in Part I, but use the bottle and hook-insert cap in place of the two cups. Record your data in the data table.

10. If you were successful in step 9 and your results reflect the conservation of mass, proceed to complete the experiment. If not, find a lab group that was successful, and discuss with them what they did and why they did it. Your group should then test the other group's procedure to determine whether their results are reproducible.

CLEANUP AND DISPOSAL

11. Clean your lab station. Clean all equipment and return it to its proper place. Dispose of chemicals and solutions in the containers designated by your teacher. Do not pour any chemicals down the drain or throw anything in the trash unless your teacher directs you to do so. Wash your hands thoroughly after all work is finished and before you leave the lab.

ANALYSIS AND INTERPRETATION—PART I

1. **Inferring Conclusions:** What evidence was there that a chemical reaction occurred?

2. **Organizing Data:** How did the final mass of the system compare with the initial mass of the system?

3. **Resolving Discrepancies:** Does your answer to the previous question show that the law of conservation of mass was violated? (Hint: Another way to express the law of conservation of mass is to say that the mass of all of the products equals the mass of all of the reactants.) What do you think might cause the mass difference?

ANALYSIS AND INTERPRETATION—PART II

1. **Inferring Conclusions:** Was there any new evidence in Part II indicating that a chemical reaction occurred?
2. **Organizing Ideas:** Identify the state of matter for each reactant in Part II. Identify the state of matter for each product.

CONCLUSIONS

1. **Relating Ideas:** What is the difference between the system in Part I and the system in Part II? What change led to the improved results in Part II?
2. **Evaluating Methods:** Why did the procedure for Part II work better than the procedure for Part I?

EXTENSIONS

1. **Designing Experiments:** How would you verify the law of conservation of mass if you were given a resealable (zippered) plastic bag, as shown in Figure B, and a twist tie instead of the bottle and hook-cap? Write your proposed procedure, step by step.

FIGURE B

2. **Predicting Outcomes:** Would you have been as successful with the resealable plastic bag and twist tie? Why or why not? If time allows, try the procedure you wrote in Extension item 1 and test your prediction. Report your results, and try to explain any discrepancies you find. (Hint: What are the differences between what happened with the bag and what happened with the bottle?)
3. **Apply Models:** When a log burns, the resulting ash obviously has less mass than the unburned log did. Explain whether this loss of mass violates the law of conservation of mass.
4. **Designing Experiments:** Design a procedure that would test the law of conservation of mass for the burning log described in Extension item 3.

EXPERIMENT 4-1

Flame Tests

OBJECTIVES

- *Identify* a set of flame-test color standards for selected metal ions.
- *Relate* the colors of a flame test to the behavior of excited electrons in a metal ion.
- *Identify* an unknown metal ion by using a flame test.
- *Demonstrate* proficiency in performing a flame test and in using a spectroscope.

MATERIALS

- 1.0 M HCl solution
- 250 mL beaker
- Bunsen burner and related equipment
- $CaCl_2$ solution
- cobalt glass plates
- crucible tongs
- distilled water
- flame-test wire
- glass test plate (or a microchemistry plate with wells)
- K_2SO_4 solution
- Li_2SO_4 solution
- Na_2SO_4 solution
- NaCl crystals
- NaCl solution
- $SrCl_2$ solution
- spectroscope
- unknown solution

OPTIONAL EQUIPMENT

- wooden splints

BACKGROUND

The characteristic light emitted by each individual atom is the basis for the chemical test known as a flame test.

To identify an unknown atom, you must first determine the characteristic colors produced by different atoms. You will do this by performing a flame test on a variety of standard solutions of metal compounds. Then you will perform a flame test with an unknown sample to see if it matches any of the standard solutions. The presence of even a speck of another substance can interfere with the identification of the true color of a particular type of atom, so be sure to keep your equipment very clean and perform multiple trials to check your work.

SAFETY

Always wear safety goggles and a lab apron to protect your eyes and clothing. If you get a chemical in your eyes, immediately flush the chemical out at the eyewash station while calling to your teacher. Know the locations of the emergency lab shower and eyewash station and the procedures for using them.

Do not touch any chemicals. If you get a chemical on your skin or clothing, wash the chemical off at the sink while calling to your teacher. Make sure you carefully read the labels and follow the precautions on all containers of chemicals that you use. If no precautions are stated on the label, ask your teacher what precautions to follow. Do not taste any chemicals or items used in the laboratory. Never return leftover chemicals to their original containers; take only small amounts to avoid wasting supplies.

When using a Bunsen burner, confine long hair and loose clothing. Do not heat glassware that is broken, chipped, or cracked. Use tongs or a hot mitt to handle heated glassware and other equipment; heated glassware does not always look hot. If your clothing catches fire, WALK to the emergency lab shower and use it to put out the fire.

Call your teacher in the event of an acid or base spill. Acid or base spills should be cleaned up promptly according to your teacher's instructions.

PREPARATION

1. Prepare a data table in your lab notebook. Include rows for each of the solutions of metal compounds listed in the materials list as well as for NaCl crystals and an unknown solution. The table should have three wide columns for the three trials you will perform with each substance. Each column should have room to record the colors and wavelengths of light. Be sure you have plenty of room to write your observations about each test.

DATA TABLE

Substance	Trial 1	Trial 2	Trial 3
1.0 M HCl			
$CaCl_2$			
K_2SO_4			
Li_2SO_4			
Na_2SO_4			
NaCl crystals			
NaCl			
$SrCl_2$			
Unknown			

2. Label a beaker *Waste*. Thoroughly clean and dry a well strip. Fill the first well one-fourth full with 1.0 M HCl on the plate. Clean the test wire by first dipping it in the HCl and then holding it in the colorless flame of the Bunsen burner. Repeat this procedure until the flame is not colored by the wire. When the wire is ready, rinse the well with distilled water and collect the rinse water in the waste beaker.

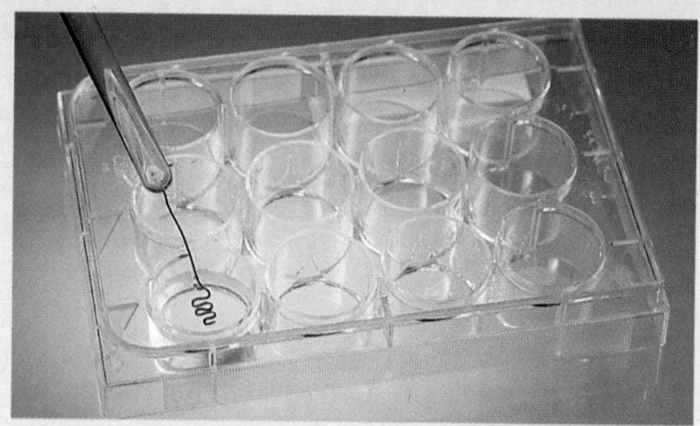

FIGURE A Be sure that you record the positions of the various metal ion solutions in each well of the well strip.

3. Put 10 drops of each metal ion solution listed in the materials list, except NaCl, in a row in each well of the well strip. Put a row of 1.0 M HCl drops on a glass plate across from the metal ion solutions. Record the positions of all of the chemicals placed in the wells. The wire will need to be cleaned thoroughly between each test solution with HCl to avoid contamination from the previous test, as shown in Figure A.

PROCEDURE

1. Dip the wire into the $CaCl_2$ solution, and then hold it in the Bunsen burner flame. Observe the color of the flame, and record it in the data table. Repeat the procedure again, but this time look through the spectroscope to view the results. Record the wavelengths you see from the flame. Repeat each test three times. Clean the wire with the HCl as you did in Preparation step 2.

2. Repeat step 1 with the K_2SO_4 and with each of the remaining solutions in the well strip. For each solution that you test, record the color of each flame and the wavelength observed with the spectroscope. After the solutions are tested, clean the wire thoroughly, rinse the plate with distilled water, and collect the rinse water in the waste beaker.

3. Test another drop of Na_2SO_4, but this time view the flame through two pieces of cobalt glass. Clean the wire, and repeat the test. Record in your data table the colors and wavelengths of the flames as they appear when viewed through the cobalt glass. Clean the wire and the well

strip, and rinse the well strip with distilled water. Pour the rinse water into the waste beaker.

4. Put a drop of K_2SO_4 in a clean well. Add a drop of Na_2SO_4. Flame-test the mixture. Observe the flame without the cobalt glass. Repeat the test again, this time observing the flame through the cobalt glass. Record the colors and wavelengths of the flames in the data table. Clean the wire, and rinse the well strip with distilled water. Pour the rinse water into the waste beaker.

5. Test a drop of the NaCl solution in the flame, and then view it through the spectroscope. (Do not use the cobalt glass.) Record your observations. Clean the wire, and rinse the well strip with distilled water. Pour the rinse water into the waste beaker. Place a few crystals of NaCl on the plate, dip the wire in the crystals, and do the flame test once more. Record the color of the flame test. Clean the wire, and rinse the well strip with distilled water. Pour the rinse into the waste beaker.

6. Obtain a sample of the unknown solution. Perform flame tests for it, with and without the cobalt glass. Record your observations. Clean the wire, and rinse the well strip with distilled water. Pour the rinse water into the waste beaker.

CLEANUP AND DISPOSAL

7. Dispose of the contents of the waste beaker in the container designated by your teacher. Wash your hands thoroughly after cleaning up the area and equipment.

ANALYSIS AND INTERPRETATION

1. **Organizing Data:** Examine your data table, and create a summary of the flame test for each metal ion.

2. **Analyzing Data:** Account for any differences in the individual trials for the flame tests for the metals ions.

3. **Organizing Ideas:** Explain how viewing the flame through cobalt glass can make it easier to analyze the ions being tested.

4. **Relating Ideas:** For three of the metal ions tested, explain how the flame color you saw relates to the lines of color you saw when you looked through the spectroscope.

CONCLUSIONS

1. **Inferring Conclusions:** What metal ions are in the unknown solution?

2. **Evaluating Methods:** How would you characterize the flame test with respect to its sensitivity? What difficulties could there be when identifying ions by the flame test?

3. **Evaluating Methods:** Explain how you can use a spectroscope to identify the components of solutions containing several different metal ions.

EXTENSIONS

1. **Inferring Conclusions:** A student performed flame tests on several unknowns and observed that they all were shades of red. What should the student do to correctly identify these substances? Explain your answer.

2. **Applying Ideas:** During a flood, the labels from three bottles of chemicals were lost. The three unlabeled bottles of white solids were known to contain the following: strontium nitrate, ammonium carbonate, and potassium sulfate. Explain how you could easily test the substances and relabel the three bottles. (Hint: Ammonium ions do not provide a distinctive flame color.)

3. **Applying Ideas:** Some stores sell "fireplace crystals." When sprinkled on a log, these crystals make the flames blue, red, green, and violet. Explain how these crystals can change the flame's color. What ingredients do you expect them to contain?

PRE-LABORATORY PROCEDURE

Gravimetric Analysis

Gravimetric analytical methods are based on accurate and precise mass measurements. They are used to determine the amount or percentage of a compound or element in a sample material. For example, if we want to determine the percentage of iron in an ore or the percentage of chloride ion in drinking water, gravimetric analysis would be used.

A gravimetric procedure generally involves reacting the sample to produce a reaction product that can be used to calculate the mass of the element or compound in the original sample. For example, to calculate the percentage of iron in a sample of iron ore, the mass of the ore is determined. The ore is then dissolved in hydrochloric acid to produce $FeCl_3$. The $FeCl_3$ precipitate is converted to a hydrated form of Fe_2O_3 by adding water and ammonia to the system. The mixture is then filtered to separate the hydrated Fe_2O_3 from the mixture. The hydrated Fe_2O_3 is heated in a crucible to drive the water from the hydrate, producing anhydrous Fe_2O_3. The mass of the crucible and its contents is determined after successive heating steps to ensure that the product has reached constant mass and that all of the water has been driven off. The mass of Fe_2O_3 produced can be used to calculate the mass and percentage of iron in the original ore sample.

Gravimetric procedures require accurate and precise techniques and measurements to obtain suitable results. Possible sources of error are the following:

1. **The product (precipitate) that is formed is contaminated.**
2. **Some product is lost when transferring the product from a filter to a crucible.**
3. **The empty crucible is not clean or is not at constant mass for the initial mass measurement.**
4. **The system is not heated sufficiently to obtain an anhydrous product.**

GENERAL SAFETY

Always wear safety goggles and a lab apron to protect your eyes and clothing. If you get a chemical in your eyes, immediately flush the chemical out at the eyewash station while calling to your teacher. Know the location of the emergency lab shower and eyewash station and the procedure for using them.

When using a Bunsen burner, confine long hair and loose clothing. Do not heat glassware that is broken, chipped, or cracked. Use tongs or a hot mitt to handle heated glassware and other equipment; heated

glassware does not always look hot.
If your clothing catches fire, WALK to the emergency lab shower and use it to put out the fire.

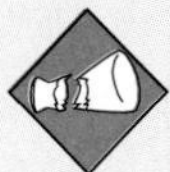

Never put broken glass or ceramics in a regular waste container. Broken glass and ceramics should be disposed of in a separate container designated by your teacher.

SETTING UP THE EQUIPMENT

1. The general setup for heating a sample in a crucible is shown in Figure A. Attach a metal ring clamp to a ring stand, and lay a clay triangle on the ring.

CLEANING THE CRUCIBLE

2. Wash and dry a metal or ceramic crucible and lid. Cover the crucible with its lid, and use a balance to obtain its mass. If the balance is located far from your working station, use crucible tongs to place the crucible and lid on a piece of wire gauze. Carry the crucible to the balance, using the wire gauze as a tray.

HEATING THE CRUCIBLE TO OBTAIN A CONSTANT MASS

3. After recording the mass of the crucible and lid, suspend the crucible over a Bunsen burner by placing it on the clay triangle as shown in Figure B. Then place the lid on the crucible so that the entire contents are covered.

4. Light the Bunsen burner. Heat the crucible for 5 minutes with a gentle flame, and then adjust the burner to produce a strong flame. Heat for 5 minutes more. Shut off the gas to the burner. Allow the crucible and lid to cool. Using crucible tongs, carry the crucible and lid to the balance, as shown in Figure C. Measure and record the mass. If the mass differs from the mass before heating, repeat the process until mass data from heating trials are within 1% of each other. This assumes that the crucible has a constant mass. The crucible is now ready to be used in a gravimetric analysis procedure. Details will be found in the following experiments.

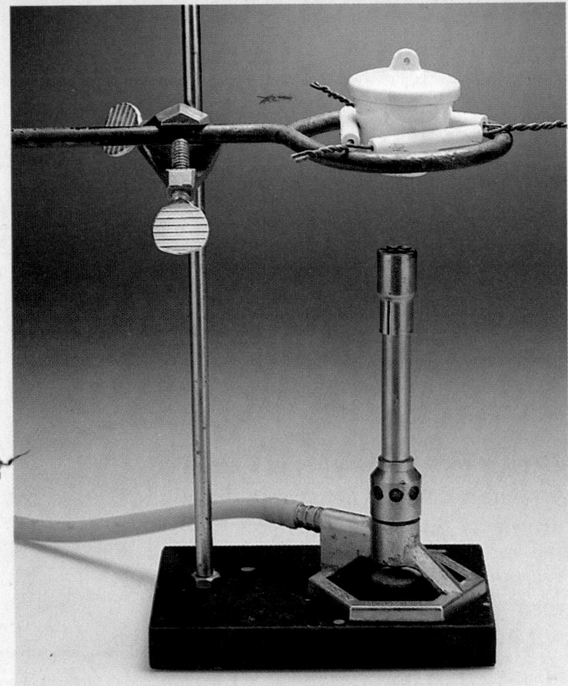

FIGURE A

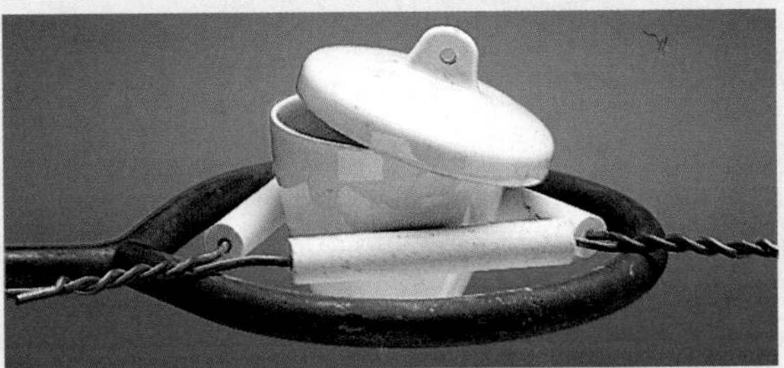

FIGURE B

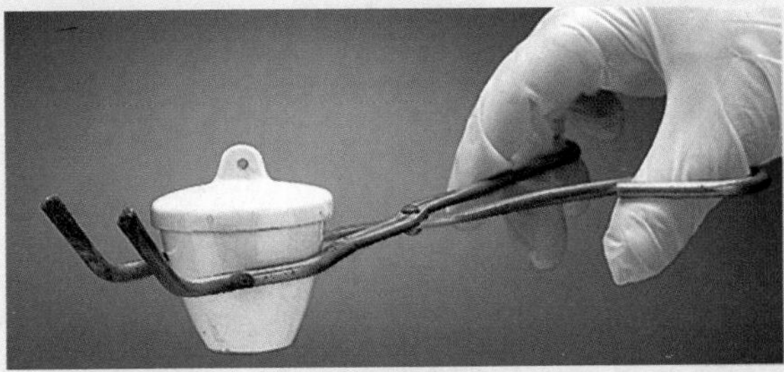

FIGURE C

Gravimetric methods are used in Experiment 7-3 to synthesize magnesium oxide, and to separate $SrCO_3$ from a solution in Experiment 9-2.

PRE-LAB • PAGE 790

Separation of Salts by Fractional Crystallization

OBJECTIVES

- ***Recognize*** **how the solubility of a salt varies with temperature.**
- ***Demonstrate*** **proficiency in fractional crystallization and in vacuum filtration or gravity filtration.**
- ***Determine*** **the percentage of two salts recovered by fractional crystallization.**

MATERIALS

- **50 mL NaCl-KNO_3 solution**
- **100 mL graduated cylinder**
- **150 mL beakers, 4**
- **balance, centigram**
- **Büchner funnel, one-hole rubber stopper, vacuum filtration setup with filter flask and tubing, or glass funnel**
- **Bunsen burner and related equipment or hot plate**
- **filter paper**
- **glass stirring rod**
- **ice**
- **nonmercury thermometer**
- **ring and wire gauze**
- **ring stand**
- **rock salt**
- **rubber policeman**
- **spatula**
- **tray, tub, or pneumatic trough**

BACKGROUND

In this experiment, you will separate a mixture of sodium chloride, NaCl, and potassium nitrate, KNO_3. Both of these substances dissolve in water, so filtering alone cannot separate them. Figure A shows that temperature does not greatly affect the amount of sodium chloride that dissolves in water, but the amount of KNO_3 that dissolves in water does vary with temperature. This difference will enable you to separate them by a technique known as fractional crystallization.

If a water solution of NaCl and KNO_3 is cooled from room temperature to a temperature near 0°C, some KNO_3 will crystallize. This KNO_3 residue can then be separated from the NaCl solution by filtration. The NaCl can be isolated from the filtrate by evaporation of the water. After drying the KNO_3 residue and the NaCl, you can measure the mass of each of the recovered substances.

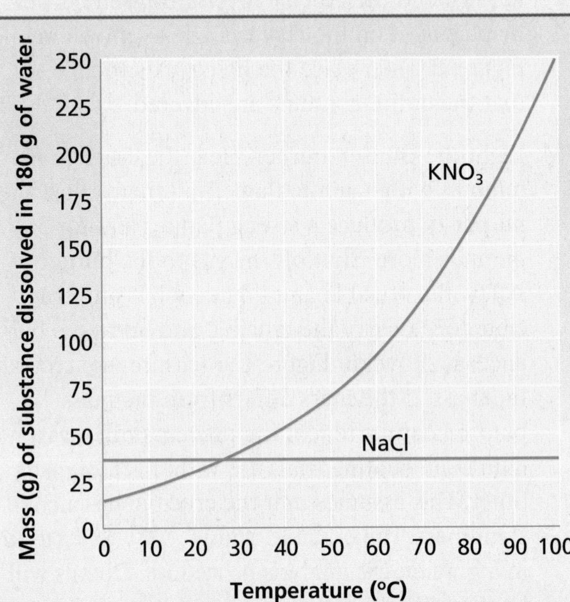

FIGURE A This graph shows the relationship between temperature and the solubility of NaCl and the solubility of KNO_3.

SAFETY

Always wear safety goggles and a lab apron to protect your eyes and clothing. If you get a chemical in your eyes, immediately flush the chemical out at the eyewash station while calling to your teacher. Know the locations of the emergency lab shower and eyewash station and the procedures for using them.

Do not touch chemicals. If you get a chemical on your skin or clothing, wash the chemical off at the sink while calling to your teacher. Make sure you carefully read the labels and follow the precautions on all containers of chemicals that you use. If no precautions are stated on the label, ask your teacher what precautions to follow. Do not taste any chemicals or items used in the laboratory. Never return leftover chemicals to their original containers; take only small amounts to avoid wasting supplies.

When using a Bunsen burner, confine long hair and loose clothing. Do not heat glassware that is broken, chipped, or cracked. Use tongs or a hot mitt to handle heated glassware and other equipment; heated glassware does not always look hot. If your clothing catches fire, WALK to the emergency lab shower and use it to put out the fire.

PREPARATION

1. Prepare a data table in your lab notebook. It should contain spaces for *Volume of salt solution added to beaker 1*, *Mass of beaker 1*, *Mass of filter paper*, *Mass of beaker 1 with filter paper and* KNO_3, *Mass of beaker 4*, and *Mass of beaker 4 with NaCl*. You will also need room to record the temperature of the mixture before and after cooling.
2. Obtain four clean, dry 150 mL beakers, and label them *1*, *2*, *3*, and *4*.

PROCEDURE

1. Measure the mass of beaker 1 to the nearest 0.01 g and record its mass in your data table.
2. Measure about 50 mL of the NaCl-KNO_3 solution into a graduated cylinder. Record the exact volume in your data table. Pour this mixture into beaker 1.
3. Using a thermometer, measure the temperature of the mixture. Record this temperature in your data table.
4. Measure the mass of a piece of filter paper to the nearest 0.01 g and record the mass in your data table.
5. Set up your filtering apparatus as described in the Pre-Laboratory Procedure on pages 790–791.
6. Make an ice bath by filling a tray, tub, or trough half full with ice. Add a handful of rock salt. The salt lowers the freezing point of water so that the ice bath can reach a lower temperature. Fill the ice bath with water until the container is three-quarters full.
7. Using a fresh supply of ice and distilled water, fill beaker 2 half full with ice and add water. Do not add rock salt to this ice-water mixture. You will use this water to wash your purified salt.
8. Put beaker 1, containing your NaCl-KNO_3 solution, into the ice bath. Place a thermometer into the solution to monitor the temperature. Stir the solution with a stirring rod while it cools. The lower the temperature of the mixture is, the more KNO_3 will crystallize out of the solution. When the temperature nears 4°C, follow step 8a if you are using the Büchner funnel and step 8b if you are using a glass funnel.

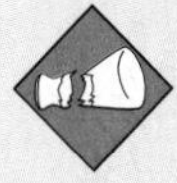

Never stir a solution with a thermometer; the bulb is very fragile.

a. Vacuum filtration

Refer to page 791 for instructions on how to set up this system. Turn on the water at the faucet that has the aspirator nozzle attached to it. Prepare the filtering apparatus by pouring approximately 50 mL of ice-cold distilled water from beaker 2 through the filter paper. After the water has gone through the funnel, empty the filter flask into the sink. Reconnect the filter flask, and pour the mixture in beaker 1 into the funnel. Use the rubber policeman to transfer all of the cooled mixture into the funnel, especially any crystals that are visible. It may be helpful to add small amounts of the ice-cold water from beaker 2 to beaker 1 to wash any crystals onto the filter paper. After all of the solution has passed through the funnel, wash the KNO_3 residue by pouring a very small amount of ice-cold water over it. When this water has passed through the filter paper, turn off the faucet and carefully remove the tubing from the aspirator. Empty the filtrate, which has passed through the filter paper and is now in the filter flask, into beaker 3. When finished, continue with Procedure step 9.

b. Gravity filtration

Refer to page 790 for instructions on how to set up this system. Place beaker 3 under the glass funnel. Prepare the filtering apparatus by pouring approximately 50 mL of ice-cold water from beaker 2 through the filter paper. The water will pass through the filter paper and drip into beaker 3. When the dripping stops, empty beaker 3 into the sink. Place beaker 3 under the glass funnel so that it can collect the filtrate from the funnel. Pour the mixture in beaker 1 into the funnel. Use the rubber policeman to transfer all of the cooled mixture into the funnel, especially any crystals that are visible. It may be helpful to add small amounts of ice-cold water from beaker 2 to beaker 1 to wash any crystals onto the filter paper. After all of the solution has passed through the funnel, wash the KNO_3 by pouring a very small amount of ice-cold water from beaker 2 over it.

9. After you have finished filtering, use either a hot plate or a Bunsen burner, ring stand, ring, and wire gauze to heat beaker 3. When the liquid in beaker 3 begins to boil, continue heating gently until the volume is approximately 25–30 mL.

Be sure to use beaker tongs. Remember that hot glassware does not always look hot.

10. Allow the solution in beaker 3 to cool, and then set it in the ice-bath. Stir until the temperature is approximately 4°C.

11. Measure the mass of beaker 4. Record the mass in your data table.

12. Repeat Procedure step 8a or step 8b, pouring the solution from beaker 3 onto the filter paper and using beaker 4 to collect the filtrate that passes through the filter.

13. Wash and dry beaker 1. Carefully remove the filter paper with the KNO_3 from the funnel and put it in the beaker. Be certain to avoid spilling the crystals. Place the beaker in a drying oven overnight.

14. Heat beaker 4 with a hot plate or Bunsen burner until it begins to boil. Continue to heat gently until all of the water has vaporized and the salt appears dry. Turn off the hot plate or burner, and allow the beaker to cool. Use beaker tongs to move the beaker, as shown in Figure B. Measure the mass of beaker 4 with the NaCl to the nearest 0.01 g and record this mass in your data table.

FIGURE B Use beaker tongs to move a beaker that has been heated, even if you believe that the beaker is cool.

15. The next day, use beaker tongs to remove beaker 1 with the filter paper and KNO_3 from the drying oven. Allow the beaker to cool. Measure the mass, with the same balance you used when you measured the mass of the empty beaker. Record the new mass in your data table.

CLEANUP AND DISPOSAL

16. Once the mass of the NaCl has been determined, add water to dissolve the NaCl and rinse the solution down the drain. Dispose of the KNO_3 in the waste container designated by your teacher. Clean up the lab and all equipment after use. Wash your hands thoroughly after all lab work is finished and before you leave the lab.

ANALYSIS AND INTERPRETATION

1. **Organizing Data:** Find the mass of NaCl in your 50 mL sample by subtracting the mass of beaker 4 from the mass of beaker 4 with NaCl.
2. **Organizing Data:** Find the mass of KNO_3 in your 50 mL sample by subtracting the mass of beaker 1 and the mass of the filter paper from the mass of beaker 1 with the filter paper and KNO_3.
3. **Organizing Data:** Determine the total mass of the two salts by addition.

CONCLUSIONS

1. **Inferring Conclusions:** Calculate the percentage by mass of NaCl in the salt mixture. Calculate the percentage by mass of KNO_3 in the salt mixture. Assume that the density of your 50-mL solution is 1.0 g/mL.
2. **Evaluating Methods:** Use the graph shown at the beginning of this experiment to estimate how much KNO_3 could still be contaminating the NaCl you recovered.
3. **Relating Ideas:** Use the graph shown at the beginning of this experiment to explain why it is impossible to separate the two compounds completely by fractional crystallization.
4. **Evaluating Methods:** Why was it important that you use ice-cold water to wash the KNO_3 after filtration?
5. **Evaluating Methods:** If it was important to use very cold water to wash the KNO_3, why wasn't the salt and ice-water mixture from the bath used? After all, it had a lower temperature than the ice and distilled water from beaker 2.
6. **Evaluating Methods:** Why was it important to keep the amount of cold water used to wash the KNO_3 as small as possible?
7. **Relating Ideas:** Your lab partner tries to dissolve 95 g of KNO_3 in 100 g of water, but no matter how well the mixture is stirred, some KNO_3 remains undissolved. Using the graph, explain what your lab partner must do to make the KNO_3 dissolve in this amount of water.

EXTENSIONS

1. **Designing Experiments:** Describe how you could use the properties of the compounds to test the purity of your recovered samples. If your teacher approves your plan, use it to check your separation of the mixtures.
2. **Designing Experiments:** How could you improve the yield or the purity of the compounds you recovered? If you can think of ways to modify the procedure, ask your teacher to approve your plan, and run the experiment again.

Naming Ionic Compounds

OBJECTIVES

- *Observe* chemical replacement reactions.
- *Use* reagent measurements to determine the formula of a chemical substance.
- *Infer* a conclusion from experimental data.
- *Apply* reaction stoichiometry concepts.

MATERIALS

- 0.1 M copper chloride
- 0.1 M iron chloride
- 0.1 M sodium hydroxide
- 8-well flat-bottom strips, 2
- fine-tipped dropper bulbs, 4
- marking pencil
- phenolphthalein indicator
- toothpicks, 10

BACKGROUND

Chemists often identify unknown substances by observing how they react with known substances. A common method involves a replacement reaction, in which a measured amount of a known substance is reacted with the unknown substance.

In this investigation, you will determine the formulas of two metallic salts. One salt is copper chloride, but you will need to find out if it is copper(I) chloride or copper(II) chloride. The other salt is iron chloride, but is it iron(II) chloride or iron(III) chloride?

The reaction to be used is a double-replacement reaction with sodium hydroxide, NaOH. Aqueous solutions of both copper and iron ions form precipitates when they are combined with a solution containing aqueous hydroxide ions.

Phenolphthalein is an indicator that turns bright pink or red in the presence of unreacted aqueous OH^- ions. It will be used to indicate when the reaction is complete. When there is no more metal left to react with the OH^- ions being added, the OH^- ions will trigger this color change.

SAFETY

Always wear safety goggles and a lab apron to protect your eyes and clothing. If you get a chemical in your eyes, immediately flush the chemical out at the eyewash station while calling to your teacher. Know the location of the emergency lab shower and eyewash station and the procedures for using them.

Do not touch chemicals. If you get a chemical on your skin or clothing, wash the chemical off at the sink while calling to your teacher. Make sure you carefully read the labels and follow the precautions on all containers of chemicals that you use. If no precautions are stated on the label, ask your teacher what precautions to follow. Do not taste any chemicals or items used in the laboratory. Never return leftover chemicals to their original containers; take only small amounts to avoid wasting supplies.

Call your teacher in the event of a spill. Spills should be cleaned up promptly according to your teacher's directions.

PREPARATION

1. Prepare a data table like the one below for the copper chloride solution. Record the number of drops of NaOH added to wells 1, 2, 3, 4, and 5 of your well strip. Prepare a similar data table for the iron chloride solution.

DATA TABLE

Copper chloride					
	1	2	3	4	5
Drops of NaOH					
Iron chloride					
	1	2	3	4	5
Drops of NaOH					

2. Obtain four dropper bulbs. Label them *Cu*, *Fe*, *NaOH*, and *In*.

PROCEDURE

1. Fill the bulb labeled *Cu* with the copper chloride solution. Fill the bulb labeled *NaOH* with the sodium hydroxide solution.
2. Using one 8-well strip, place five drops of the copper chloride solution in each of the first five wells. For best results, try to make all the drops in this experiment about the same size.
3. Using the bulb labeled *In*, put one drop of the phenolphthalein indicator in each of the five wells, as shown in Figure A.
4. Add the sodium hydroxide solution one drop at a time to the first well, mixing the solution in the well with a toothpick before adding each drop, as shown in Figure B. Continue adding NaOH until the pink or red color of the phenolphthalein just begins to show clearly. The change in color indicates that the copper chloride has reacted completely. Record the number of drops of NaOH added in your copper chloride data table. Add drops of NaOH to the other four wells in turn, and record the results. Use a different toothpick to stir the solution for each well.

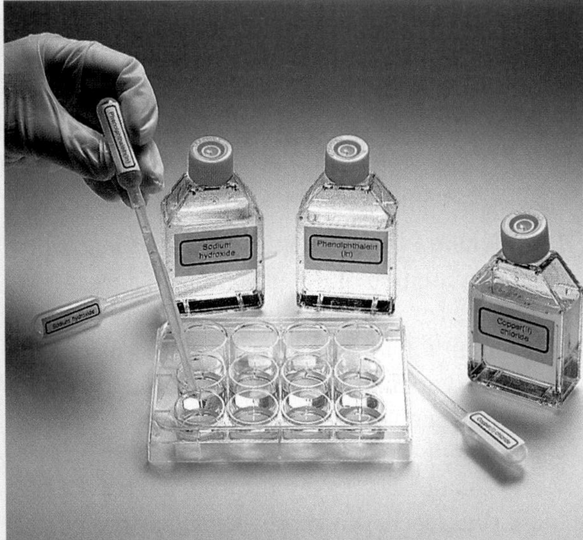

FIGURE A Add one drop of phenolphthalein to each well containing copper chloride. The size of the drops should be as uniform as possible.

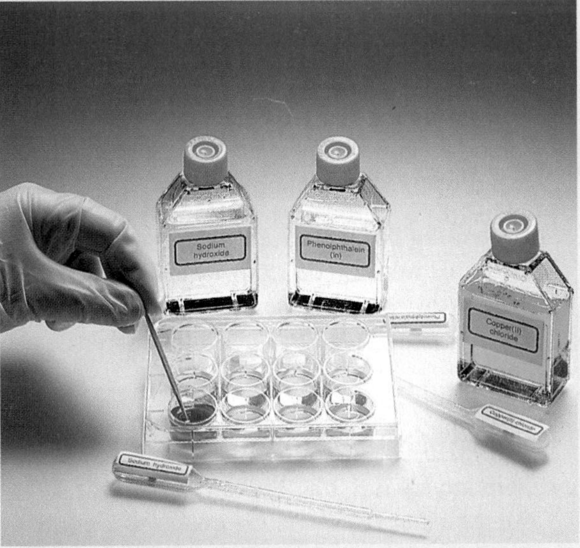

FIGURE B Add NaOH one drop at a time, stirring after each addition. A change in color to pink tells you that all the copper chloride has reacted.

5. Fill the bulb labeled *Fe* with the iron chloride solution, and repeat Procedure steps 1–4 in the second 8-well strip, using the iron chloride solution instead of the copper chloride solution. In your iron chloride data table, record the number of drops of the sodium hydroxide solution that were added to each of the five wells to cause a change in color.

CLEANUP AND DISPOSAL

1. Clean your lab station. Clean all equipment and return it to its proper place. Dispose of chemicals and solutions in the containers designated by your teacher. Do not pour any chemicals down the drain or throw anything in the trash unless your teacher directs you to do so. Wash your hands thoroughly after all work is finished and before you leave the lab.

ANALYSIS AND INTERPRETATION

1. **Organizing Ideas:** Write the balanced chemical equation for a double replacement reaction between sodium hydroxide and
   a. copper(I) chloride.
   b. copper(II) chloride.
   c. iron(II) chloride.
   d. iron(III) chloride.

2. **Organizing Ideas:** You may already have noticed that each of the solutions used has 0.1 mol of formula units for every liter of solution. Assuming that all of the drops were the same size, how do the numbers of formula units in a drop of each solution compare?

3. **Organizing Data:** On average, how many drops of NaOH were needed to react with the copper chloride? On average, how many drops of NaOH were needed to react with the iron chloride?

CONCLUSIONS

1. **Relating Ideas:** Using your answers from Analysis and Interpretation questions 2 and 3, determine how many NaOH formula units were needed to react with each copper chloride formula unit in your experiment. How many NaOH formula units were needed to react with each iron chloride formula unit?

2. **Inferring Conclusions:** Compare your answers to the previous question with the balanced chemical equations from the Analysis and Interpretations section. Which chlorides of copper and iron were in the solutions you used?

EXTENSIONS

1. **Evaluating Methods:** Share your data with other lab groups. Calculate a class average for the ratio of NaOH formula units to copper chloride formula units. Calculate a class average for the ratio of NaOH formula units to iron chloride formula units. Compare these averages with your results. Using the class average ratios as the accepted values, calculate your percent error.

2. **Designing Experiments:** What are some likely areas of imprecision in this experiment? If you can think of ways to eliminate them, ask your teacher to approve your suggestions, and run more trials.

3. **Predicting Outcomes:** How many drops of NaOH would it take to react with a solution of copper chloride (the type used in this experiment) that contains 0.5 mol/L of solution? How many drops of NaOH would it take to react with a solution of iron chloride (the type used in this experiment) that contains 0.5 mol/L of solution?
   If the solutions are available and your teacher approves, test your predictions.

4. **Designing Experiments:** What if your teacher made a solution of the same kind of copper chloride (or iron chloride) used in this experiment but forgot what the mol/L concentration is? Can you think of a way to use samples of this solution and your 0.1 mol/L solution of NaOH to find out what the concentration of the mystery solution is? If a mystery solution is available, ask your teacher to approve your suggestion, and try to determine the concentration.

Determining the Empirical Formula of Magnesium Oxide

OBJECTIVES

- ***Measure*** **the mass of magnesium oxide.**
- ***Perform*** **a synthesis reaction by using gravimetric techniques.**
- ***Determine*** **the empirical formula of magnesium oxide.**
- ***Calculate*** **the class average and standard deviation for moles of oxygen used.**

MATERIALS

- **10 mL graduated cylinder**
- **15 cm magnesium ribbon, 2**
- **Bunsen burner assembly**
- **clay triangle**
- **crucible and lid, metal or ceramic**
- **crucible tongs**
- **distilled water**
- **eyedropper or micropipet**
- **ring stand**

BACKGROUND

This gravimetric analysis involves the combustion of magnesium metal in air to synthesize magnesium oxide. The mass of the product is greater than the mass of magnesium used because oxygen bonds to the magnesium metal. Like all gravimetric analyses, success depends on attaining a product yield near 100%. Therefore, the product will be heated, cooled, and measured until two mass readings are within 0.02% of one another. When the masses of the reactant and product have been carefully measured, then the amount of oxygen used in the reaction can be calculated. The ratio of oxygen to magnesium can then be established and the empirical formula of magnesium oxide can be determined.

SAFETY

Always wear safety goggles and a lab apron to protect your eyes and clothing. If you get a chemical in your eyes, immediately flush the chemical out at the eyewash station while calling to your teacher. Know the location of the emergency lab shower and eyewash station and the procedure for using them.

Do not touch or taste any chemicals. If you get a chemical on your skin or clothing, wash the chemical off at the sink while calling to your teacher. Make sure you carefully read the labels and follow the precautions on all containers of chemicals that you use. If no precautions are stated on the label, ask your teacher what precautions you should follow. Do not taste any chemicals or items used in the laboratory. Never return leftovers to their original containers; take only small amounts to avoid wasting supplies.

When using a Bunsen burner, confine long hair and loose clothing. Do not heat glassware that is broken, chipped, or cracked. Use tongs or a hot mitt to handle heated glassware and other equipment; heated glassware does not always look hot.
If your clothing catches fire, WALK to the emergency lab shower and use it to put out the fire.

Never put broken glass or ceramics in a regular waste container. Broken glass and ceramics should be disposed of in a separate container designated by your teacher.

PREPARATION

1. Copy the following data table in your lab notebook.

DATA TABLE

	Trial 1	Trial 2
1. Mass of crucible, lid, and metal (g)		
2. Mass of crucible, lid, and product (g)		
3. Mass of crucible and lid (g)		

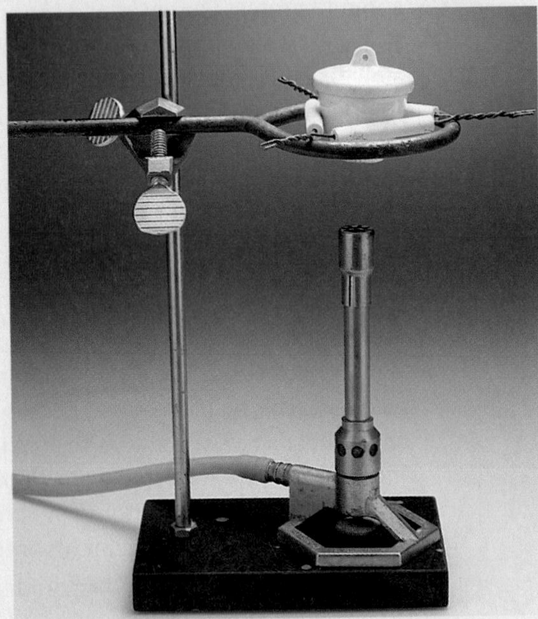

FIGURE A

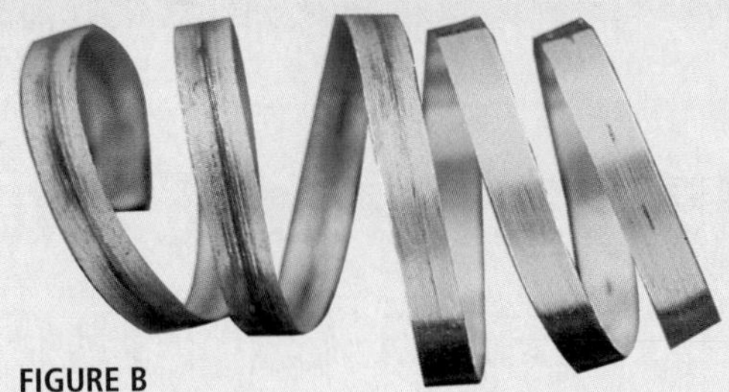

FIGURE B

PROCEDURE

1. Construct a setup for heating a crucible as shown in Figure A and as demonstrated in the Pre-Laboratory Procedure on page 805.
2. Strongly heat the crucible and lid for 5 min to burn off any impurities.
3. Cool the crucible and lid to room temperature. Measure their combined mass, and record the measurement on line 3 of your data table.

   NOTE: Handle the crucible and lid with crucible tongs. This prevents burns and the transfer of dirt and oil from your hands to the crucible and lid.
4. Polish a 15 cm strip of magnesium with steel wool. The magnesium should be shiny, as shown in Figure B. Cut the strip into small pieces to make the reaction proceed faster, and place the pieces in the crucible.
5. Cover the crucible with the lid, and measure the mass of the crucible, lid, and metal. Record the measurement on line 1 of your data table.
6. Use tongs to replace the crucible on the clay triangle. Heat the covered crucible gently. Lift the lid occasionally to allow air in, as shown in Figure C.

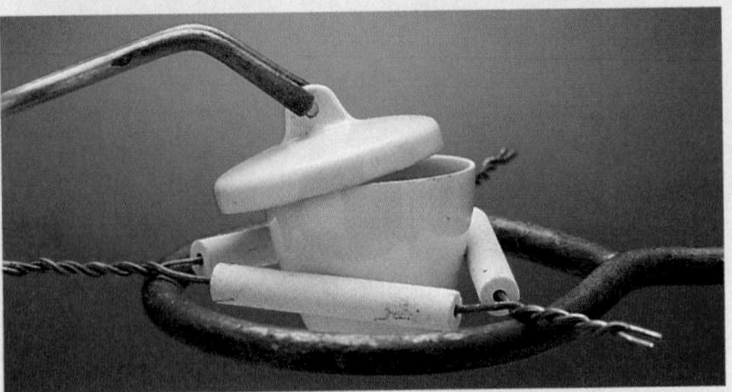

FIGURE C

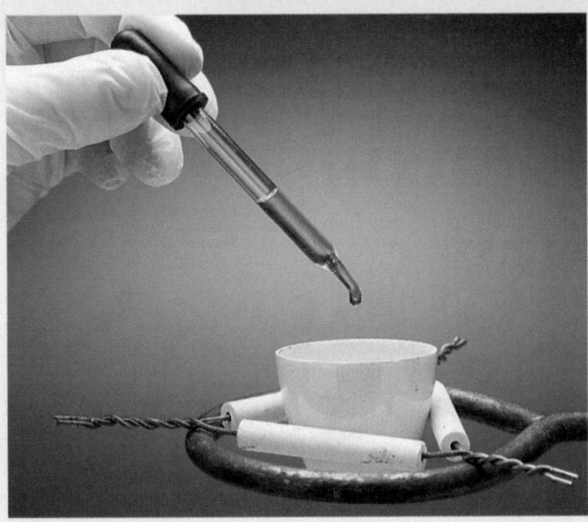

FIGURE D

CAUTION: Do not look directly at the burning magnesium metal. The brightness of the light can blind you.

7. When the magnesium appears to be fully reacted, partially remove the crucible lid and continue heating for 1 min.

8. Remove the burner from under the crucible. After the crucible has cooled, use an eyedropper to carefully add a few drops of water, as shown in Figure D, to decompose any nitrides that may have formed.

   CAUTION: Use care when adding water. Too much water can cause the crucible to crack.

9. Cover the crucible completely. Replace the burner under the crucible and continue heating for about 30–60 s.

10. Turn off the burner. Cool the crucible, lid, and contents to room temperature. Measure the mass of the crucible, lid, and product. Record the measurement in the margin of your data table.

11. Replace the crucible, lid, and contents on the clay triangle and reheat for another 2 min. Cool to room temperature and remeasure the mass of the crucible, lid, and contents. Compare this mass measurement with the measurement obtained in step 10. If the new mass is ±0.02% of the mass in step 10, record the new mass on line 2 of your data table. If not, your reaction is still incomplete. Repeat this step.

12. Clean the crucible, and repeat steps 2–11 with a second strip of magnesium ribbon. Record your measurements under Trial 2 in your data table.

CLEANUP AND DISPOSAL

1. Put the solid magnesium oxide in the designated waste container. Return any unused magnesium ribbon to your teacher. Clean your equipment and lab station. Thoroughly wash your hands after completing the lab session and cleanup.

ANALYSIS AND INTERPRETATION

1. **Applying Ideas:** Calculate the mass of the magnesium metal and the mass of the product.

2. **Evaluating Data:** Determine the mass of the oxygen consumed.

3. **Applying Ideas:** Calculate the number of moles of magnesium and the number of moles of oxygen in the product.

CONCLUSIONS

1. **Inferring Relationships:** Determine the empirical formula for magnesium oxide, Mg_xO_y. (Divide your mole ratio by the moles of magnesium since this value is derived from a measured quantity instead of a calculated quantity.)

Mass and Mole Relationships in a Chemical Reaction

OBJECTIVES

- ***Demonstrate*** **proficiency in measuring masses.**
- ***Determine*** **the number of moles of reactants and products in a reaction experimentally.**
- ***Use*** **the mass and mole relationships of a chemical reaction in calculations.**
- ***Perform*** **calculations that involve density and stoichiometry.**

MATERIALS

- **1.0 M CH_3COOH**
- **2–3 g $NaHCO_3$**
- **balance**
- **beaker tongs**
- **Bunsen burner and related equipment or hot plate**
- **dropper or pipet**
- **evaporating dish**
- **graduated cylinder**
- **ring stand and ring (for use with Bunsen burner)**
- **spatula**
- **watch glass**
- **wire gauze with ceramic center (for use with Bunsen burner)**

CH_3COOH
Acetic Acid

BACKGROUND

In this experiment, you will determine the amounts of sodium hydrogen carbonate and acetic acid needed to produce a specific amount of carbon dioxide by reacting a carefully measured mass of reactant, $NaHCO_3$, with vinegar and then measuring the mass of the product, CH_3COONa. You can then determine the number of moles of acetic acid reacted and the number of moles of CO_2 produced. Using mole relationships between reactants and products, you can calculate the mass and the number of moles of each reactant needed to produce any given volume of CO_2. To obtain the volume of CO_2 from its mass, you will need to know that the density of CO_2 is 1.25 g/L at baking temperature.

SAFETY

Always wear safety goggles and a lab apron to protect your eyes and clothing. If you get a chemical in your eyes, immediately flush the chemical out at the eyewash station while calling to your teacher. Know the locations of the emergency lab shower and eyewash station and the procedures for using them.

Do not touch any chemicals. If you get a chemical on your skin or clothing, wash the chemical off at the sink while calling to your teacher. Make sure you carefully read the labels and follow the directions on all containers of chemicals that you use. If no precautions are stated on the label, ask your teacher what precautions to follow. Do not taste any chemicals or items used in the laboratory. Never return leftover chemicals to their original containers; take only small amounts to avoid wasting supplies.

When using a Bunsen burner, confine long hair and loose clothing. Do not heat glassware that is broken, chipped, or cracked. Use tongs or a hot mitt to handle heated glassware and other equipment; heated glassware does not always look hot. If your clothing catches fire, WALK to the emergency lab shower and use it to put out the fire.

Never put broken glass or ceramics in a regular waste container. Broken glass and ceramics should be disposed of in a separate container designated by your teacher.

Call your teacher in the event of an acid or base spill. Acid or base spills should be cleaned up promptly according to your teacher's instructions.

PREPARATION

1. Prepare a data table in your lab notebook. It should contain space to record the mass of the empty evaporating dish and watch glass, the mass of the evaporating dish with the watch glass and $NaHCO_3$, and the mass of the evaporating dish with the watch glass and CH_3COONa after heating.

PROCEDURE

1. Measure the mass of a clean, dry evaporating dish and watch glass to the nearest 0.01 g. Record this mass in your data table.
2. Add 2–3 g $NaHCO_3$ to your evaporating dish. Measure the exact mass of the $NaHCO_3$ with the watch glass, to the nearest 0.01 g. Record this mass in your data table.
3. Slowly add 30 mL of the acetic acid solution to the $NaHCO_3$ in the evaporating dish. Add more acetic acid with a dropper or pipet until the bubbling stops.
4. If you are using a Bunsen burner, place the evaporating dish and its contents on a ceramic-centered wire gauze placed on an iron ring attached to the ring stand, as shown in Figure A.

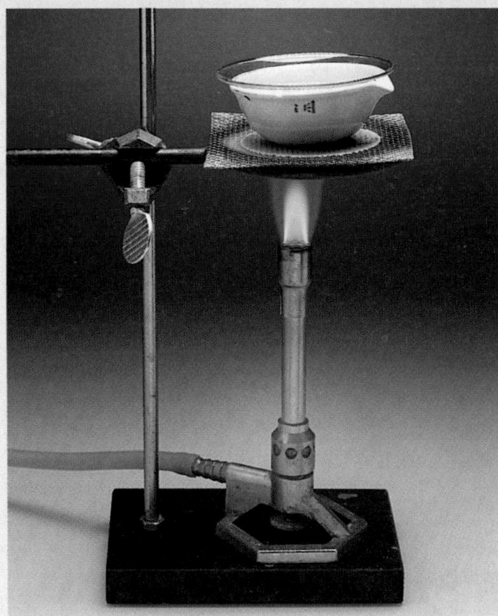

FIGURE A The watch glass is placed concave side up on the evaporating dish, which is centered on a ceramic-centered wire gauze.

Place the watch glass, concave side up, on top of the dish, making sure that there is a slight opening for steam to escape. If you are using a hot plate, position the watch glass the same way, but place the evaporating dish directly on the hot plate.

5. Gently heat the evaporating dish until only a dry solid remains. Make sure that no water droplets remain on the underside of the watch glass. *Do not heat too rapidly or the material will boil and the product will spatter out of the evaporating dish.*
6. Turn off the gas burner or hot plate. Allow the apparatus to cool for at least 15 min. Determine the mass of the cooled equipment to the nearest 0.01 g. Record the mass of the dish, residue, and watch glass in your data table.
7. If time permits, reheat the evaporating dish and contents for 2 min. Let it cool, and measure its mass again. You can be certain the sample is dry when you obtain two successive measurements within 0.02 g of each other.

CLEANUP AND DISPOSAL

8. Clean up the lab and all equipment after use. Dispose of any unused chemicals in the containers designated by your teacher. Wash your hands thoroughly before you leave the lab after all lab work is finished. Make sure to turn off all gas valves.

ANALYSIS AND INTERPRETATION

1. **Analyzing Results:** Write a balanced equation for the reaction of baking soda and acetic acid. Be sure to include the physical states of matter for all of the reactants and products.
2. **Organizing Data:** Use a periodic table to calculate the molar mass for each of the reactants and products.
3. **Analyzing Results:** Explain what caused the bubbling when the reaction took place.
4. **Analyzing Methods:** How do you know that all the residue is actually sodium acetate rather than a mixture of sodium bicarbonate and sodium acetate?
5. **Organizing Data:** Calculate the mass of $NaHCO_3$, the number of moles of $NaHCO_3$, the mass of CH_3COONa, and the number of moles of CH_3COONa.
6. **Evaluating Data:** Using the balanced equation and the amount of $NaHCO_3$, determine the theoretical yield of CH_3COONa in moles and grams.

CONCLUSIONS

1. **Analyzing Conclusions:** What is the percent yield for your reaction?
2. **Inferring Conclusions:** What is the theoretical yield of CO_2? Using the density value given, 1.25 g/L, calculate the volume of CO_2 produced in the reaction. Show your calculations.
3. **Applying Conclusions:** How many moles of $NaHCO_3$ and CH_3COONa are necessary to produce 425 mL of CO_2? Show your calculations. (Hint: Be sure to include your percent yield for this reaction in your calculations.)

EXTENSIONS

1. **Designing Experiments:** If your percent yield is less than 100%, explain why. If you can think of ways to eliminate sources of error, ask your teacher to approve your plans, and run the procedure again.
2. **Research and Communications:** Many recipes for breads use yeast, instead of baking soda, as a source of CO_2. A cake of yeast is shown in Figure B. Research the use of yeast, and explain what ingredients are necessary for the yeast to produce carbon dioxide. What is the balanced chemical equation for the reaction that yeast uses to produce CO_2?

FIGURE B Yeast can be compressed into cakes and used in baking as a source of carbon dioxide.

PRE-LAB • PAGE 804

Stoichiometry and Gravimetric Analysis

OBJECTIVES

- ***Observe*** **the double-displacement reaction between solutions of strontium chloride and sodium carbonate.**
- ***Demonstrate*** **proficiency with gravimetric methods.**
- ***Measure*** **the mass of the precipitate formed.**
- ***Relate*** **the mass of the precipitate formed to the mass of the reactants before the reaction.**
- ***Calculate*** **the mass of sodium carbonate in a solution of unknown concentration.**

MATERIALS

- **15 mL Na_2CO_3 solution of unknown concentration**
- **50 mL 0.30 M $SrCl_2$ solution**
- **250 mL beakers, 2**
- **balance**
- **beaker tongs**
- **distilled water**
- **drying oven**
- **filter paper**
- **glass funnel or Büchner funnel with related equipment**
- **graduated cylinder**
- **glass stirring rod**
- **paper towels**
- **ring and ring stand**
- **rubber policeman**
- **spatula**
- **water bottle**

BACKGROUND

This gravimetric analysis involves a double-displacement reaction between strontium chloride, $SrCl_2$, and sodium carbonate, Na_2CO_3. In general, this type of reaction can be used to determine the amount of any carbonate compound in a solution. For accurate results, essentially all of the reactant of unknown amount must be converted into product. If the mass of the product is carefully measured, you can use stoichiometry calculations to determine how much of the reactant of unknown amount was involved in the reaction. Accurate results depend on precise mass measurements, so keep all glassware very clean, and minimize the loss of any reactants or products during your lab work.

SAFETY

Always wear safety goggles and a lab apron to protect your eyes and clothing. If you get a chemical in your eyes, immediately flush the chemical out at the eyewash station while calling to your teacher. Know the locations of the emergency lab shower and eyewash station and the procedure for using them.

Do not touch any chemicals. If you get a chemical on your skin or clothing, wash the chemical off at the sink while calling to your teacher. Make sure you carefully read the labels and follow the precautions on all containers of chemicals that you use. If no precautions are stated on the labels, ask your teacher what precautions to follow. Do not taste any items used in the laboratory. Never return leftover chemicals to their original containers; take only small amounts to avoid wasting supplies.

Never put broken glass or ceramics in a regular waste container. Broken glass and ceramics should be disposed of in a separate container designated by your teacher.

PREPARATION

1. Copy the data table below in your lab notebook.

DATA TABLE	
Volume of Na_2CO_3 solution added	
Volume of $SrCl_2$ solution added	
Mass of dry filter paper	
Mass of beaker with paper towel	
Mass of beaker with paper towel, filter paper, and precipitate	

2. Clean all of the necessary lab equipment with soap and water, even if it has already been cleaned once. Rinse each piece of equipment with distilled water.

3. Measure the mass of a piece of filter paper to the nearest 0.01 g, and record this value in your data table.

4. Set up a filtering apparatus, either a vacuum filtration or a gravity filtration, depending on the equipment that is available. Use the Pre-Laboratory Procedure described on page 790.

5. Label a paper towel with your name, your class, and the date. Place the paper towel in a clean, dry 250 mL beaker, and measure and record the mass of the paper towel and beaker to the nearest 0.01 g.

PROCEDURE

1. Measure about 15 mL of the Na_2CO_3 solution into the graduated cylinder. Record this volume to the nearest 0.5 mL in your data table. Pour the Na_2CO_3 solution into a clean, empty 250 mL beaker. Carefully wash the graduated cylinder, and rinse it with distilled water.

2. Measure about 25 mL of the 0.30 M $SrCl_2$ solution in the graduated cylinder. Record this volume to the nearest 0.5 mL in your data table.

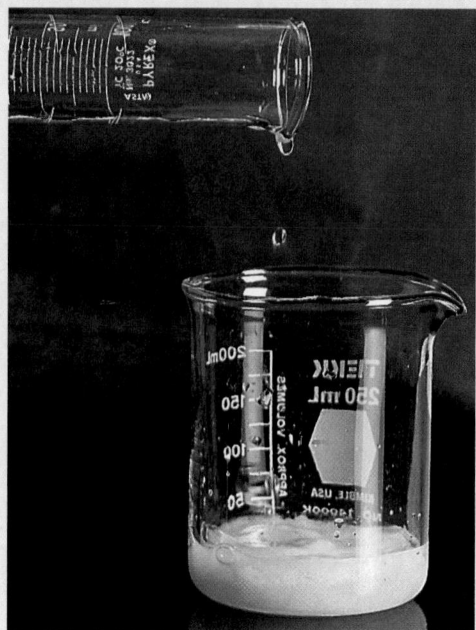

FIGURE A The precipitate is a product of the reaction between Na_2CO_3 and $SrCl_2$. Add enough $SrCl_2$ to react with all of the Na_2CO_3 present.

Pour the $SrCl_2$ solution into the beaker with the Na_2CO_3 solution, as shown in Figure A. Gently stir the solution and precipitate with a glass stirring rod.

3. Carefully measure another 10 mL of the $SrCl_2$ solution into the graduated cylinder. Record the volume to the nearest 0.5 mL in your data table. Slowly add it to the beaker. Repeat this step until no more precipitate forms.

4. Once the precipitate has settled, slowly pour the mixture into the funnel. Be careful not to overfill the funnel because some of the precipitate could be lost between the filter paper and the funnel. Use the rubber policeman to transfer as much of the precipitate into the funnel as possible.

5. Rinse the rubber policeman over the beaker with a small amount of distilled water, and pour this solution into the funnel. Rinse the beaker several more times with small amounts of distilled water, as shown in Figure B. Pour the rinse water into the funnel each time.

FIGURE B Rinse with small amounts of water, directed at all sides of the beaker, to wash all the precipitate into the funnel.

6. After all of the solution and rinses have drained through the funnel, slowly rinse the precipitate on the filter paper in the funnel with distilled water to remove any soluble impurities.
7. Carefully remove the filter paper from the funnel, and place it on the paper towel that you labeled with your name. Unfold the filter paper, and place the paper towel, filter paper, and precipitate in the rinsed beaker. Then place the beaker in the drying oven. For best results, allow the precipitate to dry overnight.
8. Using beaker tongs, remove your sample from the drying oven, and allow it to cool. Measure and record the mass of the beaker with paper towel, filter paper, and precipitate to the nearest 0.01 g.

CLEANUP AND DISPOSAL

9. Dispose of the precipitate in a designated waste container. Pour the filtrate in the other 250 mL beaker into the designated waste container. Clean up the lab and all equipment after use, and dispose of substances according to your teacher's instructions. Wash your hands thoroughly after all lab work is finished and before you leave the lab.

ANALYSIS AND INTERPRETATION

1. **Organizing Ideas:** Write a balanced equation for the reaction. What is the precipitate? Write its empirical formula.
2. **Applying Ideas:** Calculate the mass of the dry precipitate. Calculate the number of moles of precipitate produced in the reaction. (Hint: Use the results from Procedure item 8.)
3. **Applying Ideas:** How many moles of Na_2CO_3 were present in the 15 mL sample?
4. **Evaluating Methods:** There were 0.30 mol of $SrCl_2$ in every liter of solution. Calculate the number of moles of $SrCl_2$ that were added. Determine whether $SrCl_2$ or Na_2CO_3 was the limiting reactant. Would this lab have worked if the other reactant had been chosen as the limiting reactant? Explain why or why not.
5. **Evaluating Methods:** Why was the precipitate rinsed in Procedure step 6? What soluble impurities could have been on the filter paper along with the precipitate? How would the calculated results vary if the precipitate had not been completely dry? Explain your answer.

CONCLUSIONS

1. **Inferring Conclusions:** How many grams of Na_2CO_3 were present in the 15 mL sample?
2. **Applying Conclusions:** How many grams of Na_2CO_3 are present in 575 L of the Na_2CO_3 solution? (Hint: Create a conversion factor to convert from the sample with a volume of 15 mL to a solution with a volume of 575 L.)

EXTENSIONS

1. **Evaluating Methods:** From your teacher, find out the theoretical mass of Na_2CO_3 in the sample, and calculate your percent error.
2. **Designing Experiments:** What possible sources of error can you identify in your procedure? If you can think of ways to eliminate them, ask your teacher to approve your plans, and run the procedure again.

EXPERIMENT 12-1

"Wet" Dry Ice

OBJECTIVES

- *Interpret* a phase diagram.
- *Observe* the melting of CO_2 while varying pressure.
- *Relate* observations of CO_2 to its phase diagram.

MATERIALS

- 4–5 g CO_2 as dry ice, broken into rice-sized pieces
- forceps
- graduated pipet
- metric ruler
- pliers
- scissors
- transparent plastic cup

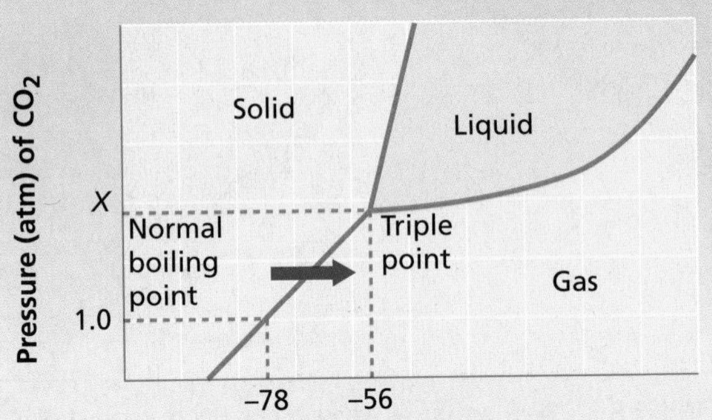

FIGURE A The phase diagram for CO_2 shows the temperatures and pressures at which CO_2 can undergo phase changes.

BACKGROUND

The phase diagram for carbon dioxide in Figure A shows that CO_2 can exist only as a gas at ordinary room temperature and pressure. To observe the transition of solid CO_2 to liquid CO_2, it will be necessary for you to increase the pressure until it is at or above the triple point pressure, labeled *X* in the diagram.

SAFETY

Always wear safety goggles and a lab apron to protect your eyes and clothing. If you get a chemical in your eyes, immediately flush the chemical out at the eyewash station while calling to your teacher. Know the location of the emergency lab shower and eyewash station and the procedure for using them.

Call your teacher in the event of a spill. Spills should be cleaned up promptly according to your teacher's directions.

Dry ice is cold enough to cause frostbite, so heat- and cold-resistant gloves should be used if you handle it.

PREPARATION

1. Organize a place in your lab notebook for recording your observations.

PROCEDURE

1. Place 2–3 very small pieces of dry ice on the table, and observe them until they have completely sublimed.
2. Fill a plastic cup with tap water to a depth of 4–5 cm.
3. Cut the tapered end (tip) off the graduated pipet.
4. Use forceps to carefully slide 8–10 pieces of dry ice down the stem and into the bulb of the pipet.
5. Using a pair of pliers, clamp the opening of the pipet stem securely shut so that no gas can escape. Hold the tube by the pliers, and lower the pipet into the cup just until the bulb is submerged as shown in Figure B. From the side of the cup, observe the behavior of the dry ice.
6. As soon as the dry ice has begun to melt, quickly loosen the pliers while still holding the bulb in the water. Observe the CO_2.

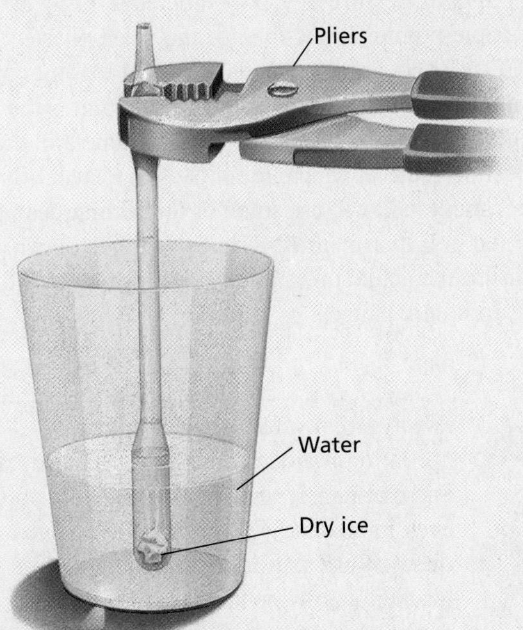

FIGURE B Clamp the end of the pipet shut with the pliers. Submerge the bulb in water in a transparent cup.

7. Tighten the pliers again, and observe.
8. Repeat Procedure steps 6 and 7 as many times as possible.

CLEANUP AND DISPOSAL

9. Clean all apparatus and your lab station. Return equipment to its proper place. Dispose of chemicals and solutions in the containers designated by your teacher. Do not pour any chemicals down the drain or in the trash unless your teacher directs you to do so. Wash your hands thoroughly before you leave the lab and after all work is finished.

ANALYSIS AND INTERPRETATION

1. **Analyzing Results:** What differences did you observe between the subliming and the melting of CO_2?
2. **Analyzing Methods:** As you melted the CO_2 sample over and over, why did it eventually disappear? What could you have done to make the sample last longer?
3. **Analyzing Methods:** What purpose(s) do you suppose the water in the cup served?

EXTENSIONS

1. **Predicting Outcomes:** What would have happened if fewer pieces of dry ice (only 1 or 2) had been placed inside the pipet bulb? If time permits, test your prediction.
2. **Predicting Outcomes:** What might have happened if more pieces of dry ice (20 or 30, for example) had been placed inside the pipet bulb? How quickly would the process have occurred? If time permits, test your prediction.
3. **Predicting Outcomes:** What would have happened if the pliers had not been released once the dry ice melted? If time permits, test your prediction.

Measuring the Triple-Point Pressure of CO_2

OBJECTIVES

- *Interpret* a phase diagram.
- *Observe* changes in pressure during a phase change.
- *Relate* pressure values to observations of air-column length.
- *Determine* the triple-point pressure of CO_2.
- *Infer* a conclusion from experimental data.

MATERIALS

- 4–5 g CO_2 as dry ice, broken into rice-sized pieces
- 7 cm tube, cut from a thin-stemmed pipet
- 15–20 cm of thread
- dark-colored water
- fine-tipped dropper bulb
- fine-tipped permanent marker
- forceps
- graduated pipet
- hot-glue gun
- metric ruler
- micropressure gauge
- pliers
- scissors
- transparent plastic cup

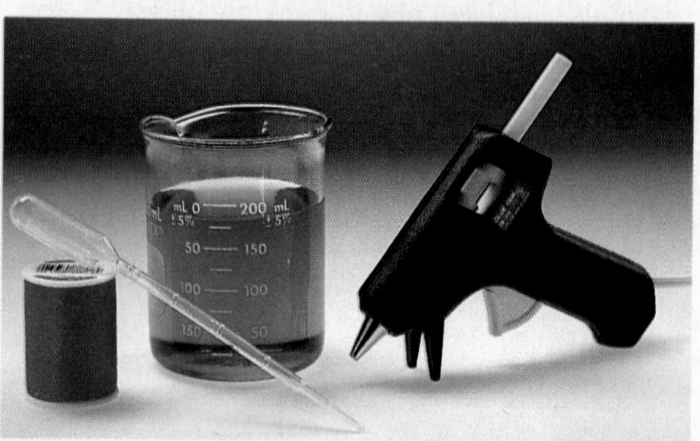

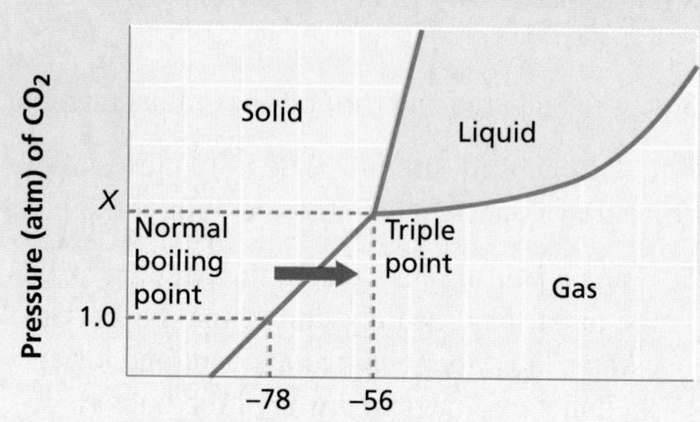

FIGURE A The phase diagram for CO_2 shows the temperatures and pressures at which CO_2 can undergo phase changes.

BACKGROUND

In the phase diagram shown in Figure A, the triple-point pressure of CO_2 is labeled *X*. Only at the triple-point is it possible for the three phases of CO_2—solid, liquid, and gas—to exist together at equilibrium. The pressure at the triple-point is the highest pressure at which CO_2 will sublime and the lowest pressure at which liquid CO_2 can exist. In this experiment, you will use some of the techniques you learned in Experiment 12–1, but this time you will measure the actual pressure at the triple point with a micropressure gauge.

SAFETY

Always wear safety goggles and a lab apron to provide protection for your eyes and clothing. If you get a chemical in your eyes, immediately flush the chemical out at the eyewash station while calling to your teacher. Know the location of the emergency lab shower and eyewash station and the procedure for using them.

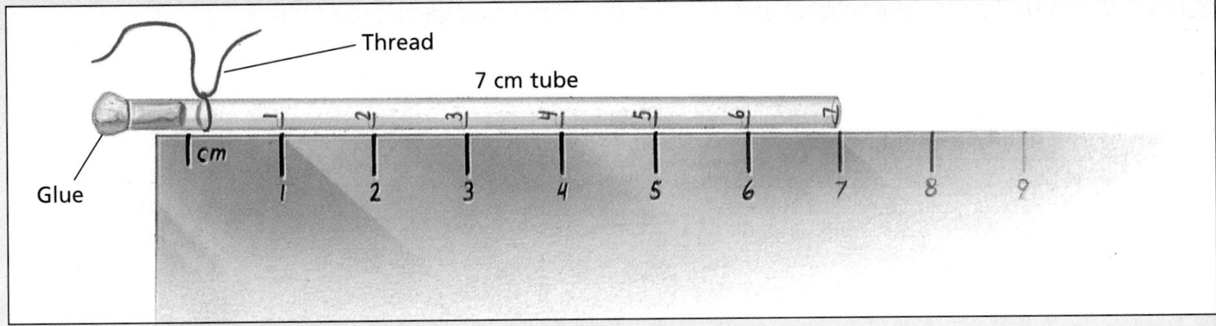

FIGURE B The sealed 7 cm tube, with thread attached, is calibrated by marking the tube at 1 cm intervals from the inside edge of the glue plug.

Do not touch any chemicals. If you get a chemical on your skin or clothing, wash the chemical off at the sink while calling to your teacher. Make sure you carefully read the labels and follow the precautions on all containers of chemicals that you use. If there are no precautions stated on the label, ask your teacher what precautions to follow. Do not taste any chemicals or items used in the laboratory. Never return leftovers to their original containers; take only small amounts to avoid wasting supplies.

Call your teacher in the event of a spill. Spills should be cleaned up promptly according to your teacher's directions.

Dry ice is cold enough to cause frostbite, so heat- and cold-resistant gloves should be used if you handle it.

The tip of the hot-glue gun is very hot, as is the glue.

PREPARATION

1. Copy the data table shown below in your lab notebook.
2. Fill a plastic cup with tap water to a depth of about 4–5 cm. Cut the tapered tip off a graduated pipet. If the micropressure gauges are already prepared, go to Procedure step 1. If not, Preparation steps 3–5 will explain how to make a micropressure gauge.
3. Take the 7 cm tube cut from a thin-stemmed pipet, and place a small drop of hot glue on one end to seal it. When the glue has cooled, tie the thread around the tube below the glue, as shown in Figure B. Trim away any excess glue so that the tube will easily pass through the end of the graduated pipet you prepared in Preparation step 2.
4. Using the metric ruler to measure from the inside edge of the glue, mark off every centimeter along the length of the tube with the fine-tipped permanent marker. Number the centimeter marks as shown in Figure B.
5. Using the fine-tipped dropper bulb, place a small drop of dark-colored water inside the open end of the tube. Record the position of the drop on the scale from the inside edge of the drop to the inside edge of the glue plug. This is a measurement of the length of the air column trapped inside the tube, as shown in Figure C. Note this measurement in your data table.

DATA TABLE

	Trial 1	Trial 2	Trial 3	Trial 4	Trial 5
Initial reading (cm)					
CO_2 melting-point reading (cm)					

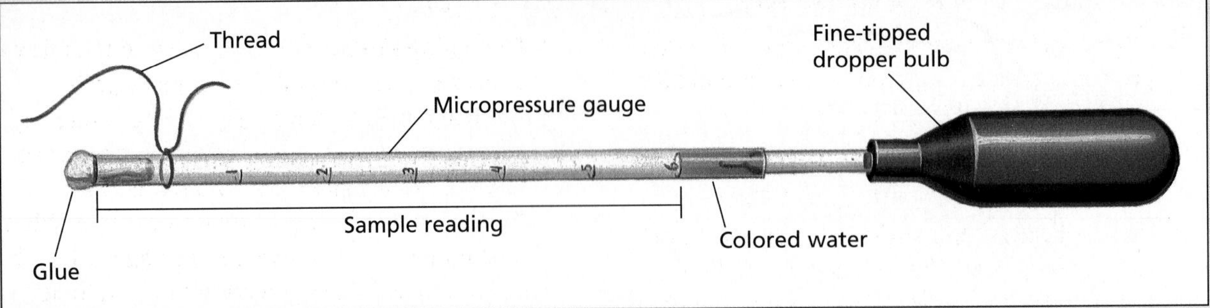

FIGURE C The pressure is equivalent to the length of the column of air measured from the inside edge of the glue plug to the inside edge of the water drop.

PROCEDURE

1. As in Experiment 12-1, carefully slide 8–10 pieces of dry ice down the stem and into the graduated pipet bulb from Preparation step 2. However, instead of clamping the graduated pipet shut, insert the micropressure gauge, open-end downward, so that the gauge hangs in the stem and NOT in the bulb of the graduated pipet, as shown in Figure D. The thread of the micropressure gauge should be hanging out of the end of the pipet. Use the pliers to clamp the pipet shut around the thread so that no gas can escape and the micropressure gauge remains suspended in the stem of the pipet.

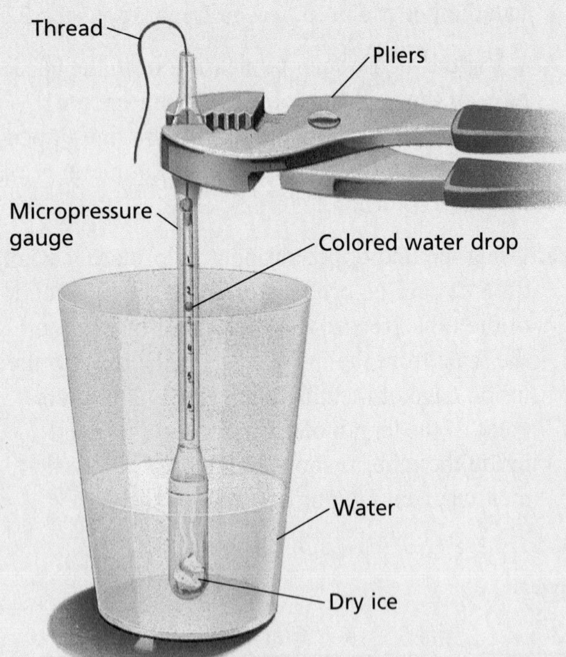

FIGURE D The micropressure gauge hangs in the stem of the pipet. The thread extends from its open end. The bulb containing the dry ice is submerged in water.

2. Holding the pipet with the pliers, lower the bulb, NOT the stem of the pipet, until it is submerged in the cup of water, as shown in Figure D.

3. From the side of the cup, observe the movement of the drop of colored water in the micropressure gauge before and while the CO_2 melts. Note the position in centimeters, as marked on the micropressure gauge, of the top of the colored-water drop the instant the dry ice begins to melt. Quickly loosen the grip on the pliers. Record the position of the drop (melting point reading) in your data table.

4. With the pliers loosened, observe the CO_2 and the drop of colored water in the micropressure gauge.

5. Tighten the grip on the pliers again and observe as before.

6. Repeat Procedure steps 3–5 as many times as possible.

CLEANUP AND DISPOSAL

1. Clean all apparatus and your lab station. Return equipment to its proper place. Dispose of chemicals and solutions in the containers designated by your teacher. Do not pour any chemicals down the drain or in the trash unless your teacher directs you to do so. Wash your hands thoroughly before you leave the lab and after all work is finished.

ANALYSIS AND INTERPRETATION

1. **Analyzing Information:** What happened to the pressure inside the bulb before, while, and after the dry ice melted?

2. **Organizing Data:** What was the initial length of the air column inside the tube? What was the length of the air column at the instant the dry ice started to melt? What was the initial pressure in atm inside the tube when you placed the drop of liquid in it?

CONCLUSIONS

1. **Inferring Relationships:** For micropressure gauges, the length of the gauge is proportional to volume. Therefore, the final pressure of the system can be calculated as follows.

$$P_2 = \frac{P_1 \times \text{length}_1}{\text{length}_2}$$

Using your answers to item 2 in Analysis and Interpretation, calculate the pressure in atm inside the pipet, P_2, when liquid CO_2 first appeared. Show your work.

2. **Relating Ideas:** How does your answer to Conclusions item 1 relate to the triple-point pressure *X*? Justify your explanation with evidence from the phase diagram for CO_2.

3. **Evaluating Methods:** The accepted value for the triple point of CO_2 is 5.11 atm. Compare this value with your experimental value, and calculate your percent error.

EXTENSIONS

1. **Evaluating Methods:** When you used the micropressure gauge, you measured the ***length*** of a trapped air column, with a glue plug at one end and a drop of water at the other. However, the calculations for gases are for a relationship between pressure and ***volume,*** not length. What assumptions are being made in using the length of the trapped air column in place of volume?

2. **Evaluating Methods and Designing Experiments:** What possible sources of error can you identify in this procedure? If you can think of a way to eliminate errors, ask your teacher to approve your suggestion. Then run the procedure again, and calculate the percent error.

3. **Inferring Relationships:** Determine the volume of the pipet bulb. Assuming that the pressure measured on the micropressure gauge is correct and that the temperature at the triple point for CO_2 is –56°C, calculate the number of moles of CO_2 gas that were present. Calculate the number of grams that this would be. What is the density of CO_2 gas under these conditions?

PRE-LABORATORY PROCEDURE

Paper Chromatography

Chromatography is a technique used to separate substances dissolved in a mixture. The Latin roots of the word are *chromato,* which means "color," and *graphy,* which means "to write." Paper is one medium used to separate the components of a solution.

Paper is made of cellulose fibers that are pressed together. As a solution passes over the fibers and through the pores, the paper acts as a filter and separates the mixture's components. Particles of the same component group together, producing a colored band. Properties such as particle size, molecular mass, and charge of the different solute particles in the mixture affect the distance the components will travel on the paper. The components of the mixture that are the most soluble in the solvent and the least attracted to the paper will travel the farthest. Their band of color will be closest to the edge of the paper.

GENERAL SAFETY

Always wear safety goggles and a lab apron to protect your eyes and clothing. If you get a chemical in your eyes, immediately flush the chemical out at the eyewash station while calling to your teacher. Know the location of the emergency lab shower and eyewash station and the procedure for using them.

PROCEDURE

1. Use a lead pencil to sketch a circle about the size of a quarter in the center of a piece of circular filter paper that is 12 cm in diameter.
2. Write one numeral for each substance, including any unknowns, around the inside of this circle. In this experiment, 6 mixtures are to be separated, so the circle is labeled 1 through 6, as shown in Figure A.
3. Use a micropipet to place a spot of each substance to be separated next to a number. Make one spot per number. If the spot is too large, you will get a broad, tailing trace with little or no detectable separation.
4. Use the pencil to poke a small hole in the center of the spotted filter paper. Insert a wick through the hole. A wick can be made by rolling a triangular piece of filter paper into a cylinder: start

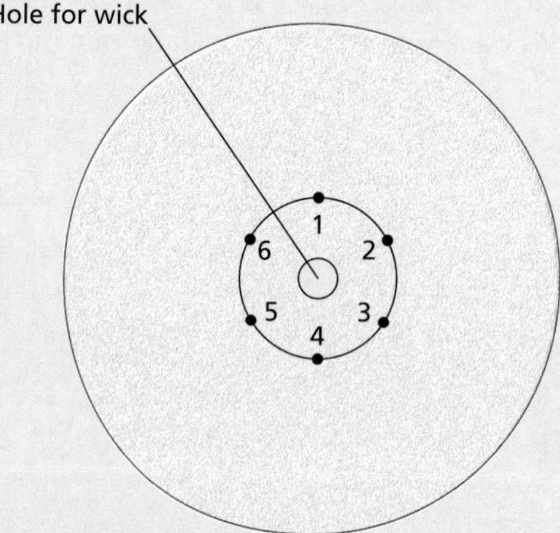

FIGURE A Filter paper used in paper chromatography is spotted with the mixtures to be separated. Each spot is labeled with a numeral or a name that identifies the mixture to be separated. A hole punched in the center of the paper will attach to a wick that delivers the solvent to the paper.

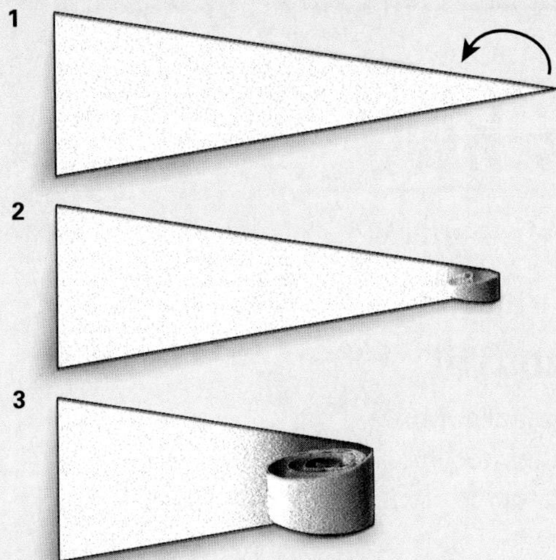

FIGURE B Cut the triangle from filter paper. Roll the paper into a cylinder starting at the narrow end.

at the point of the triangle, and roll toward its base. See Figure B.

5. Fill a petri dish or lid two-thirds full of solvent (usually water or alcohol).

6. Set the bottom of the wick in the solvent so that the filter paper rests on the top of the petri dish. See Figure C.

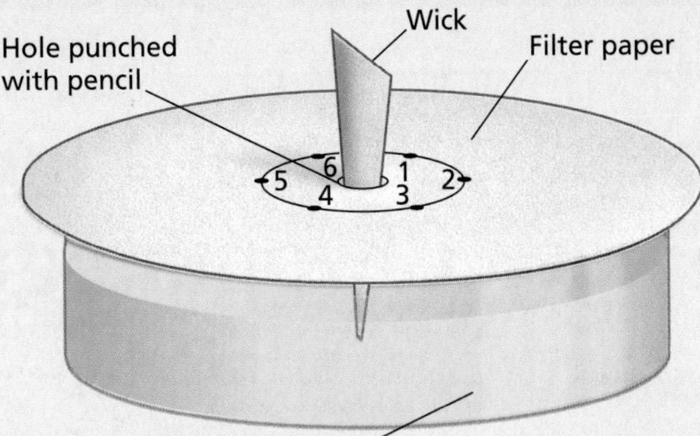

FIGURE C The wick is inserted through the hole of the spotted filter paper. The filter paper with the wick is then placed on top of a petri dish or lid filled two-thirds full of water or another solvent.

7. When the solvent is 1 cm from the outside edge of the paper, remove the paper from the petri dish, and allow the chromatogram to dry. Sample chromatograms are shown in Figures D and E.

Most writing or drawing inks are mixtures of various components that give them specific properties. Therefore, paper chromatography can be used to study the composition of an ink. Experiments 13-1 and 16-2 investigate the composition of ball-point-pen ink and marker ink, respectively.

FIGURE D Each of the original spots has migrated along with the solvent toward the outer edge of the filter paper. For each substance that was a mixture, you should see more than one distinct spot of color in its trace.

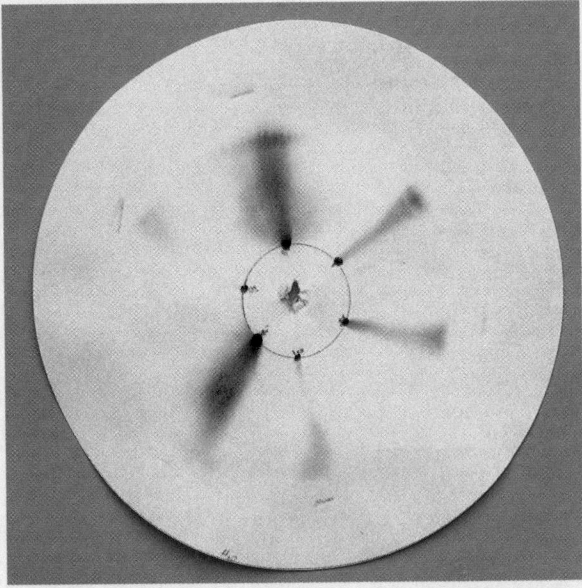

FIGURE E This chromatogram is unacceptable due to bleeding or spread of the pigment front. Too much ink was used in the spot, or the ink was too soluble in the solvent.

Separation of Pen Inks by Paper Chromatography

OBJECTIVES

- ***Demonstrate*** **proficiency in qualitatively separating mixtures using paper chromatography.**
- ***Determine*** **the R_f factor(s) for each component of each tested ink.**
- ***Explain*** **how the inks are separated by paper chromatography.**
- ***Observe*** **the separation of a mixture by the method of paper chromatography.**

MATERIALS

- **12 cm circular chromatography paper or filter paper**
- **distilled water**
- **filter paper wick, 2 cm equilateral triangle**
- **isopropanol**
- **numbered pens, each with a different black ink, 4**
- **pencil**
- **petri dish with lid**
- **scissors**

BACKGROUND

Paper Chromatography

Details on this technique can be found in the Pre-Laboratory Procedure on pages 828–829.

Writing Inks

In general, writing inks are of two types—water soluble and oil based. Inks can be pure substances or mixtures. Most ballpoint pen inks are complex mixtures, containing pigments or dyes that can be separated by paper chromatography, as shown in Figure A. Although there are thousands of different formulations for inks, the chemicals used to make them can be broadly classed into three categories: color, solvents, and resins.

An ink's color is due to dyes or pigments, which account for about 25% of the ink's mass. Black inks can contain three or more colors; the number of colors depends on the manufacturer. Each ink

FIGURE A Paper chromatography reveals the different colored dyes that black ink contains.

formulation has a characteristic pattern that uniquely identifies it.

The pigments or dyes of ink are dissolved or suspended in solvents, which comprise approximately 50% of the mass of the ink. Solvents are also responsible for the smooth flow of the ink over the roller ball in the tip of a ballpoint pen. Inks manufactured prior to 1950 have an oily base. They are usually soluble in alcohol but not in water. Inks manufactured after 1950 have a mixture of glycols as their solvent, so they are soluble in water.

The remaining 25% of the ink's mass is composed of resins. Resins control the creep, or flow, of the ink after it is applied to the surface of the paper.

In this experiment you will develop radial paper chromatograms for four black ballpoint pen inks, using water as solvent. You will then repeat this process using isopropanol as the solvent. You will then measure the distance traveled by each of the individual ink components and the distance traveled by the solvent front. Finally, you will use these measurements to calculate the R_f factor for each component.

SAFETY

Always wear safety goggles and a lab apron to protect your eyes and clothing. If you get a chemical in your eyes, immediately flush the chemical out with water at the eyewash station while calling to your teacher. Know the locations of the emergency lab shower and eyewash station and the procedure for using them.

Do not touch any chemicals. If you get a chemical on your skin or clothing, wash the chemical off at the sink while calling to your teacher. Make sure you carefully read the labels and follow the precautions on all containers of chemicals that you use. If there are no precautions stated on the label, ask your teacher what precautions to follow. Do not taste any chemicals or items in the laboratory. Never return leftovers to their original containers; take only small amounts to avoid wasting supplies.

Because the isopropanol is volatile and flammable, no Bunsen burners, hot plates, or other sources of heat should be in use in the room during this lab. Carry out all work with isopropanol in the hood.

PREPARATION

1. Determine the formula, structure, polarity, density and volatility at room temperature for water and isopropanol. The following titles are sources that provide general information on specific elements and compounds: *CRC Handbook of Chemistry and Physics, McGraw-Hill Dictionary of Chemical Terms,* and *Merck Index*.
2. Draw two data tables, one for the chromatogram made with water and one for the chromatogram made with isopropanol. Use the column headings shown below.
3. Leave room below each data table to record the distance that the solvent reaches.

DATA TABLE: Chromatogram Formed with Water

Pen no.	Dot no.	Color 1		Color 2		Color 3		Color 4	
		Distance	R_f value	Distance	R_f value	Distance	R_f value	Distance	R_f value

DATA TABLE: Chromatogram Formed with Isopropanol

Pen no.	Dot no.	Color 1		Color 2		Color 3		Color 4	
		Distance	R_f value	Distance	R_f value	Distance	R_f value	Distance	R_f value

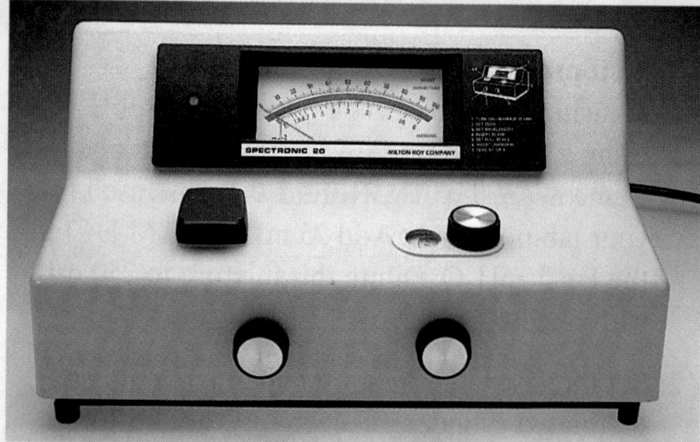

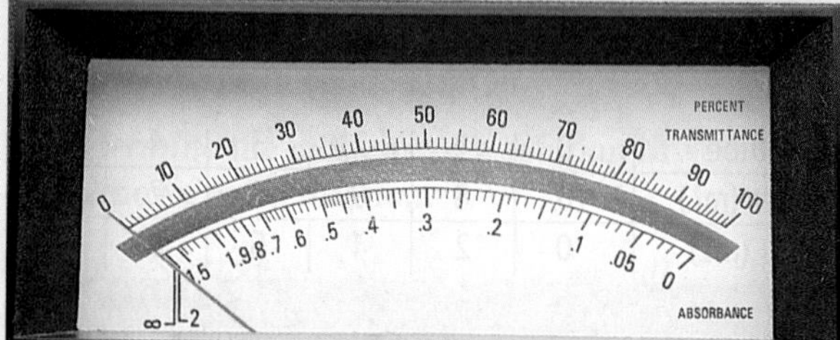

FIGURE B

e. Turn the wavelength knob to 625 nm (yellow light), and then turn the right front knob until the %T dial reads 100%.

Never wipe cuvettes with paper towels or scrub them with a test-tube brush. Use only lint-free tissues, which will not scratch the cuvette's surface. The outside of the cuvette must be completely dry, as shown in Figure C, before it is placed inside the instrument, or you will get an invalid reading.

5. Remove the cuvette and rinse it several times with distilled water, discarding the rinses in the waste beaker. Fill the cuvette approximately three-quarters full with the solution from test tube 1. Be sure the outside of the cuvette is dry by wiping it clean with a lint-free tissue. Place the cuvette in the sample compartment and place the cover over the cuvette.

6. Remove the cuvette and pour sample 1 back into the test tube. Rinse the cuvette several times with distilled water, discarding the rinses in the waste beaker. Repeat the procedure for samples from test tubes 2–5. Be sure to dry the outside of the cuvette with a lint-free tissue after each transfer. Between samples 3 and 4 check the calibration of the meter as follows: If you are measuring percent transmittance, retest the distilled water to be certain the reading is 100%; if you are measuring absorbance, retest the solution from test tube 5 to be sure the reading is 1. If the calibration test does not give the appropriate reading, start over with Procedure step 4. Otherwise, continue with Procedure step 7.

7. Test the sample of unknown concentration, and record the results in your data table.

CLEANUP AND DISPOSAL

8. Dispose of the solutions in the container designated by your teacher. Rinse the cuvettes several times with distilled water before putting them away. Clean up the lab station and all equipment after use. Wash your hands thoroughly after all work is finished and before you leave the lab.

ANALYSIS AND INTERPRETATION

1. **Analyzing Ideas:** What are the concentrations of $FeCl_3$ in test tubes 0, 1, 2, 3, 4, and 5?

2. **Analyzing Results:** What is your estimate for the concentration of the unknown? Explain your answer.

3. **Analyzing Methods:** Would you have obtained better results if, during the dilution steps, you had used a volumetric pipet that measures exactly 2.00 mL instead of using the graduated cylinder? Explain your answer.

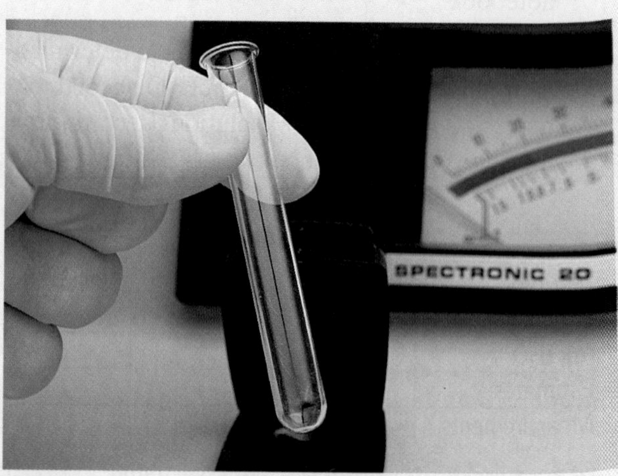

FIGURE C

4. **Organizing Data:** If you used a spectrophotometer and measured percent transmittance instead of absorbance, convert to absorbance by using the following equation. Otherwise, go on to Analysis and Interpretation item 5.

$$\text{absorbance} = \log\left(\frac{100}{\text{percent transmittance}}\right)$$

5. **Analyzing Data:** Make a graph of your data with concentration of $FeCl_3$ on the x-axis and absorbance on the y-axis. (If you did not use the spectrophotometer to make your measurements, use your values from the *Estimates* part of your data table as absorbance values.)

6. **Analyzing Data:** If your graph is a straight line, write an equation for the line in the following form:

$$y = mx + b$$

If it is not a straight line, explain why, and draw the straight line that comes closest to including all of your data points. Give the equation of this line. (Hint: If you have a graphing calculator, use the STAT mode to enter your data, and make a linear regression equation using the LinReg function from the STAT menu.)

7. **Interpreting Graphics:** Using the graph from Analysis and Interpretation item 5 or the equation from Analysis and Interpretation item 6, determine the concentration of the unknown to give the measured absorbance value.

Phenylalanine

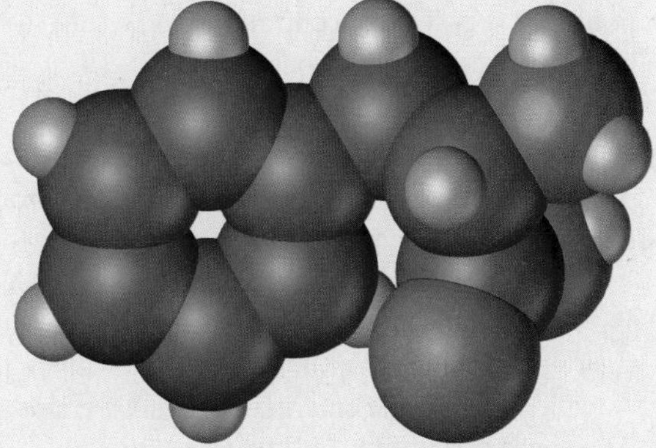

CONCLUSIONS

1. **Applying Conclusions:** A pharmaceutical company produces a 0.30 M solution of $FeCl_3$ for use in a test for the disease phenylketonuria in infants. People who have this disease are unable to break down the amino acid phenylalanine. The structure of a phenylanine molecule is shown below. The analytical department of the company found that one batch of its product had a concentration of 0.25 M rather than 0.30 M. Which of the following could have caused the problem: too much $FeCl_3$ added to the solution, too little $FeCl_3$ added to the solution, too much water added to the solution, or too little water added to the solution?

EXTENSIONS

1. **Evaluating Methods:** What are some advantages or disadvantages of colorimetry compared with other methods, such as gravimetric analysis?

2. **Relating Ideas:** Absorbance describes how much light is blocked by a sample. Transmittance describes how much light passes through a sample. As absorbance values go up, how do transmittance values change?

3. **Designing Experiments:** What possible sources of error can you identify with the procedure used in this lab? If you can think of ways to eliminate the errors, ask your teacher to approve your plan, and run the procedure again.

4. **Predicting Outcomes:** How would your absorbance or percent transmittance values for the unknown solution change if someone added an additional yellow-colored compound along with the $FeCl_3$?

5. **Evaluating Methods:** If you used a spectrophotometer, calculate the percent error for each of your estimates of absorbance compared with the values given by the equipment. (Hint: Divide each absorbance measurement by the largest measurement so that they will be on a scale of 0 to 1.00, just as your estimates are.)

EXPERIMENT 14-1

Testing Water

OBJECTIVES

- ***Observe*** **chemical reactions involving aqueous solutions of ions.**
- ***Relate*** **observations of chemical properties to the presence of ions.**
- ***Infer*** **whether an ion is present in a water sample.**
- ***Apply*** **concepts concerning aqueous solutions of ions.**

MATERIALS

- **24-well microplate lid**
- **fine-tipped dropper bulbs, labeled, with solutions, 10**
- **overhead projector (optional)**
- **paper towels**
- **solution 1: reference (all ions)**
- **solution 2: distilled water (no ions)**
- **solution 3: tap water (may have ions)**
- **solution 4: bottled spring water (may have ions)**
- **solution 5: local river or lake water (may have ions)**
- **solution 6: solution X, prepared by your teacher (may have ions)**
- **solution A: NaSCN solution (Fe^{3+} test)**
- **solution B: $Na_2C_2O_4$ solution (Ca^{2+} test)**
- **solution C: $AgNO_3$ solution (Cl^- test)**
- **solution D: $Sr(NO_3)_2$ solution (SO_4^{2-} test)**
- **white paper**

BACKGROUND

The presence of ions in an aqueous solution, even when the ions are present in small amounts, changes the physical and chemical properties of that solution so that it is no longer the same as pure water. For example, if a sample of water contains enough Mg^{2+} or Ca^{2+} ions, something unusual happens when you use soap to wash something in it. Instead of forming lots of bubbles and lather, the soap forms an insoluble white substance that floats on top of the water. This problem, created by hard water, is common in places where the water supply contains many minerals.

Aqueous solutions of other ions, such as Pb^{2+} and Co^{2+}, can be poisonous because these ions tend to accumulate in body tissues. Lead plumbing pipes in some older houses may introduce Pb^{2+} into water used for drinking and cooking because the lead metal in the pipes slowly ionizes and dissolves. The plumbing systems of such houses should be converted to another type of pipe, such as copper or polyvinyl chloride.

Because many sources of water contain harmful or unwanted substances, it is important to be able to find out what dissolved substances are present. In this experiment, you will test a variety of water samples for the presence of four ions: Fe^{3+}, Ca^{2+}, Cl^-, and SO_4^{2-}. Some of the water samples may contain these ions at very low concentrations, so be sure to make very careful observations.

SAFETY

Always wear safety goggles and a lab apron to protect your eyes and clothing. If you get a chemical in your eyes, immediately flush the chemical out at the eyewash station while calling to your teacher. Know the location of the emergency lab shower and eyewash station and the procedure for using them.

Do not touch any chemicals. If you get a chemical on your skin or clothing, wash the chemical off at the sink while calling to your teacher. Make sure you carefully read the labels and follow the precautions on all containers of chemicals that you use. If there are no precautions stated on the label, ask your teacher what precautions to follow. Do not taste any chemicals or items used in the laboratory. Never return leftovers to their original containers; take only small amounts to avoid wasting supplies.

Call your teacher in the event of a spill. Spills should be cleaned up promptly according to your teacher's directions.

PREPARATION

1. Copy the data table below into your lab notebook, and record all your observations in it.
2. Place the 24-well plate lid in front of you on a piece of white paper or other white background, and align it as shown in Figure A. Label the columns and rows as shown in Figure A. The coordinates shown will be used to designate the individual circles. For example, the circle in the top right corner would be designated 1-D, and the circle in the lower left corner would be designated 6-A.

PROCEDURE

1. Obtain labeled dropper bulbs containing the 6 different solutions from your teacher.

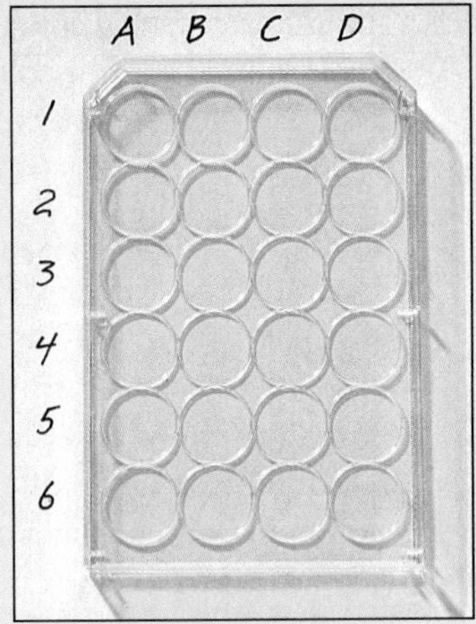

FIGURE A Label the sheet of paper that is underneath the 24-well plate to keep track of where each solution is placed.

2. Place a drop of the solution from bulb 1 in circles 1-A, 1-B, 1-C, and 1-D (the top row). Solution 1 contains all four of the dissolved ions, so these drops will show what a **positive** test for each of the ions looks like. **Be careful to keep the solutions in the appropriate circles. Any spills will cause poor results.**
3. Place a drop of the solution from bulb 2 in circles 2-A, 2-B, 2-C, and 2-D (the second row). Solution 2 is distilled water and should contain none of the ions, so it will be used to show what a **negative** test for each of the ions looks like.
4. Place a drop from bulb 3 in each of the circles in row 3, a drop from bulb 4 in each of the circles in row 4, and so on with bulbs 5 and 6. These solutions may or may not contain ions. Bulb 3 contains ordinary tap water. Bulb 4 contains bottled spring water. Bulb 5 contains local river or lake water. Bulb 6 contains solution X, which was prepared by your teacher.

DATA TABLE

Test for:	Fe^{3+}	Ca^{2+}	Cl^-	SO_4^{2-}
Reacting with:	SCN^-	$C_2O_4^{2-}$	Ag^+	Sr^{2+}
Reference (all 4 ions)				
Distilled H_2O (control—no ions)				
Tap water				
Bottled spring water				
River or lake water				
Solution X				

5. Now that each circle contains a drop of solution to be analyzed, use the solutions in bulbs A, B, C, and D to test for the presence or absence of the ions.

6. Bulb A contains NaSCN, sodium thiocyanate, which will react with any Fe^{3+} present to form $Fe(SCN)^{2+}$, a complex ion that forms a deep red solution. **Holding the tip of the bulb 1–2 cm above the drop of water to be tested,** add one drop of this solution to the drop of reference solution in circle 1-A and one drop to the distilled water in circle 2-A below it. Circle 1-A should show a positive test, and circle 2-A should show a negative test. Record in your data table your observations about what the positive and negative tests look like.

7. Use the NaSCN solution in bulb A to test the rest of the water drops in column A (the far left column) of the plate to determine whether any of them contains the Fe^{3+} ion. Record your observations in your data table. For each of the tests in which the ion was found to be present, specify whether it seemed to be present at a high, moderate, or low concentration. (If the evidence was quite visible, assume the ion was at a high concentration. If it was somewhat visible, assume the ion was at a moderate concentration. If the evidence was barely visible, assume a low concentration of the ion.)

8. Bulb B contains $Na_2C_2O_4$, sodium oxalate, which will undergo a displacement reaction with the Ca^{2+} ion to form an insoluble precipitate. **Holding the tip of the bulb 1–2 cm above the drop of water to be tested,** add one drop of this solution to the solutions in column B.

9. In your data table, record your observations about what the positive and negative $Na_2C_2O_4$ tests looked like and about whether any of the solutions contained the Ca^{2+} ion. For each of the positive tests, specify whether the Ca^{2+} ion seemed to be present at a high, moderate, or low concentration. A black background may be useful for this test and for the following tests.

10. Bulb C contains $AgNO_3$, silver nitrate, which will undergo a displacement reaction with the Cl^- ion to form an insoluble precipitate. **Holding the tip of the bulb 1–2 cm above the drop of water to be tested,** add one drop of this solution to the solutions in column C.

11. In your data table, record your observations about what the positive and negative $AgNO_3$ tests looked like and about whether any of the solutions contained the Cl^- ion. For each of the tests in which the ion was found to be present, specify whether it seemed to be present at a high, moderate, or low concentration.

12. Bulb D contains $Sr(NO_3)_2$, strontium nitrate, which will undergo a displacement reaction with the SO_4^{2-} ion to form an insoluble precipitate. **Holding the tip of the bulb 1–2 cm above the drop of water to be tested,** add one drop of this solution to the solutions in column D.

13. In your notebook, record your observations about what the positive and negative $Sr(NO_3)_2$ tests looked like and about whether any of the solutions contained the SO_4^{2-} ion. For each of the tests in which the ion was found to be present, specify whether it seemed to be present at a high, moderate, or low concentration.

14. If some of the tests are difficult to discern, place your plate on an overhead projector, if one is available. Examine the drops carefully for any signs of cloudiness. Be sure your line of vision is 10–15° above the plane of the lid, as shown in Figure B, so that you are looking at the drops from the side. Compare each drop tested with the control drops (the distilled water) in row 2. If any signs of cloudiness are detected in a tested sample, it is due to the Tyndall effect and should be considered a positive test result. Record your results in your lab notebook.

CLEANUP AND DISPOSAL

15. Clean all apparatus and your lab station. Return equipment to its proper place. Dispose of chemicals and solutions in

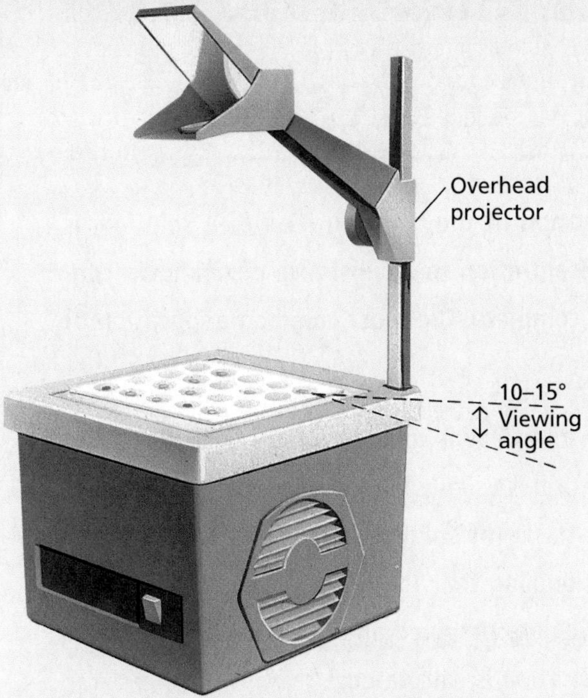

FIGURE B Look at the drops from the side of the well lid when you determine whether the drops test positive for an ion.

the containers designated by your teacher. Do not pour any chemicals down the drain or in the trash unless your teacher directs you to do so. Wash your hands thoroughly before you leave the lab and after all work is finished.

ANALYSIS AND INTERPRETATION

1. **Organizing Ideas:** Write the balanced chemical equations for each of the positive tests. Describe what each positive test looked like.

2. **Organizing Ideas:** Write the net ionic equation for each of the positive tests. (Hint: Net ionic equations are discussed in Chapter 14.)

3. **Analyzing Methods:** Why was it important to hold the dropper containing the testing solution 1–2 cm above each of the drops of water samples? Why was it important to be sure that the water samples were not spilled and accidentally mixed?

4. **Analyzing Methods:** Why was it important to use a control in the experiment? Why was distilled water used as the control?

CONCLUSIONS

1. **Organizing Conclusions:** List the solutions tested and the ions you found in each. Include notes on whether the concentration of each ion was high, moderate, or low, depending on the color or the amount of precipitate formed.

2. **Applying Conclusions and Predicting Outcomes:** Using your test results, predict which water sample would be the "hardest." Explain your reasoning.

EXTENSIONS

1. **Applying Conclusions and Predicting Outcomes:** Try to collect water samples from a variety of sources. Also collect rainwater, melted snow, swimming pool water, running water from creeks and streams, well water, and, if you're near the coast, some salt water from the ocean. Before testing the water, make a chart and try to predict which ions you will find in each sample.

2. **Relating Ideas:** Consider the substances used in the tests for this experiment and your results. What can you say about the solubilities of most sodium compounds and most nitrate compounds? Explain your reasoning without referring to a chart of solubility rules.

3. **Predicting Outcomes and Resolving Discrepancies:** Suppose you decided to determine the amount of chloride ion present in the reference solution by evaporating the water and measuring the mass of the residue left behind. If you were left with 0.37 g of white powder, could you safely assume that it was all chloride ions? Explain your reasoning.

4. **Predicting Outcomes and Resolving Discrepancies:** If you were given a sample of a solution that tested positive for sulfate ions but negative for the other three ions in this investigation and you evaporated the water and were left with 0.21 g of white powder, could you safely assume that it was all sulfate ions? Explain your reasoning.

PRE-LABORATORY PROCEDURE

Volumetric Analysis

Volumetric analysis, the quantitative determination of the concentration of a solution, is achieved by adding a substance of known concentration to a substance of unknown concentration until the reaction between them is complete. The most common application of volumetric analysis is titration.

A buret is used in titrations. The solution with the known concentration is usually in the buret. The solution with the unknown concentration is usually in the Erlenmeyer flask. A few drops of a visual indicator also are added to the flask. The solution in the buret is then added to the flask until the indicator changes color, signaling that the reaction between the two solutions is complete. Then, using the volumetric data obtained and the balanced chemical equation for the reaction, the unknown concentration is calculated.

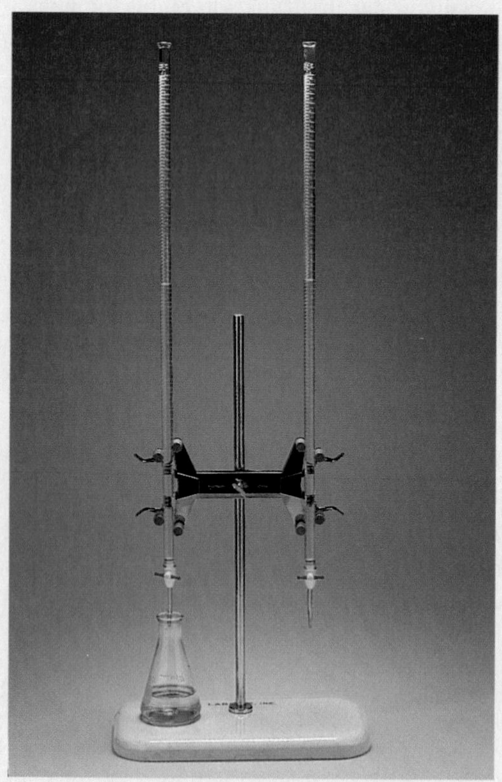

FIGURE A

GENERAL SAFETY

Always wear safety goggles and a lab apron to protect your eyes and clothing. If you get a chemical in your eyes, immediately flush the chemical out at the eyewash station while calling to your teacher. Know the location of the emergency lab shower and eyewash station and the procedure for using them.

The general setup for a titration is shown in Figure A. The steps for setting up this technique follow.

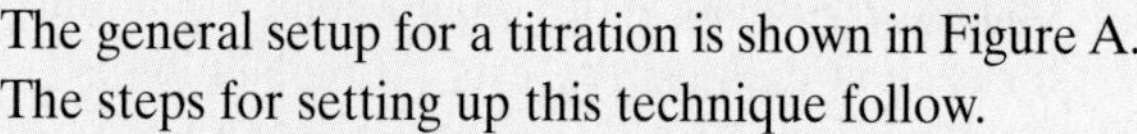

ASSEMBLING THE APPARATUS

1. Attach a buret clamp to a ring stand.
2. Thoroughly wash and rinse a buret. If water droplets cling to the walls of the buret, wash it again and gently scrub the inside walls with a buret brush.
3. Attach the buret to one side of the buret clamp.
4. Place a Erlenmeyer flask for waste solutions under the buret tip as shown in Figure A.

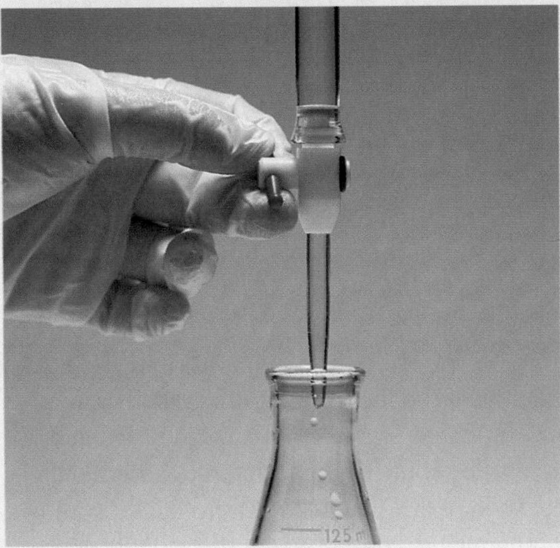

FIGURE B

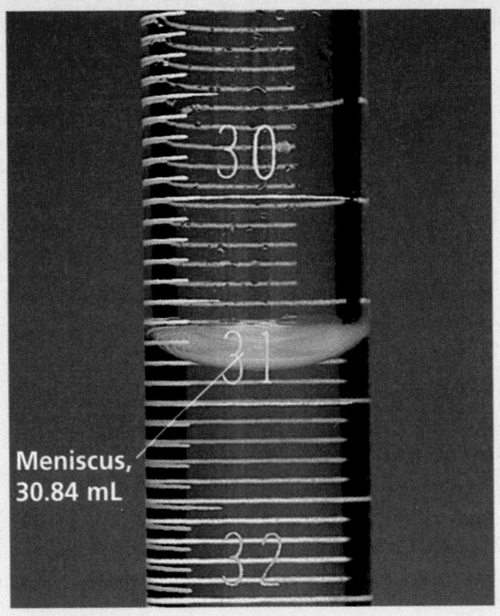

FIGURE C

OPERATING THE STOPCOCK

1. The stopcock should be operated with the left hand. This method gives better control but may prove awkward at first for right-handed students. The handle should be moved with the thumb and first two fingers of the left hand, as shown in Figure B.

2. Rotate the stopcock back and forth. It should move freely and easily. If it sticks or will not move, ask your teacher for assistance. Turn the stopcock to the closed position. Use a wash bottle to add 10 mL of distilled water to the buret. Rotate the stopcock to the open position. The water should come out in a steady stream. If no water comes out or if the stream of water is blocked, ask your teacher to check the stopcock for clogs.

FILLING THE BURET

1. To fill the buret, place a funnel in the top of the buret. Slowly and carefully pour the solution of known concentration from a beaker into the funnel. Open the stopcock, and allow some of the solution to drain into the waste beaker. Then add enough solution to the buret to raise the level above the zero mark, but do not allow the solution to overflow.

READING THE BURET

1. Drain the buret until the bottom of the meniscus is on the zero mark or within the calibrated portion of the buret. If the solution level is not at zero, record the exact reading. If you start from the zero mark, your final buret reading will equal the amount of solution added. Remember, burets can be read to the second decimal place. Burets are designed to read the volume of liquid delivered to the flask, so numbers increase as you read downward from the top. For example, the meniscus in Figure C is at 30.84 mL, not 31.16 mL.

2. Replace the waste beaker with an Erlenmeyer flask containing a measured amount of the solution of unknown concentration.

An acid-base titration is used in Experiment 16-4 to determine the concentration of acetic acid in vinegar. Experiment 16-1 is an example of a back-titration applied to an acid-base reaction; it can be performed on a larger scale if micropipets are replaced with burets. Experiment 19-1 combines a redox reaction with the titration technique to determine the concentration of Fe^{3+}.

How Much Calcium Carbonate Is in an Eggshell?

OBJECTIVES

- ***Determine*** **the amount of calcium carbonate present in an eggshell.**
- ***Relate*** **experimental titration measurements to a balanced chemical equation.**
- ***Infer*** **a conclusion from experimental data.**
- ***Apply*** **reaction-stoichiometry concepts.**

MATERIALS

- **1.00 M HCl**
- **1.00 M NaOH**
- **10 mL graduated cylinder**
- **50 mL micro solution bottle, or small Erlenmeyer flask**
- **100 mL beaker**
- **balance**
- **dessicator (optional)**
- **distilled water**
- **drying oven**
- **eggshell**
- **forceps**
- **mortar and pestle**
- **phenolphthalein solution**
- **thin-stemmed pipets or medicine droppers, 3**
- **weighing paper**

BACKGROUND

The calcium carbonate content of eggshells can be easily determined by means of an acid/base back-titration, using some of the techniques and calculations described in Chapter 16. In this back-titration, a carefully measured excess of a strong acid will react with the calcium carbonate. The resulting solution will be titrated with a strong base to determine how much acid remains unreacted. From this measurement, the amount of acid that reacted with the eggshell and the amount of calcium carbonate it reacted with can be determined. Phenolphthalein will be used as an indicator to signal the endpoint of the titration.

SAFETY

Always wear safety goggles and a lab apron to protect your eyes and clothing. If you get a chemical in your eyes, immediately flush the chemical out at the eyewash station while calling to your teacher. Know the location of the emergency lab shower and eyewash station and the procedure for using them.

Do not touch any chemicals. If you get a chemical on your skin or clothing, wash the chemical off at the sink while calling to your teacher. Make sure you carefully read the labels and follow the precautions on all containers of chemicals that you use. If there are no precautions stated on the label, ask your teacher what precautions to follow. Do not taste any chemicals or items used in the laboratory. Never return leftovers to their original containers; take only small amounts to avoid wasting supplies.

The oven used in this experiment is hot; use tongs to remove beakers from the oven because heated glassware does not always look hot.

Call your teacher in the event of an acid or base spill. Acid or base spills should be cleaned up promptly, according to your teacher's instructions.

PREPARATION

1. Remove the white and the yolk from an egg as shown in Figure A and dispose of them according to your teacher's directions. Wash the shell with distilled water and carefully peel all the membranes from the inside of the shell. Place *all* of the shell in a premassed beaker and dry the shell in the drying oven at 110°C for about 15 min. Continue with Preparation steps 2–5 while the eggshell is drying.
2. Make data and calculations tables like the ones below in your lab notebook.
3. Put exactly 5.0 mL of water in the 10.0 mL graduated cylinder. Record this volume in the data table in your lab notebook. Fill the first thin-stemmed pipet with water. This pipet should be labeled *acid*. **Do not use this pipet for the base solution.** Holding the pipet vertically, add 20 drops of water to the cylinder. **For the best results, keep the sizes of the drops as even as possible throughout this investigation.** Record the new volume of water in the graduated cylinder in the first data table under Trial 1.

Graduated Cylinder Readings (Pipet Calibration: Steps 3–5)

Trial	Initial—acid pipet	Final—acid pipet	Initial—base pipet	Final—base pipet
1				
2				
3				

Total volume of drops—acid pipet _____
Average volume of each drop _____
Total volume of drops—base pipet _____
Average volume of each drop _____

FIGURE A

4. Without emptying the graduated cylinder, add an additional 20 drops from the pipet as before, and record the new volume for Trial 2. Repeat this procedure once more for Trial 3.
5. Repeat Preparation steps 3 and 4 for the second thin-stemmed pipet. Label this pipet *base*. **Do not use this pipet for the acid solution.**
6. Make sure that the three trials produce data that are similar to each other. If one is greatly different from the others, perform Preparation steps 3–5 again. If you're still waiting for the eggshell in the drying oven, calculate and record the total volume of the drops and the average volume per drop.

Titration: Steps	
Mass of entire eggshell	
Mass of ground eggshell sample	
Number of drops of 1.00 M HCl added	150 drops
Volume of 1.00 M HCl added	
Number of drops of 1.00 M NaOH added	
Volume of 1.00 M NaOH added	
Volume of 1.00 M HCl reacting with NaOH	
Volume of 1.00 M HCl reacting with eggshell	
Number of mol of HCl reacting with eggshell	
Number of mol of $CaCO_3$ reacting with HCl	
Mass of $CaCO_3$ in eggshell sample	
% of $CaCO_3$ in eggshell sample	

FIGURE B
Use a mortar and pestle to grind the eggshell

7. Remove the eggshell and beaker from the oven. Cool them in a dessicator. Record the mass of the entire eggshell in the second table. Place half of the shell into the clean mortar and grind it to a very fine powder as shown in Figure B. This will save time when you are dissolving the eggshell. (If time permits, dry the powder again and cool it in the dessicator.)

PROCEDURE

1. Measure the mass of a piece of weighing paper. Transfer about 0.1 g of ground eggshell to a piece of weighing paper, and measure the eggshell's mass as accurately as possible. Record the mass in the second data table. Place this eggshell sample into a clean 50 mL micro solution bottle (or Erlenmeyer flask).

2. Fill the acid pipet with 1.00 M HCl acid solution, and then empty the pipet into an extra 100 mL beaker. Label the beaker *waste*. Fill the base pipet with the 1.00 M NaOH base solution, and then empty the pipet into the waste beaker.

3. Fill the acid pipet once more with 1.00 M HCl. Holding the acid pipet vertically, add exactly 150 drops of 1.00 M HCl to the bottle or flask containing the eggshell, as shown in Figure C. Swirl gently for 3 to 4 min. Observe the reaction taking place. Wash down the sides of the flask with about 10 mL of distilled water. Using a third pipet, add two drops of phenolphthalein solution.

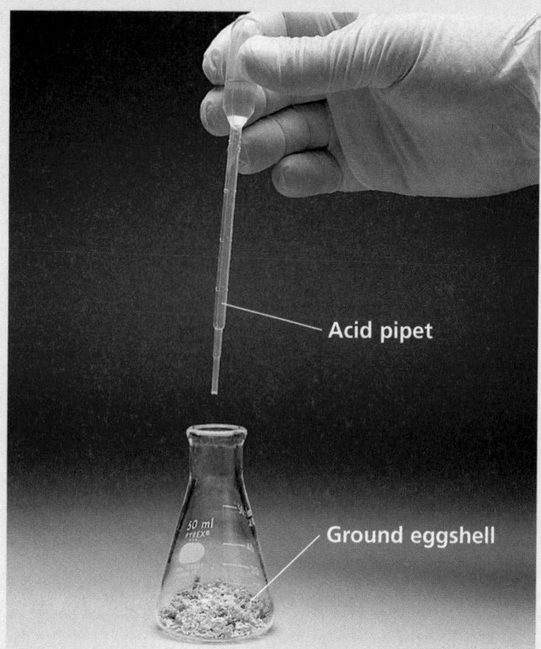

FIGURE C

4. Fill the base pipet with the 1.00 M NaOH. Slowly add NaOH from the base pipet into the bottle or flask containing the eggshell reaction mixture, as shown in Figure D, until a faint pink color persists in the mixture, even after it is swirled gently. **Be sure to add the base drop by drop, and be certain the drops end up in the reaction mixture and not on the walls of the bottle or flask. Keep careful count of the number of drops used.** Record the number of drops of base used in the second data table.

CLEANUP AND DISPOSAL

5. Clean all apparatus and your lab station. Return the equipment to its proper place. Dispose of chemicals and solutions in the containers designated by your teacher. Do not pour any chemicals down the drain or in the trash unless your teacher directs you to do so. Wash your hands thoroughly before you leave the lab and after all work is finished.

ANALYSIS AND INTERPRETATION

1. **Organizing Ideas:** The calcium carbonate in the eggshell sample undergoes a double-replacement

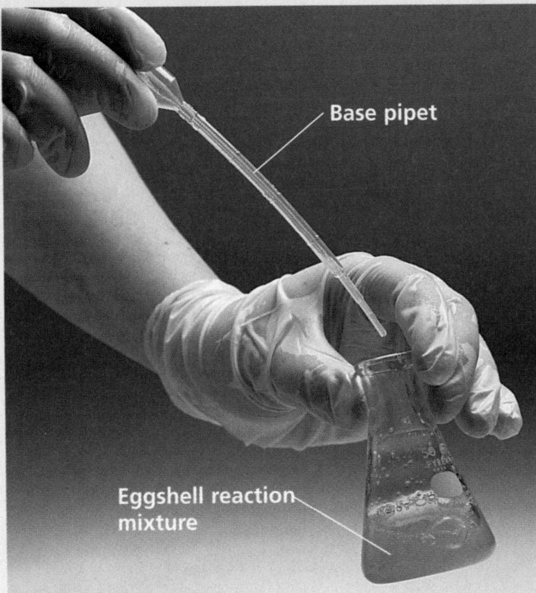

FIGURE D

reaction with the hydrochloric acid in Procedure step 3. Write a balanced chemical equation for this reaction. (Hint: The gas observed was carbon dioxide.)

2. **Organizing Ideas:** Write the balanced chemical equation for the acid/base neutralization of the excess unreacted HCl with the NaOH.

3. **Organizing Data:** Make the necessary calculations from the first data table to find the volume of each drop in milliliters. Using this mL/drop ratio, convert the number of drops of each solution in the second data table to volume in mL of each solution used.

4. **Organizing Data:** Using the relationship between the molarity and volume of acid and the molarity and volume of base needed to neutralize it, calculate the volume of the HCl solution that was neutralized by the NaOH, and record it in your table. (Hint: This relationship was discussed in Section 16-2.)

5. **Analyzing Results:** If the volume of HCl originally added is known and the volume of excess acid is determined by the titration, then the difference will be the amount of HCl that reacted with the $CaCO_3$. Calculate the volume and the number of moles of HCl that reacted with the $CaCO_3$ and record both in your table.

CONCLUSIONS

1. **Organizing Data:** Use the stoichiometry of the reaction in Analysis and Interpretation item 1 to calculate the number of moles of calcium carbonate that reacted with the HCl, and record this number in your table.

2. **Organizing Data:** Use the periodic table to calculate the molar mass of calcium carbonate. In your data table, record the mass of calcium carbonate present in your eggshell sample by using the number of moles of $CaCO_3$ you calculated in Conclusions item 1.

3. **Organizing Data:** Using your answer to Conclusions item 2, calculate the percentage of calcium carbonate in your eggshell sample, and record it in your table.

4. **Evaluating Methods:** The percentage of calcium carbonate in a normal eggshell ranges from 95% to 99%. Calculate the percent error for your measurement of the $CaCO_3$ content, using 97% $CaCO_3$ as the accepted value.

EXTENSIONS

1. **Inferring Conclusions:** Calculate an estimate of the mass of $CaCO_3$ present in the entire eggshell, based on your results for the sample of eggshell. (Hint: Apply the percent composition of your sample to the mass of the entire eggshell.)

2. **Designing Experiments:** What possible sources of error can you identify in this procedure? If you can think of ways to eliminate the errors, ask your teacher to approve your suggestion, and run the procedure again.

Investigating Overwrite Marking Pens

OBJECTIVES

- ***Demonstrate*** **proficiency in qualitatively separating marker pen inks, using paper chromatography.**
- ***Determine*** **the pH of colorless marking pens with pH paper.**
- ***Determine*** **which of the commercial pens tested are acid-bases indicators and which are not, using chromatograms.**
- ***Explain*** **the operation of overwrite marking pens in terms of acids, bases, visual indicators, and the pH scale. Suggest a method for reversing the color-change process.**
- ***Develop*** **a set of overwrite marking pens by using your knowledge of acid-base-indicator behavior in solutions with a range of pH values.**

MATERIALS

- **1 M HCl**
- **1 M NaOH (or household ammonia)**
- **alizarin red**
- **aniline blue**
- **brilliant green**
- **bromcresol green**
- **bromcresol purple**
- **bromphenol blue**
- **circular chromatography paper (or filter paper)**
- **crystal violet**
- **distilled water**
- **isopropanol**
- **gloves**
- **litmus**
- **methyl red**
- **petri dish**
- **phenolphthalein**
- **pH paper**
- **scissors**
- **set of overwrite marking pens**
- **tape or stapler with staples**
- **thymolphthalein**
- **universal indicator**

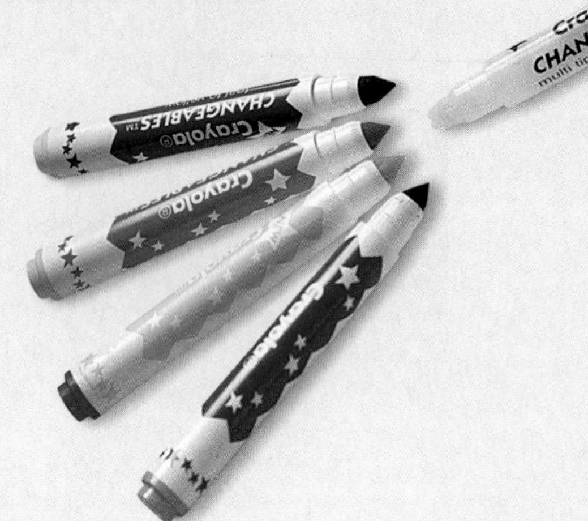

BACKGROUND

Paper Chromatography

Details on this technique can be found in the Pre-Laboratory Procedure on page 828.

Color-Change Markers

Color-change markers have been on the market for some time. The cap of each colored pen is one color, and its plug is a second color. Marks made by the pen are the same color as the cap. When these marks are overwritten by the colorless pens, they become the same color as that of the plug. The color-change properties of these markers depend on the pH values of the dyes.

Visual pH Indicators

Chapter 16 discusses the function of an acid-base indicator and the relationship between pH and H_3O^+ concentration.

Assignment

In Part A of this experiment, you will separate the inks found in marking pens by using water, isopropanol, and an acid or a base as solvents. The resulting chromatograms will help you formulate an opinion about how color changes occur in commercial overwrite marking pens. Keep a record of your data and conclusions in your lab notebook.

In Part B, you will apply your knowledge of how these pens work to develop your own set of overwrite marking pens. The colors for your ink will be derived from the acid-base indicators listed in the materials section. Record your methods, findings, and reasoning in your lab notebook.

SAFETY

Always wear safety goggles and a lab apron to protect your eyes and clothing. If you get a chemical in your eyes, immediately flush the chemical out with water at the eyewash station while calling to your teacher. Know the locations of the emergency lab shower and eyewash station and the procedure for using them.

Because the isopropanol is volatile and flammable, no Bunsen burners, hot plates, or other heat sources should be in use in the room during this lab. Carry out all work with isopropanol in an operating fume hood.

Do not touch any chemicals. If you get a chemical on your skin or clothing, wash the chemical off at the sink while calling to your teacher. Make sure you carefully read the labels and follow the precautions on all containers of chemicals that you use. Never return leftovers to their original containers; take only small amounts to avoid wasting supplies.

PREPARATION

1. Determine the formula, color changes over a pH range, and special handling requirements for each acid-base indicator in the materials list. The following handbooks, manuals, and dictionaries provide general information on specific elements and compounds: *CRC Handbook of Chemistry and Physics, McGraw-Hill Dictionary of Chemical Terms,* and the *Merck Index.*
2. Before preparing your pen set, you may wish to read about the subtractive color process in an encyclopedia or introductory physics text.
3. You will need data tables, like those in Experiment 13-1, that describe the ink separation and any color changes for the chromatograms of the commercial pens. You will need separate spaces for recording the pH of the colorless pen, for identifying the color change as acid-to-base or base-to-acid, and for identifying each pen that is an acid-base indicator.

PROCEDURE

Part A: Examining commercial pens for function and operation

1. Set up a paper chromatography apparatus, using water as the solvent to separate all of the pen dyes in your set of markers. Tape or staple the dry chromatogram to a page of your laboratory notebook. Record the composition of each ink in your data table.
2. Repeat Procedure step 1 with isopropanol as the solvent to separate all of the pen dyes. Tape or staple the dry chromatogram to a page of your laboratory notebook.
3. Determine the pH of the two colorless overwrite markers with a piece of pH paper. Record the value (or values) on your data sheet.
4. Use your result from Procedure step 3 to determine whether an acid or a base should be used as the solvent for separating the pen dyes, and repeat Procedure step 1. You should now be able to identify which dyes are acid-base indicators.
5. Evaluate your results from Procedure steps 1 through 4 by answering questions 1 through 4 under Analysis and Interpretation.

Part B: Developing a new product

6. Use the results of your evaluation to sketch out a plan in your lab notebook for making your own set of overwrite marker pens from the pH indicators in the materials list.

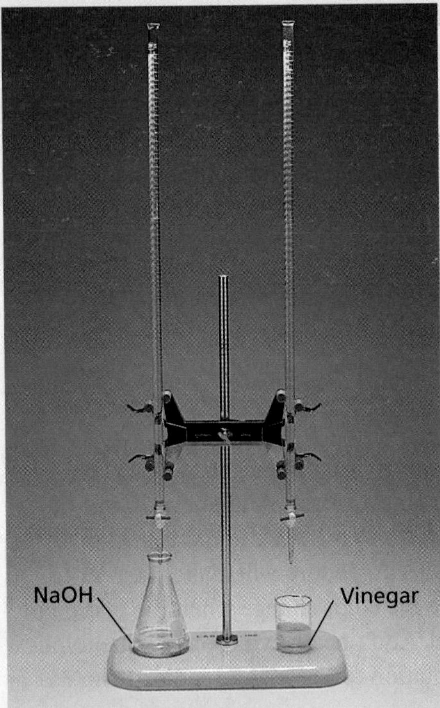

FIGURE C

your wash bottle. Add the sodium hydroxide drop by drop near the end of the titration until the last drop keeps the solution a pink color that remains after swirling.

8. Add successive quantities of both solutions, drop by drop, going back and forth from pink to colorless until the end point is clearly established. This point is indicated by the slightest suggestion of pink coloration in the flask, as shown in Figure C. You will be able to see the pink color more clearly if the flask is resting on a sheet of white paper with a beaker of distilled water next to it for comparison. Record in your data table the final buret readings of both solutions to the nearest 0.01 mL.

9. Discard the liquid in the Erlenmeyer flask in the disposal container provided by your teacher. Rinse the flask thoroughly with distilled water, and repeat the titration two more times, following Procedure steps 5–8.

CLEANUP AND DISPOSAL

10. Clean all apparatus and your lab station. Return equipment to its proper place. Dispose of chemicals and solutions in the containers designated by your teacher. Do not pour any chemicals down the drain or in the trash unless your teacher directs you to do so. Wash your hands thoroughly before you leave the lab and after all work is finished.

ANALYSIS AND INTERPRETATION

1. **Organizing Data:** Calculate the volumes of vinegar and NaOH used for each of the three trials.

2. **Organizing Data:** In your data table write the molarity of the standardized NaOH solution you used. Determine the moles of NaOH used in each of the three trials.
Hint: By definition,

$$\text{molarity} = \frac{\text{moles of solute}}{\text{1 L solution}}$$

$$\text{moles of solute} = \text{molarity} \times \text{liters of solution}$$

3. **Organizing Ideas:** Write the balanced equation for the reaction between vinegar and sodium hydroxide. (Hint: The formula for acetic acid is CH_3COOH.)

4. **Organizing Data:** Use the results of your calculations in Analysis and Interpretation item 2 and the mole ratio from the equation in Analysis and Interpretation item 3 to determine the moles of base used to neutralize the vinegar (acid) in each trial.

5. **Organizing Data:** Use the moles of base calculated in Analysis and Interpretation item 4 and the volumes of the acid used for each trial to calculate the molarities of the vinegar for the three trials.

6. **Organizing Data:** Calculate the average molarity of the vinegar.

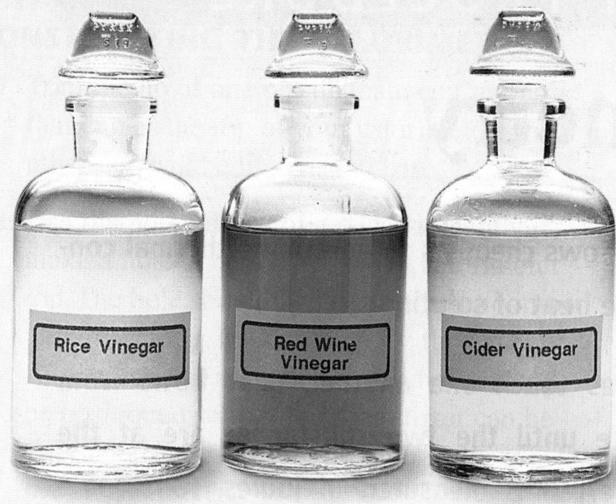

7. **Organizing Ideas:** Use the periodic table to calculate the molar mass of acetic acid, CH_3COOH.

8. **Organizing Data:** Use the average molarity for your vinegar sample to determine the mass of CH_3COOH in 1 L of vinegar.

CONCLUSIONS

1. **Organizing Conclusions:** Assume that the density of vinegar is very close to 1.00 g/mL so that the mass of 1 L of vinegar is 1000 g. Calculate the percentage of acetic acid in your vinegar sample. (Hint: The mass of acetic acid in 1 L of vinegar was calculated in Analysis and Interpretation item 8. Divide the mass of CH_3COOH in 1 L by the total mass of vinegar in a liter, then multiply by 100 to get the percentage of acetic acid in vinegar.)

2. **Applying Conclusions:** Why is it important for a company manufacturing vinegar to regularly check the molarity of its product?

3. **Analyzing Methods:** What was the purpose of using the phenolphthalein? Could you have titrated the vinegar sample without the phenolphthalein?

4. **Analyzing Methods:** At the beginning of each titration, 10 mL of vinegar was run into the Erlenmeyer flask and the vinegar was diluted with distilled water. Why was the calculated molarity of the acetic acid not affected by the water?

EXTENSIONS

1. **Evaluating Data:** Share your data with other lab groups, and calculate a class average for the molarity of the vinegar.

2. **Designing Experiments:** What possible sources of error can you identify in this procedure? If you can think of ways to eliminate the errors, ask your teacher to approve your plan, and run the procedure again.

3. **Relating Ideas:** Explain the difference between the equivalence point and the end point. Can they be the same?

4. **Resolving Discrepancies:** An industrial chemist measured the following values when titrating 10.0 mL samples from a single vat of vinegar: 15.04 mL, 16.03 mL, and 14.98 mL. What could be the source error in these titrations?

Measuring the Specific Heats of Metals

OBJECTIVES

- ***Calibrate* a simple calorimeter.**
- ***Relate* measurements of temperature to changes in heat content.**
- ***Calculate* the specific heats of several metals.**
- ***Determine* the identity of an unknown metal.**

MATERIALS

- **100 mL graduated cylinder**
- **400 mL beakers, 2**
- **balance**
- **beaker tongs**
- **boiling chips**
- **Bunsen burner, gas tubing, striker, or hot plate**
- **glass stirring rod**
- **metal samples**
- **plastic-foam cups for calorimeter, 2**
- **ring stand and ring**
- **scissors or tools to trim cups**
- **test tube, large**
- **test-tube clamp**
- **thermometer**
- **tongs for handling metal**
- **unknown metal sample**
- **wire gauze with ceramic center**
- **wire stirrer**

BACKGROUND

Changes in heat can be determined by measuring changes in temperature. When a substance is heated, the heat gained, q, depends on three factors: the mass of the substance, m, in grams; the specific heat of the substance, c_p; and the change in temperature of the substance, Δt. The following equation is used to calculate the amount of heat absorbed or lost by a substance.

$$q = m \times c_p \times \Delta t$$

In this experiment, you will measure the specific heats of several metals, but first you will need to make and calibrate a calorimeter. You will start with a known mass of water in your calorimeter. When the heated metal is added to the water, you can measure the change in temperature for the water. Using the specific heat of water (4.184 J/g•°C) and your calorimeter's calibration constant, you will be able to calculate the amount of heat gained by the water and the calorimeter. This amount is equal to the amount of heat lost by the metal. If the temperature change and mass of the metal are known, its specific heat can be determined.

$$q_{metal} = q_{calorimeter}$$

$$m_{metal} \times \Delta t_{metal} \times c_{p,\,metal} =$$
$$[(m_{H_2O} \times c_{p,\,H_2O} \times \Delta t_{H_2O}) + (C' \times \Delta t_{H_2O})]$$

SAFETY

Always wear safety goggles and a lab apron to protect your eyes and clothing. If you get a chemical in your eyes, immediately flush the chemical out at the eyewash station while calling to your teacher. Know the location of the emergency lab shower and eyewash station and the procedure for using them.

Do not touch any chemicals. If you get a chemical on your skin or clothing, wash the chemical off at the sink while calling to your teacher. Make sure you carefully read the labels and follow the precautions on all containers of chemicals that you use. If there are no precautions stated on the label, ask your teacher what precautions to follow. Do not taste any chemicals or items used in the laboratory. Never return left-overs to their original containers; take only small amounts to avoid wasting supplies.

Call your teacher in the event of a spill. Spills should be cleaned up promptly, according to your teacher's directions.

When you use a Bunsen burner, confine long hair and loose clothing. Do not heat glassware that is broken, chipped, or cracked. Use tongs or a hot mitt to handle heated glassware and other equipment because heated glassware does not always look hot. If your clothing catches fire, WALK to the emergency lab shower, and use it to put out the fire.

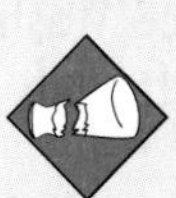

Never put broken glass in a regular waste container. Broken glass should be disposed of in a separate container designated by your teacher. Use a wire stirrer; do not use a thermometer to stir because it is fragile and can break easily.

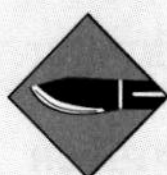

Scissors are sharp; use with care to avoid cutting yourself or others.

PREPARATION

1. In your lab notebook, you will need one data table for the calibration of your calorimeter and one table for specific heat test data for each metal you will be testing, including the unknown metal. Copy the table below in your lab notebook. You will not use the spaces for volume and mass in the *Calorimeter* column and the spaces for *Calorimeter heat capacity* in the *Cool* H_2O and *Hot* H_2O columns. For the specific heat tests, make as many data tables as you will need for the known and unknown metals you will be testing. Each table should have three columns and five rows. Label the second and third columns of the first row H_2O and *Metal*. In the first column, label rows 2 through 5 *Initial temp.*, *Final temp.*, *Change in temp.*, and *Mass.*

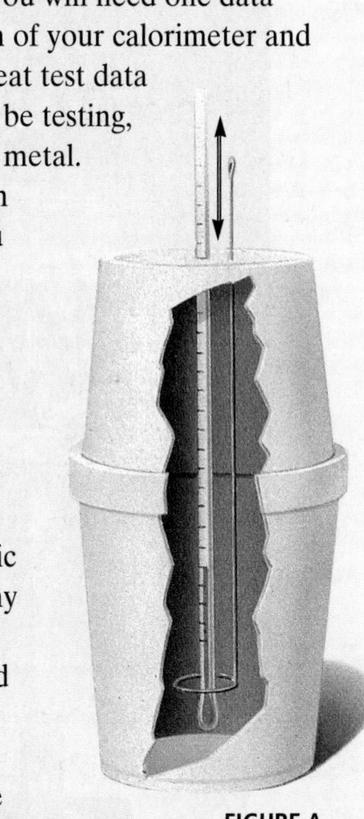

FIGURE A

2. Construct a calorimeter as in Pre-Laboratory Procedure: Calorimetery on page 858.

PROCEDURE

1. Calibrate the calorimeter as demonstrated in the Pre-Laboratory Procedure on page 858. Use the calorimeter constant in each of your calculations.

DATA TABLE

	Cool H_2O	Hot H_2O	Calorimeter
Initial temp.			
Final temp.			
Change in temp.			
Volume			
Mass			
Calorimeter heat capacity			

Calorimetry and Hess's Law

OBJECTIVES

- *Demonstrate* **proficiency in the use of calorimeters and related equipment.**
- *Relate* **temperature changes to enthalpy changes.**
- *Determine* **heats of reaction for several reactions.**
- *Demonstrate* **that heats of reactions can be additive.**

MATERIALS

- **4 g NaOH pellets**
- **50 mL 1.0 M HCl acid solution**
- **50 mL 1.0 M NaOH solution**
- **100 mL 0.50 M HCl solution**
- **100 mL graduated cylinder**
- **balance**
- **distilled water**
- **forceps**
- **glass stirring rod**
- **gloves**
- **plastic-foam cups (or calorimeter)**
- **spatula**
- **thermometer**
- **watch glass**

BACKGROUND

Hess's law states that the overall enthalpy change in a reaction is equal to the sum of the enthalpy changes in the individual steps in the process. You can infer from this statement that no matter how many steps it might take to convert reactants to products, the energy released or absorbed is the same as if the reaction had taken place in one step.

In this experiment, you will use a calorimeter to carefully measure the amount of heat released in three chemical reactions. From your experimental data, you will calculate the enthalpies of the three reactions in kilojoules per mole, and you will use the equations for the reactions and the enthalpies to verify Hess's law.

SAFETY

Always wear safety goggles and a lab apron to protect your eyes and clothing. Do not touch any chemicals. If you get a chemical in your eyes, immediately flush the chemical out at the eyewash station while calling to your teacher. Know the locations of the emergency lab shower and eyewash station and the procedure for using them.

Do not touch any chemicals. If you get a chemical on your skin or clothing, wash the chemical off at the sink while calling to your teacher. Make sure you carefully read the labels and follow the precautions on all containers of chemicals that you use. If there are no precautions stated on the label, ask your teacher what precautions to follow. Do not taste any chemicals or items used in the laboratory. Never return leftovers to their original containers; take only small amounts to avoid wasting supplies.

Never put broken glass in a regular waste container. Broken glass should be disposed of separately in a container designated by your teacher.

Call your teacher in the event of an acid or base spill. Acid or base spills should be cleaned up promptly, according to your teacher's instructions.

PREPARATION

1. Prepare a data table in your notebook like the one shown below. Reactions 1 and 3 will each require two additional spaces to record the mass of the empty watch glass and the mass of the watch glass and NaOH.
2. If you are not using a plastic-foam cup as a calorimeter, ask your teacher for instructions on using the calorimeter. At various points in steps 1 through 11, you will need to measure the temperature of the solution within the calorimeter. If you are using a thermometer, measure the temperature by gently inserting the thermometer into the hole in the calorimeter lid, as shown in Figure A. The thermometer takes time to reach the same temperature as the solution inside the calorimeter, so wait to be sure you have an accurate reading. **Thermometers break easily, so be careful with them, and do not use them to stir a solution.**

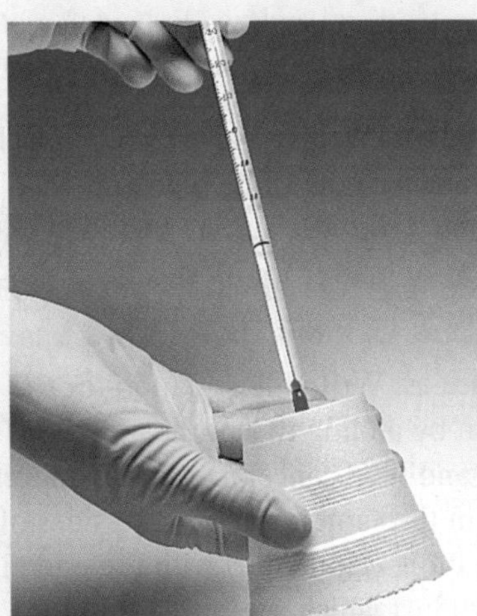

FIGURE A

PROCEDURE

Reaction 1: Dissolving NaOH

1. Pour about 100 mL of distilled water into a graduated cylinder. Measure and record the volume of the water to the nearest 0.1 mL. Pour the water into your calorimeter. Record the water temperature to the nearest 0.1°C.
2. Determine and record the mass of a clean and dry watch glass to the nearest 0.01 g. Remove the watch glass from the balance. Wear gloves and obtain about 2 g of NaOH pellets, and put them on the watch glass. Use forceps when handling NaOH pellets, as shown in Figure B. Measure and record the mass of the watch glass and the pellets to the nearest 0.01 g. **It is important that this step be done quickly because NaOH is hygroscopic. It absorbs moisture from the air, increasing its mass as long as it remains exposed to the air.**

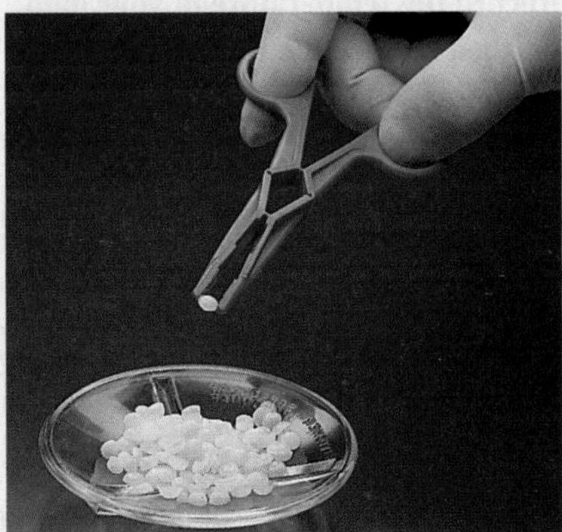

FIGURE B

DATA TABLE

	Reaction 1	Reaction 2	Reaction 3
Total volumes of liquid(s)			
Initial temp.			
Final temp.			

Rate of a Chemical Reaction

OBJECTIVES

- ***Prepare*** **and** ***observe*** **several different reaction mixtures.**
- ***Demonstrate*** **proficiency in measuring reaction rates.**
- ***Relate*** **experimental results to a rate law that can be used to predict the results of various combinations of reactants.**

MATERIALS

- **8-well microscale reaction strips, 2**
- **distilled or deionized water**
- **fine-tipped dropper bulbs or small microtip pipets, 3**
- **solution A**
- **solution B**
- **stopwatch or clock with second hand**

BACKGROUND

In this experiment, you will determine the rate of an *oxidation-reduction,* or *redox,* reaction. Reactions of this type involve a special kind of electron transfer and will be discussed in Chapter 19. The net equation for the reaction you will study is written as follows:

$$3Na_2S_2O_5(aq) + 2KIO_3(aq) + 3H_2O(l) \xrightarrow{H^+} 2KI(aq) + 6NaHSO_4(aq)$$

One way to study the rate of this reaction is to observe how fast $Na_2S_2O_5$ is used up. After all the $Na_2S_2O_5$ solution has reacted, the concentration of iodine, I_2, an intermediate in the reaction, builds up. A starch indicator solution, added to the reaction mixture, will signal when this happens. The colorless starch will change to a blue-black color in the presence of I_2.

In the procedure, the concentrations of the reactants are given in terms of drops of solution A and drops of solution B. Solution A contains $Na_2S_2O_5$, the starch indicator solution, and dilute sulfuric acid to supply the hydrogen ions needed to catalyze the reaction. Solution B contains KIO_3. You will run the reaction with several different concentrations of the reactants and record the time it takes for the blue-black color to appear.

SAFETY

Always wear safety goggles and a lab apron to protect your eyes and clothing. If you get a chemical in your eyes, immediately flush the chemical out at the eyewash station while calling to your teacher. Know the locations of the emergency lab shower and eyewash station and the procedure for using them.

Do not touch any chemicals. If you get a chemical on your skin or clothing, wash the chemical off at the sink while calling to your teacher. Make sure you carefully read the labels and follow the precautions on all containers of chemicals that you use. Never return leftovers to their original containers; take only small amounts to avoid wasting supplies.

PREPARATION

1. Prepare a data table in your lab notebook. The table should have six rows and six columns. Label the boxes in the first row of the second through sixth columns *Well 1*, *Well 2*, *Well 3*, *Well 4*, and *Well 5*. In the first column, label the boxes in the second through sixth rows *Time reaction began*, *Time reaction stopped*, *Drops of solution A*, *Drops of solution B*, and *Drops of* H_2O.

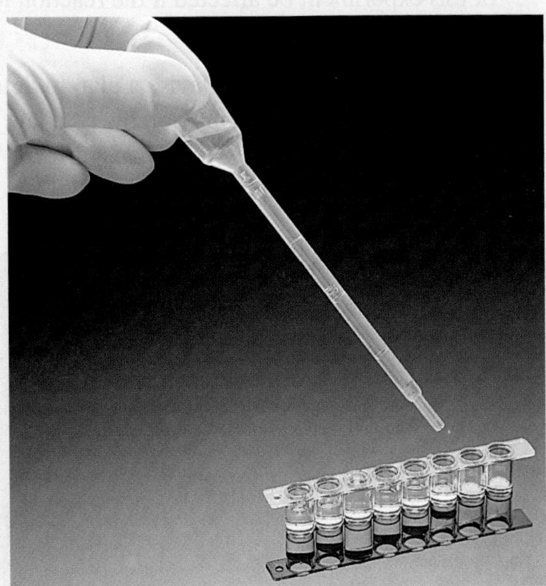

FIGURE A

2. Obtain three dropper bulbs or small microtip pipets, and label them *A*, *B*, and H_2O.
3. Fill the bulb or pipet A with solution A, the bulb or pipet B with solution B, and the bulb or pipet for H_2O with distilled water.

PROCEDURE

1. Using the first 8-well strip, place five drops of solution A into each of the first five wells, as shown in Figure A. (Disregard the remaining three wells.) Record the number of drops in the appropriate places in your data table. **For best results, try to make all drops about the same size.**
2. In the second 8-well reaction strip, place one drop of solution B in the first well, two drops in the second well, three drops in the third well, four drops in the fourth well, and five drops in the fifth well. Record the number of drops in the appropriate places in your data table.
3. In the second 8-well strip that contains drops of solution B, add four drops of water to the first well, three drops to the second well, two drops to the third well, and one drop to the fourth well. Do not add any water to the fifth well.
4. Carefully invert the second strip. The surface tension should keep the solutions from falling out of the wells. Place the strip well-to-well on top of the first strip, as shown in Figure B.

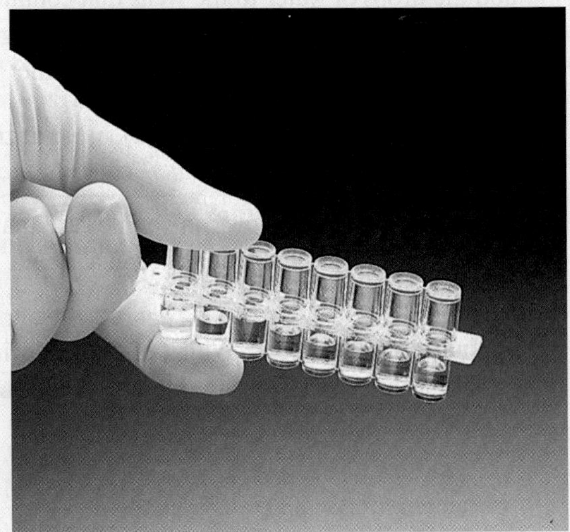

FIGURE B

while calling to your teacher. Know the locations of the emergency lab shower and eyewash station and the procedure for using them.

Do not touch any chemicals. If you get a chemical on your skin or clothing, wash the chemical off at the sink while calling to your teacher. Make sure you carefully read the labels and follow the precautions on all containers of chemicals that you use. Do not taste any chemicals or items used in the laboratory. Never return leftover chemicals to their original containers; take only small amounts to avoid wasting supplies.

PREPARATION

1. Copy the data table below into your lab notebook.
2. If you will be using a spectrophotometer, turn it on now, as it must warm up for approximately 10 min.
3. Label six test tubes *1*, *2*, *3*, *4*, *5*, and *6*.

PROCEDURE

1. Carefully measure 5.0 mL of 0.200 M $Fe(NO_3)_3$ using a 10 mL graduated cylinder. Pour this solution into test tube 1, as shown in Figure A. Record this volume and concentration in the data table.
2. Carefully measure another 5.0 mL of 0.200 M $Fe(NO_3)_3$ using the 10 mL graduated cylinder. Add 5.0 mL of 0.6 M HNO_3 to the $Fe(NO_3)_3$ solution in the graduated cylinder. Mix well with a glass stirring rod. Pour 5.0 mL of this mixture into the test tube 2. Record 5 mL as the volume in the test tube 2 and record the Fe^{3+} concentration in the data table. (Hint: The concentration of Fe^{3+} is half of what it was for test tube 1 because the $Fe(NO_3)_3$ has been diluted by an equal volume of HNO_3.)

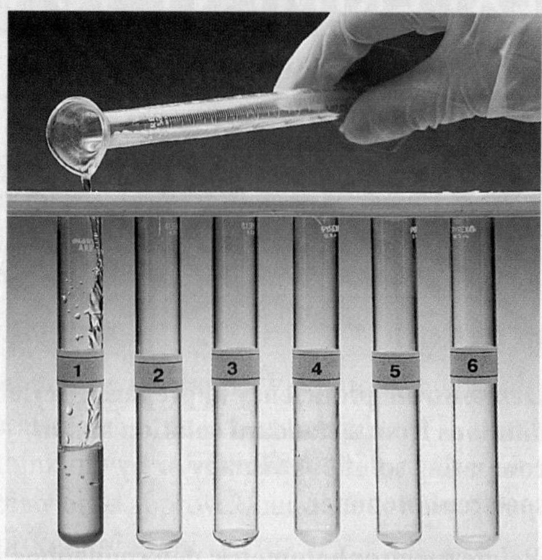

FIGURE A

3. Add 5.0 mL of 0.6 M HNO_3 to the remaining 5.0 mL of the mixture in the graduated cylinder. Mix well with a glass stirring rod. Pour 5.0 mL of this mixture into test tube 3. Record 5.0 mL as the volume for test tube 3 and record the Fe^{3+} concentration in the data table.
4. Repeat Procedure step 3 until you have filled all six test tubes.
5. Discard the contents of test tube 2, pouring it into the waste container designated by your teacher. (This dilution does not provide a measurable difference in light absorption.)

DATA TABLE

Test tube no.	Fe^{3+} conc.	mL of Fe^{3+}	mL of KSCN	Absorbance value
1				
2				
3				
4				
5				
6				
Observations				

6. Add 5.0 mL of 0.002 00 M KSCN to test tubes 1, 3, 4, 5, and 6. Mix each solution thoroughly with a stirring rod. Record the volume of KSCN used in the data table.

7. Compare the solutions while holding a piece of white paper behind the test tubes. If you are not using a spectrophotometer, estimate the intensity of the red color on a scale from 0 (clear) to 1.0 (test tube 1), and enter your estimate in the *Absorbance value* column of the data table.

8. If you are not using a spectrophotometer, go on to step 11. If you are using a spectrophotometer, allow it to warm up for 10 minutes.

Spectrophotometer

Set the wavelength to 590 nm. **Note that you should be measuring absorbance, NOT percent transmittance.** Calibrate the spectrophotometer by turning the left front knob to 0 percent transmittance while using an empty sample compartment with the lid closed. Then pour some solution from test tube 1 into the cuvette and adjust the absorbance value with the right front knob to read as close to 1.00 as possible. **Never wipe cuvettes with paper towels or scrub them with a test-tube brush. Use only lint-free tissues that will not scratch the cuvette's surface. The outside of the cuvette must be completely dry before it is placed inside an instrument, or you will get an invalid reading.**

9. With your instrument adjusted accordingly, record the absorbance value for the solution from test tube 1. If you have only one cuvette, rinse it several times with distilled water, and then make absorbance measurements for test tube 3. Record this value in your data table. Rinse the cuvette several times with distilled water, and measure and record absorbance values for test tubes 4, 5, and 6 in your data table.

10. To check your measurements, retest the solution from test tube 1 after you have finished the other measurements. Its absorbance value should still be the same, close to 1.00. If not, repeat the procedure for all solutions.

CLEANUP AND DISPOSAL

11. Dispose of all solutions in the container designated by your teacher. Wash your hands thoroughly after cleaning up the area and equipment.

ANALYSIS AND INTERPRETATION

1. **Analyzing Data:** Determine how each of the absorbance values relates to the absorbance value for test tube 1 by dividing each value by the value obtained for test tube 1. (Hint: After this calculation, the new value for test tube 1 should be 1.00, and the values for the other test tubes should be less than 1.00 because they were less concentrated than test tube 1. If you were using estimates instead of colorimetry measurements, this step can be skipped.)

2. **Analyzing Data:** Calculate the initial concentrations of SCN^- for test tubes *1, 3, 4, 5,* and *6*. Remember that each 5.0 mL of 0.002 00 M KSCN was mixed with 5.0 mL of $Fe(NO_3)_3$ solution to give a total volume of 10.0 mL. (Hint: The value will be the same for all the test tubes.)

3. **Analyzing Data:** Calculate the actual initial concentration of Fe^{3+} in the test tubes in a similar way. (Hint: The values recorded in the data table show the concentration of Fe^{3+} before it was diluted by 5.0 mL of 0.002 00 M KSCN.)

4. **Applying Ideas:** Determine the equilibrium concentrations for test tube 1. Because the initial concentration of Fe^{3+} was 0.100 M—much larger than the initial concentration of SCN^-, 0.001 M—assume that practically all of the SCN^- ions are consumed in the reaction. (Even though this is not necessarily true, the deviation from the true SCN^- concentration is so much smaller than the other factors in this equation that it can be disregarded in this case.)

5. **Analyzing Data:** Calculate the $FeSCN^{2+}$ equilibrium concentration for test tubes *3, 4, 5,* and *6* based on the absorbance data and the equilibrium concentration for test tube 1 determined in Analysis and Interpretation item 4.

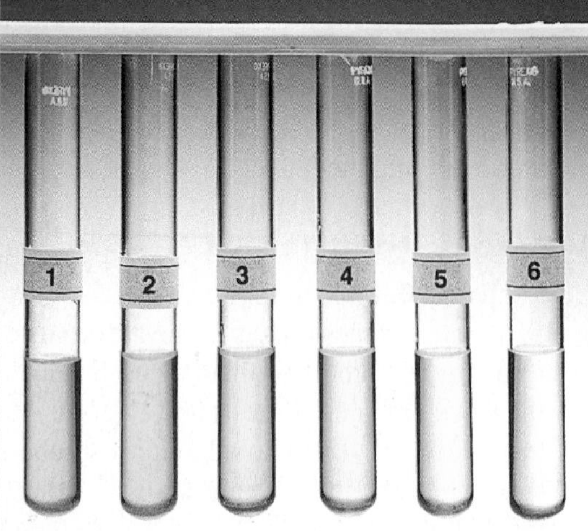

FIGURE B

(Hint: Multiply the concentration for test tube 1 by the factors calculated in Analysis and Interpretation item 1.)

6. **Analyzing Data:** Calculate the SCN^- concentration for test tubes 3–6 at equilibrium. (Hint: You know the initial SCN^- concentration from Analysis and Interpretation item 2, and you know the amount of SCN^- that has formed $FeSCN^{2+}$ from item 5.)
7. **Analyzing Data:** For test tubes 3–6, calculate the Fe^{3+} concentration at equilibrium. (Hint: You know the initial Fe^{3+} concentration from Analysis and Interpretation item 3, and you know the amount of Fe^{3+} that has formed $FeSCN^{2+}$ from Analysis and Interpretation item 5.)
8. **Analyzing Data:** Graph the adjusted absorbance values from Analysis and Interpretation item 1 against the initial concentration values for Fe^{3+}.

CONCLUSIONS

1. **Evaluating Data:** For test tubes 3–6, determine the value of the equilibrium constant, K_{eq}, for this reaction by using the equation given below. (Hint: Use the equilibrium concentrations determined in Analysis and Interpretation items 5, 6, and 7.)

$$K_{eq} = \frac{[FeSCN^{2+}]}{[Fe^{3+}][SCN^-]}$$

2. **Evaluating Methods:** Although the values for K_{eq} should be equal, describe the conditions that could cause these values to differ.
3. **Interpreting Graphics:** What general statement can you make about the absorbance values compared with the concentrations of Fe^{3+} and $FeSCN^{2+}$?

EXTENSIONS

1. **Designing Experiments:** How would you revise this procedure to determine the value of the equilibrium constant at different temperatures? Would you be able to maintain accurate data analysis for high or low temperature ranges? Explain your answers.
2. **Designing Experiments:** What possible sources of error can you identify with this procedure? If you can think of ways to eliminate them, ask your teacher to approve your plans, and run the procedure again.
3. **Applying Ideas:** Hundreds of different equilibrium reactions are taking place constantly in your body. One very important equilibrium reaction involves oxygen, O_2, and hemoglobin, a complex protein abbreviated as Hb, to form oxyhemoglobin, HbO_2.

$$Hb + O_2 \rightleftharpoons HbO_2$$

In your lungs, where oxygen is abundant, the forward reaction is favored. The oxyhemoglobin then travels in your bloodstream to your oxygen-starved cells. In the cells, the reverse reaction is favored, releasing the oxygen. In this way, equilibrium is maintained as you continue to live and breathe. Write the equilibrium expression for the reaction above involving oxygen, hemoglobin, and oxyhemoglobin.

4. **Predicting Outcomes:** At the elevation of Mexico City, 2300 m (7500 ft), the concentration of oxygen is 75% of that at sea level. Yet the same amount of oxygen needs to be delivered to the muscle cells. To compensate, does the body produce more or less hemoglobin? Explain your answer. How is your breathing rate affected at high elevations?

Measuring K_a for Acetic Acid

OBJECTIVES

- ***Compare*** **the conductivities of solutions of known and unknown hydronium ion concentrations.**
- ***Relate*** **conductivity to the concentration of ions in solution.**
- ***Explain*** **the validity of the procedure on the basis of the definitions of strong and weak acids.**
- ***Compute*** **the numerical value of K_a for acetic acid.**

MATERIALS

- **1.0 M acetic acid, CH_3COOH**
- **1.0 M hydrochloric acid, HCl**
- **24-well plate**
- **distilled or deionized water**
- **LED conductivity testers**
- **paper towels**
- **thin-stemmed pipets**

BACKGROUND

The acid-dissociation constant, K_a, is a measure of the strength of an acid. Strong acids, which are almost completely ionized in water, have much larger K_a values than weak acids because weak acids are only partly ionized. Properties that depend on the ability of a substance to ionize, such as conductivity and colligative properties, can be used to measure K_a. In this experiment, you will compare the conductivity of a 1.0 M solution of acetic acid, CH_3COOH, a weak acid, with the conductivities of solutions of varying concentrations of hydrochloric acid, HCl, a strong acid. From the comparisons you make, you will be able to estimate the concentration of hydronium ions in the acetic acid solution and calculate its K_a.

SAFETY

Always wear safety goggles and a lab apron to protect your eyes and clothing. If you get a chemical in your eyes, immediately flush the chemical out at the eyewash station while calling to your teacher. Know the location of the emergency lab shower and eyewash station and the procedures for using them.

Do not touch any chemicals. If you get a chemical on your skin or clothing, wash the chemical off at the sink while calling to your teacher. Read labels carefully and follow the precautions on all containers of chemicals that you use. If no precautions are stated on the label, ask your teacher what precautions to follow. Do not taste any chemicals or items used in the laboratory. Never return leftover chemicals to their original containers; take only small amounts to avoid wasting supplies.

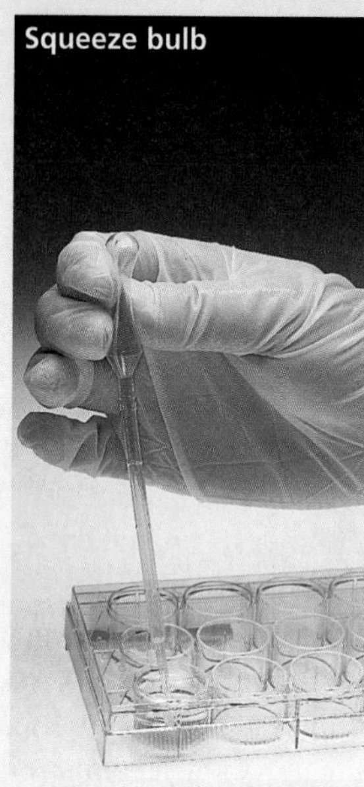

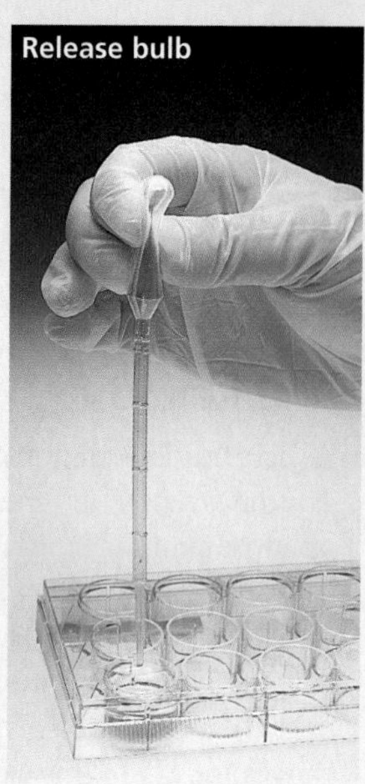

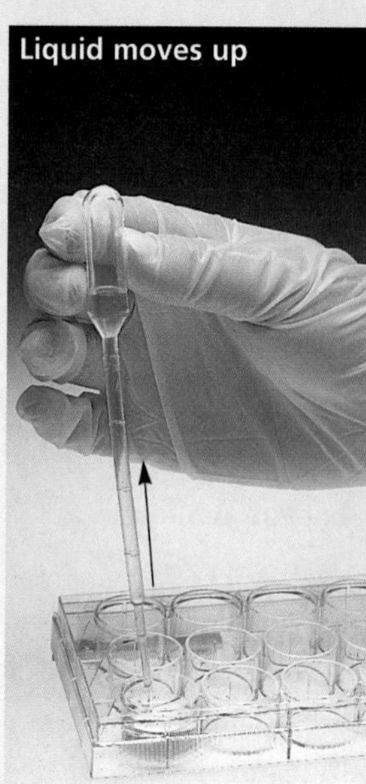

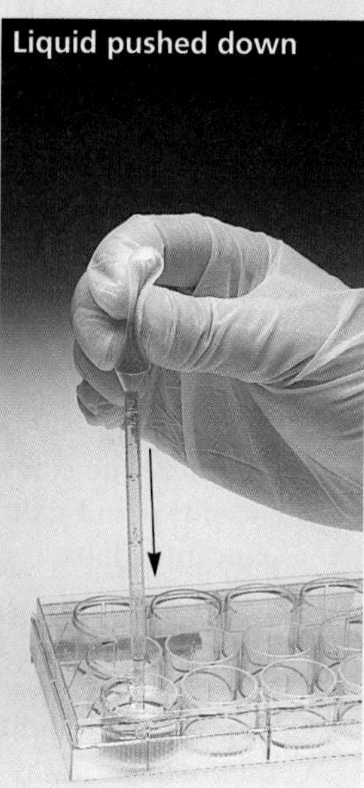

FIGURE A Use this technique for mixing the contents of a well. The pipet used for mixing can also be used to transfer a sample of the mixture to the next well, but use a clean pipet for mixing each new solution.

Call your teacher in the event of a spill. Spills should be cleaned up promptly according to your teacher's directions.

PREPARATION

1. Create a table with two columns for recording your observations. Head the first column *HCl concentration.* A wide second column can be headed *Observations and comparisons.*

PROCEDURE

1. Obtain samples of 1.0 M HCl solution and 1.0 M CH_3COOH solution.
2. Place 20 drops of HCl in one well of a 24-well plate. Place 20 drops of CH_3COOH in an adjacent well. Label the location of each sample.
3. Test the HCl and CH_3COOH with the conductivity tester. Your teacher may have prepared a conductivity tester like the one shown in Figure B. Note the relative intensity of the tester light for each solution. After testing, rinse the tester probes with distilled water. Remove any excess moisture with a paper towel.
4. Place 18 drops of distilled water in each of six wells in your 24-well plate. Add two drops of 1.0 M HCl to the first well to make a total of 20 drops of solution. Mix the contents of this well thoroughly by picking the contents up in a pipet and returning them to the well, as shown in Figure A.
5. Repeat this procedure by taking two drops of the previous dilution and placing it in the next well containing 18 drops of water. Return any unused solution in the pipet to the well from which it was taken. Mix the new solution with a new pipet. (You now have 1.0 M HCl in the well from Procedure step 2, 0.10 M HCl in the first dilution well, and 0.010 M HCl in the second dilution.)

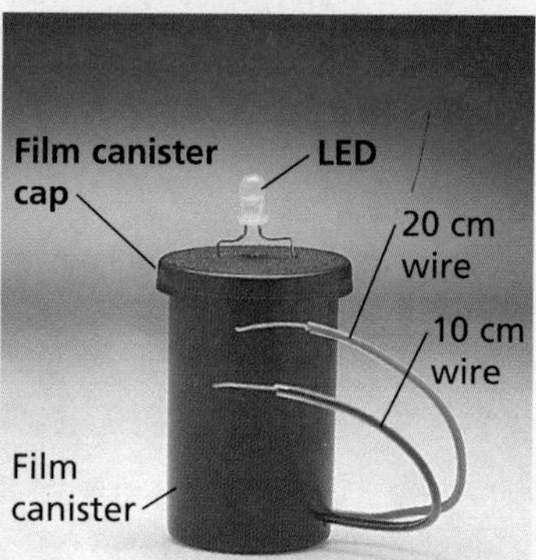

FIGURE B
Teacher-made LED conductivity tester

6. Continue diluting in this manner until you have six successive dilutions. The $[H_3O^+]$ should now range from 1.0 M to 1.0×10^{-6} M. Write the concentrations in the first column of your data table.

7. Using the conductivity tester shown in Figure B, test the cells containing HCl in order from most concentrated to least concentrated. Note the brightness of the tester bulb and compare it with the brightness of the bulb when it was placed in the acetic acid solution. (Retest the acetic acid well any time for comparison.) After each test, rinse the tester probes with distilled water, and use a paper towel to remove any excess moisture. When the brightness produced by one of the HCl solutions is about the same as that produced by the acetic acid, you can infer that the two solutions have the same hydronium ion concentration and that the pH of the HCl solution is equal to the pH of the acetic acid. If the glow from the bulb is too faint to see, turn off the lights or build a light shield around your conductivity tester bulb.

8. Record the results of your observations by noting which HCl concentration causes the intensity of the bulb to most closely match that of the bulb when it is in acetic acid. (Hint: If the conductivity of no single HCl concentration matches that of the acetic acid, then estimate the value between the two concentrations that match the best.)

CLEANUP AND DISPOSAL

9. Clean your lab station. Clean all equipment and return it to its proper place. Dispose of chemicals and solutions in containers designated by your teacher. Do not pour any chemicals down the drain or throw anything in the trash unless your teacher directs you to do so. Wash your hands thoroughly after all work is finished and before you leave the lab.

ANALYSIS AND INTERPRETATION

1. **Resolving Discrepancies:** How did the conductivity of the 1.0 M HCl solution compare with that of the 1.0 M CH_3COOH solution? Why do you think this was so?

2. **Organizing Data:** What is the H_3O^+ concentration of the HCl solution that most closely matched the conductivity of the acetic acid?

3. **Inferring Conclusions:** What was the H_3O^+ concentration of the 1.0 M CH_3COOH solution? Why?

CONCLUSIONS

1. **Applying Models:** The acid-ionization expression for CH_3COOH is the following:

$$K_a = \frac{[H_3O^+][CH_3COO^-]}{[CH_3COOH]}$$

Use your answer to Analysis and Interpretation item 3 to calculate K_a for the acetic acid solution.

2. **Applying Models:** Explain how it is possible for solutions HCl and CH_3COOH to show the same conductivity but have different concentrations.

EXTENSIONS

1. **Evaluating Methods:** Compare the K_a value that you calculated with the value found on page 570 of your text. Calculate the percent error for this experiment.

2. **Predicting Outcomes:** How would your results be affected if you tested the conductivity of a 0.10 M acetic acid solution instead of a 1.0 M acetic acid solution? What effect would it have on the value of the K_a that you calculated? If your teacher approves, try the experiment again with a different concentration of acetic acid solution.

3. **Predicting Outcomes:** Lactic acid ($HOOCCHOHCH_3$) has a K_a of 1.4×10^{-4}. Predict whether a solution of lactic acid would cause the conductivity tester to glow brighter or dimmer than a solution of acetic acid with the same concentration. How noticeable would the difference be?

Blueprint Paper

OBJECTIVES

- ***Prepare* blueprint paper and create a blueprint.**
- ***Infer* the role of oxidation-reduction reactions in the preparation.**

MATERIALS

- **10% iron(III) ammonium citrate solution**
- **10% potassium hexacyanoferrate(III) solution**
- **25 mL graduated cylinders, 2**
- **2 pieces corrugated cardboard, 20 cm × 30 cm**
- **glass stirring rod**
- **Petri dish**
- **thumbtacks, 4**
- **tongs**
- **white paper, 8 cm × 15 cm, 1**

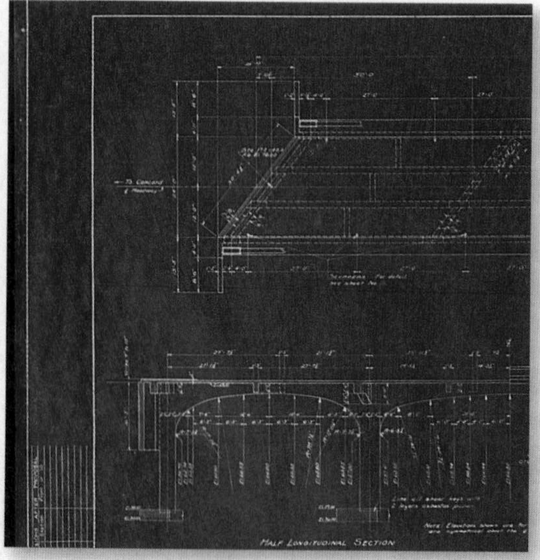

BACKGROUND

Blueprint paper is prepared by coating paper with a solution of two soluble iron(III) salts—potassium hexacyanoferrate(III), commonly called potassium ferricyanide, and iron(III) ammonium citrate. These two salts do not react with each other in the dark. However, when exposed to UV light, the iron(III) ammonium citrate is converted to an iron(II) salt. Potassium hexacyanoferrate(III), $K_3Fe(CN)_6$, reacts with iron(II) ion, Fe^{2+}, to produce an insoluble blue compound, $KFeFe(CN)_6 \cdot H_2O$. In this compound, iron appears to exist in both the +2 and +3 oxidation states.

A blueprint is made by using black ink to make a sketch on a piece of tracing paper or clear, colorless plastic. This sketch is placed on top of a piece of blueprint paper and exposed to ultraviolet light. Wherever the light strikes the paper, the paper turns blue. The paper is then washed to remove the soluble unexposed chemical and is allowed to dry. The result is a blueprint—a blue sheet of paper with white lines. Blueprints produce negative images; the part exposed to light becomes dark.

SAFETY

Always wear safety goggles and a lab apron to protect your eyes and clothing. If you get a chemical in your eyes, immediately flush the chemical out at the eyewash station while calling to your teacher. Know the location of the emergency lab shower and eyewash station and the procedure for using them.

Do not touch any chemicals. If you get a chemical on your skin or clothing, wash the chemical off at the sink while calling to your teacher. Make sure you carefully read the labels and follow the precautions on all containers of chemicals that you use. If there are no precautions stated on the label,

FIGURE A

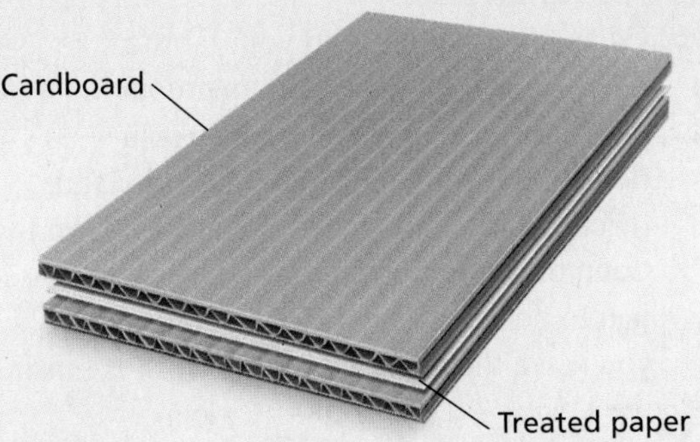

FIGURE B

ask your teacher what precautions to follow. Do not taste any chemicals or items used in the laboratory. Never return leftovers to their original containers; take only small amounts to avoid wasting supplies.

Call your teacher in the event of a spill. Spills should be cleaned up promptly according to your teacher's directions.

PROCEDURE

1. Pour 15 mL of a 10% solution of potassium hexacyanoferrate(III) solution into a petri dish. With most of the classroom lights off or dimmed, add 15 mL of 10% iron(III) ammonium citrate solution. Stir the mixture.

2. Write your name on an 8 cm × 15 cm piece of white paper. Carefully coat one side of the 8 cm × 15 cm piece of paper by using tongs to drag it over the top of the solution in the petri dish, as shown in Figure A.

3. With the coated side up, tack your wet paper to a piece of corrugated cardboard, and cover the paper with another piece of cardboard, as shown in Figure B. **Wash your hands before proceeding to step 4.**

4. Take your paper and cardboard assembly outside into the direct sunlight. Remove the top piece of cardboard so that the paper is exposed. Quickly place an object such as a fern, a leaf, or a key on the paper. If it is windy, you may need to put small weights, such as coins, on the object to keep it in place, as shown in Figure C.

5. After about 20 min, remove the object and again cover the paper with the cardboard. Return to the lab, remove the tacks, and *thoroughly* rinse the blueprint paper under cold running water. Allow the paper to dry. In your notebook, record the amount of time the paper was exposed to sunlight.

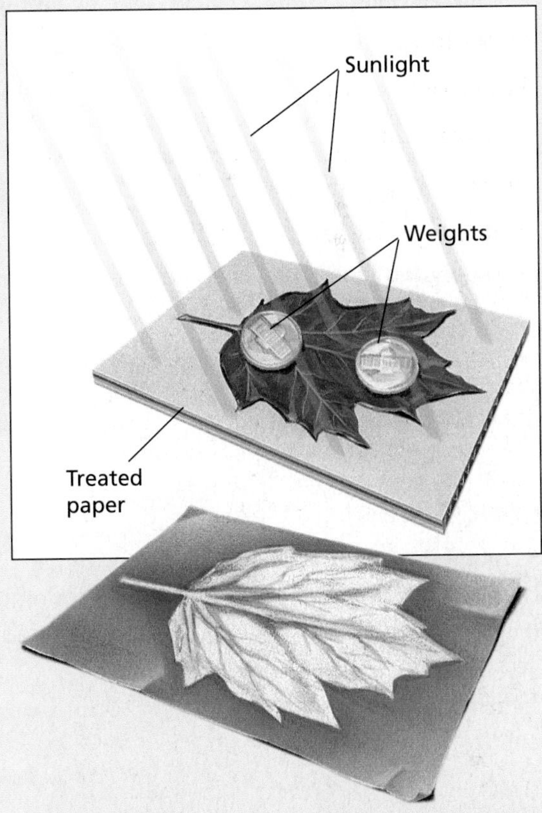

FIGURE C To produce a sharp image, the object must be flat, with its edges on the blueprint paper, and it must not move.

CLEANUP AND DISPOSAL

6. Clean all apparatus and your lab station. Return equipment to its proper place. Dispose of chemicals and solutions in the containers designated by your teacher. Do not pour any chemicals down the drain or in the trash unless your teacher directs you to do so. Wash your hands thoroughly before you leave the lab and after all work is finished.

ANALYSIS AND INTERPRETATION

1. **Relating Ideas:** Why is the iron(III) ammonium citrate solution in a brown bottle?
2. **Organizing Ideas:** When iron(III) ammonium citrate is exposed to light, the oxidation state of the iron changes. What is the new oxidation state of the iron?
3. **Organizing Ideas:** Write a chemical equation showing the reaction of the hexacyanoferrate ion, $Fe(CN)_6^{3-}$, with the iron(II) ion. Include the physical states of the reactants and the product.
4. **Analyzing Methods:** What substances were washed away when you rinsed the blueprint in water after it had been exposed to sunlight? (Hint: Compare the solubilities of the two ammonium salts you used to coat the paper and of the blue product that formed.)
5. **Analyzing Ideas:** The blueprint seems to develop as well on a cloudy day as on a clear day. Explain.
6. **Analyzing Methods:** Why was it important that you wash your hands after coating your paper and before going outside into the sunlight?

CONCLUSIONS

1. **Applying Ideas:** Insufficient washing of the exposed blueprints results in a slow deterioration of images. Suggest a reason for this deterioration.

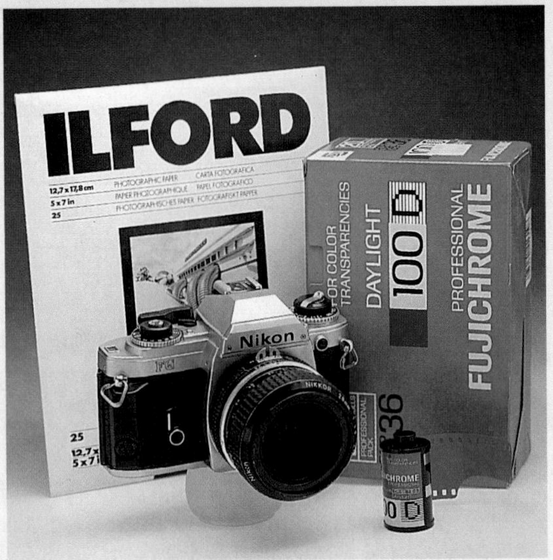

FIGURE D

2. **Relating Ideas:** Photographic paper shown in Figure D can be safely exposed to red light in a darkroom. Do you think the same would be true of blueprint paper? Explain your answer.

EXTENSIONS

1. **Applying Ideas:** How could you use this blueprint paper to test the effectiveness of a brand of sunscreen lotion?
2. **Designing Experiments:** Can you think of ways to improve this procedure? If so, ask your teacher to approve your plan, and create a new blueprint. Evaluate both the efficiency of the procedure and the quality of blueprint.
3. **Research and Communications:** The common name for the compound $KFeFe(CN)_6 \cdot H_2O$ is Prussian blue. This compound has been used for many years as a pigment because of its intense color. Write a report about how Prussian blue is manufactured and the ways in which it is used.

Reduction of Manganese in Permanganate Ion

OBJECTIVES

- ***Demonstrate*** **proficiency in performing redox titrations and recognizing end points of a redox reaction.**
- ***Write*** **a balanced oxidation-reduction equation for a redox reaction.**
- ***Determine*** **the concentration of a solution by using stoichiometry and volume data from a titration.**

MATERIALS

- **0.0200 M $KMnO_4$**
- **1.0 M H_2SO_4**
- **100 mL graduated cylinder**
- **125 mL Erlenmeyer flasks, 4**
- **250 mL beakers, 2**
- **400 mL beaker**
- **burets, 2**
- **distilled water**
- **double buret clamp**
- **$FeSO_4$ solution**
- **ring stand**
- **wash bottle**

BACKGROUND

In Chapter 13, you studied acid-base titrations in which an unknown amount of acid is titrated with a carefully measured amount of base. In this procedure a similar approach called a *redox titration* is used. In a redox titration, the reducing agent, Fe^{2+}, is oxidized to Fe^{3+} by the oxidizing agent, MnO_4^-. When this process occurs, the Mn in MnO_4^- changes from a +7 to a +2 oxidation state and has a noticeably different color. You can use this color change in the same way that you used the color change of phenolphthalein in acid-base titrations to signify a redox reaction "end point." When the reaction is complete, any excess MnO_4^- added to the reaction mixture will give the solution a pink or purple color. The volume data from the titration, the known molarity of the $KMnO_4$ solution, and the mole ratio from the balanced redox equation will give you the information you need to calculate the molarity of the $FeSO_4$ solution.

SAFETY

Always wear safety goggles and a lab apron to protect your eyes and clothing. If you get a chemical in your eyes, immediately flush the chemical out at the eyewash station while calling to your teacher. Know the locations of the emergency lab shower and eyewash station and the procedure for using them.

Do not touch any chemicals. If you get a chemical on your skin or clothing, wash the chemical off at the sink while calling to your teacher. Make sure you carefully read the labels and follow the precautions on all containers of chemicals that you use. Do not taste any chemicals or items used in the laboratory. Never return leftovers to their original containers; take only small amounts to avoid wasting supplies.

Never put broken glass in a regular waste container. Broken glass should be disposed of separately according to your teacher's instructions.

Call your teacher in the event of an acid, base, or potassium permanganate spill. Such spills should be cleaned up promptly. Acids and bases are corrosive; avoid breathing fumes. $KMnO_4$ is a strong oxidizer. If any of the oxidizer spills on you, immediately flush the area with water and notify your teacher.

PREPARATION

1. Prepare a data table in your lab notebook like the one shown below.
2. Clean two 50 mL burets with a buret brush and distilled water. Rinse each buret at least three times with distilled water to remove any contaminants.
3. Label two 250 mL beakers *0.0200 M* $KMnO_4$, and $FeSO_4$ *solution*. Label three of the flasks *1, 2,* and *3.* Label the 400 mL beaker *Waste.* Label one buret $KMnO_4$ and the other $FeSO_4$.
4. Measure approximately 75 mL of 0.0200 M $KMnO_4$ and pour it into the appropriately labeled beaker. Obtain approximately 75 mL of $FeSO_4$ solution and pour it into the appropriately labeled beaker.
5. Rinse one buret three times with a few milliliters of 0.0200 M $KMnO_4$ from the appropriately labeled beaker. Collect these rinses in the waste beaker. Rinse the other buret three times with small amounts of $FeSO_4$ solution from the appropriately labeled beaker. Collect these rinses in the waste beaker.

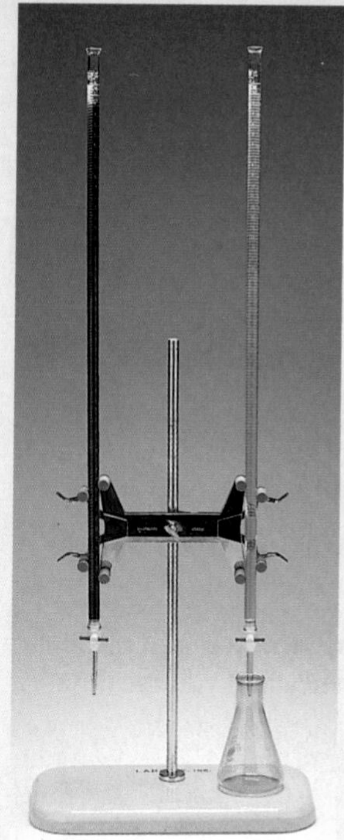

FIGURE A

6. Set up the burets as shown in Figure A. Fill one buret with approximately 50 mL of the 0.0200 M $KMnO_4$ from the beaker and the other buret with approximately 50 mL of the $FeSO_4$ solution from the other beaker.
7. With the waste beaker underneath its tip, open the $KMnO_4$ buret long enough to be sure the buret tip is filled. Repeat the process for the $FeSO_4$ buret.
8. Add 50 mL of distilled water to one of the 125 mL Erlenmeyer flasks, and add one drop of the 0.0200 M $KMnO_4$ to the flask. Set this flask aside to use as a color standard, as shown in Figure B, for comparison with the titration mixture to determine the end point.

DATA TABLE

Trial	Initial $KMnO_4$ volume	Final $KMnO_4$ volume	Initial $FeSO_4$ volume	Final $FeSO_4$ volume
1				
2				
3				

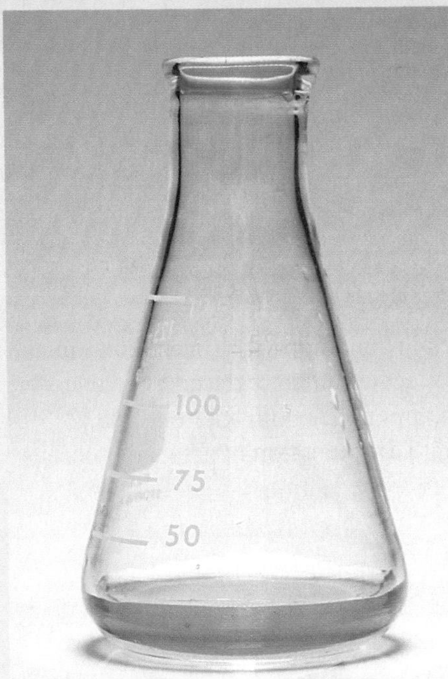

FIGURE B

PROCEDURE

1. Record the initial buret readings for both solutions in your data table. Add 10 mL of the hydrated iron(II) sulfate solution, $FeSO_4{\cdot}7H_2O$, to flask 1. Add 5 mL of 1 M H_2SO_4 to the $FeSO_4$ solution in this flask. The acid will help keep the Fe^{2+} ions in the reduced state, allowing you time to titrate.
2. Slowly add $KMnO_4$ from the buret to the $FeSO_4$ in the flask while swirling the flask. When the color of the solution matches the color standard you prepared in Preparation step 8, record the final readings of the burets in your data table.
3. Empty the titration flask into the waste beaker. Repeat the titration procedure in steps 1 and 2 with flasks 2 and 3.

CLEANUP AND DISPOSAL

4. Dispose of the contents of the waste beaker in the container designated by your teacher. Also pour the color-standard flask into this container. Wash your hands thoroughly after cleaning up the area and equipment.

ANALYSIS AND INTERPRETATION

1. **Organizing Ideas:** Write the balanced equation for the redox reaction of $FeSO_4$ and $KMnO_4$.
2. **Evaluating Data:** Calculate the number of moles of MnO_4^- reduced in each trial.
3. **Analyzing Information:** Calculate the number of moles of Fe^{2+} oxidized in each trial.
4. **Applying Conclusions:** Calculate the average concentration (molarity) of the iron(II) sulfate solution.
5. **Analyzing Methods:** Explain why it was important to rinse the burets with $KMnO_4$ or $FeSO_4$ before adding the solutions. (Hint: Consider what would happen to the concentration of each solution if it were added to a buret that had been rinsed only with distilled water.)

EXTENSIONS

1. **Designing Experiments:** What possible sources of error can you identify with this procedure? If you can think of ways to eliminate them, ask your teacher to approve your plan, and run the procedure again.
2. **Applying Ideas:** Hydrogen peroxide, H_2O_2 ,was once widely used as an antiseptic. It decomposes by the oxidation and reduction of its oxygen atoms. The products are water and molecular oxygen. Write the balanced redox equation for this reaction.
3. **Applying Ideas:** When gaseous hydrogen sulfide burns in air to form sulfur dioxide and water, the oxidation number of hydrogen does not change but that of sulfur changes from –2 to +4 and that of oxygen changes from 0 to –2. Which substance is oxidized and which is reduced? Write the balanced chemical equation for the combustion reaction.

Acid-Catalyzed Iodination of Acetone

OBJECTIVES

- ***Observe* chemical processes in the iodination of acetone.**
- ***Measure* and *compare* rates of chemical reactions.**
- ***Relate* reaction rate concepts to observations.**
- ***Infer* a conclusion from experimental data.**
- ***Evaluate* methods.**

MATERIALS

- **0.0012 M iodine solution**
- **1.0 M HCl solution**
- **1% starch solution**
- **4.0 M acetone**
- **24-well microplate**
- **distilled water**
- **stopwatch or clock with second hand**
- **thin-stemmed pipets, 5**
- **toothpicks**
- **white paper, 1 sheet**

BACKGROUND

Under certain conditions, hydrogen atoms in an acetone molecule can be replaced by iodine. The resulting compound is both a ketone and a halocarbon. The iodination of acetone proceeds according to this equation.

$$CH_3-\overset{\overset{\large O}{\|}}{C}-CH_3(aq) + I_2(aq) \xrightarrow{HCl} CH_3-\overset{\overset{\large O}{\|}}{C}-CH_2I(aq) + HI(aq)$$

Note that the placement of HCl above the arrow indicates that the reaction solution is acidic. In this experiment, HCl(*aq*) is used to provide hydronium ions. The hydronium ions act as catalysts, so the acid concentration appears in the rate equation along with the concentrations of acetone and iodine. The general rate equation for the reaction follows.

$$R = k[\text{acetone}]^x[\text{HCl}]^y[I_2]^z$$

In this experiment, you will measure the rate of the iodination reaction by using a starch indicator solution to signal the disappearance of I_2. The reaction is complete when the blue-black color disappears. By measuring the rate experimentally with differing concentrations of reactants, you can determine the values of exponents *x*, *y*, and *z*. These values can be determined only through experimentation.

SAFETY

Always wear safety goggles and a lab apron to protect your eyes and clothing. If you get a chemical in your eyes, immediately flush the chemical out at the eyewash station while calling to your teacher. Know the location of the emergency lab shower and eyewash station and the procedure for using them.

Do not touch any chemicals. If you get a chemical on your skin or clothing, wash the chemical off at the sink while calling to your teacher. Make sure you carefully read the labels and follow the precautions on all containers of chemicals that you use. If there are no precautions stated on the label, ask your teacher what precautions to follow. Do not taste any chemicals or items used in the laboratory. Never return left-overs to their original containers; take only small amounts to avoid wasting supplies.

Call your teacher in the event of a spill. Spills should be cleaned up promptly according to your teacher's directions.

Acetone solutions are flammable and the vapors can explode when mixed with air. Make sure that there are no flames or sources of sparks in the room when you are using acetone. Acetone can react violently with pure iodine and with concentrated solutions of iodine. Use only the dilute 0.0012 M solution that your teacher has provided for your use.

PREPARATION

1. Copy the data table below into your lab notebook, including the blank columns.
2. Label the five thin-stemmed pipets *Acetone*, *HCl*, *Iodine*, *Starch*, and *Water*. Label four wells on the 24-well plate *1*, *2*, *3*, and *4*.
3. Place the piece of white paper underneath the 24-well plate.

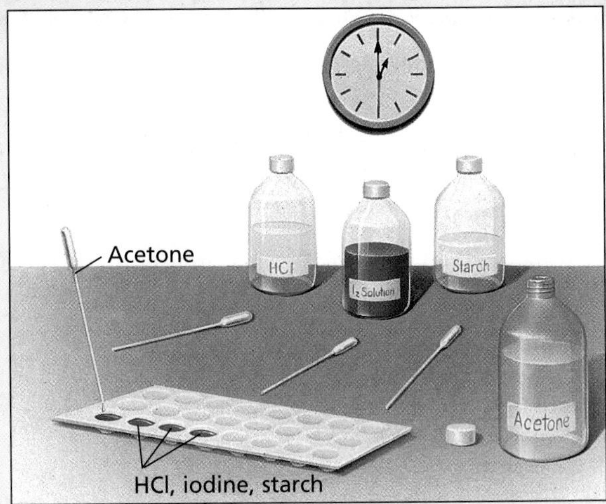

FIGURE A HCl, iodine, and starch are mixed before acetone is added. Start timing as you add the acetone

PROCEDURE

1. Using the four labeled wells of the 24-well plate, mix the starch, water, iodine, and HCl solutions in the proportions and order indicated in the data table in your notebook. **Use the appropriately labeled pipet for each solution.**
2. For reaction 1, note the time or start the stopwatch as you add 10 drops of acetone solution to well 1, as shown in Figure A. **The acetone must be added last, and you must keep track of the time elapsed.** Stir with a toothpick to thoroughly mix the reagents.
3. Continue stirring until the blue-black color disappears, as shown in Figure B on the next page. Record in your data table the time it took for the color to disappear. Also record the number of drops of acetone solution you added.

DATA TABLE

Reaction no.	Starch + H_2O (drops)	0.0012 M I_2 (drops)	1.0 M HCl (drops)	4.0 M acetone (drops)	Time (s)
1	10 + 10 H_2O	10	10	10	
2	10	10	10	20	
3	10	10	20	10	
4	10	20	10	10	

EXPERIMENT 21-2

Casein Glue

OBJECTIVES

- ***Recognize*** **the structure of a protein.**
- ***Predict*** **and** ***observe*** **the result of acidifying milk.**
- ***Prepare*** **and** ***test*** **a casein glue.**
- ***Deduce*** **the charge distribution in proteins as determined by pH.**

MATERIALS

- **100 mL graduated cylinder**
- **250 mL beaker**
- **250 mL Erlenmeyer flask**
- **funnel**
- **glass stirring rod**
- **hot plate**
- **medicine dropper**
- **baking soda, $NaHCO_3$**
- **nonfat milk**
- **paper**
- **paper towel**
- **thermometer**
- **white vinegar**
- **wooden splints, 2**

BACKGROUND

Cow's milk contains 4.4% fat, 3.8% protein, and 4.9% lactose. At the normal pH of milk, 6.3 to 6.6, the protein remains dispersed because it has a net negative charge due to the dissociation of the carboxylic acid group, as shown in Figure A below. As the pH is lowered by the addition of an acid, the protein acquires a net charge of zero, as shown in Figure B. After the protein loses its negative charge, it can no longer remain in solution, and it coagulates into an insoluble mass. The precipitated protein is known as casein and has a molecular mass between 75 000 and 375 000 amu. The pH at which the net charge on a protein becomes zero is called the isoelectric pH. For casein, the isoelectric pH is 4.6.

$H_2N - \boxed{\text{protein}} - COO^-$

FIGURE A

$^+H_3N - \boxed{\text{protein}} - COO^-$

FIGURE B

In this experiment, you will coagulate the protein in milk by adding acetic acid. The casein can then be separated from the remaining solution by filtration. This process is known as separating the curds from the whey. The excess acid in the curds can be neutralized by the addition of sodium hydrogen carbonate, $NaHCO_3$. The product of this reaction is casein glue.

SAFETY

Always wear safety goggles and a lab apron to protect your eyes and clothing. If you get a chemical in your eyes, immediately flush the chemical out at the eyewash station while calling to your teacher. Know the location of the emergency lab shower and eyewash station and the procedure for using them.

Do not touch any chemicals. If you get a chemical on your skin or clothing, wash the chemical off at the sink while calling to your teacher. Make sure you carefully read the labels and follow the precautions on all containers of chemicals that you use. If there are no precautions stated on the label, ask your teacher what precautions you should follow. Do not taste any chemicals or items used in the laboratory. Never return leftovers to their original containers; take only small amounts to avoid wasting supplies.

Call your teacher in the event of a spill. Spills should be cleaned up promptly according to your teacher's directions.

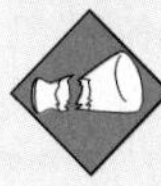

Never put broken glass in a regular waste container. Broken glass should be disposed of separately according to your teacher's instructions.

PREPARATION

1. Prepare your notebook for recording observations at each step of the procedure.
2. Predict the characteristics of the product that will be formed when the acetic acid is added to the milk. Record your predictions in your notebook.

PROCEDURE

1. Pour 125 mL of nonfat milk into a 250 mL beaker. Add 20 mL of 4% acetic acid (white vinegar).
2. Place the mixture on a hot plate and heat it to 60°C. Record your observations in your lab notebook, and compare them with the predictions you made in Preparation step 2.
3. Filter the mixture through a folded piece of paper towel into an Erlenmeyer flask, as shown in Figure C.
4. Discard the filtrate which contains the whey. Scrape the curds from the filter paper back into the 250 mL beaker.

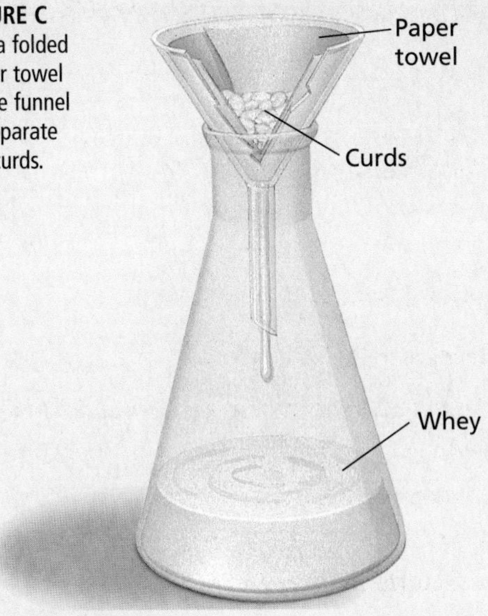

FIGURE C Use a folded paper towel in the funnel to separate the curds.

5. Add 1.2 g of $NaHCO_3$ to the beaker and stir. Slowly add drops of water, stirring intermittently, until the consistency of white glue is obtained.
6. Use your glue to fasten together two pieces of paper. Also fasten together two wooden splints. Allow the splints to dry overnight, and then test the joint for strength.

CLEANUP AND DISPOSAL

1. Clean all apparatus and your lab station. Return equipment to its proper place. Dispose of chemicals and solutions in the containers designated by your teacher. Do not pour any chemicals down the drain or in the trash unless your teacher directs you to do so. Wash your hands thoroughly before you leave the lab and after all work is finished.

ANALYSIS AND INTERPRETATION

1. **Organizing Ideas:** Write the net ionic equation for the reaction between the excess acetic acid and the sodium hydrogen carbonate. Include the physical states of the reactants and products.
2. **Evaluating Methods:** In this experiment, what happened to the lactose and fat portions of the milk?

FIGURE D
This painting was created using paints containing casein.

CONCLUSIONS

1. **Inferring Conclusions:** Figure A shows that the net charge on a protein is negative at pH values higher than its isoelectric pH because the carboxyl group is ionized. Figure B shows that at the isoelectric pH, the net charge is zero. Predict the net charge on a protein at pH values lower than the isoelectric point, and draw a diagram to represent the protein.

EXTENSIONS

1. **Research and Communications:** In addition to its use as an adhesive, casein has been used for centuries in artists' paints such as used on the painting in Figure D. Investigate the use of casein in paint—how and when it is used, its advantages and disadvantages, and its special qualities. Present your findings in a written or oral report.
2. **Relating Ideas:** Figure B represents a protein as a dipolar ion, or zwitterion. The charges in a zwitterion suggest that the carboxyl group donates a hydrogen ion to the amine group. Is there any other way to represent the protein in Figure B so that it still has a net charge of zero?
3. **Designing Experiments:** Design a strength-testing device for the glue joint between the two wooden splints. If your teacher approves your design, create the device and use it to test the strength of the glue.

EXPERIMENT 21-3

Polymers and Toy Balls

OBJECTIVES

- ***Synthesize*** **two different polymers.**
- ***Prepare*** **a small toy ball from each polymer.**
- ***Observe*** **the similarities and differences of the two types of balls.**
- ***Measure*** **the density of each polymer.**
- ***Compare*** **the bounce height of the two balls.**

MATERIALS

- **2 L beaker, or plastic bucket or tub**
- **3 mL 50% ethanol solution**
- **5 oz paper cups, 2**
- **10 mL 5% acetic acid solution (vinegar)**
- **25 mL graduated cylinder**
- **10 mL graduated cylinder**
- **10 mL liquid latex**
- **12 mL sodium silicate solution**
- **distilled water**
- **gloves**
- **meterstick**
- **paper towels**
- **wooden stick**

BACKGROUND

What polymers make the best toy balls? Two possibilities are latex rubber and a polymer produced from ethanol and sodium silicate. Latex rubber is a polymer of covalently bonded atoms.

The polymer formed from ethanol, C_2H_5OH, and a solution of sodium silicate, $Na_2Si_3O_7$, also has covalent bonds. It is known as water glass because it dissolves in water.

In this experiment you will synthesize rubber and the ethanol sodium silicate polymer and test their properties.

SAFETY

Always wear safety goggles and a lab apron to protect your eyes and clothing. If you get a chemical in your eyes, immediately flush the chemical out at the eyewash station while calling to your teacher. Know the locations of the emergency lab shower and eyewash station and the procedure for using them.

Do not touch any chemicals. If you get a chemical on your skin or clothing, wash the chemical off at the sink while calling to your teacher. Make sure you carefully read the labels and follow the directions on all containers of chemicals that you use. Do not taste any chemicals or items used in the laboratory. Never return leftovers to their original containers; take only small amounts to avoid wasting supplies.

Wear disposable plastic gloves; the sodium silicate solution and the alcohol silicate polymer are irritating to your skin.

Ethanol is flammable. Make sure there are no flames anywhere in the laboratory when you are using it. Also, keep it away from other sources of heat.

PREPARATION

1. **Organizing Data:** Copy the data table below in your lab notebook. Prepare one for each polymer. Leave space to record observations about the balls.

DATA TABLE

Trial	Height (cm)	Mass (g)	Diameter (cm)
1			
2			
3			

PROCEDURE

1. Fill the 2 L beaker, bucket, or tub about half-full with distilled water.
2. Using a clean 25 mL graduated cylinder, measure 10 mL of liquid latex and pour it into one of the paper cups.
3. Thoroughly clean the 25 mL graduated cylinder with soap and water and then rinse it with distilled water.
4. Measure 10 mL of distilled water. Pour it into the paper cup with the latex.
5. Measure 10 mL of the 5% acetic acid solution, and pour it into the paper cup with the latex and water.
6. Immediately stir the mixture with the wooden stick.
7. As you continue stirring, a polymer lump will form around the wooden stick. Pull the stick with the polymer lump from the paper cup and immerse the lump in the 2 L beaker, bucket, or tub.
8. While wearing gloves, gently pull the lump from the wooden stick. Be sure to keep the lump immersed under the water, as shown in Figure A.
9. Keep the latex rubber underwater and use your gloved hands to mold the lump into a ball, as shown in Figure B, and then squeeze the lump several times to remove any unused chemicals. You may remove the latex rubber from the water as you roll it in your hands to smooth the ball.
10. Set aside the latex-rubber ball to dry. While it is drying, proceed to step 11.
11. In a clean 25 mL graduated cylinder, measure 12 mL of sodium silicate solution, and pour it into the other paper cup.
12. In a clean 10 mL graduated cylinder, measure 3 mL of 50% ethanol. Pour the ethanol into

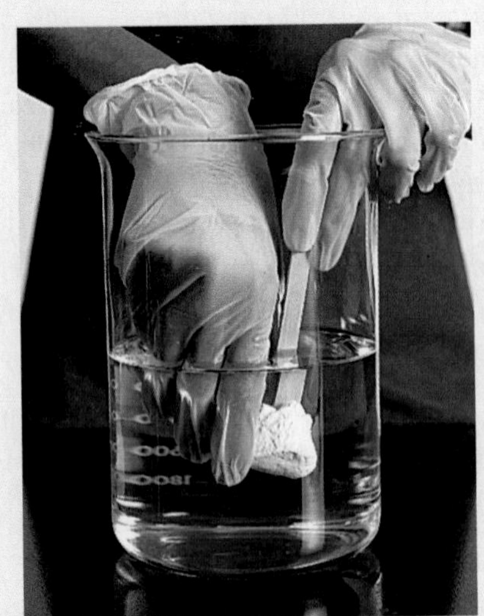

FIGURE A

FIGURE B

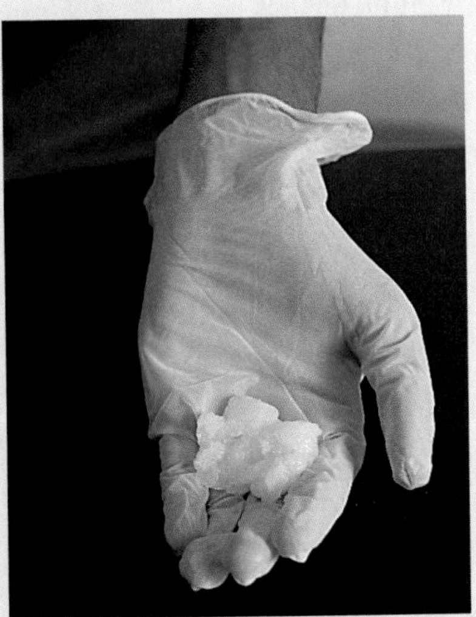

FIGURE C

the paper cup with the sodium silicate, and mix with the wooden stick until a solid substance is formed.

13. While wearing gloves, remove the polymer that forms and place it in the palm of one hand, as shown in Figure C. Gently press it with the palms of both your hands until a ball that does not crumble is formed. This takes a little time and patience. The liquid that comes out of the ball is a combination of ethanol and water. Occasionally moisten the ball by letting a small amount of water from a faucet run over it. When the ball no longer crumbles, you are ready to go on to the next step.

14. Observe as many physical properties of the balls as possible, and record your observations in your lab notebook.

15. Drop each ball several times, and record your observations.

16. Drop one ball from a height of 1 m, and measure its bounce. Perform three trials for each ball.

17. Measure the diameter and the mass of each ball.

CLEANUP AND DISPOSAL

18. Dispose of any extra solutions in the containers indicated by your teacher. Clean up your lab area. Remember to wash your hands thoroughly when your lab work is finished.

ANALYSIS AND INTERPRETATION

1. **Analyzing Information:** List at least three observations you made of the properties of the two different balls.

2. **Organizing Data:** Calculate the average height of the bounce for each type of ball.

3. **Organizing Data:** Calculate the volume for each ball. Even though the balls may not be perfectly spherical, assume that they are. (Hint: The volume of a sphere is equal to $\frac{4}{3} \times \pi \times r^3$, where r is the radius of the sphere, which is one-half of the diameter.)

4. **Organizing Data:** Using your measurements for the mass and the volumes from Analysis and Interpretation item 3, calculate the density of each ball.

CONCLUSIONS

1. **Inferring Conclusions:** Which polymer would you recommend to a toy company for making new toy balls? Explain your reasoning.

2. **Evaluating Viewpoints:** Using the table shown below, calculate the unit cost, that is, the amount of money it costs to make a single ball. (Hint: Calculate how much of each reagent is needed to make a single ball.)

Data Table

Reagent	Price (dollars per liter)
Acetic acid solution	1.50
Ethanol solution	9.00
Latex solution	20.00
Sodium silicate solution	10.00

3. **Evaluating Viewpoints:** What are some other possible practical applications for each of the polymers you made?

EXTENSIONS—ANSWERS

1. **Predicting Outcomes:** When a ball bounces up, kinetic energy of motion is converted into potential energy. With this in mind, explain which will bounce higher, a perfectly symmetrical, round sphere or an oblong shape that vibrates after it bounces.

2. **Predicting Outcomes:** Explain why you didn't measure the volume of the balls by submerging them in water.

APPENDIX A

TABLE A-1 SI MEASUREMENT

Metric Prefixes

Prefix	Symbol	Factor of Base Unit
giga	G	1 000 000 000
mega	M	1 000 000
kilo	k	1 000
hecto	h	100
deka	da	10
deci	d	0.1
centi	c	0.01
milli	m	0.001
micro	μ	0.000 001
nano	n	0.000 000 001
pico	p	0.000 000 000 001

Mass

1 kilogram (kg)	= SI base unit of mass
1 gram (g)	= 0.001 kg
1 milligram (mg)	= 0.000 001 kg
1 microgram (μg)	= 0.000 000 001 kg

Length

1 kilometer (km)	= 1 000 m
1 meter (m)	= SI base unit of length
1 centimeter (cm)	= 0.01 m
1 millimeter (mm)	= 0.001 m
1 micrometer (μm)	= 0.000 001 m
1 nanometer (nm)	= 0.000 000 001 m
1 picometer (pm)	= 0.000 000 000 001 m

Area

1 square kilometer (km^2)	= 100 hectares (ha)
1 hectare (ha)	= 10 000 square meters (m^2)
1 square meter (m^2)	= 10 000 square centimeters (cm^2)
1 square centimeter (cm^2)	= 100 square millimeters (mm^2)

Volume

1 liter (L)	= common unit for liquid volume (not SI)
1 cubic meter (m^3)	= 1000 L
1 kiloliter (kL)	= 1000 L
1 milliliter (mL)	= 0.001 L
1 milliliter (mL)	= 1 cubic centimeter (cm^3)

TABLE A-2 ABBREVIATIONS

amu	=	atomic mass unit (mass)
atm	=	atmosphere (pressure, non-SI)
Bq	=	becquerel (nuclear activity)
°C	=	degree Celsius (temperature)
J	=	joule (energy)
K	=	kelvin (temperature, thermodynamic)
mol	=	mole (quantity)
M	=	molarity (concentration)
N	=	newton (force)
Pa	=	pascal (pressure)
s	=	second (time)
V	=	volt (electric potential difference)

$\frac{1}{X} = .001$ $\frac{0.001}{X} = \frac{1}{200}$

TABLE A-3 SYMBOLS

Symbol		Meaning
α	=	helium nucleus (also ^4_2He) emission from radioactive materials
β	=	electron (also $^{0}_{-1}e$) emission from radioactive materials
γ	=	high-energy photon emission from radioactive materials
Δ	=	change in a given quantity (e.g., ΔH for change in enthalpy)
c	=	speed of light in vacuum
c_p	=	specific heat capacity (at constant pressure)
D	=	density
E_a	=	activation energy
E^0	=	standard electrode potential
E^0 cell	=	standard potential of an electrochemical cell
G	=	Gibbs free energy
ΔG^0	=	standard free energy of reaction
$\Delta G^0{}_f$	=	standard molar free energy of formation
H	=	enthalpy
ΔH^0	=	standard enthalpy of reaction
$\Delta H^0{}_f$	=	standard molar enthalpy of formation
K_a	=	ionization constant (acid)
K_b	=	dissociation constant (base)
K_{eq}	=	equilibrium constant
K_{sp}	=	solubility-product constant
KE	=	kinetic energy
m	=	mass
N_A	=	Avogadro's number
n	=	number of moles
P	=	pressure
pH	=	measure of acidity ($-\log[H_3O^+]$)
R	=	ideal gas law constant
S	=	entropy
S^0	=	standard molar entropy
T	=	temperature (thermodynamic, in kelvins)
t	=	temperature (± degrees Celsius)
V	=	volume
v	=	velocity

TABLE A-4 PHYSICAL CONSTANTS

Quantity	Symbol	Value
Atomic mass unit	amu	$1.660\,5402 \times 10^{-27}$ kg
Avogadro's number	N_A	$6.022\,137 \times 10^{23}$/mol
Electron rest mass	m_e	$9.109\,3897 \times 10^{-31}$ kg 5.4858×10^{-4} amu
Ideal gas law constant	R	8.314 L • kPa/mol • K 0.0821 L • atm/mol • K
Molar volume of ideal gas at STP	V_M	22.414 10 L/mol
Neutron rest mass	m_n	$1.674\,9286 \times 10^{-27}$ kg 1.008 665 amu
Normal boiling point of water	T_b	373.15 K = 100.0°C
Normal freezing point of water	T_f	273.15 K = 0.00°C
Planck's constant	h	$6.626\,076 \times 10^{-34}$ J • s
Proton rest mass	m_p	$1.672\,6231 \times 10^{-27}$ kg 1.007 276 amu
Speed of light in a vacuum	c	$2.997\,924\,58 \times 10^{8}$ m/s
Temperature of triple point of water		273.16 K = 0.01°C

TABLE A-5 HEAT OF COMBUSTION

Substance	Formula	State	ΔH_c
hydrogen	H_2	*g*	–285.8
graphite	C	*s*	–393.5
carbon monoxide	CO	*g*	–283.0
methane	CH_4	*g*	–890.8
ethane	C_2H_6	*g*	–1560.7
propane	C_3H_8	*g*	–2219.2
butane	C_4H_{10}	*g*	–2877.6
pentane	C_5H_{12}	*g*	–3535.6
hexane	C_6H_{14}	*l*	–4163.2
heptane	C_7H_{16}	*l*	–4817.0
octane	C_8H_{18}	*l*	–5470.5
ethene (ethylene)	C_2H_4	*g*	–1411.2
propene (propylene)	C_3H_6	*g*	–2058.0
ethyne (acetylene)	C_2H_2	*g*	–1301.1
benzene	C_6H_6	*l*	–3267.6
toluene	C_7H_8	*l*	–3910.3
naphthalene	$C_{10}H_8$	*s*	–5156.3
anthracene	$C_{14}H_{10}$	*s*	–7076.5
methanol	CH_3OH	*l*	–726.1
ethanol	C_2H_5OH	*l*	–1366.8
ether	$(C_2H_5)_2O$	*l*	–2751.1
formaldehyde	CH_2O	*g*	–570.7
glucose	$C_6H_{12}O_6$	*s*	–2803.0
sucrose	$C_{12}H_{22}O_{11}$	*s*	–5640.9

ΔH_c = heat of combustion of the given substance. All values of ΔH_c are expressed as kJ/mol of substance oxidized to $H_2O(l)$ and/or $CO_2(g)$ at constant pressure and 25°C.
s = solid, *l* = liquid, *g* = gas

TABLE A-6 THE ELEMENTS—SYMBOLS, ATOMIC NUMBERS, AND ATOMIC MASSES

Name of element	Symbol	Atomic number	Atomic mass
actinium	Ac	89	[227.0278]
aluminum	Al	13	26.981539
americium	Am	95	[243.0614]
antimony	Sb	51	121.757
argon	Ar	18	39.948
arsenic	As	33	74.92159
astatine	At	85	[209.9871]
barium	Ba	56	137.327
berkelium	Bk	97	[247.0703]
beryllium	Be	4	9.012182
bismuth	Bi	83	208.98037
bohrium	Bh	107	[262.12]
boron	B	5	10.811
bromine	Br	35	79.904
cadmium	Cd	48	112.411
calcium	Ca	20	40.078
californium	Cf	98	[251.0796]
carbon	C	6	12.011
cerium	Ce	58	140.115
cesium	Cs	55	132.90543
chlorine	Cl	17	35.4527
chromium	Cr	24	51.9961
cobalt	Co	27	58.93320
copper	Cu	29	63.546
curium	Cm	96	[247.0703]
dubnium	Db	105	[262.114]
dysprosium	Dy	66	162.50
einsteinium	Es	99	[252.083]
erbium	Er	68	167.26
europium	Eu	63	151.966
fermium	Fm	100	[257.0951]
fluorine	F	9	18.9984032
francium	Fr	87	[223.0197]
gadolinium	Gd	64	157.25
gallium	Ga	31	69.723
germanium	Ge	32	72.61
gold	Au	79	196.96654
hafnium	Hf	72	178.49
hassium	Hs	108	[265]
helium	He	2	4.002602
holmium	Ho	67	164.930
hydrogen	H	1	1.00794
indium	In	49	114.818
iodine	I	53	126.904
iridium	Ir	77	192.22
iron	Fe	26	55.847

TABLE A-6 CONTINUED

Name of element	Symbol	Atomic number	Atomic mass
krypton	Kr	36	83.80
lanthanum	La	57	138.9055
lawrencium	Lr	103	[262.11]
lead	Pb	82	207.2
lithium	Li	3	6.941
lutetium	Lu	71	174.967
magnesium	Mg	12	24.3050
manganese	Mn	25	54.93805
meitnerium	Mt	109	[266]
mendelevium	Md	101	[258.10]
mercury	Hg	80	200.59
molybdenum	Mo	42	95.94
neodymium	Nd	60	144.24
neon	Ne	10	20.1797
neptunium	Np	93	[237.0482]
nickel	Ni	28	58.6934
niobium	Nb	41	92.90638
nitrogen	N	7	14.00674
nobelium	No	102	[259.1009]
osmium	Os	76	190.23
oxygen	O	8	15.9994
palladium	Pd	46	106.42
phosphorus	P	15	30.9738
platinum	Pt	78	195.08
plutonium	Pu	94	[244.0642]
polonium	Po	84	[208.9824]
potassium	K	19	39.0983
praseodymium	Pr	59	140.908
promethium	Pm	61	[144.9127]
protactinium	Pa	91	231.03588
radium	Ra	88	[226.0254]
radon	Rn	86	[222.0176]
rhenium	Re	75	186.207
rhodium	Rh	45	102.906
rubidium	Rb	37	85.4678
ruthenium	Ru	44	101.07
rutherfordium	Rf	104	[261.11]
samarium	Sm	62	150.36
scandium	Sc	21	44.955910
seaborgium	Sg	106	[263.118]
selenium	Se	34	78.96
silicon	Si	14	28.0855
silver	Ag	47	107.8682
sodium	Na	11	22.989768
strontium	Sr	38	87.62
sulfur	S	16	32.066
tantalum	Ta	73	180.9479
technetium	Tc	43	[97.9072]
tellurium	Te	52	127.60
terbium	Tb	65	158.92534
thallium	Tl	81	204.3833
thorium	Th	90	232.0381
thulium	Tm	69	168.93421
tin	Sn	50	118.710
titanium	Ti	22	47.88
tungsten	W	74	183.84
uranium	U	92	238.0289
vanadium	V	23	50.9415
xenon	Xe	54	131.29
ytterbium	Yb	70	173.04
yttrium	Y	39	88.90585
zinc	Zn	30	65.39
zirconium	Zr	40	91.224

A value given in brackets denotes the mass number of the most stable or most common isotope. The atomic masses of most of these elements are believed to have an error no greater than ±1 in the last digit given.

TABLE A-7 COMMON IONS

Cation	Symbol	Anion	Symbol
aluminum	Al^{3+}	acetate	CH_3COO^-
ammonium	NH_4^+	bromide	Br^-
arsenic(III)	As^{3+}	carbonate	CO_3^{2-}
barium	Ba^{2+}	chlorate	ClO_3^-
calcium	Ca^{2+}	chloride	Cl^-
chromium(II)	Cr^{2+}	chlorite	ClO_2^-
chromium(III)	Cr^{3+}	chromate	CrO_4^{2-}
cobalt(II)	Co^{2+}	cyanide	CN^-
cobalt(III)	Co^{3+}	dichromate	$Cr_2O_7^{2-}$
copper(I)	Cu^+	fluoride	F^-
copper(II)	Cu^{2+}	hexacyanoferrate(II)	$Fe(CN)_6^{4-}$
hydronium	H_3O^+	hexacyanoferrate(III)	$Fe(CN)_6^{3-}$
iron(II)	Fe^{2+}	hydride	H^-
iron(III)	Fe^{3+}	hydrogen carbonate	HCO_3^-
lead(II)	Pb^{2+}	hydrogen sulfate	HSO_4^-
magnesium	Mg^{2+}	hydroxide	OH^-
mercury(I)	Hg_2^{2+}	hypochlorite	ClO^-
mercury(II)	Hg^{2+}	iodide	I^-
nickel(II)	Ni^{2+}	nitrate	NO_3^-
potassium	K^+	nitrite	NO_2^-
silver	Ag^+	oxide	O^{2-}
sodium	Na^+	perchlorate	ClO_4^-
strontium	Sr^{2+}	permanganate	MnO_4^-
tin(II)	Sn^{2+}	peroxide	O_2^{2-}
tin(IV)	Sn^{4+}	phosphate	PO_4^{3-}
titanium(III)	Ti^{3+}	sulfate	SO_4^{2-}
titanium(IV)	Ti^{4+}	sulfide	S^{2-}
zinc	Zn^{2+}	sulfite	SO_3^{2-}

TABLE A-8 WATER-VAPOR PRESSURE

Temperature (°C)	Pressure (mm Hg)	Pressure (kPa)	Temperature (°C)	Pressure (mm Hg)	Pressure (kPa)
0.0	4.6	0.61	**23.0**	21.1	2.81
5.0	6.5	0.87	**23.5**	21.7	2.90
10.0	9.2	1.23	**24.0**	22.4	2.98
15.0	12.8	1.71	**24.5**	23.1	3.10
15.5	13.2	1.76	**25.0**	23.8	3.17
16.0	13.6	1.82	**26.0**	25.2	3.36
16.5	14.1	1.88	**27.0**	26.7	3.57
17.0	14.5	1.94	**28.0**	28.3	3.78
17.5	15.0	2.00	**29.0**	30.0	4.01
18.0	15.5	2.06	**30.0**	31.8	4.25
18.5	16.0	2.13	**35.0**	42.2	5.63
19.0	16.5	2.19	**40.0**	55.3	7.38
19.5	17.0	2.27	**50.0**	92.5	12.34
20.0	17.5	2.34	**60.0**	149.4	19.93
20.5	18.1	2.41	**70.0**	233.7	31.18
21.0	18.6	2.49	**80.0**	355.1	47.37
21.5	19.2	2.57	**90.0**	525.8	70.12
22.0	19.8	2.64	**95.0**	633.9	84.53
22.5	20.4	2.72	**100.0**	760.0	101.32

TABLE A-9 DENSITIES OF GASES AT STP

Gas	Density (g/L)
air, dry	1.293
ammonia	0.771
carbon dioxide	1.997
carbon monoxide	1.250
chlorine	3.214
dinitrogen monoxide	1.977
ethyne (acetylene)	1.165
helium	0.1785
hydrogen	0.0899
hydrogen chloride	1.639
hydrogen sulfide	1.539
methane	0.7168
nitrogen	1.2506
nitrogen monoxide (at 10°C)	1.340
oxygen	1.429
sulfur dioxide	2.927

TABLE A-10 DENSITY OF WATER

Temperature (°C)	Density (g/cm^3)
0	0.999 84
2	0.999 94
3.98 (maximum)	0.999 973
4	0.999 97
6	0.999 94
8	0.999 85
10	0.999 70
14	0.999 24
16	0.998 94
20	0.998 20
25	0.997 05
30	0.995 65
40	0.992 22
50	0.988 04
60	0.983 20
70	0.977 77
80	0.971 79
90	0.965 31
100	0.958 36

TABLE A-11 SOLUBILITIES OF GASES IN WATER

Volume of gas (in liters) at STP that can be dissolved in 1 L of water at the temperature (°C) indicated.

Gas	0°C	10°C	20°C	60°C
air	0.029 18	0.022 84	0.018 68	0.012 16
ammonia	1130	870	680	200
carbon dioxide	1.713	1.194	0.878	0.359
carbon monoxide	0.035 37	0.028 16	0.023 19	0.014 88
chlorine	—	3.148	2.299	1.023
hydrogen	0.021 48	0.019 55	0.018 19	0.016 00
hydrogen chloride	512	475	442	339
hydrogen sulfide	4.670	3.399	2.582	1.190
methane	0.055 63	0.041 77	0.033 08	0.019 54
nitrogen*	0.023 54	0.018 61	0.015 45	0.010 23
nitrogen monoxide	0.073 81	0.057 09	0.047 06	0.029 54
oxygen	0.048 89	0.038 02	0.031 02	0.019 46
sulfur dioxide	79.789	56.647	39.374	—

*Atmospheric nitrogen–98.815% N_2, 1.185% inert gases

TABLE A-12 SOLUBILITY CHART

	acetate	bromide	carbonate	chlorate	chloride	chromate	hydroxide	iodine	nitrate	oxide	phosphate	silicate	sulfate	sulfide
aluminum	S	S	—	S	S	—	A	S	S	a	A	I	S	d
ammonium	S	S	S	S	S	S	S	S	S	—	S	—	S	S
barium	S	S	P	S	S	A	S	S	S	S	A	S	a	d
calcium	S	S	P	S	S	S	S	S	S	P	P	P	S	S
copper(II)	S	S	—	S	S	—	A	—	S	A	A	A	S	A
hydrogen	S	S	—	S	S	—	—	S	S	S	S	I	S	S
iron(II)	—	S	P	S	S	—	A	S	S	A	A	—	S	A
iron(III)	—	S	—	S	S	A	A	S	S	A	P	—	P	d
lead(II)	S	S	A	S	S	A	P	P	S	P	A	A	P	A
magnesium	S	S	P	S	S	S	A	S	S	A	P	A	S	d
manganese(II)	S	S	P	S	S	—	A	S	S	A	P	I	S	A
mercury(I)	P	A	A	S	a	P	—	A	S	A	A	—	P	I
mercury(II)	S	S	—	S	S	P	A	P	S	P	A	—	d	I
potassium	S	S	S	S	S	S	S	S	S	S	S	S	S	S
silver	P	a	A	S	a	P	—	I	S	P	A	—	P	A
sodium	S	S	S	S	S	S	S	S	S	d	S	S	S	S
strontium	S	S	P	S	S	P	S	S	S	S	A	A	P	S
tin(II)	d	S	—	S	S	A	A	S	d	A	A	—	S	A
tin(IV)	S	S	—	—	S	S	P	d	—	A	—	—	S	A
zinc	S	S	P	S	S	P	A	S	S	P	A	A	S	A

S = soluble in water. A = soluble in acids, insoluble in water. P = partially soluble in water, soluble in dilute acids. I = insoluble in dilute acids and in water. a = slightly soluble in acids, insoluble in water. d = decomposes in water.

TABLE A-13 SOLUBILITY OF COMPOUNDS

Solubilities are given in grams of solute that can be dissolved in 100 g of water at the temperature (°C) indicated.

Compound	Formula	0°C	20°C	60°C	100°C
aluminum sulfate	$Al_2(SO_4)_3$	31.2	36.4	59.2	89.0
ammonium chloride	NH_4Cl	29.4	37.2	55.3	77.3
ammonium nitrate	NH_4NO_3	118	192	421	871
ammonium sulfate	$(NH_4)_2SO_4$	70.6	75.4	88	103
barium carbonate	$BaCO_3$	—*	$0.0022^{18°}$	—*	0.0065
barium chloride dihydrate	$BaCl_2 \cdot 2H_2O$	31.2	35.8	46.2	59.4
barium hydroxide	$Ba(OH)_2$	1.67	3.89	20.94	$101.40^{80°}$
barium nitrate	$Ba(NO_3)_2$	4.95	9.02	20.4	34.4
barium sulfate	$BaSO_4$	—*	$0.000\,246^{25°}$	—*	0.000 413
calcium carbonate	$CaCO_3$	—*	$0.0014^{25°}$	—*	$0.0018^{75°}$
calcium fluoride	CaF_2	$0.0016^{18°}$	$0.0017^{26°}$	—*	—*
calcium hydrogen carbonate	$Ca(HCO_3)_2$	16.15	16.60	17.50	18.40
calcium hydroxide	$Ca(OH)_2$	0.189	0.173	0.121	0.076
calcium sulfate	$CaSO_4$	—*	$0.209^{30°}$	—*	0.1619
copper(II) chloride	$CuCl_2$	68.6	73.0	96.5	120
copper(II) sulfate pentahydrate	$CuSO_4 \cdot 5H_2O$	23.1	32.0	61.8	114
lead(II) chloride	$PbCl_2$	0.67	1.00	1.94	3.20
lead(II) nitrate	$Pb(NO_3)_2$	37.5	54.3	91.6	133
lithium chloride	$LiCl$	69.2	83.5	98.4	128
lithium sulfate	Li_2SO_4	36.1	34.8	32.6	$30.9^{90°}$
magnesium hydroxide	$Mg(OH)_2$	—*	$0.0009^{18°}$	—*	0.004
magnesium sulfate	$MgSO_4$	22.0	33.7	54.6	68.3
mercury(I) chloride	Hg_2Cl_2	—*	$0.000\,20^{25°}$	$0.001^{43°}$	—*
mercury(II) chloride	$HgCl_2$	3.63	6.57	16.3	61.3
potassium bromide	KBr	53.6	65.3	85.5	104
potassium chlorate	$KClO_3$	3.3	7.3	23.8	56.3
potassium chloride	KCl	28.0	34.2	45.8	56.3
potassium chromate	K_2CrO_4	56.3	63.7	70.1	$74.5^{90°}$
potassium iodide	KI	128	144	176	206
potassium nitrate	KNO_3	13.9	31.6	106	245
potassium permanganate	$KMnO_4$	2.83	6.34	22.1	—*
potassium sulfate	K_2SO_4	7.4	11.1	18.2	24.1
silver acetate	$AgC_2H_3O_2$	0.73	1.05	1.93	$2.59^{80°}$
silver chloride	$AgCl$	$0.000\,089^{10°}$	—*	—*	0.0021
silver nitrate	$AgNO_3$	122	216	440	733
sodium acetate	$NaC_2H_3O_2$	36.2	46.4	139	170
sodium chlorate	$NaClO_3$	79.6	95.9	137	204
sodium chloride	$NaCl$	35.7	35.9	37.1	39.2
sodium nitrate	$NaNO_3$	73.0	87.6	122	180
sucrose	$C_{12}H_{22}O_{11}$	179.2	203.9	287.3	487.2

*Dashes indicate that values are not available.

TABLE A-14 HEAT OF FORMATION

Substance	State	ΔH_f
ammonia	g	−45.9
ammonium chloride	s	−314.4
ammonium sulfate	s	−1180.9
barium chloride	s	−858.6
barium nitrate	s	−992.1
barium sulfate	s	−1473.2
benzene	g	+82.88
benzene	l	+49.080
calcium carbonate	s	−1207.6
calcium chloride	s	−795.4
calcium hydroxide	s	−983.2
calcium nitrate	s	−938.2
calcium oxide	s	−634.9
calcium sulfate	s	−1434.5
carbon (diamond)	s	+1.9
carbon (graphite)	s	0.00
carbon dioxide	g	−393.5
carbon monoxide	g	−110.5
copper(II) nitrate	s	−302.9
copper(II) oxide	s	−157.3
copper(II) sulfate	s	−771.4
ethane	g	−83.8
ethyne (acetylene)	g	+228.2
hydrogen (H_2)	g	0.00
hydrogen bromide	g	−36.29
hydrogen chloride	g	−92.3
hydrogen fluoride	g	−273.3
hydrogen iodide	g	+26.5
hydrogen oxide (water)	g	−241.8
hydrogen oxide (water)	l	−285.8
hydrogen peroxide	g	−136.3
hydrogen peroxide	l	−187.8
hydrogen sulfide	g	−23.9
iodine (I_2)	s	0.00
iodine (I_2)	g	+62.4
iron(II) chloride	s	−399.4
iron(II) oxide	s	−825.5
iron(II, III) oxide	s	−1118.4
iron(II) sulfate	s	−928.4
iron(II) sulfide	s	−100.0
lead(II) oxide	s	−217.3
lead(IV) oxide	s	−274.5
lead(II) nitrate	s	−451.9
lead(I) sulfate	s	−919.94
lithium chloride	s	−408.6
lithium nitrate	s	−483.1
magnesium chloride	s	−641.5
magnesium oxide	s	−601.6
magnesium sulfate	s	−1284.9
manganese(IV) oxide	s	−520.0
manganese(II) sulfate	s	−1065.3
mercury(I) chloride	s	−264.2
mercury(II) chloride	s	−230.0
mercury(II) oxide (red)	s	−90.8
methane	g	−74.9
nitrogen dioxide	g	+33.2
nitrogen monoxide	g	+90.29
dinitrogen monoxide	g	+82.1
dinitrogen tetroxide	g	+9.2
oxygen (O_2)	g	0.00
ozone (O_3)	g	+142.7
diphosphorus pentoxide	s	−3009.9
potassium bromide	s	−393.8
potassium chloride	s	−436.49
potassium hydroxide	s	−424.58
potassium nitrate	s	−494.6
potassium sulfate	s	−1437.8
silicon dioxide (quartz)	s	−910.7
silver chloride	s	−127.01 ± 0.5
silver nitrate	s	−120.5
silver sulfide	s	−32.59
sodium bromide	s	−361.8
sodium chloride	s	−385.9
sodium hydroxide	s	−425.9
sodium nitrate	s	−467.9
sodium sulfate	l	−1387.1
sulfur dioxide	g	−296.8
sulfur trioxide	g	−395.7
tin(IV) chloride	l	−511.3
zinc nitrate	s	−483.7
zinc oxide	s	−350.5
zinc sulfate	s	−980.14

ΔH_f is heat of formation of the given substance from its elements. All values of ΔH_f are expressed as kJ/mol at 25°C. Negative values of ΔH_f indicate exothermic reactions. s = solid, l = liquid, g = gas

TABLE A-15 PROPERTIES OF COMMON ELEMENTS

Name	Form/color at room temperature	Density (g/cm³)†	Melting point (°C)	Boiling point (°C)	Common oxidation states
aluminum	silver metal	2.702	660.37	2467	3+
arsenic	gray metalloid	5.727^{14}	817 (28 atm)	613 (*sublimes*)	3−, 3+, 5+
barium	bluish white metal	3.51	725	1640	2+
bromine	red-brown liquid	3.119	−7.2	58.78	1−, 1+, 3+, 5+, 7+
calcium	silver metal	1.54	839 ± 2	1484	2+
carbon	diamond graphite	3.51 2.25	3500 (63.5 atm) 3652 (*sublimes*)	3930 —	2+, 4+
chlorine	green-yellow gas	3.214*	−100.98	−34.6	1−, 1+, 3+, 5+, 7+
chromium	gray metal	7.2028	1857 ± 20	2672	2+, 3+, 6+
cobalt	gray metal	8.9	1495	2870	2+, 3+
copper	red metal	8.92	1083.4 ± 0.2	2567	1+, 2+
fluorine	yellow gas	1.69‡	−219.62	−188.14	1−
germanium	gray metalloid	5.323^{25}	937.4	2830	4+
gold	yellow metal	19.31	1064.43	2808 ± 2	1+, 3+
helium	colorless gas	0.1785*	−272.2 (26 atm)	−268.9	0
hydrogen	colorless gas	0.0899*	−259.34	−252.8	1−, 1+
iodine	blue-black solid	4.93	113.5	184.35	1−, 1+, 3+, 5+, 7+
iron	silver metal	7.86	1535	2750	2+, 3+
lead	bluish white metal	11.343716	327.502	1740	2+, 4+
lithium	silver metal	0.534	180.54	1342	1+
magnesium	silver metal	1.745	648.8	1107	2+
manganese	gray-white metal	7.20	1244 ± 3	1962	2+; 3+, 4+, 6+, 7+
mercury	silver liquid metal	13.5462	−38.87	356.58	1+, 2+
neon	colorless gas	0.9002*	−248.67	−245.9	0
nickel	silver metal	8.90	1455	2730	2+, 3+
nitrogen	colorless gas	1.2506*	−209.86	−195.8	3−, 3+, 5+
oxygen	colorless gas	1.429*	−218.4	−182.962	2−
phosphorus	yellow solid	1.82	44.1	280	3−, 3+, 5+
platinum	silver metal	21.45	1772	3827 ± 100	2+, 4+
potassium	silver metal	0.86	63.25	760	1+
silicon	gray metalloid	2.33 ± 0.01	1410	2355	2+, 4+
silver	white metal	10.5	961.93	2212	1+
sodium	silver metal	0.97	97.8	882.9	1+
strontium	silver metal	2.6	769	1384	2+
sulfur	yellow solid	1.96	119.0	444.674	2−, 4+, 6+
tin	white metal	7.28	231.88	2260	2+, 4+
titanium	white metal	4.5	1660 ± 10	3287	2+, 3+, 4+
uranium	silver metal	19.05 ± 0.02^{25}	1132.3 ± 0.8	3818	3+, 4+, 6+
zinc	blue-white metal	7.14	419.58	907	2+

† Densities obtained at 20°C unless otherwise noted (superscript)
‡ Density of fluorine given in g/L at 1 atm and 15°C
* Densities of gases given in g/L at STP

Study Skills for Chemistry

Your success in this course will depend on your ability to apply some basic study skills to learning the material. Studying chemistry can be difficult, but you can make it easier using simple strategies for dealing with the concepts and problems. Becoming skilled in using these strategies will be your keys to success in this and many other courses.

Reading the Text

- **Read the assigned material before class** so that the class lecture makes sense. While reading, one of your tasks is to figure out what information is important.
- **Select a quiet setting** away from distractions so that you can concentrate on what you are reading.
- **Have a pencil and paper nearby to jot down notes and questions** you may have. Be sure to get these questions answered in class.
- **Use the Objectives in the beginning of each section as a list of what you need to know from the section.** Teachers generally make their tests based on the text objectives or their own objectives. Using the objectives to focus your reading can make your learning more efficient.

Taking Notes in Class

- **Be prepared to take notes during class.** Have your materials organized in a notebook. Separate sheets of paper can be easily lost.
- **Don't write down everything your teacher says.** Try to tell which parts of the lecture are important and which are not. Reading the text before class will help in this. You will not be able to write down everything, so you must try to write down only the important things.
- **Recopying notes later is a waste of time** and does not help you learn material for a test. Do it right the first time. Organize your notes as you are writing them down so that you can make sense of your notes when you review them without needing to recopy them.

Reviewing Class Notes

- **Review your notes as soon as possible after class.** Write down any questions you may have about the material covered that day. Be sure to get these questions answered during the next class.
- **Do not wait until the test to review.** By then you will have forgotten a good portion of the material.
- **Be selective about what you memorize.** You cannot memorize everything in a chapter. First of all, it is too time consuming. Second, memorizing and understanding are not the same thing. Memorizing topics as they appear in your notes or text does not guarantee that you will be able to correctly answer questions that require understanding of those topics. You should only memorize material that you understand.

Concept Maps

A technique called concept mapping can help you decide what material in a chapter is important and the best way to learn that material. A concept map presents key ideas, meanings, and relationships for the major concepts being studied. A concept map for a chapter can be thought of as a visual road map for learning the material in the chapter. Using concept maps, this learning happens efficiently because you work with only the key ideas and how they fit together.

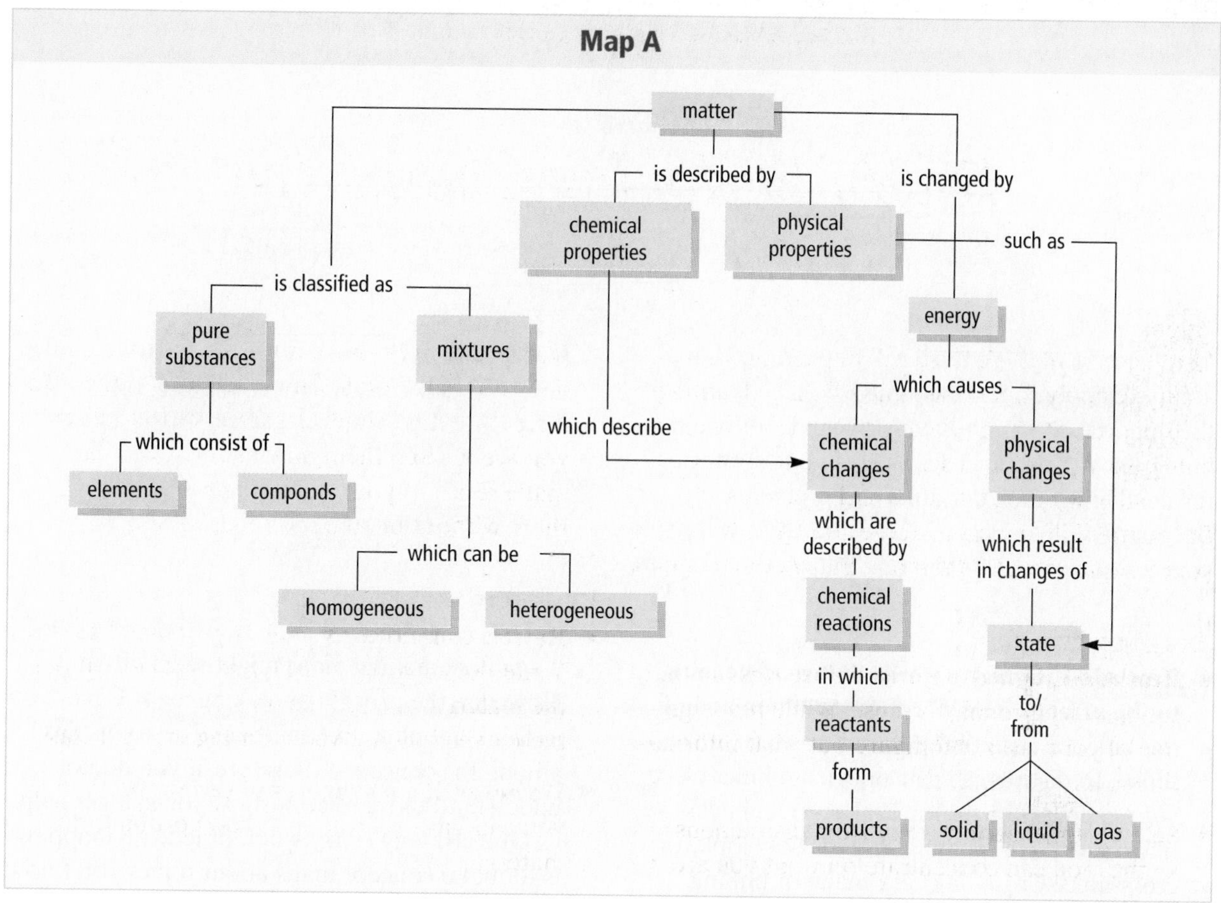

The concept map shown as **Map A** was made from many of the vocabulary terms in Chapter 1. Vocabulary terms are generally labels for concepts, and concepts are generally nouns.

Concepts are linked using linking words to form propositions. A proposition is a phrase that gives meaning to the concept. For example, on the map, "matter is changed by energy" is a proposition.

Studies show that people are better able to remember materials presented visually. The concept map is better than an outline because you can see relationships among many ideas from various sections of the outline. Read through the map to become familiar with the information presented. Look at the map in relation to all of the text pages in Chapter 1; which would be more useful to study before an exam?

How to Make Concept Maps

1. **Start by taking a text section and listing all the important concepts.** We use some terms defined in Section 1-2.

compound	mixture
element	pure substance
heterogeneous mixture	solution
homogeneous mixture	

- From this list, group similar concepts together. For example, one way to group these concepts would be into two groups—one that is related to mixtures, and one that is related to pure substances.

mixture	*pure substance*
heterogeneous mixture	element
homogeneous mixture	compound
solution	

2. **Select a main concept for the map.** We will use matter as the main concept for this map.
3. **Start building the map by placing the concepts according to their importance under the main concept, matter.** One way of arranging the concepts is shown in **Map B.**

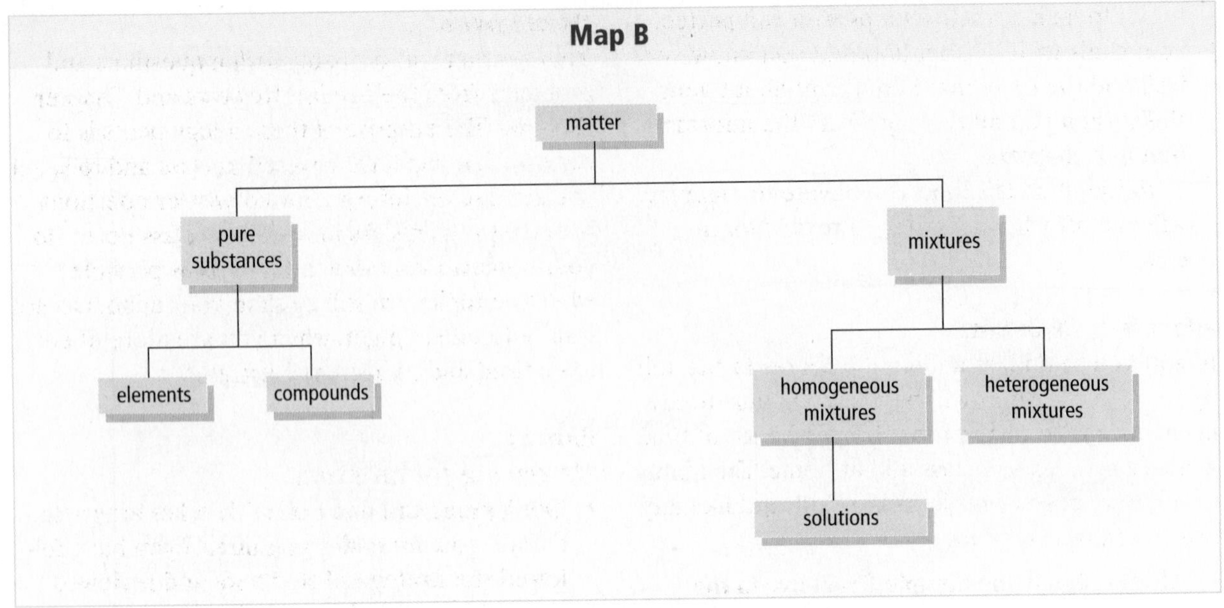

4. **Now add the linking words that give meaning to the arrangement of concepts.** When adding the links, be sure that each proposition makes sense. To distinguish concepts from links, place your concepts in circles, ovals, or rectangles—like the ones in the map shown. Now look for cross-links. Cross-links are made of linking words and arrows. **Map C** is the finished map covering the main ideas found in the vocabulary listed in step 1.

At first, making maps might seem difficult. However, the process of making maps forces you to think about the meanings and relationships among the concepts. Therefore, if you do not understand those relationships, you can get help.

One strategy to try when practicing mapping is to make concept maps about topics you know. For example, if you know a lot about a particular sport (such as basketball) or you have a particular hobby (such as music), you can use those

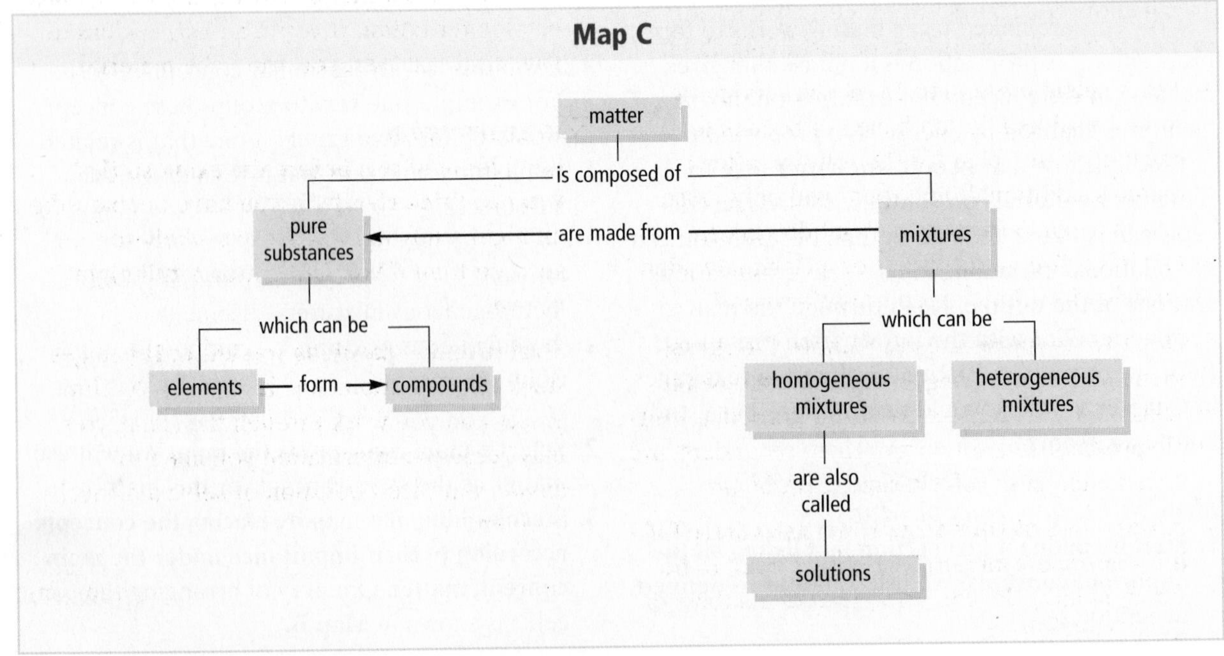

topics to make practice maps. You will perfect your skills with information that you know very well and then feel more confident about your skills when you make maps from the information in a chapter.

Remember, the time you devote to mapping will pay off when it is time to review for an exam.

Working Problems

In addition to understanding the concepts, the ability to solve problems will be a key to your success in chemistry. You will probably spend a lot of time working problems in class and at home. The ability to solve chemistry problems is a skill, and like any skill, it requires practice.

- **Always review the Sample Problems in the chapter.** The Sample Problems in the text provide road maps for solving certain types of problems. Cover the solution while trying to work the problem yourself.
- **The problems in the Chapter Review are similar to the Sample Problems.** If you can relate an assigned problem to one of the Sample Problems in the chapter, it shows that you understand the material.
- **The four steps: Analyze, Plan, Compute, and Evaluate should be the steps you go through when working assigned problems.** These steps will allow you to organize your thoughts and help you develop your problem-solving skills.
- **Never spend more than 15 minutes trying to solve a problem.** If you have not been able to come up with a plan for the solution after 15 minutes, additional time spent will only cause you to become frustrated. What do you do? Get help! See your teacher or a classmate. Find out what it is that you do not understand.
- **Don't try to memorize the Sample Problems; spend your time trying to understand how the solution develops.** Memorizing a particular sample problem will not ensure that you understand it well enough to solve a similar problem.
- **Always look at your answer and ask yourself if it is reasonable and makes sense.** Check to be sure you have the correct units and numbers of significant figures.

Homework

Your teacher will probably assign questions and problems from the Section Reviews and Chapter Reviews. The purpose of these assignments is to review what you have covered in class and to see if you can use the information to answer questions or solve problems. As in reviewing class notes, do your homework as soon after class as possible while the topics are still fresh in your mind. Do not wait until late at night, when you are more likely to be tired and to become frustrated.

Exams

Reviewing for an exam

- **Don't panic and don't cram! It takes longer to learn if you are under pressure.** If you have followed the strategies listed here and reviewed along the way, studying for the exam should be less stressful.
- **When looking over your notes and concept maps, recite ideas out loud.** There are two reasons for reciting:
  1. You are hearing the information, which is effective in helping you learn.
  2. If you cannot recite the ideas, it should be a clue that you do not understand the material, and you should begin rereading or reviewing the material again.
- **Studying with a friend provides a good opportunity for recitation.** If you can explain ideas to your study partner, you know the material.

Taking an exam

- **Get plenty of rest before the exam so that you can think clearly.** If you have been awake all night studying, you are less likely to succeed than if you had gotten a full night of rest.
- **Start with the questions you know.** If you get stuck on a question, save it for later. As time passes and you work through the exam, you may recall the information you need to answer a difficult question or solve a difficult problem.

Gook luck!

GLOSSARY

absolute zero the temperature $-273.15°C$, given a value of zero in the Kelvin scale (317)

accuracy the closeness of measurements to the correct or accepted value of the quantity measured (44)

acid-base indicator a compound whose color is sensitive to pH (493)

acid-ionization constant the term K_a (569)

actinide one of the 14 elements with atomic numbers from 90 (thorium, Th) through 103 (lawrencium, Lr) (126)

activated complex a transitional structure that results from an effective collision and that persists while old bonds are breaking and new bonds are forming (535)

activation energy the minimum energy required to transform the reactants into an activated complex (534)

activity series a list of elements organized according to the ease with which the elements undergo certain chemical reactions (265)

actual yield the measured amount of a product obtained from a reaction (293)

addition polymer a polymer formed by chain addition reactions between monomers that contain a double bond (686)

addition reaction a reaction in which an atom or molecule is added to an unsaturated molecule and increases the saturation of the molecule (682)

alcohol an organic compound that contains one or more hydroxyl groups (663)

aldehyde an organic compound in which a carbonyl group is attached to a carbon atom at the end of a carbon-atom chain (672)

alkali metal one of the elements of Group 1 of the periodic table (lithium, sodium, potassium, rubidium, cesium, and francium) (132)

alkaline a solution in which a base has completely dissociated in water to yield aqueous OH^- ions (461)

alkaline-earth metal one of the elements of Group 2 of the periodic table (beryllium, magnesium, calcium, strontium, barium, and radium) (132)

alkane a hydrocarbon that contains only single bonds (634)

alkene a hydrocarbon that contains double covalent bonds (647)

alkyl group a group of atoms that is formed when one hydrogen atom is removed from an alkane molecule (637)

alkyl halide an organic compound in which one or more halogen atoms —fluorine, chlorine, bromine, or iodine—are substituted for one or more hydrogen atoms in a hydrocarbon (666)

alkyne a hydrocarbon with triple covalent bonds (651)

alpha particle two protons and two neutrons bound together and emitted from the nucleus during some kinds of radioactive decay (706)

amine an organic compound that can be considered to be a derivative of ammonia, NH_3 (677)

amorphous solid a solid in which the particles are arranged randomly (368)

amphoteric any species that can react as either an acid or a base (470)

angular momentum quantum number the quantum number that indicates the shape of the orbital (101)

anion a negative ion (149)

anode the electrode where oxidation takes place (607)

aromatic hydrocarbon a hydrocarbon with six-membered carbon rings and delocalized electrons (652)

Arrhenius acid a chemical compound that increases the concentration of hydrogen ions, H^+, in aqueous solution (459)

Arrhenius base a substance that increases the concentration of hydroxide ions, OH^-, in aqueous solution (459)

artificial transmutation bombardment of stable nuclei with charged and uncharged particles (711)

atmosphere of pressure exactly equivalent to 760 mm Hg (311)

atom the smallest unit of an element that maintains the properties of that element (10)

atomic mass unit a unit of mass that is exactly 1/12 the mass of a carbon-12 atom, or $1.660\,540 \times 10^{-27}$ kg (78)

atomic number the number of protons in the nucleus of each atom of an element (75)

atomic radius one-half the distance between the nuclei of identical atoms that are bonded together (140)

Aufbau principle an electron occupies the lowest-energy orbital that can receive it (105)

autooxidation a process in which a substance acts as both an oxidizing agent and a reducing agent (605)

average atomic mass the weighted average of the atomic masses of the naturally occurring isotopes of an element (79)

Avogadro's law equal volumes of gases at the same temperature and pressure contain equal numbers of molecules (334)

Avogadro's number $6.022\,1367 \times 10^{23}$; the number of particles in exactly one mole of a pure substance (81)

band of stability the stable nuclei cluster over a range of neutron-proton ratios (702)

barometer a device used to measure atmospheric pressure (310)

benzene the primary aromatic hydrocarbon (652)

beta particle an electron emitted from the nucleus during some kinds of radioactive decay (706)

binary acid an acid that contains only two different elements: hydrogen and one of the more-electronegative elements (454)

binary compound a compound composed of two different elements (206)

binding energy per nucleon the binding energy of the nucleus divided by the number of nucleons it contains (702)

boiling the conversion of a liquid to a vapor within the liquid as well as at its surface; occurs when the equilibrium vapor pressure of the liquid equals the atmospheric pressure (378)

boiling point the temperature at which the equilibrium vapor pressure of a liquid equals the atmospheric pressure (378)

boiling-point elevation the difference between the boiling point of a pure solvent and a nonelectrolyte of that solvent, directly proportional to the molal concentration of the solution (440)

bond energy the energy required to break a chemical bond and form neutral isolated atoms (167)

bond length the distance between two bonded atoms at their minimum potential energy, that is, the average distance between two bonded atoms (167)

Boyle's law the volume of a fixed mass of gas varies inversely with pressure at constant temperature (314)

Brønsted-Lowry acid a molecule or ion that is a proton donor (464)

Brønsted-Lowry acid-base reaction the transfer of protons from one reactant (the acid) to another (the base) (465)

Brønsted-Lowry base a molecule or ion that is a proton acceptor (465)

buffered solution a solution that can resist changes in pH (570)

C

calorimeter a device used to measure the heat absorbed or released in a chemical or physical change (511)

capillary action the attraction of the surface of a liquid to the surface of a solid (365)

carboxylic acid an organic compound that contains the carboxyl functional group (674)

catalysis the action of a catalyst (540)

catalyst a substance that changes the rate of a chemical reaction without itself being permanently consumed (540)

catenation the covalent binding of an element to itself to form chains or rings (630)

cathode the electrode where reduction takes place (607)

cation a positive ion (149)

chain reaction a reaction in which the material that starts the reaction is also one of the products and can start another reaction (717)

change of state a physical change of a substance from one state to another (12)

Charles's law the volume of a fixed mass of gas at constant pressure varies directly with the Kelvin temperature (317)

chemical any substance that has a definite composition (6)

chemical bond a mutual electrical attraction between the nuclei and valence electrons of different atoms that binds the atoms together(161)

chemical change a change in which one or more substances are converted into different substances (13)

chemical equation a representation, with symbols and formulas, of the identities and relative amounts of the reactants and products in a chemical reaction (241)

chemical equilibrium a state of balance in which the rate of a forward reaction equals the rate of its reverse reaction and the concentrations of its products and reactants remain unchanged (554)

chemical-equilibrium expression the equation for the equilibrium constant, K (556)

chemical formula a formula that indicates the relative numbers of atoms of each kind in a chemical compound by using atomic symbols and numerical subscripts (164)

chemical kinetics the area of chemistry that is concerned with reaction rates and reaction mechanisms (538)

chemical property the ability of a substance to undergo a change that transforms it into a different substance (12)

chemical reaction a reaction in which one or more substances are converted into different substances (13)

chemistry the study of the composition, structure, and properties of matter and the changes it undergoes (5)

coefficient a small whole number that appears in front of a formula in a chemical equation (243)

colligative properties properties that depend on the concentration of solute particles but not on their identity (436)

collision theory the set of assumptions regarding collisions and reactions (532)

colloid a mixture consisting of particles that are intermediate in size between those in solutions and suspensions forming mixtures known as colloid dispersions (397)

combined gas law the relationship between the pressure, volume, and temperature of a fixed amount of gas (321)

combustion reaction a reaction in which a substance combines with oxygen, releasing a large amount of energy in the form of light and heat (263)

common ion effect the phenomenon in which the addition of an ion common to two solutes brings about precipitation or reduced ionization (567)

composition reaction a reaction in which two or more substances combine to form a new compound (256)

composition stoichiometry calculations involving the mass relationships of elements in compounds (275)

compound a substance that is made from the atoms of two or more elements that are chemically bonded (11)

concentration a measure of the amount of solute in a given amount of solvent or solution (412)

condensation the process by which a gas changes to a liquid (373)

condensation polymer a polymer formed by condensation reactions (690)

condensation reaction a reaction in which two molecules or parts of the same molecule combine (683)

conjugate acid the species that is formed when a Brønsted-Lowry base gains a proton (467)

conjugate base the species that remains after a Brønsted-Lowry acid has given up a proton (467)

control rod a neutron-absorbing rod that helps control a nuclear reaction by limiting the number of free neutrons (718)

conversion factor a ratio derived from the equality between two different units that can be used to convert from one unit to the other (40)

copolymer a polymer made from two different monomers (685)

covalent bonding a chemical bond resulting from the sharing of an electron pair between two atoms (161)

critical mass the minimum amount of nuclide that provides the number of neutrons needed to sustain a chain reaction (718)

critical point indicates the critical temperature and critical pressure of a substance (381)

critical pressure the lowest pressure at which a substance can exist as a liquid at the critical temperature (382)

critical temperature the temperature above which a substance cannot exist in the liquid state (381)

crystal a substance in which the particles are arranged in an orderly, geometric, repeating pattern (368)

crystal structure the total three-dimensional arrangement of particles of a crystal (369)

crystalline solid a solid consisting of crystals (368)

cycloalkane an alkane in which the carbon atoms are arranged in a ring, or cyclic, structure(635)

D

Dalton's law of partial pressures the total pressure of a mixture of gases is equal to the sum of the partial pressures of the component gases (322)

daughter nuclide a nuclide produced by the decay of a parent nuclide (710)

decay series a series of radioactive nuclides produced by successive radioactive decay until a stable nuclide is reached (710)

decomposition reaction a reaction in which a single compound produces two or more simpler substances (259)

delocalized electron an electron shared by more than two atoms (627)

density the ratio of mass to volume or mass divided by volume (38)

deposition the change of state from a gas directly to a solid (380)

derived unit a unit that is a combination of SI base units (36)

diamond a colorless, crystalline, solid form of carbon (626)

diatomic molecule a molecule containing only two atoms (164)

diffusion spontaneous mixing of the particles of two substances caused by their random motion (305)

dipole equal but opposite charges that are separated by a short distance (190)

dipole-dipole force a force of attraction between polar molecules (190)

diprotic acid an acid that can donate two protons per molecule (466)

direct proportion two quantities that give a constant value when one is divided by the other (55)

displacement reaction a reaction in which one element replaces a similar element in a compound (261)

dissociation the separation of ions that occurs when an ionic compound dissolves (425)

double bond a covalent bond produced by the sharing of two pairs of electrons between two atoms (172)

double-replacement reaction a reaction in which the ions of two compounds exchange places in an aqueous solution to form two new compounds (262)

ductility the ability of a substance to be drawn, pulled, or extruded through a small opening to produce a wire (182)

E

effervescence the rapid escape of a gas from the liquid in which it is dissolved (407)

effusion a process by which gas particles under pressure pass through a tiny opening (306)

elastic collision a collision between gas particles and between gas particles and container walls in which there is no net loss of kinetic energy (303)

electrochemical cell a system of electrodes and electrolytes in which either chemical reactions produce electrical energy or an electric current produces chemical change (607)

electrochemistry the branch of chemistry that deals with electricity-related applications of oxidation-reduction reactions (606)

electrode a conductor used to establish electrical contact with a nonmetallic part of a circuit, such as an electrolyte (607)

electrode potential the difference in potential between an electrode and its solution (613)

electrolysis the process in which an electric current is used to produce an oxidation-reduction reaction (610); also the decomposition of a substance by an electric current (259)

electrolyte a substance that dissolves in water to give a solution that conducts electric current (399)

electrolytic cell an electrochemical cell in which electrical energy is required to produce a redox reaction and bring about a chemical change (610)

electromagnetic radiation a form of energy that exhibits wavelike behavior as it travels through space (91)

electromagnetic spectrum all the forms of electromagnetic radiation (91)

electron affinity the energy change that occurs when an electron is acquired by a neutral atom (147)

electron capture the process in which an inner orbital electron is captured by the nucleus of its own atom (707)

electron configuration the arrangement of electrons in an atom (105)

electron-dot notation an electron-configuration notation in which only the valence electrons of an atom of a particular element are shown, indicated by dots placed around the element's symbol (170)

electronegativity a measure of the ability of an atom in a chemical compound to attract electrons (151)

electroplating an electrolytic process in which a metal ion is reduced and solid metal is deposited on a surface (611)

element a pure substance made of only one kind of atom (10)

elimination reaction a reaction in which a simple molecule, such as water or ammonia, is removed from adjacent carbon atoms of a larger molecule (684)

empirical formula the symbols for the elements combined in a compound with subscripts showing the smallest whole-number mole ratio of the different atoms in the compound (229)

end point the point in a titration at which an indicator changes color (498)

enthalpy the heat content of a system at constant pressure (515)

enthalpy change the amount of heat absorbed or lost by a system during a process at constant pressure (516)

entropy a measure of the degree of randomness of the particles, such as molecules, in a system (527)

equilibrium a dynamic condition in which two opposing changes occur at equal rates in a closed system (372)

equilibrium constant the ratio of the mathematical product of the concentrations of substances formed at equilibrium to the mathematical product of the concentrations of the reacting substances. Each concentration is raised to a power equal to the coefficient of that substance in the chemical equation (556)

equilibrium vapor pressure the pressure exerted by a vapor in equilibrium with its corresponding liquid at a given temperature (376)

equivalence point the point at which the two solutions used in a titration are present in chemically equivalent amounts (498)

ester an organic compound with a carboxylic acid group in which the hydrogen of the hydroxyl group has been replaced by an alkyl group (675)

ether an organic compound in which two hydrocarbon groups are bonded to the same atom of oxygen (669)

evaporation the process by which particles escape from the surface of a nonboiling liquid and enter the gas state (365)

excess reactant the substance that is not used up completely in a reaction (288)

excited state a state in which an atom has a higher potential energy than it has in its ground state (94)

extensive property a property that depends on the amount of matter that is present (11)

F

family a vertical column of the periodic table (21)

film badge a device that uses exposure of film to measure the approximate radiation exposure of people working with radiation (714)

fluid a substance that can flow and therefore take the shape of its container; a liquid or a gas (305)

formula equation a representation of the reactants and products of a chemical reaction by their symbols or formulas (244)

formula mass the sum of the average atomic masses of all the atoms represented in the formula of any molecule, formula unit, or ion (221)

formula unit the simplest collection of atoms from which an ionic compound's formula can be established (176)

fractional distillation distillation in which components of a mixture are separated, on the basis of boiling point, by condensation of vapor in a fractionating column (644)

free energy the combined enthalpy-entropy function of a system (528)

free-energy change the difference between the change in enthalpy, ΔH, and the product of the Kelvin temperature and the entropy change, which is defined as $T\Delta S$, at a constant pressure and temperature (528)

freezing the physical change of a liquid to a solid by the removal of heat (366)

freezing point the temperature at which a solid and liquid are in equilibrium at 1 atm (101.3 kPa) pressure (379)

freezing-point depression the difference between the freezing points of a pure solvent and a solution of a nonelectrolyte in that solvent; is directly proportional to the molal concentration of the solution (438)

frequency the number of waves that pass a given point in a specific time, usually one second (91)

fullerene a dark-colored solid made of spherically networked carbon-atom cages (626)

functional group an atom or group of atoms that is responsible for the specific properties of an organic compound (663)

G

gamma ray a high-energy electromagnetic wave emitted from a nucleus as it changes from an excited state to a ground energy state (707)

gas the state of matter in which a substance has neither definite volume nor definite shape (12)

gas laws simple mathematical relationships between the volume, temperature, pressure, and quantity of a gas (313)

Gay-Lussac's law the pressure of a fixed mass of gas at constant volume varies directly with the Kelvin temperature (319)

Gay-Lussac's law of combining volumes of gases at constant temperature and pressure, the volumes of gaseous reactants and products can be expressed as ratios of small whole numbers (333)

Geiger-Müller counter an instrument that detects radiation by counting electric pulses carried by gas ionized by radiation (714)

geometric isomers isomers in which the order of atom bonding is the same but the arrangement of atoms in space is different (632)

Graham's law of effusion the rates of effusion of gases at the same temperature and pressure are inversely proportional to the square roots of their molar masses (352)

graphite a soft, black, crystalline form of carbon that is a fair conductor of electricity (626)

ground state the lowest energy state of an atom (94)

group a vertical column of the periodic table (21)

H

half-cell a single electrode immersed in a solution of its ions (607)

half-life the time required for half the atoms of a radioactive nuclide to decay (708)

half-reaction the part of a reaction involving oxidation or reduction alone (593)

halogen one of the elements of Group 17 (fluorine, chlorine, bromine, iodine, and astatine) (137)

heat the sum total of the kinetic energies of the particles in a sample of matter (512)

heat energy the sum total of the kinetic energies of the particles in a sample of matter (512)

heat of combustion heat released by the complete combustion of one mole of a substance (519)

heat of reaction the quantity of heat released or absorbed during a chemical reaction (514)

heat of solution the amount of heat energy absorbed or released when a specific amount of solute dissolves in a solvent (410)

Heisenberg uncertainty principle it is impossible to determine simultaneously both the position and velocity of an electron or any other particle (99)

Henry's law the solubility of a gas in a liquid is directly proportional to the partial pressure of that gas on the surface of the liquid (407)

Hess's law the overall enthalpy change in a reaction is equal to the sum of the enthalpy changes for the individual steps in the process (519)

heterogeneous catalyst a catalyst whose phase is different from that of the reactants (540)

heterogeneous reaction a reaction involving reactants in two different phases (538)

highest occupied energy level the electron-containing main energy level with the highest principal quantum number (110)

homogeneous catalyst a catalyst that is in the same phase as all the reactants and products in a reaction system (540)

homogeneous reaction a reaction whose reactants and products exist in a single phase (532)

homologous series a series in which adjacent members differ by a constant unit (634)

Hund's rule orbitals of equal energy are each occupied by one electron before any orbital is occupied by a second electron, and all electrons in singly occupied orbitals must have the same spin (106)

hybrid orbitals orbitals of equal energy produced by the combination of two or more orbitals on the same atom (188)

hybridization the mixing of two or more atomic orbitals of similar energies on the same atom to produce new orbitals of equal energies (187)

hydration a solution process with water as the solvent (405)

hydrocarbon the simplest organic compound, composed of only carbon and hydrogen (630)

hydrogen bonding the intermolecular force in which a hydrogen atom that is bonded to a highly electronegative atom is attracted to an unshared pair of electrons of an electronegative atom in a nearby molecule (192)

hydrolysis a reaction between water molecules and ions of a dissolved salt (572)

hydronium ion the H_3O^+ ion (431)

hypothesis a testable statement (30)

I

ideal gas an imaginary gas that perfectly fits all the assumptions of the kinetic-molecular theory (303)

ideal gas constant the constant R, 0.082 057 84 L·atm/mol·K(342)

ideal gas law the mathematical relationship of pressure, volume, temperature, and the number of moles of a gas (340)

immiscible liquid solutes and solvents that are not soluble in each other (406)

inner-shell electron an electron that is not in the highest occupied energy level (110)

intensive property a property that does not depend on the amount of matter present (11)

intermediate a species that appears in some steps of a reaction but not in the net equation (532)

intermolecular force the force of attraction between molecules (189)

inverse proportion two quantities that have a constant mathematical product (56)

ion an atom or group of bonded atoms that has a positive or negative charge (143)

ionic bonding the chemical bond resulting from electrical attraction between large numbers of cations and anions (161)

ionic compound a compound composed of positive and negative ions that are combined so that the numbers of positive and negative charges are equal (176)

ionization the formation of ions from solute molecules by the action of the solvent (431); any process that results in the formation of an ion (143)

ionization energy the energy required to remove one electron from a neutral atom of an element (143)

isomers compounds that have the same molecular formula but different structures (630)

isotopes atoms of the same element that have different masses (76)

joule the SI unit of heat energy as well as all other forms of energy (511)

ketone an organic compound in which a carbonyl group is attached to a carbon atom within the chain (672)

kinetic-molecular theory a theory based on the idea that particles of matter are always in motion (303)

lanthanide one of the 14 elements with atomic numbers from 58 (cerium, Ce) to 71 (lutetium, Lu) (126)

lattice energy the energy released when one mole of an ionic crystalline compound is formed from gaseous ions (178)

law of conservation of mass mass is neither created nor destroyed during ordinary chemical or physical reactions (66)

law of definite proportions a chemical compound contains the same elements in exactly the same proportions by mass regardless of the size of the sample or the source of the compound (66)

law of multiple proportions if two or more different compounds are composed of the same two elements, then the ratio of the masses of the second element combined with a certain mass of the first element is always a ratio of small whole numbers (66)

Le Châtelier's principle when a system at equilibrium is disturbed by application of a stress, it attains a new equilibrium position that minimizes the stress (374)

Lewis acid an atom, ion, or molecule that accepts an electron pair to form a covalent bond (474)

Lewis acid-base reaction the formation of one or more covalent bonds between an electron-pair donor and an electron-pair acceptor (475)

Lewis base an atom, ion, or molecule that donates an electron pair to form a covalent bond (475)

Lewis structure a formula in which atomic symbols represent nuclei and inner-shell electrons, dot-pairs or dashes between two atomic symbols represent electron pairs in covalent bonds, and dots adjacent to only one atomic symbol represent unshared electrons (171)

limiting reactant the reactant that limits the amounts of the other reactants that can combine—and the amount of product that can form—in a chemical reaction (288)

liquid the state of matter in which the substance has a definite volume but an indefinite shape (12)

London dispersion force an intermolecular attraction resulting from the constant motion of electrons and the creation of instantaneous dipoles (193)

lone pair a pair of electrons that is not involved in bonding and that belongs exclusively to one atom (171)

M

magic numbers the numbers of nucleons that represent completed nuclear energy levels—2, 8, 20, 28, 50, 82, and 126 (703)

magnetic quantum number the quantum number that indicates the orientation of an orbital around the nucleus (102)

main-group element an element in the *s*-block or *p*-block (136)

malleability the ability of a substance to be hammered or beaten into thin sheets (182)

mass a measure of the amount of matter (10)

mass defect the difference between the mass of an atom and the sum of the masses of its protons, neutrons, and electrons (701)

mass number the total number of protons and neutrons in the nucleus of an isotope (76)

matter anything that has mass and takes up space (10)

melting the physical change of a solid to a liquid by the addition of heat (368)

melting point the temperature at which a solid becomes a liquid (368)

metal an element that is a good conductor of heat and electricity (22)

metallic bonding chemical bonding that results from the attraction between metal atoms and the surrounding sea of electrons (181)

metalloid an element that has some characteristics of metals and some characteristics of nonmetals (24)

millimeters of mercury a common unit of pressure (311)

miscible liquid solutes and solvents that are able to dissolve freely in one another in any proportion (406)

mixture a blend of two or more kinds of matter, each of which retains its own identity and properties (15)

model an explanation of how phenomena occur and how data or events are related (31)

moderator a material used to slow down the fast neutrons produced by fission (718)

molal boiling-point constant the boiling-point elevation of a solvent in a 1-molal solution of a nonvolatile, nonelectrolyte solute (440)

molal freezing-point constant the freezing-point depression of the solvent in a 1-molal solution of a nonvolatile, nonelectrolyte solute (438)

molality the concentration of a solution expressed in moles of solute per kilogram of solvent (416)

molar heat of formation the heat released or absorbed when one mole of a compound is formed by the combination of its elements (517)

molar heat of fusion the amount of heat energy required to melt one mole of solid at its melting point (380)

molar heat of vaporization the amount of heat energy needed to vaporize one mole of liquid at its boiling point (379)

molar mass the mass of one mole of a pure substance (81)

molarity the number of moles of solute in one liter of solution (412)

mole the amount of a substance that contains as many particles as there are atoms in exactly 12 g of carbon-12 (81)

mole ratio a conversion factor that relates the amounts in moles of any two substances involved in a chemical reaction (276)

molecular compound a chemical compound whose simplest units are molecules (164)

molecular formula a formula showing the types and numbers of atoms combined in a single molecule of a molecular compound (164)

molecular polarity the uneven distribution of molecular charge (183)

molecule a neutral group of atoms that are held together by covalent bonds (164)

monatomic ion an ion formed from a single atom (204)

monomer a small unit that joins with others to make a polymer (685)

monoprotic acid an acid that can donate only one proton (hydrogen ion) per molecule (465)

multiple bond a double or triple bond (173)

N

natural gas a fossil fuel composed primarily of alkanes containing one to four carbon atoms (643)

net ionic equation an equation that includes only those compounds and ions that undergo a chemical change in a reaction in an aqueous solution (429)

neutralization the reaction of hydronium ions and hydroxide ions to form water molecules (473)

newton the SI unit for force; the force that will increase the speed of a one kilogram mass by one meter per second each second it is applied (309)

noble gas a Group 18 element (helium, neon, argon, krypton, xenon, and radon) (111)

noble-gas configuration an outer main energy level fully occupied, in most cases, by eight electrons (112)

nomenclature a naming system (206)

nonelectrolyte a substance that dissolves in water to give a solution that does not conduct an electric current (400)

nonmetal an element that is a poor conductor of heat and electricity (23)

nonpolar-covalent bond a covalent bond in which the bonding electrons are shared equally by the bonded atoms, resulting in a balanced distribution of electrical charge (162)

nonvolatile substance a substance that has little tendency to become a gas under existing conditions (436)

nuclear binding energy the energy released when a nucleus is formed from nucleons (702)

nuclear fission a process in which a very heavy nucleus splits into more-stable nuclei of intermediate mass (717)

nuclear force a short-range proton-neutron, proton-proton, or neutron-neutron force that holds the nuclear particles together (74)

nuclear fusion the combining of light-mass nuclei to form a heavier, more stable nucleus (719)

nuclear power plant a facility that uses heat from nuclear reactors to produce electrical energy (718)

nuclear radiation the particles or electromagnetic radiation emitted from the nucleus during radioactive decay (705)

nuclear reaction a reaction that affects the nucleus of an atom (704)

nuclear reactor a device that uses controlled-fission chain reactions to produce energy or radioactive nuclides (718)

nuclear shell model nucleons exist in different energy levels, or shells, in the nucleus (703)

nucleon a proton or neutron (701)

nuclide the general term for any isotope of any element (77); another term for an atom that is identified by the number of protons and neutrons in its nucleus (701)

O

octane rating a measure of a fuel's burning efficiency and its antiknock properties (645)

octet rule chemical compounds tend to form so that each atom, by gaining, losing, or sharing electrons, has an octet of electrons in its highest occupied energy level (169)

orbital a three-dimensional region around the nucleus that indicates the probable location of an electron (100)

organic compound a covalently bonded compound containing carbon, excluding carbonates and oxides (629)

osmosis the movement of solvent through a semipermeable membrane from the side of lower solute concentration to the side of higher solute concentration (442)

osmotic pressure the external pressure that must be applied to stop osmosis (442)

oxidation a reaction in which the atoms or ions of an element experience an increase in oxidation state (592)

oxidation number a number assigned to an atom in a molecular compound or molecular ion that indicates the general distribution of electrons among the bonded atoms (216)

oxidation-reduction reaction any chemical process in which elements undergo changes in oxidation number (593)

oxidation state a number assigned to an atom in a molecular compound or ion that indicates the general distribution of electrons among the bonded atoms (216)

oxidized having experienced an increase in oxidation number (592)

oxidizing agent a substance that has the potential to cause another substance to be oxidized (602)

oxyacid an acid that is a compound of hydrogen, oxygen, and a third element, usually a non-metal (455)

oxyanion a polyatomic ion that contains oxygen (209)

P

pH the negative of the common logarithm of the hydronium ion concentration of a solution (485)

pH meter a device used to determine the pH of a solution by measuring the voltage between the two electrodes that are placed in the solution (494)

pOH the negative of the common logarithm of the hydroxide ion concentration of a solution (485)

parent nuclide the heaviest nuclide of each decay series (710)

partial pressure the pressure of each gas in a mixture (322)

pascal the pressure exerted by a force of one newton acting on an area of one square meter (311)

Pauli exclusion principle no two electrons in the same atom can have the same set of four quantum numbers (106)

percent error a value calculated by subtracting the experimental value from the accepted value, dividing the difference by the accepted value, and then multiplying by 100 (45)

percent yield the ratio of the actual yield to the theoretical yield, multiplied by 100 (293)

percentage composition the percentage by mass of each element in a compound (227)

period a horizontal row of elements in the periodic table (21)

periodic law the physical and chemical properties of the elements are periodic functions of their atomic numbers (125)

periodic table an arrangement of the elements in order of their atomic numbers so that elements with similar properties fall in the same column, or group (125)

petroleum a complex mixture of different hydrocarbons that varies greatly in composition (643)

phase any part of a system that has uniform composition and properties (373)

phase diagram a graph of pressure versus temperature that shows the conditions under which the phases of a substance exist (381)

photoelectric effect the emission of electrons from a metal when light shines on the metal (93)

photon a particle of electromagnetic radiation that has zero rest mass and carries a quantum of energy (94)

physical change a change in a substance that does not involve a change in the identity of the substance (12)

physical property a characteristic that can be observed or measured without changing the identity of the substance (12)

polar having an uneven distribution of charge (162)

polar-covalent bond a covalent bond in which the bonded atoms have an unequal attraction for the shared electrons (162)

polyatomic ion a charged group of covalently bonded atoms (180)

polymer a large molecule made of many small units joined to each other through organic reactions (685)

polyprotic acid an acid that can donate more than one proton per molecule (465)

positron a particle that has the same mass as an electron but that has a positive charge, and is emitted from the nucleus during some kinds of radioactive decay (706)

precipitate a solid that is produced as a result of a chemical reaction in solution and that separates from the solution (242)

precision the closeness of a set of measurements of the same quantity made in the same way (44)

pressure the force per unit area on a surface (308)

primary amine an organic compound in which one hydrogen atom in an ammonia molecule has been replaced by an alkyl group (677)

primary standard a highly purified solid compound used to check the concentration of a known solution in a titration (499)

principal quantum number the quantum number that indicates the main energy level occupied by the electron (101)

product a substance that is formed by a chemical change (13)

pure substance a substance that has a fixed composition and differs from a mixture in that every sample of a given pure substance has exactly the same characteristic properties and composition (17)

quantity something that has magnitude, size, or amount (33)

quantum the minimum quantity of energy that can be gained or lost by an atom (93)

quantum number a number that specifies the properties of atomic orbitals and the properties of electrons in orbitals (101)

quantum theory a mathematical description of the wave properties of electrons and other very small particles (99)

radioactive dating the process by which the approximate age of an object is determined based on the amount of certain radioactive nuclides present (715)

radioactive decay the spontaneous disintegration of a nucleus into a slightly lighter and more stable nucleus, accompanied by emission of particles, electromagnetic radiation, or both (705)

radioactive nuclide an unstable nucleus that undergoes radioactive decay (705)

radioactive tracer a radioactive atom that is incorporated into a substance so that movement of the substance can be followed by a radiation detector (715)

rate-determining step the slowest-rate step for a chemical reaction (543)

rate law an equation that relates the reaction rate and concentrations of reactants (542)

reactant a substance that reacts in a chemical change (13)

reaction mechanism the step-by-step sequence of reactions by which the overall chemical change occurs (531)

reaction rate the change in concentration of reactants per unit time as a reaction proceeds (538)

reaction stoichiometry calculations involving the mass relationships between reactants and products in a chemical reaction (275)

real gas a gas that does not behave completely according to the assumptions of the kinetic-molecular theory (306)

redox reaction any chemical process in which elements undergo changes in oxidation number (593)

reduced having experienced a decrease in oxidation state (593)

reducing agent a substance that has the potential to cause another substance to be reduced (602)

reduction a reaction in which the oxidation state of an element decreases (593)

reduction potential the measurement of the tendency for a half-reaction to occur as a reduction half-reaction in an electrochemical cell (613)

rem the quantity of ionizing radiation that does as much damage to human tissue as is done by 1 roentgen of high-voltage X rays (713)

resonance the bonding in molecules or ions that cannot be correctly represented by a single Lewis structure (175)

reversible reaction a chemical reaction in which the products re-form the original reactants (246)

roentgen a unit used to measure nuclear radiation; equal to the amount of radiation that produces 2×10^9 ion pairs when it passes through 1 cm^3 of dry air (713)

S

salt an ionic compound composed of a cation and the anion from an acid (215); an ionic compound composed of a cation from a base and an anion from an acid (473)

saturated hydrocarbon a hydrocarbon in which each carbon atom in the molecule forms four single covalent bonds with other atoms (634)

saturated solution a solution that contains the maximum amount of dissolved solute (403)

scientific method a logical approach to solving problems by observing and collecting data, formulating hypotheses, testing hypotheses, and formulating theories that are supported by data (29)

scientific notation numbers written in the form $M \times 10^n$ where the factor M is a number greater than or equal to 1 but less than 10 and n is a whole number (50)

scintillation counter an instrument that converts scintillating light to an electric signal for detecting radiation (714)

secondary amine an organic compound in which two hydrogen atoms of an ammonia molecule have been replaced by alkyl groups (677)

self-ionization of water a process in which two water molecules produce a hydronium ion and a hydroxide ion by transfer of a proton (481)

semipermeable membrane a membrane that allows the movement of some particles while blocking the movement of others (442)

shielding radiation-absorbing material that is used to decrease radiation exposure from nuclear reactors, especially gamma rays (718)

SI *(Le Système International d'Unités)* the measurement system accepted worldwide(33)

significant figure any digit in a measurement that is known with certainty plus one final digit, which is somewhat uncertain or is estimated (46)

single bond a covalent bond produced by the sharing of one pair of electrons between two atoms (171)

single-replacement reaction a reaction in which one element replaces a similar element in a compound (261)

solid the state of matter in which the substance has definite volume and definite shape (12)

solubility the amount of a substance required to form a saturated solution with a specific amount of solvent at a specified temperature (404)

solubility product constant the product of the molar concentrations of ions of a substance in a saturated solution, each raised to the power that is the coefficient of that ion in the chemical equation (578)

soluble capable of being dissolved (395)

solute the substance dissolved in a solution (396)

solution a homogeneous mixture of two or more substances in a single phase (396)

solution equilibrium the physical state in which the opposing processes of dissolution and crystallization of a solute occur at equal rates (402)

solvated a solute particle that is surrounded by solvent molecules (409)

solvent the dissolving medium in a solution (396)

specific heat the amount of heat energy required to raise the temperature of one gram of substance by one Celsius degree (1°C) or one kelvin (1 K) (512)

spectator ion an ion that does not take part in a chemical reaction and is found in solution both before and after the reaction (429)

spin quantum number the quantum number that has only two possible values, +1/2 and –1/2, which indicate the two fundamental spin states of an electron in an orbital (104)

standard electrode potential a half-cell potential measured relative to a potential of zero for the standard hydrogen electrode (614)

standard molar volume of a gas the volume occupied by one mole of a gas at STP, 22.414 10 L (335)

standard solution a solution that contains a precisely known concentration of a solute (499)

standard temperature and pressure the agreed-upon standard conditions of exactly 1 atm pressure and O°C (312)

strong acid an acid that ionizes completely in aqueous solution (460)

strong electrolyte any compound of which all or almost all of the dissolved compound exists as ions in aqueous solution (432)

structural formula a formula that indicates the number and types of atoms present in a molecule and also shows the bonding arrangement of the atoms (630); a formula that indicates the kind, number, arrangement, and bonds but not the unshared electron pairs of the atoms in a molecule (171)

structural isomers isomers in which the atoms are bonded together in different orders (631)

sublimation the change of state from a solid directly to a gas (380)

substitution reaction a reaction in which one or more atoms replace another atom or group of atoms in a molecule (682)

supercooled liquid a substance that retains certain liquid properties even at temperatures at which it appears to be solid (368)

supersaturated solution a solution that contains more dissolved solute than a saturated solution contains under the same conditions (403)

surface tension a force that tends to pull adjacent parts of a liquid's surface together, thereby decreasing surface area to the smallest possible size (365)

suspension a mixture in which the particles in the solvent are so large that they settle out unless the mixture is constantly stirred or agitated (397)

synthesis reaction a reaction in which two or more substances combine to form a new compound (256)

system a specific portion of matter in a given region of space that has been selected for study during an experiment or observation (29)

T

temperature a measure of the average kinetic energy of the particles in a sample of matter (511)

tertiary amine an organic compound in which all three hydrogen atoms of an ammonia molecule have been replaced by alkyl groups (677)

theoretical yield the maximum amount of product that can be produced from a given amount of reactant (293)

theory a broad generalization that explains a body of facts or phenomena (31)

thermochemical equation an equation that includes the quantity of heat released or absorbed during the reaction as written (515)

thermochemistry the study of the changes in heat energy that accompany chemical reactions and physical changes (511)

thermoplastic polymer a polymer that melts when heated and can be reshaped many times (685)

thermosetting polymer a polymer that does not melt when heated but keeps its original shape (685)

titration the controlled addition and measurement of the amount of a solution of known concentration required to react completely with a measured amount of a solution of unknown concentration (497)

transition element one of the *d*-block elements that is a metal, with typical metallic properties (134)

transition interval the pH range over which an indicator changes color (494)

transmutation a change in the identity of a nucleus as a result of a change in the number of its protons (704)

transuranium element an element with more than 92 protons in its nucleus (712)

triple bond a covalent bond produced by the sharing of three pairs of electrons between two atoms (173)

triple point the temperature and pressure conditions at which the solid, liquid, and vapor of a substance can coexist at equilibrium (381)

triprotic acid an acid able to donate three protons per molecule (466)

unit cell the smallest portion of a crystal lattice that shows the three-dimensional pattern of the entire lattice (369)

unsaturated hydrocarbon a hydrocarbon in which not all carbons have four single covalent bonds (647)

unsaturated solution a solution that contains less solute than a saturated solution under the existing conditions (403)

unshared pair a pair of electrons that is not involved in bonding and that belongs exclusively to one atom (171)

valence electron an electron that is available to be lost, gained, or shared in the formation of chemical compounds (150)

vaporization the process by which a liquid or solid changes to a gas (365)

volatile liquid a liquid that evaporates readily (377)

voltaic cell an electrochemical cell in which the redox reaction occurs naturally and produces electrical energy (608)

volume the amount of space occupied by an object (37)

VSEPR theory repulsion between the sets of valence-level electrons surrounding an atom causes these sets to be oriented as far apart as possible (183)

vulcanization a cross-linking process between adjacent polyisoprene molecules that occurs when the molecules are heated with sulfur atoms (688)

wavelength the distance between corresponding points on adjacent waves (91)

weak acid an acid that is a weak electrolyte (460)

weak electrolyte a compound of which a relatively small amount of the dissolved compound exists as ions in an aqueous solution (433)

weight a measure of the gravitational pull on matter (35)

word equation an equation in which the reactants and products in a chemical reaction are represented by words (243)

INDEX

Boldfaced page references denote illustrations; *t* references denote information in tables.

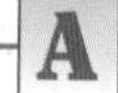

B

C

I

P

T

U

V

Abbreviated as follows: (bkgd) background; (t) top; (b) bottom; (l) left; (r) right; (c) center.

Art Credits: All art, unless otherwise noted, by Holt, Rinehart & Winston. Acknowledgments: Page iii (br), Stephen Durke/Dick Washington; iii (cr), J/B Woolsey Associates, Inc.; iv (bl), Paul Thiessen . **Table of Contents:** Page vi (bl), vii (bl), J/B Woolsey Associates, Inc. **Chapter 1:** Page 10 (bl), J/B Woolsey Associates, Inc. **Chapter 2:** Page 44 (b), David Uhl Studio. **Chapter 3:** Page 70 (bl), 71 (tr), 72 (bl, br), David Uhl Studio. **Chapter 4:** Page 92 (br,tl), 93 (cr), 95 (t), 100 (c), David Uhl Studio; 102 (b), 103 (t, b), J/B Woolsey Associates, Inc. **Chapter 6:** Page 183 (cr), 184 (tl), 185 (b), 188 (b), 189 (c), 191 (cl, t, bl), 192 (t), J/B Woolsey Associates, Inc. **Unit 4:** Page 300 (b), J/B Woolsey Associates, Inc. **Chapter 10:** Page 306 (tl), 310 (l), David Uhl Studio. **Chapter 12:** Page 371 (tr), J/B Woolsey Associates, Inc.; 373 (tl, tc, tr), 376 (tl, tr, cl, cr), Stephen Durke/Dick Washington. **Chapter 13:** Page 401 (bl, bc), 403 (tl, tc), 409 (b), Stephen Durke/Dick Washington; 411 (tr), J/B Woolsey Associates, Inc. **Chapter 14:** Page 425 (bl), 432 (tl, tr, cl, cr), 443 (bl, bc, br), Stephen Durke/Dick Washington. **Chapter 16:** Page 499 (tl, tr), Stephen Durke/Dick Washington. **Chapter 17:** Page 519 (tl), 525 (c), David Uhl Studio. **Chapter 18:** Page 560 (c), David Uhl Studio; 563 (r), Stephen Durke/Dick Washington. **Chapter 19:** Page 604 (c), 607 (t), 608 (tc), 611 (tr), 612 (tr), 614 (bc, br, tl), 620 (cr), David Uhl Studio; 623 (cr), J/B Woolsey Associates, Inc. **Chapter 20:** Page 624 (c), Paul Thiessen; 625 (bl, br), 626 (bl, t, c), 627 (cr), 628 (tc), 631 (tc), 632 (bl), J/B Woolsey Associates, Inc.; 644 (br), David Uhl Studio; 647 (b), 653 (tr), J/B Woolsey Associates, Inc. **Chapter 21:** Page 681 (c), Morgan Cain & Associates. **Chapter 22:** Page 713 (bl), 718 (b), David Uhl Studio. **Elements Handbook** Page 732 (c), Morgan Cain & Associates; 747 (b), David Uhl Studio; 753 (tr), 756 (br), 757 (b), David Uhl Studio; 758 (bl,bc,br), 759 (bl), Morgan Cain & Associates; 765 (t), Claire Booth; 766 (b), Morgan Cain & Associates; 767 (cl,cr), 775 (tr), J/B Woolsey Associates, Inc. **Laboratory Experiments:** Page 795 (bl,br), 796 (bl,br), 821 (tl), 823 (bl), 825 (t), 826 (t, bl), David Uhl Studio; 833 (tl), J/B Woolsey Associates, Inc.; 841 (tl), 858 (br), 859 (bl, tr), 861 (cr), 879 (tl, tr,cr,br), 885 (tr), 886 (tl), 889 (tr), David Uhl Studio.

Photo Credits: All photos, unless otherwise noted, by HRW Staff Photographer, Sam Dudgeon. Acknowledgments: Page ii, Frank Rossotto/The Stock Market; v, Richard Megna/Fundamental Photos. **Table of Contents:** Page vi (t), Irving Geis/Science Source/Photo Researchers; vi (cr), Lawrence Berkeley Laboratory/Science Source/Photo Researchers; vii, Alfred Pasieka/Science Photo Library/Photo Researchers; viii (t), Phil Degginger/Bruce Coleman; viii (br), Richard Megna/Fundamental Photographs; ix (tl), Irving Shapiro/The Stock Shop/Medichrome; ix (br) Breck P. Kent/Animals Animals; x (tl), Stephen Simpson/FPG International; x (bc, br), Richard Megna/Fundamental Photographs; xi (tr), Paul Chesley/Tony Stone Images; xii (t), David Parker/Science Photo Library/Photo Researchers; xii (b), Los Alamos National Laboratory; xiii (t) Tom Pantages. **Unit 1:** (bkgd/all Unit, Lab & Handbook openers), Ryoichi Utsumi/Photonica; Page 2-3 (c), Ed Bock/The Stock Market; 2 (tr), Peticolas/Megna/Fundamental Photographs; 2 (br), Howard L. Garrett/Rainbow; 3 (br), William Huber/Photonica; 3 (cr), D. Arky/Premium/Westlight. **Chapter 1:** (bkgd/all Chapter Openers), Yoav Levy/PhotoTake; Page 4, Irving Geis/Science Source/Photo Researchers; 5 (r), Volker Steger/Peter Arnold; 7, Jon Feingersh/The Stock Market; 8, Giraudon/Art Resource, NY; 17 (l), Tom Pantages; 19, James Lyle/Texas A&M University; 22 (c), Bruce Iverson; 24, Brad Simmons/Bruce Coleman. **Chapter 2:** Page 28, Lawrence Berkeley Laboratory/Science Source/Photo Researchers; 32, AP/Wide World Photos; 43, Dr. Donald Stedman & Peter Popp/University of Denver, Department of Chemistry. **Unit 2:** Page 62-63 (c), Tony Stone Images; 62 (br), IBM, Almaden Research Center/Peter Arnold; 63 (bl), Manfred Kage/Peter Arnold; 63 (br), Ken Eward/Biografx/Science Source/Photo Researchers. **Chapter 3:** Page 64, M.R. Ghadiri & D. McRee/The Scripps Research Institute; 66, Barry L. Runk/Grant Heilman Photography. **Chapter 4:** Page 90, E.R. Degginger/Color-Pic; 94, Superstock; 99 (l), Margot Bergman/Polytechnic Institute of Brooklyn; 99 (r), Courmont/Springer-Verlag; 108, Reprinted with permission from Nature (No. 1196, Vol. 46) Copyright 1892 Macmillian Magazines Limited; 109, AIP; Emilio Segŕe Visual Archives, W.F. Meggers Collection. **Chapter 5:** Page 122, Courtesy of Theodor Benfey, American Chemical Society & Instruments for Research and Industry; 123,

Kitagawa/Superstock; 132 (bl), E.R. Degginger/Color-Pic; 132 (br), Breck P. Kent/Animals Animals; 134 (l), Alfred Pasieka/Peter Arnold; 137 (l, cl), E.R. Degginger/Color-Pic; 137 (cr, r), Larry Stepanowicz. **Chapter 6:** Page 160, Alfred Pasieka/Science Photo Library/Photo Researchers; 176, Bruce Iverson; 182, 192, Photri. **Unit 3:** Page 201 (bl), Anthony Edgeworth/The Stock Market. **Chapter 7:** Page 202, Phil Degginger/Bruce Coleman; 220, Dan McCoy/Rainbow. **Chapter 8:** Page 240, Bob Burch/Bruce Coleman; 252, L.S. Stepanowicz/Visuals Unlimited. **Chapter 9:** Page 274, Richard Megna/Fundamental Photographs. **Unit 4:** Page 300-301 (c), Peter French/Pacific Stock; 301 (br), George Whiteley/Photo Researchers; 301 (cr), Breck P. Kent/Animals Animals. **Chapter 10:** Page 302, Irving Shapiro/The Stock Shop/Medichrome; 309, NASA/SIPA Press. **Chapter 11:** Page 332, Charles D. Winters/Photo Researchers, The National Audubon Society Collection; 338, Corbis/Bettmann Archive; 339 (t), James Schwabel/Panoramic Images; 339 (b), Leo L. Larson/Panoramic Images. **Chapter 12:** Page 362, Icicle Star Burst by Andy Goldsworthy. Icicles/thick ends dipped in snow then water/held until frozen together occasionally using forked sticks as support until stuck/a tense moment when taking them away/breathing on the stick first to release. Scaurwater, Dumfriesshire 12 January 1987. Courtesy of the artist and Galerie Lelong, New York; 365 (t), The Harold E. Edgerton 1992 Press, Courtesy of Palm Press; 366, E.R. Degginger/Color-Pic; 367 (br), William Huber/Photonica; 368, Dr. Dennis Kunkel/PhotoTake; 369 (tl, tc), Breck P. Kent/Animals Animals; 369 (tr, cl), Karl Hartmann/Sachs/PhotoTake; 369 (c, cr, br), Breck P. Kent/Animals Animals; 371, E.R. Degginger/Color-Pic; 383, Joanna McCarthy/Image Bank. **Unit 5:** Page 392-393 (c), Jill & Karl Wallin/FPG International; 392 (b), Phil Borden/PhotoEdit; 393 (br), Burstein Goldman. **Chapter 13:** Page 394, Stephen Simpson/FPG International; 395 (l), Dr. C.W. Dill/Texas A&M University; 397 (l), Neal Mishler/FPG International. **Chapter 14:** Page 424, J & L Weber/Peter Arnold; 435, Hulton Getty/Tony Stone Images; 446, L.S. Stepanowicz/Bruce Coleman. **Chapter 15:** Page 452, Richard B. Swint, M.D.; 465, Charles D. Winters/Photo Researchers; 474, Ray Pfortner/Peter Arnold. **Chapter 16:** Page 480, Clay H. Wiseman/Animals Animals; 482 (all), Richard Megna/Fundamental Photographs; 492, Ken Simmons/Massachusetts Division of Fisheries & Wildlife; 493 (all), Jerry Mason/Science Photo Library/Photo Researchers; 494 (b), Charles D. Winters/Photo Researchers; 498 (all), Richard Megna/Fundamental Photographs.

Unit 6: Page 508-509 (c), Myron/Tony Stone Images; 508 (br), Richard Megna/Fundamental Photographs; 509 (l, r), AP/Wide World Photos. **Chapter 17:** Page 511, Paul Chesley/Tony Stone Images; 531, Larry Stepanowicz. **Chapter 18:** Page 552, Greg Edmonds/Earth Images; 557 (all), E.R. Degginger/Color-Pic; 561, Steve McCutcheon/Visuals Unlimited. **Chapter 19:** Page 590, Frank Rossotto/The Stock Market; 593, Richard Megna/Fundamental Photographs; 596, Renee Lynn/Photo Researchers; 605 (b), Richard L. Carlton/Photo Researchers; 605 (t), Thomas Eisner and Daniel Aneshansley/Cornell University. **Unit 7:** Page 622-623 (c), Photo Researchers; 623 (br), Science Photo Library/Photo Researchers. **Chapter 20:** Page 626, Smithsonian Institution; 627, Runk/Schoenberger/Grant Heilman Photography; 628 (l), Paul Damien/Tony Stone Images; 630 (b), Gregory K. Scott/Photo Researchers; 633, Runk/Schoenberger/Grant Heilman Photography; 643, David Spindel/Superstock; 644, Keith Wood/Tony Stone Images; 645, M. Veale/Frontera Fotos; 646, David Blakenship/University of Florida Engineering Publications; 652, Kevin Anderson/Tony Stone Images. **Chapter 21:** Page 662, Ken Whitmore/Tony Stone Images; 668, Ozone Processing Team Code 916/NASA Goddard Space Flight Center; 674, Seth Joel/Science Photo Library/Photo Researchers; 678, Matt Meadows/Peter Arnold; 680, Rosalind Franklin/Photo Researchers; 688, Charles O. Cecil/Visuals Unlimited; 689, Chris Sorensen/The Stock Market; 692, Department of Chemical Engineering/North Carolina State University. **Chapter 22:** Page 700, David Parker/ Science Photo Library/Photo Researchers; 711, Fermilab Visual Media Services; 715, Custom Medical Stock Photo, 1991 Peter Berndt, M.D.,P.A.; 716, SNS/SIPA Press; 719 (l), Roger Ressmeyer/Corbis; 719 (r), Los Alamos National Laboratory; 720, 721, AIP; Emilio Segre Visual Archives.

Elements Handbook: Page 726 (bl, bc) Tom Pantages; 726 (br) Breck P. Kent/ Animals Animals; 729 (cl, cr, bc), Lester Bergman/The Bergman Collection; 729 (bl, br), E.R. Degginger/Color-Pic; 730 (all), Phillip Robbins/Phi-Lo Photo; 733, Rick Lance/PhotoTake; 734 (cl), E.R. Degginger/Color-Pic; 734 (c), Carl Frank/Photo Researchers; 734 (bl), Gary Retherford/Photo Researchers; 734 (br), Breck P. Kent/Animals Animals; 735 (bl, bc), Yoav Levy/PhotoTake; 735 (br), Larry Stepanowicz; 736, Mauritius/Superstock; 737 (b), S.L. Craig Jr./Bruce Coleman; 738, Chris Rogers/Rainbow; 739, E.R. Degginger/Color-Pic; 740 (br), Brent Cavedo/Superstock; 740 (tr), Thomas Del Brase/Tony Stone Images; 740 (cr), Larry Stepanowicz; 740 (bl), Barbara Filet/Tony Stone Images; 740 (bc), Ken Lucas/Visuals Unlimited; 741 (tl), Jeff Greenberg/David R. Frazier Photolibrary; 741 (cr), L.S. Stepanowicz/Visuals Unlimited; 743 (all), 744 (br), Tom Pantages; 744 (tr, tl), Larry Stepanowicz; 745, Dan McCoy/Rainbow; 746 (l), Hank Morgan/Rainbow; 746 (r), Dan Rubin/The Stock Shop/Medichrome; 750 (cr), Science VU-AAI/Visuals Unlimited; 750 (br), Arthur R. Hill/Visuals Unlimited; 751 (br), Tom Pantages; 752, CoCo McCoy/Rainbow; 753, Tom Pantages; 754 (bl), J & L Weber/Peter Arnold; 760, Tom Pantages; 762, Irving Geis/Science Source/Photo Researchers; 770 (cr), Peter Salek/Tony Stone Images; 771 (c, r), E.R. Degginger/Color-Pic; 772 (r), George Glod/Superstock; 772 (l), Peter Beck/The Stock Market; 774 (cl), Heather Angel/Biofotos; 774 (cr, br), E.R. Degginger/Color-Pic; 781 (bl), Tom Tracy/The Stock Shop/Medichrome.

Laboratory Experiments: Page 785, Harry Giglio/Photri; 794, Paul Berg Photography; 801, Yoav Levy/PhotoTake; 806, E.R. Degginger/Color-Pic; 818 (t), David Scharf/Peter Arnold; 878, Tom Pantages; 890, Artists of Iowa, The Magna Collections; 891, Tom Pix/Peter Arnold.

Photos by Charles D. Winters: Page viii (bl), ix (bl), xi (bl), xiii (b), xiv, 5 (l), 11, 12, 14, 16 (l, r), 17 (r), 18 (all), 22 (l, r), 23 (all), 38, 132 (t, c), 134 (c, r), 200 (br), 208, 215, 223, 241, 242 (all), 250, 257 (all), 260, 261, 268, 292, 301 (cl), 316 (all), 324, 364 (bl, bc, br), 365 (bl, bc), 367 (bc), 395 (c), 402, 407 (all), 427, 428, 434 (all), 453 (all), 454, 456 (all), 457, 458, 462, 463, 466 (all), 469, 473, 494 (t), 512 (all), 527 (all), 539 (all), 540, 553, 566, 568, 570 (all), 592 (all), 599, 603, 606 (all), 608, 665, 669, 673, 683, 684 (all), 686, 687, 690, 691, 727, 728 (l), 729 (tl, tr), 735 (t, cl, cr), 737 (t), 741 (bl, br, tr, cl), 742 (all), 750 (cl, bl), 751 (tr, bl), 754 (cr, cl, br), 755, 763, 767 (all), 770 (bl, br), 771 (l), 774 (bl), 775 (all), 780 (all), 781 (tr, br), 788, 789 (all), 790, 791 (all), 792, 799, 800, 802, 805 (all), 808, 810, 811 (all), 813, 814 (all), 815, 817, 818 (b), 822, 824, 835, 836 (all), 839, 842, 843 (all), 845, 846 (all), 847, 848, 850, 851, 852, 853, 855, 856, 857, 862, 863, 865 (all), 868 (all), 869 (all), 871, 872, 874, 876 (all), 880, 882, 883, 888. **HRW Photos by: Dennis Fagan**, Page 81 (all), 308 (all), 354, 364 (t), 572 (all), 887; **Richard Haynes**, Page 398; **John King**, Page 495 (all), 499 (all), 582 (all); **John Langford**, Page 200-201 (c), 731; **Sergio Purtell**, Page 13, 16 (c), 29, 46, 61, 245, 258, 262, 263 (all), 304, 334 (all), 367 (bl), 397 (r), 400 (all), 413 (all), 416 (l), 497, 500 (all), 501 (all), 565 (all), 571, 629 (all), 651, 676, 714 (all), 728 (c, r), 729 (tc), 820, 830, 834, 892 (all).

Periodic Table of the Elements

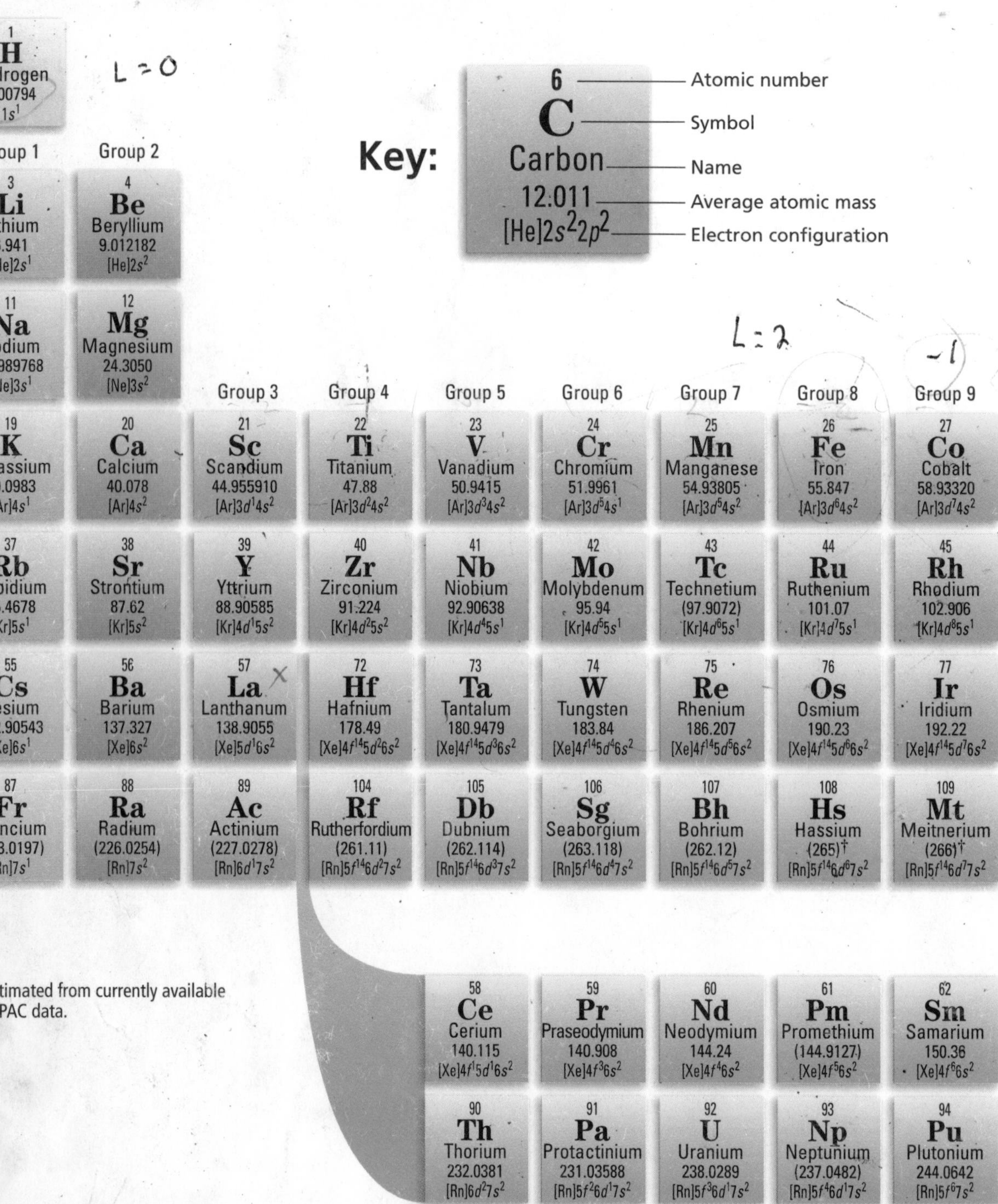

Key: 6 — Atomic number; C — Symbol; Carbon — Name; 12.011 — Average atomic mass; [He]$2s^22p^2$ — Electron configuration

Group 1	Group 2	Group 3	Group 4	Group 5	Group 6	Group 7	Group 8	Group 9
1 **H** drogen 00794 $1s^1$								
3 **Li** thium .941 e]$2s^1$	4 **Be** Beryllium 9.012182 [He]$2s^2$							
11 **Na** dium 989768 e]$3s^1$	12 **Mg** Magnesium 24.3050 [Ne]$3s^2$							
19 **K** assium .0983 r]$4s^1$	20 **Ca** Calcium 40.078 [Ar]$4s^2$	21 **Sc** Scandium 44.955910 [Ar]$3d^14s^2$	22 **Ti** Titanium 47.88 [Ar]$3d^24s^2$	23 **V** Vanadium 50.9415 [Ar]$3d^34s^2$	24 **Cr** Chromium 51.9961 [Ar]$3d^54s^1$	25 **Mn** Manganese 54.93805 [Ar]$3d^54s^2$	26 **Fe** Iron 55.847 [Ar]$3d^64s^2$	27 **Co** Cobalt 58.93320 [Ar]$3d^74s^2$
37 **Rb** bidium .4678 Kr]$5s^1$	38 **Sr** Strontium 87.62 [Kr]$5s^2$	39 **Y** Yttrium 88.90585 [Kr]$4d^15s^2$	40 **Zr** Zirconium 91.224 [Kr]$4d^25s^2$	41 **Nb** Niobium 92.90638 [Kr]$4d^45s^1$	42 **Mo** Molybdenum 95.94 [Kr]$4d^55s^1$	43 **Tc** Technetium (97.9072) [Kr]$4d^65s^1$	44 **Ru** Ruthenium 101.07 [Kr]$4d^75s^1$	45 **Rh** Rhodium 102.906 [Kr]$4d^85s^1$
55 **Cs** sium .90543 e]$6s^1$	56 **Ba** Barium 137.327 [Xe]$6s^2$	57 **La** Lanthanum 138.9055 [Xe]$5d^16s^2$	72 **Hf** Hafnium 178.49 [Xe]$4f^{14}5d^26s^2$	73 **Ta** Tantalum 180.9479 [Xe]$4f^{14}5d^36s^2$	74 **W** Tungsten 183.84 [Xe]$4f^{14}5d^46s^2$	75 **Re** Rhenium 186.207 [Xe]$4f^{14}5d^56s^2$	76 **Os** Osmium 190.23 [Xe]$4f^{14}5d^66s^2$	77 **Ir** Iridium 192.22 [Xe]$4f^{14}5d^76s^2$
87 **Fr** ncium 3.0197) n]$7s^1$	88 **Ra** Radium (226.0254) [Rn]$7s^2$	89 **Ac** Actinium (227.0278) [Rn]$6d^17s^2$	104 **Rf** Rutherfordium (261.11) [Rn]$5f^{14}6d^27s^2$	105 **Db** Dubnium (262.114) [Rn]$5f^{14}6d^37s^2$	106 **Sg** Seaborgium (263.118) [Rn]$5f^{14}6d^47s^2$	107 **Bh** Bohrium (262.12) [Rn]$5f^{14}6d^57s^2$	108 **Hs** Hassium (265)† [Rn]$5f^{14}6d^67s^2$	109 **Mt** Meitnerium (266)† [Rn]$5f^{14}6d^77s^2$

58 **Ce** Cerium 140.115 [Xe]$4f^15d^16s^2$	59 **Pr** Praseodymium 140.908 [Xe]$4f^36s^2$	60 **Nd** Neodymium 144.24 [Xe]$4f^46s^2$	61 **Pm** Promethium (144.9127) [Xe]$4f^56s^2$	62 **Sm** Samarium 150.36 [Xe]$4f^66s^2$
90 **Th** Thorium 232.0381 [Rn]$6d^27s^2$	91 **Pa** Protactinium 231.03588 [Rn]$5f^26d^17s^2$	92 **U** Uranium 238.0289 [Rn]$5f^36d^17s^2$	93 **Np** Neptunium (237.0482) [Rn]$5f^46d^17s^2$	94 **Pu** Plutonium 244.0642 [Rn]$5f^67s^2$

timated from currently available
PAC data.